Emerging Technologies in Sustainable Innovation, Management and Development

Prof. Dr. R. Udayakumar, Editor-in-Chief, completed his M.S. (Information Technology and Management) from A.V.C. College of Engineering and was awarded a Ph.D. in 2011. He has been serving in the Teaching and Research community for more than two decades, has successfully produced 5 Doctoral candidates, and is a researcher, contributing to Research in interdisciplinary areas. His work includes 256 Scopus/SCI/Wos Publications. He has an h-index of 28, with 3173 citations (Scopus). He is the Dean of the Department of Computer Science Engineering and Information Technology and Director of IPR at Kalinga University, Raipur, Chhattisgarh. Stanford recognised him among the World's Top 2% Scientists and attained 9/10 in the Vidwan Database (Government of India).

Emerging Technologies in Sustainable Innovation, Management and Development

International Conference on Emerging Technologies for Sustainable Innovation, Management and Development

Edited by

Prof. Dr. R. Udayakumar

Routledge
Taylor & Francis Group
New York London

First published 2026
by Routledge
605 Third Avenue, New York, NY 10158

and by Routledge
4 Park Square, Milton Park, Abingdon, Oxon OX14 4RN

Routledge is an imprint of the Taylor & Francis Group, an informa business

British Library Cataloguing-in-Publication Data

A catalogue record for this book is available from the British Library

ISBN: 9781041164210 (hbk)
ISBN: 9781041162797 (pbk)
ISBN: 9781003684459 (ebk)

DOI: 10.1201/9781003684459

Typeset in Times New Roman
by HBK Digital

Contents

	List of figures	xi
	List of tables	xv
	Foreword	xx
	About the Author	xxi
	Preface	xxii
	Acknowledgements	xxiii
Chapter 1	Current development status of Jizu mountain scenic area under the background of rural revitalization *Linxing L. I., Chuanchen B. I., Long Yang, and Wei Yang*	1
Chapter 2	Preference towards electric vehicles in Coimbatore district *J. Smruthymol and R. Velmurugan*	9
Chapter 3	Is there a relationship between resilience and academic burnout among school teachers? A focus on Kerala *Ranjith Somasundaran Chakkambath and M. Usha*	22
Chapter 4	AI-driven strategies for enhancing circular economy and reducing industrial waste *J. Nithyashri, S. Ramesh, R. Geetha, P. Thilagavathi, and S. Vidhya*	30
Chapter 5	A study on customer satisfaction towards e-vehicles *Smruthymol J. and R. Velmurugan*	34
Chapter 6	Urban forestry and public health studies: A case study of Tebet eco park, Jakarta *Zarina Akbar, Dian Alfia Purwandari, Alya Fajria Rahma Maulida, Irsyad Hafiz, Muhammad Fadhil Haritsah, and Vranz Damar Sophan*	43
Chapter 7	Epileptic wave identification using machine learning in EEG signals *Abdul Mahatir Najar, Raihan Anugrah, Resnawati, Juni Wijayanti Puspita, and Rais*	46
Chapter 8	Exploring the impact of customer education and environmental concern on green product purchase intentions across generational cohorts *Husna Mohamed and Meena Suguanthi G.*	51
Chapter 9	Bio-inspired optimization for sustainable smart city infrastructure *R. Geetha, S. Ramesh, P. Thilagavathi, S. Vidhya, and J. Nithyashri*	58
Chapter 10	Factors influencing soft skills acquisition among higher education faculty *S. R. Karthick and R. Velmurugan*	62
Chapter 11	Digital transformation management: analyzing the influence of new media on women's empowerment in Tamil Nadu, India *V. Muthulakshmi and R. Abdul Muthalif*	69
Chapter 12	Impact of personality traits on job satisfaction among healthcare professionals *Senthil Kumar S. and J. Sudarvel*	73
Chapter 13	Leadership attributes of elementary school teachers in Jolo I district, division of Sulu *Immir S. Juhudi*	79
Chapter 14	A study on problems of migrant workers in Coimbatore district *Jaya Prakash N. and R. Velmurugan*	83
Chapter 15	Analytical study of correlations among socioeconomic, cultural factors, and customer relationship management (CRM) in tourism management *Krishna Kumar Mishra, Kausik Mukherjee, Shinu Shukla, Prakash Kumar Sen, Seema Dwivedi, and Chandan Singh*	91
Chapter 16	Problems of tribals on availing banking services *Praveen K. P. and R. Velmurugan*	97

Chapter 17 Exploring the Blockchain: Deciphering investor preferences in urban India 105
Pradeep Chaurasia, Pramod Dwivedi, Kausik Mukherjee, Anurag Singh, Kiran Chhabariya, and Shweta Singh

Chapter 18 Consumer perception towards neuromarketing in Kerala 111
Bibin Thomas M. and J. Sudarvel

Chapter 19 Analytical study of the promotional policies adopted by cement plants in India 118
Ankit Kumar Garg, Chandan Singh, Pranjal Agrawal, Krishna Kumar Mishra, Seema Dwivedi, and Bharti Tripathi

Chapter 20 A study on work life balance of women fire workers in 124
Vairam G. and V. Mathan Kumar

Chapter 21 Empirical study of the marketing research techniques adopted by retailers in India 131
Pranjal Agrawal, Chandan Singh, Ankit Kumar Garg, Dhirendra Ojha, Kartik Soni, and Parveen Kumar

Chapter 22 A significance study on financial performance selected pharmaceutical industries in India 139
P. Easwaran, M. Kalimuthu, V. Krishnaveni, Sankar Ganesh R., J. Sudarvel, and Smruthymol J.

Chapter 23 Awareness of gender rights in Imphal west district, Manipur 145
Konika Khuraijam

Chapter 24 The impact of artificial intelligence on consumer buying behaviour: A study on online retail purchases 151
P. Easwaran, V. Krishnaveni, M. Kalimuthu, Sankar Ganesh R., J. Sudarvel, and Smruthymol J.

Chapter 25 AI-enabled waste management framework 158
Clarence S. Ordonia and Thelma D. Palaoag

Chapter 26 Day of the week effect with special reference to Nifty bank index 163
M. Sudha and J. Shanmuganandavadivel

Chapter 27 A companion guide: Smart earpiece for visually impaired 168
Honeylet D. Grimaldo, Jericho Santos, Ain Christian Panganiban, Michael Arcay, Dominic Bevere Paragoso, Nathan Michael Estonilo, Jaime M. Landingin, Jorge Z. Burgos, and Reynaldo C. Muli

Chapter 28 LIC policy holders utilization level towards online services in Tirupur district 172
D. Vishnu Vardhan, K. Jothi, E. Arulprakasam, J. Sudarvel, V. Mathan Kumar, and R. Gopi

Chapter 29 JANUS: Study and analysis of the cooling and power requirements of an NFV telecommunications data center using machine learning 180
Lyberius Ennio F. Taruc, Arvin R. De La Cruz and Nelson C. Rodelas

Chapter 30 Consumer perception on kudumbashree products 188
Mujeeb Rahiman Kattali and R. Sarojadevi

Chapter 31 Anomaly detection in induction motors through irregular sound and vibrations 194
Honeylet D. Grimaldo, Ralph John A. Dela Peña, Ruscell Ray V. Doria, Darcy Micael F. Gasgonia, Darryl Christian G. Patron, Jaime M. Landingin, Jorge Z. Burgos, and Reynaldo C. Muli

Chapter 32 A study on taxpayers' satisfaction towards e-filing system with reference to Ernakulam district 200
Thazna T. A. and Sarithamol K. K.

Chapter 33 Emotional intelligence in finance: A mediating force in cognitive bias and investment decision 209
Sreejaa G. Nair and Jegadeeswari S.

Chapter 34 A study on investment patterns followed by aided and self-financing college teachers in Ernakulam district 218
Thazna T. A. and Sarithamol K. K.

Chapter 35 Design considerations for developing a context-aware recommender system for cybersecurity 228
Jhon Zylvin M. Ramos and Arvin R. De La Cruz

Chapter 36 Competitive advantage of import of agricultural products in India with special reference to BRICS countries 235
Sarithamol K. K. and Biju P. J.

Chapter 37 Experimental study on the adversary attack vulnerabilities in IoT and CPS actuators and sensors through electromagnetic fault injection 248
Jhon Zylvin M. Ramos, Marvin O. Mallari, and Arvin R. De La Cruz

Chapter 38 Assessing India's export performance of agricultural products: A case study on China 256
Sarithamol K. K. and Biju P. J.

Chapter 39 Digital transformation strategy, management practices and challenges among academic heads of higher education institutions in Zamboanga Peninsula 266
Maricor A. Marquez and Joe-An G. Cuaresma

Chapter 40 Work life balance and job satisfaction of women working in post office 276
Sowbarani G., Shanmugapriya M., M. Manimekalai, S. Babitha, and Ambika

Chapter 41 Current situation for the sustainable cultural tourism operation management in Qianxinan prefecture under new media era 279
Aoxue Li, Chuanchen Bi, Long Yang, and Wei Yang

Chapter 42 Customer perception and problems towards solar energy products 288
Sowbarani G., M. Akilanayaki, V. Poornima, N. Omshanmugapriya, and E. Rajeswari

Chapter 43 Crowdfunding platforms as a capital formation tool for young filipino entrepreneurs: Empirical evidence and growth implications 291
Remson Mark C. Macawile and Joseph Berlin P. Juanzon

Chapter 44 Current status of food tourism development in Qianxinan prefecture 300
Sihan Chen, Chuanchen Bi, and Long Yang

Chapter 45 The current situation of the combination of health Qigong and modern health management industry 308
Xiaolin He, Chuanchen Bi, and Long Yang

Chapter 46 Digital leap: Exploring impact of digital transformation on employee performance in Philippine advertising agencies 313
Herlan Czar P. Leuterio and Marie Jeddah L. Legaspi

Chapter 47 Designing carbon-aware cloud architectures for sustainable computing 317
S. Vidhya, Ramesh S., R. Geetha, P. Thilagavathi, and J. Nithyashri

Chapter 48 Preliminary study of the implementation of PAPEDA framework based on Ethno-STEM in supporting holistic science learning 323
Henie Poerwandar Asmaningrum, Marfly, and Shraddha Verma

Chapter 49 Decentralized edge AI for real-time environmental monitoring and sustainability 331
Ramesh S., R. Geetha, P. Thilagavathi, S. Vidhya, and J. Nithyashri

Chapter 50 Building relations with common appetite: India's culinary diplomacy with South Asia and the world 336
Mouli Chakraborty, Shilpi Bhattacharya, and Ajay Kumar Shukla

Chapter 51 Skill development and empowerment of women: A case study of Nituria CD block, Purulia, West Bengal 340
Jayashri Roy and Shraddha Verma

Chapter 52 Quantum-enhanced climate models for predicting and mitigating environmental change 347
P. Thilagavathi, R. Geetha, Ramesh S., S. Vidhya, and J. Nithyashri

Chapter 53 Awareness of NEP 2020 on in-service teacher education: A professional development perspective for elementary school teachers 352
Pranati Palai and G. C. Nanda

Chapter 54 Evaluating the influence of multinational leadership on business success in global teams 356
R. K. Kasunthika, D. Kalidoss, Hemalatha Nagharajan, Sivakumar Pushparaj, and Balaji Ramalingam

Chapter 55 Advancing solar and wind power for sustainable city models 359
Sarvinoz Sayfullayevna Kasimova, Gulnora Sattorovna Mutalova, Buzakhro Marufjanovna Begmatova, Markhabo Raxmonkulovna Abdullayeva, and Doniyor Dovronovich Ergashev

Chapter 56 Harnessing emerging technologies for a sustainable future: Innovations, management, and development 364

Aravindhan Ragunathan and Nirmala Vairaganthan

Chapter 57 Synergy between creative economy and artificial intelligence for sustainable tourism 370

Putu Eka Wirawan, Ni Wayan Mega Sari Apri Yani, and I Wayan Eka Mahendra

Chapter 58 Enhancing building energy load prediction with feature engineering and deep learning: A comparative study of DNN, FCNN, and 1D CNN 375

Kanokwan Khiewwan, Narut Butploy, Phrommate Verapan, Khumphicha Tantisantisom, Pakin Maneechot, and Nivadee Klungsida

Chapter 59 A deep learning approach for real-time energy forecasting in smart buildings with NoSQL and IoT data 379

Narut Butploy, Kanokwan Khiewwan, Phrommate Verapan, Khumphicha Tantisantisom, Pakin Maneechot, and Nivadee Klungsida

Chapter 60 Adaptive load management in power systems with automatic transformers 383

Mohankumar P. and V. Ravi

Chapter 61 Innovativeness and risk-taking: A cross-cultural study of U.S., Indian, and Chinese entrepreneurs 389

Niranchana Shri Viswanathan, Delecta Jenifer Rajendren, A. Suresh Kumar, and B. Nandhini

Chapter 62 Sustainable edge AI for smart cities and renewable energy management using spiking neural networks 393

Omprakash Dewangan, Ayaz Ahmed Faridi, Ashu Nayak, Aakanksha Soy, and Kamlesh Kumar Yadav

Chapter 63 Navigating choices: Peer pressure, anxiety and decision-making in NCC cadets 398

Naveen, Saroj Nayyar, Gulsan Kumar Behera, Awtar Kishan Kaul, Jyoti Dubey, and Ekta Saxena

Chapter 64 Empowering large-scale data analysis with AutoML: Boosting processing speed and analytical adaptability 401

Sanskritee Shilpa Pandey and Kamlesh Kumar Yadav

Chapter 65 AI Integrated PBL in language classroom: Enhancing creativity, critical thinking and sustainable awareness 407

Sulagna Mohanty, Sudakshina Bhattacharya, and Ankusha Bandyopadhyay

Chapter 66 Evaluation of sustainability disclosure based on GRI standards in energy sector companies 413

Rifqi Novriyandana, Ratna Ayu Damayanti, Darwis Said, and Muhammad Rijal

Chapter 67 Exploring consumer decision-making factors for electric, hybrid, and hydrogen vehicles through focus group discussions 417

Hendri Bhirowo, Indrawati, Handrea Bernando Tambunan, Indra Ardhanayudha Aditya, Refi Rifaldi Windya Giri, Fasa Maulidan Hakim, and Fayza Najeela Pakutandang

Chapter 68 Narrative pedagogy of undergraduate students' writing literacy development and its implications for teaching writing 423

Erna Iftanti, Nani Soengkono Madayani, and Galuh Nashrulloh Kartika M. R.

Chapter 69 Skill competency test for vocational high school (SMK) students encounter labor market challenges 430

Tetty Yuliaty, Wahyu Ario Pratomo, Suhendri, and Andika Fajar Putra

Chapter 70 Implementation of the Penta Helix concept to increase climate awareness in Islamic schools through extracurricular activities 434

Efrita Norman, Arman Paramansyah, Windi Megayanti, and Nicky Rosadi, S. S.

Chapter 71 Strategic management advantages in educational organizations 442

Nur Zazin, Galuh Nashrulloh Kartika M. R., and Barsihanor

Chapter 72 Political education: Strategies and constraints for political parties and election organizers to enhance women's political literacy 448

Hariyanti, Cecep Darmawan, Rahmat, and Leni Anggraeni

Chapter 73 The role of urban forestry to enhance city sustainability: A case study of Tebet eco park in Jakarta 455

Zarina Akbar, Dian Alfia Purwandari, Irsyad Hafiz, Alya Fajria Rahma Maulida, Muhammad Fadhil Haritsah, and Vranz Damar Sophan

Chapter 74 Mastering prompt engineering: Unlocking creative potential in AI-generated art 459
Aezeden Mohamed

Chapter 75 Revolutionizing Agile project management: The role of AI in driving innovation overcoming challenges, and enhancing benefits 469
Aezeden Mohamed

Chapter 76 Antecedents of solar rooftop PV panel adoption intention: Theoretical insights and empirical evidence 479
Swathi Newashini and R. Parameswaran

Chapter 77 Analysis of marketing strategy for salted egg business in increasing sales volume: Case study of Eldona salted eggs, Banjarbaru 487
Sugiarti, Achmad Jaelani, and Siti Dharmawati

Chapter 78 A study on consumer awareness towards neuromarketing in Kannur district 491
Bibin Thomas M. and J. Sudarvel

Chapter 79 Impact of influencer marketing on tourist destination choices 495
J. Sudarvel and R. Velmurugan

Chapter 80 The influence of customer relationship management and brand image on customer satisfaction: A meta-analysis for light rail transit (LRT) transportation 502
Yunianto Aditya, Buchari Erika, Agustien Melawaty, and Ermatita Ermatita

Chapter 81 Lesson learnt from the study of the power of service quality in influencing company reputation and customer trust for LRT operator: A meta-analysis 512
Pudjiono Rode Paulus Gagok, Buchari Erika, Agustien Melawaty, and Ermatita Ermatita

Chapter 82 Analysis of the implementation of artificial neural networks in the identification and assessment of chronic kidney disease 521
R. Suhasini, N. Sandeep Kumar, V. Ramesh, Bodla Kishor, A. Vaishnavi Dali, and Rasagnya Reddy Avala

Chapter 83 A CNN's improved bio-inspired fossil image segmentation and identification for palentology 527
D. Sudha, B.Thanuja, N. Samathagoud, Katragadda Anuhya, and R. S. Sabeenian

Chapter 84 Employing a deep learning technique for image segmentation based on a level-based set method 533
M. Nagaraju Naik, Suraya Mubeen, B. Ravikumar, Swati Gupta, and S. Vijayalakshmi

Chapter 85 An effective and efficient technique that uses onvolution neural networks to identify coronary artery disease using IRIS 540
B. Venkataramanaiah, D. Srikanth, Md Yasmeen, Monika Arya, and R. S. Sabeenian

Chapter 86 An enhanced deep learning method for facial recognition built around LBP and Log-Gabor 546
Abdul Subhani Shaik, Swarna Venkatesh, B. Kishore, Mrutyunjaya S. Yalawar, and K. Manju

Chapter 87 Analysis of a quick and adaptable IOT-based approach for roadside e-vehicle charging 551
Srinivas Manikyala, K. Bharath Kumar, P. Navitha, A. Srinivasula Reddy, M. E. Paramasivam, and Rasagnya Reddy Avala

Chapter 88 An improved vertical antenna-dependent machine learning-based health diagnosis system using patient data monitoring 556
B. Premalatha, V. Mounika, K. Navya, Gadgi Sumangala, and K. Manju

Chapter 89 A programmable ALU with quantum dot cellular automata: A significantly improved fast speed multiplexer design 563
Gali Narendra, G. Srikanth, E. Aparna, Suman Mishra, and M. E. Paramasivam

Chapter 90 Comparative performance analysis for machine learning techniques on malaria detection using CNN and microscopy images 568
T. Priyanka Kumari, D. Maneiah, V. Somalaxmi, Kothapally Subramayam Chari, and R. S. Sabeenian

Chapter 91 An experienced and categorical brain tumor detection through image processing methods 574
Chandan Roy, P. Kranthi Rathan, G. Venkataswamy, Biroju Papa Chary, and T. Shanthi

Chapter 92	A real-time driver behaviour tracking system in automobiles utilizing image processing to extract features *A. Veerababu, A. Vamshidhar Reddy, S. Nagajyothi, A. Srinivasula Reddy, and R. S. Sabeenian*	579
Chapter 93	A novel ALU operation technique for parity-preserving gates driven by Xilinx and Verilog HDL *E. N. V. Purna Chandra Rao, V. Santosh Kumar, G. Srikanth, and K. Navya*	585
Chapter 94	Performance analysis of hybrid set CMOS logic for single-electron transistor using T-SPICE *B. Premalatha, E. Parvathi, Suraya Mubeen, and G. Venkat Swamy*	590
Chapter 95	Assessment of convolutional neural network-based classification on face recognition images *Y. Sailaja, P. Sandhya Reddy, Bagam Laxmaiah, and Bopparam Sampath Kumar*	594
Chapter 96	The communal effects on security access using cloud data bases in medical health records *R. Suhasini, Chandrugonda Malleshwar Rao, Voruganti Naresh Kumar, and Rapaka Usha*	600
Chapter 97	A new technique for segmenting infant brain MRI data utilizing the ridgelet and framelet transforms *Vivekanand Aelgani, Rafeeq M. D., Prasannababu J., and P. Kiran Kumar*	604
Chapter 98	Creative economic study as a center for tourism development in Banyuwangi district *Ayu Purwaningtyas, Riza Rahimi Bachtiar, Reni Nur Jannah, and Eka Afrida Ermawati*	609
Chapter 99	Beyond rationality: Overconfidence and herding on investment decisions with the moderating factor financial literacy *Sreejaa G. Nair and Jegadeeswari S.*	613

Lists of figures

Figure 3.1	Model for Burnout-Resilience moderated by Age	23
Figure 3.2	Scatter plot diagram of Burnout to Resilience moderate by Age	27
Figure 4.1	Integration of artificial intelligence in circular economy and sustainable practices	31
Figure 4.2	Integration of AI and smart sensors in waste management systems	31
Figure 4.3	Comparison of conventional and AI-enabled waste management efficiency	32
Figure 7.1	Representation of selected data	47
Figure 7.2	Comparison of accuracy of machine learning models for Epileptic wave identification	49
Figure 9.1	Smart street lighting based on Bio-inspired optimization	59
Figure 9.2	Smart parking using DELM and PSO	60
Figure 9.3	Comparison graph of parking approaches	60
Figure 19.1	Top cement manufacturers in India	119
Figure 25.1	Smart-bin prototype	161
Figure 27.1	Conceptual framework	169
Figure 27.2	Block diagram	169
Figure 27.3	System flowchart for hardware and mobile app	170
Figure 28.1	LIC'S e services with its procedures (A,B,C,D)	175
Figure 29.1	The ETSI NFV architectural framework	181
Figure 29.2	The data center dimensioning and assessment process	182
Figure 29.3	Proposed high-level model for calculating the estimated power and consumption	184
Figure 31.1	Conceptual framework	196
Figure 31.2	Block diagram	196
Figure 31.3	System flowchart for hardware and mobile app	197
Figure 31.4	The mobile application	197
Figure 31.5	The auto encoder	197
Figure 31.6	The actual and predicted value	197
Figure 31.7	The sound MSE	197
Figure 31.8	The vibration MSE	198
Figure 31.9	The model performance evaluation	198
Figure 33.1	Theoretical framework of the research	211
Figure 35.1	Proposed layered framework design for a context-aware recommender system for cybersecurity	232
Figure 36.1	WITS report for BRICS country import trading partners time series	244
Figure 36.2	Import part share percentage of Brazil, China, Russian and South Africa	244
Figure 37.1	EMFI platform setup with a programmable EMFI pulse injector sending adversary signal into two different CPS/ IoT setups for actuators and sensors applications	250
Figure 37.2	High-level view of the designed EM pulse injector circuit	250
Figure 37.3	Separate IoT sensor and IoT actuator experimental setups	251
Figure 37.4	(Top) Photocell and (bottom) LM35 temperature sensor response to EMFI injected after 10 seconds delay at 1 second and 5 second duration respectively	252
Figure 37.5	DHT11 sensor response (top) temperature and (bottom) humidity sensor response to EMFI injected after 10 seconds delay at 5 second duration	252
Figure 37.6	Relay-state response in blue to EMFI trigger in red with different pulse duration (a). Relay-state response in blue to Push button trigger in red with different pulse duration (b)	253
Figure 38.1	ARIMA stimulations (China imports from India)	260
Figure 38.2	ARIMA stimulations (Time Vs Residuals)	261
Figure 38.3	Export partner share-China	263
Figure 41.1	Relationship between stakeholder and sustainable tourism	281
Figure 41.2	Life cycle of destination	282
Figure 43.1	Conceptual framework	294
Figure 43.2	Results of hypotheses	297
Figure 44.1	2024 China food tourism development report	302
Figure 44.2	The influence of food consumption value on tourists' behavioural intention	303
Figure 46.1	Relationship of Respondents' profiles with their impact on digital transformation and job performance	313
Figure 47.1	Carbon explorer framework	318
Figure 47.2	Carbon aware cloud architecture	319
Figure 49.1	Classification of AI agents in environmental sciences	333
Figure 49.2	AI-Impact on data analysis efficiency, accuracy, and cost	334
Figure 52.1	Quantum framework for climate models	348
Figure 52.2	Comparison of classical vs. quantum models in climate simulations	349
Figure 52.3	Quantum vs. classical performance in sustainability applications	350
Figure 54.1	Impact of ethnicity on BS for ML duration	357
Figure 54.2	Mean and SD for different factors involved in hypothesis development	358
Figure 55.1	Framework for solar and wind energy application in energy sustainability	360

Figure 55.2	Sustainable building with solar/wind energy	361
Figure 55.3	Smart transport system	362
Figure 55.4	Higher quality life style with smart city	362
Figure 57.1	Percentage showing the number of mobile phone users in Indonesia	371
Figure 58.1	Deep learning-based energy load classification framework	376
Figure 58.2	The confusion matrices for DNN, FCNN, and CNN 1D	377
Figure 59.1	Plan for a zero-energy structure and system for continuous energy prediction	380
Figure 59.2	Model performance and dashboard integration	381
Figure 60.1	Block diagram of the existing system	384
Figure 60.2	Circuit diagram of the proposed system	384
Figure 60.3	Transformer circuit diagram	385
Figure 60.4	Basic relay circuit diagram	385
Figure 60.5	Protection diodes for relay	385
Figure 60.6	LCD display	385
Figure 60.7	Simulation diagram	386
Figure 60.8	Simulation result	386
Figure 60.9	Simulation result	386
Figure 61.1	Year growth	391
Figure 61.2	Bar chart country wise	391
Figure 62.1	Implementation flow diagram of the proposed model	394
Figure 62.2	Fundamental model of SNN	395
Figure 62.3	Edge computing for sustainable development	395
Figure 62.4	Performance comparison of energy efficiency model	396
Figure 62.5	Performance comparison on inference latency	396
Figure 62.6	Sustainable energy prediction of the proposed model	396
Figure 64.1	Overview of AutoML for DA	403
Figure 65.1	Student work	409
Figure 65.2	Student work	409
Figure 65.3	Student work	410
Figure 65.4	Student work	411
Figure 65.5	Student work	411
Figure 66.1	Redearch design	415
Figure 69.1	Teknik analisis data miles dan Huberman	431
Figure 70.1	Penta Helix approach	438
Figure 75.1	Flowchart	473
Figure 76.1	Conceptual framework	481
Figure 80.1	Meta-analysis forest plot	506
Figure 80.2	Funnel plot after trim-fill diagnosis	506
Figure 80.3	Meta-analysis forest plot	507
Figure 80.4	Funnel plot after trim-fill diagnosis	507
Figure 80.5	Meta-analysis forest plot	507
Figure 80.6	Funnel plot after trim-fill diagnosis	507
Figure 81.1	Meta-analysis forest plot	516
Figure 81.2	Funnel plot after trim-fill diagnosis	516
Figure 81.3	Meta-analysis forest plot	517
Figure 81.4	Funnel plot after trim-fill diagnosis	517
Figure 82.1	Framework of an detection and evaluation of chronic kidney disease by using artificial neural network	523
Figure 82.2	Use case diagram of proposed system	524
Figure 82.3	Performance analysis of various models	524
Figure 82.4	Accuracy performance of training and testing	525
Figure 83.1	Process of proposed method	529
Figure 83.2	Flowchart of proposed method	529
Figure 83.3	(a) Original ES HE (b) Quantization	530
Figure 83.4	(a) CA quantized (b) CA quantized grey	530
Figure 83.5	(a) Original image (b) ES histogram	530
Figure 83.6	Accuracy comparison of various algorithms	530
Figure 83.7	Precision comparison of various algorithms	530
Figure 83.8	Recall comparison of various algorithms	531
Figure 83.9	Precision comparison of various algorithms	531
Figure 83.10	Execution time comparison of various algorithms	531
Figure 83.11	Comparison of IoU under different noise levels	531
Figure 84.1	Block diagram	535
Figure 84.2	CNN structure model	535
Figure 84.3	Level set method partial results	536
Figure 84.4	Level set method results	536

Figure 84.5	Flow chart	537
Figure 84.6	The portrait data set's ground truth and some photos	537
Figure 84.7	Dice coefficient IoU comparison of various methods	538
Figure 84.8	Boundary IoU comparison of various methods	538
Figure 84.9	Pixel accuracy comparison of various methods	538
Figure 85.1	The coronary artery disease	541
Figure 85.2	System proposed model	542
Figure 85.3	Image pre-processing process	542
Figure 85.4	Iris image using CLAHE results of using CLAHE method on iris image	543
Figure 85.5	Architecture of CNN	543
Figure 85.6	Confusion matrix on CAD	544
Figure 85.7	Performance of training vs validation accuracy	545
Figure 86.1	Proposed system model	547
Figure 86.2	Flow chart	548
Figure 86.3	Accuracy analysis on LFW and CASIA web face	549
Figure 86.4	Training time vs inference time	549
Figure 87.1	Block diagram of a proposed system	553
Figure 87.2	Charging of battery after transaction	554
Figure 87.3	Energy transfer efficiency vs speed	554
Figure 87.4	Communication latency in IoT system	554
Figure 88.1	Representation of monitoring system with medical practitioner	558
Figure 88.2	Representation of monitoring system without medical practitioner	558
Figure 88.3	Experimental setup at patients' side	558
Figure 88.4	Experimental setup at doctors' side	559
Figure 88.5	Transmitting antenna	559
Figure 88.6	Receiving antenna	559
Figure 88.7	Temperature monitoring	560
Figure 88.8	Heartrate monitoring	560
Figure 88.9	SPO2 monitoring	560
Figure 88.10	Performance of KNN model	560
Figure 89.1	The symbol of 2:1 MUX	565
Figure 89.2	Comparison of proposed and conventional QCA MUX design	566
Figure 90.1	Proposed model structure	569
Figure 90.2	CNN architectural model	570
Figure 90.3	Flow chart	570
Figure 90.4	Input images	571
Figure 90.5	a) Detection of cells b) Infected cells	571
Figure 90.6	Performance comparison of various ML model	572
Figure 90.7	ROC curve comparison of various ML models	572
Figure 91.1	Proposed model	575
Figure 91.2	Flow chart	576
Figure 91.3	a) Input image b) Filtered image	577
Figure 91.4	a) Tumor alone b) Detected tumor	578
Figure 92.1	System model	581
Figure 92.2	Flow chart	581
Figure 92.3	System design	582
Figure 92.4	Performance evolution metrics	583
Figure 92.5	Confusion matrix	583
Figure 93.1	ALU block diagram	586
Figure 93.2	CNOT gate (a) Block diagram (b) symbol	586
Figure 93.3	Fredkin gate	587
Figure 93.4	1-bit ALU	587
Figure 93.5	Schematic of 4-bit RALU	587
Figure 93.6	Schematic of reversible 1-bit ALU	588
Figure 93.7	Design of reversible 4-bit ALU	588
Figure 93.8	Simulation of 4-bit ALU	588
Figure 94.1	Hybrid SET-CMOS inverter circuit diagram	591
Figure 94.2	Hybrid CMOS-SET half subtractor circuit	592
Figure 94.3	Simulation result of hybrid half subtractor	592
Figure 95.1	Layers of CNN	595
Figure 95.2	Deep learning workflow	596
Figure 95.3	Block illustration of face identification by means of CNN	597
Figure 95.4	Training advancement training loss versus accuracy	598
Figure 96.1	Healthcare cloud architecture	601
Figure 96.2	Frame work of PHR, HER, EMR	601

Figure 96.3	Graph showing key generation time and the amount of attributes	602
Figure 97.1	Machine learning technique for Glioma brain tumor identification	606
Figure 97.2	(a) Source brain MRI image (b) Improved brain MRI image	607
Figure 97.3	(a) Segmented image (b) Annotated image	607
Figure 97.4	Performance accuracy of various input images	607
Figure 98.1	Janger performing	611
Figure 98.2	Bamboo costume	611
Figure 98.3	Creative economy sub-sector	612
Figure 99.1	Dynamic financial	615

Lists of tables

Table 2.1	Socio-economic profile	15
Table 2.2	E Vehicle usage	16
Table 2.3	KMO and Bartlett's test	17
Table 2.4	Preference towards E-Vehicles	17
Table 3.1	Demographic characteristic	25
Table 3.2	School category and years of teaching experience of faculty	25
Table 3.3	Scheme of teaching, marital status, and employment type	25
Table 3.4	t Test results summary comparing gender with Resilience and Burnout of faculty	26
Table 3.5	Correlations test between Burnout and Resilience	26
Table 3.6	Conditional effects of facilitating conditions on expectancy in AI Chatbot at different levels of satisfaction	26
Table 3.7	Conditional effects of Burnout on Resilience at values of Age	26
Table 3.8	Conditional effect of focal predictor at values of moderator	27
Table 5.1	Socio-economic profile	38
Table 5.2	E Vehicle usage	38
Table 5.3	KMO and Bartlett's test	39
Table 5.4	Rotated component matrixa	39
Table 8.1	Descriptive statistics table	54
Table 8.2	Reliability analysis	54
Table 8.3	Mediation results for FCE → EC → CPI	54
Table 8.4	Mediation results for MCE → EC → CPI	54
Table 8.5	Dummy variable coding	55
Table 8.6	Regression model summary	55
Table 8.7	Conditional effects of FCE on CPI by age category	56
Table 8.8	Regression model summary	56
Table 8.9	Conditional effects of MCE on CPI by age category	56
Table 10.1	Demographic profile of respondents	65
Table 10.2	KMO and Bartlett's test	65
Table 10.3	Rotated component matrix	66
Table 11.1	Average access to digital technology and perception of empowerment among women in Tamil Nadu	71
Table 12.1	Descriptive statistics	75
Table 12.2	KMO and Bartlett's test	76
Table 12.3	Rotated component matrix	76
Table 14.1	Socio economic profile	87
Table 14.2	KMO and Bartlett's test	87
Table 14.3	Problems of migrant workers	88
Table 15.1	Income level and travel preferences	93
Table 15.2	Economic stability perception of participants	93
Table 15.3	Demographic influence on CRM	93
Table 15.4	Influence of cultural heritage	94
Table 15.5	Perception of balancing cultural preservation	94
Table 15.6	Correlation matrix	94
Table 15.7	Cluster analysis	95
Table 16.1	Socio-economic profile	100
Table 16.2	Banking habit	101
Table 16.3	KMO and Bartlett's test	101
Table 16.4	Problems on availing banking services	102
Table 17.1	Demographic profile of participants	107
Table 17.2	Level of awareness and understanding of blockchain	107
Table 17.3	Factors influencing investor preferences	108
Table 17.4	Investor attitudes towards regulatory frameworks	108
Table 17.5	Correlation between demographic variables and investor preferences	108
Table 18.1	Statistical analysis, frequency, and percent of the study	114
Table 18.2	KMO and Bartlett's test	114
Table 18.3	Rotated component matrix	115
Table 19.1	Cement industry with capacity of manufacturing	119
Table 20.1	Simple frequency	128
Table 20.2	KMO and Bartlett's test	129
Table 20.3	Rotated component matrix	129
Table 21.1	Demographic profile of the candidate	133
Table 21.2	Types of marketing research techniques	134
Table 21.3	Frequency of marketing research techniques usage	134
Table 21.4	Cross tabulation of techniques by retail segment	134

Table 21.5	Effectiveness ratings of marketing research techniques	135
Table 21.6	Impact of marketing research techniques on sales and customer satisfaction	135
Table 21.7	Factors influencing the adoption of marketing research techniques	136
Table 21.8	Comparison of factors by retail segment	136
Table 22.1	Five-year ROI analysis of selected pharmaceutical companies in India (FY2020–FY2024)	141
Table 22.2	Five-year ROE analysis of selected pharmaceutical companies in India (FY2020–FY2024)	142
Table 22.3	DuPont ROE salculation results	142
Table 22.4	ANOVA results: Financial performance comparison (FY2020–FY2024)	143
Table 23.1	Comparison between men and women regarding awareness of gender rights of Imphal west district	147
Table 23.2	Comparison between unmarried and married people regarding awareness of gender rights of Imphal west district	147
Table 23.3	Comparison between married men and unmarried men regarding awareness of gender rights of Imphal west district	148
Table 23.4	Comparison between married women and unmarried women regarding awareness of gender rights of Imphal west district	148
Table 24.1	Consumer awareness and perceptions of AI in online shopping	153
Table 24.2	Impact of technology on consumer buying behaviour	154
Table 24.3	Satisfaction with AI in shopping	155
Table 24.4	Challenges with AI in shopping	156
Table 26.1	Analysis and interpretation	166
Table 26.2	Dependent variable: Friday	166
Table 27.1	System testing	170
Table 27.2	Battery specification	170
Table 28.1	Analysis and interpretation	176
Table 28.2	Utilization level of e-services	177
Table 29.1	Data seed 01-a: Cabinet type power specifications	184
Table 29.2	Data seed 01-b-01: Calculated power specifications of data center C	185
Table 29.3	Data seed 01-b-02: Calculated power specifications of data center S	185
Table 29.4	Data seed 01-b-03: Calculated power specifications of data center T	185
Table 29.5	Calculated cooling requirements of data center C	185
Table 29.6	Calculated cooling requirements of data center S	185
Table 29.7	Calculated cooling requirements of data center T	186
Table 29.8	Virtual resources per data center	186
Table 29.9	Simulation results and evaluation	186
Table 30.1	Demographic profile (no. of the respondents 146)	190
Table 30.2	Perception on kudumbashree products	191
Table 32.1	Tabulation based on gender	203
Table 32.2	Tabulation based on Age	203
Table 32.3	Tabulation based on educational qualification	203
Table 32.4	Tabulation based on income level	203
Table 32.5	Testing the proximity between gender and satisfactory level on e-filing	204
Table 32.6	Chi-square test between gender and level of satisfaction	204
Table 32.7	Testing the proximity between age and satisfactory level on e-filing	204
Table 32.8	Chi-square test between age and satisfactory level on e-filing	205
Table 32.9	Testing the proximity between educational qualification and satisfactory level on e-fil	205
Table 32.10	Chi-square test between respondents qualification and satisfactory level	205
Table 32.11	Testing the proximity between income level and satisfactory level on e-filing	206
Table 32.12	Chi-Square Test between level of income and e-filing satisfactory level	206
Table 32.13	Factors identified as the hindrances of e-filing (Garrett Ranking Method)	206
Table 32.14	Calculation of mean scores based on Garrett scores	206
Table 33.1	Reliability statistics	212
Table 33.2	Demographic profile	212
Table 33.3	Correlations between variables	213
Table 33.4	Regression analysis	214
Table 33.5	Hypothesis testing	214
Table 33.6	Direct effects of biases on investment decisions	215
Table 33.7	Indirect effects of biases on investment decisions via emotional intelligence	215
Table 34.1	Data respondents in male and female count	222
Table 34.2	Frequency analysis for relevant statements	223
Table 34.3	Problems faced in understanding the importance of tax management	223
Table 34.4	T-test table – to find out the difference between the pattern of savings and the nature of employment	223
Table 34.5	Chi-square tests	224
Table 34.6	Cross-tabulation	225
Table 34.7	Correlations between investment and saving attitude	225
Table 36.1	Import product details from Brazil for the period 2019–2024	238
Table 36.2	Import product details from China for the period 2019–2024	238
Table 36.3	Import product details from Russia for the period 2019–2024	239

Table 36.4	Import product details from South Africa for the period 2019–2024	240
Table 36.5	BRICS country-based product import details for period of 2019–2024	240
Table 36.6	BRICS country-based ALCOHOLIC BEVERAGES import details for period of 2019–2024	240
Table 36.7	BRICS country-based CEREAL PREPARATIONS import details for period of 2019–2024	241
Table 36.8	BRICS country-based COCOA PRODUCTS import details for period of 2019–2024	241
Table 36.9	BRICS country-based FLORICULTURE import details for period of 2019–2024	242
Table 36.10	BRICS country-based FRUITS / VEGETABLE SEEDS import details for period of 2019–2024	242
Table 36.11	BRICS country-based MISC PROCESSED ITEMS import details for period of 2019–2024	242
Table 36.12	BRICS country-based OTHER CEREALS import details for period of 2019–2024	243
Table 36.13	BRICS country-based PROCESSED FRUITS & JUICES import details for period of 2019–2024	243
Table 36.14	BRICS country-based PULSES import details for period of 2019–2024	243
Table 36.15	Trade integration pattern index	245
Table 38.1	India China bilateral trade	260
Table 38.2	ARIMA output-goodness of fit statistics	260
Table 38.3	Export product details from India to China for the period 2020–2023	262
Table 38.4	Export partner share china	262
Table 38.5	Trade integration pattern index	263
Table 38.6	International trade analysis report	263
Table 39.1	Summary of results of the digital transformation strategy in terms of digital leadership, technology utilization and workforce capability	268
Table 39.2	Summary of results of the DT management practices in terms of digital leadership, technology utilization and workforce capability	269
Table 39.3	Academic head-participants' responses on the challenges experience in the implementation of digital transformation strategy	270
Table 39.4	Correlation: Digital transformation strategy and the DT management practices among academic heads of HEIs in Zamboanga Peninsula	271
Table 39.5	Digital transformation strategy among academic heads when respondents are grouped according to age	271
Table 39.6	Digital transformation strategy among academic heads when respondents are grouped according to educational attainment	271
Table 39.7	Digital transformation strategy among academic heads when respondents are grouped according to length of service	272
Table 39.8	Digital Transformation strategy among academic heads when respondents are grouped according to type of HEI	272
Table 39.9	DT management practices among academic heads when respondents are grouped according to age	272
Table 39.10	DT management practices among academic heads when respondents are grouped according to educational attainment	273
Table 39.11	DT management practices among academic heads when respondents are grouped according to length of service	273
Table 39.12	DT management practices among academic heads when respondents are grouped according to type of HEI	273
Table 40.1	KMO and Bartlett's test	278
Table 40.2	Rotated component matrix	278
Table 42.1	Problems of solar products	289
Table 43.1	Discriminant validity	294
Table 43.2	Factor loading, validity, and reliability of the indicators	295
Table 43.3	Blindfolding statistics for the general model	296
Table 43.4	Hypotheses testing	296
Table 46.1	The significant influence among the respondent's profile and the extent of their digitalization	314
Table 46.2	The significant influence on the respondent's profile and extent of their job performance	314
Table 46.3	Significant relationship between the extent of digitalization with their job performance	315
Table 47.1	The algorithm minimizes carbon consumption for medium switching overhead tasks by region	320
Table 48.1	Result of teacher implementation	325
Table 48.2	Result of teacher implementation	326
Table 48.3	Result of school administrators' perspective	327
Table 51.1	Role of agencies in Study area and promote skill development	343
Table 51.2	Role of stakeholders in promoting and hindering women's participation in skills development	344
Table 51.3	Participation categories	344
Table 51.4	Participation categories (age group)	344
Table 55.1	Various technology used in sustainable smart cities	360
Table 57.1	SWOT analysis	372
Table 57.2	SMART analysis	373
Table 62.1	Summary of key literature and research gaps	394
Table 63.1	Summary of two-way ANOVA for 2x2 factorial experiment of rational the style of decision-making of NCC cadets	399
Table 63.2	Showing the main effect of anxiety on rational style of decision making of NCC cadets	399
Table 63.3	Showing the interactional effect of peer pressure and anxiety on rational the style of decision making of NCC cadets	399
Table 64.1	Optimized base classifier analysis	404
Table 64.2	AutoML performance analysis	405

Table 66.1	Operational definition of variables	414
Table 66.2	Partial test results (T)	415
Table 68.1	Portrait of the undergraduate students' reading literacy development	424
Table 68.2	The supporting factors of the undergraduate students' writing literacy	426
Table 70.1	Results of interviews with principals and teachers	436
Table 70.2	Direct observation of environment-based extracurricular activities	437
Table 70.3	Documentation of environmental education materials and curriculum from academics	437
Table 70.4	Government support and community involvement in school activities	437
Table 70.5	Media publications on school environment activities	438
Table 70.6	Impact of extracurricular activities on student climate awareness	438
Table 72.1	Women's representation in the general election of the house of representatives	449
Table 75.1	Comparison of efficiency metrics between traditional and AI-enabled APM	475
Table 75.2	Comparison of cost metrics between traditional and AI-enabled APM	475
Table 75.3	Comparison of transparency and communication metrics between traditional and AI-enabled APM	475
Table 75.4	Comparison of scalability metrics between traditional and AI-enabled APM	475
Table 75.5	Comparison of risk management metrics between traditional and AI-enabled APM	476
Table 76.1	A revealed two statistically significant indirect effects on adoption intention jointly mediated by environmental concern	481
Table 76.2	Direct effects on adoption intention	482
Table 76.3	Predictor → mediator relationships	482
Table 76.4	Total effects	483
Table 76.5	Indirect effects via trust in AI	483
Table 76.6	Summary of hypothesis testing results	483
Table 77.1	Business model Canva Eldona's salted egg	489
Table 78.1	Analysis and interpretation	492
Table 78.2	Factors of respondent scores	493
Table 79.1	Analysis and interpretation	497
Table 79.2	Model summary	498
Table 79.3	Correlations coefficients	499
Table 79.4	KMO and Bartlett's test	499
Table 79.5	Rotated component matrix	499
Table 80.1	Comparison of 40 studies based on n, r, T, and R2 values, 2017–2024	505
Table 80.2	Summary effect or mean effect size	506
Table 80.3	Regression test for funnel plot asymmetry (egger's test)	506
Table 80.4	Summary effect or mean effect size	507
Table 80.5	Regression test for funnel plot asymmetry (egger's test)	507
Table 80.6	Summary effect or mean effect size	507
Table 80.7	Regression test for funnel plot asymmetry (egger's test)	508
Table 81.1	Comparison of 22 Studies based on n, r, t, F, and R2 value, 2017–2024	515
Table 81.2	Heterogeneity test	515
Table 81.3	Residual heterogeneity estimates	515
Table 81.4	Summary effect or mean effect size	515
Table 81.5	Regression test for funnel plot asymmetry (egger's test)	516
Table 81.6	Heterogeneity test	516
Table 81.7	Residual heterogeneity estimates	516
Table 81.8	Summary effect or mean effect size	516
Table 81.9	Regression test for funnel plot asymmetry (egger's test)	517
Table 85.1	Performance on training and validation set	544
Table 85.2	Performance evaluation of CNN model	544
Table 85.3	Performance comparison of various models	544
Table 86.1	Comparison of performance metrics for various models	549
Table 87.1	Energy utilization and cost effectiveness	555
Table 88.1	Display of patient data	560
Table 88.2	Results of prediction model	561
Table 88.3	Performance metric	561
Table 89.1	Truth table of 2:1 MUX	565
Table 89.2	Comparative performance analysis	565
Table 89.3	Proposed multiplexer design	565
Table 89.4	Compare QCA and CMOS based multiplexer	566
Table 90.1	Performance evaluation	572
Table 91.1	Performance comparison metrics with various models	577
Table 93.1	Truth table of 1-bit RALU	587
Table 94.1	Factors utilized in the simulation process	592
Table 94.2	Reduction in power consumption CMOS logic	592
Table 94.3	Power delay (Watt-sec) product of half subtractor	593
Table 94.4	Voltage vs. half-subtractor time delay (sec)	593

Table 95.1	Uniqueness of projected algorithm preparation on solitary CPU	598
Table 98.1	Number of tourist visits in Banyuwangi	609
Table 98.2	Population of Blimbingsari district	610
Table 98.3	Mapping of creative economy sub-sectors in Blimbingsari district	611
Table 98.4	Distribution of tourist attractions in Blimbingsari district	612
Table 99.1	Summary of reliability test	615
Table 99.2	Demographic composition of the survey participants	616
Table 99.3	Correlation analysis	616
Table 99.4	Model summary	617
Table 99.5	ANOVA	617
Table 99.6	Coefficients	617
Table 99.7	Summary of PROCESS model 1	617
Table 99.8	Summary of PROCESS model 2	617

Foreword

The book, “Emerging Technologies in Sustainable Innovation, Management and Development”, serves as the proceedings of the International Conference on Emerging Technologies for Sustainable Innovation, Management and Development (ICETSIMD, 2025), held on March 27th, 2025, in Naya Raipur, India. ICETSIMD 2025 was a vital platform for academics, researchers, and industry leaders to explore emerging technologies’ transformative potential for sustainable development. Focused on interdisciplinary collaboration and showcasing advancements in areas like AI, IoT, and green technologies, the conference addressed subject areas including Computer Science, Management, and Environmental Science. All submissions underwent a rigorous double-blind peer review process to ensure quality and relevance. This handbook compiles high-quality, impactful research, aiming to inspire impactful change and sustainable practices. Edited under the leadership of Prof. Dr. R. Udayakumar, we trust this resource will be valuable for those seeking insights at the intersection of technology and sustainability.

About the Author

Dr. R. Udayakumar is a Professor and the Director of Intellectual Property Rights at Kalinga University in Naya Raipur, Chhattisgarh. He earned his Ph.D. in 2011 and has over 25 years of experience in teaching and research., as well as Dean of the Computer Science and Information Technology department.

Dr. Udayakumar has supervised five Ph.D. scholars and has published over 261 peer-reviewed articles indexed in Scopus, Web of Science, and SCI. His Google Scholar citations exceed 5,500, with an h-index of 45 and an i10 index of 99. His Scopus citation index is 3,193, with an h-index of 28. He has also been awarded 10 patents and has organised seven international conferences and 20 national conferences. Furthermore, he was recognised as one of India's top researchers in multidisciplinary fields from 2011 to 2013 by the Career 360 survey, published on March 27, 2014. Stanford recognised him among the world's Top 2% scientist. He attained a 9/10 score in the Vidwan Database (GOI).

E-mail: rsukumar2007@gmail.com, directoripr@kalingauniversity.ac.in
Contact: +91-9940159277 / +91-9789994242

Preface

ICETSIMD 2025 was conceived as a vital platform for academicians, researchers, and industry leaders. Its primary purpose was to explore the transformative potential of emerging technologies in driving sustainable development.

To ensure the high quality, originality, and relevance of the contributions, all submissions to the ICETSIMD 2025 conference underwent a rigorous double-blind peer review process.We received 300 manuscript submissions, from which 100 were selected for inclusion in the conference proceedings. Each manuscript was evaluated by at least two independent experts based on criteria including scientific rigor, methodological soundness, clarity, and contribution to the conference themes.

Designed for academics, industry professionals, students, and NGOs, the conference, and consequently these proceedings, aim to promote networking and knowledge-sharing to inspire impactful change and sustainable practices. The contributions herein represent a meaningful effort towards shaping a sustainable future.

We acknowledge the invaluable contributions of the members of the Programme Committee and the diligent reviewers whose expertise was crucial to the peer review process.

We believe this handbook will serve as a valuable resource for anyone interested in the intersection of emerging technologies and sustainable development, offering insights and solutions for the challenges facing our world today and in the future.

Acknowledgements

The creation of this handbook, serving as the proceedings of ICETSIMD 2025, is a collaborative achievement. We express sincere gratitude to all the authors for their valuable, high-quality research that forms the core of this volume. Special thanks are extended to the members of the Programme Committee for their guidance. We are particularly indebted to the dedicated reviewers whose rigorous double-blind peer review process was crucial in ensuring the quality and relevance of the included manuscripts, upholding academic integrity. We also acknowledge Kalinga University, Naya Raipur, India, associated with the lead editor and the conference location.

1 Current development status of Jizu mountain scenic area under the background of rural revitalization

Linxing L. I.[1,a], *Chuanchen B. I.*[1,b], *Long Yang*[1,c], *and Wei Yang*[2,d]

[1]College of Graduate Study in Management, Khon Kaen University, Thailand
[2]International College, Khon Kaen University, Thailand

Abstract: In recent years, the urban tourism market has become saturated, and the cultural tourism market has gradually shifted towards county-level and rural areas. Rural tourism, on the other hand, is booming as a result of the Rural Revitalization programme. A major contributor to rural revitalization and local economic development in Dali, Yunnan, The Jizu Mountain Scenic Area is a popular tourist destination. In the wave of rural tourism development, actively exploring sustainable development paths is essential. Jizu Mountain Scenic Area has successfully attracted a large number of tourists by combining various promotional methods. From 2023, the number of tourists received will increase significantly year-on-year. However, to stand out in the preliminary list of national 5A tourist attractions, the scenic area still faces numerous challenges. The scenic area is analyzed using the PEST framework to determine how the macroenvironment affects it, and the SWOT analysis is used to evaluate the region's internal and external possibilities and threats. By carefully examining the present state of affairs and the obstacles encountered by Jizu Mountain Scenic Area, the aim is to enhance the scenic area's market competitiveness, promote its sustainable development, and contribute to the rural revitalization process of the Jizu Mountain region.

Keywords: Rural revitalization, Jizu mountain scenic area, sustainable development, current development status

1. Introduction

Unprecedented difficulties have been encountered by rural regions as a result of the fast growth of cities and the expansion of the global economy. Hindi translation arises from the phrase Outflow of the rural population, depletion of agricultural resources, along with its infrastructure lagging describes the urban rural gap leading to set-back in the potential pace for economical and social development in rural areas. In 2017, at the 19th Communist Party Congress, the 'Rural Revitalization Strategy' was initially presented in China [1]. It has strategy primarily aims to enhance agricultural productivity, deepen diversification, reinforcing a income-based economy, while placing equal focus on rural region reconstruction [2].

A new era in rural China's social and economic life is dawning with the proposal of the Rural Revitalization Strategy. The extensive reform processes accompanying the economic growth of China has strengthen the country's overall national power, but deep-seated economic disparities are still persistent in remote agrarian areas. This disparity between urban and rural communities is especially noticeable in a developing region like the one being discussed. Qiao [3] highlights that greater portions of the agricultural labor force is migrating from these rural regions to urban settings, resulting in insufficient productivity per capita as well as a pronounced void in social frameworks alongside chronic poverty afflicting rural populations. Rural Revitalization Strategy is an answer to this problem. It addresses agricultural issues from a policy perspective, making it a national concern, especially given the status of China's modernization program. This developmental strategy offers systematic mechanisms to revitalize the rural frameworks and simultaneously promote advanced living standards. Its core objectives revolved around enhanced industrial productivity, improved healthcare systems, effective governance, and ecological sustainability alongside cultural reconstruction, focusing on an all-encompassing solution to tackle the widespread structural underlining problems emerging from rural China.

This is not only a national policy: it is also a significant step toward solving problems such as the urban-rural development imbalance and the modernization lag in agriculture [4]. The context of this proposal is that the socio-economic development of rural areas has, after undergoing a phase of speedy urban industrial growth, been comparatively stagnant, marked by low income and unsatisfactory living standards among farmers. To holistically address the enduring challenges in agriculture, rural development, and the socio-economic conditions of the farmers. These challenges

[a]linxing.l@kkumail.com, [b]chuabi@kku.ac.th, [c]longya@kku.ac.th, [d]weiya@kku.ac.th

DOI: 10.1201/9781003684459-1

are brought to the forefront as part of the rural revitalization policy [5].

Rural tourism is an integral part of the Rural Revitalization Strategy, serving to preserve cultural heritage while also protecting the environment and fostering economic development in rural areas. Over the past few years, rural tourism has grown into a major force in bridging the gap between urban and rural areas, which in turn has helped boost infrastructure, employment, and farmers' earnings. Through rural tourism, natural and cultural resources in rural areas are more fully developed and utilized, further fostering comprehensive rural development.

Rural tourism not only provides farmers with diversified sources of income, but also drives local cultural and ecological preservation efforts. Many rural areas, relying on their natural environment, traditional culture, and agricultural resources, have developed unique tourism projects that attract many tourists who are eager to experience rural life. The rural economy has grown, and urban-rural exchanges and the transmission of cultural legacy have been promoted. Especially in a country like China, with a long history and abundant natural resources, Tourism in rural areas is growing, which is good for business and in line with the plan to bring back life to rural areas across the country. The rise of rural tourism has become an important part of revitalizing rural areas and getting their economies back on track in many of them. On the one hand, the growth of the tourism industry has helped the growth of related industries, such as those that sell farm goods. Handicraft production, and agricultural experience projects [6]. On the other hand, the construction of tourism infrastructure has improved rural transportation, accommodation, and other conditions, raising the living standards of rural residents. Furthermore, this tourism has also facilitated environmental improvements in rural areas, leading to the protection and restoration of natural landscapes and ecosystems in many regions [7].

Jizu Mountain, located in Binchuan County, Yunnan Province, China, is a renowned Buddhist Mountain. Its rugged terrain resembles the shape of a chicken's foot, leading to it being named 'Jizu Mountain'. Not only famous for its unique geographical features, the Mountain is also known for its profound Buddhist cultural heritage, making it one of China's sacred Buddhist sites. As a national scenic area, there are many resources that are both natural and provide a cultural heritage. In addition to its Buddhist culture, the scenic area preserves many rare species of plants, animals and unique ecosystems. Its exceptional geographical location and rich cultural history make it an ideal site for rural tourism development. Driven by the Rural Revitalization Strategy, the Jizu Mountain Scenic Area not only has the potential not only help boost local economies, but they can also be crucial in preserving cultural heritage and attracting ecotourists. Jizu Mountain has far-reaching effects on the local economy in Yunnan Province due to its importance as a tourist attraction. Nevertheless, there are a lot of obstacles and a lot of potential in its development. As the tourism industry grows, more and more people are flocking to the picturesque location, putting a strain on its infrastructure, service quality, and efforts to preserve the environment. Therefore, finding ways to promote the sustainable development of the scenic area while maintaining its ecological and cultural resources has become an urgent issue that needs to be addressed.

So as to deeply analyze the status of the Jizu Mountain Scenic Area, this paper adopted the PEST analysis method to evaluate the macro-environmental factors influencing the area. Additionally In order to better understand the difficulties Jizu Mountain encounters in all of its development and to provide theoretical support in addressing these issues. The significance of this research lies in providing a theoretical foundation, Give advice on how to make the Jizu Mountain Scenic Area a more sustainable place to visit. The study finds the area's development potential and obstacles, investigates marketing approaches that work for Jizu Mountain, and more by examining the area's internal and external contexts and provides new practical cases for enriching sustainable marketing theory. In the context of increasingly intense competition in the tourism market, sustainable marketing strategies are crucial for enhancing the market competitiveness of scenic areas. By exploring scientifically effective sustainable marketing strategies, the Jizu Mountain Scenic Area can further enhance its brand image and market appeal, strengthening its competitiveness in both domestic and international tourism markets. Better meeting the diverse needs of tourists, more satisfied tourists, healthy local tourism development, higher incomes and better quality of life for residents, integration between urban and rural areas, less urban-rural gap, and strong support for rural revitalization can all be achieved with this approach.

2. Literature Review

2.1. Rural revitalization

As much as the terminology 'rural revitalization' fresh phrase appears in studies revolving around rural development, a definition or its scope and usage is not agreed upon. Implementing the rural revitalization strategy – one of the policy 'epoch strategies' passed during China's 19th National Congress of the Communist Party of China in 2017, is a multi-faceted strategic framework aimed at the comprehensive advancement of rural areas economically, socially, culturally, and ecologically. The ultimate objectives are rural prosperity, urban-rural integration, and sustainable development, which are all considered subordinate to the achievement of farmers' living standards. Arguably, this is what is meant by 'China's ballast stone' in terms of countering globalization. This was one of the flagship tasks on the agenda of the 'rural development' as of the 19th National Congress, and it draws upon rich theoretical and practical contributions to rural building. The rural revitalization strategy serves as a locomotive for modernization in China, hosting the approach to the 'three rural issues' of the new era: agriculture, rural areas, and farmers.

Its primary goals include: 'prosperous industries, liveable ecological environments, civilized rural customs, effective governance, and affluent living standards'. To 'Accelerate the Modernization of Agriculture and Rural Areas', the guiding concept aims to 'give priority to the development of agriculture and rural areas', thereby addressing the problem of insufficient development momentum in the countryside [8]. In order to achieve the comprehensive development and modernization of rural areas, the concept of rural revitalization encompasses the economic, social, cultural, ecological, and governance dimensions. From the economic side, it underlines the importance of business diversification and the development of modern agriculture as a means to increase the income of farmers. For social reasons, it focuses on improving the lives of farmers by enhancing the quality of public services and infrastructure. The cultural objectives of the strategy focus on the advancement of rural civilization and preservation of exceptional traditional culture. Environmentally, it advocates for sustainable development to provide a healthy balance between economic and environmental benefits. On the rural governance side, it emphasizes the need to strengthen grassroots organizational construction. To close the gap in development between urban and rural areas, comprehensive policies that stimulate growth in rural areas need to be put in place [9].

2.2. *Rural tourism*

As part of plans to develop local economies and increase farmers' incomes, rural revitalization strategies also include rural tourism\/host watching as an important component. All things considered, it's an advantage from all perspectives because it helps the dual purpose of development urban and rural interconnections, as well as expands the scope of tourism, bringing attention as well as exploring unique values of rural natural picturesque, cultural heritages and traditions to urban consumers for experience consumption. Rural tourism has become an incredible phenomenon now that revitalization programmes for the regions are incessantly advertised and the culture and tourism industry is flourishing. At this time no definition is available for the term 'rural tourism.' The European Union and the Organization for Economic Cooperation and Development (OECD) define tourism to rural areas as 'rurality' which is the essential element of rural tourism [10]. Chinese scholars have provided a number of interpretations for the concept. To some, the essence of rural tourism lies in landscapes, agricultural activities and other works around the periphery of the city. Some defined it as a rural centred tourist location which offers unique rural culture, methods of rural work and scenic views. It integrates sightseeing, shopping, entertainment, leisure, and vacationing by capitalizing on the interurban contrasts in planning and design [11]. Due to the multitude of foci, viewpoints, and real-world settings, academics systematically classify the concept of rural tourism under differing perceptions as diverse in focus as the tourism itself. Scholars have defined rural tourism into three concepts: sightseeing, folk, and leisure tourism. Moreover, there is occasionally a subdivision of rural tourism into modern agricultural tourism and ancient village tourism, with studies in these areas aiding the development of rural tourism as a whole. Rural tourism is increasingly gaining attention and popularity as it has turned out to be one of the most effective ways to transform the country side in China's plan to improve the economy, employment, and income of farmers. The number of domestic tourists is reported to have increased, along with the visits to rural areas, indicating that rural tourism is becoming more prominent in the development of rural economy, as per the statistics released by the Ministry of Culture and Tourism of China. An increase in rural tourism can facilitate the revival of industries and assist in the Environmental preservation and protection of cultural heritage all contribute positively to the rural revitalization plan.

2.3. *PEST analysis*

An organization or project can benefit from doing a PEST study to determine how the external macro-environment including political, economic, social, and technological factors will affect it. In the study of scenic spot development, PEST analysis helps to identify external macro-environmental factors that affect scenic spot development. For example, government policy support or restrictions, the level of economic development, social and cultural changes, and technological progress can have a direct, or indirect, impact on the development of scenic spots [12].

2.4. *SWOT analysis*

A company's strategic planning can be informed by doing a SWOT analysis, which involves taking stock of the company's internal and external opportunities and threats. In order to establish targeted strategic measures, researchers in the field of scenic area development often utilize SWOT analysis to evaluate the internal resources and external environment of a scenic region. For example, some scenic areas may have abundant natural and cultural resources, which represent their strengths; however, simultaneously, the lack of infrastructure or the increasing number of competitors could pose weaknesses, or threats, to their development [13].

3. Methods

The primary research subject of this study is the Jizu Mountain Scenic Area, a distinctive tourist destination in Yunnan Province. This study adopts a qualitative research approach, utilizing policy documents, statistical data, academic papers, and market research reports to gather background information. Additionally, field observations and interviews are conducted to collect relevant data on the development status of Jizu Mountain Scenic Area within the context of rural revitalization. The research participants encompass managers of scenic areas, local government officials, area residents, and tourists. This study makes use of multiple groups concerning the Jizu Mountain Scenic Area which helps in understanding the area from different perspectives.

Analyzes such as PEST and SWOT are employed in the investigation. The scenic region is examined in the scenic area PEST analysis, which takes into account the various factors. Moreover, the SWOT analysis identifies the internal resources and external environment concerning the scenic area classified into strengths, weaknesses, opportunities, and threats. By doing so, the study claims it provides fundamental understanding on the development status of Jizu Mountain Scenic Area which is useful for further studies conducted on the area.

4. Results

In this chapter, the development level of Jizu Mountain Scenic Area within the revitalized countryside is studied by first applying PEST analysis to the scenic area's external environment which includes policy directives, economic indicators, shifts in societal demand, and technology consumption. Still, the effects of digital technology deployment, tourism market systems, and rural revitalization strategies on the area's growth continue to be the centre of attention. Afterwards, the SWOT analysis technique is applied to Jizu Mountain's Scenic Area to evaluate its internal assets and constraints. This includes an assessment of its inherent strengths and weaknesses in natural resources, service provision, infrastructural development, managerial practices, financial investment, scenic area competition, and regional rivalry. Moreover, incorporating the external opportunities from burgeoning tourism and the conservation challenges, this chapter addresses the sustainable development strategy for Jizu Mountain's Scenic Area within the context of rural revitalization.

At the same time, the chapter combines the external opportunities arising from the growth of the tourism market, and the pressures related to ecological environmental protection, in light of the Rural Revitalization Strategy, assess potential avenues for sustainable development and determine the area's long-term goals [14].

4.1. PEST analysis

A picturesque area's macroenvironment can be better understood by applying the PEST analysis. When analyzing the development status of Jizu Mountain Scenic Area, it helps us to comprehensively understand the key factors that impact the region's development. The analysis includes four main aspects – political, economic, social, and technological, each of which plays a significant role in regional development. The following section will discuss the impact of each of these factors on the Jizu Mountain Scenic Area. Starting with the political environment, this section will analyze the influence of relevant regulations on the region and provide a foundation for the subsequent economic, social, and technological analyzes, aiming to provide a comprehensive understanding of the external development environment of Jizu Mountain Scenic Area [15].

4.1.1. Political environment

The political environment plays a decisive role in the development of the tourism industry, providing clear guidance and strong policy support for business growth. Jizu Mountain is in Binchuan County, Dali Prefecture, Yunnan Province, and is a 4A-level tourist attraction as well as a key area supported by the Rural Revitalization Strategy. While Yunnan Province is home to numerous 4A-level scenic spots, and Dali Prefecture has several attractions of the same grade, the status of Jizu Mountain within the province, or the prefecture, is not particularly prominent. However, Yunnan Province places great importance on the development of the tourism industry and actively promotes new tourism experiences, including rural tourism, cultural tourism, and outdoor adventure, so as to attract a diverse range of tourist consumers. Rural tourism is an important initiative for Yunnan's rural revitalization, and Jizu Mountain, as a key supported area in Dali Prefecture, benefits from policy support in various forms, including tourism funds, infrastructure construction, and maintenance provided by both the provincial and prefectural governments. This policy backing offers strong support for the sustainable development of the scenic area. Moreover, the Yunnan Provincial Government is actively promoting the integration of Jizu Mountain into the Greater Western Yunnan Tourism Circuit. Leveraging cultural and tourism brands such as 'There is a lifestyle called Dali' and organizing special events, the government further enhances Jizu Mountain's visibility both domestically and internationally, helping it play a significant role in rural revitalization.

4.1.2. Economic environment

Both the external limitations placed on the market and the overall macro scale of the market are determined by the economic environment, making it an essential component in the growth of the tourism industry. The economic conditions required for tourism development include the local level of economic development, the extent of the tertiary sector's growth, transportation infrastructure, and more. In 2023, China's tourist spending will mirror the market's meteoric rise and the demand for internal travel, both of which have begun to revive. The overall value of domestic tourist excursions in China reached 4.891 billion yuan in the first half of 2023, a 93.3% rise from the previous year. The demand for inter-provincial travel has also increased significantly. An increase of 140.3% from the previous year, domestic tourists spent 4.91 trillion yuan on their travels. Jizu Mountain is situated in Binchuan County as a relatively weak economic foundation, but in recent years, with increased government investment in tourism infrastructure, the pace of cultural and tourism project construction has accelerated. Improvements in transportation conditions and visitor facilities have gradually attracted many more tourists. In addition to boosting adjacent industries like food service and lodging, the expansion of the tourism industry has boosted local governments' coffers through increased spending and ticket sales. The growth of the rural economy is a two-way street; first, tourism has increased demand for agricultural goods and local crafts, which has raised wages and improved living conditions for locals.

4.1.3. Social environment

When considering how far tourism has come in terms of cultural and social acceptance, the social environment is an important factor. In a society with advanced communication technologies, the transmission of tourism culture within the social environment is constantly evolving. Under the impetus of the rural revitalization policy, residents in the surrounding areas of Jizu Mountain have gradually become involved in the tourism industry, enhancing local services such as accommodation, dining, and cultural experiences. As the demand for nature and cultural tourism continues to grow, Jizu Mountain has attracted many tourists due to its unique Buddhist culture and beautiful natural landscapes, which have spurred the development of eco-tourism and cultural projects. The government places high importance on cultural preservation, ensuring that tourism development does not compromise the integrity of traditional culture, while also promoting the overall improvement of infrastructure and services. Rural revitalization has also encouraged many young people to return to their hometowns to start businesses, fostering the rise of agritourism, handicraft production, and other unique industries. This contributes to the long-term growth of the community's economy and way of life while also improving the experience for tourists. Additionally, since 2023, 'reverse tourism' has become a popular trend among young people, with more opting for less popular destinations to enjoy a more relaxed form of travel. This trend has contributed to the development of tourism in second and third-tier cities [16].

4.1.4. Technological environment

When it comes to marketing and developing tourist goods, the technology landscape is essential, since it determines the products' potential future markets and directions. In April 2024, the 'Smart Tourism Innovation and Development Action Plan' was issued by the Yunnan Province Department of Culture and Tourism. This plan is in accordance with the '14th Five-Year Plan for Tourism Industry Development' and 'Measures for Promoting the High-Quality Development of Tourism Consumption,' as well as the spirit of the 20th National Congress of the Communist Party of China, all of which seek to integrate the digital economy with tourism. Jizu Mountain has been progressively integrating digital management tools to carry out this strategy, building a brand that marries rural tourism with the region's distinctive agricultural cultural history. The site has introduced platforms like online booking, self-service ticketing, smart guides, and the 'One Phone to Tour Yunnan' system, which not only enhances visitor experience but also improves management efficiency. The introduction of 5G networks and smart tourism services has further supported the development of the scenic area. Additionally, to ensure the sustainability of ecotourism, the scenic area has implemented 'smart toilets,' ecological monitoring, and other environmental technologies to effectively protect the ecological environment amid increasing tourist numbers, thereby reinforcing its image as a green, eco-friendly destination.

4.1.5. Conclusion

Through the PEST analysis, a comprehensive understanding of the external macro environment of Jizu Mountain Scenic Area, in the context of rural revitalization, can be gained. From a political perspective, under the overarching framework of rural revitalization, the government has strongly supported the development of Jizu Mountain, helping it become a model for regional and rural tourism. Economically, as China's cultural and tourism market continues to shift, the tourism revenue and local economy of Jizu Mountain have been boosted. On the social level, a significant number of residents, and young people returning to their hometowns, have actively participated in the development of the regional tourism industry, enriching the services offered by the scenic area. In recent years, the significant improvement in digitalization has provided a solid foundation for the smart development of the scenic area, ensuring an enhanced visitor experience. These factors together lay a solid foundation for the sustainable development of Jizu Mountain Scenic Area.

4.2. SWOT analysis

After analyzing the external environment of Jizu Mountain Scenic Area, this study employs SWOT analysis to assess its internal strengths and weaknesses, as well as external opportunities and threats. This thorough assessment seeks to elucidate the current level of developmental progress of Jizu Mountain Scenic Area, along with possible issues it may encounter. The subsequent parts will examine the scenic area's strengths, weaknesses, opportunities, and threats individually for informed planning and resource allocation – strategic foresight and ecological engineering [17].

4.3. Strengths

Jizu Mountain Scenic Area is in Dali Prefecture, Yunan Province, and forms part of the Grand Western Yunnan Tourism Loop. The area is home to a large number of ancient temples and Buddhist relics which serves as an important site for religious cultural tourism and pilgrimage. Jizu Mountain is an ecotourism site with rich biodiversity noted for its striking natural beauty, which lends it a distinct advantage in both the ecotourism and cultural tourism markets. The scenic area has adopted a 'tourism + sports' zoning approach, utilizing its geo-sports resources to host outdoor events such as trail running and downhill cycling competitions, building further notoriety amongst sports enthusiasts.

Under the support of the rural revitalization policy, Jizu Mountain Scenic Area has received government funding for infrastructure development and cultural heritage preservation. In recent years, the area's transportation conditions, and visitor service facilities have significantly improved. Both the efficiency of scenic region management and the quality of the visitor experience have been substantially improved by the steady implementation of smart tourism technologies. Management, boosting its competitive edge in the market. In 2023, the scenic area welcomed 732,000 visitors, a 43.18% year-on-year increase, and its revenue surpassed 100 million yuan, growing by 49.12%. With the continued implementation of promotional measures, during the first day of the 2024 May Day holiday, the scenic area welcomed 5,899 visitors, marking a 58.07% year-on-year increase and setting a record, showcasing the immense potential of the Jizu Mountain Scenic Area.

4.3.1. Weaknesses

While Jizu Mountain Scenic Area's marketing strategy has achieved some success, its share of visitor numbers and revenue remains relatively low within Dali Prefecture, accounting for only 0.768% and 0.062%, respectively. Several locations that could want some attention have been highlighted by the development of the picturesque region. The region has improved in general because to the government's backing, but it still has some ways to go before it can be considered a mature 4A or 5A picturesque area. For instance, public transportation to Jizu Mountain is relatively inconvenient, particularly for visitors without private vehicles, making it difficult to access the scenic area. Additionally, transportation within the area, such as cable cars or shuttle buses, sometimes fails to meet the demand required during peak tourist seasons. Furthermore, during times of high visitor traffic, waste management and sanitation maintenance often fall behind, leading to sub-optimal environmental conditions in some parts of the area, which negatively impacts the overall visitor experience. In terms of infrastructure, the road conditions leading to Jizu Mountain are poor, particularly the mountain roads, which are steep and have uneven surfaces, thus diminishing the quality of the tourism experience. Additionally, the area may lack sufficient resting facilities and public toilets. On the management front, promotional activities during holidays are often concentrated, with clear distinctions between the peak and off-peak tourist seasons. There is also a lack of unified pricing management for internal vendors, leading to unreasonable pricing in food, shopping, and other services. This results in higher costs for visitors, impacting their satisfaction. These issues not only affect the tourist experience but also constrain the further development of the scenic area. Compared to other well-known tourist destinations in Yunnan, such as Lijiang and Dali, Jizu Mountain has a lower market profile. Limited brand promotion has hindered its ability to attract visitors and penetrate the market, affecting its long-term visitor numbers and market influence [18].

4.3.2. Opportunities

Increased investment in resources and robust policy backing are only two of the many benefits that the Jizu Mountain Scenic Area has reaped from the rural revitalization strategy's thorough execution. These factors can facilitate the improvement of infrastructure and the preservation of cultural resources. With the growing demand for ecotourism and cultural tourism among modern travelers, Jizu Mountain has the opportunity to develop more specialized tourism projects, such as religious cultural experiences and ecological hiking trails, to cater to niche markets while attracting a broader visitor base. Moreover, the progression of digital marketing offers new opportunities for promotion. Jizu Mountain can improve its marketing and draw more potential tourists through short video platforms and social media. In addition, the emergence of smart tourism allows for greater optimization in management and services provided. The implementation of 5G coverage, smart guidance, and online reservation systems can improve the experience for visitors and also the management efficiency of scenic areas.

4.4. Threats

Compared to other well-known tourist attractions in Yunnan Province, like Dali Ancient Town and Lijiang Ancient Town, Jizu Mountain Scenic Area faces stiff competition. These areas have more developed tourism infrastructures, sophisticated marketing techniques, and advanced promotional strategies which make it difficult for Jizu Mountain to stand out and capture the interest of tourists. Furthermore, the growing number of tourists also increases the ecological strain on the site. In order to prevent deterioration of the environment, achieve sustainable development, and protect biodiversity, Jizu Mountain will have to find a way to balance visitor capacity and natural resource preservation.

4.4.1. Summary

Strength (S)	1. Strategically located along the distinctive Grand Western Yunnan Tourist Loop, offering a prime geographical advantage. 2. Boasts unique natural landscapes and a will-preserved ecological environment. 3. Rich in cultural heritage, with a significant influence of Buddhist traditions. 4. Natural topographical conditions support the implementation of a "tourism + sports" strategy. 5. Enhanced infrastructure and the integration of smart tourism technologies have significantly improved scenic area management.

Weakness (W)	1. The visitor reception volume and income share in Kali Prefecture are relatively low. 2. Public transportation is inconvenient, and internal transportation arrangements within the scenic area are insufficient. 3. The maintenance and management of sanitation and hygiene require improvement. 4. Infrastructure is underdeveloped, with poor road conditions. 5. Promotional efforts are inadequate, and marketing methods are relatively simplistic.
Opportunities (O)	1. The rural revitalization strategy provides policy support and resource investment. 2. The growing demand for eco-tourism and cultural tourism presents significant market potential. 3. The development of short video platforms and other social media facilitates online promotion and publicity for the scenic area. 4. The advancement of digital technologies helps improve the service level of the scenic area. 5. The trend of regional tourism integration brings more development opportunities for the scenic area.
Threats (T)	1. The traditional tourism model is insufficient to meet the demands of visitors. 2. Intense competition from surrounding scenic areas leads to noticeable diversion of tourists. 3. Increased demands for ecological environmental protection exacerbate the conflict between development and conservation. 4. Rising expectations from tourists regarding scenic area services make it difficult to ensure consistent service quality.

4.5. Research summary

According to the study's findings, rural revitalization strategies help rural regions' economies grow over time, which opens up new possibilities for the Jizu Mountain Scenic Area's development. Politically, the scenic area benefits from government policies promoting tourism development and relevant incentives. Economically, the growth of the Dali region's economy creates more possibilities for investment in Jizu Mountain Scenic Area. From a social perspective, the increasing demand for eco-tourism and cultural tourism shows significant potential for market growth. Technologically, advancements in digital technology will bring more opportunities for the development of Jizu Mountain, helping to improve its management and service quality. As a key component in the local economic development, Jizu Mountain's unique Buddhist culture and natural conditions are advantageous for its growth. However, the weaknesses in the area of infrastructure cannot be ignored. In an increasingly competitive environment, Jizu Mountain should concentrate on building and reinforcing a locally unique unique brand while sharpening its cultural appeal and multicultural market value. In the development process, maintaining a balance with strong ecological preservation focus is crucial; in other words, there should be no contradiction between the economic development and the environment, as both must be upheld in balance along with ecology.

5. Conclusion

5.1. Main conclusions

The development of Jizu Mountain Scenic Area benefits from its favorable natural conditions and rich cultural resources, as well as national policy support and continuous technological advancements. The national government encourages the development of rural economies through tourism, and Yunnan Province has actively responded by developing the region's distinctive Great Western Yunnan Circular Route. In response to new tourism development trends, the province has focused on tourism product, model, and business innovation, aiming for the sustainable development of its tourism industry. The attention given to tourism by Yunnan Province provides Jizu Mountain Scenic Area with significant opportunities. However, as society evolves, visitor expectations continue to rise, and the scenic area faces increasing competition from other popular destinations as well as the pressure of ecological preservation. Overall, while Jizu Mountain Scenic Area has potential under the support of rural revitalization policies, there is still significant room for improvement in achieving sustainable development.

5.2. Future research directions

New possibilities for rural tourism development are emerging as a result of the rural revitalization strategy's continued implementation. Not only does rural tourism impact the growth of the tourism industry as a whole, but it is also essential to the success of rural revitalization initiatives. Since tourism is a cornerstone of Yunnan Province's economy, studying how it interacts with other sectors and developing new types of tourism products should be high priorities for academics in the years to come. Making good use of social media for promotional activities and efficiently converting potential customers will also be a crucial way to improve the economic growth of scenic locations, especially with the increasing importance of internet marketing.

References

[1] Patil, S., & Dayananda, T. B. (2016). Soft switching DC-DC converter based on a phase shift controlled active boost rectifier

using fuzzy controller. *International Journal of Advances in Engineering and Emerging Technology*, *7*(4), 01–12.

[2] Akgün, M. H., & Ergün, N. (2023). Parameters response of salt-silicon interactions in wheat. *Natural and Engineering Sciences*, *8*(1), 31–37. http://doi.org/10.28978/nesciences.1278076

[3] Qiao, J. (2008). *On the Regional Economy of Rural Areas in China*. Beijing, Science Press.

[4] Huang, Z. (2018). *Accurately grasp China's rural revitalization strategy*. China's rural economy.

[5] Priyalatha S. (2024). Effect of state of fabric on wicking characteristics of knitted fabrics suitable for sports applications. *Archives for Technical Sciences*, *2*(31), 1–8. https://doi.org/10.70102/afts.2024.1631.001

[6] Martinez, R., & Garcia, C. (2024). Integrated systems design: A holistic approach to mechanical engineering. *Association Journal of Interdisciplinary Technics in Engineering Mechanics*, *2*(4), 12–16.

[7] Nahavandi, R., Khezri, M., Rabiei, S., & Altan, Ö. (2024). Extending the shelf life of Artemia urmiana during frozen storage using Vitamin E treatment. *International Journal of Aquatic Research and Environmental Studies*, *4*(1), 101–113. http://doi.org/10.70102/IJARES/V4I1/9

[8] Han, D. (2018). *Research on the Development Model of Characteristic Town in Shaanxi Province under the Background of Rural Revitalization Strategy.* Xi'an University of Architecture and Technology.

[9] Zhang, J. (2018). *Rural Value Positioning and Rural Revitalization*. China's rural economy.

[10] Xu, X. (2024). *Evaluation of the Ecological Value of Rural Tourism Land Resources in the Context of Agricultural and Tourism Integration: A Case Study of Sanya Damao Ocean Ecological Village*. Tourism Management, Hainan Tropical Ocean University.

[11] Liu, W. (2024). *Research on the Problems and Countermeasures of Rural Tourism Development in Hebei Province*. Rural Development, Hebei Normal University of Science & Technology.

[12] Johnson, S. W. (2008). *Exploring Corporate Strategy.* University of Strathclyde.

[13] Sammut-Bonnici, T., & Galea, D. (2015). SWOT Analysis. *Wiley Encyclopedia of Management*, 1–8.

[14] Tian, T. (2009). *A Study on the Sustainable Development of Scenic Spot in Mountain—Jizu Shan Scenic Spot in Dali YunNan an Example*. Urban Planning and Design, Beijing Forestry University.

[15] LI, Y. (2023). *Research on Marketing Strategy of Scenic Area A*. Business Administration, Hebei GEO University.

[16] Wu, D. (2024). *Re-discussion on Reverse Tourism*. Journal of Northwest Minzu University.

[17] Wang, J. (2007). *The Diagnosis and Plans of Traveling Marketing about the Scenery Scenic Spot Area Based in the 4PCs Theory—take the Mt. Songshan scenery scenic spot area as the example*. Business Administration, Guangxi University.

[18] Chen, X. (2012). *Research on the Development of Buddhist Tourism in Jizu Mountain Scenic Area, Yunnan Province*. Geography, Shaanxi Normal University.

2 Preference towards electric vehicles in Coimbatore district

J. Smruthymol[1,a] *and R. Velmurugan*[2,b]

[1]Karpagam Academy of Higher Education, Coimbatore, India
[2]Associate Professor and Head, Karpagam academy of Higher Education, Coimbatore, India

Abstract: This study explores the preferences towards Electric Vehicles (EVs) in Coimbatore District, focusing on factors influencing consumer choices and adoption rates. A survey-based approach was employed, gathering primary data from 275 EV users using a structured questionnaire. The study analyzes factors such as reduced operating costs, government incentives, battery life, maintenance costs, comfort, technological advancements, traffic congestion, model availability, range anxiety, vehicle reliability, design preferences, convenience of home charging, city sustainability goals, regulatory support, charging infrastructure, environmental concerns, future outlook, horsepower, user-friendliness, vehicle price, brand image, and safety features. Data were analyzed using descriptive statistics and factor analysis to identify significant determinants influencing EV preferences. The research data shows operating cost reductions together with government benefits, technological advancement and environmental awareness as primary components that motivate EV selection. The research findings present policy measures which target cost improvement and infrastructure development together with environmental sustainability to maximize EV adoption acceptance in the Coimbatore district through market promotion initiatives.

Keywords: Consumer preferences, electric vehicles, sustainable transportation

1. Introduction

Modern technological developments have boosted Electric Vehicle (EV) popularity due to combined economic growth and environmental sustainability challenges. Traditional internal combustion engine cars generate major air pollution and emit carbon pollutions which results in environmental degradation and adverse health impacts on people while also producing global warming effects according to the European Environment Agency [1]. The immediate necessity for climate change combat along with reducing greenhouse gas emissions drives this development. EVs produce zero carbon emissions which makes them more environmentally friendly since they can diminish the total carbon output of the transportation sector [2].

Environmental factors definitely require the adoption of EVs. Internal combustion engines (ICEs) produce harmful emissions of nitrogen oxides (NOx) and particulate particles (PM) that trigger respiratory problems and smog development. Gas emissions from internal combustion engines will decrease through increasing EV adoption thus improving public health together with air quality. The environmental advantages of electric vehicles will increase because renewable energy power keeps expanding throughout the electrical grid system [3].

New research demonstrates that ICE car air pollution causes serious health problems to people. The inhalation of PM alongside NOx acts as an environmental trigger for multiple respiratory ailments that lead to premature death. Road transport in India stands as a major source of urban air pollution according to Indian Environment Agency [1] because it creates 12% of the total energy-related CO2 emissions. By transitioning to electric cars, cities may reduce harmful pollution and promote healthier living circumstances. As renewable vitality sources such as wind and solar power become more prevalent, EVs' environmental advantages will grow. Charging electric vehicles using renewable-powered electricity reduces their overall carbon impact. According to the International Energy Agency [3], as the amount of renewable energy in the energy mix grows, EV lifetime emissions will fall, making them an even more sustainable alternative than internal combustion engines (ICE cars). The need to protect the environment is reinforced by developments in battery technology, which have significantly increased the cost, efficiency, and range of electric cars. EVs are becoming increasingly appealing as a wider range of people can now afford them, thanks to declining battery costs and rising energy density [4]. One of the foremost worries of prospective EV customers is range anxiety, but the development of solid-state batteries and other cutting-edge storage technologies promises even greater gains in energy capacity and charging speed. Solid-state batteries outperform lithium-ion batteries in terms of energy capacity and also charge faster. EV batteries using this technology can travel longer distances without charging while charging durations decrease thus enabling more successful long trips.

[a]smruthymol.jayalal@kahedu.edu.in, [b]drvelsngm@gmail.com

DOI: 10.1201/9781003684459-2

The development of technologies represents an essential element that drives customers toward electric vehicles. The quick development of charging infrastructure solved one major concern of potential EV buyers known as range anxiety. The increasing number of fast-charging stations enables EV drivers to charge their vehicles anytime they need while traveling long distances according to Hardman et al. [5]. Beverly Anderson and others demonstrate that autonomous driving technologies merge perfectly with electric vehicles thus revealing how transportation systems will unite convenience and safety with sustainability [6].

The main financial gains from electric vehicles second only to environmental benefits. The natural efficiency of electric motors stems from minimal vehicle components and rapid battery-powered vehicle acceleration [7]. The power efficiency of electric vehicles allows their owners to reduced their energy costs. Electric vehicles provide better ownership affordability than regular vehicles because they need reduced fuel and maintenance costs according to Palmer et al. [8]. Electric vehicles need no oil maintenance since they utilize regenerative braking which extends their braking system performance. EVs present attractive long-term savings to consumers through their efficient features which make them attractive financial investments according to Bonges and Lusk [9].

The EV market expansion creates employment opportunities and triggers the development of three new industries which include battery manufacturing and production of electric vehicles and electric charging infrastructure. The economic transition works as a base for technological transformation which leads to lasting economic development. The adoption of EVs receives support from government incentives which include tighter emission regulations and financial benefits such as tax credits and grants [10]. These policies promote environmentally friendly transportation systems. Governments that support the EV industry through innovation promotion and green technology job creation set up a framework for sustainability alongside environmental cleanliness [11].

Current United States tax regulations allow electric vehicle buyers to receive up to $7,500 in credits from the federal government even though European Union members aim to reduce automobile CO_2 emissions by 37.5% before 2030 reaches 2021 benchmark levels. China leads global EV sales and compels vehicle manufacturers to attain specific EV sales targets as well as provides substantial financial support. EVs create significant changes in both global political systems and energy protection systems. A decreased dependence on oil creates national energy independence and shields countries from oil price volatility and raw material disruption patterns.

The EV revolution has the ability to reshape geopolitics by reducing oil-producing areas' strategic value while elevating nations having control of battery component elements including nickel, cobalt along with lithium. A prevention of new geopolitical conflicts in the oil sector requires sustainable and ethical supply chain development. The EV industry expansion creates new employment opportunities that foster industries for battery manufacturing and charging station development and EV vehicle production. This change in the economy fosters both sustainable economic growth and technical innovation. Consumers currently perceive EVs differently from how they used to several years ago. People are becoming more confident about EVs after positive experiences combined with the increasing numbers of vehicles seen on the roads. The market acceptance of electric vehicles will expand due to better perception of their performance qualities and lower operating costs. Various individuals choose EVs as their preferred solution because environmental awareness and sustainability-focused cultural changes are influencing market demand.

In the early phases of EV adoption, many buyers were cautious owing to worries regarding range, charging infrastructure, and overall car performance. However, as the technology has progressed, these problems have been mostly resolved. The range of contemporary EVs is on par with that of conventional cars, and the growing number of charging stations guarantees that drivers will always have easy access to electricity, even when traveling long distances. Many doubters have also been persuaded by the better performance of EVs, which are distinguished by rapid acceleration and a calm, comfortable ride. A network effect occurs when more customers talk about their great experiences driving electric cars, which speeds up the adoption of these cars.

It appears that electric vehicles will be the predominant mode of transportation in the prospect. This can be attributed to a number of factors, including the ongoing development of battery technology, the expansion of the charging infrastructure, and the encouragement of government policies. EVs will maintain their status as cornerstone sustainable transportation devices because renewable energy integration with the power grid strengthens their environmental benefits. The adoption of autonomous driving systems links directly to EV popularity because they will establish a new mobility paradigm focused on protecting people while also delivering maximum operational efficiency.

Transportation advances significantly because EVs have the opportunity to integrate with autonomous driving capabilities. The combination of autonomous vehicles and electric power trains generates triple improvements through higher efficiency and better traffic flow as well as reduced emissions which results in maximum energy efficiency. Such a merger between electric-powered vehicles and autonomous technology presents the opportunity to restructure whole public transportation systems. The fusion between electric buses and autonomous vehicle technology would establish self-driving electric transportation systems which give environmentally sustainable commuting options to city dwellers.

The V2G (Vehicle-to-Grid) technology shows great promise as a future advancement. Through V2G technology electric vehicles can deliver energy back to the grid which operates as power storage for unstable renewable energy systems such as solar and wind power. The ability to supply additional energy to the grid results in improved power system steadiness while maximizing solar and wind power resources plus establishing a supplemental revenue stream for EV owners. Smart grids gain more resilient and effective operation systems through the integration of electric vehicles.

Electric vehicles have gained popularity as a result of several influencing elements between environmental concerns, economic investments and technological progress together with government backing. Public understanding about fighting climate change along with decreasing pollution drives the critical adoption of EVs which leads to developing sustainable transportation systems. Technological progress in EVs combined with available economic benefits and legal incentives shapes these vehicles into multiple superior choices than conventional cars while creating increasing demand from worldwide consumers. Ehirgeous move of introducing EVs represents a transformative path toward a cleaner better and sustainable future.

EV ownership brings two major financial benefits along with reduced fuel costs and reduced maintenance expenses. Certain states together with local areas grant their EV owners financial benefits through tax incentives and rebate programs and discounted registration fees. The cheaper initial investment for EV purchase expands the market opportunity to broader groups of shoppers. Some locations have implemented two types of financial incentives for EV owners by allowing free or discounted parking privileges and permitting access to HOV lanes. The savings from using electric power instead of gasoline become highly significant. Functions of electricity as an energy source remain cheaper than fuel prices while power prices should continue to remain competitive due to growing renewable energy adoption.

2. Review of Literature

Research by Caulfield, Farrell, and McMahon [12] shows that alternative fuel and hybrid electric cars gain popularity over greenhouse gas emissions and vehicle registration taxes due to their dependability security and cost-effectiveness. Research shows that vehicle customers make purchasing decisions based on how useful the vehicles are and how financially advantageous they are. Customers seek reliable and secure transportation so safety alongside trustworthiness act as essential factors for their choice. The initial purchase price along with reduced fuel expenditure take precedence over environmental concerns such as greenhouse gas emissions and vehicle registration taxes when customers decide which models to buy.

According to Ko and Hahn [13] consumers in Korea select electric vehicles that operate using swappable batteries while paying a lump sum instead of unswappable batteries and payment terms. Swappable batteries get refueled at a faster pace which makes them the probable preference because they provide users with increased convenience and minimized downtime. Customers select lump-sum payment methods instead of installment subsidy plans because they prioritize one-time payments for ease-of-use along with potential financial advantages above the extended commitment involved in installment plans.

Seixas, et al. [14] uncovered three main factors affecting EV adoption including total cost and ownership status and regular driving activities of customers. EV adoption relies heavily on the total price customers need to invest for purchasing and running electric vehicles. Social status becomes crucial since owning an EV often indicates status along with environmental awareness. The choice to adopt an EV heavily depends on driving practices which consider regular mileage and the present availability of charging systems.

According to Morton, Anable, and Nelson [15], how well EVs perform and how open consumers are to adoption innovations both significantly influence their preferences between BEVs and PHEVs. People who easily embrace new technology because of their phenomenological style demonstrate higher preference for BEVs and PHEVs. Positive assessments of EV features including driving distance and duration of recharging and service requirements determine significantly how customers pick their choices between vehicles. Upgraded performance capabilities in EVs would likely drive their customer acceptance levels.

The study conducted by Liao, et al. [16] identifies four primary factors which determine electric automobile customer choices which include social influence along with mobility conditions and psychological aspects and socio-economic characteristics. A person's ability and interest in purchasing electric vehicles depends on their household income and degree of education and the number of family members. Representative psychological elements which impact strongly on EV decisions include personal views and abstract attitudes toward electric vehicles as well as environmental issues. The economic feasibility of EV vehicle ownership depends on factors that involve moving around from place to place and the capability to recharge batteries. Social influence shapes electric car opinions and preferences by combining elements of network relationships with prevailing cultural norms of society.

Shabanpour, Mousavi, Golshani, Auld, and Mohammadian [17] revealed that persons who possess advanced education levels, display technological sophistication and operate within young demographics and drive long distances regularly for work demonstrate a stronger preference for automated electrified solutions. People who travel frequently and actively use technology tend to accept innovative advanced car systems because they need comfort and efficiency when they drive distances.

The research conducted by Liao et al. [18] found that consumer preferences regarding various electric vehicle models determine their preference for business models. The research shows that battery-electric vehicle users most commonly use vehicle leasing but battery leasing stands as the least preferred option for conventional car and plug-in hybrid owners. Buyers demonstrate greater openness to lease entire vehicles instead of batteries alone because they probably want to minimize maintenance costs associated with battery replacement.

The decision process of Chinese EV customers is influenced by their application-specific interests combined with EV pricing alongside powertrain types and automobile classification along with aesthetics [19]. Price together with automobile type along with technical specifications (engines) and designed function and planned utilization defines Chinese consumer choices in the market.

According to Tu and Yang [20] customers decide to purchase electric vehicles based primarily on their technological item acceptance and environmental consciousness and resource control capabilities and consultation of opinions. The purchase of electric vehicles involves multiple factors which include personal financial capability (resource control) alongside expert advice (consultation opinion) and environmental concern (environmental responsibility) in combination with technological adoption readiness (technology product acceptance).

Asadi, et al. [21] in their study revealed that customers' intentions to acquire electric vehicles in Malaysia are influenced by perceptions of value, attitude, responsibility, subjective and personal standards, perceived consumer efficacy, and comprehension of the consequences. This implies that social and psychological variables play a considerable role in forming consumer intentions, underlining the necessity for encouraging favourable attitudes and raising awareness of EVs in order to stimulate their adoption.

Bansal, Kumar, Raj, Dubey, and Graham [22] found that EV features including quick charging times, long driving ranges, and low operating costs affect Indian customers' choices for EVs. This suggests that practical performance attributes and financial factors play a significant role in Indian customers' decision-making process when it comes to EVs.

Jung et al. [23] found that buyers who choose EVs are concerned about environmental pollution and give future car worth a high priority, but they give less weight to the look of the vehicle. This implies that factors other than the vehicle's instant visual appeal – such as environmental awareness and the expected long-term advantages of EVs – have a greater impact.

Lashari, Ko, and Jang [24] found that customers' intentions to buy electric vehicles are most strongly predicted by their opinions of the economy, the environment, and technology. This emphasizes how crucial it is to address the financial and environmental benefits of EVs in addition to easing technological concerns in order to increase consumer acceptance of EVs.

Jihyeok, Jung et al. [25] provided a stated preference analysis on Electric Cars in the Korean auto industry using survey information gathered in February 2018 from 1,251 participants. The study examined variables pertaining to consumer attitudes on environmental issues and characteristics of vehicles that have a major impact on prospective Korean buyers' choices for electric vehicles. A discrete automobile choice questionnaire and an attitude questionnaire were used to gather data. The survey revealed that buyers who favoured electric vehicles were cognizant of the problems caused by pollution in the environment, especially by moving cars, and gave future car worth a higher priority than looks. With the potential for a positive externality impact, economic variables including purchase price, charger subsidies, and charging infrastructure were shown to be critical for the effective adoption of EVs in Korea.

Claudia Aravena and Eleanor Denny [26], along with their co-author Denny, aimed to uncover the keys that drive Indian consumers to select electric automobiles. The research data collection depended on a well-designed questionnaire which underwent Chi-Square and percentage analysis. The results showed that consumers' purchasing decisions for electric vehicles were highly impacted by factors including 'same price as petrol/diesel,' 'fuel price,' and 'environmental consciousness.' Contrary to previous research, the study discovered that customers' choices for electric cars were not significantly influenced by their degree of awareness.

Shakil, Khan, and Hanna Maoh [27] revealed preference data from a random sample of more than 1,000 Canadian organizations gathered through an online survey, researchers examined the attitudes and perceptions of Fleet Operating Entities (FOEs) regarding EVs. The results indicated that among the studied FOEs, latent factors related to cost and technology may discourage the adoption of EVs. Several financial and non-financial incentives such as improved fuel economy and reduced maintenance costs and access to high-occupancy vehicle lanes together with free parking might facilitate adoption. The analytical findings produced two fundamental aspects about cost consciousness and early adoption behaviour. The corporate and government sectors displayed equivalent attitudinal characteristics although these groups showed minor variations in their responses to variables that discouraged including power and insurance price hikes and concerns about operational dependability and extended charging requirements. The FOEs who completed the survey considered the risks associated with adopting new technology to be greater than the total ownership costs and environmental benefits.

Siqin and his colleagues Xiong and others [28] used random coefficient logit analysis to examine electric automobile customer preferences. The random coefficient logit model served as the basis for data collection to examine electric automobile consumer choices better. The analysis of survey data specified the elements which drive customer preferences for electric cars.

The research conducted by W. Loengbudnark, Kaveh and their colleagues [29] analyzed factors which affect Australian consumers to select between hydrogen Fuel Cell Electric Vehicles (FCEVs) and Battery-powered Electric Vehicles

(BEVs). The research sample included 1,735 Australian customers. Intention to adopt EVs depended more heavily on safety concerns than on both perceived benefits and purchasing cost. The main features Australian consumers sought in BEVs consisted of long range capability and reasonable battery charging duration together with safety against hydrogen explosions. FCEVs gained favour because they offer extended driving distances together with decreased chances of battery explosion.

The research by Loengbudnark [29] examined the determining factors for Australian consumers when selecting between hydrogen FCEVs and BEVs. Research data came from 1,735 Australian customers across the nation where mean participants reached 44.9 years old and 41% identified as male. Safety issues exceeded perceived advantages and purchase expenses when it came to customers' intentions to adopt electric vehicles. The decision factors for choosing BEVs included having an adequate operating distance and short charging periods in addition to avoiding hydrogen explosion concerns. FCEVs gained popularity because they provided longer range and fewer safety risks related to battery explosions.

Joy and Dsouza [30] studied which variables motivated Indian customers to choose electric automobiles. A well-designed questionnaire was utilized to gather the data, and Chi-Square and percentage analysis were the techniques of analysis employed. The results showed that consumers' purchasing decisions for electric vehicles were highly impacted by factors including 'environmental consciousness,' 'fuel price,' and 'same price as petrol/diesel.' The study discovered that customers' choices for electric vehicles were not significantly influenced by their awareness level, in contrast to previous research.

Sumas Wongsunopparat et al. [31] discovered that the brand, government policy, at-home charger, price, battery, charge station, and at-home charger are important elements affecting inclination towards electric vehicles. This suggests that, in addition to more conventional elements like cost and brand image, policy assistance (the form of government incentives) and infrastructure (the form of charging stations and at-home chargers) also play important roles.

Udit Chawla et al. [32] shown that the adoption of EVs is moderately impacted by variables such as customer acceptability, charging infrastructure, power efficiency, and loyalty. This implies that although these issues hold significance, they do not hold as much sway as other ones like economic reasons and environmental concerns.

Wenjian, Jia, and T. Donna Chen [33] The study, which was based on a stated preference survey done in Virginia, looked at consumer preferences for Plug-in Electric Vehicles (PEVs) utilizing Mixed Logit (MXL), Latent Class (LC), and Latent Class-Mixed Logit (LC-MXL). The findings revealed that, while boosting the deployment of charging infrastructure did successfully promote PEV adoption, providing monetary incentives was the most effective strategy to increase PEV market share. The study found minimal association between battery range expansion and PEV uptake. The study's goals were to better understand preference variability in PEV adoption and provide insights into the impact of various policies on PEV adoption.

Tidmunds [34] examined the buying habits and preferences of Indian consumers with regard to electric automobiles. Online questionnaires and interviews were used to gather data. The primary goal was to comprehend how consumer purchasing intentions changed as technologies advanced. The second goal connected attitudes toward battery-electric vehicles (BEVs) and mitigating sociodemographic factors with social cognitive perception and attitude intent.

Yavuz, Selim, et al. [35] conducted research to analyze and segment consumer preferences for hybrid and electric vehicles in Turkey, the Netherlands, Germany, and Sweden. Data collection methods included online questionnaires and discrete choice exercises. The study shed light on how age, income level, fuel cost, CO2 emissions, vehicle range, and environmental concerns influence consumer decisions in the electric and hybrid automobile markets of the aforementioned countries.

Vuban, Chowdhury, et al. [36] studied the use of Electric Cars in multi-car families in the United States in terms of vehicle choice. The study devised a two-step machine-learning approach to determine the preferred trip category for EV use and to capture the effects of various factors on EV use in multi-car households. The findings revealed that households with fewer workers and lesser incomes preferred to use electric cars (EVs) for discretionary trips.

Babu, George [37] examined consumer perceptions of electric vehicles, with a particular emphasis on the socioeconomic, technical, environmental, and demographic issues that are driving the move to electric vehicles. Eighty respondents in Kerala's Kollam District provided data, which showed a favourable attitude toward electric vehicles. Nevertheless, the study found that the absence of other models, pricing, and recharging issues were barriers. According to the research, a greener environment may result from addressing these challenges through technology developments.

Reema, Bera, et al. [38] Using a stated preference framework, commuters and non-commuters in Delhi and Kolkata were surveyed on their preferences for Plug-in Hybrid Electric Vehicles (PHEVs). Data from each city's car-owning population was collected and evaluated to assess the perceived benefits of PHEV operation-specific attributes for commuters and non-commuters. The findings revealed details about consumers' willingness to pay (WTP) for factors such as fuel economy, battery range, tailpipe emissions, and warranty. The report recommended Indian authorities to implement legislative steps for subsidized pricing with extra public charging infrastructure and educational efforts to promote PHEV market expansion across India. The survey disclosed that purchasing costs combined with limited public charging infrastructure represented the biggest challenge for PHEV market acceptance.

The research by Mahjabeen, Naz et al. [39] evaluated resident opinions in Jaipur India about EVs. The research employed surveys together with questionnaires as well as face-to-face interviews to collect data. Survey participants showed willingness to convert to electric cars due to the favourable pricing along with their low fossil fuel consumption and environmental compatibility. The absence of charging infrastructure together with initial price levels negatively affected customer interest in using electric vehicles. The research indicated that electric cars could become successful in India since consumers already anticipate purchasing this technology at some point in the near future.

Reema Bera and their colleagues [38] conducted their research by allowing commuters and non-commuters in Delhi and Kolkata to rank their preferences for plug-in hybrid electric vehicles through a stated preference survey. This study examined the perceived benefits related to PHEV operation-specific features among commuters and non-commuters who owned cars in both Indian cities through collection of survey data. The research revealed consumer price readiness (WTP) information regarding expenses for gasoline along with warranty on batteries and tailpipe emissions and range of battery power and time needed for battery charging. Research findings demonstrate that India faces two key barriers to PHEV adoption because of expensive purchase costs and insufficient charging infrastructure. The research recommended altering legislation by boosting subsidies and establishing new public charging facilities while conducting public awareness initiatives.

2.1. *Research gap*

Studies about electric vehicle adoption among diverse customer types and geographical areas have not clearly established the relationship between psychological variables with technological factors that motivate EV adoption. Researchers have identified scarce evidence about the relationship between smart grid systems and ADAS technologies and psychological consumer variables such as perceived behavioural control alongside environmental awareness and technological stress. The extended effect of changing governmental policies along with their incentives on consumer adoption behaviours across multiple socio-economic conditions requires more thorough investigation. Strengthening the understanding of factors behind EV adoption would enhance sustainable transportation strategies through better knowledge of complete adoption drivers.

3. Statement of the Problem

EVs require transition as the essential element to decrease greenhouse gas emissions while creating sustainable transportation systems [3, 40]. There exists uncertainty about the specific motives which drive Coimbatore District residents to want EVs. The growing number of studies about EV adoption [41, 42] fails to provide localized research which examines the area-specific socioeconomic, psychological and infrastructure characteristics. This research analyzes the variables which determine EV customer preference in Coimbatore District based on technological acceptance [20], environmental awareness [24], economic selectivity [43] along with charging station availability [31]. The analysis of this topic would reveal what prevents EV adoption in Coimbatore thereby enabling the creation of targeted advancement programs for district electric vehicle usage.

4. Objective of the Study

4.1. *Scope of the study*

This study evaluates Electric Vehicle users' socioeconomic background together with their key decision factors that drive them toward EV adoption instead of gasoline or diesel-powered vehicles within Coimbatore District. This investigation traces target EV buying segments through analytic examination of traditional demographic factors such as age distribution and income levels and household structures and education degrees combined with occupational status. The research investigates all key elements influencing EV preference which consist of policy backing and infrastructure development and environmental concerns along with social and economic factors and technology advancements. Having a complete understanding of these variables leads to clear comprehension of how local circumstances affect the transition to electric transportation. The study achieves its goal to offer practical guidance to stakeholders by merging survey data analysis with interview-based qualitative information. This study uses comprehensive analysis to support the main objective of sustainable urban transportation by promoting focused EV adoption measures in Coimbatore District.

5. Research Methodology

5.1. *Data*

The research collects its entire data from primary sources by using a questionnaire made specifically for this study. The chosen approach provides authentic understanding of participant viewpoints and facilitates complete assessment of electric vehicle preference drivers and behavioural patterns.

5.2. *Sampling and sample size*

A total of 275 EV users received survey questionnaires under the snowball sampling method. Only 110 EV user participants failed to complete their questionnaires while 23 others left sections blank. The total number of participants available for assessment amounted to 142 people who provided complete answers to the survey questions. The research method ensured thorough collection of user insights from EV customers who participated in the study area.

5.3. *Framework of analysis*

A combination of simple percentage calculations together with factor analysis methods served to analyze the obtained data. The combination of statistical analysis and factor analysis enabled a complete investigation into numeric variables together with the detection of essential variables which affect how electric vehicle users behave.

6. Significance of the Study

The evaluation of EV preferences in Coimbatore District serves important purposes for several essential areas. The assessment of local community preferences reveals crucial variables which either speed up or slow down EV adoption among this particular socio-economic and infrastructural setting. The collected data possesses vital value because it helps policymakers and urban planners and stakeholders develop better strategies which boost sustainable transportation practices. The distinct traits of Coimbatore District consisting of industrial enterprises and urban growth patterns determine how the region will adopt EVs as opposed to other demographic areas. The research evaluates EV preferences in Coimbatore to expand understanding about sustainable urban transportation in developing regions throughout India thus providing useful insights applicable across similar worldwide locations. The research delivers essential information to help decision-making processes in order to reach environmental sustainability and energy efficiency goals for Coimbatore District and additional regions.

7. Limitations of the Study

The collection of new data through primary research methods suffers from multiple sources of possible bias that develop from survey participant perceptions and survey planning decisions as well as sampling protocols. Survey participants showed bias when they either provided responses that matched social expectations or the questions led to incorrect information due to misunderstood questions. Survey accuracy can be compromised when lead-time or unclear questions are used because such issues create biases that affect the study findings. If a sample is not representative of the total population, sampling techniques may create biases that result in the over- or underrepresentation of particular groups. It is important to use caution when extrapolating and generalizing the study's conclusions to a larger context since these biases have the potential to compromise the validity and reliability of the findings.

8. Findings

8.1. *Socio-economic profile*

The paragraph discusses the age, gender, occupation, educational qualifications, family type, familial role, monthly income, family income, and family expenditure of electric vehicle users in Coimbatore district (Table 2.1).

The data indicates that the majority of e-vehicle users are aged 26–35, comprising 81.0% (115 individuals). Users up to 25 years old account for 7.7% (11 individuals), those aged 36–45 make up 7.0% (10 individuals), and only 4.2% (6 individuals) are above 45, reflecting a predominantly young user base.

57.0% of e-vehicle users are male (81 individuals), while 43.0% are female (61 individuals), showing a relatively balanced but slightly male-dominated user base.

The data shows that e-vehicle users are predominantly employees, making up 48.6% (69 individuals). Agriculturists account for 19.7% (28 individuals), business owners comprise 12.0% (17 individuals), homemakers represent 9.2%

Table 2.1. Socio-economic profile

Variables	*Description*	*Frequency*	*Percent*
Age	Up to 25	11	7.7
	26–35	115	81.0
	36–45	10	7.0
	Above 45	6	4.2
Gender	Male	81	57.0
	Female	61	43.0
Occupation	Agriculturist	28	19.7
	Business	17	12.0
	Employee	69	48.6
	Home Maker	13	9.2
	Others	15	10.6
Educational Qualification	Diploma	6	4.2
	UG	12	8.5
	PG	82	57.7
	Others	42	29.6
Type of family	Joint	42	29.6
	Nuclear	100	70.4
Status in family	Head	20	14.1
	Member	122	85.9
Monthly income(Rs.)	Up to 20000	53	37.3
	20001–35000	52	36.6
	35001–50000	15	10.6
	Above 50000	22	15.5
Family Income (Rs.)	Upto 30000	58	40.8
	30001–50000	51	35.9
	Above 50000	33	23.2
Family Expenditure (Rs.)	Upto 20000	46	32.4
	20001–35000	50	35.2
	35001–50000	25	17.6
	Above 50000	21	14.8

Source: Author.

(13 individuals), and those in other occupations make up 10.6% (15 individuals). This indicates a diverse user base with a significant proportion of employees.

Majority of e-vehicle users have postgraduate qualifications, accounting for 57.7% (82 individuals). Those with other educational backgrounds make up 29.6% (42 individuals), while users with an undergraduate degree comprise 8.5% (12 individuals), and those with a diploma represent 4.2% (6 individuals). This suggests a highly educated user base.

70.4% of e-vehicle users come from nuclear families (100 individuals), while 29.6% belong to joint families (42 individuals). This suggests that the majority of e-vehicle users are from nuclear family structures.

The data shows that a significant majority of e-vehicle users, 85.9% (122 individuals), are members of their families, while 14.1% (20 individuals) are heads of their households. This indicates that most e-vehicle users are not the primary decision-makers within their families regarding vehicle choices.

The data reveals that among e-vehicle users, a substantial portion, 37.3%, have a monthly income up to Rs. 20,000. Additionally, 36.6% fall within the income bracket of Rs. 20,001 to Rs. 35,000 per month. A smaller proportion, 10.6%, earn between Rs. 35,001 and Rs. 50,000 monthly, while 15.5% have incomes exceeding Rs. 50,000. This distribution indicates that a significant number of e-vehicle users have moderate to lower incomes, with a notable presence in higher income brackets as well.

A significant portion of e-vehicle users have monthly family incomes below Rs. 30,000, comprising 40.8% (58 individuals). Those with family incomes ranging from Rs. 30,001 to Rs. 50,000 account for 35.9% (51 individuals), while users with incomes exceeding Rs. 50,000 represent 23.2% (33 individuals). This distribution suggests that the majority of e-vehicle users come from households with moderate to lower monthly incomes, with a notable presence in higher income brackets as well.

A significant portione-vehicle users, 35.2%, have monthly family expenditure ranges between Rs. 20,001 and Rs. 35,000. Additionally, 32.4% have expenditures up to Rs. 20,000 per month. A smaller proportion, 17.6%, spend between Rs. 35,001 and Rs. 50,000 monthly, while 14.8% have family expenditures exceeding Rs. 50,000. This distribution suggests that e-vehicle users tend to have moderate to lower monthly family expenditures, with some also falling into higher expenditure brackets.

8.2. E Vehicle usage

The following paragraph discuss about periodicity of E-Vehicles usage, Average kilometres travelled per day, Sources of information to buy E-Vehicles, Frequency of usage of E vehicles, Type of E-Vehicles used and brand of E-Vehicles used (Table 2.2).

A significant majority of users (80 respondents, 56.3%) utilize e-vehicles for periods below one year, suggesting a high turnover or possibly frequent upgrades or replacements within short intervals. A substantial but comparatively smaller group (52 respondents, 36.6%) maintains e-vehicles for 1–3 years, indicating a longer average usage period but still suggesting turnover within a few years. A minority (10 respondents, 7.0%) use e-vehicles for more than 3 years, implying a segment of users committed to longer-term ownership or less frequent upgrades. This distribution highlights varying patterns of adoption and retention of e-vehicles across different user segments, reflecting diverse consumer behaviours and preferences in the electric vehicle market.

Table 2.2. E Vehicle usage

Variables	*Description*	*Frequency*	*Percent*
Periodicity of E Vehicle Usage	Below 1 years	80	56.3
	1–3 years	52	36.6
	Above 3 years	10	7.0
Average Kilometers per day	1–10 kms	47	33.1
	11–0 kms	25	17.6
	Above 20 kms	70	49.3
Sources of Information to buy E vehicle	Self	77	54.2
	Friends	33	23.2
	Colleagues	14	9.9
	Relatives	18	12.7
Frequency of Usage of E vehicle	Frequently	47	33.1
	Occasionally	33	23.2
	Whenever Needed	62	43.7
Type of E vehicle	Two-wheeler	116	81.7
	Four-wheeler	26	18.3
Brand of E Vehicle	OLA	30	21.1
	Ather	21	14.8
	Hero	44	31.0
	TATA	26	18.3
	Others	21	14.8

Source: Author.

The data reveals the average kilometres travelled per day by users of e-vehicles across different distance ranges. A significant portion of users (70 respondents, 49.3%) cover distances above 20 kilometres daily, indicating a substantial reliance on e-vehicles for longer commutes or frequent travel. Users who travel short distances for daily activities compose 33.1% of the total respondents with 47 participants in this category. Out of all respondents a minority group totalling 25 people (17.6%) moves between 11 to 20 kilometres per day indicating moderate use patterns between short distances and longer routes. The pattern shows how e-vehicle users operate their vehicles differently because commuting distance, travel frequency and individual lifestyle choices affect their choices.

The results show that respondents who relied on self-sourced information amount to 77 participants which constitutes 54.2% of the total sample. The research shows how friends are significant information sources because 33 survey respondents (23.2%) received knowledge from personal recommendations and experiences about e-vehicles. The exchange of information between colleagues and relatives represents secondary avenues through which people become aware of e-vehicles (14 respondents, 9.9% and 18 respondents, 12.7% respectively). This distribution demonstrates how consumers utilize diverse data collection methods by taking individual action alongside getting recommendations from others thereby representing the various aspects of decision-making in the e-vehicle market.

The survey data shows that 43.7% (62 respondents) uses e-vehicles based on their needs because these users tend to drive them during particular moments. 47 of the respondents (33.1%) exhibit frequent use of e-vehicles which indicates they drive them regularly throughout their days for their daily routines and transportation requirements. The data shows 33 respondents who fall into the category of occasional e-vehicle users making up 23.2% of the sample group. The distribution reveals how users of e-vehicles operate their vehicles with different frequencies which demonstrates their diverse requirements to include electric vehicles in their everyday or irregular schedule.

The survey shows that most participants (116, 81.7%) use two-wheelers among their e-vehicle selection. This suggests a strong preference or adoption of electric motorcycles or scooters, possibly due to their affordability, practicality in urban environments, and ease of use for short-distance commuting. In contrast, a smaller segment (26 respondents, 18.3%) utilizes four-wheelers, indicating a preference for electric cars or larger vehicles which may be favoured for longer commutes, family use, or specific transportation needs. This distribution highlights the popularity of two-wheelers among e-vehicle users, potentially influenced by factors such as cost, convenience, and urban mobility requirements.

Hero emerges as the most favoured brand, with 44 respondents (31.0%) choosing it, possibly due to its established presence and reputation in the electric vehicle market, particularly in two-wheeler segments. OLA follows closely with 30 respondents (21.1%), likely reflecting its growing popularity and market penetration in electric scooters and vehicles for ride-sharing services. TATA and Ather both garner significant interest, with 26 respondents (18.3%) opting for TATA and 21 respondents (14.8%) for Ather, suggesting diverse preferences across four-wheeler and premium two-wheeler categories respectively. The category 'Others' also holds 21 respondents (14.8%), indicating a variety of lesser-known or emerging brands that also attract attention among e-vehicle consumers. This distribution underscores the competitive landscape and varied consumer choices in the evolving market for electric vehicles.

9. Preference Towards E Vehicles

The principal justifications for favouring electric cars (EVs) were established through the application of factor analysis. The data appropriateness was determined through Kaiser-Meyer-Olkin (KMO) and Bartlett's Test of Sphericity before analysis. The KMO sampling adequacy indicator at 0.878 demonstrated that the study data were appropriate for factor analysis (Table 2.3). Factor analysis of the collected data proved adequate based on the results of the statistically significant Bartlett's Test (Chi-square = 2429.896, df = 253, $p < 0.001$). The predesigned tests put in place demonstrated that the data collection was stable enough to properly examine the keys factors contributing to electric vehicle preferences (Table 2.4).

Table 2.3. KMO and Bartlett's test

Kaiser-Meyer-Olkin Measure of Sampling Adequacy.		.878
Bartlett's Test of Sphericity	Approx. Chi-Square	2429.896
	df	253
	Sig.	.000

Source: Author.

Table 2.4. Preference towards E-Vehicles

	Component			
	1	*2*	*3*	*4*
Reduced Operating Costs	.886			
Government Incentives	.875			
Battery Life	.746			
Low Maintenance Cost	.745			
Battery Capacity	.694			
Comfortable	.633			
Technological Advancements	.606			
Traffic Congestion	.554			
Model Availability	.547			
Range Anxiety		.871		
Vehicle Reliability		.713		
Outlook and Design		.609		
Convenience of charging at home		.541		
City Sustainability Goals		.501		
Regulatory Support			.896	
Charging Infrastructure			.792	
Environmental Concerns			.781	
Future Outlook			.572	
Horse Power			.569	
User Friendly				.909
Vehicle Price				.907
Brand Image				.875
Safety Features				.860
Eigenvalues	9.971	3.194	1.586	1.103
% of Variance	43.351	13.886	6.895	4.796
Cumulative %	43.351	57.236	64.131	68.927

Source: Author.

Factor analysis revealed four separate elements which exceeded unity in eigenvalues and achieved 0.5 or greater component loading values to demonstrate their essential significance. The rotated component matrix showed different factors which shape customer preferences regarding EVs. Operating costs reduction and government incentives together with extended battery life and low maintenance expenses alongside battery capacity features and comfort levels along with technological advancements and traffic congestion considerations and availability of diverse EV models constitute the main influential factors. Multiple critical factors serve to mold consumer decisions about electric vehicles because they affect how people choose EVs in the market.

The factor analysis revealed multiple important influences that affect EV preference through its second analysis component. These factors include range anxiety, which pertains to concerns about the distance an EV can travel on a single charge and the availability of charging infrastructure; vehicle reliability, which refers to the trustworthiness and performance consistency of EVs; outlook and design, encompassing the aesthetic appeal and attractiveness of EV models; convenience of charging at home, highlighting the ease and practicality of home-based charging solutions; and alignment with city sustainability goals, indicating how well EV usage supports environmental and urban sustainability initiatives. These factors demonstrate the extensive elements which determine consumer preference decisions and adoption rates toward EVs thus demonstrating both practical needs and environmental and design priorities within the purchasing process.

The third factor of the analysis shows a set of substantial elements which determine Electric Vehicle (EV) consumer preference. The support of regulations remains essential because it consists of government-made policies and incentives together with rules that foster EV adoption through both emission standard enforcement and subsidy programs. The ease of owning an EV depends directly on the availability of charging facilities and easy access to those facilities. The desire to decrease environmental impact pushes consumers toward purchasing EVs because these vehicles help reduce methane emissions as well as pollutants in the atmosphere. The way people view technological progress combined with better battery technology and market stability determines the decisions individuals will make about EV purchasing in the future. The horsepower considerations directly influence how customers perceive the power capabilities as well as acceleration performance of electric vehicles. The entrance of EVs into the market depends on various interwoven influences that span regulatory support and network growth and the preservation of nature and constant technical development.

Several vital elements which affect EVs preference choices emerge in factor number four of the factor analysis. The overall ease of EV operation combined with user-friendly interfaces determines how satisfied users become leading to better adoption rates. Vehicle price acts as a game-changer for affordability as well as cost-effectiveness which steers customer purchasing choices. Consumer preferences regarding EV manufacturers are influenced by how reliable and trustworthy and innovative they seem to customers. EV safety features consist of two elements: firstly collision avoidance systems alongside robust structural design that increase both passenger and driver safety during vehicle operation. Evaluation of EVs involves multiple influences touching every aspect of consumer behaviour from operability and cost to brand reputation and safety rating standards in market selection.

Factor one serves as the main influential element because it explains 43.351% of the total preference for EVs. The four factors that affect preferences for EVs are explained by factor one which accounts for 43.351% along with factor two at 13.886% and factor three at 6.895% and lastly factor four at 4.796%. The total variance in EV preference reaches 68.927% due to the combined effectiveness of these four factors. Business costs coupled with environmental elements, technological elements and safety opinions collectively guide the adoption rate of electric vehicles as shown in this breakdown.

10. Suggestions

- A comprehensive strategy developed to motivate people toward EVs adoption must handle various issues while catering to different population interests.
- Operating expenses of EVs remain lower than those of standard gasoline automobiles. Talking about how EVs reduce operational costs through lower fuel costs together with maintenance expenses opens up an affordable ownership option for regular commute and extended ownership time.
- The public needs to receive information about EV owner incentives including tax deductions and rebates on top of low registration fees from the government. Government incentives enable customers to buy electric vehicles at reduced prices.
- Customers require information about battery improvements that boost battery power retention periods and product longevity. The durability of modern EV batteries extends to numerous years to eliminate customer worries about battery replacement frequency.
- Highlight the low maintenance costs of EVs due to fewer moving parts and less wear and tear. The maintenance cost of electrical vehicles decreases because owners require fewer visits to mechanic shops for basic services like oil changes along with brake and engine repairs.
- People worried about battery life can see the improved range abilities through demonstrative examples. Among modern EVs consumers can now experience impressive single-charge driving distances which work effectively for city commutes and extended journey needs.

- All EVs deliver a smooth journey while delivering comfort to their users. You should describe three key features about quiet operation and how instant torque functions and the advanced temperature control systems that create enhanced comfort.
- The promotion of modern EV technologies should focus on presenting their advanced features like intelligent accessories combined with wireless connections and autonomous functionality. The combination of modern innovations integrates into a contemporary driving experience which provides convenience to users.
- The ease of commuting becomes better when EV owners can drive in carpool lanes or benefit from reduced traffic limitations within specific urban areas.
- The market offers EV models to suit various consumer preferences from luxurious to affordable options encompassing different styles. The EV market extends from basic compact cars to SUVs and luxury models so all consumers can find a suitable fit.
- Addressing customers' range anxiety should involve describing how modern EVs have longer driving ranges and how the charging station infrastructure continues to expand. The article discusses real-life scenarios where existing EV ranges satisfy day-to-day driving requirements.
- Reassure consumers about the reliability of EVs, supported by warranties and the robust performance of electric drivetrains. Share testimonials and data on the reliability and longevity of existing EV models.
- Showcase the stylish designs and aesthetics of modern EVs, appealing to consumers who value looks and brand image. Emphasize that EVs are no longer just practical but also fashionable choices.
- Explain the convenience of home charging, allowing users to start each day with a full battery. Highlight the ease of installing home charging stations and the benefits of overnight charging.
- Connect EV adoption to city sustainability goals, emphasizing how switching to EVs supports cleaner air and reduced carbon emissions, contributing to a healthier environment.
- Discuss the regulatory support for EVs, such as emissions standards and future bans on internal combustion engines. Reinforce that regulatory trends favour EVs, making them a future-proof investment.
- Update the public on the rapidly expanding charging infrastructure, including fast-charging stations and charging networks. Provide maps and resources for locating nearby charging points.
- Make an appeal to environmental concerns by emphasizing how EVs have less of an impact on the environment and emit fewer greenhouse gases than traditional cars. Stress the advantages of driving an EV for the environment, highlighting the substantial impact that opting for an EV may have on creating a cleaner, more sustainable environment.
- Present an optimistic picture of EVs' future, highlighting impending technological developments, longer range, and lower costs. Demonstrating the inventive advancements and prospective possibilities of electric cars might entice customers to join the expanding EV movement.
- Highlight EVs' remarkable performance qualities, such as their rapid torque and tremendous horsepower. Emphasize how EVs provide an exhilarating driving experience and acceleration, making them a compelling option for both daily drivers and performance aficionados.
- Highlight the ease of use of EVs, such as their simple charging procedures, user-friendly interfaces, and seamless integration with smart home technology. Every component of an EV has been specifically designed for ease of use to provide customers with a new standard of driving convenience.
- The entire ownership expense with maintenance costs and fuel savings should be considered when solving price-related concerns about the vehicle. Customers should recognize that they can choose EV models supported by various financing programs which offer reasonably priced solutions to suit different buyers.
- Take advantage of the prestige and favourable brand image that come with owning an EV. Emphasize celebrity, influencer, and environmental advocacy endorsements to increase the allure of EVs and establish them as a chic and socially conscious option.
- Highlight the cutting-edge safety features that EVs provide, such better crash test ratings, sophisticated driver assistance, and collision avoidance systems. Advise clients that electric vehicles deliver extraordinary protection benefits through their contemporary features as well as cozy riding experiences.

11. Conclusion

The collected data presents a comprehensive view of e-vehicle users throughout different characteristics. Most e-vehicle users belong to the age group between 26 and 35 but older age groups show limited adoption of these vehicles. Users show a minimal preference toward males. Users who use e-vehicles originate from diverse occupations which include workers in agriculture and business sectors as well as numerous others. Educational data shows the prevalence of postgraduate degree holders who make up most of the e-vehicle user population based on their advanced training levels. Family members make up most of users who live in nuclear family setups because heads of households currently form a minority group. Much of the Internet user population falls within the middle income bracket. Most users operate e-vehicles throughout limited time periods while searching for information independently and choose two-wheelers as their primary transportation option. The user preference for Hero demonstrates competitive conditions caused by numerous consumer preference patterns within the market. Dynamic changes between population statistics and financial standing and behavioural patterns drive how electric vehicles get adopted by customers.

Different factors intertwine to form the preferred choice for electric vehicles which dramatically affects how consumers adopt and utilize these vehicles. EVs attract consumers due to their reduced operating expenses which becomes more affordable through governmental economic incentives. Everyday vehicles find their appeal in EVs mainly because they possess longer-lasting batteries that deliver high capacity and demand little maintenance. Aside from improving performance the ongoing technological developments allow EVs to benefit from features that provide users with fast recharging capabilities and smart connectivity systems. The implementation of EVs helps reduce city traffic while supporting sustainability goals through policies which have triggered quick growth in charging stations. EVs address environmental needs by reducing carbon emissions thus helping create a sustainable future with unexploited air quality. EVs will secure more customers through future developments that enhance range capabilities alongside reliability solutions and stylish designs which also increase safety standards. Global sustainable transportation will significantly change through the prominent placement of EVs because both pricing strategies and reputations among brands are focusing on reliability and innovation.

12. Scope for Further Research

Researching electric vehicle preferences should examine additional routes as a method to improve knowledge and handle upcoming obstacles. Research should examine how changing battery solutions affect consumer preferences regarding vehicle fuel capacity and power storage length together with auto lifespan tolerance. The study of regional EV adoption patterns along with study of different socio-economic groups can reveal strategic segments that marketing teams should target through their marketing plans. Government policies together with incentives at both local and international levels hold significant importance for understanding consumer adoption behaviours and choices regarding EVs. Technical evaluations of long-term environmental consequences associated with extensive EV adoption together with studies of sustainability among infrastructure networks will assist sustainability research. Gaining insight into upcoming trends in electric vehicle design together with user interface choices and technological breakthroughs such as driverless functions will direct upcoming innovations to boost electric vehicle customer approval and usage.

References

[1] European Environment Agency. (2020). *Greenhouse gas emissions from transport in Europe*.

[2] Hawkins, T. R., Singh, B., Majeau-Bettez, G., & Strømman, A. H. (2013). Comparative environmental life cycle assessment of conventional and electric vehicles. *Journal of Industrial Ecology*, *17*(1), 53–64.

[3] International Energy Agency. (2020). *Global EV Outlook 2020: Entering the decade of electric drive?* Paris: International Energy Agency.

[4] Nykvist, B., & Nilsson, M. (2015). Rapidly falling costs of battery packs for electric vehicles. *Nature Climate Change*, *5*(4), 329–332.

[5] Hardman, S., Shiu, E., & Steinberger-Wilckens, R. (2018). Comparing high-end and low-end early adopters of battery electric vehicles. *Transportation Research Part A: Policy and Practice*, *116*, 70–84.

[6] Anderson, J. M., Nidhi, K., Stanley, K. D., Sorensen, P., Samaras, C., & Oluwatola, O. A. (2014). *Autonomous Vehicle Technology: A Guide for Policymakers*. RAND Corporation.

[7] Faria, R., Moura, P., Delgado, J., & De Almeida, A. T. (2013). Managing the charging of electric vehicles: Impacts on the electric grid and on the environment. *Energy*, *61*, 134–143.

[8] Palmer, K., Tate, J. E., Wadud, Z., & Nellthorp, J. (2018). Total cost of ownership and market share for hybrid and electric vehicles in the UK, US and Japan. *Applied Energy*, *209*, 108–119.

[9] Bonges, H. A., & Lusk, A. C. (2016). Addressing electric vehicle (EV) sales and range anxiety through parking layout, policy and regulation. *Transportation Research Part A: Policy and Practice*, *83*, 63–73.

[10] Sperling, D., & Gordon, D. (2009). *Two Billion Cars: Driving Toward Sustainability*. Oxford University Press.

[11] Lutsey, N., & Sperling, D. (2012). Regulatory adaptation: Accommodating electric vehicles in a petroleum world. *Energy Policy*, *45*, 308–316.

[12] Caulfield, B., Farrell, S., & McMahon, B. (2010). Examining individuals preferences for hybrid electric and alternatively fuelled vehicles. *Transport Policy*, *17*, 381–387.

[13] Ko, W., & Hahn, T. (2013). Analysis of consumer preferences for electric vehicles. *IEEE Transactions on Smart Grid*, *4*, 437–442.

[14] Seixas, J., Simoes, S., Dias, L., Kanudia, A., Fortes, P., & Gargiulo, M. (2015). Assessing the cost-effectiveness of electric vehicles in European countries using integrated modeling. *Energy Policy*, *80*, 165–176.

[15] Morton, C., Anable, J., & Nelson, J. (2016). Exploring consumer preferences towards electric vehicles: The influence of consumer innovativeness. *Research in Transportation Business and Management*, *18*, 18–28.

[16] Liao, F., Molin, E., & Wee, B. (2017). Consumer preferences for electric vehicles: A literature review. *Transport Reviews*, *37*, 252–275.

[17] Shabanpour, R., Mousavi, S., Golshani, N., Auld, J., & Mohammadian, A. (2017). Consumer preferences of electric and automated vehicles. *2017 5th IEEE International Conference on Models and Technologies for Intelligent Transportation Systems (MT-ITS)*, 716–720.

[18] Liao, F., Molin, E., Timmermans, H., & Wee, B. (2019). Consumer preferences for business models in electric vehicle adoption. *Transport Policy*, *73*, 12–24.

[19] Ma, S., Fan, Y., Guo, J., Xu, J., & Zhu, J. (2019). Analysing online behaviour to determine Chinese consumers'

preferences for electric vehicles. *Journal of Cleaner Production, 229*, 244–255.

[20] Tu, J., & Yang, C. (2019). Key factors influencing consumers' purchase of electric vehicles. *Sustainability, 11*(4), 1–22.

[21] Asadi, S., Nilashi, M., Samad, S., Abdullah, R., Mahmoud, M., Alkinani, M., & Yadegaridehkordi, E. (2020). Factors impacting consumers' intention toward adoption of electric vehicles in Malaysia. *Journal of Cleaner Production, 124474*, 282.

[22] Bansal, P., Kumar, R., Raj, A., Dubey, S., & Graham, D. (2021). Identifying the key attributes influencing the adoption of electric vehicles in India. *Journal of Cleaner Production*, 278.

[23] Jung, J., Yeo, S., Lee, Y., Moon, S., & Lee, D. (2021). Factors affecting consumers' preferences for electric vehicle: A Korean case. *Research in Transportation Business & Management*, 41.

[24] Lashari, Z., Ko, J., & Jang, J. (2021). Consumers' intention to purchase electric vehicles: Influences of user attitude and perception. *Sustainability, 13*(2), 497.

[25] Jung, J., Yeo, S., Lee, Y., Moon, S., & Lee, D. J. (2021). Factors affecting consumers' preferences for electric vehicle: A Korean case. *Research in Transportation Business & Management, 41*, 100666.

[26] Aravena, C., & Denny, E. (2021). The impact of learning and short-term experience on preferences for electric vehicles. *Renewable and Sustainable Energy Reviews, 152*, 111656.

[27] Khan, S., & Maoh, H. (2022). Investigating attitudes towards fleet electrification – An exploratory analysis approach. *Transportation Research Part A: Policy and Practice, 162*, 188–205.

[28] Xiong, S., Yuan, Y., Yao, J., Bai, B., & Ma, X. (2023). Exploring consumer preferences for electric vehicles based on the random coefficient logit model. *Energy, 263*, 125504.

[29] Loengbudnark, W., Khalilpour, K., Bharathy, G., Taghikhah, F., & Voinov, A. (2022). Battery and hydrogen-based electric vehicle adoption: A survey of Australian consumers perspective. *Case Studies on Transport Policy, 10*(4), 2451–2463.

[30] Dsouza, J. (2023). A study on the impact of perceived benefits on customer preference for electric vehicles. *SDMIMD Journal of Management*, 15–26.

[31] Sumas Wongsunopparat et al. (2023). Factors influencing preference towards electric vehicles: A case study in Thailand. *Transportation Research Part D: Transport and Environment, 100*, 102995. doi:10.1016/j.trd.2021.102995

[32] Chawla, U., Mohnot, R., Mishra, V., Singh, H. V., & Singh, A. K. (2023). Factors influencing customer preference and adoption of electric vehicles in India: A journey towards more sustainable transportation. *Sustainability, 15*(8), 1–15.

[33] Jia, W., & Chen, T. D. (2023). Investigating heterogeneous preferences for plug-in electric vehicles: Policy implications from different choice models. *Transportation Research Part A: Policy and Practice, 173*, 103693.

[34] Kumar, V. (2023). Analyse the consumer preferences and purchases of electric vehicles purchases and developments. In *Contemporary Studies of Risks in Emerging Technology, Part B* (pp. 87–95). Emerald Publishing Limited.

[35] Balcioglu, Y. S., Sezen, B., & İşler, A. U. (2024). Evolving preferences in sustainable transportation: A comparative analysis of consumer segments for electric vehicles across Europe. *Social Responsibility Journal, 20*(9), 1664–1696.

[36] Chowdhury, V., Mitra, S. K., & Hernandez, S. (2024). Unraveling Electric Vehicle Preference: A Machine Learning Analysis of Vehicle Choice in Multi-Vehicle Households in the United States. *Available at SSRN 4648746*.

[37] Babu, George. (2024). Consumers attitude towards electric cars with special reference to kollam district. *International Journal of Advanced Research.* doi:10.21474/ijar01/18848

[38] Sharma, R. B., Majumdar, B. B., & Maitra, B. (2024). Commuter and non-commuter preferences for plug-in hybrid electric vehicle: A case study of Delhi and Kolkata, India. *Research in Transportation Economics, 103*, 101415.

[39] Naz, M., Beniwal, A. S., & Meena, R. V. (2024). People's perception of electric vehicles in developing countries: A case study of Jaipur city. In *Smart Electric and Hybrid Vehicles* (pp. 187–197). CRC Press.

[40] Levinson, A. (2019). Electric vehicles: It's not just about the car. *Nature Energy, 4*(6), 415–416. doi:10.1038/s41560-019-0404-7

[41] Liao, F., Molin, E., & Wee, B. (2017). Consumer preferences for electric vehicles: A literature review. *Transportation Research Part D: Transport and Environment, 50*, 64–78. doi:10.1016/j.trd.2016.10.019

[42] Bansal, P., Kumar, R., Raj, A., Dubey, S., & Graham, D. (2021). Willingness to pay and attitudinal preferences of Indian consumers for electric vehicles. *Energy Economics*, 100.

[43] Ma, S., Fan, Y., Guo, J., Xu, J., & Zhu, J. (2019). Consumer preferences and choice behavior towards electric vehicles in China: An analysis based on online user-generated content. *Transportation Research Part D: Transport and Environment, 69*, 401–414. doi:10.1016/j.trd.2018.09.009

3 Is there a relationship between resilience and academic burnout among school teachers? A focus on Kerala

Ranjith Somasundaran Chakkambath[1,a] *and M. Usha*[2,b]

[1]Research Scholar and Corresponding Author, Department of Management, Karpagam Academy of Higher, Pollachi Main Road, Eachanari Post, Coimbatore, Tamil Nadu, India
[2]Associate Professor and Co-author, Department of Management, Karpagam Academy of Higher Education, Pollachi Main Road, Eachanari Post, Coimbatore, Tamil Nadu, India

Abstract: Education transforms the minds of people and is the cornerstone of any society. Teachers are the architects who facilitate this transformation. After COVID-19, there has been changes in the educational sector owing to the way in which learning has developed during the online sessions that were in force. The teachers had to recalibrate the move to a slightly different approach now which has put pressure on them. The study analyzes the interconnection between teacher burnout and their resilience capabilities. The authors collected data from 209 teaching professionals scattered throughout schools in Kerala through standardized surveys. The survey contained both Oldenburg Burnout Inventory (OLBI) and Brief Resilience Scale (BRS) evaluation items. Research using the t-test established that the female members of the target teacher audience handled burnout situations with greater efficiency and showed stronger resilience abilities. The correlation test confirmed that burnout and resilience show a meaningful relationship to each other. The role of resilience as a predictor of burnout with age as moderator generated a positive result through moderation analysis conducted using Hayee's PROCESS macro. The results indicated that as the age progressed, the teachers would be able to develop a coping mechanism to overcome burnout and be resilient.

Keywords: Burnout, resilience, moderation analysis, classroom, teachers, correlation, t-test, Hayes process macro

1. Introduction

As a profession, teaching can be quite stressful, which could result in decreased job satisfaction, burnout, and low performance at work. In the modern era, the role of imbibing knowledge to budding minds is in the hands of the teachers [1]. This also contributes to shape the culture of the society we live in. However, with the demands of the modern educational system, teachers are put through pressures that lead to burnout. Elongated exposure to adverse mental and physical stress will result is the condition of Burnout. As far as a career in the education industry goes, the professionals working here are put through different physical and mental demands along with interpersonal demands which drain these professionals of their emotional tolerance [2]. There are many reasons that teachers are put through emotional exhaustion. It has been observed that burnout among teachers is a cumulative result of contributing states of mental, emotional, and physical stress experienced for a long time. Teachers who are burned out may experience a decrease in job satisfaction, an increase in absenteeism, and a decline in their teaching performance. Along with Burnout, another important process of adaption to difficult circumstances is known as resilience. Resilience helps to overcome burnout among employees. A teacher's job is becoming increasingly challenging, demanding, and exhausting [3]. The resilience of teachers is strongly correlated with their job. Resilience and academic burnout are closely related, especially among faculty members who face a variety of stressors in their academic work. Sometimes, individuals showcase the ability to rebound from adverse situations or setbacks. This shows the resilience of the individual while burnout includes different levels of stress in an academic scenario at emotional, mental and physical levels. Interestingly, over the years, there has been some pointers that generate academic burnout among faculty, namely- emotional exhaustion and the other one is depersonalization. The capability of resilience functions as a defense mechanism which shields personnel against negative effects caused by stress and professional exhaustion. Teachers who demonstrate stronger resilience capability better handle workplace stress that sustains their job commitment as well as their work dedication and performance levels. Therefore, it is important for academic institutions to support and promote resilience among their faculty members, through interventions such as stress management training, social support, and work-life balance initiatives. By doing

[a]ranjithsc2016@gmail.com, [b]ushakarthic6969@gmail.com

DOI: 10.1201/9781003684459-3

so, institutions may be able to prevent or alleviate academic burnout and enhance the well-being and productivity of their faculty members [4].

1.1. The objective of the study

1. To look into the demographic and work-related factors that may affect resilience and academic burnout among faculty members.
2. To test the correlation between resilience and academic stress/burnout.
3 To test the association of gender and burnout of teachers and also the association between gender and resilience teachers.
4. To test the association of burnout on resilience moderated by the years of age of the teachers [5].

1.2. Limitations

- The study was limited to the geographical region of the State of Kerala, India.
- Although some ad hoc and part-time teachers were in the study, the population of this group was less and they showed less response due to fear of losing jobs.

1.3. Theoretical framework

The Conservation of Resources (COR) theory proposes that stress occurs when individuals experience a net loss of resources. Resources encompass various dimensions including material possessions like money, social aspects such as relationships and social support, as well as psychological aspects like self-esteem and mastery. The impact of challenging events on the well-being of teachers is contingent upon their resilience, specifically their ability to acquire and lose resources, as postulated by COR [6]. The COR has enriched the understanding of teacher well-being by being broader and emphasizing individual resources in addition to contextual ones than the well-known Job Demands and Resources Theory (JDR) [7]. According to the COR theory, burnout occurs when teachers experience a net loss of resources, such as time, energy, and social support. This can happen due to a number of factors. The increasing workload puts a lot of pressure on the education professionals along with a work environment where there lack of support from peers or administrators. Time and money are resources which are exhaustion contribute to burnout factors. In some situations, personal trauma can be an additional contributor to the burnout of teachers. This research aims to use COR theory to check the association between the independent variable (IV) – burnout and the dependent variable(DV) – resilience moderated (M) by age of the faculty. Based on this the following Model was developed as shown in Figure 3.1.

The model, in line with COR theory, investigates Burnout – a condition that lowers performance and motivation in any job. It disrupts the physical and mental state of the person. The following hypotheses were framed based on the model in Figure 3.1.

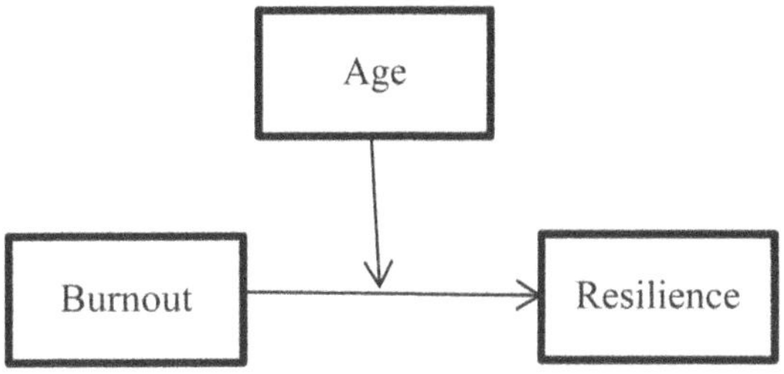

Figure 3.1. Model for Burnout-Resilience moderated by Age.

Source: Author.

1.3.1. Hypothesis 1: Burnout will have a negative effect on resilience

The negative association of Burnout with resilience among the teachers who were working in schools was investigated. This will prove that higher burnout contributes to less effort from teachers to be resilient and bounce back to their original state of performance in teaching.

1.3.2. Hypothesis 2: The effect of burnout on resilience will be moderated by age of the faculty

This will be able to highlight the impact age of the teachers who gained more exposure to teaching overs their tenure which has a strengthening or weakening effect on the association between burnout and resilience.

2. Literature Review

2.1. Teachers stress

According to published literature, it's widely accepted that teaching is a challenging career. Chronic stress exposure increases the likelihood of poor mental and physical health as well as burnout [8]. The volume of stress experienced by secondary school instructors exceeds that of teachers in primary schools according to research by Kongcharoen et al. [9]. Job responsibilities together with financial pressures serve as main stress factors for secondary educators [9]. Some research reveals that teachers encounter substantial anxiety together with stress when they integrate instructional technology into their classroom instruction [10].

2.2. Burnout

Burnout stands as a workplace phenomenon according to the World Health Organization [11] and describes a condition that develops from severe work-related stress lasting too long without appropriate support. Research indicates that emotional exhaustion influences 50% of staff members in public and service-driven work fields [12]. Understanding that teachers with high stress levels or burnout experiences negatively affect their student teaching outcomes [8, 13, 14]. It is essential to put appropriate school-based interventions into place if you want to help teachers better manage their stress, lower their risk of burnout, and generally increase their well-being [8]. Over the decades, literature pinned that teaching to be stressful career which can lead to burnout syndrome. This has been seen to impact the wellness, the relationship of the teachers with their pupils and also their quality of course delivery [15].

2.3. Resilience

The concept of resilience means overcoming hurdles in life [16]. People generally use resilience as a defensive strategy to manage both severe stress and traumatic experiences. The literature explains resilience as both 'the ability to recover and maintain a sense of normalcy after a stressful event or the initial ability to cope with such an event' [17]. The capacity to respond in a healthy and useful manner under pressure is referred to as resilience. It increases one's capacity for overcoming and adapting to challenging circumstances to the point where resilience can make the difference between success and failure in life [18].

2.4. Research gap

Although, over the years there has been research on resilience and burnout of teachers, there seems to be less studies focused on teachers working in the schools which are the foundation of educations. In this research, we also have made age a moderator to the relationship between burnout and resilience. This aspect was not much studied among the teachers working in schools in Kerala. There is a need to address the gap and shed more light on the less explored domain of burnout and resilience among education professionals.

3. Methodology

3.1. Research design

This research was used to investigate the link between resilience and academic burnout among faculty members, as well as to identify demographic and work-related factors that may affect resilience and academic burnout where the age of the Faculty acted as the moderator between burnout and resilience. Cross-sectional research design was used with a structured questionnaire. The respondents were teachers working in different categories of schools which included private, public, and aided schools from different parts of the State of Kerala, India.

3.2. Measurement scales used in the instrument

Academic burnout is a common issue among faculty members, which can lead to decreased job satisfaction, emotional exhaustion, and reduced academic performance. Research has shown that job demands and resources can significantly impact academic burnout and resilience among faculty members. Over the years, there have been many scales developed to measure burnout. In this research, we have used Oldenburg Burnout Inventory (OLBI) [19]. The OLBI consists of sixteen items, eight of which measure the exhaustion dimension of burnout (e.g., 'There are days when I feel tired before I arrive at work') and eight measuring the disengagement dimension of burnout (e.g., 'It happens more and more often that I talk about my work in a negative way'). As respondents were asked to rate their feedback on a scale ranging from 1 to 4 where former indicated strongly agree and the latter strongly disagree. The Cronbach's alpha generated for both the subscales – exhaustion (Cronbach's α = .87) and disengagement (Cronbach's α = .81), were showed higher reliability. To support the learning of students in schools, skills and knowledge of teachers along with their resilience will contribute to a better outcome [20]. To assess an individual's resilience – that is, their ability to bounce back fast from adversity – the Brief Resilience Scale (BRS) was created [21]. As Van Griethuijsen et al. [22] indicate, any value of Cronbach's alpha equal to or greater than 0.7 is deemed satisfactory for demonstrating the questionnaire's reliability.

3.3. Sampling method

Convenience sampling was used in the survey for which the major source of the data was collected through questionnaire. Additional information was collected from journals and websites.

3.4. Sample size

The sample size was 209 respondents from different parts of the State of Kerala, India. Kerala is a state known for its high literacy and cultural vibe.

3.5. Data collection method

A structured questionnaire in Google Forms, was used for the survey. The initial part of the survey focused on the demographic characteristics of the respondents. The second part was used to collect the details regarding the burnout experience by the teachers working in the schools across different schools. The third part of the questionnaire assessed the level of resilience among the respondents.

3.6. Techniques used

Demographic analysis was done to measure the characteristics of the target population. Descriptive statistics generate general picture of the sample characteristics. Inferential statistics were tested for the association between burnout and resilience among both genders was tested using a t-test. The Correlation between Burnout and Resilience was tested. Regression analysis was done using Hayee's PROCESS v4.2

to assess the ability of Burnout to predict the level of Resilience among the teachers where the relationship was moderated due to the Experience of the teachers gained over years of working in this sector [23]. 'IBM SPSS Statistics 23 was used to perform the data analysis.'

4. Results and Discussion

4.1. Demographic characteristics

The research evaluated demographic aspects which affect burnout and resilience levels among education staff members in Table 3.1. The research used self-report surveys to obtain information from 209 teachers at different schools and districts. The research collected information about participants' gender, age groups, teaching profession, personal earnings, as well as educational attainment. The outcome of the demographics interestingly had the majority of the respondents in the younger age group below 35 years while female teachers formed a sizeable portion of the response. The level of income earned by the target group was below thirty thousand rupees per month. The demographic results also prove that the respondents were highly qualified which is evident from the fact the State of Kerala is known for its high literacy rate in Table 3.2.

In this case, the majority of schools are private sector schools (67%), followed by public sector schools (29%), and aided schools (4%). This suggests that the education system in this country is largely dominated by the private sector. The findings of this research demonstrated that many of these teachers had less than five years of teaching experience. Teachers with lesser years of experience were found more likely to suffer burnout. This may be due to the challenges of new teachers, the increasing workloads and pressure to improve student performance faced by teachers with 3 to 5 years of experience, and the mid-career evaluations and increased family responsibilities faced by teachers with 5 to 15 years of experience in Table 3.3.

Table 3.1. Demographic characteristic

Age Group (Years)	%
25–29	26.9
30–34	20.1
35–39	19.6
39–43	13
44–48	11.5
Above 49 years	2.9
Gender	%
Male	43
Female	57
Occupation	%
Employed	37.9
Unemployed	4.8
Self Employed	6.8
Student	50.5
Income	%
Below 15000	14.7
15000–30000	32
30000–60000	22.3
60000–90000	15.5
Above 90000	15.5
Education	%
Graduate with B.Ed.	31
Postgraduate	23
Postgraduate with B.Ed.	12
Postgraduate with M.Ed.	11
M.Phil. with B.Ed./M.Ed.	20
Ph.D. B.Ed./M.Ed.	11
Graduate with TTC	2

Note: Sample Size n= 209

Source: Author.

Table 3.2. School category and years of teaching experience of faculty

School Category	%
Aided	4
Private Sector	67
Public sector	29
Years of Teaching Experience	%
Less than 3 years	20
Between 3 to 5 years	53
5 years to 15 years	18
15 years to 20 years	9

Note: The table presents the distribution of faculty across school categories and years of teaching experience.

Source: Author.

Table 3.3. Scheme of teaching, marital status, and employment type

Scheme of Teaching	%
CBSE	54
ICSE	7
NIOS	7
State Education Board	32
Marital Status	%
Single	18
Married	82
Employment Type	%
Ad hoc Faculty	6
Permanent	91
Visiting Faculty	3

Source: Author.

The fact that the majority of schools follow the CBSE curriculum suggests that it is the most popular and well-respected curriculum in the country. This could be due to a number of factors, such as its comprehensiveness, its focus on academic excellence, and its wide acceptance by universities and employers. Another interesting result from the survey was that the marital status of teachers are married. This shows that teaching is a stable and rewarding profession. Many of the respondents were permanent faculty members hence suggest that educational institutions and Government agencies are investing in teacher training and development. This is a positive development, as it ensures that teachers have the skills and knowledge they need to be effective in the classroom.

4.2. Independent t-test: Comparing gender with Resilience and Burnout of faculty

Thus, in this study an independent t test was used to compare the two groups: male and female school teachers in Table 3.4. According to previous literature, a score above 45 indicates high Burnout while using the OLBI scale [24]. Resilience scores were scored between 1 and 25, the higher the score the higher resilience [24].

4.3. Correlation test: Burnout and Resilience

From Table 3.5, results showed a strong positive correlation, $r(208) = .893$, $p < .05$ among the variables investigated in this research. Hence, both variables – burnout and resilience are interlinked.

4.4. Moderation analysis

Moderating Effect of Age (Age) On the Relationship Between Burnout (Burnout) And Resilience (Resilience) based on Table 3.6, the outcome of moderation analysis was conducted to assess the moderating effect of age (Age) on the relationship between burnout (Burnout) and resilience (Resilience) among school teachers. The interaction term between burnout and age was significant ($\beta = 18.24$, SE = .06, t = 294.39, $p < .05$), indicating that the relationship between burnout and resilience was moderated by age.

Table 3.7 related to conditional effects of burnout on resilience were significant at low levels of age (b = 0.42, SE = 0.0053, t = 80.75, $p < .05$) and at high levels of age (b = .44, SE = 0.0051, t = 88.49, $p < .05$).

As per the Table 3.8, conditional effects of burnout on resilience at different age level were probed. The effect of burnout on resilience was the strongest at younger ages (-0.5445, SE = 0.4387, $p < .05$) and gradually weakened with increasing age (-0.457, SE = 0.457, $p < .05$). Johnson-Neyman technique was statistically significant for the relationship between burnout and resilience was significant for different age levels of the teachers. The younger teachers were more prone to the negative effects of burnout and resilience in comparison with their senior counterparts.

The Figure 3.2 showed a linear relationship between resilience and burnout of the teachers which was moderated by the age of the teachers. Both the line are clustered together by colour which is indicative of the effect of burnout on different age groups. The slope is steeper for lower age groups

Table 3.4. t Test results summary comparing gender with Resilience and Burnout of faculty

Variable	*Gender*	*N*	*Mean*	*Std. Deviation*	*Std. Error Mean*
Resilience	Male	103	18.17	7.29	0.71
	Female	106	18.28	7.68	0.75
Burnout	Male	104	28.24	16.47	1.61
	Female	105	28.61	17.39	1.7

Note: The mean burnout score for the 209 teachers in this study was 48.43 (SD=16.90). The mean Resilience score was 18.23 (SD=7.47).

Source: Author.

Table 3.5. Correlations test between Burnout and Resilience

Variables	*n*	*M*	*SD*	*1*	*2*
1. Burnout	209	48.42	16.9	1	−0.726*
2. Resilience	209	18.23	7.47	0.893*	1

Note: M stands for Mean and SD for standard deviation respectively. n represents the number of valid cases. The mean for burnout was 48.42 with a standard deviation of 16.9 while Resilience had a mean value of 18.23 with a standard deviation of 7.47.8. * indicates p<.05.

Source: Author.

Table 3.6. Conditional effects of facilitating conditions on expectancy in AI Chatbot at different levels of satisfaction

Variable	*Effect(β)*	*SE*	*t*	*p*	*95% CI*
Intercept	18.24	0.06	294.39	<.05	[18.12,18.36]
Burnout	0.44	0.00	119.25	<.05	[.43,.45]
Age	0.00	0.01	0.25	0.802	[-0.01,0.02]
Burnout*Age	0.00	0.00	2.86	0.004	[0.00,0.00]

Note: This table presents the results of a Haye's Process Macro Model1 moderation analysis using MATRIX procedure in SPSS. The dependent variable is Resilience and the independent variables s are Burnout and Age [23]. The interaction tern Burnout*Age is significant indicating that the effect of Burnout on Resilience is moderated by Age.

Source: Author.

Table 3.7. Conditional effects of Burnout on Resilience at values of Age

Age	*Effect(β)*	*SE*	*t*	*p*	*LLCI*	*ULCI*
7.5006	0.4289	0.0053	80.7513	0.000	0.4185	0.4394
0.000	0.4394	0.0037	119.247	0.000	0.4322	0.4467
7.5006	0.4499	0.0051	88.4913	0.000	0.4399	0.46

Note: p<0.05

Source: Author.

Table 3.8. Conditional effect of focal predictor at values of moderator

Age	*Effect*	*SE*	*t*	*p*	*LLCI*	*ULCI*
-12.4493	0.422	0.0073	58.1848	0	0.4077	0.4363
-11.2588	0.423	0.0068	62.6944	0	0.4104	0.4357
-10.0683	0.4253	0.0063	67.7554	0	0.413	0.4377
-8.8778	0.4277	0.0058	73.4118	0	0.4185	0.4368
-7.6874	0.4287	0.0053	80.7932	0	0.4183	0.4391
-6.4969	0.4301	0.0048	88.5934	0	0.4205	0.4397
-5.3064	0.432	0.0045	93.9734	0	0.4229	0.4411
-4.116	0.4342	0.0042	102.7441	0	0.4265	0.4419
-2.9255	0.4366	0.004	108.1509	0	0.4287	0.4444
-1.735	0.4389	0.0038	114.539	0	0.4314	0.4464
-0.5446	0.4413	0.0036	118.359	0	0.434	0.4486
0.6459	0.4436	0.0035	126.103	0	0.4363	0.451
1.8364	0.446	0.0034	131.723	0	0.4385	0.4535
3.0269	0.4483	0.0032	138.309	0	0.4408	0.4559
4.2173	0.4506	0.0031	145.785	0	0.4431	0.4581
5.4078	0.453	0.003	151.529	0	0.4454	0.4605
6.5983	0.4553	0.0029	158.051	0	0.4477	0.4629
7.7888	0.4537	0.0028	163.82	0	0.4499	0.4651
8.9792	0.456	0.0027	169.878	0	0.452	0.4601
10.1697	0.4537	0.0026	174.875	0	0.4511	0.4665
11.3602	0.4553	0.0025	179.879	0	0.4417	0.4688
12.5507	0.457	0.007	65.1051	0	0.4432	0.4709

Source: Author.

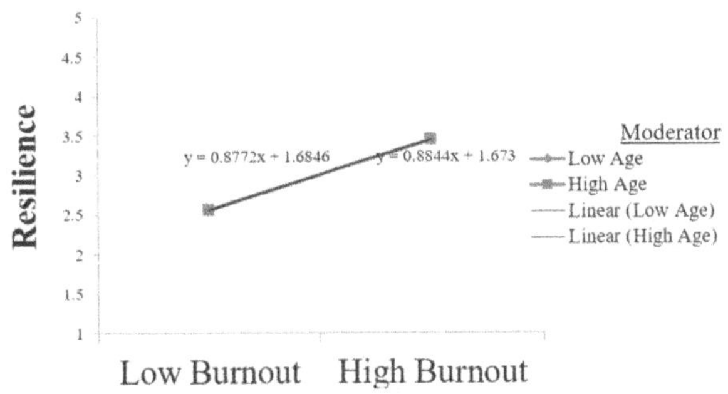

Figure 3.2. Scatter plot diagram of Burnout to Resilience moderate by Age.

Source: Author.

suggesting more burnout for this group compared to the senior teachers. Younger teachers will experience higher stress due to factors like workloads or lack of support in the work environment [25]. Studies have proven the negative correlation between resilience and burnout similar to this research [26]. This contributes to the fact that resilience would be a better predictor of burnout. The profession of teaching carries significant stress levels that may result in burnout syndrome, subsequently affecting educators' overall well-being, the quality of their instruction, and their interactions with students [15]. Nevertheless, the implementation of coping mechanisms, bolstered self-efficacy, and robust social support systems can serve as protective factors against the onset of burnout [15]. Contradictorily to the r, some studies had found that there is no significant differences between teachers' age and their levels of resilience and burnout [27]. Burnout, otherwise identified as occupational fatigue, denotes a continual strain progression that materializes in the absence of proactive measures and exerts a psychological impact on individuals [28]. Similar to the findings of this research, previous studies have noted that younger teachers or faculty members experience higher levels of burnout due to factors such as emotional exhaustion, low personal accomplishment, and depersonalization [29]. There should be an effective system to monitor and identify the teachers who are reaching the burnout stage. Use of technology tools and related awareness would be enable the faculty to reduce their workload. Similar observation were found in previous literature and also in recent publications that the use of ICT awareness can reduce the teachers burnout by empowering them to use these tools effectively [30].

5. Findings and Suggestions

Schools are the foundation of education and school teachers facilitate the building of the nation. As a result of this investigation, several interesting insights were gained regarding the relationship between resilience and burnout among the teachers. In the survey, the demographic results of this research was dominated by female respondents who were highly qualified and earned an income which was around three lakhs per annum. Private schools were the major players in the education sector in the State of Kerala under the CBSE syllabus with most of the teachers in the early stages of their careers in this sector. As the outcome of the demographic sighted, many of them were married and settled with family in their respective regions. They are permanent employees of their respective organizations. The t-test done in relation to resilience and burnout among the genders showed that females had a better mechanism to cope with burnout and were able to be more resilient than their male counterparts. The correlation test showed that burnout and resilience were interrelated among the target group. Age was found to influence the relationship between burnout and resilience among the teachers. Resilience was found to be a better predictor of burnout and young teachers were more affected by burnout. As age progresses, they become more resilient. Some suggestions that could mitigate the possible burnout among young teachers, as found in this study is to provide wellness training or workshops to improve their resilience and be ready for the challenges they will be facing in the working environment. Since the majority were highly qualified teachers, only updating their domain knowledge through attending conferences and seminars would be ideal. Another suggestion is to provide a platform or forum for open communication with the higher authorities to highlight the difficulties they face in their work.

A collaborative as well as encouraging environment through constructive feedback and recognition for efforts would empower the teachers with tremendous confidence to go the extra mile. The teachers can be provided with opportunities to take up leadership roles along with some level of autonomy which will lead to teacher self-efficacy development. Policy changes can be initiated at the Government level once these issues are brought to the public and garner their support.

6. Conclusion

The research aims to provide insight into the burnout and resilience of teachers working in Kerala. The demographics of the respondents showed that the majority of them were in the younger age groups with higher qualifications to their credentials. Many of them were working in Private Schools which followed CBSE syllabus and also the group dominated by females. The study also found that majority were married and had more than three years of teaching experience. Teaching has become one of the favourite employment domains for many people owing the fact that the State of Kerala has a high literacy rate over the decades which was evident from the former result. The independent sample t-test did not show any significant association between gender and the levels of burnout and resilience among the teachers. The correlation test was significant to establish the association between burnout and resilience of the target group. As part of the moderation analysis, the negative effect burnout has on resilience was significant which was moderated by the age of the teacher. The impact of age on the relationship between the burnout and resilience was evident. Future research can include other contributing factors like the experience of the teachers and their workload to burnout and its association to resilience which was not part of this study.

References

[1] Dewi, N. P. S. T., Wibawa, I. M. C., Parmithi, N. N., & Mahendra, I. W. E. (2024). Science education in Indonesia: A bibliometrics study in terms of science teachers' and students' perspectives. *Indian Journal of Information Sources and Services, 14*(2), 63–69. https://doi.org/10.51983/ijiss-2024.14.2.10

[2] De Vera, J. L., De Borja, J. M. A., & Orleans, A. (2020). Addressing instructional gaps in K to 12 Science Teaching through Learning Action Cell (LAC). *International Journal of Research Publications,* 46.

[3] Park, N. S., Song, S. M., & Kim, J. E. (2020). The mediating effect of childcare teachers' resilience on the relationship between social support in the workplace and their self-care. *International Journal of Environmental Research and Public Health, 17*, 1–15. doi: 10.3390/ijerph17228513

[4] Lalitha, N., & Ganesan, S. (2015). Fuzzy controlled fault tolerant converter for wind turbine application. *International Journal of Advances in Engineering and Emerging Technology, 6*(3), 24–36.

[5] Bhargavi, S. M. B., Suma, V., & Arjunan, R. V. (2024). Impact of critical success factors on productivity gain during automation testing. *Journal of Wireless Mobile Networks, Ubiquitous Computing, and Dependable Applications, 15*(3), 92–108. https://doi.org/10.58346/JOWUA.2024.I3.007

[6] Hobfoll, S. E. (2011). Conservation of resources theory: Its implication for stress, health, and resilience. *The Oxford Handbook of Stress, Health, and Coping, 127*, 147.

[7] Demerouti, E., Bakker, A. B., Nachreiner, F., & Schaufeli, W. B. (2001). The job demands-resources model of burnout. *Journal of Applied psychology, 86*(3), 499.

[8] Agyapong, B., Brett-MacLean, P., Burback, L., Agyapong, V. I. O., & Wei, Y. (2023). Interventions to reduce stress and burnout among teachers: A scoping review. *International Journal of Environmental Research and Public Health, 20*(9), 5625.

[9] Kongcharoen, J., Onmek, N., Jandang, P., & Wangyisen, S. (2020). Stress and work motivation of primary and secondary school teachers. *Journal of Applied Research in Higher Education, 12*(4), 709–723.

[10] Fernández-Batanero, J. M., Román-Graván, P., Reyes-Rebollo, M. M., & Montenegro-Rueda, M. (2021). Impact of educational technology on teacher stress and anxiety: A literature review. *International Journal of Environmental Research and Public Health, 18*(2), 548. https://doi.org/10.3390/ijerph18020548

[11] Burn-out an occupational phenomenon. (n.d.). Burn-out an Occupational Phenomenon. https://www.who.int/standards/classifications/frequently-asked-questions/burn-out-an-occupational-phenomenon#:~:text=%E2%80%9CBurn%2Dout%20is%20a%20syndrome,related%20to%20one's%20job%3B%20and

[12] Maslach, C., & Jackson, S. E. (1981). The measurement of experienced burnout. *Journal of Organizational Behavior, 2*, 99–113. http://dx.doi.org/10.1002/job.4030020205

[13] Aboorva, S., & Sowmiya, N. (2022). Student syllabus tracker. *International Academic Journal of Science and Engineering, 9*(2), 5–10. https://doi.org/10.9756/IAJSE/V9I2/IAJSE0907

[14] Alihosseini, R. (2014). How could I create the enthusiasm and willingness to pray in the students? *International Academic Journal of Innovative Research, 1*(1), 34–43.

[15] Smetackova, I., Viktorova, I., Pavlas Martanova, V., Pachova, A., Francova, V., & Stech, S. (2019). Teachers between job satisfaction and burnout syndrome: What makes difference in czech elementary schools. *Frontiers in Psychology, 10*, 2287. doi:10.3389/fpsyg.2019.02287

[16] Caycho-Rodríguez, T., Ventura-León, J., García-Cadena, C. H., Tomás, J. M., Domínguez-Vergara, J., Daniel, L., & Arias-Gallegos, W. L. (2018). Psychometric evidence of a brief measure of resilience in non-institutionalized peruvian older adults. *Psychosocial Intervention, 27*(2), 73–79.

[17] Gonzalez, S. P., Moore, E. W. G., Newton, M., & Galli, N. A. (2016). Validity and reliability of the Connor-Davidson Resilience Scale (CD-RISC) in competitive sport. *Psychology of Sport and Exercise, 23*, 31–39.

[18] Garvey, A. M., Jimeno García, I., Otal Franco, S. H., & Mir Fernández, C. (2021). The psychological impact of strict and prolonged confinement on business students during the COVID-19 pandemic at a Spanish university. *International*

Journal of Environmental Research and Public Health, 18(4), 1710.

[19] Demerouti, E., & Nachreiner, F. (1998). On the specificity of burnout for service professions: Fact or artifact. *Journal of Ergonomics, 52*, 82–89.

[20] Zhang, S., & Luo, Y. (2023). Review on the conceptual framework of teacher resilience. *Frontiers in Psychology, 14*, 1179984. doi:10.3389/fpsyg.2023.1179984

[21] Smith, B. W., Dalen, J., Wiggins, K., Tooley, E., Christopher, P., & Bernard, J. (2008). The brief resilience scale: assessing the ability to bounce back. *International Journal of Behavioral Medicine, 15*, 194–200.

[22] Van Griethuijsen, R. A., van Eijck, M. W., Haste, H., Den Brok, P. J., Skinner, N. C., Mansour, N., … & BouJaoude, S. (2015). Global patterns in students' views of science and interest in science. *Research in Science Education, 45*, 581–603.

[23] Hayes, A. F. (2013). *Introduction to mediation, moderation, and conditional process analysis: A regression-based approach.* Paris: Guilford Press.

[24] Kleinau, E. F., Lamba, T., Jaskiewicz, W., Gorentz, K., Hungerbuehler, I., Rahimi, D., Kokota, D., Maliwichi, L., Jamu, E. S., Zumazuma, A., Negrão, M., Mota, R., Khouri, Y., & Kapps, M. (2023, January 28). Effectiveness of a chatbot in improving the mental wellbeing of health workers in Malawi during the COVID-19 pandemic: A randomized, controlled trial. https://doi.org/10.1101/2023.01.24.23284959

[25] Wong, R. (2020). Job-related stress and well-being among teachers: A cross sectional study. *Asian Social Science, 16*, 19–28. https://doi.org/10.5539/ass.v16n5p19.

[26] Habibi, M., & Sadry, A. (2021). The relationship between resilience and burnout of health staff at COVID-19 dedicated hospitals: A case study of Afghan-Japan Hospital. *International Journal of Indian Psychology, 9*. https://doi.org/10.25215/0901.138.

[27] Sharifian, M. S., Hoot, J. L., Shibly, O., & Reyhanian, A. (2023). Trauma, burnout, and resilience of Syrian primary teachers working in a war zone. *Journal of research in childhood education, 37*(1), 115–135.

[28] Rossi, R., Socci, V., Talevi, D., Mensi, S., Niolu, C., Pacitti, F., Di Marco, A., Rossi, A., Siracusano, A., & Di Lorenzo, G. (2020). COVID-19 pandemic and lockdown measures impact on mental health among the general population in Italy. *Frontiers in Psychology, 11*, 790. doi:10.3389/fpsyt.2020.00790.

[29] Méndez, I., Martínez-Ramón, J., Ruiz-Esteban, C., & García-Fernández, J. (2020). Latent profiles of burnout, self-esteem and depressive symptomatology among teachers. *International Journal of Environmental Research and Public Health, 17*. https://doi.org/10.3390/ijerph17186760.

[30] Sanchez, J., Sumalinog, G., Mananay, J., Baguia, M., Goles, C., & Alejandro, I. (2023). Faculty's access to information and communication technologies in colleges and Universities in Central Visayas, Philippines. *International Journal of Information and Education Technology, 13*(3), 468–474. https://doi.org/10.18178/ijiet.2023.13.3.18

4 AI-driven strategies for enhancing circular economy and reducing industrial waste

J. Nithyashri[1,a], *S. Ramesh*[2,b], *R. Geetha*[2,c], *P. Thilagavathi*[3,d], *and S. Vidhya*[2,e]

[1]Assistant Professor, Department of Computing Technologies, School of Computing, Faculty of Engineering and Technology, SRM Institute of Science and Technology, Kattankulathur, Tamil Nadu, India
[2]Assistant Professor, Department of Computing Technologies, School of Computing, SRM Institute of Science and Technology, Kattankulathur, Chennai, Tamil Nadu, India
[3]Associate Professor, Department of Computer Science and Engineering, Aarupadai Veedu Institute of Technology, Vinayaka Mission's Research Foundation (DU), Paiyanoor, Chennai, Tamil Nadu, India

Abstract: Transformation in circular economy (CE) is essential for permanent industrial development, reduction of waste and adaptation of resource use. Artificial Intelligence (AI) has emerged as a transformative tool in increasing waste management efficiency by enabling automation, future stating analysis and real-time monitoring. This study discusses the implementation of AI based approaches to enhance significant waste regulation factors such as monitoring of wastes, waste accumulation approaches, predicting generation of wastes, recycling and regulation of landfills. The outcomes depict that approaches that incorporate AI improves functional efficacy, decreasing the chances of contamination during recycling and optimizing processing of wastes as compared to conventional approaches. Future works must explore the utilization of AI approaches along with Blockchain implementations in various industries to regulate resources and improve waste deficiency techniques.

Keywords: Artificial intelligence, circular economy, waste management, recycling

1. Introduction

In recent times, safeguarding the environment and ensuring sustainable development has increased the attention of people across the world which necessitates the adoption of an efficient CE approach which can replace the conventional economic approach which follows take-make-dispose methodology [2, 4]. This innovative ideology reduces waste, improves material durability and focuses on executing sustainable approaches which are beneficial for the environment as well as the economy [6]. A CE makes sure that the materials and products go through effective circulation by passing different stages such as design, generation, consumption as well as disposal and this develops enclosed cycles where waste is decreased and resources get recycled [16, 25]. Unlike traditional approaches, where waste acts as an unavoidable factor, the CE considers reduction of wastes as the main goal in the processes of structuring, consumption as well as manufacturing [18, 22]. As organizations as well as policymakers highly encourage this approach, incorporation of advancing technologies like AI [5] aid in implementing these approaches in real-world scenarios. Ai has the capability to bring changes in the way organizations carry out resource management, enhancing recycling methods and decrease wastes and this improves sustainability [9, 13]. Through the utilization of technologies based on AI like ML, automation as well as predictive scrutinization, organizations can decrease the intricacy of functionalities, improve efficacy of resources and deals with waste regulations in an innovative and accurate manner [7]. Predictive regulation with the aid of AI will decrease unwanted damage, increasing the durability of machineries used in industries [17]. In addition to that, waste sorting methodologies that use AI along with robotics can enhance the process of recycling [3]. As AI is emerging as a significant technology, incorporating it with CE can improve the initiatives taken by government and other organizations to decrease wastes [8, 20]. This research discusses and explores implementing AI technology along with SE regulations to detect significant applications, challenges and advantages in decreasing wastes produced by industries and establishing sustainable management of resources [10, 11].

2. Literature Review

According to Botta and Cerniglia [1], approaches that sort wastes by incorporating AI and computer vision can segregate wastes more accurately than conventional approaches, permitting the usage of significant materials which would be sent to get disposed [19]. Approaches such as these play an important role in implementing large scale automation

[a]drjnithyashri@gmail.com, [b]ramesh.isaiah@gmail.com, [c]geethar1@srmist.edu.in, [d]thiluxpriya499@gmail.com, [e]vidhyas2@srmist.edu.in

DOI: 10.1201/9781003684459-4

of sorting wastes and this is because they execute optimal recognition for detecting different materials such as metals, plastic and glasses [23].

Bashynska and Prokopenko [21], underline the fact that intelligent waste management approaches with the aid of AI are capable of predicting the generation of wastes and suggest optimal accumulation durations and pathways, decreasing the consumption of fuel and also decreases emissions of greenhouse gas. Through the modification of pathways on the basis of real-world information, these AI approaches aid in decreasing the implementation expenses [12, 24].

Oladapo et al. [15], have discussed that the ability of waste regulation approaches to decrease the number of supplies that are unsold as they turn out to be waste products, thereby supporting the aim of CE in decreasing wastes. Additionally, AI can help in identifying pathways that are effective and expense-effective for delivering returned supplies which reduces the consumption of various energies and decreases emissions [14].

3. Methodology

3.1. AI integration in circular economy methodology

The illustrated framework in Figure 4.1 highlights the integration of artificial intelligence (AI) within the circular economy, showing its role in adaptation to resource efficiency, waste management and permanent production practices. This research machine adopts a systematic approach to analyze AI-operated technologies including machine learning, deep learning, smart sensors and robotic automation, sorting in their application, sorting, life cycle evaluation and decision-making processes in their application. Taking advantage of AI, can increase industry monitoring, risk management and environmental impact assessment, facilitates data-powered infections towards a more durable economic model. The functioning involves qualitative and quantitative analysis of AI applications, supported by case studies and current literature, which is to assess their effectiveness in further circular economy principles. In addition, the study evaluates the role of AI in supporting permanent agriculture, physical recovery and emissions, contributing to the broader target of environmental stability.

3.2. AI-driven waste management for enhanced sustainability

As illustrated in Figure 4.2, AI-powered waste management integrates smart sensors, real-time monitoring and robotic automation to increase efficiency. Intelligent garbage approaches, equipped with gas, temperature, humidity, odour, sound, and infrared sensors, enabled to prevent real-time waste level monitoring, adaptation of collection and overflow. The AI-manufactured robot waste sorting is used to classify and separate materials with high precision, to reduce contamination and improve recycling efficiency. Robotic weapons automate pruning in recycling features, reduce labour costs and increase processing speed. Additionally, AI-implemented future maintenance tools help reduce failures, reducing downtime and expanding the lifetime of waste management infrastructure. By adapting the collection routes and waste processing, AI reduces operating expenses and environmental factors, which contributes to a more durable and efficient waste management system aligned with circular economy rules.

4. Results and Discussions

Integration of AI in waste management increases efficiency in various aspects compared to traditional methods. Figure 4.3 depicts a graph that compares traditional waste management with AI-competent waste management in five major aspects: monitoring wastes, waste accumulating route, prediction of waste production, recycling and pruning, and landfill management. In traditional systems, waste monitoring efficiency is 70%, while AI integration increases it by 95%. Similarly,

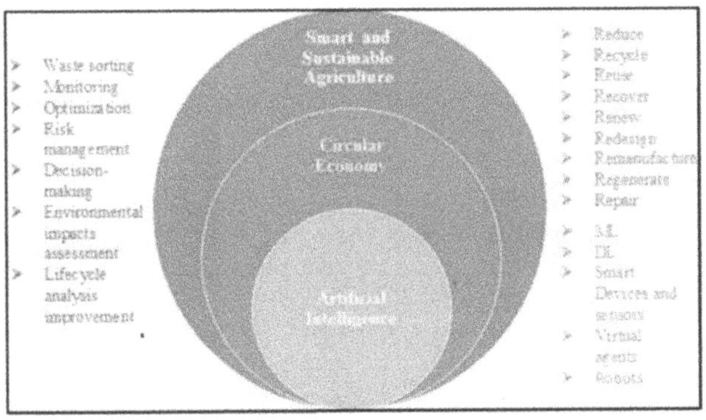

Figure 4.1. Integration of artificial intelligence in circular economy and sustainable practices.

Source: Author.

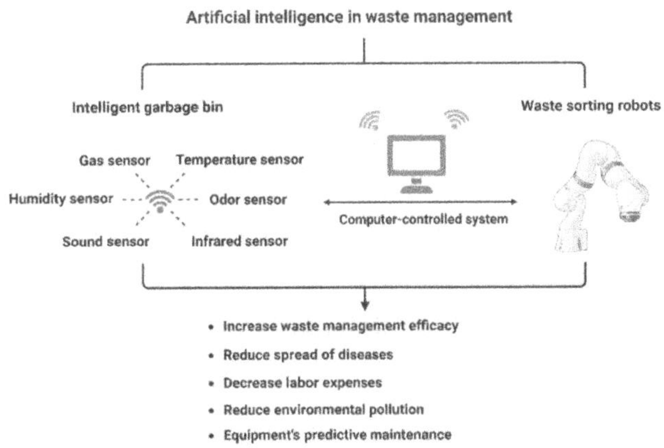

Figure 4.2. Integration of AI and smart sensors in waste management systems.

Source: Author.

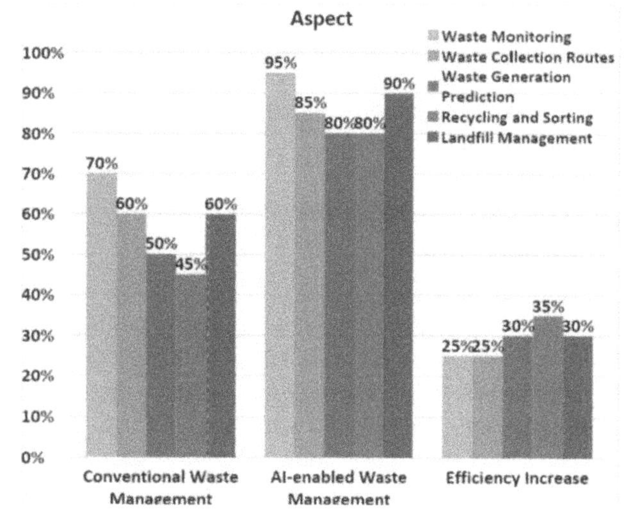

Figure 4.3. Comparison of conventional and AI-enabled waste management efficiency.

Source: Author.

AI improves waste collection passage efficiency from 60% to 85%, and waste production prediction accuracy from 50% to 80%. Recycling and pruning also show a significant improvement, with an increase in traditional systems up to 80% with AI. Landfill regulation efficiency increases from 60% to 90%, demonstrating AI's ability to enhance waste processing and disposal approaches.

The increase in overall efficiency due to AI integration is evident, with monitoring of wastes, accumulation and profiting everyone from automation and data-operated decisions. The total efficiency increase for waste management processes is 25–35% in various aspects, which underlines the adequate boons of AI in decreasing waste and improving stability efforts. These indicate AI's ability to revolutionize waste management by optimizing correction processes, reducing human error and ensuring effective approaches to deal with wastes.

5. Conclusion and Future Work

AI-operated technologies greatly increase waste management efficiency, decrease industrial waste and support circular economy regulations. With an improvement of up to 35% in major regions such as waste monitoring, collection and recycling, AI performs better by traditional methods by enabling automation and data-making decisions. Future research should focus on scaling AI applications in industries and integrating advanced technologies like blockchain for better flexibility and optimization of resources.

References

[1] Botta, S., & Cerniglia, F. (2020). AI-driven sustainable waste management: A systematic review. *Resources, Conservation and Recycling, 161*, 104915. https://doi.org/10.1016/j.resconrec.2020.104915

[2] Chatterjee, A., & Sanyal, S. (2024). From production to market: Uncovering the complexities of COVID-19's impact on fisheries and aquaculture. *International Journal of Aquatic Research and Environmental Studies, 4*(2), 37–52. http://doi.org/10.70102/IJARES/V4I2/3

[3] Muller, H., & Romano, L. (2024). An exploratory study of the relationship between population density and crime rates in Urban Areas. *Progression Journal of Human Demography and Anthropology, 1*(1), 28–33.

[4] Dubey, R., Bryde, D. J., & Fynes, B. (2017). Big data and smart technologies in the circular economy: A framework for sustainability. *Sustainable Production and Consumption, 9*, 135–146.

[5] Tharakan, J., & Qu, D. (2017). The role of AI in advancing circular economy: A framework for analysis. *Sustainability, 9*(10), 1879.

[6] Yogamadhavan, V. K., & Mannayee, G. (2024). An evaluation of various deep convolutional networks for the development of a vision system for the classification of domestic solid street waste. *Journal of Internet Services and Information Security, 14*(2), 268–283. https://doi.org/10.58346/JISIS.2024.I2.017

[7] Wainwright, C. (2019). Circular economy in practice: AI applications in waste reduction. *Journal of Cleaner Production, 229*, 1243–1252.

[8] Govindan, K., Soleimani, H., & Kannan, D. (2015). A systematic review on the applications of circular economy in supply chain management. *Journal of Cleaner Production, 102*, 289–303.

[9] Escobedo, F., Clavijo-López, R., Calle, E. A. C., Correa, S. R., García, A. G., Galarza, F. W. M., … & Flores-Tananta, C. A. (2024). Effect of health education on environmental pollution as a primary factor in sustainable development. *Natural and Engineering Sciences, 9*(2), 460–471. http://doi.org/10.28978/nesciences.1574456

[10] Su, B., Heshmati, A., Geng, Y., & Yu, X. (2013). A review of the circular economy in China: Current practices and future prospects. *Sustainable Production and Consumption, 2*, 22–32.

[11] Malešević, Z., Đeković-Šević, M., & Jovović, M. (2021). Recycling of waste sludge from WWTP 'Gorić' Valjevo using the process of stabilization and solidification. *Archives for Technical Sciences, 2*(25), 59–64. https://doi.org/10.7251/afts.2021.1325.059M

[12] Segarra-Oña, J. J., & Sanchez, A. (2019). A review of AI applications in the circular economy. *Journal of Cleaner Production, 240*, 118057.

[13] Krsmanović, M., Šušnjar, S., Golijanin, J., & Valjarević, A. (2022). GIS based vulnerability assessment of illegal waste disposal – case study East Sarajevo. *Archives for Technical Sciences, 2*(27), 63–76. https://doi.org/10.7251/afts.2022.1427.063K

[14] Sinha, R., & Sharma, R. (2020). AI-driven approaches for circular economy: A new paradigm for waste management. *Waste Management, 105*, 411–419.

[15] Oladapo, B. I., Olawumi, M. A., & Omigbodun, F. T. (2024). AI-driven circular economy of enhancing sustainability and efficiency in industrial operations. *Sustainability, 16*, 10358.

[16] Ekins, P. (2011). Green economy: The economic case for a circular economy. *Ecological Economics*, *70*(9), 1601–1603.

[17] Wang, L., & Li, M. (2020). Circular economy and artificial intelligence: A review of the literature. *Sustainability*, *12*(3), 1058.

[18] Agarwal, A., & Yadhav, S. (2023). Structure and functional guild composition of fish assemblages in the matla estuary, Indian Sundarbans. *Aquatic Ecosystems and Environmental Frontiers*, *1*(1), 16–20.

[19] Zhu, Y., & Zhang, J. (2020). A framework for integrating AI in circular economy strategies. *Journal of Cleaner Production*, *261*, 121188.

[20] Zott, C., & Amit, R. (2010). Business model design: An activity system perspective. *Long Range Planning*, *43*(2–3), 216–226.

[21] Bashynska, I., & Prokopenko, O. (2024). Leveraging Artificial Intelligence for circular economy: Transforming resource management, supply chains, and manufacturing practices. *Scientific Journal of Bielsko-Biala School of Finance and Law*, *28*(2), 85–91. https://doi.org/10.19192/wsfip.sj2.2024.13

[22] Altınışık, İ., & Yağlıoğlu, D. (2022). Age and Growth of the Bulgarian Minnow, Phoxinus strandjae (Drensky, 1926) (Actinopterygii: Cypriniformes: Cyprinidae) Living in Melen River Basin (Düzce, Turkey). *Natural and Engineering Sciences*, *7*(1), 41–49. http://doi.org/10.28978/nesciences.1098664

[23] Kathiresh, M. (2016). Multi-Biometric based User Authentication System in Banking Application. *International Journal of Advances in Engineering and Emerging Technology*, *7*(4), 250–268.

[24] Bordbar, S., & Shirazi, A. H. (2019). Satire Terminology in Persian and Arabic literature. *International Academic Journal of Innovative Research*, *6*(1), 66–76. https://doi.org/10.9756/IAJIR/V6I1/1910006

[25] Onopreychuk, D. (2018). The use of computer technological simulation for designing a Cisco hierarchical framework at the Hilton Hotel. *International Journal of Communication and Computer Technologies*, *6*(1), 19–22.

5 A study on customer satisfaction towards e-vehicles

Smruthymol J.[1,a] *and R. Velmurugan*[2,b]

[1]Karpagam Academy of Higher Education, Coimbatore, Tamil Nadu, India
[2]Associate Professor and Head, Department of Commerce, Karpagam academy of Higher Education, Coimbatore, Tamil Nadu, India

Abstract: The research looks at consumer satisfaction concerning electric vehicles (E-vehicles) since they represent a sustainable option compared to standard gasoline vehicles. Sustainable adoption of electric vehicles has increased globally because of environmental issues combined with the pressing need for decreased carbon outputs. It becomes vital to recognize and work on the main drivers of acceptance rates and customer satisfaction because this directly encourages a broader acceptance of electric vehicles. The research investigates the main factors that drive efficiency consisting of performance attributes, cost aspects, environmental effects, technical capabilities, pricing models and operational range capabilities. The sample for this study consisted of surveys and interviews from electric car owners and prospective buyers. Hence, the assessment shows that although gain in environmental protection and overall cost in the future are the main driving forces of consumer adoption, slow advances in charging facilities, reasonable battery life concerns as well as the initial cost still remain the key sources of dissatisfaction among consumers. Also, customer acceptance of batteries and related technologies depends on innovation and developments in the powertrain domain, such as battery and vehicle efficiency. From the fact it emerges that manufacturers of electric cars need to invest their efforts in enhancing battery capabilities, increasing charging stations facilities, and affordable prices with an aim to providing customers satisfaction and access that will lead to the widening of market share in the industry. There is also a call upon policy makers to offer incentives for and to support infrastructural growth in e-vehicles for more development in this technology. This paper enhances knowledge about customer perceptions of e-vehicles and offers relevant information to identify strategies that fit the customers' mindset for stakeholders.

Keywords: Customer satisfaction, electric vehicles, sustainable transportation, technology acceptance

1. Introduction

The modern automobile industry is facing an essential evolution process due to the requirement for lower CO2 emissions and the shift toward environmentally friendly transport. E-Vehicles instead of ICE vehicles have come to stand as a promising solution to emissions, pollutants and reduction of fossil fuel use [1]. The prospects of batteries, better vehicle reach, and favourable government policies have in the recent past fuel the adoption of e-vehicles [2]. Yet, there is a gap in knowledge on customer satisfaction toward e-vehicles, while satisfaction and acceptance are important indicators of the sustainable growth of this market [3].

Customer satisfaction is always important, but especially for industries as young as e-vehicles that are constantly changing with people's expectations and preferences. Therefore, the choice to acquire an e-vehicle depends on a number of factors such as; Environmental sensitivity, economic factors, technological features and perceived utility of the money [4]. Hence, knowledge of factors that influence the customer satisfaction in this sector is critical for the manufacturers, policy makers and other players in the chain to formulate strategies that are consistent with the consumer demands [5]. Perceived customer satisfaction towards e-vehicles includes performance satisfaction, reliability satisfaction, comfort satisfaction, cost satisfaction and post-purchase satisfaction including maintenance and charging facilities [6]. Happy customers are likely to remain loyal to a particular brand, endorse the brand to others, and in the process contribute to repeat patronage, which are strategic steps needed to establish oneself in such a competitive market [7]. On the other hand, the disgruntled customers can discourage potential buyers maintstream growth and damage the brand image. Thus, the evaluation and improvement of customers' satisfaction is highly essential for the continuous development of e-vehicle market [8].

Therefore, this study aims to establish the factors affecting customer satisfaction with e-vehicles with special emphasis to technological, economic, environmental, and psychological factors. Knowledge of these elements will help understand in which spheres manufacturers and policy-makers should invest more efforts and resources to increase the overall acceptance of e-vehicles by clients. Some examples of technological factors include battery technology,

[a]jayalalsmruthy@gmail.com, [b]drvelsngm@gmail.com

DOI: 10.1201/9781003684459-5

range, recharging time, and the readiness of charging stations all of which define customer satisfaction [9]. One of the biggest inhibitors of potential e-vehicle consumers is the range anxiety – the worry that an e-vehicle will be drained of energy before the next charging station is reached [10]. Some of these factors include the range of battery technology, as well as charging time; with increased improvements in these features, the anxiety experienced by owners of e-vehicles is likely to be reduced, and general satisfaction is likely to increase [11]. Moreover the other aspect includes the matters related to charging stations not only availability and accessibility but also the overall attitude of consumers. More consumers will be more willing to make the switch to e-vehicles when there are well-established charging services [12].

Another of the key factors concerning customers' satisfaction is the economic function of e-vehicles. Despite the fact that demand for e-vehicles lowers operational costs with regard to fuel and maintenance more than traditional cars, purchasing cost presents a significant hurdle for most users The study by Bubeck et al. [13] shows that public incentives with subsidies and tax breaks decrease the initial purchase prices of electric vehicles to reach a wider consumer audience [14]. Customer decisions together with satisfaction outcomes depend substantially on the total expenses that include insurance rates and maintenance fees as well as vehicle devaluation costs [15]. The long-term worth of owning an electric vehicle provides better value compared to the cost of purchase because consumers recognize this value. Most people who plan to change to electric vehicles base their decision on environmental issues.

Environmental concerns are a major driving force for many consumers considering electric vehicles. Increased awareness of climate change and the need for sustainable conveyance solutions have driven interest in EV [16]. However, consumer satisfaction is linked to the environmental benefits of vehicles. This comprises; mitigating emissions of greenhouse gases, decreasing emissions of air born particles and employing renewable energy sources for payment [27]. Any company that supports environmental friendly manufacturing processes and comes out with vehicle models that are friendly to the environment is very appealing to the conscious customer. Another factor that help in generation of consumer satisfaction with the electric vehicles includes psychological and social factors like, power of intellect, new technology and social impact.

For many consumers, electricity-driven cars are associated with advanced technologies, modernity as well as being environmentally friendly. They found that attitudinal factor such as a positive attitude toward technology, self-generated motivation and perceived social influence of friends, and media in relation to environment are the key drivers to consumer satisfaction and usage of electric vehicle. For example, the relative simplicity, comfort and safety of electric vehicles influence satisfaction. However, the use of electric vehicles has the following challenges limiting acceptance by consumers. With reference to the literature, one of the biggest issues is the absence of payment infrastructure, especially in the developing world [17]. It is also equally important also to have fast charging stations to do away with the problem of range anxiety and enhance customer satisfaction [12]. The first is the infrastructure challenge, where there is insufficient chargers for customers, which the current owners of EVs may not appreciate [10]. The next challenge is the high initial cost of the EVs irrespective of their fuel economy and maintenance costs [14]. This is significantly higher when compared with the ICE vehicles, and is beyond the reach of most middle and low income customers [13]. Furthermore, customers have issues of battery durability, replacement, and the short distance it can cover for some models of e-vehicles also reduce its usage and satisfaction level among customers [11].

This paper found out that independency cause confusion in the charging technologies and compatibility of the charging stations hence e-car owners face a number of challenges leading to frustration and disappointments. Charger time is a major barrier to use and range as some models are only good for a certain distance [9]. The approach to these challenges is filled with numerous possibilities that can be used to improve the relation with the clients and hence contribute to the development of e-car. Among those the one that is most interesting is the work on improving the battery range of the electric vehicle, the battery charging time and its life span ([11]. A crucial challenge that can increase the applicability of EVs concerns research and development of better and cheaper batteries [9]. Another area, which may have an impact on customers' satisfaction, is the extension of charging infrastructure. This involves expansion of fast charging locations, inclusion of charging infrastructure in the development of cities and towns and coordination of charging equipment between manufacturers of different car models [12]. Expanding and ensuring expenditure of effective and adequate charging stations will eradicate the range issue and make the e-vehicles discoverable and easy to use [10]. Subsidising e-vehicles through appropriate pricing models or tax incentives, and supporting affordable credit to create pool of e-vehicles that can be accessed by average consumer is another potent strategy [14]. There is support from governments through helping them to implement a suitable policy framework conducive to the implementation of e-vehicles through policies [18]. Another strategy that the manufacturers can employ is to make an effort in explaining the importance of using electric vehicles together with debunking myths of power and durability, yet providing potential customers with a test drive opportunity [19] (p. 2). According to Nguyen et al. [5], the effect on satisfaction from the perspective of enhancing consumers' knowledge and beliefs reduces skepticism about e-vehicles among manufacturers.

2. Review of Literature

The research by Rehman et al. [20] evaluated how government support and status symbol adoption influence customer satisfaction regarding the intention to buy battery electric vehicles (BEVs). Research results show environmental concerns together with social norms technophilia and self-esteem and range confidence create strong BEV adoption by customers even though their satisfaction rates are elevated.

Customer satisfaction increases in the product quality model when core functions combine with appearance along with reliability and aesthetics and cognitive quality according to Ashok Kumar and Raja [21]. Customer loyalty receives positive influence from core functions as well as reliability and consistency and cognitive quality. The relationship between core performance and reliability with cognitive quality and customer loyalty passes through the intermediary factor of customer satisfaction. The essential elements of core performance together with reliability and cognitive quality must be improved to achieve better customer satisfaction and loyalty.

According to Boonchunone et al. [22] the need fulfillment of electric vehicles depends as much on non-functional values (such as monetary, emotional, convenience, social, environmental and performance values) as functional values to create satisfaction and usage intention among customers. Thai customer preferences show non-functional value surpasses functional value because they express stronger intentions toward social value rather than environmental or emotional value.

Research by Bernarto [23] revealed that EV consumer satisfaction decreases slightly when price fairness levels increase by -0.020. Statistics show that service quality maintains an exceptionally high positive relationship (0.556) with EV consumer satisfaction. Brand image produced a positive relationship with satisfaction (0.057) although analysis failed to confirm its statistical impact on EV consumer satisfaction. The relationship between product quality and EV consumer satisfaction proved positive at 0.276.

According to customer satisfaction reached higher levels in 2023 than in 2019 due to changes in charging time toward nighttime periods. Non-first-tier city consumers demonstrated a greater preference for using charging facilities set beyond their residential area although first-tier city users lost their charging time preference influence on satisfaction ratings.

Wang and Liu [24] conducted a review analysis of new energy vehicle comments from 3,376 customers which showed positive feedback from 67.71% but 32.29% expressed displeasure. Manufacturers need to develop additional strategies to improve product quality standards and satisfy customer demands.

According to Selvi [25], operating costs emerged as the leading factor affecting e-bike user satisfaction because users rated them with the lowest score of 1.94 which demonstrated the decisive influence on customer happiness.

2.1. Research gap

The research field dedicated to electric vehicles (EVs) as well as their sustainability effects and environmental advantages and technology developments actively expands. Studies have failed to sufficiently address understanding consumer satisfaction together with the elements that affect customer satisfaction. Studies evaluating EV adoption previously conducted analysis mainly through perspectives of technology, economics and policy [9, 17]. The study of consumer satisfaction between different cultural settings and economies receives very little empirical examination and investigation Research into the sustained effects of charging infrastructure and cost along with range concerns on customer satisfaction remains limited because Hardman et al. [14], Huang and Qian [10] emphasize these aspects as critical EV adoption factors. Huang and Qian [10], comprehensive studies examining their long-term impact on customer satisfaction remain scarce. Furthermore, the psychological and social aspects of consumer behaviour – such as perceived status, social influence, and environmental consciousness – which may be very important in determining how consumers feel about e-vehicles are frequently overlooked in the research that is currently available [19, 26]. Research requires more focused attention to discover the special elements that determine customer satisfaction across diverse worldwide contexts particularly in locations which differ in electric vehicle penetration.

2.2. Statement of the problem

While e-vehicles are considered as the solution to environmental-friendly transport in the future, the role that consumers will accept and be satisfied with this shift as a key aspect of success. Several potential customers therefore have not embraced e-vehicles due to concerns on cost, accessibility, the charging infrastructure and vehicle performance, including the current government policies and technological advancement. Moreover, there is a poor understanding of the Completeness of the variables that characterize consumer satisfaction, especially in different geographies with different SES and cultural characteristics. Based on the identified consumer factors and their perception and satisfaction towards e-vehicles; technological, economic, environmental, psychological and social this research proposes to fill the above knowledge gap. Moreover, the study will seek to establish the challenges and opportunities of enhancing client satisfaction in the enduring and emergent economy hence provided a broader view on the factors that affect the global adoption of electric vehicles. It is within this context that this study will offer substantive new findings which stakeholders can use to develop more efficacious approaches for managing customers' demands or expectations toward manufacturing, regulating, and marketing acceptable and satisfying vehicles. Consumer acceptance and satisfaction determine the elements which impact customer satisfaction in e-vehicle-friendly transportation systems. Potential customers fail to purchase e-vehicles because they fear the cost together with operational ease and infrastructure access and performance capabilities. Research on consumer satisfaction factors remains insufficient for all regions especially those with different socioeconomic and cultural backgrounds. This research fills the existing knowledge gap through analysis of the elements that affect consumer perception and satisfaction regarding e-vehicles by studying technological, economic, environmental and psychological and social factors. The study investigates enhanced customer satisfaction aspects for electric vehicles through a dual examination of developed and emerging market opportunities and obstacles. This study provides stakeholders with meaningful knowledge that enables manufacturers and regulators and marketers to create alignment between consumer requirements and expectations in order to boost vehicle acceptance quality.

2.3. Objective

To identify the factors influencing customer satisfaction towards e-vehicles.

2.4. Scope of the study

This research seeks to establish factors impacting the level of consumer satisfaction with electric cars also known as EVs. The extent of consideration in the scope is not restrictive to mere technology features such as range, battery capacity, charging infrastructure, but it also considered and encompassed other features such as economic benefits including government incentives, cost of purchase and maintenance, economical efficiency inclusive of reduction of emission, and features that elicits psychological and social factors such as status, normative influence, and environmental concern among others. The aim and objectives of the survey are confirmed to the exploration of the satisfaction rates of the current consumers of the EVs and the general experience of using such cars. The study is conducted only for a specific region which gives the regional understanding of the satisfaction of the consumers with the EVs. Despite the fact that this work is mostly based on present users of EVs, the findings may present useful details about much more broader characteristics of consumer behaviour that will be useful to understand when it comes to implementing strategies which increases the option of EV adoption by different manufacturers, regulators and marketers.

3. Research Methodology

3.1. Data

The most important data in this investigation was obtained with the help of the questionnaires filled in by consumers of electric automobiles. As a result of adopting the direct approach, this study gathers pertinent primary data from participants as it evaluates the factors affecting their satisfaction with electric cars.

3.2. Sampling and sample size

Convenience sampling approach was used in the study to distribute the questionnaires among 350 owners of electric vehicles (EVs). Of these only 15 of the surveys were left incomplete and 51 users never returned the completed ones. Consequently, 284 respondents who filled the completed questionnaire formed the full sample used in the analysis. This sampling technique enabled a comprehensive assessment of customer satisfaction levels of consumers of EVs in the research area.

3.3. Tools for analysis

Factor analysis and Simple Frequency is employed.

3.4. Significance of the study

This research is important for the following reasons. First of all, it provides a deeper understanding of the determinants of electric car user satisfaction that have to become key to the growth of the market and its sustainability. Manufacturers have the potential of improving their products and, therefore, make consumers happier and loyal to their brands by attending to their grievances. This is made possible by the findings of this study. Second, using consumers' requirements and preferences information, this study helps policy makers to design better laws and incentives to bring more improvements to EVs market. In addition, improved clarification of consumers' buying behaviour by marketers assists them in the formulation of appropriate marketing strategies to target prospective EV buyers. After that, by identifying current research limitations on customer satisfaction with e-cars, especially in socioeconomic and cultural contexts, this research contributes to the pool of scholarly literature. These results can serve as a basis for further scholarly researches and presentations on sustainable means for transports.

4. Limitations of the Study

All the same this research is not without some weaknesses as highlighted below. First of all, concerning the type of the sample, it is not likely to represent all of the EV users, therefore the usage of the snowball sampling may limit the generalisability of the findings. Additionally, the study is based on a sample size of 284 participants which would limit the flexibility of recording general conclusions concerning customer satisfaction from different demographics and geographically different regions. Moreover there are a number of limitations within this research; first and foremost regarding the self-report data, which was filled by the participants some common source biases such as memory bias, and social desirable responding may have affected the results of the study. Further, the nature of using a questionnaire may reduce the extent of the responses and hide other other small details that may be very important to the customers. Subsequently, because this study is small and concentrated in one area, its findings may not accurately reflect the experiences and satisfaction levels of electric car owners in other areas with distinct socioeconomic, cultural, and structural characteristics. As a result, not everyone may find the findings applicable. Larger, more varied sample sizes and the use of qualitative techniques should help future studies yield deeper, more thorough findings.

5. Findings

5.1. Socio-economic profile

A study analyzes the population-based attributes and financial traits of electric vehicle drivers across Coimbatore district focusing on age groups, gender profiles, professions, educational backgrounds, household composition, caretaker roles, along with household salaries and costs (Table 5.1).

Table 5.1. Socio-economic profile

Variables	*Description*	*Frequency*	*Percent*
Occupation	Business	34	12
	Home Maker	26	9.2
	Employee	138	48.6
	Agriculturist	56	19.7
	Others	30	10.6
Age	Up to 25	22	7.7
	26–35	230	81
	36–45	20	7
	Above 45	12	4.2
Family Income (Rs.)	Up to 30000	116	40.8
	30001–50000	102	35.9
	Above 50000	66	23.2
Monthly Income (Rs.)	Up to 20000	106	37.3
	20001–35000	104	36.6
	35001–50000	30	10.6
	Above 50000	44	15.5
Educational Qualification	UG	24	8.5
	PG	164	57.7
	Diploma	12	4.2
	Others	84	29.6
Gender	Female	122	43
	Male	162	57
Type of Family	Joint	84	29.6
	Nuclear	200	70.4
Status in Family	Member	244	85.9
	Head	40	14.1
Family Expenditure (Rs.)	Up to 20000	92	32.4
	20001–35000	100	35.2
	35001–50000	50	17.6
	Above 50000	42	14.8

Brief Interpretation.

Source: Author.

Studies indicate that Coimbatore residents mainly select electric vehicles because they are younger adults during their 26–35 years (81%) and working employees (48.6%). Males (57%) slightly outnumber females (43%). Postgraduate education is the highest educational level for users who represent 57.7% of the population while nuclear families make up 70.4% of the respondents.

The demographic data shows that EV ownership is evenly spread among Indian residents since 37.3% earns up to ₹20,000 and 36.6% falls within the ₹20,001-₹35,000 interval. The data shows that household budgets make electric vehicles affordable because 35.2% of families allocate ₹20,001-₹35,000 for monthly expenses.

Motor vehicle adoption is rising within multiple income groups while varying demographics but elderly people together with those living in joint families still exhibit limited electric vehicle adoption tendencies. Specified incentives along with awareness initiatives would help increase the adoption of EVs among these target groups.

5.2. E Vehicle usage

Most EV users (56.3%) have been using them for 1 to 4 years, while 36.6% are recent adopters (Table 5.2). Nearly

Table 5.2. E Vehicle usage

Variables	*Description*	*Frequency*	*Percent*
Periodicity of Using e-vehicle	Below 1 year	104	36.6
	1 to 4 years	160	56.3
	Above 4 years	20	7.0
Average Kilometers per day	1 to 10 kms	94	33.1
	11 to 20 kms	50	17.6
	Above 20 kms	140	49.3
Sources of Information to buy e-vehicle	Colleagues	28	9.9
	Relatives	36	12.7
	Self	154	54.2
	Friends	66	23.2
Frequency of Usage of e-vehicle	Frequently	94	33.1
	Occasionally	66	23.2
	Whenever Needed	124	43.7
Type of e-vehicle	Four-wheeler	52	18.3
	Two-wheeler	232	81.7
Brand of using e-vehicle	Ather	42	14.8
	Hero	44	15.5
	OLA	60	21.1
	TATA	52	18.3
	TVS	44	15.5
	Others	42	14.8

Source: Author.

49.3% travel over 20 km daily, indicating frequent use. Self-research (54.2%) is the primary source of purchase decisions, followed by friends and relatives. Usage varies, with 43.7% using EVs as needed, while 33.1% use them frequently. Two-wheelers dominate (81.7%), showing affordability and convenience. Brand preferences are diverse, with OLA (21.1%) and TATA (18.3%) leading.

6. Satisfaction Towards E Vehicles

High sampling adequacy exists in the dataset because the KMO value reached 0.869. The application of factor analysis finds validation through Bartlett's Test which produced a significant result of Sig. = 0.000. These tests confirmed that the dataset was enough to conduct factor analysis (Table 5.3).

The Rotated Component Matrix breaks factors into three components that emerge from their correlation patterns (Table 5.4). The first component delivers 31.25% of the total variance through its evaluation of cost efficiency alongside support elements such as comfort, savings, incentives, maintenance, pricing, and dealership assistance – Component 2 (23.76% variance) – Emphasizes performance and sustainability, covering driving experience, environmental impact, safety, battery capacity, and range. The third component with 8.10% variance explains infrastructure aspects and model array which includes insurance costs and charging features and battery capabilities. These three components combined offer effective explanation for 63.10% of the total variation in customer satisfaction factors for electric vehicles.

The first factor explains how e-vehicle customers feel a satisfaction level of 31.247 percent. Customer satisfaction toward e-vehicles receives 23.755 percent and 8.096 percent contributions from the other factor. The percentage evaluation of e-vehicle customer satisfaction reaches 63.098 percent.

7. Suggestions

- The company should establish distribution networks through local suppliers to maintain operational readiness regarding spare part deliveries.
- A unified parts standardization approach for all EV models should be implemented to simplify inventory management systems.
- A real-time inventory tracking system should be deployed for providing instant visibility into which spare parts are available.

Table 5.3. KMO and Bartlett's test

Kaiser-Meyer-Olkin Measure of Sampling Adequacy.		.869
Bartlett's Test of Sphericity	Approx. Chi-Square	4155.000
	df	231
	Sig.	.000

Source: Author.

Table 5.4. Rotated component matrixa

	Component		
	1	*2*	*3*
Interior Space and Comfort	.867		
Cost savings	.866		
Government support and incentives	.853		
Low Maintenance Cost	.842		
Innovation and technology	.826		
Vehicle Price	.817		
Warranty and Extended Service Plans	.802		
Local Dealership Support	.763		
Convenience of charging at home	.719		
Overall Driving Experience		.805	
Environmental impact		.800	
Resilience in Extreme Weather Conditions		.784	
Safety features		.777	
Battery Capacity		.747	
Performance and Acceleration		.719	
Time taken to charge		.672	
Range on a Full Charge		.642	
Availability of Spare Parts		.633	
Insurance Costs			.835
Variety of models			.783
Availability of charging infrastructure			.652
Battery Life			.628
Eigenvalues	6.867	5.214	1.770
% of Variance	31.247	23.755	8.096
Cumulative %	31.247	55.002	63.098

Source: Author.

- The company should use mobile service vans that include critical spare parts for immediate fault correction while on the move.
- The company should team up with insurers to build special EV insurance policies which provide premium discounts.
- Insurance companies should receive implementation training about EV safety innovations to justify fair rate assessments.
- EV users who set lower mileage should benefit from usage-based insurance programming and pay-as-you-drive insurance schemes.
- The company should establish training sessions to teach insurers about EV advantages.
- The company extends warranties for EV batteries to deliver lasting peace of mind to customers.

- Free and affordable battery health assessment services are available for all customers.
- An educational program for users about proper battery care should be provided to improve both system performance and battery life.
- Customers should receive instructions to stay away from both extreme heat and charging their batteries beyond capacity because it helps prolong their lifespan.
- The accessibility of affordable battery replacements should be harmonized by implementing eco-friendly programs for battery recycling.
- The organization should team up with public as well as private partners to expand charging station networks across urban areas and suburban regions and rural locations.
- Investments into ultra-fast chargers installed at important locations should lower charging time alongside improving user comfort.
- To increase accessibility the company should provide incentives together with discounts when customers buy home charging stations.
- The development of better in-car and mobile navigation technology needs to incorporate real-time monitoring of charging station information including their sites and availability as well as their pricing structure.
- Streamline the purchase process with transparent pricing, clear communication of government incentives, and accessible financing information.
- The website should feature calculators which enable buyers to estimate the complete ownership expenses from fuel benefits together with governmental rebates and insurance packages.
- The company needs to enhance battery performance continuously to extend driving distances while speeding up charging capabilities.
- The team aims to guarantee vehicle reliability across different situations together with affordable maintenance expenses.
- The company delivers top-quality support services through expert technicians who provide effective problem-solving and maintain an easily understandable FAQ section.
- The company will maintain regular contact with customers through phone and email and it will help them resolve their problems while showing them how to optimize their vehicle performance.
- Free charging equipment installations and decreased annual maintenance costs should be offered to new EV purchasers.
- The company should establish a referral program that grants customers free vehicle services or enhanced features when they successfully refer new clients.
- The company supports green manufacturing through every destination by promoting sustainable energy use along with responsible product packaging while establishing complete battery recycling systems.
- Present the vehicle to wider customer segments by increasing the variety of interior customization features along with new colour schemes and technical functions.
- The system should allow customers to activate driving mode choices and download software improvements that adapt to their individual preferences.
- The company should extend warranty coverage together with affordable maintenance services including the battery system and drivetrain as well as charging infrastructure.
- Customers should receive free scheduled vehicle examination together with software updates that preserve peak vehicle functionality during their lifespan.

8. Conclusion

Customer satisfaction factors for electric vehicle adoption in Coimbatore district have been analyzed by taking into account socio-economic profiles and usage patterns throughout the research. The majority of participants belong to youth demographics with education backgrounds from nuclear families who possess average income and spend habits. Both data and customer behaviour patterns show increasing preferences for electric two-wheelers over other options since they remain affordable and convenient while serving extended distances that exceed 20 kilometres for daily use. Customers show the highest affinity towards OLA and TATA because they view both companies as trustworthy innovative firms.

Customer satisfaction factors exist in three dominant dimensions according to factor analysis results. The first component of satisfaction contributing 31.257% centers around cost efficiency and governmental backing and convenience aspects. This phase demonstrates 23.744% weightage which combines environmental factors and driving performance characteristics as well as vehicle features. The accessibility of fundamental services which include spare parts and insurance as well as charging infrastructure form the basis of this factor at 8.093%. These dimensions along with the others demonstrate how customer satisfaction stems from multiple economic environmental and practical factors. The 63.094% variation explained indicates a strong potential for upgrading the quality of EV ownership. The acceleration of EV adoption depends heavily on developing better affordability as well as environmental advantages and infrastructure construction.

9. Scope for Further Research

New studies must investigate difficulties EV owners encounter which include problems with available charging stations and battery performance alterations based on environment variations. Additional research based on user comparisons between cities and countryside conditions will deliver comprehensive results. Studies that monitor consumer satisfaction patterns across time significantly aid in developing policies because the industry continues to evolve. The integration of behavioural and psychological elements that affect EV adoption should be studied because it will help develop more focused marketing approaches.

References

[1] International Energy Agency (IEA). (2020). *Global EV Outlook 2020 Entering the decade of electric drive?* IEA Publications. Retrieved from https//www.iea.org/reports/global-ev-outlook-2020

[2] Hertzke, P., Müller, N., & Schenk, S. (2017). *The global electric-vehicle market is amped up and on the rise.* McKinsey & Company. Retrieved from https//www.mckinsey.com/

[3] Rezvani, Z., Jansson, J., & Bodin, J. (2015). Advances in consumer electric vehicle adoption research: A review and research agenda. *Transportation Research Part D Transport and Environment, 34*, 122–136. https//doi.org/10.1016/j.trd.2014.10.010

[4] Egbue, O., & Long, S. (2012). Barriers to widespread adoption of electric vehicles: An analysis of consumer attitudes and perceptions. *Energy Policy, 48*, 717–729. https//doi.org/10.1016/j.enpol.2012.06.009

[5] Nguyen, T., Yang, Y., Shrestha, P., & Lin, L. (2020). Factors influencing electric vehicle adoption Insights from Vietnam. *Sustainability, 12*(17), 7187. https//doi.org/10.3390/su12177187

[6] Zhou, Y., Wang, M., & Li, J. (2018). Understanding the consumer preferences for electric vehicles: A systematic review and future research directions. *Renewable and Sustainable Energy Reviews, 102*, 423–434. https//doi.org/10.1016/j.rser.2018.12.022

[7] Kim, J., Rasouli, S., & Timmermans, H. J. (2019). Expanding scope of electric vehicles and consumer satisfaction: A review of the literature. *Transportation Research Part D Transport and Environment, 70*, 225–233. https//doi.org/10.1016/j.trd.2019.04.004

[8] Liao, F., Molin, E., & van Wee, B. (2017). Consumer preferences for electric vehicles: A literature review. *Transport Reviews, 37*(3), 252–275. https//doi.org/10.1080/01441647.2016.1230794

[9] Sovacool, B. K. (2018). The neglected social dimensions to a vehicle-to-grid (V2G) transition: A critical and systematic review. *Environmental Research Letters, 13*(12), 123001. https//doi.org/10.1088/1748-9326/aaec9c

[10] Huang, X., & Qian, X. (2018). Consumer preferences for electric vehicles in lower-tier cities of China Evidence from south Jiangsu province. *Journal of Cleaner Production, 202*, 1061–1070. https//doi.org/10.1016/j.jclepro.2018.08.219

[11] Noel, L., Zarazua de Rubens, G., & Sovacool, B. K. (2019). Exploring electric vehicle adoption in a rapidly changing context: The case of India. *Transportation Research Part D Transport and Environment, 70*, 124–137. https//doi.org/10.1016/j.trd.2019.03.014

[12] Nicholas, M., Hall, D., & Lutsey, N. (2019). Quantifying the electric vehicle charging infrastructure gap across U.S. markets. *International Council on Clean Transportation (ICCT)*. Retrieved from https//theicct.org/publications/charging-gap-US

[13] Bubeck, S., Tomaschek, J., & Fahl, U. (2016). Perspectives of electric mobility Evaluation of factors influencing market penetration. *Transport Reviews, 36*(3), 405–425. https//doi.org/10.1080/01441647.2015.1100757

[14] Hardman, S., Shiu, E., & Steinberger-Wilckens, R. (2018). Comparing high-end and low-end early adopters of battery electric vehicles. *Transportation Research Part A Policy and Practice, 88*, 40–57. https//doi.org/10.1016/j.tra.2016.03.010

[15] Lee, J., & Clark, C. E. (2018). Electric vehicle development in the early stages: An assessment of consumer preferences and the impact of market structure. *Energy Policy, 115*, 436–444. https//doi.org/10.1016/j.enpol.2018.01.043

[16] Steinhilber, S., Wells, P., & Thankappan, S. (2013). Socio-technical transitions in the automotive sector: The case of the electric vehicle. *Technological Forecasting and Social Change, 80*(8), 1418–1430. https//doi.org/10.1016/j.techfore.2013.03.017

[17] Mersky, A. C., Sprei, F., Samaras, C., & Qian, Z. S. (2016). Effectiveness of incentives on electric vehicle adoption in Norway. *Transportation Research Part D Transport and Environment, 46*, 56–68. https//doi.org/10.1016/j.trd.2016.03.011

[18] Jansson, J., Nordlund, A., & Westin, K. (2017). Examining drivers of sustainable consumption: The influence of norms and opinion leadership on electric vehicle adoption in Sweden. *Journal of Cleaner Production, 154*, 176–187. https//doi.org/10.1016/j.jclepro.2017.03.186

[19] Moons, I., & De Pelsmacker, P. (2015). An extended decomposed theory of planned behaviour to predict the usage intention of the electric car: A multi-group comparison. *Sustainability, 7*(5), 6212–6245. https//doi.org/10.3390/su7056212

[20] Rehman, S. U., Bresciani, S., Yahiaoui, D., & Kliestik, T. (2024). Customer satisfaction leading the intention to adopt battery electric vehicles with the moderating role of government support and status symbol. *Journal of Cleaner Production, 456*, 142371.

[21] Kumar, M. N. K., & Raja, S. A study on customer satisfaction towards online portals with reference to Svm Kia automobiles. *Emperor Journal of Finance, 3*(5). doi:10.58394/ejf.2024.3513

[22] Boonchunone, S., Nami, M., Krommuang, A., Phonsena, A., & Suwunnamek, O. (2023). Exploring the effects of perceived values on consumer usage intention for electric vehicle in Thailand: the mediating effect of satisfaction. *Acta Logistica, 10*(2), 151–164. doi:10.22306/al.v10i2.363

[23] Bernarto, I. (2023). Effect of price fairness, service quality, brand imaged and product quality on consumer satisfaction Electric Cars (Ev) in Indonesia. *Enrichment: Journal of Management*, doi:10.35335/enrichment.v13i1.1206

[24] Wang, S., & Liu, Q. (2020, June). Analysis on the influencing elements and optimization countermeasures of new energy vehicles' consumer satisfaction—Analysis based on internet online review mining. In *IOP Conference Series: Earth and*

Environmental Science (Vol. 510, No. 2, p. 022020). IOP Publishing.

[25] Selvi, S. (2017). A study on customer satisfaction towards electric bikes with special reference to Coimbatore city. *International Journal of Applied Research, 3*(5), 355–359.

[26] Jansson, J., Marell, A., & Nordlund, A. (2010). Green consumer behavior Determinants of curtailment and eco-innovation adoption. *Journal of Consumer Marketing, 27*(4), 358–370. https//doi.org/10.1108/07363761011052396

[27] Hackbarth, A., & Madlener, R. (2016). Willingness-to-pay for alternative fuel vehicle characteristics: A stated choice study for Germany. *Transportation Research Part A Policy and Practice, 85*, 89–111. https//doi.org/10.1016/j.tra.2015.12.005

6 Urban forestry and public health studies: A case study of Tebet eco park, Jakarta

Zarina Akbar[a], Dian Alfia Purwandari[b], Alya Fajria Rahma Maulida[c], Irsyad Hafiz[d], Muhammad Fadhil Haritsah[e], and Vranz Damar Sophan[f]

Master of Environmental Education, Postgraduate School, Universitas Negeri Jakarta, Jakarta, Indonesia

Abstract: Urban forestry has been recognized as one of the solutions to overcome public health problems amid rapid urban development. Declining air quality, increasing pollution, and lack of green open space seriously challenge city residents' physical and mental health. This research examines how urban forestry contributes to enhancing public health through a case study at Tebet Eco Park Jakarta. Using descriptive qualitative methods and literature study, this study examined the contribution of urban forestry as a green open space in improving physical health through the provision of activity spaces and reducing the level of air pollution which is one of the main causes of respiratory diseases. In addition, urban forests have been proven to positively affect mental health by reducing stress and restoring mental health. The findings of the study revealed that urban forestry at Tebet Eco Park significantly contributes to public health by both lowering the risk of environmental diseases and enhancing mental well-being. This research examines the impact of urban forestry and seeksi to deepen the understanding of how urban forestry can enhance the quality of life for city residents.

Keywords: Urban forestry, public health, urban forest, urban development

1. Introduction

Population growth in urban areas such as Jakarta triggers significant changes that contribute to environmental pollution. As the population grows, the need for housing, transportation, energy, and other supporting infrastructure increases significantly. This phenomenon creates great pressure on the environment. In addition to having an impact on the environment, rapid urbanization in big cities often harms public health. Overcrowded neighborhoods caused by overpopulation, elevated air pollution from higher vehicle usage, and a shortage of green spaces all lead to a decline in the quality of life and a greater risk of mental and physical health issues such as respiratory problems, stress, and other mental health concerns. On the other hand, urban forestry as part of a green open space arranged with natural vegetation has great potential in improving the environment and the quality of urban community health.

Urban forestry provides a variety of ecoloogical and health advantages that are crucial for urban communities. From an ecological perspective, urban forestry can lower air temperature, absorb pollutants, and provide oxygen for the surrounding environment. Meanwhile, on the health side, studies showed that urban forestry can improve air quality, reduce stress, and encourage people's physical activity through the provision of space for sports or recreation. In Indonesia, the discourse on urban forestry has become part of the urban planning plan to address environmental and health issues, but its implementation still faces various challenges.

With the high population and limited green open space in urban areas such as Jakarta, the need for urban forestry is increasingly urgent. One of the urban forests, namely Tebet Eco Park, carries the theme of Connecting People with Nature. It not only plays a significant role in the environment but also can integrate overall public health through urban forestry in urban development. Therefore, this study aimed to describe the role of urban forestry, its contribution to the improvement of the physical and mental health of urban communities, and the impact of urban forestry in improving public health.

2. Theoretical Review

2.1. *Public health and environmental health*

Public health has a long history of discussing human life and the environment. The Indonesian Association of Environmental Health (IEAH) defines environmental health as the state of the environment that fosters a harmonious ecological balance between humans and their surroundings, aiming to achieve a high quality of life for individuals. [1]. According to the World Health Organization (WHO), environmental health refers to the ecological balance that needs to be maintained between humans and their environment to safeguard human health [2].

[a]zarina_akbar@unj.ac.id, [b]dian-alfia@unj.ac.id, [c]alfarama1971@gmail.com, [d]irsyaadhafzd1237@gmail.com, [e]mfh4596@gmail.com, [f]damarvranz994@gmail.com

DOI: 10.1201/9781003684459-6

2.2. Urban forestry

Urban forestry is an approach to the planning and management of vegetation (such as trees and greenery) in urban areas to provide ecological, social, and economic benefits to the people living in the area. This concept focuses on the importance of the presence of vegetation amid an urban environment dominated by concrete, asphalt, and other infrastructure buildings that tend to affect air quality, temperature, and people's welfare.

Urban forestry was first defined as a practice designed to create, manage, and protect forest resources in and around cities. Urban Forestry is 'the planning and management of all vegetation, especially trees, in urban areas to improve the quality of the environment and the lives of citizens in cities' [3]. These trees, parks, and urban forests not only enhance the visual appeal but also play a crucial role in preserving the health of urban ecosystems. In some of the world's major cities, urban forestry spaces are designed to be accessible to the public, providing green spaces essential for recreation, education, and environmental conservation.

Urban forests represent a crucial and progressively valuable element of the urban landscape. Trees and urban forests contribute to the enhancement of urban spaces, improving their suitability for living, working, and recreational activities. Studies on the preferences and behaviors of urban residents underscore the substantial influence that trees and forests have in elevating the quality of life within urban environments.

2.3. The function of urban forestry

Urban forestry is a form of green infrastructure that offers a range of environmental, economic, social, and health benefits and services in urban areas. In addition to having these functions, urban forestry also has significant benefits for the health of urban people. Urban forestry can increase people's physical activities, such as walking, jogging, and cycling. These activities contribute to heart health, obesity reduction, and fitness improvement. The urban forests encourage people to participate in physical activities, such as running, walking, or cycling. Physical activities done in a natural environment have been shown to be beneficial for heart health, improve physical fitness, and help maintain an ideal weight [4]. In addition, urban forestry has a soothing effect that helps alleviate stress, anxiety, and depression. This aligns with the claim [5] that green spaces, like urban forests, offer areas for relaxation and recreation, which can aid in lowering stress levels. Activities such as jogging in the park or just sitting under a tree can have a calming effect and improve psychological well-being.

3. Methodology

This study used a qualitative approach with a descriptive research method and was supported by literature studies. Data collection was carried out by observation and interviews at Tebet Eco Park located in Tebet, East Jakarta. Meanwhile, the analysis was deepened using literature studies, which is research conducted by collecting several books, and magazines related to research problems and objectives.

4. Results and Discussion

The rapid development of the city has brought social, economic, and environmental changes. The swift urbanization in a big city like Jakarta brings significant environmental challenges, including air, sound, water, and soil pollution, land degradation, global warming, and the lack of easily accessible green open spaces. This situation not only negatively affects the environment but also has significant implications for the physical and mental health of urban populations. The detrimental effects on the well-being of individuals living in urban areas are underscored by data from the Ministry of Environment and Forestry, which indicates that air quality in Jakarta frequently reaches unhealthy levels, primarily due to emissions from motor vehicles and industrial activities [6].

As a solution, the government and various environmental organizations encourage the implementation of urban forestry to improve air quality, reduce pollution, and improve the quality of life of people in urban environments. In this context, urban forestry emerged as a strategic solution to overcome various health problems faced by urban communities. The concept of urban forestry involves the management and preservation of trees and vegetation within urban areas to enhance the quality of life by providing functional green open spaces.

One example of the applications of urban forestry in Jakarta is Tebet Eco Park, a green open space that functions as a recreational area as well as a conservation area. The park was designed to provide various facilities, including walking paths, jogging tracks, Pilates areas, gymnastics, playgrounds, and environmental education areas that are expected to become an oasis in Jakarta's density. The presence of Tebet Eco Park shows how important green spaces are in urban areas, especially in facing health challenges related to urbanization. Research also confirmed that urban forestry in urban areas helps create a healthier environment and supports better air quality [7], which leads to a reduction in the risk of air pollution-related diseases, such as asthma and cardiovascular diseases [8].

This supports a study that highlighted the role of urban forests in filtering air pollutants, which helps mitigate some of the harmful health effects of air pollution [8]. Evidence for this includes the lower rates of lung cancer in areas with tree-covered green spaces, improved mental health in pregnant women, reduced obesity rates in toddlers, and enhanced immune function [9]. This is likely due to higher levels of physical and social activity, along with decreased exposure to air pollution in urban settings. However, it is crucial to note that some types of urban trees may release allergenic pollen, which can negatively impact the health of vulnerable individuals [10]. As such, identifying the tree species that are most likely to trigger allergic reactions is essential.

However, the benefits of urban forestry are not limited to physical health. The green environment provided by Tebet

Eco Park also provides a calming effect that can improve psychological well-being. This is crucial considering that the pressures of living in a big city can cause psychological disorders such as stress and anxiety. According to a study conducted in urban areas in Indonesia, greater access to green open spaces is positively related to reduced stress levels and increased life satisfaction [11].

According to Experience of Nature: A Psychological Perspective [12], the natural environment can help people recover from mental fatigue because the natural atmosphere provides a different experience than the artificial or urban environment. In Attention Restoration Theory (ART), it is emphasized that the natural environment can restore mental attention that is drained due to the stress of life in the city so that it functions as a recovery space for people's mental health. The natural environment that supports this element of ART can reduce stress symptoms, improve mood, improve cognitive function, and even improve overall quality of life. Research has proven that those who interact frequently with the natural environment tend to experience reduced stress, faster cognitive recovery, and better psychological well-being.

Urban Forestry creates strategic green spaces in urban areas with ecological and social functions. Its ecological functions, such as pollutant and carbon absorption, have been shown to help reduce air pollution levels and maintain environmental quality. Meanwhile, the social function of urban forestry, among others is providing places for recreation and relaxation. This certainly contributes significantly to the mental and physical health of urban people who live in stressful routines. Urban communities with access to green spaces, such as urban forests, have been shown to experience lower stress levels, greater happiness, and an improved quality of life compared to those without access to such spaces [13].

In the context of urban forestry, the ART theory is very relevant. Green spaces such as city parks, urban forests, or neighborhood parks provide opportunities for urban communities to get a restorative effect without leaving the city. These parks not only serve as an environment protector and pollution absorber but also provide a place for people to escape the pressures of city life, find peace, and reduce mental fatigue. In Tebet Eco Park, for example, the park design that includes walking paths, flower gardens, and small lakes provides a restorative experience that fits the concept of ART allowing visitors to be secluded for a moment, enjoy the soothing natural charm, and feel the vast visual depth. In this context, Tebet Eco Park is an ideal example of how the application of Urban Forestry can be a sustainable solution for urban communities' physical and mental health [14].

5. Conclusion

The findings of several studies indicate that trees and urban forests play a key role in enhancing the physical and mental health of city dwellers. Green spaces, such as urban forests, have been found to have a healing effect that helps in recovering from stress and mental fatigue. Furthermore, trees in urban environments play a crucial role in reducing pollutants and lowering air temperatures, which directly helps to improve air quality. Based on this evidence, numerous urban forestry initiatives are now specifically focused on promoting human health by providing ecosystem services like pollution control and temperature regulation. This highlights the need to incorporate urban forests into city planning to enhance the quality of life for residents.

References

[1] Marisa, A. R., Putri, F. E., Rahmasari, K., Salsabila, N. D., & Dewi, O. C. (2024, December). Like or dislike? Evaluating Jakarta's green open space through digital perception: case study of Tebet Eco Park. In *IOP Conference Series: Earth and Environmental Science* (Vol. 1419, No. 1, p. 012066). IOP Publishing.

[2] Donovan, G. H. (2017). Including public-health benefits of trees in urban-forestry decision making. *Urban Forestry & Urban Greening, 22*, 120–123.

[3] Van den Bosch, M. (2017). Impacts of urban forests on physical and mental health and wellbeing. In *Routledge handbook of urban forestry* (pp. 82–95). Routledge.

[4] Lakshmisha, A., Nazar, A. F., & Nagendra, H. (2024). Nature based solutions in cities of the Global South–the 'where, who and how'of implementation. *Environmental Research: Ecology*.

[5] O'Brien, L., Williams, K., & Stewart, A. (2010). Urban health and health inequalities and the role of urban forestry in Britain: A review. *Forest Research.*

[6] Edelman, D. J., & Gunawan, D. S. (2020). Managing the urban environment of Jakarta, Indonesia. *Current Urban Studies, 8*(01), 57.

[7] Ostoić, S. K., & van den Bosch, C. C. K. (2015). Exploring global scientific discourses on urban forestry. *Urban Forestry & Urban Greening, 14*(1), 129–138.

[8] Wolf, K. L., Lam, S. T., McKeen, J. K., Richardson, G. R., van Den Bosch, M., & Bardekjian, A. C. (2020). Urban trees and human health: A scoping review. *International Journal of Environmental Research and Public Health, 17*(12), 4371.

[9] Nowak, D. J., Hirabayashi, S., Doyle, M., McGovern, M., & Pasher, J. (2018). Air pollution removal by urban forests in Canada and its effect on air quality and human health. *Urban Forestry & Urban Greening, 29*, 40–48.

[10] Livesley, S. J., McPherson, E. G., & Calfapietra, C. (2016). The urban forest and ecosystem services: impacts on urban water, heat, and pollution cycles at the tree, street, and city scale. *Journal of Environmental Quality, 45*(1), 119–124.

[11] Yin, S., Chen, W. Y., & Liu, C. (2023). Urban forests as a strategy for transforming towards healthy cities. *Urban Forestry & Urban Greening, 81*, 127871.

[12] Kaplan, R., & Kaplan, S. (1989). *The experience of nature: A psychological perspective.* Cambridge University Press.

[13] Sanesi, G., Gallis, C., & Kasperidus, H. D. (2011). Urban forests and their ecosystem services in relation to human health. *Forests, Trees and Human Health*, 23–40.

[14] Kadhim, A. J. (2023). Media in times of crisis: A case study on the impact of media on societal awareness during the COVID-19 pandemic. *MEDAAD, 2023*, 40–44. https://doi.org/10.70470/MEDAAD/2023/006

7 Epileptic wave identification using machine learning in EEG signals

Abdul Mahatir Najar[1,a], *Raihan Anugrah*[1,b], *Resnawati*[1,c], *Juni Wijayanti Puspita*[1,d], *and Rais*[2,e]

[1]Mathematics Study Program, Tadulako University, Palu, Indonesia
[2]Statistics Study Program, Tadulako University, Palu, Indonesia

Abstract: Epilepsy is a prevalent neurological disorder in humans, typically marked by recurrent seizures resulting from brain abnormalities. This study investigated five machine learning methodologies: K-Nearest Neighbours, Random Forest, Learning Vector Quantization, Support Vector Machine, and Naïve Bayes, for the identification of epileptic waves in electroencephalogram signals. The dataset was generated using data from five channels, including one primary channel exhibiting epileptic waves and four auxiliary channels. The K-Nearest Neighbors method exhibited superior performance, particularly when the dimensionality of the dataset was reduced by the Principal Component Analysis Method. Moreover, the Random Forest technique exhibited notable performance stability, but its accuracy was lower than that of K-Nearest Neighbors.

Keywords: Electroencephalogram, machine learning, epilepsy, principal component analysis

1. Introduction

Epilepsy is a highly widespread neurological condition in humans, commonly characterized by recurrent seizures due to brain abnormalities. The World Health Organization reports that more than 50 million people globally are afflicted by this ailment. [5, 11]. Epilepsy typically does not immediately result in death; nonetheless, it induces a range of symptoms that can impact all sensory modalities, awareness, psychological states, and motor functions, hence hindering the individual's physical and social engagements [2]. One of the most frequently employed methods for substantiating an epilepsy diagnosis is the electroencephalogram (EEG). In this study, epileptic waves are the term used to describe aberrant EEG patterns that represent epileptiform activities, which are the hallmark of the epilepsy phenomenon [19].

Typically, an EEGer manually determines whether an EEG signal contains epileptic waves. To mitigate the risk of misdiagnosis, it is essential to provide EGGer with a mechanism capable of autonomously identifying epileptic waves. Numerous investigations have utilized a variety of strategies, encompassing mathematical and statistical model-based approaches as well as machine learning algorithms. Puspita et al. [12] employed the Bayesian method to classify waves obtained from the Walsh transformation into spike or sharp waves. The classification of Interictal Epileptiform Discharges (IEDs) on EEG data, which involved differentiating spike and sharp waves using a Backpropagation Neural Network, was also conducted by Puspita et al. [13]. The authors in [15] utilized three approaches for learning: Logistic Model Trees (LMTs), Support Vector Machines (SVM) and Multinomial Logistic Regression (MLR). The results demonstrate that the LMT classifier achieved greater accuracy than the others, and that the OAT is the most effective method for feature extraction. Nair et al. [9] asserted that AI-based methodologies have markedly improved the diagnosis, prediction, and management of epilepsy, hence boosting healthcare outcomes. Saminu et al. [14] examined and analyzed AI-based methods utilized in computer-aided devices (CAD) systems for the detection and characterization of epileptic seizures.

The aforementioned research primarily focus on EEG data from channels exhibiting epileptic waves. In this study we developed a profile (feature) that derived from five EEG channels, comprising one primary channel and four adjacent channels. Then, we assessed several machine learning models utilizing datasets that were previously dimensionally reduced by PCA.

2. Methods

2.1. Data acquisition

The EEG data utilized in this investigation were obtained from epileptic patients at Hasan Sadikin Hospital in Bandung, Indonesia. The EEG data was acquired using the 10–20 electrode placement method, comprising 23 electrodes organized in a bipolar configuration. The recording lasted at least 30 minutes. The acquisition of EEG data at a frequency of 500 Hz

[a]abdulmahatir@untad.ac.id, [b]raihananugrah@untad.ac.id, [c]resnawati@untad.ac.id, [d]jw.puspita@untad.ac.id, [e]rais@untad.ac.id

DOI: 10.1201/9781003684459-7

facilitated the monitoring of cerebral activity. The European Data Format (.edf) is utilized for storing data that includes brainwave recordings from both healthy individuals and those with epilepsy, together with diagnoses from healthcare specialists. The dataset's value is augmented by integrating expert diagnoses with raw EEG signals to differentiate and analyze the variances between brain functioning and epileptic events.

During manual observation, the epileptic wave generally affects both the primary channel and the neighbouring channels. Therefore, we gathered EEG data from the channel displaying the epileptic wave along with four other pertinent channels. The chosen data visualization is shown in Figure 7.1, with a focus on the EEG T2-T4 channels, which are very important as the main site of epileptiform activity. The four adjacent channels also exhibit alterations in the pattern. Each channel will be recorded for 0.5 seconds, both before and after the presence of epileptic waves, to capture the attributes of this wave.

2.2. *Profile modeling*

In the process of recording EEG data, a sample frequency of 500 Hz is utilized. Consequently, this indicates that there are 500 samples in each and every second of the EEG signal. The number of samples collected for each channel constitutes 250 samples, which is calculated by multiplying the rate of 500 samples per second by the duration of 0.5 seconds. The amplitude extraction data from the five channels is consolidated into a 5 × 250 matrix. Each row denotes a channel, while each column signifies the amplitude value.

$$A = \begin{bmatrix} a_{1,1} & a_{1,2} & a_{1,3} & \cdots & a_{1,250} \\ a_{2,1} & a_{2,2} & a_{2,3} & \cdots & a_{2,250} \\ a_{3,1} & a_{3,2} & a_{3,3} & \cdots & a_{3,250} \\ a_{4,1} & a_{4,2} & a_{4,3} & \cdots & a_{4,250} \\ a_{5,1} & a_{5,2} & a_{5,3} & \cdots & a_{5,250} \end{bmatrix} \quad (1)$$

where

$A \in \mathbb{R}^{5\times 250}$ is the matrix representing the EEG data.

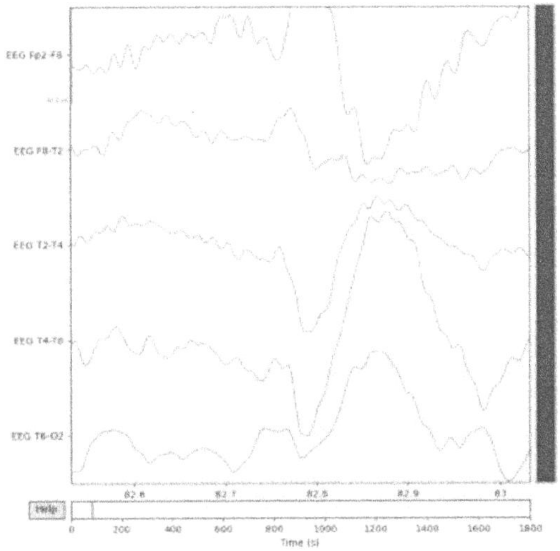

Figure 7.1. Representation of selected data.

Source: Author.

$a_{i,j}$ represents the amplitude of the signal from channel i at time j, where $i \in \{1, 2, 3, 4, 5\}$ and $j \in \{1, 2, \ldots, 250\}$.

To facilitate the machine learning training process, each data which is represented as matrix is transformed into a vector with length 1250. This process is called flattening, let B is the flattened vector from matrix A, then we can write it as:

$$B = [a_{1,1} \quad a_{1,2} \quad \cdots \quad a_{1,250} \quad \cdots \quad a_{5,1} \quad \cdots \quad a_{5,250}] \quad (2)$$

This study utilized a total of 230 data points, comprising 120 classified as epilepsy and 110 classified as non-epilepsy. The next step is to build a matrix that encapsulates the data and its attributes. Assuming each data point and its features are represented by a vector B, a matrix of dimensions 230 × 1250 is constructed with 230 datasets.

$$Dataset = \begin{bmatrix} B_1 \\ B_2 \\ \vdots \\ B_{230} \end{bmatrix} = \begin{bmatrix} b_{1,1} & b_{1,2} & \cdots & b_{1,1250} \\ b_{2,1} & b_{2,2} & \cdots & b_{1,1250} \\ \vdots & \vdots & \cdots & \vdots \\ b_{230,1} & b_{230,2} & \cdots & b_{1,1250} \end{bmatrix} \quad (3)$$

2.3. *Dimensionality reduction*

Data that is represented as a 230 × 1250 matrix demonstrates an extremely large dimension. The success rate of various machine learning techniques is reduced when the dataset possesses a high dimensionality, since it becomes progressively difficult to discern significant patterns. Furthermore, the computation of extremely large dimensions necessitates extended periods of time. Consequently, the principal component analysis algorithm was implemented to minimize the dataset's dimensions in this investigation.

A statistical technique known as principal component analysis is adaptable and can be employed to simplify a case-by-variable data table into its fundamental features, which are referred to as principal components. The most comprehensive explanation of the variance of all variables is provided by principal components, which are linear combinations of the original variables. This method approximates the original data table by employing only a small number of these principal components [1, 7]. The following is an illustration of the procedures employed in this research to reduce the dimension of dataset:

Step 1: Mean Normalization

Calculate the mean of each feature:

$$\mu_j = \frac{1}{230} \sum_{i=1}^{230} x_{i,j} \quad (4)$$

for $j = 1, 2, \ldots, 1250$. Subtract the mean of each feature from the corresponding feature values:

$$X_{norm} = X - \mu, \quad (5)$$

where μ is the mean vector of size 1 × 1250, and X_{norm} is the mean-centered dataset.

Step 2: Compute the Covariance Matrix

The covariance matrix Σ is compute as:

$$\sum \frac{1}{230-1} X_{norm}^{T} X_{norm} \quad (6)$$

Here, Σ is a symmetric matrix of size 1250 × 1250.

Step 3: Eigenvalue and Eigenvector Decomposition
Perform eigenvalue decomposition of the covariance matrix Σ,

$$\sum v = \lambda v, \quad (7)$$

where $\lambda_1, \lambda_2, \ldots, \lambda_{250}$ are the eigenvalues, sorted in descending order ($\lambda_1 \geq \lambda_2 \geq \lambda \geq \lambda_{1250}$).
$\upsilon_1, \upsilon_2, \ldots, \upsilon_{1250}$ are the corresponding eigenvectors.

Step 4: Select the Principal Components
For example, in this scenario, select the four eigenvectors that correspond to the largest four eigenvalues. Form a projection matrix W:

$$W = [v_1\ v_2\ v_3\ v_4] \quad (8)$$

where W is of size 1250 × 4.

Step 5: Project the Data
Project the normalized dataset onto the 4 principal components:

$$Z = X_{norm} W, \quad (9)$$

where Z is the reduced dataset of size 230 × 4.

2.4. Classificiation

The reduced dataset is subsequently incorporated into machine learning algorithms during the classification process to create a model capable of identifying epileptic wave patterns.

2.4.1. K-Nearest Neighbors

K-Nearest Neighbors (KNN) is a machine learning approach that is straightforward and frequently employed. This technique employs a voting mechanism to ascertain the suitable class for an unclassified object by evaluating the class of its nearest neighbors in the decision space [8, 20]. The steps of the algorithm are explained as follows:

1. Represent each data point as a vector in *Rn*, in which *n* is the number of features. For instance, a data point x_i can be expressed as:

   $$x_i = [x_{i1}, x_{i2}, \ldots, x_{in}] \quad (10)$$

   with a corresponding label $y_i \in \{1,2, \ldots, 250\}$. where C is the number of classes
2. Find the distance within x_q and each other points in the dataset. The Euclidean distance is the most frequently

   $$d(x_q, x_i) = \sqrt{\sum_{j=1}^{n} (x_{qj} - x_{ij})^2} \quad (11)$$

   Manhattan distance and Minkowski distance are additional distance metrics that may be implemented depending upon the application.
3. Select the smallest distance of the *k*-nearest neighbors of x_q. Let these neighbors be denoted as $N_k(x_q)$:

   $$N_k(x_q) = \{x_{(1)}, x_{(2)}, \ldots, x_{(k)}\}, \quad (12)$$

   where $x_{(1)}$ represents the *i*-th closet neighbor to x_q.
4. Determine The class of x_q is using majority voting among the *k*-nearest neighbors. Let these $y_{(i)}$ be the label of the *i*-th closet neighbor. The predicted class $\hat{y}_q$ is:

   $$\hat{y} = \arg \max_{c \in \{1,2,\ldots,C\}} \sum_{i=1}^{k} \mathrm{I}(y_{(i)} = c) \quad (13)$$

   where $\mathrm{I}(y_{(i)} = c)$ is the indicator function, which equals 1 if $y_{(i)} = c$, and 0 otherwise.

2.4.2. Random forest

Despite its high computational cost, random forest is still one of the most frequently employed machine learning algorithms due to its reliability and versatility to handle large and varied data. This algorithm enhances classification accuracy by integrating multiple decision trees [3, 10]. The algorithm's stages are delineated as follows:

1. Generate numerous bootstrap samples from the original dataset. Random sampling with replacement is employed to generate each sample.
2. Construct a decision tree for each bootstrap sample according to the given rules:
   - At each node, a random selection of attributes is utilized to ascertain the optimal split.
   - The tree is develop to its deepest level without pruning.
3. Integrate the predicted results of all trees to generate the outcome.
   - For classification, use majority voting among the trees.
   - For regression, take the average of the tree predictions.

2.4.3. Learning vector quantization

Learning Vector Quantisation (LVQ) is a training methodology employed for supervized learning inside a competitive layer of a single-layer network structure. The class operates as a competitive layer determined by the distance between input vectors. The competitive layer will categorize the two input vectors as belonging to the same class if their outcomes are proximate [4, 17]. The stages of the algorithm are outlined as follows:

1. Define k prototypes $\{w_j\}_{j=1}^{k}$ $\{w_j\}_{j=1}^{k}$ randomly or using clustering (e.g., k-means). Each prototype is associated with a specific class label.
2. for each training sample (x_i, y_i):
   a. Find the nearest prototype w_j to x_i using a distance metric:

   $$j = \arg\max_{j} \|x_i - w_j\| \quad (14)$$

b. Update the prototype w_j based on the class label y_i:
   - If y_i = class (w_j) (correct classification):

$$w_j \leftarrow w_j + \alpha(x_i - w_j) \tag{15}$$

   - If $y_i \neq$ class (w_j) (misclassification):

$$w_j \leftarrow w_j - \alpha(x_i - w_j) \tag{16}$$

c. Adjust the learning rate α (e.g., decay α over iterations)

2.4.4. Support vector machine

The Support Vector Machine, invented by Vapnik, is a kernel-based machine learning model employed for classification and regression tasks [6]. The exceptional generalization capability of SVM, along with its optimal solutions and discriminative strength, has garnered significant interest in the domains of data mining, pattern recognition, and machine learning in recent years. Support Vector Machines have been utilized as an efficient tool for tackling actual binary classification problems. Research demonstrates that Support Vector Machine surpass conventional supervized learning methodologies [6, 16].

For a dataset $\{x_i, y_i\}_{j=1}^{n}$ $\{x_i, y_i\}_{j=1}^{n}$, where $x_i \in \mathbb{R}^d$, and $y_i \in \{-1,+1\}$, SVM solves the following optimization problem:

$$\min_{w,b} \frac{1}{2}\|w\|^2 + C\sum_{i=1}^{n} \xi_i \tag{17}$$

Subject to:

$$y_i(w \cdot x_i + b) \geq 1 - \xi_i, \xi_i \geq 0, \forall i \tag{18}$$

With C as a regularization parameter that governs the balance between maximizing the margin and minimizing classification errors.

2.4.5. Naïve Bayes

Naïve Bayes (NB) is a prominent probabilistic classification technique. It is a straightforward yet effective algorithm with numerous practical applications, including product suggestions, medical diagnosis, and the operation of self-driving vehicles [18]. The algorithm is grounded in Bayes' Theorem, which state:

$$P(C|X)\frac{P(X|C)P(C)}{P(X)} \tag{19}$$

Where, $P(C|X)$ is the posterior probability of class C given the feature set X; $P(X|C)$ represents the likelihood of the feature set X given class C; P(C) is The prior probability of class C and P(X) equals Marginal likelihood of the feature set X.

For a given data point X, the Naive Bayes classifier predicts the class label $\hat{C}$ as:

$$\hat{C} = \arg\max_{C} P(C) \prod_{\{i=1\}}^{n} P(x_i|C) \tag{20}$$

3. Results and Discussions

In this investigation, we extracted the amplitude signal from EEG into a matrix, subsequently, a dataset was created for the development of machine learning models to detect epileptic waves. The predictions of the machine learning model were matched with the observations of a trained EEG specialist. The following formula is employed to ascertain the accuracy of each model:

$$accuracy = \frac{P}{P+N} \times 100\% \tag{21}$$

where P is the number of data that identified as epileptic wave both by the certified EEGer and the model and N is the number of data that identified as epileptic wave by the certified EEGer but not detected by the model.

One of the most important steps in this study is the reduction of the dataset's dimensionality. Employing the PCA method, we condense the dataset into three distinct scenarios, which are 3, 4, and 5 dimensions, to assess the impact of dimensionality on the effectiveness of the used machine learning algorithm. Overall, it can be clearly seen that, all five machine learning models exhibit optimal performance with a four-dimensional dataset. This suggests that condensing the information to four dimensions adequately represents the four features that are pertinent and useful for the classification process.

Figure 7.2 illustrates comparison of the developed machine learning models accuracy developed in this investigation. The KNN model demonstrated superior performance in each scenario, surpassing four other machine learning models. This is due to the fact that KNN is one of the machine learning algorithms that rely solely on distance for model training, rather than complex calculations. This results in KNN performing exceptionally well on data that is devoid of significant noise. This condition is consistent with data that has been processed using PCA to mitigate noise.

Another intriguing aspect is the random forest model's performance, which demonstrates stability in each scenario with distinct dimensions. This can be attributed to the random forest algorithm's characteristics, which generate a

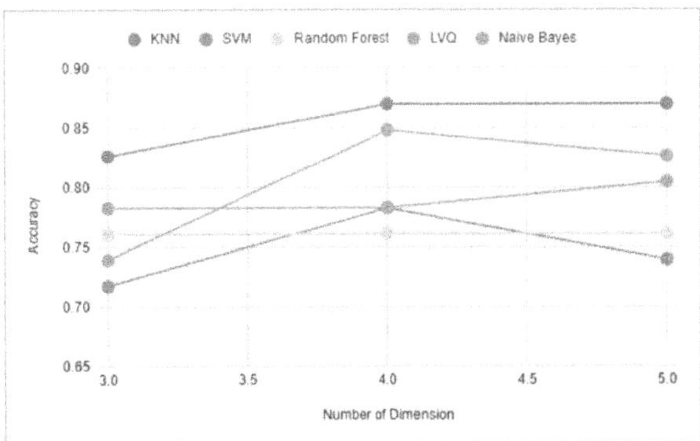

Figure 7.2. Comparison of accuracy of machine learning models for Epileptic wave identification.

Source: Author.

substantial quantity of decision trees during the model training process by utilizing random subsets of data and features. This enhances the model's resilience to modifications in the dataset. Random Forest is a dependable algorithm in dataset situations with high variation or complexity, despite the fact that it does not exhibit the same level of accuracy as KNN.

4. Conclusion

This study assessed five machine learning methods for the identification of epileptic waves. The dataset was created with data from five channels, comprising one primary channel with epileptic waves and four supplementary channels. The KNN technique demonstrated the highest performance, especially when the dataset dimensionality was reduced using PCA. Furthermore, the Random Forest method demonstrated commendable performance stability, while its accuracy was inferior to KNN. This study contributes to the development of a faster and more accurate epilepsy diagnosis method based on EEG signals.

5. Acknowledgement

This research was fully funded by Bima Grant from the Ministry of Research, Technology and Higher Education in 2024.

References

[1] Abdi, H., & Williams, L. J. (2010). Principal component analysis. *Wiley Interdisciplinary Reviews: Computational Statistics*, *2*(4), 433–459.

[2] Alalayah, K. M., Senan, E. M., Atlam, H. F., Ahmed, I. A., & Shatnawi, H. S. A. (2023). Effective early detection of epileptic seizures through EEG signals using classification algorithms based on t-distributed stochastic neighbor embedding and K-means. *Diagnostics*, *13*(11), 1957.

[3] Biau, G., & Scornet, E. (2016). A random forest guided tour. *TEST*, *25*(2), 197–227. https://doi.org/10.1007/s11749-016-0481-7

[4] Borman, R. I., Fernando, Y., Yudoutomo, Y. E. P., & others. (2022). Identification of Vehicle Types Using Learning Vector Quantization Algorithm with Morphological Features. *Jurnal RESTI (Rekayasa Sistem Dan Teknologi Informasi)*, *6*(2), 339–345.

[5] Cano, A., Fonseca, E., Ettcheto, M., Sánchez-López, E., de Rojas, I., Alonso-Lana, S., Morato, X., Souto, E. B., Toledo, M., Boada, M., & others. (2021). Epilepsy in neurodegenerative diseases: Related drugs and molecular pathways. *Pharmaceuticals*, *14*(10), 1057.

[6] Cervantes, J., Garcia-Lamont, F., Rodríguez-Mazahua, L., & Lopez, A. (2020). A comprehensive survey on support vector machine classification: Applications, challenges and trends. *Neurocomputing*, *408*, 189–215.

[7] Greenacre, M., Groenen, P. J., Hastie, T., d'Enza, A. I., Markos, A., & Tuzhilina, E. (2022). Principal component analysis. *Nature Reviews Methods Primers*, *2*(1), 100.

[8] Hamilton, D., Pacheco, R., Myers, B., & Peltzer, B. (2020). kNN vs. SVM: A comparison of algorithms. In: Hood, Sharon M., Drury, Stacy, Steelman, Toddi, Steffens, Ron (Eds.). *Proceedings of the Fire Continuum-Preparing for the Future of Wildland Fire; 2018 May 21–24; Missoula, MT. Proceedings RMRS-P-78. Fort Collins, CO: US Department of Agriculture, Forest Service, Rocky Mountain Research Station. p. 95–109.*, *78*, 95–109.

[9] Nair, P. P., Aghoram, R., & Khilari, M. L. (2021). Applications of artificial intelligence in epilepsy. *International Journal of Advanced Medical and Health Research*, *8*(2), 41–48.

[10] Najar, A. M., Abu, M., Resnawati, R., & Syahrullah, S. (2024). Rice leaf diseases classification using image processing and machine learning. *Transcendent Journal of Mathematics and Applications*, *3*(1), 11–19.

[11] Organization, W. H. (2006). *Neurological disorders: Public health challenges*. World Health Organization.

[12] Puspita, J. W., Gunadharma, S., Indratno, S. W., & Soewono, E. (2017a). Bayesian approach to identify spike and sharp waves in EEG data of epilepsy patients. *Biomedical Signal Processing and Control*, *35*, 63–69.

[13] Puspita, J. W., Soemarno, G., Jaya, A. I., & Soewono, E. (2017b). Interictal Epileptiform Discharges (IEDs) classification in EEG data of epilepsy patients. In *Journal of Physics: Conference Series* (Vol. 943, No. 1, p. 012030). IOP Publishing.

[14] Saminu, S., Xu, G., Zhang, S., Kader, I. A. E., Aliyu, H. A., Jabire, A. H., Ahmed, Y. K., & Adamu, M. J. (2023). Applications of artificial intelligence in automatic detection of epileptic seizures using EEG signals: A review. *Artificial Intelligence and Applications*, *1*(1), 11–25. https://doi.org/10.47852/bonviewAIA2202297

[15] Satapathy, S. K., Jagadev, A. K., & Dehuri, S. (2017). Weighted majority voting based ensemble of classifiers using different machine learning techniques for classification of EEG signal to detect epileptic seizure. *Informatica* (03505596), *41*(1).

[16] Sheykhmousa, M., Mahdianpari, M., Ghanbari, H., Mohammadimanesh, F., Ghamisi, P., & Homayouni, S. (2020). Support vector machine versus random forest for remote sensing image classification: A meta-analysis and systematic review. *IEEE Journal of Selected Topics in Applied Earth Observations and Remote Sensing*, *13*, 6308–6325.

[17] Sumarsono, A., & Supatman, S. (2021). Imagery identification of tomatoes which contain pesticides using learning vector quantization. *Jurnal Teknik Informatika*, *2*(1), 9–16.

[18] Wickramasinghe, I., & Kalutarage, H. (2021). Naive Bayes: Applications, variations and vulnerabilities: A Review of Literature with code snippets for implementation. *Soft Computing*, *25*(3), 2277–2293.

[19] Yogarajan, G., Alsubaie, N., Rajasekaran, G., Revathi, T., Alqahtani, M. S., Abbas, M., Alshahrani, M. M., & Soufiene, B. O. (2023). EEG-based epileptic seizure detection using binary dragonfly algorithm and deep neural network. *Scientific Reports*, *13*(1), 17710.

[20] Zhang, S., & Li, J. (2021). KNN classification with one-step computation. *IEEE Transactions on Knowledge and Data Engineering*, *35*(3), 2711–2723.

8 Exploring the impact of customer education and environmental concern on green product purchase intentions across generational cohorts

Husna Mohamed[1,a] *and Meena Suguanthi G.*[2,b]

[1]Research Scholar, Department of Management, Karpagam Academy of Higher Education, Coimbatore, Tamil Nadu, India
[2]Assistant Professor, Department of Management, Karpagam Academy of Higher Education, Coimbatore, Tamil Nadu, India

Abstract: A study examines the connection between educational customer programs such as eco-friendly and natural products and customer purchasing choices. Positive attitudes resulting from consumer product benefit comprehension lead to better purchase intentions. Research evidence shows that Millennials and Gen Z prefer green products to a greater extent than Baby Boomers who care mainly about product simplicity. A business needs to provide educational outreach and manage information gaps to succeed as this method creates effective ways to serve specific customer needs throughout different demographic groups. Consumer education stands as the primary benefit that strengthens sustainable practices alongside promoting healthy choices and therefore this research advocates for businesses to establish market strategies for future sustainable development.

Keywords: Customer education, environmental concern, green products, consumer purchase intention, generational cohorts, green consumption behaviour, sustainable practices

1. Introduction

Educational efforts directed at customers who buy green organic products provide vast advantages to the general public. Educated consumers develop improved perceptions of sustainable products which makes them more likely to purchase such items [1, 2]. Organizations can achieve consumer-based organic benefit knowledge and sustainable purchasing motivation through properly designed education programs [3]. Knowledge-empowered shoppers will choose environmentally good merchandise since they see these choices benefit their personal health along with environmental preservation [4, 5]. Educational programs designed for consumers should provide exact steps to turn awareness into actual purchase choices.

The manner in which consumers behave significantly relates to generational factors because people develop their opinions and preferences through their technological experiences and social expectations and environmental awareness [6]. Youths during the Millennial and Gen Z age brackets show enhanced commitment to environment-friendly consumption compared to Baby Boomer buyers who opt for current lifestyle practices and simple accessibility [7]. Organizations in business need to create custom market strategies for their selected demographic areas that represent the specific traits of each generation segment. To obtain better consumer acceptance and sustainability businesses need to modify their individual generational educational materials and marketing plans. According to Williams et al. [8] the sustainable practices adopted by educated consumers create joint advantages for economic development together with business-corporate society relations.

1.1. *Factors affecting consumer purchase intention*

Sustainable product purchasing from consumers hinges on three psychological elements which intertwine based on public market perceptions and societal values. The process through which consumers evaluate their purchases depends on three key components since Customer Education works as a mediator between Environmental Concern along with Generational Cohorts serving as the moderating variable [9]. Organizations must develop specific market-based plans so they can influence customer purchasing patterns and enhance purchasing motivation [10].

1.2. *Customer education*

Customer education stands as a vital factor because it supports them in making informed purchasing decisions when

[a]husna7mohamed@gmail.com, [b]meenasuganthi.govindaraj@kahedu.edu.in

DOI: 10.1201/9781003684459-8

sustainability plays an important role in their choices. These educational programs provide users access to green product advantages and informational knowledge of environmental impact while teaching accountability [11]. People who acquired knowledge about green products treat these environmentally beneficial products as premium offerings with dual benefits for preservation. Brand product education helps customers develop trust in environmentally sustainable products which reduces their skepticism toward green marketing practices. At Patagonia the company runs educational programs that explain product sustainability and transparency to prove that teaching customers leads to the purchase of environment-friendly products [12].

1.3. Environmental concern

People modify their buying decisions when they care about the environment because they consider environmental factors together with ethical manufacturing standards [13]. People who demonstrate serious environmental interest buy green products with Fair Trade certification along with carbon-neutrally labeled products. Environmentally friendly goods secure loyal customers since external observers recognize them as ethical and reliable products. Two major advantages arise from market strategies that emphasize environmental concerns because they drive higher brand choices and superior market position.

1.4. Generational cohorts

Green products generate different purchase responses from each generational segment that exists between Baby Boomers and Gen Z because they have distinctive values and consumer behaviour patterns. Due to their digital exposure to environmental activism Millennials together with Gen Z members tend to adopt ethical and sustainable buying habits [14]. People who belong to older segments of the market primarily search for dependable products having superior quality criteria. Knowing how generations differ helps companies create marketing strategies that enhance their contact with individual market segments to achieve stronger outcomes through promotional attempts.

1.5. Statement of the problem

Customers lack enough interest in purchasing organic products available in market stores even when these products appear on their shelves. Studies regarding customer education together with purchase choice decisions show limited progress due to the scarcity of available research on these fields.

The digital orientation influences the lifestyle of all members from Generation Y and Generation Z. Every piece of net-based information remains accessible to users through their associated online connections. Gen Z members process information rapidly when compared to the other generations. Each member of these modern generations follows separate ways of thinking which impacts their purchase selection methods. The study aims to identify similarities between Generation Y and Generation Z consumers in their educational receipt of customer requirements and purchase decisions.

1.6. Objectives of the study

- To investigate the effect of Market & Firm specific Customer Education on consumer purchase Intention toward organic food products.
- To examine the Mediating role of Environmental Concern towards the relationship between Customer Education and Consumer Purchase Intention.
- To analyse the Moderating role of Generational Cohorts towards the relationship between Customer Education and Consumer Purchase Intention.

1.7. Conceptual model

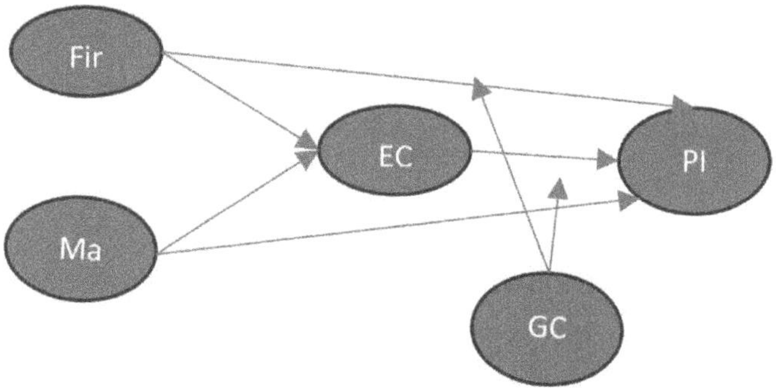

2. Literature Review and Hypothesis

2.1. Customer education and environmental concerns

Environmental knowledge education for consumers helps form better purchasing decisions for eco-friendly products by teaching them about their environmental effects to enhance their views of convenience and credibility [15–17]. Environmental education, particularly through media, raises awareness and fosters responsibility, motivating eco-friendly choices [18, 19]. A fundamental level of environmental education serves as a vital link between green products and holistic environmental issues which leads to better market success [3, 20]. According to Mehta and Chahal [21] customer education proves superior to advertising and labeling methods when promoting environmentally friendly products because it transforms consumer attitudes toward positive behaviours that boost their willingness to buy eco-friendly products [22]. Consumer environmental consciousness drives their buying decisions towards green options because environmental anxiety creates a direct connection between sustainable concern and shopping intentions [23, 24].

3. Hypothesis H1: Customer Education (CE) Correlates Positively with Environmental Concern (EC)

3.1. Environmental concern and consumer purchase intention

Environmental concern directly impacts consumer readiness to buy green products since climate change and energy generation processes and resource preservation determine what consumers think and decide [16]. The lack of product environmental qualities knowledge creates a gap between consumer concerns and effective sustainable purchasing decisions because concern does not automatically lead to intended purchases. To make informed decisions consumers need necessary information. Providing customers with essential information through education enhances environmental attitudes and purchase intentions while turning their sense of concern into effective action. Customer education enables people to obtain critical details so they can choose responsible purchases which supports sustainable consumption.

Hypothesis H2: Environmental concern (EC) mediates between customer education and the purchase intention of organic food products in Kerala.

3.2. Customer education and purchase intention

For consumers to make educated product choices particularly in regard to organic food they need proper education. Research indicates that consumers who gain knowledge about organic product benefits toward health and nature spend more on sustainable purchases [20, 25]. An educated consumer base that appreciates environmental effects of their actions tends to select eco-friendly products mainly because of their concern for nature [21, 25]. United Nations outlines how customer education enhances sustainable purchase decisions by creating environmental concern as a bridge between educational knowledge acquisition and organic product purchasing intentions. The purchasing patterns of consumers are shaped by generational differences because young people including Millennials and Gen Z members consistently demonstrate environmentally conscious behaviour. Purchasing intent responses to educational programs about organic foods show different patterns between generations because younger consumer groups prove more receptive to such educational approaches yet older population segments need strategic approaches to modify their behaviours.

Hypothesis H3: Customer education positively influences the purchase intentions of organic food products in Kerala.

Hypothesis H4: Environmental concern mediates the relationship between customer education and Consumer purchasing intention of organic food products in Kerala.

Hypothesis H5: Generational cohort differences moderate the relationship between customer education and consumer purchase intention

3.3. Research design

The research focuses on consumers living in Cochin City and takes a descriptive approach. A sample of 480 participants was selected randomly through a simple random sampling technique. Data were collected using a structured questionnaire, and scales for the study variables were adopted from previous research. The gathered data underwent regression and mediation analyses.

3.4. Tools and techniques used

The survey includes six sections using established scales and one section for demographic information. In some studies has also evaluates Food Waste Awareness with items. Environmental Concern with items and Consumer Purchase Intention with items. The analysis thoroughly observes the descriptive statistics, reliability, mediation, and moderation analysis for the study, yielding several significant insights.

4. Data Analysis

4.1. Interpretation of descriptive statistics

The descriptive statistics reveal the following insights: Firm-Specific Customer Education (FCE) has a mean of 5.52 and moderate variability, with a right-skewed distribution, indicating a higher level of education among respondents (Table 8.1). Market-Specific Customer Education (MCE) has a slightly lower mean of 5.7, with a similar range and a negatively skewed distribution, suggesting most respondents fall toward the lower end of the scale. EC has a mean of 4.13, with a smaller standard deviation and negatively skewed distribution, indicating higher concern levels among respondents. Purchase Intention (CPI) has the highest mean of 6.01, reflecting strong purchase intentions, with a strongly negative skew indicating that most respondents have high purchase intention scores. Despite some deviations, the distributions are generally well-behaved, supporting the suitability of the data for further analysis.

Table 8.1. Descriptive statistics table

Variable	*N*	*Range*	*Minimum*	*Maximum*	*Mean*	*Std. Deviation*	*Variance*	*Skewness*	*Std. Error (Skewness)*	*Kurtosis*	*Std. Error (Kurtosis)*
FCE (Firm-Specific Education)	111	5.00	2.00	7.00	5.5158	1.01650	1.033	−0.569	0.229	0.059	0.455
MCE (Market-Specific Education)	111	5.00	2.00	7.00	5.4234	1.06055	1.125	−0.422	0.229	-0.271	0.455
EC (Environmental Concern)	111	3.25	1.75	5.00	4.1329	0.66471	0.442	−0.723	0.229	0.664	0.455
CPI (Consumer Purchase Intention)	111	5.40	1.60	7.00	6.0108	1.01151	1.023	−1.914	0.229	5.255	0.455

Source: Author.

Table 8.2. Reliability analysis

Construct	*Number of Items*	*Cronbach's Alpha*	*Interpretation*
Firm-Specific Customer Education (FCE)	4	0.832	High internal consistency.
Market-Specific Customer Education (MCE)	4	0.832	High internal consistency.
Environmental Concern (EC)	4	0.832	High internal consistency.
Consumer Purchase Intention (CPI)	5	0.930	Excellent internal consistency.

Source: Author.

Reliability analysis demonstrates high internal consistency for study constructs because Cronbach's Alpha measures exceed 0.7. Research data indicates that FCE and MCE and EC possess reliability levels of 0.832. The reliability score of Purchase Intention (PI) stands at 0.930 indicating highly consistent measurement. The established reliability of these constructs can proceed to structural modeling analysis as well as hypothesis testing due to their validated suitability (Table 8.2).

4.2. *Mediation analysis results and interpretation*

Table 8.3. Mediation results for FCE → EC → CPI

Path	*Coefficient*	*SE*	*t-value*	*p-value*	*95% CI (LL)*	*95% CI (UL)*
FCE → EC	0.2932	0.0560	5.2380	0.0000	0.1823	0.4042
EC → CPI	0.8980	0.1162	7.7308	0.0000	0.6678	1.1283
Direct Effect (FCE → CPI)	0.2024	0.0760	2.6640	0.0089	0.0518	0.3529
Total Effect (FCE → CPI)	0.4657	0.0842	5.5288	0.0000	0.2988	0.6326
Indirect Effect (FCE → EC → CPI)	0.2633	0.0859	-	-	0.1139	0.4503

Source: Author.

Table 8.4. Mediation results for MCE → EC → CPI

Path	*Coefficient*	*SE*	*t-value*	*p-value*	*95% CI (LL)*	*95% CI (UL)*
MCE → EC	0.2824	0.0536	5.2697	0.0000	0.1762	0.3886
EC → CPI	0.8661	0.1144	7.5709	0.0000	0.6393	1.0928
Direct Effect (MCE → CPI)	0.2375	0.0717	3.3120	0.0013	0.0953	0.3796
Total Effect (MCE → CPI)	0.4821	0.0788	6.1154	0.0000	0.3258	0.6383
Indirect Effect (MCE → EC → CPI)	0.2446	0.0823	-	-	0.1076	0.4210

Source: Author.

4.3. Interpretation for FCE

FCE proves as a significant predictor of Environmental Concern and Environmental Concern predicts Purchase Intention in the mediation analysis (Table 8.3). The direct relationship between FCE and CPI exists independently of the mediation effect by EC. FCE demonstrates significant total impact on CPI while its relationship passes the test of partial mediation through EC. Research shows FCE produces direct and indirect effects on CPI because of its ability to boost environmental concern levels in consumers.

4.4. Interpretation for MCE

The mediation analysis for MCE shows that it significantly predicts EC, which in turn significantly influences Purchase Intention (CPI). The direct effect of MCE on CPI is significant, indicating both direct and indirect impacts through EC (Table 8.4). The total effect of MCE on CPI is also significant, with partial mediation confirmed by the significant indirect effect through EC. These findings suggest that MCE directly enhances purchase intentions while also driving an indirect effect through increased EC. Both Firm-Specific and MCE influence Purchase Intention directly and indirectly through EC, highlighting the importance of integrating environmental messaging into customer education strategies to boost purchase behaviour.

4.5. Moderation analysis results and interpretation

In the moderation analysis conducted using the PROCESS macro (Model 1), the categorical moderator variable 'AGE' was transformed into two dummy variables: W1 and W2. This transformation facilitates the examination of how different age categories influence the relationship between the independent variable (FCE) and the dependent variable (CPI) (Table 8.5).

In this coding scheme, the 18–24 years age group serves as the reference category, against which the other groups are compared.

Table 8.5. Dummy variable coding

AGE Category	*Description*	*W1*	*W2*
1	18–24 years (Reference Group)	0	0
2	25–34 years	1	0
3	35–44 years	0	1

Source: Author.

4.5.1. Interpretation

The constant term predicts the CPI score for the 18–24 age group when FCE is zero (6.0442) (Table 8.6). For this group, a one-unit increase in FCE leads to a 0.3521 increase in CPI, indicating a significant positive relationship ($p = 0.0037$). While the 25–34 and 35–44 age groups show slightly lower CPI scores compared to the 18–24 group, these differences are not sta tistically significant ($p = 0.7236$ for W1 and $p = 0.8657$ for W2). The interaction terms suggest that the effect of FCE on CPI may be stronger for the 25–34 age group (coefficient = 0.2728, $p = 0.1249$), but this effect is not statistically significant. For the 35–44 age group, the interaction term shows no significant effect (coefficient = 0.0039, $p = 0.9915$).

These results indicate that the positive effect of FCE on CPI is significant for the 18–24 and 25–34 age groups, with the effect being more pronounced in the 25–34 group. However, for the 35–44 age group, the effect is not statistically significant (Table 8.7).

The analysis suggests that age may moderate the relationship between FCE and CPI, particularly for individuals aged 25–34, where the effect appears more pronounced. However, the lack of statistical significance in the interaction terms indicates that these differences are not substantial. Further research with a larger sample size may be needed to confirm these findings.

The constant term predicts the CPI score for the 18–24 age group when MCE is zero (6.0438). For this group, a one-unit increase in MCE leads to a 0.3197 increase in CPI, indicating a significant positive relationship ($p = 0.0045$). The 25–34 and 35–44 age groups show slightly lower CPI scores compared to the 18–24 group, but these differences are not statistically significant ($p = 0.7169$ for W1 and $p = 0.5391$ for W2). The interaction term for the 25–34 age group (MCE × W1) suggests a significant increase in the effect of MCE on

Table 8.6. Regression model summary

Variable	*Coefficient*	*Standard Error*	*t-value*	*p-value*	*95% Confidence Interval*
Constant	6.0442	0.1307	46.2346	0.0000	[5.7850, 6.3034]
FCE	0.3521	0.1187	2.9666	0.0037	[0.1168, 0.5874]
W1 (25–34 years)	-0.0645	0.1819	-0.3546	0.7236	[-0.4252, 0.2962]
W2 (35–44 years)	-0.0517	0.3050	-0.1695	0.8657	[-0.6566, 0.5531]
FCE × W1	0.2728	0.1764	1.5467	0.1249	[-0.0769, 0.6225]
FCE × W2	0.0039	0.3691	0.0107	0.9915	[-0.7279, 0.7358]

Source: Author.

Table 8.7. Conditional effects of FCE on CPI by age category

AGE Category	*Effect of FCE on CPI*	*Standard Error*	*t-value*	*p-value*	*95% Confidence Interval*
18–24 years	0.3521	0.1187	2.9666	0.0037	[0.1168, 0.5874]
25–34 years	0.6249	0.1305	4.7900	0.0000	[0.3662, 0.8836]
35–44 years	0.3560	0.3495	1.0188	0.3107	[−0.3369, 1.0490]

Source: Author.

CPI (coefficient = 0.3281, p = 0.0449), while the interaction term for the 35–44 age group (MCE × W2) shows no significant effect (coefficient = 0.4724, p = 0.2370) (Table 8.8).

The results reveals that the positive effect of MCE on CPI is significant across all age groups, with the effect being more pronounced in the 25–34 and 35–44 age groups. The analysis suggests that age moderates the relationship between MCE and CPI, particularly for individuals aged 25–34 and 35–44, where the effect is more substantial (Table 8.9).

Table 8.8. Regression model summary

Variable	*Coefficient*	*Standard Error*	*t-value*	*p-value*	*95% Confidence Interval*
Constant	6.0438	0.1263	47.8556	0.0000	[5.7934, 6.2942]
MCE	0.3197	0.1101	2.9035	0.0045	[0.1014, 0.5380]
W1 (25–34 years)	-0.0639	0.1757	−0.3636	0.7169	[-0.4123, 0.2846]
W2 (35–44 years)	-0.1861	0.3020	−0.6163	0.5391	[−0.7850, 0.4127]
MCE × W1	0.3281	0.1616	2.0299	0.0449	[0.0076, 0.6486]
MCE × W2	0.4724	0.3971	1.1894	0.2370	[−0.3151, 1.2598]

Source: Author.

Table 8.9. Conditional effects of MCE on CPI by age category

AGE Category	*Effect of MCE on CPI*	*Standard Error*	*t-value*	*p-value*	*95% Confidence Interval*
18–24 years	0.3197	0.1101	2.9035	0.0045	[0.1014, 0.5380]
25–34 years	0.6478	0.1184	5.4734	0.0000	[0.4131, 0.8825]
35–44 years	0.7920	0.3816	2.0756	0.0404	[0.0354, 1.5486]

Source: Author.

5. Findings

The study carries both theoretical and practical implications. Theoretically, the content enriches existing literature by addressing identified gaps. Based on the results, marketers can leverage customer education to increase consumer knowledge and concern, thereby fostering sustainable purchase intentions.

6. Suggestions

The study focused on a few mediating variables to realize how customer education influences sustainable purchase intentions. Specifically, it only examined Market-specific and Firm-specific education within the customer education context. Future research could explore additional forms of customer education and identify other potential mediating variables for further study.

7. Conclusion

The research findings underline how customer education plays an essential part in advancing sustainable purchasing behavior through awareness creation and knowledge mix that results in ethical consumer actions. The research provides detailed instruction for marketers who need to create marketing strategies which match sustainability targets alongside consumer belief building strategies. Future research should include investigations of Knowledge & Skills Education education along with perceived value and brand trust factors along with the existing Market-specific and Firm-specific education analysis. Additional insight into sustainable consumption and global support would result from strategy adaptations across cultures and through varied generational groups as well as time-based impact studies.

References

[1] Kushwah, S., Singh, S., & Soni, P. (2019). The role of customer education in influencing purchase intentions for green products. *Journal of Consumer Research, 45*(4), 670–685.

[2] Jeganathan, S., Lakshminarayanan, A. R., Parthasarathy, S., Khan, A. A. A., & Sathick, K. J. (2024). OptCatB: Optuna Hyperparameter Optimization Model to Forecast the Educational Proficiency of Immigrant Students based on CatBoost Regression. *Journal of Internet Services and Information Security, 14*(2), 111–132. https://doi.org/10.58346/JISIS.2024.I2.008

[3] Kumar, V., & Ghodeswar, B. M. (2022). Impact of customer education on sustainable consumption and green product purchase behavior. *International Journal of Sustainable Business, 19*(2), 89–105. 76.

[4] Nguyen, T., Hoang, T., & Tran, M. (2020). The influence of customer education on green product purchasing decisions:

An empirical study. *Journal of Environmental Marketing, 14*(3), 234–249.

[5] Escobedo, F., Canales, H. B. G., Criollo, R. A. G., Romero, E. M. Y., Orellana, C. S., Ramírez, J. A. B., Vela, C. A. L., & Herrera, J. M. G. (2024). a smart crowd monitoring and management model for humanity in intelligent environments: A real-time application scenario. *Journal of Wireless Mobile Networks, Ubiquitous Computing, and Dependable Applications, 15*(4), 166–178. https://doi.org/10.58346/JOWUA.2024.I4.011

[6] Young, A., & Lim, M. (2022). Understanding consumer purchase intentions in a multi-generational context. *Sustainability and Consumer Behavior, 18*(1), 23–37.

[7] Kemp, R., DeWitt, C., & Mayer, T. (2021). Generational cohort differences and sustainable consumption patterns: An exploratory study. *International Journal of Consumer Studies, 45*(5), 678–685.

[8] Williams, A., Harris, C., & Shaw, L. (2023). A generational approach to sustainable consumption and customer education. *Journal of Sustainability and Ethics, 31*(2), 145–162.

[9] Nihlani, A., Chhabda, P. K, Ahmed, M, & Pandey, S. K. (2024). Funding and Library Resource Management in Higher Education Universities in India. *Indian Journal of Information Sources and Services, 14*(2), 34–40. https://doi.org/10.51983/ijiss-2024.14.2.06

[10] Divya, S., & Vinitha Subhashini, B. (2015). Life time enhancement in bottleneck zone of sensors of wireless sensor network using duty cycle, network coding and sink relocation technique. *International Journal of Advances in Engineering and Emerging Technology, 6*(3), 37–48.

[11] Singh, S., & Verma, P. (2020). The impact of customer education on green product purchase intentions. *Journal of Environmental Marketing, 22*(1), 52–66.

[12] Bala Krishna, M. (2021). A robust strategic theory based load balancing and resource allocation in cloud environment. *International Academic Journal of Science and Engineering, 8*(1), 19–30. https://doi.org/10.9756/IAJSE/V8I1/IAJSE0803

[13] Nguyen, T., Lee, R., & Ho, H. (2019). Environmental concern and consumer willingness to pay a premium for green products. *International Journal of Green Economics, 10*(3), 231–245.

[14] Smith, A., & Brower, T. (2012). Long-term generational trends in sustainability: An analysis of Millennial and Gen Z consumer behavior. *Journal of Consumer Marketing, 29*(3), 242–250.

[15] Laroche, M., Bergeron, J., & Barbaro-Forleo, G. (2001). Targeting consumers who are willing to pay more for environmentally friendly products. *Journal of Consumer Marketing, 18*(6), 503–520.

[16] Diamantopoulos, A., Schlegelmilch, B. B., Sinkovics, R. R., & Bohlen, G. M. (2003). Can socio-demographics help predict environmentally conscious behavior? *Journal of Marketing Management, 19*(3–4), 1–21.

[17] Wiser, R. H., Johnson, T., & Swaddle, J. (1999). The influence of environmental education on green consumer behavior. *Journal of Consumer Research, 26*(2), 121–133.

[18] McClinchie, S. (2014). Environmental education and consumer decision-making. *Environmental Communication, 8*(4), 389–404.

[19] Young, W., Hwang, K., McDonald, S., & Oates, C. J. (2010). Sustainable consumption: Green consumer behavior when purchasing products. *Sustainable Development, 18*(3), 141–157.

[20] Ottman, J. A. (1998). *Green marketing: Opportunity for innovation*. NTC Business Books.

[21] Mehta, R., & Chahal, H. (2018). The role of customer education in promoting green products. *Journal of Business Research, 73*, 201–209.

[22] Cheah, I., & Phau, I. (2011). Green marketing: A study of consumers' behavior in the green movement. *Journal of Business Ethics, 105*(3), 339–352.

[23] Sari, H. S., Anggraeni, L. S., & Rachman, A. R. (2020). The impact of customer education on the willingness to purchase green products in Indonesia. *Journal of Marketing Research, 42*(2), 89–104.

[24] Casey, T., & Scott, J. (2006). Environmental concern and individual behaviors: A study of consumers in the UK. *Environmental Politics, 15*(3), 314–329.

[25] Goldsmith, R. E., & Piscopo, S. (2014). The influence of eco-labelling on the perception of green products. *International Journal of Consumer Studies, 38*(6), 624–630.

9 Bio-inspired optimization for sustainable smart city infrastructure

R. Geetha[1,a], *S. Ramesh*[1,b], *P. Thilagavathi*[2,c], *S. Vidhya*[1,d], *and J. Nithyashri*[3,e]

[1]Assistant Professor, Department of Computing Technologies, School of Computing, SRM Institute of Science and Technology, Kattankulathur, Tamil Nadu, India
[2]Associate Professor, Department of Computer Science and Engineering, Aarupadai Veedu Institute of Technology, Vinayaka Mission's Research Foundation (DU), Paiyanoor, Tamil Nadu, India
[3]Assistant Professor, Department of Computing Technologies, School of Computing, Faculty of Engineering and Technology, SRM Institute of Science and Technology, Kattankulathur, Tamil Nadu, India

Abstract: As the population of urban areas is increasing, the requirement for sustainable and smart infrastructure in these regions has become important. Bio-inspired optimization (BO) approaches from natural methodologies like swarm intelligence as well as evolutionary algorithms provide creative solutions to enhance the architecture of smart cities. This research scrutinizes approaches of BO for establishing optimized and energy enhanced street light frameworks and smart parking. The introduced framework improves decision-making, reduces utilization of energy, and enhances distribution of resources by executing GA (Genetic Algorithm) and PSO (Particle Swarm Optimization). The outcomes illustrate that BO approaches develop the infrastructure in urban areas by decreasing congestion, enabling automation and increasing efficacy. Incorporating BO with AI (Artificial Intelligence) and IoT (Internet of Things) ensures robust and flexible smart cities. Future work will concentrate on implementing blockchain for exchanging information with higher security and for implementing reinforcement learning in real-world scenarios of urban sectors.

Keywords: Bio-inspired optimization, genetic algorithm, particle swarm optimization

1. Introduction

BO approaches, especially GA and PSO, resolve real-world urban issues. Studies depict that these techniques when incorporated with learning methodologies, can create reinforced, expense-effective, self-adaptive, and autonomous robust systems to deal with smart city implementations [1, 2]. The approaches of BO are basically developed on the basis of natural factors such as evolution, molecular activities, forging, neurology, reproduction as well as cognitivism [3, 4]. These BO approaches are highly efficient in dealing with intricate optimization issues by having nature as its source of inspiration [5]. Among the real-time industrial challenges, they address are those related to generating electricity, managing power systems, designing and controlling automobiles, communication networks, processing images and sounds, sensor networks, detecting and recovering faults, scheduling, signal processing, allocating resources, traffic control, material engineering, and chemical processing [6]. Additionally, they optimize neural networks, clustering, segmentation, classification, prediction, search operations, and pattern recognition [7]. This research offers a detailed overview of BO approaches and their flexible role in enhancing smart urban areas [8, 9]. These methods open the door for more intelligent, flexible, and ecologically responsible urban development by enhancing urban services and infrastructure's sustainability, effectiveness, and resilience [10, 11].

2. Literature Review

According to [12], establishing smart cities has acquired attention as we use technology daily. The public and commercial sectors have recently invested significantly in smart city initiatives [13]. As technology develops and urban populations increase, smart cities become crucial [14]. Systematic scrutinization of the existing condition of smart urban areas and applications has become highly important because of the public and commercial sectors' substantial global investments in smart urban area projects [15].

Del et al. [16] has discussed that many researchers from scientific associations have implemented BO approaches as computational algorithms by deriving inspiration from smart behavioural patterns from observing nature, and their goal is to enhance the self-learning abilities and flexibility of these systems to deal with intricate issues [17]. As a result, the development and design of bio-inspired approaches for

[a]geethar1@srmist.edu.in, [b]ramesh.isaiah@gmail.com, [c]thiluxpriya499@gmail.com, [d]vidhyas2@srmist.edu.in, [e]drjnithyashri@gmail.com

DOI: 10.1201/9781003684459-9

various computing challenges has traditionally taken into account an increase in inspirational resources.

According to [18], various ways concentrate on modifying the traffic light regulations to enhance road traffic. Whereas [19] utilized GA with a Traffic simulator, which works on the basis of cellular automation to optimize the Traffic in La Almozara at Saragossa, Spain, a PSO approach is executed to acquire this in an area of Malaga and Seville, Spain. Despite the encouraging outcomes they have acquired, these approaches do not provide vehicles with alternate routes; instead, they view the area as a group of traffic signals that manage all automobiles [20].

3. Methodology

3.1. Smart streetlight using BO optimization

Figure 9.1 depicts an intelligent street lighting system that uses bio-inspired adaptation to increase energy efficiency, adaptability, and intelligent decision-making. Embedded motion sensors in streetlights detect the presence of pedestrians or vehicles, triggering a customized light reaction. Instead of a simple motion-based activation, swarm intelligence (inspired by ant colonies and bird flocking behaviour) dynamically communicates streetlights, adjusting the level of brightness depending on real-time traffic flows and walking density. Using the GA, the system develops optimal light patterns by constantly analyzing historical data and environmental factors, ensuring minimum energy waste while maintaining security. Gateway transmits sensor data for a management and control center via 4G/Wi-Fi, where nerves predict traffic trends and optimize the empty lighting program. If no speed is detected, the system slows down wisely or turns off the lights, but with bio-induced intelligence, it estimates the high-mang period and suddenly prevents darkness. This advanced adaptive structure ensures a permanent and self-organizing innovative street lighting system, reducing energy consumption by increasing urban infrastructure.

3.2. Smart parking using BO optimization

The proposed innovative parking system in Figure 9.2 uses a deeper extrFeme learning machine (DELM) model, which is adapted to increase the accuracy and efficiency of parking space detection using particle herd optimization (PSO) to increase the accuracy and efficiency. The functioning consists of several layers, which ensure the process of making a structured and intelligent decision. The data acquisition layer collects real-time data from the IOT-based sensor posted in the layer parking lot and detects the presence of a vehicle and environmental conditions. This data is then processed in the pre-processing layer, where the missing values are handled, the noise is reduced using the moving average, and the normalization is applied to standardize the input values. In the prediction layer, the DELM is used to classify parking slots. To enhance learning efficiency, PSO dynamically optimizes DELM's hyperparameters, ensuring better generalization and low computational costs.

The performance assessment layer root is the perfection of predictions using matrices such as RMSE (Root Mean Square Error), MAE (Mean Absolute Error), and MAPE (Mean Absolute Percentage Error). Finally, the system reaches the decision-making phase, where the model parking slots determine the availability. If a location is found in the form of possession, it is marked as unavailable for parking; Otherwise, the system directs vehicles to available locations. This PSO with the DELM approach ensures high prediction accuracy, rapid convergence, and better efficiency in smart parking management.

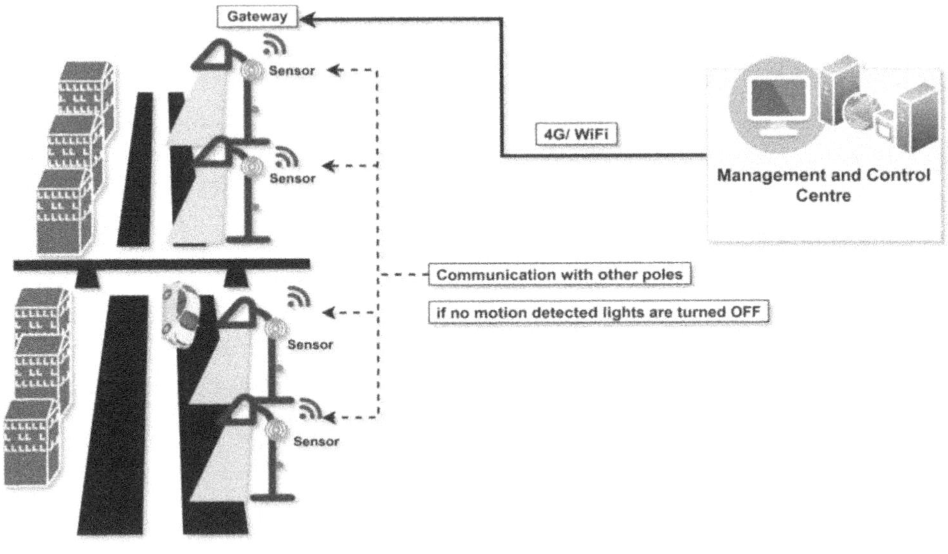

Figure 9.1. Smart street lighting based on Bio-inspired optimization.

Source: Author.

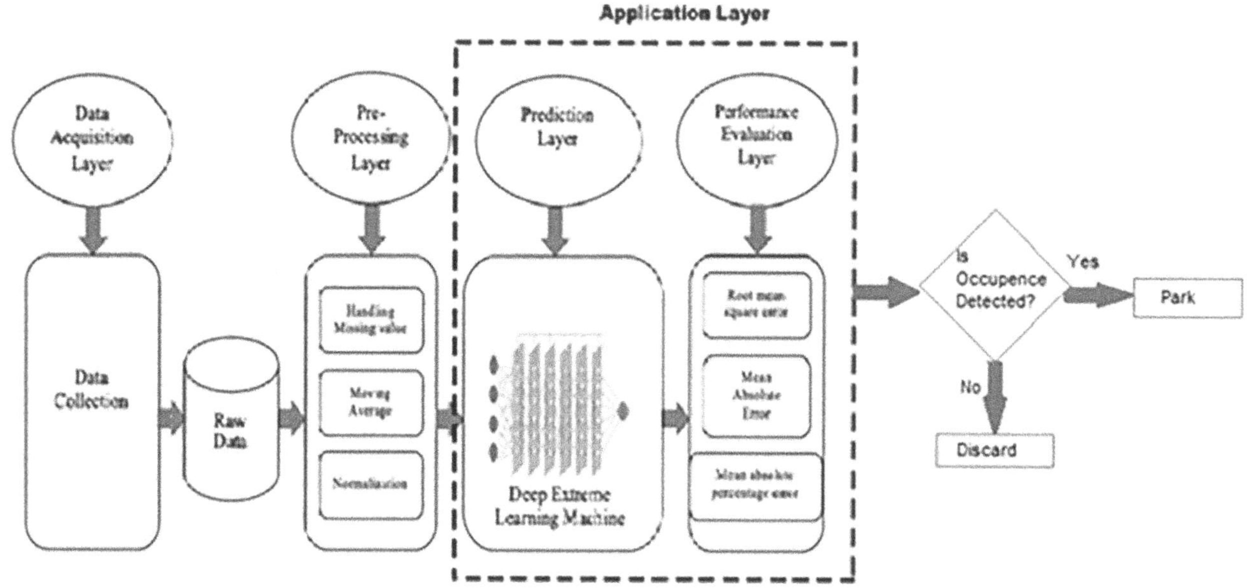

Figure 9.2. Smart parking using DELM and PSO.
Source: Author.

4. Results and Discussions

The results of this research highlight the effectiveness of bio-inspired adaptation techniques in adaptation to smart city infrastructure. The Smart Street Lighting System demonstrated important energy savings while maintaining high standards. The ability to modify brightness in a dynamic manner on the basis of real-world parameters helped in regulating energy in a sustainable and efficient manner. In addition to that, utilizing swarm intelligence in this approach increased flexibility, giving it the opportunity to optimize on itself and enhance decision making. These outcomes denote that the implementation of BO offers a creative and expense-effective remedy to enhance the infrastructure of urban regions.

Similarly, the smart parking system operated by DELM and PSO gained significant improvement in parking space detection accuracy. By decreasing the rate of incorrect segregation and improving the allocation of parking areas, this approach significantly decreases the congestion due to traffic and enhances user experience. The evaluation of performance using MAE as well as RMSE ensured efficacy and reliability of the approach as depicted in Figure 9.3. Moreover, IoT and AI-based decisions reduces the intricacy of parking methods, which decreases the consumption of fuel and the waiting time. These findings underline the capability of BO approaches to establish smart and autonomous solutions to manage smart urban cities.

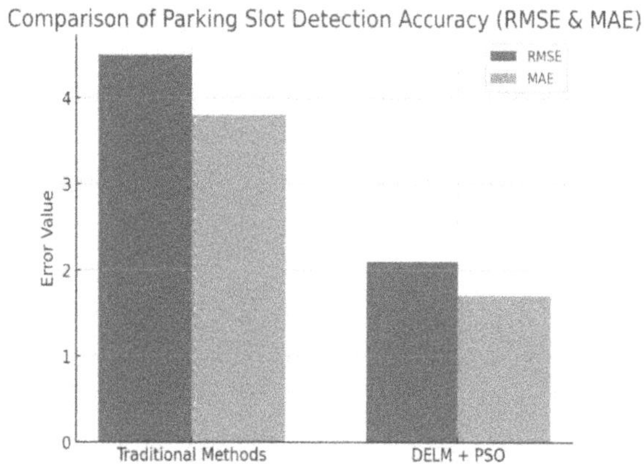

Figure 9.3. Comparison graph of parking approaches.
Source: Author.

5. Conclusion and Future Works

This study offers a comprehensive observation of bio-inspired adaptation techniques and their transformative role in shaping the future of smart cities. By contributing to stability, efficiency and flexibility in urban infrastructure and services, these approaches pave the way for more intelligent, adaptive and environmentally aware urban development. Future work will integrate BO methods with blockchain for safe data exchange, expand the adaptation framework for waste management, and will implement the model of learning reinforcement to increase decision making in real -time urban landscapes. Through continuous innovation, bio-inspired adaptation can contribute significantly to develop the next generation permanent smart cities.

References

[1] Verscheure, O. (2011). Smart cities: How data mining and optimization can shape future cities. 11. doi:10.1007/978-3-642-23780-5_10.

[2] Deshmukh, A., & Nair, K. (2024). An analysis of the impact of migration on population growth and aging in Urban Areas.

Progression Journal of Human Demography and Anthropology, *1*(1), 1–7.

[3] Smith, R. M., & Pathak, P. A., & Agrawal, G. (2018). India's 'smart' cities mission: A preliminary examination into India's newest urban development policy. *Journal of Urban Affairs*, *41*, 518–534. doi:10.1080/07352166.2018.1468221.

[4] Nagalakshmi, D., Poongodi, V., & Thangadurai, K. (2016). A systematic virtual resource allocation on dynamic data distribution in cloud environment. *International Journal of Advances in Engineering and Emerging Technology*, *7*(4), 269–276.

[5] Marrow, P. (2000). Nature-inspired computing technology and applications. *BT Technology Journal*, *18*(4), 13–23.

[6] Angiello, G., Carpentieri, G., Mazzeo, G., Pinto, V., Russo, L., & Zucaro, F. (2013). Review Pages: Smart City: Researches, Projects and Good Practices for the City. *TeMA-Journal of Land Use, Mobility and Environment*, *6*(1), 113–140.

[7] Giffinger, R., Fertner, C., Kramar, H., Kalasek, R., Pichler-Milanovic, N., & Meijers, E. J. (2007). Smart cities–Ranking of European medium-sized cities. Final report.

[8] Mohsenian-Rad, H., & Cortez, E. (2015). Smart grid for smart city activities in the california city of riverside. In *International Summit, Smart City 360°* (pp. 314–325). Cham: Springer International Publishing.

[9] Wahul, R. M., & Lomte, S. S. (2019). Smart City Project Management System Using Cloud. In *Computing, Communication and Signal Processing: Proceedings of ICCASP 2018* (pp. 225–231). Springer Singapore.

[10] Jaiswal, H., & Pradhan, S. (2023). The economic significance of ecosystem services in urban areas for developing nations. *Aquatic Ecosystems and Environmental Frontiers*, *1*(1), 1–5.

[11] Kiyomoto, S., & Martin, K. M. (2011). Model for a common notion of privacy leakage on public database. *Journal of Wireless Mobile Networks, Ubiquitous Computing, and Dependable Applications*, *2*(1), 50–62.

[12] Ismagilova, E., Hughes, L., Dwivedi, Y. K., & Raman, K. R. (2019). Smart cities: Advances in research – An information systems perspective. *International Journal of Information Management*, *47*, 88–100.

[13] Badii, A., Carboni, D., Pintus, A., Piras, A., Serra, A., Tiemann, M., & Viswanathan, N. (2013). CityScripts: Unifying web, IoT and smart city services in a smart citizen workspace. *Journal of Wireless Mobile Networks, Ubiquitous Computing, and Dependable Applications*, *4*(3), 58–78.

[14] Raj, D. S., & Dharmaraj, A. (2024). Rural women's saving and investment habits: A Study with Special Reference to Kuttampuzha Area, Ernakulam District. *Indian Journal of Information Sources and Services*, *14*(3), 39–44. https://doi.org/10.51983/ijiss-2024.14.3.06

[15] Alsafar, B. S., Alzaher, D. M., ALshwukh, R. W., Darwish, A. M., Ahmed, T. E., Alyahyan, E. A., Alboaneen, D., & El-Sharawy, E. E. (2024). Smart parking management system using IoT. *Journal of Internet Services and Information Security*, *14*(2), 284–297. https://doi.org/10.58346/JISIS.2024.I2.018

[16] Del Ser, J., Osaba, E., Molina, D., Yang, X. S., Salcedo-Sanz, S., Camacho, D., Das, S., Suganthan, P. N., Coello, C. A. C., & Herrera, F. (2019). Bio-inspired computation: Where we stand and what's next. *Swarm and Evolutionary Computation*, *48*, 220–250.

[17] Jelena, T., & Srđan, K. (2023). Smart mining: Joint model for parametrization of coal excavation process based on artificial neural networks. *Archives for Technical Sciences*, *2*(29), 11–22. https://doi.org/10.59456/afts.2023.1529.011T

[18] García-Nieto, J., Alba, E., & Olivera, A. C. (2012). Swarm intelligence for traffic light scheduling: Application to real urban areas. *Engineering Applications of Artificial Intelligence*, *25*(2), 274–283.

[19] Sánchez-Medina, J. J., Galán-Moreno, M. J., & Rubio-Royo, E. (2009). Traffic signal optimization in 'La Almozara' district in saragossa under congestion conditions, using genetic algorithms, traffic microsimulation, and cluster computing. *IEEE Transactions on Intelligent Transportation Systems*, *11*(1), 132–141.

[20] Narteni, S., Vaccari, I., Mongelli, M., Aiello, M., & Cambiaso, E. (2021). Evaluating the possibility to perpetrate tunneling attacks exploiting short-message-service. *Journal of Internet Services and Information Security*, *11*(3), 30–46.

10 Factors influencing soft skills acquisition among higher education faculty

S. R. Karthick[1,a] *and R. Velmurugan*[2,b]

[1]Ph.D. Research Scholar in Commerce, Karpagam Academy of Higher Education, Coimbatore, Tamil Nadu, India
[2]Assistant Professor of Commerce, Karpagam Academy of Higher Education, Coimbatore, Tamil Nadu, India

Abstract: Soft skills development enables higher education faculty to enhance their effectiveness because these skills strengthen communication methods combined with adaptability as well as improve their teaching leadership abilities and involvement with students. The development process of soft skills in faculty members depends on personal features together with institutional resources and educational methods and local cultural norms. The capability of individuals to enhance soft skills depends on their personal elements which include accumulated experience and motivational levels plus a clear understanding of themselves. Specialized development programs and mentorship resources and technological advancements given by organizations enable faculty members to enhance their abilities. Higher education learners obtain their competencies through direct contact with academic personnel who work with industrial affiliates and local community interaction. Regular assessments, feedback mechanisms, and exposure to global best practices also refine these competencies. The study examines the core components which impact soft skill acquisition by higher education professors since professional advancement needs structured skill development programs.

Keywords: Soft skills acquisition, higher education faculty, professional development, communication skills, teaching effectiveness, institutional support, faculty training, student engagement and workplace competencies

1. Introduction

Educational faculty in higher learning institutions need soft skills because these attributes directly shape both educational success and academic and student interactions. The combination of skills like communication alongside adaptability and teamwork combined with leadership together with emotional intelligence helps faculty members in building better student-professor interactions [1]. The current dynamic educational environment requires faculty members to develop and improve soft skills because these capabilities create positive learning environments and professional readiness among students [2].

Multiple factors influence the development of soft skills for faculty members through personal characteristics together with institutional backing. This skill set development depends on personal attributes which include motivation and self-knowledge and past learning experience as formulated by Laker and Powell [3]. Institutional factors such as faculty development programs together with mentorship alongside technological integration provide scheduled pathways for faculty to enhance their skills perpetually [4]. The ability of faculty members to enhance their soft skills depends on their student connections and their work with industries and international best practice experiences [5].

Research shows that faculty members struggle to develop and implement soft skills because of increasing recognition of their value in teaching practices. The lack of training initiatives together with institutional hesitation and resistance to change create obstacles that prevent skill development according to Clokie and Fourie [6]. Professionals need to identify the important factors that steer soft skills acquisition among higher education faculty because this knowledge supports creating interventions which advance professional development and improve learning results.

The research explores factors that influence higher education faculty member development of soft skills through personal experiences in combination with institutional and environmental influences. Research findings will enable the creation of particular faculty competency development methods which lead to enhanced learning environments.

2. Review of Literature

Academic researchers deeply analyze higher education faculty soft skill training since it produces enhanced teaching practices and improved student engagement along with increased instructor development. Research findings present an analysis of soft skills development elements for faculty personnel through investigations of personal traits and

[a]srkarthick1989@gmail.com, [b]drvelsngm@gmail.com

DOI: 10.1201/9781003684459-10

institutional backing strategies as well as teaching methodologies and societal environments.

Educational framework adaptations become easier for faculty members after they develop their soft skills which also enhances their communication skills and team capabilities. The combination of leadership competencies with emotional intelligence skills and teamwork abilities enhances technical competence for the higher education sector as per Robles [1]. Higher education institutions demand that teachers use their soft skills as educational methods to reach peak student learning outcomes and professional achievements according to Succi and Canovi [5]. Educational results will improve once teachers receive training that unites both personal development with employability skills according to Cottrell [7].

The abilities of faculty members to cultivate soft skills heavily depend on their motivational level together with their personal self-awareness and professional experience. According to Laker and Powell [3] there exists a distinction between hard and soft skills where the latter demands both practical experience and ongoing practice. The combination of strong intrinsic motivation and self-reflection in faculty members creates conditions for the improvement of their interpersonal and leadership capabilities [2]. Beausaert, Segers and Gijselaers [8] observed that faculty members use development plans to enhance their adaptability and communication and problem-solving abilities through which they progress professionally.

Research on how institutions support faculty development exists in numerous publications. O'Leary [4] explained that formal faculty training programs enable skill development through practical training activities and mentorship support and team collaboration butions. Research evidence demonstrates that mentorship programs enhance professional advancement through the avenue of experienced colleagues teaching and training faculty members to improve their teaching and communication competencies [6]. Jackson [9] stressed the significance of work-integrated learning for faculty development because real-world experience development helps their acquisition of essential soft competencies.

Student-teaching interactions are essential for faculty members who want to learn essential skills. A teaching method that bases learning on student interests leads faculty members to improve both adaptability and communication skills [10]. Daily student encounters with diverse students create positive effects on both emotional intelligence capabilities and students' abilities to resolve disagreements [2]. Digital learning environments together with virtual reality systems help faculty members develop their communication and adaptability competencies through their need for innovative methods in student engagement according to [11].

The expectations of the industry sector together with the process of globalization create new demands for soft skill training among university faculty. According to Succi and Canovi [5] faculty members who work in partnership with industry professionals encounter genuine business issues which enhances their critical thinking and negotiation capabilities. The communication approaches together with leadership tendencies of faculty members in diverse educational settings are influenced by societal customs and cultural community standards [2]. According to Knight and Yorke [12] curriculum design needs to incorporate updated professional demands because this enables teaching staff to implement soft skill training in their educational practices.

Academic teaching staff recognize soft skill importance yet encounter difficulty when developing these competencies. Organizations lag behind in skill development because they give insufficient attention along with minimal training resources and resist changes in operational methods [6]. Laker and Powell [3] discussed that soft skills differ from technical skills because they need sustained practice and reinforcement which escalates their learning difficulty. Technical faculty members such as engineers usually find soft skills development challenging according to Riemer [13] because their institutions focus primarily on subject-matter expertise instead of interpersonal competencies.

2.1. Research gap

Research shows that soft skills are essential for academic faculty yet scientists still need better understanding about which techniques lead to best teaching skill acquisition and integration results. The current body of research concentrates on development programs for faculty members and mentorship programs and institutional backing systems however it lacks evidence about their sustained effects over time. Studies are scarce about how education technology environments and business partnerships contribute to faculty soft skill development. Most university research covers student employment outcomes but does not focus on how faculty members adapt their practices to satisfy emerging educational needs. Studies about soft skills development tend to provide generalized findings which fail to address how different fields of study affect faculty training regarding these competencies. A complete solution to this problem needs to combine long-term research with fields-based investigations to develop new approaches for embedding soft skills training throughout faculty development initiatives.

2.2. Objectives of the study

- To identify the Factors Influencing Soft Skills
- Acquisition Among Higher Education Faculty.

2.3. Scope of the study

The study investigates soft skills acquisition factors within higher education faculty members by analyzing personal qualities together with institutional backing and educational practices while reviewing social influences. This investigation reveals the methods by which university educators acquire leadership competencies and communication abilities alongside adaptability and teamwork capabilities and emotional intelligence milestones through system-based instruction and guidance from colleagues and student collaboration. The research examines teaching personnel working in universities and colleges using both full-time and part-time instructors from different academic fields. The investigation evaluates how institutions create policies and implement support systems which help their staff grows their skills. The study will prioritize a certain geographical area or nation because data accessibility depends on both factors. The research uses surveys together with interviews and literature assessment to deliver a full understanding of the topic. The study does not reach beyond students nor does it target non-teaching staff members or performance assessment of faculty members in detail. Information derived from this research will help create faculty development programs and institutional approaches to improve teaching quality and education results.

3. Research Methodology

3.1. Data

The data for this study is primarily primary in nature and has been collected using surveys and interviews with faculty members in higher education institutions.

3.2. Sampling

By employing judgment sampling, data has been collected from faculty members across various disciplines in universities and colleges. The study includes responses from 249 faculty members, ensuring a diverse representation of perspectives.

3.3. Framework of analysis

The collected data has been analyzed using descriptive statistics, simple percentage analysis, and factor analysis to identify key determinants influencing soft skills acquisition among faculty members.

3.4. Significance of the study

The examination demonstrates how higher education faculty ought to learn soft skills to boost their teaching achievements and enhance both student participation and academic success. The research delivers essential information to faculty members regarding which factors impact their soft skills enhancement along with techniques to strengthen communication abilities and leadership skills and adaptability and emotional intelligence competencies. The research outcomes enable institutions to create faculty development training and establish teaching environments which support complete intellectual faculty growth along with creating policies promoting scholarly development. Educational policymakers together with administrators should include soft skills training programs into their faculty development programs in order to fulfill national and international educational requirements. Educational research enables students to interact with diverse faculty who both mentor them effectively and display leadership abilities thus delivering superior student outcomes. This research determines essential factors of soft skills learning which helps close faculty development practice gaps and develop institutional methods to strengthen higher education results.

3.5. Analysis and interpretation

Table 10.1 shows that the majority of respondents are female at 57.8% of the total while most participants fall within age group 31–40 years (45.4%) and are married (89.2%). The study population primarily includes 69.9 percent Assistant Professors holding either an MBA degree or M.Com (58.2 percent) who have accumulated more than 8 years of teaching experience (52.2 percent). Monthly earnings of respondents range from ₹20,001–30,000 (43.8%) as the second most common group after ₹10,000–20,000 (43.8%). Among families 73.1% belong to nuclear families and the population lives mostly in semi-urban (39.4%) and rural (31.3%) areas. Forty-two percent of responders need 15–30 minutes to get to their college since it is located either in an urban (41.8%) or semi-urban (27.3%) area. Additionally, 40.2% journey between 10–20 kilometers to their workplace. The collected evidence indicates workers have spent an average amount of time in their careers while earning a solid income while keeping their residence locations moderate.

3.6. Factor analysis

To identify the Factors Influencing Soft Skills Acquisition Among Higher Education Faculty factor analysis is employed.

Factor analysis demonstrates high suitability for the given dataset based on its 0.897 KMO Sampling Adequacy value since this exceeds the accepted threshold of 0.6. Bartlett's Test results indicate statistical significance at the $p < 0.001$ level thus validating the use of factor analysis because the correlation matrix deviates from identity form. The strong interrelationships within the dataset mark it suitable for identifying important factors which affect soft skill development in higher education faculty members (Table 10.2).

The Rotated Component Matrix demonstrates five main variables which control soft skills education for higher education faculty (Table 10.3). The first component analyzes the teaching techniques and adjustable practices which serve student interests and adaptability. The second factor concentrates on self-improvement together with institutional support which includes personal development along with staff adaptability and institutional collaboration support. The third factor includes professional development opportunities for conflict resolution which showcases faculty confidence

Table 10.1. Demographic profile of respondents

Particulars	*Numbers (n=249)*	*Percentage (%)*
Gender		
Male	105	42.2
Female	144	57.8
Total	249	100
Age		
Up to 30 Years	44	17.7
31–40 Years	113	45.4
Above 40 Years	92	36.9
Total	249	100
Designation		
Assistant Professor	174	69.9
Associate Professor	53	21.3
Professor	22	8.8
Total	249	100
Marital Status		
Married	222	89.2
Unmarried	27	10.8
Total	249	100
Educational Qualification		
MBA or M.Com	145	58.2
MBA With NET or SET	36	14.5
MBA With Ph.D	43	17.3
MBA With Ph.D and NET	25	10
Total	249	100
Total Service in Years		
Less than 3 Years	33	13.3
3–5 Years	32	12.9
6–8 Years	54	21.7
More than 8 Years	130	52.2
Total	249	100
Service in Present College		
Less than 2 Years	90	36.1
3–4 Years	58	23.3
5–6 Years	21	8.4
More than 6 Years	80	32.1
Total	249	100
Monthly Income (Rs.)		
Up to 20000	21	8.4
20001–30000	109	43.8
30001–40000	56	22.5
Above 40000	63	25.3
Total	249	100
Monthly Expenditure (Rs.)		
Up to 15000	82	32.9
15001–25000	82	32.9
25001–35000	51	20.5
Above 35000	34	13.7
Total	249	100

(continued)

Particulars	*Numbers (n=249)*	*Percentage (%)*
Family Size		
Up to 3 Members	26	10.4
4–5 Members	161	64.7
More than 5 Members	62	24.9
Total	249	100
Family Type		
Joint Family	67	26.9
Nuclear Family	182	73.1
Total	249	100
Area of Residence		
Urban	73	29.3
Semi-Urban	98	39.4
Rural	78	31.3
Total	249	100
Location of College		
Urban	104	41.8
Semi-Urban	68	27.3
Rural	77	30.9
Total	249	100
Distance from Residence to College		
Less than 5 km	32	12.9
5–10 km	70	28.1
10–20 km	100	40.2
More than 20 km	47	18.9
Total	249	100
Time to Reach College		
Less than 15 minutes	41	16.5
15–30 minutes	109	43.8
30–60 minutes	62	24.9
More than 60 minutes	37	14.9
Total	249	100

Source: Author.

Table 10.2. KMO and Bartlett's test

Kaiser-Meyer-Olkin Measure of Sampling Adequacy.		.897
Bartlett's Test of Sphericity	Approx. Chi-Square	3899.920
	df	300
	Sig.	.000

Source: Author.

together with mentorship programs and institutional training. Factor 4 focuses on leadership and interactive teaching methods by promoting student participation as well as leadership training and mental reasoning challenges for students. The challenges related to technical talent versus soft skill development form factor 5 which underscores the importance of teamwork advancement along with professional development and training difficulties in soft skills. The factors

Table 10.3. Rotated component matrix

	Component				
	1	*2*	*3*	*4*	*5*
I modify my communication style based on student needs and feedback.	.745				
I create a supportive and inclusive classroom environment.	.701				
I seek feedback from peers and students to improve my soft skills.	.692				
I actively listen to student concerns and respond empathetically.	.683				
I use student-centered learning approaches to enhance engagement.	.683				
I incorporate emotional intelligence strategies in my teaching approach.	.619				
I am comfortable using digital tools to improve student interaction and participation.	.608				
I actively participate in workshops and conferences related to faculty development.	.587				
I regularly reflect on my strengths and weaknesses in soft skills.		.739			
I actively seek opportunities to improve my soft skills.		.722			
Time constraints make it difficult for me to focus on developing soft skills.		.610			
My institution encourages collaboration among faculty to enhance soft skills.		.534			
I am adaptable to changes in teaching methods and institutional policies.		.519			
I effectively manage conflicts and diverse opinions in the classroom.			.782		
There are sufficient professional development opportunities available to improve soft skills.			.652		
Institutional policies support continuous faculty development in soft skills.			.560		
Mentorship programs in my institution have helped me enhance my interpersonal skills.			.516		
My institution provides adequate training programs for soft skills development.			.508		
I am confident in my ability to communicate effectively with students and colleagues.			.507		
I integrate interactive teaching methods that foster student engagement and communication.				.756	
I take initiative in developing leadership skills within my academic environment.				.610	
I use problem-solving and critical thinking exercises to develop my own and my students' soft skills.				.514	
I face challenges in balancing technical expertise with soft skills training.					.710
I encourage teamwork and collaboration among students.					.646
I believe continuous improvement in soft skills is essential for professional success.					.508
Eigenvalues	10.579	1.773	1.496	1.113	1.060
% of Variance	42.315	7.093	5.984	4.452	4.239
Cumulative %	42.315	49.407	55.391	59.844	64.083

Source: Author.

explain a major part of variance in faculty soft skill acquisition because their combined influence reaches 64.083%.

4. Suggestions

4.1. For faculty members

- Educational institutions should test soft skills regularly for enhancement through workshops and mentoring and self-assessment programs.
- Educational staff should employ interactive student-centered teaching strategies which enhance communication activities between them and their students.
- Staff members should effectively schedule their time to maintain subject mastery while improving interpersonal abilities.
- Faculty members should actively take part in programs for teacher enhancement and peer-based learning activities.

4.2. *For educational institutions*

- The organization must create formal training programs for teachers which include instructions about communication methods and leadership principles as well as teamwork techniques.
- The organization should install mentorship along with peer-learning projects which will help teams improve their performance.
- The institution should initiate policies that reward employees who engage in soft skills development while providing recognition.
- The organizations should establish collaborative approaches between different fields to spread valuable information across programs and best practices.

4.3. *For government and policymakers*

- Government institutions should develop official policies that establish mandatory soft skill training procedures for faculty development programs.
- Healthcare entities should get funding together with grant opportunities to establish institutional faculty development programs.
- Standard protocols need development to ensure consistent programs which improve soft skills operation throughout the healthcare sector.
- The organization must track training program success rates for their alignment with present industrial requirements.

5. Conclusion

Soft skills constitute vital elements which help faculty members succeed at improving their educational achievements at higher education institutions. Elementary aspects of soft skill training including adaptability and communication along with leadership functions as main factors which emerged during factor analysis. The systematic process requires faculty members to engage in continuous development chances and maintain both educational event attendance and skill self-evaluation. The power of educational professionals to influence classroom performance increases when they use student-centered instructional methods and emotional intelligence practice implementation. Educational institutions generate participative spaces to foster soft skills development along with formal tutorial sessions which create training systems led by mentors. The institution needs to create policies which allow faculty members to develop their capabilities through professional development programs and construct assessment and learning platforms for mutual growth. The government needs to support faculty soft skill training by developing policies while creating funding programs to develop capacity which includes professional development programs. Acquiring soft skills helps teachers improve both their educational performance and their speaking capabilities along with their student engagement techniques. Transformational school governing bodies as well as administrative departments should join forces to develop an active learning environment for students that delivers high-quality education for professional-ready graduates.

5.1. *Scope for further research*

Studies need to investigate how soft skills affect both faculty performance results and student educational achievements beyond short-term analysis. Studies which analyze both different academic areas and multiple academic establishments offer better understanding. Using emerging technologies for faculty training simultaneously with extended observations of their skill growth overtime proves beneficial. The understanding can reach greater depth through assessments of both student views about faculty soft skills and governmental policies that support faculty development. Researchers should study how different cultural groups and geographic regions learn and use soft skills in order to present worldwide views.

References

[1] Robles, M. M. (2012). Executive perceptions of the top 10 soft skills needed in today's workplace. Business Communication Quarterly, 75(4), 453–465.

[2] Tang, K. N. (2020). Beyond employability: Embedding soft skills in higher education. The Journal of Teaching and Learning for Graduate Employability, 11(1), 1–13.

[3] Laker, D. R., & Powell, J. L. (2011). The differences between hard and soft skills and their relative impact on training transfer. Human Resource Development Quarterly, 22(1), 111–122.

[4] O'Leary, S. (2016). Integrating employability into degree programmes: A methodology for work-based learning. Journal of Education and Work, 29(7), 751–775.

[5] Succi, C., & Canovi, M. (2020). Soft skills to enhance graduate employability: Comparing students and employers' perceptions. Studies in Higher Education, 45(9), 1834–1847.

[6] Clokie, T. L., & Fourie, E. (2016). Graduate employability and communication competence: Are undergraduates taught relevant skills? Business and Professional Communication Quarterly, 79(4), 442–463.

[7] Cottrell, S. (2015). *Skills for success: Personal development and employability*. Palgrave Macmillan.

[8] Beausaert, S., Segers, M., & Gijselaers, W. (2011). The use of a personal development plan and the acquisition of soft skills. *Career Development International*, *16*(5), 446–469.

[9] Jackson, D. (2015). Employability skill development in work-integrated learning: Barriers and best practice. *Studies in Higher Education*, *40*(2), 350–367.

[10] Bennett, N., Dunne, E., & Carré, C. (2000). *Skills development in higher education and employment*. Open University Press.

[11] Huang, H. M., Rauch, U., & Liaw, S. S. (2010). Investigating learners' attitudes toward virtual reality learning environments: Based on a constructivist approach. *Computers & Education*, *55*(3), 1171–1182.

[12] Knight, P. T., & Yorke, M. (2004). *Learning, curriculum and employability in higher education*. Routledge.

[13] Riemer, M. J. (2007). Communication skills for the 21st-century engineer. *Global Journal of Engineering Education*, *11*(1), 89–100.

11 Digital transformation management: analyzing the influence of new media on women's empowerment in Tamil Nadu, India

V. Muthulakshmi[1,a] *and R. Abdul Muthalif*[2,b]

[1]PhD Research Scholar (Full-Time), Sri Vijay Vidhyalaya College of Arts and Science, Dharmapuri, Tamil Nadu, India
[2]Assistant Professor and Research Supervisor PG and Research Department of Commerce, Dr. Zakir Husain College, Ilayangudi, Sivagangai, Tamil Nadu, India

Abstract: The digital revolution has altered the geographical landscape to the degree where promising new ways of communication, carrying of information, and social interaction has been provided. In Tamil Nadu, India, the emergence of new media has impacted positively towards women empowerment that helps in breaking the traditional constraints and embolden personal agency. This research paper seeks to investigate how digital technologies have played a nuanced role in improving the status of women in Tamil Nadu, by how digital technologies or new media are bridging the gender disparity in economic, educational and social spheres.

Keywords: Digital revolution, global landscape, media, women's empowerment, technologies, management

1. Introduction

This is the digital revolution, characterized by the all-pervasive availability of internet, smartphones and social media, which has changed the course of world and presented a brave new dawn to the people all over the globe. Amid this global transformation, Tamil Nadu, a state in southern India, serves as a vivid illustration of the profound influence of the digital age on women's empowerment [1]. In this instance, women have been using the power of new media to break the boundaries of the past and take the reins of their lives [2].

In Tamil Nadu the digital revolution has opened up new avenues for Tamul women to grow, express themselves, to be independent. Given this, this paper sets out to unravel the multifaceted influence of new media on women empowerment in this teeming and diverse Indian state. This paper attempts to study in detail the nuances of this intersection by looking into how digital technologies have significantly affected the lives of women in Tamil Nadu through their own devices [4].

This research paper is structured to provide a comprehensive examination of the subject matter, beginning with an in-depth literature review, followed by a discussion of the research methodology [6]. We will then present our findings, drawing attention to the effects of new media on women's access to information, economic empowerment, educational opportunities, social networks, and political participation. Lastly, we will synthesize these findings into a broader discussion, concluding with an outlook on future directions in this critical area of study.

The digital revolution has not only transformed a technological field but also a societal transition into one of a just, powerful, and dynamic society. Hence, understanding the impact of the above digital transformation on the lives of women in Tamil Nadu is in part an academic endeavor; and equally, it is an attempt with policy and social action implications on the larger canvas. Therefore, although the future of the topic of this research goes beyond personal interest, ultimately, it refers to the hope for how digital technologies could contribute to achieving the agenda of gender equality and social development [8].

1.1. *Strategic management of digital transformation for women's empowerment*

This explores the strategic management aspects of digital transformation in women's empowerment programs in Tamil Nadu [10]. It delves into how organizations are strategically adapting to and harnessing the power of new media to maximize the impact of their initiatives and advance women's empowerment [12].

1.2. *Inclusive management strategies for navigating the digital revolution in women's empowerment*

Focused on inclusive management practices, this content discusses how organizations are developing and implementing management strategies that ensure the digital revolution

[a]vmuthulakshmimcom@gmail.com, [b]abdulphd7@gmail.com

DOI: 10.1201/9781003684459-11

benefits all women, considering factors such as accessibility, literacy, and cultural nuances. In addition, it characterizes the importance of inclusive management in the empowerment process.

1.3. Agile management in the era of digital media: A case study of women's empowerment initiatives in Tamil Nadu

The entire case analysis would be based on the application of Agile management principles in the domain of women's empowerment in the digital revolution context in Tamil Nadu [14]. Adaptability, collaboration and fast decision making are key enablers of success for these programs in the digital dynamo.

1.4. Measuring impact: Data-driven management in women's empowerment through new media

This content focuses on utilizing data analytics for management in women's empowerment programs and how the organizations are using data driven decision making. It discusses the significance of data in evaluating the effects of initiatives, tailoring strategies, and taking decisions that contribute to a positive result for women in Tamil Nadu.

2. Literature Review

Research and policy circles in Tamil Nadu have begun to take notice of the confluence of the digital revolution and women's empowerment. Using such literature, the literature review understands the role of new media in shaping the lives of women in Tamil Nadu through five critical dimensions which include access to information, economic empowerment, educational opportunities, social networks and participation in politics.

2.1. Access to information

Empowerment requires access to information. Access to information through the digital revolution has greatly enhanced women's access to information in Tamil Nadu in terms of health, education, legal rights, and a range of other important information [3]. The internet, to boot, with mobile devices, is a gateway that provides women access to knowledge, allowing them to make informed decisions about their lives and challenge the status quo of tradition. However, studies have shown that there has been a rise in women in Tamil Nadu, who are accessing health related, financial and legal information to increase their autonomy and the control they have over their lives, through the internet.

2.2. Economic empowerment

Economic empowerment is part of women's autonomy and well being [5]. However, in Tamil Nadu, the new era of the digital revolution has made a new economic opportunity for women. E commerce, freelancing platforms, and work from the home has now let women join the current workforce and earn money as a freelancer. Nevertheless, a multitude of women in Tamil Nadu have succeeded in building upon their online businesses, which turns not only for themselves but also for the territory into an economically independent territory. The online marketplaces and platforms have provided avenues in which women showcased their skills, their products and services without financially depending on others, with a provision to support their families.

2.3. Educational opportunities

Traditionally it has been disrupted in the digital age, giving women of Tamil Nadu ample chances to gain variety of education in absence of traditional education models. Thanks to online educational platforms women have been able to gain new skills, get a degree, improve their employability. Additionally, many of these platforms also generally provide flexible learning opportunities for women to learn and carry out their studies as they also carry out their parental and professional responsibilities. Research has also shown that women in Tamil Nadu are using online courses and resources to acquire vocational skills and enhance their qualifications and thus make economic independence [7].

2.4. Social networks

In Tamil Nadu, social media platforms have played a significant role as social networks that have been facilitating women's empowerment [9]. On digital spaces they have come together, from every walk of life, exchanged views on common issues, talked and experienced what is being experienced and sought emotional support. Apart from a sense of belonging and solidarity, online communities also have within them communities of advocacy and collective action [15]. As a whole, women in Tamil Nadu use social media to create social networks, participate in discussions around gender issues and engage in gender equality initiatives [11].

2.5. Political participation

Women in Tamil Nadu have been enabled to enter into political processes and take their right to push their agendas using digital media. Social media campaigns, digital advocacy and online forums become a platform for women's voice to be shown, to participate in political discussion and to refuse traditional systems of power. Research shows that women are further taking part in digital political movements and campaigns, and taking their influence on policy agendas and the paving way for gender sensitive policies.

Therefore, in the light of this, there seems to be justification that the digital revolution impacts women's empowerment in Tamil Nadu in multiple ways; literary substantiate this. Thus, digital technologies have not only made it possible for women to access information, gain monetary independence, get opportunities for education, social connection and political engagement. Based on these results, a deeper understanding of the means by which digital revolution may bring transformational measure to gender disparities and women empowerment in Tamil Nadu is enabled.

3. Methodology

The research paper thus employs a mixed method approach with both qualitative and quantitative methods to study the impact of new media on female empowerment in Tamil Nadu represented in Table 11.1. Primarily, surveys, interviews and focus groups are used as a source of primary data to capture the personal narratives of the interviewees; whereas secondary data is provided from the appropriate reports, studies and statistical sources.

The research shows that digital technology plays a key role in empowering women in Tamil Nadu. The area has good access to digital tools for women, who also perceive these tools to be valuable for them. However, in terms of economic empowerment, this is most clear; the use of digital technology is of great importance. Besides, digital media promotes social connectivity and women participation in political processes hence promoting women empowerment. This does not seem to be a problem for accessibility to digital resources for education, which is relatively high but perhaps there is room for improvement on further empowering people in this way. The overall research converges to the fact that this digital technology has a great power of transformation to bring gender equality and development if it is used resourcefully with Tamil Nadu women [13].

4. Findings

4.1. Access to information

Tamil Nadu's women now get access to information about health, education, legal rights among other things, all thanks to Digital media. This has given them the ability to make an informed choice and rewrite the rules of the past [16].

Table 11.1. Average access to digital technology and perception of empowerment among women in Tamil Nadu

Aspect of Empowerment	*Average Access to Digital Technology (on a scale of 1.5)*	*Average Perception of Empowerment (on a scale of 1.5)*
Economic Empowerment	3.8	4.2
Educational opportunities	4.5	4.0
Social Networks	4.1	3.9
Political Participation	3.7	4.4

Sources: Primary data.

4.2. Economic empowerment

With the digital revolution, women in Tamil Nadu have many economic opportunities. They support their families and invest their futures with their financial independence thanks to freelancing, remote work and e commerce.

4.3. Educational opportunities

Online educational platforms have allowed women the power to access formal or formal education. Instead, they become able to acquire new skills, degrees and become more employable and attain greater economic independence.

4.4. Social networks

Social media is also a medium through which women in Tamil Nadu can join together to network, exchange experiences, and talk about common issues. These networks have allowed them to get a voice, increased their confidence, and created solidarity.

4.5. Political participation

Digital media enables women to discuss politics online, campaign on their rights, and confront power dynamics that may continue in the analog world. Through social media campaigns and online forums, people have been enabled to participate politically.

5. Discussion

This research paper findings, highlight the capabilities of the digital revolution to transform Tamil Nadu in India. Women's empowerment is also achieved by providing access to information, economic opportunities, education, social networks and political engagement using new media. The coming together of these factors has shifted the gender dynamics of our country and has promoted women's empowerment by helping them overcome traditional barriers.

6. Conclusion

Undoubtedly, the digital revolution has paved the path for participating women in Tamil Nadu, India. Women have been given means and window to claim their rights, acquire economic independence, be educated, form social networks and participate in political dispensation. Created as a powerful statement about the impact of new media on women empowerment, this stands out as a compelling point in how digital technology can transform for social development and bridge the gap between the sexes.

7. Future Directions

However, dynamics between digital revolution and women's empowerment are still worth researching in such an area. This includes assessing the digital literacy initiatives, the role of social media influencers in facilitating gender equality and empowerment, and other policies that encourage support of the two.

Lastly, Digital Revolution, as the last word, has an important role in bringing changes in the life of women in Tamil Nadu, and contends unarguable significance in women empowerment, both as a part of Tamil Nadu, as well as in the global context overall. This research paper is a proof of what digital technologies can do and how they can help to change the concept of dominance through gender and create an unbiased, inclusive and empowered society.

References

[1] Ahmed, N., Abulaish, M., & Anwar, S. (2019). Digital empowerment of women: A review of initiatives in developing countries. Information Systems Frontiers.

[2] Bernardos, C.J., Gramaglia, M., Contreras, L.M., Calderon, M., & Soto, I. (2010). Network-based localized IP mobility management: Proxy mobile IPv6 and current trends in standardization. *Journal of Wireless Mobile Networks, Ubiquitous Computing, and Dependable Applications (Special issue: Advances in Wireless Mobile and Sensor Technologies)*, *1*(2/3), 16–35.

[3] Chaturvedi, A., Agarwal, P., et al. (2018). Digital literacy and women's empowerment: A case study from rural India. *International Journal of Information Management*.

[4] I. You. (2021). Letter from the Editor-in-Chief. *Journal of Internet Services and Information Security*, *11*(1), 1-1.

[5] Kaimal, S., & Lingard, R. (2016). Mobile phones and women's empowerment: A review of the literature. *Gender, Technology, and Development*.

[6] Baldeón, C. P. H., Fuster-Guillén, D., Figueroa, R. P. N., Lirio, R. A. P., & Hernández, R. M. (2024). Pedagogy in Strengthening Visual Perception: A Review of the Literature. *Indian Journal of Information Sources and Services*, *14*(2), 85–96. https://doi.org/10.51983/ijiss-2024.14.2.13

[7] Dad, N., & Khan, S. (2018). Social media and women's empowerment: A global perspective. *International Journal of Communication*.

[8] Abed, F. S. (2024). Optimization of Expression and Purification of Recombinant One of Central Nervous System Enzyme (Acetylcholinesterase). *International Academic Journal of Science and Engineering*, *11*(1), 40–47. https://doi.org/10.9756/IAJSE/V11I1/IAJSE1106

[9] Singh, S., & Singh, R. (2016). Economic empowerment of women through e-commerce in India. *International Journal of E-Services and Mobile Applications*.

[10] Baggyalakshmi, N., Harrsini, M. S., & Revathi, R. (2024). Smart Billing Software. *International Academic Journal of Innovative Research*, *11*(1), 51–60. https://doi.org/10.9756/IAJIR/V11I1/IAJIR1106

[11] Rahman, M. M., & Chowdhury, A. (2019). Educational empowerment of women through online learning: A case study of MOOCs in Bangladesh. *Journal of Online Learning and Teaching*.

[12] Naveena, N., & Stalin Alex, D. (2016). Efficient Face Eyes and Mouth Detection Using Viola Jones Algorithm. *International Journal of Advances in Engineering and Emerging Technology*, *7*(4), 13–22

[13] Mitullah, W., & Kinyanjui, M. N. (2020). The role of social media in fostering women's entrepreneurship in Sub-Saharan Africa. *Information Development*.

[14] Shugars, S., & Adolphsen, P. R. (2019). Women's political participation in the digital age: A comparative analysis. *New Media & Society*.

[15] Andrade, S. S., & Moraes, M. J. S. (2017). Digital access and gender equality: A cross-national analysis. *Information Technology for Development*.

[16] AlSajri, A., & Aljanabi, M. (2024). The Impact of ChatGPT on Social Media. *MEDAAD*, *2024*, 9–14. https://doi.org/10.70470/MEDAAD/2024/002

12 Impact of personality traits on job satisfaction among healthcare professionals

Senthil Kumar S.[1,a] *and J. Sudarvel*[2,b]

[1]Research Scholar, Department of Management, Karpagam Academy of Higher Education, Coimbatore, Tamil Nadu, India
[2]Assistant Professor of Management, Karpagam Academy of Higher Education, Coimbatore, Tamil Nadu, India

Abstract: The present research evaluates how healthcare practitioners' personality types influence their work satisfaction utilizing the assessment framework of the Big Five Personality Model which includes Openness, Conscientiousness, Extraversion, Agreeableness and Neuroticism. The research implements a quantitative method through standardized questionnaires which measure both personality traits and job satisfaction scores. An analysis through factor methods detects principal factors which determine job satisfaction levels. Research outcomes demonstrate that healthcare professionals rate higher conscientiousness and agreeableness traits positively relate to job satisfaction but neuroticism produces negative effects. A factor analysis technique collects job satisfaction components into categories which uncovers detailed information about the effects of personality traits on professional work life. Healthcare organizations can leverage the research findings to improve their employee well-being and retention along with job performance through specific Human Resources policies which deliver workplace interventions.

Keywords: Personality traits, job satisfaction, healthcare professionals, big five model workplace well-being

1. Introduction

Healthcare personnel rate their job satisfaction based on its critical value for employee performance and workplace retention as well as overall well-being [1]. The evaluation of work satisfaction elements for healthcare personnel stands essential for enhancing staff engagement alongside organizational success rates. Personality traits define work satisfaction levels as essential determinants because they dictate how people experience their work environment [2]. Researchers rely on the Big Five Personality Model to examine how personality influences workplace results [3]. The research indicates that conscientious people experience better job satisfaction since they bring discipline and objective attitudes to work but neurotic individuals tend to report lower satisfaction from work because they are emotionally fragile [4]. The workplace performance of extraverted and agreeable individuals benefits from their ability to create good relationships while adapting to professional requirements [2]. Healthcare professionals face high work demands along with emotional strain which makes their personality traits play a substantial role in their ability to manage stress and their interactions at work [5]. This study performs factor analysis to discover essential personality traits that determine job satisfaction and provides knowledge to enhance workplace environment quality and promote employee well-being.

2. Review of Literature

Scholars have extensively researched how personality traits affect job satisfaction throughout different occupational fields primarily in healthcare. This section evaluates significant scholarly works about the Big Five Personality Traits and their job satisfaction impact together with behavioural understanding.

The enduring characteristics found in personality traits affect the way people think as well as the way they express emotions and carry out behaviours. The Big Five Personality Model which McCrae and Costa [3] presented has become the leading standardized method to evaluate personality traits. Multiple studies confirm that personal characteristics create substantial effects on employee perspectives of work assignments and their methods of dealing with job obstacles and their job satisfaction levels [2].

Earnest personalities manifest characteristics which include orderliness and strong accountability together with their drive to achieve goals. The evidence shows conscientious workers experience higher job satisfaction because their work efficiency rises along with a strong sense of accomplishment [2, 4]. People who enjoy pleasing others create positive interpersonal bonds thus their relationships at work improve and job satisfaction increases according to Diener, Oishi, and Lucas [6].

The social and energetic nature of extraverts creates better job satisfaction particularly when employees work

[a]vinsyed@gmail.com, [b]j.sudarvel@gmail.com

DOI: 10.1201/9781003684459-12

on teams and communicate frequently [7]. Job satisfaction impacts differently on open-minded individuals since they usually prioritize discovering new things and innovation which leads to positive workplace satisfaction [3].

People who are neurotic cope with inconsistent emotions and endure high stress levels which negatively affects their job satisfaction [8].

The positive correlation between workplace satisfaction and conscientiousness and agreeableness is the strongest according to Sutin et al. [9], but neurotic symptoms lead to decreased happiness at work [9]. The high-stress working environment of healthcare requires job satisfaction because it extends retention levels and improves performance according to research from Aiken et al. [10]. Several components which affect job satisfaction in healthcare include workload demands along with work-life balance requirements and the quality of interpersonal connection and emotional well-being [11].

The workplace responses of healthcare professionals regarding stress management and work satisfaction depend on their personality traits according to research conducted by Bakker and Demerouti [12]. The sense of personal satisfaction increases with conscientiousness and extraversion because these traits facilitate efficient responsibility management and positive interpersonal contact.

Professionals who show neurotic traits experience diminished job satisfaction and increased burnout tendencies since neurotic individuals struggle constantly with stress and emotional exhaustion [13]. Factor analysis functions as a standard statistical approach to discover the main components responsible for job satisfaction as well as personality structures. Scientists have employed Exploratory Factor Analysis (EFA) and Confirmatory Factor Analysis (CFA) to discover crucial personality factors which shape job satisfaction based on Tabachnick and Fidell [14].

Factor analysis performed by Judge et al. [2] verified through analysis that conscientiousness as well as neuroticism both significantly affect job satisfaction according to the Big Five model. Through factor analysis Sutin et al. [9] determined that intrinsic job satisfaction primarily depends on conscientiousness and openness behaviours. Bakker and Demerouti [12] conducted factor analysis in healthcare environments which demonstrated that work-related satisfaction in healthcare providers requires assessment through workload management as well as professional independence and connections with colleagues.

The satisfaction levels of healthcare workers depend on multiple elements where personality traits remain essential together with working environment conditions. A series of individual characteristics which include emotional stability and interpersonal skills determine job perception and satisfaction according to Judge, Heller, and Mount [2]. Healthcare professionals' job satisfaction improves through structured mechanisms established by organizations which focus on workforce management and leadership backing along with career development assistance [11]. Job satisfaction for healthcare professionals depends on their patient relationships and team dynamics together with their experience of new medical practice developments and global healthcare changes [12].

2.1. *Research gap*

Research has extensively studied personality traits in work satisfaction but fewer studies have researched hospital staff specifically. This research field exhibits limited use of factor analysis to study healthcare job satisfaction despite widespread employment of the method throughout general organizational studies. This research works to resolve these knowledge gaps through factor analysis for identifying fundamental personality elements which shape job satisfaction within healthcare institutions.

2.2. *Objective*

To identify the Impact of Personality Traits on Job Satisfaction Among Healthcare Professionals.

2.3. *Scope of the study*

The research investigates how personality characteristics influence the job satisfaction levels of medical practitioners. This research examines the relationship of extraversion, conscientiousness and neuroticism, agreeableness and openness as personality traits toward job satisfaction in demanding healthcare conditions. The research evaluates intrinsic motivators like motivation and emotional stability and extrinsic workload factors in combination with qualities from leadership and organizational support. A targeted group of healthcare workers from medical facilities takes part in the study to maintain real-world application. The study results will support enhancements in job satisfaction standards and employee welfare policies in healthcare facilities.

3. Research Methodology

3.1. *Data*

This research relies primarily on primary sources which healthcare professionals working in hospitals and clinics provided by means of surveys and structured questionnaires.

3.2. *Sampling*

Data collection through judgment sampling favors healthcare professionals of all specialties including medical doctors and nursing staff and allied health workers. A diverse set of responses from 176 healthcare professionals supports the research representation.

3.3. *Framework of analysis*

A factor analysis with descriptive statistics and simple percentage analysis of the collected data revealed major personality traits which impact job satisfaction among healthcare professionals.

3.4. Significance of the study

This research maintains importance because it evaluates how personality characteristics affect job satisfaction levels of healthcare personnel who play a vital role as care providers and workforce maintainers. Understanding how personality traits interact with job satisfaction will aid healthcare institutions in designing specific workplace improvement initiatives for their staff. The conclusions from the study enable policymakers along with hospital administrators and human resource managers to create specific recruitment and training programs with workplace support for diverse personality types. The study adds value to healthcare efficiency and satisfaction rates and burnout prevention within medical institutions.

3.5. Analysis and interpretation

Data from 176 healthcare professionals shows the workforce contains various profiles among its members (Table 12.1). The survey participants split into 55.7% female and 44.3% male segments and 45.5% are between 31 to 40 years old, accompanied by 34.6% who are above 40 years of age. Doctors form 40.9% of the respondents with nurses at 36.4% and allied health professionals at 22.7% as the remaining groups. A high percentage of 81.3% healthcare professionals are married to one partner and possess either bachelor's (36.9%) or master's degrees (31.3%). The working experience breakdown shows that more than one third of the respondents (36.9%) maintain over eight years of work but less than one third (30.7%) stay at their current hospital for less than two years. Among the group the highest number of individuals at 40.9% earn between Rs. 30,001 - 50,000 annually. The majority of participants (67%) live with their family nucleus in their homes while 38.1% of them choose to reside in semi-urban locations. The collected data provides valuable information about staffing distributions that affects worker satisfaction levels in healthcare settings.

3.6. Factor analysis

To identify the Impact of Personality Traits on Job Satisfaction Among Healthcare Professionals factor analysis is employed.

Factor analysis proves suitable with a Kaiser-Meyer-Olkin (KMO) Measure of Sampling Adequacy value of 0.912 because values greater than 0.80 signify meritorious adequacy for dataset analyses (Table 12.2). The significant values of Bartlett's Test of Sphericity (Chi-Square = 8833.920, df = 300, Sig. = 0.000) indicate that the correlation matrix is different from an identity matrix which shows factor analysis properly identifies latent structures in healthcare professional personality traits and job satisfaction patterns (Table 12.3).

Table 12.1. Descriptive statistics

Particulars	*Numbers (n=176)*	*Percentage (%)*
Gender		
Male	78	44.3
Female	98	55.7
Total	176	100
Age		
Up to 30 Years	35	19.9
31–40 Years	80	45.5
Above 40 Years	61	34.6
Total	176	100
Designation		
Doctor	72	40.9
Nurse	64	36.4
Allied Health Professional	40	22.7
Total	176	100
Marital Status		
Married	143	81.3
Unmarried	33	18.7
Total	176	100
Educational Qualification		
Diploma	30	17
Bachelor's Degree	65	36.9
Master's Degree	55	31.3
Professionals	26	14.8
Total	176	100
Total Service in Years		
Less than 3 Years	28	15.9
3–5 Years	34	19.3
6–8 Years	49	27.8
More than 8 Years	65	36.9
Total	176	100
Service in Present Hospital		
Less than 2 Years	54	30.7
3–4 Years	42	23.9
5–6 Years	27	15.3
More than 6 Years	53	30.1
Total	176	100
Monthly Income (Rs.)		
Up to 30,000	28	15.9
30,001–50,000	72	40.9
50,001–70,000	44	25
Above 70,000	32	18.2
Total	176	100
Family Type		
Joint Family	58	33
Nuclear Family	118	67
Total	176	100
Area of Residence		
Urban	52	29.5
Semi-Urban	67	38.1
Rural	57	32.4
Total	176	100

Source: Author.

The analysis revealed five essential dimensions that affect professional job satisfaction of healthcare workers who together explain 63.788% of the total variance. Employees experience high levels of job satisfaction due to Workplace Environment & Relationships which accounts for 43.209% of the total influence. The factor of Job Commitment & Recognition (6.642%) demonstrates that employee commitment enhances motivation through autonomy and patient care dedication as well as recognition and fair evaluation processes. The assessment of workload alongside performance metrics together with workplace referral interest and work efficiency comprise the third element affecting workplace satisfaction at 4.855%. Professional Growth & Compensation (4.659%) stands as the fourth vital component which includes workplace respect together with problem-solving capabilities and adaptability and employee financial satisfaction in its core elements. Work satisfaction stems from meaningful activities at work which deliver gratification at Meaningfulness & Fulfillment (4.424%). The research shows these variables work together to determine healthcare professionals' job satisfaction levels through their workplace associations coupled with recognition practices alongside their workload management needs and continuing development chances as well as the personal rewards they find in their job duties.

Table 12.2. KMO and Bartlett's test

Kaiser-Meyer-Olkin Measure of Sampling Adequacy.		.912
Bartlett's Test of Sphericity	Approx. Chi-Square	8833.920
	df	300
	Sig.	.000

Source: Author.

Table 12.3. Rotated component matrix

	Component				
	1	*2*	*3*	*4*	*5*
My workplace encourages teamwork and cooperation.	.742				
I have a good professional relationship with my colleagues.	.710				
My supervisor provides constructive feedback and support.	.688				
I feel comfortable communicating my concerns with management.	.677				
I feel my workplace promotes a positive and supportive environment.	.619				
I enjoy collaborating with colleagues and working in a team.	.592				
I remain calm and composed under stressful work situations.	.588				
I am confident in handling work-related challenges effectively.	.570				
I am satisfied with the level of autonomy I have in my job.		.749			
I am dedicated to providing the best possible care to patients.		.643			
I feel appreciated and valued for my contributions at work.		.596			
My job offers a fair performance evaluation and promotion system.		.553			
I feel a strong sense of belonging to my workplace.		.537			
I feel enthusiastic and motivated about my job responsibilities.		.509			
I believe my workload is reasonable and manageable.			.769		
I would recommend my workplace to others as a good place to work.			.639		
I am willing to put extra effort into my job when needed.			.619		
I take pride in completing my work with accuracy and efficiency.			.573		
There is mutual respect among staff members in my organization.				.728	
I take initiative to solve problems without waiting for instructions.				.668	
I am open to new ideas and willing to adapt to changes in the workplace.				.603	
I am satisfied with my salary and benefits.				.530	
I believe my job provides meaningful and fulfilling experiences.					.827
Eigenvalues	10.802	1.661	1.214	1.165	1.106
% of Variance	43.209	6.642	4.855	4.659	4.424
Cumulative %	43.209	49.851	54.706	59.364	63.788

Source: Author.

4. Suggestions

- The company needs to establish professional relationships between staff members together with communication channels that permit employees to interact freely with leaders.
- The organization should establish an environment through windows that strengthens staff unity while promoting cooperative work relationships.
- The evaluation of business performance and promotional practices should include free of prejudice methods that identify core duties and unbiased advancement processes.
- Employee burnout creates poor duty management that results in diminished performance capacity of staff at their workplace.
- Organizations that pay higher than market standards for compensation will prevent employees from leaving their present roles.
- Every operational unit must maintain patient-centered cultures to promote job purpose.
- The organization must regularly organize certification-related training workshops alongside seminars that help personnel gain advanced industry practices and build and sustain their skills.
- To achieve career development support the organization must develop plans for mentoring which pairs senior team members with new employees.
- The organization requires established career advancement paths to outline the necessary qualifications with job promotion eligibility prerequisites.
- The company must support student advancement through financial training assistance, educational scholarships as well as flexible work scheduling options.
- The organization should establish methods for staff to access work opportunities between departments which help them develop their expertise.
- Staff members dedicated to enhancing their professional skills through educational activities should obtain workplace rewards from their organization.
- Staff members of the organization should gain access to research discovery platforms as well as meetings which enable exchange of knowledge between employees to boost innovation.
- The workplace approach at the company will develop performance-based workplace culture that advances employee growth through enhanced organizational performance.
- Every work shift operated by the company needs flexible options which accommodate individual employee requirements.
- Workplace distribution systems should be established by lawmakers to stop creating excessive work that causes employee burnout due to stress.
- The organization needs to implement concrete procedures which grant employees access to their scheduled rest periods together with authorized mental health days and rest time to stay healthy.
- Workload problem discussion should receive space from the company for staff members to request necessary modifications.
- The company should combine mindfulness sessions with exercise and physical activities to enhance staff wellness between other duties.
- Organizations gain their highest possible workplace efficiency through digital equipment that simplifies administrative functions.
- Employees at this organization will be able to disconnect from their work duties using disabled business communication during normal working hours.
- The organization extends parental leave provisions and runs family-support programs and extended leave schemes to help workers achieve better work-life balance.
- The organizational programs develop perfect working environments that promote staff well-being while increasing workplace satisfaction.

5. Conclusion

This research identifies personality aspects in the work environment which determine essential satisfaction elements for healthcare professionals. Research data indicates employees base their work environment views primarily on team competencies as well as organization adaptability along with organizational resilience capabilities dedicated to enabling efficient communication throughout all work elements. Staff members at organizations who work in supportive communication and teamwork environments develop greater morale levels which generates superior teamwork outputs. Healthcare personnel achieve peak career fulfillment by receiving performance-driven evaluations which include personnel choice decision-making power. The management systems for workload directly impact staff job satisfaction since heavy workloads generate stress that develops into burnout patterns. Healthcare staff members become motivated to perform their duties at work after receiving strong work-life balance programs as well as health benefits programs which match their salary alongside personalized career development opportunities. High-performing satisfied staff members emerge from organizations which implement mentor programs alongside training sessions and skill development programs.

Healthcare institutions addressing these core factors will obtain increased retention rates among staff members and reduced worker burnout and simultaneously deliver better healthcare outcomes to patients. Medical organizations succeed within systemwide and healthcare service networks through satisfied staff members who improve quality performance levels through patient care delivery. Organizations addressing employee wellness through supportive settings will secure enduring long-term success in the healthcare sector.

6. Scope for Further Research

A research investigation that aims to study job satisfaction depends on analytical and conceptual analysis for understanding emotional intelligence and resilience as separate personal traits. The external assessment requires examination of both hospital cultures together with hospital leadership decisions along with hospital policies. Multi-unit healthcare laboratory research that continually monitors staff groups enables scientists to determine how job satisfaction evolves throughout time. Marking themselves as adherents to interview techniques among qualitative research methods allows scientists to extract vital information from their direct interactions with employees. Medical organizations need work satisfaction to achieve optimal operational performance in their patient care outcomes. Better workplace policies should receive a boost through increased investigation in these domains to implement employee well-being strategies.

References

[1] Locke, E. A. (1976). The nature and causes of job satisfaction. In M. D. Dunnette (Ed.), *Handbook of industrial and organizational psychology* (pp. 1297–1349). Chicago: Rand McNally.

[2] Judge, T. A., Heller, D., & Mount, M. K. (2002). Five-factor model of personality and job satisfaction: A meta-analysis. *Journal of Applied Psychology*, *87*(3), 530–541.

[3] McCrae, R. R., & Costa, P. T. (1999). A five-factor theory of personality. In L. A. Pervin & O. P. John (Eds.), *Handbook of personality: Theory and research* (2nd ed., pp. 139–153). New York: Guilford Press.

[4] Barrick, M. R., & Mount, M. K. (1991). The Big Five personality dimensions and job performance: A meta-analysis. *Personnel Psychology*, *44*(1), 1–26.

[5] Weiss, H. M., Nicholas, J. P., & Daus, C. S. (1999). An examination of the joint effects of affective experiences and job beliefs on job satisfaction and variations in affective experiences over time. *Organizational Behavior and Human Decision Processes*, *78*(1), 1–24.

[6] Diener, E., Oishi, S., & Lucas, R. E. (2003). Personality, culture, and subjective well-being: Emotional and cognitive evaluations of life. *Annual Review of Psychology*, *54*(1), 403–425.

[7] Watson, D., & Clark, L. A. (1997). Extraversion and its positive emotional core. In R. Hogan, J. A. Johnson, & S. R. Briggs (Eds.), *Handbook of personality psychology* (pp. 767–793). San Diego: Academic Press.

[8] Judge, T. A., Bono, J. E., & Locke, E. A. (2000). Personality and job satisfaction: The mediating role of job characteristics. *Journal of Applied Psychology*, *85*(2), 237–249.

[9] Sutin, A. R., Costa, P. T., Miech, R., & Eaton, W. W. (2009). Personality and career success: Concurrent and longitudinal relations. *European Journal of Personality*, *23*(2), 71–84.

[10] Aiken, L. H., Clarke, S. P., Sloane, D. M., Sochalski, J., & Silber, J. H. (2002). Hospital nurse staffing and patient mortality, nurse burnout, and job dissatisfaction. *JAMA*, *288*(16), 1987–1993.

[11] Lu, H., Barriball, K. L., Zhang, X., & While, A. E. (2012). Job satisfaction among hospital nurses revisited: A systematic review. *International Journal of Nursing Studies*, *49*(8), 1017–1038.

[12] Bakker, A. B., & Demerouti, E. (2007). The job demands-resources model: State of the art. *Journal of Managerial Psychology*, *22*(3), 309–328.

[13] Hochwälder, J. (2008). A longitudinal study of the relationship between personality and job satisfaction among nurses. *Personality and Individual Differences*, *44*(8), 1764–1774.

[14] Tabachnick, B. G., & Fidell, L. S. (2007). *Using multivariate statistics* (5th ed.). Boston: Pearson.

13 Leadership attributes of elementary school teachers in Jolo I district, division of Sulu

Immir S. Juhudi

Sulu State College, Capitol Site, Jolo, Sulu, Philippines

Abstract: This study aimed to determine the levels of leadership attributes among elementary school teachers in Jolo District I of DepEd-Sulu during the 2022–2023 school year. The research explored the differences in leadership attributes based on teachers' demographic profiles, including age, gender, civil status, and length of service. A descriptive-exploratory research design was employed, with 100 elementary school teachers purposively selected as participants. The findings revealed that the majority of teachers were middle-aged (20–30 years old), female, married, and relatively new to teaching. The teachers generally exhibited very high positive attributes in the domains of About Oneself, About Others, About Jobs, and About Life. Significant differences were observed in the domains of About Oneself, About Others, and About Jobs based on age, but no significant differences were found based on gender, civil status, or length of service. The study supported aspects of Jims Evans' leadership attributes theory, emphasizing the importance of self-awareness, knowledge gaps, a sense of humor, and a desire to be a leader. Based on the findings, recommendations were made for Jolo District I of DepEd-Sulu to establish a clear district vision, focus on leadership development programs for younger teachers, and encourage further research on leadership in education. Additionally, it was suggested that researchers in the field of education conduct similar studies, expanding the exploration to include more domains of leadership. This research contributes to the understanding of leadership attributes among elementary school teachers and provides practical recommendations for enhancing leadership development programs and improving teachers' leadership knowledge and skills.

Keywords: Leadership attributes, elementary school teachers, Jolo I district

1. Introduction

Teachers have an important role in the education sphere and lay a major contribution to the development and growth of students [9]. In order to develop a pleasant learning environment and promote student attainment, elementary school teachers will require effective leadership attributes [1]. Understanding the attributes of leadership demonstrated by teachers will therefore provide some useful insights to enhance the overall quality of education [2].

This study seeks to find out the leadership attributes of elementary school teachers of Jolo I District, Division of Sulu. The study will provide understanding about leadership characteristics exhibited by teachers in the said setting, which includes seven elementary schools on the district [3].

This research is intended to improve the present knowledge of the effective teaching methods and educational leadership by studying about the leadership attributes of elementary school teachers [4]. The results of this study can provide useful insights to the educational officials, school administrators, teachers and other educational stakeholders to improve the leadership development programs, and ultimately the quality of education in the Jolo I District [5].

Only then, it can be understood that what leadership attributes elementary school teachers possess, that on the basis of which we can establish a positive and supportive learning environment that increase the development of kids completely and also improves their future achievements [7].

2. Methodology

The level of leadership attributes of public elementary school teachers in the Jolo-I District, Division of Sulu was studied [6].

The research design used is descriptive-exploratory. It is quantitative research since the leadership qualities of elementary school teachers in the Jolo-I District, Division of Sulu, have been determined. Descriptive research deals with accurate description of educational phenomena [8].

The study was carried out in the Jolo I district, Division of Sulu, which comprises of seven elementary schools. The teachers who were at the district in the 2022–2023 school year were the respondents.

The samples used in this research were representative samples from seven elementary schools selected via purposive sampling which is a non probability sampling method. The procedure used was purposive sampling so as to collect the desired quality and quantity of data. A Quota sampling was also used to determine the number of respondents.

Various statistical treatment of data were applied which include frequency count, percentage distribution, mean,

immirjuhudi11@gmail.com

DOI: 10.1201/9781003684459-13

standard deviation, and One-Way Analysis of Variance as a way of validating the hypothesis.

3. Results and Discussion

The following parts supply the presentations, analyses, and interpretations of results derived from the comprehensive scoring and statistical procedures applied to the collected data for this study.

1. The study examined the demographic profile of elementary school teachers in Jolo I District:
   - **Age:** Among the 100 teachers surveyed, 31% fall within the age range of 20–30 years, 39% fall within the age range of 31–40 years, and 30% fall within the age range of 41–65 years. These findings suggest that around 40% of the elementary school teachers in Jolo I District fall within the middle age range. This implies that these educators possess the capacity to optimize their abilities and commitment to their occupation.
   - **Gender:** Out of the 100 teachers that responded, 49 (41.66%) are male and 51(58.33%) are female. Female teachers in Jolo I District surpass the number of male teachers.
   - **Civil Status:** Among the 100 teachers surveyed, 36% are unmarried, 59% are married, and 5% are widowed. These findings suggest that a significant proportion of elementary school teachers in Jolo I District are married and have dual responsibilities in their professional and personal lives.
   - **Length of Service:** Out of the 100 teachers surveyed, 52% have been teaching for 3–10 years, 27% have 11–20 years of experience, and 21% have 21 years or more of experience. Consequently, a majority of the elementary school teachers in Jolo I District are currently in the early phase of their teaching professions.
2. The leadership attributes of elementary school teachers in Jolo I District, Division of Sulu, are categorized into different areas, and their levels are as follows:
   - **About Oneself:** The teachers surveyed demonstrate a strong ability to reflect on and acknowledge their personal strengths and flaws. Priority is given to what others think and transparency is regarded as a good trait of leadership. They are ready to cooperate with their colleagues and open to the other's views [10].
   - **About Others:** The survey took place with teachers who are capable of leading and creating an environment that incites people to strengthen and evolve growth in themselves. They believe in the idea of 'collective intelligence' and have high regard for their subordinates. Through their actions, they demonstrate leadership and work towards making even more significant connections in the improvement of everyone involved [12].
   - **About Jobs:** Teachers surveyed develop the teaching potential as a way to advance in professional skills and progress. Continuous professional growth and a deep feeling for the importance of instructional direction and co partaking with peers is trusted to such a great extent. This is important because it convinces that priority is required on the teacher training of leadership roles and also on the systems and structures which would support the growth and collaboration of teachers.
   - **About Life:** For the teachers surveyed, life afforded the opportunity to acquire knowledge and develop what he or she knows within himself. Life is seen by them as making a voyage through their lives, where there is no ending, only progression and advancement. They aim to serve and nurture other teachers and enhance their leadership.

   Across several areas, elementary school teachers in Jolo I District in general, are regarded as having excellent leadership qualities. This includes introspection, teamwork, openness and continuous developing. They want to inspire or motivate others, show respect and consideration to others' interests, promote positive and supportive culture in the schools.

   This is in agreement with earlier studies in educational leadership that underscore personal attributes, cooperation, continuous professional development, facilitating a healthy organizational ambience as factors for effective leadership.
3. The study was able to examine whether there is a significant difference between the levels of leadership attributes of elementary school teachers in Jolo-1 District as grouped according to age, gender, civil status, and length of service [11].
   - **Difference by Age:** An analysis was conducted to know if there is any significant difference in teachers perception on teachers' leadership attributes in the category of About Oneself, About Others and About Jobs based on age. But there is no difference in the category About Life. The implication here is that teachers belonging to different age groups have different perceptions concerning their strengths, weaknesses, interactions with others, or their work performance. Teachers perceive their strengths and weaknesses, collaborate with other teachers, and do their job responsibilities differently depending on age.
   - **Difference by Gender:** Thus, elementary school teachers feel that gender has no bearing on how they perceive and evaluate their leadership abilities. Unlike some other studies that discuss gender differences in leadership styles and how women lead, this study did not show any significant differences according to gender.
   - **Difference by Civil Status:** Overall, there was no considerable difference among perceptions on

leadership qualities of all categories between single married and widowed teachers. This means that the level of civil status of elementary school teachers has no impact or is statistically inconsequential on their perception and judgment on their leadership traits. However, teachers of both marital status share similar perception about leadership power.

- **Difference by Length of Service:** The results also indicated that there is a significant difference in the perceptions of teachers' perceived leadership attributes in terms of the category of About Oneself based on length of service. However, the categories of About Others, About Jobs and About Life did not show significant difference. This is because teachers' perception and valuation of the personal leadership traits are affected by the number of years in teaching this means. It depends on the lengths of service of the teachers, how their strengths and weaknesses are different. There were no meaningful differences in relations between their perceptions of leadership attributes related to others, jobs, and life and length of service [13].

The study revealed that teachers' perceptions of their leadership traits are moderately influenced by their age and duration of service, however gender and civil status do not have a substantial impact on these perceptions. This research provides useful information regarding the components that shape teachers' views of leadership as well as ideas for the creation of leadership development programs for teachers in elementary schools.

4. Summary

The rationale of this study is to determine the levels of leadership attributes of elementary school teachers in Jolo I District, DepEd-Sulu and to discover if there are significant differences among age, gender, civil status and service in length. The study was conducted by using quantitative research method with 100 purposively selected teacher respondents. Teachers were found to have generally very high positive attributes in those domains of About Oneself, About Others, About Jobs, and About Life. The domain of About Life was not significantly different on the basis of age, but significant differences were found in the domains About Oneself, About Others and About Jobs. No gender or civil status showed significant differences regarding the facilities. However, the length of service mattered when it came to the domain About Oneself. Finally, the study concluded that teachers' perceptions of their own leadership attributes were affected by age and length of service. The district study recommends creating of clear district vision, development of the leaderships programs for young teachers and further research in the field of education.

5. Findings

1. It was found by the study that Jolo-I District teachers were predominantly middle aged, females, married, and new teachers.
2. The domains About Oneself, About Others, About Jobs, and About Life, were dominated by very high positive attributes of the teachers in terms of leadership traits.
3. Significant differences were observed based on age in the domains of About Oneself, About Others, and About Jobs, but not in the domain of About Life.
4. No significant differences were found based on gender or civil status in any of the domains.
5. There was a significant difference based on length of service in the domain of About Oneself.

6. Conclusions

1. The study concluded that age and length of service influenced teachers' perceptions of their own leadership attributes, particularly in the domain of About Oneself.
2. The findings supported Jims Evans' leadership attributes theory, which emphasizes self-awareness, a sense of humor, and a desire to be a leader.

7. Recommendations

1. The study recommends that Jolo-I District of DepEd-Sulu should establish a clear district vision, obtain leadership commitment, focus on leadership development programs for leaders and future leaders (especially younger teachers), set incentives for compliance, and promote open communication.
2. Younger teachers with less than ten years of service should be encouraged to participate in leadership development programs to enhance their leadership skills.
3. The study encourages researchers in the field of education to conduct similar studies, including more domains of leadership attributes, to further enhance teachers' leadership knowledge and skills.
4. The study suggests that Jolo-I District of DepEd-Sulu should give priority to creating mentorship programs. These would be programs that would help direct and support people who are aspiring to be teacher leaders and grow and develop in the leadership sites.
5. School administrators and district leaders, therefore, have a duty to motivate teachers aggressively to collaborate and work together as a team, exchanging their leadership experiences and best practices with each other.
6. There is a need for further research concerning the effect of leadership qualities on student outcomes, namely, academic performance, level of student participation, and a typical school atmosphere, in order to gain a more comprehensible image of the link between teacher leadership and student success.

7. The proposal for all teachers, regardless of their number of years in the field, is to be offered professional development opportunities focused on leadership qualities to promote a healthy environment where improvement and growth can continue and allow all educators to take leadership role in their schools.
8. School administrators, legislators should think through making changes to incorporate leadership development initiatives into the curriculum of teacher education programs and ensure that aspiring teachers have the appropriate leadership abilities to contribute in meaningful ways to their schools and communities.
9. To collect feedback, the study proposes to adopt regular feedback practices, for instance, through questionnaires or focus group discussions by the instructors on whether they think about regarding important leadership attributes or leadership development programs.
10. In order to develop and implement leadership development programs, it is suggested that school districts should partner with external stakeholders, for instance universities, non-profit organizations and educational associations, in the utilization of extra resources and expertise.

These recommendations are intended as practical solutions for developing leadership qualities among the elementary school teachers of Jolo 1 District in the Division of Sulu.

References

[1] Cimene, F. T. A., & Aladano, A. N. (2013). Leadership perspective from the Philippines: Its implications for theory, research and practice. In *Culture and Gender in Leadership: Perspectives from the Middle East and Asia* (pp. 48–63). London: Palgrave Macmillan UK.

[2] Nithyanandhi, S., & Arul Murugan, R. (2016). Solar photovoltaic power generating scheme with peak power tracking control using adaptive neuro-fuzzy inference system. *International Journal of Advances in Engineering and Emerging Technology*, *7*(4), 23–34.

[3] Abdurahman, N. A., & Omar, J. A. K. (2021). School heads' educational leadership practice and teachers' performance: The case of Omar District, Division of Sulu, Philippines. *Open Access Indonesia Journal of Social Sciences*, *4*(4), 432–444.

[4] Yeo, M., & Jiang, L. (2023). Resonance phenomena in planetary systems: A stability analysis. *Association Journal of Interdisciplinary Technics in Engineering Mechanics*, *1*(1), 14–25.

[5] Hafezieh, M., Seidgar, M., Alizadeh Osalou, Zh., Nekoueifard, A., Ghara, K., Mohebbi, F., & Rezaei, M. M. (2024). Mechanization impact of improvement of some quality indicators of wastewater in rainbow trout culture dualpurpose farms in Markazi Province of Iran. *International Journal of Aquatic Research and Environmental Studies*, *4*(2), 1–17. http://doi.org/10.70102/IJARES/V4I2/1

[6] Askin, W. P. (2019). Principals' Leadership Skills and Attitude towards Teaching among Public Elementary School Teachers in Sibulan District, Negros Oriental. *ACADEME University of Bohol, Graduate School and Professional Studies*, *14*(1), 41–57.

[7] Bobojonova, D., Karimov, N., Masalieva, O., Pardaev, A., Nematov, O., Sattorova, M., & Kamola, M. (2024). Traditions and history of librarianship in central Asia. *Indian Journal of Information Sources and Services*, *14*(2), 70–77. https://doi.org/10.51983/ijiss-2024.14.2.11

[8] Karia, N., & Abu Hassan Asaari, M. H. (2019). Leadership attributes and their impact on work-related attitudes. *International Journal of Productivity and Performance Management*, *68*(5), 903–919.

[9] Baggyalakshmi, N., Janani Shree Priya, M., & Revathi, R. (2024). Online Shopping System. *International Academic Journal of Science and Engineering*, *11*(1), 71–80. https://doi.org/10.9756/IAJSE/V11I1/IAJSE1110

[10] Ontai-Machado, D. (2016). *Teachers' perceptions of elementary school principals' leadership attributes and their relationship to school effectiveness* (Doctoral dissertation, Walden University).

[11] Sharma, S. (2015). Attributes of school principals-leadership qualities and capacities. In *Proceedings International Congress on School Effectiveness & Improvement.*

[12] Jailani, Raugda S. (2005). Leadership Attributes of Secondary Schools Teachers in Jolo: A self-Appraisal.

[13] David, D., & Alamoodi, A. (2023). A bibliometric analysis of research on multiple criteria decision making with emphasis on Energy Sector between (2019–2023). *Applied Data Science and Analysis*, *2023*, 143–149.

14 A study on problems of migrant workers in Coimbatore district

Jaya Prakash N.[1,a] *and R. Velmurugan*[2,b]

[1]PhD (Full-Time) Research Scholar of Commerce, Karpagam Academy of Higher Education, Coimbatore, Tamil Nadu, India
[2]Associate Professor of Commerce, Karpagam Academy of Higher Education, Coimbatore, Tamil Nadu, India

Abstract: This research investigates the multiple difficulties which affect migrant workers operating in Coimbatore district of Tamil Nadu, India. The growing industrial development of Coimbatore includes migrant workers as an essential component of its workforce despite their unexplored experiences in the region. This study bases its findings on interviews with migrant workers as well as survey data and analytical assessment to understand the primary employment challenges that workforce faces in Coimbatore's district. The study analyzes these obstacles to construct targeted interventions that will improve both welfare and living standards for migrant workers in Coimbatore district. The study enhances understanding of migrant labour patterns in urban industrial settings thus emphasizing immediate adoption of rights-based measures for labour migration governance.

Keywords: Migrant workers, work conditions, social exclusion, economic challenges, policy interventions

1. Introduction

Migrant workers contribute substantially to the industrial force in the Coimbatore district because they fill a crucial worker role within the area. Despite their essential workforce contributions the living conditions of Coimbatore's migrant workers have minimal investigation compared to the district's booming economic state. This research aims to uncover social, financial and habitation contexts of migrant workers in Coimbatore district and establish their human services connection. Migrant workers throughout India battle numerous problems according to statistical studies. These are more acute in an urban industrial hub such as Coimbatore, where accelerated economic development is ahead of a support system for migrant workers. The issue of housing is very critical because migration workers are living in overcrowded substandard housing; most residents are residing in informal settlements or labour camps lacking basic amenities. Migrant workers are subjected to unfair labour practices through wage theft, lack of contracts, or less safety measures at the workplace, which further deepens vulnerabilities in the workplace. Such workers, because of the fragile nature of their employment with limited job security, are uniquely vulnerable to shocks from the economy, which adds to the justification for stricter labour protections. After all, issues regarding housing and employment, including access to healthcare services, are also a big concern for migrant workers at Coimbatore. Many don't have an idea of healthcare resources. This paper aims to understand the precise challenges faced by the international workers in that specific region by closely examining the context of Coimbatore. We pooled together qualitative interviews with quantitative surveys for the possible discovery of critical issues such as living in very low-accorded accommodation, exploitation while working, lack of good medical facilities, or exclusion from society. By understanding these problems in the context of their community, we can then suggest targeted interventions and policy measures that can improve the welfare and living standards of migrant workers. This piece of research is based upon empirical evidence and contemporary scholarship that calls for a rights-based approach to address challenges facing the migrant workforce of Coimbatore district. A rights approach speaks to and protects the rights of all, including a migrant worker, in terms of housing, healthcare, and fair working conditions. Amplifying the voices and experiences of migrant workers will take us closer to creating more inclusive and equitable urban environments in which the dignity and rights of each and every one are respected regardless of their migration status. The systems provide quality living conditions to migrant workers in addition to creating social unity and economic stability for regions that promotes collective identity among all social populations [1].

[a]jayap9292@gmail.com, [b]drvelsngm@gmail.com

DOI: 10.1201/9781003684459-14

2. Review of Literature

Namita Mathur [2] studies the unsafe employment situations of Delhi's migrant workers to show why existing laws protecting workers fail to work effectively. The author conducted research in two Delhi slum communities where the findings demonstrated that most migrant workers experience both dismal economic situations and dangerous workplace settings without access to social protection or legal safeguards. The paper criticizes the shortcomings of present-day social security benefits because structural design limitations prevent countless migrants from accessing them despite recent expansions. Mathur explains how migrant workers have disappeared from policy frameworks during many years which requires immediate action to decrease their vulnerabilities and enhance their living situation. This study highlights the necessity of creating specific policy changes that will help resolve the problems faced by this underrepresented group.

Maksum [3]. Indonesian post-migrant workers: A challenging problem for human security. The policy examines the problems of reentry integration while it investigates economic difficulties as well as social discrimination and health problems and court problems and effects on family life. Research findings enable government officials and practitioners to understand complex post-migration worker needs thus protecting worker rights together with their well-being.

The study by Shepherd et al. [4] Challenges influencing the safety of migrant workers in the construction industry: A qualitative study in Italy, Spain, and the UK examines safety challenges faced by migrant construction workers between these three countries. Through qualitative inquiry the study examines workplace safety factors which include language issues and training inadequacies and workplace discrimination and cultural gaps as key influences. The findings will guide policy creation and intervention design for better safety standards of construction work for migrant workers.

Uddin, Uddin, and Islam [5] investigates how migrant workers experience challenges in the Malaysian construction market through their detailed analysis. The analysis investigates residential conditions alongside workplace safety matters as well as barriers under the law and worker exploitation and their difficulties adapting socially. The study produces data which will guide policy decisions aimed at protecting migrant worker rights in Malaysian construction activities.

Rasool, Hussain, and Bano [6] investigates health difficulties which affect migrant workers in their place of settlement. The research investigates safety risks at work and follows up with healthcare restrictions and mental health issues in addition to infectious illnesses and reproductive diseases. The study produces data which helps create effective policies and interventions for addressing the health needs of migrant workers.

An analysis of COVID-19 effects on migrant workers in India emerges from the report of Suresh, James, and RSj [7]. The research examines migrant worker difficulties which combine financial burden with movement barriers, social loneliness together with medical service problems and summarizes official policy responses. The research seeks to create strategies that will shape governmental policies to assist migrant workers during emergencies.

Through qualitative research Srivastava and colleagues [8] investigated both major pandemic-stressors and coping mechanisms adopted by internal migrant workers during COVID-19. The major stress factors that impact migrant workers include difficulties in obtaining employment security as well as poor living conditions and limited access to healthcare and social connections and mental health issues. Various coping mechanisms help migrants to get social support while practicing religions and seeking new sources of earnings through adapting to different situations. The research data delivers essential knowledge for designing crisis intervention initiatives to help internal migrant workers.

Ansari [9] investigates the challenges of interstate migrant workers in Kochi, Kerala, along with the perspectives of government officials. The research analyzes migrant issues from economic and social aspects as well as governmental regulations and official attitudes regarding these matters. The study delivers knowledge which can help improve social welfare initiatives that serve South Asian migrant workers.

The study by Good [10] examines how socio-political elements affect the mental health of migrant workers. This paper evaluates special migrant worker challenges alongside structural mental health risk factors and analyzes political responses and advocacy strategies for migrant support. This work exists to spread awareness and champion policy changes that will guarantee equal mental health opportunities for migrant groups.

The research by Kumar and Varsha [11] explores both economic and health outcomes resulting from COVID-19 for migrant employees in the Musiri Block at Tiruchirappalli District of India. The study investigates employment losses and wage cuts along with medical dangers faced by migrant workers as they implement different ways to handle these challenges. The research results help experts design appropriate support systems to help migrant workers through critical situations.

Wispandono, Kurniasari along with Yuherawan [12] examined international migrant workers by studying specifically the population of Madura descent. This study examines how migrants adapt their locations along with their occupational choices and also studies their financial transfers to country people while studying social integration and national and regional policies. The research results contribute valuable knowledge about Madurese migrant worker scenarios and worker welfare and migration management policy development.

The article by Baid [13] describes various difficulties faced by Indian migrant workers. The analysis covers both economic abuse alongside unstable living circumstances and social safety issues and health-related risks together with social segregation and emergency exposure. This article supports policy changes which will protect the rights and welfare of migrant workers in India.

Dwiningwarni [14] evaluated the difficulties faced by former migrant workers after they returned to the Tulung Agung District of Indonesia. The analysis studies multiple elements particularly economic reintegration together with social adjustment as well as health repercussions and policy consequences. Research findings establish the basis for programs and policies which support the return process of migrant workers and their general well-being.

Jha and Lahiri [15] investigate how domestic migrant workers in India faced health and socio-economic difficulties during their homecoming journey due to the COVID-19 pandemic. The study discusses the problems which include job loss and income instability in addition to health dangers and local population responses. The study delivers valuable information to serve both policy creation and targeted assistance for returning migrant workers while crises are ongoing.

The paper by Singh, Arun, and Chavan [1] assesses Indian government actions to minimize the economic consequences of the lockdown measures against COVID-19 upon migrant workers. This report reviews national government actions together with independent non-governmental interventions and describes current implementation challenges alongside mental health risks and provides warning system improvement strategies for future emergencies.

Loganathan, Rui, Ng, and Pocock [16] evaluate the various elements that determine medical care availability to migrant workers throughout Malaysia. The study examines how legal, financial, cultural, and structural constraints affect healthcare access together with elements of health-seeking routines for migrant workers. The investigation provides essential information that can guide lawmakers to develop better policies for enhancing healthcare services available to migrant workers in Malaysia.

The paper by Khan and Arokkiaraj [17] analyzes the difficulties facing Indian domestic and migrant laborers throughout the COVID-era economic system while studying the phenomenon of backward migration. The research examines economic harm alongside social and psychological effects after reverse migration throughout both groups then explores suitable policies for migrant worker support. The research findings discuss both the multiple aspects of reverse migration behaviour and their projected impact on India's future migration patterns.

Pathak and Purohit [18] investigates migrant worker difficulties which became worse due to the COVID-19 pandemic. This research explores the economic problems alongside health dangers and social separation and the legal and policy consequences that result from the situation. The results demonstrate why concentrated action must be taken to safeguard migrant workers both legally and physically throughout emergency situations.

Jayaram along with Varma [19] studies the informal employment situations of migrant workers in cities which restricts their ability to claim labour rights. The research explores both the root causes and social effects of self-driven arrangements among migrant workers while analyzing the impact of governmental policies on dealing with these employment impediments. This research provides essential knowledge about the operational patterns of migrant labor throughout urban areas in India.

During economic crises Vianello and Sacchetto [20] study the migration process to informal employment. The study evaluates both systemic obstacles and individual factors that motivate this labour realignment together with its effects on worker well-being in addition to necessary policy responses.

The author Siegmann and Schiphorst [21] investigates the effects of globalization which caused informal sector workers to transition into precarious employment arrangements. The study investigates both the underlying elements behind this transition as well as its workforce repercussions while discussing appropriate policy action.

Waridin [22] conducts research on approaches to help Indonesian migrant workers in Malaysia's informal economy after their placement period ends. The author examines worker difficulties while conducting research and suggests programs to develop their independence after placement.

The research by Das [23] analyzes the interplay between regional financial discrepancies and informal workers from internal migration during the COVID-19 crisis period. This research investigates the manner in which inequalities between regions further worsens migrant workers' exposure to financial difficulties because they primarily work in informal employment. The article examines solutions for coping with the situation and presents analysis of government responses while advocating for comprehensive recovery plans.

Jayaram and Varma [19] conduct research about labour migration within urban regions by studying informal employment settings and migrant workers' ability to obtain labour rights. The paper details problems such as employee mistreatment combined with inadequate security measures while suggesting governmental solutions to enhance working environments and social economic participation.

2.1. Research gap

Studies about migrant worker issues in Coimbatore require more investigation according to specific research gaps. The reporting of precarious employment together with insufficient social security systems and worker health challenges generally occurs while researchers concentrate their studies on different geographical areas and business sectors. Studies about the economic and social conditions of Coimbatore are scarce within this particular research field. This gap brings forth the requirement for targeted research on the experiences of migrant workers in Coimbatore, which would be handy in light of effective local policies and interventions.

2.2. *Statement of the problem*

Migrant workers in Coimbatore face a host of difficulties regarding health and productivity spheres. It is seen that migrant workers enjoy precarious employment conditions, inadequate living standards and scant access to social services, besides insufficient legal protection, even though the city economy is expanding due to economic growth and industrialization. Some of the specific issues they have to face are low wages, insecurity in jobs, poor conditions of staying and also poor healthcare and denial of social security benefits. These are compounded by rapid industrial development without matching improvement in worker welfare. This study attempts to examine these problems in detail, with an emphasis on experiences of migrant workers in Coimbatore. The research work intends to bring out in detail the most crucial challenges faced by this population, hence pointing out the lacunas in policies and practices at present. It will ultimately give way to a development of targeted recommendations for improving living and working conditions for migrant workers in the region, enhancing their quality of life as a whole and ensuring adequate recognition and support of their contribution to the local economy.

2.3. *Objectives*

To ascertain the problems of migrant workers in Coimbatore district.

2.4. *Scope of the study*

The present study is confined to Coimbatore district and focuses solely on the socio-economic profile and problems faced by migrant workers.

3. Research Methodology

3.1. *Source of data*

Primary data for this study is collected through interviews conducted with migrant workers in Coimbatore district. These interviews are serve as the primary source of information, allowing for direct insight into the experiences, challenges and needs of migrant workers in the region.

3.2. *Sampling method*

Convenience sampling is employed to select participants for the interviews.

3.3. *Sample size*

A total of 600 migrant workers are interviewed for this study. This sample size is determined based on practical considerations, aiming to achieve a balance between obtaining sufficient data for analysis.

3.4. *Framework of analysis*

The collected data are analysed by employing simple percentage and Factor Analysis.

3.5. *Scope of the study*

The present study narrates the challenges in the living and working conditions of the migrant workers in Coimbatore District, with particular reference to their conditions at work, living standards and utilization of basic amenities. Specific issues to be discussed are related to wage, job security, house conditions, health care utilization and legal protection. This proposed research will target migrant workers, regardless of the industry they are working in, within Coimbatore and will be based on a qualitative and quantitative approach. It covers deep understanding of the peculiar difficulties which these workers face in reaching the workers and also policy recommendations from its findings.

3.6. *Significance of the study*

The importance of the study lies in focusing on the challenges of migrant workers in Coimbatore, which is turning out to be one of the fastest-growing industrial hubs. The research shall look into aspects such as low wages, poor living standards, and access to essential services, which will reveal critical gaps in the current policies and support systems. The resultant understanding will be fundamental for devising specific interventions and policy reforms that improve migrant workers' well-being. In turn, this study will contribute to the equitable sharing of fruits from economic growth and provide useful guidance to policy makers, employers and interest groups toward a more enabling environment for migrant workers.

3.7. *Findings*

Table 14.1 discuss about socio-economic profile of migrant workers.

3.8. *Interpretation*

- Age: The majority of migrant workers fall within the age range of 21 to 35 years (34.2%), followed by those aged 35 to 50 years (32.5%). Workers above 50 years comprise the smallest proportion (6.7%).
- Gender: Female workers constitute a larger proportion (47.5%) compared to male workers (52.5%).
- Educational Qualification: The majority of workers have completed their SSLC (75.8%), while a smaller proportion are illiterate (24.2%).
- Marital Status: A larger proportion of workers are married (67.5%) compared to unmarried workers (32.5%).

Table 14.1. Socio economic profile

Particulars	*Number (n=600)*	*Percentage*
Age		
Up to 21 Years	160	26.7
21–35 years	205	34.2
35–50 years	195	32.5
above 50 years	40	6.7
Gender		
Male	285	47.5
Female	315	52.5
Educational Qualification		
Illiterate	145	24.2
SSLC	455	75.8
Marital Status		
Married	405	67.5
Unmarried	195	32.5
Status in Familhy		
Head	230	38.3
Member	370	61.7
No of Family Members		
1 member	395	65.8
2- 3 members	130	21.7
more than 3 members	75	12.5
Family Income		
Up to–15000	140	23.3
15001–25000	270	45.0
More than 25000	190	31.7
Occupation		
Mason	50	8.3
Plumber	130	21.7
Painters	75	12.5
Others	345	57.5
Languages Known		
Hindi	315	52.5
Others	285	47.5

Source: Author.

- Status in Family: Most workers are members of their families (61.7%), while a significant proportion are heads of their families (38.3%).
- No of Family Members: The majority of workers have one family member (65.8%), followed by (21.7%) having two to three family members and having more than three members (12.5%).
- Family Income: The largest proportion of workers have a family income between 15001 to 25000 (45.0%), followed by more than 25000 (31.7%) and up to 15000 (23.3%).
- Occupation: The highest proportion of workers are categorized under 'Others' (57.5%), followed by plumbers (21.7%), painters (12.5%), and masons (8.3%).
- Languages Known: A majority of workers know Hindi (52.5%), while others know languages other than Hindi (47.5%).

3.9. Problems of migrant workers

Factor Analysis was employed to identify the major problems faced by migrant workers. To ensure the suitability of the data for factor analysis, the Kaiser-Meyer-Olkin (KMO) measure and Bartlett's Test of Sphericity were conducted as preliminary checks. The KMO statistic, which assesses the adequacy of the sample size, was found to be 0.793, exceeding the acceptable threshold of 0.70. This indicates that the data was appropriate for factor analysis. Additionally, Bartlett's Test of Sphericity, which evaluates whether the correlation matrix is significantly different from an identity matrix, yielded a large chi-square value of 10,829.564 (df = 276, $p < 0.001$). This result confirms the presence of sufficient interrelationships among the variables for factor analysis. Together, the high KMO value and significant Bartlett's Test result demonstrate that the collected data is both adequate and suitable for applying factor analysis, ensuring the reliability of the findings in identifying the key challenges faced by migrant workers (Table 14.2).

Seven key factors were identified through an analysis of Eigenvalues greater than one, which signify the major problems faced by migrant workers (Table 14.3). Variables with component loadings of 0.5 or higher were considered significant contributors to these issues. A detailed examination of the rotated component matrix revealed the following findings:

- ***Factor 1: Workload and Social Exclusion***
  This factor accounts for **32.407%** of the problems faced by migrant workers, making it the most significant contributor. The issues under this factor include an overwhelming workload, social exclusion or avoidance by the local populace, an unfavourable work environment, excessive working hours, inadequate assistance from employers, and a lack of care or support from the local community. These challenges highlight the dual burden of demanding work conditions and the lack of social

Table 14.2. KMO and Bartlett's test

Kaiser-Meyer-Olkin Measure of Sampling Adequacy.		.793
Bartlett's Test of Sphericity	Approx. Chi-Square	10829.564
	Df	276
	Sig.	.000

Source: Author.

Table 14.3. Problems of migrant workers

Problems faced by migrant workers	*Component*						
	1	*2*	*3*	*4*	*5*	*6*	*7*
Overwhelming workload	.872						
Facing exclusion or avoidance from indigenous populace	.865						
Unfavourable work environment	.816						
Excessive working hours at the job.	.805						
Inadequate employer assistance	.635						
Lack of care or assistance from indigenous individuals	.588						
Employer offers significantly low wages as stated		.839					
Anxiety regarding law enforcement's distrustful behaviour		.818					
Health issues stemming from work-related factors		.769					
Mental anguish due to job hazards			.883				
Persistent salary delays			.855				
Endured negative treatment from co-workers			.815				
Salary reduction as disciplinary measure				.700			
Workplace bias				.631			
Inability to access loans from financial institutions or employer				.539			
Employee misconduct					.843		
Language Problem					.836		
Hostility from local residents						.855	
Limited support from colleagues or roommates						.675	
Work entails higher risk compared to locals							.791
Feelings of isolation or solitude							.542
Initial Eigenvalues Total	7.778	2.699	1.71	1.549	1.282	1.259	1.104
% of Variance	32.407	11.248	7.124	6.453	5.341	5.247	4.600
Cumulative %	32.407	43.655	50.779	57.232	62.573	67.820	72.420

Source: Author.

integration, significantly impacting migrant workers' well-being.

- ***Factor 2: Low Wages and Health Concerns***
The economic and psychological factors compose 11.248% of the topic. Migrant workers face three major issues that include insufficient salary compensation from employers as well as nervousness from mistrustful police interactions and health complications due to occupational conditions. Many migrant workers experience dangerous economic situations combined with mental health challenges because of these contributing elements.
- ***Factor 3: Job Hazards and Negative Treatment***
The identified problems depend on this factor to the extent of 7.124%. The principal problems affect workers through stress from hazardous work conditions together with extended payment delays and unfriendly or discriminatory employee conduct. These issues demonstrate the necessity to enhance workplace safety practices together with prompt wage payments and development of an inclusive environment at work.
- ***Factor 4: Financial and Workplace Inequities***
The 6.453% segment of problems consists of salary reductions as disciplinary punishment along with co-worker prejudice and limited loan access through banks and employers. Migrant workers face obstacles that stem from financial vulnerabilities in combination with systemic inequalities which prevent them from building financial stability.
- ***Factor 5: Employee Misconduct and Language Barriers***
The employee misconduct together with language barriers amount to 5.341% of the overall distribution. Language difficulties produce extreme communication problems which block workplace understanding and result in multiple workplace misunderstandings.
- ***Factor 6: Hostility and Limited Support***
This factor analyzes local resident hostilities plus the shortage of colleague and roommate support which makes up 5.247% of the total problems. Host

communities become less supportive because of these social challenges which create alienated and unsupported conditions for migrant workers.

- ***Factor 7: Job Risks and Isolation***
  The final considered factor demonstrates a 4.600% weight and shows both job safety challenges that migrant workers experience compared to local staff and isolation or solitude feelings among these workers. Job security differences together with the stress of being absent from home form the basis of these problems.

The seven identified factors explain 72.420% of the problems which migrant workers experience. Work conditions and social exclusion problems constitute Factor 1 which represents the dominant percentage requiring prompt attention. The subsequent factors illustrate economic, social and workplace-based challenges which demonstrate how complex migrant worker problems truly are. Research illustrates that migrant worker problems merge into unified clusters between economic difficulties while they also confront workplace problems and social isolation and psychological challenges. A solution demands an integrated method which unites government policies and industrial workplace improvements with communal support to enhance migrant workers' standard of living.

4. Suggestions

Based on the identified prevailing influential factors of the challenges faced by migrant workers, some suggestions can be targeted toward both government bodies and migrant labour societies.

- Legislate and enforce laws on workplace improvement and employers' support for the workers in providing them with a safe working environment and assuring them of fair labour practices. Extend the vigil for better facilitation by employers in terms of support and amenities to migrant workers. Design workload management-related programs and give clear guidelines to the employers.
- Promote measures of social inclusion that reduce hostility towards migrant workers. Public information campaigns should be launched for the receiving community to appreciate the contribution of migrant workers and inform them about adopting more inclusive behaviour.
- Strengthen measures against delay/delayed payment of salary and malpractices. Develop procedures on financial dispute resolution and ensure timely payment of wages to workers.
- Elaborate and apply standards that meet the needs of the workers; reduce workplace biases with anti-discrimination policies, and most importantly, train employers and employees alike.
- Training in languages and translation services would ease the communication of migrant workers to get their way into the labour market. Make basic information and resources available in multiple languages.
- Establishment and support of community-based organizations to help and resources for migrant workers. Development of partnership with NGOs and institutions at the local level for social and legal support.
- Regular evaluation of the effectiveness of policies and programs touching on or involving migrant workers is called for. Data insights shall be used to make necessary adjustments toward ensuring that interventions attain intended goals.
- Establish and strengthen the local support network that can guide, resource, and support the migrant worker. Facilitate connections of workers with local services to access support more effectively.
- Regularly advocate on behalf of workers' rights and fair treatment before policy makers, employers, and community leaders. Raise awareness among migrant workers about their legal rights and resources.
- Conduct training programs for workers that will enhance their employability. Provide language skills, financial management, and workplace safety workshops to empower the workers.
- Create programs to make possible a variety of social interactions and integrations with the local community. Cultural exchange events, experience-sharing sessions, and other such opportunities that bring migrant workers closer to each other will be highly valued.
- Several community-wide problems encompass occupational dangers and social isolation issues among community members. Your organization should deliver support systems which address these particular requirements.
- Counseling services should be accessible to workers for helping them overcome stress and mental health issues together with social isolation. Workers must have free access to specialized and confidential assistance.
- Comparable services should offer distinct pathways so workers can present their notes about workplace difficulties and service performance evaluations. Implement a system to process and enhance service delivery through feedback collection while adopting novel problems as they emerge.

5. Conclusion

Migrant workers encounter various obstacles which stem from both extreme work requirements alongside social separation and societal inequality. Workers experience two kinds of challenges because their physical demands surpass reasonable levels and their work environment remains unfavourable along with a deficit of support from their employers and local communities. Economic pressures which include insufficient pay and delayed wages and limited money access together with mental health challenges from workplace dangers and discriminatory experiences increase their exposure to harm. International workers face communication challenges and hostile treatment from native residents causing social detachment and finding it harder to assimilate while experiencing worrisome occupational hazards and systemic discrimination against them. Multiple issues work together to show the immediate requirement for broad policy changes and workplace enhancement with community-based solutions to deal with economic and social and psychological troubles facing migrant labourers.

6. Scope for Further Research

The research could be improved through additional studies which examine migration worker policy interventions in Coimbatore specifically and identify their benefits to worker health together with potential areas needing improvement. Comparative evaluations should illustrate which challenges migrant workers encounter exclusively in Coimbatore versus those present in identified industrial locations and urban areas in both Indian and international contexts. The study of employer practices alongside corporate social responsibility stands as the most effective approach to find solutions for migrant labour problems. Research conducted over time tracks migrant worker development by monitoring their condition changes specifically through economic fluctuations policy alterations and pandemic-related events. Study results enriched by psychological and social considerations concerning mental state and social relationships would advance understanding about migrant well-being. The research could advance through utilization of participant-oriented tools such as interviews to gain insights which would enable better alignment of assistance measures with migrant requirements.

References

[1] Singh, G. P., Arun, P., & Chavan, B. S. (2020). Efforts to minimize the impact of lockdown on migrant workers in India during the COVID-19 pandemic. *The primary care companion for CNS Disorders*, *22*(3), 26309.

[2] Namita Mathur (2024). *Socioeconomic Vulnerabilities of Migrant Labour Households in Delhi* (Vol. 5 No. 4 (Jul-Aug)). Indian Public Policy Review.

[3] Maksum, A. (2021). Indonesian post-migrant workers: A challenging problem for human security. *Social Sciences & Humanities Open*, *4*(1), 100223.

[4] Shepherd, R., Lorente, L., Vignoli, M., Nielsen, K., & Peiró, J. M. (2021). Challenges influencing the safety of migrant workers in the construction industry: A qualitative study in Italy, Spain, and the UK. *Safety Science*, *142*, 105388.

[5] Uddin, M., Uddin, M. S., & Islam, M. Z. (2022). Migrant workers at Malaysian construction market: A critical analysis of their facing difficulties. *Journal of Asian and African Social Science and Humanities*, *8*(4), 13–24.

[6] Rasool, F., Hussain, M. H. A. G., & Bano, S. (2023). Health problems faced by migrant workers in country of destination. *UW Journal of Social Sciences*, *6*(1), 11–19.

[7] Suresh, R., James, J., & RSj, B. (2020). Migrant workers at crossroads–The COVID-19 pandemic and the migrant experience in India. *Social Work in Public Health*, *35*(7), 633–643.

[8] Srivastava, A., Arya, Y. K., Joshi, S., Singh, T., Kaur, H., Chauhan, H., & Das, A. (2021). Major stressors and coping strategies of internal migrant workers during the COVID-19 pandemic: A qualitative exploration. *Frontiers in Psychology*, *12*, 648334.

[9] Ansari, P. A. (2019). Exploring the issues and challenges among (interstate) migrant workers and government officials: A study of Kochi, Kerala. In *Social welfare policies and programmes in South Asia* (pp. 132–146). Routledge India.

[10] Good, D. M. (2020). Migrant workers and the politics of mental health. *Mental Health and Migrant Workers*, *55*(31), 39.

[11] Kumar, S., & Varsha, U. (2022). Economic and Health Impact of Migrant Workers during Covid-19 Period in Musiri Block at Tiruchirappalli District. *International Journal of Early Childhood Special Education*, *14*(3).

[12] Wispandono, R. M., Kurniasari, N. D., & Yuherawan, D. S. (2018). Analysis of international migrant workers: Case study of worker of Madura Origin. *Management*, *6*(5), 384–394.

[13] Baid, N. (2021). Problems of Migrant Workers in India. *Supremo Amicus*, *24*, 756.

[14] Dwiningwarni, S. S. (2016). *Ex-Migrant Worker: An Assessment Problem (Study In The District Tulung Agung)*. ICEBUS.

[15] Jha, S. S., & Lahiri, A. (2020). Domestic migrant workers in India returning to their homes: Emerging socioeconomic and health challenges during the COVID-19 pandemic. *Rural and Remote Health*, *20*(4), 1–4.

[16] Loganathan, T., Rui, D., Ng, C. W., & Pocock, N. S. (2019). Breaking down the barriers: Understanding migrant workers' access to healthcare in Malaysia. *PloS One*, *14*(7), e0218669.

[17] Khan, A., & Arokkiaraj, H. (2021). Challenges of reverse migration in India: a comparative study of internal and international migrant workers in the post-COVID economy. *Comparative Migration Studies*, *9*(1), 49.

[18] Pathak, A., & Purohit, N. (2022). Migrant Workers and their Challenges: How the Crisis went from Bad to Worse. *International Journal of Law and Social Sciences*, 47–54.

[19] Jayaram, N., & Varma, D. (2020). Examining the 'labour' in labour migration: Migrant workers' informal work arrangements and access to labour rights in urban sectors. *The Indian Journal of Labour Economics*, *63*(4), 999–1019.

[20] Vianello, F. A., & Sacchetto, D. (2016). Migrant workers' routes to the informal economy during the economic crisis: structural constraints and subjective motivations. *Prakseologia*, (158/1), 299–322.

[21] Siegmann, K. A., & Schiphorst, F. (2016). Understanding the globalizing precariat: From informal sector to precarious work. *Progress in Development Studies*, *16*(2), 111–123.

[22] Waridin, W. (2020). Formulation of problem-solving design for Indonesian informal sector migrant workers towards post-placement independence: Case study in Malaysia. *Research in World Economy*, *11*(1).

[23] Das, D. (2020). Regional disparities of growth and internal migrant workers in informal sectors in the age of COVID-19. *Journal of Public Affairs*, *20*(4), e2268.

15 Analytical study of correlations among socioeconomic, cultural factors, and customer relationship management (CRM) in tourism management

Krishna Kumar Mishra[1,a], Kausik Mukherjee[2,b], Shinu Shukla[1,c], Prakash Kumar Sen[1,d], Seema Dwivedi[1,e], and Chandan Singh[1,f]

[1]Assistant Professor, Faculty of Management Studies, AKS University Satna, Madhya Pradesh, India
[2]Associate Dean, Faculty of Management Studies, AKS University Satna, Madhya Pradesh, India

Abstract: Delving into an analytical study to investigate the complex relationships between socioeconomic, cultural factors, and Customer Relationship Management (CRM) within the realm of tourism management. Given the ever-changing nature of the tourism industry and its dependence on strong customer relationships, grasping the interaction between these elements is crucial. This study utilizes a mixed-methods approach, incorporating surveys and interviews, to thoroughly examine the connections among socioeconomic indicators, cultural dimensions, and CRM strategies. Exploring theoretical frameworks related to CRM, the literature review investigates the impact of socioeconomic factors and cultural dimensions. Through a thorough analysis of previous research, the work uncovers deficiencies in the current body of knowledge, setting the stage for a unique addition to the discipline. The methodology section outlines the selected research design, data collection methods, and ethical considerations. Examining the data through statistical analysis and qualitative methods helps reveal patterns and gain deeper insights. In the following sections of the study, we delve into how economic stability, income levels, and social demographics affect CRM in tourism. In addition, the study delves into cultural aspects, highlighting the significance of heritage and cultural tourism in influencing CRM strategies. By conducting a comprehensive analysis, the study seeks to uncover the relationships between socioeconomic, economic, and cultural factors, providing valuable insights for tourism management. The results will help in developing strategies that are suitable for different cultural settings, enhancing customer relationships, and supporting the long-term development of the tourism sector. Concluding the study, the findings are summarized, the contribution to existing knowledge is emphasized, and suggestions for future research are put forward. By exploring these connections, the study aims to deepen our comprehension of the intricacies in tourism management, ultimately aiding industry professionals and scholars.

Keywords: Tourism management, customer relationship management (CRM), socioeconomic factors, cultural factors, analysis

1. Introduction

1.1. Background

As one of the most peculiar in the world, the tourism industry has become a part of global economic and cultural interactions [1]. There are a lot of tourist places – from overcrowded cities to abandoned natural wonders – with a plethora of attractions for millions of tourists who come every year [2]. It helps in boosting the economic development of the nation, generation of employment opportunities and cultural exchange along with making substantial contributions to the GDP of the other countries [3].

1.2. Overview of tourism industry in India

India's GDP (at current prices), tourism sector contributes 4.6% [4]. No other sector is like this and hence that pushes the tourism as the least important priority for the Indian government [5]. Forbes magazine listed India seventh on 'The 50 Most Beautiful Countries In The World' ranking, And in 2021, when contribution of tourism to GDP of India's was researched, it was estimated to Rs 18 lakh crore (US$170 billion) or contribute 5.8 per cent of our total GDP with employment for 32.1 million people [6]. Although the drop was compared before the pandemic, the economy of the country witnessed huge growth since the drop from 2020 to 2021. In 2014, 184,298 international patients travelled to India for medical care, which translated to a size of around US$9 billion, out of the total global market of US$44.8 billion in the same year.India is identified as the 5th largest global travel healthcare destination, worth approximately US$9 billion in 2019 within the total global travel healthcare industry estimated at US$44.8 billion in the same year.It is projected to

[a]kkmishra700@gmail.com, [b]Km466430@gmail.com, [c]Shinushukla03@gmail.com, [d]Prakashe1042@gmail.com, [e]Seemasharad2011@gmail.com, [f]Chandansingh201285@gmail.com

DOI: 10.1201/9781003684459-15

grow by 7.8% annually until 2031, amounting to an estimate of ₹33.8 lakh crore (US$ 420 billion), orBased on data from the Ministry of Tourism, there was a significant decrease in the number of foreign tourists visiting India in 2022 compared to 2019.India's large domestic tourist population means it is not heavily reliant on foreign tourists [7]. With 7.3 Million visitors in 2017, India remains as the 22nd most visited country in the world and 8th across the region of Asia and the Pacific [17]. National Geographic's Traveller magazine termed Kerala as one of 'ten paradises of the world' and the 'top spot on the list of a traveller's lifetime' in 2012 and Lonely Planet put it as their top '50 must see destinations of the a lifetime' 2012. In 2014, Tamil Nadu, Maharashtra and Uttar Pradesh had been the tourists' preferred choices [8].

1.3. Importance of understanding correlations in tourism management

The interaction of the socioeconomic, cultural and the CRM factors influences within the industry's direction, which it cannot be ignored within the complex tourism management domain [9]. Tourists tend to spend differently and make different travel decision depending on their socioeconomic factors such as income level, employment rate and economic stability [10]. Cultural factors such as traditions, customs, heritage etc enter the appeal of the places and their tourism. In the tourism industry, the strategies and practices of Customer Relationship Management (CRM) are great in building and keep the positive relationship with tourists [18, 28]. It is necessary to understand the correlation between various factors in order to develop effective CRM strategies in the competitive sector of tourism industry. Things such as cultural sensitivity, economic factors, the social dynamic among many other variables make one strategy more successful than another when it comes to people, process, and environment [12]. Knowledge of such connections is necessary for successful economic development of the destinations, protection of cultural heritage, promotion of sustainable tourism. This research seeks to analyze those various interrelations of socioeconomical, cultural and CRM factors being vital in tourism management to contribute to dynamics of this management and to assist tourism industry stakeholders [11].

1.4. Objectives of the study

1. To analyze the relationship between socioeconomic factors and CRM in tourism.
2. To examine the impact of cultural factors on CRM in the context of tourism management.
3. To identify the interplay between socio, economic, and cultural factors in influencing CRM strategies in tourism.

1.5. Literature review

Several studies have explored the effects of CRM in the tourism industry. Lin (2003) and Vogt (2011) [24, 25] emphasise the importance of CRM in enhancing customer value and loyalty, particularly focusing on high-quality hotel businesses. Providing a thorough analysis, Rahimi (2017) [26] presents a comprehensive review of CRM studies in the hospitality and tourism sector, highlighting its impact on company performance [13, 23]. In a study from 2014, the discussion delves into the concept of customer engagement, which has been proven to positively influence loyalty in the hotel and airline sectors [27]. These studies emphasise the significance of CRM and customer engagement in the tourism industry, suggesting that a deeper understanding of customer behaviour and preferences can enhance them [14]. In the present scenario it is imperative to understand the dynamics of business if you really want to sustain your customers. Small and medium sized enterprises of tourism industry also had the constraints of not enough of funding, not enough top management support towards CRM, and not enough proper coordination [15, 16]. Sigala (2005: 177 – 179) [21] in a 2005 study emphasised particularly CRM and its crucial role in attracting and retaining guests, as they are becoming more price sensitive and less brand loyal. However, while CRM in the hospitality industry has built upon the importance placed on ICT, there is currently a consensus that a good CRM implementation requires ICT to work in a way that it can be efficiently integrated and timed to support business practices. As CRM implementation framework is not widely accepted, then we propose this model that manages and integrates ICT capabilities into the CRM strategies and business processes. The main result of this research is that three main manager processes (Tail ICT, Relations both internal and external and Knowledge management) should be designed and aligned for the effective CRM implementation. The data of Greek hotels were implemented for testing the model and the results were found to be very useful from a practical point of view offering useful insights and recommendations for future studies [19].

2. Methodology

2.1. Research design

In order to deal with complexities of interrelatedness of socio-economic and cultural factors as well as in CRM in tourism management, a mixed-methods approach will be used [26]. The approach is, thus, integrated in that the models for qualitative and quantitative research are used such that a comprehensive and nuanced understanding of a subject is arrived at [25].

2.1.1. Justification for the chosen research design

For this reason, we select a mixed methodology design to triangulate the obtained data from different sources to offer a stronger and broader perspective of the correlations under study. The approach enables the validation and selfthem the quantitative findings with the qualitative insights to better understanding between the complex relationships in the tourism industry [20].

2.2. Data collection

Collected data will be secondary data from the various sources, the published research papers, government reports, organizational records and other public available datasets. For this study, secondary data can be very useful.

2.3. Data analysis

1. Quantitative Data Analysis: Appropriate statistical tools, such as regression analysis and correlation coefficients, will be used to conduct statistical analysis, which will examine the relationships among socioeconomic factors, cultural dimensions and the level of CRM effectiveness respectively. The objective of quantitative data analysis is to discover any pattern, trends and statistical significance the data has [22].
2. Qualitative Data Analysis: Both interviews will be analyzed thematically to generate key insights and extract emerging themes. Qualitative data will be coded and categorized to assist in organizing participants' view on the correlations being studied [29].

3. Socioeconomic Factors and CRM in Tourism

3.1. Economic influences

3.1.1. Income levels and travel preferences

The study explores the relationship between income levels, travel preferences, and CRM satisfaction among 200 participants.

3.1.2. Interpretations

Table 15.1 shows the association between participants' income ranges, favourite trip types, and CRM satisfaction, assessed from 1 to 10. Adventure travel is preferred by 40 low-income participants, with a CRM satisfaction rating of 7.2. This shows that low-income travellers enjoy adventurous travel. However, 80 moderate-income participants prefer cultural tourism despite a lower CRM satisfaction rating of 6.5. The data shows that moderate-income people like cultural experiences, although CRM pleasure may vary. Income, favourite travel kinds, and CRM satisfaction are positively correlated for higher-income participants. The 50-person high-income group prefers luxury travel, resulting in an 8.3 CRM satisfaction rating. Luxury travel may be more satisfying for wealthy people due to higher service expectations. Finally, 30 high-income participants favour ecotourism, reflecting their environmental awareness. This group gets the highest CRM satisfaction rating of 9.0, highlighting the link between high income, ecotourism interests, and CRM satisfaction. The data suggests a positive correlation between higher income levels and greater satisfaction with CRM, particularly evident in luxury travel and ecotourism.

Table 15.1. Income level and travel preferences

Income Bracket	*Number of participants*	*Preferred travel type*	*CRM satisfaction (1–10)*
Low	40	Adventure Travel	7.2
Moderate	80	Cultural Tourism	6.5
High	50	Luxury Travel	8.3
Very High	30	Ecotourism	9.0

Source: Author.

3.1.3. Economic stability and customer relationships

Participants were surveyed on their perception of the impact of economic stability on their travel decisions, revealing their perspectives on how economic factors influence CRM (Table 15.2).

The majority of participants express agreement on the positive influence of economic stability, suggesting a potential link between economic confidence and favorable CRM perceptions.

3.2. Social factors

3.2.1. Demographic influences on CRM practices

The study examines the impact of demographics, including age, gender, and occupation, on participants' perceptions of CRM effectiveness (Table 15.3).

The data showcases variations in CRM effectiveness based on demographic factors, with entrepreneurs and executives tending to perceive higher CRM effectiveness.

Table 15.2. Economic stability perception of participants

Economic Stability Perception	*Number of Participants*
Strongly Agree	60
Agree	80
Neutral	40
Disagree	15
Strongly Disagree	5

Source: Author.

Table 15.3. Demographic influence on CRM

Age Group	*Gender*	*Occupation*	*CRM Effectiveness (1–10)*
18–25	Female	Student	6.8
26–35	Male	Professional	7.5
36–45	Other	Entrepreneur	8.2
46–55	Male	Executive	7.0
56–65	Female	Retiree	6.5

Source: Author.

3.2.2. Heritage and cultural tourism

Influence of Cultural Heritage on CRM: The study investigates how cultural heritage influences CRM strategies, gauging participants' perceptions of the impact of heritage on customer relationships (Table 15.4).

Participants generally acknowledge a positive influence of cultural heritage on CRM, with a significant portion expressing a strong positive correlation.

3.2.3. Balancing cultural preservation and customer satisfaction

Participants were questioned on the challenge of balancing cultural preservation and customer satisfaction in tourism management (Table 15.5).

The majority of participants perceive balancing cultural preservation and customer satisfaction as a challenging aspect of tourism management, indicating the need for nuanced strategies.

4. Correlations among Socio, Economic, and Cultural Factors

That explores the interconnections between socioeconomic, economic, and cultural factors to unravel complex relationships within the context of tourism management.

4.1. Integrated analysis

4.1.1. Correlation matrix

The correlation matrix examines the relationships among income levels, economic stability, cultural dimensions, and CRM effectiveness, providing insights into the interconnectedness of these factors (Table 15.6).

Table 15.4. Influence of cultural heritage

Influence of Cultural Heritage	*Number of Participants*
Strong Positive	70
Positive	80
Neutral	30
Negative	15
Strong Negative	5

Source: Author.

Table 15.5. Perception of balancing cultural preservation

Perception of balancing cultural preservation	*Number of participants*
Challenging	100
Manageable	70
Neutral	20
Not Challenging	10

Source: Author.

4.1.2. Interpretations

The correlation matrix table shows how income, economic stability, culture, and CRM effectiveness relate. The table's cells show the correlation coefficient between the variables, ranging from −1 to 1, with 1 indicating a strong positive correlation, −1 a strong negative correlation, and 0 no association. The diagonal elements (1.00) represent the perfect connection of income levels with themselves. Off-diagonal elements show substantial positive correlations. Income has a 0.65 association with economic stability, implying that as income rises, participants perceive economic stability to increase. Income correlates positively with cultural aspects, but only 0.30. This suggests a moderate link between income levels and cultural preferences, but other factors may also influence individuals' cultural preferences. Income levels have the strongest positive association (0.75) with CRM effectiveness. Tourism CRM satisfaction increases with income, demonstrating that wealthier people are happier with CRM techniques. Economic stability positively correlates 0.60 with CRM effectiveness. This suggests that tourism CRM practices are more satisfying for participants in stable economies. Hofstede's cultural framework shows a minor positive association (0.15) with economic stability. This suggests a weak link between cultural factors and economic stability. Last, there is a moderate positive correlation between CRM efficiency and cultural factors (0.45). Additionally, participants that respect more deeply cultural diversities or particular cultural features, may be happier with tourism CRM methods.

4.1.3. Cluster analysis

Grouping of participants is done on the basis of their response to income levels, economic stability perceptions, cultural dimensions and CRM effectiveness as it is done by using the cluster analysis. Interacting with this method, will tell us the segments already exist in the participant pool (Table 15.7).

The idea is to obtain insights for each of the distinct clusters of the participants in order to perform a more in depth analysis of each segment, and to cluster participants to better understand the diversity of profiles within the study and

Table 15.6. Correlation matrix

	Levels	*Economic Stability*	*Cultural Dimensions*	*CRM Effectiveness*
Income Levels	1.00	0.65	0.30	0.75
Economic Stability	0.65	1.00	0.15	0.60
Cultural Dimensions	0.30	0.15	1.00	0.45
CRM Effectiveness	0.75	0.60	0.45	1.00

Source: Author.

Table 15.7. Cluster analysis

Cluster	*Number of Participants*	*Characteristics*
1	80	High income, strong economic stability
2	60	Moderate income, positive cultural dimensions
3	40	Low income, diverse cultural preferences
4	20	Very high income, high CRM satisfaction

Source: Author.

conduct a targeted analysis for each segment.The goal of the integrated analysis is to reveal patterns and the relationships between socioeconomic, economic and cultural data, in order to have a holistic view of the intricate interconnections that affect the relationships with the customers in the tourism sector.

5. Implications for Tourism Management

The conclusions of study on relationship between socioeconomic, economic and cultural variables and CRM in tourism industry are of several important implications for efficient tourist management.

5.1. Tailoring CRM strategies based on socioeconomic profiles

It is found that there is a strong relationship between income level and CRM satisfaction, meaning that there is a higher expectation and CRM satisfaction with higher income. This insight can be very valuable for tourism management professionals: it allows them to adapt CRM strategies with regard to socioeconomic profiles of their target audience. The high income segments in fact may provide a good opportunity for customized services and personalization of experiences to increase the overall customer satisfaction.

5.2. Balancing cultural sensitivity and economic considerations

It is demonstrated that certain cultural dimensions have an effect to CRM effectiveness, which suggests that cultural consideration is a critical aspect of tourism management. Businesses should try to understand and integrate different cultural aspects in their offering to foster positive customer relationships. Nevertheless, it is necessary to be able to reconcile cultural protection with economies. Therefore, tourism management must find a way to balance this delicate line, that these cultural experiences that are offered be authentic and minimize economic sustainability goals.

5.3. Targeted marketing strategies for different income brackets

It gives opportunities to direct the marketing for particular travel type per certain income brackets. Engaging in the creation of marketing campaigns that link to the adventure seeking tendencies of low income participants, the cultural tendencies of moderate income persons, luxury aspirations of high income segments and the ecotourism desires of very high income groups can attract a heterogeneous group of tourists.

5.4. Enhancing CRM strategies during economic instability

Based on the positive correlation between economic stability and CRM effectiveness, during periods of economic instability, it is suggested that tourism management gives assertive attention to CRM initiatives. During times of economic downturn, targeted promotional activities, loyalty programs and value added services can be implemented in order to improve customer satisfaction and loyalty so that this helps in maintaining the resilience of tourism business.

5.5. Investment in ecotourism and cultural preservation

CRM satisfaction and the preference for ecotourism are all correlated with very high income and therefore tourism managers can invest in ecotourism ventures. Such practice not only to satisfy customer's but also to support sustainable tourism activities from the point of view of growing environmental consciousness among high income tourists. In addition, preserving the cultural heritage is important because carrying out such initiatives can be recreating unique and authentic experiences, which can work on different types of tourists.

5.6. Continuous monitoring of cultural dynamics

Tourism managers should stay attuned to burgeoning cultural dynamics as there is some moderate correlation between cultural dimensions and CRM effectiveness. Periodic assessment and keeping in touch while changing cultural preference and sensibility will help tourism businesses to update their CRM strategies with what they find to be culturally valued by their sector.

6. Conclusion

This research establishes a relation between socioeconomic, economic and cultural factors of tourism industry and CRM. The study serves to indicate that income levels, economic stability, culture and CRM effectiveness strike a complex balancing act and this would be of immense use to tourism management. Those in higher income levels have been positively correlated with higher CRM satisfaction. This, then, calls for market segment specific tailoring of approaches to customers' respective market segments of different statuses of income to derive customer satisfaction so that organizations can achieve CRM goals. In addition, cultural dimensions enter into the development of CRM initiatives for the reasons that there exist, albeit only moderate, but statistically significant relationship among cultural dimensions and the effectiveness of CRM initiatives. For tourism management professionals, this study will provide implications or actionable strategies for the professionals to adopt such as targeted marketing, investment in ecotourism as well the profession's continued interest in striking a chord between preservation of culture and economic needs. It will integrate the insights into practices of the tourism industry to enhance customer satisfaction and develop sustainable cultural tourism practice and guide the way for tour operators in maintaining relationships with customers in an ever-developing industry. This research is one such contribution to a better understanding about the complexities in the tourism management and offers a basis for subsequent studies to investigate the changing dynamics of the industry.

References

[1] Bloom, L. B. (2023). *The World's 50 Most Beautiful Countries (You Won't Believe Where the U.S. Ranked)*. Forbes. Retrieved 17 August 2023.

[2] Nisha Priya, P., & Stalin Alex, D. (2016). Combining left and right palm prints for more accurate personal identification. *International Journal of Advances in Engineering and Emerging Technology*, *7*(4), 35–51.

[3] India's Travel & Tourism could surpass pre-pandemic levels by the end of 2022. (2022). *WTTC*. Archived from the original on 30 September 2022. Retrieved 30 September 2022.

[4] Barile, J., Carreño, E., & los Ríos-Escalante, D. (2024). A review of mollusks farming in Chile. *International Journal of Aquatic Research and Environmental Studies*, *4*(1), 63–69. https://doi.org/10.70102/IJARES/V4I1/6

[5] Singh, H. *India eyes to be a huge medical tourism hub in coming years*. The Times of India.

[6] Colace, F., Santo, M. D., Lombardi, M., Mosca, R., & Santaniello, D. (2020). A multilayer approach for recommending contextual learning paths. *Journal of Internet Services and Information Security*, *10*(2), 91–102.

[7] 'Medical value tourism in India: What makes the country a leading Medical Tourism Destination'. 27 September 2021.

[8] Petrova, E., & Kowalski, D. (2025). Energy-efficient microalgae filtering and harvesting using an extremely low-pressure membrane filter with fouling control. *Engineering Perspectives in Filtration and Separation*, *2*(1), 25–31.

[9] 'Promotion of Medical Tourism'. Press Information Bureau. Archived from the original on 7 May 2016. Retrieved 28 April 2016.

[10] 'India received 6.19 million Foreign Tourist Arrivals (FTAs) during 2022 as compare to 1.52 million during the same period of 2021'. pib.gov.in. Retrieved 17 August 2023.

[11] 'Foreign tourist arrivals to India 44% lesser than pre-Covid level'. Deccan Herald. Retrieved 17 August 2023.

[12] 'Most Visited Countries 2021'. worldpopulationreview.com. Archived from the original on 6 December 2021. Retrieved 22 December 2021.

[13] 'India's Domestic Tourists increase by 16% crossing 1 billion Mark'. news.biharprabha.com. Indo-Asian News Service. Archived from the original on 26 October 2014. Retrieved 21 February 2014.

[14] Aditi Sharma (7 February 2014). 'Kerala: A vacation in paradise – Times of India'. The Times of India. Archived from the original on 16 January 2023. Retrieved 16 January 2023.

[15] 'Tamil Nadu, UP pip Goa as tourist havens'. Archived from the original on 28 July 2015. Retrieved 26 July 2015.

[16] Bremner, Caroline. 'Top 100 City Destinations Ranking' (PDF). Euromonitor International. Archived (PDF) from the original on 23 December 2019. Retrieved 30 January 2017.

[17] 'Travel & Tourism Development Index 2021 Rebuilding for a Sustainable and Resilient Future' (PDF). weforum.org. World Economic Forum. May 2022. p. 13. Archived (PDF) from the original on 3 July 2022. Retrieved 24 June 2022.

[18] 'Travel and Tourism Competitiveness Report 2019' (PDF). weforum.org. Archived (PDF) from the original on 4 September 2019. Retrieved 10 September 2019.

[19] 'Rank in Tourism' (Press release). Press Information Bureau, Government of India. 7 February 2014. Archived from the original on 22 February 2014. Retrieved 8 February 2014.

[20] 'Annual Report 2017–18' (PDF). Ministry of Tourism, Government of India. p. 27. Archived (PDF) from the original on 2 July 2018. Retrieved 27 August 2018.

[21] 'e-Visa'. Government of India. 2014. Archived from the original on 19 November 2017. Retrieved 26 November 2014.

[22] Sigala, M. (2005). Integrating customer relationship management in hotel operations: managerial and operational implications. *International Journal of Hospitality Management*, *24*, 391–413.

[23] Özgener, Ş., &Iraz, R. (2006). Customer relationship management in small–medium enterprises: The case of Turkish tourism industry. *Tourism Management*, *27*, 1356–1363.

[24] Lin, Y., & Su, H. (2003). Strategic analysis of customer relationship management – a field study on hotel enterprises. *Total Quality Management & Business Excellence*, *14*, 715–731.

[25] Vogt, C. A. (2011). Customer Relationship Management in Tourism: Management Needs and Research Applications. *Journal of Travel Research*, *50*, 356–364.

[26] Rahimi, R., Köseoglu, M. A., Ersoy, A. B., & Okumus, F. (2017). Customer relationship management research in tourism and hospitality: A state-of-the-art. *Tourism Review*, *72*, 209–220.

[27] Akyol, O., Çoker, T., Toprak, H. B., & Capapé, C. (2023). Capture of a Rare Smoothback Angelshark Squatina oculata (Squatinidae) in Turkish 2 Waters, with Updated Records from the eastern Mediterranean Sea. *Natural and Engineering Sciences*, *8*(1), 38–45. http://doi.org/10.28978/nesciences.1281611

[28] So, K. K., King, C., & Sparks, B. A. (2014). Customer engagement with tourism brands. *Journal of Hospitality & Tourism Research*, *38*, 304–329.

[29] Kodati, S., Reddy, K. P., Mekala, S., Murthy, P. S., & Reddy, P. C. S. (2021). Detection of fake profiles on twitter using hybrid svm algorithm. In *E3S Web of Conferences* (Vol. 309, p. 01046). EDP Sciences.

16 Problems of tribals on availing banking services

Praveen K. P.[1,a] *and R. Velmurugan*[2,b]

[1]PhD (Part-time) Research Scholar, Department of Commerce, Karpagam Academy of Higher Education, Coimbatore, Tamil Nadu, India
[2]Associate Professor of Commerce, Karpagam Academy of Higher Education, Coimbatore, Tamil Nadu, India

Abstract: This study investigates the multifaceted barriers that tribal communities in Wayanad District, Kerala, encounter in accessing formal banking services. Primary data were collected through structured interviews with 225 tribal individuals, focusing on socio-economic characteristics and banking behaviours. The analysis, employing descriptive statistics and factor analysis, reveals that physical and infrastructural challenges – such as geographical isolation, inadequate infrastructure, limited bank branch presence, and lack of proper identification documentation – pose significant impediments. Additionally, low literacy levels and insufficient financial literacy further exacerbate these difficulties. The study also identifies cognitive and procedural barriers, including mistrust of formal financial institutions, technological and language challenges, bureaucratic hurdles, high transaction costs, and cultural differences, as well as institutional shortcomings like insufficient government outreach, reliance on informal credit systems, and the absence of tailored financial products. Factor analysis distilled these obstacles into three distinct dimensions, with the physical and infrastructural factor accounting for 41.25% of the variance, while the other factors contributed 13.55% and 9.33%, respectively, cumulatively explaining 64.13% of the total variance. The study reveals how vital it is to create strategic policies and develop improved infrastructure and educational resources and banking methods that acknowledge indigenous cultural practices to solve the financial exclusion problem among tribal communities. The study advances knowledge about financial exclusion in disadvantaged communities and gives specific suggestions to public officials and financial entities which support greater economic self-empowerment among tribal populations.

Keywords: Tribal banking, financial inclusion, rural access, socio-economic barriers, banking challenges

1. Introduction

Modern financial systems face an inclusion and equity crisis because tribal communities experience major difficulties when accessing banking sector services which serves as one of the essential economic development foundations. Tribal people living in distant geographically cut-off areas encounter multiple obstacles when trying to reach basic banking services. Kumar [1] explains that the limited availability of bank branches in tribal areas compels community members to make lengthy journeys which both raises expenses and reduces bank participation. Stringent documentation demands serve as an additional massive obstacle facing potential bank clients. Many tribal people do not possess standardized identification papers according to Patel [2] which prevents them from satisfying bank account establishment requirements. The combination of low education levels with insufficient financial training makes it worse according to Sharma [3], who analyzed how modern digital banking systems do not consider typical practices of tribal communities when describing languages.

The tribal communities maintain substantial distrust toward modern financial institutions which adds to their existing physical hurdles and institutional barriers. The author Roy [4] maintains that exploitation and historical abandonment have created an atmosphere of skepticism which results in tribal communities preferring their established informal payment methods to new banking solutions. The use of informal networking systems for financial help frequently leads tribal communities into wicked loan terms and deepens their economic risks. Government financial inclusion programs that intend to help tribes face implementation challenges because these efforts fail to properly address the specific social and cultural background of tribal populations. The programs reviewed by Kumar [5] implement standardized solutions that prove ineffective for understanding and serving tribal population requirements.

Many tribal areas remain unimproved regarding banking accessibility during the quick transition to technology-based banking. According to Singh [6] the problems with the infrastructure such as unreliable electricity and limited internet connectivity and technical support problems hinder digital banking service adoption. The digital divide makes tribal persons even more marginal to mainstream society as the financial world becomes more digital. The challenges demonstrate why tribes face such difficulty when trying to join formal banking institutions. A combination of location barriers alongside regulatory barriers and cultural misunderstandings and the lack of technological capacity forms an extensive issue which requires extensive solutions. The authors Sharma [3], Roy [4] support banking methods that

[a]praveenpulpally948@gmail.com, [b]drvelsngm@gmail.com

DOI: 10.1201/9781003684459-16

engage communities in service and product creation whereas Kumar [1] and Patel [2] focus on policy changes for straightforward documentation systems and enhanced rural financial facilities.

Tribals need solution-oriented approaches to their banking service accessibility problems which must work hand in hand with government officials and banking executives and native community leaders. Tribal communities require strategies which should combine physical and technological barrier solutions alongside trust restoration efforts and financial product improvements for their cultural and economic standards. Complete financial inclusion becomes possible only by implementing a comprehensive method that will enable sustainable economic growth in addition to establishing social equity for tribal populations.

2. Review of Literature

Kumar [1] together with Sharma [3] underline that remote physical location stands as the main hurdle. The secluded locations where tribal peoples reside with inadequate road systems make it impossible for banks to operate profitable branches nearby. The situation worsens because Singh [6] shows how insufficient equipment such as unreliable power systems and shoddy road infrastructure together with minimal communication services create more accessibility issues. According to Kumar [1] banking institutions avoid investing in remote locations because low population density and uncertain profitability prevents financial institutions from serving tribal communities adequately.

According to Patel [2] lack of identification documentation creates major obstacles for entry into services. Tribal people face difficulties opening bank accounts because they lack formal identification needed for banking transactions according to existing requirements from banks and financial institutions. The way in which formal institutions communicate through banking documents and formal communications presents challenges because these materials are often unavailable in native local dialects according to Patel [2]. Even persons who want to use the formal financial system face dismissal because of the excessive paperwork and complex procedures which form institutional barriers.

The research by Sharma [3] determines that tribal populations face significant barriers because of their low reading abilities. Unrefreshed banking service use becomes impossible when people cannot understand or read basic financial documentation. Frequent financial illiteracy among Indian tribal communities makes it harder for them to recognize the benefits and understand the procedures and risks in formal banking as explained by Sharma [3]. According to Roy [4] tribal communities maintain strong distrust toward standard financial institutions. Previous experiences of discrimination and victimization caused numerous individuals to mistrust contemporary banking operations. Traditional community practices and local customs create barriers against standard bank operating procedures due to cultural differences between formal banking institutions and tribal peoples. The cultural practices of tribal people require them to use informal credit systems because these local networks are more reliable and suitable to their traditions [4].

Banks have introduced quick digitalization in banking operations that brings efficiency to urban banking but faces new obstacles when implemented in tribal regions. According to Singh [6] weak internet connections, limited access to computing devices and insufficient digital skills among these populations make digital banking app unavailable in their areas. The existing gaps in access continue to expand creating separate groups between technology users and non-users. Economic costs such as traveling expenses together with time requirements and potentially informal fees work as deterrents against frequent banking participation among tribal people according to Kumar [5]. The dearth of suitable outreach efforts from the government prevents many tribal populations from accessing financial inclusion schemes because these strategies fail to use effective methods of communication and create programs that are not customized to tribal needs. According to Kumar [5] there remains a significant gap in specialized financial services because most banking solutions fail to recognize the distinct payment patterns, saving behaviour and risk tolerances of tribal populations making them unappealing to users.

A thorough picture emerges from the reviewed studies about the complex network of obstacles tribal people experience when trying to access banking services. Insufficient financial inclusion exists because tribal communities face problems from disconnection to limited infrastructure and poor governance and low reading skills combined with deepening suspicion and cultural communication problems and poor technology access and insufficient resources. Cultural disparities alongside geographical and economic obstacles demand collaborative support from policy designers and financial organizations and community leaders who should develop solutions that address these multitudinous barriers.

2.1. Research gap

Multiple research gaps related to tribal population banking obstacles remain evident according to Kumar [1], Sharma [3], Singh [6], Roy [4], Patel [2], and Kumar [5] with their findings about geographical constraints and inadequate infrastructure along with limited branch presence and identity problems and literacy deficiencies and technological restrictions plus cultural skepticism. A comprehensive integrated framework exists to study how financial exclusion barriers connect with one another to produce their combined effects. The shortage of thorough research about specific approaches for targeted interventions coupled with restricted assessments of digital solutions for tribal community needs stands as a critical barrier because it results in few comparative assessments between regions like insufficient analysis of gender-related obstacles. Research on policy reform as well as infrastructural development effects has been conducted through mainly short-term assessments that do not track sustained outcomes. It becomes essential to tackle these knowledge deficiencies because they enable the creation of banking solutions that suit tribal cultural preferences and persist over time.

2.2. *Statement of the problem*

Government initiatives for financial inclusion have not solved the persistent problems facing tribal communities who want to access formal banking services throughout Wayanad District in Kerala. Many tribal residents struggle to access formal banking because they need to travel extensive distances because of geographical separation and inadequate bank branches accompanied by high costs for transactions according to Kumar [1], Sharma [3] and Singh [6]. The combination of insufficient financial literacy and poor literacy skills together with language barriers and unstandardized identification documentation prevents tribal residents from effectively using banking products and services [2, 3]. Historical neglect alongside cultural differences between communities and their deep mistrust of bureaucratic institutions marginalize them even more [2, 4] besides bureaucratic hurdles. The combination of these factors creates limited economic possibilities for tribal populations while continuing to isolate them socially and economically [5]. Systematic research is essential to address the connected hurdles affecting tribal communities in Wayanad District through purposeful policy changes and interventions that will bring them into financial inclusion.

2.3. *Objective of the study*

To ascertain major barriers on tribal people on availing banking services.

2.4. *Scope of the study*

The research examines exclusively the tribal residents within Wayanad District of Kerala. The study analyzes the bank profiles alongside socio-economic information together with the particular obstacles faced by these minority communities in using banking services.

3. Research Methodology

3.1. *Data*

The research depends solely on primary data obtained by using structured interview questions. Both the tribal community socioeconomic information and their hurdles when reaching banking options appear on the set interview format.

3.2. *Sampling*

A total of 225 tribal participants took part in the research based on a snowball sampling method.

3.3. *Framework of analysis*

Simple percentages and factor analysis were used to analyze the gathered data.

3.4. *Significance of the study*

Tribal communities from Wayanad District benefit from this research because it analyzes their complete obstacles for accessing formal banking services. The research reveals important details about financial exclusion of tribal populations through its examination of geographical isolation along with infrastructural inadequacies and low literacy levels and bureaucratic challenges. The research provides essential understanding which serves as a foundation for policymakers and financial institutions together with community organizations to develop adequate interventions that boost financial inclusion and raise the socio-economic condition of these excluded groups.

This research provides evidence to help develop specific plans that will help officials overcome the weaknesses identified during the study. The research outcome emphasizes that local solutions must include setting up accessible banks and easy document processing and specific financial training for these communities. Policymakers should tackle these problems because they enable tribal communities to enter the formal financial system thus providing economic empowerment alongside savings development along with poverty reduction across the area.

The paper advances present academic understanding about financial inclusion together with rural development while specifically focusing on tribal communities who have received insufficient research attention. The research investigates one of the leading gaps in academic literature while making possible future scholarly investigations regarding social economic factors and financial service access. Additional research needs to develop these findings through new assessments of financial exclusion aspects as well as monitoring how proposed solutions impact communities in the long run.

The survey demonstrates why financial institutions must modify their services to properly serve the needs of tribal communities. The specific difficulties tribal populations encounter enable banks to create banking products which align with their economic situation and cultural needs. The approach results in improved client interaction and advanced economic literacy which produces enduring rural development.

The study holds significant value because it provides data to guide changes for better outcomes. This research reveals the complex barriers facing tribal communities in Wayanad so that it helps create comprehensive strategies which promote both social equality and sustainable economic development and financial practices within the region.

3.5. *Limitation of the study*

Primary data collection provides direct insights to the research but remains vulnerable to different types of bias.

The obtained results require cautious use due to their generalization limitations.

3.6. *Findings*

3.6.1. Socio-economic profile

This section evaluates the socio-economic aspects of tribal residents living in Wayanad district of Kerala (Table 16.1).

Table 16.1. Socio-economic profile

Variables	*Category*	*Frequency*	*Percentage*
Age	Under 20	20	8.9%
	21–30	50	22.2%
	31–40	60	26.7%
	41–50	45	20.0%
	51 and above	50	22.2%
Gender	Male	130	57.8%
	Female	95	42.2%
Educational Qualification	Illiterate	80	35.6%
	Primary	90	40.0%
	Secondary	40	17.8%
	Higher Education	15	6.7%
Occupation	Agriculture	120	53.3%
	Labour	50	22.2%
	Self-employed	30	13.3%
	Unemployed	25	11.1%
Type of Family	Nuclear	100	44.4%
	Joint	125	55.6%
Status in Family	Head	150	66.7%
	Dependent	75	33.3%
Monthly Income	< Rs. 5,000	80	35.6%
	Rs. 5,000–10,000	90	40.0%
	> Rs. 10,000	55	24.4%
Family Income	< Rs. 15,000	100	44.4%
	Rs. 15,000–30,000	90	40.0%
	> Rs. 30,000	35	15.6%
Family Expenditure	< Rs. 10,000	70	31.1%
	Rs. 10,000–20,000	100	44.4%
	> Rs. 20,000	55	24.4%

Source: Author.

The respondents predominantly belong to three age ranges including the 21–30 (22.2%) and 51 and above (22.2%) groups while the highest bracket consists of those between 31–40 (26.7%). The majority of residents within this population belong to the prime working years while adults under 20 years of age make up the remaining 8.9% segment. Working-age people are more likely to use formal financial services in banking whereas an age distribution with this demographic pattern may require specific banking services.

Males represent 57.8 percent of the subjects while females make up the remaining 42.2 percent of the collective. The limited number of female participants in this study base could either come from natural demographic trends or from traditional social expectations regarding economic involvement. The research shows male banking views appear modestly stronger than female perspectives yet the significant female participation requires dedicated gender-focused banking approaches.

The sample data shows that 76.6% of participants hold illiteracy or primary education levels only. Most bank clients have low levels of education considering that only 17.8% completed secondary education and less than half that with higher education qualifications. The limited education of the respondents may harm their financial literacy skills because it hinders their ability to unlock and fully utilize bank services.

People in this study earn their living from agriculture (53.3%) primarily although labour (22.2%) follows then self-employment (13.3%) then unemployment (11.1%) becomes the least common income source. The dominant agricultural sector influences seasonal economic cycles and variable incomes which indicates the necessity for banking institutions to create specialized financial solutions to better serve this population.

Among the respondents over 55.6 percent belong to joint family structures but nuclear family members make up the remaining 44.4 percent. Such a traditional family structure within tribal communities may impact both decision-making and financial resource pooling within the household.

The majority of 66.7% respondents lead their families while 33.3% depend on others financially. Household financial management falls mainly under the responsibilities of most respondents thus making this data central to developing effective financial inclusion plans.

A large portion of respondents receive a monthly salary under Rs. 5,000 and 40.0% obtain payment between Rs. 5,000 and Rs. 10,000 and another 24.4% earn in excess of Rs. 10,000 each month. Because of their limited economic means these respondents face barriers both when interacting with formal banking services and when purchasing various financial products.

Family income data refers to similar economic patterns as individual income with 44.4% of families taking home less than Rs. 15,000 and 40.0% earning between Rs. 15,000 and Rs. 30,000 and only 15.6% reporting earnings greater than Rs. 30,000. The data presents an overall financial condition suggesting high vulnerability of the community.

Family spending patterns show that 31.1% of households allocate fewer funds than Rs. 10,000 while 44.4% fall between Rs. 10,000 and Rs. 20,000 and lastly 24.4% spend more than Rs. 20,000 monthly. A comparison between income levels and spending data helps reveal potential saving capacities and financial wellness of domestic units in the area.

The majority of tribal community members in Wayanad consists of families with minimal income who have low levels of education and depend heavily on farming for their survival. The various barriers displayed a collective picture of considerable difficulties in accessing formal banking and suggest that specialized financial inclusion solutions should be developed to overcome these specified barriers.

3.7. Banking habit

The next section analyzes Wayanad tribal people's banking behaviour patterns (Table 16.2).

The tribal respondents demonstrated a tendency to choose Public Sector Banks as their banking institutions since they represented 53.3% of respondents possibly because of the combination of strong reliability attributes with greater accessibility across rural regions and government-directed financial initiatives for tribal areas. The data indicates that Private Sector Banks have 26.7% of user respondents while the remaining 73.3% opt for different types of banks. The data shows that Cooperative Banks maintain bank accounts for 20.0% of the respondents because the tribal population might not access or understand these financial institutions well.

Table 16.2. Banking habit

Variables	*Category*	*Frequency*	*Percentage*
Type of Bank	Public Sector Bank	120	53.3%
	Private Sector Bank	60	26.7%
	Cooperative Bank	45	20.0%
Nature of Account	Savings Account	180	80.0%
	Current Account	30	13.3%
	Recurring Deposit Account	15	6.7%
Period of Holding Account	Less than 1 year	50	22.2%
	1–3. years	100	44.4%
	More than 3 years	75	33.3%
Purpose of Holding Account	Savings & Deposits	140	62.2%
	Payment/ Transaction Needs	50	22.2%
	Receiving Government Benefits	35	15.6%

Source: Author.

Survey results demonstrate that 80.0% of tribal individuals hold Savings Accounts for storing their funds securely. Saving stands as the main purpose why the tribal population maintains bank accounts since very few of them use banks for regular transactional activities. Only 13.3% of participants have current accounts among the respondents while savings accounts served as the most popular choice among 80%. These data demonstrate that regular high-volume transacting parties tend to favour current accounts based on their economic activities or business affiliations. A small segment of bank users with Recurring Deposit Accounts totalled 6.7% and shows minimal participation in scheduled savings or long-term financial deposits.

The survey demonstrates that most respondents (44.4 percent) have used their accounts between one to three years while being part of the formal banking system. A significant number of 33.3% have kept their banking accounts active for longer periods exceeding three years. New financial inclusion efforts seem to influence the tribal population because 22.2% of respondents maintained their bank account for less than twelve months.

Bank account holders primarily use savings and deposit services as their primary account function according to 62.2 percent of respondents indicated Savings & Deposits. This demonstrates that security of funds stands as the main banking purpose for these people. Payment/Transaction Needs were the main reason for account usage among 22.2% of respondents which demonstrates their involvement with standard financial transactions. People who take their funds from the government through public aid programs compose 15.6% of the total respondents who maintain bank accounts.

3.8. Problems of tribal people on availing banking services

Factor Analysis revealed what constitutes the main obstacles tribal communities encounter while attempting to use banking services. Firstly Kaiser-Meyer-Olkin (KMO) together with Bartlett's Test of Sphericity evaluated the dataset suitability for factor analysis (Table 16.3). The KMO value showed an exceeding level of 0.70 which confirmed that the data was suitable for additional evaluation. Factor analysis modeling criteria supported the use of data collection due to

Table 16.3. KMO and Bartlett's test

Kaiser-Meyer-Olkin Measure of Sampling Adequacy.		.864
Bartlett's Test of Sphericity	Approx. Chi-Square	1786.660
	df	105
	Sig.	.000

Source: Author.

the significant values obtained from Bartlett's Test of Sphericity (1786.660, df = 105, $p < 0.001$) and the KMO statistic (0.864).

The factor analysis yielded three separate components using eigenvalues greater than one as the basis for evaluation. All elements with loadings surpassing 0.5 were considered important (Table 16.4).

The physical barriers along with infrastructure issues comprise the major component of Factor One. The first factor shows strong indications toward geographical scarcity coupled with poor infrastructure combined with few bank branches and insufficient identification documents and low reading abilities. The collected variables confirm structural limits and accessibility barriers function as primary obstacles facing tribal people when they attempt to use banking services. The challenges outlined under Factor One demonstrate that geographical distance and inadequate structure and documentation as well as low educational attainment act as fundamental obstacles to banking system participation.

Factor Two groups together cognitive and procedural obstacles. Significant variables linked to this factor include tribal participants' limited financial knowledge along with their mistrust in formal banking institutions and technical obstacles, communication problems, government red tape and high fees for banking transactions and cultural differences. Beyond physical constraints society faces additional severe barriers to comprehend and operate within banking processes. Tribal populations avoid formal banking services because financial education failure combines with natural skepticism to complex procedural requirements.

Factor Three focuses on problems created by insufficient external backing and policies which do not meet current needs. Insufficient government outreach along with dependence on informal credit systems and the absence of tailored financial products reveal institutional along with policy-level problems which form the key variables within this factor. The lack of suitable banking products which meet tribal community needs together with insufficient government initiatives both play crucial roles in why tribal populations struggle to access formal banking services.

Factor One establishes itself as the leading influence by accounting for 41.252% of the total variance that obstructs tribal community members from accessing banking services. The three principal factors account for a total variance of 41.252% followed by 13.548% from Factor Two then 9.327% from Factor Three. These three aspects explain 64.127% of the total variance between factors which proves their comprehensive ability to describe the primary service access barriers for tribes in banking.

Table 16.4. Problems on availing banking services

Barrier	*1*	*2*	*3*
Geographical Isolation	.852		
Inadequate Infrastructure	.794		
Limited Bank Branch Presence	.777		
Lack of Identification Documentation	.747		
Low Literacy Levels	.679		
Insufficient Financial Literacy		.837	
Mistrust of Formal Financial Institutions		.813	
Technological Barriers		.777	
Language Barriers		.625	
Bureaucratic Hurdles		.606	
High Transaction Costs		.522	
Cultural Differences		.516	
Insufficient Government Outreach			.827
Dependence on Informal Credit Systems			.778
Lack of Tailored Financial Products			.657
Eigen Value	6.188	2.032	1.399
% of Variance	41.252	13.548	9.327
Cumulative % of Variance	41.252	54.800	64.127

Source: Author.

4. Suggestions

A complete and simultaneous approach needs implementation to tackle banking service availability problems among tribal communities.

- The establishment of mobile banking vans together with micro-branches or banking correspondents should provide accessible services to tribal population living in remote areas.
- Tribal areas need investments to build better road networks and power systems and Internet access so digital banking steps can function efficiently.
- The banking system should accept community certificates and voter IDs to simplify account opening procedures while implementing programs to distribute standard IDs among tribal people.
- To make banking procedures more accessible the system needs streamlining of paperwork along with procedural complexities through dedicated support centers and assistance programs.
- Setting up targeted instruction programs with basic language materials and visual training material and community educational classes for improving literacy alongside financial awareness levels.
- Trust-building activities featuring community leaders should reach out through outreach programs to establish formal financial institution trust and counter common banking system misunderstandings.

- Training about digital tools along with easy-to-use interfaces for mobile banking systems will serve customers who have limited digital skills.
- The institution must provide banking materials, applications and customer support through local dialects to help customers grasp the service information with clarity.
- Rural areas should receive discounted transaction fees in order to help low-income tribal families manage financial costs.
- Pulling service offerings into match tribal cultural routines and seasonal income dispersal methods that characterise tribal customs.
- Nitin Foundation must design financial products with flexibility such as microfinance options together with seasonal loans and savings schemes that match the distinctive financial patterns of tribal populations.
- Government financial inclusion programs should increase their outreach through using local platforms and community engagement activities to deliver greater efficiency.
- Formal financial institutions should bring informal credit approaches into their product designs to make it easier for tribal people to transition from traditional to modern systems.

Improved financial inclusion and better socio-economic results for tribal people in Wayanad District become achievable by unifying various stakeholder groups which will reduce banking barriers.

5. Conclusion

The majority of Wayanad District's tribal community consists of people in their working years who practice agriculture with many recipients of limited educational and financial literacy. The population statistics show that adults between 31 and 40 years old make up most of the district yet males outnumber females since families follow traditional joint family traditions which affect their banking conduct. Most people in this demographic have low earnings together with restricted savings capabilities which drives them towards utilizing public sector banks and keeping their money in savings accounts. Factor analysis separated barriers into three fundamental groups with physical and infrastructural difficulties composing Factor One and cognitive and procedural hurdles composing Factor Two while external help and policy failures constituted Factor Three. Factor One consisting of geographical isolation together with inadequate infrastructure, limited bank branches and lack of proper identification and low literacy levels proved to be the primary barrier contributing to over 41% of the banking difficulties experienced by respondents. The analysis of these variables reveals their ability to explain 64% of all variance throughout the study which showcases how complex the obstacles are between tribal communities when they try to use banking services. The research establishes the requirement for specialized interventions that handle both system structures and local cultural obstacles to improve economic development through financial access in the area.

6. Scope for Further Research

The research field about tribal communities' banking accessibility in Wayanad District opens many possibilities for extensive investigation. Studies focusing on this subject should utilize qualitative research methods including fieldwork and interview-based approaches to fully explore tribal individuals' understanding of standard banking services. Such research would bring insight into why formal banking services receive distrust from tribal individuals based on their culture and emotions. Studies about tribal and non-tribal rural population comparisons would deliver better understanding of the special obstacles tribal groups encounter.

Research needs to study how digital transformation affects tribal population access to finance because it presents a promising field of inquiry. Research should analyze how digital training sessions work together with mobile banking solutions in helping people overcome technical challenges and network limitations because banking systems are rapidly moving toward technology-based operations. Studies following the long-term effects of government programs with microfinance options and customized financial products on tribal community social and economic development would provide beneficial results. Research needs to study if the interventions successfully create lasting improvements in the fields of financial inclusion and economic stability.

Analysis of how informal credit functions with formal banking systems would create knowledge about combining these channels into hybrid financial approaches which work better for tribal communities. Studies linking informal practices with formal financial services through integration or complementary backing will create a model for building bank products which suit cultural needs. The effectiveness of outreach and educational programs aimed at financial literacy becomes a research topic to evaluate their ability to improve banking service usage from populations who engage with traditional banking systems.

The collection of this research data will enhance our comprehension of financial inclusion barriers alongside development methods which specifically serve the diverse socio-economic transformations of tribal members across Wayanad District.

References

[1] Kumar, A. (2012). Accessing financial services in rural India: Challenges and prospects for tribal communities. *Journal of Rural Banking and Development*, *5*(2), 45–67.

[2] Patel, R. (2015). Documentation dilemmas: The impact of identification requirements on tribal inclusion in banking. *Financial Inclusion Review*, *8*(3), 112–128.

[3] Sharma, P. (2014). Bridging the gap: Literacy, technology, and financial services among tribal populations. *Journal of Economic Development and Finance*, *12*(4), 89–105.

[4] Roy, S. (2017). Trust deficit: The role of historical neglect in shaping banking practices among tribal communities. *Journal of Social and Economic Studies*, 9(1), 34–50.

[5] Kumar, A. (2019). Reforms and realities: Policy initiatives for enhancing banking services in tribal areas. *Policy and Development Journal*, *10*(2), 77–93.

[6] Singh, M. (2018). Digital barriers: Infrastructure challenges and the digital divide in tribal banking services. *International Journal of Information Technology and Development*, *7*(3), 58–75.

17 Exploring the Blockchain: Deciphering investor preferences in urban India

Pradeep Chaurasia[1,a], *Pramod Dwivedi*[2,b], *Kausik Mukherjee*[1,c], *Anurag Singh*[2,d], *Kiran Chhabariya*[2,e], *and Shweta Singh*[2,f]

[1]Associate Professor, Faculty of Management Studies, AKS University Satna, Madhya Pradesh, India
[2]Assistant Professor, Faculty of Management Studies, AKS University Satna, Madhya Pradesh, India

Abstract: To shape the financial landscape, it is crucial to understand investor preferences as Blockchain technology continues to gain popularity globally. Examining the urban investment landscape of India, this research seeks to understand investor preferences in relation to Blockchain. The study uses a quantitative approach with Likert scale instruments to examine how well-informed urban investors are about Blockchain, what factors matter most to them when making investment decisions, how they feel about current regulatory frameworks, and whether there are any correlations with demographic variables. The research aims to shed light on the future of Blockchain investments in urban India and provide policymakers and business leaders with useful information. In order to lay the groundwork for individualized investment strategies, the study postulates that demographic characteristics are significantly correlated with investor preferences in Blockchain investments.

Keywords: Blockchain investments, investor preferences, urban India, likert scale, regulatory frameworks

1. Introduction

Blockchains are decentralized ledgers that use cryptographic hashes to securely connect ever-growing sets of records, or blocks [1]. Both of the initial two Each block contains information about transactions, a timestamp, and a cryptographic hash of the previous block [3]. "This information is commonly shown as a Merkle tree, with data nodes represented as leaves. Given that each block stores data pertaining to the block before it, the structure resembles a linked list in that each block links to the previously mentioned blocks [2]. Once data of a block is added into the blockchain, for it to be modified will affect all other blocks which follow it. Thus, blockchain's transactions are irreversible [8]. Usually, peer-to-peer (P2P) decentralized computer networks keep public global ledgers, also known as blockchains. Adding and checking new transaction blocks will be done by nodes in these networks using a consensus-based system. Forks in blockchains that fight about the existence of unchangeable records could be used to argue that blockchain records are not completely immutable. Beyond that, Blockchains seem to be built with strong Byzantine fault tolerance and better security already built in. While making the Bitcoin cryptocurrency in 2008, someone or a group of people put out the first public distributed record for Bitcoin transactions. Which was based on the work of Stuart Haber, Dave Bayer, and W. Scott Stornetta later in [19]. Bitcoin was the first to use blockchain technology for double spending prevention, Getting rid of the requirement for a central computer or a reliable third party. There are additional applications for the Bitcoin blueprint. Third, blockchains that are publically accessible for use by digital currencies as well [27]. The blockchain, from one angle, may be regarded as a payment rail. It has been noted that discussion around possible value added for Business-internal blockchains is active [7]. Computerworld calls the sale of private blockchains that don't have a strong security framework 'snake oil' [4]. Yet, many believe that permissioned blockchains can, indeed, be safer due to their distributed nature if well designed. [9, 28].

1.1. History of blockchain

A protocol similar to blockchain was initially described by cryptographer David Chaum in his 1982 dissertation Computer Systems Established, Maintained, and Trusted by Mutually Suspicious Groups [10]. In 1991, Stuart Haber and W. Scott Stornetta detailed supplementary work on a cryptographically protected block chain.They aimed to establish a system that would make it impossible to alter the timestamps of documents [4, 11]. By permitting the collection of many document certificates into one block, Haber, Stornetta, and Dave Bayer enhanced the design's efficiency in 1992 by including Merkle trees [4]. Document certificate hashes have been published weekly in The New York Times since 1995 under their company Surety.in [10, 13]. In 2008, an anonymous individual or group going by the name of

[a]drpradeepchaurasia78@gmail.com, [b]pramod.dwivedi5@gmail.com, [c]km466430@gmail.com, [d]parihar_anurag@yahoo.in, [e]kiranchhabariya96@gmail.com, [f]shwetasingh.satna@gmail.com

DOI: 10.1201/9781003684459-17

Satoshi Nakamoto came up with the idea for the first decentralized blockchain. By incorporating a difficulty parameter to stabilize the number of block additions to the chain and a Hashcash-like approach to timestamp blocks without requiring their signature by a trusted party, Nakamoto significantly enhanced the concept [4]. The following year, Nakamoto integrated the architecture into Bitcoin, a cryptocurrency that functions as the network's public record for all transactions.As of August 2014, the file size of the Bitcoin blockchain—which contains records of all network transactions—had grown to 20 GB [14]. The size increased to over 30 GB in January 2015, and from 50 GB to 100 GB in the period between January 2016 and January 2017, the bitcoin blockchain expanded. Beginning in early 2020, the size of the ledger has surpassed 200 GB [15, 16]. Although Satoshi Nakamoto first used the terms "block" and "chain" independently in his paper, the term "blockchain" had already gained widespread use by 2016 [30]. In 2016, blockchains reached the early adopters' phase with an adoption rate of 13.5% inside financial services, according to Accenture's application of the diffusion of innovations theory [17]. In 2016, the Chamber of Digital Commerce launched the Global Blockchain Forum, an endeavor of industry trade groups." A mere 1% of chief information officers (CIOs) mentioned blockchain deployment in May 2018 according to Gartner, while 8% said they were "planning or [looking at] active experimentation with blockchain" in the near future [18] in In 2019, just 5% of chief information officers (CIOs) held the belief that blockchain technology would significantly impact their company [19].

2. Literature Review

The study by Tamilselvan, B. [21] aims to explore the relationship between demographic factors and investors' attitudes towards cryptocurrency investments. Primarily utilizing structured questionnaires for data collection and secondary data sourced from the internet, this research primarily focuses on Tamil Nadu, India. Target respondents include those who have invested in cryptocurrencies like Bitcoin, Ethereum, Litecoin, Cardano, etc. The analysis utilizes statistical tools to measure investors' attitudes and preferences for investment decision sources, providing awareness and suggestions to potential investors in Tamil Nadu. Agarwal [23] and Parimalakanthi [25] both highlight the risk-averse nature of Indian investors, emphasizing their preference for traditional investment avenues such as fixed deposits, life insurance policies, and gold. In their individual studies, Schuetz [24] and Muth [26] investigate blockchain technology's possible applications in combating financial marginalisation and improving the openness of city planning processes. These studies collectively underscore the need for further research on the adoption of blockchain-based technologies in India, particularly in urban areas, and the potential for these technologies to reshape the investment landscape. Jaiswal, & Joge [22] state that their paper aims to assess the potential of blockchain technology in achieving economic development goals. They emphasize the importance of development practitioners posing relevant questions to technologists and highlight challenges that innovators must overcome for the technology to be effective. The central finding suggests that blockchain-based solutions can significantly enhance efficiency and outcomes in certain use cases, with varying impact in others. However, the paper notes that the primary constraints often lie outside technological aspects. To unlock the full potential of blockchain in economic development, governments and development organizations must take steps they have traditionally resisted, such as adopting common reporting systems and establishing reliable land registry systems [5]. The positive aspect is that the enthusiasm surrounding the technology has prompted increased interest and investment from organizations in addressing these fundamental challenges. Keywords include Blockchain, Cryptocurrencies, and decentralized ledger.

2.1. Objectives of the study

- To assess the level of awareness and understanding of blockchain among urban investors in India.
- To identify and rank the factors influencing investor preferences in blockchain investments.
- To examine and gauge investor attitudes towards regulatory frameworks
- To explore the correlation between demographic variables and investor preferences.

3. Research Methodology

3.1. Study design

1. Quantitative Research Approach: The researchers in this study used a quantitative methodology to collect and analyse numerical data on the preferences of urban Indian investors in relation to Blockchain.
2. Survey Instrument Development: In order to gauge how people feel and what they desire in terms of Blockchain investments, we will create a structured survey with Likert scale questions.

3.2. Sample selection

1. Target Demographic: Recognizing the distinct features and factors related to this demographic, the research zeroes in on urban investors in India.
2. Inclusion Criteria: All participants must be at least 18 years old, live in urban regions, and have either investing expertise or an interest in investments related to Blockchain.
3. Sample Size Determination: To guarantee a representative and statistically significant dataset, we aim for a sample size of 250 participants. This will allow for robust analysis and the generalizability of findings.

3.3. Data collection

1. Survey Distribution: To guarantee a varied and geographically representative sample, the survey will be electronically distributed across several social media channels and platforms.
2. Likert Scale-Based Administration: Using Likert scales, survey takers will indicate how much they agree or disagree with claims made about Blockchain investments. A standardized and measurable evaluation of investor preferences is possible with this method.

3.4. Data analysis

1. Statistical Analysis: Quantitative data from Likert scales will be looked at using statistical methods like descriptive statistics, association analysis, and regression analysis, if necessary.
2. Thematic Analysis: If open-ended questions are included in the survey, thematic analysis will be employed to derive qualitative insights, complementing the quantitative findings.

4. Result Analysis

This survey sought to understand the preferences of urban Indian Blockchain investors, and Table 17.1 gives a quick overview of the demographic profile of the participants. The age categories of 26–35 and 18–25 make up the bulk of the participants, with 32.0 and 24.0 percent of the total, respectively. The sample is fairly balanced in terms of gender distribution, with 44.0% female and 56.0% male participation. In terms of employment status, 72.0% of participants are full-time workers, and 48.0% have a bachelor's degree or above. With 48.0% classifying themselves as intermediate investors and 32.0% as experienced, the sample reflects a wide range of financial expertise. This demographic makeup guarantees a thorough and representative dataset for the study by providing a diverse viewpoint on investor preferences.

Participants' answers to questions measuring their knowledge of Blockchain technology, their familiarity with Blockchain ideas, their practical experience with Blockchain, and their self-perceived knowledge regarding cryptocurrencies are displayed in Table 17.2. It is worth mentioning that 120 out of 280 people say that they are aware of what Blockchain is, suggesting that it is well known. When it comes to real-world experience, though, opinions are more divided, with 55 people remaining neutral. A majority of participants consider themselves educated about cryptocurrencies, demonstrating a remarkable level of confidence among themselves. Understanding Blockchain technology finds a balanced distribution. The results show that the surveyed population has a wide range of knowledge and expertise with Blockchain-related ideas.

Table 17.3 shows the average scores that participants gave to different aspects affecting their Blockchain investment preferences, on a Likert scale from 1 to 5. The participants clearly prioritize financial rewards and transaction security, as indicated by the high average ratings of 4.2 for

Table 17.1. Demographic profile of participants

Demographic Characteristic		*Frequency*	*Percentage (%)*
Age Group	18–25	60	24.0
	26–35	80	32.0
	36–45	45	18.0
	46–55	35	14.0
	56 and above	30	12.0
Gender	Male	140	56.0
	Female	110	44.0
Educational Background	High School	20	8.0
	Bachelor's Degree	120	48.0
	Master's Degree	80	32.0
	Doctoral Degree	30	12.0
Employment Status	Employed	180	72.0
	Unemployed	40	16.0
	Student	30	12.0
Investment Experience	Novice	50	20.0
	Intermediate	120	48.0
	Experienced	80	32.0

Source: Author.

Table 17.2. Level of awareness and understanding of blockchain

Statement	*Strongly Disagree*	*Disagree*	*Neutral*	*Agree*	*Strongly Agree*
1. I am familiar with the concept of Blockchain.	10	15	25	120	80
2. I understand how Blockchain technology works.	15	20	30	100	85
3. I have practical experience using Blockchain.	40	45	55	70	40
4. I consider myself knowledgeable about cryptocurrencies.	20	35	50	90	55

Source: Author.

Table 17.3. Factors influencing investor preferences

Factors	*Average Rating (Likert Scale: 1–5)*
1. Potential Return on Investment	4.2
2. Perceived Risk	3.8
3. Regulatory Clarity and Compliance	4
4. Technological Complexity and Understanding	3.6
5. Market Volatility	4.1
6. Reputation of Blockchain Projects or Companies	4.3
7. Peer Influence and Social Trends	3.7
8. Ease of Access and User-Friendly Platforms	4.2
9. Security of Blockchain Transactions	4.4
10. Government Support and Initiatives	3.9

Source: Author.

possible return on investment and 4.4 for security of Blockchain transactions. Trust and usability were emphasized by the strong average scores for Blockchain projects or firms' reputation (4.3) and platforms' ease of access and user-friendliness (4.2). Government support and initiatives (3.9) and regulatory clarity and compliance (4) both obtained above-average scores, indicating that they significantly influence investor choices. In sum, the complex factors influencing Blockchain investment decisions are illuminated by these ratings.

In Table 17.4, we can see how people felt about various statements that probed their views on regulatory frameworks as they pertain to Blockchain investments. The importance of transparent laws for fostering investor trust is highlighted by the fact that a considerable portion (110) of respondents feel that such regulations have a favourable impact on their investment choices. Out of the total number of respondents, 120 are in strong agreement that clear laws are crucial to the expansion of Blockchain investments. This shows how important regulations are seen as in promoting the growth of the business. Emphasising the detrimental impact of regulatory ambiguity, a large portion (40%) is unwilling to invest due to unclear regulations. In addition, a large portion (90) makes an effort to stay updated on regulatory developments, which shows that they understand the need of keeping up with the ever-changing regulatory landscape. The answers shed light on how investors feel about the current regulatory climate for Blockchain investments as a whole.

Investor attitudes towards regulatory frameworks and factors influencing Blockchain investment choices are two important features that are shown in Table 17.5, which demonstrates correlation coefficients between demographic characteristics. Significantly, there are positive connections between age and investment experience with characteristics that impact preferences. This indicates that individuals with

Table 17.4. Investor attitudes towards regulatory frameworks

Statements	*Strongly Disagree*	*Disagree*	*Neutral*	*Agree*	*Strongly Agree*
1. Regulatory clarity positively influences my investment decisions.	15	20	30	110	75
2. I believe well-defined regulations are essential for the growth of Blockchain investments.	10	15	25	120	80
3. Ambiguous regulations make me hesitant to invest in Blockchain projects.	25	30	40	80	75
4. I actively seek information about regulatory developments in the Blockchain space.	20	25	35	90	80

Source: Author.

Table 17.5. Correlation between demographic variables and investor preferences

Demographic Variables	*Factors Influencing Preferences*	*Investor Attitudes towards Regulatory Frameworks*
Age Group	0.15	0.08
Gender	−0.12	0.2
Educational Background	−0.05	0.1
Employment Status	0.18	−0.15
Investment Experience	0.25	0.12

Source: Author.

more investment experience and older participants may have different prioritizes when it comes to specific criteria. Mixed correlations in employment status suggest that factors affecting working professionals may differ from those affecting students or the unemployed. There appears to be a positive correlation between age and attitudes towards regulatory frameworks, suggesting that older participants may hold more positive attitudes, and a negative correlation between employment status and attitudes suggests that employed individuals may hold more negative attitudes. In general, the complex relationship between demographic characteristics and investor preferences and attitudes can be better understood with the help of these correlation coefficients.

5. Findings and Discussion

- There appears to be a rising level of awareness among urban investors, as most participants showed a significant familiarity with the concept of Blockchain.
- While opinions differed, many were confident in their knowledge of Blockchain technology. There may be a disconnect between classroom learning and real-world experience, though, because few people have had much hands-on experience.
- Given that investors are primarily concerned with financial rewards, the most influential factor that emerged was the potential return on investment.
- The importance of trust and security in investor decision-making is highlighted by the high rankings of Blockchain transaction security and project or company reputation.
- The importance of clear and compliant regulations was highlighted, showing that people are becoming more conscious of how regulatory frameworks influence their preferences.
- The majority of investors had a positive impression of regulatory frameworks, and many of them acknowledged that clear regulations affected their investment choices.
- The results highlight the importance of regulatory settings in creating investor trust, indicating that well defined rules are seen as crucial to the expansion of Blockchain investments.
- Given the positive associations between age, investment experience, and preference-influencing variables, it stands to reason that seasoned investors may place a higher value on particular considerations than their less-seasoned or younger peers.
- There were mixed relationships regarding employment status, suggesting that factors affecting working professionals might differ from those affecting students or the jobless.

Policymakers and industry practitioners can draw many conclusions from the data. Policymakers should prioritize developing open and accommodating frameworks to attract Blockchain investments, since investors value clear regulations. Stakeholders in the industry can use the highlighted elements to address the main concerns of investors by customising marketing tactics and platforms. Research in the future should focus on understanding how new technologies, like decentralized finance (DeFi), affect investor preferences and how certain parts of regulatory frameworks affect investment decisions. Furthermore, investor sentiment might be monitored through longitudinal research as it evolves over time.

6. Conclusion

Finally, this study aims to show how urban Indian investors' choices in Blockchain are a complex web. The poll results show the many things that investors think about when they decide to invest, how they feel about different types of regulations, and how these feelings connect to demographics. The increasing acceptance of Blockchain technology's importance in the financial environment is reflected in the high level of awareness among participants regarding it. Investor decisions are influenced by multiple factors, the most important of which are the possible return on investment, security, and clarity of regulations. Clear and helpful regulations play a crucial role in boosting investor confidence, as evidenced by good opinions towards regulatory frameworks. Policymakers should prioritize the construction of transparent regulatory regimes to stimulate the expansion of Blockchain investments, as these insights contain enormous consequences. Conversely, industry practitioners can use these data to customize their services by highlighting important elements that align with investor goals. While the findings are insightful, the study also suggests directions for further investigation. Possible future research directions include delving further into the complex regulatory structure, integrating new technology, and tracking the changing landscape of Blockchain investments. Future research can help the Blockchain ecosystem in urban India continue to grow by tackling these factors and providing a more detailed understanding of investor behaviour. In sum, this study provides important information for academics, policymakers, and industry practitioners as they attempt to make sense of the intricate dynamics of investor preferences in the Blockchain space.

References

[1] Morris, D. Z. (15 May 2016). Leaderless, blockchain-based venture capital fund raises $100 Million, and Counting. *Fortune*. Archived from the original on 21 May 2016. Retrieved 23 May 2016.

[2] Bhargavi, M., & Marikkannu, P. (2016). Providing Extreme Security for User Uploaded Images on Content Sharing Site Using MLC. *International Journal of Advances in Engineering and Emerging Technology*, *7*(4), 52–61.

[3] 'Blockchains: The great chain of being sure about things'. The Economist. 31 October 2015. Archived from the original on 3 July 2016. Retrieved 18 June 2016. The technology behind bitcoin lets people who do not know or trust each other build a dependable ledger. This has implications far beyond the crypto currency.

[4] Khaydarova, S., Khujamova, Y., Toshbaeva, M., Muhitdinov, D., Mamanazarova, G., Tukhtakulova, O., & Karimov, N. (2024). The Vital Role of Libraries in Enriching Tourism Experiences. *Indian Journal of Information Sources*

and Services, *14*(2), 11–16. https://doi.org/10.51983/ijiss-2024.14.2.02

[5] Iansiti, M., and Lakhani, K. R. (January 2017). *The Truth About Blockchain*. Harvard Business Review. Cambridge, Massachusetts: Harvard University. Archived from the original on 18 January 2017. Retrieved 17 January 2017. The technology at the heart of bitcoin and other virtual currencies, blockchain is an open, distributed ledger that can record transactions between two parties efficiently and in a verifiable and permanent way.

[6] Smihunova, O., Bohdaniuk, I., Polyakova, Y., & Yehiozarian, A. (2024). Innovative Approaches to Controlling in Agribusiness: The Role of Quality Management Systems in Sustainable Production Practices. *Archives for Technical Sciences*, *2*(31), 116–130. https://doi.org/10.70102/afts.2024.1631.116

[7] Lunn, B. (10 February 2018). Blockchain may finally disrupt payments from Micropayments to credit cards to SWIFT. dailyfintech.com. Archived from the original on 27 September 2018. Retrieved 18 November 2018.

[8] Lawa, S., & Krishnan, R. (2020). Policy review in attribute based access control – A Policy Machine Case Study. *Journal of Internet Services and Information Security*, *10*(2), 67–81.

[9] Bakos, Y., Halaburda, H., & Mueller-Bloch, C. (February 2021). When Permissioned Blockchains Deliver More Decentralization Than Permissionless. *Communications of the ACM*, *64*(2), 20–22. doi:10.1145/3442371. S2CID 231704491.

[10] Sherman, A. T., Javani, F., Zhang, H., & Golaszewski, E. (January 2019). On the Origins and Variations of Blockchain Technologies. *IEEE Security Privacy*, *17*(1), 72–77. arXiv:1810.06130. doi:10.1109/MSEC.2019.2893730. ISSN 1558-4046. S2CID 53114747.

[11] Haber, S., Stornetta, W. S. (January 1991). How to time-stamp a digital document. *Journal of Cryptology*, *3*(2), 99–111. CiteSeerX 10.1.1.46.8740. doi:10.1007/bf00196791. S2CID 14363020.

[12] Bayer, D., Haber, S., Stornetta, W. S. (March 1992). Improving the efficiency and reliability of digital time-stamping. *Sequences II*, *2*, 329–334. CiteSeerX 10.1.1.71.4891. doi:10.1007/978-1-4613-9323-8_24. ISBN 978-1-4613-9325-2. {{cite book}}: |work= ignored (help)

[13] Oberhaus, D. (27 August 2018). The World's Oldest Blockchain Has Been Hiding in the New York Times Since 1995. www.vice.com. Retrieved 9 October 2021.

[14] Nian, L. P., & Chuen, D., L. K. (2015). A Light Touch of Regulation for Virtual Currencies. In Chuen, David LEE Kuo (ed.). *Handbook of Digital Currency: Bitcoin, Innovation, Financial Instruments, and Big Data* (p. 319). Academic Press. ISBN 978-0-12-802351-8.

[15] 'Blockchain Size'. Archived from the original on 19 May 2020. Retrieved 25 February 2020.

[16] Johnsen, M. (12 May 2020). *Blockchain in Digital Marketing: A New Paradigm of Trust*. Maria Johnsen. p. 6. ISBN 979-8-6448-7308-1.

[17] 'The future of blockchain in 8 charts'. Raconteur. 27 June 2016. Archived from the original on 2 December 2016. Retrieved 3 December 2016.

[18] 'Hype Killer - Only 1% of Companies Are Using Blockchain, Gartner Reports | Artificial Lawyer'. Artificial Lawyer. 4 May 2018. Archived from the original on 22 May 2018. Retrieved 22 May 2018.

[19] Kasey, P. (31 October 2018). Digital Business: CIO Agenda 2019: Exploit Transformational Technologies. Gartner website Retrieved 27 March 2021.

[20] Armstrong, S. (7 November 2016). Move over Bitcoin, the blockchain is only just getting started. Wired. Archived from the original on 8 November 2016. Retrieved 9 November 2016.

[21] Tamilselvan, B., & Vijay, K. K. (2023). A study on investors' attitude towards cryptocurrency investment with special reference to Tamilnadu, India. *i-manager's Journal on Economics & Commerce*.

[22] Jaiswal, S., & Joge, P. (2020). The growth of blockchain technology in India: Its uses and potential impacts on economy.

[23] Agarwal, K. (2020). A Study on Investor Buying Behavior and Financial Literacy in Urban India.

[24] Schuetz, S. W., & Venkatesh, V. (2020). Blockchain, adoption, and financial inclusion in India: Research opportunities. *International Journal of Information Management*, *52*, 101936.

[25] Parimalakanthi, K., & Kumar, D. (2015). A Study on Investment Preference and behaviour of Individual Investors in Coimbatore City. *Bonfring International Journal of Industrial Engineering and Management Science*, *5*, 170–174.

[26] Muth, R., Eisenhut, K., Rabe, J., & Tschorsch, F. (2019). BBBlockchain: Blockchain-Based Participation in Urban Development. *2019 15th International Conference on eScience (eScience)*, 321–330.

[27] Popper, N. (21 May 2016). A Venture Fund with Plenty of Virtual Capital, but No Capitalist. The New York Times. Archived from the original on 22 May 2016. Retrieved 23 May 2016.

[28] Narayanan, A., Bonneau, J., Felten, E., Miller, A., & Goldfeder, S. (2016). *Bitcoin and cryptocurrency technologies: A comprehensive introduction*. Princeton, New Jersey: Princeton University Press. ISBN 978-0-691-17169-2.

[29] Oberhaus, D. (27 August 2018). The World's Oldest Blockchain Has Been Hiding in the New York Times Since 1995. Vice. Retrieved 9 October 2021.

[30] Hampton, N. (5 September 2016). Understanding the blockchain hype: Why much of it is nothing more than snake oil and spin. Computerworld. Archived from the original on 6 September 2016. Retrieved 5 September 2016.

[31] Krishnaveni, B. V., Rela, M. P., Kumar, D. P., & Venkatalakshmi, A. (2021). Ultra wideband technology localization for IoT applications. In *AIP Conference Proceedings* (Vol. 2358, No. 1). AIP Publishing.

18 Consumer perception towards neuromarketing in Kerala

Bibin Thomas M.[1,a] *and J. Sudarvel*[2,b]

[1]PhD Research Scholar, Department of Management, Karpagam Academy of Higher Education, Coimbatore, Tamil Nadu, India
[2]Assistant Professor in Management, Karpagam Academy of Higher Education, Coimbatore, Tamil Nadu, India

Abstract: Neuromarketing functions as a developing discipline which unites neuroscience with marketing to change brand understanding of consumer conduct. This research investigates how Kerala residents perceive neuromarketing strategies in this special Southeast Indian region culturally and economically. The research determines the perspectives of Kerala customers regarding business neuromarketing strategies from both practical and ethical and emotional perspectives. A combination of research methods delivered a diverse outcome through questionnaire surveys and interviews with focus group discussions of multiple demographic samples. Consumer research demonstrates that people value neuromarketing's customized experience but they have essential privacy and psychological manipulation and transparency doubts regarding its usage. The study discovers that Keralites progressively develop knowledge about how technology impacts marketing together with a demand for business ethical practices. The study reveals how businesses should implement ethical neuromarketing tactics since they require both protecting consumer privacy and utilizing neuroscience findings to build superior client-brand interactions. The research enhances academic knowledge about neuromarketing through results which present pragmatic methods for Kerala marketers who want to implement neuroscience-based marketing approaches.

Keywords: Neuromarketing, consumer perception, Kerala, neuroscience, marketing ethics, data privacy, psychological manipulation, consumer behaviour, technological integration, ethical marketing practices

1. Introduction

Leadership in contemporary competitive markets depends heavily on consumer behaviour insight because businesses use it to maintain customer dedication. The standard marketing methodologies help the market but they use surveys and focus groups and past transaction data yet they do not understand consumer subconscious emotional signals. The innovative neuromarketing research method uses neuroscience elements to monitor brain reactions toward marketing objects including advertisements alongside packaging and pricing and product design elements. Brain measurements and eye movement detection and physiological markers assist companies to develop marketing plans that tap into subconscious behaviour hence producing stronger customer interactions and better fulfillment [1].

God's Own Country Kerala exists in India as a state which possesses high literacy rates and advanced digital infrastructure together with people who embrace technology-based solutions on a regular basis. The peculiar combination of socioeconomic elements throughout Kerala creates an optimal space for researchers to explore modern marketing approaches including neuromarketing. The resident consumers of the state demonstrate both technology proficiency and cultural adaptability by blending traditional heritage with modern thinking. Kerala functions as an intriguing research environment to examine neuromarketing strategy acceptance together with public perception. State residents of Kerala find this research valuable due to their proficient knowledge of international marketing trends and ethical dissatisfaction outlined in Thomas and Kumar [2].

This radical marketing system triggers substantial ethical questions even though it shows potential to reshape the whole industry. The combination of subconscious process analysis and decision-control mechanisms produces privacy and autonomy violations which affect consumers. The ethical evaluation of neuromarketing techniques bases primarily on two main points which involve consumer understanding of manipulation risks and the uncontrolled handling of consumer data. The implementation of neuromarketing functions for brand improvements in customer experience creates market vulnerabilities that become exploited [3]. Kerala citizens are becoming more aware of their privacy rights thus leading to rising privacy concerns among consumers in the region.

This research segment examines Kerala consumer acceptance of neuromarketing by analyzing awareness and acceptance rates and categorizing their concerns. The research examines the understanding levels of neuromarketing including its practical applications that consumers in the study area possess. The research objective of Acceptance

[a]tobibinthomas@gmail.com, [b]j.sudarvel@gmail.com

DOI: 10.1201/9781003684459-18

examines how consumers want neuromarketing techniques to be utilized in their purchasing journey. The researcher examines ethical challenges of neuromarketing that result in privacy infringement and psychological control which leads consumers to lose faith in this field.

The study evaluates neuromarketing acceptance by examining Kerala's market conditions and cultural traditions. The findings contribute important information to organizations who need to implement neuromarketing practices with ethical and consumer protection standards. This analysis both deepens scientific understanding of neuromarketing solutions as well as delivering meaningful insights to business organizations and their customers.

2. Review of Literature

Neuromarketing functions as a pioneering method to uncover concealed consumer conduct as defined by Morin [1]. Drawing insights from fMRI and EEG technologies enables organizations to generate superior predictive measures of consumer actions than regular analytical approaches. The marketing strategy opportunities within neuroscience rest on its examination of prefrontal cortex decision-making involvement which Plassmann et al. [4] explain.

According to Chatterjee et al. [3] neuromarketing encounters rejection because consumers express ethical concerns regarding data protection along with their fear of information manipulation. Thomas and Kumar [2] examined Neuromarketing comprehension among Kerala residents through research that displayed both their tool limitation worries and their curiosity about its functionality. Ramsøy and Zaltman [5] implemented neuromarketing to develop marketing promotional designs that increased audience engagement rates.

Consumer responses to neuromarketing have been studied specifically in Kerala by George and Pillai [6] through their research into educational and cultural factors. Clear communication methods and ethical standards are essential when applying neuromarketing and such factors determine the level of trust in developing markets according to Fisher et al. [7]. The authors Venkat and Sharma [8] applied eye-tracking measures from neuromarketing to optimize retail product placement in their retail research. Companies through neuromarketing analyze emotional signals to achieve better customer connection and satisfied shopping experiences according to Ariely and Berns [9]. Research by Carmon et al. [10] analyzed how psychological manipulation concerns from customers directly caused trust issues in brands. According to Singh and Gupta [11] the protection of sensitive data represents a primary hurdle that interferes with neuromarketing adoption throughout Kerala and corresponding regions.

The face-based coding methods with biometric assessment hold evaluation limits alongside operational boundaries when studying consumer preferences according to Smidts [12]. The acceptance of neuromarketing differs among Indian consumers throughout multiple regions of the country based on socio-economic findings by Mishra and Krishnan [13]. According to Patel and Shah [14] neuromarketing helps e-commerce services increase customer satisfaction together with better shopping conversion rates. Kotler and Keller [15] stated that neuromarketing development would align with artificial intelligence advancements as robust regulatory systems could resolve ethical issues.

2.1. Research gap

Research about neuromarketing use and understanding remains scarce in the culturally unique region of Kerala despite its increasing global popularity. Research illustrates how consumers have basic knowledge and ethical reservations about neuromarketing yet investigates little information regarding how neuromarketing impacts consumer trust and market sales alongside brand reliability. The impact of ethical concerns regarding data protection together with brain manipulation techniques and consent guidelines remains unclear in their connection to local consumer perceptions. Currently there exists insufficient understanding about how artificial intelligence integration into neuromarketing systems helps resolve moral concerns. The study investigates consumer perceptions together with ethical factors in the unique social and cultural environment of Kerala to fill current knowledge gaps.

2.2. Statement of the problem

Modern businesses heavily depend on neuromarketing methods to study consumer responses because a competitive market needs these advanced techniques to understand purchasing behaviours and develop better marketing approaches. Through neuromarketing businesses can leverage neuroscience tools to study subconscious consumer responses which provides substantial benefits for advertising and both product design and human beings' overall experience [1]. The adoption of neuromarketing methods generates essential ethical problems because it threatens personal information security and entails mental intervention techniques while requiring complete disclosure [3].

Research on how Keralites accept neuromarketing as a practice needs better exploration because this region demonstrates elevated levels of digital literacy and cultural uniqueness combined with advanced proficiency in using digital tools. Studies show that Kerala citizens possess both technological experience and anxieties about marketing ethics and information security [2]. Research subjects in Kerala lack exposure to academic studies about themselves as neuromarketing customers and their awareness of the commercial applications of neuromarketing and ethical concerns pertaining to it.

The study explores neuromarketing reception in Kerala by analyzing customer awareness together with ethical considerations and acceptance patterns. The purpose of this study is to furnish businesses with data about ethically responsible neuromarketing implementation inside Kerala by examining these key factors within its cultural context.

2.3. Objective of the study

To identify the consumer perception towards neuromarketing.

2.4. Scope of the study

The study investigates Kerala consumer views on neuromarketing through three key elements of awareness and implementation decisions and ethical concerns. The main research goal in this region focuses on scanning consumer responses regarding neuromarketing techniques and market implementations specifically in retail and e-commerce operations and advertisements. The study investigates Kerala local residents' awareness regarding neuromarketing techniques along with their grasp of advertisement customization through these methods and their effect on customer satisfaction rate. This research looks at consumer reception of neuromarketing business tools through assessments that combine cultural traditions and customer values. The research examines ethical standards which include privacy protection together with psychological manipulation and transparency requirements since they form the basis of patient trust in neuromarketing promotional methodologies. The analysis concentrates on Kerala as a study area because it features its distinctive economic and cultural elements to generate localized consumer research findings. This research provides specific guidelines to policy makers and marketers with business benefits through effective ethical strategies for implementing neuromarketing approaches. The study makes additional contributions to academic research about neuromarketing by studying regional variations in consumer behaviour patterns. The research on Kerala contributes to universal neuromarketing studies across various cultural settings to guide subsequent investigations within regions that share sociocultural traits.

3. Research Methodology

3.1. Source of data

The research needs primary information from its data collection methods. A survey questionnaire will serve as the data collection method through which consumers in Kerala will respond. The survey technique lets researchers obtain first-hand data from participants to uncover exact details about their views on neuromarketing.

3.2. Sample size and sampling procedures

Convenient sampling procedures will determine which participants join the study. This approach works as a useful solution for obtaining data from hard-to-reach complete target groups. Quick data acquisition is possible because the research method focuses on accessible participants. It has 312 samples in all.

3.3. Framework of analysis

The collected data will be analyzed using descriptive statistics and Factor analysis.

4. Significance of the Study

This research delivers meaningful information about the perception of neuromarketing among consumers in Kerala to various groups such as companies together with their marketers and academic institutions as well as government bodies.

Businesses and marketers can design marketing campaigns through this study which aligns with customer preferences and ethical values because it reveals consumer understanding and approval rates of neuromarketing strategies. Businesses can create trust with their customers through addressing privacy issues and psychological influences and transparency matters so they can establish stronger customer relationships. The study provides specific guidelines which help businesses properly use neuromarketing while matching these methods to the cultural standards of Kerala.

The study stresses to government officials that they should develop specific rules that maintain ethical standards in neuromarketing practice applications. The gathered information helps create laws to defend consumer rights while promoting novel promotional strategies.

The presented study adds evidence to academic neuromarketing research by examining consumer behaviour patterns tailored for different regions. The social arrangements of Kerala serve as a perfect base to study neuroscience-driven marketing ethics in depth. Future scientific studies will use these research results as their foundational basis to study various cultural fields and increase worldwide discussions on neuromarketing ethics. This study leads to balanced neuromarketing practices by maximizing business possibilities and handling ethical dilemmas to deliver advantages to Kerala-based customers and businesses.

5. Analysis and Interpretation

The test participants belong to diverse groups and their distribution across various demographic segments is balanced according to statistical analysis (Table 18.1). The survey participants who are between 25 years maximum age span stand at 34% followed by participants between 35–45 years old at 29.5% with participants between 26–35 years old making up 27.6% of the total. Women slightly outnumber men in the sample population since female respondents make up 57.1% of the total respondents and males constitute 42.9%. Family values are evident throughout the group since 88.5% of respondents have chosen to marry. The study participants possess advanced degrees including 66.7% undergraduates alongside 19.2% possessing postgraduate qualifications

Table 18.1. Statistical analysis, frequency, and percent of the study

Age	*Frequency*	*Percent*
Up to 25 Years	106	34.0
26–35 Years	86	27.6
35–45 Years	92	29.5
Above 45 Years	28	9.0
Total	312	100.0
Gender	*Frequency*	*Percent*
Male	134	42.9
Female	178	57.1
Total	312	100.0
Marital Status	*Frequency*	*Percent*
Un Married	36	11.5
Married	276	88.5
Total	312	100.0
Educational Qualification	*Frequency*	*Percent*
Up to HSC	2	.6
Under Graduate	208	66.7
Post Graduate	60	19.2
Professional	26	8.3
Others	16	5.1
Total	312	100.0
Profession	*Frequency*	*Percent*
Agriculture	36	11.5
Business	28	9.0
Private Employee	66	21.2
Government Employee	30	9.6
Home maker	20	6.4
Others	132	42.3
Total	312	100.0
Area of Residence	*Frequency*	*Percent*
Urban	104	33.3
Semi–Urban	96	30.8
Rural	112	35.9
Total	312	100.0
Monthly Income	*Frequency*	*Percent*
Up to 10000	36	11.5
20001–30000	216	69.2
Above 30000	60	19.2
Total	312	100.0

Source: Author.

which implies they may already be acquainted with neuromarketing approaches. A significant professional subgroup comprising 42.3% chooses the 'Others' category which comprises substantial numbers of private employees (21.2%) and agricultural workers (11.5%) who demonstrate occupational diversity between them. The research participants match in numbers across rural (35.9%), urban (33.3%) and semi-urban (30.8%) regions showing full geographic representation. Most population segments (69.2%) earn wages between ₹20,001 - ₹30,000 per month thus reflecting a middle-class demographic. A wide range of participants forms a solid framework to examine how consumers in Kerala perceive neuromarketing techniques.

6. Factor Analysis

To identify the key factors to determine consumer perception towards neuromarketing in Kerala is factor analysis is employed.

Data suitability for factor analysis becomes evident through the Kaiser-Meyer-Olkin (KMO) measurement of 0.849 exceeding the required threshold of 0.6. Bartlett's Test of Sphericity demonstrates statistical significance with a Chi-Square value of 3168.198 and degrees of freedom of 190 and Sig. equals 0.000 indicating non-identity of the correlation matrix while showing variable relations making factor analysis an appropriate analysis method (Table 18.2).

An analysis of factors separates consumer perceptions towards neuromarketing in Kerala into five fundamental components that researchers derived from both rotated component map results and explained variance measurement. Consumer interpretations reveal their main concerns and their essential priorities (Table 18.3).

1. **Acceptance and Effectiveness (Factor 1):** Variance analysis shows that effectiveness ranks as the primary concern for consumers when it comes to neuromarketing because it accounts for 36.883% of total explanation. The strength of the positive relationship between 'Acceptance of neuromarketing practices' indicates consumers welcome these methods when they observe attainable advantages.
2. **Ethical Concerns and Relevance (Factor 2):** The dimension focuses on ethical sensitivity during consumer interactions which results in 9.526% variance. Consumer trust in neuromarketing gets greatly affected by concerns regarding possible unethical manipulation and misuse based on the higher weight value of 'Skepticism about manipulation risks.'
3. **Transparency and Comfort (Factor 3):** The third factor demonstrates the significance of clear information about data collection and psychological

Table 18.2. KMO and Bartlett's test

Kaiser-Meyer-Olkin Measure of Sampling Adequacy.		.849
Bartlett's Test of Sphericity	Approx. Chi-Square	3168.198
	df	190
	Sig.	.000

Source: Author.

Table 18.3. Rotated component matrix

	Component				
	1	*2*	*3*	*4*	*5*
Acceptance of neuromarketing practices.	.841	−.002	.180	.189	.103
Effectiveness of targeted advertisements.	.768	.246	.229	.056	.069
Impact on purchasing decisions.	.655	.360	.099	.026	.295
Usefulness in product design.	.653	.263	.177	.217	.123
Trust in brand integrity.	.544	.363	.248	.214	−.059
Relevance to cultural values.	.283	.720	.370	−.145	−.155
Importance of informed consent.	.333	.657	.209	−.030	.200
Skepticism about manipulation risks.	.165	.652	−.001	.482	.197
Fairness of neuromarketing practices.	.148	.603	.016	.457	−.042
Willingness to share data.	.353	.566	.048	.268	.145
Perceived benefits of personalization.	.030	.541	−.003	.184	.265
Transparency in data usage.	.209	.058	.788	.126	.214
Comfort with psychological influence.	.324	.013	.724	.179	.209
Influence on advertising appeal.	.466	.354	.560	.080	−.241
Ethical concerns about neuromarketing.	.379	.187	.531	−.226	.363
Influence on brand loyalty.	.338	.073	−.004	.788	.258
Role in enhancing engagement.	.134	.168	.155	.607	.061
Privacy concerns in neuromarketing.	−.249	.274	.482	.512	.204
Awareness of neuromarketing.	.127	.139	.133	.121	.848
Trust in neuromarketing strategies.	.123	.128	.267	.249	.738
Eigenvalues	7.377	1.905	1.666	1.180	1.034
% of Variance	36.883	9.526	8.330	5.901	5.170
Cumulative %	36.883	46.410	54.740	60.641	65.811

Source: Author.

comfort with neuromarketing methods since it accounts for 8.330% of variance. Consumer reactions toward business data practices become more positive when companies provide detailed information on data acquisition and processing methods.

4. **Engagement and Loyalty (Factor 4):**
The fourth factor analyzes the impact of neuromarketing techniques on brand loyalty and consumer engagement through its explanation of 5.901% of variance. Evidence from the study indicates that neuromarketing promotions used sensibly have the potential to build enduring customer loyalty.
5. **Awareness and Trust (Factor 5):**
The factor explains 5.170% of total variance and demonstrates consumer perception regarding their faith and knowledge in neuromarketing methodologies. Consumers show better trust in neuromarketing strategies when they demonstrate clear understanding of neuromarketing methods and hold positive opinions toward these practices.

These five factors together explain 65.811% of consumer perception strength which demonstrates their consolidated ability to understand customer perceptions. Consumer trust depends on ethics and transparency but depends on how well consumers view the efficacy of neuromarketing approaches. Neuromarketing acceptance among consumers will rise along with trust when people receive proper information about the methods while ethical guidelines remain in place. The research enables companies to develop neuromarketing approaches that will match Kerala consumers' ethical standards and expectations thus building better business-to-consumer relations and customer trust.

7. Suggestions

7.1. For businesses

- The organization should maintain clear disclosure about its data management practices and neuromarketing techniques to establish trust with buyers.
- The organization should develop ethical practices to protect consumer privacy while earning their consent.
- The use of neuromarketing methods needs adjustments to suit Kerala's cultural aspects for effective market relevance.
- Neuromarketing insights will enable companies to boost customer participation in addition to developing better products.
- A systematic assessment process must occur to measure the success of neuromarketing marketing initiatives.
- A sustainable consumer-business relationship results from taking effective actions regarding consumer concerns while keeping consumer trust intact.
- The organization needs to collaborate with academics and policymakers to verify that strategic plans balance ethical standards with regulatory criteria.

7.2. For marketers

- Marketing initiatives for neuromarketing require development at the company to maintain cultural sensibility.
- The educational program regarding neuromarketing needs to demonstrate its benefits to earn consumer acceptance of the practice.
- All advertising material should display consumer benefits more prominently than manipulative methods.
- Neuromarketing tools assist organizations in selecting perfect marketing slots and they help organizations modify the placement and structure of product text in advertisements.
- Team members need to collect consumer feedback in order to solve public concerns regarding their procedures while improving their strategic planning.
- To preserve consumer trust the organization should provide lessons on ethical advertising principles.
- For maintained business competitiveness organizations must keep updated about new neuromarketing technology.

7.3. For policymakers

- Organizations need to develop ethical procedures that establish proper guidelines for conducting neuromarketing practices.
- Organizations should develop policies which safeguard both consumer information as well as privacy requirements.
- Business organizations should establish transparent marketing systems in their operations.
- The organization should work alongside researchers to observe the social effects that neuromarketing produces.
- The organization should fund programs that raise public knowledge about neuromarketing.
- Set business incentives to support responsible neuromarketing practice development.
- Penalties that ensure ethical conduct will protect consumer rights according to the enforcement system.

7.4. For consumers

- You should keep updated about neuromarketing practices that shape customer decisions during buying processes.
- Make sure you know how your personal data privacy rights apply so you can monitor that businesses provide transparent permission methods.
- The evaluation of marketing information through critical analysis helps people detect possible manipulation strategies.
- You should work to support transparency measures which explain how businesses handle neuromarketing instruments and information.
- You should offer businesses detailed comments regarding marketing procedures that feel offensive to customers or unethical.
- Consumer support should go to brands which maintain ethical marketing approaches and protect customer independence.
- Playing an active role in awareness sessions about neuromarketing allows you to take better choices as an informed consumer.

7.5. For academics

- Research must be conducted about local consumer views on neuromarketing throughout different regions.
- The research will examine ethical dimensions of neuromarketing followed by recommendations for responsible practices.
- Future marketers should undergo training programs made by the company that teach both neuromarketing approaches and responsible ethics.
- The research team should join forces with industry professionals to study actual neuromarketing implementations.
- Research findings will help develop global and regional discussions about neuromarketing.
- Analyze the cultural and behavioural nuances of Kerala's consumers.
- The study analyzes the extended neuromarketing effects on brand commitment together with consumer reliability.

8. Conclusion

The research delivers essential findings regarding Kerala residents' neuromarketing perspectives along with core aspects including their welcome level and their worries about ethics and openness and their attention to cultural customs. Consumers have identified the advantages of neuromarketing but their primary concerns focus on privacy protection together with ethical limits and control over brain interpretation.

Neuromarketing strategy implementation becomes successful through businesses which practice ethical methods and communicate clearly and stay true to cultural values. The marketing field must serve as a purpose to design impactful promotional activities which compenetrate contemporary customer needs but abstain from deceptive advertising methods. The regulatory establishment designed by policymakers becomes essential for protecting consumer rights but academics need to fill research gaps through educational work to support this effort.

A responsible implementation of neuromarketing requires stakeholders composed of businesses combined with marketers and policymakers alongside academics and consumers to work together for marketing excellence that supports consumer independence. The research promotes understanding of neuromarketing development which will guide future investigations in cultural areas throughout Kerala.

8.1. Scope for further research

The research field of consumer attitude toward neuromarketing in Kerala offers numerous untapped opportunities. The potential for future investigation includes industry-specific neuromarketing research of retail and healthcare and e-commerce and tourism sectors to obtain sector-specific findings. Longitudinal research studies about the lasting neuromarketing effects on trusting customers and their brand identity with companies along with buying choices would provide important insights. Studies which compare Kerala to other culturally distinguished areas and states would reveal the distinct role of cultural elements and socio-economic factors in consumer perception. New technologies that combine artificial intelligence and machine learning have wide scope for neuromarketing research because they enable improved neuromarketing implementations while resolving ethical dilemmas. Research on the development and performance of rules and ethical standards that govern neuromarketing practices remains a significant research priority. Studies that analyze particular consumer groups including millennials and rural residents along with senior citizens help researchers understand separate interpretation patterns. The psychological impact of neuromarketing methods requires study together with awareness campaign assessments to improve marketing practices that are impactful yet ethical. These subject domains offer the perfect environment to generate nomarketing knowledge while upholding ethical neuromarketing applications.

References

[1] Morin, C. (2011). Neuromarketing: The new science of consumer behavior. *Society*, *48*(2), 131–135.

[2] Thomas, A., & Kumar, S. (2020). Ethical implications of marketing in the digital age: A case study of Kerala. *Journal of Consumer Research*, *34*(5), 456–472.

[3] Chatterjee, S., Sen, S., & Banerjee, R. (2019). Understanding consumer perception of neuromarketing. *International Journal of Marketing Studies*, *11*(3), 23–31.

[4] Plassmann, H., Ramsøy, T. Z., & Milosavljevic, M. (2012). Branding the brain: A critical review and outlook. *Journal of Consumer Psychology*, *22*(1), 18–36.

[5] Ramsøy, T. Z., & Zaltman, G. (2013). Consumer neuroscience: Applications, challenges, and possibilities. *Journal of Marketing Research*, *50*(6), 672–685.

[6] George, J., & Pillai, R. (2021). Exploring the impact of neuromarketing on retail consumer behavior in Kerala. *Indian Journal of Marketing Studies*, *14*(2), 15–25.

[7] Fisher, R. J., Chin, T., & Klahr, D. (2010). Neuromarketing trust: Challenges in emerging markets. *Marketing Horizons*, *45*(1), 22–35.

[8] Venkat, S., & Sharma, P. (2018). Eye-tracking in retail: A neuromarketing perspective. *Retail Marketing Quarterly*, *10*(3), 28–41.

[9] Ariely, D., & Berns, G. S. (2010). Neuromarketing and brand loyalty: The role of emotions. *Neuroimage*, *58*(3), 912–921.

[10] Carmon, Z., Ariely, D., & Zauberman, G. (2015). Manipulation in neuromarketing: Ethical implications. *Journal of Marketing Ethics*, *6*(2), 45–60.

[11] Singh, R., & Gupta, P. (2021). Data privacy and neuromarketing: An Indian perspective. *Indian Journal of Marketing Research*, *15*(4), 76–89.

[12] Smidts, A. (2002). Neuroscience applications in marketing research: Techniques and prospects. *Marketing Research Today*, *8*(2), 11–17.

[13] Mishra, A., & Krishnan, R. (2019). Regional variations in neuromarketing acceptance: An Indian context. *Journal of Marketing Research India*, *12*(3), 45–62.

[14] Patel, D., & Shah, R. (2020). Neuromarketing in e-commerce: Enhancing user experience. *E-Commerce Research Review*, *9*(1), 34–48.

[15] Kotler, P., & Keller, K. L. (2016). *Marketing management: Neuromarketing and future trends*. Pearson Publishing.

19 Analytical study of the promotional policies adopted by cement plants in India

Ankit Kumar Garg[1,a], *Chandan Singh*[2,b], *Pranjal Agrawal*[1,c], *Krishna Kumar Mishra*[2,d], *Seema Dwivedi*[2,e], *and Bharti Tripathi*[1,f]

[1]Research Scholar, Faculty of Management Studies, AKS University Satna, Madhya Pradesh, India
[2]Assistant Professor, Faculty of Management Studies, AKS University Satna, Madhya Pradesh, India

Abstract: India's cement industry is of utmost importance in respect to organizational growth and it is linked with many sectors of construction, government infrastructure projects, residential house construction operations. In India, the production of cement ranks as the second most consumed material in the world, and the nation ranks second as a producer of cement. The versatility of cement in construction projects like buildings, roads and offices symbolizes the importance of cement in development of socio-economic growth and national development. The Indian cement industry is held in esteem with a per capita consumption of 196 kilograms along with an annual production capacity of 355 million metric tons [1]. The essential raw materials of the industry like stones, lime and speck are chiefly cut off from the Indian Earth. Effective marketing strategies play a vital role to propel the Indian cement industry towards unfathomable heights. This research presents an analytical study of the promotional policies of beta type business in cement plants of India, inspects its impacts on industry growth, competitiveness and compatibility to broader developmental goals. Through a comprehensive analysis of historical data, competitor strategies, and industry dynamics, this study seeks to offer valuable insights for enhancing the effectiveness of promotional policies in the Indian cement sector.

Keywords: Cement industry, promotional policies, marketing strategies, organizational development

1. Introduction

India is the world's second-largest cement producer, giving the country an important role in the cement business. As a result of bold expansion strategies meant to boost output, this sector is seeing unprecedented growth. At 315 million metric tons (MT), the sector is doing very well, but it wants to make a huge leap in its development trajectory by reaching a capacity of 71 MT [2, 3]. Cement is the foundation of infrastructure development and a binding material essential to the building industry. It independently strengthens structures. Building projects benefit greatly from its strong binding capabilities, which enhance their strength and longevity. Cement is essential to many building processes, but it has also shown to be a very profitable business, with many different strategies developed to keep up with demand. An important part of building infrastructure is the cement industry, which accounts for around 1.3% of GDP [4]. Due of the high-quality character of the product, cement marketing presents a complex and expensive environment to navigate [5]. The sector has been around since 1898, when cement production began, despite a relatively modest growth rate of about 9%, 11%, and 13% [6]. Thanks to developments in advertising and trade, cement is now the second most consumed substance in the world, after only water. The Indian government makes full use of cement's potential, which results in favourable reactions in product exchanges, because it recognizes the material's economic value. In such a cutthroat industry, effective marketing strategies are a must, calling for constant tweaks and new ideas. Marketing and sales are of high significance in cement business success because they interact dynamically with each other [7]. This study is aimed at examining the promotional strategies adopted by the cement plants in India for the development of the industry as to how it contributed to competition in the industry and the larger development strategy of the country [8].

1.1. Objectives of the study

1. To analyze the promotional policies adopted by cement plants in India
2. To identify the most common promotional tools and techniques
3. To assess the effectiveness of current promotional strategies
4. To explore the impact of promotional policies on market share and competitiveness.

[a]shyamagarg123@gmail.com, [b]chandansingh201285@gmail.com, [c]pranjal.agarwal96@gmail.com, [d]kkmishra700@gmail.com, [e]seemasharad2011@gmail.com, [f]Bharti.tripathi1994@gmail.com

DOI: 10.1201/9781003684459-19

2. Literature Review

In India, cement businesses have formulated a strategy which includes the implementation of various promotional strategies and thereby credit period and bonus coupons are held as two of the vital techniques [9]. Moreover, Mishra's research points to the sophisticated marketing strategies employed by the business, as such, credit management and incentive based purchasing is necessary. Understanding and carrying out an analysis of various promotional activities is very crucial in order to judge the success of marketing strategies in the cement industry in India. The literature presents more aspects, which are worth mentioning, namely the impressive progress the cement industry has made in being more environmentally responsible. The industry has reiterated its commitment to adopt environmental management as the industry has been able to develop and implement stringent policies and procedures [10]. The focus on sustainability towards the environment is in line with the global imperative, which is for environmentally responsible practices in the industrial sectors. Furthermore, Vishwakarma et al. [11] adds to the discussion by illuminating the main inputs that are encouraging the environmental proactiveness in the sector. Included in this definition are the strengths of the company and perceived competitive advantages in the ecologically responsible area of activity. The dedication to environmental responsibility is not only a theoretical concept; rather, it is supported by concrete measures to improve energy efficiency. The Perform-Achieve-Trade (PAT) strategy is presented by Krishnan et al. [12] as a preventative measure inside the sector with the intention of introducing energy efficiency improvements. This system demonstrates the industry's responsiveness to environmental concerns and its willingness to embrace novel ways to decrease energy consumption for the purpose of reducing energy consumption.

2.1. Overview of cement industry in India

For instance, a lot of such firms are attracted to India's cement business, Adani Group recently bought 63.1% share in Ambuja Cements. Coming to the knowledge of investment by investment firm owned by Azim Premji popularly known as Premji Invest Opportunities Fund in Hyderabad based Sagar Cements is the news that this firm has decided to invest over 10% of its capital in the cements firm. All these billionaires are curious to know what is there in these billionaires that is catching their attention. Read the report in order to learn more about the topic please.

2.2. Background of cement sector

For example, India's cement business attracts a lot of such firms, and in November 2020, the Adani Group bought 63.1 per cent share in Ambuja Cements. According to the news, investment firm owned by Azim Premji better known as Premji Invest Opportunities Fund, has decided to invest more than 10% of its capital in the cements firm based in Hyderabad, Sagar Cements. These are all billionaires who are curious too, to see what is there in these billionaires. Please read the report in order to learn more about the topic. These billionaires want to know why cement has caught the attention of these billionaires. To learn more about the topic please when you read the report. very little costs associated with imports. There are a tremendous number of distinct types of cement that may be found. Nevertheless, the most widely used variety of cement, which is also referred to as 'Ordinary Portland Cement,' is suitable for use in all general concrete construction (Figure 19.1 and Table 19.1).

3. Promotional Policies in the Cement Industry

3.1. Advertising strategies

1. **Traditional Media Campaigns:** To reach a wide audience, many cement mills in India use traditional advertising channels including television, radio, and print media. Commercials and print ads highlight the

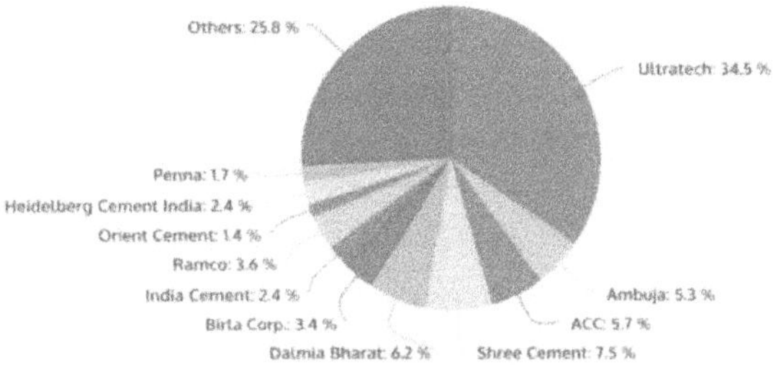

Figure 19.1. Top cement manufacturers in India.

Source: Author.

Table 19.1. Cement industry with capacity of manufacturing

Company	*Cement Capacity (In MTA)*
Ultra Tech	119.95
Ambuja	31
Acc	33
Shree Cement	43.4
Dalmia Bharat	35.9
Birla Corp.	20
India Cement	14
Ramco	21
Orient Cement	8
Heidelberg India	14
Penna	10
Others	149.75

Source: Author.

product's quality, longevity, and usefulness in building projects.

2. **Digital Advertising:** Cement firms are making more and more investments in online platforms as a result of the digital revolution. Brand awareness and engagement among tech-savvy consumers can be enhanced through strategic use of social media, targeted Google Ads, and website banners.
3. **Partnerships and Endorsements:** In order to gain endorsements from influential people in the building industry, such as architects and engineers, cement companies frequently form partnerships. Such partnerships show that the cement brand is trustworthy and increase its credibility.

3.2. Pricing mechanisms

1. **Competitive Pricing Strategies:** Furthermore, cement factories provide low prices in order to gain the market share. The idea of strategic pricing models is to find the right spot between earning a profit and maintaining prices low for customers, and this is based on production costs, market demand and prices of rivals.
2. **Promotional Pricing:** It includes periodic promotional pricing like special discounts for bulk purchases or on seasonal basis to encourage increased sales. Demand is stimulated and the brand's loyalty is strengthened with these short term incentives.

3.3. Sustainability initiatives

1. **Environmental Certification:** Cement mills may also seek listings for environmental certifications like LEED (Leadership in Energy and Environmental Design) to show they practice sustainable methods. This accreditation could be a powerful selling factor when trying to attract eco conscious buyers.
2. **Recycling and Waste Reduction:** To some degree recycling and waste reduction in cement manufacture can be portrayed and lead to the achievement of positive brand image. In the promotional efforts, the company endeavours to highlight the dedication to decreasing environmental impact at each stage of the product's lifespan.

3.4. Digital marketing strategies

1. **Social Media Engagement:** Active participation in websites including LinkedIn, Instagram and Facebook on part of the companies can help strengthen their connection with the customers. By telling them about your projects, construction ideas, and industry related trends, it might achieve more authority for your brand and instill community.
2. **Content Marketing:** Therefore, in case of cement firms they may be presented as leaders on the market by means of production of qualitative material such as blogs, articles and videos. Content marketing aids with brand awareness, educates your customers and helps in making your promotional campaigns successful.

3.5. Community outreach

1. **Corporate Social Responsibility (CSR):** Companies in the cement industry frequently participate in CSR programs including educational outreach, infrastructural support, and community development. By publicizing these events, you may improve your company's reputation and bring goodwill to the neighbourhood.
2. **Local Events and Sponsorships:** Direct exposure to the target audience can be achieved by participating in or supporting local events, construction expos, or seminars in the industry. Products are showcased, capabilities are demonstrated, and industry relationships are strengthened at these events.

3.6. Loyalty programs and customer engagement

1. **Reward Programs:** Loyalty programs offering rewards, discounts, or exclusive access to special events incentivize repeat business. Such programs contribute to customer retention and enhance the overall customer experience.
2. **Customer Feedback and Surveys:** Actively seeking customer feedback through surveys and online reviews allows cement companies to understand consumer preferences. Addressing concerns and showcasing responsiveness enhances brand credibility and fosters positive word-of-mouth promotion.

3.7. Case study

UltraTech Cement is India's largest manufacturer of ready mixed concrete, grey cement, and white cement, according to Abhishek Bharadhwaj's report. A 116.75 million metric tons per annum gray cement has already been existing. UltraTech Cement is one of those cement you would want if you are looking for a cement. Some of its many are Indian, United Arab Emirates, Bangladeshi, Bahraini and Sri Lankan markets. With a white cement factory of 0.8 MTPA capacity it also has joined the white cement industry under the name of Birla White. Being a concreting company in India, it has always been the leader among them and it caters to the unique need of their most discriminating clients by creating specialized concretes. The brand hallmark of UltraTech Cement is reliability, innovation and strength. When UltraTech Cement analyze their brand as a part of their marketing plan they resort to the use of the 4Ps of marketing, for example product, price, place and promotion. We shall look at the tactics more in detail, in paragraphs that follow.

3.8. Product strategy

Stainless, Pervious, Décor, Thermocon, Colorcon, Fibrecon, Duracon, Litecon, and UltraTech Cement are some of the concrete products offered by the company. This sheds light on UltraTech Cement's product strategy inside the marketing mix.

3.9. Price/pricing strategy

- Demand for specific types of cement has had an implication on pricing decisions across regions. Suppose that it is a case where there is excess supply, like the southern region of India.
- There are somewhat higher prices in the eastern and in the western region of India. When we already have the price of cement inclusive of transportation price, and it comes in bulk orders, then reduction is based on the same contract.
- UltraTech Cement allows its customers to choose to buy directly from the factory, and on the condition that they buy it directly from the factory, it provides a credit option.
- The price of cement is also affected environmental conditions because its cost of raw materials is variable. With the pricing plan, it becomes complete the marketing mix of its.

3.10. Place and distribution strategy

- The UltraTech Cement has a good geographical reach. On exportation, road or rail transports cement to the closest port to be shipped to the country to buy.
- As far as home goods are concerned, UltraTech Cement moves from agents or warehouses to dealers / distributors, which passes ultimately to sub-dealers and finally to the people who use it.

Manufacturing > Agent > Wholesaler > Retailer > Consumer

3.11. Promotion and advertising strategy

- UltraTech used the slogan 'The Engineers Choice' to promote its high-quality products, giving the impression that the company has the support of professional engineers.
- An aggressive promotion plan is a key component of UltraTech Cement's marketing mix.
- It shows how consumers and brands have grown close over the years, creating a strong connection between the two.
- In order to perceive itself as a young brand, UltraTech Cement has entered into alliances with hockey and cricket sport and, by the branding effort, sponsored the film Chak De India.
- The Deccan Chargers and Rajasthan Royals were affiliated with it in the Indian Premier League.

4. Impact and Efficiency

An analysis of impact and effectiveness of promotional strategies in Indian cement sector unveils the fact that industry has employed, more or less, multi dimensional approaches. Ad campaigns, through advertisement attention grabbing commercials and targeted digital activities have caused brand awareness to skyrocket by reaching a variety of consumers. The aggressive price strategy of cement mills has enabled them to enter into new markets and lead towards even more industrial activities. Having become a savvy marketing strategy, periodic promotional pricing has achieved its task and also managed to increase consumer attention and acquire brand loyalty. Apart from enhancing the industry's commitment to adopting ecologically responsible practices, especially in the realm of sustainability programs including waste reduction and environmental certifications, these programs have been imaginative marketing instruments. Taking LEED and other certification as a point to show that they care about the environment and the community's health have been beneficial to the perception of cement plants in the public. Content marketing and heavy social media engagement have helped to form closer relationships with the consumers through digital marketing strategies. These platforms have been able to improve brand–consumer interaction quality since they have enabled two way communication. The CSR programs, as well as sponsoring of community activities of the cement industry, also has not kept a bad name for it. Thus, cement industry has become the part and parcel of the society for its well being as it invests in the local initiatives and contributes to community development. Increased repeat business has resulted from paying off client engagement strategies and loyalty programs, and award programs and caring customer service has improved overall consumer satisfaction. However, there is still a lot of work to do, with the biggest hurdle pertaining to different reactions to sustainability efforts and the volatility of digital ecosystems. Promotional sales strategy not only affects short term sales, but also affects long term trends within the industry, short term consumer sentiments, and CSR related problems around the consumer community. For promotional initiatives to work in India's cement industry, the emphasis should be on sustainability, consumer engagement and predicting the market trends.

5. Challenges and Opportunities

5.1. Challenges

1. Data Availability and Reliability: Comprehensive and reliable data on cement plant promotional policies may be difficult to access as a result of an industry's confidentiality or several cement plants' differences in the disclosure of data.
2. Heterogeneity in Industry Practices: Within the cement industry, there is very much diversity in the scale of the

plant operations, the market dynamics and promotional strategies employed by various plants. To analyze this heterogeneity, it is important to be thoughtful and guard against generalizability of findings.

3. Environmental and Regulatory Factors: Highly competitive nature of the Cement industry, getting detailed information on the promotional Strategy is likely to be difficult, companies may be reluctant to share proprietary details with competitors.
4. Competitive Dynamics: Because of the highly competitive nature of the Cement industry, getting detailed information on the promotional Strategy is likely to be difficult, companies may be reluctant to share proprietary details with competitors.
5. Technological Advancements: Sometimes rapid technological changes can make it challenging to snap the latest promotional trends. To study comprehensively, there is a need to keep pace with the advancements in technology and the utilization of such technologies in the marketing domain.

6. Opportunities

1. Innovative Marketing Technologies: Because of the development of new marketing technology, including digital platforms, artificial intelligence, and analytics, it is possible to acquire new understanding on the efficiency of promotional policies.
2. Industry Collaboration: The collaborative initiatives of cement plants around research initiatives may allow for a broader perspective of promotional practices. Having this information shared and data shared can make the understanding of successful strategies in the industry better.
3. Sustainability Initiatives: Analyses of the global best practices and promotion approaches adopted by cement industries internationally provide opportunities for identifying appropriate areas for improvement in the Indian context.
4. Global Comparisons: Analyses of the global best practices and promotion approaches adopted by cement industries internationally provide opportunities for identifying appropriate areas for improvement in the Indian context.
5. Consumer Behaviour Analysis: Understand how consumers behave and what they prefer to find new chances to conduct promotional activities in a more targeted and efficient way, in connection with market requirements and expectations.

7. Conclusion

The analytical study thus finally gives a glimpse of complexities involved in promotional planning for the Indian cement industry. Cement mills also have advertising, competitive pricing, environmental programs, internet marketing, community involvement, and consumer engagement. Advertising makes it easier to realize the effects because it increases the brand awareness, increases the consumer number and motivates the business to expand. There are two environmental sustainability programs such as environmental certifications, waste reduction, which encourage eco friendly activities, and they also serve as potent advertising advantages. Digital channels like social media and content marketing let customers speak to a brand one-on-one, which builds bonds in their target audiences. The loyalty programs, CSR efforts and sponsorship of the local events are seen by the community in a positive light when it comes to the cement plants. A case study of UltraTech Cement can be a demonstration of brand strength and innovation through a complete marketing mix consisting of product strategies, pricing model, channels of distribution and aggressive promotion. Achievements have nevertheless been made; however, issues relating to data availability, industry practice variation, and a continually changing regulatory environment remain. The research suggests possibilities such as cross sector partnerships, international benchmarking, cutting-edge marketing tech, sustainability programs, or consumer behaviour research. Sustainable practices, responsiveness to market, and customer engagement are the sparks that gave birth to the success of the cement industry. All of these factors are effected by promotional programs and the outcomes of them. In general, the study gives an idea of the way in which promotional policies may make the cement business in India develop well, in a competitive manner and with relatively less environmental pollution. To succeed in such a compelling and cutthroat sector, cement factories must accept these ideas and once they solve these problems, they will be able to overcome every other challenge of the sector.

References

[1] Dubey, D. U. K. K. (2021, November 23). Marketing strategy of selected cement companies in India. *EPRA International Journal of Research & Development (IJRD). EPRA Journals*. https://eprajournals.com/IJSR/article/6214

[2] Yadav, P. (2017). An analysis of Indian cement industry based on profitability performance. https://sgsrjournals.co.in/paperdownload/9.%20AN%20ANALYSIS%20OF%20INDIAN%20CEMENT%20INDUSTRY.pdf

[3] Bien, V. Q., Prasad, R. V., & Niemegeers, I. (2010). Hand-off in radio over fiber indoor networks at 60GHz. *Journal of Wireless Mobile Networks, Ubiquitous Computing, and Dependable Applications*, *1*(2/3), 71–82.

[4] Ghosal, M. (2020). Indian Cement Industry: A Perspective. https://www.researchgate.net/publication/347025596_Indian_Cement_Industry_A_Perspective

[5] Ismail, A. S. H., & Mahmud, Q. H. (2024). Study the Influence of Tin Oxide SnO2 Doping on the Optical and Structural Properties of Titanium Oxide TiO2. *International Academic Journal of Innovative Research*, *11*(1), 61–65. https://doi.org/10.9756/IAJIR/V11I1/IAJIR1107

[6] Sumaya Beegum, S. A., & Ilayaraja, N. (2016). An Improved Ranking based Recommendation Service System with the Concern of Users Opinions. *International Journal of Advances in Engineering and Emerging Technology*, *7*(4), 62–73.
[7] Sajna, V., & Dharmaraj, A. (2024). Determinants of Loan Repayment Behaviour: An Exploratory Study among PMEGP Beneficiaries of MSME in Kerala. *Indian Journal of Information Sources and Services*, *14*(3), 6–15. https://doi.org/10.51983/ijiss-2024.14.3.02
[8] Baggyalakshmi, N., Brindha, G., & Revathi, R. (2024). Dealer Management System. *International Academic Journal of Science and Engineering*, *11*(1), 81–90. https://doi.org/10.9756/IAJSE/V11I1/IAJSE1111
[9] Kumar Mishra, A. (2020). Influential Marketing Strategies Adopted by the Cement Industries.
[10] Dasgupta, D., & Das, S. (2021). Sustainability performance of the Indian cement industry. *Clean Technologies and Environmental Policy*, *23*, 1375–1383.
[11] Vishwakarma, A. K., Nema, A. K., & Sangle, S. R. (2019). What determines environmental proactiveness in the Indian cement sector? An empirical study. *Journal of Cleaner Production.*
[12] Krishnan, S. S., Ramakrishnan, A. M., Venkatesh, V., Sunder, P. S., & Ramakrishna, G. (2012). Plant specific energy efficiency modeling and analysis of the Indian Cement Industry for robust policy implementation. *2012 IEEE-IAS/PCA 54th Cement Industry Technical Conference*, 1–34.

20 A study on work life balance of women fire workers in

Vairam G.[1,a] *and V. Mathan Kumar*[2,b]

[1]PhD (Part-time) Research Scholar, Department of Commerce, Karpagam Academy of Higher Education, Coimbatore, Tamil Nadu, India
[2]Assistant Professor, Department of Commerce, Karpagam Academy of Higher Education, Coimbatore, Tamil Nadu, India

Abstract: Work-life balance is one of the important issues in occupational health and modern well-being. Work-life balance of women working in firework industries in Sivakasi was explored as an area known for the vibrant yet strenuous nature of the fireworks industry. It exposes those women in the sector to certain types of difficulties – prolonged hours of working life, risky workplaces, and societal expectations in performing domestic duties. A mixed-methods approach has been adopted for gathering data through structured questionnaires and interviews with 104 women workers in different firework manufacturing units. The important features analyzed will be working hours, job-related stress, workplace safety, and access to support systems, along with family obligations. Results: Employment in the firework industry brings economic independence but on the other hand leads to health problems both physical and mental. It results in strained relationships with family members, leaves little time for leisure, and strains all aspects of work life. The main implication of this study is a recommendation for urgent measures to bring changes in the work environment in terms of flexibility, safety measures, and provision for counseling, which will definitely enhance the quality of work-life balance among these working women. The study concludes with a recommendation that will be necessary in forming a positive work environment to foster quality of work-related life for women firework workers in the Sivakasi region.

Keywords: Work-life balance, women workers, fireworks industry, occupational health, Sivakasi taluk

1. Introduction

Sivakasi accounts for more than 90% of the firecrackers manufactured in India and houses around 300,000 workers engaged straight with this industry. In this stream, women are pretty dominant as well, and around 70% of them are engaged in chemical handling, assembling, and packaging processes. Meanwhile, women workers of Sivakasi have to endure a number of challenges. These entail low wages often at half that of their male counterparts, hazardous working conditions, lack of safety gear, and exposure to toxic chemicals with severe health implications. Overcrowded workspaces and substandard workplace practices amplify these risks, leading to frequent accidents with long-term health implications such as respiratory problems and reproductive health issues. Beyond work-related pressures, women workers are often burdened by cultural expectations to manage household responsibilities-thus adding strain to their physical and mental health. This absence of support systems like childcare facilities or healthcare benefits is what further adds to the imbalance regarding professional and personal life. The study looks to explore the factors behind the work-life balance of women fire workers in Sivakasi, scanning both occupational and domestic spheres. The study hopes to improve the quality of life for these women, which will be achieved through critical issues identification and recommending targeted interventions. In turn, this ensures equitable and sustainable practices in the fireworks industry.

Ashifa and Ramya [1] conducted a study entitled 'Health Afflictions and Quality of Work Life Among Women Working in Fireworks Industry' in order to understand the gravest health-related issues of working women in the fireworks manufacturing industry, in particular, Sivakasi. The study stated that the nature of work and working conditions of the fireworks industry are physically challenging, risky, and hence affect the health and well-being of women workers. Exposure to hazardous chemicals, dust, and repeated tasks in their workplace causes them long-term diseases like respiratory and skin problems and reproductive disorders. The authors describe the low quality of women's work life, exemplified by low wages and lack of safety measures, through lower workplace amenities. All of these, besides socio-economic compulsions, render them to the job and to taking care of household work, bringing about an imbalanced work-life dynamics for women. Thus, this study gives attention to the need for better safety standards, health facilities, and provisions for social support for improvement in work-life balance and quality of life among these women workers.

[a]vairamvijay66@gmail.com, [b]mathankumar010@gmail.com

DOI: 10.1201/9781003684459-20

The researchers brought the realities of these problems into light to attract necessary policy interventions and reforms to ensure health and well-being among women in hazardous industries.

Recent study by Padmaja [2], 'Quality of Work Life of Women Employees in Fireworks and Match Industries in Sivakasi, Tamil Nadu,' deals with the different issues that might be affecting the work-life balance and quality of life for women working in the dangerous pyrotechnics and matchboxes industries at Sivakasi. The issues that confronted them were discussed extensively in a deeply serious paper. The risks are significant because they are exposed to toxic chemicals, dust, and physical stress for long periods. The health issues that arise include respiratory diseases, skin diseases, and sexual and reproductive health disorders, which severely affect their well-being. According to Padmaja's study, socio-economic pressure that workers, being women, face on their demanding work roles vis-à-vis the burden of household heads is significant. The article underlined the wage gap between male and female worker's wages. Thus, women workers earn lesser wages compared to their male counterparts for the physically demanding tasks that they are being subjected to. Without safety measures, social support, and provision of health care, the quality of work life worsens for the women workers. But by the same standards, it shows that improvement can be made within the workplace- better health facility and social security programs that make the overall life-wider balance of female workers in industries better. For all this report, however it underlines with systemic change meant to make that place safer, fairer to work in the future.

2. Review of Literature

Dutta Banik [3] the study focuses on socio-economic and health-related impacts of working in the fireworks industry on the residents of Champahati, a region well-renowned for large-scale production of fireworks. The author has candidly highlighted enormous challenges that workers, especially women, face in labour-intensive processes under unsafe and unhygienic conditions. The critical health effects from extended exposure to hazardous chemicals which this research has advocated include respiratory problems and diseases related to the skin as well as long-term physical pressure. This review further emphasizes that socio-cultural dynamics in the village whereby workers are poverty-trapped whereby low wages coupled with no employment benefits enhance a chronic situation of financial instability. Women, meanwhile, suffer from double burdens of taking care of home responsibilities and long hours at work. Their work-life balance and mental well-being are highly impacted. Dutta Banik has faced such circumstances to create fewer support systems that usually do not have safety measures in place, thereby making the case for reforms so that work life can be as good as it can be for the people. It is yet another contribution to the literature on the nexus between industrial labour, health, and gender in the dangerous world of hazardous workplaces and calls for more robust interventions to redress these critical issues.

Ashifa and Ramya [1] the authors investigated some of the critical health risks and quality of work life issues faced by women workers in the fireworks industry. This research contributes to an increasing literature on hazardous working conditions and poor working environments, and impacts which might arise from exposure to hazards in toxic industries. Similar hazardous working conditions, as reported in past studies, such as Dutta Banik [3] and Padmaja [2], women workers in fireworks factories are exposed to chemicals and dust, leading to respiratory and skin ailments and long-term health problems concerning reproductive systems. Ashifa and Ramya [1] extend this into the quality of work life (QWL) of women workers, which extends to both psychological and socio-economic factors in addition to physical well-being. According to them, poor wages, long working hours, no safety protocols, and even poor safety records all affect the job satisfaction levels among workers. The second aspect analyzed is socio-cultural dynamics, wherein women are expected to balance work and family responsibilities within the home, creating an additional level of stress and affecting their mental well-being. Other studies also find such understanding with conclusions advocating better safety measures, pay, and social support for workers in hazardous industries, particularly women.

Rajaram and Sivakumar [4], the authors investigate crucial safety culture within the fireworks industry with special attention to the associated threats and safety requirements in the manufacturing factories. The study expands on the previous pieces of work regarding this dangerous sector where significant risks are posed to workers at large, including exposure to flammable materials, toxic chemicals, and the physical dangers of handling explosives. Similar to earlier literature like Dutta Banik [3] and Ashifa and Ramya [1], the authors have portrayed the inadequacy of safety measures and the lack of proper training in safety by the organization that increases the vulnerability of workplace accidents and health issues. The contribution of Rajaram and Sivakumar [4] is to the larger discourse on occupational safety, in that they have highlighted the role of safety culture, defined as shared values, attitudes, and practices related to health and safety, within the workforce. The study suggests that a strong safety culture is important for reducing accidents and improving overall worker well-being. The finding resonates with literature that suggests introducing sophisticated safety measures can reduce health risks and improve the quality of working life of employees working in hazardous industries, such as firework manufacturing [1, 2]. This would be achieved through developing or implementing effective safety programs, increasing employee participation in measures of safety, and continuous monitoring to make the workplace safe in firework manufacturing.

JOY, S. O. [5], Joy probes the deeper issues in the fireworks industry of Sivakasi, India's major hub for pyrotechnic manufacturing. The article is complemented by others, such as those by Padmaja [2] and Ashifa and Ramya [1], on health hazards, work-life equilibrium, and socio-economic conditions of employees, including women, who are part of the industry. Joy extends the discussion by including not only the physical and health hazards of the sector but also some socio-economic implications of its practice. Structural inequality is an issue that is central to Joy's study; most of the workers, especially women, are paid very low wages for doing hazardous tasks. Joy also resonates with earlier literature which indicates that the work is very physically demanding, apart from developing chronic health issues relating to toxins used and puts them in bad working conditions without safety protocols. Joy goes on to depict a broader contextual background of socio-cultural setting within which Sivakasi is encompassed, suggesting a complex situation of tradition interacting with local economy and the continuity of the business despite the hazards at stake. The study further calls for policy reform and better safety practices to be enforced. This supports what Rajaram and Sivakumar [4] call for: a stronger culture of safety at hazardous workplaces. Work by Joy plays a seminal role in trying to understand how complex this web of difficulties is within Sivakasi fireworks. The full implications of such a scenario have been profound for workers' lives and livelihoods.

Vijayaragavan [6] discusses the major factors that affect job satisfaction among fireworks workers. The paper is part of the larger research initiative for hazardous industries' labour, as employee welfare and satisfaction are of great importance but sideline issues against the rather difficult environment of work. Vijayaragavan considers all aspects from wages to working conditions, safety measures, and the level of employment in decision-making processes. The results of this study reflect earlier work done by Padmaja [2] and Ashifa and Ramya [1], which indicated that low wages, long hours of work, and a lack of safety protocols cause workers to have low job satisfaction. Vijayaragavan also stresses the contribution of organizational culture and management practices to influencing employee perception. Workers who are valued by the employers and thus feel valued themselves have been known to report job satisfaction more intensely. This theory justifies the findings of previous researches, such as those of Dutta Banik [3], which established factors that include leading, communicating, and safety culture as the means to enhance job satisfaction among workers. By linking job satisfaction with productivity and overall well-being, Vijayaragavan's study adds to the literature by proposing that improved working conditions and better safety measures are essential to enhance job satisfaction and by extension the overall productivity and sustainability of the fireworks industry.

Ashifa K. M. [7] It explores health risks and general well-being for workers engaged in manufacturing firework industries, contributing to the overall debate regarding occupational hazards in hazardous industries. Ashifa's case study focuses on severe health problems caused among workers, such as respiratory problems, burns, and long-term exposure to chemical substances like nitrates and sulphides. The findings also align with earlier studies carried by Ashifa and Ramya [1] and Padmaja [2], which emphasize the physical burden that such workplaces have on the employees, especially women. The psychological and emotional effects on workers are also well discussed, with results showing extreme stress and burnout arising from the dangerous nature of the job and low job satisfaction. The findings by Ashifa are in agreement with other research studies [3, 6] since their findings show that inadequate safety regulation, insufficient compensation, and poor working conditions have great impacts on workers' mental well-being. The study further emphasizes the socio-economic challenges of workers mainly due to financial instability as most employees are forced to accept unsafe working conditions. Ashifa's research underlines the need for a holistic approach while integrating the physical, mental, and social dimensions of well-being. This may include more effective safety standards, better healthcare conditions, and increasing focus on workers' rights with calls for systemic reform in hazardous industries.

Kumar and Karunanidhi [8] A study of health problems among women workers in the fireworks manufacturing industry that deals with both physical and psychological health problems. The study points out the prevalence of respiratory diseases, skin diseases, and musculoskeletal problems through which most women are affected due to exposure to harmful chemicals, dust, and strenuous labour. This is very much in tandem with earlier studies conducted by Ashifa and Ramya [1] and Dutta Banik [3], making the point that the firecracker industry poses significant health risks to women. Kumar and Karunanidhi's research focuses on socio-economic impacts and shows how women workers are generally subject to double burdens, where they have to work out their home duties and hectic working hours simultaneously. These challenges have been linked with poor work-life balance and mental health disorders like stress and anxiety. This relates to the research findings by Padmaja [2] and Ashifa [7], which revealed a negative effect of long working hours and absence of support systems. The study then calls for even more urgent attention to improving health and safety standards in the workplace and it recommends within itself improved safety in the workplace, regularized health checks, and policies on fair wages and security of jobs for women workers. These are resonant recommendations of calls for systemic reforms seen in other studies, such as that which places emphasis upon holistic well-being at work among hazardous industries.

Karunai and Manoharan study [9], this study examines a very important lacuna in the area of information literacy: the unawareness of workers regarding safety in the workplace, health risks, and rights and benefits available to them. It is a lacuna that leaves the worker vulnerable to accidents and health hazards in the workplace and the exploitation of such. The authors argue for empowering workers by upgrading information literacy that will help the workers make rational decisions on safety and well-being matters. It fills the lacuna in existing literature on workplace health and safety by confirming findings of previous studies such as those carried out by Ashifa and Ramya [1] and Kumar and Karunanidhi [8], according to which education and learning in hazardous industries play an important role. Karunai and Manoharan's study further emphasizes that by giving workers better access to information on safety protocols, health risks, and laws, their work condition and quality of life improved significantly. This study underlines the need for targeted education programmes and sensitization activities that would actually arm workers with information they could use to fend off health-related risks, and improve their balance between work and family life, so as to lobby for better conditions of work – which is not far from several other studies suggestions of more substantial structural changes.

3. Statement of the Problem

The fireworks industry in Sivakasi, Tamil Nadu, is one of the major sectors of women's employment. The working conditions prevailing here are quite problematic for employees' health and wellbeing. Workers, specially women, face great numbers of hazards such as being exposed to toxic chemicals, facing respiratory issues, and general physical injuries through accidents and working long hours under poor conditions. These are further worsened by inadequate safety measures, low wages, and limited rights of the workers, which contribute to high stress, burnout, and health problems [1, 8]. Additionally, the double burden of handling work and family responsibilities affects the work-life balance of workers, thereby affecting their quality of life as a whole [2]. Despite such drawbacks, to this date, scarce literature exists concerning this aspect about how health and socio-economic variables have intermingled with job satisfaction. This implies that there's a lacuna in this sphere, trying to understand how this work condition overall impacts women's health in this fireworks sector in Sivakasi. The paper, therefore, hopes to bridge the gap in addressing the health problems, work-life balance, and job satisfaction issues of female employees, putting in focus systemic reformations and proper policy changes in regard to quality-of-life improvement of workers [3, 6].

3.1. Research gap

There are notable knowledge gaps about the relationship between work-life balance, gender-specific difficulties, and occupational dangers, according to study on the QWL of female employees in Sivakasi's fireworks sector. Previous research has mostly concentrated on health dangers and safety concerns, but it hasn't thoroughly examined mental health, emotional well-being, or long-term socioeconomic effects. Furthermore, little is known about how technology developments and legislative changes might allay these worries without compromising job security. Future studies should also look at how cultural elements and community support networks influence women's job experiences and quality of life.

3.2. Objectives of the study

To Ascertain in women workers perception on quality of work life.

3.3. Scope of study

This scope is set in understanding the health problems, work-life balance, and job satisfaction that affect the working women within the fireworks industries in Sivakasi in respect to the working environment with regard to its impact on both the body and mental conditions of a female worker socioeconomically. The study will critically analyze the risks of health from toxic chemicals exposure, physical injuries, and working hours, considered essential factors influencing the health of the workers as a whole [1, 8]. The socio-economic issues concerning the women also encompass wage disparities, dual household and work burden, and safety issues [2]. This research will investigate how safety culture and information literacy can boost the capacity of workers to take care of themselves and make good decisions regarding their health and safety [9]. The scope of the results is to have an in-depth view of the gaps in safety measures, education, and support structures that are robbing women of their well-being and job satisfaction. The scope extends to recommending improvements in policies, safety standards, and worker education programs to promote better health, job satisfaction, and work-life balance, thereby fostering a more sustainable and equitable working environment in the fireworks industry [4].

4. Research Methodology

4.1. Source of data

Data to be used for the study is primary in nature. Collecting primary data through interviews can prove to be an effective method of gathering detailed and specific information directly from the Respondents. The interview schedule then ensures consistency in the questions posed and standardizes the responses by maintaining the reliability of the data collection process.

4.2. Sampling and sample size

Data collected using the convenience sampling method from 104 respondents within the Sivakasi Taluk.

4.3. Framework of analysis

Basic percentage and factor analysis have been used to analyze the data that were gathered.

4.4. Significance of the study

This research will be of great importance as it can address some of the significant gaps in the understanding of health, safety, and well-being of women workers in the Sivakasi fireworks industry. The research study investigates the different challenges these workers face, including exposure to hazardous chemicals, poor working conditions, and the dual burden of household and professional responsibilities [2, 8], and aims to provide insight into how these factors impact their health, both in body, mind, and society. Better understanding of the work-life balance of such women and what contributes to low job satisfaction would, therefore enable policy-makers and industry stakeholders to forge better safety protocols, better policies on wages and support systems that undertake the creation of healthier work environments. The implications of this study are significant not only within the confines of worker safety and health but also in socio-economic female empowerment in hazardous industries [1]. The study can contribute to the design of targeted interventions enhancing both the individual well-being of employees and productivity by highlighting the need for better safety culture, education, and workers' rights, according to Karunai and Manoharan [9]. Furthermore, findings in the paper will inform broad labour and industrial policy reforms of the adoption of better practices on other hazardous sectors that will work towards a sustainable and equitable work environment [4].

4.5. Data analysis and interprepation

The demographic analysis of the 104 respondents provides valuable information regarding the socio-economic and familial context of women workers in the fireworks industry (Table 20.1). A significant majority (57.7%) are 36–45 years old, indicating that women of middle age form the principal workforce, followed by 23.1% above the age of 46 years. The study also reveals 74% of the respondents as female, which reflects reliance on women labourers within the sector. Educational level is, on the other hand low at a record 77.9%, who cannot even sign his name and requires literates as well as some type of skills up-grade courses. Nuclear family composition has taken a bigger shape as it forms 73.1% of respondents thus leading away from joint family families. Female members could assume the status of family heads to only 11.5%. The total income ranges from ₹8,001 to ₹20,000 per month, which translates to 74% and corresponds with overall household income trends. There is an evident pressure on economic performance as a huge proportion, or 73.1%, face an expenditure of more than ₹30,000. A higher proportion, 82.7% stay in a family that comprises 2 to

Table 20.1. Simple frequency

Age (Years)	*Number*	*Percentage*
Up to 35	20	19.2
36 to 45	60	57.7
Above 46	24	23.1
Total	104	100.0
Gender	*Number*	*Percentage*
Male	27	26.0
Female	77	74.0
Total	104	100.0
Education Qualification	*Number*	*Percentage*
Illiterate	81	77.9
SSLC	13	12.5
HSC	10	9.6
Total	104	100.0
Type of Family	*Number*	*Percentage*
Joint	28	26.9
Nuclear	76	73.1
Total	104	100.0
Status in family	*Number*	*Percentage*
Head	12	11.5
Member	92	88.5
Total	104	100.0
Monthly Income	*Number*	*Percentage*
Up to 8000	19	18.3
8001 to 20000	77	74.0
Above 20000	8	7.7
Total	104	100.0
Family Income	*Number*	*Percentage*
Up to 16000	13	12.5
16001 to 25000	77	74.0
Above 30000	14	13.5
Total	104	100.0
Family Expenditure	*Number*	*Percentage*
Up to 20000	28	26.9
Above 30000	76	73.1
Total	104	100.0
Number of Non-Earning members	*Number*	*Percentage*
Up to 2 members	10	9.6
2 to 4 members	86	82.7
Above 4 members	8	7.7
Total	104	100.0
Number of Earning Members	*Number*	*Percentage*
Up to 2 members	13	12.5
2 to 4 members	77	74.0
Above 4 members	14	13.5
Total	104	100.0
Number of Earning Members	*Number*	*Percentage*
Up to 2 members	14	13.5
2 to 4 members	85	81.7
Above 4 members	5	4.8
Total	104	100.0

Source: Author.

4 non-earning members. The labour force normally consists of 2–4 earning members (81.7%), balancing the expenses in the family. These findings reflect socio-economic challenges faced by women, including low education, financial strain, and household responsibilities that are likely to affect the quality of their work life. Their well-being can be significantly improved by dealing with these issues.

5. Women Fire Workers Perception on Quality of Work Life-Factor Analysis

In this study, factor analysis is carried out to unveil the most critical factors affecting the perception of the quality of work life among female Fire workers. The suitability of the full sample for factor analysis was checked prior to analysis. For this, KMO and Bartlett's Test of Sphericity have been used as pre-analysis tests. The KMO and Bartlett's Test values are greater than 0.70. Therefore, the data collected are appropriate for the factor analysis process. The large result of Bartlett's Test of Sphericity was 388.273, Df: 45, Sig.000, showing the appropriateness of factor analysis, thereby the sample being sufficient (Table 20.2).

Locating Eigen values greater than unity, two factors are identified (Table 20.3). Factors which have a component loading of 0.5 and above are said to be significant that induces influencing women Fire workers perception on quality of work life. From the rotated component matrix, it can be seen that 'necessary protective equipment,' 'enough safety training,' 'Sanitation facilities are adequate,' 'safe working in fireworks,' 'concerned about health risks,' 'The work environment is not hazardous,' 'workplace is clean and organized,' 'improves workplace conditions,' 'supportive work environment,' have a component loading of 0.5 and above. Hence, these two variables form first factor.

In the second factor, 'Long hours affect household duties,' are found to be significant. The other factors contribute, namely, 46.309 and 10.224 towards the determinants of the factors considered women Fire workers perception on quality of work life. The total cumulative percentage of determinants is the factors considered awareness among women Fire workers Perception at 46.309 per cent.

Table 20.2. KMO and Bartlett's test

Kaiser-Meyer-Olkin Measure of Sampling Adequacy.		.867
Bartlett's Test of Sphericity	Approx. Chi-Square	388.273
	df	45
	Sig.	.000

Source: Author.

Table 20.3. Rotated component matrix

	Component	
	1	*2*
Necessary protective equipment	.760	
Enough safety training	.752	
Sanitation facilities are adequate.	.751	
Safe working in fireworks	.744	
Concerned about health risks	.743	
The work environment is not hazardous.	.680	
Workplace is clean and organized.	.661	
Improves workplace conditions.	.649	
Supportive work environment	.578	
Long hours affect household duties.		.969
Total	4.631	1.022
Percentage of Variance	46.309	10.224
Cumulative Percentage	46.309	56.533

Source: Author.

6. Suggestion

- Safety at Work To reduce the risks involved with dangerous working conditions, implement more care for safety practices, provide comprehensive training, and ensure the provision of personal protective equipment (PPE).
- Health and Wellness Programs frequently conduct health camps that focus on emotional and physical health issues specific to women in the fireworks sector. Workers and their families should also be able to afford and access health care services.
- Sanitary Facilities sanitary infrastructure at workplaces should be developed to provide clean and hygienic facilities, especially women workers.
- Sanitary Facilities Workplaces should create sanitary infrastructure to provide clean and hygienic facilities, especially for women employees.
- Child Care Assistance Take use of the free and convenient daycare options available near the workplace so that women may concentrate more on their work without having to worry about their kids.
- Financial Incentives and Equality in Wages: To enhance financial security and confidence, introduce performance-related bonus schemes and ensure gender pay equality.
- Education Enhancement Plan: To offer women alternative sources of employment and to enhance their confidence levels, adult literacy and vocational skills programs must be introduced.
- Self-Emotional Nurturing Support system for helping women in handling their emotional issues, create stress-reduction support groups and also counseling support groups.
- Grievance Mechanisms enhances grievance redressal system by making the system more open and efficient towards listening to women workers and ensuring proper redressal support within their working workplace.

- Community and Family Engagement Program Encourages initiatives in communities teaching families to be supportive of working women, helping reduce the stigma attached to the dangerous sectors wherein women work.

7. Conclusion

The QWL study on women workers in Sivakasi fireworks industry points out the complex challenges they face, resulting from a combination of workplace conditions, health risks, and social responsibilities. Women in this sector contribute a lot to the output of the industry but do so under hazardous conditions. These include prolonged exposure to chemicals, fire hazards, and insufficient access to protective equipment. There are inadequate safety measures that make the situation worse, thereby exposing workers to occupational health issues, such as respiratory diseases and chronic illnesses. More to that, most of these workers have low wages, which in many cases, cannot support them to maintain their families or even to enhance their standards of living. The most significant barriers raised include poor sanitation and facilities at the workplace, thus increasing the misery and stress suffered by women at the workplace. Long working hours and stressful jobs at work do not allow one to manage their household properly, hence leading to work-life imbalance and negatively affecting their mental well-being. Many women complained of the double burden in the workplace and in their personal life, thus demanding change in workplace conditions as well as family-friendly policies. Even with all these difficulties, the study has an area of potential change. There is stringent safety measure like health insurance coverage and medical check-up twice a year; there are also good facilities for the workplace. All these could be found to drastically improve the quality of life of the workers. Wages equally distributed and tied with certain performance incentives give economic and psychological stability; flexible schedule of work helps pave ground for education and development skills training for women; this transforms the work life into being more balanced and satisfying as well. The results underline the need for collective work among policymakers, employers, and community organizations. The need of the hour is listening to women workers more for the policy to be all-inclusive and responsive towards women workers' needs in the system. Consciousness building towards safety in workplaces, redressal of grievances, and associating with healthcare service providers is significant to change long term. Empowering women workers in the fireworks industry of Sivakasi will improve the quality of their lives and, simultaneously, sustain the industry. Investments in their well-being will provide a safer and more equitable working environment, boosting productivity and retaining the workforce. This paper calls for urgent and sustained action to tackle the deeply entrenched issues faced by women in this high-risk industry.

7.1. Future scope of the study

The study on QWL of women workers at Sivakasi firework industry opens several ways for future research and pragmatic applications. Future studies could deal with developing and assessing interventions intended to enhance workplace safety and health standards. These could include, for example the effectiveness of advanced protective equipments, training programs, international safety norms compliance, thereby giving a good insight of reducing occupational hazards. Another area that can be used for exploration is where technological advancements and automation may benefit health risks but instead enhance productivity. Of particular concern to policymakers as well as the employers would be research into exactly how modern machinery would replace high risk manual processes and not reduce prospects of employment appreciably. Longitudinal studies could further trace the impacts of workplace reform on women's health, long-term economic security, and families. Yet more important are topics such as programs for education and skill development toward empowering women workers and their transformation into safer jobs with better earnings either within the fireworks industry or outside it. Comparative analysis on QWLs of women working in Sivakasi with regard to their colleagues in other hazardous industries worldwide also forms a future research focus point from which a wide context will develop and best practice identification will happen. Lastly, future participatory approaches should therefore engage the voice of workers within the policy process so that what happens can be more sustainable and just.

References

[1] Ashifa, A. S. H. I. F. A., & Ramya, P. (2019). Health afflictions and quality of work life among women working in fireworks industry. *International Journal of Engineering and Advanced Technology*, *8*(6 Special Issue 3).

[2] Padmaja, R. (2018). Quality of Work Life of Women Employees in Fireworks and Match Industries in Sivakasi, Tamil Nadu. *Asian Review of Social Sciences*, *7*(2), 50–56.

[3] Dutta Banik, P. (2019). The fireworkers' village of champahati: an analysis of their work and its impact on their life.

[4] Rajaram, S., & Sivakumar, G. D. (2023). Assessment of safety culture in the fireworks industry. *International Journal of Occupational Safety and Ergonomics*, *29*(2), 466–473.

[5] Joy, S. O. (2023). Exploring the complexities of Sivakasi's fireworks industry.

[6] Vijayaragavan, T. (2014). A study on perception of the employees relating to the job satisfaction at standard fireworks in Sivakasi. *GE International Research Journal of Management Research*, *2*(7), 363–374.

[7] Ashifa, K. M. (2022). Hazards and well-being of employees in firework industries. *Journal of Positive School Psychology*, 11798–11801.

[8] Kumar, R. S., & Karunanidhi, G. (2016). Health problems of women labourers in fireworks industries–A cross sectional study. *Interpreting*, *5*(6).

[9] Karunai, K., & Manoharan, A. (2015). Information literacy on workers in match and fireworks industries scenario in Tamil Nadu: An overview. *Indian Journal of Science*, *21*(72), 279–284.

21 Empirical study of the marketing research techniques adopted by retailers in India

Pranjal Agrawal[a], Chandan Singh[b], Ankit Kumar Garg[c], Dhirendra Ojha[d], Kartik Soni[e], and Parveen Kumar[f]

Faculty of Management Studies, AKS University Satna, Madhya Pradesh, India AKS University, India

Abstract: This empirical study examines Indian merchants marketing research methods in the dynamic retail sector. A mixed-methods study of 200 merchant's interviews and surveys examines marketing research strategies' prevalence, effectiveness, and affecting factors. Research shows a variety of methods, including customer surveys and market segmentation. Technique effectiveness ratings emphasize sales and customer pleasure. Economic variables, technical advances, customer behaviour, and competitive pressures influence marketing research adoption. Retailers should diversify, integrate technology, and react to consumer trends, according to the report. The insights aim to help retailers improve their marketing research methods and compete in the changing Indian retail marketplace.

Keywords: Retailers, marketing research, Indian retail sector, consumer behaviour

1. Introduction

The term 'marketing research' refers to the process of systematically collecting, documenting, and analyzing qualitative and quantitative data pertaining to matters of product and service marketing. The objective is to determine and evaluate the effects on consumer behaviour of modifying the marketing mix components [1]. To do this, you must first identify the problems that need solving, then plan how to gather the necessary data, and last oversee and carry out the data collection procedure. Once the data has been analyzed, the results and consequences are communicated to the people with the authority to implement them. There is a natural progression in the business world from conducting market research to conducting marketing research to actually marketing a product or service. Marketing research is a relatively new discipline compared to market research. Despite the fact that both fields focus on customers, marketing research focuses on marketing procedures including salesforce effectiveness and advertising efficacy, whereas market research is more narrowly focused on distribution and markets. The second reason people make this mistake is because they are confused with marketing research since marketing research is one of its subsets and holds some degree of similarities with market research. Top global corporations that know the two areas and have operations there certainly don't lessen the uncertainty [2].

To put matters simply, in the past several years, a number of developments in the behaviour of customers, in technology, and in the economy have brought about massive upheavals in the Indian retail sector [3]. As they look into this ever changing landscape, retailers are becoming more aware of the importance of marketing research that can help them better understand consumer tastes, market trends and the competition. The aim of this empirical study is to explicate further the tactics and strategies employed by Indian merchants through shedding light on the marketing research methodologies which they have utilized [4]. The scenario of retail sector is strong and diversified in India, which contributes towards growth of Indian economy. With the growth of digital technologies and globalization, a new wave of digital technologies and globalization have never been more present to retailers. Retailers depend on marketing research when it comes to keeping up to date and adapting to the constantly changing dynamics of industry [5].

2. Importance of Market Research for Retailers

Market research can help the retailers better adjust the product/service and marketing strategies so that they can match with the market trends by uncovering the details about customer behaviour. Retailers need this kind of information so that they can create a product, price it and design stores so they can connect with their customers as fully as possible. Bringing a new product market can always present a large financial risk. These perils can only be minimized with the aid of a thorough market research. In other words, retailers can better calculate the products that they are producing by

[a]pranjal.agarwal96@gmail.com, [b]chandansingh201285@gmail.com, [c]shyamagarg123@gmail.com, [d]dhirendraojha@rediffmail.com, [e]kartiksoni051101.ks@gmail.com, [f]parveenduhan261@gmail.com

DOI: 10.1201/9781003684459-21

researching market trends and unfulfilled customer needs. In retail, knowledge of industry and competitors is what makes the difference. In general, retailers have to be aware of what their competitors are selling and determine how they can differentiate themselves – through communication tactics, product features, customer service, etc.

Businesses can gain a better understanding on what consumers are shopping for, their taste and the reason why they are purchasing through market research. As such, this is the most valuable data in improving the overall customer service experience, meeting with more important contacts to customers, and launching more precise targeted marketing initiatives. Essential to note, therefore, means that retailers can know what their customers are in want of and need of and works on molding marketing custom to every individual to increase sales and loyalty through being on top of and even going farther than people's expectations and needs of a product; market research is an excellent manner of knowing if people really ask for a product, if it is specific to addressing their expectation. Paying attention to market trends, customer behaviour and wishes the retailers may reduce the probability of introducing products which do not sell.

2.1. *Objectives of the study*

- To identify and analyze the marketing research techniques used by retailers in India,
- To assess the effectiveness of these techniques in driving marketing strategies,
- To understand the challenges and opportunities faced by retailers in implementing marketing,

2.2. *Literature review*

A number of marketing research and analysis strategies have been implemented by retailers in India. According to Srivastava [11], in the context of developing the marketing strategies with success, regularly calculating the consumer profile, market positioning and competitor analysis is very important. Koul [6] emphasizes retail margin, customer profile and store design in the context of product selection planning for small retailers. The leading sales promotion techniques in terms of influence on sales, as per Katole [7], are price and product discounts which need the use of sales promotion strategies.

Costa et al. [8] This covers key aspects of inventory management and customer feedback in the retail sector, the requisite of differentiating products apart from price, using the sales data for trend research, and the need of differentiation of products in general. Observing all these studies, it is noteworthy that the marketing methods used in the Indian retail field would be varying, as well as quite complex in operation. Reddy [9], explains that the Indian market has undergone significant changes that include coming of Multi National Corporations, Indian companies adopting new marketing strategies, increased product variety, increasing choice diversity of consumers as well as emphasis on value and value for money. Marketers must reevaluate their strategy to provide value in the face of the evolving market environment, focusing on giving a satisfying buying experience. Marketers are resorting to price reductions to maintain consumer loyalty as competition grows more intense. Marketers want to provide significant discounts and pricing benefits on Maximum Retail Price (MRP) to surpass competition. In India, marketers strive to retain consumers and gain market share by implementing creative methods at the retail level. Loyalty is a crucial strategy for retaining consumers and encouraging their return after creating retail equity. Retailers face a significant problem in maintaining loyal customers due to the scattered distribution of retail sites and the large concentration of stores in urban areas. Ramesh Kumar highlighted that research in both Indian and Western contexts has demonstrated that loyal consumers significantly increase profitability across several categories. Atulkar and Kesari [10] found that retailers understand that improving customer happiness and retail performance helps them attract and keep customers for retail purchases. Retail in India started as unorganized but is currently expanding into organized sectors such as supermarkets, hypermarkets, and malls. The study builds upon prior research in the retail sector to investigate the significant outcomes related to customer purchasing behaviour, customer attitudes and preferences towards organized retail sectors, and how customers behave during product purchases, aiding retail stores in customer acquisition and retention. This paper presents an overview of Indian consumer behaviour focusing on prepurchase behaviour, experience, information processing, involvement, choices, and business practices such as retail space, product availability, brand communication, and retail environment. These factors are crucial for attracting target customers in both organized and unorganized retail sectors in India, enhancing retail effectiveness, and appealing to affluent customers.

3. Research Methodology

3.1. *Research design*

This empirical study adopts a mixed methods research design, combining both quantitative and qualitative approaches. The exploratory and descriptive nature of the research necessitates a comprehensive investigation into the marketing research techniques employed by retailers in India.

3.2. *Sampling*

A sample size of 200 retailers will be selected for the study. The selection will be stratified to ensure representation across various retail segments, including stores, online retailers, and hybrid models. To enhance the study's generalizability, participants will be chosen from diverse geographical regions within India. The inclusion criteria encompass retailers with a minimum of one year in operation to ensure a level of experience in the market.

3.3. Data collection

3.3.1. Surveys/questionnaires

- Structured questionnaires will be developed, focusing on gathering quantitative data related to the types of marketing research techniques utilized, frequency of usage, and perceived effectiveness.
- The questionnaires will be distributed electronically through email and online survey platforms, ensuring ease of access for respondents.
- In order to encourage participation a cover letter will give a brief summary of the importance of the research to the participant and the recruiter, documentation of confidentiality assurances, and a straightforward explanation as to the reason of the filled out questionnaire.

3.3.2. Interviews

Additionally, a subset of the sample, approximately 20% of the total sample size, will be interviewed indepth to collect qualitative thought and a nuanced point of view.

Purposive selection of retailers will be done on the basis of their variations regarding size, market presence and geographical area.

Challenges as well as success stories of adoption and effectiveness of marketing research techniques and the contextual factors that impact it are what will be explored in the interviews.

3.4. Data analysis

3.4.1. Quantitative analysis

Moreover, Statistical Software like SPSS will be used to quantify survey response.

Presentation of a wide view of the quantitative data will involve descriptive statistics like frequencies, percentages, and averages.

Comparative analyses will be done across the different segments of the retail to identify patterns and variations.

3.4.2. Qualitative analysis

- However, thematic analysis will be used to extract key themes and patterns from qualitative data obtained through interviews.
- Coding and categorization of responses will be performed, allowing for a deeper understanding of the challenges, successes, and contextual factors related to marketing research.

3.5. Ethical considerations

- Prior to participation, informed consent will be obtained from each retailer, emphasizing the voluntary nature of their involvement.
- Confidentiality of participant information will be strictly maintained, with data reported in an aggregated and anonymized form.
- The research will adhere to ethical guidelines and principles, ensuring the rights and privacy of the participants are protected.

3.5.1. Interpretation

The Table 21.1 provides a detailed demographic overview of merchants and their sample distribution across categories. Northern India has 25% of the sample's retailers, followed by Southern India with 20%. Western, Eastern, and Central India contribute 15%, 10%, and 7.5%. The sample includes 22.5% from Northeastern India, indicating a large presence. According to the distribution, 40% of merchants are brick-and-mortar, 30% are online, and 30% are hybrid, demonstrating a balanced distribution. The years in business vary, with 35% of stores operating for 1–3 years, 30% for 4–7 years, 20% for 8–10 years, and 15% for beyond 10 years. This distribution shows a mix of established and new retailers. The annual revenue breakdown shows retailers' financial size. Note that 30% of shopkeepers earn INR 1–5 million annually, whereas 22.5% earn INR 5–10 million. Retailers earning less than INR 1 million make up 12.5%, while those earning 10–50 million and over 50 million make up 20% and

Table 21.1. Demographic profile of the candidate

Demographic Characteristics		*Number of Retailers*	*Percentage of Sample*
Geographical Region	Northern India	50	25%
	Southern India	40	20%
	Western India	30	15%
	Eastern India	20	10%
	Central India	15	7.5%
	Northeastern India	45	22.5%
Retail Segment	Stores	80	40%
	Online Retailers	60	30%
	Hybrid Models	60	30%
Years in Operation	13 years	70	35%
	47 years	60	30%
	810 years	40	20%
	Over 10 years	30	15%
Annual Revenue	Less than INR 1 million	25	12.5%
	INR 15 million	60	30%
	INR 510 million	45	22.5%
	INR 1050 million	40	20%
	Over INR 50 million	30	15%

Source: Author.

15%, respectively. This distribution shows the various financial capacities of the retailers polled.

3.6. *Marketing research techniques in Indian retail*

3.6.1. Interpretation

Retailers use a variety of marketing research methods, as seen in the Tables 21.2 and 21.3. The majority of retailers (78%), use consumer surveys. Retailers seem to recognize the value of directly asking customer feedback to assess satisfaction, preferences, and areas for development. The second most common method is market segmentation, used by 65% of retailers. This highlights how important it is for merchants to segment their target customer by demographics, behaviours, and preferences. Segmentation helps marketers target specific consumer segments.

Competitor analysis is also used by 55% of retailers for marketing research. Staying informed about industry competitors' strategy and market positioning is crucial. Social media monitoring, at 48%, shows that merchants are increasingly using online platforms to understand consumer impressions. Qualitative data from focus groups, at 42%, helps retailers understand consumer thoughts and attitudes. Online analytics, mystery shopping, and in-store observations are used to collect data. At 20% and 15%, trend analysis and product testing are used, although they may not be mainstream or relevant across retail contexts.

Table 21.2. Types of marketing research techniques

Marketing research techniques	*Percentage of retailers*
Customer Surveys	85%
Market Segmentation	72%
Competitor Analysis	68%
Social Media Monitoring	60%
Focus Groups	55%
Instore Observations	48%
Online Analytics	45%
Mystery Shopping	38%
Trend Analysis	32%
Product Testing	28%

Calculation1: Average number of techniques used per retailer

$$\text{Average Techniques} = \frac{\text{Total Techniques Used}}{\textit{Number of Retailres}}$$

$$\text{Average Techniques} = \frac{356}{200}$$

$$\text{Average Techniques} = 1.78$$

Source: Author.

Table 21.3. Frequency of marketing research techniques usage

Marketing Research Techniques	*Frequently Used (%)*
Customer Surveys	78%
Market Segmentation	65%
Competitor Analysis	55%
Social Media Monitoring	48%
Focus Groups	42%
Instore Observations	35%
Online Analytics	30%
Mystery Shopping	25%
Trend Analysis	20%
Product Testing	15%

Calculation: Most commonly used marketing research technique

Source: Author.

3.6.2. Interpretations

Table 21.4 shows how merchants use marketing research methods, focusing on online and hybrid stores. These percentages show how much each retail segment uses each approach. Customer surveys dominate online retailers' marketing research, accounting for 40%. This emphasizes the importance of personally engaging with customers to gain insights. Another popular method, competition analysis, accounts for 35%, indicating online businesses' competitive awareness. Focus groups and in-store observations are similarly significant at 45% and 50%, stressing customer preferences and habits. Hybrid Models distributes marketing research techniques differently. Mystery Shopping leads with 50%, emphasizing undercover customer experience ratings. At 40% each, market segmentation and online analytics demonstrate a comprehensive approach to market dynamics and online consumer behaviour. Trend analysis is high at 55%, indicating strong awareness of Hybrid Models retail trends.

Table 21.4. Cross tabulation of techniques by retail segment

Marketing research techniques \ Retail segment		*Online retailers*	*Hybrid models*
Customer Surveys	40%	30%	30%
Market Segmentation	25%	40%	35%
Competitor Analysis	35%	20%	45%
Social Media Monitoring	20%	50%	30%
Focus Groups	45%	25%	30%
Instore Observations	50%	25%	25%
Online Analytics	15%	40%	45%
Mystery Shopping	30%	20%	50%
Trend Analysis	20%	25%	55%
Product Testing	25%	35%	40%

Source: Author.

3.7. *Effectiveness of marketing research techniques*

3.7.1. *Interpretation*

Table 21.5 shows how firms regard different marketing research methods. Customer surveys are the most effective, scoring 78%. This emphasizes the importance of actively engaging with customers to collect insights and feedback for successful marketing tactics. Market Segmentation follows closely with a moderate efficacy rating of 65%, highlighting its usefulness in identifying and targeting market categories. Competitor Analysis received a 55% rating, demonstrating the strategic importance of understanding the competition landscape. Moving forward, Social Media Monitoring and Focus Groups score moderate efficacy at 48% and 42%. This shows that firms prioritize social media and qualitative research for consumer sentiment and preferences. Instore observations obtain a moderate 35% rating, underlining their importance in retail locations. Online Analytics, however ineffective at 30%, helps firms understand online consumer behaviour. Mystery Shopping and Trend Analysis are moderately effective, rated 25% and 20%, respectively. These methods improve customer experience and market trend understanding. While valuable, Product Testing has a low effectiveness rating of 15%, suggesting that organizations may not use it to shape their marketing plans.

3.7.2. *Interpretations*

The Marketing Research Techniques Table 21.6 highlights corporate benefits and drawbacks. Customer surveys are effective, with 85% reporting favorable outcomes. Direct customer engagement for feedback and insights is essential. Another essential approach, market segmentation, improves results by 70%. Marketing to specified target categories improves strategy execution for companies utilizing this method. Market segmentation was neutral or unfavorable for 30% of respondents, implying implementation challenges or success.60% think competitor analysis impactful in marketing research. This implies firms value competitive knowledge. 40% report neutral or unfavorable outcomes, suggesting competitor analysis is difficult or needs improvement. Social media monitoring affects digital marketing 50% positively and 50% negatively. Because social media is dynamic and unpredictable, its effectiveness may vary. Focus groups, known for their qualitative insights, succeed 75%. This suggests firms value focus groups' qualitative data and depth of understanding. However, 25% report a neutral or negative effect, suggesting this strategy may have limitations. In-store observations boost marketing 55%. However, 45% indicate neutral or negative effects, highlighting the need to improve monitoring or execution. Despite being crucial to digital marketing, web analytics yield 40% favorable and 60% neutral or negative results. This means organizations may struggle to fully exploit web data or turn analytics into strategy. Mystery shopping boosts service and interactions by 45%. Unfortunately, 55% indicate neutral or negative effects, suggesting mystery shopping program performance needs improving. Trend analysis predicts market trends with 30% positive and 70% neutral or negative influence. Market trends might be hard to predict. The final product testing shows 25% positive impact and 75% neutral or negative. This suggests organizations may have trouble implementing product testing (Table 21.7).

Table 21.5. Effectiveness ratings of marketing research techniques

Marketing research techniques	*Effectiveness rating*
Customer Surveys	High (78%)
Market Segmentation	Moderate (65%)
Competitor Analysis	High (55%)
Social Media Monitoring	Moderate (48%)
Focus Groups	High (42%)
Instore Observations	Moderate (35%)
Online Analytics	Low (30%)
Mystery Shopping	Moderate (25%)
Trend Analysis	Low (20%)
Product Testing	Low (15%)

Calculation1: Overall effectiveness score

The overall effectiveness score is calculated based on the average rating of all techniques

$$\text{Overall Effectiveness Score} = \frac{\text{Total Effectiveness Ratings}}{\text{Number of Techniques}}$$

$$\text{Overall Effectiveness Score} = \frac{435}{10}$$

$$\text{Overall Effectiveness Score} = 43.5$$

Source: Author.

Table 21.6. Impact of marketing research techniques on sales and customer satisfaction

Marketing research techniques impact (%)	*Positive impact (%)*	*Neutral or negative*
Customer Surveys	85%	15%
Market Segmentation	70%	30%
Competitor Analysis	60%	40%
Social Media Monitoring	50%	50%
Focus Groups	75%	25%
Instore Observations	55%	45%
Online Analytics	40%	60%
Mystery Shopping	45%	55%
Trend Analysis	30%	70%
Product Testing	25%	75%

Source: Author.

3.8. Factors influencing the adoption of marketing research techniques

3.8.1. Interpretation

Table 21.8 covers the issues affecting retail, focusing on online retailers and hybrid models. These models are studied to see how each element affects economic, technological, consumer behaviour, and competitive landscape factors. Economic Factors: Online Retailers (60%) and Hybrid Models (50%) have high research costs and moderate return on investment (45%) and high (60%). Online Retailers (35%) and Hybrid Models (40%) face moderate economic uncertainty. Technology is important, with Online Retailers and Hybrid Models having 70% and 75% access to advanced tools and analytics, respectively. Online retailers integrate 40% of Ecommerce Platforms, while hybrid models integrate 50%. Online Retailers (55%) and Hybrid Models (60%) react well to changing consumer preferences. For Online Retailers (65%) and Hybrid Models (70%), consumer digital engagement is rising. Competitive Landscape: Online Retailers (50%) and Hybrid Models (45%) have strong industry benchmarking. Online retailers face moderate competitive pressure (45%), hybrid models high (50%), and the combined average moderate (48%).

Table 21.7. Factors influencing the adoption of marketing research techniques

	Factors	*Influence rating (%)*
Economic Factors	Cost of Research	High (60%)
	Return on Investment	Moderate (45%)
	Economic Uncertainty	Moderate (35%)
Technological Advancements	Access to Advanced Tools and Analytics	High (70%)
	Integration with Ecommerce Platforms	Moderate (40%)
Consumer Behaviour and Preferences	Shifting Trends and Preferences	High (55%)
	Increasing Digital Engagement	High (65%)
Competitive Landscape	Industry Benchmarking	High (50%)
	Competitive Pressure	Moderate (45%)

Calculation: Cumulative influence score

The cumulative influence score is calculated based on the sum of influence ratings across all factors.

Cumulative Influence Score = ∑Influence Ratings

Cumulative Influence Score = 60 + 45 + 35 + 75 + 70 + 40 + 55 + 65 + 50 + 45

Cumulative Influence Score = 540

Source: Author.

Table 21.8. Comparison of factors by retail segment

Marketing Research Techniques \ Retail Segment		*Models*	*Online Retailers*	*Hybrid*
Economic Factors	Cost of Research	High (60%)	Moderate (50%)	High (65%)
	Return on Investment	Moderate (45%)	High (60%)	Moderate (55%)
	Economic Uncertainty	Moderate (35%)	High (40%)	Moderate (45%)
Technological Advancements	Access to Advanced Tools and Analytics	High (70%)	High (75%)	Moderate (60%)
	Integration with Ecommerce Platforms	Moderate (40%)	High (50%)	Moderate (45%)
Consumer Behaviour and Preferences	Shifting Trends and Preferences	High (55%)	High (60%)	Moderate (50%)
	Increasing Digital Engagement	High (65%)	High (70%)	Moderate (55%)
Competitive Landscape	Industry Benchmarking	High (50%)	Moderate (45%)	High (55%)
	Competitive Pressure	Moderate (45%)	High (50%)	Moderate (48%)

Source: Author.

3.9. Recommendations for retailers

Drawing upon the insights gained from the empirical study on marketing research techniques adopted by retailers in India, the following recommendations are offered to enhance the effectiveness of marketing strategies and improve overall business outcomes.

3.9.1. Diversify research techniques

- Retailers should apply a combination of marketing research techniques that will enable them to understand a broader spectrum of consumer behaviour, market trends and competitors.
- By combining the traditional methods such as customer surveys and instore observations, with the latest available digital tools like online analytics and social media monitoring, we can have a wellrounded perspective.

3.9.2. Invest in technological integration

- Include investing in more advanced tools and analytics platforms to embrace technological advancements.
- Connect marketing research technologies with ecommerce platforms to make them easier to collect and analyze data for integration with the online and offline sales cycle.

3.9.3. Prioritize customer surveys and feedback

- Since customer surveys and feedback mechanisms are proven to be very effective and have a positive influence on sales and customer satisfaction, use the first.
- Collect and analyze customers' feedback regularly and identify which ones have changed and which are the new ones so as to improve product offering and increase customer experience.

3.9.4. Adapt to shifting consumer trends

- Continually monitor the evolving consumer trends and preferences in relation to the rising digital engagement that is witnessed in the Indian market.
- Make it quick to adapt the marketing strategies against the emerging trends to resonate with the target audience.

3.9.5. Benchmark against industry standards

- Engage in industry benchmarking to assess performance against competitors and industry standards.
- Regularly monitor and analyze key performance indicators (KPIs) to identify areas for improvement and innovation.

3.9.6. Manage economic uncertainty prudently

- Hence, retailers should adopt a prudent approach towards managing the research expenses, especially because economic factors, especially, the cost considerations play a vital role in influencing their decisions.
- To evaluate the return on investment for each of the research techniques and choose which techniques that will have the biggest impact on business outcomes.

3.9.7. Facilitate employee training

- Catalogue comprehensive training programs for employees who participate in marketing research activities.
- Provide training to the staff to enable them to utilize more advanced tools and technologies, which will facilitate the implementation of marketing research strategies in an appropriate and efficient manner.

3.9.8. Collaborate with industry experts

- Engage in partnerships with marketing research experts and agencies for the gain of specialized insights and guidance.
- To supplement internal capabilities and stay abreast of the latest in cutting edge research methodologies leverage the external expertise out there.

4. Conclusion

Therefore, as a conclusion, this empirical study has extensively examined the marketing research techniques pursued by the Indian retailers. The results provide useful knowledge about the types of techniques used within the industry, the extent to which they work and the variables driving their deployment. The retail landscape is evolving and data driven decision-making becomes more important.

5. Key Findings

1. Thus, the customer survey is the most commonly used method of marketing research carried out by retailers in INDIA and is ranked the highest common market research technique practiced by retailers in the country.
2. The effectiveness of Customer Surveys and Focus Groups were very highly rated, Online Analytics and Trend Analysis were seen as rather low, and Product Testing was viewed with the lowest effectiveness.
3. It had a positive impact on sales and customer satisfaction, since customer surveys, market segmentation and focus groups proved to be a positive influence.
4. The adoption of marketing research techniques was found to be affected by the economic, technological, consumer and competitive pressures.

References

[1] Hammudi, R. A., Hammudi, Y. A., & Hameed, E. M. (2024). The important role of cold atmospheric plasma in wounds healing: A literature review. *International Academic Journal of Science and Engineering, 11*(1), 91–97. https://doi.org/10.9756/IAJSE/V11I1/IAJSE1112

[2] Nazarmon, V. P., Senthil Prakash, T., & Ramesh, S. (2016). Steganography based Secured Online Payment System. *International Journal of Advances in Engineering and Emerging Technology, 7*(4), 74–81.

[3] Dulger, G., & Dulger, B. (2023). Evaluation of Antimicrobial Activities of Salvia verbenaca. *Natural and Engineering Sciences, 8*(2), 61–71. http://doi.org/10.28978/nesciences.1309249

[4] Kumar, A., & Yadav, P. (2024). Experimental investigation on analysis of Alkaline treated natural fibers reinforced hybrid composites. *Association Journal of Interdisciplinary Technics in Engineering Mechanics, 2*(4), 25–31.

[5] Sathish Kumar, M., Santhi, L., & Senthilkumar, A. (2024). Quantifying the impact of Indian virtual reality research: A scientometric study. *Indian Journal of Information Sources and Services, 14*(3), 1–5. https://doi.org/10.51983/ijiss-2024.14.3.01

[6] Koul, S., & Jasrotia, S. S. (2019). Product adoption by small retailers in India. *International Journal of Retail & Distribution Management*, 31.

[7] Katole, H. J. (2020). Effects of sales promotion campaign adopted by retailers in India. *Journal of Critical Reviews*, 39.

[8] Costa, D. D., & Hans, D. V. (2024). Opportunities and challenges for Indian retail stores and their marketing strategies. *EPRA International Journal of Multidisciplinary Research (IJMR)*.

[9] Reddy, P. N. (2004). Efficiency benefits pass on to consumers: New development in retail market environment in India. *Indian Journal of Marketing*, 34.

[10] Atulkar, S., & Kesari, B. (2014). Adoption of retailer centric philosophy in organizing buying process: a case study on Indian buying systems. *ZENITH International Journal of Multidisciplinary Research*, *4*, 204209.

[11] Srivastava, S. (2012). Marketing strategies adopted by retailers in retail store: A case study of max lifestyle storephoneix mall, *Lucknow*, 12.

22 A significance study on financial performance selected pharmaceutical industries in India

P. Easwaran[1,a], *M. Kalimuthu*[2,b], *V. Krishnaveni*[3,c], *Sankar Ganesh R.*[4,d], *J. Sudarvel*[1,e], *and Smruthymol J.*[1,f]

[1]Karpagam Academy of Higher Education, Coimbatore, Tamil Nadu, India
[2]Professor and Head, Department of Commerce with Professional Accounting, Dr. N.G.P. Arts and Science College, Coimbatore, Tamil Nadu, India
[3]Professor and Dean, Department of Management Studies, Dhanalakshmi Srinivasan College of Engineering, Coimbatore, Tamil Nadu, India
[4]Associate Professor, Department of Management Studies, Vel Tech Rangarajan Dr Sagunthala R&D Institute of Science and Technology, Chennai, India

Abstract: The present study has been carried out to identify the five leading Indian pharmaceutical organizations including Sun Pharma and Cipla with Dr. Reddy's Laboratories and Divi's Laboratories and Torrent Pharma were subjected to financial health evaluation through annual data from 2020 to 2024. The research analyzed component-wise profitability and operational efficiency and financial stability using KPIs including ROI and equity returns. Statistical tools helped identify major variations between the analyzed companies. The financial records of Dr. Reddy's demonstrated powerful performance metrics but Divi's Laboratories showed heightened performance then experienced a major decline. The financial records indicate that Sun Pharma and Cipla showed progressive growth whereas Torrent Pharma demonstrated steady medium-level returns. Research revealed that capital utilization levels differed between the pharmaceutical companies while profitability rates and earnings potentials also showed mixed results. To sustain growth in this evolving pharmaceutical market companies must pursue specialized business expansion while optimizing their financial structure together with pursuing revolutionary developments.

Keywords: Pharmaceutical industry, financial ratios, corporate finance

1. Introduction

The pharmaceutical industry plays a critical role in modern healthcare, contributing to advancements in drug development, disease management, and medical technology. As the global population grows and healthcare demands rise, pharmaceutical companies are at the forefront of scientific innovation, biotechnology integration, and regulatory adaptation. The industry's economic impact is substantial, driving employment, research investments, and healthcare accessibility worldwide.

The global pharmaceutical market has witnessed exponential growth, reaching a valuation of USD 1.48 trillion in 2022 and is projected to surpass USD 2.1 trillion by 2030 [1]. Factors contributing to this expansion include:

- Increased healthcare expenditures due to aging populations and rising chronic diseases [2].
- Technological breakthroughs such as AI-driven drug discovery and personalized medicine [3].
- Regulatory changes and patent reforms that promote innovation and competition [4].
- Globalization and market penetration in emerging economies [5].

1.1. India's position in the global pharmaceutical industry

India has established itself as a dominant force in the pharmaceutical sector, ranking among the world's largest producers and exporters of generic drugs. The Indian pharmaceutical market has reached a valuation of USD 50 billion in 2024, with a projected steady growth trajectory [6]. The country is responsible for:

- 20% of global generic drug exports, supplying affordable medicines to developing and developed nations [7].
- Over 60% of global vaccine production, making it a crucial player in the fight against infectious diseases [2].

[a]easwar83@gmail.com, [b]kalimuthusujanias@gmail.com, [c]krishnaveni.rkv99@gmail.com, [d]sankarganeshphd@gmail.com, [e]j.sudarvel@gmail.com, [f]jayalalsmruthy@gmail.com

DOI: 10.1201/9781003684459-22

- A thriving research and development (R&D) ecosystem, with increased investments in biosimilars, biologics, and advanced therapeutics [6].

Government initiatives such as the Production Linked Incentive (PLI) Scheme have further propelled India's pharmaceutical growth by encouraging domestic manufacturing, reducing import dependency, and boosting exports [8]. Policies promoting intellectual property rights (IPR) protection, infrastructure development, and public-private partnerships have strengthened India's position as a pharma hub.

1.2. Evaluating financial performance in the Indian pharmaceutical sector

To understand the financial health and profitability of the Indian pharmaceutical industry, this study focuses on the performance analysis of leading pharmaceutical companies, including:

- Sun Pharmaceutical Industries
- Cipla Limited
- Dr. Reddy's Laboratories
- Divi's Laboratories
- Torrent Pharmaceuticals

Using financial ratio analysis, key metrics such as Gross Profit Ratio, Net Profit Ratio, Operating Profit Ratio, Return on Equity (ROE), and Return on Investment (ROI) are examined to assess:

- Financial stability – How efficiently companies manage their costs and revenues [5].
- Profitability trends – The ability of firms to generate consistent earnings and maintain growth [9].
- Investment potential – The attractiveness of these companies for stakeholders and investors [10].

By analyzing these financial indicators, the study provides valuable insights into the growth potential, challenges, and sustainability of India's pharmaceutical sector within the global market. With increasing emphasis on R&D investments, technological advancements, and regulatory shifts, the future of the pharmaceutical industry remains promising, innovative, and economically influential.

2. Review of Literature

1. The authors Sharma and Kumar [11] performed a liquidity study of India's pharmaceutical sector through an assessment of current and quick ratios. The research demonstrated the pharmaceutical industry requires sufficient short-term funding resources for maintaining operational stability.
2. Through activity ratios Patel and Desai [12] analyzed the asset utilization performance of the Indian pharmaceutical sector. The research showed that proper asset management leads to higher revenue together with better financial results across the business [4].
3. Garg and Mishra [13] conducted research to evaluate macroeconomic variables as performance indicators for the financial sector of Indian pharmaceutical companies. The authors showed that inflation and interest rates play major roles in shaping financial results so organizations need strong financial planning systems.
4. The authors Das and Banerjee [14] conducted a panel data analysis to understand financial performance patterns of Indian pharmaceutical companies. According to their research panel data analysis offered superior insights into the financial evolution of this market throughout its time span.
5. The authors of Chandra and Reddy [15] conducted non-parametric analysis on Indian pharmaceutical sector financial performance evaluation. The research proved helpful when collecting data from non-normal distribution samples.
6. Research by Singh and Mishra [16] showed that effective corporate governance systems create better financial outcomes in Indian pharmaceutical firms by improving profitability together with shareholder value aspects.
7. Working capital management efficiency plays a crucial role in boosting profitability levels of Indian pharmaceutical SMEs according to Joshi and Patel [17].
8. Through their analysis Reddy and Rao [18] uncovered how digital technology adoption leads pharmaceutical firms in India to experience better operational results together with enhanced financial outcomes.
9. Menon and Iyer [19] presented findings demonstrating how strict pharmaceutical price regulations create adverse effects on profit levels and capital investment of Indian pharmaceutical organizations.
10. Banerjee and Verma [20] established that pharmaceutical companies in India achieved financial stability through robust supply chain performance during the COVID-19 pandemic.
11. Desai and Chandra [21] established that pharmaceutical firms with established ESG practices lead to improved financial performance within the Indian drug manufacturer industry.
12. Verma and Patel [22] conducted research showing that the return on investment (ROI) of pharmaceutical organizations directly links with their working capital efficiency.

2.1. Statement of the problem

The rapid growth of India's pharmaceutical industry makes it difficult for analysts to track its financial stability. Multiple challenges impede the industry such as market trends that shift and regulatory barriers and competitive competition. Decision-makers within industries and financial institutions and authorities need detailed assessments of pharmaceutical businesses' financial stability. This research investigates financial performance while collecting data about profitability and developing stability criteria to replace the current analysis gap. This study combines financial ratio analysis with market trend assessment in order to deliver important information which supports wise decision-making.

2.2. Significance of the study

The study brings essential information to comprehend financial performance within India's pharmaceutical sector. The pharmaceutical sector must maintain financial stability because its expanding influence within global healthcare needs it to succeed in sustainable growth. The assessment of financial ratios and industry patterns informs investor decision-makers who seek rewards as well as policymakers who establish rules and corporate executives who enhance their financial methods. This study conducts comparative research to reveal how well-established pharmaceutical firms in India perform financially.

2.3. Objectives of the study

The study evaluates profitability along with efficiency and financial stability of multiple pharmaceutical companies in India through quantitative financial ratio analysis.

2.4. Research methodology

The research implements a quantitative research approach that collects secondary financial data from annual reports together with stock exchange filings and industry reports. The methodology includes:

- **Data Collection:** Financial records from major pharmaceutical firms, regulatory bodies, and market analyses.
- **Sample Selection:** Leading companies based on revenue, market capitalization, and industry influence.
- **Financial Ratio Analysis:** Examination of key indicators, including Gross Profit Ratio, Net Profit Ratio, ROE, and ROI.
- **Comparative Analysis:** Yearly financial performance evaluation to identify trends and growth patterns.
- **Statistical Tools:** The study employs ANOVA and paired t-tests to measure differences in financial performance across companies.

3. Analysis and Interpretations

3.1. Return on investment (ROI) analysis

An ROI evaluation of Sun Pharma, Cipla, Dr. Reddy's, Divi's Labs and Torrent Pharma occurred during FY2020-FY2024 (Table 22.1).

Return on Investment (ROI) measures the efficiency of asset utilization that generates profits within a company. The financial performance and key trends of Sun Pharma Cipla Dr. Reddy's Divi's Labs and Torrent Pharma are displayed through five years of examination.

Table 22.1. Five-year ROI analysis of selected pharmaceutical companies in India (FY2020–FY2024)

Year	*Sun pharma*	*Cipla*	*Dr. Reddy's*	*Divi's labs*	*Torrent pharma*
FY2020	5.17%	6.55%	**11.16%**	**25.84%**	6.43%
FY2021	3.84%	9.54%	9.08%	**28.75%**	7.74%
FY2022	4.12%	10.46%	10.99%	**33.21%**	9.37%
FY2023	**10.49%**	**12.75%**	**18.33%**	23.71%	10.32%
FY2024	**11.20%**	**12.70%**	**19.95%**	**20.56%**	10.68%

Source: Author.

The ROI records of Dr. Reddy's stand as the market leader through all observed fiscal years as it reached 19.95% during FY2024. The pharmaceutical firm demonstrates outstanding industry position through its powerful financial management system and efficient asset deployment and persistent profitable operations.

During the analyzed period Divi's Labs achieved a noteworthy 33.21% ROI which stood as the highest return among all companies. The ROI ratio for this company decreased between FY2023 and FY2024 and reached 20.56% in the upcoming FY2024. During previous times the company displayed exceptional profitability; however current market dynamics and modifications to internal financial approaches have caused its performance to change.

Sun Pharma together with Cipla have demonstrated significant advancement in their company's performance. Sun Pharma commenced its year with 5.17% profit margins in FY2020 before having a short-term revenue dip in FY2021 which then generated a robust 11.20% margin in FY2024. Cipla exhibited identical financial growth as the company started at 6.55% in FY2020 and achieved 12.70% in FY2024 while implementing strategic financial investments and achieving better financial stability.

The financial return on investment at Torrent Pharma remains consistently stable but consistently modest. Through gradual improvement the company raised its financial output from 6.43% in FY2020 to reach 10.68% in FY2024. The ROI level of this company stands below those of competitors showing potential challenges in its financial strategy development.

Dr. Reddy's and Divi's Labs maintained their position at the top in ROI until Sun Pharma and Cipla reduced the difference through steady growth. The financial performance of Torrent Pharma has been stable but their return generation remains lower than other companies. These market patterns show that pharmaceutical business operations are transforming because successful companies combine strategic investments with adaptable strategies to reach profitability and achieve growth outcomes.

3.2. Return on equity (ROE) analysis

Research assesses Sun Pharma together with Cipla and Dr. Reddy's and Divi's Labs and Torrent Pharma through ROE

Table 22.2. Five-year ROE analysis of selected pharmaceutical companies in India (FY2020–FY2024)

Year	*Sun pharma*	*Cipla*	*Dr. Reddy's*	*Divi's labs*	*Torrent pharma*
FY2020	8.36%	9.76%	**12.57%**	**18.75%**	10.66%
FY2021	6.28%	**13.04%**	9.96%	**21.08%**	11.20%
FY2022	6.85%	**13.72%**	**12.37%**	**25.22%**	12.50%
FY2023	**15.20%**	**15.56%**	**19.51%**	14.23%	14.10%
FY2024	**15.10%**	**15.50%**	**19.85%**	12.04%	**15.50%**

Source: Author.

Table 22.3. DuPont ROE salculation results

Year	*Sun pharma*	*Cipla*	*Dr. Reddy's*	*Divi's labs*	*Torrent pharma*
FY2020	3.77%	**9.97%**	**12.58%**	**18.75%**	10.80%
FY2021	2.78%	**13.46%**	9.96%	**21.08%**	13.02%
FY2022	3.34%	**14.40%**	**12.37%**	**25.22%**	15.47%
FY2023	**8.26%**	**16.66%**	**19.51%**	14.23%	**17.28%**
FY2024	**8.81%**	**15.50%**	**19.85%**	11.69%	**17.34%**

Source: Author.

calculations from Financial Year 2020 to Financial Year 2024 (Table 22.2).

- Dr. Reddy's establishes itself as the leading performer in ROE by delivering sustained profitable operations. The ROE figures for the company showed an initial value of 12.57% in FY2020 followed by a temporary decrease in FY2021 before achieving 19.85% in FY2024. The company demonstrates excellent profit generation abilities and effective use of shareholder capital through its consistent growth trend.
- The five-year period has reached its highest point in FY2022 when Divi's Labs achieved an ROE of 25.22%. The corporation reported its ROE to decline progressively from FY2022 to end at 12.04% in FY2024. The financial management strategy of the organization appears to be evolving because of potential investment increases combined with profit modifications.
- Two big pharmaceutical companies Sun Pharma together with Cipla demonstrated exceptional improvements in return on equity metrics throughout this period. The utilization of equity at Sun Pharma evolved by beginning at 8.36% before a slight decrease in FY2021 then returning stronger to reach its highest point at 15.10% in FY2024. The financial stability of Cipla manifested through its strategic development as its ROE reached 9.76% in FY2020 and eventually grew to 15.50% in FY2024.
- The financial performance of Torrent Pharma has progressed steadily throughout FY2020 to FY2024 because its ROE increased from 10.66% to 15.50%. The upward trend in shareholder value indicates successful financial health and superior management practices with respect to owner value enhancement.

3.3. DuPont analysis

Financial efficiency can be found in DuPont analysis by separating ROE into three distinct parts which are net profit margin alongside asset turnover and financial leverage. Dr. The financial efficiency metrics of Reddy's showed a DuPont ROE of 19.85% during FY2024 (Table 22.3).

The financial efficiency of Dr. Reddy's reaches its highest level due to achieving a DuPont ROE of 19.85% during FY2024. Divi's Labs peaked in FY2022 (25.22%) but declined thereafter. Cipla and Sun Pharma showed steady improvement, while Torrent Pharma maintained stable growth, signaling effective asset utilization and profitability.

3.4. Paired T-Test: ROE vs. ROI

A **paired t-test** was conducted to determine if there was a significant difference between ROE and ROI among the selected companies.

Null Hypothesis (H_0): There is no significant difference between ROE and ROI.

Alternative Hypothesis (H_1): There is a significant difference between ROE and ROI.

Company	*T-Statistic*	*P-Value*	*Interpretation*
Sun Pharma	4.94	0.0078	Significant difference
Cipla	−16.26	0.00008	Significant difference
Dr. Reddy's	−3.40	0.0272	Significant difference
Divi's Labs	19.22	0.00004	Significant difference
Torrent Pharma	−12.46	0.00024	Significant difference

The paired t-test results confirm a significant difference between ROE and ROI for all companies, as all P-values are below 0.05, rejecting the null hypothesis. Divi's Labs (19.22) and Cipla (-16.26) exhibit the highest differences, possibly due to varying capital structures, investment strategies, and profit retention policies. A higher T-statistic suggests stronger deviations in how companies utilize assets versus equity for generating returns, reflecting differences in leverage, operational efficiency, and reinvestment strategies. The negative T-statistics for Cipla, Dr. Reddy's, and Torrent Pharma indicate that ROI significantly lags behind ROE, potentially due to higher asset investments relative to profits, affecting overall return efficiency. These findings highlight key differences in financial management across the pharmaceutical sector.

3.5. ANOVA test: Financial performance comparison

Analysis of Variance test was used to determine if there are statistically significant differences in the financial

Table 22.4. ANOVA results: Financial performance comparison (FY2020–FY2024)

Metric	*F-Statistic*	*P-Value*	*Interpretation*
Net Profit Ratio	12.42	0.00019	Significant difference among companies
Return on Equity (ROE)	2.82	0.0725	No significant difference at 5% level
Earnings Per Share (EPS)	11.37	0.00031	Significant difference among companies

Source: Author.

performance of Sun Pharma, Cipla, Dr. Reddy's, and Divi's Laboratories based on:

- Net Profit Ratio
- Return on Equity (ROE)
- Earnings Per Share (EPS)

The ANOVA test reveals that companies have different profitability measures through Net Profit Ratio (P = 0.00019) and EPS (P = 0.00031) levels (Table 22.4). Profit differences between companies can be explained by their cost structures and pricing strategies alongside market positioning approaches. The ROE results (P = 0.0725) demonstrate no statistical difference indicating companies use shareholder equity with comparable efficiency levels regardless of profit and earnings variations. The differences in Net Profit Ratio reveal varying approaches to controlling operational costs and earning revenues but EPS variations stem from diverse practices concerning dividend distribution and company reinvestment decisions and profit evolution patterns. Analyze of business strategies along with competitor alliances can be achieved by studying these financial insights for pharmaceutical companies.

4. Findings

- The Return on Investment (ROI) Analysis indicates that Dr. Reddy's achieved 19.95% in FY2024 as its highest level due to effective financial strategy and asset utilization. The peak value of 33.21% that Divi's Labs reached in FY2022 did not sustain and instead indicated a strategic change in business direction. The company data revealed that Sun Pharma and Cipla demonstrated consistent growth patterns but Torrent Pharma presented stable but average ROI performance because different investment choices characterized their results.
- The pharmaceutical company Dr. Reddy's maintained robust Return on Equity (ROE) between FY2020 and FY2024 because it effectively utilized equity. The company started FY2020 with 12.57% ROE and reached 19.85% ROE in FY2024. The annual financial results of Divi's Labs reached its peak of 25.22% in FY2022 but later experienced a decrease suggesting alterations in company investment strategies. Sun Pharma achieved 15.10% and Cipla recorded 15.50% in their ROE value for FY2024. The financial performance of Torrent Pharma grew consistently until FY2024 via a 15.50% ROE that displayed dependable resource usage and financial handling practices.
- The DuPont ROE calculation for Dr. Reddy's shows financial strength because it reached 19.85% in FY2024 bringing together robust profit margins and efficient asset utilization. The peak DuPont ROE performance of 25.22% occurred in FY2022 at Divi's Labs but the value decreased afterwards as the company executed different cost structure and reinvestment plans. The companies Cipla and Sun Pharma demonstrated stable growth patterns while Torrent Pharma maintained dependable returns which reflected their constant financial success.
- A paired T-Test analysis of ROE and ROI revealed declared statistical significance ($P < 0.05$) for all companies which validates the different approaches between financial strategy and capital utilization methods. The largest differences between Cipla and Divi's Labs signified major changes in their investment methods alongside their ability to generate returns. The different profitability levels combined with reinvestment plans have distinguished the chosen pharmaceutical companies from each other.
- ANOVA results for Financial Performance Comparison indicated statistically meaningful differences in both Net Profit Ratio (P = 0.00019) and EPS (P = 0.00031) across examined companies. These findings demonstrate that the selected pharmaceutical organizations have divergent profitability and earnings potential. No substantial difference appeared between the companies' returns on equity (P = 0.0725) indicating their shareholders' equity use was similar. Net profit ratios show how companies operate differently whereas EPS shows how they manage dividends and reinvestment which showcases how pharmaceutical organizations use differing financial approaches.

5. Suggestions

To enhance financial performance, pharmaceutical companies should focus on diversification and innovation, particularly those like Torrent Pharma, which exhibited lower ROI. All companies should increase R&D spending to create new products. Optimizing capital structures to balance debt and equity is essential, given the observed differences between ROE and ROI. Strengthening supply chain resilience, as emphasized by prior research, is crucial for mitigating risks from global disruptions. Continued investment in digital transformation, as highlighted in existing studies, can improve operational efficiency and reduce costs. Integrating ESG practices can enhance long-term financial performance and attract socially conscious investors. Companies must proactively adapt to evolving regulatory landscapes and maintain strong compliance. Improving working capital efficiency can directly enhance ROI. Divi's Labs should re-evaluate their investment strategies to reverse declining trends. Regular benchmarking against industry peers should be conducted to identify areas for improvement and maintain a competitive advantage.

6. Conclusion

India's pharmaceutical industry maintains robust profitability together with steady equity returns and developing investment practices. Dr. Financial efficiency data reveals lead positions to Reddy's and Divi's Labs followed by steady growth from Sun Pharma and Cipla. Torrent Pharma operated steadily yet achieved lower return optimization performance than its competitor firms. The discovered financial data enables strategic decision-making for investors and policymakers and business leaders. The pharmaceutical industry requires sustained growth that depends on optimal asset management and improved capital structure and market-responsive strategies in the future.

7. Future Scope of the Studies

Research needs to assess how long governmental initiatives like the PLI scheme impact financial results. The dynamic nature of global pharmaceutical supply chains requires evaluation of their effects on the financial resilience of Indian pharmaceutical companies. Future research must study how artificial intelligence together with biotechnology affects financial performances because this knowledge is essential. The study of both international pharmaceutical entities alongside qualitative studies on ESG factors will enhance the understanding of the Indian pharmaceutical industry's financial aspects.

References

[1] Statista. (2023). *Global pharmaceutical industry market value 2022–2030.* Retrieved from https://www.statista.com

[2] World Health Organization (WHO). (2023). *The impact of pharmaceutical advancements on global health.* Retrieved from https://www.who.int

[3] McKinsey & Company. (2023). *How AI is transforming drug discovery in the pharmaceutical industry.* Retrieved from https://www.mckinsey.com

[4] Food and Drug Administration (FDA). (2022). *Regulatory innovation and pharmaceutical advancements.* Retrieved from https://www.fda.gov

[5] Deloitte. (2023). *2023 Global Life Sciences Outlook: Reshaping the future of healthcare.* Retrieved from https://www2.deloitte.com

[6] India Brand Equity Foundation (IBEF). (2024). *Indian Pharmaceutical Industry Report.* Retrieved from https://www.ibef.org

[7] Pharmaceutical Export Promotion Council of India (Pharmexcil). (2023). *India's Role in Global Generic Drug Exports.* Retrieved from https://www.pharmexcil.com

[8] Department of Pharmaceuticals, Government of India. (2023). *Production Linked Incentive (PLI) Scheme for Pharmaceuticals.* Retrieved from https://pharmaceuticals.gov.in

[9] PwC. (2023). *The financial performance of pharmaceutical companies in India.* Retrieved from https://www.pwc.com

[10] KPMG. (2023). *Pharmaceutical industry financial analysis and investment trends.* Retrieved from https://home.kpmg.com

[11] Sharma, A., & Kumar, V. (2019). Liquidity analysis of pharmaceutical companies in India. Journal of Financial Studies, 27(1), 15–30.

[12] Patel, K., & Desai, M. (2021). Efficiency of asset utilization in the Indian pharmaceutical industry: A ratio analysis. Management Accountant, 56(3), 78–92.

[13] Garg, N., & Mishra, S. (2022). Macroeconomic determinants of financial performance in the Indian pharmaceutical industry. Economic and Political Weekly, 57(15), 39–46.

[14] Das, S., & Banerjee, R. (2023). Panel data analysis of financial performance trends in Indian pharmaceutical companies. International Journal of Finance and Accounting, 12(1), 45–62.

[15] Chandra, A., & Reddy, P. (2022). Non-parametric analysis of financial performance in the Indian pharmaceutical sector. Journal of Business Analytics, 5(2), 112–128.

[16] Singh, A., & Mishra, D. (2018). Corporate governance and financial performance: Evidence from the Indian pharmaceutical sector. *Journal of Corporate Governance*, *26*(4), 512–528.

[17] Joshi, M., & Patel, K. (2020). Working capital management and profitability of pharmaceutical SMEs in India. *International Journal of Small Business Management*, *8*(2), 88–102.

[18] Reddy, L., & Rao, S. (2021). Digital transformation and financial performance: An analysis of the Indian pharmaceutical industry. *Journal of Digital Business*, *3*(1), 45–60.

[19] Menon, N., & Iyer, T. (2022). Impact of price controls on financial performance: Evidence from the Indian pharmaceutical sector. *Regulatory Review*, *15*(3), 220–235.

[20] Banerjee, S., & Verma, A. (2023). Supply chain resilience and financial stability: Lessons from the COVID-19 pandemic in the Indian pharmaceutical industry. *Supply Chain Management Review*, *12*(1), 112–128.

[21] Desai, R., & Chandra, M. (2023). ESG factors and financial performance: A study of the Indian pharmaceutical sector. *Journal of Sustainable Finance*, *6*(2), 145–160.

[22] Verma, A., & Patel, K. (2024). Working Capital Efficiency and ROI in the Indian Pharmaceutical Sector. *Journal of Financial Management and Analysis*, *37*(1), 112–127.

23 Awareness of gender rights in Imphal west district, Manipur

Konika Khuraijam

Associate Professor, DM College of Teacher Education, Manipur University, India

Abstract: Gender rights developed from the perspective of the state have effective implementation problems in the larger consciousness of the people. Awareness makes the conscious decision of deprived genders the safe space to develop in a patriarchal setting- whether institutional or personal. The study conducted in Imphal west district of Manipur will aim to discuss the lack of awareness among the people of the state, what are the causal agencies, and the effective spatial conditions which can overcome them. Through a survey method, this paper is an attempt to signify the major barriers prevalent among inhabitants of Imphal west district. For this developing region, contemporary inquiries can help bridge the gap between gender differences through awareness of rights.

Keywords: Women's rights, political awareness, socio-economic development, gender sensititization, inclusivity

1. Introduction

Conscious awareness of women's rights and its *raison d'être* pertains to a collective sense of human rights, which exist outside of the individual. From the perspective of the people, the 'elitist' development of women's rights over the years has failed to incorporate into a diverse country like India, where several factors pertaining to an individual's community, class, caste, and religion have overplayed their part in limiting the scale of reform through negligence of it understandings on a wider, universal scale [1].

The interest in women's rights largely owes itself to the bourgeoisie democratic development of the 19th century in Europe [2]. Ever since the shifting terrains of women's rights have pushed democratic governments towards challenging the patriarchal attitudes prevalent in all areas of collective society. In the Indian state, the universality of women's rights is being pushed to the sidelines due to neglect of several complexities emerging from institutional settings and complex conservative mindsets throughout its regions and their unique compositions [3].

The argument that women's rights in India are recognized under the patriarchal lens, is to bring the fact that women's objectification, subjugation, oppression and other disadvantageous positions need to be studied individually and in their own uniqueness to fully understand the level of awareness among a particular section of society. Keeping in mind universal women's reform movements and their rights that guarantee basic human dignity and freedom, the development of women's statutes reveals the need to incorporate effective implementation of these rights. For the women of Manipur, their repute for their contributions to the state's economic, social and political affairs have been well established and have been prominent areas of study for several social scientists.

However, persistent forms of oppression in Manipur state have corroded the idea of an egalitarian society. Several social scientists have linked violence towards women with communal tensions, bringing to the forefront the negligence of state implementation of awareness of women's rights. Therefore, the paradoxical situation of women's issues that deal with occurrences of missing girl children, domestic violence, child marriages, lack of education, social exclusion, and workplace issues are contradictory to universal women's rights. Moreover, persistent campaigns to put pressure on the government to enact reforms are yet to occur, despite some real attempts to modify drafts that would safeguard women's rights.

Deprivation in all these sectors encompasses all ages, classes, and castes in the uniquely formulated state with people from different backgrounds. Stringent recognition and implementation of women's rights effectively becomes the first step towards women's development [4]. The only way to approach the level of awareness about women's rights from the tangle they are now in is to take a methodical approach to enacting small, targeted reforms through focused campaigns, and by assessing the prevalent rights and its impact through regional inquiries [5]. Through this inquiry, a new mode of rights will also be assessed- that of a mental perspective of women's lives, which will help in developing the awareness of women's rights in the region of Imphal west, Manipur [6].

2. Literature Review

Awareness of women's rights can be studied in several contextual events of time and space. Earlier, the laws that were

konikhur@hotmail.com

DOI: 10.1201/9781003684459-23

closely modelled to fit the needs of the colonial state- such as changes in consent laws, widow inheritance laws, and abortion laws took shape under the pretence of 'women's rights'. However, the formulation of rights guaranteed to serve the needs of the colonial, patriarchal state [2], and served little to no purpose for the upliftment of women. Early twentieth-century formations with strong nationalistic tones, remaining elitist and unaccessible to the larger population.

Widespread grassroots movements in the post-independence era emphasized the economic and social rights of women. The third wave emerged in the late 1980s, linking women's issues to the environment with a focus on alternative strategies of development [7].

Burns, Scholzman, and Verba [8], I identify traditional patterns of political socialisation as the reason why women perceive 'political right' as a 'man's game'. Additionally, it only further damns women to the role of equal contributor to those who are due their rights. By following certain programs of gender stereotypes, Thakur [9], provides educational interventions, economic independence and gender sensitisation as imperative measures to tackle the disconnections between people and awareness of gender rights.

The following are, therefore, the traditional patterns of political socialisation, which make women form the opinion that the political right is a 'man's game'. It also only serves to damn women once again to the roles of equal contributor to those who are due their rights.

An investigation of such gender relevant political knowledge in the state of Manipur, however, allows us to explore the debate of gender gap in Manipur in terms of gender differences in awareness of women's rights.

For their active participation in political and economic affairs in Manipur, Khuraijam [10] has provided a detailed contribution of women from various communities making conscious efforts towards collective empowerment during the colonial period.

Moreover, the diverse women of Manipur state belonging to different communities have significantly contributed to the state's development, and active political agencies gave women of Manipur significant recognition [10].

According to Babulu et al. [11], the increasing number of persons receiving assistance in Manipur state, under various programmes for empowerment and rights, is not increasing the coverage potential for these individuals and is significantly lower than what is necessary. He brings out the immediate demand for voluntary efforts in providing the services and resources required to satisfy the high demand. Khuraijam [13], further discusses the issue of the lack of basic education and technological facilities among rural women of Manipur, who even after significant contributions in agricultural markets are unable to empower themselves fully.

Therefore, an important implementation issue with welfare organisations in playing awareness roles is limited according to Babulu et al. [11] and Chhibber [7]. The stereotypical mindset of women as projections of mere 'moms and spouses', especially for those living in rural regions, is not benefiting these individuals in the State to the intended extent because of the lack of institutional assistance programs.

An important institutional effect of patriarchy on human rights has been discussed by Khuraijam [10], on the increased violation of rights on its transgender population, facing high levels of marginalisation and neglect of the state, and brings to the forefront patriarchal systems within institutions, societal perceptions, and economic deprivations. Furthermore, patriarchal norms have affected the women of the region over the years. In an early attempt to present the ideality of women's conditions in Manipur by Kipgen [12], a matrilineal culture with increased economic contribution of women in general takes place. But in Khuraijam's [13], detailed studies, we find evidence of increased work-life balance constraints among women, with several responsibilities prevalent among the selected women and responsibilities of organisations to bring about beneficiary practices for them. However, economic well-being with depreciating recognition of women's rights awareness can also be a significant factor towards this problem.

The credit given to the contemporary digital age and internet connectivity which has made it possible for women's awareness and rights to reach a significant amount of people has reached the epitome of virtual collective action. For Gurumurthy et al. [14] developments in Information and Communications Technologies (ICTs) have significantly contributed to the development of the gender justice agenda, by transforming public-political spheres of women's lives through active intervention in online debates and opinions citing concerns for women's future in the country. However, studies conducted by Mishra [15], also bring to the forefront the issue of gender-based contemporary violence to several women, citing experiences of women being subject to stalkers and online harassment. While new-age digital activism can further uplift awareness of rights for women, credibility and authentic developments on women's situations are still lacking. Therefore, while one should not ignore the role of online communication in spreading awareness of women's rights, scholarly research on the level of awareness of women's rights, becomes the prerequisite of the developing state of Manipur.

Lastly, with the provision of gender-based empirical enquiries, this study aims to map the awareness of gender and gender laws across a wide range of political and historical contexts and investigate tactics that could protect women's rights in the face of intricate personal, economical, social and political constraints.

3. Research Methodology

This study attempts to elicit and determine the levels of awareness of women's rights in the Imphal west district of

Manipur state. It contains a set of questions which measure attitudes and information on women's representation both in their political life and in the economic, social and personal spheres and in particular degrees of political efficacy, interest and participation will be assessed. A simple random sampling procedure was adopted to bring the authenticity of awareness levels amongst a generic population and to compare the level of awareness of males and females of both unmarried and married persons residing in the state of Manipur. A questionnaire was developed with 25 statements with yes/no options to determine the level of awareness of rights. Questions were primarily situated around aspects related to fundamental, political, economic and social rights.

4. Analysis

Table 23.1 it shows that there exists no significant difference between men and women regarding awareness of gender rights in Manipur.

The context of awareness of rights in its fundamental state can be associated with a political lack of knowledge. As discussed by Kalpagam [16], the simple pretexts of early feminist movements in India were associated with personal symbolic movements as motherly figures and their association with their traditional roles. Due to traditional family structures which existed within the larger caste sections of Manipuri society, there was limited awareness between both men and women of the society. The societal impact was highly affluent in the neglect of awareness of women's rights.

In contrast, progressive feminist rights awareness was limited to 'fragmented' and 'educated' sections of Imphal West society, of which, several were highly distant from popular conceptions of the existing awareness drives taking place. Lack of awareness of gender rights, also equally contributes to the oppression and subjugation of women in the state. In the survey, the absence of political positions of women in government institutes marked a major hindrance towards the lack of their rights. Studies conducted by Khuraijam [13], demonstrate the prevalence of sexual abuse, lack of education, low access to entrepreneurship opportunities and absence of digital presence for a large section of the rural women. Several of these factors are equivalent in contributing to the lack of awareness of women's rights. Contemporary studies to bring the lack of awareness amongst the population on gender rights can be encouraged to fully incorporate sustainable development into the state. To reach this goal, governments and NGOs must take full cognisance of utilising awareness mechanisms- such as local media and online communication to ensure the issues of minimal significance regarding awareness of gender rights. Moreover, state intervention in public spheres- such as schools, government offices, public centres, hospitals, etc can incorporate the role of gender rights as a means of sustainable development among people.

Table 23.2 this issue is reflected in the result between unmarried and married people on the no significant awareness of gender rights. A major noticeable change in this section found its way as married people scored lower on tests of political knowledge regarding right and opportunities. The reason why the respondents relied more often on 'no' responses was because, while the respondents had relatively high levels of risk aversion, and this therefore resulted in fear of being wrong. These results indicate that there is a psychological component to answering questions that is independent and distinct from the amount of information persons may have.

Table 23.1. Comparison between men and women regarding awareness of gender rights of Imphal west district

Category	*Total (N)*	*Mean*	*Standard Deviation (S.D.)*	*Standard Error (SE)*	*t-radio*
Men	$N_1 = 50$	$M_1 = 43.28$	$\sigma_1 = 4.693$	0.887	0.676
Women	$N_2 = 50$	$M_2 = 43.28$	$\sigma_2 = 4.q67$		

df = 98
Critical value at 0.05 = 1.99
Remark = Not significant

Source: Author.

Table 23.2. Comparison between unmarried and married people regarding awareness of gender rights of Imphal west district

Variables	*Total (N)*	*Mean*	*Standard Deviation (S.D.)*	*Standard Error (SE)*	*t-radio*
Unmarried people	$N_1 = 50$	$M_1 = 43.08$	$\sigma_1 = 4.081$	0.9067	0.2205
Married people	$N_2 = 50$	$M_2 = 42.88$	$\sigma_2 = 4.369$		

df= 98
Critical at 0.05=1.99
Remarks= Not significant

Source: Author.

A second set of explanations suggests that the gender gap exists because the determinants of political knowledge are themselves gendered: that is, they are experienced differently by women and men, which results in different knowledge levels for each. Researchers point to women's historically lower levels of education and occupation, their greater time commitment to childrearing and family life, and their lower levels of political interest and discussion as sources of the gap [17]. Therefore, the identification of gender rights awareness makes it difficult for the awareness to take cognition of, rather than focussing on prevalent rights, the respondents preferred to stay aloof given the plethora of mental strains they were suffering from.

Table 23.3 the mean awareness of gender right between married men and unmarried men in Imphal West District, Imphal did not differ.

The majority of the results were influenced by intersectional unawareness. Men of deprived sections of the society were not aware at all of women's rights but the upper sections of the male society of the Imphal west also showed equal ignorance on the subject.

The intervention of active human engagement is needed to address the implementation of the women's rights for the economically and socially weaker male sections. According to Verma et al. [18]. There is realistically not much time to discuss and reflect on women's issues, this however, is a new phenomenon, especially for those situated in rural parts of the state, as Al (2006) put it. This was often the first such occasion for them to talk and reflect on these issues. Based on reports on attitudes towards gender and sexuality, reporting of behaviour in relationships, interviews revealed that there was decline in support of inequitable gender norms and sexual harassment of girls and women. It can be useful for women in several areas of Manipuri society if young men could be engaged in discussing gender dynamics and health risks in a critical manner and if there can be change in some key gender related attitudes.

In addition, there are interventions of provision of the agency to women's agencies to recognize them to cultivate both mens and women who can rise up holistically.

Table 23.4 comparison regarding married women and unmarried women regarding their awareness of gender rights in Imphal West District was found insignificant. The Imphal west district area is under the incessant influence of the patriarchal setting which has led to the thought of women playing their subversive roles in the Manipuri society.

Identification of women as agents of change is inevitably possible through awareness programmes on education, health and economic independence of women. This can be achieved using resources which in turn have a positive impact on the problem not only in terms of women's freedom and well being but also of the well being of the whole family, society and the well being of the nation as a whole.

A child of a woman is born and she has to play her reproductive role and all this happens because of the biological differences. She requires a more tended and delicate treatment than men to truly appreciate her fundamental rights, for instance, safe transport, adaptable calendars, childcare bolster and so forth, which will lessen their weight. In reality, this will furnish the stage guidelines of the platform with the goal to empower women in developing their situation by awareness of gender rights.

Table 23.3. Comparison between married men and unmarried men regarding awareness of gender rights of Imphal west district

Variables	*Total (N)*	*Mean*	*Standard Deviation (S.D.)*	*Standard Error (S.E.)*	*t-radio*
Married Men	N_1=25	M_1=43.68	σ_1 = 4.295	1.196	0.669
Unmarried men	N_2= 25	M_2= 42.88	σ_2 = 4.159		

df = 48
Critical value at 0.05 = 2.02
Remark: Not significant

Source: Author.

Table 23.4. Comparison between married women and unmarried women regarding awareness of gender rights of Imphal west district

Variables	*Total (N)*	*Mean*	*Standard Deviation (S.D.)*	*Standard Error (S.E.)*	*t-radio*
Married Women	25	42.08	4.299	1.167	0.719
Unmarried Women	25	43.28	3.994		

df= 48
Critical value 0.05 = 1.69
Remark= not significant

Source: Author.

5. Conclusion

From the above inquiries, it is evident that the need to incorporate awareness-driven interventions in the state of Manipur is evidentiary. The lack of significant awareness between males and females, married and unmarried people, married males and unmarried males, and lastly, married females and unmarried females contribute towards gender deprivation of political, social, and economic rights for women. The worst form of these forms of awareness lack is directly attributed to the state's ongoing tensions about women's conditions. The environmental applications of women's rights programs, through means of systematic education at elementary school levels, and incorporating gender-sensitised cognition amongst the youth, can bring effective change.

Through the above inquiries, it is evident that the prevalent women's rights are incapable of delivering the desired outcome of gender equilibrium in the state. We can hence, utilize this study as a building platform for more inclusive gender rights that incorporate a diverse community such as Manipur. The diverse hilly and plain communities, people of different religions, transgenders, and economically backward sections have to be taken into account while drafting more gender-sensitized rights. Future research must be encouraged based on different sections of people to fully understand the personal issues and how awareness of gender rights can solve the problem.

The above standardisation people helps in serving the notion that marriage acts as a barrier, especially for those women who want to develop and grow economically, socially and politically. Marriages must account for the safe platform for their growth, and the transgression of it into a hindering factor in women's development must be addressed. Women's support in work, as well as their homes, can create a safe platform for them to outgrow their issues.

A significant contribution towards increased awareness of women's rights can be conducted through the development of education spheres which removes all age barriers. Special education drives that encourage more sensitisation of biological and physical surroundings to bridge the gaps of gender difference at all ages and for all people can encourage the effective role of gender awareness. In an attempt to connect gender differences, contemporary issues of social media influence, political news, and authentic information must be encouraged as an increased form of communication between males and females.

In conclusion, the globalisation and universality of gender rights can be induced to accommodate and take shape keeping in mind the region's spatial culture. Rights when discussed among a variety of different communities can bring out the existing dominant perception of gender-constructed roles of society, which can be furthered developed via increased educational drives, removal of stereotypes, workplace opportunities, and marriage support can increase the engagement of rights and discussions.

References

[1] Eshwar, L., & Selvalakshmi, G. (2024). Awareness and Perception of Gold Ornament Buyers in Bangalore: A Mixed Method Approach. *Indian Journal of Information Sources and Services*, *14*(3), 30–38. https://doi.org/10.51983/ijiss-2024.14.3.05

[2] Chitnis, V., & Wright, D. (2007). Legacy of Colonialism: Law and Women's Rights in India. *Washington and Lee Law Review*, *64*, 1315.

[3] Manosh Happy, M., & Shanmugapriya, J. (2016). Solar Photovoltaic Power Generating Scheme with DC/DC Using SEPIC Converter with PSO. *International Journal of Advances in Engineering and Emerging Technology*, *7*(4), 82–93.

[4] Jawad, M. H., & Turki, Z. T. (2024). Diagnosis of DC Glow Discharge Plasma of Krypton Gas Under Different Pressures. *International Academic Journal of Science and Engineering*, *11*(1), 98–106. https://doi.org/10.9756/IAJSE/V11I1/IAJSE1113

[5] Balaji, R., Logesh, V., Thinakaran, P., & Menaka, S. R. (2022). E-Learning Platform. *International Academic Journal of Innovative Research*, *9*(2), 11–17. https://doi.org/10.9756/IAJIR/V9I2/IAJIR0911

[6] Sun, J., Wang, Y., Si, H., Yuan, J., & Shan, X. (2010). Aggregate Human Mobility Modeling Using Principal Component Analysis. *Journal of Wireless Mobile Networks, Ubiquitous Computing, and Dependable Applications*, *1*(2/3), 83–95.

[7] Chhibber, B. (2018). Women's Rights are Human Rights. *World Affairs: The Journal of International Issues*, *22*(1), 122–135.

[8] Burns, N. S., & Verba, L. (2001). *The Private Roots of Public Action*. Cambridge, MA: Harvard University Press.

[9] Thakur, D. (2019). Gender equality and human rights in India: Issues and perspectives. *An International Peer Reviewed Open*, *2*(3): 169–275.

[10] Khuraijam, K. (2022). The Role of Women in the Freedom Struggle: The Manipur Chapter. *International Journal of Current Science*, *12*(3), 210–215.

[11] Babulu, T., & Devi, E. R. (2023). Perception of Women on Utilising Government Welfare Schemes In rural areas of Manipur. *International Journal of Research and Analytical Reviews*, *10*(2), 897–907.

[12] Kipgen, T. G. (2010). *Women's Role in the 20th Century Manipur: A Historical Study* (1st edn). Gyan Publishing House.

[13] Khuraijam, K. (2021). Issues and Challenges of Rural Women of Manipur. *International Journal of Early Childhood Special Education*, *13*(01), 1151–1158.

[14] Gurumurthy, A., Chami, N., Babbar, A., Vasudevan, M. P., & Sudharma, N. (2014). Digital technologies and gender justice in India. *IT for Change*, 202014–202011.

[15] Mishra, D. (2016). Role of Social Networking Sites on Gender Violence. *International Journal of Advanced Information Science and Technology*, *5*(11), 58–63.

[16] Kalpagam, U. (2000). The Women's Movement in India Today-New Agendas and Old Problems. *The History of Doing: Feminist Studies*, *26*(3), 645–660.

[17] Khuraijam, K., Devi, N. R., & Kangabam, A. (2023). A Study on Stereotyping Women's Leadership Roles. *Journal of Namibian Studies: History Politics Culture*, *33*, 4074–408.

[18] Verma, R. K., Pulerwitz, J., Mahendra, V., Khandekar, S., Barker, G., Fulpagare, P., & Singh, S. K. (2006). Challenging and changing gender attitudes among young men in Mumbai, India. *Reproductive Health Matters*, *14*(28), 135–143.

24 The impact of artificial intelligence on consumer buying behaviour: A study on online retail purchases

P. Easwaran[1,a], *V. Krishnaveni*[2,b], *M. Kalimuthu*[3,c], *Sankar Ganesh R.*[4,d], *J. Sudarvel*[1,e], *and Smruthymol J.*[1,f]

[1]Karpagam Academy of Higher Education, Coimbatore, Tamil Nadu, India
[2]Professor and Dean, Department of Management Studies, Dhanalakshmi Srinivasan College of Engineering, Coimbatore, Tamil Nadu, India
[3]Professor and Head, Department of Commerce with Professional Accounting, Dr. N.G.P. Arts and Science College, Coimbatore, Tamil Nadu, India
[4]Associate Professor, Department of Management Studies, Vel Tech Rangarajan Dr Sagunthala R&D Institute of Science and Technology, Chennai, Tamil Nadu, India

Abstract: AI technology is reshaping the way consumers deal with e-commerce platforms through its radical transformation of online retail sector business models. The research examines changes in AI through the analysis of predictive analytics together with conversational AI tools which are combined with customized solutions to assess their impact on customer buying behaviours. To study AI's positive effects on user satisfaction and business profitability the authors merged statistical information evaluation with feedback collected during face-to-face surveys with customers. Analysis together with survey data enables evaluation of AI advantages for enhancing customer satisfaction while simultaneously measuring its sales success. Analytical findings demonstrate that AI technology generates better customer contentment and boosts conversion ratios although it greatly affects what buyers want to purchase.

Keywords: Artificial intelligence (AI), online retail, e-commerce personalization and retail data analytics

1. Introduction

The online retail business sector experiences significant technological changes which develop individualized shopping experiences using efficient and streamlined processes. Data-driven tools implement by online retailers help them improve product suggestion systems as well as search capabilities and deliver instant customer service. The resulting innovations provide users with better ways to discover items and analyze costs thereby enabling them to buy products with full knowledge.

The influence from digital enhancements on consumer actions continues to rise based on recent statistical data. An increase of 4% from the prior year marked the $282 billion worth of online sales in the United States during the 2024 holiday season. The customer behaviour shift brings about a significant change because more than 40% of shoppers actively use automatic price comparison tools before completing their transactions. Various retailers have implemented sophisticated recommendation systems that offer taste-prediction for wines and showroom recommendations matching past purchases which enhances customer satisfaction.

Retailers embrace smart technologies to enhance customer interactions between themselves and their consumers. Community-based shopping assistants and self-operating customer support systems coupled with customized marketing initiatives have become typical retail industry functions. The implementation of these technological innovations improves consumer convenience as well as guides purchasing activities and affects customer-brand interaction.

Research will evaluate the effects of modern retail technologies on online marketplace consumer purchasing choices. Recent trends alongside statistical data and technological innovations are combined in this study to offer complete knowledge about digital tools' transformation of contemporary retail operations.

2. Review of Literature

- Smith [1] discovered that AI-driven personalized shopping produces loyal customers who frequently buy in online retail platforms.

[a]easwar83@gmail.com, [b]krishnaveni.rkv99@gmail.com, [c]kalimuthusujanias@gmail.com, [d]sankarganeshphd@gmail.com, [e]j.sudarvel@gmail.com, [f]jayalalsmruthy@gmail.com

DOI: 10.1201/9781003684459-24

- According to Brown [2] the deployment of AI chatbots produces better customer satisfaction results through timely precise solutions for online shoppers.
- According to Jones and Lee [3] consumers make their purchasing choices more readily because AI operates 24/7 customer support which promptly answers questions and addresses concerns.
- White [4] emphasizes that trust from customers toward AI technologies remains essential for helping them purchase online while pointing out their need for secure and dependable systems.
- Online retailers can predict customer preferences through behavioural analytics created by AI systems to deliver customized promotions which resulting in changed customer shopping habits according to Martinez [5].
- Wilson [6] conducted research on how AI nudges and incentives through AI mechanisms prompt customers to make impulsive online purchases.
- The use of AI-powered visual search tools that Chen [7] discusses has become increasingly widespread because they enable shoppers to search by images to find products.
- Green [8] investigated the effect of augmented reality (AR) on product depiction through immersive shopping modes that boost customer interaction and purchasing goals in online retail.

Kim and Park [9] researched how AI-based dynamic pricing affects consumer purchase decisions by showing prices change frequently leads consumers to change their buying decisions.

Lee and Choi [10] studied how AI contributes to security systems that protect online transactions while improving trust and shopping confidence of consumers.

Based on their work Nguyen and Tran [11] demonstrated how voice commerce operates as a strong determinant of online shopping ease and frequency for consumers.

Davis and Wilson [12] studied how AI can forecast product returns through its proactive recommendation system which helps customers pick better sizes for improved delivery satisfaction.

Patel and Shah [13] studied the effects of AI-powered personalized marketing advertisements which lead to substantial elevation of consumer buying engagement and conversion rates.

Sentiment analysis through AI received research attention from Rodriguez and Gomez [14] who discovered retailers could strengthen their products and support through AI-derived insights.

Through their study Thompson and Baker [15] showed that AI drives the production of virtual fitting rooms which leads to enhanced item visualization with reduced purchasing doubt.

Product availability improves and stockouts decrease when AI-based inventory management provides accurate forecasts according to Walker and Hill [16].

The study by Young and Moore [17] examined the effects of AI-powered email customization that leads to increased email interactions and customers' final purchases.

Zhao and Wu [18] conducted research which demonstrated how AI analytics on social media helps retailers identify consumer trends through their social data for improving their marketing strategy development.

2.1. Research gap

Many investigations have studied individual technological applications in online retail but no one has yet determined their complete influence over consumer behaviour. Research through time enables the assessment of how consumers change their reactions throughout various population segments and product groups. Research needs to explore the psychological consequences together with the ethical aspects created by technological progress and their resulting effects on consumer satisfaction after making a purchase. Small and medium-sized enterprises (SMEs) require a thorough evaluation of these emerging technologies because it serves to reveal their wider impacts on the retail industry.

3. Statement of the Problem

The adoption of advanced technology tools by online retail creates new consumer spending patterns which scientists still need to thoroughly study. Current research about advanced technology in online retail examines applications one at a time without exploring impacts from multiple technologies. The current research lacks a detailed examination of three areas: consumer trust building over time, behavioural understanding, racial differences between groups and product types, and practical adoption of ethical coding standards for technology devices and effects on both small business success and consumer satisfaction following purchases. Research needs to develop full perspectives about consumer preferences and ethical standards so retailers can optimize their service delivery to customers effectively.

3.1. Significance of the study

The study delivers vital information to retailers and promotes awareness about technological impact and supports digital tool development in a responsible manner. The research supports academic development and guides ethical policymaking regarding technology implementation while directing small and medium-sized enterprises through digital progress. The investigation into the intricate technology-consumer relationship helps develop an informed online retail environment which combines ethical practices with effective operations for all participating stakeholders.

3.2. Objectives of the study

- To study the AI on shaping of consumer buying habits in online retail, focusing on personalized recommendations and product offerings.
- To study the AI-powered chatbots and predictive analytics in enhancing customer experience and conversion rates.

3.3. Research methodology

The research evaluates online retail consumer behaviour changes under technology influence through quantitative survey methodology. A 5-point Likert scale measures the close-ended questions within an online questionnaire which serves as the primary source of data collection. A random sampling process identifies respondents throughout different backgrounds as part of 183 participants to gain comprehensive insight. A descriptive statistical analysis was utilized alongside interpretation of the data to understand consumer awareness and trust, satisfaction, and ethical matters about technology in online purchasing.

3.4. Data analysis and interpretation

3.4.1. Demographic profile

The data gives the demographic profile of respondents' age, gender, education, job status, and income levels, providing a clear view of their background and financial situation.

- Age: The majority of respondents are aged 25–34 years (31 responses), followed by those in the 35–44 years category (29 responses).
- Gender: More responses were received from males (97) compared to females (86).
- Education: The most common educational qualification is a Bachelor's Degree (41 responses).
- Employment Status: The highest number of respondents is privately employed (33 responses).
- Income: The most frequent income range is Rs. 50,000–Rs. 74,999 (39 responses).

3.4.2. Awareness of AI in consumer behaviour

Table 24.1 shows the consumer AI technology in online shopping, focusing on awareness, trust, impact on purchasing

Table 24.1. Consumer awareness and perceptions of AI in online shopping

Factors	*5*	*4*	*3*	*2*	*1*	*Total Score*	*Mean*	*Rank*
I am aware of AI in online shopping.	39	39	45	30	30	183	3.147541	9
	195	156	135	60	30	576		
AI influences my purchase decisions.	26	51	24	46	36	183	2.918033	3
	130	204	72	92	36	534		
I trust AI-driven product recommendations.	36	32	45	35	35	183	2.994536	5
	180	128	135	70	35	548		
AI improves my online shopping experience.	39	32	37	41	34	183	3.005464	6
	195	128	111	82	34	550		
I understand how AI personalizes my shopping.	25	34	31	50	43	183	2.715847	1
	125	136	93	100	43	497		
AI-powered ads affect my buying choices.	30	33	33	33	54	183	2.737705	2
	150	132	99	66	54	501		
I am concerned about AI and data privacy.	42	32	43	36	30	183	3.10929	8
	210	128	129	72	30	569		
AI enhances customer service (chatbots, recommendations).	44	32	46	30	31	183	3.153005	11
	220	128	138	60	31	577		
I prefer AI-driven personalized offers and discounts.	42	31	31	36	43	183	2.961749	4
	210	124	93	72	43	542		
AI influences my trust in online retailers.	48	31	34	40	30	183	3.147541	9
	240	124	102	80	30	576		
I understand the ethical issues related to AI.	43	37	25	46	32	183	3.071038	7
	215	148	75	92	32	562		
AI will continue to shape online shopping.	45	36	43	29	30	183	3.202186	12
	225	144	129	58	30	586		

Source: Author.

decisions, privacy concerns, and ethical considerations. Responses were collected using a 5-point Likert scale (5 = Strongly Agree, 1 = Strongly Disagree), with rankings based on average scores.

The findings indicate that consumer awareness of technology in online shopping varies. While many recognize its growing impact (Mean = 3.20), there is limited understanding of its role in personalization (Mean = 2.71). Trust in technology-driven recommendations and personalized offers is moderate, suggesting cautious acceptance. Concerns about data privacy (Mean = 3.10) remain significant, highlighting apprehensions regarding security. Although technology enhances customer service, targeted advertisements raise skepticism. Overall, technology contributes to a more convenient shopping experience, but issues related to trust, ethics, and privacy must be addressed to strengthen consumer confidence and ensure long-term acceptance.

3.4.3. Technology and consumer buying behaviour

Table 24.2 provides insights into consumer opinions on how technology influences their buying decisions, shopping experiences, and concerns about data privacy. Responses were gathered using a 5-point Likert scale (5 = Strongly Agree, 1 = Strongly Disagree), with rankings determined by average scores.

The study highlights the role of technology in shaping consumer buying behaviour. While it enhances convenience (Mean = 2.97) and personalizes shopping (Mean = 2.92), concerns about data privacy (Mean = 3.00) remain significant. Trust in technology-based recommendations is moderate (Mean = 3.04), indicating a cautious approach. Consumers rely on customer reviews (Mean = 2.87) but are influenced by targeted ads (Mean = 2.79). Personalized discounts (Mean = 2.97) impact purchasing decisions, yet comfort in sharing data (Mean = 3.03) varies. Overall, technology improves shopping experiences, but trust and privacy concerns must be addressed for broader acceptance.

3.4.4. Consumer satisfaction with AI in shopping

Table 24.3 represent the consumer satisfaction with technology in online shopping, covering aspects like personalized recommendations, shopping convenience, and customer

Table 24.2. Impact of technology on consumer buying behaviour

Factors	*5*	*4*	*3*	*2*	*1*	*Total Score*	*Mean*	*Rank*
Technology influences my purchase decisions.	33	49	33	32	36	183	3.060109	12
	165	196	99	64	36	560		
I trust technology-based product recommendations.	38	46	22	40	37	183	3.043716	11
	190	184	66	80	37	557		
Technology personalizes my shopping experience.	36	34	32	42	39	183	2.923497	3
	180	136	96	84	39	535		
I am concerned about data privacy.	35	35	40	41	32	183	3	8
	175	140	120	82	32	549		
Technology improves my post-purchase experience.	38	37	35	30	43	183	2.983607	7
	190	148	105	60	43	546		
I prefer online shopping due to technology.	40	39	28	40	36	183	3.038251	9
	200	156	84	80	36	556		
Targeted ads influence my purchasing choices.	31	33	35	36	48	183	2.797814	1
	155	132	105	72	48	512		
Technology makes shopping more convenient.	37	30	47	29	40	183	2.972678	5
	185	120	141	58	40	544		
I rely on customer reviews before buying.	26	36	46	39	36	183	2.874317	2
	130	144	138	78	36	526		
Chatbots improve my online shopping experience.	32	41	34	39	37	183	2.956284	4
	160	164	102	78	37	541		
Personalized discounts impact my buying decisions.	42	28	35	40	38	183	2.978142	6
	210	112	105	80	38	545		
I am comfortable sharing my data for recommendations.	42	36	31	35	39	183	3.038251	9
	210	144	93	70	39	556		

Source: Author.

Table 24.3. Satisfaction with AI in shopping

Factors	*5*	*4*	*3*	*2*	*1*	*Total Score*	*Mean*	*Rank*
I am satisfied with AI in shopping.	36	37	42	40	28	183	3.071038	3
	180	148	126	80	28	562		
AI improves my online shopping experience.	39	39	38	39	28	183	3.120219	2
	195	156	114	78	28	571		
AI-based recommendations match my preferences.	39	30	36	42	36	183	2.967213	6
	195	120	108	84	36	543		
AI makes shopping faster and easier.	37	46	27	33	40	183	3.038251	4
	185	184	81	66	40	556		
AI-powered customer service is helpful.	32	32	44	35	40	183	2.896175	8
	160	128	132	70	40	530		
AI-driven ads are relevant to my interests.	32	46	35	35	35	183	3.027322	5
	160	184	105	70	35	554		
AI helps me make better purchase decisions.	40	44	37	26	36	183	3.142077	1
	200	176	111	52	36	575		
I would recommend AI-powered shopping.	31	44	33	37	38	183	2.961749	7
	155	176	99	74	38	542		

Source: Author.

support. Responses were collected using a 5-point Likert scale (5 = Strongly Agree, 1 = Strongly Disagree), with rankings determined by average scores.

The study reveals that AI enhances consumer satisfaction in online shopping, with the highest mean score (3.14) indicating that AI helps in making better purchase decisions. AI improves shopping experiences (3.12) and makes the process faster and easier (3.03). However, AI-powered customer service (2.89) and personalized recommendations (2.96) show moderate satisfaction, suggesting room for improvement. While AI-driven ads (3.02) are seen as relevant, consumers are somewhat hesitant to fully recommend AI-powered shopping (2.96). Overall, AI contributes positively to shopping experiences, but factors like customer service and ad relevance need refinement to enhance consumer trust and satisfaction.

3.4.5. Challenges of AI in shopping

Table 24.4 presents the challenges consumers face when using technology in shopping, such as concerns about data privacy, unfair pricing, and a lack of transparency. The responses were collected using a 5-point Likert scale (5 = Strongly Agree, 1 = Strongly Disagree), with rankings based on average scores.

The study highlights several concerns regarding technology in shopping. The highest-ranked issue (3.20) is that it influences consumer choices without their awareness, raising concerns about control over decision-making. Bias in product suggestions (3.14) and chatbot inefficiency (3.08) also affect user trust. Unfair pricing (3.03) and inaccurate recommendations (3.01) further contribute to dissatisfaction. Ethical concerns (2.95) and privacy issues (2.97) indicate discomfort with data collection practices. Additionally, intrusive ads (2.90) and limited shopping options (2.90) reduce user satisfaction. Overall, while technology enhances convenience, addressing biases, transparency, and customer service quality is crucial for improving consumer trust.

3.5. Findings

- Increasing Influence of Technology: Consumers recognize the growing impact of technology on online shopping experiences (Mean = 3.20); however, there remains a limited understanding of how it personalizes recommendations (Mean = 2.71).
- Measured Trust in Digital Recommendations: While consumers acknowledge personalized suggestions and promotional offers, their acceptance remains cautious, indicating a need for greater transparency and reliability (Mean = 2.99).
- Concerns Over Data Privacy and Security: A significant proportion of consumers express apprehension regarding the collection and utilization of their personal data, underscoring the need for stricter privacy measures (Mean = 3.10).
- Convenience Versus Consumer Skepticism: Although technological advancements enhance shopping efficiency and post-purchase experiences (Mean = 2.97), concerns persist over the intrusiveness of targeted advertising (Mean = 2.79) and the perceived fairness of dynamic pricing strategies (Mean = 3.03).

Table 24.4. Challenges with AI in shopping

Factors	*5*	*4*	*3*	*2*	*1*	*Total Score*	*Mean*	*Rank*
AI-based recommendations are often inaccurate.	33	42	36	38	34	183	3.010929	5
	165	168	108	76	34	551		
AI-driven ads feel intrusive and repetitive.	35	33	37	36	42	183	2.907104	8
	175	132	111	72	42	532		
AI collects personal data without clear consent.	32	40	38	37	36	183	2.972678	6
	160	160	114	74	36	544		
AI-powered customer service lacks human empathy.	39	26	42	27	49	183	2.885246	10
	195	104	126	54	49	528		
AI-driven pricing strategies feel unfair.	31	43	40	39	30	183	3.032787	4
	155	172	120	78	30	555		
AI sometimes promotes biased product suggestions.	43	39	35	33	33	183	3.142077	2
	215	156	105	66	33	575		
AI influences my choices without my awareness.	45	43	30	34	31	183	3.202186	1
	225	172	90	68	31	586		
AI-based chatbots often fail to resolve issues.	34	48	34	33	34	183	3.081967	3
	170	192	102	66	34	564		
AI recommendations reduce my shopping options.	30	38	42	31	42	183	2.907104	8
	150	152	126	62	42	532		
AI in shopping raises ethical concerns.	34	33	48	27	41	183	2.956284	7
	170	132	144	54	41	541		

Source: Author.

- Effect of Personalized Offers: While tailored discounts and promotions influence purchasing behaviour (Mean = 2.97), consumers exhibit varying levels of comfort in sharing personal data to access these benefits (Mean = 3.03).
- Satisfaction with Technology-Driven Shopping: Technology contributes positively to consumer decision-making (Mean = 3.14) and overall shopping experiences (Mean = 3.12); however, automated customer service solutions remain inadequate and require further refinement (Mean = 2.89).
- Ethical Concerns and Consumer Autonomy: Many consumers feel that technological interventions shape their purchasing decisions without explicit awareness (Mean = 3.20), raising critical ethical concerns about transparency and user control (Mean = 2.95).
- Challenges to Widespread Adoption: Biased product recommendations (Mean = 3.14), intrusive advertisements (Mean = 2.90), and a lack of clear communication regarding technological processes continue to hinder consumer confidence.
- Need for Transparency and Ethical Standards: To foster broader acceptance, businesses must prioritize addressing algorithmic biases, strengthening data privacy protections, and enhancing digital customer service to build a more trustworthy and consumer-centric shopping environment.

4. Suggestions

Businesses should guarantee more openness in the collection and use of customer data in order to increase consumer trust. In order to accommodate customer preferences, product suggestions have to be more objective and precise. In order to properly address problems, customer support systems need to be improved to offer a more human-like experience. Maintaining ethical pricing practices will help avoid any impression of manipulation. In order to lessen intrusiveness, consumers should be able to choose which ads they see. Customers may make better selections if they are informed about how technology affects their purchasing decisions, and companies can use ethical business practices to strike a balance between innovation and morality.

5. Conclusion

The technological evolution has brought about substantial improvements to shopping according to consumer needs while providing effortless shopping solutions. The existing privacy concerns together with biased recommendations create problems that require businesses to adopt ethical principles in their operations. Businesses need to preserve customer trust through transparent practices that include fairness along with giving control to consumers regarding technological influences. The incorporation of technology in shopping with a responsible approach leads to enduring consumer satisfaction as well as loyalty.

6. Future Scope of the Study

Studies in the future should establish ethical standards to regulate shopping technology along with privacy protections. Studies should examine how businesses can achieve the right level of personalization while maintaining security with an emphasis on buying products that support sustainability. A human-touch integrated in digital customer service leads to better customer satisfaction levels. The evaluation of regulatory effects on responsible technology applications in digital marketing would build transparent marketplaces which promote trust and efficiency throughout digital shopping transactions.

References

[1] Smith, J. (2020). Personalized shopping experiences and customer loyalty in online retail. *Journal of Marketing Research, 8*(4), 120–135.

[2] Brown, A. (2019). The impact of AI chatbots on customer satisfaction in online retail. *Journal of Customer Service, 5*(2), 45–60.

[3] Jones, D., & Lee, K. (2021). 24/7 customer support: The role of AI in online retail. *Journal of Retail Management, 12*(3), 90–105.

[4] White, T. (2023). Consumer trust and AI technologies in online retail. *Journal of Technology and Society, 15*(1), 30–45.

[5] Martinez, E. (2020). Behavioral analytics and AI: Predicting consumer purchasing patterns. *Journal of Consumer Behavior, 9*(2), 65–80.

[6] Wilson, G. (2021). AI and impulse buying in online retail. *Journal of Behavioral Economics, 11*(3), 100–115.

[7] Chen, L. (2019). AI-powered visual search: Transforming online shopping. *Journal of Digital Marketing, 7*(3), 78–92.

[8] Green, S. (2022). Augmented reality and consumer engagement in online retail. *Journal of Interactive Marketing, 10*(4), 145–160.

[9] Kim, H., & Park, S. (2018). The impact of AI-driven dynamic pricing on consumer purchasing decisions. *Journal of Pricing and Revenue Management, 17*(4), 320–335.

[10] Lee, J., & Choi, Y. (2019). AI-based fraud detection and consumer trust in online retail. *Journal of Information Security and Applications, 48*, 102345.

[11] Nguyen, T., & Tran, L. (2020). The role of AI voice assistants in online shopping. *Journal of Electronic Commerce Research, 21*(2), 115–130.

[12] Davis, M., & Wilson, A. (2021). AI-driven product return prediction and consumer satisfaction. *Journal of Operations Management, 70*(1), 80–95.

[13] Patel, R., & Shah, N. (2022). Personalized AI advertising and consumer conversion rates. *Journal of Advertising Research, 62*(3), 280–295.

[14] Rodriguez, L., & Gomez, C. (2023). AI-driven sentiment analysis and customer feedback. *Journal of Marketing Analytics, 11*(1), 45–60.

[15] Thompson, K., & Baker, P. (2018). Virtual fitting rooms and AI: Enhancing online shopping experiences. *Journal of Fashion Marketing and Management, 22*(4), 510–525.

[16] Walker, S., & Hill, D. (2019). AI-driven inventory management and consumer satisfaction. *International Journal of Production Economics, 218*, 107380.

[17] Young, E., & Moore, F. (2020). AI-powered personalized email marketing and conversion rates. *Journal of Direct Marketing, 14*(2), 160–175.

[18] Zhao, Q., & Wu, X. (2021). AI-driven social media analytics and consumer trend prediction. *Journal of Social Media Marketing, 7*(3), 220–235

25 AI-enabled waste management framework

Clarence S. Ordonia[1,a] and Thelma D. Palaoag[2,b]

[1]Mariano Marcos State University, MIT, Philippines
[2]University of the Cordilleras, Philippines

Abstract: This study is to address the critical problem of school waste management situation by creating an AI based waste management system. Grounded on a synthesis of the said researches, the framework differs mainly from a flow of artificial intelligence (AI) and technologies to enhance the methods of waste in the educational institutions. Therefore, the integrated framework comprises predictive analytics, continuous monitoring and a system of the collaborative engagement with the stakeholders. Among its main objectives are to promote the efficiency of the waste collection processes and simultaneously ensure efficiency in its operations, and to be used as an education tool to spread out sustainability values to the students. Another important highlight of the study is the value demonstrated in turning AI methodology into practical application in the domain of waste management using our Smart-bin Prototype. We have Arduino Uno board, ultrasonic sensors, servo motors, batteries and jumper cables as hardware components from this prototype. The integrated package of the components together creates a system which offers real time waste management efficiency by monitoring and retrieving data about waste management activities. AI model development is confirmed by the study results, a supporting data ecosystem (which includes GPS tracking and sensors) and the use of predictive analytics to help resource allocation are all part of the study results. Further, the framework is encompassing and dynamic, as it can also be able to change with eventualities where factors like waste generation patterns and conditions in the school environment change. The framework of AI–Enabled Waste Management is capable to enhance the practices of waste management for schools in a more holistic manner. The framework also nudges students toward environmentally conscious behaviours and is a small part of a broader project of sustainable waste practices in addition to only providing operational improvements. Hence, this research will create room for cuts in schools costs, rolling up the activities in waste management at schools and establishment of the opportunity to nurture the generation of eco literate citizens.

Keywords: Artificial intelligence, predictive analytics, smart-bin prototype, sustainability, waste management

1. Introduction

As schools in some countries accumulate a massive volume of waste on daily basis, waste management in educational institutions, especially schools have become critical. Due to implications to the environment and the ability to educate students about sustainability and responsible resource management, management and disposal of this waste is important. The traditional school waste management practices, however difficult it is to integrate in real school settings and highly beneficial to the school communities, often pose challenges to efficiently sort, recycle, and minimize waste generation and the environmental footprint [1].

Every day, educational institutions, such as schools, produce a large amount of waste, including paper and electronic waste among other materials. In this context, efficient waste management is critical to reduce negative impacts on the environment and encourage recycling, as well as inspire eco-friendly behaviour among students and staff. In schools, there is already some kind of waste management system that already exists, but these systems can be updated by feeding Artificial Intelligence (AI) and other such related technologies.

Related studies have been carried out that explored the integration of AI and technology to enhance the existing practice of managing wastes in other settings, which highlights some commonalities with the objectives of this study [2].

In starting off the study, the researchers create a foundational reference for the study by comprehensively reviewing Smith's [3] view on waste management practices in educational institutions Baggyalakshmi et al. [4]. Furthermore, the challenges faced, as well as opportunities, of providing sustainable waste management in schools as outlined by Jones and Green [5] is in complete agreement with the research goal of boosting sustainability efforts within such sites.

Similarly, Brown and Taylor's [6] constructed case study is relevant to the research objectives and focuses on the role of AI in elementary school waste reduction. Garcia et al. [7] explored the IoT based smart waste management system, which is in line with the approach taken by the researchers to utilize technology in school waste management.

Chang et al. [8] also focused on the application of machine learning for sorting wastes in school cafeterias. The concept reflects that of the study's focus on AI to waste management optimization. Roberts et al. [9] already present the intelligent waste bin system in educational institutions which

[a]csordonia@mmsu.ed.ph, [b]tpalaoag@gmail.com

DOI: 10.1201/9781003684459-25

is equivalent to our framework development and the practical application of AI in this context.

In addition, Wilson et al. [10] did a pilot study on how gamification works to promote the practice of sustainable school waste management, which is complementary to the overall framework of the research. Thus, the technological orientation of the research is presented within the integration of AI and robotics in the waste collection in educational institutions, researched by Gomez et al. [11]. This research is in line with the study by Li et al. [12] in that it also adheres to the data driven school waste management optimization approach Baggyalakshmi et al. [13].

Lastly, Johnson et al. [14] considered the effect of AI based waste management in schools towards sustainability, confirming the appropriate objective orientations of our study Zhao et al. [15]. The related studies radiate a thread upon threads of efforts and thoughts converging from the different facets, pointing the importance of coupling AI and related technologies to improve waste management of educational institutions that aligns very much with the central goal of the researcher to build an AI enabled Waste Management Framework specifically for schools.

This work looks into and creates an AI designed Waste Management Framework for the case of schools towards the above mentioned challenges. It assimilates AI and related technologies to redefine school waste management practices in terms of waste minimization and recycling, and also, fits an environment friendly approach to manage waste.

Schools waste management is a complex affair which involves segregating various wastes, monitoring the amount of waste it generates and foster the responsible practice of waste disposal. Additional challenges include fluctuating school populations, changing waste volumes, and documentations within the waste categories. However, traditional approaches to waste management often fail to adapt to these dynamics – which leads to inefficiency and environmental cost. Therefore, this study concludes that AI could be a game changer in school waste management. With data analysis, machine learning and automation capabilities, AI helps to streamline the collection process for the schools, enables intelligent sorting and provides all the necessary insights to optimize school waste management strategies. In addition, AI can be incorporated into waste management practices as an educational experience that aims to educate students on sustainability and eco friendly behaviours.

This paper talks about the problem and discusses how by developing a Waste Management Framework for the educational institutions, an AI enabled Waste Management Framework can help to manage the waste problems of schools in a much better way by adapting it in schools. The scope of this framework is connected to AI based waste sorting, monitoring and data analytics, that provides an efficient green waste management system by using and integration of emerging technologies such as Internet of Things (Iot). This study additionally stresses that the implementation of waste management is not only beneficial to the operational work of such operations, but it can be converted into a new useful educational tool. The school's teachable curriculum incorporating the AI teaching can educate its students about waste reduction, recycling and environmental stewardship thus creating a sustainability culture in the school's community.

Heavy on both education and environment, schools are not just educational establishments but play a vital role in shaping generations to come and their environmental awareness; AI enabled a Waste Management Framework for schools shall, therefore, contribute substantially to both waste management as well as education on how to nurture eco conscious citizens. The purpose of this research is to contribute to the increasing body of knowledge in waste management, AI, and sustainability with the end objective of developing a practical and educational way to manage waste in schools in an efficient and effective manner, which foster an environmentally responsible generation.

2. Methodology

The methodology employed for this study entails a systematic approach to harnessing AI's potential in waste management, with a focus on the framework itself. The process begins with an exhaustive review of existing literature to establish a solid understanding of current waste management practices and AI's role within this context. This review aids in identifying gaps in knowledge and opportunities for AI integration.

Following the literature review, the study centers on collecting pertinent waste generation, collection, and disposal data within the chosen study area. This dataset encompasses historical records, geographic information system (GIS) data, and in-depth waste composition analysis.

The central point of this research is AI models and algorithms based on AI tailored to meet the focused needs of waste management. Predictive analytics to forecast the waste generation patterns, optimization algorithms for the route planning and data drive decisions support tools are defined in these models.

The paper integrates multiple data sources including sensors, GPS tracking, and citizen data reporting to create a robust data ecosystem for waste management. However, the integration of data related to nature and found within landfill goes a long way in encouraging data driven decision making which ultimately improves the overall performance of waste management operations.

Fundamentally, issues involving monitoring and evaluation come in the picture and adjustments and refinements are made on the basis of data driven insights as well as stakeholder feedback. The research also interacts with critical stakeholders such as that of the local governments, waste management entities, and environmental organizations, to help the implementation and adoption of the AIEnabled Waste Management Framework. To inform a larger audience and inspire about the potential that AI can bring to waste management, dissemination of research findings and best practices is key.

3. Results

The research has yielded significant results promising to transform school waste management practices and beyond. The study, which focused on the development of AI models and algorithms customized to meet the specific demands of waste management, has revealed several key outcomes:

3.1. Enhanced waste management efficiency

The researcher has successfully developed AI models and algorithms that optimize waste management processes.

Provided code contains algorithm of a high level of the core functionality of smart waste bin management system. Waste management process encapsulates the critical components and processes that contribute to efficient management of waste. In the initial phase, the need is to import the necessary modules and check that all the required tools and functions are ready to be used for implementation. Sensors are configured to begin collecting data by the smart bins, which come into existence and connect to sensors. It is also based on the collection plan and the routes as the basis of planned waste collection activities.

The major part of the algorithm is related to real time monitoring where data is collected from the smart bins in real time and the data includes waste type, fill level, and image data. Machine learning models are built using predictive modelling to predict the fill levels and types of the waste. Based on this data driven approach, the system is capable to decide when to pick the waste, when to service and which of the bins require immediate attention.

Further, the algorithm includes route optimization, bin priority, notification when garbage needs to be collected, dynamic scheduling of collections and a user interface for displaying data in real time. It is reasonably intended to run on a periodic basis so that resources are saved and a constant monitoring is ensured. The underlying concept of this algorithm aids to create an AI enabled smart waste bin management system which supports smooth waste collection, decreases costs and improves sustainability for schools and other similar organizations.

With predictive analytics, waste generation patterns have been anticipated and consequently, better allocation of resources and proper collection strategies have been possible. The increased efficiency and reduced operational costs would also be of advantage to schools and other education institutions. The predictive below model was used by the researcher.

```
import pandas as pd
from sklearn.model_selection import train test_split
from sklearn.ensemble import Random Forest Regressor
from sklearn.metrics import mean_ absolute_error
# Sample historical waste generation data (replace with your dataset)
data = {
'Date': pd.date_range(start='2022-01-01', end='2022-12-31', freq='D'),
'WasteGenerated': [120, 130, 140, 110, 125, 130, 150, 105, 115, 120, 140, 130, 120, 110, 150, 125, 110, 140, 130, 135, 145, 125, 130, 140, 110, 120, 130, 150, 125, 135, 140, 130, 110, 120, 125, 140, 150, 130, 135, 110, 120, 140, 130, 125, 150],
'SchoolHolidays': [0, 0, 0, 1, 0, 0, 0, 0, 0, 0, 1, 0, 0, 0, 0, 1, 0, 0, 0, 0, 0, 0, 1, 0, 0, 0, 0, 0, 0, 1, 0, 0, 0, 0, 0, 0, 0, 0, 1, 0, 0, 0, 0, 0, 0],
'SpecialEvents': [0, 0, 0, 0, 1, 0, 0, 0, 0, 0, 0, 1, 0, 0, 0, 0, 1, 0, 0, 0, 0, 0, 0, 0, 0, 0, 0, 0, 0, 1, 0, 0, 0, 0, 0, 0, 0, 0, 0, 0, 0, 1, 0, 0, 0], }
df = pd.DataFrame(data)
# Feature engineering
"df['DayOfWeek'] = df['Date'].dt.dayofweek
# Split data into training and testing sets
X = df[['DayOfWeek', 'SchoolHolidays', 'SpecialEvents']]
y = df['WasteGenerated']
X_train, X_test, y_train, y_test = train_test_split(X, y, test_size=0.2, random_state=42)
# Create and train a predictive model (Random Forest Regressor)
model = RandomForestRegressor(n_estimators=100, random_state=42)
model.fit(X_train, y_train)
# Make predictions
y_pred = model.predict(X_test)
# Model evaluation
mae = mean_absolute_error(y_test, y_pred)
print(f'Mean Absolute Error: {mae:.2f}")
```

In waste management systems, which are very important in this type of environments such as schools, this predictive model allows a more appropriate resource distribution and better waste collection strategies. The researcher gathered a synthetic dataset containing historical waste generation data and additional features like school on holidays or events. Potential influencing factors on waste generation considered by the model include the day of the week, school holidays, and special events. The data is then split into training and testing sets for purposes of having a reliable means of measuring the performance of the model.

A Random Forest Regressor, a machine learning algorithm that can capture a non linear relationship within the data is the core of the predictive model. The model is trained on historical data to find out what factors impact waste generation patterns, for that reason. It can predicts the future waste generation from given input features after training.

Mean Absolute Error (MAE) is used as a metric to assess the performance of the model because it measures the average absolute difference between actual and predicted values. The model with the lower MAE is more accurate.

This predictive analytics model yields such useful insights on when and which locations are most likely to experience waste generation. Historical data and machine learning are leveraged to help waste management teams in schools to better allocate their resources and optimally strategise waste collection. This way, it is possible to perform timely collections during the moments when we produce more waste (so that the bin does not overflow) and optimize resource usage when we produce less waste. This way also promotes sustainability because of cost reduction and waste management done more efficiently and responsibly.

3.1.1. Establishment of a robust data ecosystem

The integration of various data sources, including sensors, GPS tracking, and citizen reporting, has created a comprehensive data ecosystem. This infrastructure facilitates data-driven decision-making by providing real-time insights into waste generation, collection, and disposal. The research's focus on data integration is pivotal for waste management in schools and other settings where informed decision-making is essential. The Smart-bin Prototype is shown in Figure 25.1.

The following are the hardware components of Smart-bin:

- **An Arduino Uno board.** This is the brain of the smart bin. It controls the servo motor and reads the data from the ultrasonic sensor. It also communicates with other devices using serial or wireless protocols.
- **An ultrasonic sensor.** This is the eye of the smart bin. It emits and receives sound waves to measure the distance between the sensor and an object. If someone is approaching the bin with trash, the sensor can detect it and trigger the servo motor to open the lid.
- **A servo motor.** This is a muscle of the smart bin. The horn attached to lid can be open or close according to the signal from the Arduino board, which can make it rotating.

Figure 25.1. Smart-bin prototype.

Source: Author.

- **A battery.** These provide electricity to the Arduino board, the ultrasonic sensor and the servo motor. A 9V battery can be used for the battery or a rechargeable battery pack.
- **Jumper cables.** The wires that connect the components. There are the female-female, female-male, and male-male. And they can connect the ultrasonic sensor, the servo motor and pins on the Arduino board.

3.2. Continuous improvement

The study made an essential discovery pointing out that the school environment requires continual monitoring and evaluation. Data is always analyzed routinely and as a result data is constantly being used to tweak and improve the waste management system. The important use of this iterative method is to ensure that the framework remains effective and flexible in the face of an evolving set of waste generation patterns and environmental conditions in a school setting.

3.3. Dissemination of knowledge and best practices

One of the important aspects of the results of the research is the dissemination of the findings and best of practices. The research shares insight and best practices with regards to the use of AI in waste management, providing general awareness of this potential. Furthermore, this outreach educates the waste management industry and related industries about how AI can be used for creating sustainability and efficiency.

The work presented in this dissertation further demonstrates that the research results related to the AI Enabled Waste Management Framework hold enough potential to implement them on such a large scale. The objectives of building a complete waste management enhancement framework includes development of customized AI models and algorithms, data ecosystem building, continuous improvement, and stakeholder partnership, knowledge dissemination in order to make waste management more efficient, with more effective use of resources, environmentally responsible, and competent to deal with changes in waste requirements in educational institutions and other such domains.

4. Conclusions

Therefore, this work is a comprehensive approach of revolutionizing school waste management by an AI enabled waste management framework. This methodology demonstrates the combination of predictive analytics, real time monitoring and use of AI to involve stakeholders is important based on extensive literature review and robust data included through machine learning. Extensive research was carried out for the goal of using AI models and algorithms to increase the efficiency of waste management through the effective optimization of waste collection process and resource allocation strategies.

It has provided a solid foundation to create a solid data ecosystem with sensors, GPS, citizen reporting, etc. to aid in data based decision making. For example, the model of application of the advanced AI technology to the waste management is the Smart-bin Prototype that demonstrates the whole set of waste management complex model in real time with high operational efficiency. The continuous monitoring and evaluation of the framework, on the basis of continuous monitoring and evaluation, permits the framework to be modified where necessary and as things alter in response to the patterns of waste production and changing environmental conditions in the school.

So, the spread of AI in waste management research and best practices should be known for the application and understanding of AI. Through this study, waste management practices in educational institutions are advanced, and other sectors are encouraged to discover AI applications for sustainability and efficiency by sharing insights and process of success. Ultimately, the AI-enabled waste Management Framework represents a significant leap forward, offering a holistic, efficient, environmentally responsible, and adaptable solution, paving the way for schools to reduce costs, enhance operational efficiency, and cultivate environmentally conscious behaviours among students.

5. Acknowledgements

I am most grateful to the Commission on Higher Education (CHED), which gave me support both in terms of financial resources and other material resources which made this research possible. I would like to first give special thanks to Dr. Thelma Palaoag, my adviser, for his precious guidance and expertise that contributed greatly in the development and completion of this study.

I would like to thank the University of the Cordilleras for being the delivering institution where a hospitable environment for research is being offered, and as well as the data collection and analysis facilities.

Sending institution, Mariano Marcos State University, is thanked for caring for an academic foundation that has nurtured my research skills and intellectual growth. This research benefited from my previous academic journey at MMSU as I have been able to gain knowledge and experiences that will help me in achieving it.

I am also indebted to my family and friends for their unwavering support, encouragement, and understanding throughout the research process. Their patience and motivation were crucial in overcoming challenges and staying focused on the research objectives.

References

[1] Shyamala, R., & Shanmugapriya, T. (2016). Study on Replacement of Sand with Coarse Fly Ash in High Performance Concrete. *International Journal of Advances in Engineering and Emerging Technology*, *7*(4), 94–101.

[2] Turan, C., Yağlıoğlu, D., Doğdu, S. A., Gürlek, M., Ergüden, D., Ivanova, P. P., & Raykov, V. S. (2023). Existence of Belone svetovidovi Collette & Parin, 1970 in the Marmara Sea and Black Sea Coasts of Türkiye. *Natural and Engineering Sciences*, *8*(2), 72–81. http://doi.org/10.28978/nesciences.1331296

[3] Smith (2019), Waste management practices in educational institutions: A review. *Waste Management*, *75*, 234–245.

[4] Baggyalakshmi, N., Harshada, J., & Revathi, R. (2024). Super Market Billing Management System. *International Academic Journal of Science and Engineering*, *11*(1), 107–117. https://doi.org/10.9756/IAJSE/V11I1/IAJSE1114

[5] Jones, E., & Green, L. (2020). Sustainable waste management in schools: Challenges and opportunities. *Sustainability*, *12*(3), 1021.

[6] Brown, K., & Taylor, M. (2018). The role of AI in waste reduction: A case study in an elementary school. *Environmental Science & Technology*, *52*(11), 6236–6243.

[7] Garcia, R., et al. (2021). IoT-based smart waste management system for schools. *Journal of Cleaner Production*, *296*, 126568.

[8] Chang, J., et al. (2019). Machine learning for waste sorting in school cafeterias. *Waste Management*, *84*, 56–64.

[9] Roberts, P., et al. (2017). An intelligent waste bin system for educational institutions. *International Journal of Environmental Technology and Management*, *20*(4), 311–328.

[10] Wilson, H., et al. (2020). Gamification for sustainable waste management in schools: A pilot study. *Waste and Resource Management*, *173*(6), 156–164.

[11] Gomez, D., et al. (2019). Integrating AI and robotics for waste collection in schools. *Robotics and Autonomous Systems*, *120*, 103345.

[12] Li, H., et al. (2018). A data-driven approach to optimize school waste management. *Sustainability*, *10*(11), 3916.

[13] Baggyalakshmi, N., Keerthana, A. R., & Revathi, R. (2023). Procurement with ERP. *International Academic Journal of Innovative Research*, *10*(2), 27–35. https://doi.org/10.9756/IAJIR/V10I2/IAJIR1007

[14] Johnson, S., et al. (2019). The impact of AI-based waste management in schools on sustainability. *Journal of Environmental Management*, *240*, 292–301.

[15] Zhao, Y., Hu, N., Zhang, C., & Cheng, X. (2020). DCG: A Client-side Protection Method for DNS Cache. *Journal of Internet Services and Information Security*, *10*(2), 103–121.

26 Day of the week effect with special reference to Nifty bank index

M. Sudha[1] *and J. Shanmuganandavadivel*[2,a]

[1]Research Scholar, Department of Commerce, Karpagam Academy of Higher Education, Coimbatore, Tamil Nadu, India
[2]Under the Guidance of Dr. J. Shanmuganandavadivel, Professor, Department of Commerce, Karpagam Academy of Higher Education, Coimbatore, Tamil Nadu, India

Abstract: The Day of the Week Effect (DOWE) is a well-documented anomaly in financial markets, where stock returns exhibit systematic variations based on the trading day. This study examines the presence of this anomaly in the Nifty Bank Index, a benchmark for the banking sector in India, from January 2019 to December 2024. Using historical daily returns, we analyze whether specific days of the week show significantly different return patterns. Findings from past research suggest that Friday often records higher returns, possibly due to pre-weekend optimism, while Monday returns tend to be lower due to weekend news effects. Experts question the exact degree of this irregularity within Indian banking institutions. Price movements for the Indian market experience fundamental changes through three key factors comprised of institutional trading behaviour alongside macroeconomic events and global market influences. The research uses descriptive analysis combined with ANOVA and OLS regression modeling tests to establish if the Nifty Bank Index shows significant DOWE patterns. The study offers critical understanding about market efficiency while discovering methods for successful trading among investors and traders.

Keywords: Day of the week effect, stock market anomalies, Nifty bank index, market efficiency, trading strategies, stock returns

1. Introduction

Scholars along with investors have continuously studied financial markets to understand the patterns and unexplained events behind stock return changes. The DOWE represents an extensively documented financial anomaly because stock return patterns vary based on which trading day occurs. The DOWE undermines the Efficient Market Hypothesis proposed by Fama (1970) because it demonstrates stock prices should not reflect predictable market patterns.

Researchers worldwide have studied the DOWE because market returns tend to demonstrate favourable or unfavourable patterns during specific days. Research shows that stocks typically generate elevated returns during Fridays since investors usually display pre-weekend optimism [1]. At the same time, Mondays can produce decreased returns as weekend news influences investor sentiment and market mood [1]. Various studies in the Indian market present conflicting evidence about the anomaly because variables such as institutional transactions and macroeconomic dynamics and world market impacts might account for observed variations.

1.1. DOWE in the Indian context

The Indian stock market draws active research attention regarding anomalies especially for examining market inefficiencies in price movements. The research conducted by Sudarvel and Velmurugan [2] revealed statistical significance in NSE Nifty index daily returns which demonstrates that Indian markets possess operational inefficiencies. In another publication Sudarvel and Velmurugan [3] investigated trading prospects based on patterns in Indian stock prices through their review of market anomalies.

Representation of Indian financial systems relies heavily on the banking sector which is tracked through the Nifty Bank Index. The price activity in banking stocks gets influenced by institutional trading patterns as well as macroeconomic announcements and external market factors. According to Sudarvel et al. [3] stock returns within the BSE IT Index demonstrated systematic patterns throughout different trading days within the Indian IT sector. Further research needs to confirm if this observed phenomenon exists in the banking sector institutions.

The analyses by Thirumalaisamy et al. [4] investigated DOWE patterns between stock returns and volatility for the trading indices of MSX and Sensex. This research added new knowledge about the longevity of such anomalies within emerging markets through its detailed investigation of bank-specific sectors to determine their effect on trading behaviours.

[a]shanmuganandavadivelj.jaganathan@kahedu.edu.in

DOI: 10.1201/9781003684459-26

2. Review of Literature

Feasible research has shown the DOWE functions as a stock market anomaly because stock returns demonstrate systematic changes according to the trading day. Researchers have dedicated numerous studies to different market sectors and geographical areas to determine the existence and strength of this anomaly. This part summarizes important research about the DOWE in worldwide stock markets including emerging markets and the Indian financial sector.

2.1. Early studies on the DOWE

According to French [1] U.S. stocks demonstrated substantial decreases in market returns on Mondays before all other trading days. The study created knowledge which led to more investigations about stock market irregularities. Based on Gibbons and Hess [5] research the negative Monday effect extended its presence from equities into U.S. Treasury securities.

The research by Jaffe and Westerfield [6] showed the Monday Effect emerges within markets across the U.K., Japan, Canada and Australia as well as the United States. The Japanese stock market experienced abnormal return patterns that demonstrated a lower Tuesday performance as opposed to Monday returns.

2.2. DOWE in emerging markets

The DOWE showed different patterns when academic researchers examined new geographic regions. Researchers identified emerging markets as highly inefficient systems that easily displayed seasonal patterns including the DOWE [7].

According to Basher and Sadorsky [8] investors experienced significant negative returns during Mondays in India plus neighbouring Asian emerging economies that included China and Indonesia as well as better returns on Fridays. Researches worldwide show how investor sentiment together with institutional trading patterns create these particular patterns.

2.3. DOWE in India

Different studies analyze the presence of this anomaly in Indian stock market indices operating on NSE and BSE platforms. The NSE Nifty Index received its highest returns on Wednesdays and Fridays according to Sudarvel and Velmurugan [2] study. The researchers discovered that daily return movements depend strongly on investor sentiment alongside announcement events of macroeconomic news.

The research conducted by Sudarvel et al. [3] examined BSE IT Index data to demonstrate that technology stock prices showed high reaction to international market trends resulting in notable Friday effects. The anomaly exists to some extent because individual market factors within specific sectors play a distinct role in its formation.

According to Thirumalaisamy et al. [4] volatility levels on Mondays exceeded those of other trading days in both the Sensex Index and Muscat Securities Market (MSX) Index which tested financial market efficiency.

2.4. Institutional trading and market efficiency

Research investigations show how institutional investors create conditions which lead to the Day of the Week Effect. Lakonishok and Maberly [9] discovered that institutional investors have higher levels of Monday sales which generates the documented negative market returns. The research by Chaudhuri and Wu (2003) evaluated Asian financial markets specifically India to verify that institutional investors used systematic trading patterns which affected market returns differently on different days of the week.

The research conducted by Pandey [10] demonstrated that Indian stock markets showed reduced Monday returns following weekend pessimism and Friday returns indicated pre-weekend optimism.

2.5. Recent developments and contradictory findings

Developed financial markets show diminishing evidence of the DOWE because of increasing market efficiency along with growing algorithmic trading methods. Research from India together with other developing markets shows this effect to be significant. According to Gupta and Kumar (2020) stock market anomalies within the Indian market persist mostly in small-cap and mid-cap stocks which receive lower levels of investor institutional involvement.

Financial market experts are still actively researching the significance of the Day of the Week Effect. Research shows Indian markets along with other emerging economies still exhibit this pattern because of advanced technology and less market efficiency although it has lessened in developed countries. Available research shows that daily returns in the market suffer variations because of institutional trading as well as investor sentiment alongside macroeconomic events. An examination of the Nifty Bank Index from 2019 to 2024 adds new evidence about the continuance or disappearance of this market anomaly in India's banking sector.

2.6. Research gap

The investigation of DOWE primarily examines broad market indices (Nifty 50, BSE Sensex) while showing reduced interest in the banking sector. Studies about the Nifty Bank Index need additional exploration because it responds to institutional trade patterns alongside regulatory changes and broad economic market activities. Studies conducted before 2019 require modernization following pandemic changes and algorithmic trading and worldwide economic transformations (2019–2024). Previous research has made use of primary methods instead of the advanced OLS regression modelling approach. This research addresses the knowledge voids through updated financial data analysis with advanced statistical methods while accounting for worldwide market effects thus producing new insights about trading frameworks and market efficiency in Indian banking institutions.

2.7. Statement of the problem

Stock returns show systematic variations through the DOWE which functions as a documented market anomaly affecting trading days. Research about the DOWE in sector-specific indices mainly focusing on the Nifty Bank Index remains sparse in the academic literature.

Indian financial markets heavily depend on the banking sector because this sector operates under institutional trading regulations and is influenced by global economic events. The effect of the Day of the Week variation on banking stock movements remains unclear because doubtful market efficiency changes have stripped away such anomalies through algorithmic trading alongside emerging investor behaviour patterns.

Indian stock market studies mainly analyzed early data groups (pre-2019) with statistical methods from the past. Given post-pandemic market changes and rising financial global integration requires systematic research using OLS regression modeling to analyze (2019–2024) whether the anomaly continues in bank stocks.

The Nifty Bank Index receives evaluation for DOWE in this analysis while also investigating its market significance for investor trading approaches and policy decisions.

2.8. Scope of the study

Academics and investors together with policymakers require this research because it studies the DOWE in the Nifty Bank Index to better understand market response patterns and operational efficiency. Research is completed to address the knowledge gap by studying stock market irregularities in particular sectors with contemporary data (2019–2024) and sophisticated statistical methodologies. The day of week effect analysis assists investors to discover money-making trading opportunities while controlling risks and enhancing the effectiveness of trading algorithms. The data findings enable policymakers to track market efficiency levels thus facilitating their understanding of institutional warehouse behaviour and policy intervention requirements. Research also explores how Indian banking sector patterns match up to global finance system developments including specific effects from macroeconomic events and foreign investment activities. The research provides essential financial data for multiple stakeholders to handle their decision-making through its examination of these research areas.

2.9. Objective

To identify the existence of the Day of the Week Effect with Special Reference to Nifty Bank Index.

2.10. Research methodology

This analytical research evaluates the DOWE by analyzing secondary data from NSE web portal within the Nifty Bank Index. An updated analysis of market anomalies in the Indian banking sector occurs during January 2019 through December 2024. The research adopts quantitative methodology through Descriptive Statistics and OLS Regression Modeling to determine whether trading day effects produce significant returns modifications. This study investigates if the DOWE exists within the banking sector even as market efficiency rises combined with growing institutional trading and worldwide economic events.

2.11. Framework of analysis

The studied information undergoes statistical analysis through these methods:

Descriptive Statistics – Mean, Standard Deviation, Variance, Skewness and Kurtosis.

Shapiro-Wilk Test – Tests for normality of stock returns.

The OLS Regression Modelling method assesses stock return correlations with trading day effects through external factor adjustments.

2.12. Limitations

The investigation examines only banking sector stocks through analysis of the Nifty Bank Index. Other sectors together with their indices would fail to receive direct applicability from these research results. Stock returns might be affected by external factors consisting of global financial crises and geopolitical events and macroeconomic policies thus impacting the research outcomes' reliability. When applying these research conclusions to the entire Indian stock market readers should approach interpretation with caution.

2.13. Significance of the study

The research holds important value by analyzing market irregularities through market characteristics of the Nifty Bank Index in India's financial market. The application of statistical techniques to latest data (2019–2024) leads to improved comprehension of sectorial stock return patterns. The study enables investors along with traders to find successful trading opportunities and build risk management plans while generating data-focused investment decisions. News and trading professionals can capitalize on understanding daily return fluctuations to produce better trading algorithms as well as optimize their portfolios.

The study grants policymakers and financial regulators important knowledge about market efficiency which helps them determine the extent to which Efficient Market Hypothesis (EMH) explains price movements. This research enables a clearer view of how industry trading behaviour combines with outside financial influences on banking stock value. The research findings support academic and financial analysis work related to stock market anomalies present in emerging economies. The research allows analysts to evaluate return patterns in the Indian banking sector by studying macroeconomic indicators and foreign investment trends relative to international financial markets. The findings from this study deliver useful information for multiple stakeholders who can use it to enhance their investment approaches and regulatory policies and financial market analysis (Table 26.1).

Analysis results support the occurrence of the DOWE for the Nifty Bank Index. The negative Monday Effect appears to influence Nifty Bank Index returns because Monday has both the lowest returns (−0.1153%) and the most volatility (2.06195). The Tuesday market performance delivers the greatest average return of 0.2432% as investors seem to recover from Monday's market decline. The 0.0683% positive return on Friday indicates that investors feel optimistic before weekends. Price variations show their maximum levels during the beginning and closing days of the week when measured on Mondays and Fridays. The p-values (.000) show that returns across all days exhibit non-normal distribution patterns since stock returns deviate from normality. The statistical deviation from normality found in the NSE Nifty Index signifies that anomalous phenomena affect the index which supports the notion of the Day of the Week Effect. The analysis reveals that investors should utilize weekly return patterns along with risk management techniques that focus especially on Mondays and Fridays which exhibit elevated market volatility (Table 26.2).

2.14. *Model 1: OLS, using observations*

The OLS regression analysis evaluates how the Nifty Bank Index Friday return values react to Monday-through-Thursday returns. The research indicates Thursday returns at -0.1635 ($p = 0.0083$) display a negative significant impact on Friday returns as investors tend to sell their assets before weekend beginnings.

Friday returns remain unaffected by the measurements taken during Monday through Wednesday across the Nifty Bank Index. Through model evaluation we discovered Friday returns can only be explained by external forces which account for 4.34% ($R^2 = 0.0434$) of total variation in returns because investor sentiment and global market movements dominate the market. The F-statistic ($p = 0.0121$) shows overall model significance as indicated by its output and the Durbin-Watson statistic (2.2188) demonstrates that residual autocorrelation does not exist.

3. Suggestions

3.1. *For investors and traders*

- The negative effects of Thursday affect Friday returns so traders need to examine daily trends carefully before making weekend position adjustments.
- Strategic risk handling through stop-loss systems should apply especially to Monday and Friday periods which feature increased market volatility.
- Different industries should be included within investment portfolios to minimize sector-related performance irregularities.
- Population-based confirmation tools consisting of moving averages and RSI provide traders with signals to verify market entry points and exit points based on weekday data patterns.
- Quantitative financial traders should build trading algorithms which implement the DOWE to generate maximum financial returns.

Table 26.1. Analysis and interpretation

	N	*Mean*	*Std. Deviation*	*Variance*	*Skewness*	*Kurtosis*	*Kolmogorov-Smirnova*	*Shapiro-Wilk*
Monday	293	−.1153	2.06195	4.252	−2.114	16.391	.000	.000
Tuesday	297	.2432	1.41043	1.989	.558	13.288	.000	.000
Wednesday	295	.0880	1.40870	1.984	.565	8.035	.000	.000
Thursday	301	−.0103	1.38598	1.921	−.946	9.251	.000	.000
Friday	294	.0683	1.46271	2.140	.647	5.345	.000	.000

Source: Author.

Table 26.2. Dependent variable: Friday

	Coefficient	*Std. Error*	*t-ratio*	*p-value*	
const	0.102648	0.0859759	1.194	0.2335	
Monday	0.0551783	0.0414550	1.331	0.1842	
Tuesday	−0.0919264	0.0606242	−1.516	0.1305	
Wednesday	−0.0616372	0.0609981	−1.010	0.3131	
Thursday	−0.163548	0.0615402	−2.658	0.0083	***

R-squared	0.043424	Adjusted R-squared	0.030138
F(4, 288)	3.268451	P-value(F)	0.012138

Source: Author.

3.2. *For policymakers and financial regulators*

- Market Efficiency Assessment indicates potential biases within financial markets because of the detection of Day of the Week Effect. Regulatory entities must consider whether these irregularities affect how prices are discovered and affect investor trust.
- The government should work with institutional entities to boost their Friday trading volume since this action

will decrease market instability and minimize weekend turbulence.

- Market liquidity will strengthen through implementing measures involving additional trading hours or alternative risk management approaches to maintain price stability throughout every trading day.
- The regulations should pursue awareness programs which educate retail investors regarding market anomalies and behavioural biases for their capacity to take well-informed investment decisions.
- The tracking of international economic patterns along with tracking financial indicators allows officials to implement proactive intervention policies for Indian banking sector operations.
- The recommended actions will allow investors to strengthen their approaches and regulatory bodies to improve marketplace efficiency and stability.

4. Conclusion

Research reveals that stock prices of Nifty Bank Index show a DOWE where varying returns emerge on different trading days. Analysis results show Monday generates the least profitable returns together with elevated market volatility which validates the proven Monday Effect caused by news events between weekends and investor caution. An analysis of returns demonstrates Tuesday holds the highest value indicating market recovery then Friday reports moderate profitability because investors become optimistic before the weekend ends. The OLS regression results show that Thursday returns create significant negative pressure on Friday returns because investors tend to record profits before weekend begins. External market forces including global trends combined with macroeconomic policies and institutional trading activity explain most of the Friday return pattern based on the low R-squared value of 0.0434.

The Nifty Bank Index shows market anomalies through stock return distributions that do not follow a normal pattern according to the p-values (.000) across all trading days. The Efficient Market Hypothesis (EMH) faces challenge because traders have the chance to exploit foreseeable patterns in returns to establish better investment strategies. These practical results deliver important guidance which benefits both investors as well as traders and policymakers within their operations. Investors may modify their trading approaches using daily price fluctuations information yet policymakers alongside regulators need to track market performance to conduct required actions for lowering market volatility and enhancing market liquidity. Future investigations should study the contribution of world economic conditions along with institutional market behaviour as well as industry specific data to uncover the reasons behind DOWE permanence in Indian financial markets. This research adds to the stock market anomaly field while providing essential knowledge that benefits practice-oriented investors in the Indian banking industry.

5. Scope for Further Studies

Further research needs to review if the DOWE appears in multiple sectoral indices such as Nifty IT, FMCG and Auto to confirm its sector-based manifestations. An investigation of both macroeconomic variables and worldwide events as they relate to daily market returns can be performed. Other research should examine parameters between institutional and retail trading and how algorithmic trading affects market efficiency and should analyze results compared to global banking indices. The forecasting accuracy receives an improvement from using advanced models which incorporate GARCH alongside machine learning algorithms. The analysis of market trading behaviour from pre-COVID times to post-COVID times would show any modifications that occurred during the pandemic period. These specific market sectors yield advanced understanding of investment strategies and market irregularities.

References

[1] French, K. R. (1980). Stock Returns and the Weekend Effect. Journal of Financial Economics, *8*(1), 55–69.

[2] Sudarvel, J., & Velmurugan, R. (2016). Day-of-the-week effect in Indian stock market with reference to NSE Nifty Index. International Journal of Multidisciplinary Educational Research, *5*(9), 145–153.

[3] Sudarvel et al. (2015). DOWE in Indian IT Sector with Special Reference to BSE IT Index. International Journal for Scientific Research and Development, *3*, 706–708.

[4] Thirumalaisamy et al. (2023). Day-Of-The-Week Effect on Stock Returns. Journal of Namibian Studies, *35*, 2168–2194.

[5] Gibbons, M. R., & Hess, P. (1981). Day of the Week Effects and Asset Returns. Journal of Business, *54*(4), 579–596.

[6] Jaffe, J., & Westerfield, R. (1985). The Weekend Effect in Common Stock Returns. Journal of Finance, *40*(2), 433–454.

[7] Aggarwal, R., & Rivoli, P. (1989). Seasonal and Day of the Week Effects in Emerging Stock Markets. Financial Review, *24*(4), 541–550.

[8] Basher, S. A., & Sadorsky, P. (2006). Day-of-the-week effects in emerging stock markets. Applied Economics Letters, *13*(10), 621–628.

[9] Lakonishok, J., & Maberly, E. (1990). The Weekend Effect. Journal of Finance, *45*(1), 231–243.

[10] Pandey, I. M. (2002). The DOWE in the Indian Stock Market. Vikalpa, *27*(3), 33–42.

27 A companion guide: Smart earpiece for visually impaired

Honeylet D. Grimaldo[a], Jericho Santos[b], Ain Christian Panganiban[c], Michael Arcay[d], Dominic Bevere Paragoso[e], Nathan Michael Estonilo[f], Jaime M. Landingin[g], Jorge Z. Burgos[h], and Reynaldo C. Muli[i]

College of Engineering, FEU Alabang, Filinvest Muntinlupa City, Philippines

Abstract: The role of eyes is even important in day to day life and it is one of the precious gifts that person who have. Those who have eyesight bless them to familiarize themselves with the world and surroundings. However, there are people that do not have the capability to view. As such, a wearable device readily available for such specific people should be designed. Visually impaired people communicate with the surroundings with the help of the technology. It has sped up the communication process and dissemination of information on a far greater scale including different parts of the world that highly affected human life and in turn increased ways of entertainment and comfort and reduced causes of suffering and hardship in many things. This world has Visually Impaired people, and until today what was impossible for them became available to them with a huge impact by the technology. This paper attempts to develop a device named as A Companion Guide: Smart Earpiece for Visually Impaired intent for providing hand to people with a visually impaired disability with the assistance of technology. Therefore, they will support the visually impaired to live their daily lives independently.

Keywords: Companion, dissemination, encourage, technology, visually impaired

1. Introduction

The COVID-19 Pandemic made a huge commotion and it definitely affected the world's status quo. There are people who are not in the place to help themselves or need to rely on others in order to keep up with the current situation. Being visually impaired is now almost common when a person reaches the age 50 and over. Over 300 million–400 million people have experienced visual impairment by the age of 50. And according to references, 90% of the world's visually impaired people live with low incomes settings. When it comes to jobs, they are never in the priority or even in the regular. They should not be treated as hindrance; hence, why the researchers came up with a device that can reduce the struggles of the visually impaired. The purpose of this study will include the term visual impairment, which is used to refer to the people who have been visually impaired. Any kind of visual impairment for example, totally blind, colour blindness, night blindness, extreme light sensitivity, blurred vision, etc.

The present era of technology has been surreal in the past few decades; this generation went through living without phones, then to the time that they cannot live without one, since technology has affected the way of general living. For example, technology has greatly made life easier including things like o Smartphones, the Internet, TVs, etc., and if the researchers focus on the other side of technology, that may help the way of living other people's lives. This is where the researchers insert the people that are handicapped. Other researchers invented robotic body parts for the amputated people. This shows a bright future for the next generation because of modern technology. In this study, the researchers try to present a device that can give a hand to visually impaired people with their daily lives. This device is programmed to understand the problems the visually impaired are facing and it enhances it by assisting them to live through their daily lives independently.

2. Background of the Study

Dealing with unfamiliar situations and people is challenging. Over time, it is about adjusting. But how can the researchers improve the visually impaired at dealing with new and unfamiliar places? Even if one is gifted with being able to see, it is still very uneasy when they are in unfamiliar territory. But when one think about the visually impaired people when they are facing unfamiliar territory daily, it makes them think about the fear of unpleasant people taking advantage of their handicap.

The researchers utilized technology to enable the visually impaired sight in dealing with the environment because it is one of the most important senses in the human body.

[a]hdgrimaldo@gmail.com, [b]santos.jericho08@gmail.com, [c]ainchristianpanganiban@gmail.com, [d]michaelarcay08@gmail.com, [e]dbparagoso@gmail.com, [f]nmestonilo@gmail.com, [g]jmlandingin@gmail.com, [h]jzburgos@gmail.com, [i]rcmuli@feualabang.edu.ph

DOI: 10.1201/9781003684459-27

Since other technology, such as smart glasses, do not generally work with all types of blind people. The device itself is a vocal assistant that is intended to anticipate what is to come. This device is going to provide information on what the visually impaired may not be aware of. The researchers' solution is to create a device to communicate with the user, with technology similar to Alexa, Siri, etc. The device will be interactive so the user can easily navigate through a mobile application using speech commands. Losing a major sense in one's body would affect them physically and mentally.

2.1. General objective

To develop a companion guide device as an assistance for visually impaired people.

2.2. Specific objectives

1. To receive audio information from the application.
2. To detect and alarm the user for incoming obstacles.
3. To establish hardware connectivity between the mobile device and processing module.
4. To identify the current location.
5. To send a warning message when the battery is on critical level.
6. To obtain a 90% success rate of the system.

2.3. Scope and delimitations

The research enhanced the understanding of the problems visually impaired face daily, the processing module served as the main component of the system. When the bone conduction earpiece and processing module is turned on through a button, the ultrasonic sensor, application and battery level sensor are activated. Once the sensor has detected an obstacle, a sound output through the earpiece is sent to alarm the user. The bone conduction earpiece is connected to the mobile phone. When the visually impaired person inputs information on where his/her intended destination is, the information will be processed and will be sent to the mobile application as an input to the destination field of the application. The phone GPS determined the current location. All information such as distance from the destination and direction to take until it reaches its destination was provided by the system through the bone conduction earpiece that aided and guided the visually impaired person on their way to the destination. The button is used to turn on/off the system and since the processing module uses a battery, a sound message will be activated when the battery is in critical level. The product's input voltage is 12.6V and 11.1V as its critical level, which means 88% was set as the critical level. Based on research for 12.6V the low battery level is at 10.5V which means that the battery percentage capacity is at 0%, that is why the researchers decided to set the critical level at 11.1V for the safety of the battery at 60% capacity.

2.4. Conceptual framework

A visually impaired person would like to do things on their own, but because of the disability, it limits the activities (Figure 27.1). The bone conduction earpiece's sound feedback lets the user absorb the information. It has provided the information through sound that will be transmitted to its speaker. The obstacle detection system helped the user to navigate without the possibility of bumping into objects. When the battery is in its critical level, it will produce a sound message to inform the user.

3. Methodology

This chapter presents the methods used in an area of research. It shows the procedures and processes used to support the information about the topic. This also uses applied research as the research methodology. The process and procedures are discussed and broken down in this chapter along with the system used and its constraints.

3.1. Block diagram

The system has its own raspberry pi built in the device. The device revolved around the user's input on the bone conduction earpiece and application (Figures 27.2 and 27.3).

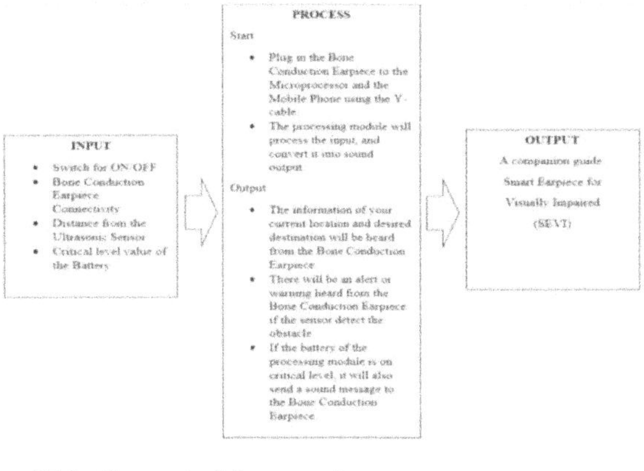

Figure 27.1. Conceptual framework.

Source: Author.

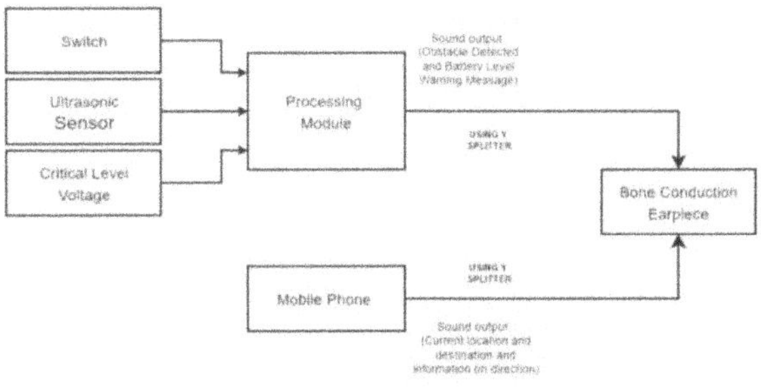

Figure 27.2. Block diagram.

Source: Author.

3.2. *System flowchart*

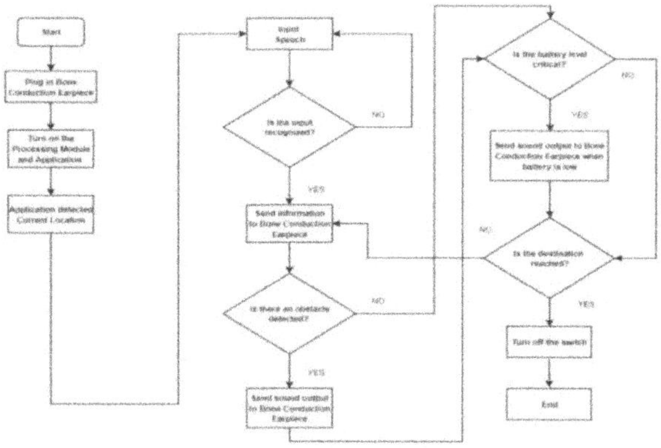

Figure 27.3. System flowchart for hardware and mobile app.

Source: Author.

4. Results and Discussion

Table 27.1. System testing

Trial	Testing if the application can recognize the command	Testing if the obstacle detection system is working	Testing if the obstacle detection system is working	Testing if the Mobile Phone is connected to the Processing Module	Testing if the identify current location works	Testing if the battery level indicator is working	Testing the functionality of SEVI Objective
1	✓	✓	✓	✓	✓	✓	✓
2	✓	✓	✓	✓	✓	✓	✓
3	✓	✓	✓	✓	✓	✓	✓
4	✓	✓	✓	✓	✓	✓	✓
5	✓	✓	✓	✓	✓	✓	✓
6	✓	✓	✓	✓	✓	✓	✓
7	✓	✓	✓	✓	✓	✓	✓
8	✓	✓	✓	✓	✓	✓	✓
9	✓	✓	✓	✓	✓	✓	✓
10	✓	✓	✓	✓	✓	✓	✓
Success Rate	100%	100%	100%	100%	100%	100%	100%
Overall Reliability	100% Reliability						

Source: Author.

Table 27.2. Battery specification

CAPACITY IN %	LEAD-BATTERY	LITHIUM BATTERY	LITHIUM AV-BATTERY
100%	12.70 V.	13.60 V.	12.60 V.
90%	12.50 V.	13.32 V.	12.10 V.
80%	12.42 V.	13.28 V.	11.60 V.
70%	12.32 V.	13.20 V.	11.35 V.
60%	12.20 V.	13.16 V.	11.10 V.
50%	12.06 V.	13.13 V.	10.80 V.
40%	11.90 V.	13.10 V.	10.70 V.
30%	11.75 V.	13.00 V.	10.60 V.
20%	11.58 V.	12.90 V.	10.45 V.
10%	11.31 V.	12.00 V.	10.25 V.
0%	10.50 V.	10.00 V.	9.00 V.

Source: Author.

The overall performance of the system depends on the functionality of each module and system (Table 27.1). The navigation system, obstacle detection system and battery level indicator play important parts on the overall performance. The researchers were able to test each system and achieved a high percent of reliability. The navigation system provides information on the designated location of the user, which is connected to the obstacle detection system that helps the user to get safely to that designated location. In order to use the prototype continuously, the battery level indicator will keep watch and alert if the battery will soon deplete. Through good location and resources provided, the system achieved a 100% success rate. The audio input and output were accurate and the user arrived at his designated location (Table 27.2).

5. Conclusion

The proposed study aimed to design a system that provides directions to the user which will also detect obstacles while going to the user's designated location. Some issues are encountered regarding the program that we used. The integration of Google Maps API became a problem to the system as it requires more programs and it is too complicated. The issue with the program was solved by replacing the Google API to Waze API as the main navigation system. The prototype's design has no problem since we assemble it in no time.

The system provides location, routes, and obstacle detection to the users, but the results are still apparent if the phone has a limited signal in the area where the user is traveling.

In the testing of the objectives, namely; receiving audio information from the application, the researchers were able to have a result of 80% reliability shown in Table 4.1. In detecting and alarming the user for incoming obstacles, the researchers obtained a 100% reliability based on testing shown in Table 4.2. In establishing hardware connectivity between the mobile device and processing module, the researchers achieved 100% reliability based on testing shown in Table 4.3. In identifying the current location, the researchers achieved 80% reliability due to internet connectivity loss, this was shown in Table 4.4. In the last objective on sending a warning message when the battery is on critical level, based on Table 4.5 warning messages were heard when battery is on its critical level from the set value of 11.1V and below at 100% reliability.

In implementing the system, trials were done to achieve every objective for the full functionality of the system. Based on the results of the trials, the researchers therefore conclude that all objectives achieved desirable results of the system's full functionality.

6. Recommendation

The objectives were successfully met with the different materials and proponents used. However, there are still some factors to consider to help with the functionality and provide efficiency to the system:

Accessibility can be improved by switching to a wireless bone conduction. This grants more convenience and freedom for the users.

The implementation of using waterproof materials can improve the longevity and the safeness of the prototype.

To improve better obstacle detection, future researchers can add a better obstacle detection device by using image processing modules. Better detection of the surroundings makes it easier for the user to perceive the information given by the prototype.

Reliability can be improved by adding more lifespan to the battery. Future researchers can also improve the capability of the prototype to handle a much higher battery voltage.

Future researchers can improve the study by improving the use of the GPS by having a separate route for the handicapped. This route should be safe and unchallenging for the users.

Future researchers may opt to design and use lighter materials to lessen the weight of the cane or may have it wearable instead of a cane.

References

[1] T. Kalaiselvi, K. Kalaivani, R. Abirami, S. Janani, V. Harshavadhani, EAR FREE LISTENING USING BONE CONDUCTION TECHNOLOGY, European Journal of Molecular & Clinical Medicine, Volume 7, Issue 4, Pages 2769–2772, 2020.

[2] Vidhya, P., Mugunthan, S. R., & Rajesh Kumar, B. (2016). An Efficient Active Contour based Brain Image Segmentation Using MRI Imaging Technology. International Journal of Advances in Engineering and Emerging Technology, 7(4), 102–110.

[3] "Voice Based Navigation System for Blind People Using Ultrasonic Sensor", Harsur & Chitra, 2017. [Online]. Available: https://ijritcc.org/download/1435907850.pd f/. [Accessed 13-November-2021]

[4] S. Real, A. Araujo, Navigation Systems for the Blind and Visually Impaired: Past Work, Challenges, and Open Problems, 2019. [Online]. Available: https://www.ncbi.nlm.nih.gov/pmc/articles/PMC6696419/. [Accessed: 13-November-2021]

[5] Zamanpoore, M., Sedaghat, F., Sorbie, M. R., & Ashjar, N. (2024). Evaluation of the primary production and maximum fish production in Salman Farsi Reservoir, Fars Province, Iran. International Journal of Aquatic Research and Environmental Studies, 4(2), 19–36. http://doi.org/10.70102/IJARES/V4I2/2

[6] W. Elmannai, K. Elleithy, Sensor-Based Assistive Devices for Visually-Impaired People: Current Status, Challenges, and Future Directions, 2017. [Online]. Available: https://www.mdpi.com/1424-8220/17/3/565. [Accessed: 14-November-2021]

[7] Makhmaraimova, S., Kurbanazarova, N., Karimov, I., Zakhidova, S., Karimov, Z., Sattorova, Z., Almamatova, S., & Amonturdiev, N. (2024). Tracing the Linguistic Journey of Geological Terms—A Philological Study of Stratigraphy and Mineralogy. *Archives for Technical Sciences*, *2*(31), 192–200. https://doi.org/10.70102/afts.2024.1631.192

[8] Mauro Avila, Katrin Wolf, Anke Brock and Niels Henze. Remote Assistance for Blind Users in Daily Life: A Survey about Be My Eyes. The 9th ACM International Conference on P Ervasive Technologies Related to Assistive Environments—PETRA'16, Jun 2016, Corfu Island, Greece. ACM, 2016.

[9] Annalakshmi, Nandhini, & Balamurugan. (2022). Reader's Dot. *International Academic Journal of Innovative Research*, *9*(2), 25–30. https://doi.org/10.9756/IAJIR/V9I2/IAJIR0914

[10] Çağıltay, F., Biçer, E., Yıldız, A., & Gökoğlu, M. (2023). Occurrence of the Scalloped Ribbonfish Zu cristatus (Lampridiformes) in the Gulf of Antalya, Türkiye. *Natural and Engineering Sciences*, *8*(2), 82–85. http://doi.org/10.28978/nesciences.1338121

[11] J. Velasco, E. Fernandez, G. Adiao, E. Alarilla, P. Cano, M. De Lara, J. Grande, J. Nuevo, I. Valenzuela, L. Tolentino, User Authentication for Computer Security Based on Sound-to-Feature Algorithm and Artificial Neural Network Using Bone Conduction, 2019. [Online]. Available:https://deliverypdf.ssrn. com/delivery.php?ID=64111707411911606808902908208409102401607302702707506210100608802207812410211212301109610006304505309800908808010310401810802805902100902303609008509602608008908412506008206306711900309702000211901408202712406908812009308312100708411109812606406702811 0&EXT=pdf&INDEX=TRUE. [Accessed:14-November-2021]

[12] C. Atienza et al, A Wearable Electronic Guidance System for the Blind Person, Cebu City, Philippines, Cebu Technological University, 2018

[13] E. Bermudo, P. Nahiwan, B. Redoble, B. Carlos, J. Reyes, J. Singzon, H. Asaba, C. Ong, Developing an Assistive Device for the Visually

[14] Impaired Using Ultrasonic Sensors for Distance an Solidity Determination, 2020. [Online]. Available: https://www.dlsu.edu.ph/wp-content/uploads/pdf/conferences/research-congress-proceedings/2020/HCT-05.pdf. [Accessed; 14-November-2021]

[15] E. Guevarra, M. Reyes, Development of Guiding Cane with Voice Notification for Visually Impaired individuals, Cavite City, Philippines, Cavite State University, 2018.

[16] Mauro Avila, Katrin Wolf, Anke Brock and Niels Henze. Remote Assistance for Blind Users in Daily Life: A Survey about Be My Eyes. The 9th ACM International Conference on P Ervasive Technologies Related to Assistive Environments—PETRA'16, Jun 2016, Corfu Island, Greece. ACM, 2016.

[17] Article on Discharge and Battery Capacity, RebelCell. https://www.rebel-cell.com/knowledge-base/the-discharge-and-capacity-of-batteries/?fbclid=IwAR0OTlJq7FR0TkxHaQT3sw-4oN8Vv3WGSxqJjM25HjVuSYQtjkv-O2WmtPc

[18] Article on Maximizing Battery Life, Century. https://www.rebel-cell.com/knowledge-base/the-discharge-and-capacity-of-batteries/?fbclid=IwAR0OTlJq7FR0TkxHaQT3sw-4oN8Vv3WGSxqJjM25HjVuSYQtjkv-O2WmtPc

[19] Article on How to Maintain Battery, Deep Cycle Battery (DBC). https://deepcyclebatterystore.com/how-to-maintain-batteries/

[20] P. Venkatapathi, Dr.Habibulla Khan and Dr.S.Srinivasa Rao "Performance of threshold detection in Cognitive Radio with improved Otsu's and Recursive One-Sided Hypothesis Testing Technique " International Journal of Innovative Technology and Exploring Engineering (IJITEE) Volume 8, Issue-9S3, July.-2019, ISSN: 2278-3075.

28 LIC policy holders utilization level towards online services in Tirupur district

D. Vishnu Vardhan[1,a], *K. Jothi*[2,b], *E. Arulprakasam*[1,c], *J. Sudarvel*[1,d], *V. Mathan Kumar*[3,e], *and R. Gopi*[4,f]

[1]Karpagam Academy of Higher Education, Coimbatore, Tamil Nadu, India
[2]Associate Professor and Head, Department of Commerce E Commerce, KPR College of Arts Science and Research, Coimbatore, Tamil Nadu, India
[3]Assistant Professor, Department of Commerce, Karpagam Academy of Higher Education, Coimbatore, Tamil Nadu, India
[4]Associate Professor, PG and Research Department of Commerce, Government Arts College, Udumalpet, Tamil Nadu, India

Abstract: The phrase 'online' has become the leading market trend during the recent years across industrial sectors including travel and retail along with banking and education in India. Indian people already conduct their banking activities online. People select fixed deposits and mutual fund investments as their primary online purchasing investments in India. People make electronic payments for money transfers as well as monthly bills through net banking methods. Insurance is not far behind. Insurance policies now gain more popularity among customers who want to buy coverage online. Prior to this period people mostly used the internet to find products but now the post office offers policy renewal and premium payment as its service portfolio. Transparency combined with effortless online processes and cost efficiency now drive customers towards purchasing policies on the internet. This research study concentrates on evaluating how rural LIC policyholders use online LIC services. Primary data collection represented the main research method in this study as it obtained responses from 166 policyholders who acquired LIC of India policies within Tirupur district. The questionnaire was used to gather the data. The research employs convenient sampling to choose the participants from the policyholder group. The researchers utilized weighted average alongside simple percentage to study and derive conclusions from the obtained data. Most rural policyholders actively use the various online services provided by LIC as the majority do not face difficulties operating these services.

Keywords: Policy holders, awareness, utilization and LIC e-services

1. Introduction

The financial sector changes through technological innovations which produce enormous efficiency improvements while creating initial periods of doubt along with uncertainty about the new ideas. The modern technological developments of recent years have generated the phenomenon known as 'FinTech.' The nature of financial services industry matches well with technological innovation because it allows for lower transaction costs as well as faster service delivery. Over financial history this evolution has occurred yet the current abundance of internet connectivity together with home computing systems and mobile devices and innovated software applications now presents financial businesses with a new prospect for market entry barriers reduction which generates rising industry competition.

Since innovation and technology naturally produce unpredictable results in the marketplace it is more accurate to describe them as disruptive than labeling them this way. Insurance joins most other industries as technology developments enable new approaches to customer service delivery while expanding data collection abilities that lead to enhanced risk assessment capabilities under the 'InsurTech' name. The InsurTech sector emphasizes enhancing individual services instead of focusing on business applications compared to the FinTech industry.

Any person searching for financial savings should consider working with an Indian insurance company. Initially people viewed insurance as simply a tax-benefit investment. The typical public opinion about life insurance moves toward a positive shift over time. Modern society holds life insurance on a par with financial investments. Since private insurance companies entered the market they introduced new standards of transparency together with more flexible options. The insurance market now has private companies controlling about 9% of total insurance despite government

[a]vishnudhanvanth07@gmail.com, [b]jothikrishnasamydr@gmail.com, [c]eapcas@gmail.com, [d]j.sudarvel@gmail.com, [e]mathankumar010@gmail.com, [f]gopigacudt@gmail.com

DOI: 10.1201/9781003684459-28

monopolization of endowment and money return products. The private insurance sector delivers better customer services at present while providing personalized treatment and complete transparency. Due to a large segment of India's insurance market being unregulated companies have many business prospects to explore in that sector.

The insurance industry of India grew at an accelerated rate after implementing liberalization processes. Life insurance represents one of India's fastest growing and developing markets at present. Insurance firms play an essential role in making available insurance products that satisfy customer needs at reasonable prices. The Life Insurance Corporation provides Term insurance along with Whole life insurance and Endowment policies and Money back plans and Individual insurance and Group insurance and Pension plans and Children's plans and Unit-linked plans. The life insurance provider offers these plans through its large monopolistic public organization for life insurance purposes and specifically designs them based on income and age group characteristics. As the government owns this firm and promises reliable returns its staff members show trustworthiness. LIC employees understand the general knowledge policyholders possess about the e-services that the company provides.

The status of policies and premium payment can only be processed through any of LIC's nationwide locations. The operational presence of LIC branches does not guarantee that customers can easily access offices for premium payments or to check insurance policy or loan statuses. LIC provides an online website accessible to policyholders which lets them perform various services through a few mouse clicks from home. The global digital tsunami has caused most insurers to transform their operations. The national insurance company LIC operates as part of the numerous organizations that deliver digital solutions to their services.

1.1. Benefits of LIC online portal

- All insurance companies operating across the nation have added their products to online platforms. Policies from LIC are now available for online viewing. LIC's online policy system delivers its best feature through a simple procedure which provides continuous post-sale assistance to policyholders. These are the main advantages which LIC's web portal provides:
- Insurance policyholders can access plan information when needed and from any place where they have internet access.
- The process of paying insurance premiums becomes possible through an online system that eliminates traditional branch requirements.
- Claim status tracking can be completed without contacting LIC customer service or police office visits by policyholders. On the LIC online portal customers can access every necessary piece of information.
- Online submission of complaints and grievances through the portal provides a service along with policy bond image viewing as well as proposal form image viewing capabilities for enrolled policies.
- The site gives customers free access to numerous policy-related information while all LIC's online services remain without charge.

1.2. Services offered by LIC's online portal

- Internet payment
  Through this option, policyholders can use LIC's online services to pay renewal premiums, loan interest, and loan repayment. Payment methods include BHIM, UPI, BHIM, debit and credit cards, as well as net banking. Prior to the due date and throughout the duration that the insurance is in effect, policyholders have one month to pay their payments.
- Policy Schedule
  Consumers can access the policy schedule on the policy bond's initial page via the portal.
- Policy Situation
  The e-services portal also shows the status of the policy and other plan information, such as the plan's duration, sum assured, and commencement date.
- Bonus Standing
  The entire bonus accumulated under the policy is also visible to policyholders.
- Loan Status
  Internet users can find loan status information such as the total loan balance of their policy and when their loan payments are due in the portal system.
- Claims Status
  The tool gives policyholders the ability to check the date of matured benefits or survival benefit dates during their policy duration.
- Revival Quotation
  The system offers revival quotation service to policy holders when their policies lapse.
- Premium Due Calendar
  The tab named Premium Due Calendar shows all payments which need to be made during the year.
- Premium Paid Certificate
  The system displays premium payment information for both individual and consolidated policies at this page. The policy area shows all claims payment information during the policy duration.
- Proposal form/Policy bond image
  Current policyholders can view policy bond and proposal form information directly from the portal through scanned image documents. Members of enrolled policies alone can access this service.
- Grievance Registration
  A policyholder also maintains the freedom of filing complaints with their insurer.

1.3. Who can avail e-services?

Through its website LIC provides all present policyholders access to these services. Registered customers of LIC India can directly create e-service accounts through filling the registration form located on the main website. Policyholders who do not hold an LIC registration need to obtain enroll- ment in their policies when signing up for e-services are shown in the Figure 28.1.

1.4. Exist a method to enroll for LIC e-services?

LIC e-services enrollment operates through three primary channels including the Website and through a Customer's visit to a LIC Branch.

- Website

  The 'Registration form' choice appears on LIC's website landing page following successful login for

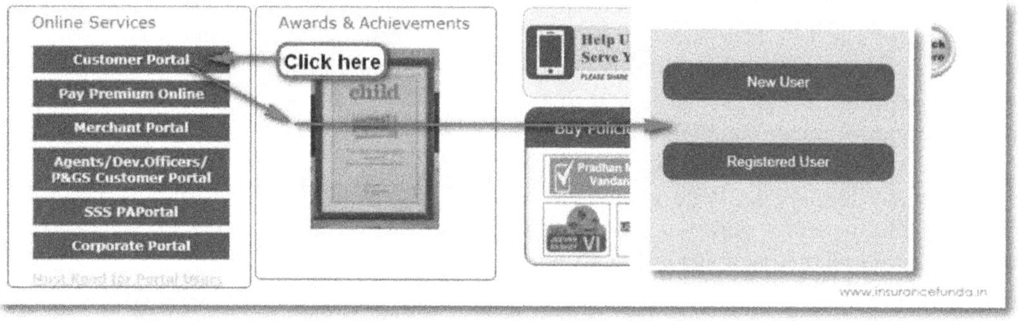

(A)

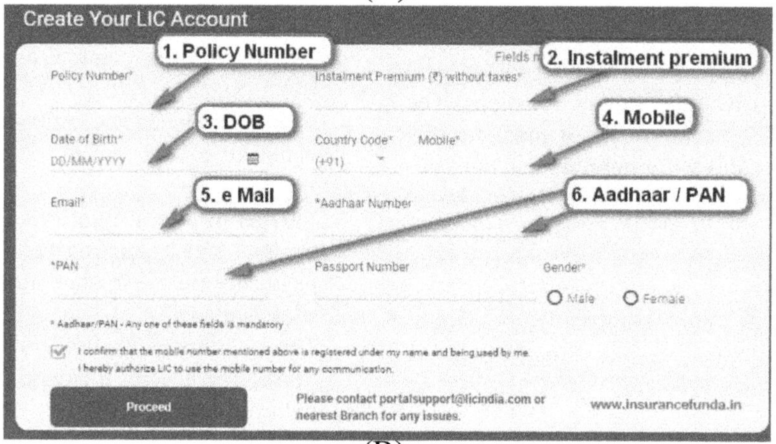

(B)

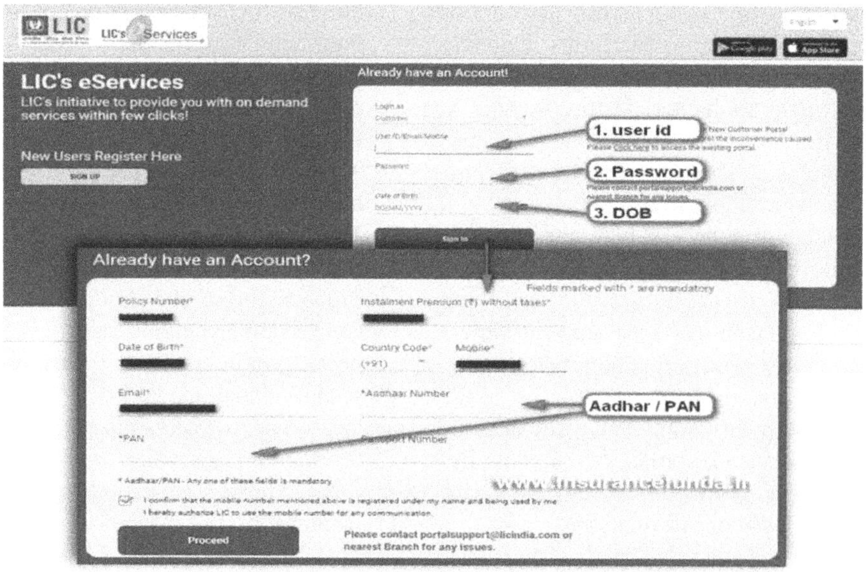

(C)

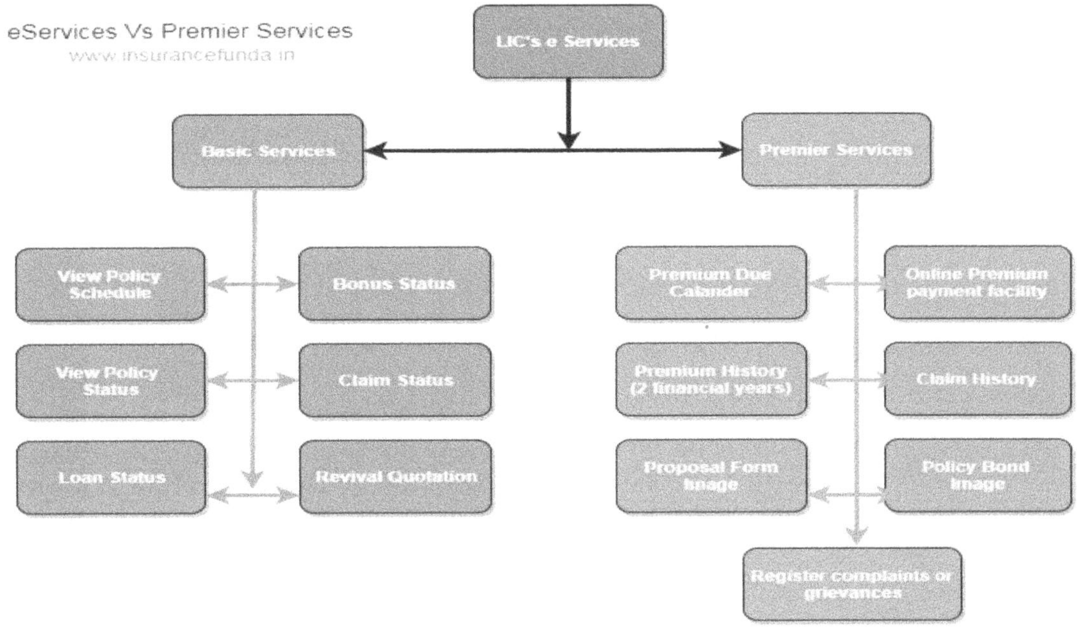

(D)

Figure 28.1. LIC'S e services with its procedures (A,B,C,D).

Source: Author.

policyholders. Users will access a fresh page which requires policy details and KYC information completion. The LIC official will verify the submitted form.

- LIC Branch

  The proposal form includes an option for policyholders who wish to enable e-services during their new insurance purchase. Both the insurance bond and proposal form scanned images become available to the consumer at the site.

2. Review of Literature

Pandemic risks produce substantial conceptual content that questions insurability by creating conditions for major insured losses such as natural events including food and earthquakes and hurricanes. The economic loss insurance protection poses substantial hurdles for insurance businesses because customers changed their priorities and needed digital platforms caused by work from home requirements.

Technological advances serve as a game-changer by delivering superior methods to generate and acquire and process insurance-relevant data so insurance organizations can transform their core risk operations. Digital Technology helps insurance industries to deliver superior customer experiences across various aspects. Digital innovation through digitalization enables the insurance industry to develop with controlled risks throughout its entire scope. Technology will serve as an enabler at an even higher level in future operations.

Digital technology serves a crucial purpose across all organizations while its implementation strengthens customer service delivery through firm collaboration and data storage and analysis per Ray et al. (2005; Fox and Connolly, 2018) along with enhancing service convenience abilities per Eisenhardt and Martin (2000; Pavlou and El Sawy, 2006; Teece et al., 1997). The connection between organizational change requirements and digital innovation was researched by Tanzi and Ancarani (2016). People who buy products online expect complete functionality during their shopping journey which extends to complete product displays in addition to top-notch service treatment. The future of customer insurance deals with digital above all else according to Hirt and Willmott (2014). The Digital world currently serves as a new consumer frontier that forces all enterprises to learn digital survival skills to prevent extinction with insurance needing adaptation (Nicoletti, 2016). The Indian insurance sector consists of 57 companies with both 24 life insurance providers and 33 non-life insurance firms. A synthesis of research studies concerning digital technology utilization in insurance businesses can be found in Table 28.1.

2.1. Scope of the study

The present study delivers information to the researcher regarding LIC's e-service utilization throughout the Tirupur district. The analysis directly addresses LIC'S E-Service challenges which exist in the current time. The research findings will support necessary changes for enhancing LIC's electronic services. The research mainly investigates the electronic service utilization rates among the public at Tirupur district levels.

2.2. Objective

To identify theutilization level towards LIC's E-Services in Tirupur District.

Table 28.1. Analysis and interpretation

Area of Residence	*Frequency*	*Percent*
Rural	128	77.1
Semi-urban	22	13.3
Urban	16	9.6
Total	166	100.0
Gender	*Frequency*	*Percent*
Male	78	47.0
Female	88	53.0
Total	166	100.0
Age	*Frequency*	*Percent*
Up to 30 Years	66	39.8
31–35 Years	58	34.9
36–45 Years	23	13.9
46–55 Years	14	8.4
Above 55 Years	5	3.0
Total	166	100.0
Marital Status	*Frequency*	*Percent*
Married	85	51.2
Unmarried	81	48.8
Total	166	100.0
Nature of the family	*Frequency*	*Percent*
Joint	90	54.2
Nuclear	76	45.8
Total	166	100.0
Education qualification	*Frequency*	*Percent*
Up to HSC	26	15.7
Under Graduate	20	12.0
Post Graduate	67	40.4
Others	53	31.9
Total	166	100.0
Occupation	*Frequency*	*Percent*
Daily wage earner	19	11.4
Agriculture	73	44.0
Business	16	9.6
Government Employee	11	6.6
Private Employee	47	28.3
Total	166	100.0
Monthly Income	*Frequency*	*Percent*
Up to 20000	86	51.8
20001–30000	50	30.1
30001–40000	17	10.2
Above 40000	13	7.8
Total	166	100.0
Savings per month	*Frequency*	*Percent*
Up to 2000	110	66.3
2001–4000	26	15.7
Above 4000	30	18.1
Total	166	100.0

Source: Author.

3. Methodology

3.1. Data

Primary data was used in this research to collect the data using a structured questionnaire and secondary data, which were also used in this study to support this study.

3.2. Area of study

Study on theutilization LIC – E Services for selected Policy holders in Tirupur District.

3.3. Sample design

A convenient sampling method was once used in this investigation. 166LIC Policy holders have been chosen in Tirupur for this research.

From the Table 28.1 clearly shows that most of the Policy holders (77.1%) are residing in the rural area, Majority of the Policy holders (53.1%) are female, Majority of the Policy holders (39.8%) are belonging to the up to 30 years of age group, majority of the Policy holders (51.2%) are married, most of the Policy holders (54.2%) belonged to the joint family, most of the Policy holders' (40.4%) educational qualification is postgraduate, most of the Policy holders (44%) belong to agriculture, majority of the Policy holders (51.8%) monthly income is up to Rs. 20000, Majority of the Policy holders (66.3%) monthly saving up to Rs. 2000.

Modern aids used	*Frequency*	*Percent*
Mobile Phone	164	98.79
Laptop	33	19.9
Desktop	33	19.9
Tablet	31	18.7
Others	6	3.6
Payment mode	*Frequency*	*Percent*
Online	84	50.6
Offline	82	49.4
Total	166	100.0
Periodicity of premium payment	*Frequency*	*Percent*
Monthly	36	21.7
Quarterly	97	58.4
Half Yearly	33	19.9
Total	166	100.0
Type of policy held	*Frequency*	*Percent*
Endowment plan	135	81.3
Whole Life plan	5	3.0
Term plan	2	1.2
Pension plan	22	13.3
Unit plan	1	.6
Special plans	1	6
Total	166	100.0

Table 28.2. Utilization level of e-services

	Low 1	Moderate 2	High 3	Population	Mean Score	Rank
Policy status	22	22	122	166	2.60	1
	22	44	366	432		
Bonus status	50	50	66	166	2.10	7
	50	100	198	348		
Loan status	38	44	84	166	2.28	4
	38	88	252	378		
Claims status	55	56	55	166	2.00	9
	55	112	165	332		
Revival Quotation	42	52	72	166	2.18	6
	42	104	216	362		
Premium due calendar	26	44	96	166	2.42	3
	26	88	288	402		
Premium paid certificate	21	34	111	166	2.54	2
	21	68	333	422		
Claim history	55	41	70	166	2.09	8
	55	82	210	347		
Grievance registration	64	62	40	166	1.86	10
	64	124	120	308		
Process of various services and online forms	29	63	74	166	2.27	5
	29	126	222	377		
Premium Payment	14	68	84	166	2.42	3
	14	136	252	402		
Insurance Plans	10	46	110	166	2.60	1
	10	92	330	432		

Source: Author.

Majority of the Policy holders (98.79%) using Mobile Phones, and majority of the Policy holders pay their premium Online portal, majority of the policy holders (58.1%) pay their premium in Quarterly, and the majority of the policy holders (81.3%) policy is Endowment plan.

It is seen from the Table 28.2 that, according to the Weighted Average test, Utilization level of LIC's e-services are namely Policy status, Insurance Plans, Premium paid certificate, Premium Payment, Premium due calendar, Loan status, Process of various services and online forms, Revival Quotation, Bonus status, Claims status and Grievance registration.

3.4. Reason for using LIC e-services

It is seen from the above table that, according to the Garrett Ranking Test, factors that influence to use the LIC's e- services for policy holders, are namely Access from Anywhere, 24*7 Service, Convenience, Organized Downloadable Reports, Time Saving, No Hidden Fees, Monitor Your Accounts Closely, User Friendly and Schedule Regular Payments.

4. Suggestions

- LIC e-services should try to launch more schemes through better marketing and innovative measures.
- LIC e-services system should be made more flexible regarding the formalities so that users feel convenience in regular use of LIC e-services.
- LIC e-services should be resolve the technical problem very quickly that the customer face during pay the premium in online.
- The company should enhance its marketing approach through digital advertising and establishing effective social media outreach.
- The organization should introduce alternative communication pathways which allow policyholders to get assistance with inquiries and claim submission needs.

Reasons	1	2	3	4	5	6	7	8	9	10	11	Sample Size	Total Score	Mean Score	Rank
	83	71	64	59	54	50	45	40	35	28	17				
Access from Anywhere	45	40	21	15	11	9	7	6	5	4	3	166			
	3735	2840	1344	885	594	450	315	240	175	112	51		10741	64.705	1
Organized Downloadable Reports	20	30	40	22	21	11	8	9	2	2	1	166			
	1660	2130	2560	1298	1134	550	360	360	70	56	17		10195	61.416	4
Schedule Regular Payments	0	2	2	2	2	2	3	3	4	50	96	166			
	0	142	128	118	108	100	135	120	140	1400	1632		4023	24.235	11
Convenience	1	60	40	39	5	4	3	2	6	4	2	166			
	83	4260	2560	2301	270	200	135	80	210	112	34		10245	61.717	3
Security Assured	2	1	1	2	3	4	18	19	22	49	45	166			
	166	71	64	118	162	200	810	760	770	1372	765		5258	31.675	10
No Hidden Fees	4	9	24	12	42	24	20	12	8	8	3	166			
	332	639	1536	708	2268	1200	900	480	280	224	51		8618	51.916	6
Monitor Your Accounts Closely	4	3	12	4	2	36	32	31	30	10	2	166			
	332	213	768	236	108	1800	1440	1240	1050	280	34		7501	45.187	7
Easy Access	0	3	8	7	4	20	40	20	48	12	4	166			
	0	213	512	413	216	1000	1800	800	1680	336	68		7038	42.398	9
24*7 Service	54	10	4	54	6	8	4	4	4	14	4	166			
	4482	710	256	3186	324	400	180	160	140	392	68		10298	62.036	2
User Friendly	8	4	5	4	12	24	13	50	35	7	4	166			
	664	284	320	236	648	1200	585	2000	1225	196	68		7426	44.735	8
Time Saving	28	4	9	5	58	24	18	10	2	6	2	166			
	2324	284	576	295	3132	1200	810	400	70	168	34		9293	55.982	5
Population	166	166	166	166	166	166	166	166	166	166	166	1826			

- The organization should adjust all written materials to decrease their insurance-specific terminology.
- The company must provide customers combined solutions consisting of artificial intelligence and human interaction alongside physical settings with digital ones as well as direct contact options and agent assistance.
- Engaging with customers daily or throughout the year instead of only at renewal time.

5. Conclusion

Indian core competencies grew because digital technologies gained wider usage during pandemic years leading to improved marketing capabilities. The approach introduces new options to expand individual awareness regarding the purchase and sale of life insurance using digital systems. This study determined the extent to which LIC's e-services are utilized by policy holders who make use of Policy status, Insurance Plans, Premium paid certificate, Premium Payment, Premium due calendar, Loan status, Process of various services and online forms, Revival Quotation, Bonus status, Claims status and Grievance registration. The insurance firm LIC maintains current digital services. The website of LIC accounts for less than one percent of its total sales transactions. Two things are happening. People are becoming progressively digital which involves both premium collections and other matters. LIC operates as the leading life insurance company throughout the Indian insurance sector. The company maintains its market leadership position because of its established brand built over a long time coupled with an extensive nationwide network. The company implements every type of marketing strategy between online platforms and offline channels and direct outreach methods.

References

[1] Chithirai Selvan.M, Velmurugan, R and S. Benjamin Christopher (2012), 'Determinants of Mediclaim Policyholder Satisfaction', Journal of Management and Science, Vol. 2 No.4, pp. 37–50,

[2] Subashini, S and Velmurugan (2016), "Policyholders Satisfaction of Life Insurance Products with Reference to

Coimbatore District", International Journal of Business and Management Invention, Vol.5, Issue 4, pp. 28–33

[3] Subashini, S and Velmurugan (2016), "Misselling – A Bane on Growth of Life Insurance", International Journal of Scientific Engineering and Technology, Vol.5, Issue 5, pp. 264–267

[4] Subashini,S and Velmurugan (2016), "A Study on Lapsation of Insurers in Life Insurance Policies in Coimbatore", International Journal of Management Research and Review, Vol. 6, Issue 5, pp. 560–566

[5] Subashini, S and Velmurugan (2017), "Lapsation of Life Insurance Policies", Journal of Advanced Research in Dynamical and Control Systems, Special issue V, pp. 33–36

[6] Subashini, S and Velmurugan (2015), "SWOT Analysis for Online Insurance India", International Journal in Management and Social Sciences, Vol.3, Issue 11, pp. 276–286

[7] Subashini, S and Velmurugan (2015), "Digital Distribution Strategy of Life Insurance in India", International Research Journal of Commerce and Law, Vol. 2, Issue 11, pp. 18–28

[8] Subashini, S and Velmurugan (2016), "The Study on Reasons for Lapsation in Life Insurance Companies in Coimbatore City", Bonfring International Journal of Industrial Engineering and Management Science, Vol. 6, Issue 1, pp.78–80

[9] Kadhim, A. J. (2023). Media in Times of Crisis: A Case Study on the Impact of Media on Societal Awareness During the COVID-19 Pandemic. MEDAAD, 2023, 40–44. https://doi.org/10.70470/MEDAAD/2023/006

[10] Prithiviraj, V., B. V. Krishnaveni, and R. Jayabharathy. "Evaluating context based indoor positioning system using Wi-Fi standards." In 2013 International Conference on Communication and Signal Processing, pp. 982–987. IEEE, 2013.

29 JANUS: Study and analysis of the cooling and power requirements of an NFV telecommunications data center using machine learning

Lyberius Ennio F. Taruc[1,2,3,a], *Arvin R. De La Cruz*[1,b] *and Nelson C. Rodelas*[4,c]

[1]Graduate School, Polytechnic University of the Philippines, Manila, Philippines
[2]Computer Engineering Department, FEU Institute of Technology, Sampaloc, Manila, Philippines
[3]Network Technical Group, Globe Telecom, Inc., Bonifacio Global City, Taguig City, Manila, Philippines
[4]Computer Engineering Department, University of the East, Caloocan, Metro Manila, Philippines

Abstract: The objective of the present study is to develop a Machine Learning (ML) model which uses Multiple Linear Regression (MLR) techniques to estimate power and cooling requirements for data centers servicing NFV enabled network elements. The study is primarily geared towards Network Planners and Architects. For the simulation, the available synthetic dataset was partitioned into three distinct sets; 70% for training, 20% for testing and 10% for validation. Then the model was trained on the dataset. It has been established that independent discrete and continuous factors function in conjunction to provide a reasonable predictive performance. However, the situation is different for continuous independent variables where a negative R-squared value is achieved along with greater volatility in MSE and MAE. The research aimed to analyze the scenario if a single data center is removed or if 33% of the data are removed. The findings observed tend to be mixed but exhibit a degree of uniformity depending on which removal is applied. The primary conclusion of the study shows inferring the scale of a data center from its volume and operational value is a more precise approach than relying entirely on machine learning computations. With a more reliable model, it can now be used in multiple data centres with disparate types of deployments because the model can be trusted to handle variated yet cohesive datasets.

Keywords: Data center planning, machine learning, multiple linear regression, network functions virtualization (NFV)

1. Introduction and Background

Network function virtualization (NFV) is a relatively new architectural component of contemporary data centres that has improved network service management in terms of efficiency, scalability, and adaptability. At the same time, however, this shift brought with it significant difficulties in estimating power consumption and cooling needs, and these are both directly related to operational efficiency and cost control. Currently, traditional data centre dimensioning heavily relies on static models or expert driven estimation which might not reflect NFV based workloads dynamics accurately. In a data driven nature, Machine Learning (ML) allows forecasting resource demands in this context, potentially aiding in more energy efficient utilization of infrastructure as a result. The conventional Multiple Linear Regression (MLR) can be used to model complex relationship between independent variables (e.g., network traffic, utilization of the hardware resource, and environmental condition) and dependent variables such as power and cooling demand [1].

This study attempts to build an MLR based predictive model for helping out Network Planners and Architects in the optimal allocation of data center resources. A fake dataset was used to train and test the model. The main goal was to see how the discrete and continuous factors used for training affected how well the model could predict the future [2]. The findings reveal that predictive performance only deteriorates with continuous variables even though both variable types are combined to prevent instability, and analyzing continuous variables alone do not produce stable results [3]. Moreover, model reliability relies also on dataset variations by eliminating a subset of data centers or reducing the amount of data. Interestingly, our results show that volume and utilization metrics are still more reliable compared to dimension based on other ML metrics, however ML models still have a role to play when trained on diverse and representative

[a]lyberiusennioftaruc@iskolarngbayan.pup.edu.ph, [b]docardelacruz@gmail.com, [c]nelson.rodelas@ue.edu.ph

DOI: 10.1201/9781003684459-29

datasets. Over the past years NFV (Network Function Virtualization) has gained momentum in telecommunications data center, however this brings the challenge of how to manage their high power consumption and cooling requirements [4]. Allocating resource in the conventional way requires static provisioning, and consequently leads to over provisioning, and thus, resource underutilization and high operational costs. Since the NFV workloads are dynamic in nature, data driven approaches need to be used to optimize the infrastructure efficiency [5].

Using Machine Learning (ML), herein to identify power and cooling requirements as a function of true time, workload, and environmental parameters in particular Multiple Linear Regression (MLR) has been used. This research pursues to give Network Planners and Architects with solutions on how to use effective forecasting models that can help to plan energy usage, minimize cost and increment sustainability in NFV – based data centers in a manner that energy utilization is optimized, costs reduced and data centers are run in a sustainable formation at a time maintaining the efficient consumption of resources and the operational reliability [6].

1.1. Network functions virtualization as a workload in data centers

NFV is a system that can change the standard, one-piece, and hardware-reliant network services (e.g., firewalls, load balancers, and routers), to software that runs as instances in virtualized software [7].

Network usage has changed from using hardware-based network appliances to using software-based network services that run on regular servers with NFV (Figure 29.1). With this change, it will be easier to offer network services like firewalls, load balancers, and attack detection services, and it will also be more cost-effective to do so. So, it's hard to tell how much power and cooling NFV-based processes will

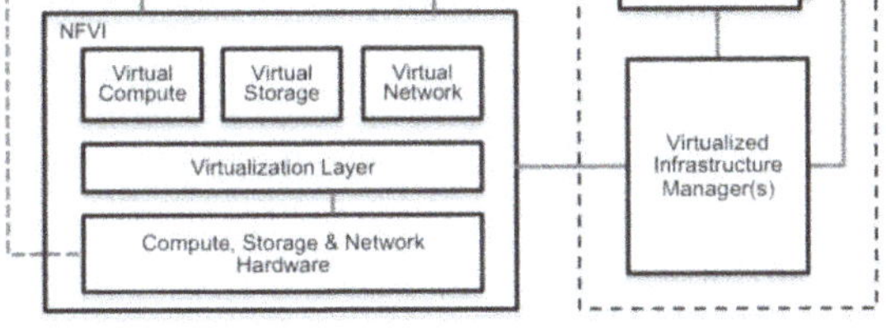

Figure 29.1. The ETSI NFV architectural framework [8].

Source: Author.

need. This makes allocating resources efficiently a difficult task. Because virtualized network functions are inherently unpredictable, accurate forecasting models become even more critical in order to simultaneously optimize energy usage and service reliability [9].

NFV workload computational loads are highly varying and rely on the transport network traffic volume, virtualization overhead and the orchestration efficiency in the data center hosting NFV workloads. Typically, such conventional IT workloads have predictable usages of resources in practice; whereas NFV workloads exhibit high sensitivity to real-time service demand and infrastructure configuration. This variability makes capacity planning hard and power and cooling management strategies must adapt. However, we can consider the deployment of NFV as a successful case only when the computation of NFV resource consumption trends is performed by an intelligent predictive model for the appropriate utilization of infrastructure. Data centre operators can use machine learning methods like MLR to boost service performance, cut down on energy costs, and improve efficiency. This will make sure that NFV workloads are run in the best possible environment.

NFV delivers several key benefits [8], including:

1.2. The challenge of data center dimensioning

Dimensioning of data center is a critical process and is concerned with determining the right capacity for power, cooling and infrastructure to meet workload demand under considerations of efficiency and economic feasibility. Dimensioning is historically composed of data from the past, fixed provisioning models and conservative estimations, producing over provisioning, with wasted resources and higher operating costs, or under provisioning and producing bottlenecks in performance and service disruption. Moreover, when data centers are hosting NFV workloads, this problem is even more pronounced due to highly dynamic and unpredictable resource consumption patterns in these environments. Unlike IT workloads in general, NFV based services run on virtualized infrastructure where network functions are provisioned dynamically, scaled/backed down or even migrated to other infrastructure as traffic (or demand for service) varies. Given the variability of the activity, it is not possible to find fixed strategies for allocation of the resources, making these approaches particularly required intelligent, and adaptive dimensioning approaches. A data driven approach for predicting power consumption and cooling requirement is provided based on real time workload parameters using ML-based models; especially Multiple Linear Regression (MLR). Nevertheless, the quality and diversity of training dataset as well as the inclusion of appropriate independent variables is crucial for the effectiveness of ML models. On the other hand, we propose to use traditional volume and utilization based metrics in the dimensioning process and a hybrid method (predictions + traditional metrics) as the ideal method to compute optimal resource allocation in data center.

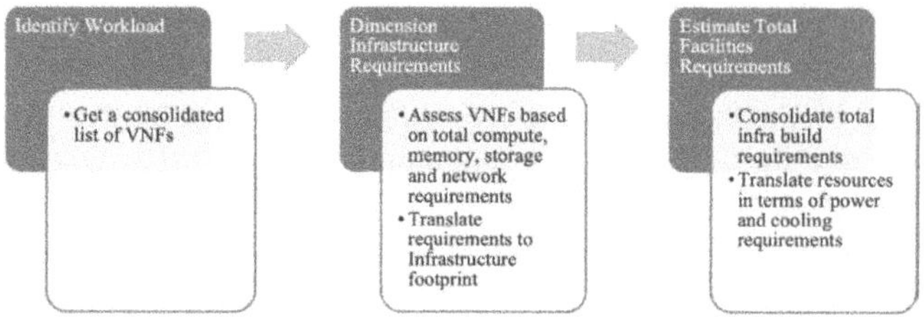

Figure 29.2. The data center dimensioning and assessment process.

Source: Author.

Figure 29.2 shows that the evaluation process usually has several steps that are done by hand.

Dimensioning of Data center includes computation of the optimal infrastructure capacity to satisfy workload demand at minimum cost, with minimum level of utilization of the infrastructure. The process of assessment includes the analysis of power consumption, cooling requirements, hardware utilization, as well as the network traffic. To determine key factors such as server density, environmental condition, and scalability needs are determined. The approach is a data driven one and ML models allow prediction of future loads with more accuracy resulting in better allocation of resources to make data center operation more sustainable and efficient.

2. Research Overview

In this part, we'll talk about how the research study was put together, as well as its goals, scope, methods, and limitations. The point of this study is to guess how much power and cooling NFV-based data centres will need by using a machine learning model and MLR. We aim to increase the amount of data allocated to resource allocation, energy efficiency and cost effectiveness in dynamically scaling virtualized environments.

2.1. Problem statement

The potential of Network Functions Virtualization (NFV) continues to change the telecommunications infrastructure and therefore keep the data centers dealing with power consumption issues as well as the likes of cooling management. The traditional methods of dimensioning have been based on static provisioning; which is often resulting into inefficiency, over-provisioning, or underutilization of resources. The resource demands of NFV-based workloads are highly dynamic and unpredictable, and conventional techniques are quite inadequate to forecast these demands with a good accuracy.

2.2. Research objective

The research's goal is to create an ML model in Python using MLR that can predict how much power and cooling data centres that house VM instances will need. Number of vCPUs, memory distribution, and instances, the given model will make sure that the forecast is accurate with the intention of optimizing resource allocation and optimizing the efficiency of NFV based data centers.

2.3. Significance of the study

This thesis, JANUS looks at Machine Learning (ML) for data center cooling and power calculations for telecommunications data centers when based on NFV. Unfortunately, traditional dimensioning models are no longer adequate for the accurate modeling of resource demand in the presence of the dynamic behaviour of virtualized network functions, given that they have widely been accepted and significantly spread in NFV. This gap is bridged by this research through development of a model based on MLR to forecast power consumption and cooling needs thereby providing the data driven insights for efficient infrastructure.

This result is highly relevant to the Network Planners and Architects as it provides a more adaptive and scaleable perspective on how to run the resources. The research shows that the prediction accuracy in ML based forecasting depends on the dataset's quality and diversity and it will assess the discrete and continuous variables effect on prediction accuracy. The results, thus, show that there is still a role for both ML and traditional metrics of volume and utilization in dimensioning strategies to achieve reliability. By all means, JANUS supports to enhance the energy efficiency, reduce the operation costs, and to make the NFV based data centers more sustainable.

2.4. Methodology

This research adopts a two step approach: Firstly, a systematic review of the literature is conducted and (secondly), model developers make use of the Machine Learning Lifecycle process to develop a MLR model.

The goal of the literature review is to find the main problems with building an NFV data centre and look into how machine learning can be used to lower power use and cooling needs. It provides comprehensive coverage of any relevant studies, industry reports, and technical papers.The MLR model development follows a structured Machine Learning lifecycle:

Data Collection and Preprocessing: By taking out noise and outliers, a fake dataset is made, cleaned, and improved.

Model Training: According to independent factors, the MLR model learns from preprocessed data to guess how much power and cooling will be needed. Model Evaluation: Performance is assessed using a holdout dataset, comparing predictions to actual values to ensure accuracy and reliability.

This methodology ensures a rigorous, data-driven approach, offering valuable insights into integrating ML for improved NFV data center planning and resource optimization.

2.5. Research scope and limitations

Researchers want to help network planners and architects make good decisions about how to allocate resources based on how accurate the predictions are. They will do this by looking at how discrete and continuous factors affect the accuracy of the predictions. A systematic literature review (SLR) and an MLR model development has been made (based on the Machine Learning lifecycle). A synthetic dataset is used for training, testing and evaluation.

However, the study did not lack some limitations. The synthetic data used to make the model predictions do not have to be representative of all potential workloads that exist in the real world NFV setting. Furthermore, only MLR is used and the other Ml techniques such as deep learning are not considered. In addition, the study does not include hardware dependent power variation as well as real time environment factors. To somewhat address these considerations, these limitations can be added as future research by incorporating of real-world datasets and more advanced predictive models.

3. A Summary of Related Literature

3.1. NFV concepts and adoption

NFV Software operating on conventional servers has mostly supplanted hardware-based network functions. such as its adoption helps realize scalability, flexibility, and cost effectiveness to achieve dynamic resource allocation and efficient operation on modern telecommunications data centers.

3.1.1. NFV and SDN: Advances and open research challenges

By separating the data plane from the control and technology planes, NFV and SDN alter the network model, making it more automated and flexible. Network function is virtualized in SFV while network is being controlled in CF. Scalability, security, a much more general question is: How to coordinate all the relevant functions of a data processing pipeline? However, problems to deal with are resource management, working together and being fault tolerant. Future research will address these issues aiming to increase efficiency and reliability of such NFV-SDN integrated environments [10].

3.1.2. Software-defined networking (SDN) and network functions virtualization (NFV) for future internet: A comprehensive survey

Technologies like NFV a as well as SDN (Software-Defined Networking) are reshaping the way networks work today. NFV enables the virtualization of network functions in a manner that does not rely on proprietary hardware, whereas SDN makes the entire network configurable through a centralized method comparable to network programmability. By working together, they increase scalability, flexibility and cost efficiency that are necessary for the future Internet. Nevertheless, there are issues like interoperability, security threats and optimally distributed resource. This survey outlines current advances in SDN-NFV integration, points out gap in research and reviews some future directions for making SDN-NFV integration optimally for next generation of networks [11].

3.1.3. Software defined networking (SDN) and network functions virtualization (NFV): A review

Both SDN and NFV make it possible to build and use modern network designs that are more flexible, scalable, and cost-effective. SDN splits the control plane from the data plane, and NFV lets you virtualize network functions to make them less reliant on hardware. This paper reviews the advantages, deployment difficulties, and integration possibilities of these solutions, including the problems regarding compatibility, security and resource management in dynamic and large scale network setting [12].

3.2. Machine learning algorithms for power and cooling forecasting

The direct use of ML algorithms in predicting power consumption and equally important is the need for cooling in data centres. These techniques MLR, Decision Trees, Random Forest, and Neural Networks – look at trends of workload and environmental factors to make prediction methods more accurate. The use of these models facilitates effective resource allocation, reduction of operating costs, and increase in energy efficiency. Although dataset quality, model generalization, and real time adaptability for dynamic NFV workloads are challenges that still remain [13, 14].

3.2.1. Predicting data center power consumption with machine learning

Machine Learning (ML) models that take workload, hardware utilization and environmental factors as input and try to predict power consumption of a data center is something I see getting increasingly predicted. With more optimal energy efficiency and lower costs, Multiple Linear Regression (MLR), Random Forest and Neural Networks algorithms are possible due to more accurate forecasting. However, the fine tuning and integration with traditional data center management to operate in an optimal way is challenging yet due to the model accuracy, dataset variability and real time adaptability [15].

3.2.2. Predicting data center power consumption with machine learning

Instead, machine learning is used to guess how much cooling the data centre will need based on changes in temperature, traffic, and how well air flows. MLR, SVM, and DL models help improve cooling systems by using algorithms to make the most of the least amount of energy used and the lowest cost of running. However, data quality, model generalization and real time prediction accuracy are issues to be solved for ensuring more efficient cooling management in NFV based data centers [16].

3.2.3. Machine learning for power and cooling optimization in data centers

ML helps data centres get the most out of their power and cooling by looking at things like workload trends, environmental conditions, and how the hardware is being used. The efficiency techniques used are MLR, Random Forest and Neural Networks that lowers power consumption and make it more environmentally friendly. It is not yet quite there and it does face challenges such as low real time adaptability, unvarying data set, and integrating with the existing infrastructure and hence, more work in that direction would be necessary for sustainable and intelligent data center resource management [17].

4. Model Overview

4.1. Overview

In the proposed model, the power and cooling consumption in the NFV based data centers are estimated by using Multiple Linear Regression (MLR). It takes into account relevant independent variables such as the CPU utilization, memory usage, network traffic of the server as well as the environmental temperature and server workload in order to predict the amount of power consumption (in kilowatts) as well as the cooling requirements (in BTU/hr). Data collection and preprocessing takes place to start the process, which includes aggregating, cleaning, and normalizing two sets of data – historical and synthetic. Next, feature selection is made to determine the factors that feature the most significant impact on power and cooling uses. To get the relationships among the input variables and energy usage, we have used structured datasets to train the MLR model. Finally, prediction accuracy is evaluated and validated on a hold out dataset. As an original and high-level model, this data driven approach represents one way to boost energy efficiency of NFV-based data centers (Figure 29.3).

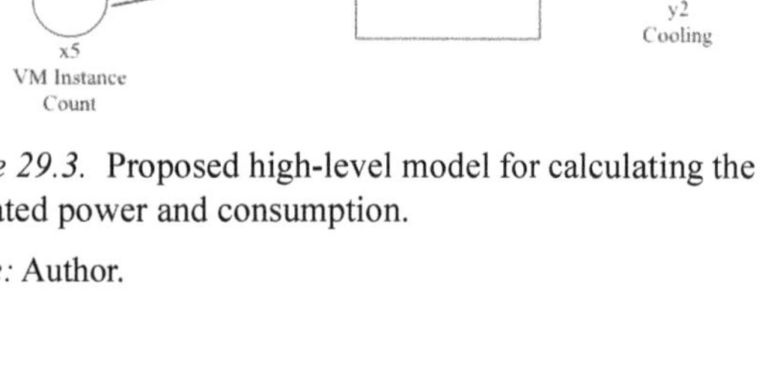

Figure 29.3. Proposed high-level model for calculating the estimated power and consumption.

Source: Author.

4.2. Data set assumptions

The training and evaluation of the Multiple Linear Regression (MLR) model are based on a dataset of synthetic and historical data about NFV-based data center workloads. It prescribes that CPU utilization, memory usage, network traffic, temperature, and workload intensity are key independent variables that work as reliable predictors for power consumption and cooling requirements. In addition, it makes the assumption that data centers run under normal workload fluctuations and not under extraordinary or atypical load conditions.

With the assumption that there are no large hardware failure or unpredictable increases in the demand, the dataset adheres to a normal distribution of resource utilization for consistency. It further assumes that cooling efficiency will not vary too much and will not be affected by external environmental factors such as humidity and airflow. Although real time to live datacenters may see spikes, they do not need to model irregularities, and this dataset can reflect what general trends are, giving a basis for predictive modeling (Tables 29.1–29.4).

Tables 29.5 through 29.7 were used to make the cooling requirements.

The following high-level formula is used to figure out BTU:

$$\text{BTU} = [\text{Area}] \times 20 + [\text{Power}] \times 3.14 + [\text{Active People}] \times 400 \quad (1)$$

where,

By examining the area unit, it is possible to determine the extent of the NFV equipment's footprint on the entire floor of the data centre.

The collective power consumption of the NFV equipment is referred to as 'power.'

At any particular moment, the term 'Active People' denotes that the number of individuals in the room is limited to that number.

Table 29.1. Data seed 01-a: Cabinet type power specifications

Cabinet Type	*Full-Width Blades*		*Subrack Blades (16-Blade Model)*		*Subrack Blades (8-Blade Model)*	
	Actual	*Max*	*Actual*	*Max*	*Actual*	*Max*
Compute	6.980	8.180	12.296	15.399	16.700	20.866
Storage	5.308	7.562	5.308	7.562	5.308	7.562
Network	7.970	10.60	7.970	10.60	7.970	10.60

Source: Author.

Table 29.2. Data seed 01-b-01: Calculated power specifications of data center C

Cabinet Type	Full-Width Blades			Subrack Blades (16-Blade Model)			Subrack Blades (8-Blade Model)		
	QTY	Actual	Max	QTY	Actual	Max	QTY	Actual	Max
Compute	0.00	0.000	0.000	0.00	0.000	0.000	4.33	72.311	90.350
Storage	0.00	0.000	0.000	0.00	0.000	0.000	1.00	5.308	7.562
Network	0.00	0.000	0.000	0.00	0.000	0.000	0.25	1.993	2.650
Total per Deployment Type		0.000	0.000	-	0.000	0.000	-	79.612	100.562
Total Site Power Consumption								79.612	100.562

Source: Author.

Table 29.3. Data seed 01-b-02: Calculated power specifications of data center S

Cabinet Type	Full-Width Blades			Subrack Blades (16-Blade Model)			Subrack Blades (8-Blade Model)		
	QTY	Actual	Max	QTY	Actual	Max	QTY	Actual	Max
Compute	0.00	0.000	0.000	0.00	0.000	0.000	4.20	70.140	87.637
Storage	0.00	0.000	0.000	0.00	0.000	0.000	1.00	5.308	7.562
Network	0.00	0.000	0.000	0.00	0.000	0.000	0.25	1.993	2.650
Total per Deployment Type		0.000	0.000	-	0.000	0.000	-	77.441	97.849
Total Site Power Consumption								77.441	97.849

Source: Author.

Table 29.4. Data seed 01-b-03: Calculated power specifications of data center T

Cabinet Type	Full-Width Blades			Subrack Blades (16-Blade Model)			Subrack Blades (8-Blade Model)		
	QTY	Actual	Max	QTY	Actual	Max	QTY	Actual	Max
Compute	4.00	27.920	32.720	9.10	111.894	140.131	0.00	0.000	0.000
Storage	5.60	29.725	42.347	0.00	0.000	0.000	0.00	0.000	0.000
Network	0.00	0.000	0.000	0.75	5.978	7.950	0.00	0.000	0.000
Total per Deployment Type		57.645	75.067	-	117.871	148.081	-	0.000	0.000
Total Site Power Consumption								175.516	223.148

Source: Author.

Table 29.5. Calculated cooling requirements of data center C

Parameter	Value	BTU Factor
Area (sq. ft.)	5705.00	114100.00
Power (kWH)	79.61	249.98
Active People (Max)	5.00	2000.00
Total BTU	116349.98	

Source: Author.

Table 29.6. Calculated cooling requirements of data center S

Parameter	Value	BTU Factor
Area (sq. ft.)	6027.00	120540.00
Power (kWH)	77.44	243.16
Active People (Max)	5.00	2000.00
Total BTU	122783.16	

Source: Author.

This formula provides a structured approach to estimating cooling requirements by incorporating key factors that influence heat generation within the data center environment.

The Virtual Resources are made by looking at Table 29.8, which assumes that the number of vCPUs, vMemory, and Active VM instances are all the same for each dataset. The Average vCPU and vMemory Utilisation Rates were taken from the 2023 data extracts and then used for extrapolation to make a fake dataset. An agreed percentage error rate per resources types was applied in order to extrapolate the variability: vCPU utilization can vary from 1% to 10% of error, whereas vMemory utilization falls between 1% to 5% of error. This method guarantees the dataset to have associated realistic variations in resource usage without a loss of statistical reliability for training and testing of the Multiple Linear Regression (MLR) model that was used in this study.

Table 29.7. Calculated cooling requirements of data center T

Parameter	Value	BTU Factor
Area (sq. ft.)	4133	82660.00
Power (kWH)	175.52	551.12
Active People (Max)	5.00	2000.00
Total BTU		85211.12

Source: Author.

Table 29.8. Virtual resources per data center

Data Centers	CPU Count	vMemory Count (GB)	Active VM Instance	Average vCPU Util	Average Memory Util
C	9812	27053	272	56.10%	53.20%
S	10238	27922	277	55.50%	53.10%

Source: Author.

5. Simulation and Testing Results

A Multiple Linear Regression (MLR) model was developed to identify the relevant parameters and their impact on CO2 capture. The quality and the effectiveness of developed MLR models are measured through the digital experiments performed through twelve test scenarios, with Python. These models were then evaluated in such scenarios to identify the superiority of two models: Separate models for vCPU Utilisation and vMemory Utilisation, as well as Total vCPU, Total vMemory, and Active VMs. To look at how accurate the predictions were, the following situations were thought about: data type (Discrete or Continuous) and dataset size (all sites and at least two sites).

To judge each case, MAE, MSE, and R-Squared numbers were used. Next, it was thought that the best model was one that had low MAE and MSE and an R-Squared value close to 1. The results show that: Continuous variables, when combined with Discrete variables make stable predictions (Table 29.9).

Negative R-Squared values and increased variance of MSE, MAE are caused by continuous variables alone.

Removing 33% of the dataset has varied but steady results based on what data is removed.

Table 29.9. Simulation results and evaluation

Test Scenario	Discrete			Continuous		Scoped Data Centers	Power Model (MLR01)			Cooling Mode (MLR02)		
	Total vCPU	Total vMemory	Active VMs	vCPU Util	vMemory Util		MSE	MAE	R-Squared	MSE	MAE	R-Squared
1	Y	Y	Y	Y	Y	C, S, T	0.0008	0.0227	0.9992	0.0054	0.0619	0.9947
2	Y	Y	Y	Y	Y	C, S	0.4460	0.5908	0.6159	0.1598	0.3533	0.8257
3	Y	Y	Y	Y	Y	S,T	0.0008	0.0233	0.9992	0.0042	0.0552	0.9957
4	Y	Y	Y	Y	Y	C,T	0.0011	0.0271	0.9989	0.0053	0.0629	0.9947
5	Y	Y	Y	N	N	C, S, T	0.0008	0.0227	0.9992	0.0054	0.0621	0.9947
6	Y	Y	Y	N	N	C,S	0.4465	0.5912	0.6155	0.1599	0.3536	0.8256
7	Y	Y	Y	N	N	S,T	0.0008	0.0233	0.9992	0.0042	0.0551	0.9958
8	Y	Y	Y	N	N	C,T	0.0011	0.0271	0.9989	0.0053	0.0631	0.9947
9	N	N	N	Y	Y	C, S, T	1.0110	0.9459	0.0027	1.0238	0.9384	0.0035
10	N	N	N	Y	Y	C,S	1.1635	0.9147	-0.002	0.9502	0.8859	-0.036
11	N	N	N	Y	Y	S,T	0.9984	0.9969	-0.01	0.9963	0.9945	-0.008
12	N	N	N	Y	Y	C,T	1.0107	1.0028	-0.026	1.0312	1.0111	-0.023

Source: Author.

6. Conclusion

ML, more especially MLR, to guess how much power NFV-based telecommunications data centres will use and how much cooling they will need. Traditional data center dimensioning was seen to be difficult, and this research showed how ML models can help in gaining data driven insights in making data centers energy efficient. Findings indicate that using both discrete and continuous independent variables lead to more stable predictions, but when using only continuous variables predictions have negative R square values along with (MSE & MAE) inconsistent errors. ML has the potential, but this study shows that volume and utilization volume based metrics are imperative for data center dimensioning. Results indicate that the most reliable prediction could be achieved with a hybrid approach by incorporating ML into traditional planning strategies. Future work should include the use of real-world datasets, advanced ML techniques and ability to adapt in real time to fine tune the predictive capabilities and improve the NFV data center efficiency.

References

[1] Han, Y. H., Kim, Kim C. M., & Gil, J. M. (2010). A Greedy Algorithm for Target Coverage Scheduling in Directional Sensor Networks. *Journal of Wireless Mobile Networks, Ubiquitous Computing, and Dependable Applications*, *1*(2/3), 96–106.

[2] Perera, T. D. P. (2018). Computer network analysis in knowledge sharing. *International Journal of Communication and Computer Technologies*, *6*(2), 5–8. s, 215, 109934.

[3] Anada, H. (2020). Decentralized Multi-authority Anonymous Authentication for Global Identities with Non-interactive Proofs. *Journal of Internet Services and Information Security*, *10*(4), 23–37.

[4] Hossain, A. B. M. A., Siddiqua, A., Islam, M. E., Kabir, S. M. H., & Mahdi, G. M. A. (2024). Economic supply chain analysis of Hilsa Fish Landing Centers (Maach Ghats) in Bangladesh: Operational insights and strategies. *International Journal of Aquatic Research and Environmental Studies*, *4*(2), 69–88. http://doi.org/10.70102/IJARES/V4I2/5

[5] Anna, K.-J. & Javid. (2017). SDN/NFV-Based Mobile Packet Core Network Architectures: A Survey. *IEEE Communications Surveys & Tutorials*, *19*(3), 1567–1602.

[6] Moises, H. (2023). Philippine Data Centers Demand for Steady Supply of Power. https://w.media/philippine-data-centers-demand-for-steady-supply-of-power/

[7] Li, Y., & Li, X. (2017). A survey on network function virtualization. *IEEE Communications Surveys & Tutorials*, *19*(1), 646–666. https://doi.org/10.1109/COMST.2016.2616295

[8] Han, B., Gopalakrishnan, V., Ji, L., & Lee, S. (2015). Network function virtualization: Challenges and opportunities for innovations. *IEEE Communications Magazine*, *53*(2), 90–97.

[9] Sivasankari, P., Uma, R. N., & Thaarrini, J. (2016). Ternary Blended Green Concrete Using Textile Effluents. *International Journal of Advances in Engineering and Emerging Technology*, *7*(4), 111–117.

[10] Bhamare, D., & Kulkarni, P. (2017). Network Functions Virtualization (NFV) and Software Defined Networking (SDN): State of the Art and Open Research Challenges. *2017 International Conference on Computing, Communication, Control and Automation (ICCUBEA)*, 1–6.

[11] Khan, M. A., & Gani, A. (2018). Software-defined networking (SDN) and network functions virtualization (NFV) for future internet: A comprehensive survey. *Journal of Network and Computer Applications*, *107*, 1–25.

[12] Kaur, A., & Rani, R. (2019). Software Defined Networking (SDN) and Network Functions Virtualization (NFV): A Review. *2019 5th International Conference on Computing, Communication, Control and Automation (ICCUBEA)*, 1–5. 16.

[13] Chen, Y., Zhang, X., & Li, Y. (2020). A hybrid machine learning approach for cooling load forecasting in data centers. *Applied Energy*, *261*, 114391.

[14] Gao, Y., Zhang, X., & Li, Y. (2020). A machine learning-based approach for cooling load forecasting in data centers. *Energy and Building*.

[15] Liu, Y., & Bash, C. E. (2018). Predicting data center power consumption with machine learning. *Applied Energy*, *228*, 1194–1204.

[16] Li, Y., & Bash, C. (2011). Machine learning for data center optimization. *Proceedings of the 2011 31st International Conference on Distributed Computing Systems Workshops*, 201–206.

[17] Kumar, S., & Bash, C. (2015). Machine learning for power and cooling optimization in data centers. *Proceedings of the 2015 IEEE International Conference on Autonomic Computing*, 11–20.

30 Consumer perception on kudumbashree products

Mujeeb Rahiman Kattali[1,a] *and R. Sarojadevi*[2,b]

[1]Research Scholar, Department of Commerce, Karpagam Academy of Higher Education, Coimbatore, Tamil Nadu, India
[2]Research Supervisor, Department of Commerce, Karpagam Academy of Higher Education, Eachanari, Coimbatore, Tamil Nadu, India

Abstract: This study aims to understand social processes underlying the perception on Kudumbashree products. It investigates a range of foods, handmade crafts, clothings, and georgraphic regions that Kudumbashree deals with. Through questionnaires, interviews and perhaps constructions of focus groups, the social perception of importance in relation to measure of social impact, product quality, and product price are examined. The study uncovers that Kudumbashree products have a busting range of uses or lack of use which profoundly differs with regarding to and not confined to, age, occupation, educational level, monthly income, how regular people buy things, and how they deal or interact with the products. It is clear that marital status and gender do not have a significant impact. These pieces of information can be beneficial for Kudumbashree's market researches, customer relationship management, and product design.

Keywords: Kudumbashree products, handmade crafts, clothings, and georgraphic regions, customer relationship management, and product design

1. Introduction

Market dynamics and consumer behaviour are greatly reliant on the customer's perception. In Kerala's economic sector, Kudumbashree products have emerged as a singular phenomenon that has garnered considerable traction. Gourmet food, handcrafted shawls, eco-friendly items, and many more industries have been actively invaded by the grassroots movement in Kerala. Kudumbashree products' associations with authenticity and local culture have a significant influence on how consumers view them. Since these products are usually made with exquisite craftsmanship that highlights local and traditional techniques, they appeal to consumers looking for an authentic experience. Additionally, the social element of Kudumbashree attracts consumers with social conscience because it helps both empower female workers and foster community advancement. Consumer perception of Kudumbashree products relies heavily on customer perceptions that their quality and value match each other excellently. Handcrafted merchandise alongside locally obtained fresh ingredients lead Kudumbashree customers to believe their products offer higher quality compared to commercial manufacturers. Kudumbashree implements ethical production practices as well as sustainable environmental practices to protect their product idea while attracting conscious environmental consumers.

Kudumbashree products receive their evaluations from consumers through the common cultural aims and future hopes of the members of society. The organization Kudumbashree together with other social enterprises holds a reputation for resolving social challenges alongside promoting economic growth. The brand story of Kudumbashree generates emotional bonds between consumers that keeps them connected to embrace the brand products.

Not only regarding its features, but the Kudumbashree product has community impact, and ethical and cultural dimensions that are often associated with it in the Kerala market. Any researcher studying the brand and consumption of kudumbashree products will be able to appreciate the effort put in the branding and the social entrepreneurship success achieved at the grassroots level in the new market.

2. Review of Literature

Thomas and George [1] looked into the loyalty and trust Keralean consumers have on Kudumbashree products is what the study investigates the relationship between customer purchase loyalty and the corporate social responsibility, business reputation, and trust in the firm. The study attempts to explain how consumers of social firms like Kudumbashree are able to develop and maintain brand trust.

Nair and Jacob [2] in their research, unlike competing products in the market, Kudumbashree's product are believed to be more costly and price elastic,. This research intends to find out how consumers evaluate Kudumbashree products in terms of quality, originality, social value and purchase decisions. The findings tend to provide insights on pricing strategy and value messaging for Kudumbashree products.

Gopinath and Menon [3] focuses on consumers' regard for the artistic and cultural value of the Kudumbashree

[a]kattalimuj@gmail.com, [b]rajendrasaro@gmail.com

DOI: 10.1201/9781003684459-30

Handicraft Products has been studied more deeply. This study quantifies the purchase intention of the consumers and the willingness to support local craftsmen within the context of Kudumbashree crafts perceived as cultural imagination and capability. The study presents results that speak to the need of sustaining cultural heritage as well as artisanal skills within the consumer market spaces.

Menon and Rajesh [4], looks that with all other modern products, the focus for is the factors that determine the purchasing intention of the consumers of Kudumbashree brand in the metropolitan region of Kerala. The study assesses the various constituents of product such as brand Kudumbashree, social and other forms of advertising or marketing and their effects on the buying behaviour towards Kudumbashree items. The document provides revealing comments concerning marketing approaches and tactics designed for Newer Adoption and increased sales of Kudumbashree products in urban areas.

Sreejith and Krishnakumar [5] examined Keralean consumer evaluations alongside their buying behaviour of Kudumbashree goods specifically through evaluation of product quality and costs and social benefits on consumer purchasing patterns. The research analyzes consumer attributes to explain why Kudumbashree products receive preference choices from consumers instead of competitor products.

Menon and Pillai [6] demonstrate how Keralean consumers hold particular preferences for items produced by Kudumbashree. This research evaluates consumer mental and psychological responses to these goods together with their fundamental cultural and responsible values and authentic qualities to explain product advantages and determining factors.

The paper by Prasad and Nair [7] studies how branding elements along with packaging affect the esteem and reliability evaluation of Kudumbashree products for local Kerala consumers. The research outcomes show that branding together with packaging elements modify consumer perception and their reception of products.

Kumar and Rajan [8] undertakes studies of Kudumbashree organic product perceptions plus Keralean customer happiness levels. The analysis includes customer satisfaction with organic food combined with their evaluation of product quality alongside health value and environmental aspects. This research indicates that consumer satisfaction with Kudumbashree's entire organic product range depends heavily on proper attention towards its eco-friendly certifications and organic practices.

2.1. Research gap

There are still a lot of possible issues that need more focus, particularly the growing awareness of Kudumbashree products in Kerala. One clear lack is the absence of in-depth research about the factors that shape consumers' perceptions toward the various categories of Kudumbashree products. Earlier works concentrate on broad generalizations rather than specific attributes of the product, as well as specific pricing and marketing communication strategies that may exist for different categories of products.

Likewise, little has been done to study the impacts of culture, society or the economy on consumer perception with the shifts in attitude among the younger and more urbanized demographics. Additionally, very little effort has been put forth regarding the ways consumers view Kudumbashree products through the lenses of branding and marketing strategies implemented on social media platforms, leading to the necessity for further research in this particular field. Closing these gaps would allow Kudumbashree and other stakeholders to properly position their products, craft appropriate marketing strategies, and capture the diverse markets in the state of Kerala.

2.2. Statement of the problem

This research analyzes the Keralean attitudes regarding the purchase Kudumbashree branded products. There is a lack of understanding about the other differences in demographic and product category that might influence the perceptions of Kudumbashree goods that are linked with social responsibility, 'real' marketing, and quasi-ethnic crafts. This research tries to explore how age, occupation, education level, income, and purchase frequency shapes consumers' perceptions and decisions towards the specific products offered by Kudumbashree. The goal this research aims to achieve is to identify any gaps or issues that consumers may experience regarding the distribution, pricing, and promotion of the goods. The research attempts to answer the question how these gaps can be filled and how best Kudumbashree can develop strong customer relationships, marketing, and appropriate product strategies appropriate for this region.

2.3. Significance of the study

This study attempts to understand the attitudes of constituents of Kudumbashree with regards to products of Kudumbashree in relation to age, occupation, income, or level of education. The conclusions highlight the need for niche marketing in marketing mix element social characteristics of the product. In addition, this suggests that more awareness can enhance consumption and loyalty to the brand across varying age and income groups. In addition, the study also emphasizes on the aspects of community and social concern reachable throw Kudumbashree's empowerment suggestive of the potential impact of such stories on customer engagement. The example of Kudumbashree motivated the researchers, employing the right marketing mix, paying attention to the social goal to suggest how Kudumbashree can improve its activities to serve better in the market, enhance customer satisfaction and increase the clientele.

2.4. Objectives of the study

To examine the consumer perception towards Kudumbashree products.

2.5. Scope of the study

Kudumbashree aims to understand how its customers perceive and choose its products, which includes buying patterns and other reasons that lead to purchase decisions. It would cover different product lines and specific areas served by Kudumbashree. The study aims to capture information that impacts the product in terms of quality, pricing, and social effects. It also involves Matters of concern. The end goal would be to provide recommendations that will reconstruct and strengthen Kudumbashree's programs towards community development and economical upliftment.

3. Research Methodology

This study is based on primary data which is collected with the help of Questionnaire. By adopting convenience sampling technique and sample size of 146 respondents from Ernakulam district were selected.

3.1. Tools for analysis

Percentage analysis and chi-square test is employed for analysis.

3.2. Limitations of the study

1. It concentrates on the Ernakulam district which may not reflect perceptions of the rest of the Kerala state. 146 respondents is a relatively small sample size considering the entire Kerala population.
2. This method violates randomness, therefore there is an issue of bias. Not all Kudumbashree products, particularly novel or highly specialized ones, may be included in the research.
3. The study encompasses consumer perceptions at a single point in time and does not longitudinally track trends. There are some intricate cultural and social aspects of consumer behaviour that exist in Kerala which are not adequately understood or explored.
4. The study takes into account demographics but has not approached the problem of purchasing decisions from psychological or emotional perspectives.

3.3. Analysis and interpretation

According to the Table 30.1, 23.3 percent of respondents are under 30, 43.8% are between 31 and 40, 23.3% are between 41 and 50, and 9.6% are over 50. In terms of gender, there are 88 male respondents and 58 female respondents; 116 respondents are married, and 30 are single; in terms of educational qualifications, 5.5 percent have completed up to HSC, 12.3 percent have completed graduate school, 68.5 percent have completed postgraduate school, and 13.7 percent have completed professional courses such as law, medicine, etc. Occupationally, six of the respondents work in agriculture, fourteen are business owners, twenty-four are employed by the government, and 102 are employed by the private sector. In terms of monthly income, 2.7 percent of respondents earn less than Rs. 5000, 5.5 percent earn between Rs. 5000 and Rs. 10,000, 11.0 percent earn between Rs. 100,000 and Rs. 15000, 12.3 percent earn between Rs. 15,000 and Rs. 20,000, and 100 percent earn more than Rs. 20,000.

Thirty-one percent of respondents had used Kudumbashree products for less than a year, twenty-eight percent had used them for one to three years, fifteen percent had used them for three to five years, and twenty-six percent had used them for more than five years (Table 30.2).

Table 30.1. Demographic profile (no. of the respondents 146)

Demographic Profile	*Classification*	*Frequency*	*Percentage (%)*
Age	Below 30 Years	34	23.3
	31–40 Years	64	43.8
	41–50 Years	34	23.3
	More than 50 Years	14	9.6
Gender	Male	88	60.3
	Female	58	39.7
Marital Status	Married	116	79.5
	Unmarried	30	20.5
Educational Qualification	Up to HSC	8	5.5
	Under Graduate	18	12.3
	Post Graduate	100	68.5
	Professional Course	20	13.7
Occupation	Agriculture	6	4.1
	Business	14	9.6
	Government Services	24	16.4
	Private Services	102	69.9
Monthly income	Below Rs. 5000	4	2.7
	Rs. 5000– Rs. 10000	8	5.5
	Rs. 10000– Rs. 15000	16	11.0
	Rs. 15000– Rs. 20000	18	12.3
	Above Rs. 20000	100	68.5
Experience of Kudumbashree products	Less than 1 Year	44	30.1
	1–3 Year	42	28.8
	3–5 Year	22	15.1
	Above 5 Year	38	26.0

Source: Primary data.

Table 30.2. Perception on kudumbashree products

Demographic variables	*Classification*	*Perception on Kudumbashree Products*				*Chi-Square value*	*Df*	*Asymptotic Significance (2-sided)*	*Association*
		Low	*Moderate*	*High*	*Total*				
Gender	Male	20	56	12	88	1.917[a]	2	.384	Not significant
		22.7%	63.6%	13.6%	100.0%				
	Female	12	42	4	58				
		20.7%	72.4%	6.9%	100.0%				
Age	Below 30 Years	4	28	2	34	39.426[a]	6	.000	Significant
		11.8%	82.4%	5.9%	100.0%				
	31–40 Years	16	44	4	64				
		25.0%	68.8%	6.3%	100.0%				
	41–50 Years	8	24	2	34				
		23.5%	70.6%	5.9%	100.0%				
	More Than 50 Years	4	2	8	14				
		28.6%	14.3%	57.1%	100.0%				
Marital status	Married	28	78	10	116	4.087[a]	2	.130	Not significant
		24.1%	67.2%	8.6%	100.0%				
	Unmarried	4	20	6	30				
		13.3%	66.7%	20.0%	100.0%				
Education Qualification	Up to HSC	4	2	2	8	19.081[a]	6	.004	Significant
		50.0%	25.0%	25.0%	100.0%				
	Under graduate	2	16	0	18				
		11.1%	88.9%	0.0%	100.0%				
	Post graduate	22	70	8	100				
		22.0%	70.0%	8.0%	100.0%				
	Professional	4	10	6	20				
		20.0%	50.0%	30.0%	100.0%				
Occupation	Agriculture	0	2	4	6	29.019[a]	6	.000	Significant
		0.0%	33.3%	66.7%	100.0%				
	Business	2	8	4	14				
		14.3%	57.1%	28.6%	100.0%				
	Government Services	8	16	0	24				
		33.3%	66.7%	0.0%	100.0%				
	Private Services	22	72	8	102				
		21.6%	70.6%	7.8%	100.0%				
Monthly income	Below-Rs. 5000	0	2	2	4	20.119[a]	8	.010	Significant
		0.0%	50.0%	50.0%	100.0%				
	Rs.5000–Rs.10000	2	6	0	8				
		25.0%	75.0%	0.0%	100.0%				
	Rs.10000–Rs.15000	4	12	0	16				
		25.0%	75.0%	0.0%	100.0%				
	Rs.15000–Rs.20000	2	10	6	18				
		11.1%	55.6%	33.3%	100.0%				
	Above Rs. 20000	24	68	8	100				
		24.0%	68.0%	8.0%	100.0%				

(continued)

Table 30.2. (Continued)

Demographic variables	*Classification*	*Perception on Kudumbashree Products*				*Chi-Square value*	*Df*	*Asymptotic Significance (2-sided)*	*Association*
		Low	*Moderate*	*High*	*Total*				
Frequency of Kudumbashree product purchase	Never	33.3%	50.0%	16.7%	100.0%	30.839ª	6	.000	Significant
		18	54	6	78				
	Rarely	23.1%	69.2%	7.7%	100.0%				
		8	36	2	46				
	Occasionally	17.4%	78.3%	4.3%	100.0%				
		2	2	6	10				
	Frequently	20.0%	20.0%	60.0%	100.0%				
		33.3%	50.0%	16.7%	100.0%				
Experience of Kuduphashree Products	Less than 1 Year	12	30	2	44	29.766ª	6	.000	Significant
		27.3%	68.2%	4.5%	100.0%				
	1–3 Year	14	26	2	42				
		33.3%	61.9%	4.8%	100.0%				
	3–5 Year	2	20	0	22				
		9.1%	90.9%	0.0%	100.0%				
	Above 5 Year	4	22	12	38				
		10.5%	57.9%	31.6%	100.0%				
		(2.6)	(33.3)	(24.4)	(100.0)				

Source: Primary data.

Gender: The results show that there appears to be no statistical significant association of the gender of the respondents with the variable of interest Kudumbashree products since the chi-square value 1.917 with 2 degrees of freedom and the p-value .384 is greater than 0.05. This means that both male and female respondents self reported their level of interaction with the Kudumbashree products did not differ.

Age: The relationship between age and the perception of Kudumbashree products seems to be statistically significant because of the p value of .000 and the chi square value of 39.426 with 6 degrees of freedom. This means that people of different age categories have different perceptions and interactions with the Kudumbashree goods.

Marital Status: There is no statistically significant relationship between p value of .130 and the variable of interest which is married status of the respondent because the chi square value is 4.087 with two degrees of freedom.

Educational Qualification: The p value is .004 and the chi square value is 19.081 with 6 degrees of freedom indicating relationship between educational level attained and the variable of interest is statistically significant. This means that the use or perception of Kudumbashree products varys with the change in educational level of an individual.

Occupation: A statistically significant correlation is indicated by the p-value of.000 and the chi-square value of 29.019 with six degrees of freedom. This indicates that people's interactions with Kudumbashree items are influenced by their line of work.

Monthly Income: A statistically significant correlation is indicated by the p-value of.010 and the chi-square value of 20.119 with 8 degrees of freedom. This implies that interactions with Kudumbashree items are influenced by monthly income levels.

Frequency of Kudumbashree Product Purchase: The p-value is.000, indicating a significant relationship, and the chi-square value is 30.839 with six degrees of freedom. This finding suggests that there are notable differences in the frequency of Kudumbashree product purchases among various demographic groups.

Experience with Kudumbashree Products: A significant relationship is indicated by the p-value of.000 and the chi-square value of 29.766 with six degrees of freedom. This implies that certain demographic factors are linked to various Kudumbashree product experiences.

4. Findings of the Study

- The majority 43.8% of the respondents are in the 31–40 years age group.
- The inference that the 60.3% of the respondents are male.
- Marital status does not significantly affect product perception, the majority of 79.5% respondents are married.
- The educational levels correlate with greater awareness of product quality and social impact, the 68.5% of respondents have completed post-graduate education and 69.9% are employed in private services.
- The majority of 68.5% earn above Rs. 20,000 per month and 30.1% have used Kudumbashree products for less than one year.

- The chi-square test reveals that demographic variables such as age, occupation, educational qualification, monthly income, frequency of purchase, and experience with the products have a significant influence on consumer perceptions of Kudumbashree products, whereas gender and marital status do not have a significant effect.

5. Suggestions

- Kudumphashree products and marketing to different age groups since age significantly impacts product preferences.
- Design products that cater specifically to the needs of various occupations.
- Adopt a tiered pricing strategy to accommodate the varying income levels of customers, making products accessible to all income brackets.
- Develop educational initiatives to enhance consumers' comprehension and product knowledge, focusing on educating them according to their educational background.
- Create a loyalty program that offers benefits to devoted clients in order to boost client retention and promote repeat business.
- Actively seek out and incorporate client feedback, especially from individuals with in-depth product knowledge, to enhance goods and services.
- To guarantee wide appeal, create marketing campaigns that are inclusive of all genders and marital statuses. Boost online visibility and effectively reach a variety of demographics by strengthening digital marketing strategies.
- Participating in local festivals and events has two advantages: it fosters direct engagement with the product and engages local people.
- By highlighting how the purchasing of these products solve larger social challenges, promote and grow corporate social responsibility (CSR) initiatives to draw in customers who are interested in ethical shopping.

6. Conclusion

Important demographic factors that are closely related to the consumption of Kudumbashree goods are captured by this specific study, including age, income, and employment. Kudumbashree could enhance its brand and customer involvement by implementing innovative marketing strategies, offering reduced prices, instructional advertising, and items tailored to specific demographics. These initiatives should also boost the company's market share and customer happiness, which will boost Kudumbashree's ability to compete.

7. Scope for Further Research

The study that follows enables the development of novel strategies and provides a thorough understanding of how customers interact with Kudumbashree products. Longitudinal studies can be used to assess the effects and focus attention on various marketing initiatives. This could involve quantitative methods, such using surveys to gather data and conducting interviews to gain a deeper understanding of the figures' context. The impact of the marketing initiatives and the cost implications should indicate how best to enhance the strategies used. Investigating new technologies and other outside influences can lead to a deeper comprehension. These research will be useful in creating new product innovations and marketing campaigns that will increase Kudumbashree's market share.

References

[1] Thomas, A., & George, L. (2021). Kudumbashree Brands in Kerala: An Empirical Study of Consumer Trust and Loyalty. *Indian Journal of Marketing*, *51*(3), 32–45.

[2] Nair, S., and Jacob, M. (2018). A comparative study of Kudumbashree products' perceived value and price sensitivity in Kerala. *Journal of Consumer Services and Retailing*, *25*, 98–106.

[3] Gopinath, R., & Menon, N. (2019). An investigation into the craftsmanship and cultural appeal of Kudumbashree handicrafts. *Journal of Cultural Heritage Management and Sustainable Development*, *9*(2), 134–148.

[4] Menon, A., & Rajesh, K. (2020). A study conducted in urban Kerala examined the factors that influence consumers' intentions to purchase Kudumbashree products. *Strategic Marketing Journal*, *28*(5), 420–435

[5] Sreejith, V. P., & Krishnakumar, C. (2018). Consumer Perception and Buying Behavior towards Kudumbashree Products in Kerala. *International Journal of Marketing and Consumer Research*, *47*(3), 56–68.

[6] Menon, S., & Pillai, S. S. (2020). A Study on Consumer Attitudes and Preferences towards Kudumbashree Products in Kerala. *Journal of Business and Management*, *15*(2), 78–92.

[7] Prasad, R., & Nair, M. (2019). Impact of Branding and Packaging on Consumer Perception of Kudumbashree Products: A Case Study in Kerala. *Journal of Consumer Studies*, *35*(4), 210–225.

[8] Kumar, R., & Rajan, S. (2017). Consumer Perception and Satisfaction towards Kudumbashree Organic Products in Kerala. *Journal of Rural Development*, *22*(1), 45–59.

31 Anomaly detection in induction motors through irregular sound and vibrations

Honeylet D. Grimaldo[a], Ralph John A. Dela Peña[b], Ruscell Ray V. Doria[c], Darcy Micael F. Gasgonia[d], Darryl Christian G. Patron[e], Jaime M. Landingin[f], Jorge Z. Burgos[g], and Reynaldo C. Muli[h]

College of Engineering, FEU Alabang, Filinvest Muntinlupa City, Philippines

Abstract: The industrial sectors, effective use in business operations due to its energy efficiency, low cost, and ability to easily embed on any kind of machines, etc. are likely to be the areas in which machine learning, and embedded systems will have the most significant impact in the near future. Around the same time, the 1770's saw the beginning of the first industrial revolution, which saw a shift away from traditional human work toward the widespread usage of steam and water. This continued until the end of the 19th century, when electricity and power generation was introduced, which prompted the second industrial revolution. At that time, the first industrial revolution was already well under way. Because of this evolution, mass production was able to begin through the use of assembly lines and automation, which further boosted the productivity of industrial processes. As a result, we now have the concept of a factory in the form that it exists today. This pattern persisted until the 1950's, when the digital revolution gave rise to the third industrial revolution. The development of computers made it possible to easily regulate and monitor industrial operations, which in turn made it possible to automate part of the work that was previously done by blue-collar workers. Industry 4.0 includes things like self-driving systems, machine learning.

This study proposal includes an important consideration for energy efficiency (EE). To maximize energy economy, we can achieve this by performing inference on-device or on the microcontroller unit (MCU) and decreasing the amount of time spent on external communications. When we perform inference on the device, we reduce latency and free ourselves from being constrained by the speed of the network. Instead, we are only bound by the hardware limitations of our embedded systems. Anomaly detection (AD), which is part of predictive maintenance (PdM), will, in the end, notice a problem early on, enable a faster reaction, and tell the appropriate people, who will then be able to follow the appropriate procedures to resolve any problems before they may result in a breakdown or downtime.

Keywords: Class-eye, tention A, machine vision, inference approach of student, attentiveness, detection

1. Introduction

Diagnosing issues in an induction motor is quite tedious, cumbersome, and time-consuming depending on how the induction motor is utilized [1]. This device can provide ease in diagnosing irregular noises and vibrations in Induction Motors that can otherwise lead to further complications within the machine [8]. One of the most prevalent causes of odd sounds coming from your electric motors is mechanical noise, which can come from a variety of sources [3, 4]. This could be caused by problems with the bearings, a stator core that is not properly attached, or even airborne noise coming from a part of the electric motor's structure. It is difficult in most industrial checks because the processes are too complex to identify alert rules, dynamic and complex. According to a DTTL article, in today's world, inadequate maintenance techniques can reduce a piece of equipment's overall productivity by 5% to 20% [12]. According to Naresh Shanker, a Forbes Councils Member, unplanned or last-minute machine downtime due to some faults costs industrial production organizations an estimated $50 billion each year [11]. The manufacturing sector is an extremely dynamic industry; every day, thousands of new products are introduced. However, it is common to observe that factories staff entire departments with individuals whose sole responsibility is to visually inspect products for flaws such as dents, scratches, or the incorrect placement of detail. It can cost hundreds of thousands of dollars per year to maintain a quality department like this one, and it still leaves room for human error, such as when workers skip defective details or goods. Machine Learning (ML) and Supervised Learning come to the rescue in situations where 1 human labour is unable to cover 100% of the cases. This is how we are able to teach the system to do the exact same thing that humans would do (sort, review, find defects), but better and faster because, unlike humans, the machine does not get tired and does not lose focus on what it is doing [9].

This paper discusses the importance of anomaly detection. An effective machine learning-based anomaly detection technique is proposed.

[a]hdgrimaldo@gmail.com, [b]radelapena@gmail.com, [c]rvdoria@gmail.com, [d]rfgasgonia@gmail.com, [e]dgpatron@gmail.com, [f]jmlandingin@gmail.com, [g]jzburgos@gmail.com, [h]rcmuli@feualabang.edu.ph

DOI: 10.1201/9781003684459-31

2. Background of the Study

The induction motor is helpful in a wide variety of applications that allow people to go about their daily lives, and it is also used in a number of essential items. Induction motors are currently the most popular choice for home, business, and factory applications [5, 6]. If there is a problem with the air compressor, the user may hear loud noises or feel strange vibrations coming from the device itself [7]. These signs indicate that there is a problem taking place within the body. When a motor fails, it is usually because something went wrong while it was supposed to be operating normally. But, what if we could identify that something was wrong with the motor's actual sound when it was operating, and hence foresee its impending failure?

Modifications on a lesser scale have the ability to prevent subsequent problems; however, this does not necessarily signal that large repairs are required due to factors such as excessive use of the device [2, 10].

According to the findings presented by Anthony M. Smith and Glenn R. Hinchcliffe in their book RCM Gateway to World Class Maintenance, human error in the maintenance process can account for up to fifty percent of all machine breakdowns [13–15]. Vibration analysis enables us to zero in on the specific component or portion that has to be addressed. It is no longer required to do extensive repairs all at once, which can sometimes result in more issues than they resolve [23].

In light of this, the research team intends to come up with a solution to the problem of it being more cost- effective to do predictive maintenance on machinery and equipment rather than to repair it after it has broken down [26]. Paired with an algorithm for machine learning to increase intelligence, as well as an embedded system to decrease power consumption and increase the likelihood of accurate prediction. Predictive Maintenance (PdM) has the potential to become a crucial part of any given sector [24].

2.1. General objective

The main objective of our system is to detect anomalies such as abnormal sound and vibrations from the vespa air compressor's motor using a machine learning model called an autoencoder (AE), which is a type of neural network [24].

2.2. Specific objectives

- To develop a system capable of diagnosing a vespa air compressor's motor. Issues caused by abnormal vibrations or sound such as bearings, loose belt or belt aging, or shafts using anomaly detection algorithms.
- To design a machine learning system that detect anomalies in induction motor using an autoencoder.
- To be able to display the results.
- To achieve at least 80% success rate of the system.

2.3. Scope and delimitations

The scope will focus on Low power Air compressors motors that are used every day in households. The aim of the study is to monitor the health of the air compressor to determine whether something was on the verge of breaking. If the error between input and the new reconstructed input (output) is greater than the set threshold, the output is flagged as an anomaly. This means that our system only learns what it is to be normal. When the behaviour deviates from expected pattern, it means that there is something off or rather a difference with the device. Then, it will issue a warning that prompts preventative measures to be taken (like routine maintenance) to forestall motor failure [16]. You can display the results via LED and Mobile app, automatically turn off the machine when an anomaly is detected, thus, a deviation is not yet a fault, so no immediate action is required. On the other hand, it is an anomaly that requires further investigation [17, 18].

For predictive maintenance, the researchers are considering using sensors and microcontrollers [19]. Machine learning (ML) may have the most significant effect in the corporate world at the national and regional levels. For instance, the smart factory, autonomous system, and machine learning will all play crucial roles in industry 4.0. By anticipating component failures and fixing them before they cause major problems.

According to the article that discusses the common causes of electric motor vibration, one condition or a combination of elements can cause electric motor vibrations. This is especially true when considering the fact that supplementary devices can also be the source of abnormalities. Motor vibration can be caused by a variety of issues, including electronic or mechanical imbalance, faulty gears, failed bearings, loose foundations, wear and tear, misalignment [20].

To narrow down the cause of a vibrating motor, you and your operators should ask a few simple questions, such as Is everything tight? Is 'soft foot' a common occurrence? Is hot alignment permitted? Does thermal compensation have an impact on cold motor alignment? Is the motor excessively vibrating? Are there any resonant motor vibrations? Do engine vibrations outnumber motor frame vibrations? Are there any loose parts on the motor shaft? Are your fan blades damaged? Is the coupling greased? [22, 25]

According to the article that discusses the most common causes of machine vibration, vibration can be an indicator of a problem and, if it is allowed to continue on its own, can cause damage or accelerate deterioration. Vibration can be caused by one or more factors at any given time, the most common of which are imbalance, misalignment, wear, and looseness. Vibration can occur at any time [21].

The scope does not apply to industrial air compressor machines because we have limited access to industrial air compressor machines; the only machine we have access to is the most commonly used air compressor, such as the vespa air compressor, which can be used in spray painting, adding air to tires, and so on [26, 27].

2.4. Conceptual framework

The researchers will be using a accelerometer (for the vibration) and microphone (for the sound) to collect data, after collecting the data and extract important features from it, we can now train, test and validate using our machine learning model if everything is working, deploy it to a microcontroller (Figure 31.1). To compare the error between the input and the output; we are able to determine if our input is similar to our normal training data. The system will characterize what normal or nominal operation of the machine looks like, we perform anomaly detection (AD) by selecting a threshold above which we classify the input as an anomaly and will display the results via LED rotating, Mobile app(GUI) and automatically turning off the machine.

3. Methodology

3.1. System requirements versus functionalities

This chapter introduces the overall method approach that will be used in this applied research. This method is further illustrated in this chapter.

3.1.1. Functionalities

- **Design and Train a model**
  The Researchers will begin to use the extracted features and use those features to design and train, validate, test a machine learning model with the help of Tensorflow 2.10.0 library.
- **Convert and optimize model**
  The Researchers will convert and evaluate/optimize the model in order to be ready for quantizing and deployment.
- **Deploy and make inferences**
  The Researchers will make an inference to deploy a converted model into a microcontroller and with the help of tensorflow lite micro library to make a much more efficient model for the MCU.

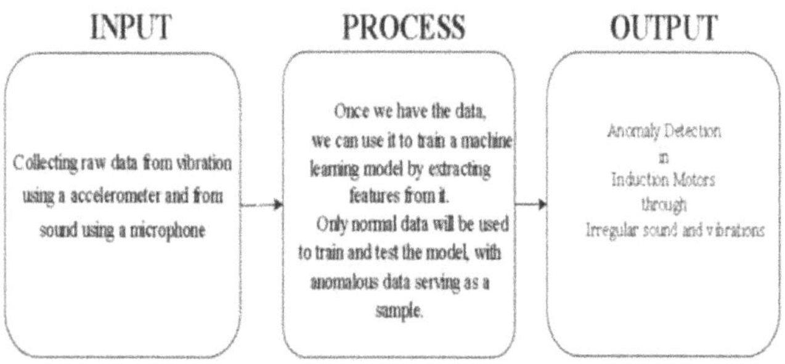

Figure 31.1. Conceptual framework.

Source: Author.

3.1.2. System Requirements

- **Collect data**
  The Researchers will collect data using accelerometer, Microphone, and Microcontroller for processing unit, for the software part arduino programming will be used.
- **Preprocess data**
  After acquiring the raw data, researchers begin to look for good features to help the model out by reducing the number of dimensions or data reduction/ transformation, for the software part, python programming will be used.

3.2. Block diagram

The system's sensors and modules control inputs and outputs. The system includes a microcontroller, microphone, accelerometer sensor, a relay, an LED with a buzzer to display the result, bluetooth module that acts as a sender or transmitter while the mobile app acts as the receiver and a program to process the inputs (Figure 31.2).

3.3. System flowchart

The system's multiple sensors and MCU control its input and output. The system includes a microphone, an accelerometer sensor, a relay, a rotating led with a buzzer to display the result, and MCU to the mobile app via the Bluetooth module. The mobile app can receive this data and process it accordingly, and a program to process the inputs. If the system detects that there is a large reconstruction error and it is greater than the threshold that was set, the system will output an anomaly and give a warning that there was an unexpected behaviour in our machinery. Since the autoencoder can learn its own data, the reconstructed input as the new (output) will be validated by the model itself. The performance of the autoencoder model was evaluated by examining the confusion matrix; a low false positive and false negative rate indicates that the model successfully reconstructed the input data (Figures 31.3–31.6).

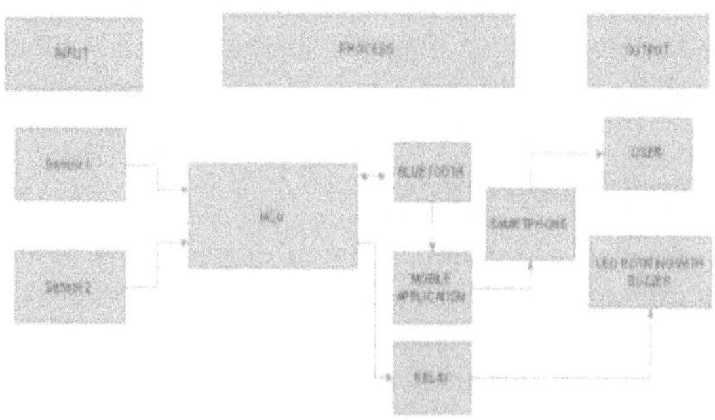

Figure 31.2. Block diagram.

Source: Author.

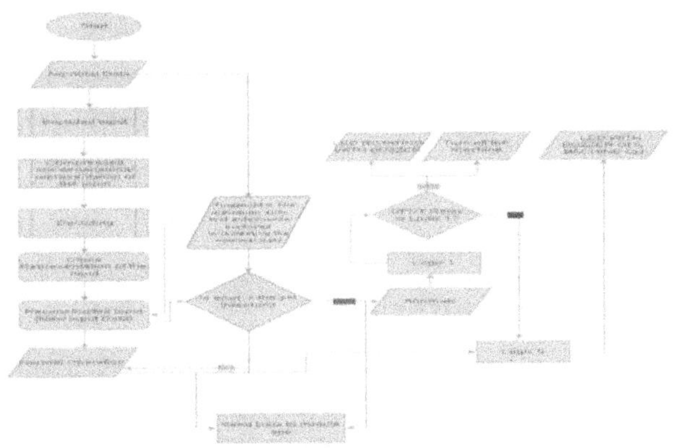

Figure 31.3. System flowchart for hardware and mobile app.

Source: Author.

3.3.1. Software system

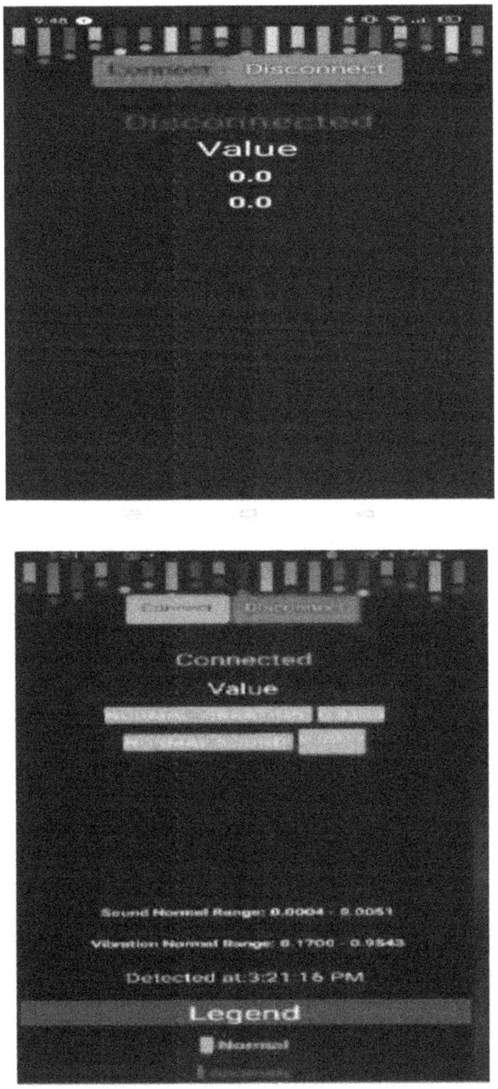

Figure 31.4. The mobile application.

Source: Author.

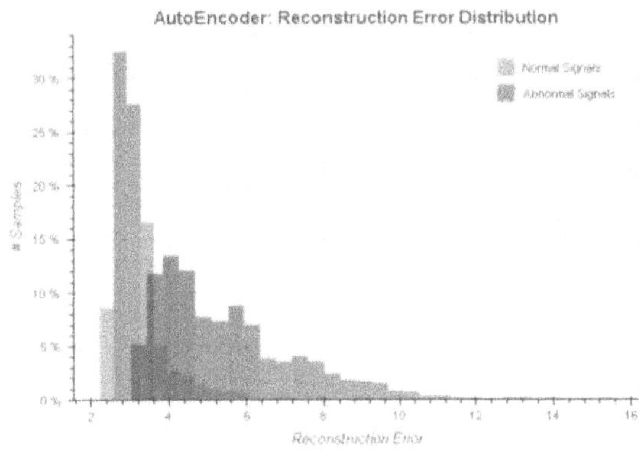

Figure 31.5. The auto encoder.

Source: Author.

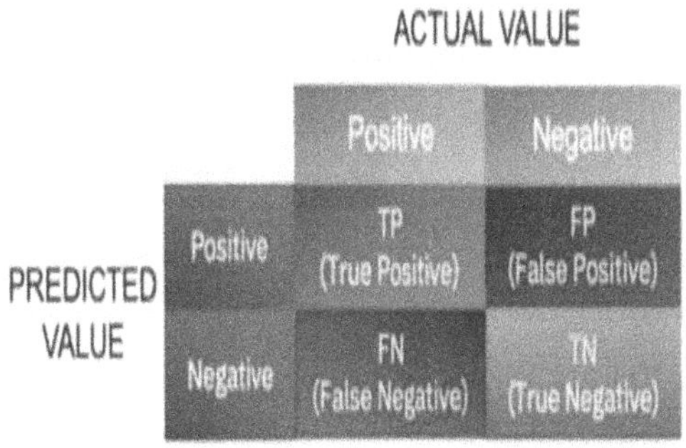

Figure 31.6. The actual and predicted value.

Source: Author.

4. Results and Discussion

The researchers conducted comprehensive tests on the entire device to assess its overall reliability and accuracy (Figures 31.7–31.9). During these tests, they observed that the device's performance was affected when it was positioned in different orientations or angles, leading to inaccurate results. However, after calibrating the device's placement and implementing training for the system, significant improvements were achieved. As a result, the system attained an overall success rate of 95%, indicating enhanced accuracy and reliability compared to the initial test.

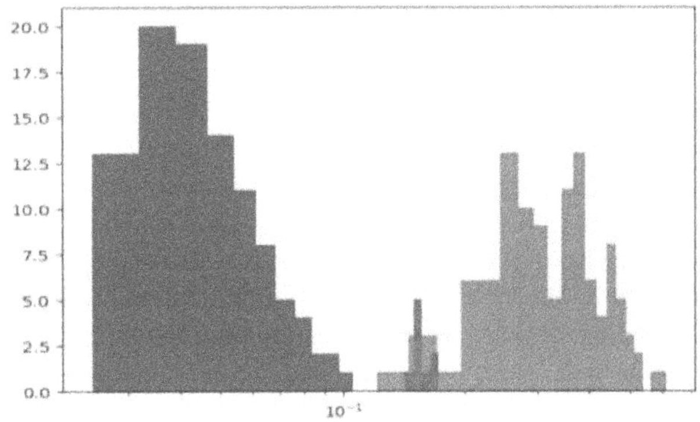

Figure 31.7. The sound MSE.

Source: Author.

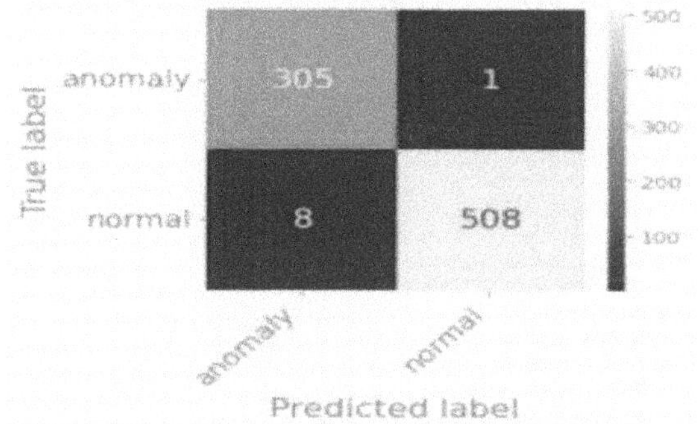

Figure 31.8. The vibration MSE.

Source: Author.

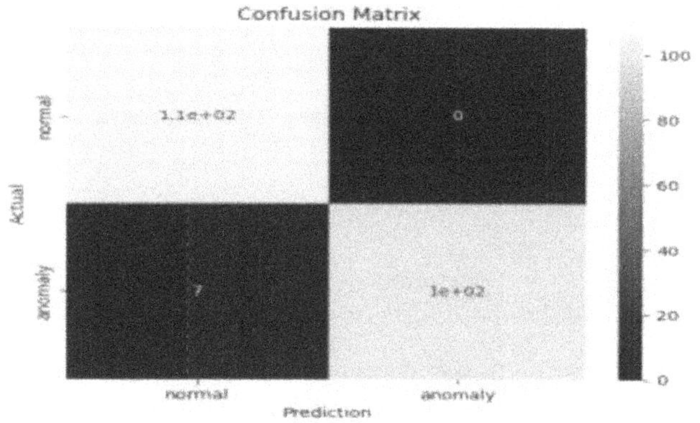

Figure 31.9. The model performance evaluation.

Source: Author.

5. Conclusion

The researchers focused on gathering the data for training the system, with an emphasis on training to recognize the normal and abnormal operations of the system. The system is based on the results of testing and training that authenticate its reliability and accuracy.

These findings are based on the actual testing of the system to distinguish between normal or abnormal sounds and vibrations caused by the bearings, loose belt, belt aging, or shaft by the use of anomaly detection, which is stated in this document, the Testing Methods were conducted on two similar air compressors, as well as trials in the results and discussion. After further trials, the researchers discovered that the prototype along with the mobile application provided sufficient results to fulfill the objectives of the study.

Designing the machine learning system that would recognize the normal and abnormal sounds and vibrations being produced by the bearings, loose or aging belt, and shaft was achieved by utilizing several programming languages, such as Python and C, through an autoencoder. Among the options for Creating the device that would collect the data produced by the air compressor is the microcontroller that would be assimilated into the system. The options are the Raspberry Pi or Teensy. The considerations for the Mobile Application are the output that it will display and being user-friendly.

Overall, the system produced the desired success rate of 95%, with outstanding results fulfilling all the objectives and the primary purpose of the system. The prototype and the mobile application corresponded to their supposed features and provided sufficient outcomes.

6. Recommendation

For the improvement of the design of this project, it is recommended for the benefit of the people who are aligned with the Computer Engineering. This study satisfied the people who benefited from the use of air compressors in the project 'Anomaly Detection in Induction Motors through Irregular Sound and Vibrations.'

To further improve the development of the mobile application and the accommodation of the hardware, the researchers have provided the following recommendations:

Improvement in the utilization of other types and brands of air compressors and to further utilize the specific parts of the air compressor. Future Researchers can design machine learning that can be transferred onto other air compressors, removing the constraint of low transferability.

Additional features can be integrated into displaying the results, such as additional features in the application being more simplified and other alternatives for the LED buzzer.

This study utilizes the AutoEncoder machine learning model and anomaly detection for the method, and it solely focuses on the general detection of the said system. The researchers recognize that it can be improved by including another model to classify specific objectives such as a loose belt, a faulty bearing, and so on.

Future studies can use the AutoEncoder model to integrate the classification model with anomaly detection. By combining classification with autoencoder-based anomaly detection, you can benefit from the advantages of both approaches. The classification model can help identify specific anomalies that have been labeled during training, while the AutoEncoder can capture more subtle or complex anomalies that may not have been explicitly labeled.

References

[1] Mohd Ghazali, M. H., & Rahiman, W. (2021, September 11). *Vibration Analysis for Machine Monitoring and Diagnosis: A Systematic Review.* [Online]. Available: from https://www.hindawi.com/journals/sv/2021/9 469318/?fbclid=IwAR2M8 Hkuomal6R87vE KIpWDtKWF5zvvz133zpB8abGmnUPx-oRdi K2B-rGQs

[2] Priyadharshini, B., & Rani, V. G. (2016). Selfish Node Detection Using Distributed Clustering Algorithm over MANET. *International Journal of Advances in Engineering and Emerging Technology*, 7(4), 118–130.

[3] Navea, R. F., Celestial, P. B., Paden, E. R., Pascual, P., & San Buenaventura, J. R. (2018). Automated Pothole Detection

and Notification System. From https://www.dlsu.edu.ph/wp- content/uploads/pdf/conferences/research-congress-proceedings/2018/hct-06.pdf

[4] Sneka, A. B., Mahajan, P., & Nithyakalyani. (2022). Shipping Management System. *International Academic Journal of Innovative Research, 9*(2), 31–34. https://doi.org/10.9756/IAJIR/V9I2/IAJIR0915

[5] Alviento, E. V., Cabais, J. B., Fajardo, J. M. M., Garcia, R. G., & Ballado, A. H. (2021). Early Detection of Single-Phase Induction Motor Failure t through the use of Accelerometer and Temperature Sensor via Arduino. *2021 IEEE 17th International Colloquium on Signal Processing & Its Applications (CSPA)*, pp. 64–6doi:10.1109/CSPA52141.2021.9377285, from https://ieeexplore.ieee.org/stamp/stamp.jsp?tp=&arnumber=9377285&isnumber=9377268

[6] Musa, K. M., & Alshemary, K. K. H. (2024). The Role of Nanoparticles in Sunscreen: UV Protection and Particle Size. *International Academic Journal of Science and Engineering, 11*(1), 153–164. https://doi.org/10.9756/IAJSE/V11I1/IAJSE1118

[7] Takahashi, Y. (2017). IoT Solution-Business Promotion Department, IoT Platform Division, ICT Business Division. *Anomaly Detection using Vibration Analysis with Machine Learning Technology for Industrial IoT System*, from https://www.oki.com/en/otr/2017/n230/pdf/ot r-230-R09.pdf

[8] Baggyalakshmi, N., Anubarathi, M., & Revathi, R. (2023). Pharmacy Management System. *International Academic Journal of Innovative Research, 10*(2), 36–55. https://doi.org/10.9756/IAJIR/V10I2/IAJIR1008

[9] Savchenko, N. (August 2020). *How Anomaly Detection Can Boost Your Business* [Online]. Available: from https://www.linkedin.com/pulse/how-anomaly-detection-can-boost-your-business- nazar-savchenko

[10] Khlefha, A. R. (2024). Comprehensive Review of Interpolation Techniques Using B-Spline. *International Academic Journal of Science and Engineering, 11*(1), 304–311. https://doi.org/10.9756/IAJSE/V11I1/IAJSE1135

[11] By IndustryWeek in collaboration with Emerson. *How Manufacturers can achieve top quartile performance* [Online]. Available: from https://partners.wsj.com/emerson/unlocking- performance/how-manufacturers-can-achieve-top-quartile-performance/

[12] Deloitte refers to one or more of Deloitte Touche Tohmatsu Limited, a UK private company limited by guarantee (DTTL), *Making maintenance smarter*, from https://www2.deloitte.com/cn/en/pages/consu mer-industrial-products/articles/making- maintenance-smarter.html

[13] Biglete, E. R., et al. (2020). Taguchi-Based Experimental Investigation of Vortex-Induced Vibration (VIV) Energy Converter under a Controlled Environment. *2020 IEEE 12th International Conference on Humanoid, Nanotechnology, Information Technology, Communication and Control, Environment, and Management (HNICEM)*, pp. 1–6, doi:10.1109/HNICEM51456.2020.9400069, from https://ieeexplore.ieee.org/abstract/document/ 9400069

[14] Tran, T., Bader, S., Lundgren, J. (2022). *Denoising Induction Motor Sounds Using an Autoencode*r, from https://arxiv.org/pdf/2208.04462.pdf?fbclid=I wAR2x8EVb86RvDm-nEUbY7- 07QzmYEaMql6_CCQcdfBqji5bghE4YUx1HZ_T4

[15] Basis for MSE [Online]. Available: from https://en.wikipedia.org/wiki/Mean_squared_ error

[16] Basis for Confusion Matrix, from https://medium.com/ibm-watson-aiops/confusion-matrix-precision-recall-and-f1-score-performance-metrics-for-log- anomaly-detection-1a187a17fca9

[17] Histogram plot for '*Autoencoder Reconstruction Error*', [Online]. Available: from https://github.com/aldente0630/sound-a nomaly-detection-with-autoencoders

[18] Frequency domain via *FFT*, [Online]. Available: f from https://github.com/aldente0630/sound- anomaly-detection-with-autoencoders

[19] Median Absolute Deviation to measure spread view in 3D scatter plot sample, [Online]. Available: from https://www.researchgate.net/figure/a-Iris-dataset-distribution-view-in-3D-space-b-Iris- dataset-clustering-view-in-3D- space_fig4_339548925

[20] Global Electronic Services, *Common Causes of Electric Motor Vibration* [Online]. Available: from https://gesrepair.com/common-causes-of- motor-vibration/

[21] Fluke Resource Center. *Most common causes of machine vibration* [Online]. Available: from https://www.fluke.com/en- ph/learn/blog/vibration/most-common- causes-of-machine-vibration

[22] Autoencoder-based Condition Monitoring and Anomaly Detection Method for Rotating Machines https://arxiv.org/pdf/2101.11539.pdf

[23] Margarit Jaile, S. (2020, July 28). *Anomaly Detection using Audio Signal*s, from https://upcommons.upc.edu/bitstream/handle/ 2117/329692/Thesis_TUD_SMargarit.pdf?se quence=2&isAllowed=y&fbclid=IwAR1Ex0 sJeQR5SiljcgM-C6bcJFbeZHZq5QXalzFY60 WJEAxvbFMo1V7ASWCg

[24] Hisashi, U., Yuma, K., Shoichiro, S., Akira, N., & Noboru, H. *Anomaly Detection Technique in Sound to Detect Faulty Equipment* [Online]. Available: from https://www.ntt-review.jp/archive/ntttechnical.php?contents= ntr201708fa5.html

[25] Cawaling, C. V., Duque, F. L. & Capricho, N. C. (2013). Noise Detector with a Warning Device. *PULSAR, 2*(1). Retrieved from https://ejournals.ph/article.php?id=7248

[26] Li, J., Liu, Y., Wang, Q., Xing, Z., & Zeng, F. (2022). *Rotating machinery anomaly detection using data reconstruction generative adversarial Networks with vibration energy analysis*, from https://aip.scitation.org/doi/10.1063/5.008535 4

[27] Purohit, H., Tanabe, R., Ichige, K., Endo, T., Nikaido, Y., Suefusa, K., & Kawaguchi, Y. (2019). Mimii Dataset: Sound Dataset for Malfunctioning Industrial Machine Investigation and Inspection, from https://arxiv.org/pdf/1909.09347.pdf?fbclid=I wAR3m6MEZBk74EUlpvD4IcTIndsO5RVo81xfOzueG0dLA84xmn-w3apzd9h8

32 A study on taxpayers' satisfaction towards e-filing system with reference to Ernakulam district

Thazna T. A.[1,a] *and Sarithamol K. K.*[2,b]

[1]Doctoral Research Scholar, Department of Commerce, Karpagam Academy of Higher Education, Coimbatore, Tamil Nadu, India
[2]Research Supervisor & Assistant Professor, Department of Commerce, Karpagam Academy of Higher Education, Coimbatore, Tamil Nadu, India

Abstract: The ideal approach to achieving large social and financial benefits is through e-documenting. It can lend considerable assistance to the public authority in its attempts to advance great administration. Annual expense records should be recorded for all evaluates whose available pay exceeds essential exception requires. A basic exclusion limit is a sum, the public authority establishes for the benefit of the tax paying community. Individuals with earnings below the specified limit are liberated from the obligation to pay personal expense.

If the pay of a singular surpasses the essential exception limit, the assessee should pay personal expense and document personal duty reports. Duty installment and return documenting ought to be done before the due date. The due date varies based on whether the assessee is a corporate or an individual. It also differs based on whether an assessee is subject to a mandatory tax audit. Previously, the procedure for recording expenditure forms was manual, which not only made it tedious but also geared towards blunders and administrative errors. Currently, this strategy of e-recording has been merged by the Personal Duty Division to manufacture a framework of duty to upgrade the personal expense evaluation method.

Keywords: E-filing, income tax returns, taxable income, exemption limit, tax audit

1. Introduction

Personal responsibility refers to how much money that a administration requires people to uphold based on their income, the worth of their property, and different factors, and is drafted to pay for taxpayer driven organizations.

This kind of documenting is done by the government through digital ways of charging readiness programming that are preapproved by the fitting duty specialists, for example, Indian Revenue Services (IRS).With energy and productivity, E-documenting has become a trend over the last ten years. The IRS assures that e-recording expedites charge concessions and helps eliminate delays.

The technological advances are not left monitored seeing as in this period existing individuals practice specific development, whereas the public authority of India creates specific probable skimming realities concerning electronic recording, for instance, the citizens presenting their government forms online through the site http://incometaxindia.gov.in.

As per the IRS, most citizens that e-document and give direct store data ought to hope to get any discount owed in somewhere around 21 days as long as their government form is precise.

One more benefit of e-recording is that an affirmation or excusal notice is shipped off the obligation filer in around 48 hours (typically inside 24 hours) of introducing the appraisal structure. Affirmation confirms that the files have been gotten and are in the system, while excusal enlightens the resident that the return has not been recognized.

The notification of excusal will likewise determine what requirements amending just when he returns to make it great. On the off chance that the obligation due date is finished, the assessee will be given a period of five extra days to correct and refile the important data. Then, need to record a remedied paper return. According to Segment 139 (1) of Annual Expense Act 1961, people whose complete pay in gossamer of the earlier year is more than the greatest sum not chargeable to sack are required to document their ITR or personal government forms. E-recording refers to the process of filing individual tax return online. Although, official cutoff time for documenting assessment structures is 31st July every year, government specialist may give a 15–30 days greatness period for online or disconnected structures to be recorded.

1.1. Some of the key benefits include

1.1.1. Prompt dealing

Your acknowledgment of the Personal Government Form (ITR) is swiftly made. More essentially, discounts, when appropriate, are handled more quickly than returns recorded via paper.

[a]thaznames@gmail.com

DOI: 10.1201/9781003684459-32

1.1.2. Improved accuracy

E-recording projects with understanding arrangements and electronic system work easily and fundamentally limit blunders. Paper entries can be prone to botching. In addition, when it moves from any paper structure to the various electronic framework there is a possibility of human mistake identified with information input.

1.1.3. Ease of purpose

Refunds on the web have neither time nor area limitations. The e-recording administration is 24 hours every day offering presenting the filings whenever and anyplace at individual comfort.

1.1.4. Secrecy

Enhanced security in contrast with paper submissions as the information is not available to anyone, either intentionally or inadvertently. Via paper sections, residents pay information could discover its way into some unacceptable hands at the bookkeeper's office, or the Individual Duty Division.

1.1.5. Access to verifiable data

Individuals may conveniently retrieve prior information while filing their returns. Most of the e-documenting programs store information securely and give favorable access while logging returns in future.

1.1.6. Confirmation of receipt

Clients will be given speedy confirmation on any accommodation, both at the time of transaction and soon after submitting, through email sent to their enlistment email address.

1.1.7. Ease of purpose

E-recording is benevolent and the settled guidelines make it simple as well, even for the individuals who are not extremely all around familiar with the web.

1.1.8. Digital banking

Since simple retail is for rebates and direct charging is for charge settlements. An individual can choose to record now and pay later – choose the day to pull out resources out of the bank for charge installment, with various different valuable highlights.

2. Objectives Framed

1. To understand the satisfaction level of e-filing among the taxpayers with respect to different demographic profile (Age, Gender, Qualification, and Level of Income).
2. To examine the hindrances taxpayers face during e-filing of returns.

3. Statement of the Problem

Income affiliations all over well-off nations have their eyes peeled to see how residents take a glimpse at the e-recording of individual indictment. This research is in the domain of e-recording of personal assessment in the city of Ernakulam. Some are entirely new to electronic exchanges and others need PC information, making it hard for residents to make use of the e-recording framework for their annual appraisal structures. Some of the populaces do not use the web and have less PC abilities. Part of the citizens are concerned about the security problem. There are very few electronic frameworks resulting in poor use. Thus, it is very basic to focus on satisfaction level and habitudes of individual citizens towards e-recording of personal duty forms.

4. Review of Literature

Ambali [1], in her report, had examined what was instructed about the e-government technique, comprehends the difficulties concerning the e-recording structure. Members were given a survey perspective, followed by an unregulated meeting to survey widely the usability, convenience, security, working with conditions, and reception of the e-filing framework. Over a two-month period of data gathering, 450 citizens were deliberately and effectively chosen to partake in the concentrate.

In the review of her, Gayathri [2] explained how citizens view it e-filing of personal tax documents. Key information was obtained using polls to survey respondents bliss and attention to the e-recording. According to the review, respondents requested progressively e-petitioned for a sooner charge markdown, 4% for a unique money refund, and simply 7% affirmed they had no reason for e-recording, elsewhere assigning constrained mental strain.

As stated in their article, if the pay earned by an individual surpasses the basic exclusion limit, an assessee is liable both to pay income tax and also file income tax returns [8]. Assessment installment and profit recording both must be done before the due date. The due date moves contingent upon whether an assessee is corporate or an individual. It also varies depending upon whether an assessee is subject to a mandatory taxaudit. Previously, the site was the most well-known strategy for recording, since the part utilized for expense forms is gigantic (and as a result was excessively genuine), and exceptionally fallible. So far, a Personal Duty Division has brought the segment of e-petitioning together to set up an expense framework through which the procedure of annual duty appraisal is made basic and easy.

As an illustration, the readers of the paper by Monika Arora and Dr. Rajni [3] focus on an inside and larger part investigation of citizen fulfillment with e-recording in India. Of the 500 who responded to the review, 367 were found to be investigative competent. The positioning procedure, ANOVA-test, and t-test were utilized for the examination of citizens fulfillment levels. You train on data until

October 2023. The information uncovers critical divisions between citizen happiness with the security and direction of their e-documents from the Individual Expense Office. It is concluded in the review that most individual subjects are gratified after using this facility. Maturity with the e-filing framework fundamentally contrasts by age, instructive level and occupation. The review recommended that the Annual Assessment Office should discover appropriate methods to educate citizens on e-recording methods. Media crusades settling issues amid duties changes can take care of e-recording administrations presentations – underlining the amount they need to make strides.

According to his review, E-recording of annual duty becomes the key venture under the e-administration framework of the annual expense division, which permits to document the profits from anyplace without an approach the division. This study intended to assess the extent of satisfaction of the salaried population with e-filing based on demographic characteristics. Respondents to a survey strategy which were drawn upon the poll consists of five main target groups: Highly Satisfied (HS), Satisfied (S), Neutral (N), Dissatisfied (DS), and Highly Dissatisfied (HDS). The review found that the segment profiles of respondents have little to no impact on citizen satisfaction.

According to them, it is basic to progress from manual to mechanized work and new innovative forward leaps benefit all organizations [4]. From this perspective, the web is simplistic. It has made an e-recording site that enables citizens to e-document their Individual Government form (ITR), on e-administrator rather than visiting the personal duty division physically. E-documenting alludes to the strategy associated with documenting, producing, and maintaining personal government forms. The predominant position of their investigation was to ascertain how fulfilled and cognizant citizens are with the Annual Duty Divisions E-documenting administration. Data was gathered through a survey of 120 participants. The review explanation advises all citizens how to enhance the frameworks ease of use and dependability.

Dr. G. Gabriel Prabhu [5], his paper is E documenting is portrayed of a few created nations documenting the assessment structure online rather than coming to it truly. This article explains the composition and cons of e-documenting among clients. Electronic recording framework enables the citizen to document the government structures online which in turn spares the time and exertion of the assessee. We selected that specific paper to analyze with the character of e-recording framework. According to the review, slip-up free exchanges, quick handling and speedy outcomes are the advantages of e-recording despite the fact that the framework conveys allowable non-human mistakes.

Amit Gupta, Shailendra Singh Bhadouria [6], their review was to find out the issues looked by the citizens while e-documenting their profits and whether it was same like the time of COVID-19 pandemic. 210 respondents were selected as the sample size reported with a near - completed organized poll. Reportedly by the review for the individuals where recorded their profits, full recounting it in reality, are quick to react in kind burdening specialists to do accordingly amidst pandemic. The complexity of e-recording throughout the pre and post pandemic was also compared.

The paper by Pritesh Thakar, Rahul Tepan, Dr. Aashka Thakkar [7], examined the responsibility specialists perspectives related to the e-filing of income forms. The paper examined the critical data concerning the e-recording framework in the country. Marketing of annual revenue is negligible; e-filing of income tax return; Outcomes of review; The outcomes of the review distinguished the helpful data, accessibility of information and strategy of e-recording of Annual assessment forms which gave to the subjects. The report made a careful reminder that there are steps that should be taken into account in recording government shapes on the web. The concentrate likewise said that a portion of the respondents are unsatisfied with e-documenting offices. This current situation is only solvable through proper awareness on e-filing among taxpayers.

Jemima Jose [9], through the review a survey of school instructors found that most of respondents have fundamental comprehension and familiarity with electronic documenting. Despite having the average knowledge of various variables, they believe in the tax computation and management. The majority of responders expressed satisfaction with the current E-recording system. However, it should be emphasized that they faced multiple challenges concurrently. Although convenient, reception e-documenting is hindered by poor organization. Lack of inability can be corrected with proper preparation.

5. Research Gap

The various literature reviews collected show that filing income tax return through the e-filing process requires immense knowledge regarding the workings of the system of e-filing. Lack of technical knowledge on e-filing may create so many issues before taxpayers while they attempt to file the tax returns online. As in different studies, the respondents face different kinds of challenges on e-filing, there is always scope for studying the drawbacks or effectiveness of e-filing among different sections of individuals. So, the study is attempting to identify the perception and problems faced by taxpayers in Ernakulam district of Kerala.

6. Hypotheses

To make a study on taxpayers perception of the e-filing system, the following hypothesis was framed:

H0: There is no remarkable association between Age, Gender, Qualification, and Income level with Level of Satisfaction on E-filing.

H1: There is remarkable association between Age, Gender, Qualification, and Income level with Level of Satisfaction on E-filing.

7. Research Methodology

This descriptive review was completed to understand the extent of satisfaction and obstacles that individual tax payers face while e-filling their income tax returns. Non-random techniques are used to draw examples, important of which is the relatively sampling approach. This study has sample size of 250 individual and core knowledge are collected using an interview schedule.

7.1. Sample size

The sample sizes of 250 taxpayers were drawn from the Ernakulam District.

7.2. Statistical tools

After the survey was finished, the scientist painstakingly backed result. Shortly after, the information was updated. The specialist performed the assessment physically, using a calculator. The citizen discernments were broken down by utilizing the factual methodologies, for example, rate, chi-square test and Garrett positioning.

8. Demographical Findings of the Study

The findings of the study are presented in different tabular forms as follows:

8.1. Classification of demographic profile

This Table 32.1 shows that out of 250 individual appraisals, 147 (58.80%) are male and the excess 103 (41.20%) are female. It is inferred that the larger part (58.80%) of the people evaluated are male.

This Table 32.2 shows that 17.2% of the 250 respondents are under the age of 30, 26% are between the ages of 30 and 40, 28.8% are between the ages of 40 and 50, and the remaining 28% are over the age of 50. According to the findings of this study, nearly the same amount of respondents are 40–50 years old and older than 50.

This Table 32.3 shows that out of 250 respondents, 87 (34.8%) have merely completed higher secondary education, 134 (53.60%) have graduated, and the remaining 29 (11.6%) have pursued professional degrees. A significant majority (53.60%) of the individuals assessed in the study are graduates.

This tabular details reveals, among 250 respondents, 56 (22.4 %) have month to month pay range under 30000, 123 (49.2%) of the full scale secure a month to month pay range some place in the scope of 30000 and 50000, 71 (28.40%) procure the month to month pay 50000 or more. An examination of the month to month pay scope of the individual evaluates has uncovered that a greater part (around half) of the individual surveys are procuring the month to month pay range somewhere in the range of 30000 and 50000.

9. Analysis and Interpretations

The analysis of the formatted findings was done through the statistical tools like Chi-square test and Garrett Ranking method. Chi-square tests were used to study the association between different demographic factors and level of satisfaction on e-filing. Garret ranking method was used to identify the hindrances come across in e-filing procedures.

9.1. Chi-Square test

Chi-Square test is a statistical method for calculating the difference among observed and expected data. This test can also figure out whether it is related to our class factors. It aids in determining whether a routine in two complete factors is due to chance or a relation between them.

Utilize a chi-square test or a practically comparable non-parametric test to test a speculation with respect to a clear-cut

Table 32.1. Tabulation based on gender

Gender	*No. of Respondents*	*%*
Male	147	58.8
Female	103	41.2
Total	250	100

Source: Primary data.

Table 32.2. Tabulation based on Age

Age category	*No. of Respondents*	*%*
Below 30 years	43	17.2
30–40 years	65	26
40–50 years	72	28.8
Above 50 years	70	28
Total	250	100

Source: Primary data.

Table 32.3. Tabulation based on educational qualification

Educational Qualification	*No. of Respondents*	*%*
Higher Secondary	87	34.8
Graduation	134	53.6
Professional Degree	29	11.6
Total	250	100

Source: Primary data.

Table 32.4. Tabulation based on income level

Income level	*No. of Respondents*	*%*
Less than 30000	56	22.4
30000–50000	123	49.2
Above 50000	71	28.4
G.T	250	100

Source: Primary data.

Table 32.5. Testing the proximity between gender and satisfactory level on e-filing

Gender and satisfactory level (observed)					*Gender and satisfactory level (expected)*				
	Yes	*No*	*Moderate*	*Total*		*Yes*	*No*	*Moderate*	*Total*
Male	70	54	23	147	Male	66.44	58.21	22.34	147
Female	43	45	15	103	Female	46.56	40.79	15.66	103
Total	113	99	38	250	Total	113	99	38	250

Source: Primary data.

variables sharing. Highlights like creatures or nations can be ostensible or ordinal; categorical features, similarly to straight out ones, can be distinct or ordinal. They can't take on a standard spread since they only contain a few specific traits.

The Chi-square *(x^2)* test of independence allows a researcher to explain whether two traits are connected or not, and the formula employed is provided below.

$$\text{Expected Frequency (E)} = \frac{\text{Row Total} \times \text{Column Total}}{\text{Grand Total}}$$

$$\text{Degree Freedom (D.F)} = (r - 1)(c - 1)$$

Where, c = Number of Columns
r = Number of Rows

After identifying the expected frequencies, chi square values are computed as follows:

$$x^2 = \Sigma\, (O\text{-}E)^2/E$$

Table 32.6. Chi-square test between gender and level of satisfaction

Cell	*O*	*E*	*O-E*	*(O-E)²*	*(O-E)²/E*
R_1C_1	70	66.44	3.56	12.65	0.19
R_2C_1	43	46.56	−3.56	12.65	0.27
R_1C_2	54	58.21	−4.21	17.74	0.30
R_2C_2	45	40.79	4.21	17.74	0.43
R_1C_3	23	22.34	0.66	0.43	0.02
R_2C_3	15	15.66	-0.66	0.43	0.03
					1.25

Source: Primary data.

Where, O = Observed Values
E = Expected values

The result obtained should be compared to the applicable table value, and conclusions can be drawn. If the calculated value exceeds the table value, the proposed hypotheses will be rejected; otherwise, they will be approved.

The below Tables 32.4–32.12 depict the association between demographic factors and pattern of satisfaction of the respondents using Chi-square analysis.

D.F = (c − 1) (r − 1) = (3 − 1) (2 − 1) = 2 × 1 = 2
T.V of x^2 at 0.05 significance level and d.f 2 = 5.99
Calculated value of x^2 = 1.25

Since the determined worth (1.25) is not exactly the table worth (5.99), the invalid speculation is acknowledged. It is observed that there is no critical relationship between citizen orientation and level of fulfillment with the e-documenting of annual assessment forms. It is presumed that orientation never influence the perspective on the e-recording framework.

D.F = (c − 1) (r − 1) = (3 − 1) (4 − 1) = 2 × 3 = 6
T.V of x^2 at 0.05 significance level and d.f 6 = 12.59
Calculated value of x^2 = 13.14

Since the calculated result (13.14) exceeds the table value, it leads to rejection of the null hypothesis. This scenario proving a substantial association among tax payers age and their satisfactory level on e-filing. It is established that age effects the view of the e-filing system.

D.F = (c − 1) (r − 1) = (3 − 1) (3 − 1) = 2 × 2 = 4
Table value of x^2 at 0.05 significance level and d.f 4 = 9.49
Calculated value of x^2 = 15.18

Table 32.7. Testing the proximity between age and satisfactory level on e-filing

Age and satisfactory level (Observed)					*Age and satisfactory level (Expected)*				
	Yes	*No*	*Moderate*	*Total*		*Yes*	*No*	*Moderate*	*Total*
Below 30 years	25	13	5	43	Below 30 years	22.36	15.31	5.33	43
30–40 years	40	20	5	65	30–40 years	33.80	23.14	8.06	65
40–50 years	30	26	16	72	40–50 years	37.44	25.63	8.93	72
Above 50 years	35	30	5	70	Above 50 years	36.40	24.92	8.68	70
Total	130	89	31	250	Total	130	89	31	250

Source: Primary data.

Table 32.8. Chi-square test between age and satisfactory level on e-filing

Cell	*O*	*E*	*O-E*	*(O-E)²*	*(O-E)²/E*
R_1C_1	25	22.36	2.64	6.97	0.31
R_2C_1	40	33.80	6.20	38.44	1.14
R_3C_1	30	37.44	−7.44	55.35	1.48
R_4C_1	35	36.40	−1.40	1.96	0.05
R_1C_2	13	15.31	−2.31	5.33	0.35
R_2C_2	20	23.14	−3.14	9.86	0.43
R_3C_2	26	25.63	0.37	0.14	0.01
R_4C_2	30	24.92	5.08	25.81	1.04
R_1C_3	5	5.33	−0.33	0.11	0.02
R_2C_3	5	8.06	−3.06	9.36	1.16
R_3C_3	16	8.93	7.07	50.01	5.60
R_4C_3	5	8.68	−3.68	13.54	1.56
					13.14

Source: Primary data.

D.f = (c − 1) (r − 1) = (3 − 1) (3 − 1) = 2 × 2 = 4
T.V of x^2 at 0.05 significance level and d.f 4 = 9.49
Calculated value of x^2 = 24.09

Since the calculated result (24.09) is greater than the value in the table, the null hypothesis is refuted. It is found that there is a substantial correlation between the taxpayers educational background and their level of satisfaction with electronically completing their income tax forms. It is found that perceptions of the e-filing system are influenced by educational background.

9.2. Identification of hindrances in e-filing – calculations using Garrett scores

GARRETT SCORE - Garrett scores analysis is a statistical ranking technique in which a respondents factor rating is converted into a numerical score. The results are then utilized to identify the most significant factors or difficulties.

Table 32.9. Testing the proximity between educational qualification and satisfactory level on e-fil

Respondents qualification and satisfactory level (observed)				
	Yes	*No*	*Moderate*	*Total*
Higher Secondary	52	31	4	87
Graduation	64	52	18	134
Professional Degree	7	15	7	29
Total	123	98	29	250

Respondents qualification and satisfactory level (expected)				
	Yes	*No*	*Moderate*	*Total*
Higher Secondary	42.80	34.10	10.09	87
Graduation	65.93	52.53	15.54	134
Professional Degree	14.27	11.37	3.36	29
Total	123	98	29	250

Source: Primary data.

Table 32.10. Chi-square test between respondents qualification and satisfactory level

Cell	*O*	*E*	*O-E*	*(O-E)²*	*(O-E)²/E*
R_1C_1	52	42.80	9.20	84.57	1.98
R_2C_1	64	65.93	−1.93	3.72	0.06
R_3C_1	7	14.27	−7.27	52.82	3.70
R^1C_2	31	34.10	−3.10	9.63	0.28
R^2C_2	52	52.53	−0.53	0.28	0.01
R_3C_2	15	11.37	3.63	13.19	1.16
R_1C_3	4	10.09	−6.09	37.11	3.68
R_2C_3	18	15.54	2.46	6.03	0.39
R_3C_3	7	3.36	3.64	13.22	3.93
					15.18

Source: Primary data.

The null hypothesis is disproved since the computed result (15.18) exceeds the value in the table. It is determined that the degree of satisfaction with e-filing income tax returns and the taxpayers educational background are significantly correlated. It is determined that educational background affects how the e-filing system is perceived.

Garrett is determined utilizing suitable Garrett positions, and the Garrett esteem is found out. The Garrett table qualities and scores of each rank are given in Tables 32.13 and 32.14. At last, by adding each line, the absolute Garrett score is gotten.

$$\textbf{Percent position} = \frac{\textbf{100(Rij-0.5)}}{\textbf{Nj}}$$

Rij is the rank given to the ith variable by the jth responder
Nj is the number of different items ranked by the jth responder

Garretts scores are shown in the Table 32.14. First, the Garret ranks are computed using the relevant Garrett ranking formula. The Garrett values for each rank are then multiplied by the respective no. of respondents opted for each rank and took the sum of these values obtained for each point of hindrance and obtained the total Garrett score for each row. The total scores thus obtained are divided by the sample size and obtained mean scores (average scores). According to the mean value obtained the ranks were allotted to different points of hindrances of e-filing.

It is seen from the table that Regular modification has been positioned as the principal factor affecting the individual surveys to approach to record charge. Security concerns

Table 32.11. Testing the proximity between income level and satisfactory level on e-filing

Income level and satisfactory level (observed)				
	Yes	*No*	*Moderate*	*Total*
Less than 30000	33	19	4	56
30000–50000	85	29	9	123
Above 50000	42	9	20	71
Total	160	57	33	250

Income level and satisfactory level (expected)				
	Yes	*No*	*Moderate*	*Total*
Less than 30000	35.84	12.77	7.39	56
30000–50000	78.72	28.04	16.24	123
Above 50000	45.44	16.19	9.37	71
Total	160	57	33	250

Source: Primary data.

Table 32.12. Chi-Square Test between level of income and e-filing satisfactory level

Cell	*O*	*E*	*O-E*	*(O-E)²*	*(O-E)²/E*
R_1C_1	33	35.84	−2.84	8.07	0.23
R_2C_1	85	78.72	6.28	39.44	0.50
R_3C_1	42	45.44	−3.44	11.83	0.26
R^1C_2	19	12.77	6.23	38.84	3.04
R^2C_2	29	28.04	0.96	0.91	0.03
R_3C_2	9	16.19	−7.19	51.67	3.19
R_1C_3	4	7.39	−3.39	11.51	1.56
R_2C_3	9	16.24	−7.24	52.36	3.22
R_3C_3	20	9.37	10.63	112.95	12.05
					24.09

Source: Primary data.

Table 32.13. Factors identified as the hindrances of e-filing (Garrett Ranking Method)

Particulars	*I*	*II*	*III*	*IV*	*V*	*Total*
Complex return	36	43	47	64	60	250
Errors in data entry	48	34	55	45	68	250
Lack of up-to-date technical expertise	48	51	54	47	50	250
Frequent revision	61	68	43	36	42	250
Security concerns	57	54	51	58	30	250

Source: Primary data.

Table 32.14. Calculation of mean scores based on Garrett scores

Hindrances of E-filing	*Total score*	*Mean*	*Rank*
Complex return	11690	46.76	V
Errors in data entry	11890	47.56	IV
Lack of up-to-date technical expertise	12490	49.96	III
Frequent revision	13295	53.18	I
Security concerns	13135	52.54	II

Source: Primary data.

has been positioned second, Absence of cutting-edge specialized skill put in the third position, trailed by Mistakes in information passage the fourth position, and Complex return positioned fifth.

10. Summary of Finding

This study is an endeavor to dissect the different variables that work on the fulfillment of the e-documenting framework in Ernakulam fully intent on presenting a few noteworthy thoughts.

To concentrate on the fulfillment of citizens towards e-recording system, 250 respondents were arbitrarily chosen from Ernakulam and their perspective were acquired.

1. Among the respondents studied 58.80 % are male and 41.20 % are females. A greater part (147) of the example respondents are men.
2. More than half of the citizens (56.8%) reviewed are in the age category of above 40 years.
3. Among the studied respondents 53.60 % are graduates.
4. Out of 250 respondents 22.4% of the citizens having a place with the pay gathering of under Rs.30000, 49.2% of the citizens are in the pay gathering of Rs.30000–50000, 28.4% of the citizens procure pay of above Rs.50000.
5. From the chi-square investigation made on segment factors and level of examination, if there should arise an occurrence of orientation as the determined worth lies beneath the table worth, it is distinguished that orientation and level of fulfillment on e-documenting does not have a critical relationship.
6. The chi square investigation made among age and level of fulfillment demonstrated that there is huge relationship exist between both the variable since the determined worth stands more prominent than table worth.
7. The chi square investigation done between staying two segment factors (instructive capability and Pay level) and level of fulfillment on e-recording likewise reflected critical relationship as the determined qualities were higher than table qualities.
8. It is seen from the table of Garrett arranged for distinguishing the variables of blocks in e-documenting process, regular correction in e-recording strategies has been positioned as the primary calculate of obstacle e-recording framework. Security concerns has been positioned

as second, Absence of state-of-the-art specialized ability set in the third position, and it is trailed by mistakes in information passage the fourth position and Complex return positioned as the fifth and last position.

11. Suggestions and Recommendations

As per the outcomes of the study, below are some of the proposals I would like to bring forward:

1. According to a survey, income taxpayers prefer eco-friendly websites that require multiple steps to function, which they find laborious.
2. The government should address income taxpayers concerns about online filing issues to improve their filing experience.
3. To improve the website, government should prioritize server maintenance to ensure optimal performance during busy months.
4. The government uses innovative technology to provide a user-friendly website for taxpayers to file income tax returns.
5. The government must protect taxpayer data from hacking, viruses, and cyber threats.
6. Individual assessors have been trained to file tax returns independently.
7. Educate assesses on deductions ranging from 80C to 80U to maximize their claims.
8. To fulfill individual assessments, the government should reduce tax rates considerably while increasing the tax cap.

Since E-filing is the initiative of the Revenue Department to reduce the burden of all the individuals dealing with it, they have to manage the system more lively, secure, and with a less complex atmosphere.

12. Conclusion

E -filing is the process of filing income tax return through online. In this new era, as all are busy in their work life, e-filing helps the individual assessee to saves their time and effort. Physically approaching a tax advisor to file the tax returns demands valuable time of the tax payers. This service demands a remarkable consultation fee too. So, this can be avoided if opting for e-filing.

There are so many benefits wrapped up with e-filing process. The prominent among them are: e-filing saves the time and effort of the tax payers, refund if applicable can be initiated quickly, human errors on paper documentation are reduced to a minimum, ease of use of application for a computer literate person, secrecy regarding the taxpayers income information will be maintained, e- filing programs will store the previous transaction history which in turn provide the data base for the assessee, confirmations and updates on filing the return will receive in mail id of the assesee, digital banking can also be enjoyed along with e-filing service.

Along with the merits provided, e-filing possess some hindrances too. Complex return procedures, Errors in data entry, Lack of up-to-date technical expertise, frequent revision of the service, and Security concerns are the notable drawbacks faced by tax payers during filing the return online.

Apart from the above-mentioned pros and cons of e-filing, taxpayers are satisfied with the services provided by the e-filing. According to the study, there is no relationship between gender and satisfaction level on e-filing services. However, the demographic factors like age, income level, and education level of the respondents show a substantial correlation to the satisfaction level on e-filing.

Throughout the paper, while studying the problems faced by the respondents during e-filing, some hindrance factors were identified and respondents ranked frequent revision on e-filing as the first problem affecting e-filing satisfaction.

13. Scope for Further Research

The study results are immensely useful for taxpayers proposing to file their tax returns through e-filing services. The paper discusses the benefits and drawbacks of e-filing services. The details thus covered in the study can be used as a guide before stepping towards the e-filing process. The opinion of 250 respondents regarding their satisfaction level on online filing of tax returns are included in the study, the above-said sample size shared their experiences and perceptions towards e-filing services. The experience thus shared will act as a user guide for the taxpayers. The study results belong to Ernakulam, a district of Kerala. So, by keeping the study results as a reference, there opened a further scope to study the topic in some other geographical areas and under some other featured individuals.

References

[1] Ambali, R. A. (2009). E-Government Policy: Ground Issues in E-Filing System. *European Journal of Social Sciences, 11*, 249–266.

[2] Gayathri, P., & Jayakumar, K. (2016). A Study on Tax Payer's Perception Towards E-Filing of Income Tax Returns. *International Journal of Trend in Research and Development*, 32–38.

[3] Monika Arora and Dr. Rajni (2017). Exploring Individual Taxpayers' Satisfaction Level for E-Filing System in Indian Context, Inspira. *Journal of Modern Management & Entrepreneurship (JMME)*, ISSN: 2231–167X, General Impact Factor: 2.3982, 07(04), pp. 259–265.

[4] Saranya, M., & Jagadeesan, P. (2024). A Study on Satisfaction Level of Individual Taxpayer Towards E-filing with reference to Chennai District, AIP Publishing. *AIP Conference Proceedings, 2965*, 060003.

[5] Gabriel Prabhu, G. (2018). Challenges and Prospects of E-Filing. *Journal of Emerging Technologies and Innovative Research (JETIR)*, (ISSN-2349-5162).

[6] Gupta, A., & Bhadouria, S. S. (2023). Challenges on the Road towards E-filing Tax Returns-Pre & Post Covid19 Pandemic, Shodh Samagam, Peer-reviewed and Referred, Quarterly, Multidiciplinary and Multilingual Research Journal.

[7] Thakar, P., Tepan, R., & Thakkar, A. (2021). The Study of Taxpayers Perception Towards E-Filing In India. *International Research Journal of Modernization in Engineering Technology and Science*, *3*(4), 460–469.

[8] Vanitha, R., & Ketharaj, M. (2022). A Study on Taxpayers Perception Towards E-Filing System. *International Journal of Mechanical Engineering*. ISSN: 0974-5823

[9] Radhakrishnan, R. P., & Jose, J. (2018). E Filing of Income Tax Returns: Awareness and Satisfaction Among Teachers of Aided Arts and Science Colleges, Thrissur Corporation. *Journal of Emerging Technologies and Innovative Research (JETIR).*

33 Emotional intelligence in finance: A mediating force in cognitive bias and investment decision

Sreejaa G. Nair[1,a] *and Jegadeeswari S.*[2]

[1]Research Scholar, Karpagam Academy of Higher Education, Coimbatore, Tamil Nadu, India
[2]Assistant Professor, Karpagam Academy of Higher Education, Coimbatore, Tamil Nadu, India

Abstract: The research investigates how emotional intelligence interacts with cognitive biases to influence investment choices of active Indian stock market traders. The research gathered data from 262 subjects who took part in a survey with an adjusted questionnaire through convenience sampling. SPSS version 26 performed the statistical analysis through correlation, regression and mediation analysis procedures. This research analyzes the primary influence and secondary mediation process of cognitive biases which include overconfidence bias, confirmation bias, gambler's fallacy bias, herding and availability bias upon investment choices while emotional intelligence functions as the main mediating element. The research shows that investment decisions experience substantial direct effects from cognitive biases which include overconfidence bias together with confirmation bias and gambler's fallacy bias. The study reveals emotional intelligence acts as a vital connecting factor which shapes the manner cognitive biases affect investment selection decisions. The direct relationship between herding bias exists but emotional intelligence establishes intricate controls over the bias's effect. The findings highlight the essential role of emotional intelligence when reducing negative impacts of cognitive biases on investment choices within Indian stock markets. Future investigation requires analysis of other factors that affect this interrelationship between emotional intelligence, cognitive biases and investment choices while studying their connection in detail.

Keywords: Cognitive biases, emotional intelligence, investment decisions, overconfidence bias, herding bias, confirmation bias, gambler's fallacy, availability bias

1. Introduction

Customers face difficulties in making investment selections which are substantially influenced by psychological and emotional elements. According to traditional financial theory everyone within the investment community needs to base their decisions on factual information. According to the Efficient Market Hypothesis rational investors need to exist without anomalies present [1]. Traditional economic models presume investors to base all their choices solely on the information they receive. Recent studies demonstrate that how information displays itself to investors majorly impacts their investment selection process. Although many investment choices follow rational lines of thinking there is contrary evidence according to behavioural finance studies [2]. Behavioural finance evaluates human behavioural effects on financial decision processes especially regarding investment decisions [3]. Behavioural finance traces the standard investor patterns by studying their irrational actions according to [4]. The use of heuristic information causes investors to develop biases according to Hayati et al. [5]. Investors let behavioural biases lead their choices even though these choices are irrational. Financial market development depends strongly on psychological aspects which affect its development [6].

The dynamic finance industry demands that people along with institutions need to make well-informed investment choices. Investors make decisions regarding investment opportunities based on cognitive biases which act as major influence factors according to research by Gargano and Rossi [7]. The behaviour of investors becomes more risky because they experience cognitive bias [8]. Behavioural biases influence choices because they result from cognitive errors and emotional influences [9]. Conclusive research finds that investor behaviour gets substantially shaped by cognitive bias in recent times. According to Chitra and Jayashree [10] cognitive biases result from mental errors during reasoning because individuals simplify how they process information. Having cognitive biases produces wrong choices that result in substantial monetary losses. People who maintain psychological stability combined with effective emotion management will decrease their behavioural biases according to Sri et al. [11].

Multiple studies investigate the destructive effects cognitive biases create for financial investment decisions. Emotional intelligence needs further research to determine its specific capabilities in counteracting such influences. The ability to understand emotions experienced by self or others and maintain control over both types of emotions is defined as emotional intelligence according to Salovey and

[a]sreejaagnair80@gmail.com

DOI: 10.1201/9781003684459-33

Mayer [12]. The four essential elements comprising emotional intelligence such as self-awareness and self-regulation and social awareness and relationship management help investors to stay emotionally stable when making decisions [13]. The paper investigates the intricate link between cognitive biases and investment choices while highlighting that emotional intelligence functions as a deciding variable that connects these elements. This research investigates how better emotional intelligence helps people avoid biases in their decision-making process which will reveal methods to boost financial decision quality.

2. Literature Review

2.1. Theoretical review

Such creations like option pricing model of Black, Scholes, and Merton, capital asset pricing model of Sharpe, Black, and Lintner, and Markowitz and Miller, Modigliani portfolio principles [14] make up a base for building conventional finance theories. One of the theories holds that markets are efficient and investors are rational in their endeavour to maximize the expected utility by understanding all available information best. But these assumptions are disputed by behavioural finance which focuses on the impact of psychology bias such as emotions, cognitive mistakes and social factors on investors [15]. The change of view draws our focus on decision-making under psychological impacts. Therefore, it encourages an investigation of investment decisions when made with biases in mind such as overconfidence, mental accounting, and heuristic approaches [16]. At the end of the day, the combination of the conventional finance principles with the behavioural finance findings will add more sense and provide an all pervasive understanding of how investment decisions come about. According to Obeng [17] this strategy relates the rational and the irrational factors for empowering individuals to make well informed monetary decision.

On the other hand, behavioural finance deals with core concepts that are highly significant in determining investment choices. Such principles include loss aversion, endowment effect, momentum effect, noisy trading, herding, overconfidence, risk propensity, heuristics, prospect theory, market factors, familiarity bias, and self attribution bias (Xu, S. 2023). All of these investment choices are herding behaviour, overconfident, risk propensity. It makes people to make different choices and overspread their capacities by following the crowd, overestimate their own capacities and evaluating the risks in a distinctive style [18]. Moreover, heuristics, prospect theory and familiarity bias play an important role in decision making in the market of cryptocurrencies by introducing cognitive biases and emotional impacts which can result in suboptimal choice of cryptocurrency [19]. These behavioural financial concepts are inevitable, and investors must know them in order to properly negotiate the knotty of the financial markets and make profitable investments.

2.2. Empirical review

[20] state that cognitive bias is a situation in which the errors of decision making stem from the application of mental shortcuts or emotions and can affect investment decision making and result in poor rate of return. It has found that cognitive bias especially cognitive dissonance bias plays a major role in how decision making in the investment [21], explored the effect of cognitive biases (like herding bias, disposition effect etc.) on investment decisions. Potentially, it will uncover how these bias have different types of impacts on investors' investment decisions: disposition effect and mental accounting bias are beneficial for investors' investment decisions, whereas herding bias plays a negative role. The study also evaluates whether financial literacy can buffer against the above effects. Cognitive biases such as herd instinct, availability, overconfidence can impact investment decisions [22].

Overconfidence bias occurs when investors underestimate the difficulty of the task. As a result, it leads to too much trading and one taking on too much risk relative to what's warranted, which tends to lead to poor investment performance [23]. Availability bias refers to the propensity for making decisions not on all available data but rather on data which is easy to obtain or date. Evidence for this can be seen where investors can overvalue the importance of recent events or trends so that when choosing options matters are not chosen that fully take into consideration the bigger picture [24]. These types of investors tend to mimic the behaviour of the majority, which can cause overpriced assets or the buckling of the market. It is this trend, based on the idea that group wisdom beats individual decision-making [25]. The way we define an 'anchor' here is tendency to base our decision on the first piece of information we get. For example, in investment situations, it can lead to bias towards initial stock prices or market trends and biased valuations as well as suboptimal investment decisions [26].

Situngkir, T. L. et al. [27] found that the investors' investment decisions are greatly inclined by confirmation bias, all the choices of investments made by the investors are affected by confirmation bias. Investors tend to look for information that support their beliefs and ignore contradictory evidence [28]. The result of this behaviour is that sometimes when investments are experiencing downward trends, people may go as far as taking speculative positions. According to [29], these cognitive biases, including optimism bias and risk perception, influence the investment acts. In fact it is highlighting the importance of emotional intelligence in making these decisions for success. Emotional Intelligence refers to capacity to identify and understand one's own feeling as well as that of others (Jamil, N., & Bashir, T. 2021). Since, emotion plays an important role in human decision making process, it is important due to the fact that it changes the connection between the investment decisions and the cognitive dissonance bias (Khilar, R. P., & Singh, S. 2020). (Hadi, F. 2017) stated that investment decision is an individual's decision to determine where and when capital is to be spent or the amount of capital to be spent in the search of making profit. According to [30], emotional intelligence significantly affects how investment decisions are made by factors including trading behaviour, the perception of risk and cognitive biases. Research provides evidence that emotional intelligence is very important in formation of investors' behavioural tendencies and preferences of stocks and tolerance to risk.

3. Statement of the Problem

Cognitive biases may influence investment choice and lead to suboptimal outcome or financial disservice to the individuals or organizations. While it is certainly the case that extensive research has been done on cognitive biases present in financial decisions, the extent to which emotional intelligence plays a mitigating role is barely touched on at all. To be emotionally intelligent is to be able to sense, understand and manage the emotions one has – which have been shown to have a huge role in how people process information and make financial decisions. But we do not know yet how exactly emotional intelligence integrates with the cognitive biases affecting investment behaviour. It renders us helpless to generate efficient strategies to enhance decision making within the finance. Thus, given the complexity of these dynamics, there is a significant need to explore the mediating role of emotional intelligence in the relation to the investment behaviour in the context of cognitive bias.

3.1. Scope of the study

This will also exhaustively examine the correlation between emotional intelligence, cognitive bias, and the behaviour of investors in this field. Secondly, it will examine how the association between cognitive biases and investment decision making is impacted by emotional intelligence that encompasses risk perception, decision making, and the outcome of investment. The study seeks to analyse with quantified data collected from investors or participants using a set of measures of emotional intelligence, cognitive biases and the investors behaviour. The objective of its examining the mediating role of emotional intelligence in financial decision making and clarifying the process that underlie. Finally, it aims to explore practical consequences in terms of improving investors' financial literacy by means of education practices concerning decision making.

3.2. Objectives

This study aims to investigate the following:

1. To determine how individual equity investors in the Indian stock market make investment decisions in light of cognitive biases.
2. To evaluate how emotional intelligence influences how cognitive biases and financial decision-making interact.

4. Methodology

4.1. Conceptual framework

The above factors are independent while the individual investor's investment decisions in the Indian stock market are dependent variable. Cognitive bias and investments decision are mediated by the Emotional Intelligence (Figure 33.1).

This study intends to examine the following alternative hypotheses based on the aforementioned review of the literature and conceptual framework:

H1: The influence of overconfidence bias is significant and positive to investment decision making process.

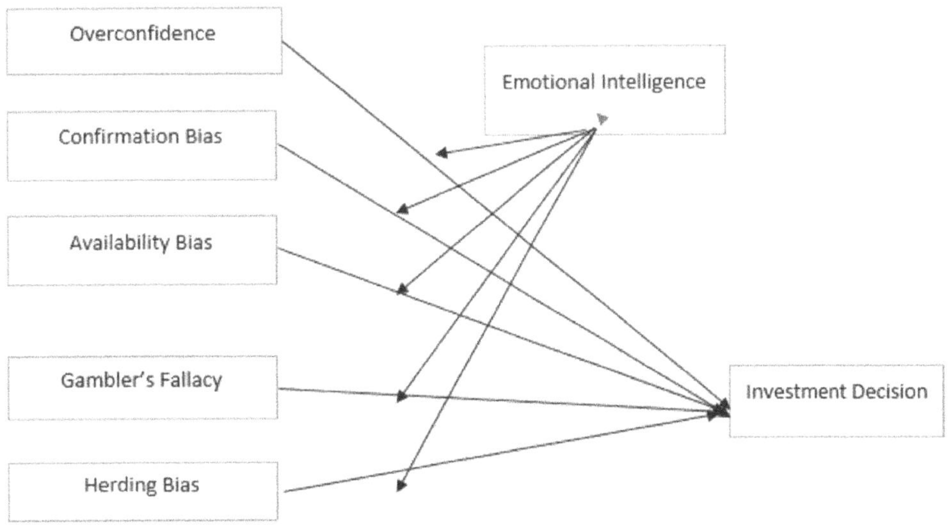

Figure 33.1. Theoretical framework of the research.

Source: Author.

H2: Herding bias is a notable and positive effect on investment decision making process.

This has narrowed my hypothesis into availability bias and it shows a notable and constructive effect on investment decision making process.

H4: There is a noteworthy and favourable impact on investment decision making processes in the case of confirmation bias.

Emotional Intelligence acts as a mediating variable in the relationship between the cognitive bias and the decision of Investment.

4.2. Data collection technique

The individual equity investors of Indian Stock market were the subjects of study. An adaptive questionnaire was used to collect data on cognitive bias's impact on investment decisions by these investors. Practical and within the reach of the investigator, convenience sampling was used to collect primary data from 262 actively engaged equity investors, to test the hypotheses. In other words, from amongst the 300 questionnaires that underwent electronic communication through emails and different online platforms, 262 responses out of 300 were able to be used for the analysis though some of them were not responded to or somehow incomplete.

Data gathering was made through a structured survey using a five point Likert scale. The survey contained 7 sections: the first section be commonly referred to as the personal information section that was to be filled in by the participants; the other 6 sections covered the topics of overconfidence, herding, confirmation, availability, investment decision, and emotional intelligence. A number of questions based on the work were used to assess investment decisions. Five, four and five items from the scale were used to measure overconfidence, herding, confirmation and availability respectively (Table 33.1). Additionally, sixteen questions were used to assess self emotion, others emotions, use of emotion and regulation of emotion and these were used to evaluate emotional intelligence. [12].

Table 33.1. Reliability statistics

Variable	*Cronbach's Alpha*	*N of Items*
Emotional Intelligence	0.954	16
Investment Decision	0.877	5
Overconfidence Bias	0.867	5
Herding Bias	0.899	4
Confirmation Bias	0.844	5
Gambler's Fallacy	0.766	2
Availability Bias	0.867	4

Source: Survey 2024 and author's calculation.

5. Significance of the Study

This study illustrates how emotional intelligence (EI) can bridge finance and human behaviour in investment making. It integrates EI as a mediating factor in refining financial theories and models and its practical applications. The findings enlighten one on ways unnecessary biases in the process of investment decision making can be minimized through targeted interventions and training programs designed to boost EI. These insights can be used by the investors, financial advisors and institutions for making better strategies. In

Table 33.2. Demographic profile

Demographic Variables		*No. of Respondents*	*Percentage*
Gender	Male	115	43.9
	Female	147	56.1
Age	< 27	14	53.4
	28–43	81	30.9
	44–59	41	15.6
	Above 60	0	0
Education	HSC	25	9.5
	Graduate	125	47.7
	Post Graduate	92	35.1
	Professional	20	7.6
Annual Income	Less than 1,50,000	0	0
	1,50,000–3,00,000	15	5.7
	3,00,000–5,00,000	206	78.6
	Above 5,00,000	41	15.6
Occupation	Private Employee	66	25.2
	Government Employee	10	3.8
	Business Person	5	1.9
	Professional	36	13.7
	House Wife	10	3.8
	Student	130	49.6
	Others	5	1.9
Investment Time	<–2 Years	141	53.8
	2–5 Years	50	19.1
	6–10 Years	35	13.4
	> 10 Years	36	13.7

Source: Based on the 2024 field survey and the authors' computations.

addition, the study shows beyond finance how psychological factors affect financial behaviour and specifically how it affects things such as consumer protection. Finally, this research can be of help to academia and practical approaches to support economic stability by the means of the more rational decision-making.

6. Results and Discussion

It was a sample of 115 males (43.9%) and 147 females (56.1%). The respondents' ages ranged, most were between 28–43 years old (30.9%), then approaching 44–59 years (15.6%), just a bit of people under 27 years (53.4%). No respondents were over 60. Of all the educational group graduates were the most represented (47.7%) with graduates of secondary education (35.1%), having of HSC (9.5%) and professionals (7.6%). On the lines of income, most of them generate between 3 and 5 lakh (78.6 per cent), 15.6 per cent earn more than 5 lakh and 5.7 per cent earn between 1 1/2 and 3 lakh. It was in occupation (49.6), then private employees (25.2), professionals (13.7), government employees and housewives (9.6), business persons (1.9) and others (1.9%). Primarily, the majority (53.8%) had invested less than 2 years (19.1%) of 2–5 years, (13.4%) of 6–10 years, (13.7%) over 10 years. In the study it turns out that the stock market investors are primarily young and with very high academic degrees (Table 33.2). Investors that are educated are more likely to participate in the stock market more often and are less likely to make more irrational, uninformed bets as characterized in the study of [31].

6.1. Inferential statistical analysis

6.1.1. Correlation analysis between variables

In the present study bivariate Pearson correlation analysis is used to find out the direction and strength of linear relationship between the variables considered. In this analysis, the dependent variable is making investment capability which is dependent on the independent factors of Overconfidence, Herding, Confirmation, Gambler's Fallacy, and Availability biases and the mediating variable is Emotional Intelligence (Table 33.3).

The emotional intelligence, investment choices, and cognitive biases are strongly linked with each other. The results indicate that investment decisions made using a decision matrix do not suffer from biases like herding ($r = .300$, $p < .01$), confirmation ($r = .391$, $p < .01$), gambler's fallacy ($r = .321$, $p < .01$), overconfidence ($r = .490$, $p < .01$), availability biases ($r = .329$, $p < .01$), emotional intelligence ($r = .712$, $p < .01$), and even do not display a negative correlation with halo effect ($r = .150$, $p < .01$). Though helpful in deciding there is no such thing as an unbiased decision; it lessens their influence. There is also a very significant correlation ($r = .524$, $p < .01$) between overconfidence, confirmation ($r = .524$, $p < .01$), and availability biases with investment decision making. The relationships between cognitive biases are characterized by strong interconnections, specifically confirmation bias with herding ($r = .700$, $p < .01$), availability ($r = .737$, $p < .01$). What this means is that decisions around money can be made with more efficacy or wisdom when our emotional intelligence is running an effective bias management system. The results of these studies are similar to the studies of [32–34].

Table 33.4 is the regression analysis of Investment Decisions as dependent variable and five biases as independent variables. The model has an Rvalue of .601, an R^2 of .361 (.361) and is able to explain 36.1% of the variance. The model is significant (ANOVA results: $F(5, 119) = 28.963$, $p < .000$). Investment decisions are affected by Overconfidence Bias ($B = 0.283$, $p = .000$), Confirmation Bias ($B = 0.252$, $p = .011$), and Gambler's Fallacy ($B = 0.575$, $p = .002$) significantly, and OCB (Beta = 0.272) and GFB (Beta = 0.242)

Table 33.3. Correlations between variables

Variables	*Emotional Intelligence*	*Investment Decision*	*Overconfidence Bias*	*Herding Bias*	*Confirmation Bias*	*Gambler's Fallacy*	*Availability Bias*
Emotional Intelligence	1						
Investment Decision	.712**	1					
Overconfidence Bias	.490**	.524**	1				
Herding Bias	.300**	.383**	.510**	1			
Confirmation Bias	.391**	.524**	.643**	.700**	1		
Gambler's Fallacy	.321**	.516**	.577**	.625**	.693**	1	
Availability Bias	.329**	.468**	.642**	.627**	.737**	.687**	1

Source: Author.

Table 33.4. Regression analysis

Model	*Unstandardized Coefficients*		*Standardized Coefficients*	*t*	*Sig.*	*Model Summary*		*ANOVA*	
	B	*Std. Error*	*Beta*			*R*	*R Square*	*F Value*	*Sig.*
(Constant)	7.334	0.874		8.391	0.000	.601a	0.361	28.963	.000b
OCB	0.283	0.073	0.272	3.898	0.000				
HB	-0.082	0.09	-0.067	-0.907	0.365				
CB	0.252	0.098	0.227	2.56	0.011				
GFB	0.575	0.182	0.242	3.16	0.002				
AB	0.002	0.1	0.002	0.021	0.983				

Source: Author.

Table 33.5. Hypothesis testing

Variables		*Impact*		
		Beta Coefficients	*p- value*	*Hypothesis Support*
Overconfidence Bias	Investment Decision	0.272	0.000 (p < 0.05)	H1 is accepted
Herding Bias		-0.067	0.365 (p >0.05)	H2 is rejected
Confirmation Bias		0.227	0.011 (p < 0.05)	H3 is accepted
Gambler's Fallacy Bias		0.242	0.002 (p < 0.05)	H4 is accepted
Availability Bias		0.002	0.983 (p >0.05)	H5 is rejected

Source: Author.

have the strongest effect. Thus, Herding Bias (B = -0.082, p = .365) and Availability Bias (B = 0.002, p = .983) were not significant. The constant term is significant (B = 7.334, p = .000). These findings agree with previous studies that Gamblers's fallacy and herding do not affect decisions significantly, while confirmation bias, availability biases and overconfidence also distort decisions inconsistently.

The results from Table 33.5 indicate the overconfidence bias (β = 0.272, p < 0.05), confirmation bias (β = 0.227, p < 0.05), and gambler's fallacy (β = 0.242, p < 0.05) are significant predictors of a decision to invest, although these are tested in support of the hypotheses H1, H3, and H4. These results echo the considerable influence that overconfidence exerts on investment behaviour [23] and the effects of confirmation bias and gambler's fallacy in dictating whether an investor takes a gamble or observes trends. On the contrary, herding bias (β = -0.067; p > 0.05) and availability bias (β = 0.002; p > 0.05), as competitors, do not form significant predictors, which rejects hypotheses H2 and H5. As studied by Waweru et al. [35] and Chen et al. [29], these biases do not always significantly affect investment choices and, therefore, such biases may be ignored without harming the usefulness of the profits motive as a means of investment allocation.

6.2. *Mediation analysis*

The PROCESS procedure was used to understand the link between Overconfidence Bias, Herding Bias, Confirmation Bias, Gambler's Fallacy Bias, Availability Bias and Investment Decisions with Emotional Intelligence (EI) as a mediating variable (Table 33.6).

The results from the mediation analysis provided evidence that Emotional Intelligence (EI) is a mediating variable between cognitive biases and the investment decisions. Involved in investment decisions are overconfidence (β = 0.2406, p < 0.001), confirmation (β = 0.3216, p < 0.001), gambler's fallacy (β = 0.7600, p < 0.001) and availability bias (β = 0.3160, p < 0.001). As opposed to its initial hypothesis, herding bias (β = 0.2290, p <0.001) also has a direct effect. Also, the indirect effects through EI further strengthen its mediating function and all the biases showed substantial mediation including herding bias (β = 0.2427, Boot LLCI =

Table 33.6. Direct effects of biases on investment decisions

Predictor Bias	*Outcome Variable*	*Direct Effect (β)*	*LLCI (Direct)*	*ULCI (Direct)*	*p-value (Direct)*
Overconfidence Bias	Investment Decision	0.2406	0.1421	0.3391	<.0001
Herding Bias		0.229	0.1218	0.3363	<.0001
Confirmation Bias		0.3216	0.2257	0.4176	<.0001
Gambler's Fallacy Bias		0.76	0.5659	0.9542	<.0001
Availability Bias		0.316	0.2134	0.4186	<.0001

Source: Author.

Table 33.7. Indirect effects of biases on investment decisions via emotional intelligence

Predictor Bias	*Outcome Variable*	*Indirect Effect (β)*	*LLCI (Indirect)*	*ULCI (Indirect)*	*p-value (Indirect)*
Overconfidence Bias	Investment Decision	0.3064	0.2139	0.403	<.0001
Herding Bias		0.2427	0.1103	0.3788	<.0001
Confirmation Bias		0.2605	0.1413	0.3672	<.0001
Gambler's Fallacy Bias		0.464	0.2338	0.6996	<.0001
Availability Bias		0.248	0.1196	0.3779	<.0001

Source: Author.

0.1103, Boot ULCI = 0.3788). Herding is directly responsible for decisions in her; however, EI moderates its effects. These findings emphasize the importance of EI to curtail biases and shape financial decision-making as it is important to understand the psychological machinery responsible for the investment behaviour (Table 33.7).

7. Suggestions

In the future research, emotional intelligence (EI) training for investors should be considered, and it should be assessed on a long-term basis, whether it can reduce cognitive biases and improve the investment outcomes. The integration of EI assessments can be included in developing decision support systems that may improve personalized recommendations. Awareness programs and workshops have been studied longitudinally in order to see what lasting effect they have. Researching an EI, cognitive bias and investment decision across demographics may provide insight about what better to tailor interventions. Furthermore, inclusion of EI within certification of financial advisors could enhance the quality of clients' interaction and outcomes on investments. Such directions will help us further grasp what EI means for finance and further about rational investing.

8. Conclusion

Finally, the study regression and mediation analyses showed how cognitive biases relate to investment decisions, since cognitive biases affect how individuals score in EI and vice versa. Evidence that cognitive biases such as, overconfidence bias, confirmation bias, and gambler's fallacy bias have significant impact on the way in which investments are chosen proved to be compelling. Their robust beta coefficients and significant p values corroborates with the prior works [36, 37] that these biases influence investment decisions. Emotional Intelligence comes out as a crucial mediator in this relationship: cognitive biases impact investment decisions through it. Although herding bias has a very strong direct effect, its effect is highly moderated by emotional intelligence suggesting that cognitive biases and behaviour decisions are intrinsically complex [38, 39]. This highlights the importance of emotional intelligence to minimize the negative effects of cognitive biases, providing clever tips for the financial practitioners, policymakers and educators. Finally, future research should study more of what influences this relationship and is further involved with the mechanisms associating emotional intelligence, cognitive biases, and investment decisions. Such efforts could contribute in refining the decision making process and develop a sounder, more rational and emotionally intelligent investing arena.

References

[1] Kafayat, A. (2014). Interrelationship of biases: effect investment decisions ultimately. *Theoretical & Applied Economics, 21*(6).

[2] Singh, A. (2021). A study of behavioral biases in investment decisions using artificial neural network. In *Rising Threats in Expert Applications and Solutions: Proceedings of FICR-TEAS 2020* (pp. 455–461). Springer Singapore.

[3] Blerina, Dervishaj. (2018). *Psychological Biases, Main Factors of Financial Behaviour - A Literature Review*. doi: 10.26417/EJNM.V1I2.P25-35

[4] Sharma, M., & Firoz, M. (2020). Do investors' exhibit cognitive biases: Evidence from Indian equity market. *International Journal of Financial Research, 11*(2), 26–39.

[5] Hayati, R., Suriyanti, L. H., & Irman, M. (2022). Bias Kognitif dalam Keputusan Investasi di Pekanbaru. *Jurnal Akuntansi Dan Ekonomika, 12*(1), 64–73.

[6] Sukanya, R., & Thimmarayappa, R. (2015). Impact of behavioural biases in portfolio investment decision making process. *International Journal of Commerce, Business and Management, 4*(4), 1278.

[7] Gargano, A., & Rossi, A G. (2018, April 23). *Does It Pay to Pay Attention?*. Oxford University Press, 31(12), 4595–4649. https://doi.org/10.1093/rfs/hhy050

[8] Fang, H. C., Siau, K. L., Memili, E., & Dou, J. (2019). Cognitive antecedents of family business bias in investment decisions: A commentary on "Risky decisions and the family firm bias: An experimental study based on prospect theory". *Entrepreneurship Theory and Practice, 43*(2), 409–416.

[9] Suresh, G. (2024). Impact of financial literacy and behavioural biases on investment decision-making. *FIIB Business Review, 13*(1), 72–86. doi: 10.1177/23197145211035481

[10] Chitra, D. K., & Jayashree, M. T. (2015). Does the influence of cognitive biases on investor behaviour differ with demographic profile? An empirical study. *International Journal of Research Publication, 5*(4), 161.

[11] Ady, S. U. (2018). The cognitive and psychological bias in investment decision-making behavior:(Evidence from Indonesian investor's behavior). *Journal of Economics and Behavioral Studies, 10*(1), 86–100. doi: 10.22610/JEBS.V10I1(J).2092

[12] Salovey, P., & Mayer, J. D. (1990). Emotional intelligence. *Imagination, Cognition and Personality, 9*(3), 185–211.

[13] Cherniss, C., Goleman, D., Emmerling, R., Cowan, K., & Adler, M. (1998). Bringing emotional intelligence to the workplace. New Brunswick, NJ: *Consortium for Research on Emotional Intelligence in Organizations*, Rutgers University, 1–34.

[14] Vaid, A. J., & Chaudhary, R. (2022). Review paper on impact of behavioral biases in financial decision-making. *World Journal of Advanced Research and Reviews, 16*(2), 989–997.

[15] Adaramola, A. O., Kayode, P. A., Adewale, O. V., & Ogiamien, O. F. (2022). Behavioural finance: an exploratory review. *Journal of Economic Research & Business Administration, 142*(4), 45–56.

[16] Sattar, M. A., Toseef, M., & Sattar, M. F. (2020). Behavioral finance biases in investment decision making. *International Journal of Accounting, Finance and Risk Management, 5*(2), 69.

[17] Obeng, G. (2019). Behavioural Antecedents Complementing Classical Financial Models for Rational Decision Making. *International Journal of Publication and Social Studies, 4*(2), 111–122.

[18] Asbaruna, L. W. B., Gorib, R. I., & Syifa, M. A. A. (2023). Behavioral Finance In *Investment Decisions. International Journal of Ethno-Sciences and Education Research, 3*(3), 95–98.

[19] Sapkota, M. P. (2022). Behavioural finance and stock investment decisions. *Saptagandaki Journal*, 70–84.

[20] Otuteye, E., & Siddiquee, M. (2015). Overcoming cognitive biases: A heuristic for making value investing decisions. *Journal of Behavioral Finance, 16*(2), 140–149.

[21] Khan, D. (2020). Cognitive driven biases, investment decision making: The moderating role of financial literacy. *Investment Decision Making: The Moderating Role of Financial Literacy* (January 5, 2020). https://doi.org/10.2139/ssrn.3514086

[22] Dhungana, B. R., Bhandari, S., Ojha, D., & Sharma, L. K. (2022). Effect of cognitive biases on investment decision making: A case of Pokhara Valley, Nepal. *Quest Journal of Management and Social Sciences, 4*(1), 71–84. doi: 10.3126/qjmss.v4i1.45868

[23] Barber, B. M., & Odean, T. (2001). Boys will be boys: Gender, overconfidence, and common stock investment. *The Quarterly Journal of Economics, 116*(1), 261–292. https://doi.org/10.1162/003355301556400

[24] Tversky, A., & Kahneman, D. (1973). Availability: A heuristic for judging frequency and probability. *Cognitive Psychology, 5*(2), 207–232.

[25] Bikhchandani, S., Hirshleifer, D., & Welch, I. (1992). A theory of fads, fashion, custom, and cultural change as informational cascades. *Journal of Political Economy, 100*(5), 992–1026.

[26] Tversky, A., & Kahneman, D. (1974). Judgment under uncertainty: Heuristics and biases. *Science, 185*(4157), 1124–1131.

[27] Situngkir, T. L., Nugraha, N., & Disman, D. (2021, June). Perilaku Bias Dalam Menentukan Keputusan Investasi. In PROSIDING SEMINAR NASIONAL EKONOMI DAN BISNIS (pp. 7–15).

[28] Cheng, C. X. (2019). Confirmation bias in investments. *International Journal of Economics and Finance, 11*(2), 50–55.

[29] Chen, G., Kim, K. A., Nofsinger, J. R., & Rui, O. M. (2007). Trading performance, disposition effect, overconfidence, representativeness bias, and experience of emerging market investors. *Journal of Behavioral Decision Making, 20*(4), 425–451. https://doi.org/10.1002/bdm.561.

[30] Siloam, Y. C., & Gunawan, L. (2023). The Influence of Emotional Intelligence Towards Investing Decision. *Jurnal Entrepreneur dan Entrepreneurship, 12*(1), 13–18.

[31] Campbell, J. Y. (2006). Household finance. *The Journal of Finance, 61*(4), 1553–1604.

[32] Lerner, J. S., & Keltner, D. (2000). Beyond valence: Toward a model of emotion-specific influences on judgement and choice. *Cognition & Emotion, 14*(4), 473–493.

[33] De Bondt, W. F., & Thaler, R. H. (1995). Financial decision-making in markets and firms: A behavioral perspective. *Handbooks in Operations Research and Management Science, 9*, 385–410.

[34] Kahneman, D., & Tversky, A. (1974). Judgment under Uncertainty: Heuristics and Biases. *Science, 185*(4157), 1124–1131. https://doi.org/10.1126/science.185.4157.1124

[35] Waweru, N. M., Munyoki, E., & Uliana, E. (2008). The effects of behavioral factors in investment decision-making: A survey of institutional investors operating at the Nairobi Stock Exchange. *International Journal of Business and Emerging Markets, 1*(1), 24–41. https://doi.org/10.1504/IJBEM.2008.019243

[36] Jain, J., Walia, N., & Gupta, S. (2019). Evaluation of behavioral biases affecting investment decision making of individual equity investors by fuzzy analytic hierarchy process. *Review of Behavioral Finance*, *12*(3), 297–314.

[37] Bihari, A., Dash, M., Kar, S. K., Muduli, K., Kumar, A., & Luthra, S. (2022). Exploring behavioural bias affecting investment decision-making: a network cluster based conceptual analysis for future research. *International Journal of Industrial Engineering and Operations Management*, *4*(1/2), 19–43.

[38] Baker, H. K., & Nofsinger, J. R. (2002). Psychological biases of investors. *Financial Services Review*, *11*(2), 97–116.

[39] Pompian, M. M. (2006). *Behavioral Finance and Wealth Management: How to Build Optimal Portfolios That Account for Investor Biases*. John Wiley & Sons.

34 A study on investment patterns followed by aided and self-financing college teachers in Ernakulam district

Thazna T. A.[1,a] *and Sarithamol K. K.*[2]

[1]Doctoral Research Scholar, Department of Commerce, Karpagam Academy of Higher Education, Coimbatore, Tamil Nadu, India
[2]Research Supervisor and Assistant Professor, Department of Commerce, Karpagam Academy of Higher Education, Coimbatore, Tamil Nadu, India

Abstract: There are different investment patterns in India, ranging from NSE, BSE, Post- office Savings Deposits, National Highway Authorities of India, chitties, stock, bond, and mutual funds certified deposits and New Pension Schemes. Sometimes investments are made to save tax; on the other hand, these are made to save money for the future. This paper focuses on the aided and self-financing college teachers' investment pattern in Ernakulam District – Kerala. This paper deals with different investment activities and examines how these respondents are preferring on the different investment schemes. Simple random sampling is used for collecting data, and 200 respondents were selected from different regions. Thus, primary data were used in this study.

Keywords: Investment schemes, stock, bond, mutual funds, certified deposits, new pension schemes

1. Introduction

Effective financial planning involves using cash or other monetary assets to manage a company or activity to generate positive returns, defined as benefits that surpass the initial inputs. It illustrates the allocation of assets, often capital (i.e., cash), aimed at producing rewards, advantages, or profits. Capital may be allocated to diverse schemes (either directly or indirectly), including starting a business or acquiring assets, such as buying real estate to produce rental income and then selling it at an appreciated value. Investing entails the procurement of assets or commodities with the objective of generating income and capital appreciation. Investments, including assets or acquired goods, are used to produce future wealth. These commodities often manifest as stocks or bonds, but they may also include real estate or other assets such as cryptocurrency or gold.

Investing capital is advantageous for several reasons. Individuals endeavour to amass wealth to sustain themselves throughout emergencies, unemployment, or future aspirations. They want to leverage compounding while considering inflation to prevent the devaluation of their money over time. Moreover, if an individual plans to cease employment and retire at a future date, prudent investments made in previous years will significantly safeguard their financial future.

The primary impetus for financial management is the anticipation of a favourable return manifested as remuneration or asset appreciation with empirical significance. The range of resources for giving and receiving returns is very extensive. Risk and return are inherently connected in astute financial planning; typically, larger returns are associated with more risk.

Individuals' interpretations of wealth differ. Wealth may denote a precise amount of money in a bank account or represent the financial goals an individual has for the future. Investment facilitates the attainment of these goals.

Diversified investing allows for the allocation of funds among many investments, therefore mitigating risk. Diversification mitigates losses from underperforming firms or asset classes by safeguarding against further investments.

Recently, investment has emerged as the cornerstone of a nation striving to develop its economy; thus, it is essential for the public to engage in prudent investment practices and to facilitate foreign investment. Investing transcends the mere allocation of present capital for returns; it inherently involves the duality of reward and risk, which varies based on the individual's perception of their savings in the now and future.

The research on the investment behaviours of Aided and Self-Financing college educators will ascertain if disparities exist in the investment options selected by these two groups, examine the correlation between the savings practices of Aided and Self-Financing sector educators, and evaluate the impact of their employment status on their savings and investment choices.

[a]thaznames@gmail.com

DOI: 10.1201/9781003684459-34

1.1. Types of investments

1.1.1. Stock

Investing in shares is a method of obtaining ownership stakes in a public entity. The small shares are designated as the agency's share capital, and by investing in this stock, investors anticipate that the organization will grow and operate effectively over time. In such instances, the shares may appreciate in value, prompting new investors to offer a higher price than what the current owners initially paid. This suggests that investors may realize gains if they choose to divest.

Investing in the stock market is a protracted endeavour. A prudent guideline is to diversify your investment portfolio and maintain engagement amid market fluctuations. A very effective approach for novice investors to acquire knowledge about stock investment is to establish an online investment account, which can thereafter be used for investing in individual stocks or mutual funds.

Due to several broker fees, investors may begin investing for the price of a single share. Certain organizations provide paper trading, enabling clients to practice buying and selling stocks using market simulators prior to execute actual monetary transactions.

Stocks, also referred to as equities or securities, are a basic and remarkable kind of investing. When investors acquire shares, they get an ownership stake in a publicly listed corporation. A diverse array of premier companies in the nation – such as General Motors, Apple, and Facebook – are publicly traded, allowing investors to own shares by them.

1.1.2. Bonds

Administration securities are the credit instruments of the central and state legislatures through which they get the cash to support their exercises and control the cash supply. The public authority regularly use protections as a mechanism to get subsidizes for foundation projects or to cover costs. For this reason, the government will issue bonds to the public, attractive to investors. At the designated recovery date, public authority pays the head and interest according to the expectations given in the bond. The public authority issues bonds overseen by the Reserve Bank of India (RBI).

The RBI or the government of India issues bonds to fill in as a solution for the services monetary imbalance. Recently, gigantic business contenders for examples, workplaces, and money-related foundations have been awarded safeguards. And recently, government bonds have become available to less well-heeled investors, such as small-time brokers and friendly banks. Financial backers that do not have to report openly are likewise extremely keen on purchasing government bonds.

In India government bonds are chiefly long haul venture instruments. The maturity of these bonds is 5 to 40 years. The images rely on the broad classification of government securities (G-secs), to get understand about government bonds. Both public and state legislatures may issue bonds. Bonds issued by state legislatures are sometimes called State Improvement Credits.

Categories of GOI Bonds: The Public authority of India (GOI) gives various classifications of bonds. These bonds also cater to the various needs of the investor. The coupon rate reflects the financing cost charged on government securities. The coupon is semi-annually distributed and can be steady or sailing. The Public authority of India powers securities of constant coupon instalments in the open market.

1.1.3. Mutual funds

A mutual fund is an investment instrument whereby a collective of investors pool their capital and choose a fund manager to invest on their behalf. The Securities and Exchange Commission (SEC) oversees the fund. In mutual funds, each participant receives a proportional return on their investment. Mutual funds may invest in shares, bonds, money market instruments, or other assets.

Investments in mutual funds may provide tax benefits contingent upon the financing and redemption approaches used. Mutual funds provide the potential for portfolio diversification across sectors, cost-effectiveness, and access to specialized expertise via fund managers. The risks associated with mutual funds include the absence of direct ownership of underlying assets, resulting in less control for traders over the fund's contents and components. Mutual funds are more expensive and riskier than index funds.

1.1.4. Exchange-traded funds (ETF)

Exchange Traded Funds, unlike conventional mutual funds, are traded on stock exchanges similar to regular equities. A licensed broker of an established stock exchange often transacts in ETF units. ETFs are traded on stock exchanges, and their Net Asset value (NAV) fluctuates with market conditions. ETF units are exclusively listed on the stock market, precluding their acquisition and sale in the manner of conventional open-end equity funds. Investors may acquire an unlimited number of units via the exchange.

Within ETFs track indices such as the CNX Nifty or the BSE Sensex. When investors acquire shares or units of an ETF, they are obtaining shares or units of a portfolio that mirrors the yield and return of its underlying index. ETFs distinguish themselves from other index funds by replicating the performance of the corresponding index instead of striving to surpass it. They aim to become the market rather than to conquer it. Financial experts may choose ETFs over mutual funds because of their superior daily liquidity and lower costs.

1.1.5. Certificate of deposit

Certificates of Deposit (CDs) are savings instruments that retain funds for a certain duration, such as six months, one year, or five years, and provide interest in compensation.

Upon redeeming their certificates of deposit, investors get their principal investment along with accrued interest. Investments in certificates of deposit are regarded as safe. But CDs have both advantages and drawbacks.

1.1.6. Retirement plans

Retirement arranging implies getting ready today to one's future lifestyles so that it keeps on gathering each without a doubt taken into consideration certainly one will achieve the goals and dreams freely and riskless. This consists of putting forth the retirement desires, assessing how much quality of cash dealing one will need, and contributing to increase the retirement reserve finances. Every retirement plan is thus remarkable. Presently, so many retirement plans are available for the employees.

1.1.7. New pension scheme

The New Pension Scheme is a governmental initiative aimed at reducing pension obligations while enabling the public to choose their investment options. It is a dual-level, contribution-based investing framework that gives people total authority over the allocation of their funds. As previously mentioned, it consists of a two-tier structure with levels I and II. The functionality of the whole pension system is rather straightforward. Tier I of the pension system forbids early withdrawal, but Tier II permits withdrawals before maturity. The minimum monthly payout for each individual is Rs. 500, equating to Rs. 6000 annually. The New Pension Scheme offers several advantages, such as cost efficiency, accessibility, investment opportunities, tax implications, and beneficiary protection.

2. Literature Reviews

Mahalakshmi Kumar and Rajesh Mankani [1] focus on how working women's understanding varies over various investment options, particularly in Mumbai. An intentional survey was used to collect urgent information, generating a deliberate cross-sectional format. Women are tutored on the importance of working, saving, and systematic saving. It caters towards their ability to analyse elective ventures, including upsides and drawbacks, empowering them to make educated speculation choices to achieve their monetary targets. It encourages people to achieve financial independence, which frees them to concentrate on other aspects of their life. Mentoring assists women with understanding the significance and importance of making monetary decisions in the quest for their monetary goals. This propels them to get information on different pioneering alternatives to build income while lessening risk.

Mak and Ip [2] stated in their work, which titled as 'An exploratory investigation of venture conduct of financial backers,' that their request intends to fill the recognized hole by utilizing direct relapse models to assess the monetary venture conduct of central area of Chinese and Hong Kong financial backers. Relapse currently focuses to show that (i) there are blaring contrasts in monetary speculation conduct between the central area of Chinese and Hong Kong financial backers, and (ii) mental, humanistic, and division qualities are major areas of strength for their venture conduct/inclinations. After that, suppliers with monetary advantage may predict similar ways of behaving and inclinations for their clients regarding handling cash, which allows them to make advertising and vital decisions like evolving the customer's portfolio of monetary ventures dependent on the models of relapse produced.

Monetary Education, Monetary Behaviour, Pay, and Individual Venture Choice Baiq Fitri Arianti [4] explored and tested the impact of monetary education, monetary way of behaving, and pay on individual venture choices. The data were collected using surveys, with a test rank of 100. The methods used include descriptive statistical analysis, data quality assessment, classical hypothesis testing, multiple linear regression analysis, F-test, t-test and coefficient of determination assessment. Monetary education, however, meaningfully influences speculation choices, even though pay is a major determinant of choices.

Research titled 'A Study of Mindfulness and Awareness Regarding Tax Saving Options Among Salaried Individuals' was conducted, which aimed at assessing the awareness and understanding of salaried employees regarding different tax saving options. The inspection was undertaken up to October 2023, to verify if there are any contrasts in compensation, hold saving, together and contributing systems in light of orientation, and thought processes engaged with choosing a specific venture. The survey additionally planned to discover the most popular expense-saving money management methodologies. A correlational study was executed to investigate the relationship between reserve funds, speculations, and remuneration, and to assess whether individuals differed in their profit, reserve funds, and speculations. According to the report, LIC is the most popular savings scheme available with the salaried class.

According to Kumari Asmita, S. Prakash and K. [3] in their article titled 'Factors Effecting Saving and Investment Behaviour of Teachers,' saving and investment is the main factor of any country's financial development and the primary targets of even state strategy is to boost the saving and capital development for financial development. As indicated by the accessible information, almost zero research is available on getting and funding propensities for auxiliary and school instructors of Samastipur locale of Bihar. The major emphasis in this study is not on the saving and putting cycles as a general rule, yet rather on understanding saving and speculation conduct, saving and speculation channels, critical saving and speculation driving forces, determinants filling in as a danger alleviation and the discouragements confronting the instructors. These findings reveal that the respondents' saving and venture propensities differ contingent upon orientation, age, work type, pay class and a few different components.

The survey was carried out through personal interviews with salaried people using a standardized questionnaire [4] 'A Concentrate on Speculation Inclinations among Representatives Regarding Kurnool City.' Using a self-appraisal test, review handles the members who preferred financial security markets. The audit hangs on huge information sources acquired through the scattering of a shut finished investigation. Data were analysed using frequencies, chi-square, and descriptive statistics. According to the report, salaried staff continue to prioritize wellbeing, strong investment returns, and ample liquidity. As indicated by the overview, the respondents are aware of the endeavour decisions accessible in India.

Diya Usman [5] undertook a study titled 'A Study on Tax Planning Strategies Adopted by the Salaried Class in Kottayam District, Kerala' to analyze the strategies followed by the salaried class, identify their limitations and offer solutions to these issues. The majority of respondents were male and employed in the public sector. As suggested by a review from Varun Sagar Singal [6], a large number of individuals are intrigued by learning about common assets and tax saving methods. The paper also looked at some of the other factors that affect investment decisions, the influence of social factors on investors, and barriers to investment.

As per their review named 'Bewilderment regarding the Different Venture Choices Accessible for Expense Arranging,' duty arranging is not just an apparatus to minimize expense responsibility but also a comprehensive technique to augment benefits from all derivations, exclusions, and reliefs approved under individual assessment laws [7]. This exposition intends to investigate levels of mindfulness among utilized women. The review looks at the different salary charge planning procedures embraced by Indian citizens, alongside their speculation propensities and inclinations.

In the survey article 'Individual Venture Arranging: An Audit of Writing,' Siddiqui, Imran, and Pratiksha [8] break down the connection between individual speculation arranging and the variables that lead to an absence of venture. The review examined the past writing on individual speculation conduct. To write down the composing audit, search for records on individual financial backers' plans aggregated from a couple of cited periodicals. Research shows that sharing information affects a person's journey of independent decision-making. Pay preparation, age, risk, and return are fabulously critical components to remember while plotting the speculations. The others' findings demonstrate that non-others are interested in low-risk, direct-reward venture participation.

3. Research Gap

Out of the articles referred, nowhere was a significant and comparative study found on the savings and investment pattern followed by the aided and self-financing college teachers from Ernakulam, a District from Kerala. So, it is found relevant to make research on the investment avenues mostly opted by the particular category of the population. The result of the study will be immensely helpful to the college teachers to make them aware of the importance of savings and investment.

4. Need and Significance of the Study

Today, in the current context, employees in the country are entering a new era in which they are not only becoming financially independent throughout their working years, but also expect a consistent return after retirement. As a result, to make an informed selection, individuals must be aware of all aspects of the numerous investment opportunities available to them. Investment decisions are also a vital aspect of life, and every worker should be aware of the many strategies for increasing savings and ways to invest in the correct avenues for a guaranteed return in the future, which will help the society to achieve economic progress. This study will aid in determining the level of awareness among aided and self-financing college professors regarding the various investing platforms available to them.

5. Statement of the Problem

Currently, it is common for most employees to spend more than their income. This issue mostly arises from inadequate management of monetary disbursements. This may stem from workers' lack of awareness of the significance of saving, ignorance of the potential consequences of insufficient reserves for future eventualities, or unawareness of available investment possibilities. In this context, it is essential to examine the investment behaviours of employees; especially, the present study concentrates on the investment patterns of both aided and self-financing college teachers. The objective of this paper is to ascertain the investment plan preferences of both aided and self-financing college teachers in the Ernakulam region.

6. Objectives of the Study

1. Identify respondents' challenges in recognizing the need for tax management.
2. To find out the difference between the pattern of savings and the nature of employment
3. To find the opinion of employees about the new pension scheme and their nature of employment.
4. The study examines the relationship between college professors' investment and savings attitudes.

7. Research Methodology

The current study aims to investigate the investment patterns of aided and self-financing college professors. To collect data, 200 experts are chosen as samples using basic random selection methods. The samples were collected from the Ernakulam District of Kerala. The data for the study were collected both from primary and secondary sources. Primary data were acquired via standardized questionnaires issued to the respondents to evaluate their investment habits and practices.

For analysing the questionnaire, statistical tools like Statistical percentages, ANOVA, Chi-square tests, Correlation analysis, etc., were used.

8. Period of Study

The period of study is from June 2023 to August 2023, which covers three months of study.

9. Limitations

The current study's examination was conducted using data acquired from selected salaried staff of aided and self-financing colleges. The following difficulties were experienced:

- The study is an opinion overview; caution may be exercised when expanding the result to other ranges.
- Due to time constraints, only 200 respondents (100 teachers each from aided and Self-Financing colleges) were considered.
- The result completely depends on the data given by the respondents, which may change from place to place and may contain individual bias.

10. Results and Discussions

For solving the stated objectives, a questionnaire was prepared and distributed among 200 college teachers from Aided and Self-Financing sectors. Samples were selected using a simple random sampling technique, and the following results were obtained using different statistical tools (Table 34.1).

Table 34.1. Data respondents in male and female count

	Male		*Female*		*Total*	
	Count	*%*	*Count*	*%*	*Count*	*%*
No. of respondents	60	30	140	70	200	100
Age classification:						
Below 35	5	8.33	15	10.71	20	10
35–45	30	50	62	44.29	92	46
45–55	15	25	35	25	50	25
Above 35	10	16.67	28	20	38	19
Year of experience:						
Below 10 years	14	23.33	16	11.43	30	15
10–20 years	40	66.67	80	57.14	120	60
Above 20 years	6	10	44	31.43	50	25
Annual income:						
Below 2,50,000	20	33.33	50	35.71	70	35
2,50,000–7,50,000	25	41.67	65	46.43	90	45
Above 7,50,000	15	25	25	17.86	40	20
Working Sector:						
Aided	35	58.33	65	46.43	100	50
Self-Financing	25	41.67	75	53.57	100	50

Source: Primary data.

10.1. Demographic data of respondents in a nutshell

Demographic data of the 200 respondents (100 Aided college teachers and 100 Self-Financing college teachers) were collected, presented, and analysed as Table 34.1.

As per the data related to relevant demographic factors of the respondents, it is observed that among the 200 employees, 46% belong to the age level of 35-45 years. When considering experience level, 60% of the respondents had a job experience level between 10 to 20 years. When comparing annual income, it is clear that 45% are between 2,50,000 and 7,50,000, and 65% of the employees earning taxable income. Regarding the nature of employment, equal proportion of the respondents are from both Aided and Self-Financing colleges.

As per the results obtained from the Table 34.2, it is found that 75% of the total respondents have monthly savings less than 25% of their income. Only 24% have savings above 25%, that too, the majority from the aided sector. Among 200, only 2 from the Aided sector are keeping above 50% of their earnings as savings.

Among the respondents, only 27.5% are following tax planning either at the beginning or end of the year, the remaining portions are not doing anything for tax planning.

From the total respondents, 61% have an emergency fund. Among that 59.02% are from the aided sector. 39% of the respondents are not maintaining any reserve fund to meet emergencies.

According to the result, 120 respondents are in the group of regular investors, of which 62.5% belong to the aided sector. Among the non-investors, 68.75% lies under the self-financing sector (Table 34.3).

Table 34.4 uncovers that 33% of respondents have absence of relevant tax knowledge, 38% of them need training and informative sessions, 14% say consistent and complex tax structure is the hindrance before them and roughly 6% express in light of a few different reasons they felt issue in figuring out the significance of tax management. So, the table

Table 34.2. Frequency analysis for relevant statements

	Aided College		*Self-Financing College*		*Grand Total (GT)*	
	Frequency	*% to GT*	*Frequency*	*% to GT*	*Frequency*	*%*
Monthly Savings of the respondents						
Less than 25%	65	43.33	85	56.67	150	75
Between 25% and 50%	33	68.75	15	31.25	48	24
Above 50%	2	100	0	0	2	1
Tax planning of the respondents						
Beginning/End of year	40	72.73	15	27.27	55	27.5
No Tax planning	60	41.38	85	58.62	145	72.5
Maintaining Emergency Fund						
Interested	72	59.02	50	40.98	122	61
Not interested	28	35.9	50	64.1	78	39
Regular Investment						
Invested	75	62.5	45	37.5	120	60
Not invested	25	31.25	55	68.75	80	40

Source: Primary data.

Table 34.3. Problems faced in understanding the importance of tax management

Response	*No. of Respondents*	*Percentage*
Lack of subjective knowledge	66	32.86
Lack of training and informative sessions	74	37.14
Continuous and complex changes in tax structure	29	14.29
Lack of time to spend	20	10.00
Others	11	5.71
Total	200	100

Source: Primary data.

discloses that lack of subjective tax knowledge, unavailability of awareness sessions and rapid and complex changes in tax structures are the major problems faced by the respondents in understanding the importance of tax management.

10.2. Analysis of Pattern of tax-saving investments using T-test

Investments are made in money-related possessions, notwithstanding substantial assets. The exact capacity of money-related resources is that you can persistently shop in little pieces. The review becomes principally founded on the ensuing money-related assets: 'monetary foundation stores', 'chitties and kuries', 'capital market securities' and 'others'.

Table 34.4. T-test table – to find out the difference between the pattern of savings and the nature of employment

Variable	*Group*	*N*	*Mean*	*Std. Deviation*	*Std. Error Mean*	*t*	*df*	*Sig. (2-tailed)*
Savings Pattern	Self-Financing College	100	0.32	0.20	0.02	18.74	198	0.000
	Aided College	100	0.16	0.13	0.01			
Bonds	Self-Financing College	100	0.08	0.28	0.03	2.93	198	0.004
	Aided College	100	1.84	0.61	0.06			
Mutual Funds	Self-Financing College	100	0.08	0.28	0.03	4.97	198	0.000
	Aided College	100	0.64	0.42	0.04			
Certified Deposits	Self-Financing College	100	0.00	0.00	0.00	0.00	198	1.000
	Aided College	100	0.48	0.70	0.07			
Retirement Plans	Self-Financing College	100	0.18	0.42	0.04	3.90	198	0.000
	Aided College	100	0.48	0.70	0.07			

Source: Primary data.

The 'others' class in money-related effects covered government monetary investment funds.

The T-test analysis presented in the table provides a comprehensive comparison of savings and investment patterns between teachers in the self-financing and aided colleges. This analysis is crucial in understanding how teachers from these two distinct sectors manage their finances and prioritize different savings instruments. Starting with the general savings patterns, the results indicate a significant difference between self-financing and aided college teachers. Self-financing teachers exhibit a higher mean score (0.32) when compared to aided college teachers (0.16). The higher standard deviation in the self-financing sector (0.20) suggests that there is greater variability in how these employees save. The t-test result, with a t-value of 18.74 and a p-value of 0.000, strongly suggests that self-financing teachers are more engaged in saving behaviours compared to their aided college teachers. This difference could be attributed to the differing financial priorities or security levels inherent in each sector. When it comes to investing in bonds, aided college teachers show a significantly higher mean score (1.84) compared to self-financing college teachers (0.08). The variability in bond investments is also higher among aided college teachers, as indicated by the standard deviation 0.61. The t-test result, with a t-value of 2.93 and a p-value of 0.004, confirms a statistically significant difference, indicating that aided college teachers are more inclined towards bond investments. This might be due to the perceived stability and lower risk associated with bonds, which aligns with the typically risk-averse nature of aided sector workers. The investment patterns in mutual funds also reveal a notable difference between the two sectors. Aided college teachers have a higher mean score (0.64) when compared to self-financing college teachers (0.08), suggesting a greater inclination towards mutual funds among the former. The t-test yields a significant t-value of 4.97 with a p-value of 0.000, indicating that this difference is statistically significant. Aided college teachers may favour mutual funds due to their balanced risk-return profile, which offers a mix of security and potential growth. In contrast, when looking at certified deposits, both self-financing and aided college teachers exhibit similar behaviour, as reflected in the identical mean score of 0.00 for the self-financing and a slightly higher score of 0.48 for the aided college teachers. The t-test result, with a t-value of 0.00 and a p-value of 1.000, indicates no significant difference between the two groups. This suggests that certified deposits are equally attractive or unattractive to employees in both sectors, possibly due to their perceived safety but relatively lower returns. The results also revealed that aided college teachers are focused on retirement plans than self-financing teachers (Table 34.5).

10.3. Acceptance of the new pension scheme and the nature of employment

The Chi-Square test results explore the relationship between the acceptance of the new pension scheme and the nature of employment (Table 34.6). The analysis was conducted on two distinct groups: those who accepted the new pension scheme ('yes') and those who did not ('no'), with a total of 200 valid cases (100 per group). For this analysis,

Table 34.5. Chi-square tests

Acceptance of New_Pension_Scheme		*Value*	*Df*	*Asymp. Sig. (2-sided)*	*Exact Sig. (2-sided)*	*Exact Sig. (1-sided)*
Yes	Pearson Chi-Square	6.250[b]	1	.012		
	Continuity Correction[c]	4.785	1	.029		
	Likelihood Ratio	10.107	1	.001		
	Fisher's Exact Test				.011	.007
	Linear-by-Linear Association	6.188	1	.013		
	N of Valid Cases	100				
No	Pearson Chi-Square	37.500[a]	1	.000		
	Continuity Correction[c]	34.440	1	.000		
	Likelihood Ratio	44.629	1	.000		
	Fisher's Exact Test				.000	.000
	Linear-by-Linear Association	37.125	1	.000		
	N of Valid Cases	100				
Total	Pearson Chi-Square	33.333[a]	2	.000		
	Likelihood Ratio	47.398	2	.000		
	Linear-by-Linear Association	.000	1	1.000		
	N of Valid Cases	200				

Source: Primary data.

Table 34.6. Cross-tabulation

Nature of Employment	*Accepted Pension Scheme (Yes)*	*Accepted Pension Scheme (No)*	*Total*
Aided College	80	20	100
Self-Financing	60	40	100
Total	140	60	200

Source: Primary data.

the Chi-Square test is appropriate, and it is observed that the acceptance of the pension scheme significantly differs based on the nature of employment, as indicated by the significance values being below the conventional acceptance limit ($p < 0.05$). For the group that accepted the pension scheme, the Pearson Chi-Square value is 6.250 with 1 degree of freedom and a significance level of 0.012, indicating a statistically significant relationship between the acceptance of the pension scheme and the nature of employment. The Continuity Correction further refines this result with a significance level of 0.029. The Likelihood Ratio test ($p = 0.001$) and Fisher's Exact Test ($p = 0.011$) confirm this relationship. Similarly, the Linear-by-Linear Association ($p = 0.013$) supports this conclusion. For the group that did not accept the pension scheme, the Pearson Chi-Square test yielded a highly significant result with a value of 37.500 ($p < 0.001$), and the Continuity Correction also showed significance ($p < 0.001$). Both the Likelihood Ratio test ($p < 0.001$) and Fisher's Exact Test ($p < 0.001$) strongly support this association. The Linear-by-Linear Association further confirms this significant relationship ($p < 0.001$). When combining both groups, the total Pearson Chi-Square value is 33.333 with 2 degrees of freedom and a p-value of less than 0.001, indicating a significant overall relationship between the nature of employment and pension scheme acceptance (Table 34.7). The Likelihood Ratio test ($p < 0.001$) also confirms this, although the Linear-by-Linear Association test shows no significant relationship ($p = 1.000$).

In conclusion, the Chi-Square test reveals that there is a significant difference in the acceptance of the new pension scheme based on the nature of employment. Specifically, aided college teachers are observed to be more accepting of the new pension scheme than self–finance college teachers, as indicated by the statistically significant results across multiple tests. This finding highlights the influence of employment sector on attitudes toward pension scheme acceptance.

10.4. Investment and savings attitude of respondents

The above reveals the correlation between investment decisions and saving attitude of the selected employees. After analysing the correlation, it is found that the employees investing in the chosen investment types seem to have a good savings attitude; since the significant value is lesser and there is an excellent positive correlation value, it is confirmed that investment and savings attitude are positively correlated.

Table 34.7. Correlations between investment and saving attitude

		Investment	*Saving attitude*
Investment	Pearson Correlation	1	.681**
	Sig. (2-tailed)		.000
	N	200	200
Saving attitude	Pearson Correlation	.681**	1
	Sig. (2-tailed)	.000	
	N	200	200

Source: Primary data.

11. Findings of the Study

- Around 46% of the respondents have a place with the age classification of 35 to 45 years. The 25% have a place in the 45 to 55 age class, and roughly 19% are over 55 years.
- A greater part of the respondents are female college teachers.
- The largest part of the respondents has experience between 10 and 20 years in their service.
- A greater part of the respondents fall in the yearly pay class of Rs.2,50,000 to Rs. 750,000. In this way, the greater part of them belongs to the taxpayer's category.
- 84% of respondents reported difficulties in recognizing the significance of tax management due to a lack of subjective knowledge, insufficient training and information, and ongoing changes in tax structure.
- Out of the total respondents, 75% are following a saving pattern of less than 25% of their earnings, among those 56% belong to self-financing sector. 24% of the respondents maintain savings above 25% of their earnings, among nearly 70% of them work in the aided sector. So, when comparing the saving habit, aided sector employees show steady and regular savings from their income.
- Results revealed that the tax planning concept is not very familiar among employees. Out of the samples selected, nearly 28% are giving importance to tax planning either at the beginning or end of the year. Responds from the remaining samples, depicting that they are not at all bothered about planning their tax payment.
- 122 teachers agreed that they are maintaining an emergency fund for meeting uncertainties, from them too, a major portion belongs to aided colleges. Rest of the 39% of the sample size are doing nothing for meeting their future emergencies.

- When 60% of the total teachers regularly invest their savings in secured investment avenues, the rest of the 40% are not interested in making any investment. Aided sector employees show more interest in regular investment of their part of savings.
- Both in aided and self-financing sector college teachers, investments in mutual funds and retirement plans are mostly opted by the respondents as their safe and secured investment avenues.
- Certificate of Deposit and investment in Bonds are treated more or less the same by both aided and self-financing sector teachers.
- It is also disclosed that the aided sector college teachers are more accepting of the New Pension Scheme than the self-financing teachers.
- The results obtained from the study confirmed that investment and savings attitude of the employees are positively correlated. Which revealed that a saving habit improves the investment possibilities among the employees.

12. Suggestions and Recommendations

The following suggestive measures can be recommended to improve the saving and investment attitude of employees.

- Self-financing college teachers should give a little more importance to inculcating a saving habit.
- Self-financing colleges should arrange regular awareness programs for the employees to make them knowledgeable regarding saving and spending habits.
- In aided sector organizations too, employees should be provided with informative classes on the need for building a saving habit to safeguard the future.
- Organizations should arrange various investment avenues before the interested employees and make them aware of different schemes.
- Instead of spending the whole amount of salary within a month, self-financing teachers should also find risk-free investment schemes which will match with their salary slab.
- Providers of different investment schemes should arrange the schemes in such a way as to match with the income level of employees from different sectors.
- Employers should take initiative to make arrangements in the bank accounts (salary account) of the employees for automatic and regular spendings for retirement plans and investment with guaranteed return.
- All the teachers should arrange their savings and investments in such a way as to secure the health, education, and the marriage of the children, and the medical needs of parents.

13. Conclusion

Tax management is essential as it indicates a reduction in residents' tax rates and an increase in investment options. The survey indicates that most assesses are oblivious to the advantages of paying personal taxes; the majority have opted to evade taxation instead of managing their taxable income. Tax avoidance is more common than tax management; thus, it is essential to enhance evaluations by providing sufficient tax information to effectively lower tax rates. The revenue agency should conduct suitable awareness activities to educate aided and self-financing college instructors about tax matters. Tax preparation is a crucial component of financial planning; nevertheless, it should not predominantly influence investment decisions. After establishing a financial plan, it will be easier to save funds to capitalize on tax benefits.

Upon analysis and interpretation of the data, it is determined that investors lack awareness of available investment alternatives; yet, they exhibit a preference for allocating their savings into bank deposits, real estate, equities, bonds, mutual funds, certificates of deposit, retirement plans, and the New Pension Scheme.

The study's data analysis indicates that safety is a crucial factor in investing decisions, rendering other options less relevant to investors. This is a reasonable investment option, particularly for college educators, whose mutual funds and retirement plans consistently perform at or near their maximum potential. Nonetheless, there is no certain return, and the risks linked to mutual fund investments are considerable; also, the benefits of retirement plans are accessible only during post-retirement. It is essential to maintain earnings, formulate a plan, and refrain from expenditures above one's means. Consequently, if compensated educators, such as college professors, are equipped with enough tax and investment knowledge, they will be capable of lawfully preserving a substantial portion of their earned income.

References

[1] Mahalakshmi Kumar, & Rajesh Mankani. (2017). A Study of Level of Awareness Regarding Investment Avenues among Educated Working Women with Special Reference to Mumbai City. *International Journal of Research in Economics and Social Sciences (IJRESS)*, *7*(9), 271–277.

[2] Mak, K. Y. M., & Ip, W. H. (2017). An exploratory study of investment behavior of investors. *International Journal of Engineering Business Management*, *9*, 1847979017711520.

[3] Kumari Asmita, S., & Prakash, K. (2018). Factors Affecting Saving and Investment Behavior of the Teachers-Economics. *International Journal of Current Microbiology and Applied Sciences*.

[4] Arianti, B. (2018). The influence of financial literacy, financial behavior, and income on investment decisions. *Economics and Accounting Journal*, *1*(1), doi:http://dx.doi.org./10.32493/eaj.vlil.y2018.pl-10.

[5] Diya Usman. (2019). A Study on Tax Planning Measures Adopted by Salaried Class Based on Kottayam District in Kerala. *Think India Journal*, *22*(10).

[6] Singal, V. (2019). Factors affecting investment in mutual funds. *Journal of General Management Research*, *5*(1), 210–224.

[7] Pankhuri Agarwal, Vipin Jain, & Surabhi Goel (2020). Awareness of the Various Investment Options Available for Tax Planning. *Wesleyan Journal of Research*, *13*(41).

[8] Mishra, P., & Siddiqui, I. N. (2020). Individual Investment Planning: A Review of Literature. *International Journal of Management, Technology and Engineering*, *10*, 237–241.

35 Design considerations for developing a context-aware recommender system for cybersecurity

Jhon Zylvin M. Ramos[a] *and Arvin R. De La Cruz*[b]

Graduate School, Polytechnic University of the Philippines, Philippines

Abstract: The complexity and sophistication of cyber-attacks on the one hand and, on the other hand, to the rise in the need for efficient and effective cybersecurity solutions. Additionally, recommender systems (RS) based on Artificial (AI) intelligence have demonstrated great promise in improving the cyber defender's capability to detect and, more importantly, to counter the cyber-attack. This paper proposes layered holistic approach for building an AI based RS for cybersecurity in the light of major research challenges and opportunities in the process. The design elements such as user interface design, data privacy and security, scaling, and integration and orchestration and contextual awareness are among the recommended. With consideration of these design considerations, the proposed RS can provide a more efficient and effective method in the real time detection, prevention or response to cyber threats, in turn increase the effectiveness and efficiency of incident response as well as security compliance to incorporate a stronger resilience and robustness of security posture.

Keywords: Context-aware recommender system, network security, security recommender systems, cybersecurity

1. Introduction

However, as cyberattacks become more frequent and intricate, organizations need to employ advanced cybersecurity solutions to protect their assets, whereby, such as data, networks, and applications [1, 2] recommend systems offer an alternative cyber solution to analyze huge volumes of network data and give each company a customized suggestion [3]. Machine learning and artificial intelligence algorithms can ease the decision making by applying these systems [4]. Then RS can be used as a useful tool that can be implemented in the real time threat predictions, vulnerability identification, and suggestions for prevention and correction [5].

Research has also been conducted earlier trying to implement recommender systems within cybersecurity arena but there is some gap in structured framework that covers the specific requirement to implement AI within cybersecurity arena so that it can be executed in real time to identify, avoid, and counteract cyber attacks [6].

The development of Artificial Intelligence (AI) has created new cybersecurity concerns and possibilities [7–11]. AI systems can enhance identity management and access control, give predictive security analytics, execute autonomous and semi-autonomous planning, coordinate and orchestrate numerous cybersecurity tools, and increase human-machine teaming to reduce negative repercussions [12]. However, including an RS into AI systems for cybersecurity has issues that must be addressed [13].

This paper aims to investigate the existing RS implementations applied in cybersecurity, identify the design considerations, and propose a framework for an RS that can address the unique requirements of AI systems in cybersecurity [14].

2. Background

In order to provide assistance to users in making decisions and find out what is most suitable for them, without spending time and effort to look for it [15], recommender systems have been widely used in nowadays in many fields such as e-commerce, social media and entertainment. In 2021, Das and Sandhane [7] conducted a systematic study to examine some of the applications of recommender systems in cybersecurity and they concluded that indeed, the systems have been applied to cybersecurity.

Three main areas of applications for existing RS implementations in cybersecurity can be identified.

2.1. *Vulnerability and anomaly detection*

RS can evaluate data for vulnerabilities in terms of flaws using the data stored in Knowledge Base and the data sourced from multiple sources such as vulnerability database (e.g., Common Weakness Enumeration (CWE) and Common Vulnerabilities and Exposures (CVE)), and hence the vulnerabilities in hardware, software applications, networks, and systems can be detected using RS. (CVE). Subsequently, the system is able to suggest remediation actions (patch, upgrade, replace) for the exploited components to reduce the risk.

[a]ramosjhonzylvin@gmail.com, [b]docardelacruz@gmail.com

DOI: 10.1201/9781003684459-35

2.2. *Threat prevention through prediction and analysis*

RS can analyze huge amounts of data from a variety of sources, including network traffic, logs, and system events, to detect abnormalities and potential threats. The system can use machine learning techniques to predict and assess the possible effect of these threats, as well as suggest relevant mitigation strategies.

2.3. *Incident response and security compliance*

RS can be implemented to improve incident response to various cyber security threats and cybersecurity compliance by providing proactive measures to prevent cyber-attacks or remedial actions to mitigate their impact. These systems can suggest security tools, best practices, standard and improved configurations or policies to protect against possible threats, and response actions such as quarantine, isolation, etc. in the face of a threat detected. The response will give both cyber defenders and analy1sis more ways, and more insights, to know their operating environment and to better understand situational awareness as well as three ways in which to evaluate the security and the robustness of potential defense mechanisms. Using machine learning algorithms and data analysis, recommender systems can improve the incident response and security compliance efforts thereby improving the overall security posture.

3. Related Works

Several research have been carried out on RS for cybersecurity usage, yet mostly within respect to the individual and combined applications outlined above. As an example [16], suggested an RS- based vulnerability detection. Specifically, the vulnerability information is extracted from the publicly available data of software vendors and CVE Naming Authorities using a pipeline with techniques including natural language processing, fuzzy matching, machine learning and others. The RS system then automatically matches the vulnerabilities to the software packages in an organization's inventory using machine learning algorithms. The main purpose of the RS is to detect the most important product software vulnerabilities for an organization, the reliable and versions of software products to which it belongs, as well as provide minimization of software vulnerabilities.

A more advanced implementation of RS in Cybersecurity calls for Lyons [17] who recommends the use of a recommender system as a tool to predict an attack and cyber defense conjointly with an Intrusion Detection System (IDS). The recommender system is fed with the input of IDS alerts to predict the likelihood of events for some node in the network. The system is capable of producing a list of actions ranked according to effectiveness in reducing the upcoming predicted events. The action recommended to the cybersecurity defenders for implementation is the top ranked.

Polatidis et al. [18], in 2017 also proposed a recommendation system for future attack steps in a network with a multi-level collaborative filtering method which utilize attack graphs to predict future attack steps in a network infrastructures of maritime supply chain. In the implementation in this paper: Business Partners need to have expertise in information technology (IT) and security and have knowledge of what the infrastructure is required to configure the RS with to input direct conditions such as the attacker's location, capabilities, and the entry and target points. The inputs comprise of the open source datasets such as Common Weakness Enumeration (CWE) and Common Vulnerabilities and Exposures (CVE). Subsequently, the generated attack paths are used as an input in the attack prediction algorithm which can predict attack paths, vulnerable assets, and a chain of sequential vulnerabilities as recommendations. The design of RS provided the security analysts with the ability to proactively detect a cyber attack path and to give the analysis and prediction of a potential cyber attack path.

In the study of Ayala et al. [19], in 2021 involved the development of a recommender system designed to aid security analysts in identifying the most appropriate recommendation from a list of expert recommendations using ratings relevant to the inputted anomaly. It uses collaborative and knowledge based techniques to provide ratings for the worst cases as well as the best recommendations that are available. A graphical interface was also created for the security analyst to use to facilitate interaction and recommendation viewing.

A similar study was carried out to build a cloud security recommender system that suggests suitable security standards according to the given security threat [28]. Cloud consumers can use the system to identify and enforce security and compliance policies within their organization as the system supplies the information on cloud threats, security and compliance models.

In the research by [20] a recommendation system assisted by cloud based was created intended for multimedia data filtering and correlation for Cyber Situational Awareness. In addition, cyber threat understanding and risk assessment improved significantly, and analysts' productivity enhanced. The methodology entailed transition from analysts' preferences, contextual information, and multimedia data type, and performed experiments using different data sources to evaluate the system. In all, the research provided a concise and effective framework to Cyber Situational Awareness domain.

Abuhussein et al. [21], proposed a security recommender for Cloud Computing (CC) environments (CCRS) in 2016, as a recommender of defense actions against possible attack vectors according to the deployment and service models of the CC. The Attack Surface Taxonomy is predefined and the recommender system expects as input CC model consumption scenarios from the client for reusing the Attack Surface Taxonomy to predict classes of vulnerabilities arising from the CC model. Separating the categories makes it analysable, comprehensible, expandable and structured recommendation algorithm. A visualization tool included in a web portal is also made available to display a complete list of security and privacy issues which need to be taken care of as controls to reduce the CC attack surface.

Another study proposed to develop a security investigation recommender system based on a knowledge based approach that takes hash of the user's query and domain knowledge as inputs [22]. One of the datasets the project uses is a dataset provided by DARPA however this dataset consists of descriptive data about a number of Advanced Persistent Threat (APT) attacks. The machine learning algorithm used in this system serves as a visual analysis tool to propose an exploration path for cybersecurity investigators. From the logs of the investigator, the system would provide contextual information and dynamically present potentially useful parts of the dataset to investigate. The system's output is a list of recommended exploration paths that contain, for each path, relevant parts of the dataset for that path. The system also supplies visualizations to help security analysts understand and explain it what happened in the compromised system.

The MENTOR recommender system [23], suggest offers of offsite protection services (a list of products and services) to the customers based on their specific needs. Users can access a customizable dashboard, on which end users can specify their profile, define their required budget and their protection requirements. Recommending the cheapest service that is sufficient for the end-user according to their demands, distance based algorithms were the most applicable.

Husák [24] in 2021 Therefore, we developed a recommender system for responding to cybersecurity incidents in particular (i.e., ransomware and similar incidents). System then receives this IP address from a device infected with a RAT and does graph traversal algorithm to discover devices in proximity and fetches their profiles. In order to achieve this, the system leverages the CRUSOE data model for cyber situational awareness. [30], which allows for the data collection [29] and processing of heterogeneous data from an organization's network. The system also incorporates existing network scanning and traffic monitoring tools, as well as local environmental knowledge, to find hosts with similar physical and logical network topologies. The system then sorts the identified devices by their similarity, which is determined by factors such as their behavioural profile, fingerprint, or common history.

In 2017, Sayan et al. [25] proposed the architecture of the Intelligent Cyber Security Assistant (ICSA), which aims to empower a human cyber defender in the attack detection task (with the defense suggestion) through the use of the machine learning. In this, the system uses the existing monitoring tools to capture network traffic data and then analyzes it using an anomaly based intrusion detection approach. The Intelligent Recommender Assistant (IRA) module uses a knowledge based, feature engineering and cycle on an existing machine learning algorithm approach that recommends defenses for real time attacks. In addition, the authors expect future improvement on the system that may allow for automated or semi automated cyber defense actions of similar systems.

Overall, the existing research and implementation of RS in cybersecurity have demonstrated their potential for improving threat understanding, predicting threats, assessing risks, detecting vulnerabilities, and facilitating prevention and remediation. RS implementations has been found to enhance security analyst productivity, incident response, and compliance efforts, ultimately resulting in a stronger and more resilient security posture.

However, there is still a need for more research to develop more effective and efficient RS [21] that can better address the specific cybersecurity requirements holistically [26]. In order to address this gap in knowledge, the researcher was motivated to investigate and formulate a layered approach to the design of a recommender system.

4. Design Considerations

Although cybersecurity has become much more developed, cyber-attacks are becoming more complex and challenging for the field [2]. Mcdaniel [31] in 2020 reported the key research challenges and opportunities in the area of cybersecurity and artificial intelligence. Based on this research the report stresses out role of AI based systems in detection, prevention and response to cyber attacks.

Consequently, it is very important that an AI based recommender system, which encompasses these key research challenges and opportunities delineated in the report, is designed. These two design considerations are used by the researcher to address this need by proposing a layered approach to a recommender system for cybersecurity.

4.1. User interface design

The system should be easy to use and have a friendly interface so that recommendations are easy to understand and easy to follow for human cyber defenders. The security posture (dashboard) of analysis and visualization, forecasts, analysis and recommendation should be available via this interface. It should also be possible to involve humans in the interface and get feedback from them while operating it. That could be achieved by letting the human operator provide contextual data of the cybersecurity scenario to the machine for it to suggest.

Addition of these characteristics into the system's interface allows human cyber defenders to utilize the system's suggestions quickly and successfully. It can contribute to an improvement of the system's global efficiency in detecting and fighting cyber-attacks.

4.2. Contextual awareness

In order to improve suggestion process, contextual information needs to be included in the RS [27]. The system should be able to respond to changes in settings and adapt to appropriate suggestion depending on how the current cyber security scenario may be, in which there are knowledge-based multimedia data from different forms of modalities that will need to be studied and analyzed to find cyber-attacks.

Moreover, for the system to be operational, it should smooth the way for passive as well as active continuous collection of contextual information from network or cybersecurity domain.

4.3. Explainability and interpretability

The system has to provide its explanations as to why the provided recommendations are correct, so that the human cyber defenders are able to understand the reasoning that goes into it. It is necessary to build trust in the recommendations made by the system. In this case, the system should be able to give the user an explanation of why a certain recommendation has been done, and what factors have had the influence on the recommendation.

4.4. Flexibility

The system should be adaptable and able to give advice in respect to the individual situation based on the type of cybersecurity risk it faces. The system must be adaptable on three major areas. First, it should be able detect vulnerabilities, abnormalities, patterns and trends, security threats in real time, be able to classify and analyse that and accordingly present it with recommendations of what needs to be done for which it can replace the incident response capability of an operator. Second, it should be able to carry out other advanced cyber situation awareness activities, threat analysis and synthesizing automated and semi-automated procedures to mitigate risks. In the third place, the system should be ready to ascertain, react to, and analyze probable and actual threat occurrences while ensuring compliance with security requirements. Effective recommendation algorithms that perform well and help them to reach to their goals, have to be flexible enough to adapt to changing situations and provide a fast and true analysis of preprocessed data and suggestion creation.

4.5. Human intervention and human-machine teaming

A Recommender System is expected to handle threats with minimum human intervention, but with human supervision. They also suggest that the system should allow for human intervention at different stages of the decision–making process because the human cyber defenders should be able to override the system recommendations if appropriate. It will increase the humans and machines' collaboration and coordination. The important role of human – machine teaming for human–AI system outcomes in a diverse environment with feedback interactions, to reduce human error, ensure safety and provide accountability is also discussed.

The output generated by the RS should be recommendations that consider the human operator's expertise, and that turn into actionable insights for the human operators to make decisions. The system should also be able to accept human operator feedback on the recommendations that they are given and adapt its recommendations based on this feedback.

4.6. Data privacy and security

The system should protect the privacy and security of both the data used to produce suggestions and the recommendations themselves. The RS guidelines should consider access control measures and guarantee that only authorized users have access to critical resources and data.

4.7. Scalability

For effective recommender system in cybersecurity context, it is crucial to have the ability to scale to massive datasets. However, as the threats grow and become more complex over time, the system needs to make recommendations based on a large number of up to date data. As such the system will be a dynamic learning agent able to periodically update and retrain its underlying knowledge base as soon as there is new available attack vector or threat intelligence. It makes the system up to date and useful giving consideration to varying threats. To stay current with the ever changing threat landscape in cybersecurity, the recommender system focuses on scalability to enable it to continue to provide timely & accurate recommendations to the human cyber defenders. This may therefore allow organizations to ascertain a possible security incident and act on it before such an incident becomes costly.

4.8. Integration and orchestration

In order to make the recommender system aware of the additional context and enhance its real time detection and response ability, it is necessary to integrate the system with present cybersecurity tools and systems. This will allow for extracting live and historical data as inputs or triggers for recommendations to human cyber defenders with seamless user experience. Besides integration, the system should also be able to coordinate and orchestrate by multiple and existing cybersecurity tools. This includes interactions with other systems, cybersecurity tools, and with human actors to support autonomous and semi-autonomous cybersecurity incident response, also referred to as cybersecurity automation. This enables the system to thwart and circumvent cyber security threats. Overall, by improving contextual awareness thru collecting and processing data from various sources including data collected from the existing cybersecurity tools and systems, the recommender system can provide a more efficient and effective means of detecting and responding to cyber threats. These are the proposed design considerations for a context-aware cybersecurity recommender that can better detect, prevent, and respond to cyber threats in real time, improve identity management and access control, provide predictive analytics for security, perform autonomous and semi-autonomous planning, coordinate and orchestrate multiple cybersecurity tools, and maximize human-machine teaming to minimize negative consequences.

5. Proposed Framework

With the design considerations above, a layered framework design in Figure 35.1 for a context-aware recommender system for cybersecurity was proposed and details each layer. Each layer has a specific function in the RS framework, and the layers work together to generate relevant and actionable recommendations for cybersecurity.

5.1. Application interface layer

The application layer is the top layer of the proposed framework, and the security administrator provides interface to RS. It allows to see the processed data by the user, to configure the RS, and see the output suggestions, through the user interface provided at this layer. The application layer listens to inputs from the security administrator, like security policies, feedback, rules and others thresholds, to decide whether a particular behavioural vector constitutes a threat of fall under the category of an Application Error. In this layer, the human interventions are performed.

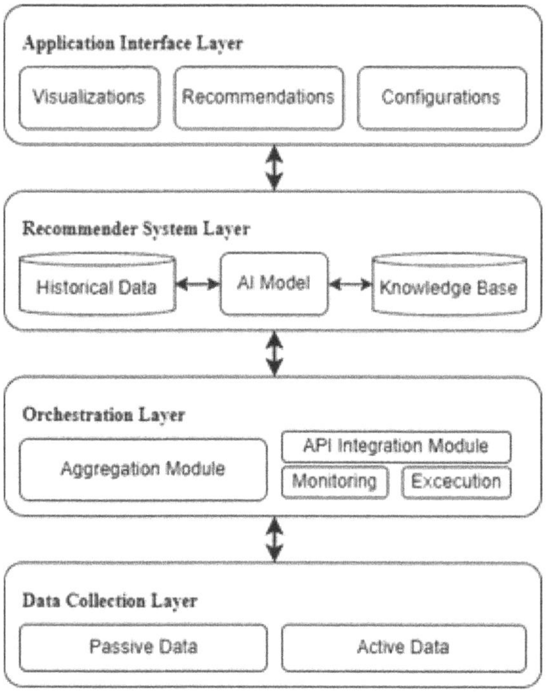

Figure 35.1. Proposed layered framework design for a context-aware recommender system for cybersecurity.

Source: Author.

5.2. Recommender system layer

The first layer is the recommender system layer which is of critical importance. The function is to analyze cybersecurity data and suggest account specific mitigation and remediation practices for vulnerabilities. To do this, the layer design is twofold: it is focused on cyberattack, error detection, prediction and analysis. However, it was desirable that it should use a large amount of historical data and knowledge base data for identifying anomalies and detecting errors, and at the same time, should not be susceptible to adversarial attacks. Within this layer, machine learning algorithms used should be optimized both for speed and accuracy. In addition, they must have the ability to analyze large volumes of data and come up with appropriate recommendations based on contextual factors in a short and precise manner.

5.3. Orchestration layer

The layer has two components that are central, the Aggregation, and the API integration modules. These components allow the Recommender System to quickly aggregate and process data of importance and tie into other cybersecurity tools easily. It helps to develop the system's capabilities as a whole and allows to offer a better and more complete recommendations for users.

5.3.1. Aggregation module

The Data Collection Layer is responsible for organizing and executing the collection data from numerous data sources, and the Aggregation is the module which carries out this operation. The recommender system has a corresponding agent sub-module for each of the data types in which the collected data is pre-processed and converted into a consumable format for the recommendation system layer.

5.3.2. API integration module

In this module, the API Integration is responsible for the coordination and integration of multiple cybersecurity tools. This module is made of two sub modules: monitoring and execution. The sub-module uses API to collect information about the current state of this cybersecurity tool that is connected. It differs from the Execution sub module which is responsible for managing all configuration change and command execution to the connected tools.

5.4. Data collection layer

The area of responsibility that pertains to the Data Collection Layer is that of collecting and aggregating contextual information pertaining to the time domain. It encompasses topology, application and system OS fingerprints, services, ports, logs, IDS and IPS data, threat landscape, etc. The layer must be collecting the passive data, the data that doesn't change infrequently and the active data, which is time based data.

The collected data must be classified and categorized to ensure that there is a proper adapter in the Orchestration Layer that can collect and prepare the data before inputting it into the Recommender System. This layer is the foundation of the RS framework, and the quality of the recommendations generated by the RS depends on the accuracy and quality of the data collected in this layer.

The Data Collection Layer is essential as it leads to thorough insights of domain and facilitates the RS to make suggestions based on the data available. This is why it is crucial to ascertain that the data being collected is brand new and relevant and appropriately labeled to allow for maximizing performance of the RS.

6. Conclusions

In this research, we propose a layered approach to context aware, cybersecurity recommender system that can be used to address key research challenges in context of cybersecurity and Artificial Intelligence. In order to address the need for efficient and effective cybersecurity in the presence of more and more complex and sophisticated cyber attacks, design considerations in terms of user interface design, contextual awareness, explainability and interpretability, flexibility, human intervention and human machine teaming, data privacy and security, scalability, and integration and orchestration are proposed. The proposed RS can enhance the effectiveness and efficiency of incident response and security compliance efforts through the more efficient and effective means of detecting, preventing and responding to cyber threat in real time, and in turn, they ensure a more robust and resilient security posture. The proposed design considerations can serve as a groundwork for future research in the field of cybersecurity and AI-based recommender systems and help advance the development of more effective and efficient cybersecurity measures. This study also designs and implements elements on how the design considerations of this study can be applied to build a recommender system in the future work. The RS can serve as a cybersecurity measure using an innovative architecture built using real or live network environment to detect threat, predict threat and provide mitigation recommendation.

7. Acknowledgements

The researchers express their deepest gratitude to their families for the unwavering support and patience shown throughout this research endeavour. The love and encouragement from their wives, and relatives fuelled the researchers' determination to persevere and attain their goals, even amid challenges.

The researchers are also thankful for the spiritual strength and wisdom bestowed upon them by almighty God, which provided inspiration and hope during trying times. This work would not have been possible without these profound blessings guiding them along the journey.

References

[1] Pawlicka, A., Pawlicki, M., Kozik, R., & Choraś, R. S. (2021). A systematic review of recommender systems and their applications in cybersecurity. In *Sensors* (Vol. 21, Issue 15). MDPI AG. https://doi.org/10.3390/s21155248

[2] Segal, E. (2022). The Impact of AI on Cybersecurity. *IEEE Computer Society*. https://www.computer.org/publications/tech-news/trends/the-impact-of-ai-on-cybersecurity

[3] Srinivasan, B., & Mallika, N. M. (2016). Artificial Neural Network based Data Integrity for Multi Cloud Computing. *International Journal of Advances in Engineering and Emerging Technology*, *7*(4), 143–153.

[4] Gadepally, V. N., Hancock, B. J., Greenfield, K. B., Campbell, J. P., Campbell, W. M., & Reuther, A. I. (2016). Recommender Systems for the Department of Defense and Intelligence Community. In *74 LINCOLN Laboratory Journal n Volume* (Vol. 22).

[5] Turan, F., & Ergenler, A. (2023). The Genotoxic Damage in Cyprinus carpio Exposed to Abamectin. *Natural and Engineering Sciences*, *8*(2), 119–128. http://doi.org/10.28978/nesciences.1338147

[6] Zhang, J., Chen, C., & Cohen, R. (2010). A Scalable and Effective Trust-Based Framework for Vehicular Ad-Hoc Networks. *Journal of Wireless Mobile Networks, Ubiquitous Computing, and Dependable Applications*, *1*(4), 3–15.

[7] Das, R., & Sandhane, R. (2021). Artificial Intelligence in Cyber Security. *Journal of Physics: Conference Series*, *1964*(4). https://doi.org/10.1088/1742-6596/1964/4/042072

[8] Sarker, I., Kayes, A. S. M., Badsha, S., Alqahtani, H., Watters, P., & Ng, A. (2020). Cybersecurity data science: an overview from machine learning perspective. *Journal of Big Data*, https://doi.org/10.1186/s40537-020-00318-5

[9] Shaukat, K., Luo, S., Varadharajan, V., Hameed, I. A., & Xu, M. (2020). A Survey on Machine Learning Techniques for Cyber Security in the Last Decade. *IEEE Access*, *8*, 222310–222354. https://doi.org/10.1109/ACCESS.2020.3041951

[10] Sun, N., Zhang, J., Rimba, P., Gao, S., Zhang, L. Y., & Xiang, Y. (2019). Data-Driven Cybersecurity Incident Prediction: A Survey. *IEEE Communications Surveys & Tutorials*, *21*(2), 1744–1772. https://doi.org/10.1109/COMST.2018.2885561

[11] Thuraisingham, B., Khan, L., Masud, M. M., & Hamlen, K. W. (2008). Data Mining for Security Applications. *2008 IEEE/IFIP International Conference on Embedded and Ubiquitous Computing*, *2*, 585–589. https://doi.org/10.1109/EUC.2008.62

[12] Zhang, Z., Ning, H., Shi, F., Farha, F., Xu, Y., Xu, J., Zhang, F., & Choo, K.-K. R. (2022). Artificial intelligence in cyber security: Research advances, challenges, and opportunities. *Artificial Intelligence Review*, https://doi.org/10.1007/s10462-021-09976-0

[13] Soy, A., & Balkrishna, S. M. (2024). Automated detection of aquatic animals using deep learning techniques. *International Journal of Aquatic Research and Environmental Studies*, *4*(S1), 1–6. https://doi.org/10.70102/IJARES/V4S1/1

[14] Kitana, A., Traore, I., & Woungang, I. (2020). Towards an Epidemic SMS-based Cellular Botnet. *Journal of Internet Services and Information Security*, *10*(4), 38–58.

[15] Recommender System with Machine Learning and Artificial Intelligence. (2020). In *Recommender System with Machine*

Learning and Artificial Intelligence. Wiley. https://doi.org/10.1002/9781119711582

[16] Huff, P., McClanahan, K., Le, T., & Li, Q. (2021). A Recommender System for Tracking Vulnerabilities. *The 16th International Conference on Availability, Reliability and Security*, 1–7. https://doi.org/10.1145/3465481.3470039

[17] Lyons, K. B. (2014). AFIT Scholar Theses and Dissertations Student Graduate Works A Recommender System in the Cyber Defense Domain. https://scholar.afit.edu/etd/612

[18] Polatidis, N., Pimenidis, E., Pavlidis, M., & Mouratidis, H. (2017). Recommender Systems Meeting Security: From Product Recommendation to Cyber-Attack Prediction. https://doi.org/10.1007/978-3-319-65172-9_43

[19] Ayala, C. S., Jiménez, K., Loza-Aguirre, E., & Andrade, R. O. (2021). A Hybrid Recommender System for Cybersecurity Based on a Rating Approach. *Advances in Security, Networks, and Internet of Things*.

[20] Alnusair, A., Zhong, C., Rawashdeh, M., Hossain, M. S., & Alamri, A. (2017). Context-aware multimodal recommendations of multimedia data in cyber situational awareness. *Multimedia Tools and Applications*, *76*(21), 22823–22843. https://doi.org/10.1007/s11042-017-4681-2

[21] Abuhussein, A., Shiva, S., & Sheldon, F. T. (2016). CSSR: Cloud Services Security Recommender. *2016 IEEE World Congress on Services (SERVICES)*, 48–55. https://doi.org/10.1109/SERVICES.2016.13

[22] Brisse, R., Boche, S., Majorczyk, F., & Lalande, J. F. (2021, November). KRAKEN: a knowledge-based recommender system for analysts, to kick exploration up a notch. In *International Conference on Information Technology and Communications Security* (pp. 1–17). Cham: Springer International Publishing

[23] Franco, M. F., Rodrigues, B., & Stiller, B. (2019). MENTOR: The Design and Evaluation of a Protection Services Recommender System. *2019 15th International Conference on Network and Service Management (CNSM)*, 1–7. https://doi.org/10.23919/CNSM46954.2019.9012686

[24] Husák, M. (2021). Towards a Data-Driven Recommender System for Handling Ransomware and Similar Incidents. https://doi.org/10.1109/ISI53945.2021.9624774

[25] Sayan, C. M., Hariri, S., & Ball, G. L. (2017). Cyber Security Assistant: Design Overview. *2017 IEEE 2nd International Workshops on Foundations and Applications of Self* Systems (FAS*W)*, 313–317.

[26] Husák, M., & Cermak, M. (2022). SoK: Applications and Challenges of using Recommender Systems in Cybersecurity Incident Handling and Response. https://doi.org/10.1145/3538969.3538981

[27] Adomavicius, G., Mobasher, B., Ricci, F., & Tuzhilin, A. (2011). Context-aware recommender systems. *AI Magazine*, *32*(3), 67–https://doi.org/10.1609/aimag.v32i3.2364 19.

[28] Kalaiprasath, R., Elankavi, R., & Udayakumar, R. (2017). Cloud security and compliance-a semantic approach in end to end security. *International Journal on Smart Sensing and Intelligent Systems*, *10*(5), 482.

[29] Husák, M., Laštovička, M., & Tovarňák, D. (2021). System for Continuous Collection of Contextual Information for Network Security Management and Incident Handling. https://doi.org/10.1145/3465481.3470037

[30] Komárková, J., Husák, M., Laštovička, M., & Tovarňák, D. (2018). CRUSOE: Data Model for Cyber Situational Awareness. 1–https://doi.org/10.1145/3230833.3232798

[31] Mcdaniel, P. (2020). Artificial intelligence and cybersecurity: Opportunities and challenges technical workshop summary report. *The National Science Technology Council's NITRD and MLAI Subcommittees gratefully acknowledge*. http://www.whitehouse.gov/ostp.

36 Competitive advantage of import of agricultural products in India with special reference to BRICS countries

Sarithamol K. K.[1,a] *and Biju P. J.*[2,b]

[1]Assistant Professor, Department of Commerce, Karpagam Academy of Higher Education, Coimbatore, Tamil Nadu, India
[2]Assistant Professor, Loyola College, Chennai, India

Abstract: This research focuses into the beneficial performance of imports of Indian agricultural products, paying particular attention to the BRICS nations. There are a lot of motivation in evaluating the BRICS countries' agricultural product import performance. Brazil, Russia, China, and South Africa's agricultural commodity imports are the focus of this study. For the years 2017–2021, it also examines trade analyses with each bilateral partner. The goal is to find potential BRICS nations for agricultural product imports with analysing trade pattern and trends in agricultural products import. Moreover, top agricultural product which are imported from BRICS nations over the past five years is also examined. Secondary sources will be gathered by the Agricultural Export Processed Development Authority (APEDA) and WITS (World Integration Trade Solutions) for this research analysis. This report should also include the fact that agricultural imports from the BRICS nations especially China have increased significantly over the past five years.

Keywords: Import, agricultural product, India, BRICS, APEDA

1. Introduction

The five emerging and newly industrialized nations are referred to as BRICS. South Africa, China, India, and Russia. All five members of the BRICS have large and rapidly expanding economies and considerable influence over local government. Over fifty percent of the world's growth since the start of the global financial crisis in 2007 had come from the BRICS nations by 2014. With a combined population of around 3.6 billion in 2015, each of the five BRICS nations account for almost 41% of the global total. Not only that, four of the five countries are in the top ten in terms of population. There is an approximate US$4 trillion in combined foreign reserves and an average nominal GDP of US$16.6 trillion, which is equal to around 22% of the total global revenue. The combined GDP (PPP) is over US$37 trillion. After an estimated expansion of 3.9% in 2015, the BRICS are projected to increase by 4.6% in 2016. In 2017, the World Bank expects a growth rate of 5.3% from the BRICS. Numerous critics have offered both positive and negative evaluations on the BRICS. Principles of equality, mutual benefit, and non-interference have always guided bilateral ties among BRICS countries [1].

The primary objective of the BRICS nations is to trade agricultural products with one another [12]. Anything cultivated or produced from the fields of agriculture, horticulture, floriculture, viticulture, or vegetables is considered an agricultural product [4]. All crops, farm animals, and farm animal products are considered agricultural goods. This includes fruits, vegetables, horticulture specialties, livestock, poultry, fur-bearing animals, eggs, milk, and furs. One of the many precious commodities and raw resources used to make food components is agricultural product. Governments should pay close attention to agricultural production since it is a key factor in the well-being of farmers, the agro-based sector, and all of mankind [2].

Above all else, BRICS is very important to India's economic ties with other countries. In addition to being a growing regional trading organization, it serves as a major market for Indian exports, imports, strengthening ties between India and the BRICS countries. From 8% in 2002 to over 14% in 2018, the proportion of India's overall trade that is attributed to the BRICS countries has increased significantly. That India is serious about strengthening commercial ties inside the BRICS is evident here. A trade imbalance exists between India and the other BRICS countries. The trade deficit between India and the rest of the globe including the BRICS nations has increased dramatically since 2006. From $8.73B in 2006 to $65B in 2018, the trade gap has increased significantly. A lack has been increasing higher over the last several years. It is clear that India's economy isn't as competitive as the BRICS countries from the country's massive

[a]sarithamol.kamalasanan@kahedu.edu.in, [b]bijupathayil@gmail.com

DOI: 10.1201/9781003684459-36

trade deficit with them. In addition, the export/import ratio of India has been steadily decreasing over the last several years, falling from 0.475 in 2010 to 0.234 in 2017 [14, 15]. Since the growth rate of India's exports and imports to and from BRICS nations has been almost constant since BRICS establishment, the significance of BRICS for India cannot be adequately justified by the size of its commerce with BRICS countries. Commodity imports from India have not improved and the country's trade gain with the BRICS has not led to major structural shifts, in contrast to China's trade imbalance. The consequent reduction in India's import inter-dependence relative to imports within the BRICS has been noticeable. So, it seems very clear that India's commercial potential in the BRICS nations has gone unexplored [10].

To fill that research gap, this study analyses trade across BRIC countries and focuses on agricultural commodity import data. This study aims to examine the effects of importing agricultural goods from BRICS countries to India. The main contribution is detailed below:

1.1. Objectives

1. To determine the import partner share of agricultural product from BRICS countries.
2. To examine patterns and trends in the import of agricultural goods
3. In order to determine which imported agricultural items are most important

Below is the outline for the remainder of the article. Section II delves into the study that relates to the import and export of agricultural products in comprehensive way. In Section III, the study's paramilitaries are detailed. The findings and comments based on secondary data and analysis are presented in Section IV. The article concludes in section V with a discussion of the study's limitations.

2. Literature Survey

According to a study titled 'Study of Impact of Provisions of WTO on Indian Agricultural Exports with Special Focus on Rice: A Causal Research' [3], Basmati rice exports were relatively stable between 2013 and 2015, and they showed growth in quantity and value addition from 2010 to 2017. From 1995 to 2018, the World Trade Organization (WTO) supported a variety of policies, including covert import substitution and outward export promotion. Before the World Trade Organization (WTO), exports of Basmati rice were stable, but after the WTO, there was an overall increase in the amount of rice shipped. There are non-tariff obstacles, such as rising local demand, which reduces the export capacity of Basmati and reduces its share of the global market [9].

According to the article 'Export and Import Performance of Agriculture in India' (p. 4), this country has had a rough go of it recently due to expansion stemming from the worldwide financial crisis of 2008–2009 and its weak growth rate. Imports of these cereals increased between fourteen percent and twenty percent between 2009 and 2010. Imports of products followed a tendency of fluctuation from 15% in 2010–11 to 12% in 2011–12 to 9% in 2012–13 to 8% in 2013–14 to 5% in 2014–15. In terms of agricultural exports, it was a resounding success between 2010 and 2018, with a rising trend in export value. The government of India should help its exporters by taking effective measures to increase the country's agricultural exports.

According to the publication 'An Analysis of India's Trade Relationship with SAARC Countries: A Review' [7], India has been concerned that its business with the SAARC members has seen a general increase in trade recently. It is emphasized that SAARC has failed to progress to the same level as more mature organizations such as the European Union (EU). It should also be noted that gross capital creation (as a percentage of GDP) was 27% in 1990 and 34.3% on average between 2007 and 2016. The conclusion is that India should increase trade with its neighbours, particularly Sri Lanka, Bangladesh, and Nepal.

According to the article 'Rice Export from India: Trends, Problems and Prospects' [5], the vast majority of Indians make an income in the agricultural sector. When it comes to rice exports, India is one of the world's most significant nations. The global market for Indian rice has hit a new peak. Notably, the global market presents a promising opportunity for some kinds of Indian rice, notably basmati rice. Conclusion: Some types of Indian rice still have a strong market, and India continues to rely on her conventional export commodities.

According to a study named 'An Analytical Study on The Export Performance of Dairy Industry in India' [6], despite being one of the world's top milk producers, India's exports do not account for a significant portion of the total. Since there are many undiscovered aspects to the area of exporting dairy products, it is considered that it contains brilliant projection for future researchers. The ever-increasing demand for dairy products, together with the relatively modest exports and imports of the industry, has made it imperative to understand the current state of the dairy industry's commerce.

2.1. Problem statement

Significant opportunities and challenges arise when agricultural products are imported into India, particularly through trade relations with the BRICS countries (Brazil, Russia, India, China, and South Africa). India is facing a growing demand for agricultural products that surpasses its domestic production capacity in certain areas, despite being one of the world's largest agricultural economies. The domestic agricultural sector in India faces risks from imports, which can help with supply-and-demand, food security, and consumer choice at competitive prices. The challenge is in understanding the complexities of BRICS agricultural imports, which include variables impacting tariff policies, trade patterns, and access to markets. The long-term viability of domestic agriculture, price volatility, and competition from cheaper imports are some of the negative impacts that local farmers are worried about, despite the potential benefits. Furthermore, India's ability to develop its agricultural infrastructure and its level of self-sufficiency in essential food staples could be impacted by the import of agricultural products.

2.2. Research gap

Imports of agricultural products are often compared in terms of their relative merits, but the economic effects of these imports on domestic agriculture, food safety, and consumer pricing are rarely considered. The effects of imports on local output, market dynamics, and agricultural employment require additional research. Studies that look at patterns and trends in agricultural imports are common, but there aren't many that look at how these patterns change over products. Previous studies have attempted to determine which agricultural imports are most crucial, but little is known about the interdependencies among these products, their relationship to domestic demand, or their overall effect on food systems.

3. Research Methodology

Based on secondary sources, this research provides both descriptive and analytical information. They originated from Agricultural and Processed Food Products Export Development Authority (APEDA) and WITS (World Integration Trade Solutions) and online resources made up the bulk of the secondary materials.

3.1. Sampling details

The research examines trade statistics between India and the BRICS nations between 2019 and 2024 using a thorough enumeration method. The comprehensive dataset seeks to pinpoint significant patterns and changes in trade throughout time, including a broad spectrum of agricultural goods. The longitudinal data is analysed using sophisticated methods such as trend analysis and ARIMA, which provide a thorough understanding of the dynamics of agricultural trade between India and the BRICS countries.

3.2. Period of the study

The research team has chosen the five-year span between 2019–20 and 2023–24 to examine agricultural imports. Our goal here has been to examine the items traded between India and the BRICS countries between 1996 and 2018. This data on trade flow is derived from the UN Common Format for Transient Data Exchange (UN COMTRADE) and the International Trade Centre (ITC Trade Map).

3.3. Index of trade integration pattern

A country's trade volume and composition change over time. The idea of comparative advantage sheds light on international trade patterns and is one of the drivers of this transformation. Consequently, we used data on product comparative advantages to build an indicator of trade integration trends [13]. We aimed to measure India's trade integration pattern (similarity/dissimilarity) with other BRICS countries by establishing an index of trade integration patterns. To be more specific, it operates on the assumption that trade patterns between parties are more comparable when there are more commodities in which they share a comparative advantage. We estimate that this index will be in the range of 0 to 1. A greater proximity to zero indicates more inter-industry trade and a more different trading pattern. When the value is around 1, however, intra-industry trade is more common and the trading pattern is more comparable. Here are the details of the index is given in Equation (1):

$$M = \frac{2C}{N_1 + N_2 + 2C} \tag{1}$$

The variables C, N1, and N2 represent the number of commodities in which the two trading countries share a comparative advantage, the number of commodities in which one trading country gets a comparative advantage, and the total number of products in which the second trading country gets a comparative advantage, respectively. The stated comparative advantage index was calculated using the following equation [2].

$$RCA^k_{ab} = (X^k_{ab}) \,/\, (X_{ab})/(X^k_{wb}\,/X_{wb}) \tag{2}$$

Here X^k_{ab} imports from country a to country *b* for product *k*; X_{ab} represents the total imports from country a to country *b*; X^k_{wb} represents imports of product *k* from world w to country *b*; X_{wb} = total imports to country *b*.

4. Result and Discussion

Table 36.1 displays the specifics of Brazilian imports from 2019–20 to 2023–24. We may infer that, with 232.32 Cr values, pulses will be the most heavily imported commodity in the 2019–20 timeframe. The number of pulses imported is considerable, reaching a maximum value of 169.29 Cr, even in the 2020–21 timeframe. The largest import value for pulse goods in 2021–22 is 489.1 Cr, while alcoholic drinks take the top place at 56184500. With 52,595,000 metric tons, pulse imports reach a peak value of 461.23 Cr in 2022–23. Once again, the value of alcoholic drinks reaches a peak of 541.75 Cr in the 2023–24 period, with a quantity of 88823613.

4.1. Product wise details

Details of imports from China over the five years starting from 2019 to 2024 are shown in Table 36.2. In 2019, the total value of agricultural imports was 1051.23 Cr, with 49984000 number of pulses being the top import commodity. The highest-value import commodity in 2020 was pulses, valued at 345.12 Cr, while the total import value dropped to 932.84 Cr. After falling to 899.88Cr in 2020, the total import value

Table 36.1. Import product details from Brazil for the period 2019–2024

	2019–20		*2020–21*		*2021–22*		*2022–23*		*2023–24*	
ProductName	*Qty*	*Rs. Crore*	*Qty*	*Rs. Crore*	*Qty*	*Rs. Crore*	*Qty*	*Rs. Crore*	*Qty*	*Rs. Crore*
Alcoholic Beverages	93000	0.57	198276	0.69	56184500	268.63	25452551	165.75	88823613	541.75
Pulses	40409000	232.32	24206000	169.29	53503000	489.7	52595000	461.23	45261000	406.05
Fresh Fruits	5801000	40.36	7162000	50.44	23899000	159.76	11525000	81.53	13654000	97.49
Processed Fruits & Juices	1116038	9.62	725273	9.86	2698530	32.23	4274537	56.1	6697501	92.45
Misc Processed Items	0	18.77	0	24.56	0	36.03	0	14.17	0	23.55
Other Cereals	11061000	35.98	2554000	9.43	1427000	6.99	304000	1.54	837000	4.15
Other Meat	0	0	0	0	0	0	33000	0.92	51000	1.59
Cocoa Products	23217	0.43	28680	0.34	20808	0.53	19758	0.5	654	0.16
Cereal Preparations	2000	0.04	0	0	1000	0.04	0	0	2000	0.01
Poultry Products	0	0.15	0	0.43	0	0.69	0	1.18	0	0.01
Fruits / Vegetable Seeds	112	0.04	4	0	28	0	259	0.04	1	0
Non-Basmati Rice	0	0	0	0	0	0	0	0.01	0	0
Fresh Vegetables	0	0	0	0	0	0.01	0	0	0	0
Floriculture	0	0	0	0	0	0	722	0.48	0	0
Processed Vegetables	0	0	0	0	0	0	14000	0.3	0	0
Total		338.28		265.04		994.61		783.75		1,167.21

Source: CIS.

Table 36.2. Import product details from China for the period 2019–2024

	2019–20		*2020–21*		*2021–22*		*2022–23*		*2023–24*	
ProductName	*Qty*	*Rs. Crore*	*Qty*	*Rs. Crore*	*Qty*	*Rs. Crore*	*Qty*	*Rs. Crore*	*Qty*	*Rs. Crore*
Misc Processed Items	0	207.96	0	169.4	0	242.68	0	341.6	0	377.35
Pulses	49984000	364.56	38688000	345.12	15603000	182.68	4475000	57.4	20666000	251.67
Fresh Fruits	8769000	85.44	10173000	75.99	7573000	73.88	12315000	115.09	28011000	232.35
Cereal Preparations	9216000	86.86	8515000	85.68	9967000	112.64	8516000	132.75	12749000	175.2
Processed Fruits & Juices	14064185	144.82	13677609	137.99	14307524	137.01	11302301	118.8	10073306	100.66
Floriculture	1517263	26.54	1056413	26.04	2321253	61.75	3193171	74.33	4612513	77.81
Fruits / Vegetable Seeds	41165	10.08	333489	21.63	111663	27.09	65646	33.32	123990	50.14
Processed Vegetables	17426293	111.74	7212212	57.44	5084619	51	3931974	49.48	2219553	35.36
Cocoa Products	1027309	8.22	565047	5.42	879935	8.29	722553	8.68	479462	5.94
Fresh Vegetables	250000	1.14	9000	0.1	0	0.03	99000	1.13	246000	3
Poultry Products	0	0.14	0	0.17	0	0.18	0	0.58	0	1.68
Guargum	36000	1.79	61000	3.29	45000	2.11	49000	2.86	41000	1.47
Alcoholic Beverages	162252	1.44	399393	4.51	55611	0.52	1309698	21.63	31362	0.29
Groundnut	0	0	0	0	0	0	0	0	5000	0.08
Milled Products	5922	0.1	2800	0.04	0	0	1400	0	2800	0.01
Other Cereals	74000	0.39	0	0	0	0	56000	0.15	0	0
Cashew Nut Shell Liquid	100	0.01	101	0.02	100	0.02	0	0	0	0
Total		1,051.23		932.84		899.88		957.8		1,313.01

Source: CIS.

will rise to 957.8Cr in 2022 and 2023. Fresh fruits were the most valuable farm commodity imported in 2023, totalling 28011,000 metric tons and 251.67 crore rupees.

Table 36.3 displays the information of agricultural products imported from Russia throughout the last five years. Pulses, with a value of 414.11 Cr and a quantity of 17,442,4000, are the most expensive import in 2019–20. The highest number of pulses in the 2020–21 period is 72,199,000, and their value is 229.63. During the years 2021 and 2022, alcoholic drinks accounted for 2.97Cr of the total import value. While cereal preparation goods had the highest imported value of 7.38 Cr in 2022, pulses ranked first with a quantity of 911000 as an imported agricultural commodity. With a quantity of 40,177,1000 and a value of 1820.3 Cr, India's pulse imports were the most in the last four years, in the 2023–24 fiscal year. Import product value as a whole was 1841.63Cr in 2023 and 2024 period.

Table 36.4 displays the specifics of imported products from South Africa during the last five years. For 2019–2020, the total import value is 302.89 crore rupees. During 2020 and 2021, its value climbed to 310.5 Cr. Over the course of 2021–2022, the total value rises from 508.72 to 744.56 Cr, and in 2023 it reaches to 768.75 Cr. In 2019, 2020, 2021, 2022, and 2023, the fresh fruits import quantity as 28990000, 28220000, 53159000, 81653,000, and 85967,000 of the total imported goods, respectively.

4.2. Country wise details

Table 36.5 displays the information of import products by country from 2019–20 to 2023–24. Russia has the most quantity at 17,5751222, while China has the highest value at 10,51.25 Cr for the 2019–2020 period. The total import value in 2020 was 1750.79Cr, with China accounting for 932.84Cr, the highest quantity, at 80693089. Brazil accounted for the highest value of 994.63Cr with 137733866 quantity out of a total import goods worth of 2414.01Cr in 2021–22. subsequently estimates a total value of 2506.38Cr for the fiscal year 2022–23, with China accounting for maximum value of 957.81Cr and South Africa for maximum quantity of product at 102470667. Compare with previous years, 2023–24 is having the maximum value of 5090.61 Cr, in which the maximum quantity by Russia at 40495912 with the cost of 1841.65Cr.

Details of India's imports of alcoholic beverages from the four BRICS nations – Brazil, China, Russia, and South Africa – from 2019 to 2024 are shown in Table 36.6. Both the amount (in kilos) and the value (in rupees) of alcoholic drinks imported annually are included in the statistics. In terms of alcoholic beverage imports to India, Brazil leads the BRICS countries, and imports will increase significantly after 2020. China exhibits unstable patterns, particularly in the latter years, while South Africa and Russia maintain

Table 36.3. Import product details from Russia for the period 2019–2024

	2019–20		*2020–21*		*2021–22*		*2022–23*		*2023–24*	
ProductName	*Qty*	*Rs. Crore*	*Qty*	*Rs. Crore*	*Qty*	*Rs. Crore*	*Qty*	*Rs. Crore*	*Qty*	*Rs. Crore*
Pulses	174424000	414.11	72199000	229.63	209000	0.67	911000	4.09	401771000	1820.3
Cereal Preparations	115000	1.23	848000	7.59	161000	1.93	453000	7.38	2560000	11.76
Alcoholic Beverages	159865	2.12	126663	1.74	271543	2.97	272158	7.18	196902	5.27
Other Meat	11000	0.23	26000	0.65	0	0	0	0	50000	1.47
Non-Basmati Rice	786000	52.55	156000	0.53	249000	0.83	0	0	260000	1.08
Groundnut	108000	0.82	57000	0.51	34000	2.68	0	0	89000	0.99
Misc Processed Items	0	0.63	0	1.14	0	1.59	0	0.95	0	0.38
Cocoa Products	491	0.03	0	0	350	0.01	0	0	32194	0.35
Poultry Products	0	0	0	0	0	0	0	0	0	0.03
Dairy Products	0	0	6	0	0	0	0	0	25	0
Fruits / Vegetable Seeds	37	0.12	19	0.04	1	0	5	0	7	0
Floriculture	0	0	0	0	640	0.08	0	0	0	0
Fresh Vegetables	80000	0.4	100000	0.33	0	0	84000	0.65	0	0
Processed Fruits & Juices	46824	0.33	0	0	0	0	0	0	0	0
Processed Vegetables	5	0	16000	0.26	0	0	0	0	0	0
Other Cereals	20000	0.06	0	0	0	0	0	0	0	0
Total		472.63		242.42		10.76		20.25		1,841.63

Source: DGCIS.

Table 36.4. Import product details from South Africa for the period 2019–2024

	2019–20		*2020–21*		*2021–22*		*2022–23*		*2023–24*	
ProductName	*Qty*	*Rs. Crore*	*Qty*	*Rs. Crore*	*Qty*	*Rs. Crore*	*Qty*	*Rs. Crore*	*Qty*	*Rs. Crore*
Fresh Fruits	28990000	204.1	28220000	202.6	53159000	358.83	81653000	569.61	85967000	599.21
Fruits / Vegetable Seeds	19158	10.57	25436	4.3	34819	22.66	36000	28.26	49932	73.25
Other Cereals	11240000	33.03	7677000	29.07	15491000	64.18	13314000	64.13	8841000	48.71
Processed Fruits & Juices	1556360	16.69	896815	10.2	714598	8.06	1196265	20.29	1111685	22.24
Pulses	4770000	33.13	6491000	55.15	6710000	49.13	5919000	45.43	1368000	16.14
Alcoholic Beverages	190741	3.25	109922	2.93	133796	4.8	317064	9.58	183236	5.62
Floriculture	26746	0.98	17487	1.19	3469	0.19	5120	2.21	22691	1.82
Misc Processed Items	0	0.33	0	0.38	0	0	0	4.49	0	1.53
Cereal Preparations	26000	0.44	32000	0.63	44000	0.86	30000	0.54	8000	0.18
Cocoa Products	23494	0.33	5	0	210	0.01	218	0.02	585	0.05
Cashew	1000	0.04	480000	4.05	0	0	0	0	0	0
Total		302.89		310.5		508.72		744.56		768.75

Source: GCIS.

Table 36.5. BRICS country-based product import details for period of 2019–2024

	2019–20		*2020–21*		*2021–22*		*2022–23*		*2023–24*	
Country name	*Qty*	*Rs. Crore*	*Qty*	*Rs. Crore*	*Qty*	*Rs. Crore*	*Qty*	*Rs. Crore*	*Qty*	*Rs. Crore*
Russia	175751222	472.65	73528688	242.42	925534	10.76	1720163	20.25	404959128	1841.65
China	102573489	1051.25	80693089	932.84	55948825	899.89	46036743	957.81	79260987	1313.01
Brazil	58505367	338.29	34874233	265.03	137733866	994.63	94218827	783.76	155326769	1167.21
South Africa	46843502	302.89	43949665	310.5	76290892	508.73	102470667	744.56	97552129	768.74
Total		2,165.08		1,750.79		2,414.01		2,506.38		5,090.61

Source: GCIS.

Table 36.6. BRICS country-based ALCOHOLIC BEVERAGES import details for period of 2019–2024

Country/Region Name: Brazil, China P Rp, Russia, South Africa
Product Name: ALCOHOLIC BEVERAGES
Qty In Kg; Value in Rs. Crore

Year	*2019–20*		*2020–21*		*2021–22*		*2022–23*		*2023–24*	
CountryName	*Qty*	*Rs. Crore*	*Qty*	*Rs. Crore*	*Qty*	*Rs. Crore*	*Qty*	*Rs. Crore*	*Qty*	*Rs. Crore*
Brazil	93000	0.57	198276	0.69	56184500	268.63	25452551	165.75	88823613	541.75
South Africa	190741	3.25	109922	2.93	133796	4.8	317064	9.58	183236	5.62
Russia	159865	2.12	126663	1.74	271543	2.97	272158	7.18	196902	5.27
China P Rp	162252	1.44	399393	4.51	55611	0.52	1309698	21.63	31362	0.29
Total		7.38		9.87		276.92		204.14		552.93

Source: DGCIS.

stable import levels. A rising demand for alcoholic drinks is reflected in the overall value of imports from these nations, with Brazil leading in this field.

Table 36.7 lists the specifics of India's imports of cereal preparations from the four BRICS nations (South Africa, China, Russia, and Brazil) between 2019 and 2024. The amount (in kilos) and value (in rupees) of grain preparations imported from each nation during the designated years are shown in the table. China is by far the greatest supplier of cereal preparations – both in terms of quantity and value – to India. While South Africa and Brazil make up a relatively tiny percentage of overall imports, with South Africa's share decreasing over time, Russia has seen significant increase in recent years, especially in 2023–2024. Overall, the trend shows that China and Russia are the main importers of cereal preparations.

Table 36.8 shows the information of cocoa goods imported by India from four BRICS nations (South Africa, Brazil, China, and Russia) from 2019 to 2024. Information is provided for each year on the amount of cocoa products imported, measured in kilos, as well as their value, expressed in Rs. Crore. China is the primary source of BRICS nations' cocoa product imports to India, however its share has been shrinking over the years. Imports from Russia increased significantly in 2023–24, whereas those from South Africa and Brazil had a little impact and had been decreasing in previous years. The general trend indicates that the total import value of cocoa goods from BRICS nations is decreasing, with China maintaining its dominance even though this trend is happening.

Table 36.9 illustrates the floriculture imports made by India from four BRICS nations (South Africa, Brazil, China, and Russia) between 2019 and 2024. The data features both the amount and the quality. China's continuous increase in both quantity and value is the primary source of floriculture imports into India. In terms of overall floriculture imports, Russia has maintained a small but steady presence, while Brazil and South Africa both make limited contributions; South Africa's imports for 2022–23 were particularly small. As overall, things are looking up for floriculture items, with the majority of those goods coming from China.

India imported fruit and vegetable seeds from four BRICS countries – South Africa, Brazil, China, and Russia – from 2019 to 2024, as shown in Table 36.10. Information is provided on the value (in Rs. Crore) and amount (in kg) of seeds imported from each nation. China has been the most constant

Table 36.7. BRICS country-based CEREAL PREPARATIONS import details for period of 2019–2024

Country/Region Name: Brazil, China P Rp, Russia, South Africa
Product Name: CEREAL PREPARATIONS
Qty In Kg; Value in Rs. Crore

Year	*2019–20*		*2020–21*		*2021–22*		*2022–23*		*2023–24*	
CountryName	*Qty*	*Rs. Crore*	*Qty*	*Rs. Crore*	*Qty*	*Rs. Crore*	*Qty*	*Rs. Crore*	*Qty*	*Rs. Crore*
China P Rp	9216000	86.86	8515000	85.68	9967000	112.64	8516000	132.75	12749000	175.2
Russia	115000	1.23	848000	7.59	161000	1.93	453000	7.38	2560000	11.76
South Africa	26000	0.44	32000	0.63	44000	0.86	30000	0.54	8000	0.18
Brazil	2000	0.04	0	0	1000	0.04	0	0	2000	0.01
Total		88.57		93.9		115.47		140.67		187.15

Source: DGCIS.

Table 36.8. BRICS country-based COCOA PRODUCTS import details for period of 2019–2024

Country/Region Name: Brazil, China P Rp, Russia, South Africa
Product Name: COCOA PRODUCTS
Qty In Kg; Value in Rs. Crore

Year	*2019–20*		*2020–21*		*2021–22*		*2022–23*		*2023–24*	
Country Name	*Qty*	*Rs. Crore*	*Qty*	*Rs. Crore*	*Qty*	*Rs. Crore*	*Qty*	*Rs. Crore*	*Qty*	*Rs. Crore*
China P Rp	1027309	8.22	565047	5.42	879935	8.29	722553	8.68	479462	5.94
Russia	491	0.03	0	0	350	0.01	0	0	32194	0.35
South Africa	23217	0.43	28680	0.34	20808	0.53	19758	0.5	654	0.16
Brazil	23494	0.33	5	0	210	0.01	218	0.02	585	0.05
Total		9.01		5.76		8.84		9.2		6.5

Source: DGCIS.

supplier of fruits and vegetable seeds to India among the BRICS nations, although Russia's contribution has increased significantly in recent years. The contributions from South Africa and Brazil are small, and their market shares have either decreased or remained low over the time frame. China and Russia are the primary sources of the overall pattern of a significant growth in seed demand.

For the years 2019–2024, Table 36.11 details the processed products that India imported from the four BRICS nations (South Africa, China, Russia, and Brazil). The number of imports (in kilos) and their value (in Rs. Crore) are both shown in the table. China is the primary contributor in terms of both quantity and value to India's imports of various processed goods. The contributions of Russia

Table 36.9. BRICS country-based FLORICULTURE import details for period of 2019–2024

Country/Region Name: Brazil, China P Rp, Russia, South Africa
Product Name: FLORICULTURE
Qty In Kg; Value in Rs. Crore

Year	*2019–20*		*2020–21*		*2021–22*		*2022–23*		*2023–24*	
Country Name	*Qty*	*Rs. Crore*	*Qty*	*Rs. Crore*	*Qty*	*Rs. Crore*	*Qty*	*Rs. Crore*	*Qty*	*Rs. Crore*
China P Rp	1517263	26.54	1056413	26.04	2321253	61.75	3193171	74.33	4612513	77.81
Russia	26746	0.98	17487	1.19	3469	0.19	5120	2.21	22691	1.82
South Africa	0	0	0	0	0	0	722	0.48	0	0
Brazil	0	0	0	0	640	0.08	0	0	0	0
Total	1517263	26.54	1056413	26.04	2321253	61.75	3193171	74.33	4612513	77.81

Source: DGCIS.

Table 36.10. BRICS country-based FRUITS / VEGETABLE SEEDS import details for period of 2019–2024

Country/Region Name: Brazil, China P Rp, Russia, South Africa
Product Name: FRUITS / VEGETABLE SEEDS
Qty In Kg; Value in Rs. Crore

Year	*2019–20*		*2020–21*		*2021–22*		*2022–23*		*2023–24*	
Country Name	*Qty*	*Rs. Crore*	*Qty*	*Rs. Crore*	*Qty*	*Rs. Crore*	*Qty*	*Rs. Crore*	*Qty*	*Rs. Crore*
China P Rp	19158	10.57	25436	4.3	34819	22.66	36000	28.26	49932	73.25
Russia	41165	10.08	333489	21.63	111663	27.09	65646	33.32	123990	50.14
South Africa	112	0.04	4	0	28	0	259	0.04	1	0
Brazil	37	0.12	19	0.04	1	0	5	0	7	0
Total		20.81		25.97		49.75		61.62		123.39

Source: DGCIS.

Table 36.11. BRICS country-based MISC PROCESSED ITEMS import details for period of 2019–2024

Country/Region Name: Brazil, China P Rp, Russia, South Africa
Product Name: MISC PROCESSED ITEMS
Qty In Kg; Value in Rs. Crore

Year	*2019–20*		*2020–21*		*2021–22*		*2022–23*		*2023–24*	
Country Name	*Qty*	*Rs. Crore*	*Qty*	*Rs. Crore*	*Qty*	*Rs. Crore*	*Qty*	*Rs. Crore*	*Qty*	*Rs. Crore*
China P Rp	0	207.96	0	169.4	0	242.68	0	341.6	0	377.35
Russia	0	18.77	0	24.56	0	36.03	0	14.17	0	23.55
South Africa	0	0.33	0	0.38	0	0	0	4.49	0	1.53
Brazil	0	0.63	0	1.14	0	1.59	0	0.95	0	0.38
Total		227.69		195.48		280.3		361.21		402.81

Source: DGCIS.

and South Africa have increased, particularly in the last few years, while Brazil's contribution has remained low. As a whole, the trend is indicative of rising demand for various processed goods, with most of those goods coming from China, which maintains its dominance in this import category.

Table 36.12 shows the yearly amount (in kilos) and value (in Rs. Crore) of 'Other Cereals' imported by India from four BRICS nations (South Africa, China, Russia, and Brazil) from 2019 to 2024.Other cereals imported from these nations had a dynamic total value, reaching a high of Rs. 71.17 crore in 2021–22 and falling to 52.86 crore in 2023–24. The table shows that each BRICS nation imports grain to varied degrees, with China being the biggest source.

Table 36.13 shows the monthly quantities (in kg) and values (in Rs. Crore) of 'Processed Fruits & Juices' imported by India from the BRICS nations (South Africa, Brazil, China, and Russia) from 2019 to 2024. From 2019–20 to 2023–24, the overall value of imports from these nations increased consistently, rising from Rs. 171.46 crore to Rs. 215.35 crore. This expansion shows that processed fruit and juice consumption is on increasing, especially in Russia and China.

Table 36.14 shows the annual amount (in kg) and value (in Rs. Crore) of 'Pulses' imported by India from the BRICS

Table 36.12. BRICS country-based OTHER CEREALS import details for period of 2019–2024

Country/Region Name: Brazil, China P Rp, Russia, South Africa
Product Name: OTHER CEREALS
Qty In Kg; Value in Rs. Crore

Year	*2019–20*		*2020–21*		*2021–22*		*2022–23*		*2023–24*	
Country Name	*Qty*	*Rs. Crore*	*Qty*	*Rs. Crore*	*Qty*	*Rs. Crore*	*Qty*	*Rs. Crore*	*Qty*	*Rs. Crore*
China P Rp	11240000	33.03	7677000	29.07	15491000	64.18	13314000	64.13	8841000	48.71
Russia	11061000	35.98	2554000	9.43	1427000	6.99	304000	1.54	837000	4.15
South Africa	74000	0.39	0	0	0	0	56000	0.15	0	0
Brazil	20000	0.06	0	0	0	0	0	0	0	0
Total		69.46		38.5		71.17		65.82		52.86

Source: DGCIS.

Table 36.13. BRICS country-based PROCESSED FRUITS & JUICES import details for period of 2019–2024

Country/Region Name: Brazil, China P Rp, Russia, South Africa
Product Name: PROCESSED FRUITS & JUICES
Qty In Kg; Value in Rs. Crore

Year	*2019–20*		*2020–21*		*2021–22*		*2022–23*		*2023–24*	
Country Name	*Qty*	*Rs. Crore*	*Qty*	*Rs. Crore*	*Qty*	*Rs. Crore*	*Qty*	*Rs. Crore*	*Qty*	*Rs. Crore*
China P Rp	14064185	144.82	13677609	137.99	14307524	137.01	11302301	118.8	10073306	100.66
Russia	1116038	9.62	725273	9.86	2698530	32.23	4274537	56.1	6697501	92.45
South Africa	1556360	16.69	896815	10.2	714598	8.06	1196265	20.29	1111685	22.24
Brazil	46824	0.33	0	0	0	0	0	0	0	0
Total		171.46		158.05		177.3		195.19		215.35

Source: DGCIS.

Table 36.14. BRICS country-based PULSES import details for period of 2019–2024

Country/Region Name: Brazil, China P Rp, Russia, South Africa
Product Name: PULSES
Qty In Kg; Value in Rs. Crore

Year	*2019–20*		*2020–21*		*2021–22*		*2022–23*		*2023–24*	
Country Name	*Qty*	*Rs. Crore*	*Qty*	*Rs. Crore*	*Qty*	*Rs. Crore*	*Qty*	*Rs. Crore*	*Qty*	*Rs. Crore*
China P Rp	174424000	414.11	72199000	229.63	209000	0.67	911000	4.09	401771000	1820.3
Russia	40409000	232.32	24206000	169.29	53503000	489.7	52595000	461.23	45261000	406.05
South Africa	49984000	364.56	38688000	345.12	15603000	182.68	4475000	57.4	20666000	251.67
Brazil	4770000	33.13	6491000	55.15	6710000	49.13	5919000	45.43	1368000	16.14
Total		1,044.12		799.19		722.18		568.15		2,494.16

Source: DGCIS.

nations (South Africa, Russia, China, and Brazil) from 2019 to 2024. A significant rise in imports from China increased up the overall value of imports in 2023–24, which reached Rs. 2,494.16 crore, a significant increase over the previous years.

4.3. Import partner share from BRICS country

The proportion of India's total imports that originate from a particular BRICS partner nation is known as the import partner share. Figure 36.1 provides the detailed import % from 2017 to 2021. When compared to the other BRICS nations, China's partner share % from 2017 to 2021 is the most significant, as shown in the Figure 36.1.

The proportion of South Africa imports going to India is seen in Figure 36.2(a). It reveals that from 2017 to 2021, South Africa's import partner proportion increased significantly. There will be a 1.99% rise in imports in 2021 from 2017's 1.5%. Figure 36.2(b) shows the specifics of the imports from Brazil. In 2019, the import proportion is lower than in 2020

Indicators by Trading Partner Timeseries - Print

Country : India Year : 2017-2021 Trade Flow : Import Indicator : Import Partner Share (%)

Partner Name	2017	2018	2019	2020	2021
China	16.20	14.63	14.28	15.98	15.35
Russian Federation	1.80	1.38	1.30	1.61	1.52
South Africa	1.55	1.33	1.39	1.81	1.94
Brazil	1.15	0.91	0.64	0.87	0.86

Figure 36.1. WITS report for BRICS country import trading partners time series.

Source: Author.

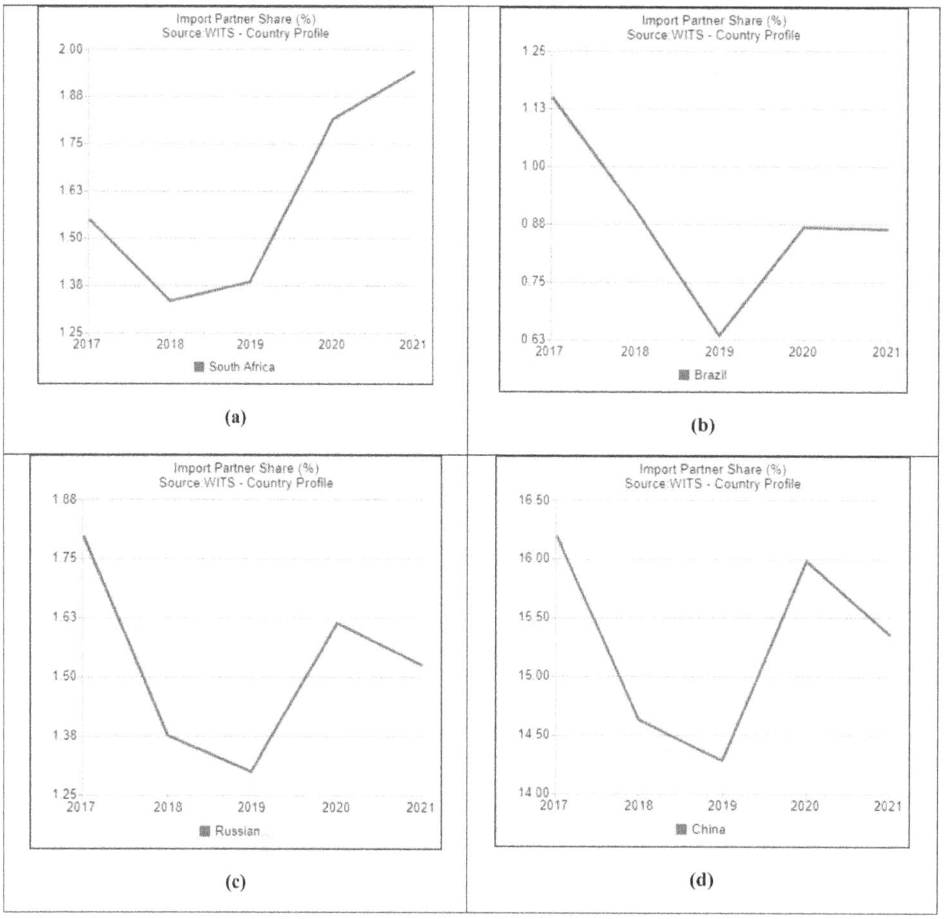

Figure 36.2. Import part share percentage of Brazil, China, Russian and South Africa.

Source: Author.

and 2021, when it will rise linearly. Figure 36.2(c) shows that the percentage of Russian imports was 1.8% in 2017, but it will drop to 1.42% in the period of 2021. China has the highest percentage of import partners among all nations from 2017 to 2021. The import percentage dropped from 16.2% in 2017 to 15.35% in 2021. The specifics of the imports from China are shown in Figure 36.2(d).

4.4. *Trade integration pattern index*

A systematic methodology for evaluating and classifying the different trade integration patterns across worldwide marketplaces is the Trade Integration Pattern Index (TIPI). Businesses, politicians, and economists rely on it as an analytical tool for analyzing the interplay of global business and market dynamics. The global economy is shaped by trade flows, alliances, regulations, and technology breakthroughs. The index gives a structured approach to analyzing these factors. With its insightful data on trade integration across regions and industries, the TIPI is a helpful tool for navigating this complicated world. Economic agreements, cross-border partnerships, logistics innovation, and digital platform innovation are identified as significant drivers of trade. Both assessment and strategic planning can benefit from this index. Market entry, investment possibilities, and supply chain management decisions can be better made by stakeholders who have a good grasp of trade integration trends. The TIPI offers a comprehensive perspective on the effects of trade flows on economic performance and competitiveness, and it is useful for a wide range of stakeholders, including large corporations, small businesses, and government agencies. The goal of the data-driven Trade Integration Pattern Index is to promote market alignment and cooperation for the purpose of easing trade connections and spotting growth possibilities in today's globally interconnected economy.

The Trade Integration Index (TII) values in Table 36.15 reveal a low trade integration within the BRICS nations, indicating a unique pattern of trade interdependence. These countries have diverse but complementary endowments in terms of capital, labour, and raw commodities, leading to unique trading patterns [8]. Products that depend on more available resources, like labour or natural resources, are exported by these countries, while those relying on scarce components, such as advanced technology, rare materials, or specialized skills, are imported by these countries. Trade integration on a country-by-country basis becomes more complex, with India's trading ties with South Africa and Russia being stronger than those with China and Brazil. This is indicative of distinct economic priorities or comparative advantages [10]. The overall BRICS market is not very similar to the trading relationship between Russia and South Africa, suggesting different trade structures and economic interests.

The BRICS countries have made significant strides in economic development by playing to their strengths, particularly in using their large pools of low-skilled workers for industrialization and growth driven by exports.

Table 36.15. Trade integration pattern index

Country	*1*	*2*	*3*	*4*	*5*	*6*
Brazil	-	0.36	0.51	0.29	0.42	0.30
China	0.29	0.25	0.43	-	0.30	0.30
India	0.51	0.33	-	0.43	0.24	0.39
South Africa	0.42	0.29	0.24	0.30	.	0.28
Russia	0.36	-	0.33	0.25	0.29	0.24
BRICS	0.30	0.24	0.39	0.30	0.28	-

Source: Author.

However, most BRICS nations have not fully capitalized on their advantages in industrial sectors that require manual labour and low levels of expertise, hindering their economic growth. To reach their maximum economic potential, BRICS nations should focus on enhancing domestic manufacturing capacities, capitalizing on their comparative advantage in labour-intensive industries, investing in capital, developing workforces, and enhancing technology to diversify trade patterns and guarantee long-term economic growth. This process of dynamic comparative advantage allows economies to adapt and evolve their capabilities, meeting the changing demands of global markets [11].

4.5. *Limitations and future research*

In order to get into the core of India-BRICS trade, our research focused on a handful of crucial, quantitative macroeconomic indicators. Many factors, including those of a political, legal, technical, etc. nature, impact the commerce between India and the BRICS countries. The COVID-19 epidemic and the political tension between Russia and Ukraine are only two examples of the recent external shocks that have altered the economic flows between India and the BRICS countries. The trade patterns between India and the BRICS countries might be better understood with the help of such external shocks. Better insights may also be gained by using other measures of transport costs and multilateral trade barriers. The variables that impact India's desire to import from the BRICS nations should be the subject of future studies. Finally, for more accurate results, additional research should compare India's import goals to those of the other BRICS nations.

4.6. *Significant of the study*

The contribution of agricultural imports from BRICS countries to India's agricultural sector can be better understood by conducting a thorough analysis of agricultural imports, which includes evaluating imports, analysing trends in imports, and identifying critical imported items. The study's findings will guide trade policies that will allow India to take advantage of the benefits of these imports to the maximum level possible, while simultaneously protecting its local agricultural sector and guaranteeing its users access to food in the future.

4.7. Suggestions

We could make the following suggestions on India's BRICS (Brazil, China, Russia, and South Africa) trade patterns for several commodities from 2019 to 2024:

1. Take Advantage of Brazil's Competence in the Alcoholic Beverage Market – With Brazil being the top importer of alcoholic beverages, India can benefit from strengthening economic ties with the country to secure a steady supply. To meet the growing demand in the United States, businesses can work with Brazilian providers to negotiate better trade terms or form partnerships.
2. Expand the Range of Cereal Preparation Suppliers – Since China and Russia are the leading importers of cereal preparations into India, the country should look into forming relationships with other BRICS countries or even non-BRICS countries to increase supply chain stability, diversify its sources, and balance South Africa's decreasing share.
3. Maximize Cocoa Imports – Since China's share of India's cocoa trade is going down and Russia's share is going up, India may be able to deal with Russia for better prices and a steady supply. One way to reduce our dependence on China as a source for cocoa products is to establish ties with different suppliers.
4. Assist the Domestic Floriculture Sector – India is heavily dependent on China for floriculture imports, which are seeing continual expansion. To lessen its need on imports and encourage local production, India may strengthen its domestic floriculture sector by implementing development programs or offering incentives.
5. Support Local Seed Producers – A domestic seed production program would be an excellent venture for India given the country's rapidly expanding seed imports, especially from Russia and China, for fruits and vegetables. This strategy has the potential to reduce dependence on unpredictable market changes and increase agricultural self-sufficiency in seed supplies.
6. Strengthen Collaborations on Processed Goods – Imports of processed commodities are mostly from China, with growing contributions from Russia and South Africa. India may sustainably supply the increasing demand for processed products by encouraging local manufacturing capacity and exploring more collaborations with these nations.
7. Maintain Stable Pulse Supply – Pulses may be in short supply in the United States in 2023 and 24 due to an enormous increase in imports from China. In addition to promoting local pulse production to fulfill increasing demand, India should think about forming strategic relationships with Russia and China.
8. Monitor the Market for Manufactured Fruits and Juices – In order to secure stable pricing and a steady supply, India should seek out long-term contracts with producers of processed fruits and juices, especially in Russia and China, where demand is on increasing. Domestic processing businesses may also get assistance as part of this plan to help them keep up with rising consumer demand.
9. Improve Import Storage and Infrastructure – India has to upgrade its import infrastructure and storage facilities to keep up with the increasing imports of different types of food and drink, including processed fruits and juices, alcoholic drinks, and pulses. Potential supply disruptions might be mitigated and seasonal demand spikes could be better managed with this.
10. Observe the Market's Volatility – India must keep an eye on global pricing trends and diversify its supplier base in light of the observed volatility in cocoa product and other grain imports. Risks associated with price fluctuations should be mitigated by strengthening commercial connections with BRICS countries in order to secure stable import conditions.

In order to effectively satisfy future demand, these ideas seek to establish a balanced import strategy that makes use of significant relationships while promoting the expansion of India's local manufacturing capabilities.

5. Conclusion

According to the study's findings, India might potentially increase its imports of agricultural goods to other trading nations in the BRICS, which are free from tariffs. According to the study, from China imports the most agricultural goods to India, totalling 5154 Cr values for the period of 2019–2023. Its foreign currency worth is boosted by imports. When India imports goods from other BRICS nations, it may build strong and positive relationships with them. Additionally, the amount of agricultural product imports traded across BRICS nations has been considered. China also has a high partner percentage index, rising from 14.28% to 16.59% during the last five years.

References

[1] Savanur, K. P. (2019). Economics Research Output in BRICS Countries: A Scientometric Dimension. *Indian Journal of Information Sources and Services*, *9*(1), 14–21.

[2] Kumar, M. A., Ramki, R., & Devi, R. S. (2021). Agricultural Products Export from India–With Reference to Brics Countries. *The Journal of Oriental Research Madras*. XCII. 34–40.

[3] Patel, P. G. (2019). Study of Impact of Provisions of WTO on Indian Agricultural Exports with Special Focus on Rice: A Causal Research. *Research Review International Journal of Multidisciplinary*, *4*(5), 2186–2189.

[4] Reena, S. R. (2019). Export and Import Performance of Agriculture in India. *International Journal of Innovative Technology and Exploring Engineering*, *8*(12S), 502–504.

[5] Ramakrishna, B., Degaonkar, C. K. (2016). Rice Export from India: Trends, Problems and Prospects. *International Journal of Research - GRANTHAALAYAH, 4*(7), 122–136.

[6] Shiv Kumar, Md.Kashif Ansari (2016). An Analytical Study on the Export Performance of Dairy Industry in India. *International Journal of Research Granthaalayah, 4*(1), 153–157.

[7] Nandeesha, K., & Kadam, R. N. (2018). An Analysis of India's Trade Relationship with SAARC Countries: A Review. *International Journal of Electronics, Electrical and Computational System (IJEECS), 7*(4), 631–638.

[8] Saji, G. T. (2019). Does Economic Integration Really Improve Trade Competitiveness? The Case of BRICS. *Economic Affairs, 64*(3), 469–480.

[9] Maryam, J., & Mittal, A. (2019). An empirical analysis of India's trade in goods with BRICS. *International Review of Economics, 66*(4), 399–421. https://dx.doi:org/10.1007/s12232-019-00328-7

[10] Myeongjoo, K. (2024). A Study on the Effect of Information and Communication Technology (ICT) on Trade in Services in the United States. *Journal of Internet Services and Information Security, 14*(1), 266–281.

[11] Chen, L. (2012). The BRICS in the Global Value Chains: An Empirical Note. *Cuadernos de Economia, 31*(57), 221–239.

[12] Dessy, A., Ratna, D., Leni, S., Yadi Heryadi, D., Fatma, S., I Gede I. S., & Robbi, R. (2023). Using Distance Measure to Perform Optimal Mapping with the K-Medoids Method on Medicinal Plants, Aromatics, and Spices Export. *Journal of Wireless Mobile Networks, Ubiquitous Computing, and Dependable Applications, 14*(3), 103–111.

[13] Beyene (2014a). Observes that the current reality of a country's or region's comparative advantage has an impact on the ease with which economic integration in that region can be facilitated.

[14] Tharik, M., Saraswathi, S., & Arumugam, K. (2021). Uncommon Mass Beaching of Porpita porpita (Linnaeus, 1758) in the Gulf of Mannar, Tamil Nadu, India. *Natural and Engineering Sciences, 6*(3), 256–260.

[15] Kansra, P., Kumar, P., Mahapatra, S. K., & Oberoi, S. (2021). Economic concert, cooperation and trade specialisation of India in comparison with rest of BRICS nations. *International Journal of Trade and Global Markets, 14*(6), 580–602.

37 Experimental study on the adversary attack vulnerabilities in IoT and CPS actuators and sensors through electromagnetic fault injection

Jhon Zylvin M. Ramos[1,a], *Marvin O. Mallari*[2,b], *and Arvin R. De La Cruz*[1,c]

[1]Graduate School, Polytechnic University of the Philippines, Philippines
[2]School of Engineering, Computer and Library Studies, Holy Cross College, Philippines

Abstract: In Internet of Things (IoT) and Cyber-Physical Systems (CPS) applications, actuators or sensor data integrity plays a paramount role. We created and validated, in this experimental work, a cheap programmable Electromagnetic Fault Injection (EMFI) pulse injection circuit on several experimental IoT/CPS sensor and actuator set ups, testing at various EMFI parameters with respect to the EMFI fault model technique, to investigate susceptibilities to adversary fault injection attacks. Specifically, the results expose convincing attack scenario, exemplifying the adversarial ability to corrupt the readings of sensors and actuation control signals by exploiting the EMFI. In addition, this study stresses the importance of the attack accuracy for the EMFI attack success, especially communicating with secure sensors and actuators. Communication protocols and vulnerability profiles that we are talking about are extremely complex and this points out how important is to understand the complex protocols. This research adds to the existing knowledge base of the topic by highlighting adversarial EMFI as a vulnerability in IoT/CPS systems and helps to explore building more robust and resilient systems in the future.

Keywords: Cybersecurity, electromagnetic fault injection, second order attack, IoT, CPS

1. Introduction

Over the years, as the IoT and CPS applications keep broadening, it brings new problems in the field of cybersecurity [1]. Sensors and actuators are viewed as important parts of the IoT/CPS application since they can interconnect the physical processes with to cyber network [2]. When adversaries attack the sensors and actuators [2–5], as in the continuous expanding and heterogeneity of the IoT/CPS applications, they may take advantage of the vulnerabilities in these devices.

Glaringly, EMFI attacks pose formidable challenge [1, 6, 7] to IoT/CPS security because these attacks are both cost effective and easy to carry out [8–10]. Adversaries manipulate EM signals to create local and often controlled transient magnetic fields near target devices to induce currents in wires and may disrupt the normal operation of the device noninvasively (i.e., without physically contact and direct access) [9, 11–13]. The integrity compromise at the IoT/CPS data or operation level can link to EMFI attacks that change the value that is delivered to an actuator or read by a sensor [14, 15]. This work aims to use EMFI Fault Model approach to investigate, and to expose vulnerabilities in Electromagnetic Interference (EMI) injection attacks on IoT/CPS actuators and sensor applications as a form of adversary attack.In particular, the researchers aim to design and develop a programmable [16] EMFI circuit that would allow the varying EM signal parameters like varying the duration of the pulse can be produced into transient or continuous pulse for a specific duration in different Device Under Test (DUT) [17]. For investigating the separate effects of EMFI in CPS/IoT sensor and actuator circuits, two different IoT/CPS systems Device Under Test (DUTs), an IoT actuator and an IoT sensor, will be developed. A better understanding of possible security issues can be obtained by exploring different attack scenarios and measuring the effects of electromagnetic injection in each of the setup [18].

2. Background

Background on electromagnetic signal injection attacks on embedded IoT/CPS applications in this section.

2.1. *From interference to exploitation*

EMI is a noise coupling mechanism which can lead to varying magnetic flux in a digital circuit board and generate unwanted voltage and currents called Eddy currents. Malfunctions or even immediate destruction of the circuit can occur with these kind of fluctuations [19].

[a]ramosjhonzylvin@gmail.com, [b]mallarimarvin022@gmail.com, [c]docardelacruz@gmail.com

DOI: 10.1201/9781003684459-37

Electromagnetic fault injection (EMFI) was first introduced by Quisquater et al. in 2002. In addition, they demonstrated that EMI can be used to inject faults on secure circuits, specifically on cryptographic devices. This threat has been subsequently confirmed and used to attack a device such as to extract encryption key of a device performing AES encryption [20], to bypass secure boot [11], exploit firmware security protection [21], clear set or reset these DFF D-type Flip-Flops (DFFs) [22] and even introduce instruction skips [23]. In 2019, Dumont et al. [9] analyzed how injecting electromagnetic signals could induce faults to circuit. Here they demonstrated that the type of fault that is created depends on frequency and amplitude of the signal and, more importantly, its location.

2.2. *EMFI techniques and platforms*

There are two ways to carry out EMFI: continuous (harmonic) EM wave injection and single transient EM pulse injection [24, 25].

2.3. *Security of embedded IoT sensors and actuators*

Sensors and actuators are required in cyber physical systems (CPS) and Internet of Things (IoT) applications. It is their duty to collect data from physical world and process that data, as well as to control physical systems. For the normal functioning of the PC, its input and output are important that must be in integrity [25, 26].

The sensors and actuators are also subject to attacks however. The False Data Injection (FDI) is one of the common attacks, where the attacker has access to the communication channel and has the ability to change the value read by a sensor or sent to an actuator [27]. Two types of FDI attacks are classified in this study:

Information that is received from a sensor can be corrupted or hidden, or non-existent information can be created. Incorrect decisions can be taken by the CPS or IoT system with this.

The actuator attacks can transmit any command to the actuators making them malfunction or operate in an unintended way. Such damage or harm to people can result from this.

In 2019 Khoumsi discussed how smart meter data could be used to launch an attack (sensor attack) to manipulate this information and have a major impact on the way power grids operate. In 2019, Hasan and Mohan [4] demonstrated how such attacks can be used to prevent a robotic arm from gripping an object – what would be a very dangerous loss of functionality in a manufacturing or healthcare setting.

In this regard, the EMFI attacks on the actuators and sensors for the same embedded application in the electric vehicles are also possible, see [26]. The causes are such attacks that inject an electromagnetic signal to the system so as to disrupt it while working as it should.

In 2019, Tu et al. [28] show how spoofing EMI signals can be used to evade the temperature readings of a sensor, resulting in a control system making bad decisions based on the evaded sensor. In diverse scenarios, no longer is the injecting EMFI attacks into the sensors and actuators shown by Selvaraj et al. [25] proven demonstrated in efficacious.

2.4. *First and second order attack*

In 2017's study from Cui and Housley [11], emphasize the difference between first order and second order EMFI attacks. However, due to high temporal and spatial resolution requirements, conducting first order EMFI attacks against modern processors is challenging [24]. Second order EMFI attacks rely on the fact that interdependent components and interconnects can reduce the temporal and spatial resolution requirements in a meaningful manner.

3. Literature Review

Previously, modeling of EM fault injection has highlighted that it can expose security vulnerabilities [21] in cryptographic operations, corrupt data, skip instructions, and represents a potential threat to microcontrollers from electromagnetic interference. It has been shown to be useful in academic and industrial research [29].

The majority of previous research [7, 23, 24, 29, 30] into attack models for first order EMFI had confirmed the vulnerability of microcontroller components to localized and repeatable EM fault injection. Nevertheless there are theoretical and practical constraints in applying these fault models in real world.

A previous argument has also been made by authors [31, 32] that these fault models should not require strong assumption of triggering an attack, intimate knowledge of the system, unlimited access to the system, or the use of expensive, sophisticated equipment to trigger EMFI attacks.

4. Contributions

Following previous research on the generic issue of embedded devices' EM susceptibility to their operation [33], we focus using first order attacks. Yet, to the author's best knowledge, little research has been done to characterize resilience of sensors and actuators in CPS/IoT applications to FDI using EMFI attacks, which are the second order attacks.

The contribution of this research is to provide an EMFI fault model that evaluates its effectiveness and detects vulnerabilities of interest. Unlike previous research using first order EMFI attacks on chip, this research will prove the direct attack using the IO signal to induce the attack [32]. The main difference is that injected EMFI signal will be inserted onto its data cable of sensor and actuator in order to alter sensor data readings or even trick an actuator to actuate [26]. The main purpose of this study is to examine EMFI fault models that inject EM parameters according to [9] in various combinations. In order to achieve this, a custom built EM pulse generator which is based on previous works [11, 21, 34, 35] will be developed. For validating the potential threats to IoT/CPS systems, this generator would be used using sensors and actuators.

The provided results on the impact of EMFI on actuators and sensors would bring enhanced knowledge about sensing and actuating IoT/CPS platforms, the foundation building blocks of sensing applications.

5. Methodology

In the following section, the experimental setup and methodology are described followed by empirical results on the use of EMFI to attack the chosen IoT/CPS applications using False Data Injection based on manipulating and tampering with the output.

The experimental setup for EM fault injection is depicted in Figure 37.1. The setup features several components: the EMFI Pulse Injector, the two IoT Device Under Tests (DUT) – an actuator (DUT1) and a sensor (DUT2) type, and an oscilloscope to visualize EM effects to input/output signal. The EM Pulse Injector is controlled by a laptop to vary EM parameters configuration. A Universal Foldable Phone Holder was modified to hold the EM Pulse injector circuit. The holder will enable the injector probe tip to be adjusted to a variable distance up to 1mm.

5.1. A programmable electromagnetic pulse injector circuit

For this experiment, a custom EM Pulse injector circuit was designed and developed. Like previous research [31], due to it practicality a simple commercial mosquito killer was adapted and modified to create the pulse generator circuit.

The EMFI circuit can deliver a voltage pulse generator delivering pulse of amplitude ranging between 50V and 1.5kV AC to a target device. Figure 37.2 shows the high-level setup. The setup consists of 3 stage circuits.

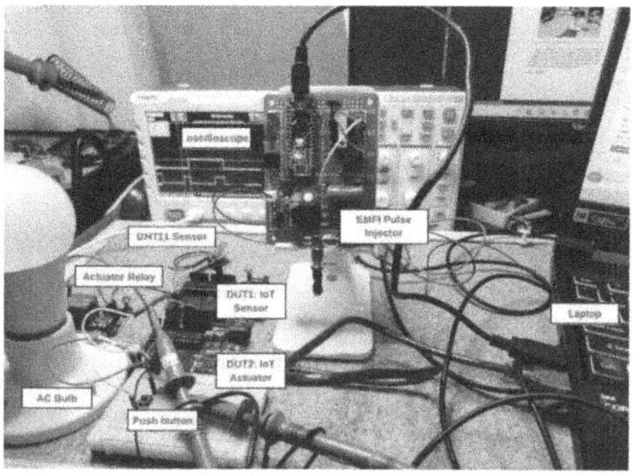

Figure 37.1. EMFI platform setup with a programmable EMFI pulse injector sending adversary signal into two different CPS/IoT setups for actuators and sensors applications.

Source: Author.

The first stage is a Pulse Controller circuit which is composed of Arduino Nano microcontroller (MCU) and FQP47P06, a p-channel MOSFET, that operates in its ohmic region to act as a variable resistor to vary the pulse intensity in voltage to be amplified in the next stage.

The MCU was configured to vary the intensity of the pulse voltage using Pulse Width Modulation (PWM). By altering the duty cycle in the MCU code, the PWM can deliver 0-5V to drive and vary the MOSFET resistance to vary the output voltage from the connected 5V battery.

MCU was also configured to set programmable parameters. A push button was included to act as a trigger button to initiate the pulse. Every pulse has a delay of 10 seconds for safety and to avoid unwanted high voltage pulse triggering.

The second stage is the Boost Converter circuit which is mainly consist of the adapted design commercial mosquito killer. It takes the input voltage from the first stage then it amplifies the input DC voltage into a high voltage AC [36].

The circuit consists of a joule thief circuit that acts as an inverter circuit. The oscillations created from an NPN transistor that is connected to a small EE-type ferrite core transformer AC square wave this then amplified in a Cockcroft-Walton voltage multiplier circuit [37] to step up the AC voltage up to a max 1.5kV AC. This stage is a high voltage circuit, so a safety switch was included to avoid unintentional shock.

The last stage is the High Voltage Pulse Injector circuit, which takes the High Voltage AC input and a pulse trigger signal to switches high-voltage and high-current loads through an EM Injector Probe to emit EMFI to a target circuit. The combination of an Insulated Gate Bipolar Transistor (IGBT) and a protection diode for switching was used similar to [21]. To provide isolation for the low voltage pulse trigger from the MCU and the High Voltage and high-current load to emit EMFI.

The IGBT gate was controlled by a drive signal from optocoupler powered by another 9V external power-supply which is connected to the MCU for pulse trigger signal. When a control pulse signal from the MCU is applied, the optocoupler will allow voltage to the gate of the IGBT which discharges the stored energy into the EM probe generating strong electromagnetic field.

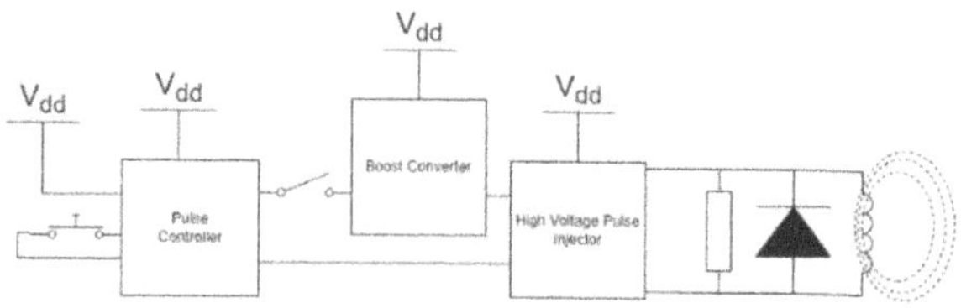

Figure 37.2. High-level view of the designed EM pulse injector circuit.

Source: Author.

The EM Injector Probe is a custom coil probe connected to a female 50-Ohm SMA connector. For this setup, it is fabricated from a coil of 10 mm diameter with 9 turns like [21].

5.2. Device under test (DUT) IoT device setup

For the experimental study, Arduino Uno R3 microcontrollers based on the ATmega328P as the Device Under Test (DUT) was selected for electromagnetic fault injection analysis. These microcontrollers operate at 16MHz and support a maximum I/O voltage of 5V for General-Purpose Input/Output (GPIO) pins.

5.2.1. Setup 1: IoT sensor application

The investigation on how EMFI attacks affect IoT/CPS sensor applications was lead by this setup. An Arduino Uno R3 microcontroller (Data Unit Test 1, DUT1) was made to collect data from some sensors. Among the sensors used were digital and analog temperature sensors and an analog photocell sensor for measuring light levels. To perform a thorough investigation of the effects of EMFI, the selection of these sensors was made.

The components and configurations of the setup were as follows:

- Arduino Uno R3 microcontroller
- DHT11 temperature with humidity sensor (Digital)
- LM35 temperature sensor (Analog)
- Photocell sensor (Analog)
- Power supply

5.2.2. Setup 2: IoT actuator application

This setup assessed the impact of EM injection attacks on IoT actuator applications. An Arduino Uno R3 microcontroller as DUT2 was used to control an AC light bulb via a DC relay (Figure 37.3).

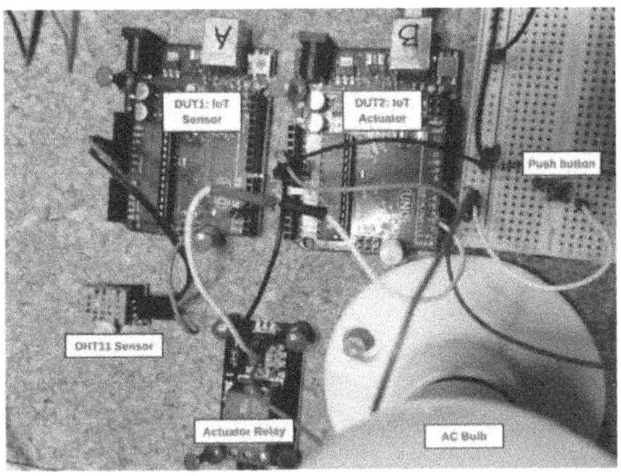

Figure 37.3. Separate IoT sensor and IoT actuator experimental setups.

Source: Author.

The components and configurations of the setup were as follows:

- Arduino Uno R3 microcontroller
- Push Button
- Pull down resistor
- AC light bulb
- 5V DC relay
- Power supply

5.3. EM fault injection attacks

To maximize the effectiveness of EM pulse injection for each Device Under Test (DUT), the EM Injector Probe tip were positioned approximately 1mm above the exposed data cable for each sensor and actuator connected to the microcontroller. Controlled experiments are conducted to introduce electromagnetic faults into two distinct target devices.

Unlike traditional first-order attacks, this research explores an indirect approach to compromising the microcontroller by altering the sensor and actuator inputs/outputs through induced perturbations.

EMFI is conducted on each DUT at varying EMFI settings of the EMFI Pulse Generator microcontroller. The frequency of the injected signal remained fixed at 120MHz due to the design of the mosquito zapper circuit, while the pulse intensity and duration were adjusted based on the potential threats identified in the literature reviews.

5.4. Data collection

Baseline sensor readings and actuator behaviour were initially recorded for both DUTs and for each sensor type without any EM injection. An oscilloscope was used to visualize and a laptop computer to collect the data. Trigger pulses were also visualized, collected, and compared against the sensor and actuator signals, both with and without EMFI perturbations. Data for various settings and responses were systematically collected.

The microcontroller (MCU) was programmed to incorporate a 10-second delay before injecting the EM into the device for a specific amount of time. Sensor/actuator response were collected for 26 seconds.

For the actuator data collection, the actuating signal and the current state of the relay will be collected and visualized. Data and visualization from a normal switch action will be collected as baseline data. When perturbations are applied, the actuating signal with perturbation will also be visualized.

5.5. Statistical analysis

The acquired data is statistically analyzed using the Independent Paired T-Test. The Null Hypothesis (H_0) proposes that there are no significant changes between the EMFI-exposed and control groups, and that any differences are due to chance. The Alternative Hypothesis (H_a) implies, on the other hand, that EMFI causes significant changes in sensor readings and actuator control behaviour.

By comparing paired data before and after EMFI exposure, the Independent Paired T-Test evaluates these assumptions. A p-value (Sig.) of less than 0.05 supports H_a, suggesting that EMFI has a statistically significant impact. A p-value greater than indicates no significant influence, confirming H_0.

6. Results and Discussion

The acquired dataset produced significant insights into the response of individual sensors when subjected to EMFI perturbations. The statistical analysis and data visualization revealed that DUT1 with Analog sensors exhibited the highest susceptibility to alterations in their reading.

This susceptibility is primarily ascribed to the presence of the Analog-to-Digital Converter (ADC) integrated into the Analog GPIO pins of the Arduino Uno MCU. The ADC plays a pivotal role in directly converting analog voltage readings into numerical data. It is noteworthy that, when the data cable was subjected to EMFI attacks, the resultant electrical noise amalgamated with the authentic sensor readings, leading to substantial variations in the recorded data. This direct voltage-to-numerical conversion process renders Analog sensors particularly vulnerable to fluctuations in the input voltage levels, thereby contributing to the observed alterations in the sensor readings. Figure 37.4 provides a visual representation of the effects of introducing EMFI to LM35 and Photocell analog sensors.

In contrast, the Digital sensor DHT11, equipped with both temperature and humidity sensing capabilities, displayed a more resilient response to EMFI injection. Specifically, the analysis revealed that only the humidity readings were susceptible to tampering by EM perturbations as shown in Figure 37.5, and this vulnerability occurred at specific voltage levels. Further investigation revealed that increasing the voltage beyond 500V caused the DHT11 to display corrupted readings and could even lead to sensor malfunction.

The communication protocol associated with DHT11 is notably different from the microcontroller, hence its difficult to handle data. In order to alter the digital signal and get past the checksum, the data had to be tampered with in a precise timing and with an accurate pattern of perturbations. These papers illustrate just how important it is to understand both the specific vulnerabilities and communication protocols

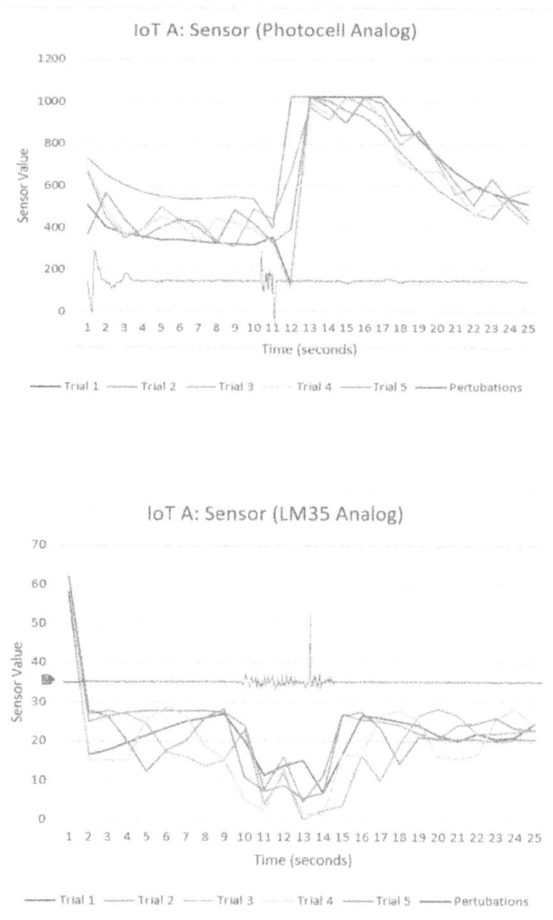

Figure 37.4. (Top) Photocell and (bottom) LM35 temperature sensor response to EMFI injected after 10 seconds delay at 1 second and 5 second duration respectively.

Source: Author.

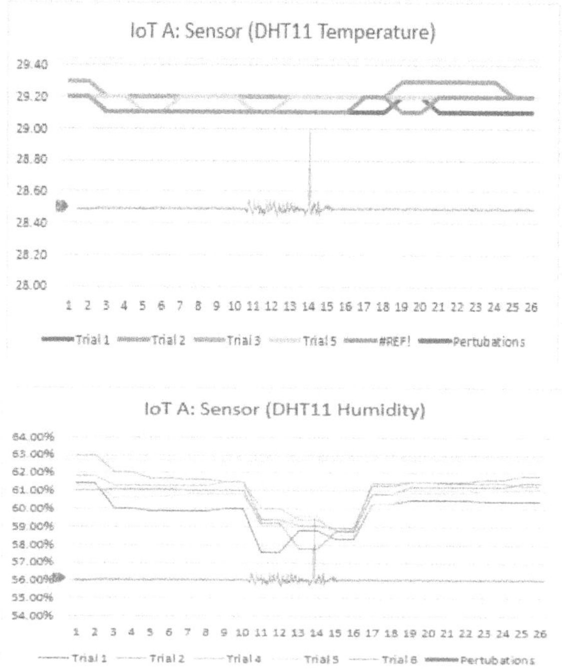

Figure 37.5. DHT11 sensor response (top) temperature and (bottom) humidity sensor response to EMFI injected after 10 seconds delay at 5 second duration.

Source: Author.

inherent to the digital sensors you target when trying to evaluate their susceptibility to EMFI attacks.

Furthermore, in viewing the option of attacking both analog and digital sensors, the research indicated that continuous EMFI attack signal should be applied at a set voltage level to manipulate the reading. These observations highlight the necessity of ongoing disturbance for the successful data manipulation under the assumption of IoT/CPS applications with sensors.

The response of DUT2 with actuator switch relay over EMFI perturbations is given in Figure 37.6. This demonstrated the crucial insights into EMFI to manipulate the relay states. We observed that relaxing the EMFI consumption bound to include an adversarial EMFI signal, could indeed alter the state of a relay. It was then found that the success of flipping the relay actuator state depended on the pulse duration of the EMFI pulse.

Particularly, more lengthy EMFI injections caused the relay to be turned on and off erratically, which may cause continuous erratic on and off condition. It is due to the prolonged exposure of EMFI which might have caused the abnormal functioning and also breaking the relay. However, it was shown that short transient EMFI pulse were more effective to produce the desired variation of the system state. Though it must be kept in mind that the duration of these pulses should be greater than the charging time of the pulse capacitor and switching time of the IGBT.

The results of this work suggest that for EMFI use when actuating binary state actuators in IoT/CPS devices, there is an optimal time to inject the EMFI pulses, as well as an optimal duration of the EMFI pulse sequence. They also point out that EMFI attacks require fine grain control and precision in order to have expected outcomes and to minimize side effects.

7. Conclusions

To sum up, this work utilizes EMFI fault model to reveal and uncover new vulnerabilities that result from Electromagnetic Interference (EMI) injection attacks on IoT/CPS actuator and sensor application involving an important platform of adversarial attacks.

The results show that Analog sensor, whose voltage to reading conversion mechanism is Analog to Digital Conversion (ADC), are highly susceptible to reading tampering due to electromagnetic interference (EMFI). On the other hand, Digital sensors(DHT11) were more resolute since it had specific vulnerabilities in humidity read at some voltage levels. For sensors that have built in security, such as the DHT11, digital encoding pattern must be recognized and timing must be set just right to successfully manipulate data via the digital signal and pass the checksum.

The study also showed that it is possible to manipulate binary actuation states of an actuator via relays through EMFI attacks. While EMFI injection duration was important, longer pulses caused erratic relay behaviour while shorter pulses produced more reliable results if within certain timing bounds.

Collectively, these findings emphasize the need to identify and understand the actuator and sensor vulnerabilities as well as communication protocols of an IoT/CPS when defending IoT/CPS devices from EMFI based adversarial attacks.

Furthermore, the research also highlights the necessity for an enhanced security measure in countering an EMFI attack in the same technology application because IoT/CPS technology is expanding at a rapid pace.

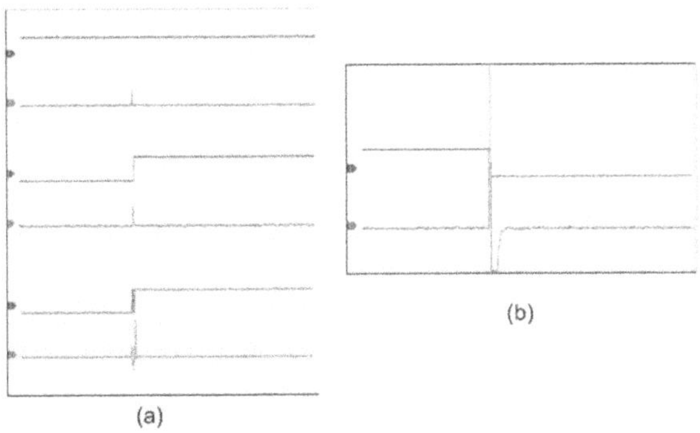

Figure 37.6. Relay-state response in blue to EMFI trigger in red with different pulse duration (a). Relay-state response in blue to Push button trigger in red with different pulse duration (b).

Source: Author.

8. Acknowledgements

As the research proceeded, the researchers want to take the opportunity to sincerely express their heartfelt gratitude to their wives, children and families for their continuous support and understanding. The support and affection of the researchers fuelled and encouraged them to survive over challenges and accomplish their goals. They also thank the Almighty God for His gift of strength, fortitude and wisdom in exactly completing this journey they embarked on together. It was during those tough times that his guidance and blessings inspired and hoped me.

References

[1] Gangolli, A., Mahmoud, Q. H., & Azim, A. (2022). A Systematic Review of Fault Injection Attacks on IoT Systems. In *Electronics (Switzerland)* (Vol. 11, Issue 13). MDPI. https://doi.org/10.3390/electronics11132023

[2] Gaur, V. S., Sharma, V., & McAllister, J. (2023). Abusive adversarial agents and attack strategies in cyber-physical systems. *CAAI Transactions on Intelligence Technology*, *8*(1), 149–165. https://doi.org/10.1049/cit2.12171

[3] Cybersecurity for embedded system devices | Arrow.com. (n.d.). Retrieved July 22, 2023, from https://www.arrow.com/en/research-and-events/articles/cybersecurity-for-embedded-devices

[4] Hasan, M., & Mohan, S. (2019). Protecting Actuators in Safety-Critical IoT Systems from Control Spoofing Attacks. http://arxiv.org/abs/1908.09444

[5] Khoumsi, A. (2019). Sensor and actuator attacks of cyber-physical systems: A study based on supervisory control of discrete event systems. *2019 8th International Conference on Systems and Control, ICSC 2019*, 176–182. https://doi.org/10.1109/ICSC47195.2019.8950627

[6] Jinteng, J., Li, H., Feng, Y., Qian, C., & Liu, Q. (2022). In-depth Analysis of the Effects of Electromagnetic Fault Injection Attack on a 32-bit MCU. https://doi.org/10.1109/SOCC56010.2022.9908097

[7] Liao, H., & Gebotys, C. (2019). Methodology for EM Fault Injection: Charge-based Fault Model. https://doi.org/10.23919/DATE.2019.8715150

[8] Cruz, L. A. A., Griño, M. T. T., Tungol, T. M. V., & Bautista, J. T. (2019). Development of a Low-Cost Air Quality Data Acquisition IoT-based System using Arduino Leonardo. *International Journal of Engineering and Manufacturing*, *9*(3), 1–18. https://doi.org/10.5815/ijem.2019.03.01

[9] Dumont, M., Lisart, M., & Maurine, P. (2019). Electromagnetic fault injection: How faults occur. *Proceedings - 2019 Workshop on Fault Diagnosis and Tolerance in Cryptography, FDTC 2019*, 9–16. https://doi.org/10.1109/FDTC.2019.00010

[10] Trouchkine, T., Bukasa, S., Escouteloup, M., Lashermes, R., & Bouffard, G. (2021). Electromagnetic fault injection against a complex CPU, toward new micro-architectural fault models. *Journal of Cryptographic Engineering*, *11*, 1–15. https://doi.org/10.1007/s13389-021-00259-6

[11] Cui, A., & Housley, R. (2017, August). BADFET: Defeating Modern Secure Boot Using Second-Order Pulsed Electromagnetic Fault Injection. *11th USENIX Workshop on Offensive Technologies (WOOT 17)*. https://www.usenix.org/conference/woot17/workshop-program/presentation/cui

[12] CWE - CWE-1319: Improper Protection against Electromagnetic Fault Injection (EM-FI) (4.12). (n.d.). Retrieved July 22, 2023, from https://cwe.mitre.org/data/definitions/1319.html

[13] Kazemi, Z., Hely, D., Fazeli, M., & Beroulle, V. (2020). A Review on Evaluation and Configuration of Fault Injection Attack Instruments to Design Attack Resistant MCU-Based IoT Applications. *Electronics*, *9*, 1153. https://doi.org/10.3390/electronics9071153

[14] Zhang, Y., & Rasmussen, K. (2022). Detection of Electromagnetic Signal Injection Attacks on Actuator Systems. http://arxiv.org/abs/2203.07102

[15] Sredić, S., Knežević, N., & Milunović, I. (2024). Effects of Landfill Leaches on Ground and Surface Waters: A Case Study of A Wild Landfill in Eastern Bosnia and Herzegovina. *Archives for Technical Sciences*, *1*(30), 97–106. https://doi.org/10.59456/afts.2024.1630.097S

[16] Koçak, C., Aydin, C. M., & Ünlüoğlu, A. (2023). Occurrence of Homola barbata (Fabricius, 1793) (Decapoda, Brachyura) in Finike Bay (Türkiye, Eastern Mediterranean Sea). *Natural and Engineering Sciences*, *8*(3), 159–167. http://doi.org/10.28978/nesciences.1404924

[17] Wong, S. K., & Yiu, S. M. (2020). Identification of device motion status via Bluetooth discovery. *Journal of Internet Services and Information Security*, *10*(4), 59–69.

[18] Kumar, S., & Ramesh, C. (2024). Mechanical Component Design: A Comprehensive Guide to Theory and Practice. *Association Journal of Interdisciplinary Technics in Engineering Mechanics*, *2*(2), 1–5.

[19] Quisquater, J.-J. (2002). Eddy current for magnetic analysis with active sensor. https://www.researchgate.net/publication/239065886

[20] Qiu, P., Wang, D., Lyu, Y., & Qu, G. (2019). VoltJockey. Proceedings of the 2019 ACM SIGSAC Conference on Computer and Communications Security, 195–209. https://doi.org/10.1145/3319535.3354201

[21] Abdellatif, K. M., & Hériveaux, O. (n.d.). *SiliconToaster: A Cheap and Programmable EM Injector for Extracting Secrets*.

[22] Ordas, S., Guillaume-Sage, L., & Maurine, P. (2017). Electromagnetic fault injection: the curse of flip-flops. *Journal of Cryptographic Engineering*, *7*(3), 183–197. https://doi.org/10.1007/s13389-016-0128-3

[23] Dutertre, J.-M., Menu, A., Potin, O., Rigaud, J.-B., & Danger, J.-L. (2021). Experimental Analysis of the Electromagnetic Instruction Skip Fault Model and Consequences for Software Countermeasures. https://www.elsevier.com/open-access/userlicense/1.0/

[24] Beckers, A., Kinugawa, M., Hayashi, Y., Fujimoto, D., Balasch, J., Gierlichs, B., & Verbauwhede, I. (2020). *Design Considerations for EM Pulse Fault Injection* (pp. 176–192). https://doi.org/10.1007/978-3-030-42068-0_11

[25] Selvaraj, J., Dayanıklı, G. Y., Gaunkar, N. P., Ware, D., Gerdes, R. M., & Mina, M. (2018). Electromagnetic Induction Attacks Against Embedded Systems. *Proceedings of the 2018 on Asia Conference on Computer and Communications Security*, 499–510. https://doi.org/10.1145/3196494.3196556

[26] Dayanikli, G. Y., Hatch, R. R., Gerdes, R. M., Wang, H., & Zane, R. (2020). Electromagnetic Sensor and Actuator Attacks on Power Converters for Electric Vehicles. *2020 IEEE Security and Privacy Workshops (SPW)*, 98–103. https://doi.org/10.1109/SPW50608.2020.00032

[27] Sousa, Á. E., Messai, N., & Manamanni, N. (2022). False Data Injection Detection in Cyber-Physical System. *IFAC-PapersOnLine*, *55*(6), 420–426. https://doi.org/10.1016/j.ifacol.2022.07.165

[28] Tu, Y., Rodriguez, A., Rampazzi, S., Fu, K., Hao, B., & Hei, X. (2019). Trick or heat? Manipulating critical temperature-based control systems using rectification attacks. *Proceedings of the ACM Conference on Computer and Communications Security*, 2301–2315. https://doi.org/10.1145/3319535.3354195

[29] O'flynn, C. (n.d.). PicoEMP: A Low-Cost EMFI Platform Compared to BBI and Voltage Fault Injection using TDC & External VCC Measurements. https://github.com/newaetech/ChipSHOUTER/blob/master/

[30] Kumar, R., Jovanovic, P., & Polian, I. (2014). Precise fault-injections using voltage and temperature manipulation for differential cryptanalysis. *Proceedings of the 2014 IEEE 20th International On-Line Testing Symposium, IOLTS 2014*, 43–48. https://doi.org/10.1109/IOLTS.2014.6873670

[31] Delarea, S., & Oren, Y. (2022). Practical, Low-Cost Fault Injection Attacks on Personal Smart Devices. *Applied Sciences (Switzerland)*, *12*(1). https://doi.org/10.3390/app12010417

[32] Lim, H., Lee, J., & Han, D.-G. (2020). Novel Fault Injection Attack without Artificial Trigger. *Applied Sciences*, *10*(11), 3849. https://doi.org/10.3390/app10113849

[33] Maistri, P., Leveugle, R., Bossuet, L., Aubert, A., Fischer, V., Robisson, B., Moro, N., Maurine, P., Dutertre, J.-M., & Lisart, M. (2014). Electromagnetic analysis and fault injection onto secure circuits. *2014 22nd International Conference on Very Large Scale Integration (VLSI-SoC)*, 1–6. https://doi.org/10.1109/VLSI-SoC.2014.7004182

[34] Balasch, J., Arumi, D., & Manich, S. (2017). Design and validation of a platform for electromagnetic fault injection. https://doi.org/10.1109/DCIS.2017.8311630

[35] Selvaraj, J., Mina, M., Song, J., Somani, A., Lu, M., & Tuttle, G. (n.d.). Intentional electromagnetic interference attack on sensors and actuators.

[36] Makama, J., Alpha, M., Kure, N., Bello, I. A., Daniel, T. S., I., & Adoyi, E. (2016). Design and Construction of a Tripler Circuit for a Mosquitor Zapper. *American Journal of Engineering Research*, *5*, 256–260.

[37] Thakare, A. R., Urkude, S. B., & Argelwar, R. P. (2015). Analysis of Cockcroft-Walton Voltage Multiplier. https://api.semanticscholar.org/CorpusID:45270286

38 Assessing India's export performance of agricultural products: A case study on China

Sarithamol K. K.[1,a] *and Biju P. J.*[2,b]

[1]Assistant Professor, Department of Commerce, Karpagam Academy of Higher Education, Coimbatore, Tamil Nadu, India
[2]Assistant Professor, Loyola College, Chennai, India

Abstract: This study evaluates India's agricultural goods export performance to China with an emphasis on trade trends, obstacles, and prospects. With China as one of its main trading partners, exports of agriculture are important to India's economic growth. The paper also offers new prospects for growing agricultural exports, emphasizing possible avenues for market diversification, logistics advancements, and policy action. Through an analysis of agrarian export patterns to China, this study seeks to offer a strategy. This study thoroughly investigates the China-India agribusiness energy exchange between 2002 and 2021. It utilized (RCA) and (RSCA) lists and different factual examinations to survey the exchange execution and give an improved understanding of the concentration design. The findings uncovered that India has a more expanded trade portfolio than China, with a more critical quantity of cultivation sections illustrating a comparative advantage. On the other hand, China's trade load is more focused, highlighting fewer sections with a virtual advantage but advanced values of RCA and RSCA.

Keywords: Agricultural exports, RCA, RSCA, India, China

1. Introduction

Agriculture is the main source of the Indian economy, it provides a significant portion along with population contributed to a large share of the country's GDP. In numerous sectors of the agriculture budget, exports should play a substantial role in boosting financial growth and also enhancing foreign exchange earnings, and improving the farmer's income. China is the main economy in the world, represented by the critical market along with the Indian agricultural products. To compare the population, to rising the demand for diverse agricultural commodities, china has to emerge through a key trading partner along with India, specifically for 'rice, tea, spices and fruits.'

Despite the growing significance of agricultural exports in China, India was faced with both challenges and opportunities in tapping the full potential of this market. Trade analysis between India and China contains geopolitical, economic, and policy factors that are influenced by volume, variety, and agricultural exports. China played a dominant worker in global agricultural trade. India country performed to the limited factors along with tariff barriers, non-tariff measures, and political dynamics. It is the understanding factor for formulating the strategies along with India's export performance in the agriculture sector. The aim of this study is to India's export performance of agricultural products to China by analyzing the key values followed by 'trade flows, identifying challenges, exploring the potential opportunities for enhancing the export volumes' [1]. This study mainly discusses the comprehensive analysis of India's agriculture export strategy to China followed by the various framework related to optimizing trade between the two countries.

On October 30, 1947, 23 countries engaged in the (GATT or GATT-47), including China and India. However, China's financial course was altered in 1949 when it was removed from the GATT system. On April 1st, 1950, India became the first non-socialist country to initiate political negotiations with the PRC. China's change from a centrally planned to a market-oriented economy began in 1978 when it began enforcing market regulations. Agricultural productivity was greatly increased under the Household Accountability System, which gave farmers more autonomy. These policies helped China's economy rebound by ensuring food safety and making a surplus of workers for urban industrialization [5]. China's history depended on the parallel adoption of these changes. More than twelve years later, in 1991, India implemented economic reforms with the primary objective of advancing economic liberalization and change. However, agricultural reform was not the main focus of India's 1991 economic reform, in contrast to China [6]. India continued to participate in the GATT-47 contract at the same time, and on January 1st, 1995, it joined the World Trade Organization (WTO) as one of its founding members. On December 11, 2001, over five years later, China became a member of the WTO, marking a significant step in its integration of international trade.

[a]sarithamol.kamalasanan@kahedu.edu.in, [b]bijupathayil@gmail.com

DOI: 10.1201/9781003684459-38

To fill out the research gap, this study discusses the trade across China and India focusing on agricultural commodities across the import data. This research study examines the importing of agricultural goods from China to India. Here discussed the main contributions below.

1.1. Objectives

1. To identify key export products from India to China
2. To evaluate export trends and patterns of agricultural products to China

These basic ideas are discussed in the remaining article, Section I describes about import and export of agricultural products comprehensively. Section II describes export trends and patterns of agricultural products to China. In the final section, we concluded the discussion with the study limitations.

2. Literature Survey

Based on Keynesian theory, increased exports lead to an analysis of heightened income growth in the short term followed by foreign exchange multiplier effects. Additional thing as exports generate a greater inflow of foreign exchange, which is utilized to acquire manufactured agricultural products. This study examined how China and India's markets may be evaluated in light of various NEER (Nominal Effective Exchange Rate) approaches. The findings imply that the different changes are absorbed in opposite directions by India's exports. Further discussion to check decisions followed by the export market through particular products correlated to PTM decisions. The correlation estimates to analyze the exchange rate pass-through (ERPT) will biased. Outcomes revealed that a variety of biases persists in the exports of affluent markets, and the pricing of transfers to low-income marketplaces operates independently of the choice to export.

The research study mainly describes the two important areas related to identifying the key export products from India to China and evaluating export products and patterns of agricultural products to China. Shukla et al. [4] describes that The RCA and RSCA indices play a central role in analyzing and evaluating the comparative advantages in agricultural trade, along with trade trends, patterns, and the efficiency of producing exact agricultural merchandise [4]. The RCA index helps in measuring trade specialization within the agricultural sector by highlighting the products that a nation exports in larger quantities relative to its total exports. Meanwhile, the RSCA index provides a balanced view of comparative advantage by taking into account both exports and imports. These indices allow investigators or academic people to identify clear specialization patterns, indicating areas where a country holds a distinct comparative advantage.

A further significant factor is that, although numerous studies have examined the comparative advantage in agricultural trade between China and India, relatively few have concentrated on the two countries' bilateral agricultural commerce. This study attempts to provide a contemporary and thorough viewpoint on the parallels and discrepancies in China's and India's agricultural trade comparative advantage structures over time, as well as the consequences for trade trends and market dynamics, by evaluating and synthesizing previous research [9]. Through careful examination, this study aims to offer insightful information that will help researchers and policymakers create plans that will improve trade cooperation and make use of China's and India's comparative advantages in agricultural commerce.

Using a range of quantitative techniques, Wang et al. [12] empirically examined the comparative superiority and complementarity of Sino-US agricultural trade with respect to sixteen major agricultural products in 1997 [12]. Sharma SK and Bugalya K used RCA, Bilateral RCA, and TCI to evaluate the agricultural sector's competitiveness in India's cotton production and exporting. Narayan S. and Bhattacharya P. investigated India's relative export competitiveness in four agricultural commodities and their determinants between 1961 and 2012 using time-varying competitiveness indicators based on the RCA index. Using RCA and other quantitative techniques, Tao Z. assessed the agricultural trade between Thailand and China to ascertain the benefits and drawbacks of this sector on a complementary and competitive spectrum. Using descriptive statistics and visual aids, Investigated and evaluated the competitiveness of the agricultural product trade between China and the countries along the 'Belt and Road' initiative. Investigated the dynamics of China's export relative gain from 2001 to 2020 [10]. The QAP regression technique and data from various sources to perform a network study on international agricultural commerce between 1991 and 2021.

Agricultural exports played an important part in the Indian economy, making substantial contributions to national revenue, employment, and foreign currency earnings. China, as one of India's principal trading associates, presents both problems and forecasts for Indian agricultural exports. A comprehensive analysis of the existing literature on these subjects highlights several aspects, such as trade policy frameworks, product-dependent outcomes, logistical challenges, and geopolitical factors, that impact India's agricultural export performance to China. Numerous investigations have recorded the advancement of agricultural exports from India over the past few years. India has experienced a steady state of increase in exports, predominantly influenced by commodities such as rice, spices, tea, and seafood. Increasingly varied in recent years, with China rising as a significant participant in promoting trade, particularly for items like rice and spices. The growth of export volumes is described to elements such as India's expanding production capacity, heightened competitiveness, and enhanced marketing approaches. Observe that India's agricultural exports continue to encounter obstacles due to infrastructural limitations, elevated production expenses, and challenges regarding the quality standards mandated by the Chinese market.

India's agricultural exports to China are shaped by both domestic and global trade policies. Research shown by Ghosh [8] to observes how China's agricultural import policies, encompassing tariff and non-tariff barriers, affect the movement of agricultural goods from India [7]. For example, India's rice exports to China have encountered stringent phytosanitary regulations and tariffs, which have at times obstructed the growth of trade in this domain underscores the significance of trade contracts, such as the China-India Free Trade Agreement (FTA), in promoting more efficient trade flows. Nonetheless, the contend that the absence of a complete and updated bilateral trade agreement concentrated on agriculture may have resulted in lost chances for enhancing export presentation [11].

Product-specific analyses offer an enhanced understanding of India's export capabilities to China that indicate that Indian rice exports to China have consistently increased in recent years, particularly attributed to the competitive pricing and superior quality of non-basmati rice [12]. Nonetheless, trade obstacles, such as China's rigorous quarantine regulations and inspection protocols, are regarded as constraining elements for the growth of rice exports. Likewise, Bhatia [3] underscores that Indian tea exports to China have fared relatively well, with Assam and Darjeeling tea gaining traction. However, Indian exporters encounter difficulties in competing with China's extensive domestic tea sector, which is both substantial and intricately connected to the global supply chain. [14]. Regarding spices, India continues to be the leading supplier to China, and research conducted by Singh & Kumari (2021) highlights the rising demand for Indian spices owing to the increasing acceptance of Indian cuisine in China. However, logistical issues, quality uniformity, and market accessibility remain aspects that necessitate enhancement for Indian exporters [12]. Patel (2018) points out that China's demand for high-quality spices is on the rise, which provides an opportunity for India to increase its market share, provided that it complies with China's quality and safety requirements.

The body of work regarding the potential for enhancing India's agricultural exports to China is expanding, with numerous research efforts concentrating on emerging patterns and growth areas. Bhat and Kapoor (2021) propose that India could boost its agricultural exports to China by concentrating on organic farming and sustainable production methods, which correspond with the rising consumer demand in China for organic and nutritious food items [13]. Likewise, Singh et al. [16] underscore the necessity for India to investigate unexploited markets for agricultural products such as processed foods, herbs, and medicinal plants, which are gaining popularity in China. Additionally, Mitra (2020) supports the advancement of bilateral trade relations through policy modifications and improved alignment of agricultural standards to satisfy Chinese consumer preferences.

2.1. Problem statement

India's agricultural sector played an important role in the national economy which contributed to the significant effects of foreign exchange earnings and financial growth. China is the main economy in the world, it represents the critical market through Indian agricultural products. Based on that India's agricultural exports to China faced various challenges for realizing the full market potential, here to understand key issues along with the agricultural products from India most successful in the Chinese market through evaluating the trends and patterns followed by the exports. It's for identifying the factors that influenced these dynamics. This research study discusses India's export performance of agricultural products through China, specifically focusing on identifying the key products, evaluating the export trends, and understanding the trade patterns among both countries. To analyze through an examination of the performance of key agricultural commodities such as rice, spices, tea, fruits, and seafood, this study will emphasize the determinants influencing their achievement or constraints within the Chinese market. Additionally, the research will explore wider trade dynamics, encompassing the effects of trade regulations, tariffs, logistical obstacles, and geopolitical factors on India's agricultural exports to China. The major issue of this research addresses the comprehensive understanding of ideas related to India's agricultural exports to China including the identification of exported products, evaluate performance over time, to the analysis of the relationship between critical barriers and trade. The main findings of this study as valuable information are formulated to the strategies, and also enhanced the India agriculture export performance and strengthened the trade between with China.

2.2. Research gap

The different advantages of agricultural product imports are frequently discussed, but the economic impacts of these imports on native agriculture, food safety, and consumer prices are rarely taken into account. More research is needed to find out how imports affect local output, market dynamics, and agricultural employment. Many studies examine many studies that examine trends and patterns in agricultural imports, but few that examine how these trends vary across different goods. Past studies have tried to identify the most important agricultural imports, but little is known about how these items are interdependent, how they relate to domestic demand, or how they affect food systems as a whole.

3. Research Methodology

Based on this research followed by both descriptive and analytical methods, these data sets were collected from the APFPEDA and WITS. These are all the online resources made of bulk information of secondary data. Quantitative analysis is used to track the growth and identify agricultural exports over the years. The research methodology discusses a systematic approach to analysis the of India's agricultural export performance to China. To combine primary and secondary data used by relevant frameworks. The main goal of this research is to provide information related to policymakers, stakeholders, and exporters.

3.1. Sampling details

Using a comprehensive counting method, the study looks at trade numbers between China and India from 2019 to 2024. The comprehensive dataset, which covers a wide range of agricultural products, aims to identify major patterns and shifts in trade across time. Modern methods like trend analysis and ARIMA are used to analyze the longitudinal data, giving a comprehensive picture of the dynamics of agricultural commerce between China and India.

3.2. Period of the study

The five years between 2019–20 and 2023–24 have been selected by the research team to investigate agricultural exports. Our aim has been to examine the goods exchanged between China and India from 1996 to 2018. The International Trade Center (ITC Trade Map) and the UN Common Format for Transient Data Exchange (UN COMTRADE) are the sources of this trade flow data.

3.3. Index of trade integration pattern

The index of trade integration pattern between India and China describes the extent of nature and provides trade relationships and also examines the both quantity and quality of their needs. Both countries have major economies globally, providing the relationship between an integral part of trade dynamics. India primarily imports and manufactures goods, electronics, and other machinery from China [11]. To export raw materials like cotton, ores, and agricultural products to China. It also reflects the asymmetric relationship that India won runs up to the trade deficit with China. The bilateral trade to GDP ratio is used to measure the intensity of the trade relationship. Both India and China have high trade-to-GDP ratios. For India, it reflects the dependence on trade to some extent, it's had lower than China. The ratio between the India-China trade gives some sense related to the trade relationship in the global context.

Calculating Trade Integration Pattern:

Some few methods are used to calculate the Bilateral Trade Intensity Index,

Trade Intensity Index=Bilateral Trade (India-China) / Combined GDP of India and China

To integrate the two countries followed by interns of total global trade related to economic size.

Trade Intensity Index:

To estimate the trade intensity index used to measure how well the export patterns of India and China are with each other. TII measures the value of employment between the two countries that would be expected from importance in world trade. It is also defined as the share of one country's exports to another country separated by the share of the world's exports with the same partner.

$$TII= (Xij/Xit) / (Xwj/Xwt) \tag{1}$$

From the above Equation (1) we describe X_{ij} as the country's value, I noted as exports to country j, Xwj as world exports to country j, and X_{it} means country i total exports. To calculate the average intensity index value as one, greater than the computed values. It also indicated the higher degrees of trade intensity between the two countries India and China.

3.4. Modified trade intensity index

The given formula calculates the trade intensity index modified and used by rest of the world (RoW) exports instead of the used world shown as below, RoW is defined as the world export excluding the country value of 'I'.

$$T''II= (Xij/Xi\ t)/X''wj/X''wt) \tag{2}$$

J = Xwj-Xij, X'wt=Rest of the World exports=Xwt-Xit, and X'wj=Rest of the World exports to the nation. In this case, we defined Tij as altered by TII. The values of Tij and T'ij are nearly equal for nations with a negligible global export share. This section discusses the differences between the important indicators for the two nations whose share values are connected to global exports [9]. The bilateral trade flow should be used to predict the index's value. According to the anticipated level of trade flow, the index's value should be equivalent to the percentage of global exports to the same partner. The extent of the expansions relates to export trade in the partner country if the country is unable to take advantage of its market share. The guide value is less than indicating the bilateral trade flow between the larger and expected to analyze the partner country of the world trade importance.

3.5. Revealed comparative advantage index

This research paper identifies the revealed option through the comparative advantage index value export from data. Balassa index value is measured to use for to assess the country's export as potential. To calculate the index value calculated through sector and commodity levels used by the SITC-2-digit level of classifications. The RCA value indicates whether the country's process expands to the products that have trade potential. On the other side, products should be followed by new partners. Both countries have similar RCA profiles with have potential trade between the industry trade. To measure RCA estimate and calculate the high level of production disaggregation focused on attention to the non-traditional products, it might be positively exported. RCA index of the country I for the product j is often slow by the product share along with the country export relation share through the world trade economy,

$$RCA_{ij} = (X_{ij}/X_{it}) / (X_{wj}/X_{wt}), \quad (3)$$

Where, X_{ij} and X_{wj} are mentioned as the values of republic export products through j, X_{it}, and X_{wt} represent the exports of world total exports. The value should be less unity, it is implied to the country and reveals the comparative disadvantage in the products.

3.6. Trade pattern between India and China

The quick growth of India's China trade is the starting stage of the century for China to connect through India's largest goods trading partner by 2008. In this situation, china continues to hold it. In the last decades, the bilateral trade between the two countries has been recorded and followed by exponential growth. From Table 38.1 2015 to 2022, India-China bilateral trade grew by 84.44%, with an average yearly growth of 12.89%. Overall trade, compared to both countries, china increased by 8.47% over reached USD 136.26 billion crossing through the USD 100 billion mark for the second time in a row. Followed by trade shortage comes under the USD of 101.28 billion India imports from China and increased by 118.77% reached through USD of 118.77 billion. Based on India exports to China reduced by 37.59% year by year reaching 17.49 billion compared to last year's remaining exports of USD 28.03 billion.

Based on ARIMA Output Table 38.2 analysis, evaluate how the model was fit and data will provide accuracy and reliability. To interpret AIC, 49.899 indicates a better fitting model, compared to goodness of fit helps to avoid overfitting. Followed by this interpretation to taken as observation as 8, SSE value of 144.225, which indicates a better fit and also represents the model that was predicted. MSE as

Table 38.1. India China bilateral trade

INDIA CHINA BILATERAL TRADE (Figures in USD Bn)							
Year	*China Export to India*	*% Change*	*China imports from India*	*% Change*	*Trade Deficit*	*Total Trade*	*% Change*
2016	11.75	−12.29	59.43	2.01	47.68	71.18	−0.67
2017	16.34	39.11	68.1	14.59	51.76	84.44	18.63
2018	18.83	15.21	76.87	12.89	58.04	95.7	13.34
2019	17.97	−4.55	74.92	−2.54	56.95	92.9	−2.93
2020	20.87	16.15	66.78	−10.87	45.91	87.65	−5.64
2021	28.03	34.28	97.59	46.14	69.56	125.62	43.32
2022	17.49	−37.59	118.77	21.7	101.28	136.26	8.47
2023	16.49	−32.59	116.77	20.7	105.28	142.26	10.47

Source: General administration of customs, China.

Table 38.2. ARIMA output-goodness of fit statistics

Goodness of fit statistics:	
Observations	8
DF	6
SSE	144.225
MSE	18.028
RMSE	4.246
WN Variance	18.028
MAPE (Diff)	16.870
MAPE	16.870
-2Log(Like.)	45.899
FPE	23.179
AIC	49.899
AICC	52.299
SBC	50.058
Iterations	16
Convergence	Yes

Source: Author.

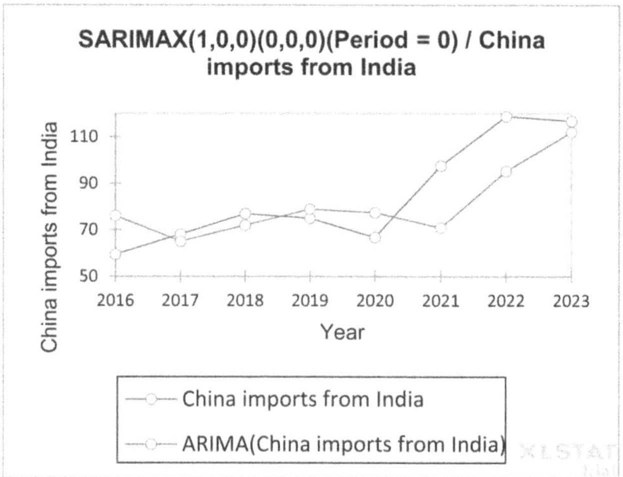

Figure 38.1. ARIMA stimulations (China imports from India).

Source: Author.

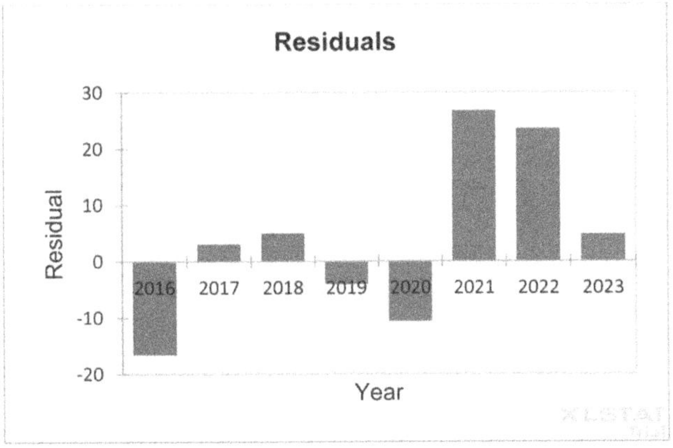

Figure 38.2. ARIMA stimulations (Time Vs Residuals).

Source: Author.

18.028, which indicates a better fit. RMSE as 4.246, which provides the ideas followed by to predict the errors. 2log like as 45.899, FPE as 23.179, which provides the model followed by the better performances. AICC as 52.299, indicates the model was a good fit and contains minimal complexity. Based on this analysis taken as iteration as 16. Figure 38.1 shows the ARIMA stimulation along with Time and China imports the products to India. This graph clearly shows the SARIMAX value under the period interval. Figure 38.2 shows the stimulation among the Time and Residuals followed by year, China imports to India, and % of changes.

4. Trade Deficit

This analysis provides results as Indian imports shifted from $66.78 in 2020 to 97.59 in 2021, which is the leading trade deficit among contains more than 23%. The limited range of primary commodities we export to China and the barriers to market access for the majority of our agricultural products and industries in which we compete, like pharmaceuticals, IT, etc., are the two main causes of the widening trade deficit with China. Iron ore, cotton, copper, aluminum, diamonds, and other natural gems have been our top exports [7]. These commodity markets based on raw materials have gradually lost ground to Chinese exports of fertilizers, organic chemicals, telecom equipment, power-related equipment, and machinery. We are still working with the Chinese side to resolve issues related to market access. The expansion in the two countries' trading volume has outpaced the growth in bilateral investment. Mutual investment flows have not yet caught up, even though both nations have become popular global investment destinations. Chinese investment in India in 2021 was USD 28.03 million, down 17.49% year over year, and by the end of 2021, the total amount of Chinese investment in India was USD 97.59 billion, according to China's Ministry of Commerce. Indian investment in China in 2021 was USD 125.62 million, down 87.65% from the previous year. By the end of 2021, cumulative Indian investment in China had reached USD 136.26 million.

From the Table 38.3 describes the imported products from China from 2020 to 2023, all information collected from Un Comtrade Database-Trade data. From the data analysis, the agricultural commodities are mentioned in code format. For example (HS 07 is vegetables, fresh, frozen, chilled, or dried vegetables), and HS-08 is fruits such as apples, bananas, and grapes. In the year 2020, China exported agricultural commodities to the world at 2,589,098,353,298. In India, this was noted at 66,719,471,588. From the year 2021, the total export commodities all over the world is 3,316,022,133,985, and In India 96,367,053,567. In the year 2022, china exports import products through the world level as 3,593,601,435,602. In India 118,501,522,799. To compare all export levels of China, most of the commodities are exported to India.

5. Export Partner Share China

Figure 38.3 and Table 38.4 provide information on the value of export partner shares in China. Approximately 3.2976 percent of China's total exports in 2019 came from exports to China. This suggests that at this time, China was a sizable but not particularly important export market for India. India's export percentage to China rose to 6.9% in 2020, a significant increase from the previous year. China's rapid return to industrial activity following the first wave of the COVID-19 pandemic is responsible for the increase [15]. China's demand for new raw materials, like other products, is another factor. The percentage of China exports to India dropped marginally to 2.9061% in 2021 (3.2976). Increased competition from other countries for China's attention in terms of export markets may be one of the reasons for this decline. Trade relations may be impacted by the ongoing political tensions between China and India, particularly in the wake of the border clashes in 2020. In 2022 (3.2976), the export share to China fell sharply to 3.33%, which was a discernible decrease. This might indicate a change in India's trade strategy, with an emphasis on lessening reliance on China. Global supply chain problems and political tensions may be additional factors. Push for greater reliance on particular sectors in India. Furthermore, the demand for China exports may be impacted by India's economic difficulties.

According to the overall trend analysis, China's sharp increase in reliance in 2020 probably reflects its increased need for specific Indian products and raw materials as it started to recover from the pandemic. The decline in India's export share to China from 2021 to 2022 after 2020 suggests that the country may be attempting to diversify its export markets or lessen its dependency on China. Political tensions, shifts in global demand, and India's economic policy priorities may also have an impact on this. From 2019 to 2022, India's exports to China fluctuated somewhat, reaching a peak in 2020 before falling in 2022. The steep decline in 2022 might indicate that the political and economic

Table 38.3. Export product details from India to China for the period 2020–2023

Period	*Reporter*	*Partner*	*Commodity Code*	*Trade Value (US$)*	*Net Weight(kg)*	*Qty*	*Alternate Quantity*
2023	China	World	7	$10,985,708,536	0	0	0
2023	China	India	7	$29,097,139	0	0	0
2023	China	World	8	$5,929,751,831	0	0	0
2023	China	India	8	$34,285,237	0	0	0
2023	China	World	80111	$286,962	278,542	278,542	278,542
2023	China	World	803	$28,232,714	26,370,052	26,370,052	26,370,052
2023	China	World	81110	$71,782,791	44,480,825	44,480,825	44,480,825
2023	China	World	TOTAL	$3,379,747,834,884	0	0	0
2023	China	India	TOTAL	$117,678,762,947	0	0	0
2022	China	World	7	$10,166,535,635	0	0	0
2022	China	India	7	$8,823,403	0	0	0
2022	China	World	8	$5,480,917,770	0	0	0
2022	China	India	8	$16,641,805	0	0	0
2022	China	World	80111	$583,354	480,394	480,394	480,394
2022	China	World	803	$24,764,380	22,087,989	22,087,989	22,087,989
2022	China	World	81110	$76,871,653	43,181,981	43,181,981	43,181,981
2022	China	World	TOTAL	$3,593,601,435,602	0	0	0
2022	China	India	TOTAL	$118,501,522,799	0	0	0
2021	China	World	7	$10,032,607,561	0	0	0
2021	China	India	7	$26,720,537	0	0	0
2021	China	World	8	$6,332,805,669	0	0	0
2021	China	India	8	$13,405,204	0	0	0
2021	China	World	80111	$319,645	261,891	261,891	261,891
2021	China	World	803	$18,439,716	20,342,383	20,342,383	0
2021	China	World	81110	$76,936,189	47,556,608	47,556,608	47,556,608
2021	China	World	TOTAL	$3,316,022,133,985	0	0	0
2021	China	India	TOTAL	$96,367,053,567	0	0	0
2020	China	World	7	$9,671,900,149	0	0	0
2020	China	India	7	$47,687,412	0	0	0
2020	China	World	8	$7,063,595,973	0	0	0
2020	China	India	8	$20,999,520	0	0	0
2020	China	World	80111	$307,873	292,729	292,729	292,729
2020	China	World	803	$15,938,545	19,753,515	19,753,515	0
2020	China	World	81110	$78,155,483	54,636,382	54,636,382	54,636,382
2020	China	World	TOTAL	$2,589,098,353,298	0	0	0
2020	China	India	TOTAL	$66,719,471,588	0	0	0

Source: UN comtrade database.

Table 38.4. Export partner share china

Reporter Name	*Partner Name*	*Trade Flow*	*Indicator*	*2019*	*2020*	*2021*	*2022*
India	China	Export	Export Partner Share (%)	5.345	6.9	5.835	3.332

Source: Author.

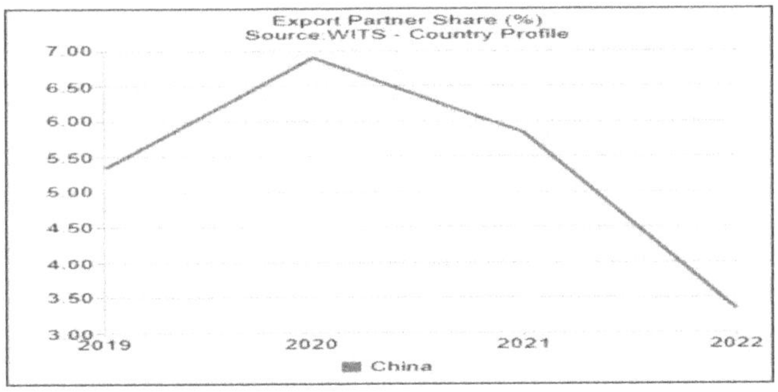

Figure 38.3. Export partner share-China.

Source: Author.

landscape of the two nations is changing. This trend may also be impacted by India's changing trade policies, diversification of its export markets, and initiatives to lessen reliance on China.

6. Trade Integration Pattern Index

Table 38.5 describes the trade integration pattern index followed by the various key aspects, both countries have the strength specific for growth driven by exports. To reach the maximum economic, China and India should be focused on manufacturing products, to invest in the automobile industry shares. Enhancing the technology to diversify the trade pattern means it has grown through long economic growth.

To analyze the Table 38.6 requires several types of potential factors related to geopolitical tensions, trade policy changes, economic slowdown in China, supply chain issues, and climate factors. The analysis report for the year 2020–23 maintains a strong performance, India consistently exports products through China from 2020–21 to 2022–23 with a positive growth rate of 19.4% and 8.6% in the first two years. In years 2023–24 suddenly dropped in year 2023 to 24, which is a major shift trade dynamic between the two countries. In future exports, to decline year 2023–24 followed by the long-term effects on India's agricultural export performance to India. Finally, we suggest that India's agricultural exports to China have strong growth from 2020 to 2023, declining through 2023–24.

Table 38.5. Trade integration pattern index

Country	*1*	*2*	*3*	*4*	*5*	*6*
China	0.28	0.24	0.42	0.41	0.32	0.32
India	0.54	0.32	0.43	0.44	0.25	0.40

Source: Author.

7. Significance of the Study

A comprehensive analysis of agricultural imports, including import evaluation, import trend analysis, and critical import item identification, can help better understand the contribution of agricultural imports from China to India's agricultural sector. The results of the study will inform trade policies that give India the best chance to benefit from these imports while preserving its domestic agriculture industry and ensuring future food supply for its citizens.

8. Suggestions

Geopolitical tensions, shifting global market dynamics, and India's continuous attempts to increase domestic agricultural production and exports have all influenced India's agricultural commodity trading patterns with China between 2019 and 2024. Based on patterns from 2019 to 2022 and anticipated changes through 2024, the following recommendations are for India's agricultural trade with China, with an emphasis on important agricultural commodities.

Trends for rice, especially Basmati, in 2019–2022, China is one of India's top export destinations for rice, especially Basmati rice. However, because China prefers to produce its rice and prefers shorter-grain types over Basmati, the amount of rice exported has been restricted. Despite sporadic shipments of Basmati rice in recent years, India's rice exports to China have not increased much.

Market Diversification refers to the high value of Basmati rice in these regions, India should keep concentrating on markets like the Middle East, Africa, and Southeast Asia. Exports could be increased, though, if deals are pursued with China to get more access to its rice market.

To increase its competitiveness, India could investigate stricter quality certification procedures, such as fulfilling China's requirements for food safety and pesticide residues.

Increase Non-Basmati Rice Exports: Given China's substantial and expanding rice demand, India may want to look into the potential to export non-Basmati rice types to that country.

Table 38.6. International trade analysis report

Country		*Value in USD Million*				
Exporting Country	*Importing Country*	*2020–21*	*2021–22*	*2022–23*	*2023–24*	*Growth(%)*
India	China	968.33	1,155.87	1,255.72	218.72	–82.58
Total		968.33	1,155.87	1,255.72	218.72	

Source: UN comtrade.

Trend 2019–2022: China's internal production and erratic demand have contributed to India's restricted trade with China, despite India being a major exporter of sugar. Brazil and Thailand are two nations that have fiercely competed with China for sugar exports in recent years.

Address Trade obstacles: In order to increase Indian sugar's competitiveness in the Chinese market, China and India should cooperate to lower trade obstacles such as import bans and tariffs.

Diversify Markets: India should concentrate on expanding its sugar exports to other countries, such as Southeast Asia, Africa, and the Middle East, where demand is growing, in light of China's robust domestic sugar production.

Develop Sugar-Based items, since these items may be more appealing in Chinese markets, India should concentrate on exporting sugar-based goods including molasses, jaggery, and value-added processed sugar goods.

Trend 2019–2022: India exports only a handful of fruits to China, including grapes, pomegranates, and mangoes. Market preferences for specific fruit kinds, logistical obstacles, and adherence to China's stringent quarantine regulations are the primary obstacles.

Boost Fruit Exports to China, by making sure that premium fruits like pomegranates, grapes, and mangoes (especially Alphonso) satisfy Chinese phytosanitary and safety regulations, India should try to boost exports of these products. By fortifying procedures and guaranteeing prompt clearance, the Indian government can aid in facilitating access.

Expand Logistics and Cold Chain Infrastructure: India should improve its transportation and cold storage facilities to guarantee freshness throughout lengthy transit periods and increase fruit shipments to China.

To increase the demand for its fruits in the Chinese market, India should look into forming regional partnerships with other Asian nations for cooperative marketing campaigns.

Trend 2019–2022, India is one of China's biggest suppliers of raw cotton, making it a significant exporter of cotton to China. However, China's growing domestic output and competitive imports from the United States and other nations have reduced India's proportion of cotton exports.

The establishment of long-term supply agreements with China for raw cotton should be India's goal to guarantee steady shipments even in the face of changes in the price of cotton globally.

Expand Export of Cotton Yarns and Fabrics: Considering the rising demand in China's textile industry, India should prioritize expanding its exports of cotton yarns and textiles to China in addition to raw cotton.

Diversify Markets, Although China is a significant importer of cotton, India should keep expanding its export market base by concentrating on Southeast Asia and Africa.

The Chinese market has a robust domestic tea sector, but India has been a major supplier of tea to China from 2019 to 2022. However, there are certain niche markets for Indian tea in China, especially for Assam and Darjeeling kinds.

Examine Specialty and Premium Teas, India should concentrate on marketing its high-end teas, including those from Assam, Darjeeling, and the Nilgiris. The increasing number of Chinese consumers who are drawn to premium imported teas may find these appealing.

Target Health Trends, Due to China's growing health consciousness, Indian teas, such as organic, herbal, and green teas, can be promoted for their health advantages (such as antioxidant qualities).

Enhance Branding and Marketing: By working with Chinese distributors and fortifying its position in important retail locations and e-commerce sites, India might concentrate on enhancing the branding and marketing of Indian teas in China.

India and China have a lot of potential for agricultural trade, especially when it comes to premium goods like cotton, rice, spices, and fruits. To lessen reliance on any one market, however, the emphasis should be on expanding exports, removing trade restrictions, and enhancing market access. India can improve its standing in the Chinese agricultural market while looking at prospects in other parts of the world by making investments in infrastructure, quality, and market diversification. Based on statically analysis India's export performance of agricultural products to China, is essential for applying some methods due to the agricultural commodities. From the statically point of view, to identify which types of commodities are dominated by India's agricultural export portfolio to China, understand the export performance and growth. Based on the analysis, we provide statically results as in the years of 2020 to 2023, export products are having strong performances. In the year 2023–24 suddenly dropped export value (–82.58) %, which indicates trade dynamics between the two countries. In the future declined in 2023–24, India's agricultural export performance to China. The exporter must be focused on addressing the trade and enhancing bilateral relationships between the positive growth trend.

9. Conclusion

India's export relationship with China was multifaceted with the key products reflecting both India's agricultural and industrial strength. To analyze the various trends from 2019 to 2022, it is the potential development through 2024 highlights some critical sectors for export growth. India's key export to China from the above years is likely to continue with agricultural commodities such as rice, spices, fruits, and raw materials. It also strengthens to sustain export growth, India country must be focused on diversifying the product types, and it is also enhanced through quality and compliance, reducing the dependence on China by discovering new global markets. Although there is a lot of room for expansion in India's agricultural exports to China, a few tactical changes are required to take full advantage of the market. Meeting China's need for high-quality agricultural products, emphasizing organic and health-conscious products, and resolving logistical issues are key opportunities. To guarantee steady growth in the Chinese market, India must also remove trade restrictions, raise quality standards, and develop its logistics system. Targeting China's expanding health-conscious consumer base, shifting toward value-added products, and broadening the export portfolio of agricultural goods should be the main priorities. India should simultaneously seek to improve its economic ties with China, lessen its dependency on single markets, and guarantee price and quality competitiveness. By making these attempts, India can increase its agricultural exports to China and succeed in the long run with this important trading partner. The scope of this study is followed by the trade policies and agreements, and China bilateral relations. The historical trade agreement is the relationship between India and China related to agricultural exports. Any changes in trade policies, tariffs, and non-tariff barriers affect the agricultural sector. To explore the influence of the political and economic relationship between China and India on agricultural products. To expand this, develop the more comprehensive ideas related to India's agricultural export performance to China. Incorporating the economic, political, historical, and environmental factors provides various challenges and opportunities in the trade relationship. Finally, the study's conclusion suggests that India may boost its imports of tariff-free agricultural products to China. This analysis finds that with $66,719,471,588 values from 2022 to 2023. China is India's top importer of agricultural products. Imports increase their foreign exchange value. India may establish solid and fruitful ties with other countries when it imports commodities from them.

References

[1] Alam, K., Uddin, G., & Alam, M. M. (2009). Trade patterns of Bangladesh with India and China: An empirical evidence of the PPP theory. *Journal of Regional Economic Studies*, *2*, 26–34.

[2] Akram, H. W., Ahmad, A., Dana, L. P., Khan, A., & Akhtar, S. (2024). Do Trade Agreements Enhance Bilateral Trade? Focus on India and Sri Lanka. *Sustainability*, *16*(2), 582.

[3] Bhatia, J. K., Mehta, V. P., Bhardwaj, N., & Nimbrayan, P. K. (2021). Export-import performance of major agricultural commodities in India.

[4] Shukla, S., Sadykova, K., & Tian, J. (2023). Research on India-China agriculture trade dynamics: A comparative advantage analysis. *Plos One*, *18*(11), e0294561.

[5] Sudhagar, S., Gurupandi, M., & Eswaran, M. S. (2022). Export performance of the agricultural products in India–with special reference to Brics Countries. *Economic Empowerment of Tiny Agricultural Sector in India*, 174.

[6] Kolte, A., Sangvikar, B., Pawar, A., & Lirong, Q. (2023). Socio-economic relations between India and China: understanding and examining the significance of trade. *International Journal of Business and Globalisation*, *33*(4), 387–424.

[7] Raju, S. (2024). Impact of imports from China on Indian manufacturing performance: an analysis of trade competitiveness. *International Journal of Emerging Markets*, *19*(12), 4503–4525.

[8] Ghosh, M. (2023). Growth and Development under Alternative Policy Regimes in India: A Political Economy Perspective. *Journal of Asian and African Studies*, *58*(7), 1134–1155.

[9] Montalbano, P., & Nenci, S. (2011). Trade Patterns and Trade Clusters: China, India, Brazil and South Africa in the Global Trading System. *India, Brazil and South Africa in the Global Trading System (February 2011)*.

[10] Naik, P. K. (2020). Exploring Factors Shaping India's Trade Patterns: Evidence from Major Trading Partners. *Journal of Business and Economic Options*, *3*(4), 150–157.

[11] Raghuramapatruni, R. (2014). A study on the trade patterns between India & China: with special focus on services trade. *Indian Ocean Digest*, *33*(1), 56–81.

[12] Wang, C., Shuai, C., Chen, X., Sun, J., & Zhao, B. (2024). Linking local and global: assessing water scarcity risk through nested trade networks. *Sustainable Development*.

[13] Veeramani, C. (2006). India and China: Changing patterns of comparative advantage?. *Georgia Institute of Technology*.

[14] Panda, R., Sethi, M., & Kumaran, M. (2016). A study of bilateral trade flows of China and India. *Indian Journal of Science and Technology*, *9*(15).

[15] Reddy, P. K., Kuchay, M. S., Mehta, Y., & Mishra, S. K. (2020). Diabetic ketoacidosis precipitated by COVID-19: a report of two cases and review of literature. *Diabetes & Metabolic Syndrome: Clinical Research & Reviews*, *14*(5), 1459–1462.

[16] Singh, S., Arya, V., Yadav, M. P., & Power, G. J. (2023). Does financial development improve economic growth? The role of asymmetrical relationships. *Global Finance Journal*, *56*, 100831.

39 Digital transformation strategy, management practices and challenges among academic heads of higher education institutions in Zamboanga Peninsula

Maricor A. Marquez[1,a] *and Joe-An G. Cuaresma*[2,b]

[1]Colegio de la Ciudad de Zamboanga, Ayala and Vitali, Zamboanga City, Philippines
[2]Commission on Higher Education, Baliwasan, Chico Rd., Zamboanga City, Philippines

Abstract: The study aimed to determine digital transformation strategy (DTS) and management practices among academic heads of higher education institutions (HEIs) in the Zamboanga Peninsula during the school year 2022-2023. The digital transformation strategy and management practices were measured in terms of three areas namely; digital leadership, technology utilization, and workforce capability through a Survey Questionnaire (SQ) and the challenges experienced by academic heads through a semi-structured interview. This study made use of mixed research design involving 108 teacher-respondents and eight academic heads-participants. Analysis of results was conducted using mean and standard deviation, parametric tests such as Pearson-r Coefficient of Correlation and t-independent test, and Analysis of Variance (ANOVA). The findings revealed that the faculty digital transformation strategy among academic heads was moderately evident in technology utilization and highly evident in terms of digital leadership and workforce capability. Moreover, their DT management practices were highly evident. With regard to challenges that were identified, academic heads experienced the challenges of managing incapable teachers, retaining qualified and competitive faculty, lack of structures and technological resources and poor internet connection. Thus, analysis on the digital transformation strategy and DT management practices in HEIs by academic heads showed there was a significant relationship in the two.

Keywords: Digital transformation strategy, management practices, digital leadership, technology utilization, workforce capability

1. Introduction

Digital transformation then is described as significant change which includes the use of digital technology to alter or to create new teaching and learning models and processes as a way of increasing the efficacy and efficiency of learning. Four, which are teaching and learning, communication, governance, planning and decision making, and process and operations are the four main areas that use of information and communication technology (ICT) falls on [1]. As a route guide and a tool, it can be used to guide the entire business through the process of how an organization should change from where there is a business demand and is calling for serious reasons for change [2]. Limited in their access in the past, digital technologies are now quickly and rapidly progressing with the necessity in the educational facilities, changing the learning environment to become digital learning, that is why the strategy of digital transformation in education – known as DTS – is recognized as equally necessary as the inevitable digital transformation in education. In universities, students are forced by DTS because of the global issues such as internet use, information interchange, digitalization, virtualization, and social media use which in one way or the other affect how students learn.

With the DTS, the theory is put into practice, especially when universities use digital capabilities to revolutionize the way that teaching and learning are delivered and when they use digitally oriented management to give students an exceptional experience [3].

On the other hand, management practices relate to the strategies employed by managers to raise the standard of their work, such as staff empowerment and training, the introduction of quality-improving programs, and the use of new technologies to boost employee performance in the direction of organizational objectives. Similar to other groups, Afridi et al. [4] noted that in order for knowledge management to function well and support the success of a company, Higher Education Institutions (HEIs) must adhere to proper management practices, including information storage and accessibility [5].

In addition to above, the HEIs during the digital transformation have to develop 'digital pedagogies' and make use of technology as a process to analyse needs of stakeholders, demands and students' needs to be able to provide world

[a]maricormarquez1206@gmail.com, [b]joeangcuaresma@gmail.com

DOI: 10.1201/9781003684459-39

class learning experience. DTS requires an intense alteration in the admission, form, and instruction that deliver the highest quality of education to make the world-class education experience, transform the students to accommodate the demands of the globalized industry with little capital expenditure, resource saving and zero regular budgetary constraints with high standard set and a competitive edge that sustains the competition in the global education landscape [6].

The DTS when being practiced in HEIs shall improve the learning process and automate plenty of essential operations. There is an increasing concern about how education handles its place in the knowledge society, along with the development of several digital tools and technology. By aligning the HEI management in preparation of implementation of changes and adaptation to the digital transformation strategy to the rapid technological advancement of the world, the HEI mission will be satisfied and be relevant as HEIs enter the fourth industrial revolution. By instilling innovative, adaptive models and by making good for the means of learning and development, delivery and continuous improvement, HEIs will have to learn to cope with the disruptive digital technology. In this light, it also serves as the answer to all these concerns observed regarding the significant changes and challenges that HEIs in Zamboanga Peninsula had to face in the acquisition of digital transformation including in the management, in adopting digital tools and technology in the teaching and learning process, human resources, in modifying the existing systems, processes, communication modes and channels, and in all the other academic and administrative activities that such institutes have to perform. In the report, the management, HEIs processes, and technology providing it with HEIs will be discussed [7].

The purpose of this study was to find out the digital transformation strategy, management practices and challenges in Hels in Zamboanga Peninsula during the school year 2022-2023 regarding digital leadership, technology utilization and workforce utilization and the challenges among academic heads, particularly the deans, department heads, and program chairs/coordinators of the offered education programs of Hels. This study also includes the profile of the respondents as age, educational attainment, length of service and type of HEI.

In addition, the significant difference in the digital transformation strategy and management practices as hypothesized based on profile was established. The findings of this study were also used as baseline data for the proposed Digital Leadership Technology Utilization Workforce Capability Model.

2. Materials and Methods

The study employed the mixed method research design. Raosoft sample calculator was utilized resulted to 108 teacher-population in eight higher education institutions (HEIs). Furthermore, eight academic heads-participants particularly one per HEI were interviewed about the challenges experience in the implementation of digital transformation strategy.

Digital transformation strategy and management practices were measured by using a modified survey questionnaire based on digital leadership, technology utilization and workforce capability of five indicative statements with very high reliability coefficients of 0.8605 meaning the survey questionnaire is reliable. The faculty of a college of the school year 2022- 2023 were pilot tested as respondents.

The parts of the survey questionnaire are four. In Part 1 of the questionnaire, the profile of teacher-respondents in respect of their age, educational attainment, length of service and type of higher education institution (HEI) was described. Part 2 and 3 were 4-point Likert Scale survey question with description as 4-Highly Evident; 3-Moderately Evident; 2-Fairly Evident; 1-Not Evident. The study, further, looked into the digital transformation strategy and management practices in form of digital leadership, technology utilization and workforce capability respectively using survey questionnaire with five indicative statements. In Part 4, the challenges confronted by academic heads in the implementation of the digital transformation in HEI were thematically analysed and arranged in the form of common themes.

3. Data Collection

I then sought permission from the school officials of the selected higher educational institutions to gather data. Finally, research instrument was used to get approval from validators and experts. The teachers were requested to answer the survey questionnaire introducing digital transformation strategy and the management practices of the digital leadership, the technology utilization, capability of workforce in their-faculty rooms and classrooms respectively. Eight academic heads were arranged to be interviewed to find out the challenges they face in the implementation of DTS. Thirdly, the treated data were analyzed with SPSS and Microsoft Excel for a statistical analysis and interview data was transcribed while analysis was done through thematic analysis.

4. Data Analysis

Mean and Standard Deviations were measured in order to find out the impact of digital leadership and technology utilization strategy and workforce capability management practices in case of digital transformation among academic heads in terms of teacher – respondents. Additionally, the common themes from the result of the interview with the academic heads on challenges in the implementation of DTS in HEIs were done by using Thematic Analysis. A significant relationship between the digital transformation strategy, and DT management practices was determined among the academic heads of HEIs using the Pearson-r Correlation Coefficient Test. The use of the T independent test was to help find the hypothesized significant differences between digital transformation strategy and DT management practices among academic heads of HEIs based on type of HEI, Private and Public.

5. Results and Discussion

5.1. What is the digital transformation strategy among academic heads of HEIs in Zamboanga Peninsula in terms of digital leadership, technology utilization and workforce capability?

The digital transformation strategy in relation to digital leadership, technology utilization, and workforce capability was Moderately Evident with an average weighted mean of 3.20 with the description was interpreted moderately evident as indicated in Table 39.1. This implies that teachers in teacher education programs agreed to some areas of strategies implemented in the DTS by academic heads such as on their leadership style and in managing their workforce capabilities but not to all indicators in technology utilization. These teacher-respondents were exposed to the digital transformation strategy of academic heads when academic heads led the team towards the attainment of goals, on how they managed their existing people in the department and how tasks were assigned to them based on their knowledge and skills but needed improvement in terms on technology utilization. Conceivably, academic heads utilized the existing technologies but since technologies innovate now and then, not all technologies were being highly utilized. The availability of these digital tools in offices and learning environment which could also be considered factors in technology utilization was not just the responsibility of academic heads since there were higher management, who has the highest authority over purchasing it and prioritizing it to be available in the educational setting.

The efforts of academic heads in leading the team towards the attainment of its goals in digital transformation support Wooldridge [8] that a key component of digital transformation is cultivating a digital attitude among staff members by top executives and other influential figures. Alenezi [25] also proposes in the Google Framework that to support an innovative culture, the university leadership should be ready to motivate people to attempt and fail as a means for learning. For the institution to achieve the academic and operational objectives, the leadership must establish the technology (tools and procedures) suitable to the institution and later, solicit for team support to adopt it. Administrators should proceed to develop a long term budget, establish which funding options are available, and seek out cost saving and relief of resources directly tied to student driven objectives.

However, the findings are consistent with [10] idea that the DTS in HEI is a movement or a change process that begins with adopting and using digital technologies. When it was carried out using Porter's 1985 architecture of building advantages and when DTS in virtual learning and virtual program management capabilities delivered a top-tier educational experience, it raises the idea that competitiveness and DTS are inextricably linked. Similarly, Hess et al. [11], DTS is the digital technology at an organizational level, developed strategies and incorporated changes are needed to improve the overall performance and productivity of the organization.

All of these pointed to the necessity of having a clear vision and creating a supportive culture for effective digital leaders, according to Vuyisile [12]. It demonstrates that leaders who employ digital leadership are the ones who will prepare people to adopt digitization by helping followers grasp the chaos brought on by digital transformation. To close the digital divide, every leader also needs to be inclusive in their leadership style and be able to communicate the institution's goal to the workforce.

5.2. What is the digital transformation (DT) management practices among academic heads of HEIs in Zamboanga Peninsula in terms of digital leadership, technology utilization and workforce capability?

Table 39.2 shows that the digital transformation management practices in terms of digital leadership, technology utilization, and workforce capability were Highly Evident with an average weighted mean of 3.31 described strongly agree and interpreted highly evident. This tells those teachers in teacher education programs highly agreed with some aspects of the leadership style, technological use, and workforce management practices of academic heads. By observing how these academic heads used their digital skills to lead the teachers toward the achievement of high-quality education through making sound decisions, responding appropriately to various situations, respecting diversity, taking risks, and learning from experiences. This further means that academic heads supported teachers in the use of digital learning platforms, interactive multimedia, and other digital tools to support teaching

Table 39.1. Summary of results of the digital transformation strategy in terms of digital leadership, technology utilization and workforce capability

Digital transformation strategy indicator	*Weighted mean*	*Description*	*Verbal interpretation*
Digital Leadership	3.29	Strongly Agree	Highly Evident
Technology Utilization	2.98	Agree	Moderately Evident
Workforce Capability	3.33	Strongly Agree	Highly Evident
Average Weighted Mean	3.20	Agree	Moderately Evident

Source: Author.

in the learning environment. They also strongly agreed with how their academic heads used technology for online meetings, data storage, collaboration, and communication. When the teacher respondents strongly agreed on how academic heads had planned the teaching load, evaluated them for their teaching strategies, put policies and procedures into place, and came to an understanding of the teachers' evaluation for their IPCR assessment reflects that academic heads perform proper management of workforce capability.

The good strategic management of academic heads in managing their leadership style, technology utilization, and workforce capability improves performance and supports Langat et al. [13] idea that strategic management is a key aspect to enhance the performance of an organization performance and failure of strategic management is due to a lack of knowledge, coordination, and poor strategy execution. The execution of strategies improves performance however several local institutions have failed to leverage strategic institutionalization. Similarly, Ruloff and Petko [14] indicated that while incorporating technology in the schools, all of the principals have a clear commitment to student-centered instruction. One principal who focused on transformational leadership reported more quick and significant improvements than all the principals who were focused on a more transactional style of leadership. Moreover, the faster adoption of digital technology and a stronger focus on educational objectives rather than tools are both indicators of transformative leadership. However, this finding contradicts Najeeb and Muhammad [15] results that teachers' reflections highlighted their extreme mental stress as a result of the rough and harsh behaviour of school heads. As a result, school heads' general practices are discovered to be autocratic, which is directly related to the low morale of instructors.

5.3. *What are the challenges that academic heads experienced in the digital transformation in HEIs?*

The descriptions of the challenges that academic heads experienced in the Implementation of digital transformation strategy involved the experience of dealing with incapable teachers in using technologies (specifically senior teachers), the lack of structures and some technological resources (facilities, tools, and equipment), the poor or weak internet connections caused by unstable signals and intermittent power interruptions and retention of competitive and qualified teachers.

Table 39.2. Summary of results of the DT management practices in terms of digital leadership, technology utilization and workforce capability

DT management practices indicator	*Weighted mean*	*Description*	*Verbal interpretation*
Digital Leadership	3.40	Strongly Agree	Highly Evident
Technology Utilization	3.27	Strongly Agree	Highly Evident
Workforce Capability	3.27	Strongly Agree	Highly Evident
Average Weighted Mean	3.31	Strongly Agree	Highly Evident

Source: Author.

The incapabilities of teachers in using technologies were generated because there were different age groups that they were managing, teachers' age limits them to explore more on the use of computers and their level of willingness to learn differs from younger ones. In the 21st century, there was difficulty in teaching senior teachers because they belong to the non-computer era, who used the conventional method of instruction since they lacked skills in the use of Google classrooms and other technologies which led them to retire because of digitalization. However, the situation was managed by determining the series of training and mentoring strategies implemented by young teachers (who were competent and very willing to help) and students who were technology savvy. Also, some teachers were difficult to convince because they could think and made decisions on their own.

The HEIs prioritized making structures and some technological resources (facilities, tools, and equipment) in their respective universities/colleges. However, it was challenging for academic heads due to several factors, including the growing student population that the existing facilities could no longer support, the destruction of existing technological resources by unforeseen circumstances or termite attacks, the inadequate physical infrastructure, and the information and communication technologies already in place, but needs to be upgraded with air-conditioned rooms to safeguard the available technologies.

Another challenge was the poor internet connections brought on by erratic signals and unpredictable power outages which academic heads were powerless to address that they needed to request help from the government on this. The academic heads in HEIs faced difficulties implementing their digital transformation strategy because of poor internet access. Internet connectivity had a significant impact on using online platforms for teaching and learning, communicating with students and professors, and taking online classes. It disrupted online apps, halted online learnings, and caused no immediate response in online platforms, which had an impact on how the digital transformation strategy was being implemented, as a whole. Poor internet services were being paid for through internet plans, which gave academic heads the impression that they were paying service providers for poor services. Some HEIs had the technological capacity to digitally modify the teaching-learning environment and to promote online learning, but these technological resources would be ineffective in the absence of signals, low speed or bandwidth, or internet connectivity. Neither the academic heads nor those in senior management have any control over this challenge, thus, it cannot be managed. The inadequate internet connection had an effect on the bigger plans of the

HEIs in placing all the necessary infrastructure, digital tools, and equipment.

Lastly, the academic heads in HEIs found it challenging to retain competitive and qualified teachers. When these teachers found better employment opportunities and more favorable conditions in other HEIs in the middle of semesters, they were free to go and the academic heads faced the challenge of assigning the subjects left to the remaining teachers and the finding of a new one.

In conclusion, managing incapable teachers (particularly senior or elderly teachers), keeping qualified and competitive faculty, a lack of structures and some technological resources (facilities, tools, and equipment), and poor or weak internet connections that affected the digital transformation strategy were the challenges faced by academic heads in HEI. It requires dedication and perseverance to be a capable leader with decision-making abilities, responsibility for making technological tools available, and the capacity to manage workforce capabilities in the digital transformation strategy.

Academic heads of these HEIs experienced challenges in the digital transformation which for Mohamed et al. [6] are the evolution of strategic management practices in universities. According to Aljanazrah et al. [16] stated that a variety of barriers to digital transformation, such as those relating to social context, software usability, digital pedagogy, and online evaluation. IGI Global [17] claim that when DTS is rolled out, it affects the way employees perform their tasks, and leaders face more challenges that they need to prepare the workforce and identify the tasks and projects because transformation does not happen immediately, it takes time, consistency, and diligence to achieve the desired outcomes and it is having the right business strategy, data, process, and people with flexible workflow, a decentralized decision-making process, learning, and a greater reliance on business ecosystems and not just on technology. Thus, leaders need to be more aware and able to adapt and practice the changes that digitization brings (Table 39.3).

Table 39.3. Academic head-participants' responses on the challenges experience in the implementation of digital transformation strategy

Common themes	*Participant identification*	*Number of participants/ Total number*	*Percentage*
1. Incapabilities of teachers in using technologies	P1, P2, P3, P5	4/8	50
2. Lacking of structures and technological resources (facilities, tools and equipment)	P1, P4, P7	3/8	37.5
3. Poor or weak internet connections caused by unstable signals and intermittent power interruptions	P2, P4, P5, P6, P7, P8	6/8	75
4. Retention of competitive and qualified teachers	P3, P6	2/8	25

Source: Author.

5.4. *Is there a significant relationship between the digital transformation strategy and management practices among academic heads of HEIs in Zamboanga Peninsula?*

Table 39.4 shows the correlation between the academic heads' level of digital transformation strategy and DT management practices in the implementation of digital transformation. The coefficient of correlation is 0.870 with a probability value of 0.000 which is significant at alpha = 0.05. Hence, the null hypothesis that states there is no significant relationship between the digital transformation strategy and DT management practices among academic heads of HEIs is hereby rejected. This means that there is a significantly strong correlation between the academic heads' digital transformation strategy and DT management practices among academic heads. This implies that the views of teachers on the academic heads' digital transformation strategy affect their management practices in HEIs. Based on the findings of this research, the digital transformation strategy among academic heads is significantly related to their DT management practices in the implementation of digital transformation. When academic heads applied digital transformation strategy can be seen in their managerial practices which was highly evident through the teachers' responses. Conversely, when academic heads implemented a digital transformation strategy, it was highly obvious in their administrative practices to their teachers. Their management practices in terms of digital leadership, technology utilization, and workforce capability were indeed the set digital transformation strategy on how to manage for the attainment of set goals and targets through being a transformational leader who considers managing the workforce in the process and its culture. This finding relates to Teng et al. [18] result that revealed small- and medium-sized enterprises (SMEs) concluded that digital technology and employee digital skills are positively correlated with performance, and it is the moderator of DTS on performance to maintain their sustainable development. Moreover, McCarthy et al. [19] recognized the critical components for system change leaders in the context of digital transformation in education. Leaders are aware that for information, services, and human relationships to be efficient and effective for stakeholders, a digital revolution is required. Since leaders in education and decision-makers have faced pressure to establish a logical strategy for the transition brought on by digital transformation, decision-makers in the educational system aspire to go beyond technological developments and adopt a transformational mindset, employing technology as an enabler.

5.5. Is there a significant difference in the digital transformation strategy when respondents are grouped according to age, educational attainment, length of service, and type of HEIs?

Table 39.5 shows the digital transformation strategy among academic heads when the respondents are grouped according to age. The data indicate that regardless of the respondents' age, their perceived digital transformation strategy among academic heads does not vary. Whether the teacher-respondents were young or old, they had similar level of perceived digital transformation strategy among academic heads. The age was not a factor to the perceived digital transformation strategy among academic heads because whatever were the age, the DTS was not affected. This finding has some connection to Chanin et al. [20] research findings in which it identified six components of the digital transformation factor: (1) Strategy, (2) Process, (3) Product/Service, (4) People, (5) Data, and (6) Technology, and they are consistent with empirical data (p-value = 0.860). The data was gathered from 100 higher institutions, with 40.67% of the population being between the ages of 31 and 40.

Table 39.4. Correlation: Digital transformation strategy and the DT management practices among academic heads of HEIs in Zamboanga Peninsula

X	*Y*	*r-value*	*p-value*	*Interpretation*
Academic Heads' Digital Transformation Strategy	Academic Heads' Management Practices	0.870	0.000	Significant

Correlation is significant at 0.05 level.

Source: Author.

Table 39.6 shows the academic heads' digital transformation strategy grouped according to educational attainment. With an alpha of 0.05, the F value is 3.367, with a probability value of 0.012 is significant. The data show that apart from the educational attainment of the respondents, perceived digital transformation strategy among academic heads is perceived differently. This means that regardless of educational attainment, the teacher-respondents had different levels of perceived digital transformation strategy among academic heads. Educational attainment was a factor in the DTS among academic heads in terms of digital leadership, technology utilization, and workforce capability. It does matter whether they were with master's degrees or doctorates. They were not the same in the level of perceived digital transformation strategy among academic heads. This finding somewhat relates to Su et al. [21] results that there is a positive correlation between education (managers' and employees' degrees) and the management practice gap, but a negative relationship between competition and the management practice gap.

Table 39.7 shows the data indicate that regardless of the respondents' length of service, their perceived digital transformation strategy among academic heads does not vary. This means that regardless of the length of service, the teachers had a similar level of perceived digital transformation strategy among academic heads. The length of service was not a factor in the level of DTS among academic heads in terms of digital leadership, technology utilization, and workforce capability. And programs to witness academic heads' digital transformation strategy. They were all the same in the level of perceived digital transformation strategy among academic heads. This finding somehow contributes to Marquez [22] conclusions that the perceived level of implementation of digital transformation among new junior high school

Table 39.5. Digital transformation strategy among academic heads when respondents are grouped according to age

Variable	*Source of variance*	*Sum of Squares*	*Mean Square*	*F-Value*	*P-Value*	*Decision*
Digital Transformation Strategy of Different Age Groups	Between	1.043	0.209	0.866	0.507	Not Significant
	Within	24.569	0.241			
	Total	25.612				

Source: Author.

Table 39.6. Digital transformation strategy among academic heads when respondents are grouped according to educational attainment

Variable	*Source of variance*	*Sum of Squares*	*Mean Square*	*F-Value*	*P-Value*	*Decision*
Digital Transformation Strategy of Different Educational Attainment	Between	2.962	0.740	3.367	0.012	Significant
	Within	22.650	0.220			
	Total	25.612				

Source: Author.

teachers in the teaching profession moderately implemented DT in public high schools. Moreover, the result was caused by the unavailability of ICT resources or limited digital tools in classrooms, internet connectivity for Augmented Reality (AR) based learning tools, and not enough ICT training for teachers.

Table 39.8 shows the digital transformation strategy among academic heads when the respondents are grouped according to type of HEI. The T value of 0.331 with the corresponding probability value of 0.566 is not significant at alpha = 0.05. The data indicate that regardless of the respondents' type of HEI, their perceived digital transformation strategy among academic heads does not vary. This means that regardless of type of HEI, the teachers had similar level of perceived digital transformation strategy among academic heads. The type of HEI was not a factor in the DTS among academic heads in terms of digital leadership, technology utilization and workforce capability. It does not matter whether they belong to public or private HEI in the education programs. They were all the same in the level of perceived digital transformation strategy among academic heads. This result contradicts Alshammri and Alenezi [9] result that there is a statistically significant differences between the mean rankings of educational university leaders-participants on the flexibility domain according to the type of university (public or private), with public universities favoring educational leaders.

5.6. *Is there a significant difference in the digital transformation (DT) management practices among academic heads when respondents are grouped according to age, educational attainment, length of service, and type of HEIs?*

Table 39.9 shows the DT management practices among academic heads when the respondents are grouped according to age. The F value of 0.804 with the corresponding probability value of 0.550 is not significant at alpha = 0.05. The data indicate that regardless of the respondents' age, their perceived DT management practices among academic heads do not vary. This means that regardless of age, the teachers had similar levels of perceived management practices among academic heads. Whether the teacher-respondents were young or old, they had similar levels of perceived DT management practices among academic heads. Age was not a factor in the perceived management practices among academic heads because, whatever the age, the management practices were not affected. This finding somehow adds to Semerci and Aydin [24] findings that there is no significant difference between teachers' perceived Information and Communication Technology (ICT) willingness by age, gender, teaching experience, ICT experience, ICT skills, and ICT training.

Table 39.7. Digital transformation strategy among academic heads when respondents are grouped according to length of service

Variable	*Source of variance*	*Sum of Squares*	*Mean Square*	*F-Value*	*P-Value*	*Decision*
Digital Transformation Strategy of Different Length of Service	Between	1.482	0.371	1.582	0.185	Not Significant
	Within	24.130	0.234			
	Total	25.612				

Source: Author.

Table 39.8. Digital Transformation strategy among academic heads when respondents are grouped according to type of HEI

Variable	*Type of HEI*	*Mean*	*T-Value*	*P-Value*	*Decision*
Digital Transformation Strategy	Private	3.30	0.331	0.566	Not Significant
	Public	3.16			

Source: Author.

Table 39.9. DT management practices among academic heads when respondents are grouped according to age

Variable	*Source of variance*	*Sum of Squares*	*Mean Square*	*F-Value*	*P-Value*	*Decision*
DT Management Practices of Different Age Groups	Between	0.993	0.199	0.804	0.550	Not Significant
	Within	25.207	0.247			
	Total	26.200				

Source: Author.

Table 39.10 shows the DT management practices among academic heads when the respondents are grouped according to educational attainment. The F value of 2.086 and their perceived digital transformation strategy among academic heads does not vary. This means that regardless of educational attainment, the teacher-respondents had the same level of perceived management practices among academic heads. Educational attainment was not a factor in the management practices among academic heads in terms of digital leadership, technology utilization, and workforce capability. It does not matter whether they were graduates of bachelor's degree, master's degree, doctorate's degree, or even with earned master's degree units or earned doctorate's degree units in the education programs. They were all the same in the level of perceived management practices among academic heads.

Table 39.11 shows the digital transformation strategy among academic heads when the respondents are grouped according to the length of service. This means that regardless of the length of service, the teachers had a similar level of perceived digital transformation strategy among academic heads. The length of service was not a factor in the level of DTS among academic heads in terms of digital leadership, technology utilization, and workforce capability. It does not matter whether they were experienced teachers or less experienced teachers or novice teachers in teaching education programs to witness academic heads' digital transformation strategy. They were all the same in the level of perceived digital transformation strategy among academic heads.

This finding somehow contributes to Marquez [22] conclusions that the perceived level of implementation of digital transformation among new junior high school teachers in the teaching profession moderately implemented DT in public high schools. Moreover, the result was caused by the unavailability of ICT resources or limited digital tools in classrooms, internet connectivity for Augmented Reality (AR) based learning tools, and not enough ICT training for teachers.

Table 39.12 shows the DT management practices among academic heads when the respondents are grouped according to the type of HEI. The T value of 0.001 with the corresponding probability value of 0.982 is not significant at alpha = 0.05. The data indicate that regardless of the respondents' type of HEI, their perceived management practices among academic heads do not vary. This means that regardless of the type of HEI, the teachers had similar levels of perceived management practices among academic heads. The type of HEI was not a factor in the DT management practices among academic heads in terms of digital leadership, technology utilization, and workforce capability. It does not matter whether they belong to public or private HEI in the education programs. They were all the same in the level of perceived DT management practices among academic heads. This result supports Alshammri and Alenezi [9] result that there were no statistically significant differences between the means of ranks of participants' scores in the remaining domains such as problem sensitivity, fluency, and originality

Table 39.10. DT management practices among academic heads when respondents are grouped according to educational attainment

Variable	*Source of variance*	*Sum of Squares*	*Mean Square*	*F-Value*	*P-Value*	*Decision*
DT Management Practices of Different Educational Attainment	Between	1.963	0.491	2.086	0.088	Not Significant
	Within	24.237	0.235			
	Total	26.200				

Source: Author.

Table 39.11. DT management practices among academic heads when respondents are grouped according to length of service

Variable	*Source of variance*	*Sum of Squares*	*Mean Square*	*F-Value*	*P-Value*	*Decision*
DT Management Practices of Different Length of Service	Between	0.745	0.186	0.753	0.558	Not Significant
	Within	25.455	0.247			
	Total	26.200				

Source: Author.

Table 39.12. DT management practices among academic heads when respondents are grouped according to type of HEI

Variable	*Type of HEI*	*Mean*	*T-Value*	*P-Value*	*Decision*
DT Management Practices	Private	3.37	0.001	0.982	Not Significant
	Public	3.30			

Source: Author.

of the creative leadership scale and its total score according to university type (public-private). In addition, Costley [23] added that technology integration is becoming more common in public and private schools.

5. Conclusion

The digital transformation strategy among academic heads is moderately evident in their digital leadership, technology utilization, and workforce capability and the digital transformation management practices among academic heads were Highly Evident in their digital leadership, technology utilization, and workforce capability. The identified challenges that academic heads experienced were managing incapable teachers (particularly senior or elderly teachers), retaining qualified and competitive faculty, lack of structures and some technological resources (facilities, tools, and equipment), and poor or weak internet connections in the implementation of digital transformation strategy in HEI. There was a correlation between the digital transformation strategy and digital transformation management practices among academic heads of HEIs in Zamboanga Peninsula. There were no significant differences on the digital transformation strategy among academic heads of HEIs when respondents are grouped according to age, length of service and type of HEI however, there is a significant difference when grouped according to educational attainment. Lastly, there were no significant differences on the digital transformation management practices among academic heads of HEIs when grouped according to age, length of service, educational attainment and type of HEI.

Based on the foregoing findings, it is safe to conclude the academic heads utilized the existing technologies but not all since technology innovation occurs periodically. However, they were leading the team towards goal attainment and managing teachers' capabilities as the basis for task assignment. The academic heads used their digital skills to lead the teachers toward the achievement of high-quality education through making sound decisions, responding appropriately to various situations, respecting diversity, taking risks, and learning from experiences, recommending the use of digital learning platforms and other digital tools for collaboration, and communication and more effective digital learning and they performed proper management in assigning teacher's assignments, evaluation, and adjustments in policies and procedures. Furthermore, the challenges that academic heads experience requires them to be more dedicated and persevere more to be the capable digital leader in managing the workforce, making decisions, taking responsibility to make technological tools available, and being capable to handle diverse people in HEIs. The views of teacher-respondents to the academic heads' digital transformation strategy affect their management practices in the implementation of digital transformation in HEIs. The academic heads apply digital transformation strategy seen in their managerial practices as highly evident in the teachers' responses. Additionally, the perceived digital transformation strategy among academic heads in HEIs are equally or similarly evident regardless of the respondents' age, length of service, and type of HEI clustering. However, when they are clustered according to educational attainment, the digital transformation strategy among academic heads varies.

It is recommended that Digital Leadership Technology Utilization Workforce Capability Model may be used by digital Leaders in digital transformation. Digital transformation strategy refers to the leaders' ability to direct the team, which involves digital technologies and their utilization in the learning environment and the workforce's capability in using digital technologies for digital learning. It involves the leaders, the technologies, the process, and the people of the educational institutions. Effective digital leaders consider the technology utilization (usage and availability of digital tools) and the capability of the workforce in the transformation process.

References

[1] The Malaysian Online Journal of Educational Technology (2014). www.mojet.net UNESCO (2018). Mahatma Gandhi Institute of Education for Peace and Sustainable Development (UNESCO MGIEP). UNESCO, ICT in education (2019). Retrieved at https://en.unesco.org/themes/ict-education.

[2] Storey, J. (2023). The Only Digital Transformation Framework You Will Need. https://expert360.com/articles/perfect-digital-transformation-framework#:~:text=A%20digital%20transformation%20framework%20is,the%20organisation%20through%20the%20journey.

[3] Shankar, R., & Karthikeyan, E. (2016). Modified BAT based Handoff Scheme and Verifiable Secret Sharing Security with M-M Scheme for VOIP in on-demand Video Streaming Applications. *International Journal of Advances in Engineering and Emerging Technology*, *7*(4), 154–174.

[4] Afridi, O., Gul, S., and Naeem, M. (2019). The Role of Management Practices of Higher Education Institutions. *Journal of Management and Research (JMR)*, *6*(1), 2019.

[5] Perera, K., & Wickramasinghe, S. (2024). Design Optimization of Electromagnetic Emission Systems: A TRIZ-based Approach to Enhance Efficiency and Scalability. *Association Journal of Interdisciplinary Technics in Engineering Mechanics*, *2*(1), 31–35.

[6] Mohamed H. M., Tlemsani, I., & Matthews, R. (2022). Higher education strategy in digital transformation. *Education and Information Technologies*, *27*, 3171–3195. https://doi.org/10.1007/s10639-021-10739-1

[7] Chitra, Sasidharan, Ahmed, S., & Balamurugan. (2022). Loan Approval Prediction Model. *International Academic Journal of Innovative Research*, *9*(2), 35–38. https://doi.org/10.9756/IAJIR/V9I2/IAJIR0916

[8] Wooldridge, S. (2022). The Role of Digital Transformation in Workforce Management. *MHR Blog*. Retrieved at https://mhrglobal.com/uk/en/blog/role-digital-transformation-workforce-management-0

[9] Alshammri F. S., & Alenezi, A. K. (2021). Creative Leadership and its Relationship to Thinking Styles among Saudi University Leaders. *International Journal of Education and Practice, Conscientia Beam*, *9*(2), 340–353. https://archive.conscientiabeam.com/index.php/61/article/view/710/1022

[10] TeacherPh (2014). DepEd Computerization Program (DCP) Guide. Vial, G. (2019). Understanding digital transformation: A review and a research agenda. *Journal of Strategic Information Systems Review*. Retrieved at https://www.sciencedirect.com/science/article/pii/S0963868717302196

[11] Hess, T., Matt, C., Benlian, A., & Wiesböck, F (2020). Options for formulating a digital transformation strategy. In *Strategic Information Management*. Routledge: London, UK, pp. 151–173.

[12] Vuyisile, M. (2022). Higher Education Leadership in a Time of Digital Technologies: A South African Case Study. *International Journal of Information and Education Technology*, *12*(10). Retrieved at http://www.ijiet.org/vol12/1728-IJIET-5005.pdf

[13] Langat, M., Kirimi, H., & Wainaina, L. (2019). Strategic Management Practices and Organization Performance of DTS. LAP Lambert Academic Publishing. Retrieved at https://www.lap-publishing.com/catalog/details//store/gb/book/978-3-659-56210-5/strategic-management-practices-and-organization-performance-of-dts

[14] Ruloff, M., & Petko, D. (2021). School principals' educational goals and leadership styles for digital transformation: results from case studies in upper secondary schools. *International Journal of Leadership in Education*. doi:10.1080/13603124.2021.2014979

[15] Najeeb E. K., & Muhammad U. H. (2021). How and to What Extent Managerial Practices of School Heads are Influencing the Morale of Primary School Teachers? *Bulletin of Education and Research*, *43*(2), 85–Retrieved from https://files.eric.ed.gov/fulltext/EJ1338292.pdf

[16] Aljanazrah, A., Yerousis G., Hamed G., & Khlaif, Z. (2022). Digital transformation in times of crisis: Challenges, attitudes, opportunities and lessons learned from students' and faculty members' perspectives. https://doi.org/10.3389/feduc.2022.1047035

[17] IGI Global (2023). *What is Management Practices*. https://shorturl.at/hikx7

[18] Teng, X., Wu, Z., & Yang, F. (2022). Research on the Relationship between Digital Transformation and Performance of SMEs. *Sustainability, 14*, 6012. https://doi.org/10.3390/su14106012

[19] Mccarthy, A. M., Maor, D., Mcconney, A., & Cavanaugh, C. (2023). Digital Transformation in Education: Critical Components for Leaders Of System Change. *Social Sciences & Humanities Open*, *8*(1), 1004Issn 2590-2911. https://Doi.Org/10.1016/J.Ssaho.2023.100479.

[20] Chanin T., Prachyanun N., & Panita W. (2022). Factors Influencing Digital Transformation Adoption among Higher Education Institutions during Digital Disruption. *Higher Education Studies*, *12*(2), 2022 ISSN 1925-4741 E-ISSN 1925-475X. https://doi.org/10.5539/hes.v12n2p9

[21] Su, Y., Liu, B., & Li, D. (2018). What Causes the Differences in Management Practices between US and UK? An Application of Decomposition Analysis on Firms across Countries. *Theoretical Economics Letters*, *8*, 1684–1700. doi: 10.4236/tel.2018.810109.

[22] Marquez, Maricor A. (2020). Digital Transformation among Public Secondary School Teachers: Basis for Instructional Enhancement. Master's Thesis.

[23] Costley, K. C. (2014). The Positive Effects of Technology on \ Teaching and Student Learning. DepEd Order No. 78, s. 20Guidelines on the Implementation of the DepEd Computerization Program. Department of EducationDepEd empowers digital transformation in schools nationwide distribution of annual Adobe licenses. DepEd Page. https://www.deped.gov.ph/2022/01/26/deped-empowers-digital-transformation-in-schools-through-nationwide-distribution-of-annual-adobe-licenses/

[24] Semerci, A., & Aydin, M. K. (2018). Examining High School Teachers' Attitudes towards ICT Use in Education. *International Journal of Progressive Education*, *14*(2), 93–105.

[25] Alenezi, M. (2021). Deep Dive into Digital Transformation in Higher Education Institutions. *Educational Sciences*, *11*, 770. https://doi.org/10.3390/educsci11120770

40 Work life balance and job satisfaction of women working in post office

Sowbarani G.[1,a], *Shanmugapriya M.*[2,b], *M. Manimekalai*[3,c], *S. Babitha*[3,d], *and Ambika*[4]

[1]Assistant Professor in Commerce, Karpagam Academy of Higher Education, Coimbatore, Tamil Nadu, India
[2]Assistant Professor of Commerce, NGM College, Pollachi, Tamil Nadu, India
[3]Assistant Professor of Commerce, Arulmigu Pazhani Andavar Arts College for Women, Palani, Tamil Nadu, India
[4]Assistant Professor of Commerce, GVG College of Arts and Science, Tamil Nadu, India

Abstract: Postal services introduce particular problems to hinder female employees in post offices from attaining work-life balance. People encounter different obstacles when they work to fulfill commitments between their household responsibilities and work duties. Woman postal workers could achieve work-life balance through the provision of child care services with flexible hours and flexible rules. Opportunities for professional advancement, a supportive work environment, and acknowledgment of women's achievements all impact post office workers' job satisfaction. Work-life balance and job satisfaction in this industry improves when suitable work environments emerge simultaneously with ensuring adequate support for women and paying attention to their needs.

Keywords: Work, women, environment, challenges

1. Introduction

A balanced equilibrium between professional duties and private dedication remains essential for both workplace happiness and employee health in the workplace according to professional requirements. Employer systems and employee satisfaction levels strongly rely on the essential alignment between professional responsibilities and personal free time marked as work-life balance. Female postal workers encounter unique difficulties since strict operational guidelines make it difficult to achieve work-life balance at their jobs. This research examines female postal workers to build supportive workplaces which maintain employee loyalty and minimize turnover rates through studies on balance-related staff satisfaction. The research objective utilizes the assessment of work-life balance barriers to satisfaction growth to deliver actionable solutions for enhancement. The study focuses on demonstrating how creating suitable work-life boundaries ensures better job satisfaction and workplace well-being by directly studying on-the-job experiences of women postal staff members. This study aims to open the door for future research by highlighting the difficulties women encounter in juggling their home and work life.

2. Review of Literature

Amit Verma [1] Businesses across a variety of industries employ a wide range of tactics, policies, procedures, practices, and initiatives to support their employees in achieving a healthy work-life balance. Certain policies are imposed by legislation, while others are implemented at will. The industrial sector has witnessed a surge in the significance of work-life balance due to several factors, such as summary of the future work-life balance situation for women in the Indian business in light of technology developments, increased work pressure, remote work, and changing demographics (gender, experience, income, inflation, and living standards). Given the growing amount of household and professional duties that women are bearing, this investigation becomes even more important.

TarunaYadav, Sushma Rani [2] The topic of this study is 'Work life balance: opportunities and challenges.' This essay focuses on the opportunities and problems associated with work-life balance in Indian businesses. Employers are being forced to reconsider their people practices by a confluence of authoritative trends in the early 21st century. It is unquestionably appropriate that businesses and policymakers prioritize work-life excellence as a strategic aim. Employers now have new opportunities to meet employees' personal ambitions and achieve organizational performance goals in tandem with the obstacles posed by an aging workforce, an increasingly competitive labour market, information technology, and rising benefit expenses. Many workers are also feeling that their quality of worklife has been diminished. Job stress and an unbalanced work-life schedule are the two main

[a]Sowba13@gmail.com, [b]mail2priya.sanju@gmail.com, [c]Gdhanisha2019@gmail.com, [d]yesbabithavikash@gmail.com

DOI: 10.1201/9781003684459-40

indicators of this. The two halves of the same coin are the work and personal lives. Uneven work-life balance between job and family obligations has become a big problem in the workplace in this troubled globe.

Girija, Kalaivani [3] In their analysis, the authors evaluated the need for numeracy education for rural women, as well as the benefits of learning about compound interest, inflation, time value of money, and illusion. These topics would enhance the women's standard of living, encourage long-term savings, and lessen unconventional lending.

No job satisfaction and consistency in life, it can create a dilemma for working women. Work life balance requires attaining equilibrium between professional work and personal work, so that it reduces friction between official and domestic life. The ultimate performance of any organization depends on the performance of its employees, which in turn depends on numerous factors. These factors can be related to job satisfaction or family or both. A comparative study is conducted among the working women of some Govt. colleges and national institutes of Bhopal in education sector and nationalized banks, running their in Bhopal (M.P.) for banking sector. The objective of this research is to study the working environment and women's perception about the work life balance and job satisfaction, who are working in banking and education sector. Apart from it, another significant objective is to study effects of work life balance on job satisfaction and initiatives taken by the organizations for effective work life balance and its relation with the job satisfaction. Standard Deviation is applied in this paper to check the authenticity of data given by the respondents of both the sectors. Finding suggests that WLB can be achieved by the factors responsible for job satisfaction such as: supportive colleagues, supportive working conditions, mentally challenging work, equitable rewards and employee oriented policies etc.

2.1. Statement of the problem

In order to achieve a satisfactory work-life balance, women employed in post offices encounter major obstacles. Women face substantial obstacles when trying to balance postal work with their personal needs because of the demanding work schedule at postal services that includes extended irregular shifts. Female employees struggle to reach work-family balance due to strict postal service working schedules and inadequate supportive infrastructure and policies. The inequality between job hours in postal services produces negative effects on women's professional growth and employment contentment that reduce their life satisfaction and operational effectiveness as workers. Resolving these matters constitutes a vital requirement for postal industry development of a better working environment.

2.2. Objective of the study

To evaluate the balance between women working in the postal service and their personal lives.

2.3. Scope of the study

The purpose behind this study is to analyze how postal industry women handle their professional commitments together with their personal responsibilities. The research examines shift-based operations alongside work area adaptability and job-related duties and personal responsibilities to determine their impact on female employees' ability to handle workplace and personal domains. The analysis based on surveys and interviews together with proper data evaluation provides understanding concerning the barriers women face in achieving work-life balance within postal services as well as potential solutions to address these obstacles.

2.4. Statistical tools used for the study

The research examined 129 of 130 collected responses using statistical computations of Simple Frequency along with factor analysis, ANOVA and Chi-Square and weighted average methods.

2.5. Limitation

Research studying work-life balance and job happiness of post office female employees requires understanding several research and treatment constraints. The ability to generalize study findings becomes limited because post offices have different organizational policies and structures. Jobs within the postal industry represent another problem because various positions require different qualifications and produce varying impacts on work-life equilibrium.

3. Factor Analysis

An approach to statistical analysis known as factor analysis helps to identify underlying latent variables, or factors, that underlie patterns of correlations among the observable variables and help interpret the relationships between the variables as they are observed. Finding the important components that explain the variance in the data set helps to simplify the data. In order to simplify the interpretation of data by identifying underlying dimensions that account for observed patterns, this technique is frequently employed in research to investigate the structure of interactions among variables.

4. Worklife Balance of Women Working in Post Office

4.1. Factor analysis

Factors the influence the work life balance of women working in post office.

The data may be somewhat sufficient for performing a factor analysis, according to the Kaiser-Meyer-Olkin Measure of Sampling Adequacy (KMO) score of 0.601. KMO values above 0.5 are generally regarded as acceptable, meaning that factor analysis can move on because the variables in the dataset have a sufficient amount of shared variance (Table 40.1). Moreover, a statistically significant result ($\chi^2 = 114.905$, df = 21, $p < 0.001$) from Bartlett's Test of Sphericity demonstrated that the correlation matrix is not an identity matrix, confirming the data's eligibility for factor analysis. All things considered, these findings imply that the dataset is quite appropriate for additional exploratory component analysis to find underlying structures or patterns within the variables being studied.

Two separate components pertaining to work-life balance in the context of post offices are shown by the rotational component matrix (Table 40.2). Component 1 variables include time flexibility for women's working hours, the efficacy of post office policies in supporting women, workload conducive to a healthy work-life balance, the availability of support services for women, equal opportunities for career development, and the provision of remote work options for work-life balance. These variables load significantly, indicating that they are interconnected and contribute to a comprehensive understanding of work-life balance initiatives. The second component strongly focuses on meeting the needs of female employees according to analysis. The eigenvalues demonstrate Component 1 assumes 30.632% of the variation and Component 2 captures 18.465% while their cumulative variance amounts to 49.097%.

5. Conclusion

The postal service faces substantial challenges regarding women's ability to maintain work-life equilibrium that requires detailed problem solving. While advances in gender equality and workplace flexibility have emerged in recent times several gender-based problems persist which include excessive work hours together with inflexible schedules and insufficient social support systems. Decisions about flexible schedules together with gender-inclusive workplace environments and proper childcare help and supportive leadership must be established to create the full solution needed to solve these problems. Promoting these activities will help the post office develop an environment where women gain control over both their personal life and professional success.

Table 40.1. KMO and Bartlett's test

Kaiser-Meyer-Olkin Measure of Sampling Adequacy.		.601
Bartlett's Test of Sphericity	Approx. Chi-Square	114.905
	Df	21
	Sig.	.000

Source: Author.

Table 40.2. Rotated component matrix

Work-life Balance	*Component*	
	1	*2*
Time flexibility for working hours for women	.800	
Effectiveness of post office policies in supporting women	.704	
Workload in the post office is conducive to healthy work-life	.640	
Support services available to Women	.625	
Providing equal opportunities for career Development		.698
Providing remote work option for work life balance		.692
Supporting in accommodating and needs of women employees		.673
Eigen Values	2.144	1.293
Percentage of Variance	30.632	18.465
Cumulative Percentage of Variance	30.632	49.097

Source: Author.

References

[1] Amit Verma et al. (2022). A Study on Work-Life Balance of a Female Employee in Indian Industry. *International Journal of Special Education, 37*(3).

[2] TarunaYadav, Sushma Rani (2019). This study is conducted on "Work life balance: challenges and opportunities".

[3] Girija, P., & Kalaivani, R. (2018). The study on saving habits of rural women's with special reference to Pollachi. *International Journal of Academic Research and Development, 3*(1), 50–52.

41 Current situation for the sustainable cultural tourism operation management in Qianxinan prefecture under new media era

Aoxue Li[1,a], *Chuanchen Bi*[1,b], *Long Yang*[1,c], *and Wei Yang*[2,d]

[1]College of Graduate Study in Management, Khon Kaen University, Khon Kaen, Thailand
[2]International College, Khon Kaen University, Khon Kaen, Thailand

Abstract: The cultural and tourism industries are continuously evolving, with the marketing environment undergoing constant change. The use of new media marketing to promote rural tourism has become a prevailing trend within the contemporary cultural and tourism sectors. Abundant tourism resources serve as a fundamental prerequisite for the development of a tourist destination, while effective tourism marketing strategies are equally indispensable. Qianxinan Prefecture, located in Guizhou Province, leverages its favorable climate, rich natural tourism resources, and ecological health industries to advance the coordinated development of a diverse range of 'tourism +' industries. The region aims to establish itself as an international mountain tourism destination and a leading domestic holiday health care destination. This study focuses on the sustainable cultural tourism in Qianxinan Prefecture. In the context of new media, the cultural tourism industry in Qianxinan Prefecture faces fresh opportunities for transformation and growth. This paper employs a PEST analysis to examine the current state of cultural tourism in Qianxinan Prefecture, investigates the existing challenges in the marketing and management of cultural tourism operations.

Keywords: New media era, cultural tourism operation, sustainable development, qianxinan prefecture

1. Introduction

With the rapid development of Internet technology, there has emerged a new type of communication media based on Internet technology, 'new media', with its massive, interactive, portable, timeliness and global characteristics, which are deeply loved by everyone. In the era of continuous development of new media, various short video platforms came into being. Users can obtain information of interest through short videos ranging from a few seconds to a few minutes on the platform [1]. From the barbecue in Zibo, the 'Village Super League[1]' in Guizhou, to the Ice and Snow Festival in Harbin and the Malatang in Tianshui, these successful hits have given full play to the role of new media and the market in mass large-scale cultural and sports tourism activities, making the local characteristics widely spread and attracting many tourists to try the experience [2].

The karst topography and geomorphology in Xingyi City of southwest Guizhou Province are obvious. Here, mountains and rivers sit close to each other, field upon field face each other, city scenery is integrated, and culture and production embrace each other, forming a beautiful outlook of 'six sceneries in one city', all within 25 kilometers around the city center. The short video 'Wanfeng Mountain Residence Map', shot in Qianxinan Prefecture with Xingyi as the background swept the network, enabling various scenic spots in Xingyi to receive traffic, and local tourism to retain its popularity. From the new cultural tourism hot spots in China, the net red cities, net red scenic spots and net red cultural phenomena in recent years, two characteristics are both the same: one is to open the model of tourism development in the way young people like, and the other is the great impact of digitalization and technology achieving [3]. It can be concluded that if the young generation wins to the tourism, the future wins too. Young people can obtain information from the internet to a certain extent; at the same time, it is also the main body of China's tourism market and the natural main consumer body of rural tourism. The continuous deepening of the integration of new media and cultural tourism is no longer in the era of passive play, but is now more 'I like it, so I'll visit,' and 'I want to experience it, so you build it' [4]. Although the development of the tourism economy in Qianxinan Prefecture is obvious to all, the research on tourism operation management and marketing strategy under the new media environment is still a little weak. In the face of increasingly mature

1The Village Super League, officially known as the Rongjiang (Sanbao Dong Village) Hemei Rural Football Super League, is a football tournament held in Rongjiang County, Qiandongnan, Guizhou Province.

[a]aoxue.l@kkumail.com, [b]chuabi@kku.ac.th, [c]longya@kku.ac.th, [d]weiya@kku.ac.th

DOI: 10.1201/9781003684459-41

consumer groups and new tourism market demands, the marketing concept of tourism operation management, under the new media environment, must be changed [5].

The investigation and field study show that there are some problems in the operation and management of Qianxinan cultural tourism under the new media environment:

1. New media marketing is single channel and the content is old and homogenized. The Qianxinan tourist area has unique cultural resources with national characteristics, but many short video accounts use the same subject matter, the same editing method, the same hash-tags, and similar content soundtrack, resulting in a lack of vitality and innovation within the short video content. This may lead to audience aesthetic fatigue, and inevitably result in the loss of a large number of audience groups, over time.
2. The coverage of new media is not extensive enough to arouse and gain the support of the people. Short video platform construction is in a fragmented state, the main body of short video is mostly the agriculture, rural areas, and farmers network celebrities or media practitioners, and official intervention is insufficient. Most short video works are dominated by static natural scenery, and the fragmentation of the narrative text is obvious, which is not conducive to improving the audience's sense of resonance [6].
3. The team is not professional enough and lacks a scientific and clear marketing plan. In the process of new media marketing, there is the dilemma of a lack of excellent operational talents. Most short video creators are local farmers, self-employed people and ordinary tourists, who lack experience and skills in short video shooting, resulting in a lack of professionalism in regard to topics, editing, shooting and soundtrack choice of the short video creation.
4. Most of the new media audiences are local and 'quite aged'. The audience group is mainly in the province, which is weak and does not radiate to the national audience. Relying only on local media and Guizhou provincial level operations to carry out external communication results in the effect not being distinct.

As a new type of tourism marketing method, new media has shown great potential and value in promoting the cultural tourism industry and stimulating people's impulsive travel intention [7]. Researching this phenomenon can further enrich the tourism marketing theory system, while at the same time effectively enhancing the brand awareness and influence of cultural tourism enterprises in better operation and effective use of cultural tourism resources. A successful operation model is the basic prerequisite for the sustainable development of cultural tourism projects. By developing sustainable tourism, we can achieve the dual goals of improving tourism economic benefits and coordinating local social interests, provide practical cases for the construction of local cultural tourism operation management and marketing models, and provide research experience for rural revitalization [8].

2. Literature Review

2.1. New media era

2.1.1. Concept of the new media era

The new media environment mainly relies on digital technology to efficiently process various types of information, ensuring that users can participate in content creation through comments, shares, likes, and other forms of engagement. This environment can provide personalized content recommendations based on user interests and needs, thereby enhancing the accuracy of information and user satisfaction. New media is characterized by interactivity and immediacy. Through social media platforms, users can connect with others and share diverse types of information – such as text, images, audio, and video – anytime and anywhere. Combined with the fragmented nature of information dissemination, this environment allows users to access large amounts of brief, simple, and fast information. This feature accelerates the speed of information transmission, enabling users to obtain the latest news and the information they need in real time [9].

2.2. Sustainable development

2.2.1. Concept of sustainable development

According to the book World Conservation Strategy; Living Resource Conservation for Sustainable Development, the term 'sustainable development' was first mentioned there. As of 1992, the United Nations Conference on Environment and Development brought in a widely prescribed international meaning, 'development that is meeting the needs of the present without compromising the capability of future generations to meet their own needs [10]'. There are two significant international definitions of sustainable tourism development. The first was proposed by the World Tourism Organization in 1993: 'Sustainable tourism development is aimed at meeting people's growing economic, social, and aesthetic needs while preserving cultural integrity and maintaining ecological balance. It provides livelihoods for today's hosts and guests while promoting and protecting the interests of future generations, granting them the same opportunities.' This definition highlights the goal of tourism development and establishes the principle of equitable regional development between hosts and guests. This definition emphasizes that the essence of sustainable tourism is to view tourism, nature, culture, and the human environment as an interconnected whole [11].

2.2.2. Dimensions of sustainable development

The goals of sustainable development aim to achieve a balance and harmony among economic, social, and environmental factors, promoting mutual growth in economic advancement, social equity, and environmental protection.

Economic Sustainability: This involves achieving economic growth through the rational use of resources and efficient production methods to ensure the long-term availability of resources and the sustainability of economic development. It emphasizes avoiding short-sighted approaches and excessive exploitation or wastage of resources.

Social Sustainability: This aims to meet basic human needs, promote social equity, reduce poverty, and improve quality of life. It stresses equality across different social classes and groups, with a focus on fair distribution in areas such as education, healthcare, and employment, to foster social stability and growth.

Environmental Sustainability: This requires protecting natural resources and ecosystems in economic and social activities, reducing pollution, conserving biodiversity, and addressing climate change. The goal is to maintain the health and stability of natural ecosystems.

2.2.3. Sustainable development stakeholder relationship

Stakeholders are those who have an economic relation to the enterprises. The stakeholder theory stresses the equitable balance of interests of all the components to attain 'win win' results. We will focus on the comprehension and analysis of the relationship among stakeholders and build up an effective communication and cooperation method between the organization and the environment in the new media environment, so as to promote a sustainable development of enterprises in the long term. As the social economy grows, it is not only necessary to consider the interest of the stakeholders, but it is also obliged to undertake social responsibilities. Enterprises can adapt to the changes, enhance the ability for resisting risk and achieve better strategic goals of the organization through interests and risks sharing. The thesis on the basis of preliminary research by the tools of observation and literary methods can conclude that the 6 stakeholders in tourism sustainability management. The relationship between stakeholders are presented in Figure 41.1.

In Figure 41.1, each indicator of sustainable development is addressed by a model involving six stakeholders. Visitors

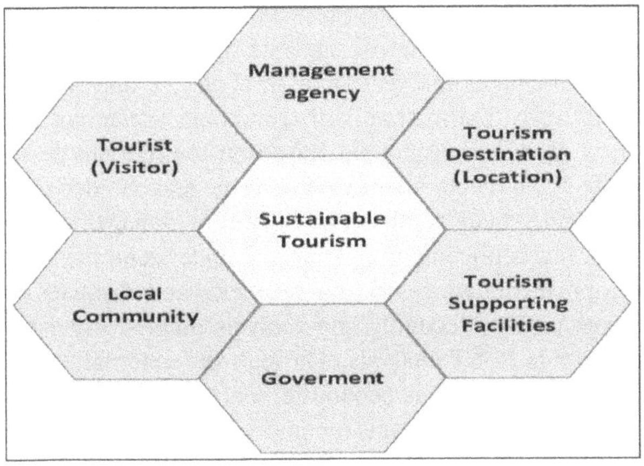

Figure 41.1. Relationship between stakeholder and sustainable tourism [12].

Source: Author.

focus on satisfaction, safety, cultural understanding, their role in nature conservation, efforts to minimize environmental impact, and energy use. The local community is concerned with tourism's economic impact, including income, market promotion, local product sales, job opportunities, residents' responsibilities, and control over pollution and resource consumption. Management agencies are responsible for destination management, advertising, public relations, providing facilities for visitors, water and energy consumption, and opportunity costs of the location. The government emphasizes public sector investment in facilities such as accommodation and sanitation, increasing tourism revenue, and managing the socioeconomic effects on the local population, energy consumption, and waste management. Tourism support facilities include accommodations (hotels), dining, and existing tourism-related businesses. Support facilities are not always physical; they also include social aspects, such as NGOs, which advocate for infrastructure protection, cultural promotion, pollution reduction, rural environment conservation, and participation in events and event tourism [12].

2.2.4. Tourism life cycle theory

In 1980, Butler proposed the S-curve and the six-stage model, introducing the Tourism Life Cycle Theory. This theory outlines the life cycle of a tourist destination, progressing through stages from inception, development, and maturity to eventual decline. Butler divided the evolution of a tourist destination into six stages: Exploration, Involvement, Development, Consolidation, Stagnation, and either Decline or Rejuvenation [13].

The Tourism Life Cycle Theory is a well-established theory in tourism research and serves as an important tool for guiding the high-quality and sustainable development of tourist destinations. Researchers can use the Tourism Life Cycle Theory to determine which stage a destination is currently in, describe the characteristics of that stage, and predict potential issues that may arise. This enables timely interventions and adjustments to identify development trends for the next phase and to develop feasible strategies, allowing the tourism life cycle to progress into a new stage of growth. By effectively applying the Tourism Life Cycle Theory, not only can the development and market changes of a destination be explained and described, but the future trends of its growth and evolution can also be forecast (Figure 41.2). This predictive capability is precisely what makes the Tourism Life Cycle Theory valuable in the practical application of tourism development [14].

2.3. Sustainable operations management

2.3.1. Concept of operations management

Operations management refers to the control and management of the operational process, encompassing the entire process related to products and services. This includes organizing, planning, executing, and controlling processes to ultimately achieve a company's goal of profitability [16].

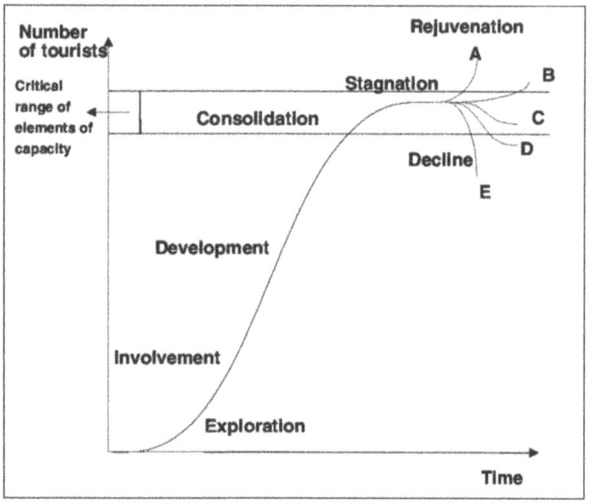

Figure 41.2. Life cycle of destination [15].

Source: Author.

2.3.2. Objectives and essence of operations management

The goal of operations management in a company is to increase both value and utility, thereby achieving sustainable economic benefits and long-term development. The essence of operations management is for a company to achieve value-added outcomes through effective management. Under optimal conditions, it integrates company resources to deliver products and services that satisfy customers [17].

2.3.3. Analysis of key success factors in tourism operations management

The study of key success factors began in the 1860s and was first used in the field of information systems, but has been expanded to a wider range of management in the past 20 years. Among the key success factors, it is necessary to pay attention to its limiting areas, characteristics, environments, situations, events, activities, strategic factors, variables, etc. If these factors are used properly and maintained for a long time, they will lead an event, individual, department, organization, and company to success and generate profits. Key success factors can be identified and controlled. The core idea of key success factors is that several key factors play a decisive role in the success or failure of a company or project. After leaders screen out a few key influencing factors from a large number of influencing factors, they regard them as the key control points for project success, and formulate project management goals around them to determine the priorities of management [18].

2.4. PEST analysis

In the late 20th century, Johnson and Scholes of the United States established the PEST model based on their summary of environmental analysis. PEST analysis has been widely used in recent years and can analyze the external macro-environment of an enterprise, including various external factors related to the operation and management of the enterprise. Based on the characteristics of the research object, the four key external environmental factors, namely the political environment, economic environment, social environment and technological environment, are mainly used to explore their impact on corporate activities and decisions [19].

1. Political environment, mainly analyzing factors that may have an impact on the long-term development of the enterprise, such as political strategies, systems, laws, etc.;
2. Economic environment, analyzing the internal and external economic environment, economic policies and other factors faced by the enterprise, and understanding their role in the formulation of corporate goals and strategies;
3. Social environment, including factors affecting the operation of the enterprise from national culture, beliefs, values, etc.;
4. Technological environment, including factors affecting the operation of the enterprise from technological capabilities, policies, dynamics, etc.

2.5. Summary

This chapter reviews the new media environment, sustainable development, operation management theory, PEST analysis method and related literature. This article will use the PEST analysis method to deeply study the cultural tourism industry and build a three-dimensional media operation development system based on new media and sustainable development, and provide a basis and practical guidance for the sustainable operation management of the cultural tourism industry in Qianxinan Prefecture.

3. Method

3.1. Research design

This study takes Qianxinan Prefecture as the research object. This paper uses qualitative analysis method. First, through field research and observation, it sorts out the tourism statistics and news text materials of Qianxinan Prefecture under the new media environment. Based on the tourism destination life cycle theory, combined with the stakeholder-related theory, understand its impact on operation management. And through marketing situation analysis, understand the current development situation and what problems and obstacles exist in development. Secondly, the analysis method involved in this paper is PEST analysis. Through the external environment analysis (political, economic, social, and technological) based on the PEST analysis method for the cultural and tourism operation management of Qianxinan Prefecture, finally, suggestions are made for sustainable cultural and tourism operations, in order to provide certain theoretical references for the development and operation of tourism projects and increase their possibility of success.

3.2. Data collection methods

3.2.1. Observation method

This study adopts a non-participatory observation method. During the observation process, in order to effectively record the observed information and obtain detailed and comprehensive data, systematic observation will be further used to ensure the scientificity and objectivity of the observation results. Through in-depth observation, this study aims to provide a deeper understanding of the research issues and make substantial contributions to related research in this field.

3.2.2. Interview method

This study will adopt an in-depth interview method. The in-depth interview method can deeply understand the expectations, acceptance psychology and specific behaviours of the actual interviewees in this research field. Achieve a deep understanding of the current status of the local cultural tourism industry. The interview method can provide a favorable basis for authenticity and reliability in the research field. Through sorting and analysis, corresponding optimization paths and countermeasures can be proposed later.

3.3. Determine the sample size

The data collection for this study will be derived from the analysis of stakeholder theory, and structured in-depth interviews will be conducted with the local tourism bureau, employees of local cultural and tourism enterprises, foreign tourists, and local people. The thoughts and expressions of the interviewees will provide powerful data. Knowing their needs, interests, and preferences through interviews can help the study to have a deeper understanding of their behaviour and decision-making process, as well as to determine their areas of interest and the factors affecting consumer behaviour.

Finally, the reliability is guaranteed in the following way: the researcher improves the validity and accuracy by using triangulation in data collection, which can reduce bias or the limitations of a single source. Therefore, reliable data can be provided for the direction of its research.

4. Result

4.1. Current status of sustainable cultural tourism development in qianxinan prefecture under the new media environment

4.1.1. Qianxinan prefecture

The Qianxinan Prefecture in Guizhou Province is located at the junction of Guizhou, Yunnan, and Guangxi. Its capital, Xingyi City, connects westward to Yunnan and southward to Guangxi. With its pleasant climate, beautiful landscapes, rich cultural heritage, abundant natural resources, and strong ethnic customs, the prefecture has been recognized by the Chinese Meteorological Society with the title 'China's Capital of Year-Round Health and Wellness.' Qianxinan Prefecture, as a region inhabited by ethnic minorities, is home to more than 40 ethnic groups, including the Bouyei, Miao, and Yi peoples. The hard-working, sincere, hospitable, and supportive nature of the local people has cultivated a vibrant and diverse ethnic culture and customs.Qianxinan boasts a wealth of intangible cultural heritage, with numerous items recognized across different levels. It currently has 14 items on the national intangible cultural heritage list, including the Yi people's 'Amei Qituo[2],' covering 20 sites, as well as 110 items on the provincial list, including Buyei cultural elements, covering 128 sites, and 244 items on the prefecture-level list [20]. These rich intangible cultural resources provide excellent opportunities for tourism development in scenic areas and villages, empowering the growth of the local cultural and tourism industries.

On December 29, 2023, the 13th meeting of the Standing Committee of the Ninth People's Congress of Qianxinan Prefecture reviewed and approved a proposal to position 'Health and Wellness Destination, Cultural Xingyi' as the city's official brand [21]. This new city positioning is aimed at promoting high-quality tourism development in the prefecture. Qianxinan continues to promote the integration of culture and tourism, fully leveraging its unique geographical and ethnic advantages. The strategic development plan, 'Cultural Tourism to Revitalize the Prefecture,' focuses on cultivating market players, upgrading industry segments, improving service quality, and revitalizing underutilized projects. Xingyi City, as the permanent host of the International Mountain Tourism and Outdoor Sports Conference, has increasingly capitalized on the mountain tourism and outdoor sports sectors. By focusing on 'resources, customers, and services,' the region actively explores multi-industry integration of cultural tourism, wellness tourism, sports tourism, and rural tourism, strengthening the 'Health and Wellness Destination, Cultural Xingyi' city brand, enhancing its international reputation.

4.1.2. Current status of sustainable cultural tourism development in qianxinan under the new media environment

The Guizhou Provincial Government has vigorously implemented the 'Tourism Industrialization' strategy, focusing on the three key elements of 'resources, customers, and services,' and has been continuously advancing four major

2Amei Qituo, meaning "Maiden's Marriage Dance," is a traditional dance of the Yi ethnic group. Commonly known as the "Foot-Stomping Dance," This lively and jubilant dance form conveys heartfelt blessings for the bride.

actions: cultivating market players, upgrading industry segments, improving service quality, and enhancing project quality and efficiency. The aim is to accelerate the transformation and upgrading of the tourism industry, striving to build a world-class tourism destination and boost the development of a strong tourism province. In 2023, Guizhou Province received a total of 636 million tourists and achieved a tourism income of more than 740 billion yuan [22]. New forms of tourism continue to emerge, 'Huang Xiaoxi, eat dinner[3]' has become a new tourism word, 'village super' and 'village BA' are popular in the circle, 'small cars and small groups[4]' and 'Zhizhichuanfei[5]' and other new products have formed new explosive models, and the travel market entities have a net increase of 19,000 households, reaching a total of 340,000 households.

Qianxinan Prefecture has taken full advantage of the tourism hotspots across the province by leveraging platforms like TikTok, WeChat Video, REDnote, and others to strengthen online marketing and promote tourism. The prefecture has also collaborated with online travel agencies such as Tongcheng and Ctrip to increase visibility and boost marketing efforts. By integrating Qianxinan's tourism resources under the 'Wanfeng Mountain Forest' and 'Sunshine Qianxinan' tourism brands, it has enhanced offline promotional efforts and focused on making the 'off-peak seasons lively.' Initiatives like 'Travel with the Event' and events like the 'Legendary Challenge Race (Guizhou Xingyi)' have been held to attract more visitors to Qianxinan for tourism and cultural experiences. In 2024, the prefecture will focus on promoting intangible cultural heritage through various travel incentives and six types of summer special products, adding new play styles, business models, and experiences in the field of intangible heritage tourism, thereby enriching the tourism market supply. Qianxinan's scenic areas have launched a variety of immersive, interactive, and experiential cultural tourism products based on the region's rich intangible cultural heritage, allowing visitors to experience the unique charm of the intangible heritage. Over the past few years, Qianxinan has targeted domestic markets, frequently conducting promotional activities in regions like the Guangdong-Hong Kong-Macao Greater Bay Area, Chengdu-Chongqing Economic Zone, Central China, East China, and North China. With its strategic location as a gateway in the 'Guizhou-Anshun Huangguoshu-Xingyi Wanfeng Mountain-Kunming Yunnan-Southeast Asia' golden tourism line, the prefecture has been actively promoting connectivity while also expanding the international tourism market. As the consumer market matures and new tourism demands emerge, Qianxinan's challenge lies in how to ensure sustainable development and promote more 'hotspots.' This is crucial not only for the high-quality development of the tourism industry but also for meeting the growing demand for a better quality of life among the population. However, new media-driven tourism development also reveals areas for improvement in China's tourism industry. The current status of sustainable cultural tourism development in Qianxinan can be summarized as follows:

1. Emergence of New Travel Trends: Recently,lesser-known, off-the-beaten-path destinations have become popular. During the 2023 'May Day' holiday, Qianxinan saw significant growth in tourist numbers, marking a new development opportunity for the region's tourism industry. However, this surge in tourism also brought challenges, such as congestion at certain scenic spots and difficulties in finding parking during peak periods.
2. Mismatch between Expectations and Experience: Qianxinan offers stunning natural landscapes such as dramatic mountain scenery, cloud-covered peaks, and misty valleys, but some visitors, attracted by promotional materials, feel disappointed upon arriving. This issue stems from poorly planned tourism routes, poorly located viewpoints, cable car stations, and key information points, leading to underwhelming visitor experiences due to inadequate development work.
3. Lack of Comprehensive Planning and Service Diversification: At present, many of the cultural tourism projects in Qianxinan lack overall planning and design, leading to a narrow range of products and services. The existing offerings fail to fully attract tourists, and there are issues with limited visitor demographics and types of tourism. The goal of 'attracting customers, retaining customers, and serving customers' has yet to be fully realized, which hampers the broader development of related industries.
4. Seasonal Imbalance in Visitor Demographics: Currently, cultural tourism projects primarily target middle-aged and young adults, whose travel patterns are concentrated during peak seasons. This creates an imbalance in visitor numbers, with some projects being overcrowded during peak periods but experiencing low turnout during off-peak times. This seasonal fluctuation leads to wasted resources and increased operational costs as facilities such as hotels, accommodations, and sports venues remain underutilized.
5. Limited Geographic Reach and Audience: Due to the region's geographic and transportation challenges, current cultural tourism and wellness projects have limited impact and small outreach, mainly attracting visitors from neighboring provinces and districts. Weekend getaways and day trips account for a significant portion of the visitor base, further limiting the potential for broader tourism expansion.

3"Huang Xiaoxi, eat dinner" This is the homophony of the first word of the six most representative scenic spots in Guizhou.
4"Small cars, Small groups" is a personalized travel model where small groups use private vehicles with agencies managing all services.
5"Zhizhichuanfei" is a tourism product that connects various branch airports and scenic spots in Guizhou Province through air travel.

4.2. PEST analysis of sustainable cultural tourism operation and management in Qianxinan prefecture under the new media environment

4.2.1. Political environment

It is an inevitable trend for the high-quality development of tourism industry in the new era to integrate health and wellness tourism and cultural tourism. The decision is necessary for promoting the innovative development of culture and tourism industry. 'Tourism and the cultural industry cannot be separated, and tourism should be designed by our culture,' stressed General Secretary Xi Jinping. We must insist on using culture to shape tourism and using tourism to highlight culture, promoting the integration of culture and tourism development, allowing people to experience the beauty of culture and the beauty of nature while appreciating the beauty of nature and cultivating their hearts and souls [23]. He further stated, 'We must develop the tourism industry, enrich the ecological and cultural connotations of tourism, and actively promote health care, elderly care, leisure, and entertainment services, making these sectors important emerging industries.' This important speech by General Secretary Xi provides a clear direction for promoting the integration of health and wellness tourism with cultural tourism.

Currently, the Chinese government is implementing strategies such as 'Healthy China,' 'Cultural Power,' and 'Rural Revitalization.' Against this backdrop, Qianxinan Prefecture is positioning itself as a 'health and wellness destination' for the city and actively promoting the development of cultural tourism and wellness industries, fostering a diverse integration of rural industries. In 2020, the 12th Plenary Session of the 7th Qianxinan Prefecture Committee made a strategic decision to develop the prefecture through culture and tourism, taking the lead in raising the tourism industry to a strategic level in the prefecture. During the 14th Five-Year Plan period, we will actively promote the development of integrated industries such as leisure and vacation, health tourism, and mountain sports tourism based on the '14th Five-Year Plan for Cultural and Tourism Development in Guizhou Province' [24]. In 2023, Qianxinan fully launched the 'Health and Wellness Destination, Cultural Xingyi' initiative, outlining its vision to develop the prefecture into a 'health and wellness product supply region in the border area of Guizhou, Yunnan, and Guangxi,' an 'ecological wellness backyard for the Greater Bay Area,' and an 'international tourism wellness destination [25].' A series of related documents have been issued to promote the development of the cultural tourism and wellness industries, providing important support for the development of foundational, related, and derivative wellness industries in Qianxinan Prefecture. Through policy-level support, it is possible to attract more inter-provincial and international tourists, promote regional tourism linkages, and create tourism clusters with complementary economic effects.

4.2.2. Economic environment

In 2023, Qianxinan Prefecture welcomed 42.49 million domestic tourists, generating a total tourism revenue of 41.213 billion yuan, demonstrating a rapid growth trend in the tourism market [26]. Through the development of the 'Health & Wellness Destination, Cultural Xingyi' brand, Qianxinan has actively guided the development of cultural tourism resources and the creation of economic benefits. Qianxinan Prefecture is rich in intangible cultural heritage resources, combined with its natural and cultural assets, they have provided a solid economic foundation for the development of the cultural tourism industry. The growth of tourism has driven the transformation and diversification of the local economy, helping to alleviate the constraints imposed by a single economic model. However, as the number of tourists continue to grow, Qianxinan faces challenges related to transportation, infrastructure, and service support. In particular, road congestion and a shortage of parking spaces during peak seasons have led to a decline in the overall visitor experience. This phenomenon has somewhat limited the further enhancement of economic benefits.

4.2.3. Social environment

Qianxinan Prefecture is home to more than 40 ethnic minorities, including the Bouyei, Miao, Yi, and others. The traditional cultures, arts, festivals, and customs of these ethnic groups provide rich tourism resources. For example, the Yi ethnic group's 'Amei Qituo' dance, Bouyei wedding customs, and Miao embroidery are part of the region's intangible cultural heritage, offering visitors the opportunity to experience the unique ethnic flavors of the area. Modern tourists are increasingly inclined toward experiential and immersive tourism, demanding higher levels of interactivity and participation. By integrating intangible cultural heritage with tourism, Qianxinan allows tourists to deeply engage with ethnic cultures and experience the region's diverse cultural atmosphere. This meets the needs of the new generation of tourists who prefer personalized travel experiences, and the shift in consumption patterns has encouraged the local tourism industry to move from traditional sightseeing to interactive, experience-based tourism. This has also driven the development of more innovative cultural tourism products. However, due to the lack of a comprehensive planning and layout, the current cultural tourism projects in the region are not sufficiently diverse, and the goal of 'attracting, retaining, and serving customers' has not been fully realized.

In recent years, Xingyi City in Qianxinan has been continuously promoting mountain tourism, leveraging its geographical advantages and using new media to develop folk customs and sports such as road running, mountain biking, cave exploration, car camping, and river rafting. The region has hosted various domestic and international mountain tourism and outdoor sports events, contributing to the integration of cultural tourism, sports, wellness, and scientific education into a multi-dimensional industrial development model. However, despite these efforts, Qianxinan's tourism industry faces challenges in terms of sustainable development. There is a significant seasonal disparity in the tourism market, with the middle-aged and young groups traveling mostly during holidays, while many homestays and tourist facilities remain idle during the off-peak season. This imbalance creates inefficiencies and limits the full potential of the local tourism sector.

4.2.4. Technological environment

Following the decision by the Guizhou Provincial Committee and the Provincial Government to develop the big health industry and promote wellness tourism, Qianxinan Prefecture has focused on leveraging its unique local resources and on implementing the 'Industry-Strong State' strategy. The region has made significant strides in expanding its health and wellness tourism sector. There are currently 2,307 medical institutions at various levels, and traditional Chinese medicine therapies, such as acupuncture, medicinal baths, foot therapy, and massage, are thriving. The local wellness and health product industry is also expanding, with notable products such as 'Xingren Job's Tears' and 'Anlong Dendrobium' gaining prominence. A wellness-focused atmosphere is gradually being established [27]. The combination of 'wellness + tourism,' 'wellness + sports,' and 'wellness + agriculture' has emerged as a new trend, and mountain sports and hot spring wellness have gradually become the region's tourism highlights. Ten national and provincial forest wellness pilot bases have been established, laying a solid foundation for the future development of cultural tourism and wellness industries.

Additionally, new media technologies, such as big data and social media, are empowering the region's wellness tourism industry. Through big data analysis, businesses and administrators can better understand market demand and optimize marketing strategies. Qianxinan Prefecture has actively utilized social media platforms such as TikTok, WeChat Channels, and REDnote for online promotion, while also partnering with online travel agencies like Ctrip and Tongcheng to promote its tourism resources. This comprehensive digital marketing strategy has significantly enhanced the visibility of Qianxinan's tourism brand, allowing more potential visitors to easily access travel information. By leveraging online platforms for precise marketing, bookings, and inquiries, Qianxinan has greatly expanded its tourism service coverage, attracting a growing number of young domestic and international tourists. At the same time, the intelligent reservation and service system can also improve the overall experience of visitors.

4.2.5. Summary of PEST analysis under the new media environment

The development of Qianxinan's cultural tourism industry benefits from policy support, favorable natural conditions, abundant social and cultural resources, and technological innovation. National policies that encourage the integration of health and wellness tourism and cultural tourism provide strong support for Qianxinan's wellness tourism industry. The region's unique subtropical climate, excellent air quality, and abundant ecological resources create a natural advantage for health tourism. Folk sports activities, rich ethnic culture and red revolutionary history also are unique cultural attractions to wellness tourism. In addition, the use of big data and new media technologies have beefed up market analysis, brand visibility and visitor experience, and provided a sound barometer for the sustainable and quality development of Qianxinan's cultural tourism industry; and Qianxinan is also facing more opportunities and challenges in this regard.

5. Discussion and Conclusion

5.1. Discussion

Cultural tourism operations not only involve the development and utilization of tourism resources, but also require a deep understanding of the practices of regional culture. The key to success lies in accurately integrating the unique local culture with tourism resources, while emphasizing the three elements of culture, innovation, and emotional value. In cultural tourism operations, the operational model is the fundamental prerequisite for the sustainable development of tourism projects. By improving aspects such as operational management, content integration, marketing strategies, and innovation, and applying relevant marketing strategies, it is possible to promote sustainable development of the cultural tourism industry in Qianxinan. This will be an area of further study for researchers moving forward.

5.2. Conclusion

The findings reveal that the era of new media offers unprecedented opportunities for innovation and growth, while also presenting challenges that require adaptive strategies. Politically, Qianxinan benefits from favorable policies and initiatives by the government to promote cultural and eco-tourism, aligning with national strategies for rural revitalization and sustainable development. Economically, the region's stable growth provides a foundation for investments in tourism infrastructure, while its abundant natural landscapes, such as the Maling River Canyon and Wanfenglin, serve as unique selling points. Culturally, Qianxinan's diverse ethnic traditions and heritage offer rich opportunities for storytelling in the digital age. Technologically, the rapid development of new media platforms enables innovative marketing strategies, real-time interaction with tourists, and data-driven decision-making, enhancing visitor experiences.

Despite these advantages, challenges remain. Ecological preservation is crucial as increasing tourism activities put pressure on the region's fragile environment. Moreover, fierce competition in the tourism market demands that Qianxinan differentiate itself with unique products and branding.

Consumer preferences are also shifting toward experiential and personalized travel, necessitating flexibility in product design and service delivery. The role of new media in amplifying both positive and negative feedback further underscores the importance of maintaining high-quality services and effectively managing public perceptions. To achieve sustainable growth, Qianxinan must adopt integrated measures that leverage its strengths while addressing key challenges. Others would be to develop eco tours that are in conformity with global development goals and encouraging community engagement so as to ensure that the local residents benefit both economically and culturally there by ensuring the sustainability of the tourism facility while at the same time developing new media that will support immersive and interactive tours that are appealing to diverse communities. The region's natural and cultural heritage needs to be preserved within a balance between development and protection through facilitating the collaboration of government agencies, private enterprises and local communities.

In summary, Qianxinan is well-positioned to become a model for sustainable cultural tourism development in the era of new media. By effectively harnessing its political support, economic potential, cultural richness, and technological advancements while proactively addressing ecological, competitive, and consumer-related challenges, a sustainable tourism ecosystem can be built in the region.

References

[1] Xiaoyu, W. (2023). Research on short video marketing strategies of tourist attractions under the background of new media—A case study of Weizhou Island, Beihai, Guangxi. *Trade Show Economy,* (15), 64–67.

[2] Uchida, N., Sato, G., & Shibata, Y. (2019). Device-to-Device Communication based DTN for Disaster Information System by using Emergent User Policy and Locational Information. *Journal of Internet Services and Information Security*, *9*(3), 41–51.

[3] Vranješ, B., Vajkić, M., Figun, L., Adamović, D., & Jovanović, E. (2024). Analysis of Occupational Injuries in an Iron Ore Mine in Bosnia and Herzegovina in the Period from 2002 to 202 *Archives for Technical Sciences*, *1*(30), 33–44. https://doi.org/10.59456/afts.2024.1630.033V

[4] Ren, L. (2023). 'Two-way travel': the concept and model of marketing innovation under the new trend of cultural travel development. *Theoretical Monthly*, (12), 106–116.

[5] Shimazu, S. (2023). Maximizing Employee Satisfaction Through Wellness Initiatives. *Global Perspectives in Management*, *1*(1), 49–65.

[6] Abbedllah, H. A., & Kori, H. A. (2023). On Two -Way Nested Classification Random Model. *International Academic Journal of Science and Engineering*, *10*(2), 71–79. https://doi.org/10.9756/IAJSE/V10I2/IAJSE1010

[7] Shengyuan, W., & Yanqiang, W. (2023). Study on the influence of short video promotion of cultural travel on users' impulse travel intention based on SOR theory. *New Media Studies*, *9*(21), 43–46.

[8] Thirunavukkarasu, E. S., & Karthikeyan, E. (2016). Security Enhancement Using Hierarchical Threshold Secret Sharing with M-M Scheme and Quantum Particle Swarm Optimization (QPSO) for Congestion Control in Skype VOIP. *International Journal of Advances in Engineering and Emerging Technology*, *7*(4), 175–193.

[9] Pingjing, M. (2024). Development strategy of network marketing of small and medium-sized enterprises under new media environment. *Modernization of Shopping Malls,* (06), 53–55.

[10] Yinxing, H., & Futian, Q. (2023). Sustainable development economics. The Commercial Press.

[11] Yanfen, J. (2016). Research on sustainable development of rural tourism based on life cycle theory of tourist destination.

[12] Purwaningsih, R., et al. (2020). Assessment sustainable tourism: A literature review composite indicator. *E3S Web of Conferences*, *202*, 03001.

[13] Mei, H. (2023). Study on tourism sustainable development of Dihua Ancient Town in Shaanxi Province.

[14] Dan, M. (2023). Research on the strategy of secondary development at Horse's Mountain Scenic spot based on the life cycle theory of tourist destination.

[15] Reindrawati, D., et al. (2022). Threats to Sustainable Tourism in National Parks: Case Studies from Indonesia and South Africa. *African Journal of Hospitality, Tourism and Leisure*, *11*, 919–937.

[16] Tinggui, S. (2020). Jingxiu Company Chenzhou animation Water World project operation management improvement research.

[17] Youtao, Y. (2023). Research on Optimization of M enterprise operation management strategy based on value chain perspective.

[18] Changhong, D. (2019). Research on key success factors of tourism real estate projects.

[19] Yuting, Z. (2023). Research on marketing strategy of HQC Group AR Guzhen Cultural Tourism real estate project.

[20] Daily, G. (2023). Humanistic charm and living heritage: Southwest Guizhou promotes the in-depth integration of intangible cultural heritage and tourism. From http://gz.news.cn/20231129/f10d5611b48e428bb3d32e706f9ec148/c.html.

[21] Congress, Q. P. P. S. (2023). The 13th meeting of the Standing Committee of the 9th People's Congress of the Prefecture. From https://qxnrd.gov.cn/#/news/aticle/778.

[22] Guizhou, C. T. i. S. (2024). Guizhou Cultural Tourism Industry Investment Promotion Conference. From https://mp.weixin.qq.com/s?__biz=MzI1NjY2MDg0Mw==&mid=2247605168&idx=2&sn=e68ea6da1e142b5432442c029ad6564d&chksm=eb497a29e4ae6f95bf70fd686ebc29beb11bada27794c05206d664aeb9f9128bd12974a773bd&scene=27.

[23] Ke, W. (2024). Promote the integration of culture and tourism, enrich market supply, optimize tourism services, and promote high-quality development of tourism to a new level. From https://www.gov.cn/yaowen/liebiao/202405/content_6951426.htm.

[24] Bureau, Q. P. T. (2022). Qianxinan Prefecture Culture, Radio, Television and Tourism Bureau's reply to Proposal D1 of the 1st Session of the 9th CPPCC. From https://www.qxn.gov.cn/ztzl/bljytagz/202208/t20220802_75924398.htmlQianxinan

42 Customer perception and problems towards solar energy products

Sowbarani G.[1,a], M. Akilanayaki[2,b], V. Poornima[3], N. Omshanmugapriya[4], and E. Rajeswari[5,c]

[1]Assistant Professor, Karpagam Academy of Higher Education, Coimbatore, Tamil Nadu, India
[2]Assistant Professor and Head, Department of Commerce Business Process Services, Nallamuthu Gounder Mahalingam College, Pollachi, Tamil Nadu, India
[3]Assistant Professor, Department of Commerce Business Process Services, Nallamuthu Gounder Mahalingam College, Pollachi, Tamil Nadu, India
[4]Assistant Professor, PG and Research Department of Commerce, Arulmigu Palaniandavar Arts College for Women, Chinnkalayamputhur, Tamil Nadu, India
[5]Assistant Professor of Commerce, Karpagam Academy of Higher Education, Coimbatore, Tamil Nadu, India

Abstract: Accessibility, cost, and dependability are some of the aspects that influence customer perception and issues with solar energy solutions. Some consumers may be worried about the expense of the initial investment, upkeep, and efficiency of solar energy, while others may view it as a sustainable and environmentally good option. Frequently encountered obstacles encompass insufficient comprehension of solar technology, sporadic energy provision, and deficient support infrastructure. To improve customer confidence and encourage the wider adoption of solar energy solutions, addressing these misconceptions and problems calls for thorough education, easily accessible financing choices, quality assurance, and strong after-sales service.

Keywords: Investment, eco-friendly, sustainable, technology

1. Introduction

Products that harness solar energy are a key component of the worldwide energy shift. These gadgets, which range from solar water heaters to solar panels, utilize the abundant and sustainable power of sunlight. Their importance stems from their ability to lessen environmental damage by decreasing dependency on fossil fuels, but also from their promise to democratize access to energy, especially in isolated or underdeveloped areas. In addition, solar energy products generate jobs in the renewable energy industry, lower long-term energy costs, and encourage clean technology innovation. Notwithstanding these benefits, problems like sporadic nature, initial expenses, and technology constraints still exist, requiring continuous study, governmental backing, and funding to maximize effectiveness, affordability, and scalability. The growing development and popular acceptance of solar energy products become essential elements of a sustainable future as society comes to understand the gravity of halting climate change and attaining energy security.

2. Review of Literature

Easwaran and Sudarvel [1]. Energy from the sun, either as heat or light, can be used to generate clean, green power. Solar energy can be sustained and is completely limitless, unless fossil fuels become scarce. When it creates electricity, it is also a clean energy source that emits no greenhouse gases. Energy suppliers and the government are pushing hard for solar product usage in order to fulfill growing demand and finish the shift to a sustainable economy. The different solar products need to be known to consumers in order to encourage them to buy them.

Shakeel, Juntunen, and Rajala [2]. Despite recent growth, the global distribution of solar energy, which is essential for meeting the world's sustainable energy needs, is uneven. Value-aligned solutions provided by solar photovoltaic (PV) firms are a key factor in boosting adoption. The business model that a company chooses to use dictates how well it can meet market demands. Numerous facets of business models are explored through research, including ownership, financing, the impact of regulations, industry practices, and models for distributed and utility-scale generating.

Agarwal, Bhadauria, Kaushik, Swami, and Rajwanshi [3]. India's rapidly growing energy needs, driven by both rapid economic and demographic growth, underscore the significant potential of solar energy in the nation's energy portfolio. The aim of this study is to identify and evaluate the factors that propel the adoption of solar renewable

[a]sowba13@gmail.com, [b]akilalingesh@gmail.com, [c]rajiguna84@gmail.com

DOI: 10.1201/9781003684459-42

energy products in India. We found 19 significant parameters influencing the adoption of solar renewable energy goods by using a nominal group technique and analyzing 40 ABDC-indexed articles. We created a driving force-dependency quadrant graph using the cross-impact matrix multiplication (MICMAC) technique to clarify the relationships between these variables.

2.1. *Statement of the problem*

Numerous obstacles, such as the erratic nature of sunshine, the high initial cost of installation, the need for ongoing maintenance, grid integration problems, and the limited energy storage capacity of solar energy products, affect users' experiences and acceptance of these technologies. Investing in solar energy solutions may be discouraged by these difficulties, which provide major obstacles to its efficient application. Many factors influence consumer decisions while making purchases, such as product affordability, dependability, performance efficiency, financing alternatives available, government incentives, environmental awareness, and projected long-term cost savings. To overcome adoption hurdles and encourage the general uptake of solar energy products among consumers, it is imperative to recognize and address these concerns.

2.2. *Objective of the study*

To study the various problems encountered by the solar energy product users.

2.3. *Scope of the study*

Solar energy is useful for many things, such as heating, lighting, and power generating. diminishing reliance on fossil fuels. The government may lessen its reliance on fossil fuels by utilizing solar energy more frequently. Which will only get more costly over the next few years; technological advancements will guarantee that solar power stays even more affordable. It's possible that throughout a significant portion of the world, solar energy will have supplanted other energy sources as the primary method for producing power by 2030. Climate change and the environment will benefit from this as well.

2.4. *Statistical tools used for the study*

106 of the 110 respondents submitted data, which were thereafter thoroughly reviewed, tallied, and analysed using statistical techniques such factor analysis, weighted average, chi square test, Simple Frequency, and so on.

2.5. *Limitation*

Due to sample size and demographic limitations, potential biases in survey responses, reliance on self-reported data, inability to account for regional variations in solar energy adoption, and difficulties capturing nuanced aspects of customer perceptions and experiences are some of the limitations of this study.

2.6. *Weighted average*

Using a weighted average analysis technique, various data are given varying weights in order to generate a more representative average. Usually, the weights are determined by how significant or important each value is.

Table 42.1 displays participant rankings of solar product challenges as evaluated by survey average responses. Survey results indicate that people primarily fear solar products are unaffordable thus presenting a significant barrier to adoption. Real-life space restrictions and solar resource-interrupting

Table 42.1. Problems of solar products

S.No	*Problems of solar products*	*7*	*6*	*5*	*4*	*3*	*2*	*1*	*Total*	*Mean score*	*Rank*
1	Takes up a lot of space	33	27	21	10	6	6	3	106	5.3867	2
		231	162	105	40	18	12	3	571		
2	Works only when the sun shines	17	41	22	11	9	0	6	106	5.2075	3
		119	246	110	44	27	0	6	552		
3	Too Expensive	9	30	26	10	5	2	4	106	5.4339	1
		203	180	130	40	15	4	4	576		
4	Least energy at night	22	23	31	11	7	5	4	106	4.9905	5
		154	138	155	44	21	10	7	529		
5	More batteries are needed to save solar power	18	28	22	14	9	8	7	106	4.8113	7
		126	168	110	56	27	16	7	510		
6	After Sales Service	20	34	17	7	5	6	7	106	4.9150	6
		140	204	85	28	45	12	7	521		
7	Leakage, Glass breakage or physical damage	20	37	17	13	6	6	7	106	5.0566	4
		140	222	85	52	18	12	7	536		

Source: Author.

factors emerge as the second most significant concerns with solar energy solutions according to survey respondents. The rating of 'leakage, glass breakage, or physical damage' at position four points to customer worries about product endurance and reliability. Two important concerns expressed by customers relate to nighttime power storage capacity as well as post-sale service support. The assessment of solar power battery storage options as a challenge for effectiveness appears despite being the designated bottom-ranking statement. The listed customer difficulties while buying solar energy goods serve as important data points for both enhancing satisfaction and expanding adoption levels. The evaluation determines essential improvement zones that should focus on product design alongside lowering costs and guaranteeing dependability and quality of after-sales assistance.

3. Conclusion

The adoption and application of solar energy products mainly depend on how customers perceive them as well as technical obstacles associated with these products. Various tests demonstrate that consumers view solar products through the lens of price and space concerns as well as reliability of solar power and post-sale assistance. Improved acceptance rates for solar energy solutions require effective solutions to the detected problems. The solar energy sector requires players to develop research and development priorities which address these issues and enhance product affordability and prevent negative reliability incidents. The solar energy sector can meet customer requirements and serve the energy transition by resolving the identified issues and maintaining high standards of product performance.

References

[1] Easwaran, P., & Sudarvel, J. (2024). A study on impact of demography factor on consumer awareness on solar energy product. In *E3S Web of Conferences* (Vol. 477, p. 00060). EDP Sciences.

[2] Shakeel, S. R., Juntunen, J. K., & Rajala, A. (2024). Business models for enhanced solar photovoltaic (PV) adoption: Transforming customer interaction and engagement practices. *Solar Energy*, *268*, 112324.

[3] Agarwal, R., Bhadauria, A., Kaushik, H., Swami, S., & Rajwanshi, R. (2023). ISM-MICMAC-based study on key enablers in the adoption of solar renewable energy products in India. *Technology in Society*, *75*, 102375.

43 Crowdfunding platforms as a capital formation tool for young filipino entrepreneurs: Empirical evidence and growth implications

Remson Mark C. Macawile[1,a] *and Joseph Berlin P. Juanzon*[2,b]

[1]Polytechnic University of the Philippines, Philippines
[2]Pamantasan ng Lungsod ng Maynila, Philippines

Abstract: This essay discusses young Filipino entrepreneurs' perceptions that influence their decision-making regarding crowdfunding as a financing option for their small businesses. It also examined how well behavioural intention could predict actual behaviour in using empirically developed models of intention and is based on sociocognitive assumptions. We employed structural equation modeling (SEM) using partial least squares (PLS) with data from 270 young Filipino entrepreneurs. It was combined with UTAUT, which had three additional constructs incorporated. The findings indicate that the Blockchain technology has little to no impact on the perceived ease and usefulness of a system. However, social influence and performance expectancy indeed have a significant and positive impact on perceived ease and usefulness. It was also discovered that perceived usefulness and perceived ease of use would correlate in a strong way with intention to use a crowdfunding platform in the future. This study is one of several that consider the relevant literature with the aim of analyzing the available works for potential future use by other researchers and practitioners.

Keywords: Blockchain technology, performance expectancy, platform quality, intention to use crowdfunding platforms

1. Introduction

There are many factors responsible for growth and The main focus of this study is on how small and medium-sized businesses (SMEs) and entrepreneurship can help an economy grow. Small and medium-sized businesses (SMEs) are very important to the growing economy because they help new businesses get started and find other ways to make money [1]. This helps create jobs and keep unemployment low. But they are almost always condemned to failure because of lack of expertise, lack of technical background, absence of strong formal and informal institutional environment in many developing countries and informational asymmetries that either trap the aspiring entrepreneur or prevent him from venture capital. [2]. One such hurdle which is more urgent than the other ones is a shortage of funds – it can hinder the absolute foundation of an enterprise long before anything to do with workforce or capital comes into the equation [4].

In the Philippines, start-ups and small businesses face many challenges in securing essential capital [3]. Many new entrepreneurs turn to traditional financial institutions for funding, yet they often lack sufficient collateral or strong credit histories. As a result, nearly half of their credit applications are declined [41].

This pushed many entrepreneurs to invest on alternative sources of funding, for instance, crowdfunding platforms, venture capital or business angels. As a result this study is based on the perspectives of young Filipino entrepreneurs or qualified students on the use of crowdfunding platform in financing their small businesses. As a financial technology tool to support startups of Southeast Asia, crowdfunding platforms have been proven as a promising solution [5]. Moreover, crowdfunding is closely linked to the rise of Fintech, which has rapidly gained popularity, reshaping the financial landscape [6].

For many entrepreneurs, crowdfunding has become a reliable alternative financing source, which is helpful especially for those entrepreneurs who have problems accessing traditional sources of funding. The intention to use crowdfunding platforms is driven by different aspects and this is well explained by several theories that capture the notion of technology adoption and user behaviour. The Unified Theory of Acceptance and Use of Technology (UTAUT) is a well-known model for how people adopt new technologies. It says that things like performance expectations, effort expectations, social impact, and facilitating conditions all play a big role in getting people to use new technologies. For crowdfunding, these traits could be expanded to include platform-specific factors like how well people think the platform works, how easy it is to use, and how much they trust the platform [42]. The blockchain technology, the quality of the platform, and social impact all play a big part in how users see the value and trustworthiness of the platform, which in turn affects their

[a]remsonmarkmacawile@gmail.com, [b]jbpjuanzon@plm.edu.ph

DOI: 10.1201/9781003684459-43

plans to use it. Framing these foundations, we identify incentives for adopting crowdfunding among young entrepreneurs and discuss implications for the design of platforms as means to reach more users and stimulate entrepreneurship [43].

2. Literature Review and Hypothesis Development

2.1. *Blockchain technology, perceived usefulness and perceived ease of use*

Blockchain technology makes crowdfunding platforms seem more helpful because they can provide a clear, safe, and quick way to handle transactions. According to [7], one of the biggest issues with crowdfunding platforms for young entrepreneurs and investors is the lack of trust, as they have to rely on the platform for sensitive financial transactions and personal data. The inherent properties of blockchain, such as transparency and tamper proofing, reduce fraud risk and provide verifiable data records [8, 9]. This further security makes the platform appear far more trustworthy and allows for enhancement in perceptions that the platform can be really useful both for funding and investment. Furthermore, blockchain enables decentralized, peer-to-peer transactions with minimal intermediaries most of the time [10]. Combined with less time, cost, and complexity than their traditional methods of funding, these aspects have huge potential to improve user appeal on the platform.

H1a. *People will see blockchain technology as being valuable.*

Blockchain also makes crowdfunding more perceived as easy to use; the process of transactions is smooth and automated [11]. Therefore, it reduces the need for any type of manual verification processes. It is evident in the prior literature that smart contracts are a set of self-executing orders that direct written code of an agreement [12]. This brings it into effect when predefined rules are satisfied. This also reduces administrative steps; since the entire cycle of funding agreements is automated, it becomes less cumbersome to operate on the platform for both the investors and the entrepreneurs. In recent research, [13] stated that blockchain user experience innovations, such as digital wallets and processes for easy identity verification, ease use because one can trace, receive, and manage their investments with reduced effort. To a young entrepreneur with no previous experience in crowdfunding, such blockchain-driven features make the process of onboarding for fundraising, along with interaction with potential investors, more intuitive and smoother, fostering an overall more accessible platform experience.

H1b. *Blockchain technology is going to affect the impression of how easy it is to use positively.*

2.2. *Platform quality, perceived usefulness and perceived ease of use*

Looking at it through the lens of the crowdfunding site being built for aspiring young business owners, perceived usefulness for a consumer will be influenced by the design and features, including functionality that a certain platform will have [14]. Typical characteristics of crowdfunding platforms include searching for projects, following funding, and communication with project creators [15]. The proposed platform is informative, useful, and navigates a user efficiently with effective communication features looked upon by all the end- users as an aiding tool in making the right funding decisions. Since the platform smooths out the process by streamlining project information, thereby smoothing out the entire process of engagement and investment-the perceived usefulness increases as one of the most useful mediums to reach out to potential backers for young entrepreneurs.

H2a: *The perceived usefulness of a platform is positively correlated with its quality.*

As a term, 'perceived of ease use' describes how much work people think goes into using the platform [16]. The facility should be user-friendly and easy to manage for young entrepreneurs creating their profiles and managing campaigns without deep technical knowledge to engage with backers [17]. Facilites are more likely to be viewed as user-friendly when they include straightforward instructions, easy-to-navigate interfaces, and few obstacles to participation. A well-designed platform reduces the learning curve, enhancing overall experience that will encourage both young entrepreneurs and backers to participate in crowdfunding. Ease of use mainly pertains to security in a competitive market, whereby user experience could be a direct contributor to the success of the campaign.

H2a: *Users will rate the platform's ease of use higher if it is of high quality.*

2.3. *Social influence, perceived usefulness and perceived ease of use*

Social influence can improve perceived usefulness of augmented reality features for online shopping by influencing users' perceptions of how useful this technology will be to improve their shopping experience [18]. When people around them start using AR features, or when they get encouraging feedback about the features from friends, family, and influencers, they would think such features would be helpful in making better decisions and evaluating the product for increased satisfaction of the shopping experience [19]. This also resonates with TAM, stating that social influence increases PU since it supports perceptions of the technology as worthwhile; therefore, the consumers will be well likely to view AR as useful in their health and wellness if the social network values or supports its use.

H3a. *Hypothesis 3a: Perceived utility will be positively affected by social influence.*

Second, one's social circle can also play a role in how easy they find AR to use. when shopping online. Similar to how people who experience no problem in navigating and reaping the benefit from AR features foster an understanding that the technology is straightforward and user-friendly, so too will others change based on what they perceive happening to others [18]. Social influence has an indirect influence on learning, in that users' experiences often breed through other people; this decreases uncertainty and increases confidence in how easily the AR features can be used [20]. This runs parallel to findings related to TAM showing how social influence can decrease the perceived complexity of new technologies by normalizing their use amongst peer groups. It creates in users the feeling that AR is something quite easy to communicate with, since people see how others interact with it without much hassle. In light of the foregoing, we postulate the following theory.

H3b. *The degree to which people feel a product is easy to use is positively impacted by social influence.*

2.4. *Performance expectancy, perceived usefulness and perceived ease of use*

In the context of online shopping, social influence has the greatest impact on users' perceptions of digital technologies. If friends, family, or members of one's social circle recommend a certain technology, it's more likely to be considered useful by the user [21]. In this regard, the hypothesis may now be developed for AR in online shopping: The higher the social influence, the more useful AR features are perceived to be [22]. If people perceive that others find AR helpful and easy to use to get views of products, they too may find AR a useful technology, for this enhances one's shopping experience through improved information quality about the product on sale with reduced uncertainties. Therefore, the authors confirmed that users would find AR more useful when significant others around them endorse its benefits.

H4a. *Perceived utility will rise as a result of social influence.*

In a similar vein, when people in a user's social circle show signs of being comfortable using augmented reality technology, it could lead to the user believing that the technology is easy to use themselves [23]. Users have been found to be more open to finding a system that is easy to use if they observe their peers navigating it effectively. For instance, with online shopping using AR, people would feel assured about navigating AR interfaces if others are showing ease in using features of that nature. Put simply, this will alleviate any personal concerns about the technology and make it seem easier to use. As stated in [24], the perceived ease of use of an augmented reality technology is positively affected by social reassurance, which gives the idea that the tool is intuitive and easy to use.

H4b. *Perceived ease of usage is positively impacted by social influence.*

2.5. *Perceived usefulness and intention to use crowdfunding platform*

Crowdfunding platforms are designed to enable funding, networking, and visibility of projects for creators and backers alike. Users' belief in the ability of the platform to enhance project success or provide access to key resources is one of the important drivers of use intention in evaluating the platform [22]. In the context of crowdfunding, perceived usefulness may mean raising funding, soliciting feedback from the public, building a supportive community-all of which suggest that use of such platforms has perceived utility for potential users [25]. It is argued by [15] that when potential users perceive crowdfunding as a means to serve their practical needs effectively, interest in it is more likely to follow. In that respect, creators consider crowdfunding useful if it allows them immediate access to cash and exposure, they will not have from other funding options. In regard to backers, those would be believers in contributing to the realization of ideas or taking part in the birth of new firms; they see in the platform an efficient tool for personal or social returns, which raises their intention to participate.

H5. *Easy of use is positively impacted by perceived usefulness.*

2.6. *Perceived ease of use and intention to use crowdfunding platform*

The Technology Acceptance Model states that consumers' expectations of a new technology's usability are a key factor in whether or not they will actually utilize it. This deeply relates to crowdfunding, where people are often looking forward to a fast and easy way of getting funds without the hassle that has always accompanied traditional financing. According to [22], crowdfunding platforms perceived as easy to use reduce cognitive load, making core functions like setting up a campaign, processing the transactions, and reaching the backers much easier. This is going to be crucial, particularly for first-time users or casual users who might not necessarily be familiar with the minutiae of online funding [25]. If this crowdfunding platform requires less effort to learn and use, the users will be less frustrated and more unlikely to give up using it, which in turn again reinforces their continued intention to use it in the future [9]. Ease can also support a smooth user experience that affects user satisfaction and perceived value. First, for example, with a navigable platform, users can focus more on the achievement of major goals such as visibility and the collection of funds rather than getting in the way of technical issues or confusing interfaces. (See Figure 43.1).

H6. *The impact of perceived ease of use on perceived ease of use is favourable.*

3. Methods

3.1. *Sampling and data collection*

This study examined 270 young entrepreneurs in the Philippines. The participants were either recent college grads or in their last year of community college or university. All participants demonstrated an interest in applying the crowdfunding

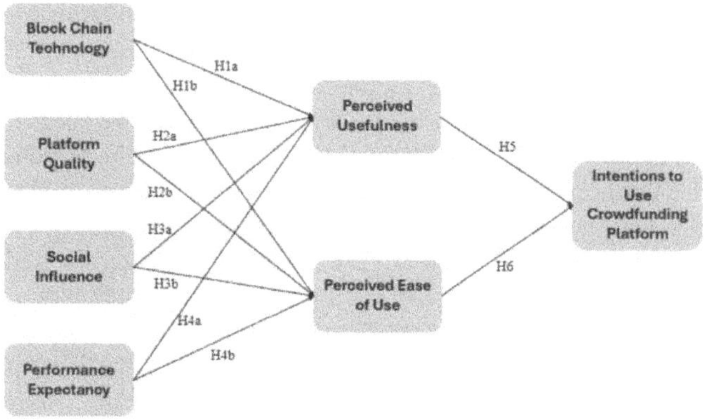

Figure 43.1. Conceptual framework.

Source: Author.

platforms to start their own small businesses. Non probability purposive sampling technique has been employed and the primary data was collected from online questionnaire which was transmitted via electronic mail and social media in order to reach the intended audience within an acceptable amount of time. A preliminary trial was conducted before the full rollout was performed with a sample of fifteen young business owners drawn from the target demographic in order to evaluate the clarity of the survey, reliability, and validity. Since there were no significant issues arising from this test, the survey was then distributed to a main sample of 285 young entrepreneurs. To maximize the number of responses, the Web questionnaire was kept online for one month.

4. Measurement

Our assessment was based on a survey we created based on scientific works; however, a few questions were modified in order to match our research needs.

We utilized a 7-point Likert scale to code the survey findings, where 1 indicates a strong disagreement and 7 indicates a strong agreement. We built this questionnaire on top of previous research. Secondly, the measure was adapted from the prior literature [10] and scales on five items scale. Furthermore, four items of platform quality has also been adapted from the research [26]. Additionally, it was adapted from the research of [26] and comprised five items of the scale of social influence construct. The fourth construct is adapted from [21] The items for the construct performance expectancy are from the research and consist of four items. All items pertaining to both perceived ease of use (PEOU) as well as PU, which stands for perceived utility, were used to gauge how people felt about Davis' Technology Acceptance Model (TAM) [27]. The TAM is a well-known framework in the field of information systems studies. As it uses an up to date method in evaluating user acceptance of technology. Our dependent variable in last is our intention to use crowdfunding platform and it was adapted from [16] and consist of five items.

5. Results

5.1. Measurement model validation

Interrelationship among the variables was assessed by correlation analysis. When the model was subjected to statistical analysis, a very significant relationship was results for the examined factors (Table 43.1). The rectangular root of AVE was used to examine the construct validity. This aspect in other words means that there are evidences in support of validity based on discrimination, as the square of AVE is greater than the correlations of this construct with others constructs [28]. The else, the validity of discriminant may be tested by comparing AVE and MSV value with all the factors. The MSV is greater than the AVE for achieve the discriminant validity [29]. As it is recommended [29] for the case of discriminant validity model selection, a construct's square root of its AVE should be greater than its correlation with others constructs. Table 43.2 also shows that the composite reliability (CR) value of every variable exceeding 0.70 [30]. The second was to conduct a convergent validity assessment to assess whether there is a relationship between these factors as quantified by AVE and item loadings [31]. In the end, it is decided that all variables meet the criteria and show more than 50% fluctuation most of the time, with an AVE greater than 0.5. The full test results are shown in Table 43.2.

Table 43.1. Discriminant validity

Constructs	*1*	*2*	*3*	*4*	*5*	*6*	*7*
Blockchain Technology	**0.855**						
Intention to use Crowdfunding Platform	0.390	**0.836**					
Perceived Ease of Use	0.614	0.541	**0.790**				
Perceived Usefulness	0.573	0.500	0.749	**0.813**			
Performance Expectancy	0.640	0.551	0.684	0.739	**0.819**		
Platform Quality	0.781	0.488	0.718	0.714	0.709	**0.770**	
Social Influence	0.666	0.613	0.747	0.765	0.774	0.795	**0.770**

The bold values are the √AVE.

Source: Author.

5.2. Reliability analysis

We used the Cronbach's alpha method to do reliability analyses for all categories, as suggested by [32]. The findings showed that Cronbach's alpha values for all constructs were higher than the 0.70 level, which means that the data can be trusted. To figure out the level of each construct's items.

Along with the Composite Reliability (CR) numbers, we found the internal consistency. Based on what [33] says, the CR values were higher than the minimum level of 0.70. Table 43.2 shows more information about these studies.

5.3. Common method variance

Numerous statistical and methodological approaches have been used to check for common method variation (CMV). Priority one was giving the questions as few words, specific details, and simplicity as possible. Partially completed study was done to check how well the instruments fit [34]. Furthermore, Harman's single factor test shows that CMV exists if one factor explains at least half of the total variation [34]. [35] was also used to assess how CMV impacted on the results. The main significant component in research accounted for

Table 43.2. Factor loading, validity, and reliability of the indicators

Constructs	*Items*	*Loadings*	*VIF*	*α*	*CR*	*AVE*
Blockchain Technology				0.816	0.891	0.732
	BCT1	0.853	1.883			
	BCT2	0.902	2.162			
	BCT3	0.808	1.633			
Intention to use Crowdfunding Platform			0.857	0.902	0.699	
	INTC1	0.884	2.544			
	INTC2	0.862	2.284			
	INTC3	0.872	2.373			
	INTC4	0.715	1.650			
Perceived Ease of Use			0.799	0.869	0.624	
	PEU1	0.755	1.546			
	PEU2	0.799	1.754			
	PEU3	0.865	2.006			
	PEU4	0.736	1.535			
Perceived Usefulness			0.828	0.886	0.661	
	PUS1	0.749	1.574			
	PUS2	0.770	1.524			
	PUS3	0.871	2.461			
	PUS4	0.856	2.395			
Performance Expectancy			0.754	0.859	0.670	
	PEX1	0.818	1.518			
	PEX2	0.859	1.926			
	PEX3	0.777	1.475			
Platform Quality			0.773	0.853	0.593	
	PLQ1	0.762	1.391			
	PLQ2	0.793	1.645			
	PLQ3	0.826	1.809			
	PLQ4	0.693	1.526			
Social Influence			0.769	0.853	0.594	
	SIN1	0.704	1.398			
	SIN2	0.726	1.515			
	SIN3	0.783	1.523			
	SIN4	0.860	1.931			

Source: Author.

36.89% of the variation and was less than 50%, in order to indicate that common method variance (CMV) was not present. Furthermore, analyzes the relationship between latent variables in order to analyze CMV. All of the variable correlations were less than 0.90. According to our statistical studies, therefore, there appears to be no CMV in the data.

5.4. *Multicollinearity*

It was used to find the benchmark values, the variance inflation factor (VIF), and multicollinearity through a regression study. This means that VIF shouldn't be greater than 0.3 [37]. All of the variables' VIF scores and thresholds fall within the recommended ranges. This means that there is no multicollinearity in this model [38].

5.5. *The predictive power of the model (Q2)*

The Stone and Geisser test was used to judge our structure model on SmartPLS. If the Q2 number for a theoretical framework is greater than zero (>0), it means that the model can predict things [39]. In the path model, this means that all of the dependent factors have Q^2 values greater than zero, confirming the validity of the path model (see Table 43.3).

Table 43.3. Blindfolding statistics for the general model

Construct	*SSO*	*SSE*	*($Q^2 = 1 - SSE/SSO$)*
Blockchain technology	800	635.12	0.206
Intention to use crowdfunding platform	800	689.25	0.138
Perceived ease of use	800	611.58	0.236
Perceived usefulness	1000	947.23	0.053
Performance expectancy	800	694.77	0.132
Platform quality	800	685.225	0.143
Social influence	1000	648.514	0.189

Source: Author.

5.6. *Structural model and hypothesis outcomes*

The hypothesized relationships were tested through use of regression analysis [40]. Therefore, the results support the first hypothesis, which says that blockchain technology has a big positive impact on how useful something is (H1a–2 = 0.052, $p > 0.01$) and how easy it is to use (H1b–^{2} = 0.075, $p < 0.01$). The second hypothesis (H2a–β = 0.252, $p < 0.01$; H2b–β = 0.253, $p < 0.001$), which assesses the quality of the platform on PU and PEOU, was also confirmed as significantly positively affecting both PU and PEOU. The results also proved the third hypothesis, because there was a positive and a strong link between how useful something is thought to be and social influence (H3a–2 = 0.360, $p<0.001$), as well as a link between how easy something is thought to be to use and social influence (H3b–β=0.366, $p<0.001$). These findings contribute to the significant effect of blockchain technology, platform quality on perceived usefulness and ease of use and on social influence (Table 43.4).

6. Discussion

This looked at the main things that make young Filipino businesspeople want to use crowdfunding sites to get money for their small businesses. The goal was reached with the help of the extended UTAUT model (Figure 43.2). As we discuss below, the results supported some constructs and hypothesized relationships, and did not support other constructs and hypothesized relationships.

Table 43.4. Hypotheses testing

	Hypotheses	**Beta**	**S.D**	***t*-values**	***p*-values**	**Decision**
H1a	Blockchain Technology -> Perceived Ease of Use	0.052	0.114	0.460	0.646	Not Supported
H1b	Blockchain Technology -> Perceived Usefulness	0.075	0.091	0.822	0.411	Not Supported
H2a	Platform Quality -> Perceived Usefulness	0.253	0.103	2.467	0.014	Supported
H2b	Platform Quality -> Perceived Ease of Use	0.252	0.109	2.313	0.021	Supported
H3a	Social Influence -> Perceived Usefulness	0.360	0.102	3.521	0.000	Supported
H3b	Social Influence -> Perceived Ease of Use	0.366	0.103	3.558	0.000	Supported
H4a	Performance Expectancy -> Perceived Usefulness	0.329	0.075	4.408	0.000	Supported
H4b	Performance Expectancy -> Perceived Ease of Use	0.189	0.078	2.436	0.015	Supported
H5	Perceived Usefulness -> Intention to use Crowdfunding Platform	0.216	0.051	4.235	0.010	Supported
H6	Perceived Ease of Use -> Intention to use Crowdfunding Platform	0.379	0.117	3.236	0.001	Supported

Source: Author.

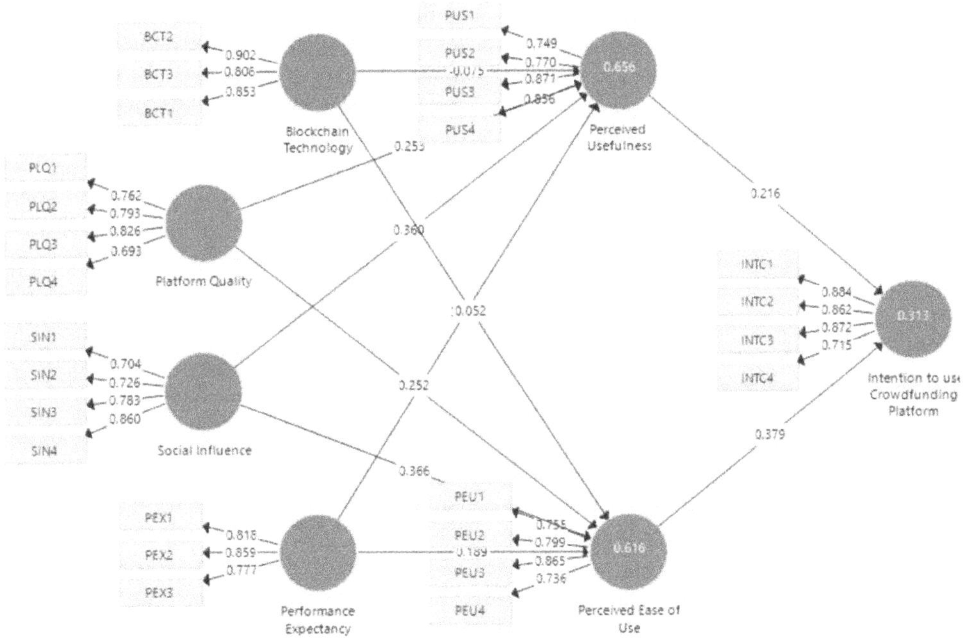

Figure 43.2. Results of hypotheses.

Source: Author.

Firstly, our it turns out that the quality of the platform has a good effect on how easy and useful it is thought to be. There is a claim that the more straightforward way the platform makes this by showing all the details about the project, similar results are found. thereby making this the entire engagement and investment procedure smooth, the greater the perceived usefulness increases as a noteworthy means of approaching potential backers for young v underwriters.

Will use simple-to-navigate interfaces, clear instructions, and minimal barriers to participation are offered. A well-designed platform reduces the learning curve, enhancing overall experience that will encourage both young entrepreneurs and backers to participate in crowdfunding.

Secondly, it was also found that social influence has a good and significant effect on how easy and useful something is thought to be. Along with [18], these results make sense which indicate that social influence can improve perceived usefulness of augmented reality features for online shopping by influencing users' perceptions of how useful this technology will be to improve their shopping experience. When people around them start using AR features, or when they get encouraging feedback about the features from friends, family, and influencers, they would think such features would be helpful in making better decisions and evaluating the product for increased satisfaction of the shopping experience [19]. Moreover, this runs parallel to findings related to TAM showing how social influence can decrease the perceived complexity of new technologies by normalizing their use amongst peer groups. It creates in users the feeling that AR is something quite easy to communicate with, since people see how others interact with it without much hassle.

Thirdly, performance potential has a big effect on how useful and easy something is thought to be to use. The results are in line with with [21], who suggested that If friends, family, or members of one's social circle recommend a certain technology, it's more likely to be considered useful by the user. If people perceive that others find AR helpful and easy to use to get views of products, they too may find AR a useful technology, for this enhances one's shopping experience through improved information quality about the product on sale with reduced uncertainties. Moreover, users have been found to be more open to finding a system that is easy to use if they observe their peers navigating it effectively. For instance, with online shopping using AR, people would feel assured about navigating AR interfaces if others are showing ease in using features of that nature. In other words, this will reduce one's subjective apprehensions about the technology and enhance perceived ease of use.

Lastly, our research showed that how useful and easy something is thought to be has a big effect on how likely someone is to use it. crowdfunding platform. The findings are aligned with the prior research which claim that users' belief in the ability of the platform to enhance project success or provide access to key resources is one of the important drivers of use intention in evaluating the platform [22]. In the context of crowdfunding, perceived usefulness may mean raising funding, soliciting feedback from the public, building a supportive community-all of which suggest that use of such platforms has perceived utility for potential users [25]. It is argued by [15] that when potential users perceive crowdfunding as a means to serve their practical needs effectively, interest in it is more likely to follow. Moreover, if this crowdfunding platform requires less effort to learn and use, the users will be less frustrated and more unlikely to give up using it, which in turn again reinforces their continued intention to use it in the future [9]. Ease can also support a smooth user experience that affects user satisfaction and perceived value.

6.1. Practical implication

Firstly, developers of platforms should pay attention on improving the design and usability of the crowdfunding platforms that can increase participation by raising the impression that they are useful and simple to use. By removing some UX features, suggested directions, and copy that reads like a how- to guide, platforms can open up their services to young entrepreneurs, especially those without much experience in digital finance. Such a move can motivate entrepreneurs to take up crowdfunding as a potential financing means for their businesses. *Secondly*, incorporating and It is important to promote blockchain technology so that people think it is easier to use crowdfunding sites. For young entrepreneurs who are hesitant to adopt new financial tools, blockchain can offer a secure and transparent transaction environment that will reduce uncertainty and help build trust. We must reinforce what is achieved through blockchain, because if we explain the prevention of fraud and preservation of data integrity, the investor will have more confidence in the product, and consequently more entrepreneurs will seek these platforms for capital formation. *Thirdly*, platform providers need to invest in platform quality (i.e., good platform functionality, reliability, and security). You can do this by reducing downtime, offering quick customer support, and good data security. A good platform is better able to fulfill the needs and expectations of your young entrepreneurs, resulting in higher satisfaction and a proclivity to adopt. *Finally*, the social influences also contribute massively to the young entrepreneur's intention to use the crowdfunding platforms. As a solution, platform providers may want to work with successful entrepreneurs or create partnerships with influencers that used crowdfunding and have them relate their experience or success stories. Providers can also take advantage of social influence to generate interest and get more potential users to adopt by fostering a positive community where clans form naturally, and users compete with each other which results in word- of-mouth marketing.

6.2. Limitations and future research direction

This study has some flaws that make it possible for more research to be done in the future. However, the study has two flaws: One, the group is made up of only young Filipinos entrepreneurs, thus, the findings may not be applied or generalized to other cultural or economic context. This limitation could be resolved for future work, by carrying out cross-cultural studies to identify if the observed indicators impact the adoption of crowdfunding in different or similar ways in different areas. There was also a chance that the results might not have been accurate because they were based on self-reported data, which can be affected by various types of bias. In the future, continuous studies might also give us a way to observe how behavioural intention changes over time and determine if crowdfunding has a lasting effect on entrepreneurs in the long run. Furthermore, this study only examines blockchain technology, platform quality, social influence and performance expectancy as variables while other variables (i.e., perceived risk, financial literacy and regulatory support) can be important factors affecting crowdfunding adoption as well. Including these factors would create a more thorough model of adoption influencing variables. Lastly, emerging technologies like AI can create personalized experiences for every entrepreneur and further improve user engagement. Future research is needed to study capabilities and techniques of AI to contribute to better experiences of entrepreneurs on crowdfunding platforms.

References

[1] Troise, C., Battisti, E., Christofi, M., van Vulpen, N. J., & Tarba, S. (2023). How can SMEs use crowdfunding platforms to internationalize? The role of equity and reward crowdfunding. *Management International Review*, *63*(1), 117–159.

[2] Stanier, J., Naicken, S., Basu, A., Li, J., & Wakeman, I. (2010). Can We Use Trust in Online Dating?. *Journal of Wireless Mobile Networks, Ubiquitous Computing, and Dependable Applications*, *1*(4), 50–61.

[3] Canare, T., Francisco, J. P., & Price, N. A. (2017). An empirical analysis of SME and large business linkages: Evidence from the Philippines. *RSN-PCC Working Paper*, 17-007.

[4] Anđelko, C., & Radomir, F. (2023). Time Dependent Deformations of A Coupled Bridge: A Case Study. *Archives for Technical Sciences*, *2*(29), 23–34. https://doi.org/10.59456/afts.2023.1529.023C2009.

[5] Rao, P. (2007). Greening of the supply chain: An empirical study for SMES in the Philippine context. *Journal of Asia Business Studies*, *1*(2), 55–66.

[6] Ergüden, D., Gürlek, M., Kabaklı, F., & Turan, C. (2022). The First Occurrence of Xanthochromic Fish, Diplodus sargus (Family: Sparidae) in the Eastern Mediterranean. *Natural and Engineering Sciences*, *7*(1), 34–40. http://doi.org/10.28978/nesciences.1098658

[7] Brem, A., Bilgram, V., & Marchuk, A. (2019). How crowdfunding platforms change the nature of user innovation–from problem solving to entrepreneurship. *Technological Forecasting and Social Change*, *144*, 348–360.

[8] Zhu, H., & Zhou, Z. Z. (2016). Analysis and outlook of applications of blockchain technology to equity crowdfunding in China. *Financial Innovation*, *2*, 1–11.

[9] Nunes, R., Alturas, B., & Fernandes, A. L. (2021, June). Creating value in equity crowdfunding platforms using blockchain technology. In *2021 16th Iberian Conference on Information Systems and Technologies (CISTI)* (pp. 1–6). IEEE.

[10] Behl, A., Sampat, B., Pereira, V., Jayawardena, N. S., & Laker, B. (2024). Investigating the role of data-driven innovation and information quality on the adoption of blockchain technology on crowdfunding platforms. *Annals of Operations Research*, *333*(2), 1103–1132.

[11] Mutharasu, J. M., Pandey, U., Rethick, B., Kulkarni, B., & Pawar, M. (2022). A framework for crowdfunding platform

using Ethereum blockchain technology. In *Cyber Security in Intelligent Computing and Communications* (pp. 99–113). Singapore: Springer Singapore.

[12] Zkik, K., Sebbar, A., Fadi, O., Kamble, S., & Belhadi, A. (2024). Securing blockchain-based crowdfunding platforms: an integrated graph neural networks and machine learning approach. *Electronic Commerce Research*, *24*(1), 497–533.

[13] Shelke, P., Zanjal, S., Patil, R., Desai, D., Chavan, H., & Kulkarni, V. (2022, November). Blockchain technology based crowdfunding using smart contracts. In *2022 International Conference on Augmented Intelligence and Sustainable Systems (ICAISS)* (pp. 939–943). IEEE.

[14] Lin, W. S., Chen, H. R., & Yang, M. L. (2020). How do we learn to get success together through crowdfunding platform? From the perspectives of system learning and multi-motivations. *Telematics and Informatics*, *52*, 101428.

[15] Lacan, C., & Desmet, P. (2017). Does the crowdfunding platform matter? Risks of negative attitudes in two-sided markets. *Journal of Consumer Marketing*, *34*(6), 472–479.

[16] Baber, H. (2021). Examining the intentions to use crowdfunding platform-An extended technology acceptance model. *International Journal of Services, Economics and Management*, *12*(2), 149–163.

[17] Febrianti, N. M. N., & Darma, G. S. (2023). Millennials' intention to invest through securities crowdfunding platform. *Aptisi Transactions on Technopreneurship (ATT)*, *5*(1), 19–30.

[18] Islam, M. T., & Khan, M. T. A. (2021). Factors influencing the adoption of crowdfunding in Bangladesh: A study of start-up entrepreneurs. *Information Development*, *37*(1), 72–89.

[19] Rodriguez-Ricardo, Y., Sicilia, M., & López, M. (2018). What drives crowdfunding participation? The influence of personal and social traits. *Spanish Journal of Marketing-ESIC*, *22*(2), 163–182.

[20] Li, Y. Z., He, T. L., Song, Y. R., Yang, Z., & Zhou, R. T. (2018). Factors impacting donors' intention to donate to charitable crowd-funding projects in China: a UTAUT-based model. *Information, Communication & Society*, *21*(3), 404–415.

[21] Alshebami, A. S. (2022). Crowdfunding platforms as a substitute financing source for young Saudi entrepreneurs: Empirical evidence. *Sage Open*, *12*(3), 21582440221126511.

[22] Baah-Peprah, P. (2023). Explaining reward crowdfunding backers' intentions and behavior. *Baltic Journal of Management*, *18*(2), 262–281.

[23] Friederich, F., Palau-Saumell, R., Matute, J., & Sánchez-Torelló, J. L. (2024). Investors' acceptance and use of investment-based crowdfunding platforms: an integrated perspective. *Journal of Financial Services Marketing*, *29*(4), 1291–1303.

[24] Algharabat, R., Alalwan, A. A., Rana, N. P., & Dwivedi, Y. K. (2017). Three dimensional product presentation quality antecedents and their consequences for online retailers: The moderating role of virtual product experience. *Journal of Retailing and Consumer Services*, *36*, 203–217.

[25] Husin, M. M., Haron, R., & Aziz, S. (2020). Adoption of financial technology in islamic crowd-funding: predicting small and medium-sized enterprises' intention to use the investment account platform. In *Impact of financial technology (FinTech) on Islamic finance and financial stability* (pp. 12–35). IGI Global.

[26] Panitkulpong, K., Saengnoree, A., & Teerawatananond, T. (2024). Financial Innovation and Crowdfunding: Influencing Investment Decisions in Tech Startups. *International Journal of Financial Studies*, *12*(4), 103.

[27] Davis, F. D., Bagozzi, R. P., & Warshaw, P. R. (1989). User acceptance of computer technology: A comparison of two theoretical models. *Management science*, *35*(8), 982–1003.

[28] Ahmad, M., Zhao, Z. Y., Irfan, M., Mukeshimana, M. C., Rehman, A., Jabeen, G., & Li, H. (2020). Modeling heterogeneous dynamic interactions among energy investment, SO 2 emissions and economic performance in regional China. *Environmental Science and Pollution Research*, *27*, 2730–2744.

[29] Fornell and D. F. Larcker, Structural equation models with unobuervable variables and measurement error: Algebra and statistics. *J. Mark. Res.*, vol. 18, no. 3, p. 382, 1981, doi: 10.2307/3150980.

[30] Hair Jr, J. F., Anderson, R. E., Tatham, R. L., & Black, W. C. (1995). *Multivariate data analysis with readings*. Prentice-Hall, Inc.

[31] Wong, K. K. K. (2013). Partial least squares structural equation modeling (PLS-SEM) techniques using SmartPLS. *Marketing Bulletin*, *24*(1), 1–32.

[32] Nunnally, (1994). *Psychometric theory 3E*. New York, NY: Tata McGraw-Hill Education.

[33] Hair Jr, J. F., Matthews, L. M., Matthews, R. L., & Sarstedt, M. (2017). PLS-SEM or CB-SEM: updated guidelines on which method to use. *International Journal of Multivariate Data Analysis*, *1*(2), 107–123.

[34] Podsakoff, P. M., MacKenzie, S. B., Lee, J. Y., & Podsakoff, N. P. (2003). Common method biases in behavioral research: a critical review of the literature and recommended remedies. *Journal of Applied Psychology*, *88*(5), 879.

[35] Harman, (1976). *Modern factor analysis, 3rd rev. ed.*

[36] Bagozzi, R. P., Yi, Y., & Phillips, L. W. (1991). Assessing construct validity in organizational research. *Administrative Science Quarterly*, 421–458.

[37] Strupeit, L., & Palm, A. (2016). Overcoming barriers to renewable energy diffusion: business models for customer-sited solar photovoltaics in Japan, Germany and the United States. *Journal of Cleaner Production*, *123*, 124–136.

[38] Hair, J. F., Hult, G. T. M., Ringle, C., & Sarstedt, M. (2016). A Primer on Partial Least Squares Structural Equation Modeling (PLS-SEM). *SAGE Publication Inc*, 2016. https://us.sagepub.com/en-us/nam/a-primer-on-partial-least-squares-structural-equation-modeling-pls-sem/book244583 (accessed Oct. 02, 2020).

[39] Roxas, B., & Chadee, D. (2016). Knowledge management view of environmental sustainability in manufacturing SMEs in the Philippines. *Knowledge Management Research & Practice*, *14*(4), 514–524.

[40] Rao, P., Singh, A. K., la O'Castillo, O., Intal Jr, P. S., & Sajid, A. (2009). A metric for corporate environmental indicators… for small and medium enterprises in the Philippines. *Business Strategy and the Environment*, *18*(1), 14.

[41] Kim, M. J., & Hall, C. M. (2020). What drives visitor economy crowdfunding? The effect of digital storytelling on unified theory of acceptance and use of technology. *Tourism Management Perspectives*, *34*, 100638.

[42] Rajeswari, J., & Gunasundari, R. (2016). Single Pass Attribute Value Frequency based Outlier Detection for Large-Scale Categorical Data. *International Journal of Advances in Engineering and Emerging Technology*, *7*(4), 194–207.

[43] Al Barazanchi, I. I., & Hashim, W. (2023). Enhancing IoT Device Security through Blockchain Technology: A Decentralized Approach. *SHIFRA*, *2023*, 10–16.

44 Current status of food tourism development in Qianxinan prefecture

Sihan Chen[a], *Chuanchen Bi*[b], *and Long Yang*[c]

College of Graduate Study in Management, Khon Kean University, Khon Kaen, Thailand

Abstract: Food is a reflection of culture, and food tourism is considered a branch of cultural tourism. It is a type of travel that focuses on the enjoyment of food, emphasizing the use of food-related resources as attractions. Through these experiences, tourists can enjoy unique, unforgettable, and culturally rich journeys. Food tourism also plays a crucial role in differentiating tourism development strategies. The ability of food to attract tourists to various places – whether streets, malls, or scenic spots – brings significant economic benefits, making it one of the primary economic drivers in the tourism industry. Consequently, promoting food has become an essential component of tourism marketing strategies. Qianxinan Prefecture in Guizhou Province boasts a distinctive ethnic minority food culture. However, its tourism marketing strategy tends to copy those of other cities, failing to capitalize on its own unique cultural strengths. This has led to a homogenized approach and lower visibility. This study focuses on the food tourism of Qianxinan Prefecture, employing methods such as observation and interviews, and uses the STP (Segmentation, Targeting, Positioning) framework to clarify market positioning and explore the current state of food tourism in the region.

Keywords: Situation of development, food tourism, qianxinan prefecture

1. Introduction

Countries around the world focus on developing unique, locally distinctive tourism products in the growth of culinary tourism. By offering a range of one-of-a-kind culinary experiences, they aim to attract visitors seeking authentic, original flavors that can only be found in specific regions [1]. Many places are renowned worldwide for their culinary specialties, with popular destinations including Bangkok in Thailand, Provence in France, and Munich in Germany. Numerous studies have shown that local cuisine can be enhanced during travel. At the same time, tourist experience Countries have also emphasized branding in culinary tourism, working to build high-value tourism brands. For instance, in 2023, Thailand's Tourism Authority launched the 'Amazing Thailand Culinary City' project to increase global awareness of Thai cuisine. France, moving beyond simply hosting international wine festivals, saw its tourism agency in 2022 promoting an immersive culinary travel route called the 'Valley of Gastronomy,' allowing visitors to experience local culture and customs more deeply. In the United States, as a culturally diverse country, the National Tourism Board introduced a multilingual version of The Culinary Guide: Tasting Your Way Through the USA in 2014 to showcase regional American flavors to international visitors and promote culinary tourism across the country [2].

The International Mountain Tourism and Outdoor Sports Conference, which has been held seven times, is a national and international summit focusing on mountain tourism that was approved by the State Council to be held in Guizhou. The first six conferences were all held in Qianxinan Prefecture, making it the only region in China to host this event. Through organizing this conference, tourism in Qianxinan saw significant growth in the four years from 2016 to 2019 (before the COVID-19 pandemic), with tourist numbers increasing from 28.66 million to 70.90 million visitors, and tourism revenue rising from 22.6 billion yuan to 73.2 billion yuan. By the end of 2022, the prefecture received a total of 33 million tourists throughout the year, generating a comprehensive tourism revenue of 29.777 billion yuan [3]. However, the development of culinary tourism in the region faces some inevitable challenges. First, the market positioning is unclear, with insufficient focus on specific target groups. Second, there is a high degree of homogeneity, and the region's unique strengths are not highlighted. Third, outdated business practices in the food and beverage industry no longer meet modern consumer demands. Lastly, commercialization and lifestyle shifts have altered some traditional ethnic cuisines [4, 5].

In light of the information shown above, this paper examines the current state of culinary tourism in Qianxinan, revealing the essential characteristics of culinary tourism, its influencing factors, and its role within tourism marketing strategies. By using Qianxinan as the case study, the paper analyzes its culinary tourism development to provide academia and practitioners with a more comprehensive, in-depth understanding. This analysis aims to lay a foundation

[a]sihanchen317@.gmail.com, [b]chuabi@kku.ac.th, [c]longya@kku.ac.th

DOI: 10.1201/9781003684459-44

for future culinary tourism marketing strategies in Qianxinan and offers insights that may serve as references for other cities developing their own tourism marketing strategies [6].

2. Literature Review

2.1. Food tourism

2.1.1. The definition of food tourism

Food tourism was originally defined as a special form of tourism activity, stemming from the concept of 'culinary tourism' introduced by Long L. in 1998. He considered food tourism as tourists' cross-border participation in a series of activities including the consumption, preparation, and presentation of exotic foods [7]. However, after research showed that most English speakers associated the term 'culinary' specifically with professionally trained chefs, the World Food Tourism Association changed the term to 'Food Tourism' in 2012 [8]. This type of tourism emphasizes food-related resources as attractions, allowing travelers to gain distinctive, unforgettable, and culturally rich experiences through food, combining social and recreational aspects. Food is an essential part of tourism products, with food-related expenses making up about 25% of total travel spending [9]. In the 1950s, culinary tourism evolved from French wine tourism, farm tourism, and rural tourism. With the rise of culinary tourism and governments' recognition of its economic contributions, food tourism has flourished in the global tourism market. This is particularly evident in two key areas: 'intangible cultural heritage' and 'internet popularity' [10]. This information indicates that tourists often travel to destinations that are renowned either for their currently trending cuisine or rich cultural culinary traditions, such as Provence in France, Chengdu in China, and Bangkok in Thailand.

2.1.2. The characteristics of food tourism

As early as 2001, Erik Wolf, the president of the World Food Travel Association, categorized food tourism into three types based on consumer behavioural intentions, cooking classes, food tours and food festivals. This categorization was also validated by Shenoy in their thesis [11]. In a journal published by Xiang Li. in 2019, these three types of food tourism were further elaborate [12]. According to the 2024 analysis report by Value Market Research [13], food tourism involves exploring local and regional culinary experiences as the primary motivation for travel. Travelers engage in food tourism to discover and appreciate different cuisines, local cooking traditions, and food-related cultural practices, including visiting food markets, participating in cooking classes, and dining at local restaurants. It contributes to the destination's economic development, promotes cultural exchange, and allows travelers to immerse themselves in the culinary heritage of different regions.

1. Cooking Classes. Culinary education is an important part of European gastronomic tourism [14]. Europe cooking schools typically offer flexible short-term training courses lasting a week or a month to meet tourists' needs. Food Tours for the purpose of learning and experiencing culinary skills, focusing more on food preparation and cooking styles [15]. For example, the culinary training programs at 'Top 5 Culinary Schools In The World' reflect people's enthusiasm for learning culinary skills [16].
2. Food Tours. Ashleigh and other scholars believe that this kind of travel emphasizes the actual use, preparation style and food context of food. 'Food tourism' not only pays attention to the consumption of the food itself, but also pays attention to the culture behind food consumption and the communication with local residents [17]. Chen Duoling believes that food tourism is a series of activities carried out by tourists who leave their daily residence or living place and are attracted by any tourism resources related to food, such as food tasting, consumption, production and display, and take this as their main purpose [18]. In the new media era, Zhou believes unique cuisines have gained visibility through short mobile videos, creating a new marketing model of 'food check-ins.' Governments are increasingly emphasizing various unique local cuisines in their tourism marketing strategies, making them city landmarks [19].
3. Food Festivals. These are celebratory events that showcase and sell food from specific regions or areas [20]. In a report by Mastercard Economics [21], benefiting from the European Football Championship, Munich, Germany topped the list of popular cross-border travel destinations from June to August 2024. During the Rio Carnival, international tourists' spending on food and groceries increased by 156%. Additionally, restaurants within 4 kilometers of Taylor Swift concert venues saw a 68% increase in sales during event periods compared to a regular day. Similar food festivals are held all over the world. According to Mehmet Kaback research, food festivals have three commonalities: first, the time is the national holiday of each country; second, the activity method is combined with the local ethnic culture; third, the food is of mainly local characteristics [22].

2.2. Development status of food tourism in China

China is one of the world's three major culinary nations, with a rich and long-standing food culture. In Lin Wang's research on the development history of Chinese food, Chinese food culture has developed into a complete and unique cultural system after thousands of years of continuous development and improvement [23]. As people's living standards have improved, food culture has entered an era of diversification, with dining demands shifting from 'satisfying hunger' to 'pursuing gastronomic pleasure,' and from 'physiological needs' to 'psychological satisfaction.' People's dietary demands have gradually become more diverse and personalized – they not only want to eat, but also want to experience distinctive flavors, refined taste, and cultural significance through their dining experiences [24].

As an important branch of niche tourism, culinary tourism not only enriches tourism products but also showcases local history and culture. In Yunyun Li's research [25], it is found that the proportion of 'food' consumption in the major elements of tourism is getting higher and higher, and it is second only to shopping consumption in international tourism consumption. In a report released by the China Tourism Academy [26], over 40% of post-90s generation experienced rural tourism at least once in 2021. Young people particularly favor ancient villages and towns with cultural backgrounds, where diverse experiences such as culinary tastings, fruit picking, and intangible cultural heritage activities, often centered around guesthouses, have become new engines for rural tourism development. This clearly illustrates the strong appeal of culinary experiences to travelers during their trips. According to data from the China Tourism Academy cited in a Zhao news report [27], 93.1% of tourists consider experiencing local cuisine as a major factor in tourism, and 82.5% of tourists are interested in the cultural connotations inherent in the food, indicating that food culture has a tremendous appeal to people. In a specialized survey [10], 21.8% of tourists consider food as a journey to deeply experience local culture. 59.3% of tourists express a willingness to further understand the destination's food culture, if given the opportunity. 26.2% of tourists are willing to learn the cooking techniques and methods of the destination's specialty dishes.

Dongxiang Li cited the rich raw materials and variety of Chinese cuisine [28]. Yang believes that China has a systematic set of craftsmanship and techniques in food preparation. The cooking methods are diverse, with different techniques varying in details, all pursuing the perfect combination of 'color, aroma, and taste' [29]. Maria believes that the reason why Chinese food culture stands unique in the world and remains enduring is primarily because China has perfectly combined food with culture, forming a complete and independent theoretical system that can also include medicinal cuisine or therapeutic food [30].

2.3. STP

Three essential elements of marketing strategy – Segmentation, Targeting, and Positioning (STP) were proposed by American marketing scholar Wendell R. Smith in 1956 [31]. Segmentation is the practice of dividing the market into different customer groups. The variables that should be considered during segmentation have been a subject of debate among scholars. American scholar Philip Kotler proposed incorporating geographical, demographic, psychological, and behavioural factors into the consideration [41]. Targeting is determined after Segmentation is completed. It is the process where businesses use undifferentiated marketing strategy, differentiated marketing strategy, and concentrated marketing strategy to satisfy one or several sub-markets with corresponding products and services. Positioning is about creating a product position and marketing mix that is most likely to attract the selected target audience. Jack Trout believes that market positioning should offer more differentiated products or services than competitors and communicate this difference to the consumers effectively [32]. Sales entities, targeting the psychology of potential customers, establish certain personality characteristics of the products and services in the minds of target customers through functional positioning, symbolic positioning, and experiential positioning.

2.4. Consumer behaviour research

2.4.1. Maslow's hierarchy of needs

Human needs appear in a hierarchical form, starting from basic needs and gradually developing upward to higher-level needs. When one set of needs is satisfied, that set no longer serves as a motivating factor. The hierarchy of needs theory is widely applied in the field of sociology and serves as an assessment tool [33], especially in areas such as education, social work, and marketing. Similarly, the demand for food culture experiences stems from tourists' elevated demands for food tourism shown in the Figure 44.1. Fenglin and his group based on Maslow's Hierarchy of Needs theory, categorizes food tourism consumer motivations into three types [10]:

The first level of food tourism needs is basic needs, corresponding to the physiological and safety needs in Maslow's hierarchy, including appetite satisfaction and food safety assurance. The second level of food tourism needs is sense of belonging needs, corresponding to the social and esteem needs in Maslow's hierarchy, including social interaction needs and the need for respect. The third level of food tourism demand is growth demand, corresponding to self-realization in Maslow's demand theory, that is, the experience of food culture.

2.4.2. Serial mediation model about food consumption value

The consumption value theory (CVT) was first put forward by Sheth [34]. Consumers are influenced by underlying

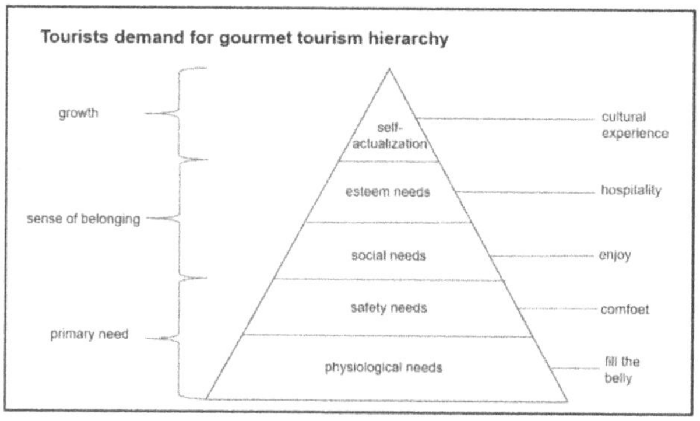

Figure 44.1. 2024 China food tourism development report [10].
Source: Author.

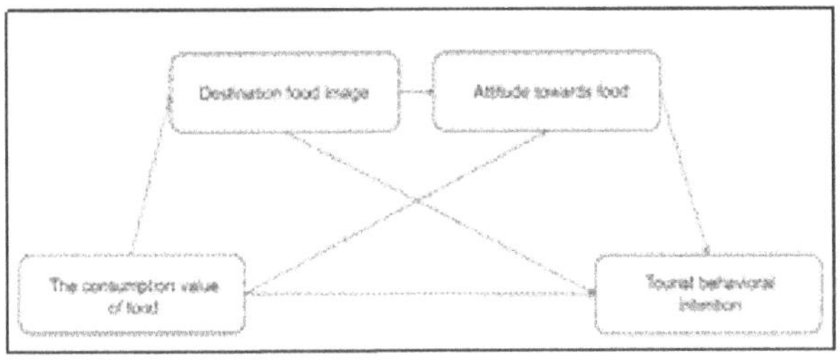

Figure 44.2. The influence of food consumption value on tourists' behavioural intention [37].

Source: Author.

functional, conditional, social, emotional, and cognitive values when making purchasing decisions, but the importance of each value depends on the context in which they consume [35]. However, food consumption differs from general product purchases, as it also needs to focus on multiple value factors such as health and quality. Therefore, Choe [36]. Proposed the seven-dimensional extended consumption value theory based on the theory of consumption value, which includes the quality value, health value, economic value, emotional value, prestige value, interactive value, and cognitive value of the food [37]. Constructed a chain mediation model to study the impact of food consumption values on tourists' behavioural intentions. The results showed that food consumption values not only have a significant positive impact on tourists' behavioural intentions, but also indirectly influence their behavioural intentions through three pathways: the mediating effects of destination food image, food attitudes, and the chain mediation effect of both factors.

2.4.3. Summary

With the shift in tourism space from scenic spots to experiences, the role of food in the construction and development of tourist destinations has become increasingly prominent. In the 2024 China Tourism Innovation Development Press conference [27], marketing cases that highlight local food culture, such as the 'Successful Family Feast' in Nanan, the Tangshan Banquet, Liaoning 'Red Flag Grand Market', and 'Jiangnan's No.1 breakfast'[1], have achieved significant success. In the tourism industry, consumer behaviour has changed, and food has now been incorporated into the strategic vision of local tourism development, becoming a new driving force for promoting tourism consumption, advancing the high-quality development of the tourism sector. The development of tourism is no longer just about the commercialization of scenic areas but should focus more on meeting consumer needs.

3. Method

This study takes China's Qianxinan Prefecture in Guizhou Province as the observation subject. It employs qualitative research methods, through observation and interviews, to record and analyze the current state of food tourism development in Qianxinan Prefecture, and thereby identify existing issues through its development status. The aim is to use the current development status as an entry point to leverage the distinctive advantages of ethnic minority cuisine in Qianxinan Prefecture and enhance its position in cultural tourism promotion. Using STP theory, this study is divided into three steps:

First, based on the chain mediation model of food consumption values' impact on tourist behavioural intentions, this study categorizes food tourism visitors into food tourists, experience tourists, and general tourists. Interviews are conducted with these three tourist groups, and the collected information is organized and summarized using triangulation to derive market segmentation in Qianxinan Prefecture's food tourism development.

Second, based on the established market segmentation, determine the status of these markets and evaluate the commercial attractiveness of each segment, while analyzing the food consumption values of Qianxinan Prefecture to observe whether each market segment has corresponding food service providers targeting them as their audience.

Finally, define the market positioning of Qianxinan Prefecture's food tourism industry within the overall tourism sector, including its current product positioning strategy and marketing mix choices. This will ultimately yield a comprehensive understanding of the current state of food tourism development in Qianxinan Prefecture.

4. Result

4.1. Development status of gourmet tourism in Qianxinan

4.1.1. Qianxinan prefecture food culture

Qianxinan is located at the junction of Guizhou Prefecture, Yunnan and Guangxi provinces, it consists of the Buyi and Miao[2]. Its unique geographical features, mild climate, and the integration of diverse ethnic cultures have shaped the distinctive culinary culture of Qianxinan. In recent years, the exploration of local culinary traditions has given Qianxinan cuisine a vibrant life, allowing visitors to savor local flavors while experiencing the unique nostalgia and cultural essence of the region. On major travel platforms in China, 'food tourism' has already become an appealing 'cultural emblem' of Qianxinan [38], Buyi Badawan[3], Anlong banquet of lotus[4], Zhenfeng Grey rice dumpling[5], Qinglong spicy chicken, carrying rich cultural heritage and ethnic customs, has deeply attracted tourists from all over the world.

1A traditional banquet organized by various Chinese provinces and open to tourists.

2Two ethnic minorities in China.

3Eight rather large bowls each containing eight different Bouyei specialties.

4A feast of lotus roots, seeds, etc. at the edge of a pond full of lotus flowers.

5The rice is soaked in lye water and wrapped in reed leaves together with various ingredients. The whole color is gray.

Food culture is the embodiment of a national culture and is primarily composed of three elements. Ingredients, the region's proximity to mountains and rivers provides a unique variety of local delicacies, including rare mountain ingredients and fresh river fish, which lay a solid foundation for its culinary traditions. Character, the local ethnic minorities, who often call themselves 'children of the mountains,' have long nurtured a culture of respect and reverence for nature. This has given rise to a distinctive culinary identity, with cooking techniques and methods unique to each ethnic group. Openness, while preserving their own culinary heritage, the people of Qianxinan have also embraced elements from neighboring Sichuan, Yunnan, Hunan, and Guangxi cuisines, creating a richly diverse and ever-evolving food culture.

4.1.2. Development status of food tourism in Qianxinan

With the collaborative efforts of government departments in culture, tourism, commerce, and broadcasting, the development of the Qianxinan food tourism brand has entered a new phase, led by government initiatives. Cultural promotion events such as the International Mountain Tourism and Outdoor Sports Conference and the 'August 8' (Lunar calendar) Ethnic Festival demonstrate that the strategic vision for tourism development in Qianxinan is no longer limited to scenic beauty; food, as a cultural medium, has become a top priority. Additionally, coverage on domestic media platforms like China News and News Live, alongside the influence of food bloggers on TikTok, REDnote, and Bilibili[6], has brought Guizhou's cultural heritage into the public eye, attracting numerous visitors.

In terms of impact, food tourism has successfully established a foundational concept; however, food festivals and tourism promotion still rely heavily on local specialties and traditional dishes. Innovations in cuisine to cater to the contemporary tastes of modern travelers, particularly with regard to restaurants and culinary creators, remain limited and there is room to further explore the 'food + tourism' business model.

4.2. Food consumption value

1. Quality and value. In terms of quality and value, the ingredients in Qianxinan cuisine mainly consist of mountain delicacies and river-fresh produce, reflecting the region's hot and humid climate, with a flavor profile that leans toward spiciness. Traditional dishes often feature, not only common farm-grown vegetables, but also wild greens like fiddlehead ferns, Chinese toon, and Qingming vegetables[7] [39]. These wild plants are highly valued in traditional Chinese medicine for their medicinal properties, though their taste may not be universally appreciated, especially by first-time visitors, who might be deterred by their unique aroma. Take houttuynia cordata as an example: in China, only Guizhou commonly serves this herb as a dish, incorporating it frequently as an ingredient or garnish. This distinct usage often prompts visitors to ask restaurant owners whether any local dishes include houttuynia cordata, as its strong flavor is unfamiliar and sometimes challenging for newcomers.
2. Health value. The Qianxinan Culinary Association has signed cooperation agreements with local farming and breeding bases, turning the region's green products into high-quality ingredients and ensuring the healthiness of the sourced raw materials. Each year, Qianxinan hosts the 'Guizhou Skills Competition—Tourism Star-Rated Hotel Industry Staff Competition,' a comprehensive event aimed at enhancing the skills of culinary chefs and service staff. This competition contributes to upgrading tourism dining services and guarantees healthy cooking practices. Traditional Chinese medicine (TCM) incorporates an extensive understanding of herbal remedies, many of which grow in secluded mountainous areas. Surrounded by mountains, Qianxinan is rich in these native herbs. Although many medicinal herbs are now cultivated on a larger scale, the local tradition of using herbs in cooking remains. This practice, steeped in TCM knowledge, is integrated into a variety of dishes, from stews to cold appetizers, and involves the use of numerous herbs and fungi, enhancing both the flavor and health benefits of the cuisine.
3. Economic value. According to the report [40]. In 2022, the government boosted cultural and tourism consumption by distributing discount coupons. From the date of issuance until the end of the year, a total of 722,355 coupons were distributed across various mobile app platforms, amounting to 25.765 million yuan in discounts, with 17.827 million yuan already redeemed. This initiative stimulated a total consumption of 68.831 million yuan. According to a report by the Guizhou Provincial Department of Culture and Tourism, from January to November 2022, the average spending per tourist in Qianxinan was 905.35 yuan, with 5.2018 million overnight stays recorded.
4. Emotion and cognition value. Qianxinan has adopted an 'agriculture + culture + tourism' model, utilizing agritourism to offer visitors hands-on experiences such as fruit and vegetable picking and family-friendly farm activities. These experiences help tourists learn how to identify various crops and produce. Additionally, with an 'open kitchen' design, visitors can participate in preparing traditional ethnic foods, such as dyeing sticky rice with natural plant pigments to make coloured rice or using wooden mallets to pound cooked sticky rice into mochi. At the 2022 'Cultural and Natural Heritage Day' in Guizhou Province, Qianxinan organized two traditional craft enterprises to participate in demonstrations,

6TikTok, REDnote, and Bilibili is Chinese New Media Platforms.
7Fiddlehead ferns, Chinese toon (Toona sinensis), and Qingming vegetables (affine cudweed), houttuynia cordata, In traditional Chinese medicine, they have high medical value.

providing hands-on teaching in pastry making and tea roasting techniques for attendees.

5. Prestige value. Since hosting the inaugural International Mountain Tourism and Outdoor Sports Conference in 2015, the Qianxinan Department of Culture and Tourism has introduced local specialty dining services in tourist attractions and ethnic minority settlements. The focus has been on developing high-end dining experiences that highlight ethnic minority cultures. In a select few upscale restaurants themed around local minority cultures, not only are traditional ethnic dishes served, but the dining environment also reflects the rich cultural heritage of the region. For example, a special hospitality etiquette of the Miao nationality is 'Gaoshan Liushui[8]'. In the taste of ethnic food and wine, you can also feel the national culture.
6. Interactive value. During special festivals or peak tourist seasons, local ethnic groups in Qianxinan organize traditional banquets in their villages. These events allow visitors to savor specialty dishes, especially the 'Eight Bowls', while enjoying 'Bayin Zuochang' (a seated musical performance featuring eight kinds of traditional instruments). Before the banquet, locals warmly welcome tourists by demonstrating the preparation and cooking processes of the ingredients, often inviting guests to join in. Carefully designed scenes create interactive spaces where visitors can fully engage in the cultural experience.

4.3. STP analysis of food tourism development

4.3.1. Segmentation

Since this research discusses food tourism using Qianxinan as a starting point, local tourists from Qianxinan Prefecture were not included as samples in the discussion. Based on interviews about travel motivations and behavioural intentions of non-local tourists, visitors were segmented by behaviour and psychology into three categories: food tourists, experience tourists, and general tourists. Through the feedback received from these three types of tourists after their travels, the study also determined the food consumption value of culinary tourism in Qianxinan Prefecture.

1. Food tourists. Qianxinan governs eight main cities, each separated by at least 80 kilometers, and each city is famous for distinct local delicacies. This means that food tourists seeking to try all the region's specialties often come with well-researched plans, clear motivations, and detailed itineraries. With the high prevalence of new media, many of these tourists are drawn by video content on platforms like TikTok, REDnote, and Bilibili, where local cuisine inspires them to make food exploration their primary reason for travel.
2. Experience tourists. As an autonomous prefecture for ethnic minorities, Qianxinan hosts various cultural festivals, such as the 'Torch Festival' celebrating ethnic customs and the 'Colored Sticky Rice Festival' highlighting traditional foods. Additionally, each year on San Yue San (the third day of the third lunar month) and Ba Yue Ba (the eighth day of the eighth lunar month), the entire region observes a one-day holiday with celebratory events. In April 2023, Wangmo County, under Qianxinan jurisdiction, organized a parade with 55 teams comprising around 3,000 members from the Buyi and Miao ethnic groups across nearby towns. The event featured tastings of traditional ethnic foods, performances of ancient Buyi songs, a tea tasting for Babu purple tea, and a showcase of Wangmo unique agricultural products.
3. General tourists. Influenced by the traditional 'sightseeing + group tour' model, developing tourist attractions in Qianxinan often rely on designated restaurants as focal points for food tourism. These attractions typically use the standardized rating systems of star-rated hotels and A-level scenic spots to classify and grade dining options. For group tours, travel agencies usually arrange for tourists to stay in hotels that combine dining and lodging for efficiency, allowing travelers to complete their daily itinerary more quickly. Hotels which partner with travel agencies, primarily serve tourists who have no time to enjoy food. The food they cook has been modified in taste to cater to foreign tourists and has lost local characteristics. If the travel agency's itinerary lacks opportunities for traditional local food tastings, tourists may leave with little understanding of the unique culinary identity of Qianxinan.

4.3.2. Targeting

In the Wanfenglin[9], the dining industry centers around a pastoral experience, with 'agritourism' as the main operating model. These establishments often retain traditional Chinese rural tools, which intrigue urban visitors eager to try them. Examples include wood-fired clay stoves for cooking, stone mills for grinding soybeans to make tofu or soy milk, and grooved stone vats for mashing sticky rice.

Each agritourism site typically has its own theme, such as fishponds for fishing or gardens where visitors can pick vegetables or fruit. After gathering their ingredients, guests have them prepared by the host for their evening meal. Agritourism in this area combines hands-on cooking experiences with food culture. However, as the hosts are usually local farmers, they may lack in-depth knowledge of the cultural meanings behind the dishes they prepare and can mostly explain only the cooking techniques, leaving the cultural stories behind the food largely unexplored.

1. Enjoy delicacy. Traditional cuisine in Qianxinan is primarily represented by street food and long-established eateries, often tucked away in inconspicuous corners or small storefronts with minimal promotion, catering mainly to locals. However, with the rise of new media,

8The wine flowing like a mountain stream.

9A famous tourist attraction in Qianxinan, Mainly pastoral scenery.

online exposure has made it easier for these hidden gems to gain visibility at a lower marketing cost. While egg fried rice isn't exclusive to Qianxinan, pairing it with local side dishes—such as pickled radish, water celery, and cold shredded potatoes, has created a regional twist. In Wanfenglin Scenic Area, there is even a 'Egg Fried Rice Street,' where over 30 restaurants specialize in this dish. During peak tourist season, they collectively serve more than 30,000 portions daily.

2. Visit food culture. As an autonomous prefecture for ethnic minorities, Qianxinan celebrates the traditional festivals of San Yue San (the third day of the third lunar month) and Ba Yue Ba (the eighth day of the eighth lunar month) each year. During these festivals, locals wear traditional attire and hold a 'Hundred Family Banquet.' Ba Yue Ba, known as the Miao New Year, retains several traditional customs despite increasing modernization and features a three-day celebration. During this period, Wangmo County saw an average daily visitor flow of over 90,000 people, marking an increase of over 500% year-on-year. On the peak day of Ba Yue Ba, the number of visitors reached 100,000, a growth of over 600% compared to the previous year. By September 11, the scenic area had welcomed approximately 300,000 visitors overall.

4.3.3. Positioning

Popular food tourism in Qianxinan promotes local culinary culture in distinct ways. Popular food tourism highlights authentic local street culture, allowing visitors to experience the vibrant local scene. High-end food tourism, on the other hand, emphasizes quality ingredients, elegant settings, and refined service to convey a sense of sophistication and unique appeal. Both approaches showcase Qianxinan culinary culture in different formats.

For popular food tourism, trendy 'internet-famous' dishes and modernized presentations of local specialties are often introduced. Utilizing marketing models like the Hotelling model, Qianxinan has established food streets and food cities. For instance, Maling Night Consumption Cluster, a night economy hub integrating nighttime scenery, performances, tours, and shopping, has gained popularity, especially during summer. Since its opening in January 2023, this area has attracted nearly 400 vendors. From January to August, the region saw a 9.76% year-on-year increase in tourist arrivals and an 11.41% growth in total tourism revenue.

High-end food tourism in the region emphasizes health and wellness. Najv·Heyuan Wellness Tourism Town, located near the Wanfenglin, is a comprehensive service center built upon a 600-year-old Buyi village, integrating Buyi culture with tourism, wellness, conferences, accommodation, and dining. According to the town's manager, Peng Ning, from January to August, the town welcomed over 2.8 million visitors, generating approximately 24 million yuan in revenue. Additionally, Yuntun Ecological Sports Park in Lutun, offers unique accommodations such as treehouse hotels, hot spring resorts, and luxury stargazing hotels, alongside outdoor barbecue and camping experiences.

In terms of service quality, Qianxinan has been actively enhancing its hospitality standards. At the 2022 Guizhou Skills Competition in the Tourism and Hospitality category, the region achieved notable accolades, winning one individual second prize, two individual third prizes, one individual excellence award, and a group second prize.

5. Discussion and Conclusion

In the progression of modern tourism, traditional tourism faces issues like severe homogenization, lack of effective marketing, and unclear target markets. Developing food tourism requires a cultural tourism marketing approach that highlights the unique attributes of different regions, creating differentiation in the tourism market based on distinct cultural characteristics. The growth of culinary culture relies heavily on building and refining a modern dining industry framework, which involves modernizing various aspects such as ingredient cultivation, product development, culinary innovation, the preservation of traditional skills, supply chain management, and sustainable practices. Through cross-industry and cross-sector collaboration, food tourism can draw from advanced concepts and technologies in other fields, enhancing its own innovation and competitiveness. This integration also encourages the transformation and upgrading of other industries, introducing new business models and revenue streams.

This study uses the development status of food tourism in Qianxinan Prefecture, Guizhou Province, China, as its research focus. A chain mediation model is applied to segment customer groups, followed by an analysis of the current food industry target markets based on these segmented groups. The goal is to determine the market positioning of Qianxinan food tourism and construct an STP model to understand the current state of its development.

Qianxinan has actively promoted mountain tourism festivals and Guizhou cuisine; however, some tourist restaurants and food providers have yet to prioritize enhancing the value of food consumption, leading to persistent challenges. These issues underscore the strategic importance of food tourism within tourism development. Only by continually enhancing the value of food experiences and exploring local culinary culture can a positive tourism environment be created.

References

[1] Ikechukwu, O., Ikpe Aniekan, E., & Paul, S. (2019). Design of a Beam Structure for Failure Prevention at Critical Loading Conditions. *International Academic Journal of Innovative Research*, *6*(1), 53–65. https://doi.org/10.9756/IAJIR/V6I1/1910005

[2] Boyne, S., et al. (2003). Policy, Support and Promotion for Food-Related Tourism Initiatives. *Journal of Travel & Tourism Marketing*, *14*, 131–154.

[3] Maoli, W., et al. (2023). Research on the integrated development of mountain tourism and high-end tourism service industry in Qianxinan. 2023 Qianxinan Forum, Xingyi, Guizhou Province, China.
[4] Zhao, F. (2023). Research on Evaluating the High-Quality Development of Mountain Tourism in Qianxinan Prefecture.
[5] Rukunuddin Ghalib, M., Singh Kuntal, H., & Garg, V. (2016). Vehicular Anti-Theft Protection and Real Time Intelligent Tracking System. *International Journal of Advances in Engineering and Emerging Technology*, *7*(4), 208–225.
[6] Hossain, S., Bakhshi, S. I., Raihan, M. M., & Zaffar, H. (2024). Gastrointestinal Impact of Flatulence-Causing Compounds in Foods: A Scientometric Study. *Indian Journal of Information Sources and Services*, *14*(3), 110–114. https://doi.org/10.51983/ijiss-2024.14.3.15
[7] Long, L. (2004). Culinary tourism: A folkloristic perspective on eating and otherness. *Culinary Tourism*, *55*, 20–50.
[8] Association, W. F. T. (2020). What Is Food Tourism?. Retrieved 16-07, 2019, from https://web.archive.org/web/20190716162714/https://www.worldfoodtravel.org/cpages/what-is-food-tourism.
[9] Erik, W. (2006). Culinary Tourism: the Hidden Harvest.
[10] Fenglin, W. (2024). *Culture-Driven High-Quality Development of Food Tourism: Practices and Pathways — 2024 China Food Tourism Development Report*. China Tourism Academy.
[11] Shenoy, S. S. (2005). *Food Tourism and The Culinary Tourist*. Clemson University.
[12] Xiang Li, X. H., & Shiyong Liu (2019). Chinese food tourism development trend. *Tourism Research*, *11*(4), 5–9.
[13] Research, V. M. (2024). Global Culinary Tourism Market Research Report – Industry Analysis, Size, Share, Growth, Trends and Forecast 2024 to 2032.
[14] Zhen, G., et al. (2019). The historical evolution and other judgments of food and tourism research. *Tourism Research*, *11*(04), 1–5.
[15] Henderson, J. C. (2009). Food tourism reviewed. *British Food Journal*.
[16] Ireland, S. (2022). Top 5 Culinary Schools in The World. *CEOWORLD Magazine*.
[17] Ellis, A., et al. (2018). What is food tourism? *Tourism Management*, *68*, 250–263.
[18] Duling, C., & Yixian, X. (2017). Overview of food tourism research. *Tourism Research*, *9*(2), 77–87.
[19] Rui, Z. (2016). Research on Food Culture Tourism Image Communication Strategies in the New Media Era — A Case Study of Chengdu as an International 'City of Gastronomy'. *Gastronomy Research, 33*(04), 26–31.
[20] Hall, C. M., Sharples, L., Mitchell, R., Macionis, N., & Cambourne, B. (Eds.) (2003). *Food Tourism Around The World (1st ed.)*. London.
[21] Martinez, Q. (2024). 2023–2024 Cross-border Tourism Consumption Trend Research Report.
[22] Kabacık, M. (2024). Examining gastronomy festivals, a type of event tourism, within the scope of the United Nations sustainable development goals. *Worldwide Hospitality and Tourism Themes*, *16*(4), 519–530.
[23] Ling, W. (2023). Value coupling and path innovation to enhance China's national image: A case study of the development of Chinese food culture resources. *Journal of Central South University for Nationalities*, *43*(09), 71–77+184.
[24] Xuelian, W., et al. (2007). Market demand analysis of food tourism – taking Guilin World Food Expo Park as an example. *Journal of Leshan Teachers College*, (05), 55–58.
[25] Yunyun, L., et al. (2019). Characteristics and influencing factors of tourists' food consumption in plateau tourism cities: A case study of Lhasa City. *Resource Science, 41*(03), 494–508.
[26] Bin, D. (2022). Global Free Travel Report 20From https://www.ctaweb.org.cn/index.php?m=home&c=View&a=index&aid=6682.
[27] Shan, Z. (2014). 'Food + tourism' has spawned a new business format. People's Daily Online: 012.
[28] Dongyang, L. (1998). *Food culture*. China Building Materials Industry Press.
[29] Liu, Y. (2007). Research on the competitiveness of China's catering industry.
[30] Yani, M. (2012). Traditional Chinese food therapy in Chinese culinary culture.
[31] Moutinho, L. (1956). Segmentation, targeting, positioning and strategic marketing. *Strategic Management in Tourism*, 121–166.
[32] Jack Trout, A. R. (2011). *Positioning-The Battle for Your Mind*. China Machine Press.
[33] William Kremer, C. H. (2013). Abraham Maslow and the pyramid that beguiled business. From https://www.bbc.com/news/magazine-23902918.
[34] Sheth, J. N., et al. (1991). Why we buy what we buy: A theory of consumption values. *Journal of Business Research*, *22*(2), 159–170.
[35] Rousta, A., & Jamshidi, D (2020). Food tourism value: Investigating the factors that influence tourists to revisit. *Journal of Vacation Marketing*, *26*(1), 73–95.
[36] Choe, J. Y., & Kim, S. (2019). Development and validation of a multidimensional tourist's local food consumption value (TLFCV) scale. *International Journal of Hospitality Management*, *77*, 245–259.
[37] Wenli, Z., et al. (2022). The influence of food consumption value on tourists' behavioral intention: A chain mediation model. *Gastronomy Research*, *39*(01), 35–43.
[38] Jinjin, W. (2024). Cuisine of Qianxinan in Guizhou Province. From https://www.imsilkroad.com/news/p/522566.html.
[39] Lei, L., et al. (2008). Investigation and collection of wild vegetable resources in Chongqing, Sichuan and Guizhou. *Southwest Agricultural Journal,* (04), 1054–1058.
[40] Government, Q. P. S. (2022). Summary of tourism industrialization work in Qianxinan in 2022.
[41] Philip, K. (1997). *Marketing management: Analysis, planning, implementation and control*. Prentice Hall International.

45 The current situation of the combination of health Qigong and modern health management industry

Xiaolin He[a], Chuanchen Bi[b], and Long Yang[c]

College of Graduate Study in Management, Khon Kaen University, Khon Kaen, Thailand

Abstract: With the increase in global health awareness and the rapid development of the health industry, consumers have displayed an increasing demand for health healing and traditional Chinese health preservation. As one of the traditional Chinese exercise and health methods, Health Qigong has shown remarkable effects in regulating the body and mind, enhancing immunity, and has gradually attracted the attention of the modern health management industry. However, due to factors such as cultural differences and ambiguous market positioning, Health Qigong still faces many challenges to its promotion within the industry. Based on a literature review, this study systematically analyzes the application status, potential advantages and obstacles of Health Qigong in the health management industry and proposes a commercialization model that combines innovative paths, such as cultural tourism and digital communication. The research results show that Health Qigong has broad potential in the global health management market, and its modernization and international promotion will provide new ideas for the health management industry.

Keywords: Commercial model, digital communication, health qigong, modern health management industry

1. Introduction

As global health awareness continues to increase, consumer demand for health treatments and traditional health care is growing year by year. The modern health management industry is developing rapidly under this background, and continues to integrate diversified health services and healing methods to meet people's comprehensive concerns regarding their physical and mental health. In this context, traditional Chinese health-preserving methods have gradually entered the international perspective along with their unique theoretical system and curative effects. Among these effects, as an important representative of traditional Chinese exercise health care, Health Qigong shows significant physical and mental regulation and health-promoting potential with the holistic concept of 'regulating the body, regulating the breath, and regulating the mind'. Extensive research supports its effectiveness in improving mental health, improving chronic disease management, and enhancing body balance. However, the promotion of Health Qigong in the modern health management industry faces many practical challenges. First, because Health Qigong is based on Eastern culture, its connotations and practical methods are relatively unfamiliar to some international audiences, and these cultural differences have become a major obstacle to its promotion. Secondly, Health Qigong has not yet formed a systematic path in terms of market positioning and promotion strategy, it lacks the support of a business model that meets the needs of modern health management. The article aims to research the question: What is the current application status of Health Qigong in the modern health management industry? What are the main obstacles it faces? This study aims to analyze the current application status of Health Qigong in the modern health management industry, to identify its potential obstacles, and propose targeted commercialization paths. Furthermore, the study aims to explore the application model of combining Health Qigong with the modern health management industry, and to provide theoretical support and practical guidance for its commercialization path [1].

Based on the above, this study takes Health Qigong as an entry point, focuses on its application status and development potential in the modern health management industry, explores the commercialization path of Health Qigong in the global health management market, and provides a basis for the modernization and internationalization of traditional sports and health projects [2]. It will provide theoretical support and practical reference for cultural promotion [3].

2. Literature Review

2.1. *Theoretical basis of health Qigong*

2.1.1. *Definition and classification of health Qigong*

In 1996, the Chinese government issued the 'Notice on Strengthening the Management of Health Qigong in Society', officially using the term 'Health Qigong'. The term was introduced to facilitate the standardization of Health Qigong practice, to distinguish it from Health Qigong and Medical

[a]xiaolin.h@kkumail.com, [b]chuabi@kku.ac.th, [c]longya@kku.ac.th

DOI: 10.1201/9781003684459-45

Health Qigong, and to emphasize that the main purpose of Health Qigong is to promote physical health, rather than to treat diseases. In 2000, the State Sports General Administration issued the 'Health Qigong Management Measures', defining Health Qigong as: 'A traditional national sports project that combines physical activities, breathing, and psychological adjustment as the main form of exercise, and is an integral part of China's long-standing culture' [4]. Health Qigong is a mind-body exercise that integrates traditional Chinese medicine and Taoist philosophy. It achieves harmony and balance between body and mind through the organic combination of 'adjusting the body' (movement), 'adjusting the breath' (breathing), and 'adjusting the mind' (consciousness). Common Health Qigong exercises include Ba Duan Jin, Wu Qin Xi, Liu Zi Jue, Yi Jin Jing, Da Wu, Ma Wang Dui Dao Yin Shu, etc. Each exercise has its own unique characteristics and is suitable for people of various ages and health needs. Studies have shown that Health Qigong is not only a form of physical exercise, but also contains profound cultural and philosophical connotations. It helps to regulate Qi and blood, balance Yin and Yang, and promote the overall development of the practitioner's physical and mental health through positive psychological guidance and self-regulation.

2.1.2. Physiological and psychological benefits of health Qigong

The mechanism of Health Qigong's effect on health has been confirmed by many studies. Physiologically, it improves cardiopulmonary function, enhances immunity, and improves symptoms of chronic diseases through slow and gentle movements, deep breathing, and focused mental activities. For example, Health Qigong can significantly lower blood pressure, relieve chronic pain, and improve the balance of practitioners. Psychologically, Health Qigong helps to reduce stress, improve anxiety and depression symptoms, and enhances overall well-being. Some systematic literature reviews have further confirmed that Health Qigong is particularly effective for patients with chronic diseases (such as diabetes, hypertension and arthritis) and can serve as an important auxiliary means of comprehensive treatment [5].

2.2. Current status and shortcomings of the commercialization of health Qigong

Although the potential of Health Qigong in the field of health is widely recognized, its commercialization direction faces many challenges. On the one hand, current research focuses on the health and wellness effects of Health Qigong, while relatively few studies are conducted on commercialization models, operating mechanisms and market demand analysis. Moreover, the promotion of Health Qigong is also limited by cultural differences and market acceptance. Existing literature shows that although Western countries are increasingly accepting traditional Chinese sports, their understanding of their cultural background and theoretical basis is still limited. Furthermore, the relatively boring and single practice content, as well as the lack of a standardized curriculum system and training certification mechanism, also lead to limiting the development potential of Health Qigong in the international health market.

In short, although the theoretical basis and health benefits of Health Qigong have been fully recognized by the academic community, its application and commercialization model within the modern health management industry still need to be explored further. This study will be dedicated to filling these research gaps and proposing innovative paths and strategies for the integration of Health Qigong within the health management industry.

3. Methods

A qualitative research method is adopted by this study, with a systematic literature review as the main tool, to integrate and analyze the research results in the fields related to Health Qigong and the modern health management industry. By focusing on the key issues in existing research, the core advantages of Health Qigong and its potential path to combine with the industry are summarized. This study relies mainly on the collection and analysis of secondary data, and provides a solid theoretical and practical foundation for the research by integrating existing literature, industry reports and market data. The sources of literature data include international authoritative journals, academic databases (such as Scopus, Web of Science) and related books, which are used to analyze the application theory and development trend of Health Qigong in the field of health management.

4. Results

4.1. Overview of the development of the modern health management industry

4.1.1. Concept and trend of health management

Health management is a systematic management process with disease prevention and health promotion at its core. Its goal is to improve the health level of the population through early intervention, personalized services and comprehensive health monitoring. In recent years, with the aging of the global population and the increase of chronic diseases, the health management industry has seen rapid development. Market trends show that digital, personalized and integrated health services are gradually becoming the key development direction of the industry. For example, the integration of health big data and artificial intelligence technology makes personalized health intervention and prediction possible, further promoting the refinement and intelligence of health management [12].

Advantages of Combining Health Qigong with Modern Health Management: As an important part of traditional Chinese health culture, Health Qigong has significant physical and mental adjustment effects, especially in chronic disease management and mental health promotion.

Physiological level: Through the practice of Health Qigong, the cardiovascular function of practitioners can be improved and the immune system can be enhanced. For example, a randomized controlled study showed that Health Qigong was significantly effective in reducing blood sugar levels in patients with type 2 diabetes. The gentle movements and breathing adjustments of Health Qigong have significant advantages in improving the balance ability of the elderly, thereby reducing the risk of falls.

Psychological level: By 'regulating the mind,' Health Qigong can significantly reduce stress levels and improve symptoms of depression. Research shows that the anxiety scores of patients in the Health Qigong group were reduced by more than 25%. Combined with the uniqueness of modern health management, Health Qigong not only emphasizes the coordination of body and mind, but is also extremely similar in concept to Western mindfulness training (Mindfulness), which helps to integrate with international health management methods [6].

4.2. *The current status of health management in the world and China*

4.2.1. The current status of global health management

The global health management industry is experiencing rapid growth. According to Statista, the global health management market size equated to around US$390 billion in 2023 and is expected to exceed US$600 billion by 2028, with an average annual growth rate of 7.8% [7]. This trend has driven the growth in demand for chronic disease management, preventive medical care and mental health services. Digital and personalized health management has become the main direction of future development. For example, Livongo in the United States provides personalized health management solutions for diabetic patients through big data and artificial intelligence. The increasing global aging population and surge in chronic diseases have also further promoted the development of the health management industry, especially in regard to chronic disease intervention and health management services for the elderly.

4.2.2. The status of health management in China

The Chinese health management industry is developing rapidly. In 2023, the market size exceeded RMB 600 billion, with an average annual growth rate of around 10%. The Outline of the Healthy China 2030 Plan has promoted the construction of a health management service system, covering multiple fields such as disease prevention, health promotion and chronic disease intervention. The integration of traditional Chinese medicine and traditional culture is a major feature of China's health management industry. The rapid development of digital health management has also brought new opportunities to China's health industry. For example, platforms such as Ping An Health and Ali Health have gained a broad user base through remote consultation and health management services [8]. Common trends in global and Chinese health management.

The global and Chinese health management industries are facing common trends of digitalization, personalization and growing demand for chronic disease management. China's health management industry has obvious advantages in the unique advantages of traditional culture, especially the integration of traditional Chinese medicine provides a good foundation for the promotion of Health Qigong. In the future, by combining traditional culture with modern technology, China is expected to play a greater role in the global health management industry.

4.3. *Successful cases of combining traditional sports with the health industry*

4.3.1. Enlightenment from the commercialization path of yoga

As a traditional way of healing the body and mind, yoga has successfully achieved global commercialization. Its successful experience includes:

Branded operation: Yoga has established international recognition through the establishment of branded institutions such as the Yoga Alliance. Health Qigong can learn from this model to promote the construction of standardized courses and certification systems.

Diversified communication channels: Yoga has attracted a large number of international users through multi-channel promotion such as video platforms and social media. Health Qigong can expand its market influence through short videos, live courses and other means.

Combination of culture and health: Yoga combines Indian culture with the needs of modern health concepts to form a brand image of a healthy lifestyle. Health Qigong can combine the concept of 'adjusting the body, regulating the breath, and regulating the mind' to attract users around the world who are interested in physical and mental balance [9].

Health tourism: Yoga is integrated into the health tourism industry and attracts tourists from all over the world. Health Qigong can also be promoted in health tourism destinations to enhance brand awareness and market share [10].

4.3.2. Experience in the development of Tai Chi industrialization

The international promotion experience of Tai Chi also provides valuable reference for Health Qigong. Its successful experience includes:

Dual-track development of competition and health preservation: Tai Chi not only participates in competitions, but also enters the field of health management through health preservation courses. Health Qigong can learn from this idea and provide personalized services and courses for various groups of people.

Government support and cultural promotion: Tai Chi entered the international market with the help of government support and cultural promotion. Health Qigong can also strive for similar policy support to promote the process of globalization [11].

Enterprise operation and cross-border cooperation: Tai Chi expands its influence by cooperating with fitness brands, Health Qigong can also cooperate with international fitness chains and be incorporated into regular courses.

Research support and scientific certification: The health benefits of Tai Chi have been widely recognized, and Health Qigong should also strengthen scientific research, provide data support, and enhance market trust [12].

4.4. Challenges in the commercialization of health Qigong

4.4.1. Cultural cognition gap

The Western market still has limited knowledge of traditional Eastern sports forms, which leads to cultural cognition barriers regarding the international promotion of Health Qigong. Compared with widely accepted sports forms such as yoga, the concepts, methods and health benefits of Health Qigong have not yet become familiar to most Western consumers. The lack of easy-to-understand popular science materials and communication strategies for the target market makes it difficult for Health Qigong to quickly occupy the market [13].

To meet this challenge, it is necessary to use digital communication models to develop multilingual promotional content, and convey the core concepts and practical values of Health Qigong to audiences from various cultural backgrounds through channels such as short videos, social media and health bloggers. At the same time, it can be combined with immersive cultural experience projects to allow consumers to feel the unique charm of Health Qigong as an actual experience, thereby enhancing awareness.

4.4.2. Lack of international standardization

At present, there is a lack of unified international certification standards for curriculum setting and the coaching qualifications of Health Qigong. This not only affects its speed of promotion in the global market, but also easily forms a sense of distrust among consumers [14]. In addition, the lack of a standardized system may lead to uneven teaching quality, affecting its brand reputation and market competitiveness.

To solve this problem, we can learn from the successful experience of industries such as yoga and establish an international certification system covering curriculum outlines, coach qualifications, and exercise effect evaluation. For example, we can cooperate with the International Health Management Association to develop a universal Health Qigong certification framework to provide authoritative qualification endorsements for coaches and institutions. This will not only help to standardize industry standards, but also enhance consumer trust and lay the foundation for Health Qigong to open up the national market development.

4.4.3. Fierce market competition

When entering the international market, Health Qigong will inevitably face competition from mainstream sports forms such as yoga and Pilates. Yoga, in particular, has occupied a large market share worldwide, has a mature curriculum system and a broad user base. If Health Qigong cannot find a differentiated entry point, it may be difficult to make a breakthrough in the health and fitness industry.

5. Conclusion

5.1. Research summary and conclusion

Through systematic review and analysis, this study clarified the academic value of Health Qigong in the field of health promotion and the practical significance of its commercialization path. Research shows that Health Qigong provides significant effects in regulating physical and mental health, especially in improving immunity, reducing stress, and improving sleep, offering a scientific basis for its use as a health intervention method. At the same time, Health Qigong has significant potential in the modern health management industry. Through the comprehensive effects of regulating the body, breath, and mind, combined with cultural communication and digital promotion, it can build a competitive commercial model and achieve globalization. promotion. This research provides a new perspective and theoretical support for the integration of traditional Chinese medicine health theory with modern health management systems, and provides specific guidance for industry practice.

5.2. The innovation of commercialization of health Qigong on the health management industry

The commercialization path proposed in this study provides a new perspective for the global promotion of Health Qigong. By drawing on the successful business cases of yoga and Tai Chi, Health Qigong can be combined with innovative models such as cultural tourism and digital communication, so that Health Qigong can be better integrated into the modern health management industry. The study suggests that when promoting Health Qigong globally, digital technology and health management platforms can be used to improve its accessibility and user experience; at the same time, cultural tourism resources can be used to carry out immersive experience courses to enhance its global popularity and appeal. Through this integration, Health Qigong can not only expand its market, but also promote the mutual penetration and integration of traditional culture and modern health concepts.

5.3. Recommendations for future research

Future research should further explore the cultural communication strategies in the commercialization process of Health Qigong, especially in terms of communication effects and acceptance in the Western market. It is necessary to strengthen research on the adaptability and popularity of Health Qigong, in different cultural backgrounds, to promote global development. Future research can focus on the integrated development of Health Qigong and modern health management science, as well as its application effects in regard to chronic disease treatment and health management in an aging society, to enhance its scientific nature and international recognition.

References

[1] Khudhur, T. M. (2024). Effect of Preparation Method on the Physical Properties of SnO2 Thin Films. *International Academic Journal of Science and Engineering*, *11*(2), 22–27. https://doi.org/10.71086/IAJSE/V11I2/IAJSE1145

[2] Moore, A. P., Cappelli, D. M., Caron, T. C., Shaw, E., Spooner, D., & Trzeciak, R. F. (2011). A Preliminary Model of Insider Theft of Intellectual Property. *Journal of Wireless Mobile Networks, Ubiquitous Computing, and Dependable Applications*, *2*(1), 28–49.

[3] Maheswari, D., & Preethi, T. (2016). A Novel Approach for Global Pattern Analysis in Dermoscopic Images. *International Journal of Advances in Engineering and Emerging Technology*, *7*(4), 226–238.

[4] Guo, Y., Xu, M., Wei, Z., Hu, Q., Chen, Y., Yan, J., & Wei, Y. (2018). Beneficial effects of qigong wuqinxi in the improvement of health condition, prevention, and treatment of chronic diseases: evidence from a systematic review. *Evidence-Based Complementary and Alternative Medicine*, *2018*(1), 3235950.

[5] Toneti, B. F., Barbosa, R. F. M., Mano, L. Y., Sawada, L. O., Oliveira, I. G. D., & Sawada, N. O. (2020). Benefits of Qigong as an integrative and complementary practice for health: a systematic review. *Revista Latino-Americana De Enfermagem*, *28*, e3317.

[6] McCaffrey, R., & Fowler, N. L. (2003). Qigong practice: a pathway to health and healing. *Holistic Nursing Practice*, *17*(2), 110–116.

[7] Statista. (2023). Market Size of the Wellness Industry Worldwide in 2023, by Segment. https://www.statista.com/statistics/270720/market-size-of-the-wellness-industry-by-segment/.

[8] Ye, P., Wang, J., Wang, L., Sun, D., & Wang, L. (2024). Development of TCM Health Industry: Current Situations, Difficulties and Breakthroughs. *Journal of Jiangxi University of CM 2023*, *36*(5).

[9] Jingwei, L. (2023). Between Wellness and Elegance: Yoga Consumption in China. *Journal of Consumer Culture*, *23*(1), 104–21.

[10] Sutarya, I. G. (2020). The Potentials and Prospects of Yoga Pilgrimage Exploration in Bali Tourism. *International Journal of Religious Tourism and Pilgrimage*, *8*(8), 11.

[11] Oh, B., Yeung, A., Klein, P., Larkey, L., Ee, C., Zaslawski, C., Knobf, T., et al. (2018). Accreditation Standard Guideline Initiative for Tai Chi and Health Qigong Instructors and Training Institutions. *Medicines*, *5*(2), 51.

[12] Wang, M. (2024). The spread and influence of Chinese martial arts in the United States. *Marital Arts Research*, *9*(5), 63–68.

[13] Munir, K., Ansari, S., & Brown, D. (2021). From Patañjali to the 'gospel of sweat': Yoga's remarkable transformation from a sacred movement into a thriving global market. *Administrative Science Quarterly*, *66*(3), 854–899.

[14] Boaventura, P., Jaconiano, S., & Ribeiro, F. (2022). Yoga and Qigong for Health: two sides of the same Coin?. *Behavioral Sciences*, *12*(7), 222.

46 Digital leap: Exploring impact of digital transformation on employee performance in Philippine advertising agencies

Herlan Czar P. Leuterio[1,a] and Marie Jeddah L. Legaspi[2,b]

[1]Manuel L. Quezon University, Philippines
[2]Pamantasan ng Lungsod ng Maynila, Philippines

Abstract: This research observes the effect of digitalization on the work performance of advertising employees in the Philippines' private firms. Through a strategic blend of convenience and random sampling methods, data from 408 employees were analysed. The study revealed a positive link between digital adoption and job effectiveness. However, factors like age, digital experience, or education showed no significant correlation with their use of digital technology or job performance, indicating a uniform impact of digital tools across different demographics. The key finding is the strong association between the extent of digital transformation and enhanced job performance, highlighting the critical role of digital integration in the competitive and dynamic advertising industry. This insight is vital for professionals and organizations globally, advocating for embracing digital change in advertising.

Keywords: Digital transformation, job performance, demographic influence

1. Introduction

As digital transformation is changing the way businesses are done and hence, also affects many sectors such as advertising, especially in the Philippines as the digital platform will replace the traditional media. This shift not only changes operational techniques but also profoundly affects the agents in advertising agencies [1]. Despite the focus on technical aspects of this transformation, there's a lack of research on its human dimension, especially regarding agents' professional efficacy, task execution, and overall well–being. This study endeavors to bridge this lacuna by exploring how digital transformation affects job performance in Philippine advertising agencies [2]. It combines technological and human behavioural insights, emphasizing the importance of understanding digital transformation's broader implications on the workforce.

Figure 46.1 shows the conceptual framework that examines how the socio–demographic and technological profiles of 408 respondents affect their digital transformation levels and job performance. Independent variables include personal details and digital utilization metrics, which are posited to impact the extent of digitalization as well as the degree of job efficacy, through engagement and adaptation mechanisms [3].

2. Hypotheses

Ho1: There is no significant influence among the regarding the respondents' profile and their extent of digital transformation.

Ho2: There exist no significant relationship regarding the respondents' profile and their level of job performance.

Ho3: There exist no significant relationship regarding the digital transformation level with their extent of job performance.

Research method: This study applied a descriptive–correlational research method, using a validated questionnaire to assess 408 respondents' profiles in personal, technological, digital transformation, and job performance aspects. The method effectively correlates these profiles with their digital

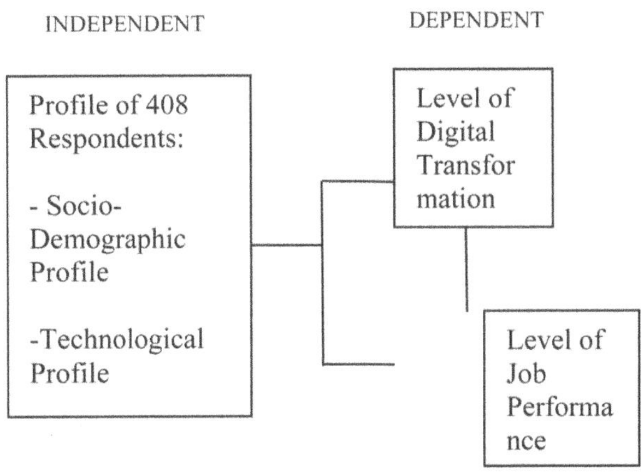

Figure 46.1. Relationship of Respondents' profiles with their impact on digital transformation and job performance.
Source: Author.

[a]leuterio.hcp@gmail.com, [b]jeddahlegaspi@gmail.com

DOI: 10.1201/9781003684459-46

transformation and job performance, providing precise, valuable insights.

Respondents of the study: The study's 408 respondents were advertising agents from the top advertising agencies in the Philippines, encompassing diverse regions. These agents were randomly selected and participated in the survey over a month, providing a comprehensive representation [4].

Sampling procedure: Data were collected via a three–part, self–created four–point Likert–scale online questionnaire. It evaluated the respondents' socio–demographic and technological profiles, their preparedness for embracing digital transformation, and their job performance. This survey was conducted on a limited timeframe from September 15 at 8:00AM to October 15 at 8:00PM using Google Forms, covering various aspects including socio–demographics, technology usage, and job performance.

Procedure for collecting data: The researcher disseminated survey questionnaires across various websites on the internet on September 15 to October 15, 2023. This involved using the web–based survey application Google Forms to distribute the questionnaires on social media platforms such as Facebook, Twitter, and LinkedIn. Additionally, electronic mail was dispatched to various advertising agencies. The survey reached 616 advertising agencies, various Facebook groups, and a Facebook ad campaign, resulting in 85,700 impressions, 1,273 engagements, and 1,255 survey link clicks. Ultimately, 408 responses were collected from advertising agents in the Philippines.

Table 46.1 shows the analysis of correlations among various variables within the advertising sector, particularly focusing on digital transformation. It reveals a spectrum of correlations from weakly positive to negative. This indicates diverse relationships between demographic variables and digital engagement.

A slight positive correlation is observed between 'College Course' and 'Digital Process', suggesting that the type of college course has a minor influence on digital processes. While the relationship exists, it is rather weak, meaning that other factors may be more important factors in digital engagement. In contrast, a negative correlation between 'Employment status' and 'Digital culture' indicates that there is no relation; a change in employments status does not associate with a change in digital culture. However, this lack of significant correlation shows how digital transformation is not dependent on any of these demographic variables.

Beyene et al. [5], however, showed that the demographic effects affect the job satisfaction. This analysis, on the other hand, discovers no significant correspondence between demographic factors and the advancement of digital transformation in the advertising sector. According to Motte–Muñoz [6], such means that digital transformation in the context of technical services does not seem to be based on the conventional demographic features, as opposed to distinguished technical knowledge and competencies.

The analysis also shows the intricate relation between these two variables and digital transformation in advertising, which requires a service force conversant with digital literacy and technical skills. Understand these dynamics is fundamental if organizations wish to take full advantage of the opportunities that emerge in the now rapidly changing digital world.

Table 46.2 offers insightful findings on the relationship between digital transformation and job performance, alongside the impact of educational and personal factors. The result shows some minimal yet significant positive correlation between an individual's course of college and his personal traits which indicate that intellect has a delicate positive effect on the formation of personality. The observation on this aspect furnishes odious effects of education on individual attributes.

The table also reveals strong positive correlations between elements of digital transformation and performance indicators related to job (Output, Job Knowledge, Work Management, Concern for the Organization, and Personal Qualities), all of these elements they must possess in order to achieve the various stages of the digital transformation. These findings clearly demonstrate the strong relationship between organizations' digital transformation efforts and

Table 46.1. The significant influence among the respondent's profile and the extent of their digitalization

Profile	*Digital Mindset*	*Digital Processes*	*Digital Knowledge and Skills*	*Digital Culture*
Age	−.016	.074	.039	.056
Gender	.049	.112	.028	.058
Education	.012	.032	−.016	.025
College Course	.093	.118	.057	.044
Position	−.049	−.135	−.03	−.065
Employment Status	−.057	−.079	−.09	−.157
Region	.11	.069	−.022	.007

Source: Author.

Table 46.2. The significant influence on the respondent's profile and extent of their job performance

Profile	*Output*	*Job Knowledge*	*Work Mngmt*	*Concern for the Org*
Age	.043	.069	.013	.041
Gender	.077	0.04	.053	.074
Education	−.007	.073	.012	.038
College Course	.064	.1	.062	.079
Position	−.064	−.086	−.09	−.11
Employment Status	−.108	−.12	−.11	−.13
Region	.045	.014	.048	.097

Source: Author.

improved job performance given consistent rejection of null hypothesis on all the variables, and therefore highlight the statistical significance of the relationships.

On the other hand, correlation of employment status with its organization concern proved to be a negative correlation which is a reason to take into account that there is no direct link between demographic variables and organizational attitudes.

The relevance of specialized academic courses needed to acquire those necessary skills in the digital sector will resonate Joaquin et al.'s [7] work, enabling students to develop required skills to join the workforce. Therefore, Trenerry et al. [8] emphasize the usefulness of creating targeted training and development programs to facilitate employees and make them good using digital technologies, similar to the workforce skills and organizational goals.

Essentially, In the epoch of digitalization, the relations between education (on the one hand) and personal qualities and employment status (on the other hand), as well as the relations between education and job performance, are very intricate (see Table 46.2). As such, it highlights the importance of both academic training and ongoing professional development for developing employees who can manage challenges and exploit opportunities in the digital landscape.

Table 46.3 presents a revealing picture of how digital elements are tied to the workplace and how they relate to job performance with correlations ranging from moderate to high. It should be noted that 'Digital Culture' positively correlates with 'Output,' meaning that when Company A improves a digital culture, Company A will naturally produce an output. The findings of this study also support that digital technologies at the organizational level should be embraced as a force for driving efficiency.

On the other side, we can see that there is a moderate negative correlation between 'Digital Processes' and 'Personal Qualities', meaning that development of digital tools usage can solve the problem of employees' flexibility and digital literacy. Although the negative coefficient appears to describe the inverse type of relationship, in reality, there are positive outcomes because digital proficiency is a crucial competency for a contemporary workforce.

Table 46.3. Significant relationship between the extent of digitalization with their job performance

Job Performance	*Digital Mindset*	*Digital Processes*	*Digital Knowledge and Skills*	*Digital Culture*
Output	.535	.637	.682	.725
Job Knowledge	.519	.584	.652	.68
Work Management	.524	.666	.563	.617
Concern for the Organization	.573	.659	.635	.671
Personal Qualities	.509	.488	.49	.56

Source: Author.

The correlations between digital transformation and the variables of Manager Effectiveness, Innovation, and Customer Focus are significant and support these findings. This resonates with the McKinsey [9] paper that established that digital innovation boosts productivity. Moreover, Anton et al. [10], through studying other aspects, reminds of the larger effects of technological developments, e.g. increase in wages and job satisfaction attributable to efficiency gains.

Gnana Sanga Mithra et al. [11] further explains in this discussion, as deep learning models show a promise in combating fake news and that all contributions should be continuous in digital technology innovations. Through the comparison of both these two scenarios, we emphasize the significance of continuous adaptation to the challenges of the time as the only way to take advantage of maximal benefits of the digital transformation not only for short–term improvements of workplace, but also the society itself.

In short, the digital transformation data restates about the magnitude of the advantages brought by digital transformation in the business, in relation to the organization's personal qualities of the workplace and output. This reflects a broader trend that companies now have to cope with: living in a digital world where they need to be digitally adaptable – being successful in a world of perpetual innovation with digital evolution, where greater economic efficiency is no longer enough to reach the level of societal well being, but requires productive application of digital technology, itself evolving over time, from which greater economic efficiency can be achieved.

3. Conclusion

This study shows that there is great correlation between digital transformation and job performance of Philippine advertising agencies; that digital transformation positively affects the productivity and efficiency of personnel working in Philippine advertising industry. The studies show clearly that using of digital tools and technologies helps not only in performing tasks in a better way but also leads to the creation of an innovative and adaptive culture key in the fast changing and competitive advertising milieu.

These outcomes highlight the need for a whole treatise of digital transformation. Of course, just introducing technology does not mean that the establishment will perform promptly, it is also about improving the digital competencies of the employees. Those advertising agencies who invest in digital literacy and/or foster a culture of continuous learning will probably become more successful at several levels, organization outputs and employee satisfaction. Additionally, targeted training programs need to be spent on training up the workforce on the skills that they are going to need to manage in this new fast paced digital environment.

The research also paves the way for investigation of similar questions in the future. Although the present study provides indications about the relationship between digital transformation and job performance, other studies could focus on determining the effects of individual digital tools and platforms on job performance. As well, it would also be advantageous to explore the long term effects of digital transformation on an employee's satisfaction, agency's performance as well as the entire advertising industry in the Philippines.

Additionally, elucidating ways in which digital transformation affects creativity and innovation in building advertising campaigns, how it changes client relationship and business strategies, would present a more clear notion of its reach. Those that can change quickly to adapt to digitalization and use these tools in a way makes them effective are not only going to survive, but they are going to do very well in this industry.

To conclude, the research rests on the fact that advertising agencies must seriously pursue digital transformation. Through the adoption of new technologies, encouraging a digital first culture, and developing employee competencies, agencies will be able to meet the needs of the future, and therefore maintain success in the currently rapidly evolving digital landscape.

References

[1] Talegaon, S., & Krishnan, R. (2020). Administrative models for role based access control in android. *Journal of Internet Services and Information Security*, *10*(3), 31–46.

[2] Sajna, V., & Dharmaraj, A. (2024). Awareness of Central Sector Scheme among the Entrepreneurs in MSME Sector: An Investigative Study. *Indian Journal of Information Sources and Services*, *14*(3), 115–122. https://doi.org/10.51983/ijiss–2024.14.3.16

[3] Sakthivel, S., Manikandan, R., & Arjunan, T. V. (2016). Evaluation on Mechanical Properties of LM25/FLY ASH/SIC Hybrid Metal Matrix Composites (MMC). *International Journal of Advances in Engineering and Emerging Technology*, *7*(4), 239–249.

[4] Saidova, K., Abdullayeva, S., Yakubova, D., Gudalov, M., Abdurahmonova, K., Khudoykulova, H., Mukhammadova, G., & Zokirov, K. (2024). Assessing the Economic Benefits of Climate Change Mitigation and Adoption Strategies for Aquatic Ecosystem. *International Journal of Aquatic Research and Environmental Studies*, *4*(S1), 20–26. https://doi.org/10.70102/IJARES/V4S1/4

[5] Beyene, T., & Gituma, M. (2017). The influence of employee demographic factors on job satisfaction: A case study of Segen Construction Company, Eritrea. *African Journal of Business Management*, *11*(21), 608–618.

[6] Motte–Muñoz, H. (2020). Empowering Filipino youth through technology and community. In Anticipating and preparing for emerging skills and jobs. *Education in the Asia–Pacific Region: Issues, Concerns and Prospects. Springer Link*, *55*, 247–254.

[7] Joaquin, J. J. B., Biana, H. T., & Dacela, M. A. (2020). The Philippine higher education sector in the time of COVID–1*Frontiers in Education*, 5.

[8] Trenerry, B., Chng, S., Wang, Y., Suhaila, Z., Lim, S., Lu, H., & Oh, P. (2021). Preparing workplaces for digital transformation: An integrative review and framework of multi–level factors. *Frontiers in Psychology*, 12.

[9] McKinsey. (2017). *What's now and next in analytics, AI, and automation. Executive Briefing*. Mckinsey & Company.

[10] Anton, J., Muñoz–de–Bustillo, R., Macias, E., & Esteve, F. (2023). *Measuring more than money: The social economics of job quality*. Edward Elgar Publishing.

[11] Bhavana, S. (2024). Navigating the Truth: Advanced Deep Learning Strategies Against the Spread of Digital Misinformation. *SPAST Reports*, *1*(2).

47 Designing carbon-aware cloud architectures for sustainable computing

S. Vidhya[1,a], *Ramesh S.*[1,b], *R. Geetha*[1,c], *P. Thilagavathi*[2,d], *and J. Nithyashri*[3,e]

[1]Assistant Professor, Department of Computing Technologies, School of Computing, SRM Institute of Science and Technology, Kattankulathur, Chennai, India
[2]Associate Professor, Department of Computer Science and Engineering, Aarupadai Veedu Institute of Technology, Vinayaka Mission's Research Foundation (DU), Paiyanoor, Chennai, Tamil Nadu, India
[3]Assistant Professor, Department of Computing Technologies, School of Computing, Faculty of Engineering and Technology, SRM Institute of Science and Technology, Kattankulathur, Tamil Nadu, India

Abstract: The increasing demand for cloud computing has made ICT energy efficiency a significant concern. The number of companies spending money on big data centres to run Cloud services is growing. Infrastructure-wise, these data centres are very complicated, and they use a lot of power. Using technologies like virtualization and streamlining could help this data centre use less energy, according to many studies. Nevertheless, these answers are mainly based on how much they cost, so they don't really address the serious issue of CO2 emissions that pose a long-term health risk to the environment. This research proposed a user-centered approach that contains cloud architecture, also known as carbon-aware cloud architecture. It mainly focuses on environmental issues related to various types of problems. We concluded by discussing future studies associated with the help of cloud computing about carbon as an efficient.

Keywords: Cloud computing, green IT, resource management, green border, environment

1. Introduction

Data centers use a lot of electricity, and that amount is growing quickly. In 2018, data centers worldwide used 205 TWh, more than what countries like Ireland and Denmark use in a year. Up until 2030, IT and communications used 24 hours a day, seven days a week, which is about 7% to 20% of the world's power. Currently, tech companies invest in green energy to compensate for the energy data centers use [2]. We have put 22 GW of renewable energy into our plans to reach Nett Zero, along with Google, Amazon, and Meta. The plans for these projects are part of bigger ones. The US wants to have no nett emissions by 2050. which means that renewable energy generation will rise from 20% in 2020 to 42% by then. 47% of this green energy comes from the sun and 34% from the wind [3]. However, due to fluctuations in renewable energy sources, it remains challenging to operate carbon-free around the clock. As more solar and wind farms are built, more significant supply changes occur every hour and season [4]. Figure 47.1 shows changes that happen when green energy makes up 33% of the total. If the amount of renewable energy the grid supplies exceeds that consumed, wasteful curtailments may be imposed to bring the supply and demand levels back into balance. Sometimes there isn't enough sun and wind power [26]. In these situations, data centers use energy high in carbon, even though they have reached Net Zero [5]. It is still difficult to run carbon-free all the time because renewable energy sources are unpredictable. Data centers claim to be Net Zero yearly by matching points with their computing energy. Without an adequate carbon-free supply, data centers will continue emitting carbon with the use of fossil fuels that are high in carbon, such as coal and gas. Therefore, in a sustainable design context, additional measures are required to eradicate the daily carbon emissions generated by computers. Maintaining carbon-free computing around the clock, old data center design and management questions must be looked at again [6, 25].

Architects have researched power infrastructure to find ways to utilize it most efficiently while still meeting performance objectives. They've made schedulers that let jobs ask for resources and then give them to them, maybe with fairness and sharing rewards in mind. These plans for carbon-free energy need to be rethought by architects [27]. Architects must first devise parameterized ways to deal with green energy that doesn't always work. They also need tools that show how to combine these answers to save carbon [7]. Rearranging computer schedules to align with the energy demands of data centres and renewable sources is one

[a]vidhyas2@srmist.edu.in, [b]ramesh.isaiah@gmail.com, [c]geetharl@srmist.edu.in, [d]thiluxpriya499@gmail.com, [e]drjnithyashri@gmail.com

DOI: 10.1201/9781003684459-47

method called demand response [28]. The second is energy storage, which makes it less vulnerable to changes in supply. In the end, carbon-free energy is produced by increasing investments in renewable energy sources like wind and solar [8]. These ideas will change the way infrastructure works. In the future, data centers will have to invest in energy generation that takes into account their where you are and how readily wind and solar electricity are available, have servers that can do more processing when carbon-free energy is plentiful, and use batteries that are significantly larger than what is currently used in the power grid. This equipment has embodied carbon costs, an essential but little-studied part of making data centers more sustainable [9]. To find the carbon-optimal combination of remedies, designers should aim for incorporated creation and administration within data center structures, networked software, and hardware architecture. Carbon Lookup. With Carbon Explorer, construction companies may configure data centres to operate continuously without emitting carbon dioxide [10]. The number ten. It makes use of data regarding worldwide access to renewable sources of energy and the amount of power required by data centers. Also, it guesses what will happen if you put money into green energy Using data from the grid, storing energy with physically accurate battery models, and the market-responsive routing number of servers needed using production data center traces. In its comprehensive analysis, Carbon Explorer considers data center architecture in light of the increased carbon levels resulting from the production of servers, batteries, wind and solar farms, and practical measures to reduce carbon emissions [11, 29–30].

2. Carbon Cloud Architecture

Our approach is shown in Figure 47.2, which notes cloud architecture followed by various requirements. These requirements include various kinds of services with cloud CO2 emissions. This proposed approach consists of various components, such as a directory, a carbon emission directory, and a list of cloud services. Requests for Cloud services (like software, platform, or infrastructure) are sent to Green Broker, which then chooses the greenest Cloud source. Source: Green Middleware makes it possible to operate the most carbon-efficient cloud [12]. This software's components vary according to the cloud service (SaaS, PaaS, or IaaS).

3. Third-Party: Directory for Green Offers and Carbon Emissions

Two additional sections, the GOD and the CED, are recommended to be implemented. These components are indispensable for the enforcement of environmentally friendly cloud computing practices [13]. Governments have already put limits on how much energy datacenters can use by rating them and passing laws. Also, people are becoming more aware of how greenhouse gases affect climate change. As a result, people will probably choose to use Cloud services from companies that leave the smallest carbon impact. Cloud service companies can also use these directories to get more users by promoting them. For example, Google has shared information about its data center's energy efficiency. Adding these kinds of folders makes sense in the current Cloud Computing environment. A Green Offer Directory is where cloud service providers list their services, and Green Brokers can view them. These deals specify the kind of service, its price, and the optimal time to utilize it to reduce CO2 emissions. The next step in this process is to calculate the network cost, CO2 emissions, power consumption efficiency, and cooling impacts of cloud data centres in providing the services. These all help to measure the electricity among the various

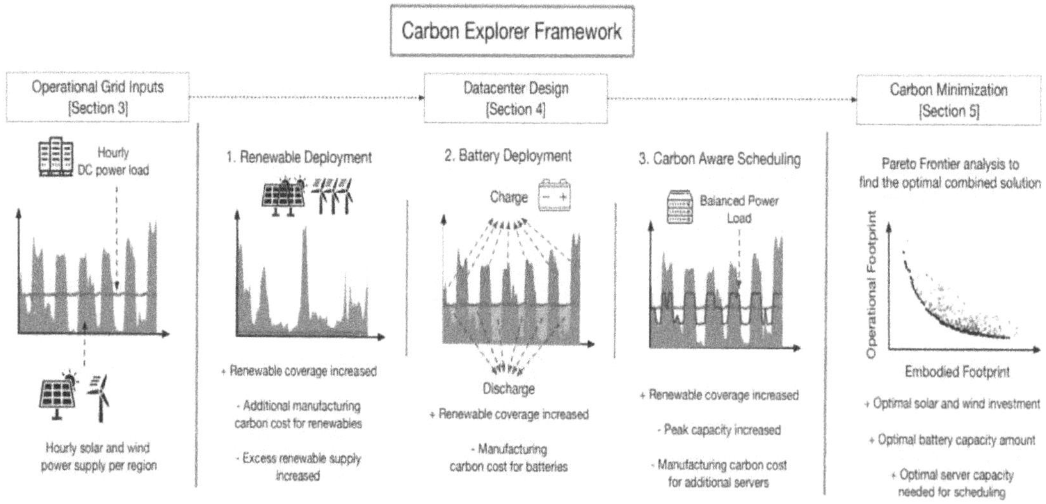

Figure 47.1. Carbon explorer framework.

Source: Author.

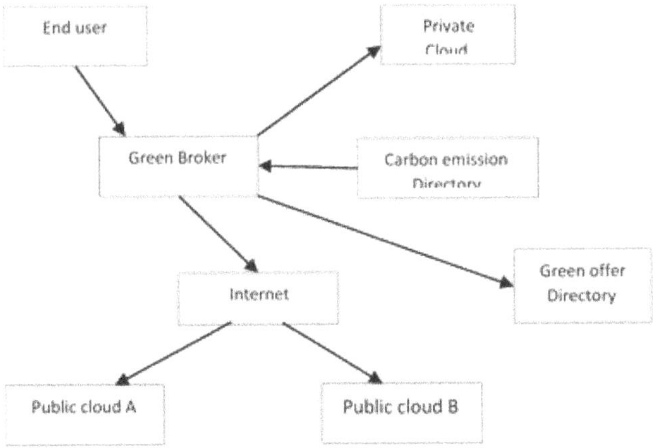

Figure 47.2. Carbon aware cloud architecture.

Source: Author.

metrics in the carbon pollution directory. It includes various information connected to the efficient use of energy by cloud services. Using energy parameters, Green Brokers can get the most recent data and make use of the various cloud services listed in CED.

3.1. User: eco-friendly broker

Similar to a standard Cloud trader, a Green trader (Figure 47.2) assists individuals in renting cloud services and strategizing their applications. Cloud request services comprise its first layer, analyzing requests to determine the type of quality of service required. The second layer uses information on different cloud providers along with up-to-date CO2 emission parameters culled from the Carbon Emission Directory (CED) and the Green Offer Directory (GOD) to figure out how much it costs to lease certain Cloud services and how much it impacts the environment [14]. Green Policies determines whether to lease cloud services by analyzing these figures. Cloud Ask Accounts will present the user with additional "Green Offers" if an exact match cannot be discovered for a request. Here are the various types of services, such as Saas, Paas, and Iaas. It also determines the various requests related to carbon footprint, followed by the computation analysis of CO2 emission. Data should be transferred through various execution services, considering it is also called a data center. SaaS and PaaS requests are also defined as the long-term analysis of CO2 emission as per Co2. Iaas should request data transfer used predominantly with one time [15].

3.2. SAAS and PAAS request (CO_2) emission per second

$$Co2PS_{SaaS/Paas} = (r_{dT}^{co2} E_{dT} * a_{dT}) + \left(r^{CO2} * \frac{1}{DCiE} * E_{serv}\right) \quad (1)$$

The rate of CO_2 emissions per joule of energy expended by user computers on their way to the data centre, as determined by the aforementioned equation, is r sub d cap T to the c o 2. A cloud service's overall power consumption is comprised of processing, lighting, wiring, HVAC, and power conditioning, among other things. Therefore, while evaluating cloud service providers, DCiE is the most appropriate metric to utilize [16].

3.3. IAAS request (CO_2 emission)

$$CO2_{Iaas} = (r_{dT}^{CO2} E_{dt} * IOdata) + \left(r^{Co2} * \frac{1}{DCiE} * E_{serv} * Vtime\right) \quad (2)$$

According to the previous equation, Vtime is the amount of time an active IO data transfer takes place when using a virtual machine (VM) rented from the cloud [1].

4. Cloud Architecture and Sustainable Computing

New developments in cloud technologies show that ideas, architectures, and tools for integrating clouds are being created. These can facilitate the transition of on-premises and infrastructure-related services to the cloud and the integration of cloud-based infrastructure with preexisting business and campus infrastructures. For businesses to use cloud infrastructure services in more complex ways, new models for setting up services and keeping them safe will need to be made [17]. These approaches may make Project and group-based infrastructures set up on demand and across several providers easier. Data and computing centre compute unit counts and system integration levels are skyrocketing to meet the surging need for faster processors cited as [18]. Power management has thus become a highly challenging topic for developing and operating large computer systems. Yongpeng Liu, Hong Zhu, Kai Lu, and Xiaoping Wang's study "Self-Adaptive Power Management of Idle Nodes in Large Scale Systems" introduced a novel approach to the management of inactive node sleep states, which ensures outstanding efficiency and low power consumption [19]. The self-adaptive mechanism of the proposed technique adjusts the size of reserve pools according to the degree to which the study pools satisfy the computational resource requirements during system operation. The trials demonstrated that the proposed method may significantly improve power efficiency compared to current power management strategies for large-scale systems [2].

5. Baseline Algorithms

The single Threshold Method contains the theoretical approaches of literature in an online search [4]. It's used for the threshold to manage the job executions related to carbon intensity. It checks if the carbon content is below a certain level at each time step. Task execution is dependent on the intensity level; if too high, it will stop. Not knowing how long a job takes or how much extra effort is involved in moving is not assumed by this strategy. Variations of this technique are denoted as ST[p], with p representing different threshold values. Algorithm with two thresholds: The author describes a Double-Threshold algorithm for temporal shifting, motivated by the single-threshold technique's failure to account for switching overhead. The technique sets low and high limits, with a space between them (α). If the previous time step interrupted the action, continuing execution requires reducing the carbon strength to at least v below the high threshold. This plan ensures that the advantages of running at a lower carbon intensity are more significant than the emissions caused by fixing the job. By reducing the need to repeatedly stop and restart the job, this strategy helps to reduce the total carbon footprint caused by checkpointing and restoration. Various variations of this method are written as DT[L, H], $s.t.H - L = \Omega$, where L and H are different low and high thresholds [20].

WaitAWhilePred: Both the job duration and the carbon intensity are estimated using this method. After acquiring these estimates, it chooses a period of time with the lowest anticipated carbon intensity within the given time frame (T). This type of job should be scheduled to execute the selected types of times. Here also mentioned the basic version of this algorithm accounts for the job's actual duration and carbon intensity estimates [21]. An agnostic approach to carbon: This program starts working on the job as soon as it is received and doesn't stop until it's done, regardless of how much carbon is being created now or in the future. The method consists of the start execution to meet the various deadline for time slot, its also less than or equal to time required for assuming the Cmax [22].

6. Motivating FTL

Task attributes, switching overhead, task length, local context (such as the accuracy of carbon intensity projections), and the unique power supply configuration are some of the major elements that impact carbon-aware scheduling algorithms (specific to a geographical location). For instance, jobs with high switching overhead perform worse with methods that overlook this aspect, such as Single Threshold and WaitAWhilePred, compared to the Double-Threshold algorithm, which determines when to pause and resume based on significant variations in carbon intensity [23]. Here, it is also mentioned that the duration of job security is influenced by the threshold of settings related to the single- and double-threshold algorithms. The tighter threshold values are used to minimize the carbon footprint and the periods of low cabor intensity. Conversely, longer jobs with strict deadlines may need to be executed at the last moment when carbon intensity is high. Thus, we adjust the threshold parameters to raise awareness about carbon emissions. WaitAWhilePred accurately predicts carbon intensity and job length, particularly when the cost of switching is low [25]. An ideal solution is virtually attained by aggressively selecting execution slots with low carbon emissions. In contrast, scenarios with minimal fluctuations in carbon intensity, large forecasting errors, or high switching overheads are more suited to the carbon-agnostic algorithm, which operates independently from changes in carbon intensity. Here, we evaluate the effectiveness of ST, DT Carbon Agnostic, and WaitA WhilePred algorithm from 2020 to 2022. Which is contains the 10% of forecasting error in the carbon intensity and job length. To analyze the above, Table 47.1 should note the frequency of each algorithm related to the least types of carbon. To analyze the various performances in the specific location to reveal the method that outshines others. Consider that AU-NSW has the top performance method, which contains 25% of the time. Furthermore, the rankings of algorithms vary by region emphasizing how they respond to specific regional circumstances. In comparison to France's Carbon-Agnostic Party, which received 46% of the vote, DT [31, 32] is most successful in Ontario, where it received 18. approach, which claims 20. 62%. We explore in greater depth the performance (carbon savings) implications of these methodologies. This variable performance necessitates a strategy capable of instantly selecting the most effective algorithm for each task, ensuring more significant and more reliable reductions in carbon emissions [25].

Table 47.1. The algorithm minimizes carbon consumption for medium switching overhead tasks by region

Algorithm	*Ontario*	*California*	*AU-NSW*	*France*
DT	18.46%	22.38%	24.15%	14.77%
Carbon	9.92%	4.08%	13.02%	20.62%
ST	11.92%	11.00%	9.92%	9.00%
DT	7.45%	5.62%	8.54%	13.02%
Other Algorithms	45.69%	11.38%	15.46%	27.54%

Source: Author.

7. Conclusion

This paper summarizes the main concepts related to cloud-aware cloud architecture, which helps mitigate the carbon impacts of cloud computing, followed by the global environmental impacts. Both users and service providers are encouraged to use and offer the "greenest" services by how our system is set up. Thus, it incorporates user-facing components like Green Broker to ensure that programs are implemented with minimal carbon footprint. Likewise, we suggest capabilities for the future generation of cloud providers, such as sharing the environmental impact of their services in public databases and providing "Green Offers" to lower energy usage generally. We also suggest that green brokers establish a carbon-efficient green policy (CEGP). Under this policy, user applications with short turnaround times are assigned to cloud data centers with lower energy consumption and carbon emissions. Additionally, simulations in various IaaS Cloud provider scenarios validate our design. Here, we compare the various scheduling methodologies to demonstrate the architecture related to carbon and energy footprint related to cloud computing. Our proposed architecture combines energy efficiency and a green policy among CEGP. It provides the potential difference among the carbon emissions as 25% and 23% as energy. These are all indicated in the results of performance evaluations.

References

[1] Garg, S. K., Yeo, C. S., & Buyya, R. (2011, August). Green cloud framework for improving carbon efficiency of clouds. In *European Conference on Parallel Processing* (pp. 491–502). Berlin, Heidelberg: Springer Berlin Heidelberg.

[2] Hsu, C. H. (2013). Theory and Algorithms for Emerging Cloud Computing and its Sustainability. *International Journal of Next-Generation Computing*, *4*(2).

[3] Anderson, T., Belay, A., Chowdhury. M., Cidon, A., & Zhang, I. (2022). Treehouse: A Case for Carbon-Aware Datacenter Software. arXiv preprint arXiv:2201.02120.

[4] Andrae, A. S., & Edler, T. (2015). On global electricity usage of communication technology: Trends to 203*Challenges*, *6*(1), 117–157.

[5] Barroso, L. A., & Hölzle, U. (2007). The case for energy-proportional computing. *Computer*, *40*(12), 33–37.

[6] Bashir, N., Guo, T., Hajiesmaili, M., Irwin, D., Shenoy, P., Sitaraman, R., Souza, A., & Wierman, A. (2021). Enabling Sustainable Clouds: The Case for Virtualizing the Energy System. arXiv preprint arXiv:2106.08872.

[7] Chien, A. A. (2020). Characterizing Opportunity Power in the California Independent System Operator (CAISO) in Years 2015–2017. *Energy and Earth Science*.

[8] Culpa, E. M., Mendoza, J. I., Ramirez, J. G., Yap, A. L., Fabian, E., & Astillo, P. V. (2021). A Cloud-Linked Ambient Air Quality Monitoring Apparatus for Gaseous Pollutants in Urban Areas. *Journal of Internet Services and Information Security*, *11*(1), 64–79.

[9] Emilsson, E., & Dahllöf, L. (2019). Lithium-ion vehicle battery production-status 2019 on energy use, CO2 emissions, use of metals, products environmental footprint, and recycling.

[10] Balaji, R., Deepakkumar, A., Prabhu, G., Thinakaran, P., & Gowtham, S. (2023). Enhancing Network Security by using SDN Algorithm in Cloud Computing. *International Academic Journal of Science and Engineering*, *10*(1), 14–19. https://doi.org/10.9756/IAJSE/V10I1/IAJSE1003

[11] Liu, Z., Liu, I., Low, S., & Wierman, A. (2014). Pricing data center demand response. *ACM SIGMETRICS Performance Evaluation Review*, *42*(1), 111–123.

[12] Masanet, E., Shehabi, A., Lei, N., Smith, S., & Koomey, J. (2020). Recalibrating global data center energy-use estimates. *Science*, *367*(6481), 984–986.

[13] Rajaeifar, M. A., Ghadimi, P., Raugei, M., Wu, Y., & Heidrich, O. (2022). Challenges and recent developments in supply and value chains of electric vehicle batteries: A sustainability perspective. *Resources, Conservation and Recycling*, *180*, 106144. https://doi.org/10.1016/j.resconrec.2021.106144

[14] Sakalkar, V., Kontorinis, V., Landhuis, D., Li, S., DeRonde, D., Blooming, T., et al. (2020). Data center power oversubscription with a medium voltage power plan and priority-aware capping. In *Procedings International Conference on Architectural Support for Programming Languages and Operating Systems (ASPLOS)*.

[15] Yemunarane, K., Chandramowleeswaran, G., Subramani, K., ALkhayyat, A., & Srinivas, G. (2024). Development and Management of E-Commerce Information Systems Using Edge Computing and Neural Networks. *Indian Journal of Information Sources and Services*, *14*(2), 153–159. https://doi.org/10.51983/ijiss-2024.14.2.22

[16] Achar, S. (2022). Cloud computing: toward sustainable processes and better environmental impact. *Journal of Computer Hardware Engineering (JCHE)*, *1*(1).

[17] Imanov, E., Asojo, J. O., & Imanova, G. (2024, April). Green Cloud Computing Strategies for Carbon Footprint. In *International Conference on Smart Environment and Green Technologies* (pp. 439–446). Cham: Springer Nature Switzerland.

[18] Mahadasa, R., & Surarapu, P. (2016). Toward Green Clouds: Sustainable practices and energy-efficient solutions in cloud computing. *Asia Pacific Journal of Energy and Environment*, *3*(2), 83–88.

[19] Kamil, A. A., Mousa, A. J., Aljuboury, A. S., & Abdul-Rahaim, L. A. (2024). Control System Design for Failure Starting of Diesel Power Block for Cell on Wheels Communication Tower based on Cloud Service System. *Journal of Internet Services and Information Security*, *14*(3), 275–292. https://doi.org/10.58346/JISIS.2024.I3.017

[20] Buyya, R., & Gill, S. S. (2018). Sustainable cloud computing: foundations and future directions. arXiv preprint arXiv:1805.01765.

[21] Laurent, A., Olsen, S. I., & Hauschild, M. Z. (2012). Limitations of carbon footprint as indicator of environmental sustainability. *Environmental Science & Technology*, *46*(7), 4100–4108.

[22] Correa, J. P., Montalvo-Navarrete, J. M., & Hidalgo-Salazar, M. A. (2019). Carbon footprint considerations for biocomposite materials for sustainable products: A review. *Journal of Cleaner Production*, *208*, 785–794.

[23] Li, B., Basu Roy, R., Wang, D., Samsi, S., Gadepally, V., & Tiwari, D. (2023, November). Toward sustainable hpc: Carbon footprint estimation and environmental implications of hpc systems. In *Proceedings of the international conference for high performance computing, networking, storage and analysis* (pp. 1–15).

[24] Ranjetha, K., Alengaram, U. J., Alnahhal, A. M., Karthick, S., Zurina, W. W., & Rao, K. J. (2022). Towards sustainable construction through the application of low carbon footprint products. *Materials Today: Proceedings*, *52*, 873–881.

[25] Bala Krishna, M. (2021). A Robust Strategic Theory based Load Balancing and Resource Allocation in Cloud Environment. *International Academic Journal of Science and Engineering*, *8*(1), 19–30. https://doi.org/10.9756/IAJSE/V8I1/IAJSE0803

[26] Issa Zadeh, S. B., & Garay-Rondero, C. L. (2023). Enhancing urban sustainability: unravelling carbon footprint reduction in smart cities through modern supply-chain measures. *Smart Cities*, *6*(6), 3225–3250.

[27] Rahman, S., & Begum, A. (2024). Analysis of Structural Integrity in High-Rise Buildings Under Dynamic Load Conditions Using AI: A Computational Perspective. *Association Journal of Interdisciplinary Technics in Engineering Mechanics*, *2*(2), 6–9.

[28] Beula Princy, S., & Duraisamy, S. (2016). A Comparative Study of Different Algorithms for Detection and Segmentation Blood Vessels from Retinal Image. *International Journal of Advances in Engineering and Emerging Technology*, *7*(4), 277–291.

[29] Khyade, V. B. (2019). Efficiency of Mulberry, Morus alba (L) as fodder for cattle. *International Academic Journal of Innovative Research*, *6*(1), 77–90. https://doi.org/10.9756/IAJIR/V6I1/1910007

[30] Ristono, A., & Budi, P. (2025). Next-gen power systems in electrical engineering. *Innovative Reviews in Engineering and Science*, *2*(1), 34–44. https://doi.org/10.31838/INES/02.01.04

[31] Liang, H., Ahmad, A., & Khan, A. (2024). A Systematic Review of Effective Models in School-Enterprise Cooperation in China in the English and Chinese Literature. *Indian Journal of Information Sources and Services*, *14*(3), 138–144. https://doi.org/10.51983/ijiss-2024.14.3.19

[32] Srividhya, E., P, J., Anusuya, V., Deepthi, K. J., Gopalsamy, P., & Gopalakrishnan, S. (2024). Deep Learning-Driven Disease Prediction System in Cloud Environments using a Big Data Approach. *EDRAAK*, *2024*, 8–17. https://doi.org/10.70470/EDRAAK/2024/002

48 Preliminary study of the implementation of PAPEDA framework based on Ethno-STEM in supporting holistic science learning

Henie Poerwandar Asmaningrum[1,a], *Marfly*[2,b], *and Shraddha Verma*[3,c]

[1]Mathematics and Science Education Department, Universiti Malaya, Kuala Lumpur, Malaysia
[2]Chemistry Department, Gadjah Mada University, Yogyakarta, Indonesia
[3]Faculty of Education, Kalinga University, NayaRaipur, India

Abstract: This research investigates the use of the PAPEDA Framework by utilizing an Ethno-STEM approach in the context of holistic science learning outcomes. Data was collected using a mixed-method design with teachers (N=10), students (N=50), and school administrators (N=6) corresponding to the Papua Region in Indonesia. The findings from the data collected indicate a successful implementation of the framework – with teacher comments indicating teacher readiness in the Provide (4.2/5) and Performance (4.1/5) components, and students reporting elevated experiences of science engagement (84%) and school administrators reporting a high commitment to adjusting resources (100%) to accommodate the framework. Student results indicated an increase in the use of scientific thought in the classroom, with average scores of 84% in scientific understanding and 76% in cultural appreciation. All administrators reported better family and community relationships as part of the increased engagement. Key challenges identified included time (90% of teachers), resources (80% of teachers), and standardization of assessments to the research objectives (70% of teachers). The authors offer recommendations focusing on professional development, the need for modifications to assessment approaches, and future work with community organizations to aid implementation of the framework. Findings from this research contribute to the dialogue of culturally responsive science education, while providing a replicable model of holistic learning that integrates inductive knowledge and imperial science learning to improve educational outcomes.

Keywords: PAPEDA framework, Ethno-STEM, holistic science education, cultural integration

1. Introduction

The National Education Policy (NEP) of 2020 indicates a change in India's educational landscape, particularly with respect to science education. The policy emphasizes holistic learning that combines cultural knowledge and contemporary scientific thinking. This change is a direct response to the calls for education based on the notion that education can be culturally relevant [1]. This paradigm of learning intersects with global and Western educational paradigms that also recognize the value of Indigenous knowledge systems and their role in creating science knowledge [2, 3].

In response to these educational shifts, the PAPEDA Framework (Provide, Analyze, Perform, Evaluate, Disseminate, and Assess) creates a theoretically and methodologically sound approach to bridge traditional knowledge with traditional science education [4]. The PAPEDA framework provides a procedural model to enact integrated solutions to address the diverse learning needs related to NEP 2020, specifically, to create holistic science learning that integrate cultural perspectives. The integration of Ethno-STEM education, which combines ethnoscience with science, technology, engineering, and mathematics (STEM), contributes to the creation of one of the goals of NEP 2020 to promote multidisciplinary learning about India's unique and rich cultural understanding.

The intersection of the NEP 2020 objectives with the PAPEDA framework and an Ethno-STEM experience provide a basis for innovative and integrated education model that responds to the philosophy of policy while ensuring science education remains relevant and accessible for students from diverse cultural backgrounds and prior learning experiences [5, 6]. Jaudinez and Joaquin [7] provide examples of traditional tools and practices that exemplified ideas about ethnoscience, ethnomathematics, and ethnoengineering informed their instruction. It will be important for the education community to be cognizant of pedagogical principles, culturally relevant knowledge and backgrounds, and comprehensive assessment to combine traditional knowledge and experiences with the standards for modern science knowledge.

[a]s2138457@siswa.um.edu.my, [b]marfly1995@mail.ugm.ac.id, [c]shraddha.verma@kalingauniversity.ac.in

DOI: 10.1201/9781003684459-48

2. Literature Review

2.1. National education policy 2020

The National Education Policy (NEP) 2020 indicates a significant transformation in India's education architecture because it provides clear articulation of reforms that will transform education [8, 9]. The policy promotes a comprehensive orientation that will combine traditional knowledge systems with contemporary learning [10, 11]. In the realm of science education, for example, NEP 2020 will promote interdisciplinary learning that transcends subjects and emphasizes cultural context of scientific concepts learned and practiced in the world [12, 13]. It calls for the incorporation of traditional knowledge into mainstream courses, enriching a learning landscape that respects local knowledge as much as it does science – an idea that is celebrated internationally and is also firmly rooted in India's various traditions of knowing and situated knowledge within its socio-cultural context [10, 41].

2.2. Ethno-STEM

Ethno-STEM is a novel viewpoint to education that integrates ethnoscience and mainstream STEM content [14, 15, 44–45]. It recognizing indigenous understandings as a contribute to scientific understanding while developing opportunities to integrate it with contemporary scientific methods [16, 17]. It promotes the value of context to learning by recognizing the development of scientific knowledge to occur within, and influenced by, cultural contexts [6, 18]. Ethno-STEM aims towards building significant links between traditional knowledge and modern scientific principles [7, 43], validating indigenous knowledge systems while offering students cultivable knowledge of scientific principles, approaches of application with the familiar perspectives and examples of their culture [19]. Ethno-STEM practice in a context of integration presents strong evidence illustrating effectiveness of a combined approach towards engaging learners and improving students' understanding of scientific content [20–22, 42].

2.3. PAPEDA framework

The PAPEDA Framework is a systematic approach to educational implementation that is based around six central components: Provide, Analyze, Perform, Evaluate, Disseminate, and Assessment [4]. The philosophical grounding provides an immersive learning experience connected to both theory and practice while being culturally situated. Each of the components serves a purpose: Provide refers to resourcing mobilization and organizing content; Analyze is to explore learning needs and contextual aspects; Performance implementation is conducting the actual learning experience; Evaluation is the ongoing monitoring of learning progress; Dissemination is sharing and adopting effective practice; and Assessment is collecting full feedback about the learning outcome. As the components are designed to work together, the PAPEDA Framework provides a holistic educational implementation that incorporates both curricula delivery and assessment related to the learning.

2.4. Integration conceptual framework

The combination of NEP 2020, PAPEDA Framework and Ethno-STEM provides a cohesive conceptual framework to guide educational implementation. This approach fulfills NEP 2020 policy aims by providing a structured implementation process, while being culturally relevant through Ethno-STEM [23]. The framework also considers the holistic interrelationships amongst the three elements to facilitate comprehensive educational aims [13]. By emphasizing the relationship between policy, implementation process, and cultural relevance, this conceptual framework enables a structured response to realizing the NEP 2020 vision for holistic, culturally relevant, and scientifically sound education [24–26]. As such, the combined elements provide a basis for implementing culturally relevant science education, while addressing NEP 2020 desired outcome of respect for traditional knowledge systems in science education. The integration of the three elements provides possible pathways for implementation that consider opportunities for resource use, teacher preparation, and assessment that balances culturally relevant instruction and scientific accuracy [8, 9].

3. Methodology

This research employed a mixed-method approach with a sequential explanatory design, to investigate the implementation of the PAPEDA Framework in the Ethno-STEM context in selected schools in Papua Region, Indonesia. In this research context, the research population comprised science teachers implementing the framework, students who experienced the integration, and educational administrators overseeing the implementation, with school selection occurring based on diversity within the local cultures, availability of local resources and connections, and the extent of integration of traditional knowledge within schools. As part of the data collection phase, multiple methods were employed to gather data: quantitative methods (structured surveys, tests, and performance measures) and qualitative methods (semi-structured interviews, observation, focus group discussion, document review, and ethnographic considerations). Instruments that were included as part of the research included structured questionnaires about the teachers' preparedness and measures of student engagement, observation protocols, interview protocols, and rubrics for assessment based on the objectives of the PAPEDA Framework, all validated through a review process with an expert in the field, and pilot testing for reliability and cultural relevance. The analysis drew from quantitative methods (descriptive statistics, correlation analysis, comparative studies) and qualitative methods (thematic analysis) with triangulation of data being used to triangulate against validity and reliability measures. There was a specific focus on analysing how traditional knowledge was integrated with scientific concepts, with the contextual analysis examining the cultural contexts, resources and access to resources, and stakeholder involvement as significant variables influencing the success of the implementation.

4. Result and Discussion

The comprehensive findings from our mixed-methods investigation of the PAPEDA Framework implementation through an Ethno-STEM approach in selected schools in Papua Region, Indonesia. Data collected from three key stakeholder groups – teachers (N=10), students (N=50), and school administrators (N=6) – reveals multifaceted insights into the implementation process, effectiveness, challenges, and outcomes of this integrated educational approach.

4.1. Teacher's perspectives

Teachers demonstrated varying levels of understanding and implementation confidence across PAPEDA Framework components as shown in Table 48.1.

Table 48.1. Result of teacher implementation

Component	*Aspect*	*Averge Score / Average Percentage*
Top Strenghth	Hands-on activities	90%
	Project-based learning	90%
	Local practices integration	80%
Implementation Success	Enhanced student engagement	90%
	Improved scientific understanding	80%
	Increased cultural awareness	70%
Key Challenges	Time management	90%
	Resource limitations	80%
	Assessment complexity	70%
Priority Needs	Professional development	90%
	Teaching materials	80%
	Technology training	70%
Weakest Areas	Disseminate component	3.5
	Evaluation methods	3.4
	Assessment tools	40% prepared
Strongest Areas	Provide component	4.2
	Classroom execution	4.2
	Performance component	4.1

Source: Author.

Teachers exhibited different levels of understanding and implementation confidence across components of the PAPEDA Framework. The most preparedness among teachers was shown in Provide (4.2) and Performance (4.1), while the lowest preparedness was indicated for Disseminate (3.5) and Evaluate (3.7). The component that teachers indicated the most confidence in implementing in their classrooms was in performance (4.2), while the lowest confidence was seen in the evaluation strategies (3.4).

Further analysis of implementation experiences revealed significant positive impacts on teaching and learning with 90% of teachers indicating successful implementation of hands-on activities and project-based learning strategies. Considerable rates of adoption of local practices (80%) and the use of examples from the local culture (70%) were also reported by teachers. Teachers reported having moderate to major improvements in student engagement (80%), understanding about science (80%) and cultural awareness (70%). The teachers identified time management as the most challenging (90%), followed by resource limitations (80%), and difficulty assessing students (70%). The reported support needs were for professional development (90%), teaching materials (80%), and technology training (70%).

The pattern of teacher readiness illustrates the progressivity of the implementation of the PAPEDA Framework, with higher levels of readiness demonstrated for the first components (Provide, Performance) and challenges arising from the later components (Disseminate, Evaluate). The data obtained from this study are aligned with Rahayu's [27, 28] determinates of teacher preparedness for integrated learning approaches. The high frequency of implementation of hands-on activities and project-based learning is indicative of the successful application of the Ethno-STEM pedagogical strategies, which is corroborated by other research Hiqmah et al. [29] and Marufi et al. [16] on integrated learning ethnic-STEM approaches.

The challenges teachers reported about managing time and limits on resources is aligned with Aslam et al.'s [30] research that identifies common challenges in implementing educational reforms. The noted interest in professional development (90%) suggested teachers recognize they require specialized training for successfully integrating traditional knowledge with STEM education. This aligns with Darling-Hammond et al.'s [31, 32] research agenda pertaining to ongoing professional development in school and educational reforms, and education transformation. Teachers presented significant positive student experiences and engagement, despite implementation challenges.

Even with limited resources, this would suggest the implementation of the PAPEDA Framework meaningfully engages students and enhances experiences when implemented well. 60% of teachers reported modified assessment approaches to sufficiently address the issues related to the rated assessment processes indicating practitioner inventiveness regarding overcoming the challenges of implementation; an important consideration and component for educational innovations.

4.2. Students' perspectives

Data on student engagement showed high satisfaction levels with the PAPEDA Framework implementation (Table 48.2), with 84% at least satisfied with the learning approach. In-class activities were by far the highest-rated, at 4.3 on a scale of 1 to 5 satisfaction – cultural integration came in lower at an overall 3.8 satisfaction rating. Scientifically meaningfully ways (84%) and appreciation for local wisdom (76%) were both reported significantly improved, with an associated context to every day ways of living (74%).

The learning outcomes showed varying levels of engagement with aspects of the learning experience where localized learning experiences showed the highest engagement ratings (56% Always and 28% often), whereas 'learning about local wisdom' rated lower engagement levels overall (32% Always and 30% Often). Of the interviewed students, local experience reflected strong preferences for hands-on experiences, where 80% students preferred hands-on learning (e.g., experiments) and 75% of behalf of students stated they understood better when learning to connect components of cultural and scientific components together.

Table 48.2. Result of teacher implementation

Component	*Aspect*	*Averge Score / Average Percentage*
Top Strenghth	Scientific understanding	84%
	Program satisfaction	84%
	Local wisdom appreciation	76%
Implementation Success	Class activities satisfaction	4.3
	Learning materials satisfaction	4.2
	Daily life application	74%
Key Challenges	Cultural knowledge sharing	2.9
	Local context application	3.0
	Cultural connection	3.1
Priority Needs	More hands-on activities	70%
	Field trips	60%
	Tech integration	56%
Weakest Areas	Cultural knowledge sharing	2.9
	Simplified assessment	40%
	Knowledge sharing	64% success
Strongest Areas	Scientific understanding	84%
	Class activities	4.3
	Assignment completion	3.6

Source: Author.

Future recommendations of students supported greater levels of hands-on activities (70%), field trips (60%), and technology integration (56%), with 40% of students seeking simplification of assessment methods. The high levels of student satisfaction and improved scientific understanding and cultural appreciation provides evidence that the PAPEDA Framework successfully provides engaging, culturally relevant learning experiences. This aligns with the research of Boutte et al. [33] and Johnson & Elliot [34] which reported positively on the benefits of culturally integrated science learning. The strong engagement with hands-on activities, specifically supports Ariyatun [35], who found scaffolded culturally contextual learning practices helps with participation, understanding of concepts and student motivation.

The somewhat lower ratings for sharing cultural knowledge indicates an area to enhance the framework, likely through better learning collaboratively practices or community involvement practices. This supports the work of Izzah et al. [36], which focused on establishing richer connections between scientific principles and cultural experiences. Student suggestions for more field trips and hands-on activities as their preferred types of learning opportunities supports the use of project-based and context-rich learning elements to the Ethno-STEM model.

The strong improvement in scientific understanding (84%) and cultural appreciation (76%) shows the PAPEDA framework achieves its dual purpose of enhancing STEM education while strengthening cultural identity. This balance supports the work of Todd [37], which emphasized the significance of nurturing culture in education, and demonstrates the framework's effectiveness in providing traditional wisdom to scientific learning.

4.3. Administrators' perspectives

School leaders indicated that there were significant shifts in the structure of their institutions to facilitate the application of the PAPEDA Framework, such as 67% indicating alignment with policies, 83% indicating curriculum adjustments, and 83% indicating adjustments in assessment systems. All school leaders identified adjustments to resources to support implementation in the school and 83% indicated an improvement in academic results, while 100% reflected an increase in relations with their community (Table 48.3).

Support systems primarily consisted of conducting training workshops (100%), providing resources (83%), and having mentoring systems (67%). The primary monitoring systems were through classroom observations (100%) and measuring student performance (83%), with collecting student feedback and parent feedback and was reported by 67% of school leaders, and an indication for utilizing both student feedback and parent feedback for the development of academic policies.

Table 48.3. Result of school administrators' perspective

Component	*Aspect*	*Averge Score / Average Percentage*
Top Strenghth	Community relations	100%
	Cultural awareness	100%
	Academic performance	83%
Implementation Success	Resource adjustment	100%
	Curriculum modification	83%
	Student engagement	83%
Key Challenges	Technology integration	100%
	Resource allocation	83%
	Time management	83%
Priority Needs	Technology integration	100%
	Teacher training	100%
	Community partnerships	83%
Weakest Areas	Assessment standardization	67%
	Technical support	67%
	Resource development	67%
Strongest Areas	Community relations	100%
	Cultural awareness	100%
	Training workshops	100%

Source: Author.

Future considerations indicated 100% wanted to incorporate technology, 100% practitioner learning, 83% establish community partners, and 67% ongoing resource development. The major challenges noted included 100% in implementing technology,83% equity in allocating resources, 83% time management, and 67% instructional agree upon assessment.

The reported complete institutional changes shown by the administrators illustrate the significant organizational commitment needed for the successful implementation of the PAPEDA Framework. The overall focus on policy alignment and resource reallocation supports Anand [38] and McClain and Neri [39] in their finding on the role of institutional support in the implementation of educational reforms. Overall, the reported improvement in academic achievement (83%) and community relations (100%) indicate that the framework provides both an increase in learning achievement along with enhanced school and community relationships.

The focus given by the administrators to technological support (100%) and teacher training (100%) for future growth aligns with teachers' identified challenges and needs, and would indicate a level of administrator awareness related to implementation. The substantial support given to expand community partnerships (83%) is notable acknowledgement of the critical role of local knowledge holders in the Ethno-STEM approach, and supports research on effective community-school partnerships for culturally relevant education.

The reported challenge of standardizing assessments by 67% of the administrators aligns with teachers' identified challenge of the complexities of assessments as a significant barrier, indicating a systematic and collective challenge. The administrators' engagement of a thorough monitoring process especially in terms of student and parent inputs (67%) is a representation of a more responsible assessment process, accounting for more stakeholders' involvement perspectives.

4.4. *Integrated discussion: Cross-stakeholder perspectives*

Analysis of stakeholder data from teachers, students, and administrators shows consensus in perceptions of implementation successes and tensions. All groups identified the use of hands-on, culturally relevant activities as highly successful, and saw improvement both in scientific understanding and cultural appreciation. This is validating in showing the PAPEDA Framework's work to support the combination of traditional wisdom with science or STEM education.

All stakeholder groups reported similar challenges regarding time, the need for resources, and the need for standardization in assessments, demonstrating a shared understanding and acknowledgment that these challenges represent systemic issues requiring collective systematic efforts toward solutions, rather than issues of implementation. The agreement between the developments of identified needs for teachers (professional development, resources), designed into administrators' future plans (training teachers, supporting teachers resources), represents a promising opportunity for solutions as institutional support continues.

The demonstrated implementation improvement reported by administrators regarding community engagement (100%) supporting teachers' implementation of community experts (70%) and student enjoyment of cultural connections (76%) are supporting the success of the PAPEDA Framework in developing school and community partnerships. This finding is also aligned with the findings of Eden et al. [40] regarding the value of fostering an inclusive education through cultural relevance.

4.5. *Future directions*

In light of the key findings from this preliminary study, there are at least two major areas for future research. First, all studies should focus on developing comprehensive professional development programs aimed at the Disseminate and Evaluate components, where teachers indicated a lower sense of preparedness around effective knowledge sharing, specifically workshops focused on assessment design that bridges cultural knowledge with the scientific principles, and effective strategies for knowledge sharing. Since 70% of teachers indicated they faced problems with the assessment's standardization as a challenge, future studies should focus on the development and validation of culturally relevant evaluation frameworks, that honor indigenous epistemologies while adhering to academic rigor. This could involve a collaborative study with assessment experts, knowledge-holders from local communities, and science educators in the same space to develop standardized assessments that are culturally relevant.

The second major area for future research is resource development and use, as 80% of teachers indicated a lack of culturally relevant resources. Future studies should examine effective models of creating, distributing, and using culturally relevant STEM resources into a variety of educational contexts, including the use of technology due to the geography factor in school location. This may include an open-access repository of Ethno-STEM learning resources organized by cultural context and scientific concept. Additionally, the teachers (70%) and student (56%) response to the use of technology was an important theme they identified. Therefore, there is a possibility for future studies of how to effectively integrate appropriate technology into the PAPEDA Framework implementation, specifically in low-resourced environment.

5. Conclusion

The examination of the implementation of the PAPEDA Framework through an Ethno-STEM approach shows its clear potential for developing science education that is culturally responsive, and bridges the gap between traditional understanding and formal STEM education. All stakeholders reported positive results of the framework implementation (i.e., 84% of students reported enhanced scientific understanding; 76% of students reported increased appreciation for their culture; 90% of teachers successfully delivered hands-on learning experiences; 100% of school administrators reported stronger community relationships) confirming the framework's influence to improve educational experiences with teacher support. Given the barriers surrounding time and resources (i.e., 90% of teachers; 80%), and standardized assessment (i.e., 70%), the framework made possible learning experiences that were meaningful and respectful of students' cultural contexts, while still being scientifically rigorous. Supports or strategies need to be identified for effective implementation and scaling (i.e., professional development, assessment practices, access to resources, and access to technologies) of the framework. The study furthers the conversation of culturally responsive science education with a replicable model that provides academic excellence and cultural relevance. The study highlights that the integration of traditional ways of knowing and modern science education can build student learning experiences, while maintaining values of cultural identity.

References

[1] Sharma, M., & Solanki, P. K. (2024). A new dawn in education: The promise and potential of NEP 2020 for students and teachers. *International Journal of Science and Research Archive*, *13*(1), 1864–2070.

[2] Choudhuri, S. S. (2024). National Education Policies and the Impact on Overall Development. *Journal of Educational Research and Policies*, *6*(9), 189–191.

[3] Pathak, J. S., & Maity, S. (2024). Innovative Assessment Practices Aligned with India's National Educational Policy-202*Asian Journal of Education and Social Studies*, *50*(9), 241–253.

[4] Asmaningrum, H. P., Sathasivam, R. V., & Shahali, E. H. B. M. (2025). Cultural Roots, Scientific Fruits: Exploring the PAPEDA Framework in Ethno-STEM Teaching. *Academic Journal of Interdisciplinary Studies*, *14*(1), 2025.

[5] Edelen, D., & Bush, S. B. (2021). Moving Toward Inclusiveness in STEM With Culturally Responsive Teaching. *Kappa Delta Pi Record*, *57*(3), 115–119.

[6] Nugraheni, F. S. A., Sari, M. W., Wati, I. K., Suciati, Widyastuti, A., & Kamaliah, K. (2023). Indigenous Knowledge and Its Potential for Junior High School Ethno-STEM Learning. *2023*, 11001. doi: 10.1063/5.0106474.

[7] Jaudinez, A., & Joaquin, Ma. N. (2024). Ethno-Science, Technology, Engineering, and Mathematics (EthnoSTEM) Ideas in the Sama Mat-Weaving. *Southeastern Philippines Journal of Research and Development*, *29*(2), 161–190.

[8] Kulal, A., N. A., Dinesh, S., Bhat, D. C., & Girish, A. (2024). Evaluating the Promise and Pitfalls of India's National Education Policy 2020: Insights from the Perspectives of Students, Teachers, and Experts. *Sage Open*, *14*(4). doi:10.1177/21582440241279367.

[9] Sardar, S., Saha, S., & Akhand, M. H. (2024). The Challenges and Opportunities of NEP-2020 in the Reference of Inclusive Education in India: An Overview. *International Journal for Multidimensional Research Perspectives*, *2*(10), 45–56.

[10] Sharma, S. V. (2024). Learning the Concepts of Magnetic Effects of Current through Experiential Learning Approach as Envisaged in National Education Policy 20*International Journal of Science and Research (IJSR)*, *13*(11), 35–41.

[11] Sharma, S., & Rai, R. (2024). Role of Holistic Education Practices to Enrich the Indian Knowledge System in Contemporary Education. *International Journal of Advanced Research (Indore)*, *12*(05), 659–666.

[12] Prabhakar, A. K. (2023). The Intersection of Traditional Wisdom and Modern Education: Unpacking the Potential of Folk Pedagogy in the Context of India's National Education Policy 20*The Creative Launcher*, *8*(2), 115–121.

[13] Roy, M. (2022). A Critical Study on the Holistic and Multidisciplinary Approach of National Educational Policy 2020 (NEP 2020) in India. *International Journal For Multidisciplinary Research*, *4*(6). doi: 10.36948/ijfmr.2022.v04i06.971.

[14] Asmaningrum, H. P., Sudarmin, & Irianto, O. (2022). Development of Syntax for Learning Chemistry Based On Etno-STEM to Build Scientific Literacy Skills and Critical Thinking Skills. *6*(6), 275–283.

[15] Gumilar, N. M. A. R., Sudarmin, S., Marwoto, P., & Wijayati, N. (2022). Ethno-STEM Research Trends Through Bibliometric Analysis on Science Learning in Elementary School. *Unnes Science Education Journal*, *11*(3), 166–172.

[16] Marufi, M., Ilyas, M., Winahyu, W., & Ikram, M. (2021). An Implementation of Ethno-STEM to Enhance Conceptual Understanding. *Al-Jabar : Jurnal Pendidikan Matematika*, *12*(1), 35–44.

[17] Sudarmin, S., Pujiastuti, R. S. E., Asyhar, R., Prasetya, A. T., Diliarosta, S., & Ariyatun, A. (2023). Chemistry project-based learning for secondary metabolite course with ethno-STEM approach to improve students' conservation and entrepreneurial character in the 21st century. *JOTSE*, *13*(1), 393–409.

[18] Sumarni, W., Mursiti, S., & Sumarti, S. S. (2020). Students' Innovative and Creative Thinking Skill Profile in Designing Chemical Batik After Experiencing Ethnoscience Integrated Science Technology Engineering Mathematic(Ethno-STEM) Learnings. *Journal of Physics: Conference Series*, *1567*(2), 0–7.

[19] Awaliyah, N. A., Widiyawati, Y., & Nurwahidah, I. (2022). Development of HOTS Science Test: Ethno-Science Technology Engineering and mathematics (STEM) based on Javanese Gamelan. *Journal of Science Education Research*, *6*(2), 81–90.

[20] Agussuryani, Q., Sudarmin, S., & Sumarni, W. (2021). STEM Integrated Ethnoscience-Based Vocational School Science Teaching Materials in Improving Student's HOTs. *Berkala Ilmiah Pendidikan Fisika*, *9*(2), 185.

[21] Bramastia, B., et al. (2023). Effectiveness of EthnoSTEM-Based Science Learning to Improve Junior High School Students' Science Literacy Ability. *Jurnal Penelitian Pendidikan IPA*, *9*, no. SpecialIssue, 332–337.

[22] Batarchuk, D. S. (2019). The Role Of Ethno-Cultural Knowledge In The Development Of Multicultural Student's Personality. *Cognitive-Crcs*, 9–20. doi: 10.15405/epsbs.2019.12.02.2.

[23] Vaithianathan, V., Shastri, D. S., Nandakumar, V., Misba, M., Kumar, P. S. V. V. S. R., & Boopathi, S. (2024). NEP Policy Implementation Strategies in Education in India. 325–353. doi: 10.4018/979-8-3693-3587-1.ch013.

[24] Sethi, J. (2024). Implications of Nep2020 for Education. *International Journal of Innovative Science and Research Technology (IJISRT)*, 424–429. doi: 10.38124/ijisrt/IJISRT24NOV283.

[25] Mwansa, B., et al. (2022). Research Methodological Choice: Explaining Research Designs; Qualitative and Quantitative Sample Size Determination, Sampling, Data Collection, and Analysis Techniques. *Journal of Marketing and Consumer Research*, Nov. 2022, doi: 10.7176/JMCR/87-06.

[26] Toyon, M. A. S. (2021). Explanatory sequential design of mixed methods research: Phases and challenges. *International Journal of Research in Business and Social Science (2147–4478)*, *10*(5), 253–260. doi:10.20525/ijrbs.v10i5.1262.

[27] Sichel, C. E., & Connors, E. H. (2022). Measurement feedback system implementation in public youth mental health treatment services: a mixed methods analysis. *Implement Sci Commun*, *3*(1), 119.

[28] Rahayu, W. E., & Sudarmin, S. (2015). Pengembangan Modul Ipa Terpadu Berbasis Etnosains Tema Energi Dalam Kehidupan Untuk Menanamkan Jiwa Konservasi Siswa. *Unnes Science Education Journal*, *4*(2), 2015, doi: 10.15294/usej.v4i2.7943.

[29] Hiqmah, N., Rienovita, E., Al-Latief, I. S., Sholehuddin, S., & Santosa, T. A. (2023). Effectiveness of Ethno-STEM Based Chimistry to Improve Students Critical Thinking Skills. *Jurnal Penelitian Pendidikan IPA*, *9*, no. SpecialIssue, 72–79. doi:10.29303/jppipa.v9ispecialissue.6422.

[30] Aslam, S., Alghamdi, A. A., Abid, N., & Kumar, T. (2023). Challenges in Implementing STEM Education: Insights from Novice STEM Teachers in Developing Countries. *Sustainability (Switzerland)*, *15*(19). doi:10.3390/su151914455.

[31] Darling-Hammond, L., Hyler, M., & Gardner, M. (2017). Effective Teacher Professional Development. doi:10.54300/122.311.

[32] Darling-Hammond, L., Flook, L., Cook-Harvey, C., Barron, B., & Osher, D. (2020). Implications for educational practice of the science of learning and development. *Applied Developmental Science*, *24*(2), 97–140. doi: 10.1080/10888691.2018.1537791.

[33] Boutte, G., Kelly-Jackson, C., & Lee Johnson, G. (2010). Culturally Relevant Teaching in Science Classrooms: Addressing Academic Achievement, Cultural Competence, and Critical Consciousness. *International Journal of Multicultural Education*, *12*(2), 1–20.

[34] Johnson, A., & Elliott, S. (2020). Culturally Relevant Pedagogy: A Model To Guide Cultural Transformation in STEM Departments. *Journal of Microbiology & Biology Education*, *21*(1). doi: 10.1128/jmbe.v21i1.2097.

[35] Ariyatun, A. (2021). Analysis of Ethno-STEM Integrated Project Based Learning on Students' Critical and Creative Thinking Skills. *Journal of Educational Chemistry (JEC)*, *3*(1). doi:10.21580/jec.2021.3.1.6574.

[36] Izzah, S. N., Wiyanto, S., & Wardani, S. (2023). Analysis of Science Concept Mastery, Creative Thinking Skills, and Environmental Attitudes After Ethno-STEM Learning Implementation. *International Journal of Instruction*, *16*(3), 777–796.

[37] Todd, W. F., Towne, C. E., & Clarke, J. B. (2023). Importance of centering traditional knowledge and Indigenous culture in geoscience education. *Journal of Geoscience Education*, *71*(3), 403–414. doi:10.1080/10899995.2023.2172976.

[38] Anand, N. (2020). A Need to Support Teacher Diversity Through Culturally Relevant STEM Education at the Elementary School. In *The IAFOR International Conference on Education* [Online]. Available: www.iafor.org

[39] McClain, J., & Colina Neri, R. (2022). 'A Tale of Two Classrooms': Designing Culturally-Relevant Hip Hop Curriculum to Support STEM Identity of Underrepresented Students. *Northwest Journal of Teacher Education*, *17*(1). doi: 10.15760/nwjte.2022.17.1.3.

[40] Eden, C. A., Chisom, O. N., & Adeniyi, I. S. (2024). Cultural Competence in Education: Strategies for Fostering Inclusivity and Diversity Awareness. *International Journal of Applied Research in Social Sciences*, *6*(3), 383–392.

[41] Nobles, P., Ali, S., & Chivers, H. (2011). Improved Estimation of Trilateration Distances for Indoor Wireless Intrusion Detection. *Journal of Wireless Mobile Networks, Ubiquitous Computing, and Dependable Applications*, *2*(1), 93–102.

[42] Moriyama, H., Yamauchi, T., Sato, M., & Taniguchi, H. (2022). Improvement and Evaluation of a Function for

Tracing the Diffusion of Classified Information on KVM. *Journal of Internet Services and Information Security*, *12*(1), 26–43. https://doi.org/10.22667/JISIS.2022.02.28.026

[43] Palash, P. S., & Dhurvey, P. (2024). Analysis of Flyash Aggregate Behavior in Geopolymer Concrete Beams Using Method of Initial Functions (Mathematical Programming). *Archives for Technical Sciences*, *2*(31), 168–174. https://doi.org/10.70102/afts.2024.1631.168

[44] Renjith, K. V., & Sudhan, P. (2016). Fairness and Network Central Locations based Cooperative Caching Scheme for Efficient Data Access in Disruption Tolerant Networks. *International Journal of Advances in Engineering and Emerging Technology*, *7*(4), 292–303.

[45] Baggyalakshmi, N., Bava Harini, G., & Revathi, R. (2023). Internal Sales Management of Heating Materials and Heating Systems. *International Academic Journal of Science and Engineering*, *10*(2), 140–152. https://doi.org/10.9756/IAJSE/V10I2/IAJSE1018

[46] Ibrahim, A. A. A. (2024). Features and functional components in the use of artificial intelligence in human capital development (higher education as a model). *ESTIDAMAA*, *2024*, 1–6.

49 Decentralized edge AI for real-time environmental monitoring and sustainability

Ramesh S.[1,a], *R. Geetha*[1,b], *P. Thilagavathi*[2,c], *S. Vidhya*[1,d], *and J. Nithyashri*[3,e]

[1]Assistant Professor, Department of Computing Technologies, School of Computing, SRM Institute of Science and Technology, Kattankulathur, Chennai, Tamil Nadu, India
[2]Associate Professor, Department of Computer Science and Engineering, Aarupadai Veedu Institute of Technology, Vinayaka Mission's Research Foundation (DU), Paiyanoor, Chennai, Tamil Nadu, India
[3]Assistant Professor, Department of Computing Technologies, School of Computing, Faculty of Engineering and Technology, SRM Institute of Science and Technology, Kattankulathur, Tamil Nadu, India

Abstract: Revolutionizing data processing and intelligence at the network edge, decentralized edge artificial intelligence (AI) minimizes latency, bandwidth utilization, and privacy threats while enabling real-time processing. The incorporation of decentralized learning is the subject of this article. Techniques, such blockchain technology and federated learning, to make Edge AI systems more secure, scalable, and robust. Our suggested method leverages distributed computing structures to ensure data sovereignty during model training on edge devices effectively. Its implications on various domains, including healthcare, smart cities, and industrial automation, the article evaluates the feasibility of decentralized Edge AI. The viability and advantages of a decentralized solution to latency-sensitive inference, improved model performance, and increased security are established through trial outcomes. The results of this research bring out the potential of distributed AI and the transformative ability of decentralized Edge AI in the world of smart systems.

Keywords: Artificial intelligent, blockchain, environment, decentralized, real time, monitoring, privacy

1. Introduction

This study seeks to understand how artificial intelligence (AI) can very much enhance the precision of environmental data. Further, the investigation aims to find out how far modern technology can go into attaining the following issues on privacy, correctness, and use of data in real-time tracking and predictive analysis towards achieving ecological management. The current environmental monitoring situation is analyzed with secondary data, reviewing published academic papers, case studies, and company reports focusing on artificial intelligence. For example, though, some breakthroughs in remedying environment problems have been possible with technology development towards making environmental monitoring data more open, accessible, and precise. Environmental data such as air quality indices, water pollution, deforestation, and carbon emissions are important in formulating policies, sustaining development, and climate change mitigation [1, 2]. According to, dealing with vast data sets, ensuring that the data is accurate, combating fraud, and all the real-time monitoring all make the trust and verification of the data more difficult. Therefore, it is necessary to develop solutions that offer reliable, immutable, and audit-proof environmental data. At the core of environmental monitoring is the act of observing, measuring, and assessing the natural world or its systems and constituent parts. Traditional environmental monitoring methods involve statistical analysis, laboratory testing, or manual sampling, assess environmental conditions and identify trends that could pose risks to ecosystems or human populations. However, these approaches are often costly, time-consuming, and lack accuracy. The conventional environmental monitoring systems have some advantages but are very inefficient in some aspects. One of the major challenges is that these processes are very costly. Laboratory testing and manual sampling are also relatively costly, as it requires specialists, advanced equipment, and expensive chemicals [3]. Following a natural disaster or an incident of pollution, it may take months or even weeks to get laboratory tests, which can significantly hinder decision-making and response efforts in an emergency. With the likelihood of errors during manual sampling and laboratory work, the uniformity of data gathering and processing can actually be compromised.

Numerous environmental issues confronting modern society pose a threat to public health, ecosystems, and the global economy [4]. Global warming, water poisoning, deforestation, and biodiversity loss are just a few of the environmental disasters that are becoming worse due to

[a]ramesh.isaiah@gmail.com, [b]geethar1@srmist.edu.in, [c]thiluxpriya499@gmail.com, [d]vidhyas2@srmist.edu.in, [e]drjnithyashri@gmail.com

DOI: 10.1201/9781003684459-49

urbanization, industrialization, and irresponsible resource extraction. These problems are occurring at a rate that has never been observed previously. For instance, as global temperatures rise, weather patterns are changing, increasing the frequency of extreme weather events including floods, droughts, and storms. The impact on agricultural productivity and water security is multiplied by these catastrophic events. Decisions in the environmental sciences must be made promptly and accurately if we are to tackle these complicated and interconnected problems. Accurate trend forecast, proactive reactions to new risks, and real-time environmental monitoring are the three pillars upon which effective mitigation and management initiatives rest. The exponential growth of both environmental change and the quantity of data generated is making traditional monitoring techniques, such as spotty data collecting and manual processing, inadequate. Therefore, cutting-edge technical solutions are desperately needed to enhance the precision, effectiveness, and extensibility of environmental data collection and processed.

2. Literature Review

In order to investigate how AI is being used in environmental research, namely in water quality and climate monitoring, a comprehensive literature analysis was conducted. Review findings include prominent directions in the field, practical uses, current obstacles, and promising new directions. future innovations that could take place. To round out the search, we looked through conference proceedings, industry reports (WHO, EPA, IEA), and electronic databases such as IEEE Xplore, SpringerLink, Science Direct, Wiley Online Library, and PubMed. Search terms such as 'AI agents,' 'IoT,' 'environmental monitoring,' and 'predictive analytics' were utilized, along with Boolean operators, to narrow down the outcomes [5]. With a focus on peer-reviewed research, we included studies published between 2010 and 2024 that discussed the integration of AI and the Internet of Things in the field of environmental sciences. Research that lacked empirical validation, was published in an obsolete format, or was not based on scholarly principles were not considered. To guarantee a comprehensive view, research diversity from around the world was taken into account. Prior to doing full-text analysis, titles and abstracts were evaluated using selection criteria. The data extracted covered the study's aims, methods, AI-IoT technologies, important results, and difficulties. We believe that AI can assist in developing culturally appropriate organizational procedures and individual habits to minimize the resources and energy consumption of human operations, as proposed by in their method for stimulating revolutionary research and practical AI solutions for environmental sustainability [6]. The fundamental value of AI will not lie in its capacity to help society reduce its water, land, and power consumption levels, but rather in its promotion and facilitation of environmental management. Research on artificial intelligence for environmental projects has been extensively documented in the scientific literature.

Despite the promising potential of blockchain and AI solutions for environmental monitoring, there is still need for improvement in both research and implementation [7]. Few research have investigated the possibility of combining Blockchain with AI to address issues with environmental data monitoring and integrity. Research showing a scalable system that successfully integrates the two technologies to provide reliable data and practical experience is noticeably lacking. There doesn't appear to be much use for AI-driven blockchain systems in addressing environmental concerns, and it's not obvious how to apply these technologies without reducing their efficacy and environmental impact [8].

AI model to monitor the environment, it's important to think about the needed computer power, skills, and data. Some AI models can effectively learn from smaller data sets without being too expensive, but others require a lot of data and computer power [9]. Support Vector Machines (SVMs) are particularly good for managing complex data with many layers and understanding how different factors work together [10]. They are versatile and can handle both text and image data. For example, SVMs have been successful in sorting images to recognize animals and detect forest destruction [11]. Support Vector Regression (SVR), a specific type of SVM, is an effective tool for predicting environmental factors when dealing with regression problems in environmental monitoring [12].

3. Methodology

AI is widely used by now in a wide range of areas, from ecological monitoring and has radically altered the way of processing and analyzing data. As a result, AI acts as an efficient tool for the problems arising from the acquisition of environmental data from diverse sources and their characteristics which are, at their core, their size, diversity, and the collective issue. Automated analysis of this kind could be a source of tremendously useful trends, patterns, and insights when it comes to the analysis of very large datasets. The quantity and variability of data make monitoring the environment challenging. Environmental datasets composed of citizen science platforms, water, air quality, and remote sensing feed into ecological datasets that could be numerical, graphical, textual, or geospatial in nature. Processing and analyzing large diverse datasets by hand takes time and often leads to mistakes. AI systems might automate this task, as they can handle complex multi-layered information. Machine learning programs can look at past data to find links between factors like temperature, humidity, and pollution levels helping to predict future events. AI-driven data analysis makes it easier to spot patterns and unusual events in environmental data as it comes in. Machine learning models use both guided and unguided learning methods to spot weather changes or sudden pollution increases that don't fit the usual pattern. AI models might catch air quality issues that show pollution allowing for quicker cleanup [13]. AI also helps protect wildlife by noticing changes in animal or environmental groups. Spotting these unusual events could help officials stop or lessen the impact of environmental problems.

AI plays a key role in watching land use, forest loss, water, and climate change. It can boost our ability to work with satellite images and data from remote sensors. Environmental risks like pollution, urban sprawl, and deforestation can be more easily detected, through high-resolution images and the use of convolutional neural networks (CNNs). This facilitates the opportunity for global environmental monitoring and provides important ecological data to local and global policymakers. The application of AI to the processing of environmental data is still in its early stages, but it shows great promise. Problems with data availability, standardization, and quality arise when ecological data must be more efficient, comprehensive, and consistent. Understanding the decision-making or prediction processes of complex algorithms is challenging due to the interpretability of AI models. To find solutions, data scientists, ecologists, and lawmakers will need to collaborate. Through the automation of data processing, identification of abnormalities, anticipation of trends, and optimization of monitoring systems, artificial intelligence can enhance environmental data analysis. AI has the potential to help environmental stakeholders gain a better understanding of the environment and make data-driven decisions to combat climate change, pollution, and biodiversity loss. Environmental monitoring systems might be enhanced with the help of artificial intelligence and blockchain technology, which would make data more secure, trustworthy, and usable [14]. The field of environmental research has recently seen the rise of AI agents as crucial tools for progress. These agents optimize decision-making processes, enhance monitoring, and permit analysis by using their different capabilities. All of these are critical for solving complicated environmental concerns. The definitions, classifications, functionality, and environmental applications of artificial intelligence agents are detailed in this chapter (Figure 49.1).

3.1. AI applications in environmental monitoring

In the fields of ecosystem management, habitat preservation, and animal monitoring, artificial intelligence has been an invaluable tool in the fight to protect biodiversity. explores the many uses of machine learning in conservation, highlighting how it might improve approaches to managing resources. There is a shortage of empirical case studies that demonstrate the efficiency of AI in real-world conservation contexts, while the paper does highlight practical uses. To enhance its impact on biodiversity research, AI must overcome obstacles connected to data accessibility, ethical considerations, and AI acceptance.

4. Results and Discussion

Based on the Figure 49.2 data, a triple bar graph was formed, with the X-axis indicating the soil health parameters, biodiversity, air quality, and water quality of environmental monitoring. Three bars are used to show the influence of each of these parameters on accuracy, efficiency, and cost reduction on a scale of 0 to 100. This infographic thus propagates possible applications for AI, including enhancing monitoring processes and reducing operational costs.

Several real-world applications, including smart cities, environmental sensing, and factory automation, were used to evaluate the proposed decentralized Edge AI architecture. In order to gauge the efficacy of decentralized learning methodologies, metrics like energy consumption, model correctness, latency, and security resilience were measured. An improvement of up to 92% in real-time environmental data classification was achieved by federated learning, which allowed for effective model training in a distributed setup of edge nodes. While maintaining data privacy and excellent model performance, decentralized learning reduced data transmission

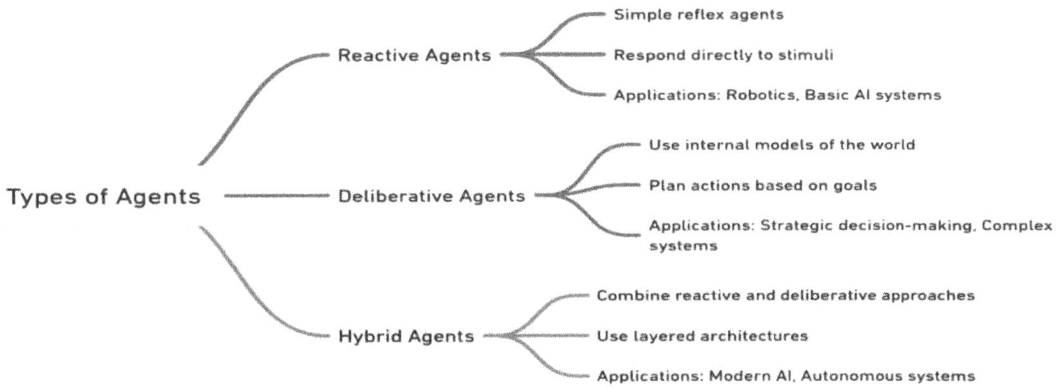

Figure 49.1. Classification of AI agents in environmental sciences.

Source: Author.

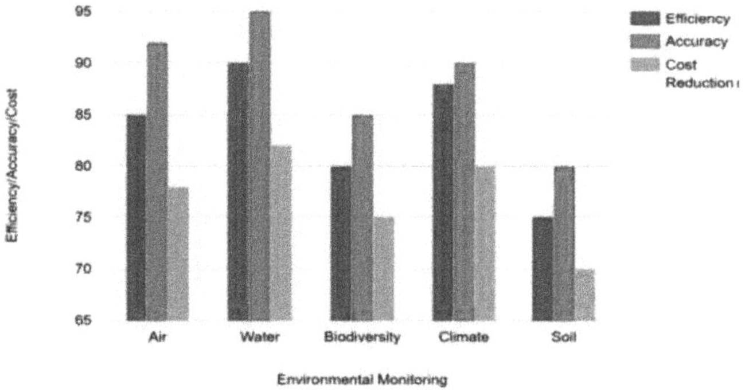

Figure 49.2. AI-Impact on data analysis efficiency, accuracy, and cost.

Source: Author.

overhead by 60% compared to traditional centralized AI systems. Compared to cloud-based processing, decentralized inference at the edge significantly reduced decision-making latency, improving it by 35%. Dynamic task scheduling and power-aware optimization are two examples of energy-efficient edge computing technologies that contributed to a 40% reduction in energy consumption, which in turn extended the operational life of IoT sensors. Integrating blockchain technology enhanced security on edge devices and allowed for tamper-evident data management. A decentralized ledger system increased system dependability in the face of cyberattacks by protecting data from illegal access and manipulation. Data integrity and traceability are improved by 95% compared to standard cloud-based systems, according to experimental results. Decentralized Edge AI has the potential to create intelligent systems that are both sustainable and scalable, as shown by the outcomes [15]. Because the methodology doesn't rely as much on centralized cloud infrastructure, it has a smaller carbon footprint, lower operating expenditures, and better real-time decision-making capabilities for different applications. Despite its benefits, decentralized Edge AI has a number of obstacles, such as computer resource limitation, model convergence difficulty in federated learning, and edge device heterogeneity. Optimization of distributed AI algorithms, enhancement of edge node interoperability, and development of low-power hardware should be the emphasis of future effort to significantly improve system performance.

5. Challenges and Future Directions

In that regard, the future of artificial intelligence associated with environmental monitoring seems bright. With the latest improvements in AI algorithms, more efficient data collection methods, and increased processing capabilities, the tendency is for deployment into practice to enhance the efficacy and accuracy of AI models. Future research efforts should be aimed at making AI models clearer, solving issues related to ethics and safety, and developing robust AI models that can handle complex and diverse environmental data. Researchers must resolve and work closely with legislators and agents of commercial interest to consider ethical innovations and ensure that fair use is guaranteed for artificial intelligence in environmental monitoring. Finally, this paper excels because of its depth and practical treatment of AI in environmental monitoring, its consideration regarding the present and future scenarios in the subject area, and its technical and ethical issues coverage in balanced ways. This publication is basically a treasure trove of information to researchers, policymakers, and practitioners who have in mind strategies that could guide the ethical use of AI for environmental monitoring. Decentralized Edge AI is challenged in terms of computational resource limitation, edge device heterogeneity, and model convergence problem in federated learning. Emphasis optimizing distributed AI algorithms, maximizing interoperability of edge nodes, and designing low-power hardware for further improving the system performance these are the feature work.

6. Conclusion

AI could be highly beneficial in improving environmental monitoring. It will be help us keep on our personal carbon footprints as well as the quality of the air, water, and land around us. These innovations substantially improve public health, environmental protection, and the management of resources. Analyzing data in real-time helps make better choices and boosts sustainability efforts with the help of AI. But there are some problems that come with using AI. For example, data centers use a lot of energy, and making AI can hurt the environment, particularly when it comes to mining rare earth materials. These issues might lead to more greenhouse gas emissions, deforestation, and pollution. On the bright side, pairing AI with blockchain can address concerns about data security and trustworthiness. In order to facilitate proactive decision-making, analytics driven by AI can sift through enormous environmental datasets, identify anomalies, and offer accurate predictions. Before AI and blockchain are widely employed, we still have a ways to go in terms of reducing energy usage, fixing integration issues, and improving computing speed. Implementation of requirements, privacy of data, and development of energy-efficient AI models all demand modifications in regulation. To create sustainable and scalable environmental monitoring systems using blockchain and AI, collaboration across the public, corporate, and academic sectors is crucial. With the development of these interconnected technologies, a more sustainable and environmentally friendly future. Decentralized and self-adaptive monitoring systems are necessary for effective administration of edge environments. But because they are based on a centralized architecture, existing methods don't work in very uncertain edge cases. Because of this setup, data storage and retrieval will be sluggish and there will be bottlenecks when an error occurs.

References

[1] Addimulam, S., Rahman, K., Karanam, R. K., & Natakam, V. M. (2021). AI-Powered Diagnostics: Revolutionizing Medical Research and Patient Care. *Technology & Management Review*, *6*, 36–49.

[2] Muralidharan, J. (2024). Innovative materials for sustainable construction: A review of current research. *Innovative Reviews in Engineering and Science*, *1*(1), 16–20. https://doi.org/10.31838/INES/01.01.04

[3] Dressing, S. A., Meals, D. W., Harcum, J. B., Spooner, J., Stribling, J. B., Richards, R. P., & O'Donnell, J. G. (2016). Monitoring and Evaluating Nonpoint Source Watershed Projects, 4United States Environmental Protection Agency, Washington, DC, USA, 2016-06.

[4] Jaiswal, H., & Pradhan, S. (2023). The Economic Significance of Ecosystem Services in Urban Areas for Developing Nations. *Aquatic Ecosystems and Environmental Frontiers*, *1*(1), 1–5.

[5] Shih, C. S., Hsieh, W. Y., & Kao, C. L. (2019). Traceability for Vehicular Network Real-Time Messaging Based on Blockchain Technology. *Journal of Wireless Mobile Networks, Ubiquitous Computing, and Dependable Applications*, *10*(4), 1–21.

[6] Surendar, A. (2024). Emerging trends in renewable energy technologies: An in-depth analysis. *Innovative Reviews in Engineering and Science*, *1*(1), 6–10. https://doi.org/10.31838/INES/01.01.02

[7] Kommineni, H. P. (2020). Automating SAP GTS Compliance through AI-Powered Reciprocal Symmetry Models. *International Journal of Reciprocal Symmetry and Theoretical Physics*, *7*, 44–56. https://upright.pub/index.php/ijrstp/article/view/162

[8] Cheng, L. W., & Wei, B. L. (2024). Transforming smart devices and networks using blockchain for IoT. *Progress in Electronics and Communication Engineering*, *2*(1), 60–67. https://doi.org/10.31838/PECE/02.01.06

[9] Zhou, X., Chai, C., Li, G., & Sun, J. (2020). Database meets artificial intelligence: A survey. *IEEE Transactions on Knowledge and Data Engineering*, *34*(3), 1096–1116.

[10] Manikandan, G., & Abirami, S. (2021). An efficient feature selection framework based on information theory for high dimensional data. *Applied Soft Computing*, *111*, 107729.

[11] Zagajewski, B., Kluczek, M., Raczko, E., Njegovec, A., Dabija, A., & Kycko, M. (2021). Comparison of random forest, support vector machines, and neural networks for post-disaster forest species mapping of the krkonoše/karkonosze transboundary biosphere reserve. *Remote Sensing (Basel)*, *13*(13), 2581.

[12] Ye, Z., Yang, J., Zhong, N., Tu, X., Jia, J., & Wang, J. (2020). Tackling environmental challenges in pollution controls using artificial intelligence: A review. *Science of the Total Environment*, *699*, 134279. https://doi.org/10.1016/j.scitotenv.2019.134279

[13] Kothapalli, S., Manikyala, A., Kommineni, H. P., Venkata, S. G. N., Gade, P. K., Allam, R., Sridharlakshmi, N. R. B., Boinapalli, N. R., Onteddu, A. R., & Kundavaram, R. R. (2019). Code Refactoring Strategies for DevOps: Improving Software Maintainability and Scalability. *ABC Research Alert*, *7*(3), 193–204. https://doi.org/10.18034/ra.v7i3.663

[14] Hui, H., An, X., Wang, H., Ju, W., Yang, H., Gao, H., & Lin, F. (2019). Survey on Blockchain for Internet of Things. *Journal of Internet Services and Information Security*, *9*(2), 1–30.

[15] Ziwei, M., & Han, L. L. (2023). Scientometric Review of Sustainable Land Use and Management Research. *Aquatic Ecosystems and Environmental Frontiers*, *1*(1), 21–24.

50 Building relations with common appetite: India's culinary diplomacy with South Asia and the world

Mouli Chakraborty[1,a], Shilpi Bhattacharya[2,b], and Ajay Kumar Shukla[3,c]

[1]Assistant Professor in Political Science, Department of Arts and Humanities, Kalinga University, Raipur, India
[2]Professor in English, Dean in Department of Arts and Humanities, Kalinga University, Raipur, India
[3]Professor in Hindi, Department of Arts and Humanities, Kalinga University, Raipur, India

Abstract: Culinary diplomacy is a less spoken form of diplomacy in the International forum. Indian cuisine is historically diverse and rich and therefore has huge potential to showcase as soft power in international relations. India's diet plan matches with other South Asian countries due to the geographical proximity, cultural and religious similarities. Over the years, it has been observed that food acted as a medium of bridging the gap between countries. Be it the mango diplomacy of Sheikh Hasina or PM Bajpayee hosting dinner during the Agra summit. The choice of food and dishes also can create deep cracks within nations such as the consumption of beef in Islamic countries and the right wing Indian government opposing it. Further the paper discussed the true identity of Indian cuisine and the relevance of local identity in upholding the traditions of these innovative dishes.

Keywords: Culinary, diplomacy, South Asia, cuisines, summits

1. Introduction

Food and Diplomacy are the two words that somehow don't secure a mainstream place in international relations [1]. But somehow culinary diplomacy has huge potential to reflect on the heterogeneous culture and traditions across the world [2]. Both economically and culturally, culinary diplomacy can facilitate trade and foster stronger cultural ties [3].

Defining South Asia as a region has stronger multicultural roots and thereby opens the gate towards diversity in cuisines. From ancient times, gastro-diplomacy has remained part and parcel of international relations. Such as the use of the Silk Route that connected Europe and Asia in probable trade of exports and imports of herbs and spices [4].

The Indian Subcontinent during the medieval era have practised experimentation with diverse styles of cooling [5]. Particularly, the Mughal Empire opened the gate for diverse routes and grandeurs, which led to the ultimate food legacy [6]. The arteries of Trade of commerce, such as the Grand Trunk Roads, Silk Roads, and other Maritime routes connecting the Arabian Sea and the Bay of Bengal, have served as the main links for food exchange within and outside the landscape of South Asia [7].

1.1. *South Asia and rich diversity of meals*

South Asia as a region is among the most populated and diverse regions. Present-day South Asia comprises India as the largest and geopolitically dominant country, Pakistan, Bangladesh, Nepal, Bhutan, Sri Lanka, Maldives, and Afghanistan. The culinary practices of the region are particularly dominated by the physical and religious preferences of the region.

Each country is situated in a strategic geographical position and their diet depends on the available food resources. Like the coastal Indian subcontinent, including the regions of Odisha, Bengal, Districts of Ratnagiri, Sindhudurga, Mumbai in Maharashtra, Andhra, And Southern provinces are dependent on variations in fishes, crabs, and other marine species. Even Bangladesh and Sri Lanka are hard-core fish trading and consuming States. In the Northern regions of the country, due to the rich production of wheat, the population widely consumes *Chapati* (thin layered bread). But in the eastern part of South Asia, the soil is rich for paddy cultivation, and thereby, people prefer rice-made dishes.

1.2. *Religion, culture, and food choices*

South Asia as a region has diverse religious beliefs and practices and so are the food choices. Major Religions, including different sects of Hinduism like Shaivism, Vishunism, Shaktism, and Tantrism; Islamic practices, Buddhism, Jainism, Sikhism, and Christianity, have an immense influence on local inhabitants.

Particularly, belief systems like Vaishnavism, Jainism, and Buddhism follow a vegetarian diet abiding with the ideology of *Ahimsa* (non-violence). The reflection of their foundational beliefs is followed in major north and central Indian landscapes. They emphasize lacto-vegetarian meal options. Dairy products such as *Paneer* (Cottage cheese),

[a]mouli.chakraborty@kalingauniversity.ac.in, [b]shilpi.bhattacharya@kalingauniversity.ac.in, [c]ajay.shukla@kalingauniversity.ac.in

DOI: 10.1201/9781003684459-50

Butter, and *Ghee* (clarified butter) are consumed daily as a part of a protein-based diet. Also, lentils, cereals, and chickpeas with vegetables are cooked to inculcate protein in diet charts [8].

The believe and practise of Shaktism and Tantrism spread from eastern part of India comprising Bihar, Bengal, towards the north east and later spread towards countries like Nepal, Tibet and part of Bhutan with interaction between Buddhism and Tantrism (Tantric-Buddhism). Along with the religious practices, the dietary choices got intermingled. People within the belief systems consume a purely non-vegetarian diet and majorly intake animal fat or meat. Shakti devotees even devote fishes during special religious or cultural festivals, as fishes like *Hilsa* are considered pious for the purpose.

Islam, predominant in South Asian countries like India, Pakistan, Bangladesh, Maldives, and Afghanistan, adheres to the concept of *Halal*. The Quran outlines the dietary guidelines with a list of food allowed for Muslims to consume and whatnot. Even the way of slaughtering an animal is given attention, so the process should not be inhumane or less suffering. So, traditional Islamic recipes from aromatic *biryanis* to grilled *Kebabs* only permit halal meat as exclusive cuisine from the region.

Geography, along with religious and historical traditions, influence the major food items in these regions. Be it the indulging Kulfis in dairy-rich Punjab and Lahore region or *Hilsa Paturi* (steamed Hilsa) in Bengal and Dhaka region.

1.3. The grandeur of India in food diplomacy

When it comes to soft diplomacy, India has always remained an esteemed nation. The richness of Indian cuisines has as much as 5,000 years of expanded history. Indians incorporate the cooking recipes of Persian, Mughals, Rajputs, and Britishers to form distinguishing meals. In today's time at international dialogues and world forums, India's gastro-diplomacy is gaining prominence [9].

1.4. India's G20 summit with sumptuous meals

The most talked about discussions regarding the G20 summit hosted by India in 2023 are not about the impact of the meeting but more about the royal dishes. India diplomatically hosted the Global leaders with meals in gold and silver plated utensils thereby showcasing the rich heritage and culture of the country [10]. The menu was a medley of dishes from different regions across India. Dishes like *Vanavarnam* (Jackfruit galette served with forest mushrooms), *Mumbai Pao* (soft bun), *Kashmiri Kahwa*, etc were served. The common tie between these dishes is the ingredients. Firstly, all these dishes are vegetarian, and secondly, they are made with different varieties of Millet. Therefore, all these dishes are purposefully made with millets to commemorate the year 2023 as the International Year of Millets (IYOM) as part of the UNGA resolution on India's initiative.

1.5. India hosting global leaders with traditional dishes

For modern state crafting, food diplomacy works as an antipode of hard-core power exercise. The Indian government hosted foreign ministers, diplomats, and political heads of State with Indian touches. This takes us back to the time when the Union government hosted the Namaste Trump event. Donald Trump, while traveling to cities like Delhi, Agra, and Ahmedabad, has tried side dishes like corn-stuffed Samosa and Gujarati snack Khaman. But as speculated by Kenzie Bryant, a Western Journalist, whether he has half eaten the food or enjoyed the vegetarian snacks is an uncertain comment. Rastrapati Bhaban, considering Trump's love for meat, has also served Mutton Gosht Biryani for dinner (Beef and Mutton servings).

Not only Trump but other successor Presidents, including Barack Obama, also hosted by Indian Prime Minister Modi with diverse Indian menus. The visit of former President Obama not only signaled a positive sign ending the uncertainty over nuclear trade but can also be seen as a way of representing regional meals like *mustard fish* and *Kashmiri Gustaba* securing global spotlights [11].

Recently, Indian Street foods like Golgappa also gained international fame when Modi Japanese Prime Minister Fumio Kishoda cherished the Indian snack together. Other local street foods are supposedly served at the grand G20 summit [12]. Dignitaries discussing over tea and samosa may not be the headline of any newspaper, but it does tacitly represent the local Indian Street food culture to the forefront.

1.6. India making global investments in food delicacies

Propagandizing about a country's culture is not always through movies and mainstream means. This further becomes evident when the Ministry of Food Processing Industries organizes grand events like World Food India to welcome global food business hubs to look for opportunities here. Indian cuisine dominated the World stage with culinary diplomacy when restaurants with Government support participated in the Global 'Taste of India' food festival in 2003. Such initiatives foster cross-cultural understanding and strengthen the country's diplomatic efforts [13].

1.7. India and South Asia over food and discussions

Other South Asian countries have intricate culinary ties with India as their traditions follow deeper roots. In this region, food has acted as a medium to both enhance cordial bonds as well as have worsened the ties. Before the Agra summit, when the relations between India and Pakistan were suffering from major rifts, nobody thought about the role of food in softening their relations. Consequently Parvez Musharaf visited India and held talks with PM Bajpayee over full-course meals. At the dinner invitation in 2004, the meal included traditional dishes from both countries. Even though the Summit talk did not lead to any progress, the dinner menu somehow managed to fill the widening gap between their relations.

The *Mango Diplomacy* signaled the betterment of relations between India and Bangladesh when former PM Sheikh Hasina sent baskets of mangoes for Prime Minister Modi, President Draupadi Murmu, and Sonia Gandhi. Such exchange of gifts is a known classic example of India-Bangladesh diplomatic relations where both countries exchange fish during the monsoon times [14].

The Indian Culinary of Kerala and Tamil Nadu have common roots with Sri Lanka and so follow similar ingredients for cooking. Mostly spices: cardamom, cloves, and cinnamon are grown in abundance in both regions. The exchange of fisheries and spices is most common between India and Sri Lanka. Infact the common Sri Lankan drink *Hoppers* (fermented with toddy) is similar to Kerala's *Appam* [15].

But food does not always serve as a source of commonality and pleasure, as it also acts as a medium of disdain. Mainly in the case of South Asia where food consumption coincides with religious practices. For instance, the region has countries like Pakistan, Bangladesh, India, Afghanistan, etc., with people from the Islamic faith. For them beef or meat consumption is common. But that negates the Vaishnavite idea of devoting cows or cattle. Even Buddhist predominant States like Bhutan follow a strict vegetarian diet and reject the consumption of meat. So, any policies or actions banning the consumption of beef hurts the true Islamic sentiments.

Also there are instances when food acted as an indicator of the relationship status of two countries. In 2017, a right-wing legislator commented on the ban of *'Momos'* as a life-threatening street food. The Himalayan countries of Nepal and Bhutan are the biggest sellers of Momos which hinted towards a weakening tie with Nepal at that point of time [16].

Another example is when the interim government of Yunus in Bangladesh banned the export of Hilsa in September 2024. However, due to public backlash and other strategic reasons, such a ban was lifted within the next two weeks [17].

So, does culinary diplomacy have the potential to strengthen relations among South Asian Nations! The answer will be on grey lines. The taste palette of these regions are quite familiar because of the proximity in geography and availability of similar resources. Majorly, South Asian regions share traditional and cultural roots; and food being an important aspect of passing on tradition can glue the region together.

However, amidst such potentials, the religious politics played by the Government at different points of time is failing to bridge the long drawn boundaries within these Nations. At a time when South Asian rich dishes can dominate the world market, such narrow political interests are interrupting the common motive of bringing prosperity within the region.

1.8. National or regional food identity?

Speaking of Indian gastronomic diplomacy [18], the question remains, is there a true Indian national food that exists or rather this culinary diplomacy is based on the regional meals! Unlike Western nation-states, India is not a country with homogenous historical, traditional or cultural identity. Accordingly, the diet plan among Indians spread across regions are not the same. So, when it's about culinary diplomacy between India and foreign states, there's no homogenous identity as such. Rather such gastronomic interactions highlight the heterogeneous cooking styles, ingredients and spread of food resources across the Indian subcontinent. The taste palette among Indians also varies depending on their local conditioning. In place of overshadowing such traditional dishes with a rhetorical national eminence at global stages, there are talks about also emphasizing on the regional or local identity of the dishes.

2. Conclusion

The beauty about India is the heterogeneity and that must be sustained to maintain the true 'Indianness' in character. Cultural affinity, preserving the traditional dishes like the locals become important at this point when there are drives going on for mainstreamization or westernizing the dishes. Indian cuisine is not based on a common taste and that should be preserved when practising culinary diplomacy with the world.

References

[1] Rajalakshmi, A. R., Jayakani, S., Shanmugam, H., & Ramakrishnan, P. R. (2024). A Study on Green Entrepreneurship and Electronic Word of Mouth: Special Reference with Kanchipuram District. *Indian Journal of Information Sources and Services*, *14*(3), 270–276. https://doi.org/10.51983/ijiss-2024.14.3.35

[2] Selvakumar, K. (2016). Improved Software Fault Detection System Using Hybridized Genetic with Artificial Bee Colony Algorithm. *International Journal of Advances in Engineering and Emerging Technology*, 7(4), 304–308.

[3] Muhammad, A., & Adilbekova, K. (2023). Culinary diplomacy: Unveiling the palate as a pathway to stronger international relations. Zenodo (CERN European Organization for Nuclear Research). https://doi.org/10.5281/zenodo.8382293

[4] Aghababaei, F., Jouki, M., & Mooraki, N. (2024). Evaluating the quality of fried whiteleg shrimp (Litopenaeus vannamei) fillets coated with quince seed gum containing encapsulated cinnamon extract. *International Journal of Aquatic Research and Environmental Studies*, *4*(2), 99–115. http://doi.org/10.70102/IJARES/V4I2/7

[5] Khyade, V. B., & Jagtap, S. G. (2019). Diversity of butterflies (Order: Lepidoptera) in Mayureshwar Wildlife Sanctuary of Baramati Tehsil Dist. Pune (India). *International Academic Journal of Innovative Research*, *6*(1), 91–118. https://doi.org/10.9756/IAJIR/V6I1/1910008

[6] Çiftçi, N., Cicik, B., & Ayas, D. (2023). First Report on the Elemental Composition of the Bigeye Thresher Shark Alopias superciliosus Lowe, 1841 from the Mediterranean Sea. *Natural and Engineering Sciences*, *8*(2), 106–118. http://doi.org/10.28978/nesciences.1338143

[7] Lal, C., & Saha, M. (2024). History of Food in Mughal Empire. *International Journal of Creative Research Thoughts (IJCRT)*. ISSN: 2320-2882.

[8] Adil. (2024, March 5). South Asian cuisine: religious influence and festive significance. Usmania: Fine Dining Restaurant Dallas. Retrieved February 3, 2025, from https://usmaniarestaurant.com/dallas/south-asian-cuisine/#:~:text=In%20addition%20to%20dietary%20restrictions,victory%20of%20light%20over%20darkness

[9] Gilada, P. (2024, September 27). Savouring Diplomacy: How India uses culinary heritage to build global bridges. Samvada World. Retrieved February 3, 2025, from https://samvadaworld.com/analysis/savouring-diplomacy-how-india-uses-culinary-heritage-to-build-global-bridges/#:~:text=India%20practices%20food%20diplomacy%20through,in%20the%20global%20culinary%20narrative.

[10] Roy, S., & Ojha, A. (2023, September 9). G20 dinner menu celebrates Millets, Diversity: 'Taste Connects Bharat.' India Today. Retrieved February 3, 2025, from https://www.indiatoday.in/amp/india/story/g20-summit-dinner-menu-narendra-modi-bharat-millets-2433475-2023-09-09

[11] President Barack Obama in India: Here's what he ate. (2017, June 23). NDTV News. Retrieved February 3, 2025, from https://food.ndtv.com/news/president-barack-obama-in-india-heres-what-he-ate-733887

[12] The Cuisine Diplomacy: Culinary statecraft. (2023, November 18). The Geostrata. Retrieved February 3, 2025, from https://www.thegeostrata.com/post/the-cuisine-diplomacy-culinary-statecraft#:~:text=Cuisine%20diplomacy%2C%20a%20time%2D%20tested,cross%2D%20culture%20understanding%20among%20nations

[13] Bhatt, Dr. A. (2013). The culinary diplomacy: Enhancing India's soft power. *International Journal of Research in Social Science*, *3*(1), 820–828.

[14] Bhattacharjee, K. (2023, June 13). Mango Diplomacy: Sheikh Hasina sends gifts to PM Modi, President Murmu. The Hindu. Retrieved February 3, 2025, from https://www.thehindu.com/news/national/mango-diplomacy-sheikh-hasina-sends-gifts-to-pm-modi-president-murmu-cm-mamata/article66963428.ece

[15] Food Diplomacy: India's role in shaping Sri Lankan cuisine. (2020, October 25). The Florida Star, from https://www.thefloridastar.com/articles/food-diplomacy-indias-role-in-shaping-sri-lankan-cuisine/

[16] Khajuria, R. (2017, June 23). BJP legislator wants momos banned, says they cause life-threatening diseases by Ravi Krishan Khajuria, Hindustan Times. Retrieved February 3, 2025, from https://www.hindustantimes.com/india-news/bjp-legislator-wants-momos-banned-says-they-cause-life-threatening-diseases/story-RaiSxYPhiZytNzWK3FYjgJ.html

[17] Bose, S. (2024, October 5). Bangladesh's Hilsa export ban: a hasty decision reversed. Observer Research Foundation. Retrieved February 3, 2025, from https://www.orfonline.org/expert-speak/bangladesh-s-hilsa-export-ban-a-hasty-decision-reversed

[18] Chakraborty, S. (2023). Gastrodiplomacy, conflict, and culinary debate. *Economic and Political Weekly*, *58*(33).

51 Skill development and empowerment of women: A case study of Nituria CD block, Purulia, West Bengal

Jayashri Roy[1,a] *and Shraddha Verma*[2,b]

[1]Assistant Professor in Geography, Vidyasagar Teachers' Training College, Midnapore Paschim, Medinipur, West Bengal, India
[2]Dean faculty of Education, Kalinga University Naya, Raipur, Chhattisgarh, India

Abstract: This study explores the impact of skill development initiatives aimed at empowering women in the rural area of Nituria CD Block, Purulia district in West Bengal. Characterized by rural isolation and limited access to economic opportunities the region faces significant challenges in terms of gender equality and women participation in the workforce. However various skill development programmes led by both government and private sector organizations have played a significant role in transforming the economic potential of rural women. Different Government schemes and local initiatives such as self help groups introduce disadvantaged women in Nituria CD Block to vocational skills such as sewing, embroidery, food preparation digital literacy and others. These programmes not only impart technical skills but also emphasize entrepreneurship which facilitates women to generate income and improve their standard of living. This study highlights both the successes and challenges of the above initiatives. Several positive outcomes include increased economic independence as well as decision making capacity building within the household, development and growth of small scale businesses. However women working outside the home face cultural resistance lack of infrastructure and other challenges that rural women face. Overall this study highlights the potential of skill development programs in promoting social and economic change in rural areas as well as identifying areas for further improvement in awareness rising. By analysing the experiences of women who are disadvantaged by the Nituria CD block this study aims to contribute to understanding of how vocational training programs can empowering women in this study area reduce poverty and contribute to rural development in marginalized areas.

Keywords: Skill development, rural area, women, social and economic change

1. Introduction

Empowering women in rural areas is an important approach to improving their economic and social status and developing skills for sustainable development. Skill development programmes play a transformative role in rural India where women face multiple challenges in terms of education, health, clothing and economic opportunities [1]. These initiatives not only enhance their technical and vocational skills but also increase their self-confidence, enable financial independence and significantly improve their overall quality of life [2].

Nituria, a rural area in Purulia district of West Bengal is an ideal place to study the impact of skill development programmes on women [3]. The district is predominantly an agricultural region facing various challenges such as low literacy rate limited employment opportunities and economic dependency. However in recent years there has been increasing focus on empowering rural women through skill development, vocational training and entrepreneurship support [4, 5].

This study examined the existing skill development programmes in Nituria CD Block to explore their impact on the socioeconomic status of women and to assess how these programmes can be further scaled up to meet the needs of rural women. This paper highlights the role of Government and Non-Government Organisations in this rural area and the success stories of women who have benefited from these initiatives. This research study aims to provide valuable insights for developing more effective and inclusive skill development strategies for rural women [6, 7].

Skill development in rural areas like Nituria Block of Purulia District has paved the way for gender equality and socio economic progress, while also empowering women to take leadership roles in their families and communities ultimately contributing to the achievement of the larger national development goals [8].

2. Objectives

- To assess the current status of development programs for women in Nituria CD Block

[a]jayashriroy19@gmail.com, [b]shraddha.verma@kalingauniversity.ac.in

DOI: 10.1201/9781003684459-51

- To identify the challenges faced by women in Nituria CD Block in participating in skill development programs
- To assess the roles of agencies in promoting skill development in Nituria CD Block.

3. Methodology

In depth interviews were conducted with local community leaders, trainers and representatives of NGOs, women and different Government agencies involved in the skills development programme. These interviews provided qualitative insights into the experiences, challenges and outcomes of the programme. A discussion was held with women who participated in the skills development programme to discuss their work experiences, challenges and the impact of the skills development programme on their lives. Data was collected from different local Government offices, NGOs and other organisations that have helped implement the skills development initiative for women in the rural areas of Nituria CD Block. 100 women from the rural areas of Nituria CD Block who are participating in the skills development programme were recruited for the study. The data is obtained from the survey and questionnaires conducted on the sample and the data was analysed using statistical tools including descriptive statistics.

4. Study Area

Nituria is a CD block which is situated in Purulia district of West Bengal, India. The study area is situated in the western part of West Bengal and it is also known for its hilly terrain and underdeveloped areas. Nituria CD Block is one of the rural administrative Blocks with several small villages scattered across a geographical area. The region is characterized by its rugged terrain and agricultural land where agriculture is the main source of livelihood. The region is mainly populated by people from other socio economically marginalized groups which including Santals are the main caste. The literacy rate in the region is relatively low for women (45% according to Census 2011) compared to urban areas and there is a significant gender gap in education, health and employment opportunities.

5. Current Status of Skill Development Programs for Women in Nituria CD Block, Purulia District in West Bengal

The Purulia districts Nituria CD Block is a largely rural community distinguished by a largely agricultural economy, low literacy rates and inadequate infrastructure. However since the necessity for womens economic empowerment and independence has become more widely acknowledged, skill development programs especially those targeted at women have become more significant in the region in recent years [6].

Among the most important skill building initiatives for women in Nituria CD Block are:

- ***Rural Self Employment Training Institutes:*** RSETIs offer instruction in subjects like computer literacy, handicrafts, food processing, dairy farming and tailoring. To encourage womens entrepreneurship several programs have been made available in collaboration with nearby banks and non-governmental organisations [2].
- ***State and Local NGO Initiatives:*** A number of state and local NGOs in Purulia District including Vikas Bharti and Pradip Foundation have been focussing on enhancing womens abilities in fields like weaving, agriculture, healthcare and the manufacturing of regional items. These NGOs frequently collaborate with government programs to offer thorough assistance and training [4].

Skill development programs in Nituria CD Block, Purulia District is relatively limited due to following factors:

- ***Awareness Issues:*** Due to cultural or communication barriers, many women are not completely aware of the programs that are available. Some of them lack access to information or are not aware that they qualify for specific programs.
- ***Mobility and Accessibility:*** Due to a lack of transport options and the distance to the closest training facilities, many rural women find it difficult to go to training facilities. Participation rates are lowered as a result especially for people living in rural areas.
- ***Community Support:*** Social norms and conventional roles frequently limit womens participation in skill development programs despite the encouragement of some local leaders and community influencers. In Nituria CD Block women may encounter social opposition to deviating from their customary responsibilities, particularly when doing so entails achieving economic or professional empowerment [4].
- ***Low participation in Formal Training Programs: Despite:*** the fact that many women in the region participate in informal skill development activities, family responsibilities, childcare shortages and home duties frequently result in lower participation in formal training programs.

5.1. Categories of skill development of the study area

- ***Vocational Skills:*** These comprise abilities like embroidery, handloom weaving, sewing and tailoring. These abilities are frequently closely related to earning money through independent employment or small scale local enterprises.
- ***Entrepreneurial Skills:*** Training programs also emphasise entrepreneurship, assisting women in joining self help groups

launching their own small enterprises. The development of entrepreneurial skills frequently focuses on industries including dairy farming, cosmetics and food manufacturing.

- ***IT and Digital Literacy:*** As technology becomes more prevalent in rural regions efforts are being made to teach women the fundamentals of computers and digital literacy so they can access information, financial services and online employment prospects [9].
- ***Health and Beauty Care:*** Beauty therapy, skincare and healthcare courses are well liked in Nituria where many women use their education to launch neighbourhood beauty salons or provide in home services.
- ***Agriculture and Horticulture:*** Given the regions agrarian economy, training in organic farming, sustainable farming methods and horticulture is also a key priority. Women receive training in household earning pursuits like poultry, dairy farming and kitchen gardening.

6. Identify the Challenges Faced by Women in Accessing Skill Development Programs in the Study Area

The following are the main obstacles that women in Nituria, Purulia district of West Bengal experience when trying to access skill development programs:

Limited Program Knowledge: Rural women of Nituria, Purulia District are not aware of the range of skill development programs that are accessible to them. This lack of awareness might be ascribed to inadequate outreach by NGOs and Government organisations to poor information distribution through local channels. Women might not be aware of their eligibility for different programs or how or where to obtain training.

Language and Communication Barriers: Women may find it difficult to comprehend the specifics of the programs if program facilitators speak in local dialects and languages they do not understand.

Gender standards and Social Expectations: Traditional gender standards frequently restrict women's mobility and their capacity to engage in extracurricular activities in rural places such as Nituria CD Block. It may be expected of women to put household responsibilities, including caring for young children and ageing family members ahead of their own growth and involvement in the economy. This restricts the amount of time and effort they can devote to skill building exercises.

Family Opposition: Many women encounter resistance from family members particularly men who do not encourage them to enrol in skill development courses. Women chances for advancement may be further restricted by cultural beliefs that prevent them from leaving the home.

Child Marriage and Family Responsibilities: In Nituria CD Block child marriage is a common occurrence and as a result women often face a social barrier to family responsibilities.

Program Cost and Related Expenses: Although many skill development programs are subsidised women are frequently prevented from fully participating due to hidden costs. These could include the price of materials, transportation and other ancillary expenditures. These expenses may be unaffordable for women from low-income families particularly if the programs advantages are not immediately apparent.

Cash Loss During Training: Women may have to take time off from jobs that provide cash such small businesses or housework in order to participate in training programs. For women who might not have enough financial support to cover the training period, this little salary loss is a major disincentive.

Geographical Barriers: The infrastructure in Nituria CD Block is restricted due to its rural location. Women find it challenging to get to training facilities particularly those who live in rural areas due to inadequate roads public transport and connectivity. In order to access skill development programs women frequently have to travel great distances which is expensive and time consuming.

Limited Employment prospects: Although women may successfully finish skill development programs they frequently have few options for using their newly obtained abilities in Nituria CD Block because to a lack of employment prospects. Because agriculture makes up the majority of the regions economy there might not be enough need for skills like computer literacy, tailoring and beauty care.

Inadequate Post-Training Support: Many women find it difficult to obtain jobs or start their own enterprises after completing training because they lack networks, mentorship and financial support. Women find it challenging to move from training to income generating activities when there are no resources for career counselling and company incubation.

Lack of experienced Trainers: The quality and scope of skill development programs may be constrained in rural areas like Nituria CD Block by the lack of experienced trainers in specific fields, such as IT, healthcare and other technical skills. Program effectiveness degrades if trainers lack the most recent information or are not properly trained.

Gender Bias Among Trainers: In conservative societies women may feel better at ease participating in training programs led by female trainers. But in rural skill development programs there is frequently a lack of female trainers which makes it hard for women to connect with the trainers or feel completely supported during the learning process.

7. Evaluate the Role of Agencies in the Study Area Promoting Skill Development

The role of agencies in promoting skill development programmes for women in Nituria CD Block under Nituria Police Station of Purulia District is very important (Table 51.1). Various agencies both government and private, have always played a vital role in ensuring the success of skill development programmes in this area by providing training and financial assistance [6].

West Bengal State Rural Development Department: Implements state level programmes to help rural women develop their skills and is also doing a good job in the development of this study area.

Purulia District Administration: Under its supervision, local training centres are monitored and ensured to ensure that all programmes are in line with government policies and women in the region are empowered to develop their skills.

Self Help Groups: Play an important role in mobilising local women for skill development providing adequate support to peers and improving various financial services.

Non-Governmental Organisations: Notable NGOs in the region are Pradeep Foundation, Vikas Bharati. They provide training resources and financial assistance to the local people and help them in various ways.

Training Centres: Local training centres affiliated to the government and NGOs such as Nituria Training Club, Bhamuria Pragati Sangha and various skill development centres provide vocational training and support to women.

A survey was conducted on 100 women 10 local government representatives and 5 representatives of local NGOs to assess the role of l agencies in promoting women's skill development programs in Block under Nituria Police Station in Purulia District, the summary of which is given Table 51.1.

8. Findings

The survey suggests that access to transportation and training canters needs to be improved. More local training centres need to be set up in remote areas. Many feel that while training is effective the quality is not always in line with industry standards. There is a need to upgrade the skills of trainers, modernize training facilities and improve content to meet the needs of the job market day after day. This disparity suggests that government support is not always visible to women at the grassroots level. More visible efforts by local governments such as workshops, awareness campaigns and direct engagement with women are greatly needed. While government financial support is available on paper it does not always reach the women who need it most. Many feel that small loans, transportation subsidies and other financial assistance programs need to be strengthened. Lack of local employment opportunities, training and job opportunities are significant barriers.

According to the findings of the study, there is a greater need to expand local industry or self employment options. A better link should be established between training programs and employers [6].

The study found that although local institutions and government agencies play an important role in promoting womens skill development programs in Nituria police station of Purulia district they are facing many challenges (Table 51.2). Such as awareness, financial support and employment after training [10].

Findings from exploring womens attitudes towards skills training programs and the factors that motivate or discourage their participation.

The Chi-square test results are as follows:

- **Chi-square:** 53.33
- **p-value:** 7.26×10−117.26/times 10^{-11}7.26×10−11

A low p-value which is less than 0.05 indicates that the null hypothesis which assumes no relationship between the variables is to be rejected. It means that there is a statistically very much significant relationship between the variables and their categories in the data as shown here (Tables 51.3–51.4).

Table 51.1. Role of agencies in Study area and promote skill development

Activities	*Evaluate the Role of agencies in the study area Promoting Skill Development*			
	Women Participants (n=100)	*Government Representatives (n=10)*	*NGO Representatives (n=5)*	*Average Rating (1–5)*
Awareness of skills development programs	65%	80%	90%	78%
available in the program	60%	70%	80%	70%
Quality of Training Provided	3.6%	4.1%	4.3%	4%
Role of local administration in support programs	55%	90%	80%	75%
Provision of financial support for women's participation	45%	70%	60%	58%
Employment opportunities after training	50%	60%	75%	61%
Agreement between local institutions and organizations	50%	80%	70%	66%
Satisfaction with support from government and organizations	3.5%	4%	4.2%	3.9%

Source: Field Survey, 2024.

Table 51.2. Role of stakeholders in promoting and hindering women's participation in skills development

Stakeholder	*Role in Promoting Participation*	*Role in Hindering Participation*
Government Agencies	Provide funding and infrastructure for training programs.	Bureaucratic delays in program implementation.
	Implement policies supporting women's skill development.	Insufficient monitoring and evaluation of programs.
Local NGOs	Deliver tailored training programs focused on women's needs.	Limited outreach due to resource constraints.
	Offer mentorship and financial support.	Occasional misalignment with community needs.
Community Leaders	Mobilize community participation and encourage women to join programs.	Reinforce traditional gender roles that limit women's mobility and participation.
Family Members	Provide moral support and allow time for women to attend training sessions.	Oppose participation due to cultural norms or household responsibilities.

Source: Field Survey, 2024.

Table 51.3. Participation categories

Variables	*Category 1*	*Category 2*	*Category 3*	*Category 4*	*Participation (Yes)*	*Participation (No)*
Age Group	18–25	26–35	36–45	46+	50	30
Education	Primary	Secondary	Higher Secondary	College level	40	40
Family Support	Yes	No			60	20
Motivating	Yes	No			70	10
Discouraging Factor	Yes	No			30	50

Source: Field Survey, 2024.

Table 51.4. Participation categories (age group)

Category (Age Group)	*Participation (Yes)*	*Participation (No)*	*Total*
18–25	20	10	30
26–35	15	15	30
36–45	10	20	30
46+	5	15	20
Total	50	60	110

Source: Filed Survey, 2024.

The results suggest that factors such as age and family support are closely linked to women's participation in skills training programs in the region.

This insight will help to design programs to address specific barriers and motivations for different demographic groups.

Women's attitudes toward skills training programs and their motivation to participate should include attitudes toward empowerment, career advancement, family responsibilities, financial independence and societal expectations.

9. Challenges Faced by Women in the Region in Participating in Skills Development Programmes

- Women in the region are often discouraged from participating in programmes by family members who prioritize traditional roles [11].
- Many programmes in the region are subsidised, yet financial barriers such as costs such as transport, course fees and other expenses prevent some women in the region from engaging in skills development [10].
- Lack of adequate infrastructure such as training centres, trainers and materials in the region hampers the effectiveness of programmes. [6].

10. Both Government and Private Sector Organizations Play a Key Role in Delivering Skill Development Programmes For Women in the Region

NGOs are often effective in reaching out to rural women in the region, conducting awareness campaigns and providing additional support services such as financial literacy, childcare and post-training support, but more needs to be done at the grassroots level, according to the survey report.

Programmes tailored to local needs and contexts, such as tailoring, food processing and handicrafts, are successful in engaging women, but more needs to be done, the survey report found.

Strategies to scale up skill development programmes for women in rural areas of Nituria block of Purulia district include addressing the barriers faced by women in accessing these programmes such as awareness, infrastructure, financial support and employment opportunities.

11. The Detailed Strategies

Training modules should be customized to suit the local economy, cultural context and available resources to pave the way for improvement. For example, focus on skills related to agriculture, handicrafts, tailoring and food processing that are in demand by rural women in the region.

Provide hands-on training in areas such as tailoring, embroidery, pottery, organic farming, livestock management and digital literacy to empower them and create livelihood opportunities [10].

The survey report highlights that overall empowerment of women should include leadership, financial literacy, communication and entrepreneurship training.

Women in the region should be facilitated to partner with banks and NGOs to provide microfinance or grants to start their own small businesses.

Community centres providing access to tools, raw materials and internet connectivity to support skill development and business initiatives should be considered to empower women in the region.

Women who have been successful in similar ventures should be identified and employed as mentors for new students.

Women in the region should form self help groups that can empower women by providing peer support, knowledge sharing and collaborative business opportunities.

Linkages should be established with local, regional and online markets so that women can sell their products and services.

Awareness campaigns should be actively conducted to reduce the social stigma surrounding women working outside the home in the region.

Success stories should be documented, shared and the impact of the program should be demonstrated to stakeholders to inspire other women [6].

By adopting these strategies it is expected that skill development programs can effectively empower women in Nituria CD Block improve their socio-economic status and contribute to the overall development of the region.

12. Impact of Skill Development Programs on Women in the Region

Some of the key outcomes observed include:

***Economic empowerment*:** Many women who completed skill development programs in the region reported increased household income due to their ability to generate income from their new skills. The most significant impact was seen in tailoring, weaving and handicrafts where women were able to sell their products locally or in nearby towns, which was a boon to them.

Social empowerment: These programs have increased the confidence of women in the region, which is helping them to participate in decision making and leadership roles in many places according to the field survey.

Entrepreneurship growth: The survey found that some women in the region are contributing to their economic development by using the skills they have trained in to start their own small businesses.

Improved literacy and education: As part of many skills development programs, women in the region have also gained access to basic education, literacy and numeracy programs which are further enhancing their ability to participate in the workforce and effectively manage businesses, the survey found [11].

13. Conclusion

This research survey of women in the Nituria police station of Purulia district of West Bengal, particularly in the area of skill development particularly in the areas of limited access to education for women, financial dependence, social constraints and traditional gender roles, shows that skill development programs have considerable potential to advance economic and social empowerment despite the numerous challenges faced by women in these areas [11]. The survey report found that just as skill development programs have paved the way for women in Nituria to achieve financial independence women who were previously dependent on their families for financial support have been able to secure employment or start small businesses. As women in this region acquire skills and start contributing to the family income perceptions of their role in society are gradually changing [12]. The current status of skill development programmes for women in Nituria police station, Purulia District presents a mixed picture, according to the survey report. Although these programmes have contributed significantly to womens empowerment in terms of economic independence and social status, challenges such as cultural barriers and financial constraints have hindered their full potential. Addressing these challenges and improving the reach and quality of skills development initiatives can help further empower women in the region, which is expected to help contribute to broader socio-economic development [10].

14. Acknowledgment

We would like to express our sincere gratitude to all those who have contributed to the successful completion of this study, *Skill Development and Empowerment of Women: A Case Study of Nituria CD Block, Purulia, West Bengal.*

We did this work ourselves by conducting field surveys, it is completely our own work. We relied on field surveys to do this work. We are deeply grateful to the women of Nituria CD Block who shared their experiences, challenges and aspirations with me. Their voices and stories form the foundation of this study and without their cooperation, this research would not have been possible.

This research is a small step towards understanding and contributing to women's empowerment and we hope it inspires further efforts in this field.

References

[1] Jude Lavanya Kumar, X., & Gopinath, T. (2016). A New implementation of Employee Engagement Model and Retention Strategies in BPO Companies in India. *International Journal of Advances in Engineering and Emerging Technology*, *7*(3), 01–08.

[2] Rao, P., & Saha, S. (2017). Impact of skill development programs on rural women in Purulia district. *Indian Journal of Rural Development*, *35*(6), 54–68. https://doi.org/10.7890/ijrd.2017.356

[3] Sujatha, S. (2024). Strategic Management of Digital Transformation: A Case Study of Successful Implementation. *Global Perspectives in Management*, *2*(1), 1–11.

[4] Kumari, S. (2020). Challenges and opportunities for skill development among rural women in Purulia district, West Bengal. *Journal of Women's Studies*, *22*(3), 91–106. https://doi.org/10.3456/jws.2020.223

[5] Crescenzo, G. D., Ghosh, A., Kampasi, A., Talpade, R., & Zhang, Y. (2011). Detecting Anomalies in Active Insider Stepping Stone Attacks. *Journal of Wireless Mobile Networks, Ubiquitous Computing, and Dependable Applications*, *2*(1), 103–120.

[6] Kittur, P. (2014). Development of Rural women Entrepreneurs through workshop training. *Research Journal of Management Science*, *3*(2), 15–18.

[7] Seo, M. (2022). Authentication for V2X Communication in an Open-Source Plug-and-Play Platform. *Journal of Internet Services and Information Security*, *12*(2), 1–20. https://doi.org/10.22667/JISIS.2022.05.31.001

[8] Mohanty, S., Sahoo, S., Dhal, S., & Swain, S. C. (2024). Expanding LGBTQ+ Inclusivity: An Ecological Model of School Library Practice in Addition to Collection Development. *Indian Journal of Information Sources and Services*, *14*(3), 169–174. https://doi.org/10.51983/ijiss-2024.14.3.22

[9] Unni, J., & Uma, R. (2004). Technical change and workforce in India: Skill Based Growth? *Indian journal of Labour Economics*, *47*(4).

[10] Gupta, P (2016). Study of women empowerment through skill development & vocational education in India. *International Journal of Education & Applied Science Research*, *3*(07).

[11] Gupta, P., & Sengupta, T. (2019). Women's participation in skill development programs in rural areas of West Bengal: A study on Nituria block. *International Journal of Rural Economics*, *15*(1), 45–59.

[12] Mokta, M. (2014). Empowerment of women in India: A critical Analysis. *Indian Journal of Public Administration*, *474*(3).

52 Quantum-enhanced climate models for predicting and mitigating environmental change

P. Thilagavathi[1,a], *R. Geetha*[2,b], *Ramesh S.*[2,c], *S. Vidhya*[2,d], *and J. Nithyashri*[2,e]

[1]Associate Professor, Department of Computer Science and Engineering, Aarupadai Veedu Institute of Technology, Vinayaka Mission's Research Foundation (DU), Paiyanoor, Chennai, Tamil Nadu, India
[2]Assistant Professor, Department of Computing Technologies, School of Computing, SRM Institute of Science and Technology, Kattankulathur, Chennai, Tamil Nadu, India

Abstract: Global challenge of climate change demands precise predictive models to grasp its effects and create sensible mitigating plans. Based on classical computing methods, traditional models suffer in processing complicated climate data, which results in lower accuracy in long-term forecasts and extreme weather prediction. With its capacity to do huge parallel computations via superposition and entanglement, quantum computing offers a potential fix for improving climate models. In addition to focused on its ability to increase weather forecasting accuracy, long-term climate projections, and extreme weather event predictions, this work investigates the integration of quantum computing in climate science. Quantum-assisted climate models demonstrate better computational efficiency, higher forecast accuracy, and better resource management for mitigating techniques including sustainable agriculture, carbon capture, and solar energy deployment. However, obstacles, developments in quantum technology and multidisciplinary cooperation between climate researchers and quantum physicists are opening the path for more dependable and scalable climate solutions.

Keywords: Climate, weather, efficiency, prediction, environmental conditions, quantum, ecosystem, mitigation

1. Introduction

Rising temperatures, more frequent extreme weather events, and broad disturbances of ecosystems, businesses, and human livelihoods follow from climate change's major worldwide problem. Precise and effective climate modelling is therefore essential to grasp the direction of climate change, project possible hazards, and create data-driven mitigating solutions given its growing speed [5].

Although they depend on classical computers, traditional climate models which have been quite helpful in analyzing climate patterns have trouble digesting the enormous and complicated data needed for reliable forecasts. Numerous interdependent variables such as air circulation, ocean currents, emissions of greenhouse gases, changes on the land surface, and human activity all of which interact very nonlinearly are all components of climate systems. Especially in long-range projections and simulations of extreme weather events, classical models may require simplifications under computational requirements, and this leads to uncertainty and reduced predicted accuracy. Applying concepts such as superposition, entanglement, and quantum parallelism, quantum computing presents a revolutionizing potential in climate modeling. In contrast to traditional computers that handle information sequentially, quantum computers can potentially study multiple climate parameters at the same time, thus allowing for faster and more precise simulations of complex climate systems. High-resolution climate modeling, severe weather prediction, and real-time data analysis for disaster preparedness all rely on this ability in particular advantage. Moreover, quantum optimization methods can be used in attempts to mitigate climate change by means of better resource management, carbon capture technology, and optimal location and efficiency of renewable energy infrastructure, so strengthening agricultural sustainability [1]. The possible benefits of quantum-enhanced climate models to raise forecasting accuracy, maximize environmental resource management, and propel sustainable climate solutions are investigated in this study [3]. This looks at how conventional climate models and quantum algorithms might be integrated, talks about current developments in quantum-assisted simulations, and notes practical uses of quantum computing in climate science. It also offers understanding of future paths of study and addresses the difficulties and restrictions of quantum computing in this domain. Climate scientists and legislators can make more informed decisions by using quantum computing, therefore guiding more efficient climate adaption and mitigating techniques that support a sustainable and strong future [7].

[a]thiluxpriya499@gmail.com, [b]geethar1@srmist.edu.in, [c]ramesh.isaiah@gmail.com, [d]vidhyas2@srmist.edu.in, [e]drjnithyashri@gmail.com

DOI: 10.1201/9781003684459-52

2. Literature Survey

Discovering and forecasting climate change by simulating the interactions among the atmosphere, oceans, land, and ice depends on conventional climate models such General Circulation Models (GCMs) and Earth System Models (ESMs). These models do, however, confront major difficulties like computational complexity, data constraints, and predictive uncertainty. Using the ideas of superposition, entanglement, and quantum parallelism, quantum computing presents a revolutionary way to address computer problems. Solving high-dimensional problems, complicated differential equations, and large-scale optimization chores is appropriate for quantum computers since they can process enormous volumes of data in simultaneously. Recent developments in quantum computing show promise in materials research, encryption, and optimization among other domains. Quantum computing can boost data analysis, raise simulation accuracy, and maximize climate mitigating techniques in climate science. More effectively analyzing large and complicated climate information, quantum-based machine learning (QML) methods can find trends that would be challenging to detect with conventional models. Furthermore, the capacity of quantum computing to precisely replicate molecular interactions has resulted in discoveries in material science and chemistry, therefore facilitating the identification of improved materials for carbon capture, energy storage, and environmentally friendly manufacturing [8].

Including quantum computers into climate science could transform prediction, modeling, and mitigating techniques in general. Differential equation solving quantum techniques can greatly speed up climate simulations, hence enabling faster and more accurate climate predictions. Better predictions of extreme weather occurrences result from enhanced pattern identification and anomaly detection in large-scale climate datasets made possible by quantum machine learning (QML) for climate data analysis. By means of optimal distribution of resources, carbon capture techniques, and renewable energy deployment, quantum-enhanced optimization can strengthen climate mitigating efforts. By means of real-time climate conditions, energy demand, and grid capacity, quantum algorithms can ascertain the most optimal sites for solar and wind farms, thereby fostering a more resilient and sustainable energy infrastructure [14]. By hastening the hunt for new compounds that can increase carbon capture efficiency or battery storage, quantum computing can also help to identify materials.

3. Methodology

Employing quantum-enhanced methods to increase prediction accuracy and mitigating strategies, the approach investigates how quantum computing may be included into climate modelling [6]. Four main elements make up the approach; these are shown in Figure 52.1.

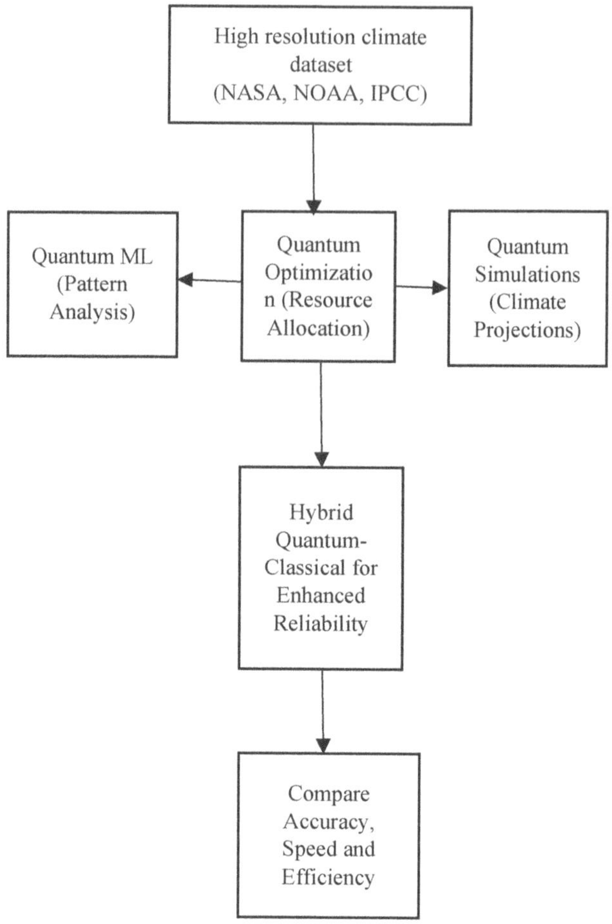

Figure 52.1. Quantum framework for climate models.

Source: Author.

3.1. Data collection

A crucial step in climate modeling is Obtaining high-resolution climate information from trustworthy worldwide sources such NASA (National Aeronautics and Space Administration), NOAA (National Oceanic and Atmospheric Administration), and the IPCC (Intergovernmental Panel on Climate Change) is a fundamental step in climate modeling. These databases cover several environmental factors that support correct climate forecasts [9]. Understanding climate dynamics and future weather changes depends on knowing temperature variations, greenhouse gas concentrations, humidity levels, and wind patterns – all of which are included in atmospheric data. Covering sea surface temperatures, ocean currents, salinity levels, and changes in marine ecosystems, oceanic data offers understanding of the ocean's part in climate control [11]. Terrestrial data emphasizes human-created environmental changes by concentrating on land use changes, patterns of deforestation, and urban heat island effects. Extreme weather records including historical data on hurricanes, droughts, floods, and wildfires also assist to better forecast and manage climate-related calamities [15]. This research guarantees that quantum-enhanced climate models are constructed on thorough and varied real-world datasets by combining satellite imaging, remote sensing data, and sensor network readings, so producing more accurate and dependable climate forecasts.

3.2. Quantum algorithm implementation

The research focuses on using quantum algorithms that, by means of data processing, pattern recognition, and optimization techniques, can improve climate modeling and mitigating activities. Climate pattern analysis benefits much from quantum machine learning (QML), which addresses the restrictions of conventional machine learning approaches struggling with high-dimensional climate datasets due to computational constraints. Using quantum superposition and entanglement, QML helps to efficiently handle vast amounts of climate data, hence improving detection of climate patterns, anomaly detection that is, early warning of extreme events and long-term climate projections. Apart from developments in machine learning, quantum optimization offers major advantages in resource allocation for techniques of climate control. While quantum optimization methods, such the Quantum Approximate Optimization Algorithm (QAOA), provide more efficient decision-making tools for renewable energy deployment, carbon capture strategies, and sustainable land-use planning, classical computers sometimes struggle with solving high-dimensional optimization problems efficiently [2]. These systems guarantee improved resource allocation for climate adaption measures, increase energy grid efficiency, and help solar panel and wind farm installation to be ideal [12]. Higher accuracy and efficiency attained by including these quantum-driven solutions would help to promote more successful environmental policies and sustainability projects by means of climate models.

3.3. Hybrid quantum-classical models

Since quantum computing has great promise, present quantum hardware is still in its early phases of development hence it is not feasible to run extensive climate simulations just on quantum computers. Leveraging the capabilities of both computational paradigms, a hybrid quantum-classical technique is used to solve this restriction. Under this paradigm, quantum algorithms are used to speed particular sub-tasks including optimization, machine learning, and differential equation solving while classical computers continue to handle the majority of large-scale climate simulations where they remain more efficient and dependable [10]. By means of this focused application of quantum computing, the general performance of climate models is improved without requiring complete quantum execution. In addition, developments in quantum algorithms can be applied to real-world problems through the use of cloud-based quantum computing platforms like Google Quantum AI and IBM Quantum, which offer scalable environments for testing and refining quantum-enhanced climate models. This research bridges the gap between developing more efficient mitigating strategies and emerging quantum technology by embedding quantum computing inside conventional climate modeling frameworks, so adopting a pragmatic and scalable approach to improve climate predictions.

3.4. Evaluation metrics

The present research compares the performance of quantum-enhanced climate models to classical models using several important evaluation criteria. Primary attention is on prediction accuracy, which compares observed data with projected climatic trends to ascertain whether quantum-assisted models increase dependability. Measuring the time needed for quantum-enhanced simulations relative to conventional calculations helps one to evaluate computational speed and highlight possible efficiency increases. Furthermore, investigated is energy efficiency to see whether quantum models, providing a more sustainable form of climate modeling, use less computational energy than conventional supercomputing techniques [13]. As quantum technology develops, scalability is ultimately tested by looking at how well quantum algorithms can manage ever complicated climate datasets. This research attempts to show the useful advantages of quantum computing in climate modeling by methodically assessing these elements, therefore providing insightful analysis of both its present capacity and future directions.

4. Results and Discussion

Emphasizing the higher accuracy of quantum computing in three main areas weather forecasting, climate predictions, and extreme storm prediction. Figure 52.2 shows a

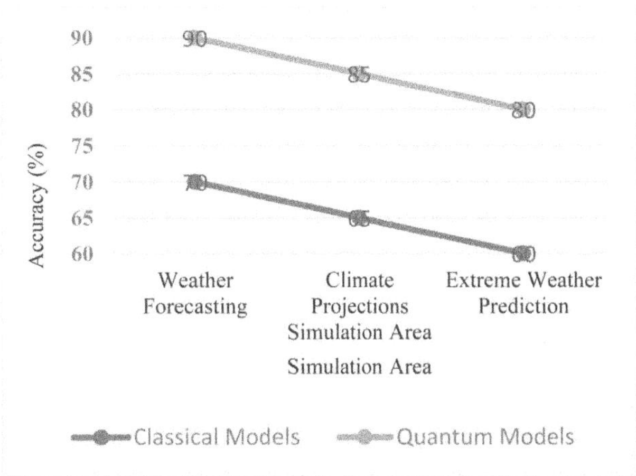

Figure 52.2. Comparison of classical vs. quantum models in climate simulations.

Source: Author.

comparison between classical and quantum models in climate simulations.

Classical models have about 70% accuracy in weather forecasting; quantum models greatly exceed them with 90% accuracy, therefore revealing their capacity to more precisely comprehend intricate atmospheric patterns. Analogously, for long-term climate forecasts, quantum models preserve a higher degree of accuracy at 85% whereas classical models show a drop to 65% accuracy. Critical weather event prediction shows the most noteworthy improvement; classical models decrease to 60% accuracy while quantum models continue to perform strongly at 80% accuracy, proving their capacity to identify anomalies and more consistently anticipate severe weather events. Moreover, it underlines generally the possibilities of quantum computing to improve climate models by means of faster processing, enhanced prediction accuracy, and greater understanding of climate variance. The growing accuracy difference between quantum and conventional models implies that quantum computing may transform environmental sustainability initiatives, disaster readiness, and climate research.

Figure 52.3 varies in three important sustainability applications carbon capture efficiency, energy grid optimization, and agricultural predictions the performance gains of quantum and conventional approaches [4]. Classical methods show approximately 60% improvement in the efficiency of carbon capture; quantum methods achieve 100%, thus showing the potential of quantum computing to identify superior CO_2 absorbing materials. In comparison, in energy grid optimization, quantum methods achieve 90% while classical methods achieve a 70% improvement, thus showing their potential to maximize energy distribution and reduce system failures. With an 85% increase over 65%, quantum computing outperforms conventional techniques in agricultural predictions, thus enhancing water resource management and crop production forecasting precision. In comparison to traditional methods, the graphic highlights the principal

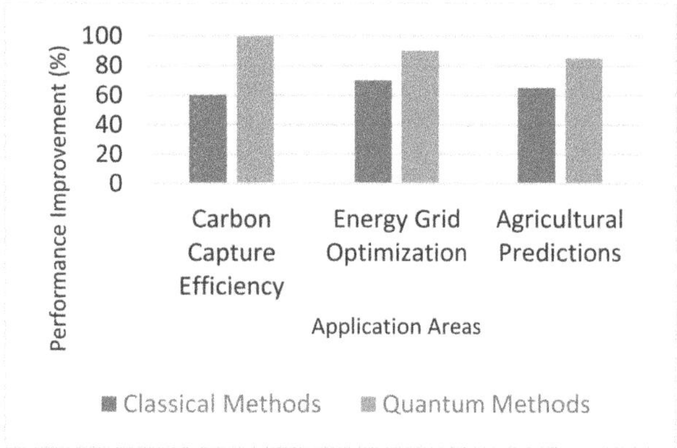

Figure 52.3. Quantum vs. classical performance in sustainability applications.

Source: Author.

advantages of quantum computing in sustainability solutions by emphasizing more efficiency, improved optimization, and superior predictive capabilities. Such findings suggest that quantum technologies could play a significantly crucial role in enhancing environmental sustainability and tackling problems related to climate change with more success.

5. Strengths and Challenges

In the field of climate science, quantum computers have three main benefits. Its higher computing power is one of its most significant benefits because it allows for quicker and more accurate climate simulation compared to conventional supercomputers. This added processing power for climate-related events like hurricanes, droughs, and heat waves makes earlier warnings and improved preparedness possible, thus improving prediction accuracy. In addition, quantum algorithms achieve maximum energy and resource optimization, thus making it possible to design sustainable agriculture practices and more efficient renewable energy systems. Quantum computing does, however, still pose some challenges despite these advantages. One primary limitation is the current state of quantum hardware, particularly with regards to issues of qubit stability, error rates, and scalability. These technological limitations hinder the universal application of quantum solutions. Additionally, hybrid quantum-classical solutions are typically needed because current quantum systems cannot now single-handedly handle complex climate models. Interpretation of data in quantum-boosted climate models is also an area of challenge, as it is still a complex and emerging science to make meaningful conclusions from outcomes delivered by quantum computations.

6. Conclusion

Climate-modeling revolution through quantum computing increases accuracy and efficacy in projections relative to environmental change. Quantum computing provides more accurate long-term forecasts of climate, enhanced identification of extreme weather events, and streamlined processing of complex climate data relative to traditional models. While the traditional computing method is still fundamental, adding the enhanced quantum methodologies helps to cross limitations in processing and generate higher precision, evidence-based climate inferences. This advancement provides powerful tools for carbon sequestration, grid optimization, and farm resource management, thus revolutionizing sustainability efforts. How quantum technology evolves will render its role in mitigating climate problems and advancing global sustainability solutions increasingly important.

6.1. Future scope and recommendation

A number of crucial steps should be placed at the top of the agenda if we are to make complete use of quantum computing for sustainability and climate modeling. Improve quantum algorithms and enhance hardware stability by way of investment in quantum research, thereby addressing issues like qubit errors and scalability limits. Making practical applications possible will be hastened through the encouragement of collaboration between climate scientists and quantum professionals. This will make advancements in quantum computing result in tangible solutions for global warming. In an attempt to outsmart the limitations of current quantum technology, scientists are shifting their attention towards hybrid quantum-classical methods, which blend the most superior characteristics of the two schools of quantum theory to optimize computational efficiency. Additionally, government support for quantum-powered climate solutions is important because governments and businesses ought to promote initiatives involving quantum computing to tackle carbon capture, energy efficiency, and climate resilience strategies. These recommendations will allow quantum computing to design more precise climate forecasts, improved sustainability strategies, and more ecological resilience, therefore altering our method of dealing with climate issues.

References

[1] Sahoo, S., & Timmann, P. (2023). Energy Storage Technologies for Modern Power Systems: A Detailed Analysis of Functionalities, Potentials, and Impacts. *IEEE Access*, *11*, 49689–49729. doi:10.1109/ACCESS.2023.3274504.

[2] Varshavardhini, S., & Rajesh, A. (2023). An Efficient Feature Subset Selection with Fuzzy Wavelet Neural Network for Data Mining in Big Data Environment. *Journal of Internet Services and Information Security*, *13*(2), 233–248. https://doi.org/10.58346/JISIS.2023.I2.015

[3] Chandravanshi, N., & Neetish, K. (2023). Diurnal Variations in Greenhouse Gas Emissions from a Macrophyte-Covered River. *Aquatic Ecosystems and Environmental Frontiers*, *1*(1), 11–15.

[4] Quantum Machine Learning in Climate Change and Sustainability: A Short Review A. Nammouchi, A. Kassler, A. Theocharis DOI: 10.1609/aaaiss.v2i1.27657

[5] Manfredi, S., Ceccato, M., Sciarretta, G., & Ranise, S. (2022). Empirical Validation on the Usability of Security Reports for Patching TLS Misconfigurations: User-and Case-Studies on Actionable Mitigations. *Journal of Wireless Mobile Networks, Ubiquitous Computing, and Dependable Applications*, *13*(1), 56–86. https://doi.org/10.22667/JOWUA.2022.03.31.056

[6] Dahale, G. R. (2023). Quantum Simulations for Carbon Capture on Metal-Organic Frameworks. *2023 IEEE International Conference on Quantum Computing and Engineering (QCE)*. Bellevue, WA, USA, 89–93. doi:10.1109/QCE57702.2023.10189.

[7] Hu, X., & Sinniah, S. (2024). The Role of Green Risk Management Approaches in Promoting Green and Sustainable Supply Chain Management. *Natural and Engineering Sciences*, *9*(2), 33–54. https://doi.org/10.28978/nesciences.1569144

[8] Jaiswal, H., & Pradhan, S. (2023). The Economic Significance of Ecosystem Services in Urban Areas for Developing Nations. *Aquatic Ecosystems and Environmental Frontiers*, *1*(1), 1–5.

[9] Surendiran, B., Dhanasekaran, K., & Tamizhselvi, A. (2022). A Study on Quantum Machine Learning for Accurate and Efficient Weather Prediction. *2022 Sixth International Conference on I-SMAC (IoT in Social, Mobile, Analytics and Cloud) (I-SMAC)*. Dharan, Nepal, 534–537. doi:10.1109/I-SMAC55078.2022.9987293.

[10] Assegid, W., & Ketema, G. (2023). Assessing the Effects of Climate Change on Aquatic Ecosystems. *Aquatic Ecosystems and Environmental Frontiers*, *1*(1), 6–10.

[11] Takao, T., Arisaka, G., Sonoda, S., Fukano, K., Katayama, T., & Shirotani, Y. (2022). Increase of Thermal Stability Caused by High Thermal Conduction Fiber and Sheet in Conduction-Cooled HTS Coils. *IEEE Transactions on Applied Superconductivity*, *32*(6), 1–5. doi:10.1109/TASC.2022.3171186.

[12] Ziwei, M., & Han, L. L. (2023). Scientometric Review of Sustainable Land Use and Management Research. *Aquatic Ecosystems and Environmental Frontiers*, *1*(1), 21–24.

[13] Iyengar, S., & Bhattacharya, P. (2024). Assessing the Effects of Climate Change on Population Displacement and Migration Patterns in Coastal Communities. *Progression Journal of Human Demography and Anthropology*, *1*(1), 15–21.

[14] Radhakrishna, K., & Sreeja, T. K. (2025). Cluster Formation for Lane and Road Based Stability in Infrastructure-Based Vehicular Ad Hoc Networks (LRSC). *Cybernetics, Human Cognition, and Machine Learning in Communicative Applications* (pp. 1–13). Singapore: Springer Nature Singapore.

53 Awareness of NEP 2020 on in-service teacher education: A professional development perspective for elementary school teachers

Pranati Palai[1,a] *and G. C. Nanda*[2,b]

[1]Research Scholar, Department of Education, Ravenshaw University, Cuttack, Odisha, India
[2]Professor (Retd.), Department of Education, Ravenshaw University, Cuttack, Odisha, India

Abstract: The National Education Policy (NEP) 2020 significantly reformed India's educational system, emphasizing teacher education and professional development. This study explores the state of awareness of NEP 2020 for in-service educators in primary school and its possible impact on professional development activities. The strategy lays out a plan for raising teaching standards through localized, context-sensitive teaching strategies, technological integration, and ongoing professional development. The paper describes the introduction of new modules, the overall comprehensive approach to teacher development, and the focus on CPD in line with international standards through an analysis of NEP 2020. It also addresses the obstacles to raising awareness, including lack of infrastructure, resource constraints, and resistance to change, and offers solutions. We have considered case studies from various Odisha regions to raise awareness and demonstrate the use of NEP-aligned training programs. It illustrates how such initiatives give rise to creative practices. The study outlines potential directions for creating technology-based and feedback-mechanism-based sustainable frameworks to improve teacher awareness and professional growth. Raising awareness of NEP 2020 may alter teacher in-service training, which could increase the standard of primary education in India.

Keywords: National Education Policy (NEP) 2020, in-service teacher education, professional development, elementary school teachers, teacher training reforms

1. Introduction

The National Education Policy (NEP) 2020 is a paradigm change in the Indian education system towards creating an all-around, inclusive, and equitable learning environment [1]. It focuses on high-quality teaching, reimagined curricula, and the use of technology to promote lifelong learning. It acknowledges teachers as the pillars of a successful education system [2]. It places professional development and teacher training as necessary change agents at the forefront of its agenda [3]. Along with innovative pedagogic approaches and rigorous training drills, the NEP seeks to raise awareness among teachers regarding the policy's goals to address the needs of their students more effectively [4]. These needs of the NEP 2020 apply to elementary teachers, as they will define the foundational years of learning. In-service training courses under the policy require giving teachers knowledge and skills that will empower them to manage the complexities of the modern classroom. To achieve this, the programs promote a culture of reflection, collaboration, and responsiveness [5].

This article examines in-service primary school teachers' understanding of NEP 2020 and its implications for professional development [6]. It examines the policy guidelines to show how awareness-raising could enhance teacher preparation. Through a discussion of reforms, opportunities, and challenges, the debate shows how NEP 2020 may transform the teaching-learning ecosystem at the elementary school level [7].

2. Contextual Background

Indian teacher education, especially at the elementary level, grapples with serious challenges that impact foundational literacy and numeracy. Several teachers receive poor pre-service training, which translates into pedagogical ability and subject matter weaknesses. In-service training programs for teachers are also theoretical and lack practical classroom conditions. Imbalances between rural and urban teaching setups, like resource allocation and mobilization, further intensify disparities in learning achievements. Sudden advancements in digital technology in the education sector have further expanded this gap, with most teachers failing to use digital tools properly, especially in schools with resource constraints.

Ongoing professional development programs are critical to solving these problems, providing teachers with current instructional methods, critical thinking, creativity, effective

[a]pranatipalai@gmail.com, [b]gcnanda2005@yahoo.co.in

DOI: 10.1201/9781003684459-53

communication, problem-solving abilities, and carrying out Action Research projects. NEP 2020 prioritizes specialized Foundational literacy and numeracy training through innovative, child-centric methods. Accessible, context-specific, and pedagogy-oriented training will fill skill gaps, make teaching more effective, and enhance learning outcomes for young children, building a stronger and more equitable elementary education system.

3. Key Provisions of NEP 2020 Related to Teacher Education

NEP 2020 brings revolutionary modifications in teacher training during service to reinforce professional growth. It focuses on fresh training modules with skill-based, technology-supported, and pedagogy-centric learning. The modules utilize digital media such as MOOCs and address emerging technologies such as AI and augmented reality to involve the learners. They also make experiential, child-centered pedagogies possible [8].

The policy incorporates an integrated teacher development plan interlinking subject matter and application. It promotes teachers to appreciate local contexts, culture, and morality and encourages critical thinking and creativity in children [5].

There is an emphasis on Continuous Professional Development (CPD), and 50 hours of professional development each year are made compulsory. CPD follows international best practices, developing pedagogic capabilities, classroom management, and inclusivity. Peer learning and lifelong professional development are promoted by NEP 2020, and quality education in the primary segment is guaranteed [3, 9].

4. Implications for In-service Teacher Training

NEP 2020 has far-reaching implications for in-service teachers' training in making teachers aware of digital platforms, integrating local context, and foundational literacy and numeracy (FLN).

The policy encourages digital learning platforms, such as AI-based and MOOCs, to offer teachers adaptive, flexible training. They facilitate cooperative learning via webinars and forums, although issues such as restricted internet connectivity and digital literacy need to be resolved [3, 9].

NEP 2020 also focuses on the inclusion of local context in teacher education, and it supports the utilization of local languages and culturally appropriate instructional materials. Through this, there is increased engagement and meaning made to learning by the students, especially in rural and tribal regions [5, 11].

A central emphasis is on FLN, with specialized teacher training in play-based and experiential learning approaches. Embedded assessment and data-driven practices enable monitoring of student progress and personalized learning. These strategies enhance education equity and lifelong learning [8, 12].

5. Challenges and Barriers

Despite its transformative vision, NEP 2020 faces several challenges in implementing in-service teacher education, including resource constraints, teacher resistance, and inadequate evaluation mechanisms.

Resource constraints, especially rural, are hindrances to accessing digital infrastructure, training budgets, and technological development tools. Digital divide, unstable internet, and inadequate technical support limit the extent to which teachers can take advantage of online learning platforms [3, 11]. Investment by the government and private institutions would bridge these gaps.

Resistance to change by teachers and administrators hinders the implementation of new teaching practices. Teachers are reluctant because they are not familiar with digital tools and are doubtful about their efficacy. Increasing awareness, offering mentorship, and creating teacher-focused training programs can facilitate this shift [5, 8].

Challenges in evaluation stem from the lack of formal mechanisms to evaluate the effect of training. NEP 2020 demands stringent monitoring, encompassing self-assessments, classroom observations, and data analytics, to make professional development worthwhile [10, 12].

6. Strategies for Effective Implementation of NEP 2020

For NEP 2020's in-service teacher training to be impactful, practical implementation strategies are crucial. This involves **policy recommendations, stakeholder collaboration, and continuous review mechanisms** to ensure professional development is relevant and effective.

6.1. Policy recommendations

- **Increased Budgetary Allocation:** Sufficient funding is necessary for digital tools, training centers, and teacher incentives, particularly in rural areas [3].
- **Contextualized Training Modules:** Training should incorporate regional languages, cultural contexts, and local challenges to enhance relevance [11].
- **Digital Inclusion:** Bridging the digital divide requires affordable internet, subsidized devices, and technical support [10].
- **Mandatory Professional Development:** Regular skill enhancement should be enforced with certification-based incentives [8].

6.2. Stakeholder collaboration

- **Teacher Involvement:** Active participation ensures training aligns with classroom realities. Peer networks foster a professional learning culture.
- **Institutional Support:** Schools should provide resources, dedicated training time, and platforms for knowledge-sharing.

- **Public-Private Partnerships (PPPs):** Collaborations with private organizations can introduce innovative digital training solutions [12].
- **Community Engagement:** Involving local communities connects training to students' real-world needs.

6.3. Periodic review and feedback

- **Impact Assessments:** Classroom observations, self-assessments, and student data gauge effectiveness [10].
- **Data-driven Decision-making:** Analytics track participation and identify improvement areas.
- **Regular Feedback Loops:** Surveys and discussions ensure dynamic, need-based training.
- **Third-Party Evaluations:** Independent reviews improve transparency and effectiveness [5].
- **Recognition and Rewards:** Acknowledging innovative teaching boosts motivation and participation.

By implementing these strategies, NEP 2020 can enhance in-service teacher training and promote sustainable professional development.

7. Case Studies and Best Practices in In-Service Teacher Training

Real-life examples of in-service teacher training under NEP 2020 highlight effective strategies for raising awareness and translating policy into practice. Odisha has pioneered several innovative approaches, significantly impacting teacher education.

7.1. Successful in-service teacher training programs

- **DIKSHA Platform:** A digital tool enabling continuous teacher training, DIKSHA has reached over 10,000 teachers in Odisha, improving engagement and learning outcomes. Government-NGO collaborations ensured smooth implementation [10].
- **Peer Learning Circles in Tribal Areas:** Initiatives in Mayurbhanj and Rayagada integrated local culture into lesson plans, enhancing teacher morale and student participation [13].
- **FLN Program by NGOs:** Using activity-based learning in Ganjam and Kalahandi, FLN scores improved by 20% within a year, demonstrating the effectiveness of contextual teaching [14].

7.2. Insights from innovative schools

- **Activity-Based Learning (Bhubaneswar):** Hands-on experiments improved student engagement and teamwork.
- **Multilingual Education (Kandhamal):** Teaching in mother tongues reduced dropout rates and improved comprehension.
- **Digital Learning (Cuttack):** Gamified applications enhanced literacy and numeracy teaching.

8. Future Directions in In-Service Teacher Training

The future of in-service teacher training under NEP 2020 must be adaptive, inclusive, and aligned with evolving educational needs. A sustainable teacher development ecosystem will ensure educators are equipped for future challenges.

8.1. Sustainable teacher training ecosystem

A structured lifelong learning framework is essential, focusing on:

- **Continuous Development:** Micro-credentials, modular courses, and career-long training pathways.
- **Incentives for Upskilling:** Financial rewards, recognition, and career growth opportunities.
- **Teacher Learning Communities:** Peer networks for collaboration, innovation, and support.

8.2. Role of technology

- **AI-Driven Personalization:** Tailored training based on performance analysis.
- **Machine Learning Evaluation:** Data-driven program optimization.
- **Immersive Learning:** VR/AR simulations for practical teaching experiences.

8.3. Bridging the digital divide

- **Offline & Low-Bandwidth Solutions:** Radio-based and preloaded digital content.
- **Infrastructure Development:** Affordable devices and internet expansion.

8.4. Alignment with vision 2030 and SDGs

- Global Standards with Local Relevance
- Inclusive Classrooms & Diversity Sensitivity
- SDG-Focused Training for Sustainability & Social Justice

Future teacher education must integrate technology, equity, and global best practices to create effective, future-ready educators.

9. Conclusion

NEP 2020 has the potential to revolutionize elementary education in India by the focus on in-service training of teachers. Its success is contingent upon mass awareness among teachers, institutions, policymakers, and the society at large. In addition to simply apprising teachers, measures should be taken to make them feel a sense of ownership and motivation towards professional and technological development. Specific attention should be drawn to the backward regions to provide even access to training facilities. Stakeholder collaboration, peer learning, and common best practices will drive continued momentum. Educating teachers is crucial to making NEP 2020's vision for quality education a success.

References

[1] Sarvi, B., & Pourmozaffar, S. (2024). A review of the reproduction and rearing of clownfish, Amphiprion clarkii, in captivity. *International Journal of Aquatic Research and Environmental Studies*, *4*(2), 145–158. http://doi.org/10.70102/IJARES/V4I2/9

[2] Rabet, F., & Mousavi, S. A. (2017). Performance Evaluation of Contractors of Shiraz Municipality using Balanced Scorecard. *International Academic Journal of Innovative Research*, *4*(1), 20–29.

[3] Ministry of Education. (2020). *National Education Policy 202*Government of India.

[4] Kabasakal, H., & Özbek, E. Ö. (2022). Length-weight relation of the angular rough shark, Oxynotus centrina (Linnaeus, 1758) in the Mediterranean Sea. *Natural and Engineering Sciences*, *7*(2), 97–107. http://doi.org/10.28978/nesciences.1159204

[5] National Council for Teacher Education (NCTE). (2021). *Guidelines for teacher education reforms under NEP 2020.* Government of India.

[6] Ramya, S., & Sudha, N. (2016). Performance of Bacterial Foraging Algorithm (BFO) to Generate Prioritized Optimal Path in Software Testing Environment. *International Journal of Advances in Engineering and Emerging Technology*, *7*(3), 09–25.

[7] Rahmah, N. M. (2023). Heat Exchangers Using Multi Types of Nanofluids: A Review. *International Academic Journal of Science and Engineering*, *10*(2), 51–61. https://doi.org/10.9756/IAJSE/V10I2/IAJSE1008

[8] Singh, R. (2021). Transforming teacher education: Insights from NEP 202*Journal of Pedagogical Innovations*, *14*(2), 67–78.

[9] Sharma, A. (2021). Continuous professional development in the NEP 2020 framework. *Educational Research and Reviews*, *29*(3), 45–52.

[10] Sharma, A. (2021). Digital platforms in teacher professional development: Opportunities and challenges under NEP 20*Educational Innovations Journal*, *12*(1), 34–50.

[11] Kumar, S. (2021). Local contexts and teacher training: Aligning NEP 2020 with cultural realities. *Indian Educational Review*, *59*(3), 67–80.

[12] Kapur, R. (2021). Foundational literacy and numeracy: A cornerstone of NEP 20*Journal of Early Childhood Education Research*, *15*(2), 45–58.

[13] Patnaik, B. (2021). Peer learning for professional development: An Odisha perspective. *Journal of Innovative Pedagogies*, *8*(4), 56–70.

[14] Mishra, A. (2021). Tribal education and teacher training in Odisha: Successes and lessons learned. *Odisha Education Journal*, *10*(1), 34–49.

54 Evaluating the influence of multinational leadership on business success in global teams

R. K. Kasunthika[1,a], *D. Kalidoss*[2,b], *Hemalatha Nagharajan*[3,c], *Sivakumar Pushparaj*[4,d], *and Balaji Ramalingam*[5,e]

[1]Lecturer, Edith Cowan University, Sri Lanka
[2]Associate Professor, Kalinga University, Raipur, Chhattisgarh, India
[3]Data Engineering Lead, London Stock Exchange Group, Bangalore, India
[4]Assistant Professor, Panimalar Engineering College, Chennai, Tamil Nadu, India
[5]Professor, Saveetha School of Engineering, Saveetha Institute of Medical and Technical Sciences, Saveetha University, Chennai, Tamil Nadu, India

Abstract: Due to advancements in innovation, globalization, and technological advances, Global Teams (GT) have become essential for several businesses. Advancements in computer technology have given businesses novel ways to establish and oversee GT. In GT, Multinational Leadership (ML) operates via digital tools, distinguishing it from conventional styles of leadership seen in in-person contexts. ML is crucial in this context, and based on an empirical study that emphasizes multiculturalism, it has been suggested that leaders' exposure to diverse cultures positively influences the link between societal variation and Business Success (BS). This work recognized the ML's ethnicity (native vs foreign to the nation's culture) and the duration of the ML as significant determinants. The impact of GT's cultural variation on BS is more favourable with an ML than with a domestic leader and an ML possessing an extended GT term. Furthermore, it has been anticipated that one moderating effect compensates for the other, indicating that the impact of a leader's ethnic origin is more pronounced for ML with less duration with the GT.

Keywords: Global teams, multinational leader, ethnicity, business success, native, foreign, culture

1. Introduction and Review of Literature

With the swift progression of science and technology, communication has evolved, enhancing the quality of interaction via digital interaction tools. In today's world, everyone, irrespective of their status, can access and utilize the information they desire due to technological advances [1]. The rise in digitization globally has enhanced communication among persons across many locations. In the latter stages of the twentieth century, personal computing devices prompted businesses to embrace technological advancements, yielding benefits such as establishing global enterprises, time efficiency, removing geographical constraints, and reducing spatial distances [8].

Businesses must navigate dynamic, ever-evolving, and intricate environments, presenting numerous problems. Commercial considerations in the altered economic landscape, characterized by modified marketplace dynamics, prompt a trend toward developing diverse organizational structures that may address the requirements of the contemporary day [2]. Research indicates that firms must assemble the appropriate individuals at the optimal moment to achieve specific objectives. At this juncture, GT has arisen to meet the demands of the contemporary period [9].

Multinational corporations and ML are utilizing GT more extensively than in previous periods. Although a centralized management system exists for every team within such corporations, these divisions must adhere to interrelated regulations and for organizations to oversee smaller corporate units in foreign nations [3]. ML and business entities possess subsidiaries, affiliates, and partnerships distinct from the centralized leadership [12]. Affiliates and small firm structures in several locations operate in unison with those at the corporate office [4]. The production and sale of goods and amenities occur within the same company, while connected entities conduct marketing and shipment [11]. ML, regarded as pivotal in the evolving global landscape, is responsible for economic endeavours. Their continual expansion and augmentation of power are attributable to the necessity for manufacturing, consumption, and dominating capital to proliferate throughout nations [5]. The proliferation of technical advancements has enabled a significant rise in transnational corporate structures. Such organizations, which minimize

[a]k.ratiyalage@ecu.edu.au /r.k.kasunthika@gmail.com, [b]dr.kalidoss@kalingauniversity.ac.in, [c]Hemalatha.nagharajan@lseg.com, [d]sivakumarpushparaj@gmail.com, [e]balaji_2410@yahoo.co.in

DOI: 10.1201/9781003684459-54

interaction and delivery expenses, have segmented their manufacturing process through technological advancement [6, 7]. Consequently, the global proliferation of this sort of organization has been achieved by facilitating the execution of all conceivable manufacturing and service-related processes across various countries to achieve BS [10].

2. Proposition Development

2.1. ML's ethnicity and GT's multiculturalism

In global enterprises, even with an organization's operations in other countries, the predominant scenario is that the host culture or ethnicity is significantly more prevalent in GT members. A crucial aspect of a ML's ethnic origin is to consider whether the ML is 'regional' or 'foreign'; specifically, whether the leader originates from the country of residence and generally shares the heritage of a large proportion of GT members within that country, or hails from a different nation, thereby representing a minority group relative to most GT members in the country where they are staying. This element is becoming increasingly significant as the proportion of ML in international company rises [13].

Increased awareness among ML regarding cultural disparities and responses to radically different individuals in their home nation may be crucial in alleviating cross-cultural tensions among GT members, thereby enhancing BS in diverse groups and mitigating the possible negative consequences of diverse cultures. Simultaneously, due to an enhanced understanding of the various viewpoints linked to cultural variations and an improved capacity to assist culturally distinct GT members in acclimating to the local culture and their roles, they may also play a crucial role in directing the GT towards learning possibilities and knowledge advantages that diverse cultures can provide, thereby fostering the BS derived from GT's diversity. Consequently, the following hypothesis has been developed:

Proposition 1: GT's cultural variation has a more favorable correlation with BS under a ML compared to a regional leadership.

2.2. ML duration and GT ethnic variation

Managing a GT may provide valuable intercultural expertise, and the longer a ML remains in their position, the more expertise they gather. Consequently, it has been anticipated that ML duration in the GT will positively moderate the association among GT ethnicity and BS. Accumulating expertise leads to the enhancement of one's learning. Increased expertise in managing culturally varied GT, along with an extended duration, should correlate with an enhanced comprehension of cultural variation. The idea that individuals progressively enhance their ability to navigate cultural variety aligns with existing diversity studies.

Higher ML duration is associated with increased confidence and awareness of differing values and views within the GT, leading to enhanced collaboration and BS. Consequently, by managing GT, executives can acquire an understanding and proficiency in managing cultural diversity within their GT. The duration of a ML's stay with the GT may correlate with an enhanced capacity to mitigate the disruptive impacts of ethinicity and to foster synergistic BS.

Proposition 2: The favourable correlation between GT's ethnicity and BS increases with longer ML duration.

2.3. The ML duration mitigates the impact of a ML's ethnicity

The foreign ethnicity and duration with the GT are recognized as moderating factors in the association between GT ethnicity and BS, as ML is anticipated to have acquired insights into ethnic variation through their international experiences or their duration of working within a multicultural group. The result that ML duration with the GT contributes less to the comprehension of ethnic variety for ML compared to local administrators suggests a triadic interaction including ML ethnicity, ML duration, and GT's cultural variation.

Proposition 3: The moderating influence of a ML's ethnicity on the link between cultural variation and BS is more pronounced with shorter ML duration.

3. Results and Discussion

Questionnaire have been distributed to 275 individuals across 65 GT and their respective team leaders, located in 35 different units throughout 20 nations. GT were engaged in several technical sectors, including field operations, sales, production, branding, and recruitment. Replies have been obtained from 63 GT. The GT had individuals from 20 nations, with 128 male and 135 female GT members.

Figure 54.1 depicts the impact of ethnicity on BS for ML duration. It suggests that people from diverse ethnic origins achieve differing degrees of BS based on the length of their leadership tenures. One group has higher values than the other, indicating a possible association between ethnicity and ML efficacy in international contexts. Furthermore, BS seems to be elevated for shorter ML durations and diminishes with

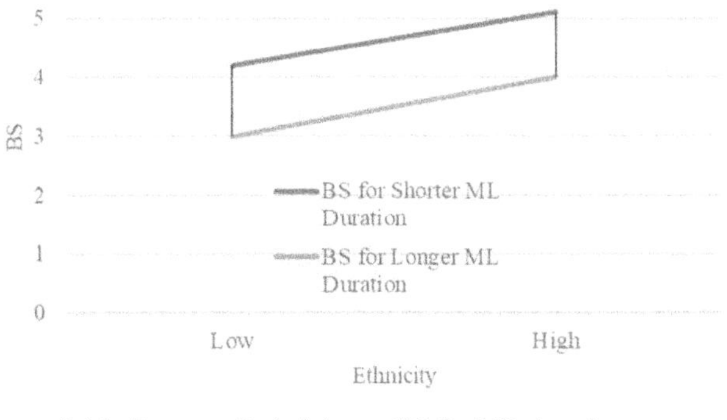

Figure 54.1. Impact of ethnicity on BS for ML duration.

Source: Author.

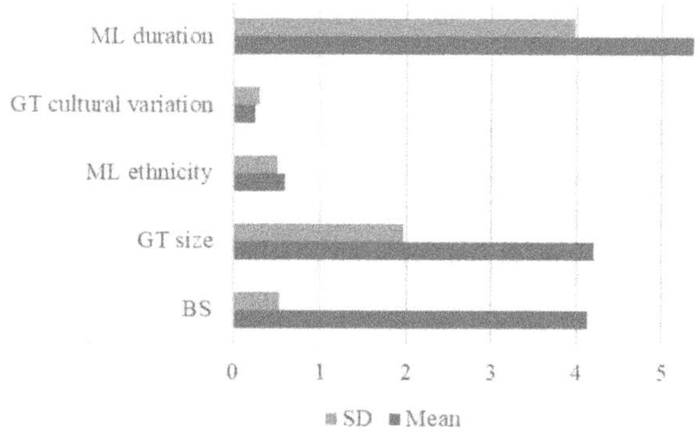

Figure 54.2. Mean and SD for different factors involved in hypothesis development.

Source: Author.

longer durations in both groups, suggesting that ML length may potentially influence BS.

Figure 54.2 illustrates the Mean and SD for different factors involved in hypothesis development. BS has a consistently steady mean with little fluctuation. The size of the GT exhibits more diversity, reflecting discrepancies in team sizes across situations. ML ethnicity and GT cultural variance have reduced mean values and standard deviations (SD), indicating constrained variability in these dimensions. The duration of ML exhibits the greatest variety, indicating substantial disparities in leadership duration across instances. These insights provide a fundamental comprehension of the principal factors affecting the hypothesis framework.

4. Conclusion

This study acknowledged the ethnicity of the ML and the duration of the GT as critical factors. The influence of GT's cultural diversity on BS is more advantageous with an ML than with a domestic leader, particularly when the ML has a lengthy GT tenure. Moreover, it is predicted that one moderating effect offsets the other, suggesting that the influence of a leader's ethnic background is more significant for ML with shorter time with the GT. Results suggests that people from diverse ethnic origins achieve differing degrees of BS based on the length of their leadership tenures. BS seems to be elevated for shorter ML durations and diminishes with longer durations in both groups, suggesting that ML length may potentially influence BS. ML ethnicity and GT cultural variance have reduced mean values and SD, indicating constrained variability in these dimensions. The duration of ML exhibits the greatest variety, indicating substantial disparities in leadership duration across instances. These insights provide a foundational understanding of the key factors influencing the hypothesis framework.

References

[1] Batırlık, S. N., Gencer, Y. G., & Akkucuk, U. (2022). Global virtual team leadership scale (GVTLS) development in multinational companies. *Sustainability*, *14*(2), 1038.

[2] Lu, J. G., Swaab, R. I., & Galinsky, A. D. (2022). Global leaders for global teams: Leaders with multicultural experiences communicate and lead more effectively, especially in multinational teams. *Organization Science*, *33*(4), 1554–1573.

[3] Soon, C. C., & Salamzadeh, Y. (2021). The impact of digital leadership competencies on virtual team effectiveness in MNC companies in Penang, Malaysia. *Journal of Entrepreneurship, Business and Economics*, *8*(2), 219–253.

[4] Rickley, M., & Stackhouse, M. (2022). Global leadership effectiveness: A multilevel review and exploration of the construct domain. *Advances in Global Leadership*, *14*, 87–123.

[5] Cuypers, I. R., Patel, C., Ertug, G., Li, J., & Cuypers, Y. (2022). Top management teams in international business research: A review and suggestions for future research. *Journal of International Business Studies*, *53*(3), 481–515.

[6] Hung, S. W., Cheng, M. J., Hou, C. E., & Chen, N. R. (2021). Inclusion in global virtual teams: Exploring non-spatial proximity and knowledge sharing on innovation. *Journal of Business Research*, *128*, 599–610.

[7] Talebi Fard, P., & Leung, V.C.M. (2011). Context-Aware Mobility Management in Heterogeneous Network Environments. *Journal of Wireless Mobile Networks, Ubiquitous Computing, and Dependable Applications*, *2*(2), 19–32.

[8] Indrawan, P. E., Parwati, N. N., Tegeh, I. M., & Sudatha, I. G. W. (2024). Trends in the Use of Augmented Reality in Character Development within Local Wisdom in Schools: A Bibliometric Study. *Indian Journal of Information Sources and Services*, *14*(4), 7–15. https://doi.org/10.51983/ijiss-2024.14.4.02

[9] Oblomurodov, N., Madraimov, A., Palibayeva, Z., Madraimov, A., Zufarov, M., Abdullaeva, M., Pardaev, B., & Zokirov, K. (2024). A Historical Analysis of Aquatic Research Threats. *International Journal of Aquatic Research and Environmental Studies*, *4*(S1), 7–13. https://doi.org/10.70102/IJARES/V4S1/2

[10] Khlefha, A. R. (2023). An Efficient Method for Solving Nonlinear Boundary Value Problems Via the Non-Polynomial Tension Spline Function. *International Academic Journal of Science and Engineering*, *10*(2), 176–183. https://doi.org/10.9756/IAJSE/V10I2/IAJSE1022

[11] Hari Shankar, S. (2016). Moving Object Detection Using Integration Frame Work-SILTP Color and Texture Information with Foreground Extraction Methods. *International Journal of Advances in Engineering and Emerging Technology*, *7*(3), 26–38.

[12] Bakari, S. (2018). Impact of Exports and Imports on Economic Growth in Canada: Empirical Analysis Based on Causality. *International Academic Journal of Innovative Research*, *5*(1), 17–29. https://doi.org/10.9756/IAJIR/V5I1/1810003

[13] Rangarajan, S., & Al-Quraishi, T. (2023). Navigating the future of the internet of things: Emerging trends and transformative applications. *Babylonian Journal of Internet of Things*, *2023*, 8–12. https://doi.org/10.58496/BJIoT/2023/002

55 Advancing solar and wind power for sustainable city models

Sarvinoz Sayfullayevna Kasimova[1], Gulnora Sattorovna Mutalova[2], Buzakhro Marufjanovna Begmatova[2], Markhabo Raxmonkulovna Abdullayeva[2,a], and Doniyor Dovronovich Ergashev[3,b]

[1]Associate Professor, University of Journalism and Mass Communications of Uzbekistan, Uzbekistan
[2]Associate Professor, Tashkent State University of Oriental Studies, Uzbekistan
[3]English Teacher, Faculty of Tourism, Alfraganus University, Uzbekistan

Abstract: The transition among sustainable development in urban areas to increase the advancement technologies of renewable energy such as solar and wind power. Recently, cities have grown through the demands of energy rises and integrated solar and wind power into city infrastructure, which has been achieved through a long-term basis of economic sustainability. This research study mainly discussed the overall development of the smart city, which contains renewable energy resources in solar energy applications with sustainable developments. Solar and wind power help reduce carbon emissions and improve the environment. This research paper explains the technological advancements in adopting solar and wind energy systems related to city models in various areas. This also explains that the sustainable city model consists of a healthy lifestyle, Green economic growth, sustainable building, and Transportation.

Keywords: Wind power, solar power, renewable energy, smart building, transportation, healthy lifestyle

1. Introduction

Solar and Wind energy are essential renewable energy resources. Different types of technologies follow both energies. Solar energy should be contributed to achieving energy solutions, followed by sustainable developments. This result shows that solar energy is produced explicitly through daily data collection. Solar PV cells are applied through various technologies developed to meet energy needs. Based on the result, large capacity should be used for global solar energy applications, which contain benefits along with the energy sectors and help to develop sufficiently. This type of analysis included economic and environmental development. Solar energy applications contribute to sustainable development to meet various requirements such as energy needs, employee modeling, and improved environmental protections. The common usage of renewable energy sources includes wind power, which helps to reduce gas emissions and is currently followed by sustainable alternatives to fossil fuels. Wind energy should be followed by various resources, which are considered primary in urban areas. The financial and engineering bases follow the result. Most of the issues are resolved by wind energy systems in urban areas. Additionally, the large buildings have the standard developments of sustainable development with unlimited space and contain 70% of the people who should be predicted with urban areas in 2050.

2. Role of Solar and Wind Energy in Sustainable Development

Several principles, such as energy generation, utilization, and distribution, follow sustainable development in the energy sector [1]. Energy systems directly impact the environment. The needs of a global sustainable energy system include the following: to minimize emissions and maximize efficiency [2]. Following the economic point, the sustainable development process uses common types of long-term energy availability and climate benefits. This is followed by the IEA recovery plan, which helps to analyze and improve the economic level quickly. It depends upon the energy infrastructure with the short-term specifications [3]. The sustainable development process includes innovative grid technology, energy pricing, and renewable energy applications [4]. To meet the electrical system's various needs, reduce the CO2 emission, and preserve the system stability [5].

High-quality infrastructure must adhere to international standards. Participation in a global PV market will be possible by establishing legislative convergence, boosting competition, and encouraging innovation [6]. Many nations could gain from playing a more proactive part in establishing international solar PV standards. Countries at the forefront of solar photovoltaic production and installation have embraced these global standards, greatly enhancing the growth of clean energy. Additional funding and capacity-building to improve

[a]A.Marxabo.R@Mail.Ru, [b]D_Ergashev@Afu.Uz

DOI: 10.1201/9781003684459-55

infrastructure quality in developing nations can promote wider acceptance and compliance with international solar PV standards. Consequently, such assistance can harmonize legislative requirements and frameworks, supporting safe and high-quality solar PV product marketing [7]. In addition to being a crucial part of renewable energy, wind energy has several benefits, such as being economical, dependable, and ecologically beneficial. These days, power is generated using it. Considering Wind's vast potential, it is essential to examine how much energy it consumes. Using renewable energy in sustainable development offers a strong basis for global cooperation in pursuing a sustainable future for the earth. It is challenging to lead a respectable life without access to energy services. Low-cost, renewable energy can be produced and supplied by more than only technology [8]. Renewable energy not only serves as an end goal but also facilitates the achievement of additional objectives, ensuring inclusivity for all individuals.

Energy systems can be independently implemented to create local value with the right funding source [9]. Renewable energy sources are simple to implement and oversee because of their ease of installation, adaptability, and modularity. Every person or community can produce energy, which might lead to the development of new industries, encourage innovation and ecologically conscious industrialization, and create job opportunities. The disparities between rural and urban people can be lessened by renewable energy [10]. Since women and children usually suffer the most from a lack of basic amenities in their homes, renewable energy enables attention to be paid to their needs [11]. Using renewable energy to cook can help protect our wildlife and ecosystems. Furthermore, renewable energy reduces greenhouse gas emissions and air pollution, making it one of the most crucial strategies for halting climate change [23].

The framework of energy sustainability helps achieve the goal. The large availability of solar energy resources for PV and CSP in energy applications is moved through energy sustainability. Figure 55.1 describes the plan for solar energy sustainability as economic development, solar energy applications, environmental development, and sustainable development. The various applications, such as environmental and operating conditions, follow environmental considerations [24]. It is straightforward and economically used for energy savings. This type of technology does not have removable parts and has low maintenance and longevity [25]. Economic and social development are followed by various job opportunities to provide cheap energy options. Solar energy should be used with clean energy technology to increase the energy transition in sustainable energy systems [12].

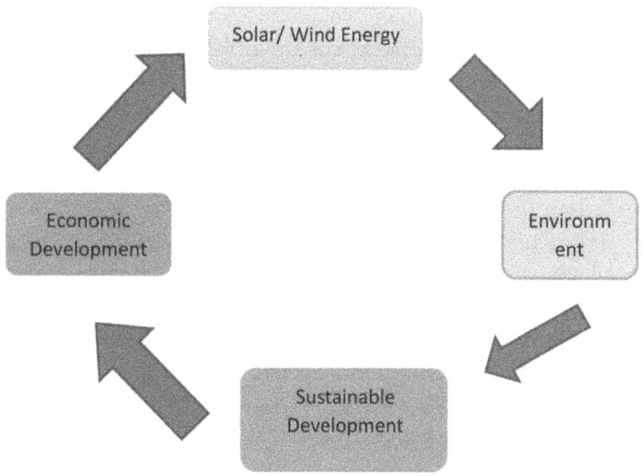

Figure 55.1. Framework for solar and wind energy application in energy sustainability.

Source: Author.

3. Literature Review

Table 55.1 discusses the various advanced solar and wind power technologies and their advantages and applications. An describes PV Solar systems as highly efficient and long-lived. This technology can be integrated with smart buildings [13]. Thermal System describes the advantages as energy efficient with low maintenance and applicable as a commercial heating system [14]. The Hybrid system describes enhancing reliability and mentions the application as power building with remote locations [15]. Notes that Smart Grid technologies have various advantages, such as real-time monitoring with energy-efficient applications and energy efficiency with smart cities [16].

4. Sustainable Development Models

4.1. Green economic growth

The top solar and Wind markets are considered conventional sources to promote sustainable economic growth with affordable and reliable power. The city's municipal utility supported

Table 55.1. Various technology used in sustainable smart cities

Ref No	*Technology*	*Advantages*	*Application in Smart Cities*
[13].	PV Solar System	Its efficiency was very high, and it had a long lifespan.	Installation in Rooftop, Solar parks integrated with Smart Buildings.
[14].	Thermal System (Solar)	Energy Efficient and Low maintenance.	Residential and commercial heating systems.
[15].	Hybrid System	To enhance reliability	Backup through power buildings in remote locations.
[16].	Smart Grid Technologies	Real-time monitoring system and energy efficient.	Energy management in smart cities.

Source: Author.

the renewables aim because it would lower fuel prices and regulatory risk, save money on power, and use less water. 14 The renewable power purchase agreement (PPA) has a 25-year tenure. In contrast, natural gas has a limit of seven years. This adds reliability. The complementary nature of Wind and solar energy also helped Georgetown by lowering the demand for storage. Here, we also discussed the applications of the green economy in plant-based homes, recycling batteries, and organic farming. Eco-friendly investments are the main emphasis of the green economy. Government investment in geothermal, hydroelectric, Wind, and solar energy is balanced and at its best in nations that use the green economy model. According to this economic paradigm, organic food production and consumption are prioritized. In regions where the green economic model has been adopted, it shapes people's everyday lives. For example, green energy is used, organic farming is carried out, green buildings and homes made of plants are built, and electric cars, bicycles, scooters, and non-motorized transit options that recycle lithium batteries are selected as modes of Transportation. Plant-based homes are healthy and sustainable, are designed for smart cities, and have the proper ventilation system to make them better than traditional homes. Recycling lithium batteries contributes to global warming. It is not sustainable, but it helps to make it easy to solve the existing problems. Organic farming means increasing crop production without using pesticides and fertilizers. Some technologies are used to make a smart city green [17].

4.2. Sustainable building and transportation

Evaluating the influence of multinational leadership on business success in global teams:

Integrating different energy sources in green buildings offers several financial and ecological advantages. One example of this strategy is combining photovoltaic (PV) and wind energy systems. Figure 55.2 this hybrid setup makes the most of the energy resources available, irrespective of the time of day or the weather. Photovoltaic (PV) systems mainly produce energy during the day, while wind turbines provide a steady electricity supply at night or during overcast conditions. Using energy management systems, which track and optimize a building's energy usage to increase its effectiveness, illustrates an integrated strategy [18]. These systems may control energy production and enable efficient distribution and storage to preserve the building's energy stability. In conclusion, contemporary energy sources and their combined use may be crucial to the building industry's shift to sustainability. We have a chance to not only satisfy the energy needs of buildings but also lessen their adverse effects on the environment and improve the lives of their occupants by utilizing cutting-edge technology and their interactions [19].

To interpret Figure 55.3, smart Transportation consists of advanced technologies for improved efficiency, sustainability, safety, and convenience. EVs play an essential role in reducing carbon emissions and promoting energy efficiency. Infrastructure contains the essentials for various components of a smart transport system to integrate with change. Real-time data analytics played a transformative role in improving the traffic flow and reducing accidents [20]. It also helps to improve mobility for those who cannot drive.

An intelligent transport system plays the role of monitoring traffic and controlling real-time traffic analysis. A smart traffic management system helps integrate various technologies to improve transportation efficiency.

4.3. Higher quality of life

SRC offers high-quality amenities like initiative and healthier, more powerful living spaces. Giving lower-income households access to renewable energy promotes inclusion.

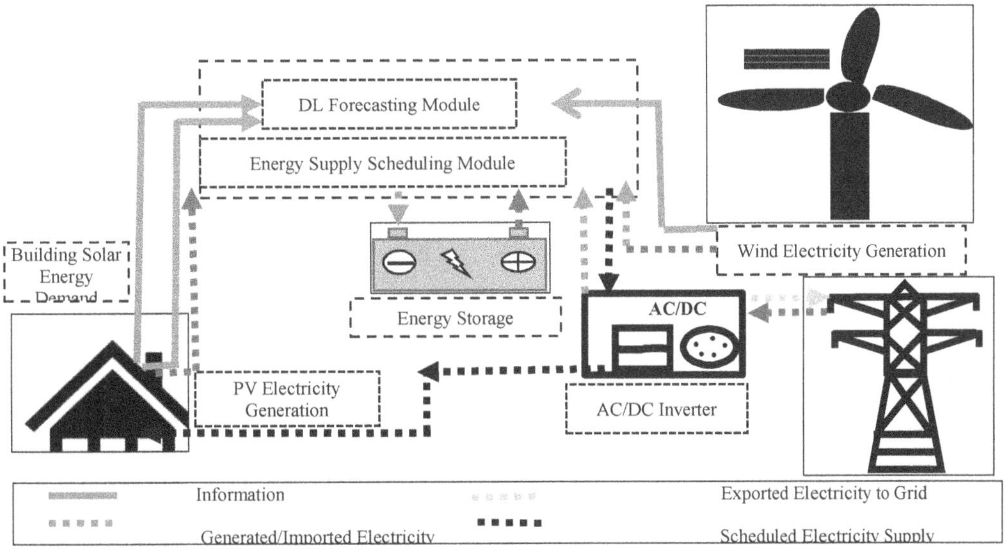

Figure 55.2. Sustainable building with solar/wind energy.

Source: Author.

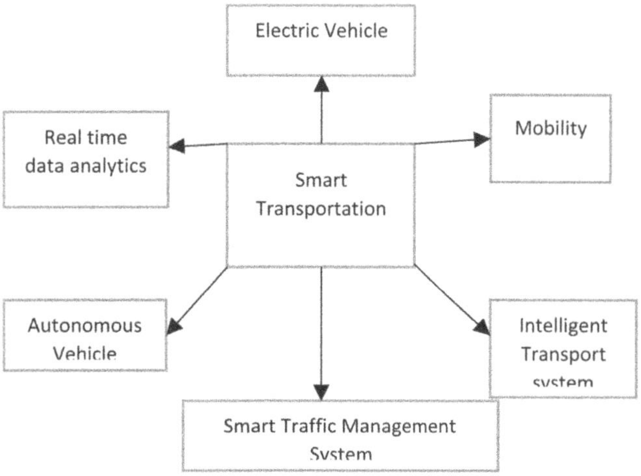

Figure 55.3. Smart transport system.

Source: Author.

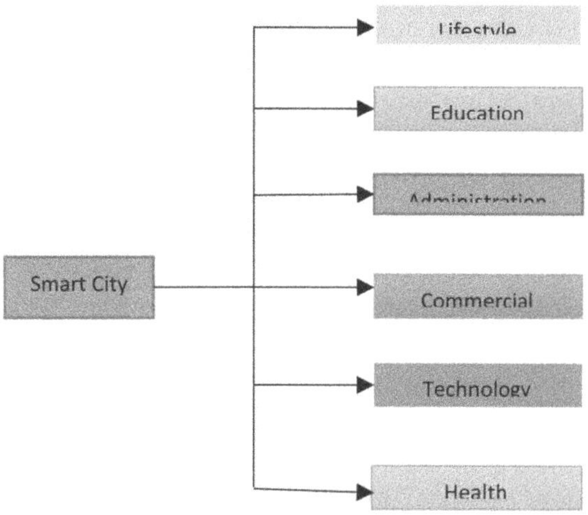

Figure 55.4. Higher quality life style with smart city.

Source: Author.

Adopting 'green' projects is frequently undermined by the false association of elitism. A widening green divide between wealthy families, in which solar panel and electric vehicle ownership are still prevalent, and the remaining households, who are left to pay for legacy energy infrastructure, is a cause for concern. Many SRCs have responded by starting programs to increase access. 35 Although rooftop solar has become more popular with SRC Bangalore's fewer wealthy residents, due to net metering, many utilities are not entirely sure how it will affect their business lines. Other SRCs, including Adelaide, have implemented initiatives to reduce the upfront costs of solar systems. By limiting vehicle access to their city cores, SRCs are increasingly adopting the strategy of further improving air quality. The ultra-low emission zone in London will go into effect in April 2019; SRC Paris said that internal combustion engines (ICE) will be prohibited by 2030, and SRC Copenhagen banned diesel vehicles from the city and established car-free zones.

To analyze the above, Figure 55.4 describes the quality of lifestyle that contributes to promoting balance between personal fulfillment. This means it is sustainable and includes various key aspects such as work-life balance, personal growth, maintaining a positive relationship, and engaging in recreational activities. Quality education provides various tools for improving socioeconomic status and developing critical thinking skills [21, 22]. It mainly focuses on interconnecting through lifestyle, administration, education, health, technology, and commercial activities.

5. Conclusion

This research also highlighted the significance of developing sustainable energy. There are several social, environmental, and financial advantages to using solar and Wind energy, which would also help stabilize energy prices. The contribution of Wind and solar energy to attaining sustainable development by supplying energy needs, generating employment, and safeguarding the environment has demonstrated this. Therefore, looking into a crucial aspect of long-term sustainability is imperative. Combining solar and wind hybrid systems offers a practical way to help diverse communities become resilient and achieve sustainable energy independence. This study highlights the frequent benefits of the hybrid system, including load management, high efficiency, minimal maintenance, reliable power supply, and effective resource utilization. In light of the declining fossil fuel reserves, finding innovative methods to adopt clean energy technology is crucial and anticipated.

References

[1] Radovanovic, M., Popov, S., & Dodic, S. (2012). *Sustainable Energy Management*. Cambridge, MA: Academic Press.

[2] Salvarli, M. S., & Salvarli, H. (2020). For sustainable development: future trends in renewable energy and enabling technologies. In: M. Al Al Qubeissi, A. El-kharouf, H. S. Soyhan (Eds). *Renewable Energy-Resources, Challenges and Applications*. London: IntechOpen.

[3] Cozzi, L., Gould, T., Bouckart, S., et al. (2020). *World Energy Outlook 202*Paris: International Energy Agency.

[4] Ellabban, O., Abu-Rub, H., & Blaabjerg, F. (2014). Renewable energy resources: Current status, future prospects and their enabling technology. *Renewable and Sustainable Energy Reviews*, *39*, 748–764.

[5] Rourke, F. O., Boyle, F., & Reynolds, A. (2009). Renewable energy resources and technologies applicable to Ireland. *Renewable and Sustainable Energy Reviews*, *13*(8), 1975–1984.

[6] Maka, A. O., Salem, S., & Mehmood, M. (2021). Solar photovoltaic (PV) applications in Libya: Challenges, potential, opportunities and future perspectives. *Cleaner Engineering Technology*, *51*, 100267.

[7] Gahrens, S., Alessandra, S., & Steinfatt, K. (2021). *Trading Into a Bright Energy Future. The Case for Open, High-Quality Solar Photovoltaic Markets*. Abu Dhabi: IRENA, 1–44. https://irena.org/-/media/Files/ IRENA/Agency/

Publication/2021/Jul/IRENA_WTO_Trading_ Energy_Future_2021.pdf (21 April 2022, date last accessed).

[8] Mustafa, M., Ghazali, F., & Yuniza, M. E. (2022). Governance interventions in providing affordable, reliable, sustainable, and modern energy in the ASEAN region (Goal 7). In *Good Governance and the Sustainable Development Goals in Southeast Asia*. Routledge: Abingdon, UK. Available online: https://www.taylorfrancis.com/chapters/edit/10.4324/9781003230724-8/governance-interventions-providing-affordable-reliable-sustainable-modern-energy-asean-region-goal-7-maizatun-mustafa-farahdilah-ghazali-mailinda-eka-yuniza (accessed on 1 November 2022).

[9] McCollum, D., Gomez Echeverri, L., Riahi, K., & Parkinson, S. (2022). Sdg7: Ensure Access to Affordable, Reliable, Sustainable and Modern Energy for All. Available online: https://www.un.org/en/chronicle/article/goal-7-ensure-access-affordable-reliable-sustainable-and-modern-energy-all (accessed on 1 November 2022).

[10] Carley, M., & Spapens, P. (2017). Sharing the World: Sustainable Living and Global Equity in the 21st Century; Routledge: Abingdon, UK. Available online: https://www.taylorfrancis.com/books/mono/10.4324/9781315087993/sharing-world-phillipe-spapens-michael-carley (accessed on 1 November 2022).

[11] Raman, A., Balakrishnan, R., Arokiasamy, A. R., Pant, M., & Sukirthanandan, P. (2024). Design of Quality of Experience-based Green Internet Architecture for Smart City. *Journal of Internet Services and Information Security*, *14*(3), 157–166. https://doi.org/10.58346/JISIS.2024.I3.009

[12] Maka, A. O., & Alabid, J. M. (2022). Solar energy technology and its roles in sustainable development. *Clean Energy*, *6*(3), 476–483.

[13] Liu, J., Zhang, Y., & Wang, F. (2020). Recent advancements in photovoltaic solar energy technologies: A review. *Renewable and Sustainable Energy Reviews*, *120*, 109505. https://doi.org/10.1016/j.rser.2020.109505

[14] Fong, S., & Tso, H. (2020). A review on solar thermal systems and their applications in residential and commercial buildings. *Applied Thermal Engineering*, *171*, 115045. https://doi.org/10.1016/j.applthermaleng.2020.115045

[15] Chowdhury, M., Khosravi, A., & Hossain, M. (2021). Hybrid solar-wind systems for urban microgrids: A review of design and optimization strategies. *Renewable Energy*, *164*, 1269–1285. https://doi.org/10.1016/j.renene.2020.08.111

[16] Zhao, P., Lin, J., & Xie, F. (2020). Smart grid technologies and their integration with renewable energy systems in smart cities: A review. *International Journal of Electrical Power & Energy Systems*, *120*, 106377. https://doi.org/10.1016/j.ijepes.2020.106377

[17] Celikyay, H. H., & Celikyay, M. (2022). A brief review on green economy in smart city applications. *Journal of Global Strategic Management*, *16*(2).

[18] Nwagu, C. N., Ujah, C. O., Kallon, D. V., & Aigbodion, V. S. (2024). Integrating solar and Wind energy into the electricity grid for improved power accessibility. *Unconventional Resources*, 100129.

[19] Akin, O., Gulmez, U. C., Sazak, O., Yagmur, O. U., & Angin, P. (2022). GreenSlice: An Energy-Efficient Secure Network Slicing Framework. *Journal of Internet Services and Information Security*, *12*(1), 57–71. https://doi.org/10.22667/JISIS.2022.02.28.057

[20] Kwok, K. C. S., & Hu, G. (2023). Wind energy system for buildings in an urban environment. *Journal of Wind Engineering and Industrial Aerodynamics*, *234*, 105349.

[21] Energy, S. (2019). Renewable energy for sustainable development in India: Current status, future prospects, challenges, employment, and investment opportunities.

[22] Kadhim, M. J. H., Alradha, R. M., Jawad, H. K., Al-Dabbagh, B., & Al-Khafaji, Z. (2024). Investigating the Effect of the Yellow Chlorophyll on the Characteristics of Liquid Polyethylene Glycol for Liquid Electrolyte Solar Cells. *Natural and Engineering Sciences*, *9*(2), 244–256. https://doi.org/10.28978/nesciences.1479785

[23] Van Vuuren, D. P., Stehfest, E., Gernaat, D. E., Doelman, J. C., Van den Berg, M., Harmsen, M., de Boer, H. S., Bouwman, L. F., Daioglou, V., & Edelenbosch, O. Y. (2017). Energy, land-use and greenhouse gas emissions trajectories under a green growth paradigm. *Glob. Environ. Change*, *42*, 237–250. [Google Scholar] [CrossRef] [Green Version]

[24] Olabi, A. G., Obaideen, K., Abdelkareem, M. A., AlMallahi, M. N., Shehata, N., Alami, A. H., … & Sayed, E. T. (2023). Wind energy contribution to the sustainable development goals: case study on London array. *Sustainability*, *15*(5), 4641.

[25] He, B. J. (2022). Green building: A comprehensive solution to urban heat. *Energy and Buildings*, *271*, 112306.

56 Harnessing emerging technologies for a sustainable future: Innovations, management, and development

Aravindhan Ragunathan[1,a] and Nirmala Vairaganthan[2,b]

[1]Senior Full Stack Software Developer, Digital Factory, Instrumentation Laboratory, Bedford, MA, USA
[2]Architect-Technical, Wesco, Bengaluru, India

Abstract: Driven by the need to balance environmental preservation with economic growth, sustainability has become a central issue in worldwide innovation, management, and development. By means of resource management, waste minimization, and efficiency enhancement, emerging technologies including artificial intelligence (AI), blockchain, and the Internet of Things (IoT) are revolutionizing corporations. Through improved transparency, automated decision-making, and real-time environmental metric monitoring, these technologies help to meet sustainability goals. For intelligent energy distribution, sustainable urban development, and climate modeling, artificial intelligence enhances predictive analytics. Blockchain advances ethical finance for sustainability projects, carbon credit tracking, and supply chain openness. Effective water and waste management is improved by the Internet of Things; it also drives smart city development and intelligent transportation systems. Notwithstanding its disruptive power, challenges include high implementation costs, cybersecurity concerns, legal restrictions, and scalability limits hinder general use. The contributions of IoT, Blockchain, and artificial intelligence to management, development, and sustainable innovation are investigated in this work. It looks at their applications in many different sectors, evaluates present challenges, and provides viewpoints on future advancements. Emerging technologies can enable notable advancement towards a sustainable and resilient future by means of legislative support, technological advancements, and international cooperation overcoming of these challenges.

Keywords: Sustainable innovation, emerging technologies, AI, blockchain, IoT, renewable energy, green technology, smart management

1. Introduction

As countries try to reduce environmental damage and boost economic growth, sustainability has become a top worldwide objective. Influence of sustainable practices depends on emerging technologies. Resource management is being changed, waste is being reduced, and efficiency is being raised via artificial intelligence (AI), blockchain, and the Internet of Things (IoT). This paper investigates how these technologies affect development, management, and sustainable innovation [1].

Global institutions – most notably the United Nations (UN) – have realized the need of sustainability and created Sustainable Development Goals (SDGs) stressing technical advancements in mitigating climate change and advancing ethical manufacturing. This study investigates how technology can help to reach these goals [51–53].

As the world's population keeps growing, so does the need for energy, natural resources, and good waste control. Because traditional sustainable projects haven't been able to solve these growing problems, technology needs to be used [2, 3]. Real-time tracking made possible by the Internet of Things (IoT), openness in blockchain technology, and systems run by AI are starting to make economic growth and environmental duty work together. Authorities, businesses, and groups are using new technologies to build long-term systems that use energy more efficiently, make the environment cleaner, and make circle economies stronger [4]. Artificial intelligence could help make better decisions about environmental policy by analyzing huge amounts of data, predicting environmental trends, and suggesting changes to the law. The Internet of Things (IoT) makes it easier to build smart infrastructure that wastes less, and blockchain technology makes sure that carbon trade is honest, that materials are sourced ethically, and that businesses run in a way that is good for the environment [5]. Even with these improvements, there are still problems that make it hard to use them widely. Some of these problems are the digital gap, very high operational costs, and legal restrictions [6]. It is very important for scientists, business leaders, and lawmakers to work together to find solutions to these problems. In order to get the most out of the positive effects that new technologies have on sustainable development, it is important to work on improving digital literacy, finding cheap technology solutions, and putting in place global sustainability laws [7]. A full look at how the Internet of Things (IoT), blockchain technology, and AI might help with sustainability is given in this piece. The article also talks about how these technologies can be used, the problems they've had, and the chances that lie ahead. Using these technologies speeds up the move to a stronger

[a]contact2aravind@gmail.com, [b]nimmicsegct@gmail.com

DOI: 10.1201/9781003684459-56

and more stable future. This ensures that economic growth won't hurt the environment or people's well-being.

2. Emerging Technologies in Sustainable Innovation

The main reason for the changes happening in sustainable innovation is how quickly new technologies are getting better. Recent technology advances, like bitcoin, AI, and the Internet of Things (IoT), are changing the way sustainability is done by making things more efficient, reducing waste, and managing resources better. Businesses, cities, and communities may be able to help the environment and grow the economy by putting these tools together in programs for sustainable innovation.

A. **AI-Driven Sustainable Solutions:** Artificial intelligence greatly advances sustainability by automating decisions and best use of resources. Advanced AI-driven models are improving energy economy in urban planning, transportation, and industry [10]. Forecasting of climate changes is made easier by AI-driven predictive analytics, therefore enabling proactive measures against environmental damage and natural disasters. Furthermore, artificial intelligence-driven automation in agriculture is significantly lowering water and fertilizer use, improving food security by means of this minimum of environmental effect [11].

B. **Blockchain for Transparency and Accountability:** Blockchain technology guarantees data integrity and security, therefore improving openness in sustainability projects and accountability. In supply chain management especially, it is quite helpful since it allows one to track product lifetime and ensure moral raw material procurement [12]. By allowing carbon credit trading, decentralized blockchain platforms help businesses to participate in verifiable and safe carbon offset programs. Blockchain eliminates fraud and improves openness, therefore promoting sustainable economic practices.

C. **IoT-Enabled Smart Infrastructure:** By means of real-time monitoring and automation made possible by IoT technology, infrastructure gains sustainability. IoT sensors compile and evaluate information on waste management, water use, and energy consumption, thereby enabling businesses to improve operations with less influence on the surroundings. The Internet of Things (IoT) is being used in smart cities to control traffic flow, ease congestion, and improve air quality. Moreover, IoT-enabled smart grids maximize energy distribution, therefore reducing transmission losses and enabling the more effective integration of renewable energy sources [15].

D. **Innovations in Green Energy:** The change toward renewable energy depends much on newly developing technology. AI systems improve the efficiency of solar and wind energy generation, hence increasing the predictability of energy output [16]. Peer-to-peer exchange of renewable energy is enabled by blockchain technology, therefore allowing consumers to sell extra solar power. IoT-enabled energy storage devices concurrently enhance grid resilience by bettering energy distribution and enabling seamless integration of renewable sources into power systems [17].

E. **Sustainable Manufacturing and Industry 4.0:** Industry 4.0 technologies – which include blockchain, IoT, and artificial intelligence – are revolutionizing production by improving efficiency and lowering environmental impact [18]. Intelligent factories use IoT-enabled automation and AI-driven analytics to cut energy use and waste. Monitoring supply chains provided by blockchain guarantees ethical labor standards and sustainable sourcing. These technologies help to reduce carbon footprints and promote circular economy concepts whereby waste products are turned into new items rather than being thrown away [19].

F. **The Function of Digital Twin Technology in Sustainability:** Digital twin technology, which creates virtual versions of real-world assets, is used to improve infrastructure and industrial operations in order of sustainability. Before they are carried out, AI-driven digital twins can predict maintenance needs, improve energy efficiency, and replicate the environmental effects of suggested projects [20]. In the building and transportation sectors, this approach greatly improves resource economy and reduces pollution.

G. **Emerging Trends in Sustainable Innovation:** Future technologies include quantum computing, 5G, and edge computing will see increased integration of artificial intelligence, blockchain, and IoT with emerging trends in sustainable innovation. By processing vast environmental data at hitherto unheard-of speeds, quantum computing has the ability to change climate modeling. By enabling faster and more consistent connectivity among IoT devices, 5G networks will improve real-time sustainability monitoring. Edge computing helps data processing close to the source, therefore reducing energy consumption and increasing operational efficiency [22].

Including these technologies into sustainable innovation will continue to advance and provide fresh approaches to solve environmental issues while building financial resilience. Adopting these technologies would help businesses significantly lower their carbon emissions, improve resource economy, and assure long-term environmental sustainability [23].

3. Management of Sustainable Technologies

Maximizing the benefits of sustainable technologies and minimizing any issues depend on effective application of them. Governments and companies have to create strategic plans that enable the seamless acceptance, integration, and governance of IoT, blockchain, and artificial intelligence into environmental initiatives.

A. **Strategic Planning and Policy Execution:** Effective management of sustainable technology depends on well stated rules that enable long-term application. It is important for governments and regulatory bodies to come up with standard procedures for the safe and fair use of AI, the completion of blockchain transactions, and the installation of energy-efficient Internet of Things devices. To make sure that technology is used in a good way, these guidelines need to cover things like data privacy, hacking, and environmental compliance.
B. **Spending money on research and development:** Research and development (R&D) needs to be constantly funded in order for lasting technology solutions to move forward. Companies must pay for tools to improve climate models that use AI, create blockchain systems that use less energy, and expand smart grids that can connect to the internet of things. When universities, study centers, and tech companies work together, new ideas can be used on a larger scale. This also speeds up the creative process.
C. **Training and building up the skills of the workforce:** It is very important to have a staff that is knowledgeable and skilled in how to use and manage eco-friendly technologies like the Internet of Things (IoT), blockchain, and AI. To make sure that people can use technology in a way that is good for the earth, businesses and states should work together to fund training programs. Schools should teach technology with an eye toward sustainability so that the next generation is ready for the green economy.
D. **Lifecycle Assessment and Performance Monitoring:** The effectiveness of sustainable technologies depends on a methodical approach for evaluating their environmental impact. Performance monitoring and lifetime assessment are thus crucial. Instantaneous assessments of resource use, carbon emissions, and operational efficiencies enabled by artificial intelligence can [28]. Blockchain technology guarantees accountable and visible sustainability projects, hence enhancing traceability. IoT-enabled monitoring systems could be able to accurately track environmental indicators, therefore guaranteeing the achievement of sustainability goals.
E. **Cross-Sector Collaboration:** Effective implementation of sustainable technology depends on cooperation across several governmental agencies, business sectors, and non-profit organizations. Extensive sustainability initiatives including smart city efforts, renewable energy transitions, and carbon offset programs [29] can be facilitated by public-private partnerships (PPPs). Through worldwide alliances, companies can share best practices, therefore encouraging creativity in sustainable technology management.
F. **Scalability and Infrastructure Development:** Scalability and Infrastructure Development: Resilient infrastructure is absolutely necessary for the great use of sustainable technologies. Governments and businesses have to distribute funds to digital infrastructure enabling large-scale IoT, blockchain, and artificial intelligence projects. Smooth integration of these technologies across several sectors depends on edge computing, cloud platforms, and secure data networks.
G. **Sustainability Metrics and Reporting:** Organizations must create important performance indicators (KPIs) to evaluate the effect of sustainable technologies. The Global Reporting Initiative (GRI) and the Carbon Disclosure Project (CDP) among other standardized reporting systems help businesses assess and communicate their improvements in sustainability. AI-powered analytics provides actionable insights for continuous improvement and helps to simplify sustainability reporting.

Strategic management ideas allow businesses and government agencies to help sustainable technologies to be adopted effectively. In line with world environmental and economic objectives, an anticipatory strategy including research, workforce training, intersectoral cooperation, and infrastructure enhancement will generate long-lasting sustainability benefits.

4. Sustainable Development through Technology

Promoting sustainability depends on technology since it increases efficiency, lowers resource consumption, and helps to enable the change to environmentally friendly economies. AI, blockchain, and the Internet of Things (IoT) taken together have accelerated efforts at world sustainable development. These developments improve environmental monitoring, decision-making, and industrial operations, so guaranteeing continuous ecological and financial stability [33].

A. **AI for Sustainable Development:** AI for Sustainable Development: AI-powered analytics let companies assess environmental effect, maximize energy resource management, and use predictive modeling to help to prevent disasters. In sectors, AI-driven automation reduces waste, improves productivity, and guarantees compliance to environmental regulations [34]. By optimizing traffic management, enhancing garbage disposal efficiency, and therefore reducing carbon emissions, artificial intelligence helps smart city projects in urban planning [35].
B. **Blockchain for Environmental Accountability:** Blockchain technology advances openness and confidence in environmental responsibility projects. It assures companies follow ethical sourcing policies, therefore enabling the confirmation of environmental claims and the

tracking of carbon credits. By means of the decentralization of financial transactions, blockchain promotes green funding and helps companies to participate in renewable energy projects with better responsibility [36].

C. **IoT in Sustainable Infrastructure:** IoT in Sustainable Infrastructure: Real-time environmental variable monitoring facilitated by IoT-connected sensors guarantees best energy efficiency and waste minimising effect. Using IoT technology, smart grids improve the supply of electricity, hence lowering reliance on fossil fuels and helping the growth of renewable energy sources [37]. Furthermore, IoT-enabled water saving systems examine consumption trends and spot leaks, therefore enabling efficient water management in rural as well as urban environments [38].

D. **Circular Economy and Green Technologies:** Combining IoT, blockchain, and AI improves recycling processes, reduces emissions, and supports sustainable manufacturing techniques thereby enabling the shift to a circular economy. While blockchain assures openness in waste management and upcycling initiatives [39], AI-enhanced material sorting in recycling facilities improves processing efficiency. IoT-enabled smart manufacturing systems improve resource economy, hence reducing industrial waste and encouraging environmentally friendly manufacturing techniques [40].

E. **Future Prospects in Sustainable Development:** The continuous merging of artificial intelligence, blockchain, and IoT with emerging technologies such 5G, quantum computing, and bioengineering will determine how sustainable development via technology unfolds. These developments will improve climate modeling, streamline farming practices, and hasten the shift to a global economy carbon-neutral [41]. To ensure the moral and fair application of new technology, governments, businesses, and research institutes have to work together, therefore promoting long-lasting environmental and financial benefits.

5. Challenges and Future Prospects

Though IoT, blockchain, and artificial intelligence have transforming power to advance sustainability, many obstacles prevent their wide acceptance and efficiency. Overcoming these obstacles calls on governments, businesses, and technology developers working together [43].

A. **Elevated Implementation Expenses:** One major challenge to the acceptance of sustainable technology is the large initial money needed for infrastructure, research, and implementation. Blockchain verification solutions, IoT-enabled monitoring systems, and AI-driven smart grids all need significant financial investment, therefore limiting access particularly in developing nations [44]. Accelerating adoption will depend on reducing costs by means of scalable and open-source technologies.

B. **Data Privacy and Security Risks:** The massive data gathered by IoT devices, artificial intelligence-powered analytics, and blockchain networks exposes security and privacy issues. Trust and regulatory framework adherence are called into question by cybersecurity threats, illegal access, and probable data exploitation both personally and commercially [45]. Reducing these threats will help by improving encryption methods, using distributed security systems, and keeping data protection laws followed.

C. **Regulatory and Policy Barriers:** Regulatory and Policy Obstacles: Their inclusion into sustainability initiatives is hampered by the lack of generally consistent regulations for IoT, blockchain, and artificial intelligence. Governments and regulatory bodies have to work together to create coherent rules that support creativity while handling security, ethical, legal, and legal challenges. Extensive adoption of these technologies depends on strong models for data governance, digital asset validation, and suitable use of artificial intelligence.

D. **Energy Consumption of Emerging Technologies:** Emerging technologies' energy consumption remains a concern even if blockchain and artificial intelligence support sustainability. Blockchain proof-of- work (PoW) consensus methods and artificial intelligence training models both demand significant processing capability, which drives higher energy consumption [47]. Reconciling technology advancement with environmental stewardship requires using energy-efficient substitutes including low-power AI models and proof-of- stake (PoS) consensus for blockchain.

E. **Scalability and Integration Challenges:** Technical challenges of scalability and interoperability must be addressed if present industrial systems effectively incorporate IoT, blockchain, and artificial intelligence into them. Many historical systems show conflicts with modern digital solutions, which calls for significant changes for flawless integration [48]. Adoption of technology can be improved by cooperation among sectors and the development of uniform policies.

Future Outlook: Research and development of sustainability via IoT, blockchain, and artificial intelligence will continue to improve these technologies, so offering promise. Improvements in the Internet of Things will maximize energy grids and waste management; blockchain technology will improve the recording and transparency of carbon credits; artificial intelligence is expected to enhance environmental monitoring systems.

Combining these technologies with innovative ideas like 5G networks and quantum computing will expose fresh chances for sustainability. With its unmatched speed, quantum computing can examine large environmental datasets, hence improving climate modeling and disaster readiness. Concurrent with this will be 5G connectivity enhancing IoT capabilities, enabling industrial automation and real-time data transfer for smart cities.

Legislators, corporations, and technology specialists must work together to solve obstacles and promote sustainable innovations. To maximize the beneficial effects of artificial intelligence, blockchain technology, and the Internet of Things on sustainability, research money, innovation legislation, and ethical technology usage are needed.

6. Conclusion

The landscape of sustainability is undergoing a transformation as a result of the introduction of new technologies that encourage transparency, improve resource management, and reduce waste. AI, blockchain technology, and the internet of things are bringing about breakthroughs that are beneficial to a variety of industries, including waste management, energy, agriculture, and city planning. Despite the fact that these technologies are not yet recognized by the majority of people, recent developments in them point to a future with a less influence on the environment.

The increasing number of companies that are working toward carbon neutrality will lead to an increase in the significance of new technologies. In order to improve energy efficiency, the infrastructure of smart cities will be operationally optimized with the assistance of the Internet of Things (IoT) and optimization models that are powered by AI. While this is going on, the immutable ledger that blockchain technology provides will assure compliance with global sustainability requirements by measuring various carbon emissions. For the purpose of accelerating the transition away from fossil fuels and toward renewable energy sources, investments in green finance that are based on blockchain technology and renewable energy projections that are supplemented by artificial intelligence may be of assistance.

When these technologies are integrated with cloud computing and big data to construct predictive sustainability models, businesses and politicians are able to make better decisions that are more informed. As these concepts grow, they will provide scalable solutions to global issues including waste management, resource depletion, and climate change. Governments, corporations, and research institutions must collaborate to achieve these outcomes while protecting the environment and boosting economic growth.

The constraints include high implementation costs, regulatory restrictions, and data security issues. These strategies may increase research and legislative support to address these issues. By encouraging innovation, raising public awareness, and establishing standard frameworks for sustainable technology deployment, societies can maximize the benefits of AI, blockchain, and the IoT while minimizing their drawbacks. Sustainability will depend on how well these technologies create an ecologically friendly and equitable society.

References

[1] IBM Institute for Business Value. (2022). *Driving Sustainability with AI.*

[2] UN Environment Programme. (2021). *Blockchain for Sustainable Development.*

[3] World Economic Forum. (2020). *AI and Climate Change.*

[4] Tesla Energy. (2023). *The Future of Renewable Energy Storage.*

[5] International Renewable Energy Agency. (2022). *Advancing Green Hydrogen.*

[6] European Commission. (2021). *IoT and Smart Cities.*

[7] NASA. (2019). *Machine Learning in Climate Science.*

[8] FAO. (2021). *AI in Precision Agriculture.*

[9] Smart Cities World. (2022). *AI-Driven Waste Management.*

[10] The World Bank. (2021). *Carbon Credit Markets and Blockchain.*

[11] Toyota Research Institute. (2022). *Hydrogen Fuel Cell Technology.*

[12] MIT Technology Review. (2020). *Blockchain for Transparent Supply Chains.*

[13] IEEE Spectrum. (2022). *IoT for Disaster Management.*

[14] Google DeepMind. (2022). *AI for Energy Optimization.*

[15] International Energy Agency. (2023). *Global Renewable Energy Outlook.*

[16] European Environment Agency. (2022). *Renewable Energy Trends in Europe.*

[17] US Department of Energy. (2023). *Hydrogen and Fuel Cells Strategy.*

[18] International Telecommunications Union. (2021). *IoT Standards for Sustainable Development.*

[19] IBM Blockchain Team. (2022). *Blockchain for Ethical Sourcing.*

[20] Deloitte Insights. (2023). *Digital Twin Technology for Sustainability.*

[21] Quantum Computing for Environmental Science. (2022). *Nature Climate Change.*

[22] 5G Alliance. (2023). *Next-Generation Networks for Smart Cities.*

[23] Edge Computing World. (2023). *Enhancing Sustainability Through Edge AI.*

[24] United Nations Environment Programme. (2022). *Sustainability and Technology Integration.*

[25] World Economic Forum. (2021). *Policy Frameworks for Emerging Technologies in Sustainability.*

[26] European Commission. (2020). *Investment in Green Technology Research.*

[27] International Energy Agency. (2023). *Developing Skills for the Green Economy.*

[28] MIT Technology Review. (2022). *AI in Environmental Monitoring.*

[29] World Resources Institute. (2021). *Public-Private Partnerships for Sustainability.*

[30] Smart Cities Council. (2022). *Digital Infrastructure for Smart Sustainability Initiatives.*

[31] Carbon Disclosure Project. (2023). *Global Standards for Sustainability Reporting.*

[32] International Institute for Sustainable Development. (2023). *The Role of AI in Environmental Conservation.*

[33] The World Bank. (2022). *Technology for a Sustainable Future.*

[34] European Environment Agency. (2021). *Artificial Intelligence for Smart Cities and Climate Resilience.*

[35] United Nations Climate Change Secretariat. (2023). *AI-Powered Solutions for Carbon Neutrality.*

[36] Blockchain Research Institute. (2022). *Blockchain for Sustainable Development Goals.*

[37] IEEE IoT Journal. (2023). *Smart Grid and IoT: Enhancing Renewable Energy Utilization.*

[38] IoT Applications in Water Conservation. (2021). *Water Resources Research.*

[39] Ellen MacArthur Foundation. (2022). *Blockchain and AI in the Circular Economy.*

[40] World Economic Forum. (2023). *Smart Manufacturing and Green Industry Innovation.*

[41] Quantum Computing for Climate Modeling. (2022). *Nature Sustainability.*

[42] Global Sustainability Initiative. (2023). *The Future of AI, Blockchain, and IoT in Sustainable Development.*

[43] United Nations Environment Programme. (2023). *Challenges in Emerging Technologies for Sustainability.*

[44] International Energy Agency. (2022). *Investment Barriers in Green Technology Adoption.*

[45] Data Privacy and Risks in AI and IoT. (2023). *Cybersecurity Journal.*

[46] World Economic Forum. (2021). *Regulatory Frameworks for AI and Blockchain in Sustainability.*

[47] MIT Energy Initiative. (2023). *Energy Consumption in AI and Blockchain Systems.*

[48] IEEE Transactions on Industrial Informatics. (2022). *Challenges in Integrating AI and IoT for Sustainability.*

[49] Future Trends in AI and Blockchain for Sustainability. (2023). *Nature Communications.*

[50] 5G Alliance. (2023). *Enhancing Sustainability Through 5G and Quantum Computing.*

[51] Shiraishi, Y., Mohri, M., & Fukuta, Y. (2011). A Server-Aided Computation Protocol Revisited for Confidentiality of Cloud Service. *Journal of Wireless Mobile Networks, Ubiquitous Computing, and Dependable Applications*, *2*(2), 83–94.

[52] Hadi, B. (2023). Convergence Problems with Ansys's Solid 65 Finite Element in Concrete-filled Tubular (CFT) Columns as A Case Study. *Archives for Technical Sciences*, *1*(28), 29–38. https://doi.org/10.59456/afts.2023.1528.029B

[53] Radhakrishna, K., Satyaraj, D., Kantari, H., Srividhya, V., Tharun, R., & Srinivasan, S. (2024). Neural Touch for Enhanced Wearable Haptics with Recurrent Neural Network and IoT-Enabled Tactile Experiences. *2024 3rd International Conference for Innovation in Technology (INOCON)*. Bangalore, India, pp. 1–6. doi:10.1109/INOCON60754.2024.10511642.

57 Synergy between creative economy and artificial intelligence for sustainable tourism

Putu Eka Wirawan[a], *Ni Wayan Mega Sari Apri Yani*[b], *and I Wayan Eka Mahendra*[c]

The International Institute of Tourism and Business, Indonesia

Abstract: Efforts to increase local business productivity and encourage economic growth can be done through the development of the tourism sector. In this context, technological advances, especially artificial intelligence (AI), play an important role by providing a significant impact on the creative economy and tourism. The synergy between the creative economy, which includes art, design, and culture, with AI can create innovations that enrich the tourist experience while maintaining the sustainability of the environment and local culture. This collaboration not only contributes to economic growth, but also focuses on preserving culture and the environment. By utilizing AI to optimize resource management and understand market trends, tourism destinations can increase their competitiveness and relevance in an increasingly competitive global market. Therefore, the integration of the creative economy and artificial intelligence is the key to creating sustainable and innovative tourism.

Keywords: Intelligence artificial, creative economy, tourism sustainable, technology, synergy

1. Introduction

Effort increase productivity business local and push economic growth can through sector tourist [1]. Along with development era, technological progress does not can unavoidable. One example of technology that currently developed is artificial intelligence (AI) [2]. This technology has a significant impact on various sectors, including the creative economy and tourism. In the tourism sector and the creative economy, there is a significant impact from the presence of artificial intelligence (AI) [3]. The collaboration of the Creative economy and *AI* can create innovations that bring new experiences to tourism. This is in the form of improving the tourist experience and of course maintaining the sustainability of the environment and local culture.

Individual creativity expressed via industry enhances the economy, referred to as the Creative Economy. The creative economy, which includes art, design, and culture [4]. This scope has great potential to attract tourists by offering unique and authentic experiences. The importance of this collaboration lies not only in improving the economy, but also in preserving culture and the environment. Thus, the synergy between the creative economy and artificial intelligence is the key to building sustainable and highly competitive tourism. In addition, this synergy offers new opportunities to develop more attractive and environmentally friendly products and services [5].

Acceleration of economic growth can be implemented through the development of tourism areas. Sustainable tourism seeks to enhance sustainability, safeguard local culture and the environment, and offer advantages in fulfilling community needs [6]. Basically, sustainable tourism development is an activity that is continuous and minimizes other negative impacts. Sustainable tourism can be associated with vacations, business trips or events, but also trips to visit friends and family too [7]. Aspects that influence sustainable tourism include economic, social, and environmental. These three aspects are interrelated with each other.

Rapidly growing tourism includes increasing accommodation capacity, growing local population, and preserving the environment. These three elements create a synergy known as sustainable tourism [8]. The main principle of sustainable tourism is to involve local communities in every aspect of development [9]. Thus, a balance will be created between the needs of the community and the desires of tourists. This balance can only be achieved if all parties, including the government, entrepreneurs, and the community, work together with the same goal. This harmonious cooperation will not only improve the quality of the tourist experience, but also ensure that the benefits of the tourism sector can be felt by all layers of society without damaging the environment or local culture. With an inclusive approach, sustainable tourism has the potential to become a sustainable economic driver, providing long-term benefits for all parties involved.

The impact of artificial intelligence (AI) technology is crucial in increasing the competitiveness of the tourism sector as a whole. Collaboration between the creative economy

[a]wirawanputu@gmail.com, [b]yanimegasari4@gmail.com, [c]ekamahendra@ipb-intl.ac.id

DOI: 10.1201/9781003684459-57

and AI technology not only improves operational efficiency but also expands market reach [10]. This condition allows industry players to reach a larger and more diverse audience. Artificial intelligence also has the ability to monitor and manage sensitive tourism resources, such as natural parks and historical-cultural sites, in real-time. By utilizing AI systems, managers can regulate access based on capacity, and better respond to surges in visitor numbers [11]. This helps prevent environmental damage and minimize the negative impacts of over-tourism. In addition, by optimizing the tourist experience, this technology ensures that visitors can enjoy a more authentic and satisfying experience. Generally, the integration of AI in the tourism sector has the capacity to create a more sustainable and competitive ecosystem, providing benefits to all stakeholders, including tourists, local communities, and the environment.

2. Theory and Concept

The creative economy is a sector driven by individual creativity, skills, and talents generating prosperity and job opportunities [12]. Furthermore, the creative industry encompasses various knowledge-based activities centered on arts and culture, with the ability to generate revenue through commerce and intellectual property rights [13]. The outputs can take the form of tangible and intangible goods or artistic services that hold economic value and serve a clear market purpose. This industry spans multiple sectors, integrating elements of art, services, and commerce, making it a dynamic player in global trade.

From the above definition, it can be concluded that the creative industry is a sector driven by innovation and fresh ideas from talented human resources. The creative economy in Indonesia has great potential, especially if assisted by three main components: creative knowledge, skilled labor, and labor-intensive. This creates vast opportunities for the continuous growth of creative products, including crafts, advertising, publishing, television, radio, architecture, music, design, and fashion. With its diverse sectors, the creative economy is playing an increasingly vital role in driving national economic growth.

In the context of regional development, there are two conceptual frameworks that are often used to understand the dynamics of the creative economy. The fundamental economic principle suggests that input demand will rise only when there is an expansion in demand for output from both the base (export) and non-base (local) sectors [14]. The difference in return rates is influenced by the environment or available infrastructure, where underdeveloped areas are not caused by bad luck, but by reduce efficiency. Nevertheless, there is still little experiential research that explores this idea, mainly due to limited data. The government plays a crucial role in overseeing and supporting the creative economy sector. The regulatory function is connected to the character of a modern state as a state of law, while the service function is related to welfare institutions [15]. Therefore, the development of tourism as part of the creative economy is very important, because it has the capacity to strengthen regional development and become a significant revenue stream. With the right support, the tourism sector can harness the power of the creative economy to support sustainable and inclusive growth.

3. Artificial Intelligence (AI) in Tourism

Artificial intelligence (AI) is a branch computer science dedicated to creating systems that replicate human cognitive abilities, including learning, reasoning, and decision-making. In the context of tourism, AI might be divided within several categories, including Weak AI is developed to handle specific tasks, while strong AI aims to replicate human-like intelligence across various functions that has the ability to learn and adapt in a wider context [16]numerous hotels are adopting AI-based service technologies, but it is still in its early stage to provide guidelines for the overall service strategy for this technology. Therefore, this study conducted a study on the failure of AI based self-service technologies (SSTs. These technologies include machine learning, where algorithms learn from data, and natural language processing, which allows for more natural interactions between users and systems [17]. With these different types and capabilities, AI can have a significant impact on many aspects in the tourism sector.

AI applications in the tourism sector are diverse and continue to grow. One of its main uses is in booking systems, where AI algorithms can examine user preferences and behaviors to deliver more tailored recommendations [18]. It is essential to recognize that artificial intelligence technology might be applied to both mobile phones and computers (Figure 57.1).

According to telecommunications statistics in Indonesia, in 2023 the total count of mobile phone users in Indonesia will increase. significantly within a period of 4 years [19]. This is started from existence covid-19 pandemic, society doing activities use telephone mobile at home. However on

Figure 57.1. Percentage showing the number of mobile phone users in Indonesia.

Source: Author.

2023 usage phone cellular in Indonesia is around 67.29% experience decrease from year to year previously. Things This because of expenditure public diverted to needs others. The correlation with tourist sustainable is telephone mobile This connected to AI technology that can support economic activities creative on sector tourist sustainable.

4. Synergy between Creative Economy and AI

The synergy model between creativity and technology creates an innovative ecosystem in the tourism industry. By combining human creativity capabilities, such as art and design, with artificial intelligence (AI), industry players can develop more attractive and relevant products and services [20]. For example, in marketing content creation, the use of AI to analyze trends can help create more targeted campaigns, while creativity in creating visual and narrative content can add to the appeal. This synergy also allows for the development of more personalized tourism experiences, where technology supports the creation of deeper and more meaningful interactions between tourists and the destinations they visit.

Several destinations around the world have successfully implemented this synergy model with impressive results. For example, in China, AI technology is being used to analyze visitor data and design engaging public art programs. These programs not only enhance the tourist experience, but also strengthen the local cultural identity [21]. In Bali, several tourism businesses have adopted VR (Virtual Reality) technology to offer interactive experiences for tourists, allowing them to explore local culture and traditions before they visit [22]. These case studies show that the integration of creativity and technology can produce innovations that support tourism growth and enrich the tourist experience.

Collaborative synergy between AI as well as creative economy possesses the capability to have a significant positive impact on sustainable tourism [23]. By leveraging AI to manage and monitor natural and cultural resources, destinations can avoid *over-tourism* and preserve the environment. For example, *AI-based monitoring systems* can regulate access to sensitive tourist sites, ensuring that visitor numbers do not exceed the capacity that can be handled. In addition, creative approaches to promoting sustainable tourism can raise recognition of the significance of environmental conservation and local culture. Technology This implemented to in mobile phone smart. Refer to sub chapter previously.

5. Results and Discussion

5.1. *SWOT analysis*

SWOT analysis is effort know factor in a way structured For formulate the expected strategy at a time road best in solve problem [24]. In the case of implementation *AI* (intelligence artificial) synergize with the Creative Economy, the hope is can maximize strengths, and opportunities at a time minimize impact the bad in the form of weaknesses and threats. The following SWOT analysis in implementation technology Synergistic *AI* with the Creative Economy that supports tourist sustainable (Table 57.1).

5.2. *SMART Analysis*

SMART analysis aims for formulate a specific, measurable, achievable goals achievable, relevant and of course there is

Table 57.1. SWOT analysis

Strengths (Strengths)	• Innovation and Creativity: The combination of creative economy and AI creates unique and attractive products and services for tourists. • Personalized Experiences: AI technology enables the offering of experiences tailored to individual preferences, increasing traveler satisfaction. • Operational Efficiency: AI can improve efficiency in resource management and operational management, reducing costs and increasing profitability.
Weaknesses	• Technology Limitations: Reliance on technology can be a constraint if the necessary infrastructure is inadequate. • HR Skills: Specific training and skills are required in the use of AI and creative technologies, which may not be possessed by all industry players. • Privacy Risks: Collecting and analyzing traveler data through AI can raise privacy and trust issues.
Opportunities	• Growth of Sustainable Tourism: Growing awareness of the importance of sustainability can be an opportunity to develop responsible tourism practices. • Technology-Based Innovation: Opportunities to create new products and services through the integration of emerging new technologies. • Multisectoral Collaboration: Opportunities to collaborate with various sectors, such as government, local communities, and creative industries, to strengthen the tourism ecosystem.
Threats (Threats)	• Global Competition: Competition from other destinations that are also utilizing technology and creativity to attract tourists. • Regulatory Changes: Changes in government policies regarding data privacy and technology may impact operations. • Resistance to Change: Some industry players may be reluctant to adapt to new technologies, which can stifle innovation.

Source: Author.

prone to time [25]. In the Table 57.2 this will present SMART analysis in effort synergy between potential economy creative and intelligent artificial in tourism sustainable.

5.3. Synthesis from SWOT and SMART analysis

The synergy between the creative economy and artificial intelligence for sustainable tourism offers significant power in creating unique, personalized products and experiences, while also improving operational efficiency. However, challenges such as technological limitations and privacy risks must be addressed through proper training and adequate infrastructure development. Growth opportunities in sustainable tourism and multi-sectoral collaboration can be leveraged to develop more responsible practices. By setting SMART goals, industry players can effectively measure the success of these initiatives, ensuring that the steps taken are relevant to market and environmental needs. Overall, this synergy has the potential to not only enhance the attractiveness of tourist destinations, but also contribute to long-term sustainability in the tourism industry [26].

6. Conclusion

Artificial intelligence can help optimize resource management, predict market trends, and improve visitor engagement.

Table 57.2. SMART analysis

Specific (Specific)	• Enhancing collaboration between the creative economy and AI in the development of sustainable tourism products, such as personalized tourism experiences and the use of technology to manage resources.
Measurable	• Determine indicators of success, such as increased tourist satisfaction (e.g., through surveys), growth in tourism revenue, and reduced environmental impact (e.g., waste reduction).
Achievable (Can be achieved)	• Ensuring that training for the workforce is conducted so that they have the skills to use new technologies, and that supporting infrastructure is available for implementation.
Relevant (Relevant)	• The focus on developing sustainable tourism is highly relevant to current global trends, where tourists are increasingly seeking environmentally friendly and sustainable experiences.
Time-bound (Time Limit)	• Set short-term targets (1–2 years) for initial implementation and testing of new products, and long-term targets (3–5 years) for impact evaluation and further development.

Source: Author.

By integrating these two sectors, tourism destinations can increase their competitiveness and relevance in an increasingly competitive global market. The creative industry is driven by innovation and fresh ideas from skilled human resources. The creative economy holds great potential if supported by three key factors: creative knowledge, skilled workers, and labor-intensive efforts, enabling broader opportunities for creative industry products.

References

[1] Tahir, T., Hasan, M., & Dinar, M. (2023). Tourism Development Strategy in Improving the Community Economy at Tanjung Bira Beach, Bonto Bahari District, Bulukumba Regency. *Pinisi Journal of Entrepreneurship Review*, *1*(1), 1–15.

[2] Hirsch-Kreinsen, H. (2024). Artificial intelligence: A 'promising technology'. *AI & Society*, *39*(4), 1641–1652.

[3] United, N. (2024). Digitalization and artificial intelligence in the creative economy. *Creative Economy Outlook 202*Geneva: United Nation, pp. 51–7doi:10.18356/9789211065558c006.

[4] Setiadi, N. J., & Boediprasetya, A. (2012). Development of the Creative Economy In The City of Bandung: Measuring Creative Behavior among Indonesian Creative Workers. In *the third seminar of the Regional Studies Association Research Network, 453/SP2H/P* (Vol. 9).

[5] Kačerauskas, T., Streimikiene, D., & Bartkute, R. (2021). Environmental sustainability of creative economy: Evidence from a Lithuanian case study. *Sustainability*, *13*(17), 9730.

[6] Xing, W. (2024). Sustainable tourism: Pathways to environmental preservation, economic growth, and social equity. *Applied and Computational Engineering*, *66*, 166–171.

[7] Rauf, J. (2015). Sustainable Tourism | Sustainable Tourism. *Novat. Publ.*, 32–36 [Online]. Available: https://sustainable-tourism.net/sustainable-tourism/

[8] Baloch, Q. B., Shah, S. N., Iqbal, N., Sheeraz, M., Asadullah, M., Mahar, S., & Khan, A. U. (2023). Impact of tourism development upon environmental sustainability: a suggested framework for sustainable ecotourism. *Environmental Science and Pollution Research*, *30*(3), 5917–5930.

[9] Amerta, I. M. S., Sara, I. M., & Bagiada, K. (2018). Sustainable tourism development. *International Research Journal of Management, IT and Social Sciences*, *5*(2), 248–254.

[10] Pabilando, S. R., TLT, J. R., Carreon, J. S., Gupta, D. A., & Naseem, S. (2022). *Artificial Intelligence In Tourism Management*. Tamil Nadu: Edurights Global #102.

[11] Jaelani, A. K., Luthviati, R. D., Siboy, A., Al Fatih, S., & Hayat, M. J. (2024). Artificial Intelligence Policy in Promoting Indonesian Tourism. *Volksgeist: Jurnal Ilmu Hukum dan Konstitusi*, 109–137.

[12] Hartono, I. B., Tjiptoherijanto, P., Chotib, C., Wulandari, S. L., & Sidiq, F. F. (2023). Potential And Challenges In Creative Economy Development In Depok City: SWOT Analysis And Strategic Recommendations. *Journal of Governance*, *8*(4), 541–562.

[13] Martinaityte, E., & Kregzdaite, R. (2015). The factors of creative industries development in nowadays stage. *Economics & Sociology*, *8*(1), 55.

[14] Pratomo, S., Ashar, K., & Satria, D. (2021). Role of creative economy on local economic development. *Journal of Indonesian Applied Economics*, *9*(2), 27–35.

[15] Faur, D. L. (2010). Regulation & Regulatory Governance. *Jerusalem* [Online]. Available: http://regulation.huji.ac.il

[16] Tuo, Y., Ning, L., & Zhu, A. (2021). How Artificial Intelligence Will Change the Future of Tourism Industry: The Practice in China. *Information and Communication Technologies in Tourism 2021*, *1*, 83–94. doi:10.1007/978-3-030-65785-7.

[17] Goyal, A. A. (2023). The Role of Machine Learning in Natural Language Understanding. *International Journal for Research in Applied Science and Engineering Technology*, *11*(7), 280–284.

[18] Mishra, D., Das, S., & Patnaik, R. (2024). Application of AI Technology for the Development of Destination Tourism towards an Intelligent Information System. *Economic Affairs (New Delhi)*, *69*(2), 1083–1095.

[19] Sutarsih, T., Sari, E., Syakilah, A., & Maharani, K. (2023). Indonesian telecommunications statistics. *Jakarta.*

[20] Ahmed, M. R., & Ahmed, B. (2023). Artificial Intelligence and Product Development. *SSRN Electron J*, no. September. doi:10.2139/ssrn.4534559.

[21] Ying, L., Ziyi, Q., Shizhuan, H., Yan, L., & Tongqian, Z. (2024). The development of digital tourism in China. *Cogent Social Sciences*, *10*(1), 20doi:10.1080/23311886.2024.2347013.

[22] Subawa, N. S., Widhiasthini, N. W., Astawa, I. P., Dwiatmadja, C., & Permatasari, N. P. I. (2021). The practices of virtual reality marketing in the tourism sector, a case study of Bali, Indonesia. *Current Issues in Tourism*, *24*(23), 3284–3295.

[23] Irawan, B. (2023). Impact of artificial intelligence on the tourism industry: A literature review. *Innovative: Journal Of Social Science Research*, *3*(6), 4865–4878.

[24] Sasoko, D. M., & Mahrudi, I. (2023). SWOT Analysis Techniques in Activity Planning. *Jurnal Perspektif Jayabaya Journal of Public Administration*, *22*(1), 8–19.

[25] Sudarmanto & Faizal, E. (2023). Tourism Potential Analysis Using Smart Method Based on Community-Based Tourism Approach. *J. Inform. Computer, Business and Manaj*, *17*(3), 1–15.

[26] Wirawan, P. E. (2024). Turtle Conservation and Education Center (TCEC) As a Digital Promotion Strategy to Increasing the Number of Tourist Visits and Sustainability. *Acta Innovations*, (52), 43–50.

[27] Akshayaa Raaja Shri, I. T., Kamalee, S., Rahul, B., & Menaka, S. R. (2022). Fine Motor Therapy Using Avatar-based AI for Autistic Children (Virtual Reality). *International Academic Journal of Science and Engineering*, *9*(2), 11–17. https://doi.org/10.9756/IAJSE/V9I2/IAJSE0908

[28] Villacis, X. L. R. V., Zuta, E. R., Lozano, S. M., Ramírez, S. V. L., Lozano, D. A. R., & Vela, J. R. (2024). Analysis of the Scientific Production on Direct Consumer Behavior. *Indian Journal of Information Sources and Services*, *14*(4), 86–91. https://doi.org/10.51983/ijiss-2024.14.4.14

[29] K. Radhakrishna, D. Satyaraj, H. Kantari, V. Srividhya, R. Tharun and S. Srinivasan, "Neural Touch for Enhanced Wearable Haptics with Recurrent Neural Network and IoT-Enabled Tactile Experiences," 2024 3rd International Conference for Innovation in Technology (INOCON), Bangalore, India, 2024, pp. 1–6, doi: 10.1109/INOCON60754.2024.10511642.

58 Enhancing building energy load prediction with feature engineering and deep learning: A comparative study of DNN, FCNN, and 1D CNN

Kanokwan Khiewwan[1,a], *Narut Butploy*[1,b], *Phrommate Verapan*[2,c], *Khumphicha Tantisantisom*[2,d], *Pakin Maneechot*[3,e], *and Nivadee Klungsida*[3,f]

[1]Department of Computer Technology, Faculty of Industrial Technology, Kampaeng Phet Rajabhat University, Kampaeng Phet, Thailand
[2]Department of Information Technology, Faculty of Science and Technology, Kampaeng Phet Rajabhat University, Kampaeng Phet, Thailand
[3]Department of Energy, Faculty of Industrial Technology, Kamphaeng Phet Rajabhat University, Kamphaeng Phet, Thailand

Abstract: Precise forecasting of energy demands for buildings is essential for optimizing energy use and promoting sustainability. In this study, we employ three deep learning models, which include a Fully Connected Neural Network (FCNN), a 1D Convolutional Neural Network (CNN 1D), and a Deep Neural Network (DNN), to forecast heating and cooling requirements based on the features of the architecture. An 8-featured dataset of 768 samples from the UCI Machine Learning Repository belongs to 8 architectural design features associated as a class attribute of the building samples. The research thus sets up the problem as a matter of classification. The result demonstrates that CNN 1D has a higher classification performance compared to FCNN and DNN with an accuracy of 0.8214. Its superiority comes from its ability to capture spatial dependencies among architectural features. Class imbalance and whether it can apply to datasets beyond simple artificial examples need further investigation. This result highlights the role of deep learning approaches in thermal load prediction, which informs sustainable building design, HVAC optimization, and the integration of sensors in real time.

Keywords: Deep learning, CNN 1D, thermal load classification, energy efficiency, heating and cooling load

1. Introduction

The International Energy Agency (2023) is information showed that buildings account for 30% of carbon dioxide emissions and 40% of global energy consumption [1]. The buildings' architecture is designed to reduce energy and greenhouse gas emissions, promoting environmental sustainability. Even though physics-based simulation and statistical models might provide some answers, the nonlinear reactions plus the high computational expenses usually stand in their way respectively [2, 3]. Another notable thing, stochastic adjustments like changes in shading, will impact the peak loads greatly, therefore the need for adaptive modeling techniques, as pointed out by Yao [4]. Among these strains, machine learning and deep learning have emerged as better alternatives, unearthing complex relationships between architectural parameters and energy consumption to enhance the prediction of using energy [5, 6]. Performance-wise, deep learning, specifically neural networks, has shown far better results regarding energy load forecasting. Even up to the short-term prediction level, better results have been illustrated with the integration attention mechanisms in hybrid modeling [7]. However, comparison challenges to benchmark the architectures, feature engineering optimization, and improving generalization for real-world deployment remain somewhat tough. This study evaluates FCNN, 1D CNN, and DNN classifiers for heating and cooling loads to observe the potential they share in capturing complex architectural relationships with the challenges faced for their practical deployment feasibility.

2. Methodology

2.1. *Study overview and data preparation*

In this research, we use deep learning models to predict loads of heating and cooling, mainly studying the architectural characteristics of buildings. The data for this study is derived

[a]kanokwan_kh@kpru.ac.th, [b]narut@kpru.ac.th, [c]phrommate_v@kpru.ac.th, [d]khumphicha_t@kpru.ac.th, [e]pakin@kpru.ac.th, [f]nivadee_k@kpru.ac.th

DOI: 10.1201/9781003684459-58

from the UCI Energy Efficiency Dataset, which includes 768 samples of buildings with eight architectural characteristics, including compactness, surface area, wall area, roof area, total height, orientation, glazing area, and glazing distribution. In principle, these factors determine the energy demand and how strongly it impacts the heating and cooling loads (Y1, Y2). Feature scaling is done using StandardScaler meant for numerical stability while KBinsDiscretizer makes heating and cooling loads categorized into five groups Therefore, the problem has been translated into a classification task. Nevertheless, it increases the interpretability of the model; it may have some class imbalance. Further exploration of alternative discretization techniques and class weighting is warranted. The polynomial feature expansion creates feature interaction terms to help facilitate model learning [8]. The data was split into 60% training and 40% testing data in an attempt to have sufficient generalization. Unlike conventional methods of feature selection, deep learning models have the ability to autonomously extract key features thereby improving its predictive accuracy and, hence, its applicability toward scaling up for real-world applications.

2.2. *Model training and evaluation*

This paper compares the three deep learning architectures including the Fully Connected Neural Network (FCNN), 1D Convolutional Neural Network (CNN 1D) and Deep Neural Network (DNN). While FCNN operates independently on input features, the quest for spatial dependencies hampers its ability to relate the architectural parameters meaningfully [9]. In contrast, 1D CNN applies convolutional operations in the extraction of local feature patterns, and this is the reason leading to its efficiency – structured architectural data wherein spatial dependencies mean much [10]. DNN builds upon the structure of FCNN by incorporating additional hidden layers and techniques to enhance learning stability and reduce overfitting, including batch normalization and dropout [11]. Each model is implemented with the Adam optimizer, utilizing a learning rate of 0.001 and a loss function defined as sparse categorical cross-entropy. The model was trained to 120 epochs with early stopping implemented on validation loss. This prevented overfitting. The data was divided into 60% used for training and 40% for testing. In such an evaluation, the model was assessed for accuracy, precision, recall, and F1-score, with a particular emphasis on the potential class imbalances in the heating and cooling loads. The model was further evaluated using a confusion matrix; this not only gave the misclassifications but also presented a complete comparison of model effectiveness. Results though indicated that CNN 1D is the model with the highest accuracy that can easily capture spatial dependencies within architectural features, further optimization remains an area of interest, and maximal performance is achievable, especially for underrepresented energy load categories.

The energy load classification pipeline based on deep learning is illustrated in Figure 58.1. In the pipeline,

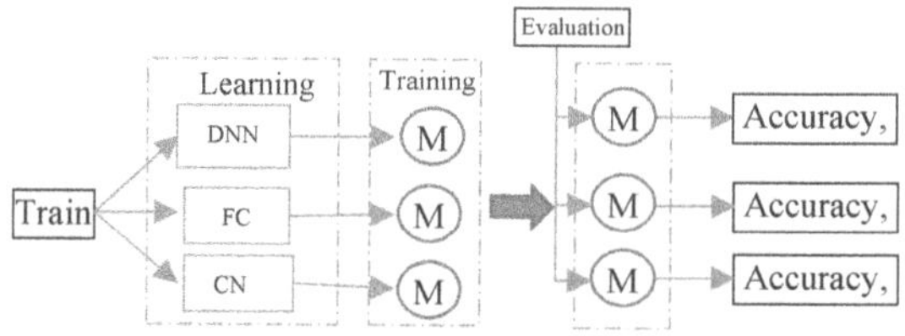

Figure 58.1. Deep learning-based energy load classification framework.

Source: Author.

architectural parameters are extracted from preprocessed data and feature engineering is subsequently conducted, which is an essential component of the neural model architecture. Three deep learning architectures, namely DNN, FCNN, and CNN 1D, are trained simultaneously to learn pattern mapping concerning energy load classification. After training, the models (M1, M2, and M3) are evaluated on unseen data using accuracy, precision, recall, and F1-score to ensure a correct comparison of performance.

3. Results and Discussion

The three deep learning models, DNN, FCNN, and CNN 1D, have been reported regarding accuracy, precision, recall, and F1-score. Among the models, CNN 1D dominated the event with 82.14% accuracy followed by FCNN at 77.62% and DNN at 75.65%, respectively. The best performance of CNN 1D is due to its ability to identify spatial dependencies in architectural features, which improves its capacity to forecast heating and cooling loads. The results shed light on the sometimes drastically different impacts of different deep learning architectures. CNN 1D will be able to detect meaningful patterns in structured data thus making it much more appropriate for energy load classification. FCNN and DNN, being very predictive, are plagued by the complicated spatial relationship and, therefore a lower rate of accuracy. This shows that convolutional structures dominate other feed-forwards in energy modeling, proving itself strong in improving the predictability of building energy management.

3.1. *Analysis of correlation matrix*

The analysis of the correlation matrix offers understanding into the models' ability to differentiate among various heating and cooling load categories.

The matrices of confusion for DNN FCNN and CNN 1D are presented in Figure 58.2; they compare predicted and true labels. The correct classifications are indicated by diagonal values while the off-diagonal values represent misclassifications. The 1D CNN achieves the maximum accuracy, implying good generalization of performance across different classes. The FCNN emerges relatively stronger, but the difficulty is indeed faced at mid-level energy loads. More

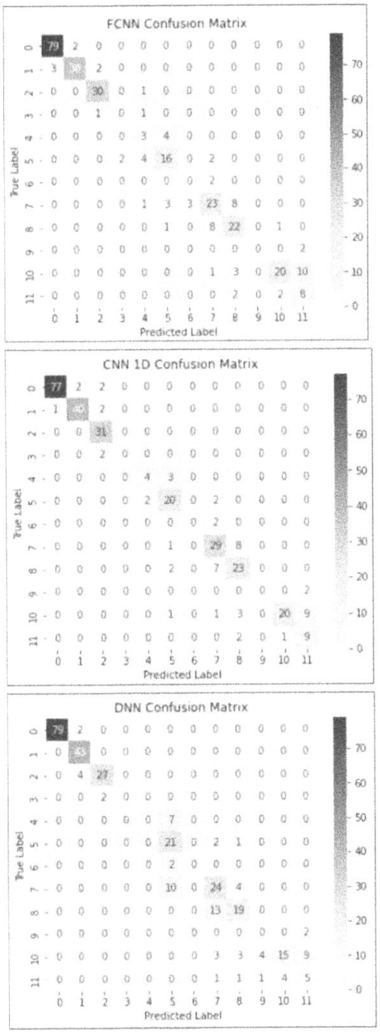

Figure 58.2. The confusion matrices for DNN, FCNN, and CNN 1D.

Source: Author.

misclassifications are indeed in the DNN model, and that too in the low level of frequency categories.

3.2. Discussion

The results confirm the better performance of CNN 1D over FCNN and DNN. Since FCNN works independently with each input feature, in contrast, CNN 1D captures effectively the interactions between the features and, thus, is ideal for energy load prediction [12]. DNN FCL eventually overfits and does not generalize well. The problems are, however, countered by the process of dropout and batch normalization, requiring some careful hyperparameters to determine the number of layers [13]. There is also a presence of KBinsDiscretizer, whose presence is useful to make things interpretable but creates the presence of class imbalance. Hybrid regression-classification models and data augmentation might be interesting avenues of research with a threat from the imbalance. Real-world datasets should be sensor-based energy monitoring and with that way may prove interesting to improve the generalizability of the model.

4. Conclusion

This research analyzed DNN, FCNN, and 1D CNN regarding the prediction of heating and cooling loads for buildings. One finding indicates that 1D CNN achieves superior classification accuracy compared to the other models, attributed to its capacity to understand spatial relationships. While deep learning models thrive in pattern recognition, more attention is needed in the application for energy prediction because the model should account for the uncertainties driven by building occupants and keep track of the evolving real-time data. Future work may concentrate on hybrid AI-optimization which involves Teaching–Learning–Based Optimization (TLBO) to enhance the productivity machinery further [14]. Real-time sensor data with a transfer learning approach to model explainability would further establish the robustness of our models enabling the deep learning models to be more suitable for real-world energy management systems.

References

[1] Özkan, V., & Özkan, A. (2022). Adsorptive Desulfurization of Crude Oil with Clinoptilolite Zeolite. *Natural and Engineering Sciences*, *7*(3), 284–293. http://doi.org/10.28978/nesciences.1222495

[2] Gil-Ozoudeh, I., Iwuanyanwu, O., Okwandu, A. C., & Ike, C. S. (2022). The role of passive design strategies in enhancing energy efficiency in green buildings. Engineering Science & Technology Journal, 3(2), 71–91.

[3] Chinju, C., Madathil Sasi, S., & Sankar, P. (2016). Symmetric Cross Switched Multilevel VSI with Reduced Number of Circuit Components Using Artificial Bee Colony (ABC) Algorithm. *International Journal of Advances in Engineering and Emerging Technology*, *7*(3), 63–77.

[4] Yao, J. (2020). Minimum number of simulation runs for reliable building energy and peak load prediction at different building scales: A study on stochastic shade adjustment. Journal of Asian Architecture and Building Engineering, 19(5), 527–540.

[5] Bourdeau, M., Zhai, X. Q., Nefzaoui, E., Guo, X., & Chatellier, P. (2019). Modeling and forecasting building energy consumption: A review of data-driven techniques. Sustainable Cities and Society, 48, 101533.

[6] Fakiha, B. (2023). The Role of Raspberry Pi in Forensic Computer Crimes. *Journal of Internet Services and Information Security*, *13*(4), 76–87. https://doi.org/10.58346/JISIS.2023.I4.005

[7] Ji, X., Huang, H., Chen, D., Yin, K., Zuo, Y., Chen, Z., & Bai, R. (2023). A hybrid residential short-term load forecasting method using attention mechanism and deep learning. Buildings, 13(1), 72.

[8] Li, J., Ren, L., Wang, B., & Li, G. (2020). Probabilistic load forecasting of adaptive multiple polynomial regression considering temperature scenario and dummy variables. Energy Reports, 6, 157–167.

[9] LeCun, Y., Bengio, Y., & Hinton, G. (2015). Deep learning. Nature, 521(7553), 436–444.

[10] Fawaz, I., Forestier, G., Weber, J., Idoumghar, L., & Muller, P.-A. (2019). Deep learning for time series classification: A review. Data Mining and Knowledge Discovery, 33(4), 917–963.

[11] Schmidhuber, J. (2015). Deep learning in neural networks: An overview. Neural Networks, 61, 85–117.

[12] Zhang, C., Yi, M., Ye, F., Xu, Q., Li, X., & Gan, Q. (2022). Application and evaluation of deep neural networks for airborne hyperspectral remote sensing mineral mapping: A case study of the Baiyanghe uranium deposit in Northwestern Xinjiang, China. Remote Sensing, 14(20), 5122.

[13] Song, X., Deng, L., Wang, H., Zhang, Y., & He, Y. (2021). Deep learning-based time series forecasting. Journal of Machine Learning Research, 22(1), 55–67.

[14] Zheng, S., Xu, H., Mukhtar, A., Yasir, A. S. H. M., & Khalilpoor, N. (2023). Estimating residential buildings' energy usage utilizing a combination of Teaching–Learning–Based Optimization (TLBO) method with conventional prediction techniques. Engineering Applications of Computational Fluid Mechanics, 17(1), 2276347.

59 A deep learning approach for real-time energy forecasting in smart buildings with NoSQL and IoT data

Narut Butploy[1,a], *Kanokwan Khiewwan*[1,b], *Phrommate Verapan*[2,c], *Khumphicha Tantisantisom*[2,d], *Pakin Maneechot*[3,e], *and Nivadee Klungsida*[3,f]

[1]Department of Computer Technology, Faculty of Industrial Technology, Kampaeng Phet Rajabhat University, Kampaeng Phet, Thailand
[2]Department of Information Technology, Faculty of Science and Technology, Kampaeng Phet Rajabhat University, Kampaeng Phet, Thailand
[3]Department of Energy, Faculty of Industrial Technology, Kamphaeng Phet Rajabhat University, Kamphaeng Phet, Thailand

Abstract: A significant advance in controlling smart building energy is applying real-time updates from machine learning algorithms to IoT-based energy monitoring. By combining information gathered from IoT sensors with LSTM networks and NoSQL databases, this research presents an online energy forecasting model that can predict energy use for dynamic 10-minute intervals. Operational and environmental data include weather, humidity, occupancy, solar power output, and precipitation which can all be found in Database powered by Firebase in real-time. Proactive energy management is made possible with the use of real-time data and short-term forecasts from the LSTM model. The results suggest the model does succeed in capturing energy demand trends, given the 0.29 kW RMSE, 11.56% MAPE, and 0.78 R^2. Facility managers can monitor off-console energy forecasting and access real-time data, enabling better identification of inefficiencies and more error-corrected adjustments. Evidence supports that smart building energy forecasting is overly simplified and stands to get the most out of deep learning when used with IoT and NoSQL databases.

Keywords: A NoSQL database, machine learning, energy forecasting, internet of things, smart building

1. Introduction

Smart energy management systems, when integrated with building energy demand management systems, allow for the optimization of power consumption through real-time monitoring and forecasting [3, 12]. IoT technologies can be utilized for energy monitoring to gather the required data to improve efficiency. Such a system facilitates the observing environmental variables including occupancy, temperature, and humidity in real-time [14]. Most current systems, however, remain primitive because they are based on historical data analysis, which stagnates their predictive capability for demand bursts [8]. This is why deep learning models, especially LSTM networks, used for short-term energy forecasting that seek to unravel the interdependence of sequential data are increasingly gaining attention [6]. Compared to ARIMA and other classic counterparts, LSTM networks excel in time-series forecasting, particularly among nonlinear and long-term dependent variables [9]. Our relational database has improved, it continues to falter in efficiently managing the incessant high-frequency data by IoT [4]. Firebase and other NoSQL databases are more suitable owing to their high expandability and low data retrieval latency, which are more advantageous for systems that operate in real-time [5, 15]. For time-series forecasting SQL databases, real-time data dealing necessitates fixed schemas that are prone to real-time crashes caused by high-velocity queries. However, these NoSQL databases are perfect for smart constructing time-series forecasting because of their horizontal scalability and dynamic schema adaptation. Sensor readings with missing values are an example of the kind of unstructured or semi-structured data that can be analysed using NoSQL. Mofradi and Nasab [13] state that SQL databases have to go through multi-stage preparations that disqualify them from being used in real-time applications.

2. Forecasting Methodology

We are aiming for more efficient energy management and resource optimization. One NoSQL database that the system

[a]narut@kpru.ac.th, [b]kanokwan_kh@kpru.ac.th, [c]phrommate_v@kpru.ac.th, [d]khumphicha_t@kpru.ac.th, [e]pakin@kpru.ac.th, [f]nivadee_k@kpru.ac.th

DOI: 10.1201/9781003684459-59

makes use of is Firebase, which is real-time, to control access to data which is available in real time. Firebase has three main parts: To facilitate model training, the smart-building-train collection in the Firebase database contains historical data in three separate collections. In 'smart-building-predict,' projected values meant for validation are stored while 'smart-building-test' has real-time data from sensors. The system retrieves the ten most recent timestamps, from the smart-building-test, with a structured data pipeline to generate LSTM model forecasts for the subsequent six timestamps an hour ahead. The system-predicted energy values are compared with real-time data from sensors. The system logs its predictions in smart-building-predict which helps revise the energy consumption against actual sensor measurements. To optimize retrieval time for frequently updating sensors, document-oriented storage is employed by NoSQL services such as Firebase. The use of these systems enables responsive and proactive learning. The easy addition of more IoT devices without requiring extensive alterations is made possible through the use of NoSQL's horizontal scaling. This research circumvents issues related to traditional forecasting by leveraging the real-time capabilities of Firebase and the flexibility of NoSQL. With the proposed system's persistent updates, facility managers will have access to current data pertaining to energy consumption and using it optimally to increase efficiency.

2.1. Forecasting model development

The Anaconda configuration uses Jupyter Notebook to implement the forecasting model in Python. Some useful libraries to include are scikit-learn for pre-processing, NumPy for data handling, and TensorFlow for creating the LSTM models with pandas. Energy consumption (kWh) is the relevant variable, and every ten minutes, 10,000 documents are collected from the Internet of Things devices. Solar power output (kw), humidity, temperature, several time indicators, and occupation are the input features. Using 'train test split,' we divided the dataset in half, setting aside 70% for training and 30% for testing. The model was then run using the provided architecture. By adding ten timestamps over the past 100 minutes to every training sample, the LSTM model is able to identify sequential dependencies, leading to more accurate energy usage estimates.

Figure 59.1(a) depicts the Zero Energy Building's off-grid solar PV system outfitted with real-time monitoring. A real-time energy analysis dashboard, an LSTM backend, Internet of Things (IoT) sensors, and a Firebase database make up the forecasting architecture shown in Figure 59.1(b).

2.2. Model training, evaluation, and deployment

Train with 70% of the data and validate with 30% to train the LSTM model for 100 epochs with a batch size of 32. Early pausing is employed to prevent overfitting. The model has two stacked LSTM layers, each with 50 units. Results are evaluated using RMSE, MAPE, and R^2. Using ten

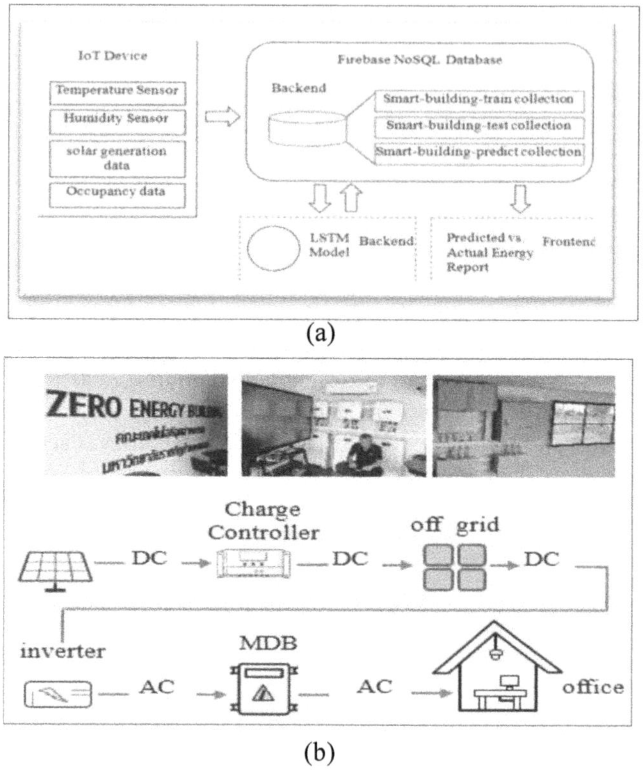

Figure 59.1. Plan for a zero-energy structure and system for continuous energy prediction.

Source: Author.

timestamps retrieved from Firebase, the model is deployed after training to forecast energy use for the following hour, divided into six 10-minute periods. In order to facilitate real-time monitoring, a dashboard is updated every 10 minutes. Upcoming updates are expected to include enhanced capabilities for detecting anomalies and trend analysis.

3. Results and Discussion

3.1. Model performance and dashboard integration

Analysing the training and validation losses allows one to test the model's learning capability over 100 epochs. This is useful for checking whether the model can track energy consumption patterns without either over- or under-fitting the data. When the loss keeps going down, it means the model is learning effectively and can better generalize to new data.

Actual and projected energy use are displayed in Figure 59.2(a). Figure 59.2(b) shows the web-based dashboard that uses Firebase data fetching to monitor energy in real-time; the demand forecasting model is very accurate, with only small deviations caused by occupancy and environmental conditions. This dashboard allows both technical and non-technical users to analyze trends, evaluate errors, and change parameters for data-driven energy management.

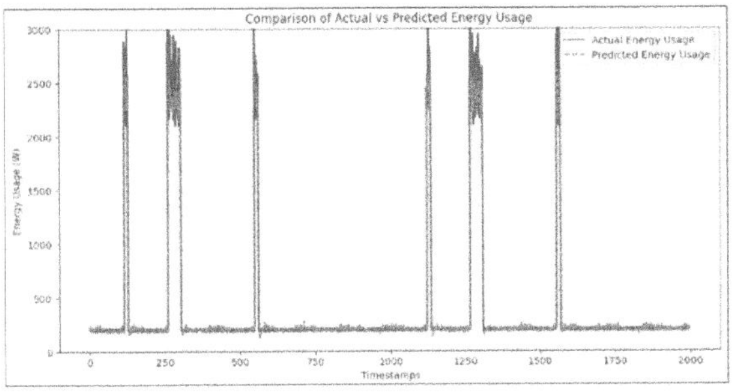

(a)

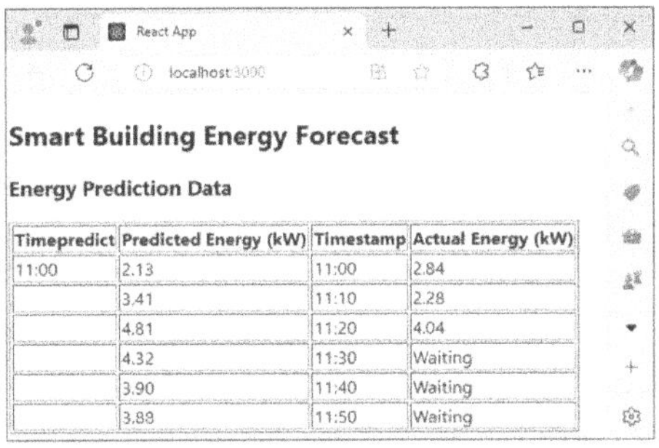

(b)

Figure 59.2. Model performance and dashboard integration.

Source: Author.

3.2. Validation measures against the ground truth of the model

R^2=0.78, MAPE=11.56%, and RMSE= 0.29 kW were used to validate the model. The model's excellent performance, with minor discrepancies in energy consumption estimates and an explained variance of 82.56%, is predicted by the error values mentioned above. These findings confirm the model's dependability and highlight areas that should be improved upon.

3.3. Real-world implementation and practical implications

Through web-based energy monitoring, the forecasting model is implemented into a dashboard. To support adjustments based on facts gathered for energy efficiency measures, Firebase guarantees that projected and actualized energy values are met at ten-minute intervals. This aligns with Chawla's [2] study exploring the significance of real-time energy visualization for consumption optimization. The model is capable, indirectly, of aiding HVAC efficiency, and renewable energy future planning from a forecasting perspective. Zhou et al. [10] and Zhu et al. [11] as illustrated deep learning's efficacy in energy forecasting, aside excluding a handful of cases involving modifications in the actual world. Creating models that can represent these actions is crucial, as Shi et al. [7] noted. considering HVAC scheduling and occupancy behaviour capture the most forecast error. Matrix of correlation show that temperature and energy consumption had a moderate to strong positive correlation, and further emerged as positively associated predictors. This, together with Abdalla et al.'s [1] findings which showed some degree onsite renewable energy use significantly reduces grid dependency, showcases once again the impactful role of HVAC on energy demand. Moreover, the accuracy of the model may only be adjusted for real-world applications through the use of adaptive learning and integration processes that draw upon multiple organizational resources at the organizational level.

4. Conclusion

This research utilizes Internet of Things (IoT) systems in conjunction with smart buildings for LSTM energy load forecasting. For optimal energy management, IoT sensors monitoring energy consumption retrieve data from Firebase, which allows for prediction every ten minutes per the written code on top. Model accuracy is shown with R^2=0.78, MAPE=11.56%, and RMSE=0.29 kW. As described previously, the web-based dashboard provides real-time prediction monitoring for automated managerial decision-making, enabling comparison against actual figures for enhanced optimization. Through the use of Firebase, data synchronization is effortless both online and offline – filling any existing data gaps. Energy generation has outstripped consumption, revealing opportunities for energy storage and redistribution. Further work will include improvements to optimization techniques, data integration from multiple sources, and adaptive learning methods.

References

[1] Abdalla, M. A. A., Wang, M., Wang, B., Ishag, A. M., & Saleh, B. (2023). Double-layer home energy management strategy for increasing PV self-consumption and cost reduction through appliances scheduling, EV, and storage. *Energy Reports*, *10*, 3494–3518.

[2] Chawla, P., Gao, J. Z., Gao, T., Luo, C., Li, H., & We, Y. (2023). An interactive web-based solar energy prediction system using machine learning techniques. *Journal of Management Analytics*, *10*(2), 308–335.

[3] Chen, Z., Sivaparthipan, C. B., & Muthu, B. (2022). IoT-based smart and intelligent smart city energy optimization. *Sustainable Energy Technologies and Assessments*, *49*, 101801.

[4] Fu, Q., Han, Z., Chen, J., Lu, Y., Wu, H., & Wang, Y. (2022). Applications of reinforcement learning for building energy efficiency control: A review. *Journal of Building Engineering*, *50*, 104165.

[5] Li, W.-J., Yen, C., Lin, Y.-S., Tung, S.-C., & Huang, S. (2018). JustIoT: Internet of Things based on the Firebase

real-time database. In *Proceedings of the 2018 IEEE International Conference on Smart Manufacturing, Industrial & Logistics Engineering (SMILE)* (pp. 43–47).

[6] Pushpavalli, M., Dhanya, D., Kulkarni, M., Rajitha Jasmine, R., Umarani, B., RamprasadReddy, M., & Rajaram, A. (2024). Enhancing electrical power demand prediction using LSTM-based deep learning models for local energy communities. *Electric Power Components and Systems*, 1–18.

[7] Shi, J., Yu, N., & Yao, W. (2017). Energy-efficient building HVAC control algorithm with real-time occupancy prediction. *Energy Procedia, 111*, 267–276.

[8] Wang, L. (2022). Research on the impact of energy price fluctuations on regional economic development based on panel data model. *Resources Policy, 75*, 102484.

[9] Zhang, Y., Li, Y., Liang, B., & Ma, R. (2022). A prediction method for fuel cell degradation based on CNN-LSTM hybrid model. In *Proceedings of the 2022 25th International Conference on Electrical Machines and Systems (ICEMS)*.

[10] Zhou, H., Liu, Q., Yan, K., & Du, Y. (2021). Deep learning enhanced solar energy forecasting with AI-driven IoT. *Wireless Communications and Mobile Computing, 2021*, 1–12.

[11] Zhu, Z., Chen, W., Xia, R., Zhou, T., Niu, P., Peng, B., Wang, W., Liu, H., Ma, Z., Gu, X., Wang, J., Chen, Q., Yang, L., Wen, Q., & Sun, L. (2023). Energy forecasting with robust, flexible, and explainable machine learning algorithms. *Innovative Applications of Artificial Intelligence, 44*(4).

[12] Kathiresh, M. (2016). Biometric based Voice Security System for Defence Sector. *International Journal of Advances in Engineering and Emerging Technology*, *7*(3), 78–86.

[13] Mofradi, R. F., & Nasab, M. S. (2017). Using of the Vernier frequencies method to resolve problem of the ambiguity in range of the pulsed radars. *International Academic Journal of Innovative Research*, *4*(2), 10–21.

[14] Roy, J., Nihlani, A., Talwar, R., & Gautam, S. (2024). Assessing the social impacts of fisheries decline on fishing communities. *International Journal of Aquatic Research and Environmental Studies*, *4*(S1), 77–82. https://doi.org/10.70102/IJARES/V4S1/13

[15] Gunalan, Kumar, K., Surya, & Chelvi, K. (2022). Stock Market Prediction. *International Academic Journal of Science and Engineering*, *9*(2), 18–22. https://doi.org/10.9756/IAJSE/V9I2/IAJSE0909

60 Adaptive load management in power systems with automatic transformers

Mohankumar P.[1,a] *and V. Ravi*[2,b]

[1]M.E. Power Electronics and Drives, K.S.R. College of Engineering, Tiruchengode, Tamil Nadu, India
[2]Professor, Department of EEE, K.S.R. College of Engineering, Tiruchengode, Tamil Nadu, India

Abstract: The Automatic Load Sharing Transformer is designed to split the power demands in control systems that utilize Arduino-based methods between multiple transformers. By employing real-time dynamic load management, problems related to energy waste or overloading are dealt with effectively. An Arduino microcontroller processes the readings from voltage and current sensors to decide the work allocation. When any one of the additional transformers reaches a set limit, it is further turned on to share the extra load. In periods of low demand, unneeded transformers are deactivated to conserve energy. The system directly imposes other functions such as automated load shedding, fault detection, and over-current protection. Its control unit, which is based on Arduino, can be easily linked with IoT platforms, offering remote monitoring and diagnostics and allowing seamless integration to expand functionalities [27–28].

Keywords: Power, arduino, transmission, distribution, transformers, sensor data, microcontrollers, electrical loads, energy, voltage

1. Introduction

Power distribution systems require increasing reliability because electrical systems have become increasingly complex over the years. Power transformers, which distribute energy, are also at risk of damage through overheating or becoming inefficient due to improper load balancing. One approach developed to solve these problems is an Arduino microcontroller automatic load-sharing transformer [6]. An overload condition is avoided by dynamically redistributing the load on each transformer through the current and voltage readings. This increases the system's optimal efficiency, extends the operational life of the transformers, and maximizes their utilization. The system also supports alarm and monitoring interface integration.

1.1. Objective

This paper aims to design and implement a self-adaptive mechanism for load sharing between transformers. The system will:

- Keep track of the load on each transformer at all times
- Distribute the load across the transformers automatically to ensure fair sharing of the burden
- Turn off or switch off transformers as needed to prevent overloading.

2. Literature Survey

Increasing energy needs pose a threat of overloading or overheating transformers, which are costly and massive pieces of power system equipment. Multiple transformers should be used in tandem to prevent overloading [3, 26]. When the primary transformer's load increases, an auxiliary transformer shares the load. When running in parallel, fuzzy logic can avoid power outages caused by overloaded or overheated transformers. By introducing a load share circuit, automated current balancing can be achieved using commercially available general power supplies, doing away with the requirement for current sensing and additional control loops in parallel DC power supplies [1]. Extensive testing and analysis of the circuit's operation and performance are conducted. A method has been suggested for isolated DC-DC converters to share the load current equally with non-isolated DC-DC converters connected in parallel [4]. We evaluated the technique for output voltages up to 700W using Average current mode control [7]. It offers redundancy in a failure and can be utilized for ABRs in Satellite Power systems. A high-power anode supply for a HET thruster can also be incorporated within the design [8]. A load-sharing controller integrated circuit was suggested to solve the problems of thermal stresses and uneven distribution of load current in paralleled power modules [2]. Fault protection, under-voltage detection, automatic master-slave control, start-up and adjustment techniques, and short circuit prevention are all features it employs [9]. Our 1.5mum BCD technique allowed us to optimize and realize the IC. This paper presents a control scheme for frequency regulation in inverter-based islanded microgrids to address steady-state deviations in active power sharing caused by clock drift in digital processors [5]. The study includes a

[a]pmk441993@gmail.com, [b]ravi@ksrce.ac.in

DOI: 10.1201/9781003684459-60

theoretical analysis, a design procedure, and experimental tests on a laboratory microgrid with three independent processors [25].

3. Methodology

3.1. Existing method

The features and quality of the AC power source differ according to location and demand. The transformer allows voltage transformation and minimizes power losses during long-distance transmission, making it crucial for efficient electrical energy transmission and distribution. Maintaining power system component compatibility is its primary function. Figure 60.1 shows a relay, a critical power system control component that regulates the current flow, separates or connects components according to preset criteria, and so on. Equipment failure prevention, enhanced reliability of the power distribution network, and overall system stability are equally important. Relays fulfilling these other tasks dramatically improve the efficiency of the process. Ensuring the proper condition of the power system requires being able to measure voltage and current. Current meters measure the amount of electrical current passing through a complete circuit, while voltage meters measure the change of potential difference between two points in a circuit. These devices are essential in monitoring the load, fault detection, and system protection.

3.2. Proposed method

The safe and consistent delivery of electric power is based upon an alternating current (AC) power system. Figure 60.2, shows Circuit Diagram of the proposed system. Alternating current (AC) input is typically provided through the grid. To change voltages, minimize losses of power, and ensure that everything cooperates, a transformer is essential. Relays control the current flowing through them by making or breaking

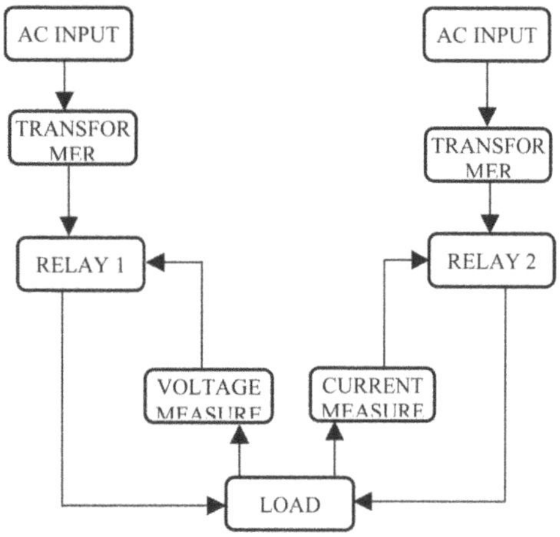

Figure 60.1. Block diagram of the existing system.

Source: Author.

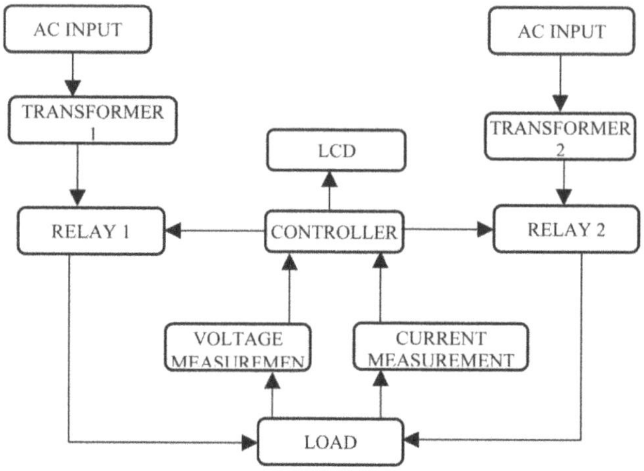

Figure 60.2. Circuit diagram of the proposed system.

Source: Author.

different system components [10]. Their capacity to sense and respond to faults or abnormal conditions, prevent equipment from being damaged, and maintain the power supply ensures much higher reliability and safety. Essential for controlling and monitoring electrical characteristics in the system are devices that measure voltage and current. The difference between potential and current is measured by voltage meters, whereas current meters quantify current flow. System stability, device voltage accuracy, and network health assessment depend on these measurements.

3.3. Components required

- AC INPUT
- TRANSFORMER
- RELAY
- LOAD
- CURRENT MEASUREMENT
- VOLTAGE MEASUREMENT
- CONTROLLER
- LCD

- **AC input:** Electricity for homes and businesses is often provided via alternating current (AC), the initial power source in a system [24]. A power supply is an apparatus that converts one kind of energy into another and feeds loads with electrical current [11–12]. Unregulated power supplies that run on alternating current (AC) use a chopper to reduce the voltage from the wall outlet, a rectifier to transform the AC voltage into pulsed direct current, and filters to eliminate the undesirable pulsation.
- **Transformer:** Figure 60.3, displays a circuit diagram for transformer. An electrical energy transfer device known as a transformer uses inductively coupled conductors to move current from one location to another. It converts alternating current (AC) voltage levels into AC utilizing the principle of electromagnetic induction.

Transformer:

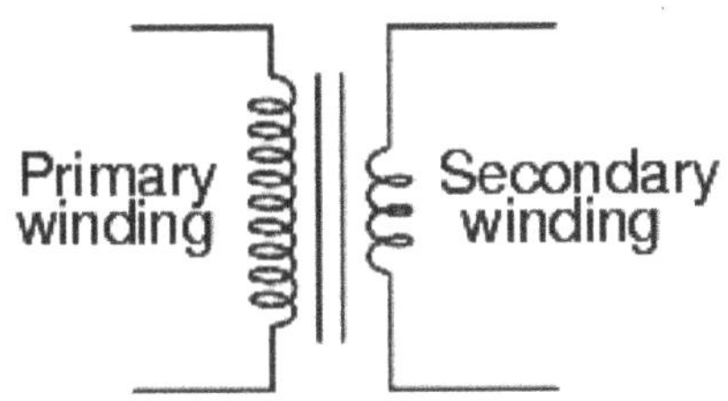

Figure 60.3. Transformer circuit diagram.

Source: Author.

Voltage and current in alternating current (AC) circuits can be easily multiplied or divided by this mechanism, which is called mutual induction. To efficiently transmit and distribute electrical energy, transformers play a key role in power systems [13]. While step-down transformers lower voltage for household or commercial usage, step-up transformers raise voltage for long-distance transmission, minimizing energy losses.

- **Relay:** Electrical switches, called relays, automate tasks like separating broken parts of a system and safeguarding it from unusual circumstances by controlling the flow of electricity within the system [14–17]. They improve the system's dependability and safety by reacting to events and taking the necessary steps in response [18]. Relays are electrical devices that enable one circuit to switch to another without requiring an internal electrical connection.

  A significant current is passed via a relay coil, as shown in Figure 60.4. Usually, 30mA is used for 12V, and 100mA is used for lower voltages. A transistor is necessary since most integrated circuits cannot supply this amount of current.

  To power high-current devices, such as lights or relays, the tiny output current from a logic IC must be amplified by a transistor, as shown in Figure 60.5. The level of current amplification is represented by current gain (HFE), and they can either act as switches or amplifiers.
- **Load:** All the electrical appliances and machinery in the house and beyond make up the load, drawing power from the system. Power system design and operation must prioritize efficient management and supply of power to various types of loads. A capacitive, inductive, or resistive load measures measures the power or current flowing through a circuit or equipment.

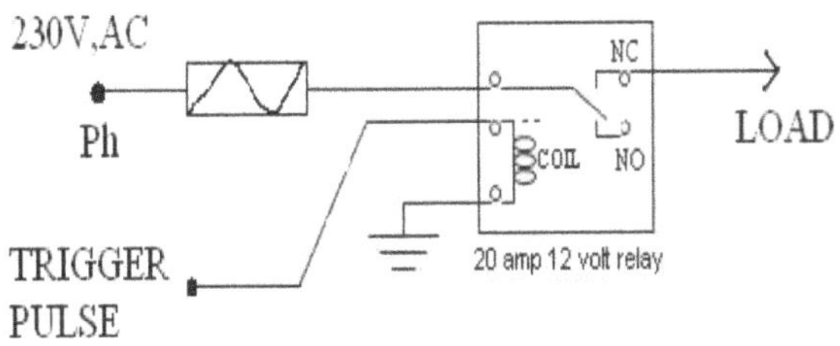

Figure 60.4. Basic relay circuit diagram.

Source: Author.

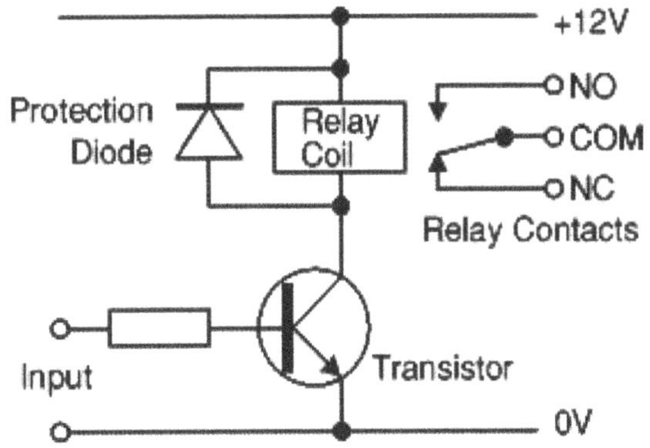

Figure 60.5. Protection diodes for relay.

Source: Author.

- **Current Measurement:** Transformers and other current-measuring devices check the levels of electric current in circuits, identify problems, and maintain them at appropriate levels. Modern sensors produce digital or analog signals for analysis and control.
- **Voltage Measurement:** Transformers and other voltage-measuring equipment are essential for maintaining stable power distribution systems, accurate voltage readings, and healthy networks. Precise readings are also crucial for ensuring device safety, designing circuits, and servicing.
- **Controller:** Sensors, measurements, and user inputs processed using the open-source electronics platform Arduino can optimize power system performance. The Arduino software and programming language allow it to read and convert inputs into outputs. Thousands of projects have been built on top of Arduino, and the platform's global maker community constantly adds to an easily navigable body of information.
- **LCD:** In Figure 60.6, the user interface is an LCD. This lets operators see the system's status, parameters, and diagnostic data in real-time to make educated decisions about the system's dependability and efficiency.

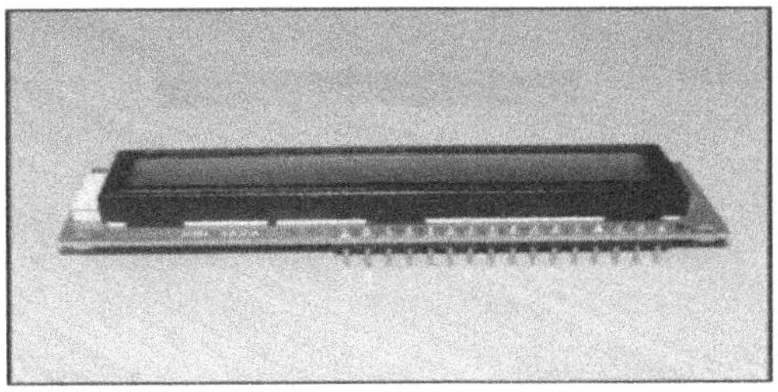

Figure 60.6. LCD display.

Source: Author.

4. Software Description

Lab Center Electronics Ltd.'s Proteus Design Suite is a comprehensive software package for making PCB layouts, modeling embedded circuits and electronics, and drawing schematics. Users can choose from English, Chinese, Spanish, or French as their language options. Working under perfect circumstances, Proteus is forgiving regarding circuit design. Likewise, Proteus ARES is utilized for PCB layouts in the PCB design process. Arduino, PIC, and 8051 are just a few microcontrollers that may have their programming codes created using this tool.

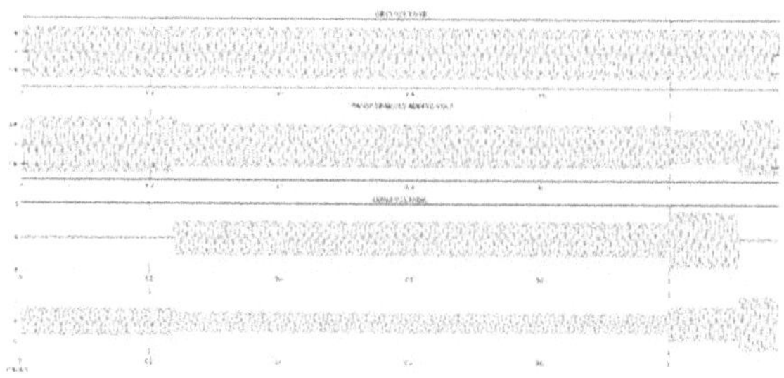

Figure 60.8. Simulation result.

Source: Author.

5. Results and Discussion

Embedded system design, simulation, automatic code generation, and continuous testing and verification are all made possible with Simulink, a block diagram environment for multi-domain simulation and Model-Based Design [20].

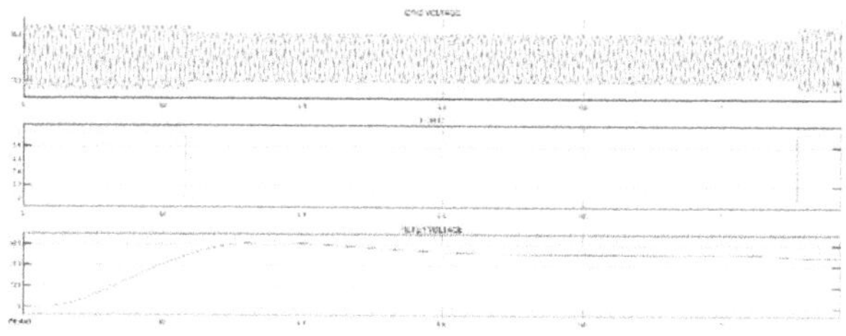

Figure 60.9. Simulation result.

Source: Author.

5.1. Simulation diagram

Simulink makes embedded system design, simulation, automated code generation, and continuous testing possible by Simulink, a block diagram environment for multi-domain simulation and Model-Based Design shown in Figure 60.7 [21]. In addition to being compatible with MATLAB for additional analysis, it has a graphical editor, block libraries that may be customized, and solvers.

Figure 60.8, illustrates that substations require supplementary transformers to accommodate increased load during peak hours. However, there will be no need for either under normal loads, particularly during off-peak hours, because our project would automatically link transformers under critical loads [19].

To avoid overloading, improve efficiency, and prolong the operating life of transformers, the idea of automatic load sharing between transformers is presented in Figure 60.9. The parts of an Arduino simulation and how it works will be covered in this article [22]. Transformer for Automatically Sharing Loads An easy and scalable solution to current power distribution problems is using an Arduino [23]. It sets the groundwork for further investigation and improvement in transformer load control and is in line with the increasing need for effective energy management in grid-connected systems. This marks another step toward building a more sustainable and reliable energy distribution infrastructure and, in turn, opens the doors to more sophisticated energy distribution systems.

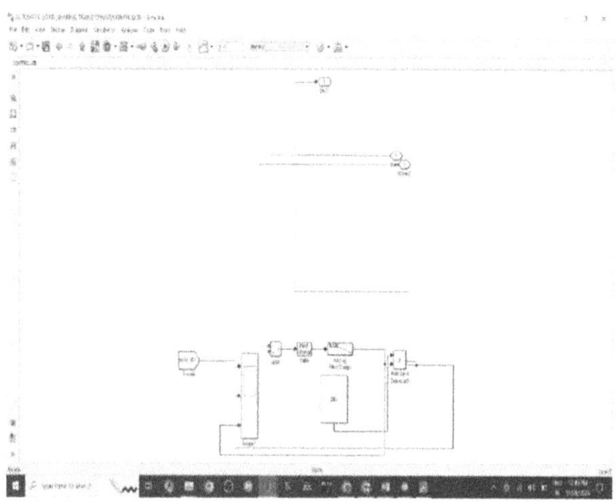

Figure 60.7. Simulation diagram.

Source: Author.

6. Conclusion

Automated load sharing using Arduino is an inexpensive way to enhance power distribution in systems with many transformers. It spreads electrical load evenly, avoids overloading, and maximizes equipment lifespan through real-time monitoring and control via the Arduino platform. For system stability, sensors monitor voltage and current. Dynamic load balancing algorithms enhance performance by dynamically balancing the load on each transformer. Modifying and improving the system to support more advanced control techniques is easy. Sustainable electrical infrastructure can take advantage of this scalable and realistic solution, which addresses the growing demand for efficient energy management in grid-connected systems.

6.1. Future scope

More efficient power supply and less energy loss are the objectives of the Arduino-based automatic load sharing transformer project. As the integration of renewable energy sources continues to rise, the Arduino-based load-sharing system can seamlessly integrate renewable energy inputs, ensuring grid stability and maximum energy utilization. This method assists in decentralized power distribution, lowering overloading, energy loss, and transformer wear and tear. Besides the scalability of Arduino platforms, machine learning algorithms can potentially change transformer settings proactively according to anticipated patterns of load demand.

References

[1] On-Road Automated Driving (ORAD) Committee. (2023). *Taxonomy and definitions for terms related to driving automation systems for on-road motor vehicles*. SAE international.

[2] Blanco, M., Atwood, J., Vasquez, H. M., Trimble, T. E., Fitchett, V. L., Radlbeck, J., … & Morgan, J. F. (2022). *Human factors evaluation of level 2 and level 3 automated driving concepts* (No. DOT HS 812 182).

[3] Abe, G., Sato, K., & Itoh, M. (2015, October). Driver's Trust in Automted Driving when Passing Other Traffic Objects. In *2015 IEEE International Conference on Systems, Man, and Cybernetics* (pp. 897–902). IEEE.

[4] Sonoda, K., & Wada, T. (2017). Displaying system situation awareness increases driver trust in automated driving. *IEEE Transactions on Intelligent Vehicles*, *2*(3), 185–193.

[5] De Winter, J. C., Happee, R., Martens, M. H., & Stanton, N. A. (2014). Effects of adaptive cruise control and highly automated driving on workload and situation awareness: A review of the empirical evidence. *Transportation Research Part F: Traffic Psychology and Behaviour*, *27*, 196–217.

[6] Beller, J., Heesen, M., & Vollrath, M. (2013). Improving the driver–automation interaction: An approach using automation uncertainty. *Human Factors*, *55*(6), 1130–1141.

[7] Yamashita, T., & Itoh, M. (2016). Driver involvement in lane-change decision making for maintaining situation awareness during automated driving. *SICE Journal of Control, Measurement, and System Integration*, *9*(6), 257–263.

[8] Pandey, V., & Gupta, N. (2024). Mechanical Engineering Design: A Multidisciplinary Approach. *Association Journal of Interdisciplinary Technics in Engineering Mechanics*, *2*(4), 6–11.

[9] Lu, Z., & De Winter, J. C. (2015). A review and framework of control authority transitions in automated driving. *Procedia Manufacturing*, *3*, 2510–2517.

[10] Reddy, S., & Mohan, P. (2024). Optimizing Energy Efficiency in Wireless Power Transmission Systems for Industrial Applications. *Association Journal of Interdisciplinary Technics in Engineering Mechanics*, *2*(2), 19–23.

[11] Mok, B., Johns, M., Lee, K. J., Miller, D., Sirkin, D., Ive, P., & Ju, W. (2015, September). Emergency, automation off: Unstructured transition timing for distracted drivers of automated vehicles. In *2015 IEEE 18th International Conference on Intelligent Transportation Systems* (pp. 2458–2464). IEEE.

[12] Telpaz, A., Rhindress, B., Zelman, I., & Tsimhoni, O. (2015, September). Haptic seat for automated driving: preparing the driver to take control effectively. In *Proceedings of the 7th International Conference on Automotive User Interfaces and Interactive Vehicular Applications* (pp. 23–30).

[13] Chandragupta Mauryan, K. S., & Purrnimaa Shiva Sakthi, R. (2020). Amelioration of Hybrid Power System Using IOT. *International Academic Journal of Science and Engineering*, *7*(2), 01–05. https://doi.org/10.9756/IAJSE/V7I2/IAJSE0703

[14] Petermeijer, S. M., Cieler, S., & De Winter, J. C. (2017). Comparing spatially static and dynamic vibrotactile take-over requests in the driver seat. *Accident Analysis & Prevention*, *99*, 218–227.

[15] Payre, W., Cestac, J., & Delhomme, P. (2016). Fully automated driving: Impact of trust and practice on manual control recovery. *Human Factors*, *58*(2), 229–241.

[16] Merat, N., Jamson, A. H., Lai, F. C., Daly, M., & Carsten, O. M. (2014). Transition to manual: Driver behaviour when resuming control from a highly automated vehicle. *Transportation Research Part F: Traffic Psychology and Behaviour*, *27*, 274–282.

[17] Nilsson, J., Falcone, P., & Vinter, J. (2014). Safe transitions from automated to manual driving using driver controllability estimation. *IEEE Transactions on Intelligent Transportation Systems*, *16*(4), 1806–1816.

[18] Kaustubh, M., Willemsen, D., & Mazo, M. (2016, June). The modeling of transfer of steering between automated vehicle and human driver using hybrid control framework. In *2016 IEEE Intelligent Vehicles Symposium (IV)* (pp. 808–814). IEEE.

[19] Russell, H. E., Harbott, L. K., Nisky, I., Pan, S., Okamura, A. M., & Gerdes, J. C. (2016). Motor learning affects car-to-driver handover in automated vehicles. *Science Robotics*, *1*(1), eaah5682.

[20] Wada, T., Sonoda, K., Okasaka, T., & Saito, T. (2016, October). Authority transfer method from automated to manual driving via haptic shared control. In *2016 IEEE International Conference on Systems, Man, and Cybernetics (SMC)* (pp. 002659–002664). IEEE.

[21] Sheridan, T. B. (2011). Adaptive automation, level of automation, allocation authority, supervisory control, and adaptive control: Distinctions and modes of adaptation. *IEEE Transactions on Systems, Man, and Cybernetics-Part A: Systems and Humans*, *41*(4), 662–667.

[22] Inagaki, T. (2003). *Adaptive Automation: Sharing and Trading of Control*. Boca Raton, FL, USA: CRC Press.

[23] Flemisch, F. O., Adams, C. A., Conway, S. R., Goodrich, K. H., Palmer, M. T., & Schutte, P. C. (2003). *The H-Metaphor as a guideline for vehicle automation and interaction* (No. L-18448).

[24] Fathima Sapna, P. (2021). Load Frequency Control of Thermal Power System by using Extended PI & FLC. *International Academic Journal of Innovative Research*, *8*(2), 01–05. https://doi.org/10.9756/IAJIR/V8I2/IAJIR0803

[25] Karimov, N., Kalandarova, M., Makhkamova, N., Asrorova, Z., Saydamatov, F., Ablyakimova, R., Karshiboeva, M., & Odilov, B. (2024). Impact of Mobile Applications on Tourism Development in Uzbekistan. *Indian Journal of Information Sources and Services*, *14*(4), 175–181. https://doi.org/10.51983/ijiss-2024.14.4.27

[26] Ghate, A. D., Sandilya, R., Verma, M., & Chakraborty, P. (2024). Developing a framework for inclusive fisheries governance in India. *International Journal of Aquatic Research and Environmental Studies*, *4*(S1), 89–94. https://doi.org/10.70102/IJARES/V4S1/15

[27] Abhinaya, J., & Kathirvel, C. (2016). A Novel Method of DC Current Suppression and Maximum Power Point Tracking for GRID Connected Photovoltaic Inverter. *International Journal of Advances in Engineering and Emerging Technology*, *7*(3), 87–95.

[28] Lukić, A. (2023). GIS Analysis of the Vulnerability of Flash Floods in the Porečka River Basin (Serbia). *Archives for Technical Sciences*, *1*(28), 57–68. https://doi.org/10.59456/afts.2023.1528.057L.

[29] Reddy, K. P., Kodati, S., Veeranna, T., & Ravi, G. (2022). Machine Learning-Based Intelligent Video Analytics Design Using Depth Intra Coding. In *Big Data Management in Sensing* (pp. 77–86). River Publishers.

61 Innovativeness and risk-taking: A cross-cultural study of U.S., Indian, and Chinese entrepreneurs

Niranchana Shri Viswanathan[1,a], *Delecta Jenifer Rajendren*[2,b], *A. Suresh Kumar*[3,c], *and B. Nandhini*[4,d]

[1]Assistant Professor, School of Management Studies, Sapthagiri NPS University, Bengaluru, India
[2]Assistant Professor, Department of Management Studies, Saveetha Engineering College, Chennai, Tamil Nadu, India
[3]Professor and Head, Department of MBA, Vel Tech High Tech Dr. Rangarajan Dr. Sakunthala Engineering College, Avadi, Tamil Nadu, India
[4]Assistant Professor, Department of MBA, S. A. Engineering College, Thiruverkadu, Chennai, Tamil Nadu, India

Abstract: This study analyzes the innovation and risk-taking attitudes of U.S-India and China – two key cultural and economic factors forming entrepreneurial dispositions among entrepreneurs. The U.S. alone innovates more than the rest of the world, combined, due to capitalist and individualistic culture with high risk tolerance, venture capital and massive technology jumps. Record levels of starts in India (thanks to 'Startup India' and the emergence of a middle-class in one of the world's fastest growing information technology ecosystems!) as far as commerce is concerned! AI consumers and e commerce in the middle of hastily, and this government and the wait on the society is 'the back-path trail adapted to technology in China' in what is? It explores a broader underlying question: how do a variety of other forces market conditions and government influence and even social norms shape not just inventions in general but risk taking. By capturing the dynamics at play in so many kinds of entrepreneurial ecosystems, the study doesn't just shed light on the way that context shapes entrepreneurship it offers potential lessons for policymakers and business leaders looking to ignite the imaginations of innovators in their communities, and spur economic growth.

Keywords: Innovativeness, risk taking, entrepreneurship, cross cultural analysis, entrepreneurial ecosystems

1. Introduction

As a torchbearer of economic growth, disruption and jobs in the domestic economy and in the global marketplace, every thriving industrialized nation on the planet relies on the small businesses and entrepreneurs as the lifeblood of its prosperity. As a last note having recent years [1] literally post globalization and phase of transforming world order, this is a very essential part of economic competitiveness. Entrepreneurship flows differently between these all of which are cultural contexts. America is a venture society on steroids, the land of can-do entrepreneurialism. Indian entrepreneurship has been riding high on the back of a rising middle class, initiatives like Startup India aimed at enabling innovation, and an overall knowledge incentive. China has a hugely successful startup industry, despite issues like a lack of property protections, repression and the like, thanks to a massive consumer market, huge government support and a pretty decent ecosystem in general [2]. Whilst the rest of the world relies on entrepreneurship as the primary driver of the latest tech and job creation, we must embrace this mindset [18]. but you are specimens of unpursued opportunity. US, Indian and Chinese entrepreneurial ecosystem have different nuances in terms of the senses it drives businesses to: [3] the driving forces of innovation and risk taking in US as drive to adopt new technology;; in India strong government support mechanisms that enable speed of huge business and in China immense commitment to push the edge of what tech can do [4]. For the many players, the future of capitalist firms, both large and small, is so based on the need to challenge normativity, with risk-taking and innovation being one of the best differentiators in a more volatile global economy [4]. The experience of the Royal Family across cultural contexts provides important lessons for policies and programs aimed at encouraging entrepreneurial activity in these cultures [17].

2. Literature Review

Specifically it covers academic literature that summarizes relations between innovativeness and risk-taking related to entrepreneurship from a cultural perspective. Unlike risk-taking which is the buzzword when it comes to entrepreneurship in American soil, entrepreneurship in traditional country of Indian where everything is about stability tends

[a]niranchanaphd@gmail.com, [b]delectajenifer@saveetha.ac.in, [c]dr.sureshkumararaju14@gmail.com, [d]nandhinijohnsundarmba@gmail.com

DOI: 10.1201/9781003684459-61

to follow boundary lines which proves them more cautious (read: conservative) in their professional Decision Making [5]. I´d say that our Chinese entrepreneurs in our portfolio are usually funded and mix risk-taking, and strategy and they don´t interest only to push moment on moonshots innovations. Modern [6, 15] Hofstede Cultural Dimensions profile levels of individualism vs collectivism, uncertainty-active, power distance within various regions can thus delineate regional attitudes around risk, entrepreneurship and innovation that entrepreneurs intending to enter global markets must be cognizant of, as reflected in these differences [7]. There's been a lot of ink spilled describing our risk appetite, and it translates well here too, whether it's risk attitude or innovation appetite these are fundamentally culture factors as much as practicalities and in-mind reactivity factors; the entrepreneur is going to have to identify themselves as part of a culture, and society whole, as the best determinant of best practices for success. The US a melting pot of execution and progress through individualism scribbles in the book, India a hybrid model of entrepreneurialism together with a cautionary colour palette, and China the hybrid model of state-pushed evolution and tech with statistic hammer in the direction of well-calculated dangers. It is the regulatory environment, the ability to access funding resources and supporting ecosystems, more than the specifics of a successful startup that defines a startups eventual commercial viability [8, 16] whether these include U.S. inspired initiatives like Startup India China and its state backed facilitators. Cross-culture comparative studies on the other hand are relatively rare in the literature also cite our ignorant knowledge on how the risk awaiting behaviour of the entrepreneurs or factors like entrepreneurial capabilities which would differ among [9] USA, UK and India due to cultural and socio-economic diversity. Their research answers this question, revealing new insights into the ways entrepreneurs act in different parts of the world [15].

3. Methodology

The data was collected through a mixed method approach that combined qualitative and quantitative data from extensive surveys and detailed interviews with entrepreneurs in a wide range of industries. The analysis also accounts for variation in individual characteristics and business sizes within all countries. Quantitative surveys will evaluate risk-taking propensities, innovative behaviours, and business outcomes, while semi-structured interviews will offer qualitative insights into the influence of cultural values and economic considerations in shaping entrepreneurial behaviours. Examples of successful ventures will illuminate important patterns. Using advanced statistical techniques to analyze survey data so that differences in trends of innovation-driven versus large-enterprise activity can be described, while also applying qualitative techniques to explore the cultural influences that underpin entrepreneurial choices. These principles define all those who are assuming the fundraiser the research field that involves the confidentiality of the participants or obtaining the informed consent of the participants and minimization of the researcher bias in terms of data collection and data analysis.

4. Comparative Analysis Across Cultures

The counties following table demonstrate some interesting things about the end- preprocedural mindset of each country. And the Amer- can propensity for risk- taking has also made the U.S. the world leader in technology, and in doing so made us one of but three countries in the annals of human history to ensnare the imagination of those pining for freedom and thus prosperity. All over the world. The rounds of fundings will not be the one which we hear a lot about yet, but the Indian IT startup landscape is getting a law and order in the terms of global scenario. This is a fraction of a much, much larger storyline: [10] Left to its own devices, China has grown due to a variety of government programs and the ferocious efficacy of the market in AI, e-commerce and elsewhere. Trends in such countries are rooted not just in their culture and regulation, or in the size of their market, but also in their innovations.

Venture capital and a culture that is risk tolerant and technology focused equals innovation in America. While as mentioned earlier India remains a developing economy, with a fast growing middle class and an unambiguous governmental push to the startup explosion pushing core innovation leverages businesses at the core of action in the subcontinent whereas where the government encourages at the fastest pace possible by [exercising control over] near all aspects of life of citizens and businesses from marrying to having children, from doing sports to writing books [11].

The U.S. the unrivaled world technology superpower, and that let us be one of the only handful of civilizations in all human history who managed to win the hearts and minds of people around the world who themselves crave freedom. All over the world. Due to that the entrepreneur's habitat in India is at a stage of immaturity but growing as the core of IT-startups [10]. On top of that, along with the large market size, there has formed an efficient innovation ecosystem in China, which has done great wonders for everything from AI to e-commence [11]. On the other hand, India is developing countries that are capable of emerging rapidly in the growth of the middle class and a government that is very supportive in the creation of a start-up that can bring extraordinary changes while the China success than the drive (both for business and citizen) by the culture of the collective success of by the positive spirit of the environment, where the government control an approximate everything you do through business and life (Figure 61.1).

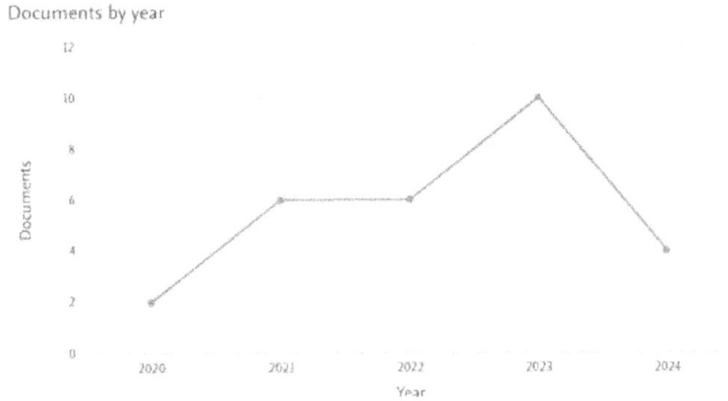

Figure 61.1. Year growth.

Source: Author.

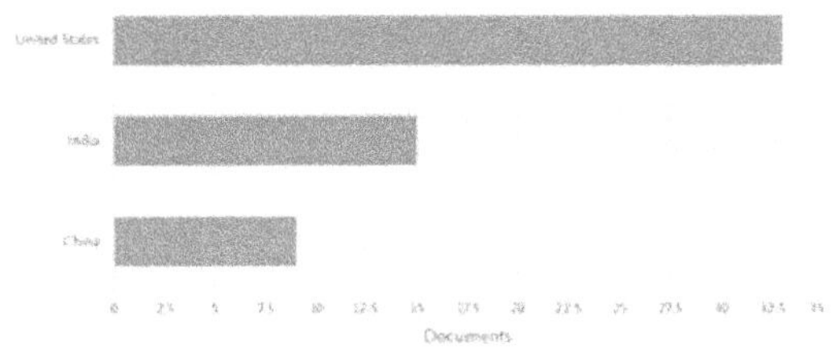

Figure 61.2. Bar chart country wise.

Source: Author.

5. Discussion

Using near the translated trends of advancement for a comparative examination of three nations (U.S.A and India, China) indigenously researched onto and fundamental realities of the pre procedural mindset of each country were discovered. Ends with Summary 2: The American proclivity to take risks has made us the world's leader in technology, and in so doing has made the U.S. (one of) only(s) countries in human history who have captured the imagination of those desiring freedom and therefore prosperity. All over the world. Although these rounds of fundings are yet not too common, Indian IT startup scenario is being constructed around the universal setting. And much else: [10] With its rough tumble very much on its own device, China has grown off an eclectic mix of government interventions with mind-numbing success in A.I., e-commerce and more and the hammer of the market. Those trends arise not just from cultural norms and regulations, or the size of their markets, but from opportunities for innovation. Combine venture capital with a risk tolerant, technology-oriented culture and you have a recipe for American innovation. Not to mention the fact that China is such a homogenous population that halo investors could be considered the government (because the masses are winning literally forever, a company can flourish in China with some government capital and mass market penetration), whereas India on the other hand is still very, very nascent with an appending middle class and a government that, on the face of it, actually propagating a growth in the startup ecosystem [11].

The U.S. the one indisputable tech superpower of the world, and, it has allowed us to be one of only a handful of civilizations in human history whose ideals are held in higher regard by people who value freedom. All over the world. As for different entrepreneur's culture of India, it is maturing at its own pace, yet establishing itself as nucleus for IT-startups. All of these plus the mammoth market size have enabled an innovative ecosystem to thrive in China that has worked wonders for industry from AI to e-commence but by an ability to innovate. That patient capital is American venture capital, nurtured by a risk-taking, tech-friendly culture, in America. These are two different sides, China is a country that the government controls almost everything you do to the extent of the business and your life but the climate is of the encouragement because the government helps you to achieve a success (as well in the case of business and the citizen) through a culture of the sine qua non success, other side there is India a developing country where the middle class is still growing on an astonishing rate and with the government being very supportive in the idea of start-up homes which can bring the revolution (Figure 61.2).

References

[1] Altnaa, U., Vasa, L., & Neszmélyi, G. I. (2021). The role of network ties on SMEs internationalization process in the developing countries/emerging markets: A comparison between Mongolia, China, and Malaysia. *International Journal of Economics & Business Administration*, *9*(1), 274–291.

[2] Lovitz, M., & Easterbrooks, M. A. (2025). Parent-provider relationships in home visiting: An introduction to relational-cultural theory and critical race theory frameworks. *Children and Youth Services Review*, 108177.

[3] Ayoade, O. (2025). Integrating risk management with financial Ab Rahim, M. S., Reniers, G., Yang, M., & Siwayanan, P. (2025). Integrating Process Safety and Process Security Risk Management: Practitioner Insights for a Resilience-Oriented Framework. *Processes*, *13*(2), 392.

[4] Evans, K., & O'Donnell, K. (2025). Relational Cultural Theory and Intervention Approaches with Adolescent Girls: An Integrative Review. *Affilia*, *40*(1), 83–102.

[5] Kraude, Richard, Sriram Narayanan, Srinivas Talluri, Prakash Singh, and Takehisa Kajiwara. (2018). Cultural challenges in mitigating international supply chain disruptions. *IEEE Engineering Management Review*, *46*(1), 98–105.

[6] Wren, C. D., Romanowska, I., & Riede, F. (2025). Bad year econometrics: Agent-based modeling of risk management strategies under varying regimes of environmental change. *Science Advances*, *11*(3), eadr0314.

[7] Messikh, A. (2022). Does entrepreneurial risk-taking affect the business performance of Micro-enterprises? Evidence

from Skikda in Algeria. *Nase Gospodarstvo: NG*, *68*(2), 65–77.

[8] Pidgeon, N. F. (1991). Safety culture and risk management in organizations. *Journal of cross-cultural psychology*, *22*(1), 129–140.

[9] Sauer, B. A. (1996). Communicating risk in a cross-cultural context: A cross-cultural comparison of rhetorical and social understandings in US and British mine safety training programs. *Journal of business and technical communication*, *10*(3), 306–329.

[10] Tsohou, A., Karyda, M., Kokolakis, S., & Kiountouzis, E. (2006). Formulating information systems risk management strategies through cultural theory. *Information Management & Computer Security*, *14*(3), 198–217.

[11] Cunha, A. B., & Canen, A. G. (2008, July). Requirements gathering in information technology: a Cross-cultural perspective. In *2008 IEEE International Professional Communication Conference* (pp. 1–8). IEEE.

[12] Ali, Q. M., Ahmed, Q., Mubarak, S., Baloch, A., Bat, L., Qazi, H., & Shaikh, I. (2023). Two New Records of Eunicidae (Annelida, Errantia) Along the Makran Coast of Pakistan, Northern Arabian Sea. *Natural and Engineering Sciences*, *8*(3), 168–182. http://doi.org/10.28978/nesciences.1405169

[13] Lee, J. H., & Teraoka, F. (2010). Guest editorial: Advances in wireless mobile and sensor technologies. *Journal of Wireless Mobile Networks, Ubiquitous Computing, and Dependable Applications*, *1*(2/3), 1–2.

[14] Nabeesab Mamdapur, G. M., Hadimani, M. B., Sheik, A. K., & Senel, E. (2019). The Journal of Horticultural Science and Biotechnology (2008–2017): A Scientometric Study. *Indian Journal of Information Sources and Services*, *9*(1), 76–84. https://doi.org/10.51983/ijiss.2019.9.1.589

[15] Roozbahani, Z. A., & Ghafari, A. O. (2014). Poly Phenol & flavanoid Contents of Lime Juice. *International Academic Journal of Innovative Research*, *1*(2), 29–37.

[16] Nandhini, C., & Nandhini, B. (2016). GSM based Automatic Smart Irrigation System Using Fuzzy Logic. *International Journal of Advances in Engineering and Emerging Technology*, *7*(3), 115–120.

62 Sustainable edge AI for smart cities and renewable energy management using spiking neural networks

Omprakash Dewangan[1,a], *Ayaz Ahmed Faridi*[2,b], *Ashu Nayak*[3,c], *Aakanksha Soy*[3,d], *and Kamlesh Kumar Yadav*[3,e]

[1]HoD, Faculty of CS & IT, Kalinga University, Raipur, Chhattisgarh, India
[2]Assistant Professor, Kalinga University, Raipur, Chhattisgarh, India
[3]Assistant Professor, Faculty of CS & IT, Kalinga University, Raipur, Chhattisgarh, India

Abstract: A Spiking Neural Network (SNN) model that is based on artificial intelligence has been suggested in this research study for the purpose of enhancing the effectiveness of the renewable energy system in conjunction with smart city applications. The approach that has been presented is an excellent choice for managing power consumption at a lower level and allowing it to be idle for edge computing applications. The architecture that has been suggested has the capability to include real-time data from edge devices, which allows for decentralized decision-making, a reduction in latency, and an improvement in efficiency across all relevant regions. SNN that incorporates edge AI and a sustainable model outperforms traditional AI in terms of a variety of measures, including scalability, efficiency, and real-time adaptability. On the basis of his findings, an intelligent and sustainable management model has been put into place with the purpose of achieving sustained economic and environmental advantages.

Keywords: Spiking neural networks, sustainable edge AI, smart cities, renewable energy management, neuromorphic computing, energy efficiency, real-time data processing, adaptive load balancing, edge computing, sustainable urban infrastructure

1. Introduction

Smart cities are play a crucial role in contributing to the sustainable development of urban areas as the population in urban regions is rising along with technological advancements. In factors such as sustainability of environment, energy efficacy and real-word decision-making, technologies such as AI (Artificial Intelligence), IoT (Internet of Things) and RES(Renewable Energy Systems) provides opportunities and drawbacks [2]. Conventional AI approaches have significant limitations such as maximized latency, energy-related evaluations and congestion of networks [1], because of being dependant on Cloud Computing (CC). These drawbacks necessitate the adoption of AI systems which have minimal latency, are energy efficient and process local information for enhancing sustainability and real-world adaptability.

An efficient solution to these issues is Edge Artificial Intelligence (Edge AI), it permits the algorithms used in AI to be executed on edge devices directly than being dependant on CC technology. This technique reduces the flow of data in networks and increases energy efficiency, latency and privacy [16] Edge AI aids in decision-making processes in crucial sectors such as traffic regulation, distribution of energy, regulation of environments as well as predictive maintenance in and around smart cities. Additionally, it enables balance of loads, demand-response procedures in the management of RESs, as well as optimization of intelligent grids to get better sustainability and energy-related outcomes [3].

Traditional edge AI approaches that function based on deep learning (DL) are power-intensive and have high intricacy despite of being beneficial, particularly in real-world scenarios where there are minimal resources. This approach encourages the research of neuromorphic computing models such as SNNs (Spiking Neural Networks) which is an efficient substitution for implementations of edge AI [4]. Smart cities can enhance their functional efficacy with the use of these innovative computing approaches and solve critical problems that occur due to technological enhancement and urbanization [6].

Compared to traditional ANN (Artificial Neural Network), SNNs can process neurons in the brain with the aid of unique electrical impulses called as spikes. The computing methodology enables SNN to acquire higher efficiency power when utilized on IBM's TrueNorth or Intel's Loihi which are neuromorphic hardware [8, 14]. When it crosses the required threshold, neurons are set to generate asynchronous as well as sparse evaluations which reduces the consumption of energy. Contrast this with traditional deep learning models, which need substantial parallel processing capability [10].

[a]omprakash.dewangan@kalingauiversity.ac.in, [b]ayazahmed.faridi@kalingauniversity.ac.in, [c]ashu.nayak@kalingauniversity.ac.in, [d]aakansha.soy@kalingauniversity.ac.in, [e]kamlesh.yadav@kalingauniversity.ac.in

DOI: 10.1201/9781003684459-62

1.1. *Key contribution*

This research aims to create an efficient SNN framework for sustainable intelligent edge applications, such as smart cities and renewable energy management.

- It aims to enhance neuromorphic computing effectiveness by developing lightweight SNN models for deployment on edge devices with limited energy resources.
- The study also aims to enhance interoperability by combining artificial intelligence based on SNN with existing renewable energy infrastructures and ecosystems for smart cities.
- This research contributes to the expanding area of sustainable artificial intelligence by addressing difficulties and paving the way for intelligent, energy-efficient urban infrastructure.
- The results will be useful to politicians, urban planners, and academics in artificial intelligence, providing a scalable and sustainable foundation for future cities.

The organization of the paper is as follows: Section II explains the recent paper on the relevant field and the need for the proposed model. The proposed model architecture and working details are illustrated in Section III. The Section 4 describes the comparisons result and evaluation of the proposed model. All the details are conclusion in Section V.

2. Literature Review

This section provides a comprehensive evaluation of previous research on artificial intelligence at the edge, spike neural networks, smart cities, and renewable energy management, highlighting key discoveries and potential research needs. It also provides a comparative analysis of previous efforts and their limitations [12].

Although Edge Artificial Intelligence and Spiking Neural Networks (SNNs) have great possibilities for smart cities and renewable energy management, more work has to be done [11, 18]. Though conventional AI approaches are intensive in the aspect of computation and resources, SNNs present in Edge AI are not scrutinized entirely to be utilized in predictive maintenance, trading of energies and grid optimization applications (Table 62.1). This research attempts to resolve these issues through the creation of a Sustainable Edge AI architecture based on SNN for developing smart cities. This approach helps in enhancing decision-making, reduces computational expenses and increases the usability of AI-driven applications in smart cities.

3. Proposed Methodology

SNNs are widely known for its employment in AI development applications as they are capable of managing data in a way that is more robust than traditional neural networks [17]. Additionally, SNNs are efficient in detecting nonlinear

Table 62.1. Summary of key literature and research gaps

Study	*key findings*	*Limitations / research gap*
Li et al. [1]	Edge AI improves smart city real-time decision-making.	High power consumption of traditional deep learning models.
Zhou et al. [11]	AI-driven smart grids optimize energy distribution.	Cloud-based AI remains computationally expensive.
Zhang et al. [3]	AI enhances solar energy forecasting and management.	Deep learning models have high computational overhead.
Chen et al. [5]	SNN-based predictive maintenance for wind farms is effective.	SNNs require better optimization for accurate predictions.

Source: Author.

patterns and rarely deal with overfitting which makes it appropriate to be used in applications that use AI.

The AI approach discussed for Sustainable Edge which utilizes SNNs offer significant enhancements in the optimization of RESs and management of smart cities [15]. DL as well as other traditional AI approaches have higher cost of computation and consume more amount of energy, hence this makes them inappropriate for Edge uses. In contrast, SNNs are capable of replicating brain's event-driven processes which aids them to make decisions at the Edge while consuming a very low amount of power. In smart cities, this results in increased energy efficiency, decreased costs associated with infrastructure, and the guarantee of decentralized, low-latency operations. A contribution to sustainable urban development is made by the model via the incorporation of artificial intelligence-driven energy management, the reduction of power junk, and the optimization of resource allocation, which eventually results in a reduction in carbon footprints and an improvement in system integrity (Figure 62.1).

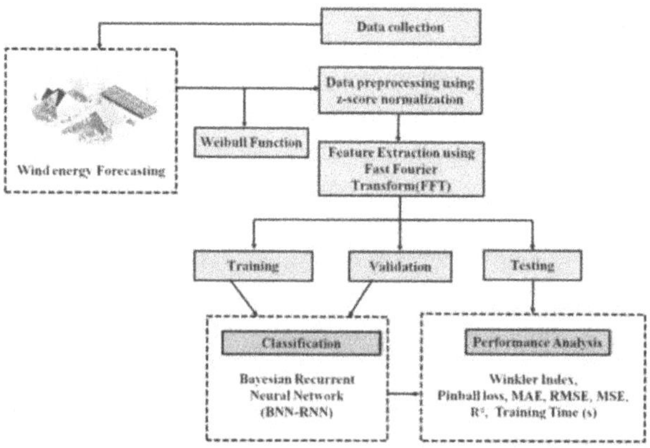

Figure 62.1. Implementation flow diagram of the proposed model.

Source: Author.

Figure 62.2 depicts a basic and fundamental SNN model that may be represented as a directed graph consisting of a sequence of linked layers. This kind of representation is shown in the figure. There is also the possibility that the model may integrate inter-layer connections, which will improve the network's capacity for learning and connection. The ability of a neural network to learn from data is enabled by the fact that each neuron in an SNN is linked to a certain number of other neurons, and the connection weights may be adjusted. An SNN is a neural network that processes weighted input spikes from neighbouring neurons during computation. These spikes are then generated and sent to output neurons by each neuron in the network. As long as the output layer is responsible for producing the final results, this iterative process will continue.

The edge-fog-cloud architecture, represented in Figure 62.3, is a popular distributed computing paradigm with several practical applications owing to its numerous advantages in controlling computational resources and reducing communication latency [7]. This design has three levels: the fog layer connects the previous two, which are the edge and cloud layers; the third is the edge layer. Edge computing provides processing and storage closer to the consumers, therefore lowering the pressure on cloud computing centers, unlike traditional cloud computing, which mostly depends on data inputs from edge devices such images, audios, and videos [8]. In contrast, even with more connection overheads and latency, moving from the edge to the cloud usually results in improved data processing and machine learning capability [13]. Sometimes modern building designs use a fog layer, which balances the respective advantages and drawbacks by serving as a grid between the cloud and the edge [9, 10]. Data sharing, communication overhead, load balancing, and dependability are all handled at this layer. The fog layer is responsible for collecting data at the edge layer and filtering it close to the cloud layer. It streamlines the flow of data in near real-time so that processing, compression, and transformation may be done efficiently.

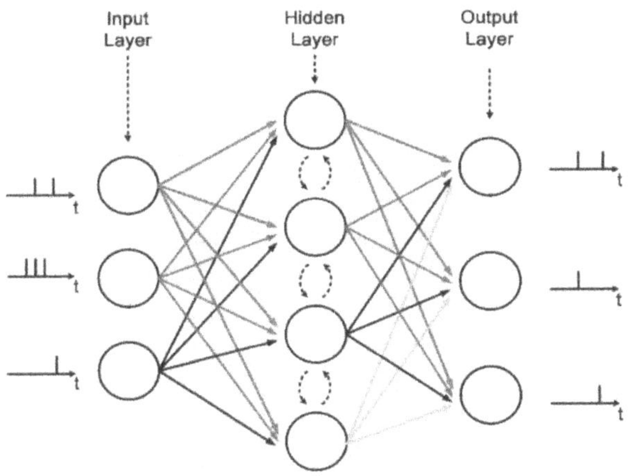

Figure 62.2. Fundamental model of SNN.

Source: Author.

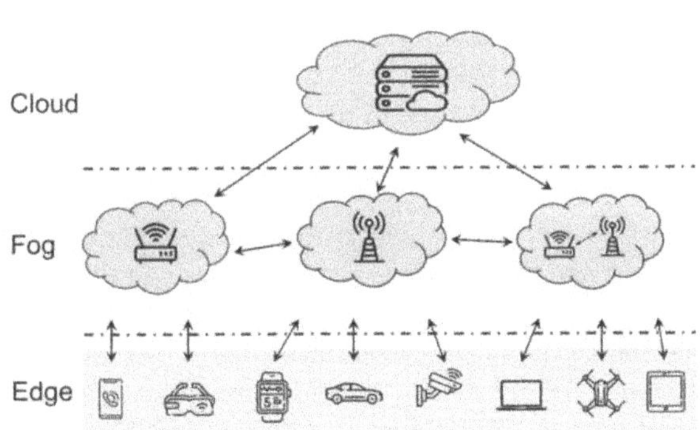

Figure 62.3. Edge computing for sustainable development.

Source: Author.

For the purpose of making decisions in real time and with maximum efficiency, this model is based on the integration of Spiking Neural Networks with Edge Computing as its underlying principle of operation [5]. Information is processed by SNNs via the use of discrete spikes, in which neurons only become active when a certain threshold is achieved. This results in a low amount of power usage. The capability of the model to handle and interpret real-time data from Internet of Things (IoT) sensors, smart grids, and energy storage devices with high accuracy and low latency is made possible by the fact that it is event-driven. The edge artificial intelligence framework makes it possible to make decisions at the local level, which lessens reliance on cloud-based systems and amplifies security. In the context of the management of renewable energy, the model forecasts the demand for energy, stabilizes variations in supply, and maximizes storage, so guaranteeing that smart cities maximize the consumption of energy in a sustainable manner. The implementation of this innovative strategy improves the autonomous control of energy grids, traffic management, and public infrastructure, while enhancing the intelligence, efficiency, and environmental sustainability of urban areas.

4. Result and Discussion

We discuss the study's findings in this section. The hardware setup for the experiment included 8 GB of RAM, a 100 GB C-drive, and a 64-bit version of Windows 10. Every single test procedure was executed in Python. In this part, we use the performance metrics of the proposed model to define its performance requirements.

The Energy Efficiency Comparison of AI Models in Figure 62.4 reveals that deep neural networks (DNNs) are not practical for energy-saving AI systems due to their high-power consumption of around 50 watts. Convolutional neural networks (CNNs) use about 40 watts less and perform better, while models based on transformers use around 35 watts less. Spiking neural networks (SNNs) are the most energy-efficient, with a power consumption of just 15 watts. SNNs' low power consumption is due to their event-driven processing, allowing efficient calculations only when needed. This makes them suitable for AI applications that use minimal power in smart city employment and RES management approaches. These AI approaches are energy efficient and can optimize the utilization of power, enhancing its operation, reducing operational expense and increasing the system's environmental sustainability.

Inference Latency Comparison of AI Models is illustrated in Figure 62.5 representation of the processing speed, in milliseconds (ms), of several AI systems. The inference latency of traditional deep neural networks (DNNs) is about 120 ms, making them the slowest. With a latency of around 100 ms, convolutional neural networks (CNNs) do somewhat better, while models based on transformers further decrease the latency to about 80 ms. With an inference delay of just around 20 ms, spiking neural networks (SNNs) accomplish the best performance increase. The event-driven and asynchronous processing of SNNs allows for quicker and more efficient decision-making, which in turn significantly reduces latency. Rapid reaction times are critical for optimizing resource allocation, traffic management, and power grid monitoring in smart cities and renewable energy management. The findings emphasize the promise of SNNs for these

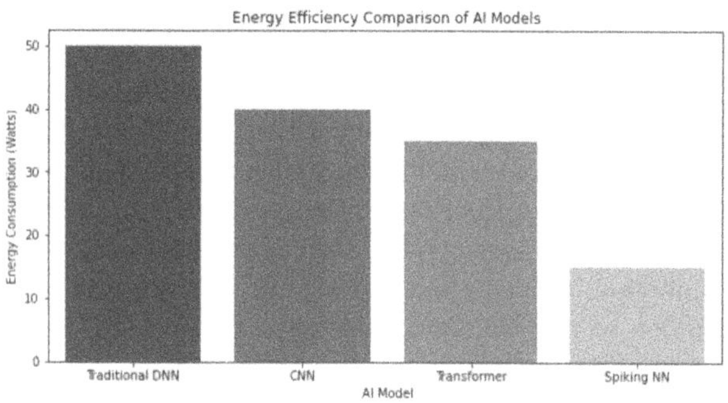

Figure 62.4. Performance comparison of energy efficiency model.

Source: Author.

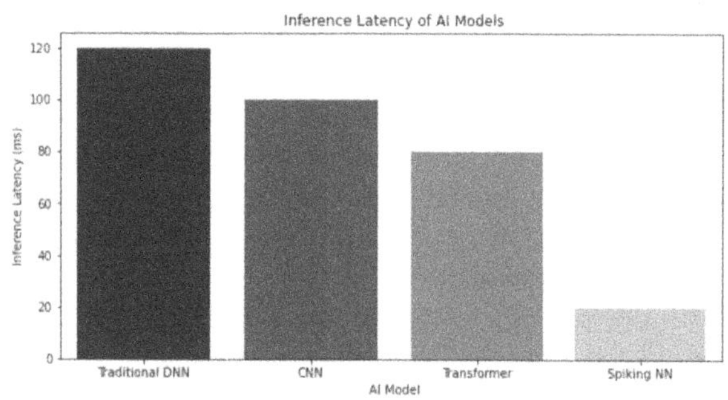

Figure 62.5. Performance comparison on inference latency.

Source: Author.

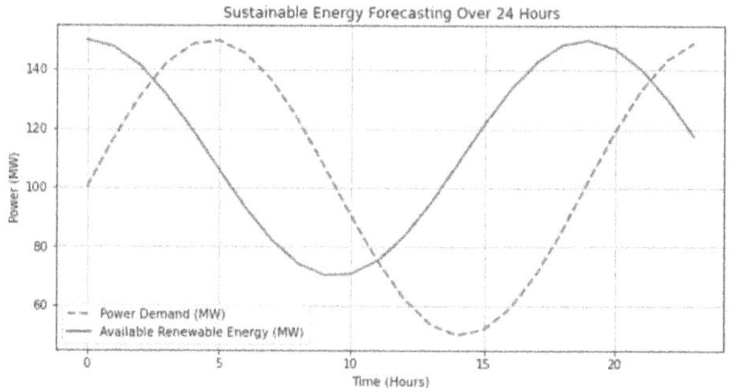

Figure 62.6. Sustainable energy prediction of the proposed model.

Source: Author.

real-time applications. The results highlight the fact that edge AI systems may be made much more efficient and environmentally friendly by using low-latency AI models like SNNs, which are both energy-efficient.

The Figure 62.6 illustrates the availability and demand for sustainable energy over the next 24 hours, highlighting patterns in renewable energy production and consumption. The green solid line represents available renewable energy, while the red dashed line represents power demand. The graph reveals that renewable energy supply is inversely proportional to electricity demand, with renewable generation peaking in the morning and decreasing during the day. Residential and commercial energy use tends to spike in the evening. This graph also depicts supply-demand processes, especially during mornings and evenings as the generation of RESs loses electricity and demands for more. This emphasises the necessity of demand-response approaches and energy storage applications to regulate the supply of power in a consistent manner. Solar power is a crucial factor which has an impact on the market of RESs. This research discusses the significance of smart grid approaches, hybrid energy models, predictive energy regulation, energy distribution optimization and sustainable urban energy approaches.

5. Conclusion

The aim of this research is to analyse the ability of SNNs in the employment of AI approaches in smart cities and applications that manage RESs which have Edge sustainability. As SNNs have the ability to process real-world data while utilizing minimal energy, they are a best choice for edge devices which have minimal resources. Enhancements can be done for adaptive balancing of loads, demand-response approaches and the utilization of RESs through the incorporation of SNNs with Edge AI. Due to their ability of consuming less power and maximized inference speed, SNNs are a great choice for enhancing urban architectures. The optimization of interoperability and scalability of SNN approaches in smart city applications should be the crucial goal of future research.

References

[1] Li, Y., Sun, J., & Zhang, P. (2021). Edge AI for Smart Cities: Challenges and Future Directions. *IEEE Internet of Things Journal, 8*(3), 1980–1992.

[2] Vakhguelt, V., & Jianzhong, A. (2023). Renewable Energy: Wind Turbine Applications in Vibration and Wave Harvesting. *Association Journal of Interdisciplinary Technics in Engineering Mechanics*, *1*(1), 38–48.

[3] Zhang, Q., Li, R., & Chen, Y. (2023). Advancements in Renewable Energy Optimization Using Edge AI. *Energy Reports*, *9*(1), 112–126.

[4] Tan, W., Sarmiento, J., & Rosales, C. A. (2024). Exploring the Performance Impact of Neural Network Optimization on Energy Analysis of Biosensor. *Natural and Engineering Sciences*, *9*(2), 164–183. https://doi.org/10.28978/nesciences.1569280

[5] Cheng, X., Hao, Y., Xu, J., & Xu, B. (2020). LISNN: Improving Spiking Neural Networks with Lateral Interactions for Robust Object Recognition. In *Proceedings of the Twenty-Ninth International Joint Conference on Artificial Intelligence, IJCAI.* Yokohama, Japan, 11–17 July 2020; pp. 1519–1525.

[6] Sharifzadeh, M. (2015). The Survey of Artificial Neural Networks-Based Intrusion Detection Systems. *International Academic Journal of Science and Engineering*, *2*(1), 11–18.

[7] Firouzi, F., Jiang, S., Chakrabarty, K., Farahani, B. J., Daneshmand, M., Song, J., & Mankodiya, K. (2023). Fusion of IoT, AI, Edge-Fog-Cloud, and Blockchain: Challenges, Solutions, and a Case Study in Healthcare and Medicine. *IEEE Internet Things J*, *10*, 3686–3705.

[8] Sofiazizi, A., & Kianfar, F. (2015). Modeling and Forecasting Exchange Rates Using Econometric Models and Neural Networks. *International Academic Journal of Innovative Research*, *2*(1), 49–65.

[9] Yang, S., Gong, Z., Ye, K., Wei, Y., Huang, Z., & Huang, Z. (2020). EdgeRNN: a compact speech recognition network with spatio-temporal features for edge computing. *IEEE Access*, *8*, 81468–81478.

[10] Surendar, A. (2024). Emerging trends in renewable energy technologies: An in-depth analysis. *Innovative Reviews in Engineering and Science*, *1*(1), 6–10. https://doi.org/10.31838/INES/01.01.02

[11] Zhou, X., Wang, C., & Zhao, L. (2022). AI-Driven Sustainable Energy Management in Smart Cities. *Renewable Energy*, *183*, 501–512.

[12] Akin, O., Gulmez, U. C., Sazak, O., Yagmur, O. U., & Angin, P. (2022). GreenSlice: An Energy-Efficient Secure Network Slicing Framework. *Journal of Internet Services and Information Security*, *12*(1), 57–71. https://doi.org/10.22667/JISIS.2022.02.28.057

[13] Hazan, H., Saunders, D. J., Khan, H., Patel, D., Sanghavi, D. T., Siegelmann, H. T., & Kozma, R. (2018). Bindsnet: A machine learning-oriented spiking neural networks library in python. *Frontiers in Neuroinformatics*, *12*, 89.

[14] Indriyani, I., Giriantari, I. A. D., Sudarma, M., & Widyantara, I. M. O. (2023). Facial Skin Type Detection for Race Classification using Convolutional Neural Network and Haar Cascade Method. *Journal of Wireless Mobile Networks, Ubiquitous Computing, and Dependable Applications*, *14*(2), 41–58. https://doi.org/10.58346/JOWUA.2023.I2.004

[15] Bilal, K., & Erbad, A. (2017, May). Edge computing for interactive media and video streaming. In *2017 Second International Conference on Fog and Mobile Edge Computing (FMEC)* (pp. 68–73). IEEE.

[16] Yemunarane, K., Chandramowleeswaran, G., Subramani, K., ALkhayyat, A., & Srinivas, G. (2024). Development and Management of E-Commerce Information Systems Using Edge Computing and Neural Networks. *Indian Journal of Information Sources and Services*, *14*(2), 153–159. https://doi.org/10.51983/ijiss-2024.14.2.22

[17] Zhang, H., Jolfaei, A., & Alazab, M. (2019). A face emotion recognition method using convolutional neural network and image edge computing. *IEEE Access*, *7*, 159081–159089.

[18] Bing, Z., Meschede, C., Röhrbein, F., Huang, K., & Knoll, A. C. (2018). A survey of robotics control based on learning-inspired spiking neural networks. *Frontiers in Neurorobotics*, *12*, 35.

63 Navigating choices: Peer pressure, anxiety and decision-making in NCC cadets

Naveen[1,a], Saroj Nayyar[2,b], Gulsan Kumar Behera[2], Awtar Kishan Kaul[3], Jyoti Dubey[4,c], and Ekta Saxena[5,d]

[1]Research Scholar, Amity University, Haryana, India
[2]Assistant Professor, Faculty of Education, Kalinga University, Raipur, Chhattisgarh, India
[3]Assistant Professor, Faculty of Arts and Humanities, Kalinga University, Raipur, Chhattisgarh, India
[4]Assistant Professor, Faculty of Education, Dev Sanskriti College of Education and Technology, Khapri, Chhattisgarh, India
[5]Assistant Professor, Faculty of Education, St. Thomas College, Bhilai, Chhattisgarh, India

Abstract: The National Cadet Corps (NCC) instills responsibility, commitment, discipline, and moral values in youth. This study examines the influence of peer pressure and anxiety on the rational decision-making style of NCC cadets using a descriptive survey method. The research involved 100 final-year NCC cadets randomly selected from 20 units under Ambala and Rohtak NCC group headquarters in Haryana. Data was collected using the Peer Pressure Scale, Anxiety Scale, and General Decision-Making Style Scale, with statistical analysis conducted through two-way ANOVA. Findings revealed that while peer pressure had no significant effect, anxiety influenced the rational decision-making process. Additionally, an interactional effect between peer pressure and anxiety was observed. The study underscores the need for interventions and support programs to help NCC cadets develop stronger decision-making skills.

Keywords: Peer pressure, rational, decision making, NCC cadets, anxiety, adolescents

1. Introduction

The National Cadet Corps (NCC) was initially formed to address the shortage in the Army and aims to inspire young individuals toward a career in the Armed Forces while fostering discipline, commitment, and leadership without mandating military service [1, 2]. Adolescents, particularly college students, often experience anxiety and depression due to various stressors, including emotional, academic, social, and economic challenges [3]. Spielberger [4] suggests anxiety is a stable trait with neural underpinnings influencing behavior, including decision-making [5, 6]. Emotions significantly impact cognition, affecting higher-order processes like decision-making [7]. Decision-making is a structured process involving problem identification, information gathering, evaluation of options, and action [8, 9]. A rational decision-making style is systematic, logic-driven, and considers all pros and cons. Peer pressure, defined as the influence of same-age individuals, can significantly impact behavior, thoughts, and actions, sometimes leading to harmful consequences [10].

1.1. Objective

To assess the effect of peer pressure, anxiety on rational the style of decision making of NCC cadets [11].

1.2. Hypothesis

H_0: There exists no noteworthy effect of peer pressure, anxiety on rational the style of decision making of NCC cadets.

2. Methodology

The study employed a purposive sampling technique to select 100 final-year NCC cadets. A correlational research design was used to examine the relationship between anxiety and peer pressure (independent variables) concerning rational decision-making (dependent variable). Tools Used are Peer Pressure Scale, Anxiety Scale, General Decision-Making Style (GDMS) Scale.

2.1. Procedure

Telephonic consent was obtained from college principals before data collection. The investigator visited selected colleges, introduced the study to participants, and assured confidentiality. Instructors assisted in data collection. Respondents were given clear instructions, and doubts were addressed. On average, participants took 30 minutes to complete the survey. Gratitude was expressed to respondents and authorities. Collected data were coded and analyzed using SPSS software, with two-way ANOVA employed for statistical analysis.

[a]naveenjakhar1987@gmail.com, [b]saroj.nayyar@kalingauniversity.ac.in, [c]jd1084644@gmail.com, [d]saxenae76@gmail.com

DOI: 10.1201/9781003684459-63

3. Results

Additionally, the interactional effect of peer pressure and anxiety on rational decision-making was found to be significant ($F1,89 = 5.438$, $p < .05$), as the obtained value exceeded the standard threshold of 3.92 at the 0.05 significance level. This indicates that while peer pressure alone did not significantly impact decision-making, its combined effect with anxiety played a crucial role. Therefore, the null hypothesis stating no interactional effect of peer pressure and anxiety on rational decision-making is rejected. These findings highlight the importance of addressing anxiety and peer pressure collectively when developing interventions to enhance NCC cadets' decision-making abilities (Table 63.1).

It was observed from Table 63.2 Rational the style of that Decision making of NCC cadets with high Anxiety (M= 20.72) differ significantly from Rational the style of Decision making of NCC cadets with low Anxiety (M=21.11). Therefore, NCC students with high Anxiety have the low rational quality whereas NCC students with low Anxiety have greater rational attitude. Anxiety together influence the rational decision-making style of NCC cadets. The interactional effect was found to be significant ($F1,89 = 5.438$, $p < .05$), as the obtained value exceeded the standard threshold of 3.92 at the 0.05 significance level. This indicates that while peer pressure alone did not have a significant impact, its combined effect with anxiety plays a crucial role in shaping decision-making.

Table 63.1. Summary of two-way ANOVA for 2x2 factorial experiment of rational the style of decision-making of NCC cadets

Source of Variance	*Type III Sum oof Squares*	*df*	*Mean sum of squares*	*F-ratio*
Peer Pressure	928.32	1	928.32	3.172
Anxiety	746.88	1	746.88	8.784*
Peer Pressure **x** Anxiety	432.86	1	432.86	5.438**
Error	874.22	85	874.22	
Total	8963.11	89		
Corrected Total	45982.96	88		

Note. *Significant at 0.01 level, **Significant at 0.05 level

Source: Author.

Table 63.2. Showing the main effect of anxiety on rational style of decision making of NCC cadets

Anxiety	High	N = 40 M = 20.72 SD = 2.19
	LOW	N = 49 M = 21.11 SD = 2.02

Source: Author.

Consequently, the null hypothesis stating no interactional effect of peer pressure and anxiety on rational decision-making is rejected. Anxiety produced interactional effect on Rational the style of Decision making of NCC cadets.

From the Table 63.3 it can be concluded that mean of low anxiety (M=21.98) is the highest mean among high and low Peer Pressure and high and low Anxiety as well and mean of high Anxiety (M=20.01) is the lowest mean overall. Therefore, it concluded that Peer Pressure and Anxiety influences interactional impact on Rational the style of decision making of NCC cadets.

4. Discussion

The results of this research produced a number of noteworthy findings, including:

- **Main Effect of Peer Pressure:** The analysis demonstrated no any significant main effect of peer pressure on the Rational the style of decision-making process of NCC cadets.
- **Main Effect of Anxiety:** The study found significant main effect of anxiety on Rational the style of decision-making process. NCC cadets with high levels of anxiety showed a lower capacity for Rational the style of decision making compared to those with lower anxiety levels.
- **Interactional Effect of Peer Pressure and Anxiety:** The most noteworthy result of the study was the interactional effect of peer pressure and anxiety on Rational the style of decision making. When both high peer pressure and high anxiety were present, NCC cadets exhibited the lowest level of Rational the style of decision-making ability. On the other hand, those with low peer pressure and low anxiety displayed the highest rational quality the style of decision-making capability.

These findings suggest that anxiety have individual and combined effects with peer pressure on Rational the style of the decision-making process of NCC cadets. It is essential for educators, mentors, and policymakers to understand these influences, as they can significantly impact the development of young individuals in programs like the NCC.

Overall, this research contributes to our understanding of the psychological factors that influence the Rational the

Table 63.3. Showing the interactional effect of peer pressure and anxiety on rational the style of decision making of NCC cadets

	High	*Low*
Peer Pressure	M = 20.21 SD = 2.37	M = 21.62 SD = 2.06
Anxiety	M = 20.01 SD = 2.79	M = 21.98 SD = 2.13

Source: Author.

style of decision making among youth in organizations like the NCC. It highlights the need for interventions and support programs that address peer pressure and anxiety to enhance the rational skill which is the style of decision-making of cadets and ensure their holistic growth and development as responsible citizens and potential leaders of the nation.

5. Suggestions and Educational Implications

For NCC students to demonstrate their ability, interest, and success in life, strong decision-making skills and good mental health are essential. Mentally stable students can overcome challenges, achieve academic success, and secure a bright future. Educational institutions should take measures to enhance NCC students' decision-making abilities, including: (1) motivating their learning process through teachers, (2) establishing guidance and counseling cells for NCC cadets, (3) conducting regular diagnostic tests to identify academic and personal struggles, and (4) ensuring teachers provide individual attention to students.

References

[1] Chen, H. H., Lin, Y. B., Sung, Y. C., & Liou, R. H. (2011). Direction-based Wireless Remote Controller: A Smartphone Application. *Journal of Wireless Mobile Networks, Ubiquitous Computing, and Dependable Applications*, *2*(2), 33–45.

[2] Turan, C. (2022). Estimation of CPUE and CPUA of pufferfish (Tetraodontidae) caught by the Bottom Trawl Fishery in the eastern Mediterranean Coasts. *Natural and Engineering Sciences*, *7*(2), 108–119. http://doi.org/10.28978/nesciences.1159210

[3] Arslan, G., Ayranci, U., Unsal, A., &Arslantas, D. (2009). Prevalence of depression, its correlates among students, and its effect on health-related quality of life in a Turkish university. *Upsala Journal of Medical Sciences*, *114*(3), 170–177.

[4] Spielberger, C. D. (2013). Anxiety: Current trends in theory and research. Elsevier. https://books.google.com/books?hl=en&lr=&id=UmhaBQAAQBAJ&oi=fnd&pg=PP1&ots=9xn3fTMsFI&sig=N3_SYljOTHv6Z1rLr5eqOvqgAXo

[5] Khyade, V. B., & Wanve, H. V. (2018). Review on Use of Mathematics for Progression of Biological Sciences. *International Academic Journal of Innovative Research*, *5*(1), 30–38. https://doi.org/10.9756/IAJIR/V5I1/1810004

[6] Majstorović, S., Tošić, D., & Torbica, D. (2023). Choice of Excavation Method of the Ore Deposits. *Archives for Technical Sciences*, *1*(28), 1–10. https://doi.org/10.59456/afts.2023.1528.001M

[7] Cabra, J. L., Parra, C., & Trujillo, L. (2022). Earprint touchscreen sensoring comparison between hand-crafted features and transfer learning for smartphone authentication. *Journal of Internet Services and Information Security JISIS*, *12*(3), 16–29. https://doi.org/10.22667/JISIS.2022.08.31.016

[8] Daabool, F. S., & Hussein, F. H. (2022). Synthesis; Characterization and Photocatalytic Activity Studying of TiO*International Academic Journal of Science and Engineering*, *9*(1), 18–25. https://doi.org/10.9756/IAJSE/V9I1/IAJSE0903

[9] Rahman, F., & Lalnunthari. (2024). Development of an image processing system for monitoring water quality parameters. *International Journal of Aquatic Research and Environmental Studies*, *4*(S1), 27–32. https://doi.org/10.70102/IJARES/V4S1/5

[10] Sahoo, S., Mohanty, S., Barik, S., & Swain, S. C. (2024). Beauty Industry Trends and Library Collections: A Perspective on Curation and Economic Impact on Beauty Parlour Workers. *Indian Journal of Information Sources and Services*, *14*(3), 251–257. https://doi.org/10.51983/ijiss-2024.14.3.32

[11] Ashok, K., & Daniel, D. (2016). DC Current Injection to the Grid for PV Inverter Using Artificial Neural Network. *International Journal of Advances in Engineering and Emerging Technology*, *7*(3), 39–48.

64 Empowering large-scale data analysis with AutoML: Boosting processing speed and analytical adaptability

Sanskritee Shilpa Pandey[1,a] *and Kamlesh Kumar Yadav*[2,b]

[1]MCA Student, Kalinga University, Raipur, Chhattisgarh, India
[2]Assistant Professor, Department of CS & IT, Kalinga University, Raipur, Chhattisgarh, India

Abstract: The proliferation of sensors and electronic gadgets in recent years has significantly accelerated the data generation rate of Internet of Things (IoT) networks. In IoT systems, substantial data must be handled, converted, and analyzed regularly to facilitate many IoT offerings and capabilities. Machine Learning (ML) methodologies have demonstrated their efficacy in IoT Data Analysis (DA). Applying ML models to Large-scale DA encounters numerous obstacles and hurdles, particularly in the practical choice of model, layout, tuning, and updating, resulting in a significant demand for skilled DA. The characteristics of IoT data lead to idea drift, resulting in diminished model performance. To minimize human labour, Automated ML (AutoML) has emerged as a prominent domain that seeks to autonomously select, develop, optimize, and refresh ML algorithms to attain optimal performance on designated tasks. This study reviews current methodologies in choosing models, tweaking, and upgrading processes within AutoML to determine the best solutions for each stage of applying ML algorithms to Large-scale DA. This paper presents a case study on using AutoML to identify anomalies and challenges, aimed at substantiating the results and assisting industry users and academics in effectively using AutoML methodologies. Finally, the research examines and categorizes this field's problems and research trajectories.

Keywords: Data analysis, automated machine learning, adaptability, internet of things

1. Introduction

By utilizing fast-advancing communication technology, Internet of Things (IoT) [1] platforms facilitate exchanging significant knowledge and data among IoT gadgets and networks to provide value for humanity. The IoT is a network of interconnected devices and endpoints that directly engage to gather, trade, and analyze essential data via the cloud. Common IoT applications encompass smart grids [2], autonomous vehicles, interactive homes [3], precision agriculture, and advanced healthcare. IoT nodes and gadgets can establish several subnets within IoT systems. A smart city system, typically overseen by a municipality, exemplifies a prevalent type of IoT system comprising numerous subcategories, including innovative residences, smart utilities, manufacturing facilities, Intelligent Transportation Systems (ITS) [4], and medical facilities [11].

In IoT analytics, while human professionals proficiently identify fundamental data trends when dimensionalities are below three, the dimensions of most real-world IoT data significantly exceed three, rendering manual Data Analysis (DA) [5] very challenging [12]. A comprehensive study comparing machine research and human analysis results from another abstract classifying investigation indicates that Machine Learning (ML) [6] classifications outperform human classifications in classifying efficacy overall. Most ML classifiers attain F1 scores exceeding 0.8, whereas just 2 of 65 human classifications surpass an F1 score of 0.5. ML models exhibit resilience to alterations and discrepancies between training and testing databases. Human classifications require considerable energy and time to attain a considerably elevated F-score [13]. The ML classifiers categorized 245 abstracts in under 10 seconds, whereas the quickest human classifications took over 2.5 hours to complete the task in this study. ML models can attain superior accuracy when trained on massive databases (e.g., thousands or millions of observations). Yet, it is impractical for humans to retain numerous intricate patterns within such databases. This is chiefly attributable to human brains' restricted memory capacity and perceptual latencies. ML models surpass human classifiers in accuracy, efficiency, and dependability. ML techniques have emerged as essential components in Large-scale DA, facilitating the quick and precise analysis of substantial data volumes generated by IoT devices to discern patterns necessary for IoT services [14].

Despite the extensive application of ML methods to Large-scale DA, implementing ML algorithms frequently necessitates substantial domain expertise and human intervention. Numerous DA and ML scientists have been investigating Automated ML (AutoML) [7] technological advances,

---

[a]isha.sanskritee@gmail.com, [b]ku.kamleshkumaryadav@kalingauniversity.ac.in

DOI: 10.1201/9781003684459-64

which seek to execute DA jobs via ML techniques with little human oversight. AutoML methodologies represent cutting-edge methods for automating Large-scale DA operations and minimizing human intervention. AutoML allows individuals to conserve essential resources, such as time, financial assets, and human capital, by autonomously rendering precise decisions [15].

Human-In-The-Loop (HITL) [8] is implemented to harness human skills and knowledge in developing ML models by integrating Artificial Intelligence (AI) with human intellect. HITL denotes the method by which human specialists oversee the ML procedure to mitigate prediction errors. Human specialists can continue to engage in ML pipelines to generate innovative selections. At the same time, AutoML methodologies can assist with monotonous, repetitive, and labour-intensive DA chores with enhanced accuracy and reduced human exertion. In the context of AutoML applied to Large-scale DA [9], specific processes inside AutoML cables, such as data collection and feature extraction, are influenced by HITL methodologies to effectively manage the substantial volumes of IoT information and facilitate innovative decision-making. Human experts can assist in identifying potential ML model applicants, and their hyperparameter varies for preliminary model choosing and expedited parameter tuning, thereby minimizing unneeded assessments by AutoML methods and facilitating more effective analysis of high-volume and high-velocity IoT information [10]. Human specialists can assist in validating DA results, labeling information for model updates, and maintaining model efficacy in activities involving significant variability in statistical analysis. Although some ML techniques necessitate HITL involvement, the majority can be fully automated, enabling human experts to engage in innovative ML processes.

2. AutoML Overview

Despite the prevalence of numerous ML methods utilized in Large-scale DA for data analysis and decision-making, a randomly chosen ML model with default construction or hyperparameter settings typically fails to attain optimal analytical outcomes and render precise decisions. Proficient DA is essential at various stages of ML pipes, encompassing data preparation and cleansing, algorithm selection, hyperparameter tuning, and assessing the necessity for model updates. Data scientists frequently perform experiments utilizing diverse ML techniques and hyperparameter configurations to identify the most effective combination. These techniques are labour-intensive, time-consuming, and require specific ML and data processing skills. The automation of the ML development and tuning process is termed AutoML. AutoML denotes the entirely automated ML implementation procedure in actual, practical uses. AutoML applies to novices and specialists who want to implement ML models effectively. It can enhance the efficiency and efficacy of ML algorithms by reducing work cycles, improving the efficiency of models, and obviating the need for DA. AutoML is a viable remedy for the scarcity of data scientists and elevated labour expenses. Three notable advantages of utilizing AutoML are:

- It enhances productivity and diminishes computational expenses by automating repetitive ML procedures.
- It aids in preventing errors attributable to human effort.
- It reduces the barrier to applying ML models by necessitating less experience and knowledge in the field.

Manual selection, creation, and calibration of ML methods for Large-scale DA typically take longer than AutoML. AutoML aims to facilitate the automation of the laborious and time-intensive processes of selecting, creating, and refining ML models by machines rather than humans.

Compared to manually constructed ML models, AutoML-based systems can attain higher standards regarding precision and effectiveness of execution time. Employing AutoML methods will not prolong the IoT analysis duration relative to standard ML models, as they simplify and streamline the selection, layout, and tuning processes of conventional DA modeling. Figure 64.1 summarizes the AutoML pipelines for IoT time-series data processing. It comprises four phases: automated information pre-processing, automatic feature design, automatic model training, and automatic model update. Automatic model learning can be categorized into automated choice of models and Hyperparameter Optimization (HPO). AutoML commences with information pre-processing, which seeks to convert the initial information into a refined version. This method is both time-consuming and crucial, significantly impacting learning effectiveness. The subsequent phase involves engineering features, encompassing extraction of features and choice. This phase retains significant database structures while improving learning generalization. The following phase requires model selection, employing a method of optimization to determine the most effective ML algorithm that yields the highest accuracy in predictions. Hyperparameter tuning seeks to improve the model's learning efficiency. AutoML solutions frequently require various ways to optimize and execute multiple phases within the workflow.

3. Experimental Design

The primary objective of this empirical investigation is to evaluate the AutoML architecture compared to a humanly trained Super-Learning (SL) ensemble using real-world databases from risk assessment, claims for insurance, and advertising.

3.1. Databases

The present study utilizes three publicly accessible databases to enhance repeatability and comparison. The selected databases are available for download from the ML Archive or the free ML contest platform, Kaggle. The databases comprise

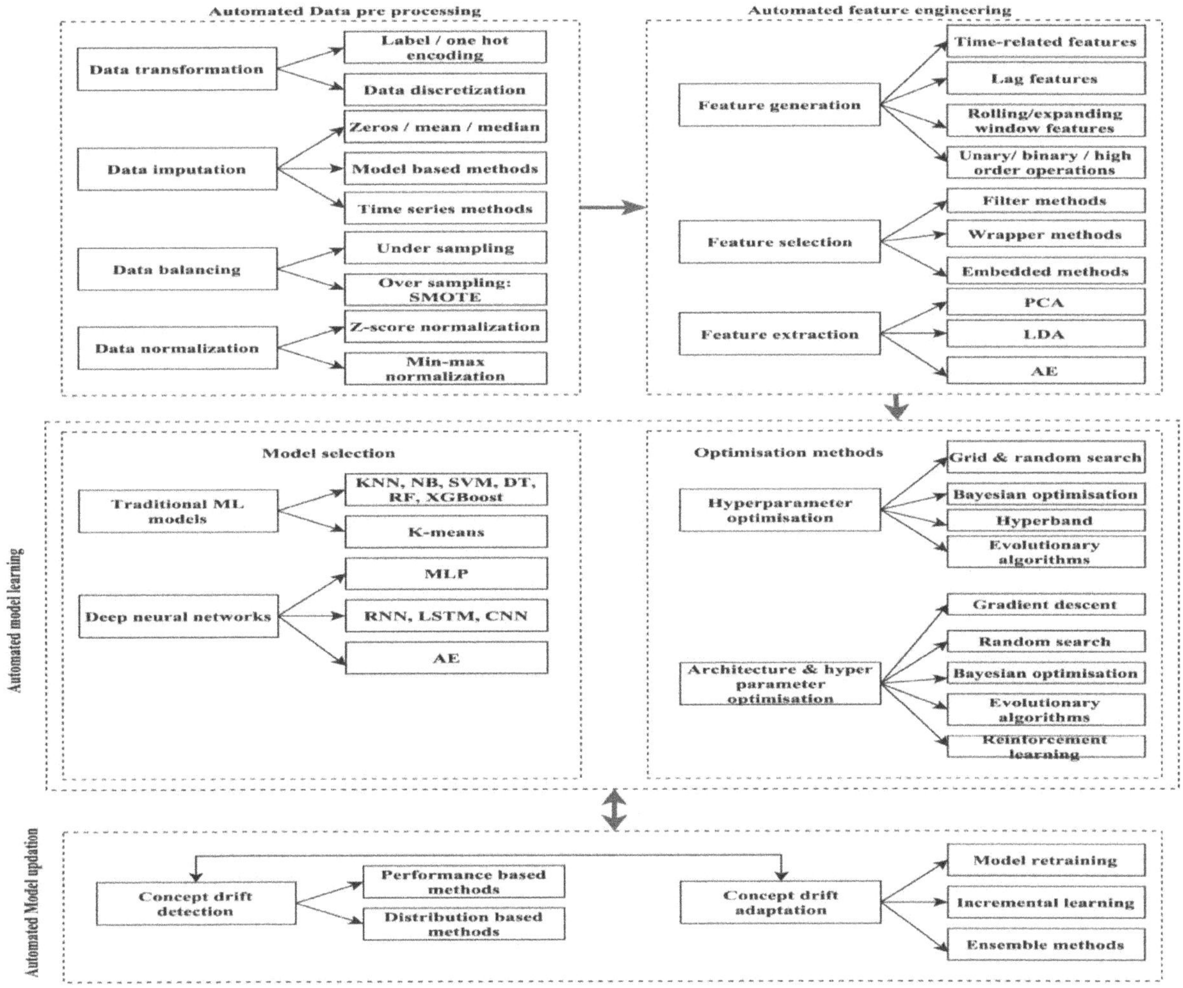

Figure 64.1. Overview of AutoML for DA.

Source: Author.

25, 55, and 15 attributes, accordingly representing the database's properties (e.g., previous client information), and will function as predictor variables for determining the categorized group of each occurrence. Each database includes a binary answer column indicating whether the client went bankrupt, initiated a claim with an insurer, or if a marketing campaign resulted in a sale.

3.1.1. Credit risks

The first information originates from the credit risk sector and comprises payment data from credit card customers. The complete database shall consist of 30k events. There are 23.4k classified as cases of positive significance and 6.7k as negative instances, indicating instances of client failure. The database consists of 24 attributes, including an answer column that indicates binary default or non-default status. This database includes concrete elements such as previous payment information and demographic details, including gender, age, marital position, and schooling.

3.1.2. Prediction of claiming

The additional database originates from the insurance sector and comprises data regarding vehicle insurance customers. The database contains 598k occurrences. A significant proportion of 575k records comprises non-filed claims, while 21.7k represents filed claims. The database includes 56 attributes plus an answer column that denotes the claim state of each insurer.

3.1.3. Promotion

The third database contains marketing and sales information. The document comprises client data for an advertising initiative in the financial services sector. The database includes a total of 45.4k events. Of those, 39.3k were failed, while 5.3k were victorious. Effective denotes achieving a conversion or concluding a sale. Each observation comprises 15 attributes and a reaction column denoting the binary outcome – whether the marketing activity has led to a final transformation or sale.

3.2. Data preprocessing

Multiple preprocessing processes were essential before conducting the study:

3.2.1. Random undersampling

A balanced proportion of favorable and adverse observations is essential; there would be an inherent bias towards the class with the most significant findings, particularly in databases with a ratio of 80:20 or above. This scenario would distort the results since the precision of predictions stems not from the classification algorithm but from the information set's imbalances. This is accomplished through random undersampling. The objective is to calibrate the available data and achieve balance by removing observations from every other category. Undersampling leads to the loss of specific data; this is disregarded in this context as the primary objective is to evaluate AutoML against the fully optimized SL ensemble. Another option is to oversample the minority group; this would inflate the database and prolong the training period, particularly concerning databases that already include significant data.

3.2.2. Encoding

A necessary preprocessing step for ML systems is the compression of categorical information. Models in ML necessitate numerical variables as inputs. Category information must be converted to numerical format before model training and assessment. If a ranking already exists, this is often accomplished using one-hot or numeric encoding. It includes an optional setting termed one_hot clear, which generates N + 1 more rows for categorical characteristics with N degrees.

3.2.3. Division of training and testing data

The selected division is 75:25, indicating that 75% of the raw data will be utilized for training, while the other twenty percent will assess the ability to generalize the trained classification. The identical cross-verification framework is essential for integrating the base students into the meta-student. In the model development phase, 75% of the raw data will be partitioned into distinct sets for training and checking by cross-verifications.

3.3. Configuration and assessment

The AutoML system is provided by foundational classifiers: Random Forest (RF), Gradient Boosting (GB), and Deep Learning (DL). In the next phase, it employs the ensemble technique of stacking to integrate all pre-trained contender models into a superclassifier to enhance accuracy rates. The optimal model is dynamically identified according to a specified assessment criterion. To evaluate the robustness of this configuration, the system replicated the internal mechanisms of the AutoML solution by physically learning the foundational models and integrating them through stacking into SL. The contrasts involve two distinct configurations of SL. One is produced mechanically by the AutoML remedies, while the other is manually adjusted and customized. The efficacy was evaluated using four approaches: Area Under Curve (AUC), Accuracy, F-score, and LogLoss.

4. Results

This section presents the experimental outcomes. The dynamically optimized Stacking Ensemble (SE) learners were evaluated against the AutoML solution. Stacking was essential to replicate the internal mechanics of the AutoML process, which depends on training various base classifications and combining those pre-trained systems to form the ultimate ensemble structure through stacking. The study utilized three real-world instances of risky credit, claim handling, and advertising. The four assessment metrics were employed to compare the AutoML answer with manually adjusted SL. The Accuracy and F-score are presented at an acceptable threshold of 0.5. The experiment was organized in the following manner: Initially, the three foundational models – RF, GB, and DL – were meticulously developed. Hyperparameter optimization of the basis models was conducted using grid searching and random searches within a specified variable range, together with hand changes during the learning process.

Table 64.1 presents the numerical outcomes for the foundational classifications corresponding to each database. GB achieved superior efficiency in general and was succeeded by RF. DL exhibits the poorest indicators of performance. This is uniform across all three databases. In the subsequent phase, the potential models were integrated with SL through the ensemble technique of arranging them, demonstrating the capacity to yield asynchronously optimal enhancements over a collection of basic classifiers. All three foundational models (RF, GB, DL) were employed to construct the SL for every study. All three permutations of the initial models for the SE were evaluated, with optimal performance attained by

Table 64.1. Optimized base classifier analysis

Case Study	*Method*	*AUC*	*Accuracy*	*F-score*	*Logloss*
Credit Risk	RF	77.29	71.75	68.88	57.77
	GB	78.45	72.34	69.75	58.02
	DL	76.07	70.65	69.14	61.3
Claims Prediction	RF	64.1	60.41	58.64	66.79
	GB	64.87	60.59	59.54	66.39
	DL	63.61	60.68	53.99	67.28
Marketing	RF	94.67	88.31	88.56	31.82
	GB	95.46	87.85	89.46	30.34
	DL	93.72	87.15	87.46	33.07

Source: Author.

utilizing RF, GB, and DL as inputs for the SL across all three instances. This is not universally applicable. In the last stage, the stacked SL developed in the second step functions as a standard for the AutoML system to assess its efficiency, durability, and dependability. Table 64.2 presents the conclusive comparison between the AutoML answer and the SL. The results are notably constant, with the stacking SL surpassing the AutoML model across all three databases, with an AUC differential of 0.003.

Although the results differentials for the other indices are not uniform, the SE consistently surpassed the AutoML approach in most instances. The discrepancies in the financial risk investigation are 0.004 for Accuracy, 0.004 for F-score, and 0.003 for LogLoss. The efficiency disparity in the insurance database is 0.035 for Accuracy, 0.018 for F-score, and 0.002 for LogLoss. The results discrepancies for the advertising investigation are -0.001 for Accuracy, -0.003 for F-score, and 0.009 for LogLoss. AutoML marginally surpassed the SE just in the advertising investigation for Accuracy and F-score. The personally optimized SE demonstrates improved efficiency relative to the AutoML approach across all three case instances.

5. Conclusion

ML and DL methods have attained significant success in analyzing information for IoT-related uses, including autonomous vehicles, home automation, e-health, and security of IoT devices. Creating efficient ML models for specific IoT applications necessitates significant human skill, constraining their practicality. AutoML has emerged as a viable alternative for developing ML models with little human involvement. This study thoroughly examines the processes of the conventional AutoML pipeline, encompassing automated information pre-processing, automated feature design, automatic model selection, HPO, and automatic model update with drifting concept adaption. The research has examined the tasks associated with large-scale DA and the ML and DL models that are frequently utilized in this domain. This study presents current libraries and instruments to enable AutoML and Large-scale DA. The research includes a case study on IoT detection of anomalies to illustrate the processes involved in AutoML implementations. Experimental findings have demonstrated the advantages of employing AutoML systems in Large-scale DA challenges. The research will examine the unresolved issues and prospective research avenues about current AutoML and IoT analytics jobs. This research offers an extensive analysis and case study on the use of AutoML technologies in Large-scale DA; due to limitations in time and resources, a more thorough and equitable standard of AutoML methodologies should be established for users in the future. The automated process of information pre-processing and feature development presents significant opportunities for enhancement; thus, improved automation approaches should be developed to optimize these two ML processes. In future endeavors, the research will investigate the utilization of AutoML in additional domains, such as sentiment evaluation, and examine alternative methodologies to enhance DA efficiency, including distributed ML.

Table 64.2. AutoML performance analysis

Case Study	*Method*	*AUC*	*Accuracy*	*F-score*	*Logloss*
Credit Risk	SE	77.92	72.27	70.39	56.63
	AutoML	77.64	71.57	70.06	57.63
Claims Prediction	SE	65.13	60.81	59.74	66.98
	AutoML	64.45	60.25	59.19	66.75
Marketing	SE	95.19	88.44	89.56	30.39
	AutoML	94.76	89.03	89.36	30.82

Source: Author.

References

[1] Laghari, A. A., Wu, K., Laghari, R. A., Ali, M., & Khan, A. A. (2021). A review and state of the art of the Internet of Things (IoT). *Archives of Computational Methods in Engineering*, 1–19.

[2] Moreno Escobar, J. J., Morales Matamoros, O., Tejeida Padilla, R., Lina Reyes, I., & Quintana Espinosa, H. (2021). A comprehensive review on smart grids: Challenges and opportunities. *Sensors*, *21*(21), 6978.

[3] Lal, B., Dwivedi, Y. K., & Haag, M. (2023). Working from home during COVID-19: doing and managing technology-enabled social interaction with colleagues at a distance. *Information Systems Frontiers*, *25*(4), 1333–1350.

[4] Lv, Z., & Shang, W. (2023). Impacts of intelligent transportation systems on energy conservation and emission reduction of transport systems: A comprehensive review. *Green Technologies and Sustainability*, *1*(1), 100002.

[5] Paganelli, A. I., Mondéjar, A. G., da Silva, A. C., Silva-Calpa, G., Teixeira, M. F., Carvalho, F., ... & Endler, M. (2022). Real-time data analysis in health monitoring systems: A comprehensive systematic literature review. *Journal of Biomedical Informatics*, *127*, 104009.

[6] Khan, T., Tian, W., Zhou, G., Ilager, S., Gong, M., & Buyya, R. (2022). Machine learning (ML)-centric resource management in cloud computing: A review and future directions. *Journal of Network and Computer Applications*, *204*, 103405.

[7] Singh, V. K., & Joshi, K. (2022). Automated machine learning (AutoML): an overview of opportunities for application and research. *Journal of Information Technology Case and Application Research*, *24*(2), 75–85.

[8] Kumar, S., Datta, S., Singh, V., Datta, D., Singh, S. K., & Sharma, R. (2024). Applications, challenges, and future directions of human-in-the-loop learning. *IEEE Access*.

[9] Ligero, M., Jordi-Ollero, O., Bernatowicz, K., Garcia-Ruiz, A., Delgado-Muñoz, E., Leiva, D., ... & Perez-Lopez, R. (2021). Minimizing acquisition-related radiomics variability by image resampling and batch effect correction to allow for large-scale data analysis. *European Radiology*, *31*, 1460–1470.

[10] Bzai, J., Alam, F., Dhafer, A., Bojović, M., Altowaijri, S. M., Niazi, I. K., & Mehmood, R. (2022). Machine learning-enabled Internet of Things (IoT): Data, applications, and industry Perspective. *Electronics*, *11*(17), 2676.

[11] Haval, A. M., & Afzal, M. (2024). Aquatic object detection using YOLO (you only look once) algorithm. *International Journal of Aquatic Research and Environmental Studies*, *4*(S1), 52–57. https://doi.org/10.70102/IJARES/V4S1/9

[12] Sungur, Ş., & Jobasi, D. (2022). Determination of biogenic amines in some cheese consumed in Hatay region. *Natural and Engineering Sciences*, *7*(2), 120–130. http://doi.org/10.28978/nesciences.1159217

[13] Papadopoulos, G., & Christodoulou, M. (2024). Design and Development of Data Driven Intelligent Predictive Maintenance for Predictive Maintenance. *Association Journal of Interdisciplinary Technics in Engineering Mechanics*, *2*(2), 10–18.

[14] Sujatha, S. (2024). Strategic Management of Digital Transformation: A Case Study of Successful Implementation. *Global Perspectives in Management*, *2*(1), 1–11.

[15] Sachdeva, L., & Upadhyay, N. (2024). Digital Transformation and Sustainability: A Study of how Firms Use Digital to Achieve Sustainable Goals. *Indian Journal of Information Sources and Services*, *14*(4), 42–47. https://doi.org/10.51983/ijiss-2024.14.4.07

65 AI Integrated PBL in language classroom: Enhancing creativity, critical thinking and sustainable awareness

Sulagna Mohanty[1,a], *Sudakshina Bhattacharya*[2,b], *and Ankusha Bandyopadhyay*[2,c]

[1]Assistant Professor, Department of English and Foreign Languages, SRM Institute of Science & Technology, Kattankulathur, Chennai, India
[2]Assistant Professor, Department of English & Humanities, Amrita School of Arts, Humanities & Commerce, Amrita Vishwa Vidyapeetham, Coimbatore, India

Abstract: The National Education Policy (NEP) 2020 calls for an innovative educational pedagogy that encourages skill development, educational technology, holistic education, experimental learning, creativity, critical thinking, digital infrastructure and integration of the Indian knowledge system to prepare students for an ever-evolving global platform. In the context of the inclusion of novelty and ingenuity in education, Project-Based Learning (PBL) can be a powerful and transformative educational tool. This study tries to experiment with the use of AI-integrated PBL in Language teaching. This research is targeted especially for the Communicative English classes for first-semester students of a technical private university in Chennai. This study intends to examine the effectiveness of PBL if it is infused with AI. It shall scrutinise if such a unique pedagogy boosts creativity, critical thinking and sustainable awareness among engineering students which would align with the policies of NEP 2020.

Keywords: Project-based learning (PBL), english language teaching (ELT), NEP 2020, AI integration, sustainability

1. Introduction

The National Education Policy (NEP) 2020 calls for an innovative educational pedagogy that encourages skill development, educational technology, holistic education, experimental learning, creativity, critical thinking, digital infrastructure and integration of the Indian knowledge system to prepare students for an ever-evolving global platform. Hence, NEP 2020 emphasises 'an education system that contributes directly to transforming our nation sustainably into an equitable and vibrant knowledge society'. This new perspective aims to reform the Indian educational system into a groundbreaking interdisciplinary framework [1].

In the context of the inclusion of novelty and ingenuity in education, Project-Based Learning (PBL) can be a powerful and transformative educational tool. Affirms that Project Based Learning can be helpful for students to develop skills and practical knowledge. PBL usually focuses more on student collaboration, creativity, critical thinking and active participation and engagement with relevant contemporary issues. Hence, by introducing PBL into the teaching pedagogy, Indian education can shift from conventional teacher-centric to student-centric learning. Similarly, if PBL is infused with artificial intelligence (AI) it can further enhance the potential of Indian students and make them future-ready. It will also bridge the gap between theoretical knowledge and practical application [2] which has been an enduring problem faced by the Indian education system [3].

This study tries to experiment with the use of AI-integrated PBL in Language teaching. This research is conducted especially for the Communicative English classes for the first-semester students of a technical private university in Chennai [4]. This study intends to examine the effectiveness of PBL if it is infused with AI. It shall scrutinise if such a unique pedagogy boosts creativity, critical thinking and sustainable awareness among engineering students which would align with the policies of NEP 2020 [5].

2. Literature Review, Research Gap and Research Objectives

Thomas in his work (2000) emphasises that PBL can introduce students to an unconventional yet effective way of learning. Likewise, Barron and Darling-Hammond in their work [6] assert that PBL in education can improve the student learning outcomes. Similarly, Luckin et al. in their work

[a]sulagnam@srmist.edu.in, [b]s_bhattacharya@cb.amrita.edu, [c]b_ankusha@cb.amrita.edu

DOI: 10.1201/9781003684459-65

[2] affirm that using AI in education can refine the personalised learning of students. Ruiz Viruel et al.'s work [7] do a comparative study between the traditional PBL and the AI-integrated PBL and check how the AI tools can help students improve their efficiency. Likewise, Leicht, Heiss and Byun in their work [8] emphasise the need to integrate SDGs with education to make students more aware of sustainable development. Richards in his work [9] focuses on using literature in a language class can bring more creativity and critical thinking [10].

However, limited research can be seen on the integration of AI in PBL for language learning which would enhance creativity, critical thinking, and sustainable awareness in students. Hence, this study aims to fill this gap by providing a pedagogical model for integrating AI and PBL in the communicative English classroom.

This study aims to scrutinise how the student's creativity, critical thinking, and sustainability awareness have been impacted through these projects. It further seeks to examine the effective integration of AI tools and digital platforms and how they have helped students enhance their understanding of the practical application of technology in an educational setting. Moreover, this study highlights how, through these projects, students perceive the benefits and challenges posed by AI-integrated project-based learning and how the students handle them and overcome them.

3. Methodology

3.1. Participants

The participants in this research are two classes of first-semester engineering students (n=140) from a technical institution in Chennai, Tamil Nadu. Each class has 70 students and they are divided into 10 groups having 10 members each. So in total, the sample involves 20 groups.

3.2. Materials

Five novels are assigned to the students for this study. The texts are *The Palace of Illusions*, *1984*, *To Kill a Mockingbird*, *The Hungry Tide*, and *Brave New World*. The students are supposed to choose one text out of the assigned literary texts and develop a project based on it over one semester.

3.3. Procedure

This study spans over 16 weeks (one semester). The students are asked to come up with unique and innovative ideas related to the chosen novel and bring forth imaginative projects with the help of AI and various digital tools and platforms. Here, in this research paper, we discuss some specific projects that are highly original, creative, and inventive. The findings are based on the multiple review presentations given by students and the observations and assessments by the teachers and external evaluators.

Examples of these projects include: **A Graphic Novel on Bhishma** (based on *The Palace of Illusions*), a **Comparative Study of Autocratic Governance** (based on *1984 vs. 2024*), **a Diary Entry of All Major Characters Present During Draupadi's Disrobement** (based on *The Palace of Illusions*), **a Documentary on Falsely Accused Individuals** (based on *To Kill a Mockingbird*) and **a Podcast on Gender Equality** (based on *To Kill a Mockingbird*).

4. Data Collection

We collected this data through 3 reviews where students presented their ideas through an extensive report, PPT, and then a final presentation of the project. The most important rubrics for the assessment included creativity, critical thinking, integration of AI, and the awareness of sustainability.

5. Discussion and Findings

In this discussion, the study showcases some selected projects by the students. These projects highlight a strong integration of creativity, critical thinking, and sustainability awareness while using advanced technological tools. A closer and more detailed look at these projects accentuates how the students have utilised their skills to address contemporary issues and challenges. The core of their work reflects their diverse approach, which according to NEP 2020, is crucial to making the Indian education framework updated, efficient, and restructured.

5.1. Five selected projects

5.1.1. The graphic novel project based on the text the palace of illusions

This first group of 7 students came up with a graphic novel based on the character of Bhisma who is an integral part of the text *Palace of Illusions* by author Chitra Banerjee Divakaruni. Bhisma was the grandsire of both Pandavas and Kauravas and the eldest of the great Kuru dynasty during the era of the Mahabharata War. With the help of technology and narrative storytelling, the students decided to create this graphic novel which would innovatively explore the life of Bhisma.

The first step towards the project was to understand the character of Bhisma, previously known as Devabrata. The discussion with the students during the review process revealed that the biggest challenge to this project was the younger generation's unfamiliarity with the Indian Knowledge System and ignorance about ancient Indian epics like the *Ramayana* and the *Mahabharata*. The text *Palace of Illusions* is based on the epic the *Mahabharata,* where Divakaruni writes from the perspective of Draupadi and outlines her innermost complex emotions. Interestingly, the students shifted the POV from Drupadi to Bhisma in their graphic novel, which provided them with a better understanding of the character of Bhisma. Moreover, it also aligned with the NEP 2020 policy, which profoundly encourages Indian students to explore the Indian Knowledge System.

Regarding technology, the students used Gen AI like Chat GPT, Gemini and Copilot and created an abbreviated narrative of Bhisma's eventful life. They had numerous brainstorming sessions to integrate AI and generate the narrative productively. Each student provided valuable input to make the narrative less complex and more engaging. After the rigorous brainstorming, they included all the significant events and divided them into chapters. Some of the important chapters were called The Curse of Silence, Bhishma Pratigya: The Vow of Sacrifice, Amba's Vow and Revenge, The Game of Dice etc.

Once the storyline was finalised, the students created visual graphics using DELL-E, Canva and Copilot. They produced high-quality vibrant graphics which brought the mythical characters to life. These images were relevant to the narrative and the students made sure that the images evoke intense emotions in the readers for the characters.

While brainstorming about the narrative of the condensed plot, students made the most use of their critical thinking. To understand the complicated life of this great Kuru warrior, students analysed all the ethical dilemmas faced by Bhisma. In his overextended extraordinary life, Bhisma was frequently tried and tested for his duty, loyalty and conscience. Sometimes, we won in this battle, and sometimes, he lost like he did on the traumatic day of the disrobing of Draupadi. Digging deeper into the life of Bhisma, the students were able to gain a profound understanding of the inner conflict of Bhisma. It made them contemplate how these timeless tales can open a discourse on the ethical challenges faced by leaders even in the contemporary setting (Figure 65.1).

With the incorporation of creativity, critical thinking and technical tools, this project provided an innovative and non-conventional method for the students to learn. Moreover, through this project, the students could align with sustainable awareness as they explored the theme of the horrendous consequences of the Mahabharata War. It made them realise the importance of essential requirements for peace, justice and strong institutions in contemporary society; the traits which are closely related to SDG 16 as proposed by the UN.

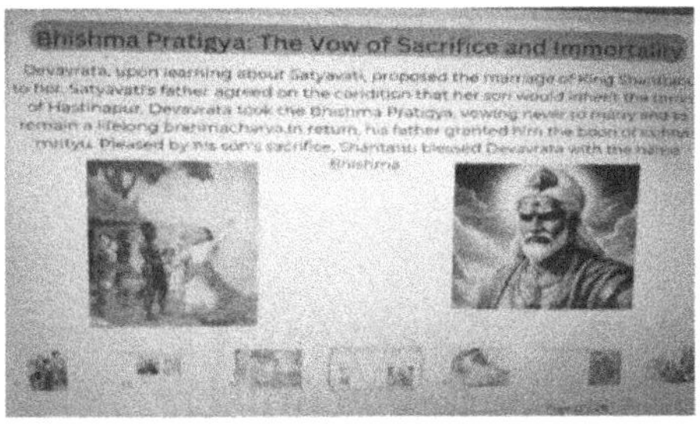

Figure 65.1. Student work.

Source: Author.

5.1.2. The comparative study project on autocratic governance based on the text 1984

This third project under discussion is a thought-provoking comparative study based on the text *1984* written by George Orwell. The students of this group came up with this innovative idea by comparing this text with the contemporary, relevant context rather than discussing the plot and narrative of the novel.

1984 is a futuristic novel that deals with various dangerous themes like power, control, surveillance, censorship, propaganda and totalitarianism. Orwell designates a futuristic society where the government controls not only the state but every aspect of its citizens' lives no matter personal or professional. The text also delineates how the state controls the memory of its people by erasing the past and rewriting the present by creating an alternative reality. Interestingly, the students could relate such a futuristic world with the contemporary, where many countries can be labelled as autocratic governments. Even various countries with democracy are autocratic and they do not hesitate to execute suppression and surveillance over their citizens.

This engaging project motivated the students to use creativity and critical thinking as they started their research with rigorous brainstorming. First, they explored the government depicted in *1984* and further resonated with contemporary states like North Korea, China, Russia, etc. The students analysed how these states execute their power for surveillance, manipulation of truth, propaganda and spread of misinformation. They also criticised democratic countries like America, which is known for surveying its citizens. The students also explored how AI tools like Siri and Alexa are also accused of selling the personal conversation data of customers to third parties, and this kind of surveillance happens with the knowledge of the government (Figure 65.2).

Another very important aspect the student discussed was the erosion of memory. The concept of rewriting past and present resonated with the students because they could see how the totalitarian government tries to change the past which is uncomfortable or unacceptable for them. The students gave

Figure 65.2. Student work.

Source: Author.

examples of how such governments try to change history by distorting it, spreading rumours and changing academic history textbooks in many countries. Likewise, in many such countries, media is used as a profound tool to spread lies and create an alternative reality for people. These governments control the media and breach the fundamental rights of the freedom of speech of citizens.

The students made this project innovative with a good use of AI as they did not want to submit a conventional comparative report. Hence, they made the project visual by using comic graphics. They created stunning professional colourful images using Canva and DELL-E. These relevant images represented the findings of complex issues like social control and autocratic power structure.

As the students worked on themes like suppression, control and breach of personal freedom, it made them realise the importance of concepts like equality, justice and fairness. This is how this process aligns with SDG 6 supports reduced inequalities and SDG 16 which encourages peace, justice and strong institutions. Hence, this project taught the importance of democracy, and social equality which helped the students to become better global citizens in future. It also taught them to combat against various forms of autocracy.

5.1.3. The diary entry project of the major characters of the mahabharata based on the text palace of illusions

The 'Diary Entry of Major Characters of the *Mahabharata*' project by the students was based on the text *The Palace of Illusions* by author Chitra Banerjee Divakaruni. This creative project was an exploration of the deep psychological complexities of the major characters of the *Mahabharata* who were present during the harrowing event of the disrobing of Draupadi in the Kaurava court.

Interestingly, this project was a handwritten diary and the creativity was evident from the way the students had adopted the paper aging technique. The students submerged the papers in coffee water and then dried them to give the paper an older look.

The students tried to put themselves in the shoes of characters such as Draupadi, Arjuna, Bhima, Yudhhisthir, Bhisma, Duryodhana, Karna, Dushashana, etc. The students assigned one specific character to each of the members of the groups Then they tried to understand the perspective of each of these characters and crafted the diary entry. They tried to comprehend the multifaceted emotions and ethical dilemmas these characters underwent in that particular fateful event. In this way, the diary was an introspective piece of writing that brought these mythical characters back to life. The innermost thoughts, emotions, fears, justifications, and retrospection of these characters became the most important focus of these diary entries.

This creative process enabled the students to reimagine these characters from a fresh point of view where they could be situated in a contemporary context. By doing so, the students could comprehend universal themes like power dynamics, injustice, patriarchal structure, and gender politics. The students minutely assessed this gender power imbalance and realised how it continues to manifest in contemporary society. The students could understand that the treatment of women in these historical and mythological contexts can be educational about contemporary gender equity (Figure 65.3).

These diary entries were not simply retelling of the event. Rather, these entries echoed the voices of these characters as the students investigated the motivations behind the character's actions and decisions. For example, what emotions Draupadi might have gone through as she was humiliated in the Kaurava court? How might she have perceived the men around her including her abuser Dushashan and her five husbands? How could the great grandsire Bhisma let such a terrible event happen in the court despite being a man of *dharma* and justice? How could King Dhritarastra blind his ethics, conscience, and integrity? How could Duryodhana, Karna, and Dushashan justify their horrible actions and rationalise their actions on that particular day? Such questions help the students demonstrate their ability of critical thinking and emotional intelligence.

This project helped the students to connect with ancient Indian mythological narratives. By doing so, the students could explore the Indian knowledge system, a significant motif of NEP 2020. Moreover, this project was helpful from the perspective of SDGs such as SDG 5 and SDG 16. SDG 5 invites our focus on gender equality and SDG 16 brings forth concepts like peace, justice, and strong institutions. Exploring gender issues and comprehending the importance of gender equality seeking a fairer and more just world is one of the most important outcomes of this particular project.

5.1.4. The documentary project on falsely accused individuals and b. the podcast project: both based on the text To Kill a Mockingbird

Analysing the text *To Kill a Mockingbird:* These two innovative projects, a documentary and a podcast were based on the classic text *To Kill a Mocking Bird.* This text is a thought-provoking novel on the theme of racial prejudice and

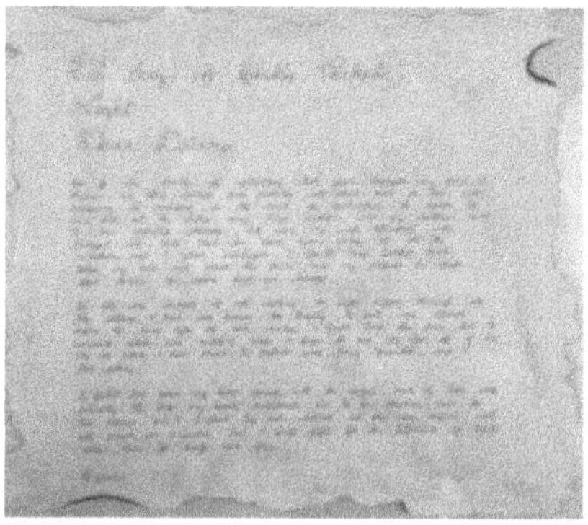

Figure 65.3. Student work.

Source: Author.

marginalisation. However, the students could connect this concept with various types of prejudiced behaviour prevalent in contemporary society. The 'mockingbird' symbolises innocence, so the students could see all the falsely accused men in society symbolising innocence and harmlessness.

5.2. The documentary project on falsely accused individuals

In this innovative project, the student came up with a documentary based on the falsely accused individuals in Indian society. In India, every year many men are accused of crimes like rape, dowry and domestic violence. But in some cases, men are accused falsely and just because of their gender, they fall victim to bias and prejudice. The students made this documentary to highlight this issue and raise their voices against it (Figure 65.4).

Students showed their creativity by compiling the documentary with existing information clips, footage and real-life interviews. They added value to the project by conducting interviews in and around the university to understand what people think about the victims who are falsely accused in such scenarios. In their documentary, the critical thinking of the students is displayed when they point out that whenever it comes to gender equality, everyone gives their opinion about female rights, but rarely in India, do people raise their voices for men's rights. India does not even have an official Men's commission nor does the Indian law support these falsely accused male victims. It creates a gap between the existing law and India's desire to achieve a gender-equal society.

This project gave the students ample chance to use their creativity and use AI tools while making the documentary. They made extensive use of video editing tools like Adobe Premier Pro. They used AI tools like Gemini and Chat GPT for brainstorming, writing the script and adding subtitles. These tools helped them to refine and polish their project and make it visually appealing and captivating.

5.3. The podcast project based on the text to kill a mockingbird

This group of students chose to come up with a more effective way of expressing their perceptions and hence they chose to make a podcast. In recent years, both audio and video podcasts have become extremely popular and they have been reviving the art of storytelling. Hence, this group created a powerful podcast to tell their story based on the classic text *To Kill a Mockingbird.*

This classic text remains the most-read novel related to racism and injustice in America. The story highlights the racial heroism of lawyer Atticus who fights the case for an African American man Tom Richardson who is falsely accused of raping a white girl named Mayella Ewell. Even after Atticus tries hard, Richardson is still convicted by the jury because of racial bias against coloured people (Figure 65.5).

Through this podcast, the students discussed the inner turmoil of characters like Atticus, Scout and Tom Robinson as discussed in the text. They put themselves in the shoes of the victims of racial prejudice and realised the relevance of a just and equal society. They used their critical thinking when they discussed subjects like marginalisation, intolerance and discrimination and understood how these issues are still relevant and prominent in contemporary society. Further, the students used their creativity by using AI tools to make the podcast more professional. They used Gen AI like Gemini and Chatgpt for their script and subtitling. Through video tools, they edited their recordings and inserted relevant audio clips making it more engaging. AI tools like Grammarly help them to polish their writing. By making this podcast, the students learnt to discard the conventional learning method and move beyond it.

5.4. Aligning with SDGs: The documentary and the podcast

Lastly and most importantly, these two projects namely the documentary and the podcast helped the students to comprehend the relevance of sustainable development and how to achieve social sustainability. By working on such sensitive topics as injustice related to race and gender, their attention was immediately drawn to SDG which promotes reduced equality and SDG 16 which promotes peace, justice and strong institutions. Through these two digital projects, the students tried to appeal to the audience to raise their voices for a most just, fair and equal society while advocating for justice and equality.

Figure 65.4. Student work.

Source: Author.

Figure 65.5. Student work.

Source: Author.

5.5. Conclusion, limitations, and future scope

This study validates that AI-integrated Project-Based Learning (PBL) can significantly encourage students to be creative and innovative while adopting critical thinking. Carefully chosen texts can significantly enhance the student's awareness of sustainability and Sustainable Development Goals proposed by the UN like gender equality, reduced inequalities, climate preservation, and quality education. Moreover, integrating AI tools in their education setting can increase their practical acquaintance and engage them in deeper discussions about complex and relevant issues where they go beyond the mundane conventional modes of learning. This powerful hybrid method brings education into the same line as the NEP 2020 policy which focuses more on a flexible and holistic way of learning pedagogy. While the results of this study are quite encouraging, this study has some limitations too. This study has been conducted in only one institution and the sample size is also relatively small. Moreover, this study deals only with engineering students which might limit the findings. So, future researchers can broaden the scope of this research by taking it further with a broader sample size. Additionally, this research framework can be adopted and conducted across diverse institutional settings with students from varied disciplines like arts, commerce, architecture, law, hotel management, etc. By broadening the scope, more applicability and effectiveness of this research can be reassessed and it could be a significant contribution toward the rapidly

References

[1] Daabool, F. S. (2022). Green Synthesis of Nanoparticles Selenium. *International Academic Journal of Science and Engineering*, *9*(1), 26–29. https://doi.org/10.9756/IAJSE/V9I1/IAJSE0904

[2] Luckin, R., Holmes, W., & Griffiths, M. (2016). *Intelligence Unleashed: An Argument for AI in Education*. Pearson.

[3] Dhivyalakshmi, R., & Marikkannu, P. (2016). Intelligent Bus Location Tracking System Using Android Smart Phones. *International Journal of Advances in Engineering and Emerging Technology*, *7*(3), 49–56.

[4] John, B., Rana, N. S., Nijhawan, M., & Sharma, G. (2024). Developing a framework for ecosystem-based fisheries management in India. *International Journal of Aquatic Research and Environmental Studies*, *4*(S1), 64–70. https://doi.org/10.70102/IJARES/V4S1/11

[5] Shahriar, A. Z. M., Atiquzzaman, M., & Ivancic, W. D. (2011). Evaluation of the Route Optimization for NEMO in Satellite Networks. *Journal of Wireless Mobile Networks, Ubiquitous Computing, and Dependable Applications*, *2*(2), 46–66.

[6] Barron, B., & Darling-Hammond, L. (2008). *Teaching for meaningful learning: A review of research on inquiry-based and cooperative learning*. The Metiri Group.

[7] Ruiz Viruel, S., Sánchez Rivas, E., & Ruiz Palmero, J. (2025). The role of artificial intelligence in project-based learning: Teacher perceptions and pedagogical implications. *Education Sciences*, *15*(2), 150. https://doi.org/10.3390/educsci15020150

[8] Leicht, A., Heiss, J., & Byun, W. (2018). Sustainability in education: An overview of current research. In A. Leicht, J. Heiss, & W. Byun (Eds.), *The role of education in the SDGs* (pp. 1–12). Springer. https://doi.org/10.1007/978-3-319-76251-7_1

[9] Richards, J. C. (2013). The role of creativity in language teaching. *English Teaching Forum*, *51*(3), 18–24.

[10] Eliyas, S., & Ranjana, P. (2022). Gamification: Is E-next Learning's Big Thing. *Journal of Internet Services and Information Security*, *12*(4), 238–245. https://doi.org/10.58346/JISIS.2022.I4.017

66 Evaluation of sustainability disclosure based on GRI standards in energy sector companies

Rifqi Novriyandana[1,a], Ratna Ayu Damayanti[2], Darwis Said[2], and Muhammad Rijal[1]

[1]Faculty of Economics and Business, University of Lambung Mangkurat, Indonesia
[2]Doctoral Program of Accounting, Faculty of Economics and Business, Hasanuddin University, Indonesia

Abstract: This research is to identify the impact of business activities, company size, and GCG on sustainability disclosure in energy sector The foundation for data processing this study is secondary data. This study utilizes purposive sampling according to specific criteria. This study utilizes a descriptive analytic approach using SPSS 26 software. Findings from the study demonstrate that both firm size and company activity factors affect sustainability disclosure. Concurrently, institutional ownership, independent boards of commissioners, and audit committees have minimal impact on sustainability disclosure. This signifies that GCG has insufficient control over sustainability disclosures.

Keywords: GCG, firm size, company activities, sustainability disclosure

1. Introduction

In recent years, stakeholders have placed greater emphasis on corporate sustainability performance in their investment decisions. They observe that an effective sustainability strategy not only reflects corporate social responsibility, but also reduces long-term business risks associated with ESG issues. Globally, regulations requiring companies to provide Sustainability Disclosures are increasing. In Indonesia, the Financial Services Authority (OJK) mandates, in accordance with POJK No. 51/POJK.03/2017, that public corporate entities disseminate sustainability reports as part of their corporate social responsibility obligations.

In a competitive industry, Sustainability Disclosure also serves as a differentiating tool. Companies that proactively communicate sustainability information are often perceived as having added value by consumers, especially in market segments that care about environmentally friendly and ethical products. Despite progress in Sustainability Disclosure practices, there are differences between large and small companies in the quality and completeness of disclosures. Large companies are generally better equipped with resources and reporting systems to support comprehensive disclosure, while smaller companies often face challenges related to cost and limited knowledge of sustainability reporting.

Organizations often choose the Global Reporting Initiative as their standard for sustainability performance reporting. In 2021, GRI amended its regulations to tackle modern sustainability issues by improving transparency and stakeholder engagement. The adoption of sustainability disclosure based on the 2021 GRI criteria is becoming important as society, governments, and investors demand transparency in company sustainability performance. This aids firms in methodically identifying, measuring, and disclosing the social as well asenvironmental impacts of their activities. GRI 2021 guidelines call for organizations to acknowledge their substantial effects and offer comprehensive disclosures, facilitating stakeholders' assessment of their contributions to global sustainability goals.

In 2020, there were 66 energy sector companies, but only 19 companies disclosed sustainability reports. By 2023, there were 83 energy sector companies and 79 companies disclosed sustainability reports. This shows that every year more and more companies realize the importance of sustainability reports.

Institutional ownership is considered to play an important role in corporate sustainability disclosure. With high institutional ownership, management may face additional pressure to report activities that demonstrate sustainable performance in an effort to maintain the institution's reputation in the eyes of investors. This theory is supported by research Alsaeed [1] and Bushee [2], says that companies with significant institutional ownership would be more tends to be report sustainability since shareholders want transparency. However, research conducted by Reverte [3] show that different institutional shareholders have different interests in sustainability. For example, many short-term institutional shareholders concentrate on short-term profits rather than sustainability.

Independent commissioners undertake various roles to oversee business management, including improving

---

[a]rnovriyandana@ulm.ac.id

DOI: 10.1201/9781003684459-66

transparency and accountability in sustainability reporting. Their neutral position, apart from the company's interests, requires the delivery of unbiased monitoring. Research conducted by Michelon and Parbonetti [4] and Mallin et al. [5] demonstrate that companies with a greater percentage of independent directors had improved sustainability reporting. Khan et al. [6] discovered that the issuance of a sustainability report by businesses was not improved by an independent board of commissioners.

Audit committees have evolved beyond just examining financial statements to supervising comprehensive governance procedures, including sustainability disclosures and reporting. Assessing audit committees allows businesses to improve openness and accountability in sustainability reporting. Moreover, research conducted by Al-Shaer and Zaman [7] and Kiliç et al. [8] shows that having an independent board of commissioners does not make it easier for businesses to provide sustainability reports. The audit committee's impact on sustainability disclosure is not always evident, particularly when the committee just performs ostensibly related tasks without placing a specific focus on reporting on sustainability [9].

The comprehensive sustainability disclosure of a corporation is often associated with its scale. This is because large firms have higher public visibility and are more likely to widely share sustainability information. Furthermore, large firms possess a greater aptitude for sustainability disclosures than smaller enterprises. Major firms have more resources to dedicate to environmental management and sustainability efforts, as evidenced by Clarkson et al. [10] and Hackston and Milne [11]. Roberts [12] found that business size is not the principal factor influencing sustainability disclosure, since tiny enterprises with a robust commitment to sustainability may nonetheless provide extensive reports.

Corporate activities affect sustainability reporting or sustainability disclosure, especially considering the pressure from stakeholders who expect transparency and social responsibility from companies. Companies that frequently engage in environmental activities often face pressure from regulators and the public to be more open about their environmental impacts. Gray et al. [13] and Chen and Roberts [14] supports this idea by saying that businesses that engage in environmentally friendly and sustainable activities are motivated to report them. However, according to research conducted by Cho and Patten [15], certain companies that carry out activities that have environmental and social consequences tend to avoid reporting in their entirety so as not to draw more attention to controversial activities.

This study looks at how institutional ownership, an independent board of commissioners, an audit committee, company size, and company operations affect the level of sustainability disclosure in energy sector companies listed on the Indonesia Stock Exchange for the 2020–2023 period. It is anticipated that this study will help identify the elements that motivate businesses to publish their sustainability initiatives more openly, given the increasingly strict rules and stakeholder expectations.

2. Methods

2.1. Research place/location

This research was performed using the official website of the Indonesia Stock Exchange, www.idx.co.id, and the official websites of the sampled firms.

2.2. Population and sample

Purposive sampling was the method used in this study to screen the samples to be tested. After screening the samples, a total of 44 samples were found to be suitable for testing.

2.3. Data collection technique

This study uses data collection techniques with the documentation method, so the data source used is secondary data.

2.4. Data Analysis techniques

Multiple linear regression analysis is used in this research to analyze the data. Descriptive statistics, traditional assumption tests, and hypothesis testing are among the several tests that are carried out. The most latest SPSS program, SPSS version 26, was utilized in this study (Figure 66.1 and Table 66.1).

3. Results and Discussions

3.1. Discussion of the effect of institutional ownership on sustainability disclosure

Test score indicates that institutional ownership has no impact on sustainability disclosure (Table 66.2). One of the main reasons that is often cited is the short-term focus of most institutional investors. According to Bushee [2] most institutional investors are more focused on short-term

Table 66.1. Operational definition of variables

Variable	*Alat Ukur*
Institusional Ownership	Institutional ownership shares / Total number of shares
Independent Board of Commissioners	Total number of independent commissioners / Total number of commissioners
Audit Committee	Total number of audi committee
Company size	Ln=(total asests)
Company Activity	Sales / Total Assets
Sustainability Disclosures	Number of disclosures / expected number of items GRI 2021

Source: Author.

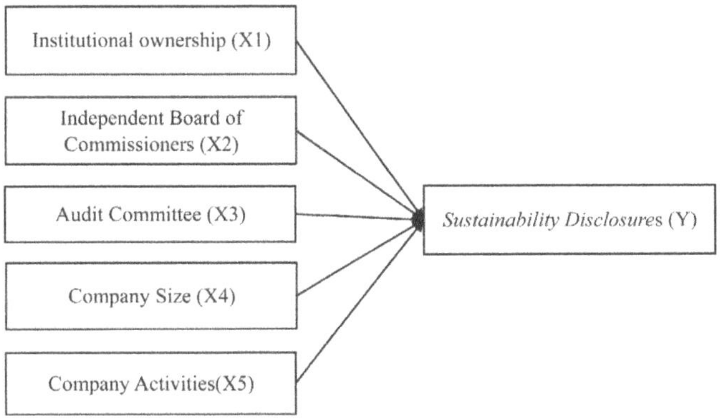

Figure 66.1. Redearch design.

Source: Author.

financial benefits than sustainability challenges that need long-term consideration. Businesses may not feel compelled to increase sustainability disclosures in these circumstances since institutional investors don't seem to be very interested in doing so. When making investment choices, many institutional investors in developing countries have not ESG factors. Institutional ownership has little effect on sustainability disclosure as investors don't need information on social and environmental aspects.

3.2. Discussion of the effect of independent board of commissioners on sustainability disclosure

Test results demonstrad that sustainability disclosure is unaffected by the independent board of commissioners. Because non-financial issues like sustainability disclosures are often ignored by the board of independent commissioners, which is typically entrusted with making sure that accounting rules and regulations are followed [4] Klik atau ketuk di sini untuk memasukkan teks. Independent boards concentrate on financial reporting and internal controls to guarantee the accuracy of financial statements rather than creating corporate sustainability initiatives. In practice, they often face limitations in exerting significant influence on the company's strategic policies and decisions, especially those related to sustainability issues.

3.3. Discussion of the effect of audit committee on sustainability disclosure

Test score indicates that audit committe has no impact on sustainability disclosure. Instead, they concentrate on the quality of external and financial audit reports [16] Klik atau ketuk di sini untuk memasukkan teks. As a result, sustainability disclosure becomes less important. They focus more on evaluating financial and audit risks, while sustainability is often considered outside the committee's primary responsibilities. Although audit committees have oversight responsibilities, they tend to focus on operational and financial issues. They may not be active enough in advocating or encouraging management to increase transparency in sustainability reports, especially if there are no external or regulatory pressures requiring such disclosures.

3.4. Discussion of the effect of company size on sustainability disclosure

It is known that company size has an influence on sustainability disclosure. Large companies often provide more information on sustainability. This is because they are more visible to the public than small enterprises [11] Klik atau ketuk di sini untuk memasukkan teks. Large corporations could invest more money in environmental management and sustainability projects. An organization's size has a significant impact on how much sustainability disclosure it provides, especially for businesses in ecologically sensitive sectors. More financial, human, and technical resources are often available to larger companies. Large corporations often have more powerful stakeholders, including the media, governments, customers, and large investors. International sustainability reporting standards, legislation, and regulations – like OJK's requirements on sustainability report disclosure – are more likely to apply to large businesses.

Table 66.2. Partial test results (T)

		Coefficients[a]				
Model		*Unstandardized Coefficients*		*Standardized Coefficients*	*t*	*Sig.*
		B	*Std. Error*	*Beta*		
1	(Constant)	−3.169	1.197		−2.647	.012
	Institutional Ownership	.226	.166	.197	1.366	.180
	Independent Board of Commissioners	.129	.270	.072	.479	.635
	Audit Committee	−.006	.039	−.024	−.155	.878
	Company Size	.110	.036	.508	3.084	.004
	Company Activities	.307	.094	.532	3.277	.002

a. Dependent Variable: Sustainability Disclosure

Source: SPSS 26 Output, 2024.

3.5. Discussion of the effect of company activities on sustainability disclosure

It is known that company activity has an effect on sustainability disclosure. Companies that carry out significant environmental activities often face pressure from social institutions and regulations to make them more open about the consequences of their actions Chen and Roberts [14]. Companies that participate in sustainability and social initiatives are incentivized to report on these initiatives. This will increase their public trust and attract institutional investors who have a particular interest in ESG (Environment, Social, Governance) elements.

4. Conclusions

The study's findings demonstrate that neither the audit committee nor the independent board of commissioners nor institutional ownership have an impact on sustainability disclosure. In conclusion, independent boards of commissioners, audit committees, and institutional ownership levels own little impacts on sustainability disclosure.

The size of the business affects sustainability disclosure. Bigger companies are more possibility to provide sustainability reports that detail their social and environmental responsibilities in sequence to preserve a favorable images. Business operations also affect sustainability disclosure. Businesses with a lot of activity often publish sustainability reports to meet their social and environmental responsibilities.

Given the study's limitations and findings, further research is urged to include other variables that may affect sustainability disclosure openness, like liquidity and management ownership. Additionally, it is expected that data from other businesses, such manufacturing or industry, would be used by future scholars.

References

[1] Alsaeed, K. (2006). The association between firm-specific characteristics and disclosure: The case of Saudi Arabia. *Managerial Auditing Journal*, *21*(5), 476–496.

[2] Bushee, B. J. (1998). The influence of institutional investors on myopic R&D investment behavior. *The Accounting Review*, *73*(3), 305–333.

[3] Reverte, C. (2009). Determinants of corporate social responsibility disclosure ratings by Spanish listed firms. *Journal of Business Ethics*, *88*(1), 351–366.

[4] Michelon, G., & Parbonetti, A. (2012). The effect of corporate governance on sustainability disclosure. *Journal of Management and Governance*, *16*(3), 477–509.

[5] Mallin, C., Michelon, G., & Raggi, D. (2013). Monitoring intensity and stakeholders' orientation. *Journal of Business Ethics*, *15*(3), 29–43.

[6] Khan, A., Muttakin, M. B., & Siddiqui, J. (2013). Corporate governance and corporate social responsibility disclosures: Evidence from an emerging economy. *Journal of Business Ethics*, *114*, 207–223.

[7] Al-Shaer, H., & Zaman, M. (2018). Board gender diversity and sustainability reporting quality. *Journal of Contemporary Accounting & Economics*, *14*(3), 266–287.

[8] Kiliç, M., Kuzey, C., & Uyar, A. (2015). The impact of ownership and board structure on Corporate Social Responsibility (CSR) reporting in the Turkish banking industry. *Corporate Governance*, *15*(3), 357–374.

[9] García-Sánchez, I. M., Martínez-Ferrero, J., & García-Meca, E. (2019). Independent directors, CEO career concerns and CSR reporting in the European banking sector. *Journal of Business Ethics*, *151*(3), 853–875.

[10] Clarkson, P. M., Li, Y., Richardson, G. D., & Vasvari, F. P. (2008). Revisiting the relation between environmental performance and environmental disclosure: An empirical analysis. *Accounting, Organizations and Society*, *33*, 303–327.

[11] Hackston, D., & Milne, M. J. (1996). Some determinants of social and environmental disclosures in New Zealand companies. *Accounting, Auditing & Accountability Journal*, *9*(1), 77–108.

[12] Roberts, R. W. (1992). Determinants of corporate social responsibility disclosure: An application of stakeholder theory. *Accounting, Organizations and Society*, *17*(6), 695–612.

[13] Gray, R., Owen, D., & Adams, C. (2001). *Accounting & Accountability: Changes and Challenges in Corporate Social and Environmental Reporting*. Prentice Hall.

[14] Chen, J. C., & Roberts, R. W. (2010). Toward a more coherent understanding of the organization-society relationship: A theoretical consideration for social and environmental accounting research. *Journal of Business Ethics*, *97*(4), 651–665.

[15] Cho, C. H., & Patten, D. M. (2007). The role of environmental disclosures as tools of legitimacy: A research note. *Accounting, Organizations and Society*, *32*(7–8), 639–647.

[16] Al-Shaer, H., & Zaman, M. (2016). Board gender diversity and sustainability reporting quality. *Journal of Contemporary Accounting & Economics*, *12*(3), 210–222.

67 Exploring consumer decision-making factors for electric, hybrid, and hydrogen vehicles through focus group discussions

Hendri Bhirowo[1,a], Indrawati[2,b], Handrea Bernando Tambunan[1], Indra Ardhanayudha Aditya[2,c], Refi Rifaldi Windya Giri[2,d], Fasa Maulidan Hakim[2,e], and Fayza Najeela Pakutandang[2,f]

[1]Power Generation Research Division, lPT PLN (Persero), South Jakarta, Indonesia
[2]School of Economics and Business, Telkom University, Bandung, Indonesia

Abstract: Indonesia's transportation system faces significant challenges due to its reliance on fossil fuel-based vehicles, which contribute heavily to greenhouse gas emissions. To combat this, the government has initiated policies to encourage the adoption of green vehicles, such as electric, hybrid, and hydrogen-powered models. Despite these efforts, adoption rates remain low. This research explores the reasons behind this slow adoption by conducting Focus Group Discussions (FGDs) with representatives from academia, industry, and policymakers. The findings reveal key barriers. High vehicle prices and operational costs during the early stages of green vehicle adoption deter consumers. Additionally, inadequate charging and refueling infrastructure, especially in remote areas, hampers widespread use. Concerns about vehicle performance and safety also influence consumer hesitation. These factors underline the need for strategic interventions by policymakers and industry stakeholders. The research integrates these findings into an Analytical Hierarchy Process (AHP) model to rank the critical factors influencing adoption. This model suggests that effective policies, such as imposing higher taxes on polluting vehicles while providing substantial subsidies for green vehicles, could drive adoption. Furthermore, addressing consumer education, financial support, and infrastructure development is essential. By identifying these barriers and opportunities, this study aims to assist in formulating targeted strategies for increasing green vehicle adoption. Achieving this goal will not only advance Indonesia's transition to sustainable transportation but also contribute to the nation's net-zero emissions target by 2060.

Keywords: Consumer decision-making, electric vehicles, focus group discussion, hybrid vehicles, hydrogen vehicles

1. Introduction

Indonesia, being one of the densely populated countries faces numerous challenges when it comes to resolving the environmental issues with respect to transportation and the most concerning greenhouse gas emissions and air pollution. As reported by the Ministry of Environment and Forestry [1], transport is among the foremost reasons of carbon emissions in Indonesia whereby motorized means of transport in primary cities such as Jakarta account for over 25% of total carbon dioxide emissions. The use of more fossil fueled cars makes the quality of air worse and increases the burden on the environment as well.

One of Indonesia's firm commitments in carbon reduction is reflected in policies that support low-emission vehicles such as electric, hybrid, and hydrogen-powered vehicles. The previous government released Presidential Regulation No. 55/2019 on the acceleration of the Battery-based Electric Motor Vehicle (BEV) program for Road Transportation in 2020 [2]. This action also aligns with the SDG 7 (ensuring access to affordable, reliable, sustainable and modern energy for all) by electrifying transportation, thus facilitating the transition to clean energy. In addition, this policy is a contributively action to SDG 12 (Sustainable consumption and production patterns) by increasing the environmental efficiencies using ecological efficient crops. Indonesia has outlined 3 strategies, to attain net zero carbon emission by 2060, eventually aiming to create a long term vision until the year 2060 including increasing energy efficiency, sustainability, and electrification of the automotive sector.

Electric vehicles (EVs), hybrids and hydrogen can be effectively used to control carbon emissions and reduce dependencies on fossil fuels [3–5]. When EV is running there is no direct emissions and the working mechanism of EV with smart charging solutions as well as implement vehicle to grid strategies can work towards supporting the energy grid which means that even during times of non-use, the EV can serve in generating energy. Thus, cars that run on electricity are indeed helpful to modern economies in combating greenhouse gas emissions and ensuring that worldwide climate

[a]hendri.bhirowo@pln.co.id, [b]indrawati@telkomuniversity.ac.id, [c]indra.aditya@pln.co.id, [d]rifaldi@telkomuniversity.ac.id, [e]fasamaulidan@gmail.com, [f]najeela.fayza2088@gmail.com

DOI: 10.1201/9781003684459-67

targets are achieved [6]. When it comes to emissions, hybrid cars which have a conventional engine and electric one is at a better place as they emit less than others that are solely driven by fossil fuels [3]. Also important to note is the fact that vehicles that run on hydrogen have an immense promise of becoming environmentally friendly seeing as their only byproduct is water. Most emissions related to hydrogen come from its production especially when hydrogen is sourced using natural gas. Nevertheless, when hydrogen is produced via electrolysis with renewable sources such as solar or wind, the emissions associated with its production can be greatly reduced [5, 7].

However, remaining, the rate at which Indonesian consumers turn to eco-friendly automobiles remains quite slow, despite nearness of the government support and clear benefits to the environment. One factor that can affect this phenomenon include awareness of the consumers, price of the vehicles, existence of supporting infrastructure such as electric or hydrogen charging stations, and even the expectations on the performance and reliability of the vehicles themselves [2].

2. Literature Review

2.1. Consumer awareness and perceptions

When consumers want to adopt a new technology, awareness of the importance of the technology is needed. Consumer knowledge of the environmental benefits they will get if they use EV, HEV, or HFCV vehicles can increase their desire to adopt the technology. This is in line with the findings of Rezvani [8] which states that if consumers know about the ecological benefits of environmentally friendly vehicles, they will be more likely to consider buying them. Then another finding from Wu et al. [9] states that increased awareness of the dangers of motor vehicle emissions, as well as fuel efficiency will directly affect consumer preferences for hybrid and electric vehicles.

2.2. Financial considerations

Financial factors such as purchase price, operating costs, and government incentives are important factors in consumer decision making. As stated by Xia et al. [10] that although consumers realize that using electric vehicles can save long-term expenses, the initial purchase cost remains the main obstacle. Incentives provided by the government such as tax breaks and fuel subsidies can reduce consumer spending and can stimulate demand. Furthermore, research by Lévay et al. [11] found that consumers also consider the total cost of ownership (TCO) when deciding to buy a vehicle.

2.3. Infrastructure and convenience

The availability of charging and refuelling infrastructure is another factor that can influence consumer decisions. As stated by Ledna et al. Ledna et al. [12] that consumers tend to be reluctant to buy electric vehicles because they are worried that there will be no charging stations available. Especially for HFCVs, because charging stations are not yet widely available. However, Hassan et al. [13] emphasized that there needs to be an increase in adequate charging infrastructure to ease consumer anxiety and increase the use of environmentally friendly vehicles.

2.4. Environmental impact

Another factor that can influence consumer preferences is the perceived environmental benefits of using EVs, HEVs, and HFCVs. This is evidenced by several studies stating that consumers tend to choose vehicles with lower emissions if they know that this technology can support sustainable development. As stated by Shahrier & Habib [14] that consumer awareness of the positive environmental impacts of environmentally friendly vehicles can increase consumer desire to adopt them. Furthermore, Xia et al. [10] found that consumers also consider choosing environmentally friendly vehicles because they want to reduce their personal carbon footprint.

2.5. Policy support

Another factor is government policy that can encourage consumer confidence in adopting environmentally friendly vehicles. With strong policies such as regulations on emissions and promotion programs, it can increase consumer desire to buy environmentally friendly vehicles. According to Langbroek et al. [15] effective government policies can encourage the adoption of electric vehicles and make consumers more confident in the technology. Furthermore, Langbroek et al. [15] explained that government incentives provided for the purchase of environmentally friendly vehicles can significantly influence consumer decisions because they can reduce the financial burden, especially related to initial costs.

3. Research Methodology

This study uses a qualitative approach to identify factors that influence consumer preferences in adopting Electric, Hybrid, and Hydrogen vehicles. The Focus Group Discussion (FGD) method is used in this study, and is the most suitable method for collecting in-depth information about consumer preferences for environmentally friendly vehicles. In this study, there are several stages in conducting a focus group discussion as follows.

3.1. Participants and sampling

This study used purposive sampling to select participants with expertise in the field of environmentally friendly vehicle development. Non-random sampling was applied to select participants based on predetermined criteria, and ensure that all participants had direct involvement and adequate experience in the field of environmental vehicle development for at least two years. A total of five experts from various disciplines were obtained in order to obtain broad knowledge and diversity of perspectives. Participants who took part in this FGD included academics who would provide theoretical knowledge; researchers who had insight into the development of environmentally friendly vehicle technology; automotive industry players, who would provide practical insight into vehicle production and marketing; and policy makers, who would provide insight into government policies that support the adoption and development of environmentally friendly vehicles. By involving experts from various fields, it is hoped that a holistic discussion with results that can discuss various aspects including policy, technology, economics, and environmental issues. This approach is also in accordance with the FGD method which aims to explore various perspectives from knowledgeable individuals. Krueger et al. [16] stated that a diverse group composition can produce more in-depth discussions and have diverse perspectives. For this reason, this study is expected to provide a comprehensive understanding of what factors can be challenges and opportunities for developing and adopting environmentally friendly vehicles in Indonesia.

3.2. Data collection

In this study, the FGD session was conducted in a semi-structured format that combines directed discussions with participants' freedom to express their opinions. Open-ended questions used in this FGD covered key topics such as consumer awareness of environmentally friendly vehicle technology (EV, HEV, HFCV), perceived environmental benefits, financial barriers, readiness of charging infrastructure, and policies that support the development of environmentally friendly vehicles. This is also supported by the opinion of Creswell & Creswell [17] who stated that open-ended questions can encourage a diversity of perspectives. The facilitator ensures that the participation of each expert is balanced and is actively and equally involved. This is very important to create a dynamic discussion, as stated by Morgan [18] and supported by Krueger et al. [16] who suggest managing the discussion evenly to avoid dominant voices and inactive participants. This FGD session was conducted for 90 minutes each session, which is enough to explore every idea from the participants. Due to distance and time constraints, this FGD was conducted online by recording the discussion sessions, both audio and video, for further use in compiling transcripts and thematic analysis. These data will be transcribed and analysed according to the process outlined by Braun & Clarke [19], which allows for the identification of themes and recurring patterns that explain the factors that influence consumers to adopt environmentally friendly vehicles.

3.3. Data analysis

To analyse the transcript from the focus group discussions (FGDs), content analysis is performed in order to categorize carefully different elements systematically that emerge from qualitative data. The researcher will start by reading and thoroughly becoming familiar with the transcript. During this step, the researcher identifies keywords, phrases, and concepts related to consumer attitude and decision making. After the initial categorization, the analysis shifts to organizing the categories into broader themes that highlight the most significant insights from the data, like consumer awareness around green vehicle capabilities and issues related to infrastructure. Once this is done, a review is conducted to ensure that the categories are logical, accurate and can reflect the data around the identified pattern. After a comprehensive analysis, the findings of the data will be reported in a structured narrative, with quotations from participants to support and illustrate the main insight. Content analysis, according to scholars like Krippendorff [20], can provide a consistent and quantitative meaning to qualitative data that permits reliable conclusions related to the consumer behaviours and preferences.

3.4. Validity and reliability

To strengthen the validity, reliability and credibility of the findings, the FGD results are compared with secondary data from existing literature. This process involves reviewing literature on consumer behaviour and vehicle adoption patterns through a technique also known as triangulation. The data is then checked by members by sharing initial findings with participants to ensure accuracy. According to Espedal et al. [21], both triangulation and member checking are crucial techniques to ensure validity in qualitative research. After the triangulation process, the identified factors were finally validated with experts who had earlier participated in FGDs. Finally, face validity was employed to ensure that the analysed factors aligned with each expert's perceptions on what is relevant and suitable for the development of the AHP questionnaire.

3.5. Ethical consideration

Prior to the discussion, consent was obtained from the participants to ensure that they were in agreement with this study's purpose before discussing it. Participants are also allowed to withdraw their consent at any time. This step is essential for protecting participants' independence and fostering trust during the research process. Wiles [22] highlighted that ethical considerations are crucial when dealing with personal opinions and attitudes in qualitative research because participants may share sensitive or subjective insights. This made the whole process anonymous to maintain confidentiality and ensure participants' privacy. Maintaining ethical integrity is essential to protect participants' rights and enhance the credibility and reliability of research results. Being respectful and transparent when engaging with the data collection process ensures that researchers adhere to standards of ethics in qualitative research, which is advantageous to both the participants and the overall research.

4. Research Findings

This section presents the results of the analysis derived from a series of FGD conducted with key stakeholders in the automotive sector, academia, researcher and policymakers. It is also possible through these findings to establish the primary determinants of the decision-making process of the consumers in Indonesia regarding the adoption of hybrid, electric, and hydrogen vehicles, providing a comprehensive overview of the criteria generated by experts.

4.1. FGD participants and dates

In this FGD, the participants were individuals who had at least two years of experience in the sustainable vehicle industry. The first participant was HR, a professional scholar in technology of sustainable transportation. Next was OA, representing the automotive sector focusing on Electric car technology. Next was RN, a researcher and policymaker engaged in green energy transition policies in Indonesia. Following that, AS participated as an industry consultant with expertise in hybrid vehicle technology. Lastly, HD was a technologist specializing in hydrogen fuel cell development. The first FGD involving HR, OA, and RN was conducted on August 16, 2024. The second FGD involving AS and HD was held on August 19, 2024 FGD included follow-up questions based on initial feedback.

During this focus group discussion, participants were presented with a series of questions aimed at identifying the key factors that influence consumer acceptance of environmentally friendly vehicles. The questions ranged from problems that consumers may have when adopting hybrid, electric, and hydrogen vehicles in Indonesia to other issues. Participants were also asked to rate how consumer decisions are made in terms of what internal purchase and ongoing operational costs entail. Additionally, they discussed the role of government policies and incentives in promoting green vehicle adoption, as well as technological concerns like battery lifespan, charging times, and vehicle range. The session also covered safety considerations and the availability of infrastructure in the decision-making process, along with the influence of environmental awareness on consumer behaviour. These discussions provided comprehensive observations that were identified as factors to influence the implementation of green vehicle technology in Indonesia.

4.2. Key criteria derived from FGD

Eight main criteria were identified through the in-depth interviews and FGDs. The prioritization of the eight criteria was then determined using content analysis. The key factors influencing consumer adoption of hybrid, electric and hydrogen vehicles in Indonesia are reflected in these criteria. The eight criteria were also analysed using the Analytic Hierarchy Process (AHP) method. The following criteria were prioritized for further analysis based on the frequency and level of importance discussed by participants.

4.3. Initial cost

One of the most significant barriers to consumers considering adoption of hybrid, electric and hydrogen vehicles is the high initial purchase price. OA supports this statement by pointing out 'When consumers see the price of an electric or hydrogen vehicle, they immediately change their minds. They don't consider the savings or the long-term benefits.' In addition, AS also supported this statement by opining 'In Indonesia, affordability is a key factor, and unless the initial price decreases, adoption will remain slow.' This represents that initial cost is an important factor in influencing consumers to adopt hybrid, electric, and hydrogen vehicles.

4.4. Operational cost

While often overshadowed by upfront cost considerations, low fuel costs and maintenance costs summarized in long-term cost savings are also among the most important factors in consumer decision-making. RN emphasized that awareness of long-term savings is critical 'The challenge is to make consumers aware of the long-term fuel and maintenance savings offered by these vehicles.' AS made a similar point 'Despite lower running costs, consumers are still more focused on upfront costs, which are currently a significant barrier.' These statements highlight the economic benefits that accrue from green vehicles, even as consumers continue to focus on the initial price.

4.5. Safety

Consumer confidence in green car technology is mainly affected by concerns about the long-term safety, stability of batteries, and security of hydrogen storage. In this regard, OA voiced its doubts about hydrogen storage: 'Consumers will continue to hesitate unless we can ensure 100% safety, as hydrogen storage is still a sensitive issue.' 'Battery safety, especially regarding long-term stability, is one of the main concerns in the electric vehicle market,' HR said, referring to the possible danger of overheating batteries in electric cars. These opinions emphasize how important it is to address safety issues to increase customer confidence in green car technology.

4.6. Government subsidies and incentives

Consumer financial burdens such as high initial costs can be helped by the government through subsidies and tax incentives so that consumers are more encouraged to adopt green vehicles. RN advocated for stronger subsidies, stating 'We need stronger incentives from the government, as we see in countries like Norway or Japan, where subsidies are quite significant.' HR continued 'Subsidies can make a big difference, but right now, they are not enough to offset the high initial costs.' Both opinions highlight the important role of the government in driving adoption.

4.7. *Charging speed*

In traveling long distances using electric vehicles, practicality and convenience are supported with the availability of fast charging infrastructure. OA has concerns regarding inconvenience due to slow charging times, 'Charging time is a major issue for consumers. No one wants to wait for hours to charge their vehicle, especially if they are traveling long distances.' HD also had similar concerns, 'We need a fast-charging solution, or adoption will stagnate, especially outside urban areas.' These opinions emphasize the urgency of an efficient charging infrastructure so that people can be more confident about adoption.

4.8. *Maintenance cost*

Consumers often question the maintenance costs of electric and hydrogen vehicles, especially in rural areas. They still consider maintenance costs associated with advanced technologies, such as service availability and battery replacement. HD emphasized the possibility of high battery replacement costs saying, 'Battery replacement is a major concern for consumers. They know that while maintenance costs are generally lower for electric vehicles, when the battery needs to be replaced, it can be expensive.' In addition, RN emphasized the importance of service center availability, stating, 'It's not just about the initial cost, but also how easy it is to maintain these vehicles over time and whether there is sufficient infrastructure to support that.' In this regard, a strong support system is needed to address consumers' concerns on adoption.

4.9. *Resale value*

While most participants were not concerned about the resale value of electric and hydrogen vehicles, this uncertainty affects consumer confidence, especially when it comes to long-term investments. AS stated, 'We don't yet have sufficient data on how electric and hydrogen vehicles perform in the used car market, and that makes people cautious.' HD added 'Resale value may not be a major concern for everyone, but for some, it is a deciding factor.' This shows how important it is to collect reliable resale data to increase customer confidence in green vehicles.

4.10. *Environmental impact*

Considerations regarding reducing emissions and contributing to sustainability are often sidelined by consumers due to financial concerns in the Indonesian market despite their importance. These issues are often ranked lower than cost considerations. HR stated, 'Consumers appreciate the idea of reducing emissions, but without strong financial incentives, environmental benefits alone will not drive adoption.' RN reinforced this by stating, 'Sustainability is important, but right now, it is not enough to make consumers choose these vehicles over cheaper options.' This suggests that stronger financial incentives are needed to drive the adoption of green vehicles in Indonesia.

5. Conclusion

This study provides extensive insights into the key factors influencing consumer adoption of hybrid, electric and hydrogen vehicles in Indonesia. High initial costs, limited infrastructure and concerns about vehicle safety were the most significant barriers identified through focus group discussions. Consumers tend to focus on financial considerations, especially the high initial cost, rather than the environmental benefits of these vehicles in their decision-making. Government incentives, such as subsidies and infrastructure development, are essential to encourage wider adoption. In supporting Indonesia's transition to cleaner transportation and achieving its long-term environmental goals, it is essential to address these barriers, especially in terms of affordability and infrastructure.

References

[1] Ministry of Environment and Forestry. (2019). *Inventory of Greenhouse Gas Emission Profile of DKI Jakarta Province*. Ministry of Environment and Forestry.

[2] Kementerian Lingkungan Hidup dan Kehutanan. (2019). Inventarisasi Profil Emisi Gas Rumah Kaca Provinsi DKI Jakarta. Kementerian Lingkungan Hidup dan Kehutanan.

[3] Pambudi, I., & Juwono, V. (2023). Electric vehicles in Indonesia: public policy, impact, and challenges. *Asian Journal of Social and Humanities*, *2*(2), 347–360.

[4] Searle, S., Bieker, G., & Baldino, C. (2021). Decarbonizing road transport by 2050: Zero-emission pathways for passenger vehicles.

[5] Subekti, R. A. (2014). *Peluang dan Tantangan Pengembangan Mobil Listrik Nasional*. Lipi Press.

[6] Union of Concerned Scientists. (2014). How Clean Are Hydrogen Fuel Cell Electric Vehicles? California Leading the Way on Clean-Hydrogen Policies.

[7] Environmental Protection Agency. (2024). Electric Vehicle Myths. Environmental Protection Agency. Accessed: Oct. 04, 2024. [Online]. Available: https://www.epa.gov/greenvehicles/electric-vehicle-myths

[8] Environmental Protection Agency. (2024). Hydrogen in Transportation. Environmental Protection Agency. Accessed: Oct. 04, 2024. [Online]. Available: https://www.epa.gov/greenvehicles/hydrogen-transportation

[9] Rezvani, Z., Jansson, J., & Bodin, J. (2015). Advances in consumer electric vehicle adoption research: A review and research agenda. *Transportation research part D: transport and environment*, *34*, 122–136.

[10] Wu, J., Liao, H., Wang, J. W., & Chen, T. (2019). The role of environmental concern in the public acceptance of

autonomous electric vehicles: A survey from China. *Transportation Research Part F: Traffic Psychology and Behaviour*, *60*, 37–46.

[11] Xia, Z., Wu, D., & Zhang, L. (2022). Economic, functional, and social factors influencing electric vehicles' adoption: An empirical study based on the diffusion of innovation theory. *Sustainability*, *14*(10), 6283.

[12] Lévay, P. Z., Drossinos, Y., & Thiel, C. (2017). The effect of fiscal incentives on market penetration of electric vehicles: A pairwise comparison of total cost of ownership. *Energy Policy*, *105*, 524–533.

[13] Ledna, C., Muratori, M., Brooker, A., Wood, E., & Greene, D. (2022). How to support EV adoption: Tradeoffs between charging infrastructure investments and vehicle subsidies in California. *Energy Policy*, *165*, 112931.

[14] Hassan, Q., Azzawi, I. D., Sameen, A. Z., & Salman, H. M. (2023). Hydrogen fuel cell vehicles: opportunities and challenges. *Sustainability*, *15*(15), 11501.

[15] Shahrier, H., & Habib, M. A. (2024). Electric vehicle (EV) type choice model: Latent class modelling (LCM) approach. *International Journal of Transportation Science and Technology*. doi: 10.1016/j.ijtst.2024.08.004.

[16] Langbroek, J. H., Franklin, J. P., & Susilo, Y. O. (2016). The effect of policy incentives on electric vehicle adoption. *Energy Policy*, *94*, 94–103.

[17] Newcomer, K. E., Hatry, H. P., & Wholey, J. S. (Eds.). (2015). *Handbook of practical program evaluation* (Vol. 864). Hoboken, NJ: John Wiley & Sons.

[18] Creswell, J. W., & Creswell, J. D. (2014). *Qualitative, Quantitative, and Mixed Methods Approaches*, Fifth Edit. California: SAGE Publications, Inc. doi: 10.4324/9780429469237-3.

[19] Morgan, D. (2012). Focus Groups as Qualitative Research. *Focus Groups as Qualitative Research*, SAGE Research Methods, pp. 32–46. doi:10.4135/9781412984287.

[20] Braun, V., & Clarke, V. (2006). Using thematic analysis in psychology. *Qualitative Research in Psychology*, *3*(2), 77–101.

[21] Krippendorff, K. (2019). *Content Analysis: An Introduction to Its Methodology*, Fourth Edition. SAGE Publications, Inc. doi: 10.4135/9781071878781.

[22] Espedal, G., Løvaas, B. J., Sirris, S., & Wæraas, A. (2022). *Researching Values Methodological Approaches for Understanding Values Work in Organisations and Leadership*. Springer Nature Switzerland AG. doi:10.1007/978-3-030-90769-3.

68 Narrative pedagogy of undergraduate students' writing literacy development and its implications for teaching writing

Erna Iftanti[1,a], *Nani Soengkono Madayani*[1,b], *and Galuh Nashrulloh Kartika M. R.*[2,c]

[1]English Education Program, UIN Sayyid Ali Rahmatullah Tulungagung, Indonesia
[2]PGMI Uniska MAB, South of Kalimantan, Indonesia

Abstract: Some undergraduate students from both Language and Science Departments of a state Islamic university in Tulungagung Indonesia have successfully developed their writing skills. This article aims at unveiling the development of their writing literacy and its implication for teaching writing. The results of this study demonstrate three key findings: the portrait of students with strong writing literacy skills, the four key factors contributing to their good writing literacy development, and its pedagogical implication in writing instructional practices. This research provided valuable insights for preparing students to develop good writing literacy skills to pursue academic, professional, social, and personal benefits.

Keywords: Narrative pedagogy, undergraduate students, writing literacy development

1. Introduction

The rapid advancement of information and technology occurs concurrently with the evolution of literacy. This dual development plays a crucial role in shaping people's reading and writing. The Indonesian government has a big concern in literacy as regulated in the Regulation of the Ministry of Education and Culture Number 23 year 2015 aiming to ensure that Indonesian students attain literacy in various forms. Being literate or illiterate is defined by possessing the ability for critical thinking, which includes basic reading and writing skills, considered one of the essential 'life skills.' Literacy is regarded as a fundamental right for every individual since it enables people to comprehend information presented in written or spoken form. Consequently, the Indonesian government places significant emphasis on enhancing literacy among the population from elementary to tertiary education levels.

Literacy in writing has been examined from three angles including writing skills and Literacy, digital environment and reading-writing literacy skills, and reading-writing relationship. Firstly, writing skills can be the predictive factor of literacy. Children deemed susceptible to future literacy challenges fall behind their classmates in a variety of early writing abilities [1] and preschool writing interventions enhanced children's early literacy outcomes [2]. Secondly, their digital environment can positively affect reading and writing literacy skills [3]. Thirdly, beyond the impacts of early literacy skills, the spelling that preschoolers invented provided a distinctive variance to later reading outcomes [4]. Moreover, reading and writing are inter-related and develop together because they share several skills and knowledge [5], and the emergent literacy skills presented significant positive correlations with reading and writing performance [6]. However, a comprehensive study on how undergraduate students develop their good writing literacy skills is hardly found.

Despite the importance of writing literacy, the portrait of literacy development and support of students in both language and science departments are under-researched. Moreover, this study was also done in response to the gap identified by [7] stating that instructional designs and implementations, particularly for the development of language and literacy are urgently needed. Thus, to portray those having good writing literacy is worthy not only to map out their literacy competence but also beneficial to set instructional programs for teaching writing that can help undergraduate students to build good writing literacy skills. Accordingly, this article intends to portray the development of the undergraduate students' writing literacy, what factors affect the development of their writing literacy skills, and propose instructional techniques for teaching writing according to their successful stories of building and developing writing literacy skills.

[a]erna.iftanti@uinsatu.ac.id, [b]Nany_Soengkono@uinsatu.ac.id, [c]galuh.nashrulloh@uniska-bjm.ac.id

DOI: 10.1201/9781003684459-68

2. Methods

2.1. Research design

This study employed a qualitative approach through three steps, namely subject selection through a questionnaire test, in-depth interviews with the selected subjects, and observation. The population was students from the Language and Science Departments of a state Islamic University in Tulungagung with good writing literacy. The questionnaire test items were developed based on relevant theories about indicators, methods, and factors affecting writing literacy development and the result of the validity and reliability test using Cronbach Alpha is 0,873.

2.2. Subject selection

To identify students from the Language and Science Departments who were potential to have good writing literacy, a questionnaire test was conducted to investigate their portrait in writing skills. This survey was distributed to 147 students of either gender enrolled in various programs within these two departments, including Biology, Mathematics, Physics, Chemistry, English Education, Arabic Education, and Indonesian Language Education. The distribution of the test was facilitated with the assistance of the Study Program Coordinators. They took the test voluntarily. This ensured that the survey was completed by individuals who were genuinely committed to sharing their writing literacy experiences. The survey test resulted in eight potential subjects who have good writing literacy. Then the in-depth interview carried out to five of them who are genuinely interested in sharing and providing insights into their writing literacy development.

2.3. Data collection method

The primary method employed for gathering data in this research involved conducting in-depth interviews with the chosen participants in the survey test. The interviews were conducted following mutual agreements between the interviewers and interviewees, which encompassed decisions regarding the mode of the interview and the timing and location. Consequently, interviews were conducted in a face-to-face manner or remotely via online methods such as video calls, accommodating the respondents' availability and preferred settings. This data collection phase spanned from the end of June to July 2023. The observation toward their writing skills was also done to find the credible and dependable data. Therefore, these data are also considered as the primary dataset for this research.

2.4. Data analysis

The data gathered through questionnaire tests were subject to coding and categorization. Irrelevant or redundant data were eliminated. This process of coding and categorization yielded potential subjects for further interviews and observations. The data obtained from these subsequent interviews and observations underwent qualitative analysis, incorporating relevant theories on literacy development. Following data analysis, the research was to provide insights into the literacy development of undergraduate students in the Language and Science Departments. These insights would be instrumental in shaping appropriate instructional programs that can support the academic vision of these study programs, positioning them as the basis to propose the instructional techniques of teaching writing for helping students in building and developing good literacy skills.

3. Results and Discussions

3.1. The portrait of undergraduate students' writing literacy development

This study reveals that students' development of writing literacy can be characterized by five fundamental components as seen in Table 68.1, namely good writing habits, nurturing writing habits from a young age, possessing the capacity for personal writing skill development, understanding the relationship between technology and writing, and engaging in collaborative writing.

One of the important aspects of the portrait of students' writing literacy involves having strong writing habits. This is demonstrated by their ability to write for various purposes essential in today's diverse and multifaceted communication landscape such as for personal or educational goals as

Table 68.1. Portrait of the undergraduate students' reading literacy development

No	*Components*	*Sub-Components*
1.	Having good habits of writing	Keeping writing for various purposes
2	Nurturing Good habits of writing since childhood	Writing as a means of self-expression Like to take notes Writing reading summary Doing process of writing: Drafting, writing, and revising
3.	Having a good capacity for personal writing skill development	Joining some writing clubs to support the development of writing literacy
4.	Understanding the relationship between technology and writing	Making use of technology to write Using smartphone /technology to write)
5.	Engaging in collaborative writing	Students' collaboration for writing Teachers' invitation to collaboratively write

Source: Author.

explicitly stated in a student's quoted interview: '*I keep writing for my purpose and for doing school assignments, but then I believe that through writing I can more easily feel and express my thoughts, opinion, and perception mainly through online media.' Ajg Sri.* The ability to write effectively for various purposes demonstrates one's versatility as a writer and reflects their capability to excel in a dynamic and interconnected world.

Nurturing the good habit of writing from an early age is another crucial indicator of students with good literacy in writing. Such a component can be built through four sub-components. The first is putting down anything like expressing daily experience, actualizing happiness or sadness, and taking some notes in writing '*..., the habit that I have applied from an early age is writing. Since childhood, I have been fond of writing whatever I can put in writing or narrative form.' Lina.* Writing from an early age equips students with the adaptability and versatility needed to excel in an information-rich, communication-driven world, making it a valuable predictor of their overall writing literacy. Similarly, a study indicated that children at risk for later literacy difficulties lag behind their peers in a range of early writing skills, including name writing, letter writing, invented spelling, and story composition in preschool [1]. The third, the students' preference to take notes and summaries from reading built good writing habit: '*...indeed I like to write or take notes even though my writing is not very good and neat.' Ulfah.* Taking notes after reading has been the regular activity of students with good literacy in writing. This portrait supports a study by Kim and Graham [8] revealing that reading comprehension was related to writing quality and those with strong reading habits have strong writing habits [9]. The forth, following the process of writing like drafting, writing, and revising is found to be significant to build good writing literacy as clearly stated in this excerpt: '*When I write, of course, sometimes there are still words that are not following the rules. Therefore, I usually make a draft every time I finish writing and revise it according to the linguistic dictionary.*' (Putra). These four sub-elements indicate that good literacy in writing and the continuity of the writing exercises should be developed in a comprehensive process of writing nurtured from an early age.

The other portrait of students with good writing literacy is having a good capacity for personal writing skills by joining a writing club. They gain access to a supportive and collaborative environment that fosters further development of their writing skills. Students with good literacy in writing is also portrayed as having an understanding the relationship between technology and writing. They can utilize online applications and social media platforms like Wattpad for various purposes, such as writing, expanding horizons, staying updated with technological advancements, facilitating educational endeavours, and expressing knowledge. This is explicitly stated in a student's statement: '*I learned to write for the first time while trying to enter the world of oranges. Or what is commonly known as wattpad. After several attempts, I finally wrote a youth novel on the orange lapak. ... I think this is a great development in terms of my writing.' (Nia).* A study proved that Wattpad is useful to increase the students' motivation in giving better engagement to reach the literary competencies [10]. Online applications and social media have transformed the way they connect, access information, write, and share their writings like novels. This finding is parallel to Alneyadi et al. [3] who proved that the use of digital learning platforms (Teams, platform LMS, Nahla and Nahil) led to an improvement in the students' levels of achievement in reading and writing. These digital tools have revolutionized the way students build their talent and skills of writing. communication, enabling global connections with ease and offering invaluable support in the digital era.

The last portrait of students with good writing literacy is writing collaboration. Their habit of regular writing in several media like notebooks has been built through collaborative writing short stories, book chapters, and articles for publication as explicitly stated in the student's quoted interview below:

'The development of my writing literacy began when I was in kindergarten... Apart from handwriting, during my college days I received encouragement from my lecturers to collaborate with my lecturers and started trying to make various scientific articles and book chapters and then publish them.' Risma.

'...Then in 2018 I wrote short stories in collaboration with an Indonesian teacher when I was in high school.' (Nia)

Their interest and courage in collaborative writing with their teachers and lecturers indicate their good literacy in writing.

In short, this study found that students who have good literacy in writing are characterized with having and building good habit in writing since childhood, having good capacity of writing skills, making use of technology as a means of writing, and having courage to do collaborative writing. The efforts to keep improving writing skills have been carried out over time and writing has been his habit. It has also been felt like nothing without writing.

3.2. The supporting factors in the development of the undergraduate students' writing literacy

This study revealed four distinct factors that influence students' writing literacy development. These factors encompass internal factors, external elements, literary works, and technology (see Table 68.2).

There are five essential points within the realm of internal factors. The first is the inspiration from self-reading. It is found that self-reading like reading stories is claimed to contribute to build the students' love of writing which can further improve their good literacy in writing. This is clearly stated in the students' quoted interview: '*I tried to write because*

Table 68.2. The supporting factors of the undergraduate students' writing literacy

No	*Components*	*Sub-components*
1.	Internal Factors	inspiration from Self-reading
		Self-awareness to develop writing literacy
		The belief that writing can have positive impacts on readers
		Willingness to integrate education and technology through writing
		Active participation in training on academic writing
2.	External Factor	Educational achievement
		Writings on media social
		Lecturers' support for collaborative writing
		Joining to some writing clubs
3.	Literary works	Literary works like narratives
4	Technology	Technology as media to write
		Writing by using technology media for various purposes

Source: Author.

I was inspired by several stories I read.' (Nia). This finding shows that self-reading and self-awareness play a crucial role in fostering the development of strong writing literacy that support a study by [11] revealing that self-regulated Reading to Write (R2W) strategy use had positive impacts on writing competence. Reading literacy also has a very strong correlation with writing literacy [12]. Self-reading entails a conscious effort to engage with various written materials that serve as a wellspring of inspiration for budding writers. Immersing in various texts can broaden their knowledge and expose them to different writing styles and techniques. Complementing this is self-awareness, a pivotal component in honing writing skills. Self-awareness as reflected in a student's quoted interview: '*at first it was just for fun, but over time I attended several writing webinars ...which made me realize that writing will be eternal. That's why I try to develop my writing literacy to benefit readers and perpetuate every knowledge I seek.' (Nia),* involves a deep understanding of their willingness, strengths, and weaknesses as a writer, enabling the use of their personal experiences, emotions, and perspectives to craft authentic and compelling narratives. Camacho, Alves, and Boscolo [13] in their study also found that motivation and writing performance have moderate positive associations. Moreover, the belief that writing can have a positive impact on readers contributes to developing good writing literacy habits as reflected in the student's quoted interview: '*..., I am also learning to continue to develop my writing so that it can have a positive impact and motivation for my readers and audience.' Nia.* The willingness to integrate education and technology through writing is also another important point that can develop good writing literacy as clearly stated by the student: '*The supporting factor for developing my writing literacy is because I want to collaborate education with technology.' Putra.*

In terms of external factors, some subfactors like educational achievement, writing on social media, and lecturers' support to write collaboratively have built good writing literacy. Students who are eager to make a positive impact on their readers through their work are increasingly aware of the educational achievement as stated in a quoted student's interview: '*The first is for the benefit of my own education.' Ajg Sri.* Writings on social media have also inspired them to get ideas for writing: '*But not infrequently when I go on social media, there I find lots of beautiful and very deep meaningful writings which in the end also become an inspiration for me.' Ajg Sri.* Such an external factor contributes to develop students' good writing literacy. Lecturer's support for collaborative writing also contributes to developing their writing literacy. This is clearly stated in the student' quoted interview: '*...because I was a teaching assistant, I was invited to carry out various research and then make products in the form of scientific articles to be published in journals and make book chapters...' Risma.* Studies also revealed that the collaborative writing technique produces better writing ability [14], and improved the EFL students' writing performance, writing self-regulation, and writing self-efficacy [15]. Another contributing external factor is the writing club which can create a good writing atmosphere. By joining the clubs, students do not only get knowledge for writing but also experience of writing either fiction or nonfiction which can be published on their social media. '*Meanwhile, when I started joining the Indonesian Writing Institute, East Java Province, I gained writing knowledge related to the correct vocabulary for writing non-fiction novels. Apart from that, I also wrote on websites related to biology...and I channeled this knowledge (a shorter amount) on my Instagram social media related to embryo development, mushrooms in tempeh and so on.' Lina.* They are encouraged to write because their surroundings have made them write whether because of completing the writing assignment given by the tutors or because of their own willingness to write.

Furthermore, literary works, such as stories or narratives, also serve as instrumental elements in the development of students' writing literacy. Stories and novels have long been acknowledged as essential tools in nurturing students' writing skills as claimed in the student's quoted interview: '*The supporting factors for writing literacy that I applied from childhood to school were only limited to stories or narratives.' Lina.* These narratives offer a rich tapestry of language, style, and creativity, furnishing students with both inspiration and practical guidance in honing their own written compositions. As proved by Graham [16] that reading and writing have a reciprocal relation. Additionally, reading habits in common [17] and reading literary works specifically [18] have a significant influence on students' writing skills. By immersing themselves in the world of storytelling, students not only expand their vocabulary and grasp of grammar but also gain valuable insights into the nuances of narrative structure, character development, and thematic exploration.

The final supporting factor is technology, that is utilized as a medium for writing across various purposes. Within the educational landscape, an array of media platforms- like Instagram and X- and tools like smartphones have emerged as potent aids in fostering students' writing skills. This finding is reflected in the student's excepts: '*I used a smartphone for the first time in 2016, initially only using a chat application to communicate with my parents and friends. Then I made a social media account...' Nia. The supporting factors for writing literacy that I applied from childhood to teenage have been ...from learning to write using technology media. Lina.* These quotations indicate that technology like chat applications has made them get used to writing because they express and put their messages down through their social media. Wen and Walters [19] revealed that technology has influenced students' writing by integrating many learning modes (visual, auditory, and reading/writing). What they have written with their social has proved to be able to help them improve their writing literacy. Social media platforms and interactive websites engage students in education communication management [20], enhance their educational experience [21], and have positively impacted students' writing styles [22], because through social media like Facebook, they can engage in practice writing [23] and are encouraged to share knowledge and improve their learning performance [24]. Technology is crucial in disseminating information and enhancing learning experiences in today's digital age.

3.3. The proposed instructional techniques for writing

Following the portraits of the development of undergraduate students' good writing literacy and factors affecting its development, some instructional programs for writing literacy can be proposed, namely Process-Product-based Writing Instruction (PPWI), Digitally Project-based Writing Program (DPWP), Writing Clubs, and Collaborative Writing.

Process-Product-based Writing Instruction (PPWI) is a structured approach where the focus is on creating a final written product, such as an essay, report, or research paper, through a systematic and guided process. Studies proved that a process-oriented approach to writing effectively develops writing skills [25, 26]. Meanwhile, Product-based Writing is also proven to trigger classroom dynamics and a positive atmosphere [27] and can yield success for each student who is taught writing in a short time [28]. This instructional method emphasizes planning, drafting, revising, and editing, which helps students develop writing literacy skills step-by-step. By focusing on the end product, students are encouraged to analyze and emulate exemplary writing structures, grammar, and style. El Ouidani and Madaoui [29] in their study revealed that even though process approach is more effective than the product writing approach in enhancing students' writing, both approaches can enhance L2 learners' writing performances. So, through combining Process and Product approaches, they build a strong foundation in organizing thoughts and ideas coherently, enhancing both technical writing skills and critical thinking.

Digitally Project-based Writing Program (DPWP) is another instructional technique that can be proposed to help students improve their writing literacy skills. As revealed in this study that students with good writing literacy skills make use of technology tools to naturally train themselves in writing for various purposes like self-expression through writing novels, short stories; academic purposes like writing assignment completion; and communication with their relatives make them. A study by Nobles and Paganucci [30] uncovered that the use of digital tools helps students compose quality writing and that writing in online environments fosters enhanced writing skill development. Technology-mediated writing instruction resulted in improving students' composing processes and writing skills as well as their knowledge and use of new literacies [31]. Accordingly, DPWP can be considered as meaningful technique to develop students' writing literacy and is parallel with the characteristics of the z generation who are native digital.

A Writing Club is another instructional technique for writing classes that can provide a dynamic platform to build and enhance students' writing literacy. In such a technique, students engage in regular, diverse writing activities, ranging from personal essays to creative storytelling and peer editing. These activities encourage students to explore their writing styles, practice new techniques, and receive constructive feedback from peers and mentors. As they work on varied projects, students improve their vocabulary, grammar, and comprehension skills, essential components of writing literacy. Additionally, attending English Clubs fosters motivation and accountability [32] and joining National Journal club improves students' confidence with the critical appraisal process [33]. Through consistent participation, students not only strengthen their writing abilities but also develop a deeper understanding of the writing process, preparing them for both academic and personal writing endeavours.

Collaborative Writing (CW) is a powerful strategy for developing writing skills. In this instructional technique, teachers work closely with students through guided-writing sessions, helping them construct ideas, organize content, refine grammar, proofread, and review draft. This interaction can foster a hands-on learning experience where students actively participate in each phase of the writing process. Such collaboration not only addresses students writing challenges but also builds students' confidence and motivation to write as being revealed in this study. CW strategy has helped students generate their writing ideas and activate the students' background knowledge [34] and had great effects on students' writing fluency [35]. Teacher-student collaborative writing, for example, provides a supportive framework that encourages transforming writing from a solitary task into an interactive, skill-building activity that strengthens literacy foundations for specifically the success in writing.

4. Conclusion and Suggestion

This research study uncovered three key findings, which include the portraits of undergraduate students' writing development, factors affecting writing literacy development, and the proposed instructional techniques of writing. The results of this study portrayed that those with strong writing literacy development are indicated by having good habits of writing for various purposes that are nurtured from an early age, taking writing as a means of self-expression, writing reading summaries, using technology as means to writing, and engaging in collaborative writing. Additionally, it explored the factors that support their writing literacy development, that is, internal factors covering inspiration of self-expression, self-awareness to develop their writing literacy, positive belief that writing can enhance their reading skills, and active participation in training on writing; external factors like educational achievement, writing on social media, teacher's support to collaborative writing, and joining writing clubs. Literary works and technology used to write for various purposes contribute to students' writing literacy development as well. Finally, referring to these findings, some instructional techniques to teach writing can be proposed: product-based writing instruction, digitally project-based writing programs, writing clubs, teacher-student collaborative writing, writing clubs, writing competitions, and writing rewards. In summary, this study provided insights into pedagogical implications of how to help learners have good writing literacy for seek of preparing them with living in Society 5.0.

References

[1] Thomas, L. J., Gerde, H. K., Piasta, S. B., Logan, J. A., Bailet, L. L., & Zettler-Greeley, C. M. (2020). The early writing skills of children identified as at-risk for literacy difficulties. *Early Childhood Research Quarterly*, *51*, 392–402. https://doi.org/10.1016/j.ecresq.2020.01.003

[2] Hall, A. H., Simpson, A., Guo, Y., & Wang, S. (2015). Examining the effects of preschool writing instruction on emergent literacy skills: A systematic review of the literature. *Literacy Research and Instruction*, *54*(2), 115–134. https://doi.org/10.1080/19388071.2014.991883

[3] Alneyadi, S., Abulibdeh, E., & Wardat, Y. (2023). The impact of digital environment vs. traditional method on literacy skills; reading and writing of Emirati fourth graders. *Sustainability*, *15*(4), 3418. https://doi.org/10.3390/su15043418

[4] Hand, E. D., Lonigan, C. J., & Puranik, C. S. (2024). Prediction of Kindergarten and First-Grade Reading Skills: Unique Contributions of Preschool Writing and Early-Literacy Skills. *Reading and Writing*, *37*(1), 25–48. https://doi.org/10.1007/s11145-022-10330-1

[5] Kim, Y. S. G. (2020). Interactive Dynamic Literacy Model: An Integrative Theoretical Framework for Reading-Writing Relations. In: Alves, R. A., Limpo, T., Joshi, R. M. (Eds.), *Reading-Writing Connections. Literacy Studies* (Vol 19). Springer, Cham. https://doi.org/10.1007/978-3-030-38811-9_2

[6] Castro, D. A. S., & Barrera, S. D. (2019). The contribution of emergent literacy skills for early reading and writing achievement. *Trends in Psychology*, *27*(2), 509–522. https://doi.org/10.9788/TP2019.2-15

[7] Li, J., Brar, A., & Roihan, N. (2021). The use of digital technology to enhance language and literacy skills for Indigenous people: A systematic literature review. *Computers and Education Open*, *2*, 100035. https://doi.org/10.1016/j.caeo.2021.100035

[8] Kim, Y.-S. G., & Graham, S. (2022). Expanding the Direct and Indirect Effects Model of Writing (DIEW): Reading–writing relations, and dynamic relations as a function of measurement/dimensions of written composition. *Journal of Educational Psychology, 114*(2), 215–238. https://doi.org/10.1037/edu0000564

[9] Quadir, B., & Chen, N. S. (2015). The effects of reading and writing habits on learning performance in a blog learning environment. *The Asia-Pacific Education Researcher*, *24*, 635–644. https://doi.org/10.1007/s40299-014-0210-5

[10] Al Falaq, J. S., Suprayogi, S., Susanto, F. N., & Husna, A. U. (2021). Exploring The Potentials of Wattpad for Literature Class. *Indonesian Journal of Learning Studies (IJLS)*, *1*(2), 98–105.

[11] Bai, B., & Wang, J. (2023). Conceptualizing self-regulated reading-to-write in ESL/EFL writing and investigating its relationships to motivation and writing competence. *Language Teaching Research*, *27*(5), 1193–1216. https://doi.org/10.1177/1362168820971740

[12] Shara, A. M., Andriani, D., Ningsih, A. W., & Shinoda, K. (2022). Correlating reading and writing literacy in jurior high school Pematangsiantar: Correlating reading and writing literacy. *International Journal of Curriculum and Instruction*, *14*(2), 1457–1470. doi: http://dx.doi.org/10.31327/jee.v5i2.1249

[13] Camacho, A., Alves, R. A., & Boscolo, P. (2021). Writing Motivation in School: A Systematic Review of Empirical Research in the Early Twenty-First Century. *Educ Psychol Rev*, *33*, 213–247. https://doi.org/10.1007/s10648-020-09530-4

[14] Risman, K., & Rozimela, Y. (2019, August). The Effect of Collaborative Writing and Reading Habit toward Students' Writing. In *1st International Conference on Education Social Sciences and Humanities (ICESSHum 2019)* (pp. 310–315). Atlantis Press. https://doi.org/10.2991/icesshum-19.2019.50

[15] Rahimi, M., & Fathi, J. (2022). Exploring the Impact of Wiki-Mediated Collaborative Writing on EFL Students' Writing Performance, Writing Self-Regulation, and Writing Self-Efficacy: A Mixed Methods Study. *Computer Assisted Language Learning*, *35*(9), 2627–2674. https://doi.org/10.1080/09588221.2021.1888753

[16] Graham, S. (2020). The sciences of reading and writing must become more fully integrated. *Reading Research Quarterly*, *55*, S35–S44. https://doi.org/10.1002/rrq.332

[17] Linuwih, E. R., & Winardi, Y. K. (2020). The influence of reading habit on students' writing skill. *Anglo-Saxon: Jurnal Ilmiah Program Studi Pendidikan Bahasa Inggris*, *11*(1), 37–47.

[18] Abdalrahman, K. K. (2021). Teaching and learning writing skills through literature. *Canadian Journal of Language and Literature Studies*, *1*(2), 1–10. https://doi.org/10.53103/cjlls.v1i2.11

[19] Wen, X., & Walters, S. M. (2022). The impact of technology on students' writing performances in elementary classrooms: A meta-analysis. *Computers and Education Open*, *3*, 100082. https://doi.org/10.1016/j.caeo.2022.100082

[20] Ohara, M. R. (2023). The Role of Social Media in Educational Communication Management. *Journal of Contemporary Administration and Management (ADMAN)*, *1*(2), 70–76. https://doi.org/10.61100/adman.v1i2.25

[21] Ramzan, M., Javaid, Z. K., & Fatima, M. (2023). Empowering ESL students: Harnessing the potential of social media to enhance academic motivation in higher education. *Global Digital & Print Media Review, VI*, *2*, 224–237. http://dx.doi.org/10.31703/gdpmr.2023(VI-II).15

[22] Muftah, M. (2024). Impact of social media on learning English language during the COVID-19 pandemic. *PSU Research Review*, *8*(1), 211–226. https://doi.org/10.1108/PRR-10-2021-0060

[23] Putri, N., & Aminatun, D. (2021). Using Facebook To Practice Writing Skill: What Do the Students Think?. *Journal of English Language Teaching and Learning*, *2*(1), 45–50. https://doi.org/10.33365/jeltl.v2i1.852

[24] Hosen, M., Ogbeibu, S., Giridharan, B., Cham, T. H., Lim, W. M., & Paul, J. (2021). Individual motivation and social media influence on student knowledge sharing and learning performance: Evidence from an emerging economy. *Computers & Education*, *172*, 104262. https://doi.org/10.1016/j.compedu.2021.104262

[25] Khaki, M., & Tabrizi, H. H. (2021). Assessing the Effect of Direct and Indirect Corrective Feedback in Process-Based vs Product-Based Instruction on Learners' Writing. *Language Teaching Research Quarterly*, *21*, n36–53. https://doi.org/10.32038/ltrq.2021.21.03

[26] Kadmiry, M. (2021). The comparison between the process-oriented approach and the product-oriented approach in teaching writing the case of Moroccan EFL students in preparatory classes for the grandes ecoles. *Arab World English Journal (AWEJ)*, *12*. https://dx.doi.org/10.24093/awej/vol12no1.14

[27] Rungwaraphong, P. (2021). Inductive teaching approaches in business English writing in an EFL context: Paper-based and product-based instructions. *English Language Teaching Educational Journal*, *4*(1), 34–48. doi.org/10.12928/eltej.v4i1.3227

[28] Kara, S., & Abdulrahman, S. A. (2022). The Effects of Product Approach on Language Preparatory School StudentsWriting Score in an Academic Writing Course. *Canadian Journal of Language and Literature Studies*, *2*(4), 45–65. https://doi.org/10.53103/cjlls.v2i4.57

[29] El Ouidani, Y., & Madaoui, R. (2024). Enhancing L2 Learners' Writing: Impact of Product and Process Approaches. *European Journal of Foreign Language Teaching*, *8*(1), 134–153. http://dx.doi.org/10.46827/ejfl.v8i1.5313

[30] Nobles, S., & Paganucci, L. (2015). Do digital writing tools deliver? Student perceptions of writing quality using digital tools and online writing environments. *Computers and Composition*, *38*, 16–31. https://doi.org/10.1016/j.compcom.2015.09.001

[31] Williams, C., & Beam, S. (2019). Technology and writing: Review of research. *Computers & Education*, *128*, 227–242. https://doi.org/10.1016/j.compedu.2018.09.024

[32] Rahmiani, N., & Malena, N. M. (2020). The Model of Students English Club Contribution to Writing Motivation and Punctuation Awareness Improvement. *Journal of English Educational Study (JEES)*, *3*(1), 71–80. https://doi.org/10.31932/jees.v3i1.672

[33] Gokani, S. A., Sharma, E., Sharma, T., Moudhgalya, S. V., Selvendran, S. S., & Aggarwal, N. (2019). Impact of a national journal club and letter writing session on improving medical students' confidence with critical appraisal. *Advances in Medical Education and Practice*, 1081–1087. https://doi.org/10.2147/AMEP.S235260

[34] Anggraini, R., Rozimela, Y., & Anwar, D. (2020). The effects of collaborative writing on EFL learners' writing skills and their perception of the strategy. *Journal of Language Teaching and Research*, *11*(2), 335–341. http://dx.doi.org/10.17507/jltr.1102.25

[35] Pham, V. P. H. (2021). The effects of collaborative writing on students' writing fluency: An efficient framework for collaborative writing. *Sage Open*, *11*(1), 2158244021998363. https://doi.org/10.1177/2158244021998363

69 Skill competency test for vocational high school (SMK) students encounter labor market challenges

Tetty Yuliaty[1,a], *Wahyu Ario Pratomo*[1,b], *Suhendri*[2,c], *and Andika Fajar Putra*[1,d]

[1]Universitas Sumatera Utara Medan, North Sumatra, Indonesia
[2]Universitas Dharmawangsa, Indonesia

Abstract: To ascertain whether or not graduates from vocational high schools have market-related challenges after undertaking a competency test, this study seeks to look into the policy as it is being executed in the schools. This descriptive qualitative research utilized triangulation as a multi-method data collection approach which included observation, interviews, and documentation. The subjects of this study included students from vocational high schools in 33 districts and cities in North Sumatra Province, along with teachers, external examiners, and test administrators. Based on this study, it was found that the schools set criteria that are easy to implement. Vocational school graduates can find employment even with a sluggish economy as long as they are prepared for the market from the onset.

Keywords: Expertise competency test, expertise competency test, vocational school labor market

1. Introduction

According to Kamdhi and Jedun [5], the Specialized Vocational High Schools SMK have the singular aim of producing effective workers in accordance with the requirements of the industry. As noted by Rahmadhani et al. [7], these schools offer curriculum modules that prepare students for a specific occupation. Sandra and Supriyati [8] state that SMK is a subdivision of secondary education which develops the requisite knowledge, skills, and work-ready disposition aimed at particular career pathways. During industrialization, there is increase in demand for skilled workers in Indonesia and their ability to embrace changes. As the qualification requirements in the labor market rise, so should the benchmark set for students [1]. Among the competencies for students to possess today are: (1) capability of critical and problem-solving skills, (2) creativity, (3) communication, and (4) ability to work with others [6]. In 2022, the BPS data suggested Vocational High School (VHS) graduates as the highest unemployed demographics, with the Central Bureau of Statistics (2024) reporting a marginal decrease in the educational based unemployment rate from 9.42% in 2022 to 9.3% in 2023. Then there is [7], who points out the glaring disparity on the goals attributed to vocational high schools of creating an immediately employable population. The employment options are scarce, and the level of aptitude among the graduates is dismally low relative to the requirements of their field of study. The lack of quality in SMK graduates stems from multiple factors rather than just the students themselves. There is need to focus on the competency of educators and how relevance their competencies are to the students' graduates [7]. Consequently, as mandated in Government Regulation Number 74 of 2005, all students as well as teachers must possess relevant academic and standard competency qualifications along with teaching certificates [2]. In addition, education that is responsive to training requirements for skills needed from the students is necessary to ensure that they can adapt to changes in the industry [9]. The current implementation of the skills competency test continues to have material and technical gaps. Teacher preparation, curriculum design, and even evaluation processes still pose issues that must be addressed [2]. From field observations, there appear to be several issues regarding the policy of defining competency assessors from BNSP not fully implementing assessing competencies. There are incomplete information sets which give rise to a rather careless approach to test materials, test questions, and er for assessment resources. The guidelines provided have yet to be adhered to, thereby making the implementation of the Expertise Competency Test (UKK) lack the requisite quality. There are additional roadblocks whereby verifiers supervising the field lack requisite qualifications, which are essential in determining the TUK (tempat uji kompetensi or competency test site for TUK) and other peripheral areas. Although North Sumatera is host to a myriad of with multiple vocational schools (SMKs), they have set a regional standard requiring all SMKs to conduct Skills Competency Tests (UKK). Unfortunately, the SMKs' level of preparedness and overall implementation of the Skills Competency Test is subpar. The geographical position of the vocational schools determines

[a]tettyyuliaty@usu.ac.id, [b]wahyu@usu.ac.id, [c]Medan Suhendri165@dharmawangsa.ac.id, [d]andikafajarputra@usu.ac.id

DOI: 10.1201/9781003684459-69

the location and infrastrucuture of the SMKs while internal conditions and support from stakeholders also play a role. The SMKs are still lagging behind in infrastructure development when it comes to districts as opposed to provincial cities. Furthermore, a majority of the schools do not design a curriculum which is appropriate to the contours of the labor markets competency benchmarks. This study is relevant in order to ascertain whether or not the policy on competency tests have been implemented appropriately, particularly for vocational school graduates in North Sumatra who are expected to be competent in their skillset vis-a-vis the current labor market demands.

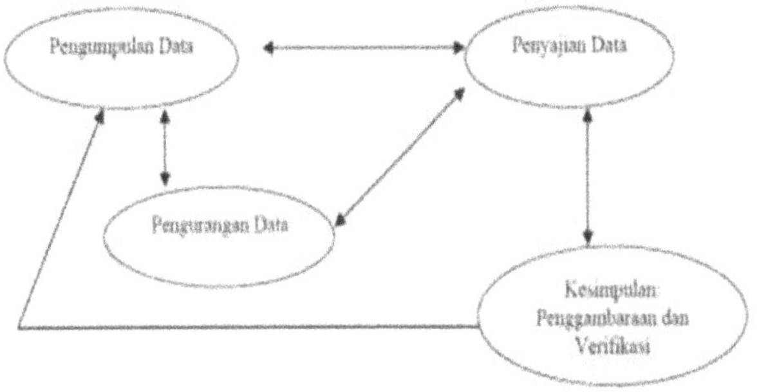

Figure 69.1. Teknik analisis data miles dan Huberman.

Source: Author.

2. Method

Methods such as participant observation, in-depth interviews, and meticulous record-keeping comprise this descriptive qualitative study. The pre-survey was conducted in April 2024, and the research in August at SMK schools in North Sumatra that have conducted the Office Management Automation Expertise Competency Test in 2024. Determination of the subject of this research using purposive technique, namely the technique of determining the subject using certain considerations [10] (p. 85). The criteria in this research subject are: (1) Teacher as internal examiner, (2) Principal, (3) External examiner assessor (4) and Students In order for this research to be scientifically accountable, researchers need to ensure that the data is valid by employing triangulation procedures. The term 'triangulation' refers to a method of data collection that uses multiple methods and sources to verify and validate information. Three methods were employed to gather data for this investigation, notably frank observation, structured interviews and unstructured interviews and documentation as data sources [10] (p. 241). This study will employ a triangulation technique that makes use of sources by comparing interview results with documentation, data collected from one informant with another, and results obtained via observations. By comparing results from multiple sources using diverse methods, or 'triangulation,' researchers can determine how reliable a set of data is. In addition, a triangulation test is used to verify the reliability of the data, namely by comparing it to other sources, in order to produce more accurate study results (Figure 69.1).

Qualitative and quantitative information were gathered for this study using a mixed-method approach. with a sequential explanatory design, where quantitative collection and analysis were carried out in the first stage, then continued with the collection and analysis of qualitative data in the second stage in order to be able to collect quantitative data. Qualitative research was taken from interview and documentation instruments, while quantitative research was taken from observation and questionnaire instruments. Quantitative Data Analysis, for this study, the data was analyzed using descriptive statistics, namely the mode, median, and mean, on information gathered via interviews, documents, observations, and questionnaires. The Analysis of Qualitative Data Using the data analysis method developed by Miles and Huberman, we examined the interview and documentation results. The following are the procedures for data analysis: (1) Reducing data levels; (2) Presentation of data; (3) Drawing conclusions.

3. Results and Discussion

Based on the responses to surveys sent to principals, internal examiners, and classroom teachers, students and external examiners or assessors from 32 city districts in North Sumatra and conducting in-depth interviews with random regional representatives, it can be seen that respondents are dominated by groups with a young age range, where most have completed undergraduate degrees, and have quite extensive teaching experience (range 5–16 years). Based on the data obtained, the following data were obtained: (1) 95.6% of the respondents agreed that the UKK is effective in measuring students' competencies in SMK; (2) 50% of the total respondents considered that the facilities and infrastructure in the implementation of UKK are inadequate and not standardized; (3) 50% stated that the UKK question materials provided by the center are not appropriate; (4) 66.2% agreed that the UKK assessment sheet needs to be improved. The survey data obtained shows that there is a strong urge to evaluate the implementation of the UKK; (5) 72.1% of students were unable to answer UKK questions without guidance or training. These results indicate the need for proper guidance to students before facing the competency test; (6) 55.9% of respondents stated that changes are needed to the UKK implementation schedule; (7) The majority of respondents chose the priority for creating competent graduates is competent teachers, followed by practical facilities and an appropriate curriculum; (8) 67.2% of respondents said that the UKK system needs to be changed to more effectively follow student competencies. Meanwhile, the results of interviews with external examiners revealed that: (1) External examiners invited by SMKs in carrying out the UKK are mostly from industry, and do not have relevant work experience in accordance with the field tested at school, so the quality of

UKK implementation is still in doubt; (2) An official government verification team has conducted the review, and the individuals conducting the review are partially in line with the findings, which may involve the participation of independent experts in the subject; (3) Competency test venues have been provided so that the implementation of the UKK can be carried out but seem less ready; (4) Some schools still have inadequate facilities and infrastructure; (5) The question material provided by the center seems inappropriate and has many mistakes, where the question material is almost the same every year with minor modifications; (6) This year's quality question packages are lower than the quality of previous years; (7) The assessment format is not provided with complete information. In addition, it is clear from the interviews with internal school inspectors that:

(1) Implementation of competency tests carried out in accordance with applicable regulations, where internal examiners and external examiners must have in-depth information and comprehension of the content included in the OTK competency tests. (2) In order to enhance the reliability and validity of the UKK, it is necessary to confirm that competency testing facilities and infrastructure are available and conducted in compliance with relevant rules. (3) The competency test venue is good and is carried out in the OTKP laboratory in several schools that have office laboratories. (4) Facilities and infrastructure are complete so that the implementation of the Expertise Competency Test (UKK) runs well but there are still schools that are not adequate. (5) The question material provided must be revised because there are several mistakes in the questions and they seem original. (6) The UKK question package for 3 years is almost the same, the school has never changed the questions and only revised them. (7) The assessment format is good enough but needs to be arranged more appropriately.

Given the above, it should come as no surprise that the UKK is an assessment system put in place as part of the state's endeavours to guarantee that vocational education is still lacking in quality. In accordance with government policy in the guidelines or regulations regarding the skills competency test, it is known that the objectives of the Competency Test are: (1) Measuring the competency achievements of vocational students who have completed the learning process in line with the competency-based instruction received. (2) Make it easier for students enrolled in vocational programmes to earn a certificate of competency when they finish their coursework. (3) In line with the Indonesian National Qualification Framework, optimise the implementation of competency accomplishment for SMK graduates who are aiming for certification. (4) Encouraging SMK collaboration with business and industry to provide competency assessments that meet the demands of both sectors (DUDI) [4]. So, it is clear from the survey and interview data that SMK in North Sumatra has not been able to accomplish the competency test's objectives since implementing UKK. It is well-known that there has been the deployment of competency test activities for experts in various locations of North Sumatra. are still many crucial aspects that need attention and need improvement both from examiners who still use from industry, and there are still minimal examiners who come from externally who have competency certificates from the National Professional Certificate Agency (BNSP) with the golden Garuda Bird logo and are still valid and active, according to government guidelines. Not to mention the question material that is not suitable or outdated so that it can be updated and adjusted to the UKK standard. The implementation of verification has not been very appropriate in the appointment of personnel. So that the implementation of the UKK at this time can be said to have not been able to provide good news for SMK graduates, it needs to be changed to be better at measuring student abilities.

Vocational schools have a mission to form workers with qualified skills and are able to compete, for this reason collaboration from various parties is needed [5]. The collaboration of schools and government, industry and all parties to guide students and provide adequate facilities is an urgency that must be taken immediately. This must be done immediately to maintain a high standard of education so that students leave with the skills necessary to succeed in today's competitive employment environment. Good collaboration in the implementation of UKK will create a good system as well. The poor UKK system found in the vocational schools studied can be caused by each party not having the same vision [2], thus having an impact on the implementation of UKK that is not in accordance with standards. The implementation of UKK that meets the standards should be able to guarantee four competencies that must be possessed by vocational school students, in particular, the capacity for in-depth analysis and resolution of complex problems, original thought, effective expression of ideas, and teamwork.

4. Conclusion

There are a number of ongoing issues with the Expertise Competency Test (UKK) at SMK North Sumatra, including a lack of suitable infrastructure and facilities and deteriorating question material quality. External and internal examiners who are not yet BNSP certified are a major problem. Including aspects of assessment, the availability of infrastructure, and the involvement of the education office as a coach is still not optimal, such as selecting and supervising external examiners in ensuring the competence of vocational students encounter labor market challenges. With this condition, graduates of SMK students who have taken the Fearing that it will be difficult to compete in the job market with an Expertise Competency Test, all parties involved are being asked to take this matter very seriously and fix the issues that still arise. According to the results, educational institutions must be able to choose effective and efficient requirements to be implemented, vocational school graduates are able to face a tight job market, if they have prepared themselves and are used to being tested comprehensively from the beginning. The implementation of quality Vocational Competency Tests for Vocational Students is highly expected by all parties, both teachers, examiners and especially industry in order to be able to create competent, skilled and competent vocational school grads who can hold their own in the job market.

References

[1] Abdurrahman, P., & Muryanto, S. (2022). Evaluastion on the automotive skill competency test through 'discontinuity' model and the competency test management of vocational education school in Central Java, Indonesia. *Helivon*, 1–10.

[2] Basahil, A., Sabarun, S., & Oamariah, Z. (2023). Evaluating Competence- Based Vocational Education in Indonesia. *Jurnal Pendidikan dan Sastra Inggris*, *3*(3), 1–13.

[3] Direktorat Sekolah Menengah kejuruan, Kementrian Pendidikan dan Kebudayaan, Pedoman Penyelenggaraan Uji Kompetensi Keahlian Sekolah Menengah Kejuruan, 2023

[4] Ikram, S. (2017). Influence of the Average Value of Competency Test of Manual Accounting on the Avarage Value of Myob Competency Test. *International Journal of Scientific & Technology Research*, *5*(10), 367–371.

[5] Kamdhi, W., & Jedun, A. (2022). Evaluasi Pelaksanaan Uji Kompetensi Siswa SMK Kompetensi Keahlian Desain Pemodelan dan Informasi Bangunan pada Lembaga Sertifikasi Profesi P-1 di Daerah Istimewa Yogyakarta. *Jurnal elektronik Pendidikan Teknik Sipil*, *10*(2), 138–149.

[6] Koloway, R., Ronga, W., Sumual, H., olii, D., & Takaredase, A. (2024). The Influence of Vocational Theory Competency Test Results on Vocational Practice Competency Test Results. *Jurnal Pendidikan Teknik Elektro*, *4*(1), 1–8.

[7] Rahmadhani, S., & Suryati, L. (2022). Vocational High School Students Competency Needs to the World of Work. *Jurnal Mimbar Ilmu*, *27*(2), 349–355.

[8] Sandra, D., & Supriyati, Y. (2019). Evaluasi Program Uji Kompetensi Peserta Didik Keahlian Teknik Elektronika Industri di Kabupaten Bekasi dengan Model Evaluasi Discrepancy. *Jurnal evaluasi Pendidikan*, *10*(1), 38–46.

[9] Suminar, T., Wardani, S., Yusuf, A., Siswanto, Y., & Islahuddin, I. (2024). Determinant Model of Graduate Competence Partnership-based Life Skills Training in the Industry 4.0 Era. *Journal of Nonformal Education*, *10*(1), 195–209.

[10] Sugiyono, Metode Penelitin Pendidikan Pendekatan Kuantitatif, Kualitatif dan R & D, Bandung: Alfabbeta, 2012.

70 Implementation of the Penta Helix concept to increase climate awareness in Islamic schools through extracurricular activities

Efrita Norman[1,a], *Arman Paramansyah*[2,b], *Windi Megayanti*[3,c], *and Nicky Rosadi, S. S.*[4,d]

[1]Doktoral UI BBC, Bungan Bangsa University Cirebon, Cirebon, Indonesia
[2]STIT Al-Marhalah Al-'Ulya Bekasi, Bekasi, Indonesia
[3]Indraprasta PGRI University, Indonesia
[4]Univesitas Indraprasta PGRI, Indonesia

Abstract: The research investigates the integration of climate awareness programs in Islamic schools through the Penta Helix concept, which encompasses collaboration among academia, government, community, business, and media. This approach addresses the global challenge of climate change by utilizing Islamic values as a foundation to enhance ecological responsibility. The study aims to analyze how this concept can be effectively implemented in extracurricular activities to foster climate awareness among students. Employing a qualitative case study methodology, the research involves interviews, observations, and documentation in selected Islamic schools actively engaged in environmental programs. Academia contributed to curriculum design and teacher training, while government support included infrastructure and policy development. Community participation enriched program implementation, media amplified awareness campaigns, and business sectors provided financial and technological resources. The findings reveal that Penta Helix collaboration successfully enhances students' understanding of climate issues and promotes environmentally friendly behaviors such as recycling and reducing waste. Activities integrated Islamic teachings like stewardship, gratitude, and balance, aligning spiritual values with environmental practices. The study demonstrates significant improvements in students' ecological awareness and their ability to act as agents of sustainable change. This research highlights the importance of integrating multi-sectoral partnerships in education to address global environmental challenges. By aligning Islamic values with environmental education, the study provides a replicable framework for other institutions. It concludes that the Penta Helix approach offers a sustainable and impactful solution to fostering ecological responsibility and achieving long-term environmental goals.

Keywords: Climate awareness, environmental responsibility, extracurricular activities, islamic education, multi-sectoral collaboration, Penta Helix

1. Introduction

Climate change is a global issue that affects various aspects of life, including education. In this context, education in Islamic schools has a very important role in developing students' awareness of sustainability and environmental responsibility. This education is not only academic but also moral, using Islamic values as a foundation to strengthen the relationship between environmental education and ecological responsibility. Islamic values have proven to be effective in building students' ecological awareness, which is expected to strengthen their participation in environmental conservation. Islamic religiosity plays a role as a moderator that strengthens the relationship between environmental education and environmentally friendly behavior, increasing students' commitment to nature conservation.

Research shows that Islam teaches the importance of protecting nature as part of the mandate given by Allah to humans, so that these religious values can be a strong foundation in environmental education Islamic religiosity acts as a moderator that strengthens the relationship between environmental education and environmentally friendly behavior, increasing students' commitment to nature conservation (Begum et al., 2021). This is evident in the implementation of the concept of eco-pesantren in several Islamic boarding schools, such as in Nurul Haramain NW Lombok. This pesantren uses the teachings of the Quran and Hadith to form a sustainable lifestyle that focuses on conservation and greening, which is also an example for students to be actively involved in environmental conservation activities (Quddus, 2020).

Islamic Religious Education (PAI) in schools also teaches the role of human beings as caliphs (keepers) on earth.

[a]efritanorman@gmail.com, [b]Paramansyah.aba@gmail.com, [c]megayantiwindi@gmail.com, [d]nickyrosadi@gmail.com

DOI: 10.1201/9781003684459-70

Islamic Religious Education teachers play an important role as role models that can form social and environmental awareness in students, teaching Islamic values that support environmentally conscious behavior in daily life (Sahri & Aziz, 2023). Through the PAI curriculum, students are taught to understand the importance of protecting and caring for nature as a form of spiritual and social responsibility. This approach helps shape an environmentally conscious school culture, where Islamic values about nature conservation are integrated into various school activities, both through intra-curricular and extra-curricular activities (Muzadi & Mutholingah, 2019).

However, environmental education in Islamic schools has not been optimally integrated in the curriculum and extracurriculars. Environmental materials are almost absent in PAI books, so additional materials are needed to increase student awareness (Kosim, 2022). Studies show that environmental education in Integrated Islamic Science Schools relies heavily on teacher creativity and formal support, but there is no systematic approach that ensures sustainable implementation. This makes environmental education inconsistent among Islamic schools (Rahman et al., 2023). Some schools implement environmental education with monotheistic values, but real practice is still limited to the example of teachers and the teachings of the Qur'an (Rahman et al., 2020). The implementation of environmental education in Islamic schools requires a strategic management approach for effective integration (Pramudita, 2023). Environmental extracurricular programs without a standard curriculum hinder their consistency and effectiveness (El-Batri et al., 2019).

Many extracurricular programs in neighborhood clubs still lack a standard curriculum, so activities tend to be inconsistent and have an immeasurable impact on increasing students' environmental awareness (El-Batri et al., 2019). Environmental education based on monotheistic values is often limited to teaching values through the advice of teachers and role models in Quranic verses, but direct practice for student involvement in environmental conservation is still lacking. This suggests that students need more structured practical experience in environmental education (Rahman et al., 2020). Although there are extracurricular programs that emphasize energy conservation, the integration of religious and scientific values is still not comprehensive and structured enough to build students' environmental awareness in depth (Rohmatulloh et al., 2023). Although communities are often involved in environmental education programs, their involvement has not been fully integrated into the Islamic school education system. This hinders full synergy between schools and communities in building a culture of environmental care (Irawan et al., 2021).

This study introduces the pentahelix model as a collaborative framework for developing comprehensive climate awareness programs in Islamic schools. The model involves five main sectors: academia, government, community, business, and the media, which are specifically integrated to support extracurricular activities based on Islamic values. The *pentahelix* concept, which involves this collaboration offers an innovative approach to facilitate climate awareness through multi-sectoral engagement The program creates direct action activities that raise children's awareness of climate change through active participation that the pentahelix model can adopt in environmental education (Trott, 2019).

Previous research has shown that collaboration between various sectors can strengthen the implementation of environmental education. The *pentahelix* concept is a form of collaboration between five main stakeholders: Academics: Developing relevant curricula and teaching materials. Government: Provide policy and regulatory support. Community: Participate in school activities as a form of social awareness. Business: Providing financial and technological support for environmentally friendly activities. Media: Disseminating information and supporting climate awareness campaigns. · Pentahelix collaboration involving the government, academia, communities, businesses, and the media in building public awareness through educational programs that can be applied in the context of environmental education and climate awareness (Setiawan & Darmawan, 2020).

The research not only focuses on raising climate awareness, but also integrates Islamic teachings and values that teach environmental responsibility as part of faith. This novelty lies in a unique approach that aligns sustainability goals with the perspective of Islamic education, which has rarely been found in previous environmental education research. in addition, the concept of pentahelix applied in the context of Islamic schools has not been widely explored, especially with a focus on extracurricular activities specifically on climate awareness. Extracurricular activities have not been carried out much.

With the increasing urgency of the issue of climate change, Islamic schools have the opportunity to play an active role in environmental education in accordance with Islamic values. This urgency arises from the need to have an environmental education model that not only focuses on knowledge transfer but also builds the character of students who care about the environment, making it an integral part of the faith. climate awareness.

2. Research Methods

The approach in this study is a qualitative approach to explore in depth how the concept of Penta Helix is implemented in extracurricular activities in Islamic schools. This approach allows researchers to understand phenomena holistically and interpret data according to their context. This type of research is a Case Study, focusing on Islamic schools that apply the concept of Penta Helix in extracurricular activities related to climate awareness. This case study helps in understanding the process, challenges, and outcomes of the implementation of Penta Helix specifically in an Islamic school environment (Hamilton & Finley, 2019).

Location This research was conducted in several Islamic schools that were selected based on the criteria of active involvement in environmental activities or climate awareness-based extracurricular programs. The subject of the research includes, the principal as the leader of the institution who directs policy. Teachers who act as facilitators of extracurricular activities. Students who are directly involved in environmental activities. Representatives from academia, government, community, business, and media involved in the Penta Helix collaboration at the school.

The Data Collection Instrument uses In-Depth Interviews, semi-structured interviews will be conducted with school principals, teachers, students, and representatives from Penta Helix sectors. This interview aims to gain a detailed understanding of each party's contribution to climate awareness programs. Participatory Observation: Researchers will be directly involved in extracurricular activities to observe the interactions and learning processes that occur. This observation will record how to implement the Penta Helix concept in daily activities. Documentation: The researcher will collect a variety of related documents, such as extracurricular curriculum, activity notes, program reports, and media-supported publication materials, to supplement the data obtained from interviews and observations.

Data Collection Procedure in the Preparation Stage: The researcher will conduct initial coordination with the school and related stakeholders to explain the research objectives and methods and ask for permission to conduct interviews, observations, and documentation. Interview Implementation: Interviews will be conducted in stages, starting with the principal and teachers, then continuing with students and representatives from other Penta Helix sectors. Interviews will be recorded to facilitate data analysis Observation of Extracurricular Activities: Researchers will conduct participatory observations during extracurricular activities, recording interactions between students, teachers, and the Penta Helix parties involved. Documentation Collection: Researchers will collect supporting documents related to programs and activities relevant to the Penta Helix concept and climate awareness.

Data analysis is carried out by Data Reduction, Data from interviews, observations, and documentation will be reduced by selecting information relevant to the focus of the research, Coding: Data that has been reduced will then be categorized through a coding process, grouping information based on main themes such as the contribution of each Penta Helix sector, the application of Islamic values in activities, and the effectiveness of activities in increasing climate awareness, Thematic Analysis: The categorized data will be analyzed thematically to identify key patterns in the application of the Penta Helix concept. The analysis also includes an in-depth understanding of the role of Islamic values in building environmental awareness in schools, Data Presentation: The results of the data analysis will be presented in a narrative form complemented by direct quotes from the research subjects, tables, or diagrams depicting the Penta Helix collaboration and its impact on students' climate awareness.

The Data Validation technique is carried out by Source Triangulation, Data validation is carried out by comparing the results of interviews, observations, and documentation. This ensures that the information obtained is accurate and consistent. Member Checking: The researcher will recheck the results of interviews and observations with the research subject to ensure that the interpretation carried out is in accordance with the subject's intent. Peer Debriefing: Discussions with peers who have knowledge in this field of research to get input and ensure that data analysis has been carried out objectively and systematically.

3. Research Results

Interviews with principals and teachers show that the implementation of the Penta Helix concept has helped schools develop more holistic climate awareness programs (Table 70.1). The principal revealed that collaboration with external sectors such as academics, government, and the community allows for more varied and relevant extracurricular activities. The teacher also mentioned that the support from these stakeholders not only enriches teaching materials but also expands environmental learning methods, such as through recycling, greening, and climate education projects (Table 70.2).

Observations of extracurricular activities show high student involvement in climate awareness programs. Activities

Table 70.1. Results of interviews with principals and teachers

Collaboration aspects	*Explanation*
Collaboration with Academics	Academics help develop a curriculum based on environmental education that is in line with Islamic values, using verses of the Quran and Hadith.
The Role of the Government in Supporting Infrastructure	The government provides physical facilities such as recycling bins, greening areas, and provides technical training to teachers.
Community Engagement in School Programs	Local communities contribute through activities such as mutual cooperation, mentoring students in recycling projects, and the establishment of green parks.
Support in Enrichment Materials and Learning Methods	Stakeholders enrich teaching materials and expand learning methods such as project-based learning and field studies.
Varied and Relevant Extracurricular Activities	Group discussions, recycling workshops, and second-hand creative competitions increase students' active participation.
Effects of Collaboration on Climate Awareness Formation	This combined support results in a holistic education that encourages students to adopt eco-friendly behaviors

Source: Author.

Table 70.2. Direct observation of environment-based extracurricular activities

Observation aspect	*Explanation*
Varied and Innovative Extracurricular Activities	Various activities such as tree planting, recycling creative competitions, and group discussions on climate change help students understand environmental issues.
Application of Islamic Values in Activities	Each activity is designed to practice Islamic values such as trust, responsibility, and gratitude, integrating religious teachings in activities.
Increasing Student Awareness and Behavior	These activities influence changes in student behavior such as reducing the use of plastic, recycling waste, and creating a small garden at home.
The Role of Teachers and Stakeholders	Teachers act as facilitators, supported by a community that provides training and technical assistance for students in environmental activities.
Relationship with the Penta Helix Concept	Penta Helix involves academics, governments, communities, and media who each contribute to providing materials, facilities, and publication of activities.
Long-Term Impact	This practical activity creates students who are more environmentally conscious and are projected to become agents of change in society.

Source: Author.

such as tree planting, creative competitions from recyclables, and discussions on the impacts of climate change provide an in-depth, hands-on learning experience for students. Through this activity, students not only gain knowledge but also practice Islamic values in protecting the environment. This observation shows that the Penta Helix approach plays an important role in facilitating activities that actively involve students, so that students' climate awareness is increasing (Table 70.3).

Documentation of Environmental Education Materials and Curriculum from Academics: The documentation obtained shows that academics have contributed to compiling curriculum materials that integrate environmental education with Islamic values. This material focuses on Islamic teachings on ecological responsibility and includes relevant verses of the Quran and Hadith. These materials help students understand the importance of their role as caliphs on earth who are in charge of preserving nature. The document also shows that teachers are trained to teach this material in an interactive way, allowing students to better understand the relationship between religion and the environment (Table 70.4).

The results of the documentation show that the local government and local communities also support the implementation of climate awareness programs in Islamic schools. The government provides facilities and technical guidance for environmental activities such as waste management and

Table 70.3. Documentation of environmental education materials and curriculum from academics

Documentation aspects	*Explanation*
Preparation of Integrated Curriculum Materials	Curriculum materials include Quranic verses, Hadith, and scientific studies to build an in-depth understanding of ecological responsibility.
Teacher Training for Material Delivery	Teachers receive training in interactive methods, the use of technology, and contextual approaches to convey environmental materials.
Strengthening Islamic Values in Environmental Education	The material emphasizes trust, gratitude, and tawazun in protecting the environment as relevant Islamic values.
Impact on Student Understanding and Action	Students demonstrate a better understanding of environmental issues and begin to implement eco-friendly practices and participate in collaborative projects.
The Relationship Between Religion and the Environment	The integration of religion and the environment helps students understand the relevance of religious teachings in social life and increases motivation to act.
Long-Term Benefits	This material builds the foundation of sustainable education that produces a generation that cares for the environment and strengthens value-based education.

Source: Author.

Table 70.4. Government support and community involvement in school activities

Support aspects	*Explanation*
Government Support in Infrastructure and Technical Guidance	Provision of facilities such as recycling bins, tree seedlings, cleaning tools; technical training for teachers and students; and incentive programs for environmentally friendly schools.
Community Engagement in School Activities	The community is involved in mutual cooperation, assisting students in recycling activities, and strengthening local wisdom in protecting the environment.
Increased Collaboration Between Schools, Government, and Communities	Collaboration includes joint planning, overseeing program implementation, and sharing resources between governments, schools, and communities.
The Impact of Government Support and Community Engagement	Support creates more active students, greener school environments, and increased collective awareness of environmental issues.
Manfaat Jangka Panjang	This program creates community-based education, produces a generation that cares about the environment, and serves as a model for other schools.

Source: Author.

school greening. The community is also actively involved through mutual cooperation programs and assists students in various environmental projects, such as the creation of green gardens and recycling systems. This collaboration shows that government and community support is very important in creating a learning environment that supports environmental education (Table 70.5).

Mass media plays an important role in supporting climate awareness programs by publicizing various school activities related to the environment. This publication not only raises the profile of the school as a school that cares about the environment but also inspires other schools and the community to contribute. Media publication documentation shows how the media helps expand the impact of environmental education by making school activities a more widely applicable example. Local media and social media used also amplify student engagement, encouraging them to feel proud and responsible for their environmental actions (Table 70.6).

Based on the results of interviews, observations, and documentation, it can be seen that extracurricular activities supported by the Penta Helix approach significantly increase students' awareness of climate issues. Students are beginning to adopt eco-friendly habits in their daily lives, such as reducing the use of plastic, recycling, and maintaining a clean environment. This positive impact shows that cross-sector collaboration is not only successful in improving students' knowledge, but also encourages behavior change that is in line with Islamic values in protecting the environment (Figure 70.1).

Table 70.5. Media publications on school environment activities

Media publication aspects	*Explanation*
The Role of the Media in Supporting Climate Awareness Programs	Mass and social media publish environmental activities such as tree planting, recycling competitions, and reforestation.
The Impact of Media Publications on School Profiles	Publications improve the school's image as an institution that cares for the environment, motivate teachers and students, and attract external support.
Inspiration for Other Schools and the Community	The publication of activities motivates other schools and the community to contribute to environmental action.
Use of Local Media and SocialMedia	Local and social media are used to disseminate information in real-time, reaching a wider audience.
Strengthening Student Engagement through Media	Publications make students proud of their contributions, increase responsibility, and provide a platform for sharing ideas.
Long-Term Benefits of Media Publications	Publications build collective awareness, improve access to information, and create a positive digital footprint for future learning

Source: Author.

The results of this study show that the application of the Penta Helix concept effectively supports a comprehensive climate awareness program in Islamic schools. Each sector in Penta Helix plays a significant role in providing support, both in terms of providing materials, training, facilities, and publications. This entire collaboration proves that

Table 70.6. Impact of extracurricular activities on student climate awareness

Impact Aspects	*Explanation*
Increasing Student Awareness of Climate Issues	Students understand the impact of global warming, the importance of waste management, and the concept of sustainability through extracurricular activities.
Changes in Student Behavior	Students demonstrate eco-friendly behaviors such as reducing plastic, planting trees, and being active in school cleanliness.
Integration of Islamic Values in Environmental Awareness	This activity instills the values of trust, gratitude, and tawazun as an Islamic responsibility in protecting the environment.
The Role of the Penta Helix Approach	Penta Helix involves academics, government, community, media, and business in supporting students' environmental activities.
Impact on School Communities	Impact on teachers, parents, and the relationship between schools and communities through collective awareness and collaboration.
Long-Term Benefits	Long-term benefits include environmentally conscious generations, schools as environmental models, and a reduction in carbon footprint.

Source: Author.

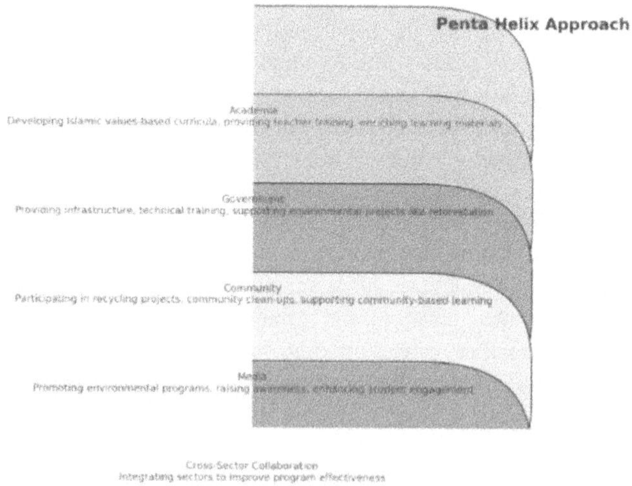

Figure 70.1. Penta Helix approach.

Source: Author.

environmental education based on Islamic values can be strengthened through synergy involving various stakeholders, so as to be able to create a long-term positive impact on students' environmental awareness.

4. Discussion

The Penta Helix concept has had a significant impact in the school's efforts to integrate climate awareness programs into the learning system. This concept involves cross-sector collaboration that includes academics, government, communities, businesses, and the media, which allows for the development of more comprehensive and relevant programs, cross-sector collaboration can increase the effectiveness of environment-based programs (Caraka et al., 2021). Collaboration with, Academics The Principal explained that collaboration with academics has brought new insights in the development of teaching materials related to the environment (Yasir et al., 2021).

Academics help develop an environmental education-based curriculum that is aligned with Islamic values, involving the use of relevant verses of the Quran and Hadith. This approach enriches the curriculum and provides a strong spiritual dimension in environmental education. The study by Turnbull emphasizes that the contribution of Islamic values in environmental education can provide a unique approach to ecological justice. The implementation of Islamic sharia such as fiqh law in education can strengthen spiritual-based ecological awareness (Turnbull, 2021). Hayat et al. emphasized the importance of Islamic environmental ethical principles rooted in the Qur'an and Hadith, such as the concepts of Tawhid, Caliphate, and Hima, to build a holistic understanding of ecological responsibility as part of worship (Hayat et al., 2023). Research by Rahmawati et al. shows the effectiveness of the integration of Islamic values in environment-based education to form ecological awareness. This approach includes problem-based learning that connects science issues with religious teachings (Rahmawati et al., 2018).

Local governments have a strategic role in supporting environmental education through the provision of physical infrastructure and technical training. One of the main supports is the provision of physical facilities that include recycling bins and greening areas. According to Agarwal et al. (2020), the provision of special bins and waste management facilities makes it easier for the community to participate in recycling, thereby increasing the effectiveness of domestic waste management. In addition, Jamal et al. (2019) pointed out that local authorities also provide composting facilities and organic waste recycling sites to manage waste segregation at the community level.

On the other hand, the government also provides technical training to teachers to improve their ability to teach environmental materials with innovative methods. Salazar et al. (2022) highlight the importance of training through environmental education programs, such as the 'Sustainable School' in Chile, which integrate schools into sustainable initiatives. This training not only improves the technical skills of educators but also strengthens their environmental awareness. Waheed et al. (2023) underline that this kind of training allows teachers to implement environment-based learning and relevant school projects.

In addition, the government plays a role in encouraging cross-sector collaboration to support sustainable programs. Altassan (2023) emphasizes that collaboration between schools, communities, and recycling agencies can create programs such as tree planting and waste processing into compost. Harrington and Hsu (2018) also highlighted the importance of cooperation between governments and non-governmental organizations in strengthening the implementation of green infrastructure and increasing local capacity in environmental management.

The involvement of local communities in school environment programs has proven to make a significant contribution to the development of climate awareness among students. Teachers revealed that active community participation, such as mutual cooperation in cleaning the school environment, assisting students in recycling projects, and the establishment of green parks, has expanded the scope of environmental programs (Trott, 2020). A study by Salazar et al. (2022) supports these findings by showing that communities play an important role in facilitating student involvement in school-based environmental programs, such as 'Sustainable Schools' in Chile, which strengthens the relationship between schools and local communities (Salazar et al., 2022).

Community support also plays a role in improving student learning outcomes. Careemdeen (2022) revealed that a supportive community environment directly affects the success of environmental education programs in secondary schools. This support includes participation in educational activities, improvement of school infrastructure, and provision of social and financial resources to support school programs (Careemdeen, 2022).

In addition, community involvement not only enriches teaching materials but also expands learning methods. Gladwin et al. (2022) emphasized that stakeholder support, including the community, can introduce innovative learning methods that involve students in hands-on activities, such as environment-based projects oriented towards real solutions (Gladwin et al., 2022). Community participation in school-based environmental education programs also facilitates interactive learning, where students can develop interpersonal skills and a deeper understanding of environmental issues, as found by Selby et al. (2020) in community-based programs in Costa Rica (Selby et al., 2020).

The media, both traditional and digital, has a significant role to play in raising public awareness of environmental issues and climate change. The media is the main channel to disseminate information, shape opinions, and motivate actions that support sustainability. Research by Jiménez-García et al. (2023) shows that the media has a greater influence than formal education in increasing climate change awareness among the younger generation. This study highlights the importance of the media in conveying information that can build understanding and encourage environmentally friendly behavior (Jiménez-García et al., 2023).

In the context of school programs, the media also plays a role as a tool to promote and disseminate the positive impact of environmental initiatives. This supports students' increased sense of pride in their contributions, as described in a study by Ahmed (2022), which emphasizes the need for media to accelerate climate awareness campaigns through special segmentation and inspirational stories involving various community groups (Ahmed, 2022).

Further, digital media provides a platform that allows for direct interaction and wider public participation. Rousell et al. (2021) show how young people are using digital media to build climate awareness through creative activities, such as educational apps, speculative fiction, and social media activism. Digital media not only conveys a message, but also creates opportunities for individuals to become agents of change in environmental issues (Rousell et al., 2021).

Another study by Junsheng et al. (2019) also underscores the role of mass media in shaping attitudes and behaviors towards climate change. The media acts as a mediator between science and the general public, allowing the dissemination of complex information in an easily understood manner by the public. The results of this study show that the media can increase public awareness, positive attitudes, and understanding of climate change, which ultimately encourages environmentally friendly behavior (Junsheng et al., 2019).

Collaborations involving various parties, such as teachers, communities, and educational institutions, create a more holistic and effective environmental education program in increasing students' awareness of climate issues. The program includes activities such as group discussions, recycling simulations, and second-hand creative competitions. Research by Kamanga and Shumba (2022) shows that a collaborative learning approach that integrates the concept of environmental education increases students' understanding of climate change issues and positive attitudes towards environmental conservation (Kamanga & Shumba, 2022).

In addition, this collaboration allows the integration of Islamic values such as amanah, gratitude, and tawazun in environmental education activities. Trott (2019) highlights that a participation-based approach involving students in environmental projects, such as tree planting and community gardens, can strengthen climate awareness while giving students a sense of responsibility as agents of change (Trott, 2019).

Furthermore, technology-based activities can also support innovative collaborative learning. Salas-Rueda et al. (2021) found that the use of technologies such as 'collaborative walls' in climate learning increases students' active participation and their motivation to understand global issues related to climate change (Salas-Rueda et al., 2021).

Other research by Barrón et al. (2022) shows that collaboration-based activities, such as research projects and environmental awareness campaigns, not only deepen students' understanding of climate change but also encourage them to engage in real action, such as environmental policy advocacy and local community engagement (Barrón et al., 2022).

Environment-based activities in education have a significant positive influence on changes in student behavior. Students began implementing concrete actions such as reducing the use of single-use plastics, separating waste for recycling, and creating small gardens in their homes. Research by Mullenbach and Green (2018) shows that environmental education can improve students' eco-friendly behaviors, such as waste reduction and awareness of the ecological impact of daily activities (Mullenbach & Green, 2018).

The practical approach in environmental education also helps students not only understand the importance of taking care of the environment theoretically, but also apply it in daily life. Kosta et al. (2022) found that activities such as outdoor education increased students' knowledge and connection with nature, thereby encouraging more environmentally conservation-oriented behaviors (Kosta et al., 2022).

Another study by Suarlin and Ali (2020) noted that well-designed environmental education encourages the formation of positive habits, including energy-saving behaviors and concern for the surrounding ecosystem. The program has a significant impact on students' attitudes towards the environment, especially through relevant hands-on experiences (Suarlin & Ali, 2020).

A values-based approach also plays an important role. Begum et al. (2021) found that environmental education integrated with religious values, such as Islam, strengthens students' environmental responsibility and their eco-friendly behaviors. This integration of spiritual values provides a more holistic dimension to environmental education (Begum et al., 2021).

Collaboration-based environmental education provides significant long-term benefits, including the formation of a generation that cares more about the environment and a reduction in the carbon footprint within the school community. Research by Schmäing and Grotjohann (2022) shows that environmental education can produce a sustainable positive effect on students' attitudes and connections with nature. This study highlights the importance of integrating environmental materials in learning to influence students' values and actions in the long run (Schmäing & Grotjohann, 2022).

In addition to the direct benefits to students, schools that successfully implement this program also serve as a model for other institutions. A study by Gal and Gan (2018) observed how long-term environmental programs in Israel have survived for nearly two decades, thanks to a combination of leadership, human and economic resources, and community engagement. The program creates a sustainable school culture and serves as an inspiration to other schools in the region (Gal & Gan, 2018).

The place- and community-based education model has also proven effective in building collective awareness of the importance of environmental conservation. Cruz et al. (2018) proposed a social capital-based approach that connects schools with local community resources. This approach enhances social, academic, and environmental benefits by involving students in locally relevant community-based projects (Cruz et al., 2018).

In Indonesia, the Adiwiyata program has been a concrete example of how collaboration between schools, local governments, and communities can create a holistic model of environmental education. Research by Aeni et al. (2020) noted that the success of this program involves integrating an environment-based curriculum that combines theory with real practice, such as maintaining school cleanliness and doing greening (Aeni et al., 2020).

5. Conclusion

The implementation of the Penta Helix concept has proven to be an effective and innovative framework for integrating climate awareness programs into Islamic schools. By involving academia, government, community, business, and media, this approach creates a holistic, multi-sectoral collaboration that enriches both the curriculum and extracurricular activities.

Academia contributes to the development of environmentally-focused, Islamic values-based curricula, while the government provides critical infrastructure and training. Communities enhance program reach and effectiveness through direct involvement, media amplifies awareness and engagement, and businesses support sustainability initiatives. These combined efforts not only equip students with theoretical knowledge but also foster practical, environmentally friendly behaviors, guided by Islamic principles such as stewardship (khalifah), gratitude (shukr), and balance (tawazun).

Moving forward, this study highlights the potential of such a collaborative framework to transform education into a powerful tool for addressing global challenges like climate change. The success demonstrated in these Islamic schools offers a replicable model for other institutions, both within and outside Islamic educational contexts. This approach fosters not only a deeper ecological awareness among students but also positions them as agents of sustainable change within their communities.

To build on this foundation, future initiatives should focus on scaling the Penta Helix model, ensuring consistent application across diverse schools, and incorporating advanced technologies and innovative teaching methods. This strategy will reinforce the alignment of education with sustainable development goals, creating a generation of leaders who are environmentally conscious, spiritually grounded, and socially responsible.

References

[1] Niu, Y., Abdullayev, V., Alyar, A. V., & Kamran, A. T. (2023). Resilience and Adaptation to Climate Change: Community-Based Strategies in Coastal Regions. *ESTIDAMAA, 2023*, 36–43. https://doi.org/10.70470/ESTIDAMAA/2023/005

[2] Pan, H., Wang, R., Wang, H., & Jia, Z. (2023). Sustainable Supply Chain Management: Best Practices for Reducing Environmental Footprints in the Global Apparel Industry. *ESTIDAMAA*, 2023, 18–26. https://doi.org/10.70470/ESTIDAMAA/2023/003

71 Strategic management advantages in educational organizations

Nur Zazin[1,a], *Galuh Nashrulloh Kartika M. R.*[2,b], *and Barsihanor*[3,c]

[1]Islamic Education, STIT Darul Ulum, Kotabaru, Indonesia
[2]Islamic Education, Universitas Islam Kalimantan Muhammad Arsyad Al Banjari, Banjarmasin, Indonesia
[3]Madrasah Ibtidaiyah Teacher Education, Universitas Islam Kalimantan Muhammad Arsyad Al Banjari, Banjarmasin, Indonesia

Abstract: To aid educational organizations in accomplishing their objectives, this study evaluated the advantages and disadvantages of strategic management, which is a relatively new phenomenon within the field. The study team performed a literature review to collect relevant information from various sources which included books, journals, and other media. They reached conclusions using content analysis. Each one of these actions, namely: data reduction, data presentation, data collecting, and data interpretation are parts of the whole process which constitutes data processing. This study claims that educational organizations can attain high profitability and productivity, favorable competitive position alongside advantages in technology and human resources, a good work environment, and enhanced organizational ethics and social responsibility, by shifting to the new paradigm of strategic management. With regard to the organization, one can expect a more dynamic organization, efficient and effective control function, increased consensus on the changes or new strategies to be adopted, greater proactive behaviour, heightened sense of belonging, active participation, and augmented responsibility.

Keywords: Strategic management, educational organization

1. Introduction

An organization is a place where one or more people gather to achieve a goal [9]. In order for the organization to be effective in achieving its goals, a certain strategy or tactic is needed that is considered capable of providing convenience and its own advantages in compiling, implementing, and evaluating the institution or organization [10]. Initially, strategic management was used more in profit organizations and organizations engaged in the economic sector, but along with the development of the times, strategic management began to penetrate non-profit organizations including education [1], even becoming a new paradigm that is popular in the effectiveness of achieving the goals of educational organizations.

As a new paradigm, if applied in the educational organization environment [6], it is impossible to do it as an activity to take over all its activities as applied in the profit organization environment (business), because the two organizations differ from each other in many ways, especially in terms of the philosophy behind it and the goals to be achieved.

The application of Strategic Management in the business organization environment is based on a philosophy that contains the values of free competition between similar business organizations, through the use of resources to achieve strategic goals. The goal is to maintain and develop each other's existence for the long term, through the ability to achieve competitive profits sustainably. Meanwhile, educational organizations are based on a philosophy that contains values of devotion and humanity to educate the nation's life.

The research team behind this project set out to talk about how educational organizations are embracing a new model of strategic management, namely the benefits and advantages of strategic management, so that it can be a reference in adopting it, facilitating the achievement of educational goals. This study is expected to be useful for observers, thinkers, educational leaders, lecturers, teachers, students both in the fields of economics and education, as enlightenment, adding to the treasury of knowledge in the field of strategic management.

2. Method

Through literature review, researchers collect various information relevant to the topic of strategic management excellence in educational organizations from books, journals, or media that have objectivity or can be verified for their truth. Data sources are sorted into categories that are relevant to the research topic, then reviewed and used as references. This data analysis technique uses descriptive analysis. The process involves processing data by selecting and sorting various ideas and thoughts from educational management figures. After that, the selected data will be described and analyzed [8]. The author's data analysis was carried out using the content analysis technique after the data was collected properly. The author uses this content analysis to obtain accurate results in

[a]nurzazin@gmail.com, [b]galuh.nashrulloh@uniska-bjm.ac.id, [c]barsihanor@uniska-bjm.ac.id

DOI: 10.1201/9781003684459-71

understanding the context of sentences in the manuscript [3]. Data analysis occurs both during and after data gathering; in other words, after data collection, one must write, edit, categorize, reduce, and present the data [2]. According to Huberman [5], data reduction, presentation, collecting, and analysis are the four steps that make up data processing.

3. Discussion

3.1. Definiton of strategic management

Words like 'management' and 'strategic' have different meanings when used alone, but when combined they form a new term with a different meaning altogether: strategic management. According to Hadari Nawawi [4] (pp. 148–149), there are 4 (four) definitions of strategic management. Strategic Management is first described as 'strategy is the process through which an organization's upper management makes important decisions and then everyone works together to put those decisions into action so that the organization may reach its objectives.'

Several crucial points can be derived from this definition, such as:

1. Making decisions is an integral part of strategic management.
2. Decisions are both broad and essential, meaning they touch on key areas of an organization's existence, such as its objectives and the means by which to attain them.
3. Top management, or the principal, is ultimately responsible for the organization's success or failure, so they must be involved in the decision-making process.
4 The implementation of the decision as an organizational strategy to achieve its strategic goals is carried out by all levels of the organization (school community), all of whom must know and carry out their roles according to their respective authorities and responsibilities.
5. In order to achieve the organization's long-term objectives, all school communities must carry out the decisions made by the principal.

Strategic management is defined as 'management's endeavours to cultivate organisational strengths in order to take advantage of new chances and accomplish its intended objectives' in the second definition. From this understanding, there is a relatively broad concept from the first understanding which emphasizes that 'strategic management is a managerial effort to develop organizational strength,' which requires the principal with or without the help of subordinate managers (Vice Principal, Student Council Advisor, Head of Administration), to identify and cultivate the organization's strengths in line with its mission so that it can accomplish its specified strategic objectives. We must make the most of every available opportunity.

Strategic management, according to the third definition, is a series of choices and activities that, when put into motion, assist an organization reach its objectives. This understanding emphasizes that the flow of decisions from organizational leaders (Head of Service, Principal) and actions in the form of implementing decisions, need to come up with a strategy or strategies, then choose the best one or ones to help the organization reach its objectives.

The fourth summary, 'Strategic management entails long-term planning that is future-focused (Strategic Planning), made as a fundamental and principled decision by top management (Top Management Decision), to facilitate effective interaction within the organization (Mission), and to achieve its goals (Strategic Objectives) and operational targets (Operational Objectives) to the best of its ability (Operational Planning).' With this wide-ranging concept in mind, we can see that Strategic Management is best understood as an integrated system whose parts work together, impact one another, and advance in tandem. A company's vision, mission, and strategic objectives make up the first part of strategic planning. Operational planning is the second part of the process. It includes operational objectives or targets, management functions like organizing and budgeting, situational policies, internal and external networks, control and evaluation methods, and feedback.

From the various definitions, it can be seen that Strategic Management has the following characteristics:

a. Strategic management is put into practice through extensive planning that encompasses all aspects of an organization's environment. This planning is then poured into a Strategic Plan (RENSTRA), which is referred to as an Operational Plan (RENOP), and later on, into work programmes.
b. Looking 25–30 years into the future, the Strategic Plan is future-focused. In contrast, the Operational Plan is revised annually or semiannually.
c. The RENSTRA is based on the main strategy (parent) and the organization's long-term strategic objectives, vision, and mission; however, the placement approach incorporates all of these references as a documented decision by top management.
d. The operational programmes provided by RENSTRA are referred to as RENOP.
e. Because it is crucial to the execution of the complete mission of the organization, the determination of RENSTRA and RENOP must involve Top Management (Leaders).
f. Management functions such as planning, executing, monitoring, and controlling make it possible for programmes to achieve their goals by putting strategies into action.

Based on the characteristics and components of Strategic Management as a system, many factors can be seen that influence the seriousness and formality with which it is put into practice in an educational setting that is not for profit. Some of these factors include:

a. The magnitude of the organization,
b. The leader's management style,

c. The intricacy of the demographic, socioeconomic, political, and cultural environment, and
d. Government rules, among other external factors.

Similarly, internal obstacles impact the degree of formality and intensity. These include, for example, the capacity to transform strategy into a work implementation process or series that provides effective, efficient, and high-quality public service (in the realm of education, for instance, identifying instructional models/systems, learning resources, learning media, etc.).

3.2. Dimentsions of strategic management

Strategic management is multidimensional, according to the perspective of its dimensions. These are the measurements at issue:

3.2.1. Time dimension and future orientation

Strategic Management in maintaining and developing the existence of an organization looks far into the future, and behaves proactively and anticipatively towards future conditions that are predicted to be faced [7]. Anticipation of the future is formulated and determined as the Vision of the organization which will be realized 25–30 years in the future. According to Hadari Nawawi [4] (p. 155), Vision can be interpreted as 'the ideal condition that is desired to be achieved in the existence of the organization in the future.' In this regard, Lonnie Helgerson, quoted by J. Salusu in his book Hadari Nawawi, said that 'Every member of an rganization can read the vision statement, which paints a picture of the organization's future state that is not yet evident. The ability to see into the future and summon others to join you is a hallmark of visionary leadership. The organization's senior leadership is responsible for developing the vision of the company.' Still according to J. Salusu who quoted Naisibit's opinion: 'Vision is a clear picture of what will be achieved along with details and instructions for each step to achieve the goal. A vision is said to be effective if it is very necessary it brings contentment, values history as a springboard to the future. J. Salusu cites Kotler, who says the following about Hadari Nawawi [4]: A company's vision statement articulates the mission of the company as it relates to the goods and services it provides the needs that can be addressed, the community groups served, the values obtained, and aspirations and ideals for the future.

The vision of the organization can be interpreted as a perspective on the future in realizing the strategic goals of the organization, which directly affects its mission now and in the future. In this regard, the mission of the organization basically means all the main tasks outlined from the strategic goals to realize the vision of the organization.

3.2.2. Internal and external dimensions

A non-profit organization's (an educational institution's) internal dimension is its present state, which includes its strengths, shortcomings, opportunities, and challenges. That is why it's important to do things like engage in self-evaluation, which may involve quantitative analysis statistical calculations, using quantitative data available in the Management Information System (MIS). However, quantitative data is often inadequate, due to the weakness of the MIS in recording, searching, conducting research and developing data in the past. Therefore, Self-Evaluation should not depend entirely on quantitative data, because it can also be done with Qualitative Analysis using various qualitative information or some quantitative data and some qualitative data.

Using SWOT Analysis is one way to conduct qualitative analysis. An examination of the surrounding environment of the organization (school) is the main focus of the external environmental dimension. This environment encompasses the operational, national, and global levels and encompasses a wide range of factors, such as political climate, economics, culture, religion, and the advancement and development of science and technology. To put Strategic Management into action, one must first determine the organization's strengths and work to build upon them, while simultaneously addressing its deficiencies.

3.2.3. Dimension of utilization of resources

Strategic management, as a management activity, is inseparable from the ability to make effective use of the various resources at one's disposal in order to carry out management functions in a way that meets the goals outlined in each RENOP. This, in turn, allows one to realize the school's vision and accomplish the organization's strategic objectives. Existing resources include physical assets, such as buildings and infrastructure, monetary assets, such as programme budgets, human resources, technological assets, and informational resources. It is essential to have a thorough understanding of all of these assets, which fall under the category of internal resources, in order to conduct an accurate self-evaluation (Internal Analysis).

3.2.4. Dimension of top management participation (Leadership)

The future of the organization can be controlled by strategic management that starts with creating a strategic plan, which allows it to exist in line with its vision. The power and obligation for ensuring that the Strategic Plan can adapt to all parts of the organization's life that impact its future existence rests with top management. Top management is responsible for proactively carrying out all operations to realize the strategic plan, which is a major decision that is not only taken by involving but also falls under their purview.

3.2.5. Multi-field dimension

Strategic Management as a System, its implementation must be based on placing the organization as a system. Thus, it means that an organization will be able to prepare RENSTRA and RENOP if it does not have any attachment or dependence as a subordinate to another organization as a superior. The school does not have complete control over the selection and determination of the vision, purpose, objectives, and tactics because of its subordinate status (it is subservient to the Education Office).The school only plays a role as the compiler of RENOP and annual programs.

From the description, it is clear that RENSTRA and RENOP are multidimensional, especially if RENSTRA is primarily implemented in a wide range of non-profits, especially those involved in higher education. Because there are so many aspects, it's simple for some of them to be overlooked.

3.3. Advantages and benefits of strategic management for educational organizations

The implementation of Strategic Management through the formulation of RENSTRA and RENOP using certain strategies in carrying out management functions, and realizing the main tasks in the educational organization environment must be measured and assessed for its excellence. From these measurements and the entire implementation process, everyone knows that Strategic Management is good for the company.

Strategic management has several positive uses in educational organizations, including the following:

3.3.1. *Benefits of putting strategic management into practice*

The following criteria can be used to assess the benefits of implementing strategic management:

a. **Profitability:** This benefit demonstrates that all tasks are executed with precision and efficiency, making good use of a limited budget to ensure that no effort is wasted.
b. **High Productivity:** This benefit demonstrates that the quantity of labour that can be accomplished typically rises. Students and the community are able to be satisfied with the process and results of providing public services, as there are less mistakes or errors and higher quality results.
c. **Competitive Position:** The presence of well-respected and essential educational institutions is indicative of this benefit. This competitive nature lies in its products (e.g., quality of graduates) that satisfy the community served.
d. **Technological Advantage:** All main tasks run smoothly in the sense that public services are carried out quickly, on time, according to quality based on the level of uniqueness as well as the simplicity of the duties that are required to be fulfilled, such they are able to adapt to technological developments and progress.
e. **Human Resource Advantage:** In the educational organization environment, an organizational culture is developed that places humans as a central factor, or a resource that determines the success of the organization. Consequently, the human resources that are owned are constantly being trained and bettered in what they know, how they do their jobs, and how they feel about helping pupils. Concurrently, the capacity to address current school challenges and foresee future challenges brought about by globalization is cultivated.
f. **Work Climate:** An organization's culture based on human values develops via the cultivation of formal and informal work connections, as demonstrated by this benchmark. According to their roles, responsibilities, and authority both during and outside of work hours, employees in educational organizations are expected to maintain cordial formal and informal relationships with one another.
g. **Ethics and Social Responsibility:** This benchmark shows that in working, high ethics and social responsibility are implemented and developed, by always prioritizing the interests of society, nation and state above personal, group and/or organizational interests. The aforementioned standard of excellence is crucial to the present and future of any society, country, or state. For this reason, cooperation and community support are needed in developing organizations in implementing Strategic Management optimally, so that the advantages above can be realized, the results of which will also benefit society.

In reality, for educational organizations (schools), the conditions for realizing these advantages still face various dilemmas. Educational organizations that exist today are relatively consumptive, while to implement Strategic Management relatively requires a lot of funds/budget. In conditions like this, it is very necessary to have the ability to create a balance between the openness of the government to allocate sufficient funding and oversee the utilization of alternative resources like community members, parents, and loans/assistance.

3.3.2. *Benefits of strategic management*

In light of the aforementioned benefits, it follows that there are numerous reasons to actively pursue the efficient and effective application of Strategic Management within the context of educational organizations. The advantages that can be gained are: 'strategic management can reduce uncertainty and complexity in preparing planning as a management function, and in the process of implementing work by using all resources that are actually owned through a process that is integrated with other management functions and the results can be assessed based on organizational goals.' In detail, the benefits of strategic management for non-profit organizations (education) are:

a. Educational organizations (schools) as work Adapting RENSTRA and RENOP to the ever-changing internal (organizational) and external (environmental) conditions is essential for organizations to remain dynamic, particularly in the face of the ever-increasing impact of globalization. In other words, Strategic Management as management and control that works realistically in its dynamics, will always be directed at Strategic Goals and Missions that are also realistic.

b. Implementation of Strategic Management through the realization of RENSTRA and RENOP functions acting as a supervisor to ensure the efficient and effective execution of management responsibilities by making full use of all available resources. Thus, it means that Strategic Management is able to support the control function, so that the entire process of achieving Strategic Goals and realizing the Vision takes place in a controlled manner.
c. Choosing and establishing a strategy as a reasonable, methodical, and logical approach is the first step in implementing strategic management. The strategy then serves as a guide for creating and carrying out work programmes. The chosen and agreed strategy can minimize and even eliminate differences and conflicts of opinion in realizing excellence that is directed at achieving strategic goals.
d. Strategic Management can function as a means of communicating ideas, creativity, initiatives, innovations and new information as well as how to respond to changes and developments in the operational, national and global environment, to all parties in accordance with their authority and responsibilities. As a result, reaching consensus on future strategy development or adjustments – with or without modifying the benefits to the organization – will be less of a challenge.
e. With strategic management as a new standard in educational organizations, everyone can take the initiative to contribute in line with their job descriptions. Therefore, in order to bolster the organization's quality, every unit and/or work unit will strive to achieve perfection in their respective fields.
f. All parties engaged are expected to take part in strategic management in educational organizations, which increases the feeling of belonging, responsibility, and engagement. Put simply, strategic management serves to foster a collective mindset that views success as benefiting not only upper management, but the entire organization and the society it serves. Despite the many positives mentioned above, it is important to remember that implementing strategic management in a school setting does not ensure a certain level of performance. Strategic management is just a means to an end – people and the people who put it into action are what really matter for success. Human resources, in their role as implementers, should be composed of professionals who are well-informed, have strong moral convictions, and, most importantly, will not utilize strategic management to further their own or their group's narrow interests.

4. Conclusion

The new model for the positive effects of strategic management on educational institutions, including:

1. Advantages of Implementing Strategic Management
   By implementing Strategic Management, educational organizations (schools) will have advantages, including: financial success, increased productivity, market dominance, technology edge, human resource advantage, ethical practices, and a more congenial work environment.
2. Benefits of Strategic Management
   - With strategic management in place, you can expect your organization to be more dynamic,
   - Your control function to run smoothly and efficiently,
   - Disagreements and conflicts over how to achieve excellence to be minimized,
   - Consensus on changes or new strategies to be developed to be more easily reached,
   - Proactive behaviour from all parties involved in the pursuit of excellence to be encouraged,
   - An overall increase in a sense of belonging, involvement, and responsibility across the board.

References

[1] Alzghoul, A., Algraibeh, K. M., Khawaldeh, K., Khaddam, A. A., & Al-Kasasbeh, O. (2023). Nexus of Strategic Thinking, Knowledge-Oriented Leadership, and Employee Creativity in Higher Education Institutes. *International Journal of Professional Business Review*, *8*(4), e01107. https://doi.org/10.26668/businessreview/2023.v8i4.1107

[2] Bungin, B. (2017). *Analisis Data Penelitian Kualitatif (Jakarta: RajaGrafindo Persada, 2003).* Jakarta: RajaGrafindo.

[3] Galuh NKMR, Barsihanor, Sari Kumala, Abdul Hafiz. (2024). Epistimology of Eco-Friendly Character Education in Islam. *Proceeding Islamic University of Kalimantan (UNISKA_ICET 2024).* Banjarmasin: Publikasi Uniska MAB. 692–713. Diakses April 4, 2024. https://ocs.uniska-bjm.ac.id/index.php/icet/picet/paper/viewFile/60/60.

[4] Hadari Nawawi (2005); *Manajemen Strategik.* Gadjah Mada Pers: Yogyakarta.

[5] Huberman, M., & Matthew B. Milles (2014). *Qualitative Data Analysis (3th Edition An Expunded Sourcesbook).* USA: Sage Publication.

[6] Mercedes, M. R. (2023). Decision in Crisis Using Strategic Thinking. *International Journal of Faculty and Leadership Development*, *1*(1), 38–41. https://journal.dia-ny.org/ijfld/article/view/8

[7] Zazin, N., & Mr, G. N. K. (2018). Manajemen Organisasi tanpa Korupsi Perspektif QS. Al-'Ashr (103): 1–3. *Manageria: Jurnal Manajemen Pendidikan Islam*, *3*(1), 195–213.

[8] Sembiring, B., Br, T., & Galuh, N. K. (2024). *Buku ajar metodologi penelitian kuantitatif dan kualitatif.* Karawang: CV Saba Jaya Publisher.

[9] Amason, A., & Ward, A. (2020). *Strategic Management: From Theory to Practice (2nd ed.)*. Routledge. https://doi.org/10.4324/9781003000594

[10] Asobee, M. S. (2021). Exploring the importance of strategic thinking to strategic planning in the strategic management process. *Journal of Business and Management Sciences*, *9*(2), 68–70. doi:10.12691/jbms-9-2-2.

72 Political education: Strategies and constraints for political parties and election organizers to enhance women's political literacy

Hariyanti[a], Cecep Darmawan, Rahmat, and Leni Anggraeni

Civic Education, Indonesia University of Education, Bandung, West Java, Indonesia

Abstract: The present study focused on socio-cultural and political problems, highlighting the disparity between the high rate of women's political participation in elections and the lack of a corresponding increase in women's representation in parliament, alongside persisting gender stereotypes that hinder women's involvement in the public sphere. The study employed a qualitative design with a descriptive methodology and was conducted from 2022 to 2023, involving informants from the National Democratic (Nasional Demokrat) Party and the General Election Commission of Pekanbaru City. Data collection methods included interviews, documentation analysis, and a review of relevant literature. The results showed that (1) political parties have started political education programs for women, mainly through the party's women's wing, but these programs have not been carried out consistently or systematically; and (2) the general election commission teaches women about politics in several ways, including letting women's groups know about elections and hiring women as community-based election volunteers to teach other women. However, these activities typically only occur during the election stages; women's political education struggles to function effectively outside the process. In the future, a clear model of political education for women should be created, consisting of sustainable and collaborative strategies (political parties, election organizers, and women's organizations), comprehensive and integrative materials, and media that can attract women's interest in knowing political information.

Keywords: Political education, political literacy, women, political parties, general election commission

1. Introduction

Political education is an initiative undertaken by both formal and informal educational institutions aimed at cultivating a political identity that aligns with the prevailing political culture of their constituents. It seeks to enhance political consciousness among citizens at all levels, enabling them to develop self-awareness and the capacity for political engagement. This process encourages active participation in addressing societal issues through various forms of involvement that contribute to positive change. Political education should be grounded in and aligned with the ideology of Pancasila [1]. This term encompasses political personality, political awareness, political engagement, educational institutions, political culture, and individuals or citizens. Educating people in a planned, directed, and systematic way is meant to make them politically aware and able to become morally and ethically responsible political actors who help achieve political goals [2].

Political education seeks to enhance citizens' awareness and engagement with political events within society and the state rather than merely producing experts in political science. Its objective is to cultivate informed and participatory citizens knowledgeable about political concepts and theories [3, 4]. Political education asks for enhancing citizens' political literacy, thereby fostering a participatory political culture, which is essential for an effective democratic process. Therefore, political education is important for citizens [5]. Multiple parties possess both moral and legal duties in the area of political education for citizens, which has been addressed through various laws and regulations. Political parties are responsible for one of these duties.

A political party is an organization that advocates for specific values and ideologies by seeking control of the power structure, which is achieved through participation in general elections [6]. Law No. 2 of 2011, which amends Law No. 2 of 2008 regarding political parties, defines a political party in Article 1 as a national organization established by a group of Indonesian citizens who voluntarily unite based on shared will and ideals. Its purpose is to advocate for and protect the political interests of its members, society, the nation, and the state while upholding the integrity of the Unitary State of the Republic of Indonesia in accordance with Pancasila and the 1945 Constitution. Political education provided by parties is crucial in developing citizens, particularly party cadres, to cultivate a political identity, enhance political awareness, and engage actively and responsively, as these cadres are integral participants in the dynamics of public power. Political education provided by parties serves as a crucial mechanism for

[a]hariyanti2023@upi.edu

DOI: 10.1201/9781003684459-72

enhancing the political participation of female cadres, both within the party and in governmental roles [7, 8].

Other than political parties, election organizers, particularly the general election commission, are responsible for conducting political education. While the general election law specifically refers to this as voter education, the content covered is largely similar, focusing on elections, politics, and democracy. However, there is a greater emphasis on enhancing participation in the electoral process and the technical execution of elections. The General Election Commission (KPU) implements political education through various methods, including election socialization, democracy volunteer programs, public hearings, invitations to KPU's election smart house, and collaborations with women's organizations for seminars, training, and workshops [7]. Currently, the issue at hand is the inadequate political representation of women in parliament. Ironically, this situation persists despite women's higher political participation in voting during general elections compared to men. The following Table 72.1 illustrates this point:

A more focused analysis of the research location, Riau province, reveals the following regarding women's political participation in the 2019 elections: (1) Women's political participation in the presidential election was 50.2%, compared to 49.8% for men; (2) in the Regional Representative Council (DPD) election, women's participation reached 86.08%, while men's was 82.74%; (3) in the Regional House of Representatives (DPRD) election, women's participation was approximately 85.94%, whereas men's was 82.96%. The Permanent Voters List (DPT) recorded 1,900,018 women and 1,963,287 men.

Table 72.1. Women's representation in the general election of the house of representatives

Terms (s)	*Women*	*Men*
1950–1955 (DPRS)	9 (3.8%)	236 (96.2%)
1955–1960	17 (6.3%)	272 (93.7%)
1956–1959 (The Constitutional Assembly)	25 (5.1%)	488 (94.9%)
1971–1977	36 (7.8%)	460 (92.2%)
1977–1982	29 (6.3%)	460 (93.7%)
1982–1987	39 (8.5%)	460 (91.5%)
1987–1992	65 (13%)	500 (87%)
1992–1997	62 (12.5%)	500 (87.5%)
1997–1999	54 (10.8%)	500 (89.2%)
1999–2004	46 (9%)	500 (91%)
2004–2009	61 (11.8%)	489 (88.9%)
2009–2014	101 (18.10%)	459 (82%)
2014–2019	97 (17.3%)	463 (86,3%)
2019–2014	21%	79%

Source: Compiled from various sources.

The analysis of women's representation indicates that the proportion of female representatives in parliament has consistently remained below 30%. Despite the government implementing a 'special' policy to support women politicians through *affirmative action*, the situation remains unchanged. Relevant legislation mandates that political parties must incorporate a minimum of 30% women in their management structures and adhere to a 30% quota for female candidates in legislative elections at the regency/city, provincial, and governmental levels. Nonetheless, the effectiveness of the quota system in enhancing women's representation relies upon the *political will* of party administrators [9]. Suppose these administrators actively promote the regeneration of women cadres through specialized and intensive political education programs, empowering them to become competitive and trustworthy community members while also minimizing financial barriers typically faced by women. In that case, an increase in women's representation is likely to occur. The quota system for women politicians has been implemented in nearly all democratic nations globally, as gender and political issues have gained prominence, often being strictly regulated in state legislation [10].

Several factors are believed to contribute to the exclusion of women from political institutions. The analysis begins with cultural constraints and religious interpretations, followed by considerations of economics, family support, education level, mental health, and the political system, which is often unfriendly to women. There are two main problems in the area of religion: (a) people in the community do not know or understand enough about the religious values that define the role and function of women; and (b) different interpretations of religious teachings keep happening that hurt women's status and role [11]. In the socio-cultural domain, cultural norms continue to categorize political activity as predominantly male-dominated. Women, paradoxically, contribute to the notion that power is incompatible with femininity. Many women tend to reject power as an aspect of their identity [12].

Another issue is (2) the limited awareness among women regarding voting for female representatives, primarily influenced by their low political literacy. This statement aims not to encourage voters to elect representatives based on gender but to suggest that women legislators may possess distinct perspectives, instincts, and ideas compared to their male counterparts, particularly concerning gender, children, and family issues. Representation by women is preferable, particularly given that Indonesian regulations mandate political parties to nominate 30% of women among legislative candidates.

Further analysis identifies the following factors contributing to women's low electability among female voters: (1) Women continue to perceive women's legislative candidates as supplementary, as demonstrated by the serial numbers typically positioned at the bottom of the list. This underscores the importance of the party in promoting women's representation in parliament. The public continues to perceive female candidates as supplementary [13]; (2) Additionally, the recruitment process for female legislative candidates remains inadequate, as parties often aim merely to meet the requirement of nominating 30% of women, resulting in candidates viewed as less suitable by the community. Political party administrators possess significant internal authority regarding the recruitment patterns and the assignment of serial numbers to female legislative candidates, which impacts the promotion of *affirmative action* [14, 15]. It is essential to examine the party's initiatives in delivering political education to cadres and election organizers within the broader women's community. The aim is not to influence political preferences but to furnish political knowledge for electing representatives in general elections, particularly those committed to gender equality. This paper examines the efforts made by relevant parties, particularly political parties and the General Election Commission (KPU), to enhance women's political literacy.

2. Methods

This study employed a qualitative descriptive method to comprehensively describe the strategies and challenges encountered by parties in enhancing women's political literacy. Qualitative research methods, often referred to as naturalistic research methods, are conducted in natural conditions or settings (Abdussamad, 221 C.E.). The study focused on the National Democratic (Nasional Demokrat) Party and the Pekanbaru City General Election Commission, with data collection occurring between 2022 and 2023. Data collection methods include interviews, document analysis, and literature reviews of relevant studies. The analysis employed the Miles and Huberman model alongside data source triangulation.

3. Results and Discussion

3.1. Political education of women cadres of the National Democratic (Nasional Demokrat) political party

The political education of female cadres aims to cultivate loyalty and pragmatism regarding national issues and foster an attachment and sensitivity to the surrounding community [16]. The National Democratic (Nasional Demokrat/Nasdem) Party implements political education via its political and cultural education division. The party conducts political education both systematically and opportunistically. The party implements an annual political education program, as outlined in the board's work program; however, its full realization has not been achieved. The party's political education activities are funded by budget allocations from government assistance funds, with 60% of these funds designated for political education initiatives. The party has conducted various political education activities, including (1) legislative schools; (2) legislative meetings; and (3) faction consolidation activities. The purpose of these activities is to enhance the quality and capabilities of cadres.

The chairman of the Regional Leadership Council (DPW) of the National Democratic Party (Nasdem) stated that the party has implemented various political education strategies to enhance female cadres' political literacy. However, these efforts have not yet reached their full potential, including: (1) programs conducted through the political and cultural education division of the Regional Leadership Council. Political education is carried out in a planned and incidental manner. As outlined in the board's work program, the party implements an annual political education program but has not fully realized its potential. The party's political education activities are funded by budget allocations from government assistance funds for political parties, with 60% designated for such educational initiatives. The party has conducted various political education activities, including: (a) legislative schools, (b) legislative meetings, and (c) faction consolidation activities. The purpose of these activities is to enhance the quality and competence of cadres; (2) the party collaborates with other socio-political organizations through planned political education programs for party cadres, as outlined in the party management work plan. The political education activities encompass training sessions, public discussions, and seminars featuring invited resource persons. The event invites experts from government agencies and socio-political, economic, and religious organizations to promote women's creativity in entrepreneurship and enhance their understanding of religious values. Nonetheless, this type of cooperation has not been formalized through an official agreement or Memorandum of Understanding (MoU); it remains merely an invitation for participation as a resource person during activities organized by the Nasdem Party. The implemented cooperation strategy lacks a systematic approach. The resource person is identified according to the requirements prior to the execution of the activity. The establishment of a specialized women's organization, GARNITA MALAHAYATI, aims to empower women. Party-wing organizations are a prevalent feature within the party political system. Article 12 letter j of Law Number 2 of 2008 concerning Political Parties states that 'Political parties have the right to form and own political party wing organizations,' thereby establishing the legal basis for the existence of such organizations. Garnita has not formally partnered with community organizations focused on women's needs or mass organizations aimed at enhancing women's skills to implement sustainable political education. However, due to its designation as a wing organization, its nature lies outside the internal structure of party management. This indicates that it serves solely as a supplementary system to fulfill the objectives of the party program. Garnita operates as a wing organization focused on women's issues, aiming to enhance the capacity and quality of female cadres to prepare them for advancement and election in legislative positions. Creating a specialized women's wing within the party is an inadequate method for empowering women, as this wing operates outside the party's structural framework and functions merely as a supplementary system. Wing organizations lack the capacity to influence party decisions and policies, resulting in Garnita's inability to promote and impact the formulation of gender-responsive party policies. The research findings indicate that Garnita Malahayati has not been significantly active in providing political education for women cadres. Additionally, the State Defense Academy (ABN) program, initiated by Surya Paloh, chairman of the National Democratic (Nasdem) Party, aims to deliver structured and centralized political education to cadres. The program enables ABN cadres to provide political education to fellow cadres and the local community, positioning them for nomination as legislative candidates.

The National Democratic (Nasdem) Party provides the following political education materials. The initial focus is on instilling militancy and loyalty among cadres to the party, ensuring that the materials provided encompass the party's vision, mission, platform, jargon, ideology, and line of struggle. Secondly, it is essential to possess knowledge specific to the region, particularly regarding West Sumatra. Cadres are local candidates registered as Nasdem members. Therefore, debriefing is essential to develop the intelligence to explore and advance regional interests through practical political platforms, specifically within party activities. Third, briefings are provided on specific examples of community activities that cadres can implement. Fourth, impart knowledge and comprehension of political package legislation, including laws governing political parties, regional head and deputy regional head elections, and legislative elections. The laws and regulations are the foundation for cadres to participate in practical politics. A cadre aspiring to become a politician must engage in political activities following relevant laws to ensure their power exercise is not arbitrary. Fifth, the material concerning political norms and ethics is equally significant. This is underscored to govern the behaviour of Nasdem party cadres. Sixth, the material presented serves as a strategy for cadres to engage the broader community in joining and supporting the Nasdem Party. Political education is delivered in a comprehensive manner. Seventh, the need for practical politics arises from the party's objective of providing cadres to earn the trust and acceptance of the populace. The outcome of political education is that cadres can serve as representatives of the people.

The implementation of political education for women party cadres faces several obstacles from both internal and external sides. The first obstacle political parties face in promoting political education for women cadres originates from within the women themselves, specifically their low self-confidence in participating and engaging actively in practical politics. Historically, women have been socialized to experience insecurity and discomfort regarding political engagement. The second obstacle arises from men who are similarly resistant to including women at parties. From the perspective of desire, willingness, and participation, these elements appear to be present in women. This represents a fundamental issue. The involvement of women in practical politics is not aimed at usurping power from men or exerting dominance over them. Misconceptions can result in societal misunderstandings and hinder women's political engagement. The third obstacle pertains to obtaining support and permission from family members. Women participating in various socio-political activities and employment are required to obtain family consent. The challenge of obtaining family permission is also connected to the factor of time. Female cadres often struggle to balance family responsibilities with active political engagement within the party, consequently impacting their involvement in party activities and programs [17]. When considering the relationship between time and economic factors, it appears that female cadres with adequate financial resources are more likely to be active compared to those with constrained economic circumstances. The fourth obstacle is the enduring strength of patriarchal political culture within society. The societal culture restricts women's involvement in political parties, thereby limiting their participation in politics. The fifth obstacle arises from internal political parties. Funding for political education is still a big problem for political parties that want to teach their cadres about politics. This makes it harder for the parties to plan and carry out programs that will give their cadres more power.

The sixth problem is that Indonesia's laws do not have harsh punishments for (1) parties that do not teach their cadres about politics, especially women cadres, and (2) parties that do not support affirmative action policies that give women party cadres more power to increase women's political participation and representation in parliament [18].

Political parties offer political education to female cadres to improve their political literacy. The hope is that the rise of knowledgeable, proactive, and responsive cadres will increase the public's trust in electing women as representatives. An increase in the number of women representatives in parliament enhances the effectiveness of gender-sensitive policy-making, as the representation of women directly influences their voice and position within the legislative body [19]. An increase in the number of women in strategic government positions correlates with a rise in policies aimed at empowering and protecting women.

3.2. Women's political education by the general election commission

Voters are individuals who fulfil the legal criteria necessary to use their voting rights. In Indonesia, voting eligibility requires being an Indonesian citizen, at least 17 years of age, or married. Election organizers must ensure that voters are adequately facilitated to cast their votes. Women voters represent a significant segment of voter education, as outlined in Article 8 of PKPU No. 8 of 2017 [20]. In the 2019 elections in Riau Province, the Permanent Voter List (DPT) recorded a total of 3,863,305 votes, comprising 1,963,287 male voters and 1,900,018 female voters. Women's political participation rate significantly exceeds that of men. During the 2019 presidential election, the participation rate for females was 50.2%, compared to 49.8% for males. In the Regional Representative Council (DPD) election, the participation rate for females was 86.08%, compared to 82.74% for males. In the 2019 Regional House of Representatives election, the participation rate for females was 85.94%, compared to 82.96% for males. However, an analysis of women's representation in local legislative institutions reveals a decline. In the 2014 elections, the count of female politicians in the Regional House of Representatives (DPRD) was 18, whereas in the 2019 elections, it declined to 12 [21].

Women voters necessitate political education due to multiple considerations, including: (1) the persistent dichotomy between women and politics. Politics is often viewed as a cruel and competitive activity, contrasting with characteristics typically associated with women, such as meekness, conflict aversion, and a lack of firmness in upholding principles [22]. This perception requires clarification through political education for women, emphasizing that politics is a collective concern and a significant social dimension of human existence, as it influences various life fields [23]. Additionally, (2) women remain regarded as objects within the political sphere. Female voters are characterized by their loyalty and subjectivity, making them frequent targets for pragmatic election candidates [24]. Also, (3) women hold a strategic position due to their dual roles as wives and mothers and their involvement in the community and other associations, equipping them with the knowledge and skills necessary for providing political education in these environments [25].

The Pekanbaru General Election Commission conducts political education for female voters through various activities, including serving as resource persons in seminars and socializations organized by other agencies and institutions and utilizing social media and YouTube. However, these activities can be characterized as incidental rather than sustainable. This paper provides a detailed account of women's political education initiatives. Specifically, (1) on July 18, 2023, Mrs. Yelly Nofiza, a commissioner of the Pekanbaru General Election Commission, served as a resource person during capacity-building activities focused on law and politics, organized by the Women's Empowerment, Child Protection, and Community Empowerment Office of Pekanbaru City. This event was attended by representatives from various women's organizations in Pekanbaru and aimed at enhancing women's political and legal literacy. Additionally, (2) podcasts were produced on the Pekanbaru General Election Commission's YouTube channel, focusing on the role of women in politics. The speakers included Mrs. Yelly Nofiza and Gabby Vionalinsyah, the heads of a women's organization in the Pekanbaru area. This podcast seeks to inform women about their potential involvement in politics, emphasizing the importance of political literacy and active citizenship through participation in elections. In addition, (3) it features a podcast on the YouTube channel centered around Kartini's legacy as a heroine of democracy in Pekanbaru city, commemorated on April 21. This activity seeks to identify contemporary figures in Pekanbaru who exemplify the movement for women's emancipation, akin to Kartini's legacy.

The General Election Commission of the Republic of Indonesia has implemented a democracy volunteer program to empower women in political activities, targeting various segments of voters, including women voters. The Democracy Volunteer Program, henceforth termed 'Relasi,' is a social initiative aimed at enhancing voter participation and the quality of voting rights exercised by individuals. This program incorporates community participation in its recruitment, positioning individuals as pioneers of democracy within their communities. Relasi seeks to (1) enhance the quality of the electoral process; (2) boost voter participation; (3) foster public trust in the democratic process; and (4) promote civil society volunteerism in the election and democratization agenda [26]. The *Relasi* program was launched by the General Election Commission of the Republic of Indonesia, engaging community groups from 11 strategic voter bases, including the women's voter base, during the 2014 and 2019 elections.

Relasi's voter education initiatives aimed at the female voter base are considered strategic for several reasons. First, Relasi is a national program designed to disseminate electoral and democratic information on a large scale effectively. A broader dissemination of information will enhance the political education of women voters. Second, Relasi is a voluntary initiative by voters aimed at educating their communities about elections and democracy. The objective is to effectively communicate electoral information using accessible language the community can easily understand and comprehend. Third, the community hosts numerous women's associations, including formal organizations like Family Empowerment and Welfare (PKK) and Dharma Wanita (the organization for the wives of Indonesian civil servants), as well as informal groups such as arisan (social gatherings) and yasinan (communal Quran readings). Relasi may originate from these groups to ensure that the electoral and democratic information presented is effectively understood in a conducive environment.

The findings of a study regarding democratic relations in Pekanbaru and Dumai support the abovementioned statement [27]. The study indicates that the political education model implemented by women's organizations is effective in enhancing women's political participation. The strategies employed in women's relations include: (1) the mapping of groups of women voters affiliated with specific communities. Women voters typically participate in Quran recitation groups, Family Empowerment and Welfare (PKK), Dharma Wanita (the organization for the wives of Indonesian civil servants), health exercise communities, and neighborhood social gatherings known as arisan. The primary women's association for political socialization is the women's Quran recitation/wirid group, alongside the gymnastics community, arisan (social gatherings), and Family Empowerment and Welfare (PKK). Additionally, (2) the association engages in interpersonal communication with the management of the target community. This request seeks permission from community management to use the association's activities to conduct voter education initiatives. The selection of the target community relies on personal introductions from democracy volunteers to community management. Additionally, (3) women in the vicinity are approached informally, including neighbors and members of the local mosque congregation. Democracy volunteers may engage in informal discussions in residential outdoor spaces or at mosques while awaiting prayer times. This strategy is implemented individually by volunteers rather than through group activities like those planned in the base program for voter education. This strategy is also relatively simple to implement without incurring significant costs. Furthermore, Relasi has the personal responsibility to educate the surrounding community regarding electoral information and democracy.

The materials provided by women involved in Relasi are as follows: (1) The significance of democracy, electoral processes, and civic engagement; (2) Schedule and timing of voting; (3) Voting procedures during elections; (4) Overview of election candidates (presidential and vice-presidential candidates, political parties, and regional representatives); (5) Additional topics relevant to the voter demographic. The primary weakness is the lack of specific materials the General Election Commission provides for women voters. Women involved in Relasi are expected to create their own political materials even though many do not have backgrounds in political science, women's organizations, or government work. Most women involved are housewives seeking to participate in electoral activities, which results in limited knowledge regarding specific strategies to enhance women's political capacity [20]. In the 2024 simultaneous elections, the Indonesian General Election Commission did not continue the Relasi Program due to various considerations. This program is recognized for its effectiveness in empowering women as volunteers within their groups and facilitating their engagement with women's groups to educate them on politics, democracy, and electoral issues.

Obstacles encountered by election organizers in implementing political education for women can be identified as follows. First, the lack of funding for voter and political education is notable. Funding for political education activities is available annually; however, it must be allocated across various programs targeting different voter segments, limiting its effectiveness if focused solely on one group within the community. The voter education budget remains limited outside the election period, as indicated by an interview with a commissioner from the General Election Commission. Secondly, there is a lack of focused and effective collaboration between the General Election Commission and women's organizations or associations. The commissioners of the General Election Commission are infrequently invited to serve as resource persons in seminars that are conducted. Currently, there is a lack of collaboration in which the General Election Commission can serve as a partner and resource for ongoing studies conducted by women's organizations or joint initiatives to enhance women's political literacy. Third, the General Election Commission lacks a political education curriculum specifically designed for women, which would address the roles and positions of women in politics, their political participation, and the importance of women's involvement in the political sphere. This initiative is designed to enhance women's political awareness and knowledge.

From the analysis and comparison of the research findings presented, the following conclusions can be drawn concerning the women's political education programs implemented by the General Election Commission. First, women's political education occurs primarily during election periods, lacking a sustainable framework. This is marked by the lack of women's political education programs outside the electoral period organized by the General Election Commission. Secondly, women's political literacy includes not only general election issues but also broader aspects of politics and democracy. Political education should promote women's active participation in voting (electoral participation) and encourage engagement in non-electoral activities, such as joining political parties, participating in community forums, and becoming members of women's organizations that contribute input to the government. To achieve this goal, intense, extensive, and ongoing political education for the women's base is essential. Relying on social media, government, and political parties for women's political literacy comes with risk.

4. Conclusions

The rate of women's political participation exceeds that of men in every election; however, this does not correlate directly with the level of women's representation in parliament. Disparities persist despite the government's implementation of affirmative action regulations to support women politicians. This situation is unfortunate. The law mandates the nomination of 30% of women in legislative elections; however, there are no penalties for parties that contravene this requirement, nor are there rigorous rules concerning the parties' duty to promote regeneration through political education. Political education serves as a fundamental asset for women in the electoral process, influencing their abilities to communicate, lobby, express opinions, formulate visions and missions, and develop programs. It also enables them to gather community aspirations and translate these into effective and acceptable policies. Election organizers exhibit limited concern for women's political education, primarily conducted during the election phase. This is hindered by insufficient funding for political education and a lack of collaboration with women's organizations. Currently, it can be observed that the political education conducted by political parties and election organizers lacks direction and sustainability, often appearing to be incidental. It is essential to develop a precise model of political education for women in the future to achieve practical political literacy. This model should serve as a guideline for implementation and include a comprehensive political education curriculum along with appropriate media support.

References

[1] Arif Prasetyo Wibowo, N. Y. (2022). Pendidikan politik sebagai usaha partai politik menyemai nilai-nilai pancasila dalam kehidupan keorganisasiannya. *Jurnal Ilmiah Pendidikan Pancasila Dan Kewarganegaraan*, *7*(Nomor 2), 458–467. http://journal2.um.ac.id/index.php/jppk/article/view/7820/3749

[2] Rafni, A., & Suryanef. (2012). *Pengembangan Model Pembinaan Perempuan Kader partai oleh Partai Politik Sebagai Upaya Peningkatan Kualitas Keterwakilan Perempuan di Lembaga Legislatif Daerah*.

[3] Handoyo, D. E., & Lestari, P. (2017). Pendidikan Politik. In *Percetakan Pohon Cahaya* (Issue November).

[4] Yusriati, Y., & Amrizal, D. (2020). Efektivitas Model Pendidikan Politik Dalam Pelaksanaan Sosialisasi Pemilu di Kabupaten Deli Serdang. *Warta Dharmawangsa*, *14*(3), 500–507. https://doi.org/10.46576/wdw.v14i3.831

[5] Hariyanti. (2022). Political education pocket book for pre-electors as an effort to prepare rational beginning voters. *Jurnal Ilmiah Pendidikan Pancasila Dan Kewarganegaraan*, *7*(November), 682–691.

[6] Jeffs, R. A., Hayward, J., Roach, P. A., & Wyburn, J. (2016). Activist model of political party growth. *Physica A: Statistical Mechanics and Its Applications*, *442*, 359–372. https://doi.org/10.1016/j.physa.2015.09.002

[7] Hariyanti, Cecep Darmawan, & Masyitoh, I. S. (2018). Peran partai politik dalam meningkatkan partisipasi politik kader perempuan melalui pendidikan politik. *Jurnal Civics: Media Kajian Kewarganegaraan*, *15*(1), 74–85. https://doi.org/10.21831/jc.v15i1.17659

[8] Roviana, S. (2015). Pendidikan Politik Koalisi Perempuan Yogyakarta Wilayah Yogyakarta 2000–2008. *Jurnal Sosiologi Reflektif*, *9*(2). https://www.neliti.com/publications/131795/pendidikan-politik-koalisi-perempuan-yogyakarta-wilayah-yogyakarta-2000-2008

[9] Susanti, S., & Lubis, A. (2015). Partisipasi Politik Perempuan pada Partai Keadilan Sejahtera Kota Medan. *JPPUMA: Journal of Governance and Political Social UMA*, *3*(1), 1–13. http://ojs.uma.ac.id/index.php/jppuma/article/view/906

[10] Rosen, J. (2017). Gender quotas for women in national politics: A comparative analysis across development thresholds. *Social Science Research*, *66*, 82–101. https://doi.org/10.1016/j.ssresearch.2017.01.008

[11] Warjiyati, S. (2016). Partisipasi Politik Perempuan Perspektif Hukum Islam. *Al-Daulah: Jurnal Hukum Dan Perundangan Islam*, *6*(1), 1–27. https://doi.org/10.15642/ad.2016.6.1.1-27

[12] Oktaviani, A., & Megawati, M. (2021). Quo Vadis Kebijakan Affirmative Action dalam Sistem Demokrasi di Indonesia. *Ahmad Dahlan Legal Perspective*, *1*(2), 90–111. https://doi.org/10.12928/adlp.v1i2.4730

[13] Masitoh, N. (2016). Gerakan Dan Representasi Politik Perempuan Di Kota Tasikmalaya. *PALASTREN Jurnal Studi Gender*, *9*(1), 215–232. https://doi.org/10.21043/palastren.v9i1.1924

[14] Suhaimi, E. (2021). Prinsip-Prinsip Umum Rekrutmen Politik Sebagai Landasan Idil Penyusunan Pola Rekrutmen Dalam Ad/Art Partai Politik Di Indonesia. *Jurnal Hukum Tri Pantang*, *7*(1), 27–51. https://doi.org/10.51517/jhtp.v7i1.295

[15] Zamroni, M. (2013). Perempuan dalam kajian komunikasi politik dan gender. *Jurnal Dakwah*, *14*(1), 103–132.

[16] Afrianty, I., & Afrianty, I. R. A. (2017). Rekrutmen Calon Anggota Legislatif Perempuan Partai Demokrat dan Partai Nasdem Tahun 2014. *Politeia: Jurnal Ilmu Politik*, *7*(2), 64–73.

[17] Aisyah, N. (2013). Relasi gender dalam institusi keluarga (Pandangan Teori Sosial Dan Feminis) | MUWAZAH: Jurnal Kajian Gender. *MUWAZAH: Jurnal Kajian Gender*, 203–224. http://e-journal.iainpekalongan.ac.id/index.php/Muwazah/article/view/346/0

[18] Rafni, A. (2013). Kesetaraan Jender Dalam Politik: Pembinaan Kader Perempuan Oleh Partai Politik, Upaya Menuju Penguatan Kapasitas Legislatif Daerah. *Jurnal Ilmiah Kajian Gender*, 144–171.

[19] Mendelberg, T., & Karpowitz, C. F. (2016). Women's authority in political decision-making groups. *Leadership Quarterly*, *27*(3), 487–503. https://doi.org/10.1016/j.leaqua.2015.11.005

[20] Hariyanti, & Hambali. (2021). Identifikasi Kendala Relawan Demokrasi Basis Perempuan sebagai Fasilitator Pendidikan Pemilih di Kota Pekanbaru dan Kota Dumai. *Journal of Moral and Civic Education*, *5*(1), 1–11. https://doi.org/10.24036/8851412512020490

[21] Ahmad Eddison, Hariyanti, H. (2023). Kader desa peduli pemilu dan pemilihan sebagai fasilitator pendidikan pemilih guna meningkatkan partisipasi politik masyarakat 1,2,3. *JCES (Journal of Character Education Society)*, *6*(1), 46–57.

[22] Kumar, S., & Prakash, N. (2017). Effect of political decentralization and female leadership on institutional births and child mortality in rural Bihar, India. *Social Science and Medicine*, *185*, 171–178. https://doi.org/10.1016/j.socscimed.2017.04.013

[23] Koyuncu L., B., & Sumbas, A. (2016). Discussing women's representation in local politics in Turkey: The case of female mayorship. *Women's Studies International Forum*, *58*, 41–50. https://doi.org/10.1016/j.wsif.2016.06.003

[24] Mukarom, Z. (2015). Strategi Komunikasi Politik Perempuan di Lembaga Legislatif. *Ilmu Dakwah: Academic Journal for Homiletic Studies*, *5*(18), 599–624.

[25] Homan, P. (2017). Political gender inequality and infant mortality in the United States, 1990–2012. *Social Science and Medicine*, *182*, 127–135. https://doi.org/10.1016/j.socscimed.2017.04.024

[26] Suryanef, S., & Rafni, A. (2020). First-Time Voter Education Through the Democracy Volunteer Movement. *Journal of Moral and Civic Education*, *4*(2), 60–68. https://doi.org/10.24036/8851412422020499

[27] Hambali, H., Hariyanti, H., & Eddison, A. (2021). The model of voter education of women-based democracy volunteers. *Jurnal Civics: Media Kajian Kewarganegaraan*, *18*(1), 48–60. https://doi.org/10.21831/jc.v18i1.33224

73 The role of urban forestry to enhance city sustainability: A case study of Tebet eco park in Jakarta

Zarina Akbar[a], Dian Alfia Purwandari[b], Irsyad Hafiz[c], Alya Fajria Rahma Maulida[d], Muhammad Fadhil Haritsah[e], and Vranz Damar Sophan[f]

Master of Environmental Education, Postgraduate School, Universitas Negeri Jakarta, Jakarta, Indonesia

Abstract: Jakarta as the largest metropolitan city in Indonesia faces various environmental challenges, including air pollution, green space degradation, and climate change which are generally caused by population growth and rapid urbanization in Jakarta. Urban forestry has emerged as a strategic solution in maintaining environmental balance, providing green spaces that not only improve air quality but also support biodiversity and mitigate the impacts of climate change. The main focus of this study is to assess the positive impacts of urban forests on air quality, the Urban Heat Island (UHI) phenomenon, rainwater management, and biodiversity conservation, while also looking at the social benefits for urban communities. Through literature analysis and field case studies in one of the urban forests in Jakarta, this study found that urban forests play a significant role in absorbing carbon dioxide and pollutants, reducing environmental temperatures through the evapotranspiration process, and providing effective shade to reduce the impact of UHI. In addition, urban forests contribute to rainwater absorption, which helps reduce the risk of flooding and become an important habitat for various flora and fauna species that are threatened by urbanization. This study recommends a sustainability-based approach in urban spatial planning which strengthens the role of urban forests as a sustainable ecosystem solution and supports the quality of life of the community.

Keywords: Urban forest, urban ecosystem, environmental sustainability, ecological balance, environmental degradation

1. Introduction

The development of big cities around the world, including Jakarta, has resulted in various significant impacts on the environment. Rapid and uncontrolled urbanization has triggered serious challenges for urban ecosystems, ranging from air pollution, and degradation of green open spaces, to increasing temperatures. The chances of environmental damage will increase along with increasing population density [1]. Rapid population growth especially in developing countries will simultaneously increase environmental degradation. Urbanization in Jakarta continues to increase every year, and this exacerbates the decline in environmental quality in this city. One of the consequences is the loss of important ecosystem functions previously provided by natural areas, such as carbon absorption and microclimate regulation.

In the midst of this urban environmental crisis, urban forestry has emerged as a sustainable solution that is increasingly relevant in overcoming the increasingly degraded environmental balance in urban areas. Urban forest is not only function as the lungs of the city, but also provide various vital ecosystem services, including improving air quality, providing space for biodiversity, and reducing the impact of climate change. In the Jakarta area itself, as the center of government and economy, it has faced environmental problems for quite a long time, so a systematic solution is needed to restore its ecological balance. Urban forests are developed with an approach that focuses on the rehabilitation of green open spaces, which can be an important instrument in maintaining environmental sustainability.

The rapid development in Jakarta has brought real pressure on environmental aspects where the reduction of green spaces, increasing air pollution, and the increasing risk of flooding are problems that have been going on for a long time but have not received real solutions to overcome them. Based on data from the Ministry of Environment and Forestry (KLHK), the coverage of green open spaces in Jakarta is currently only around 9% of the total area, far below the national target of 30% as regulated in Law No. 26/2007 concerning Spatial Planning. With the lack of Green Open Space (RTH), not only are urban environmental conditions deteriorating, but the visual aesthetics and comfort of living for the community are also decreasing.

The Forest Development offers a solution that can be implemented in Jakarta, with an approach that not only improves environmental conditions but also improves the

[a]zarina_akbar@unj.ac.id, [b]dian-alfia@unj.ac.id, [c]irsyaadhafzd1237@gmail.com, [d]alfarama1971@gmail.com, [e]mfh4596@gmail.com, [f]damarvranz994@gmail.com

DOI: 10.1201/9781003684459-73

quality of life of its citizens. As regulated in Jakarta Regional Regulation No. 1/2012 concerning Regional Spatial Planning, urban forests are an integral part of green open spaces that aim to maintain ecological balance amidst rapid urban development. Increasing urban forests can contribute to reducing air pollution, providing healthy recreational spaces, and increasing the city's resilience to climate change and natural disasters, such as flooding. However, there are many limitations in the application of policies on urban forestry, such as limited land, lack of public awareness of the importance of environmental sustainability, and regulatory constraints. This study aims to analyze the role of urban forestry in maintaining environmental balance in Jakarta, evaluate the benefits it can provide, and identify challenges that need to be faced in its management and development. Using a case study of Jakarta provides a localized perspective, this study will explore the significant role of urban forests in supporting environmental sustainability in Indonesian context.

2. Theoretical Review

Urban forests are areas that are located in or around urban areas that are covered by trees that are allowed to grow naturally resembling forests and are not arranged like parks. The benefits of having this green area are not only to reduce urban environmental degradation but also to improve the environment and aesthetics, making good and healthy environmental information easily accessible. The existence of urban forests is needed to balance the increasingly dense physical development of cities and limit the green open spaces. In addition, green spaces such as urban forests also provide ecological compensation as watersheds and places to recycle carbon dioxide into urban oxygen [2]. Urban Forests are stretches of land where trees grow compactly (densely) in urban areas, both on state land and private land, and states that they are designated as urban forests by authorized officials [3].

Urban forests have tangible benefits such as wood, medicinal plants, biomass, recreation, and so on, and intangible benefits such as hydrological functions, carbon assimilators, and so on. Urban forests need to be maintained and developed considering their function for the environment [4]. Urban forests are one type of green open space in urban areas, urban forests are areas in which there are dense and compact trees and are important conservation in urban environments because they have benefits such as regulating urban air heat, preventing dust solid pollution, providing O2, habitat centers for wild flora and fauna, and providing urban aesthetics [5].

The urban ecosystem theory states that urban ecosystems are all ecological systems in a city and around the city area that are densely populated by people it. This ecosystem is a mixture of man-made elements and natural elements that interact with each other such as the behaviour, culture, and socio-economic conditions of the community [6]. In the urban ecosystem, the problems of population density and environmental degradation are two things that are interconnected, where the more a city grows, the less green open space there is in the city [5]. Urban development is something that cannot be avoided, accessibility to and from the city encourages people to migrate in search of a better life. The development of city infrastructure to facilitate the needs of its citizens often takes up green space, reducing the number of green open spaces in cities [7].

Green Open Space (RTH) in the current global development environment is needed to maintain the balance of the quality of the environment in an area, especially in urban areas that have various problems related to such complex spatial issues. As a part of a strategic step in urban ecosystem restoration, related to ecological functions, for example, as a climate controller, namely as an oxygen producer, noise reducer, and as a visual control, namely by blocking sunlight or reflected light [8]. Other ecological potentials of the City Forest are also related to the UHI (urban heat island) phenomenon, which causes microclimate changes in urban areas and contributes to the phenomenon of global warming [9]. In providing various environmental services, the City Forest reflects a commitment to environmental sustainability in metropolitan areas [10].

3. Research Methods

This research used literature study and field observation research methods. Literature Study is research which was conducted by the researcher by collecting a number of books, and magazines related to the problem and research objectives. Field observation was carried out by conducting direct observations on urban forests in the DKI Jakarta area.

4. Discussion

Urban forests play an important role in improving air quality in urban areas by absorbing carbon dioxide (CO_2) and other pollutants such as sulfur dioxide (SO_2), nitrogen dioxide (NO_2), and dust particles. Clean natural air is often polluted by dust, both from natural activities and human activities [11]. One of the factors causing the decline in air quality is the increasing number of vehicles. As is known, the city of Jakarta is an area with a high population density, which is directly proportional to the number of vehicles used by its residents. Urban forests play a role in absorbing high motor vehicle emissions in Jakarta, especially in dense traffic areas. According to [10] this opinion is supported that the function of the Urban Forest is an area within the city that functions as a regulator of the microclimate in the surrounding environment. Urban forests not only play a role in absorbing pollutants but also act as a provider of oxygen for humans and other living things.

In urban conditions experiencing massive urbanization such as Jakarta, the Urban Heat Island (UHI) phenomenon, which is a significant increase in temperature in urban areas compared to the surrounding areas, has become a serious issue in big cities. UHI is a phenomenon where the surface temperature in urban areas is higher than the temperature in the surrounding areas due to various community activities in it. This often occurs in urban areas because the conversion of green land into hard surfaces reduces the soil's ability to absorb water and increases surface temperatures. Human activities such as industry, transportation, and the use of emitting energy also cause an increase in the effects of UHI. Urban forests act as part of mitigation that functions to reduce local temperatures through evapotranspiration and shade provision. In carrying out its function, urban forests play a role in lowering surface temperatures by several degrees, which is important for reducing the effects of UHI in urban areas.

Jakarta is a low-lying area, so flooding at the beginning of the rainy season is an annual event that occurs in Jakarta. This is reinforced by data that topographically Jakarta is located in the lowlands [12]. Almost 40% of its area is below sea level. This causes Jakarta to often be hit by floods. Urban forests play an important role in absorbing rainwater and reducing surface water flow, thereby reducing the risk of flooding. Tree roots in urban forests increase water infiltration into the soil, which slows the flow of water into drainage channels and rivers. From the perspective of the DKI Jakarta city government itself, the existence of conservation urban forests is an effort by the DKI Jakarta Provincial Government to increase the area of green open space which plays a role in mitigating climate change, including flooding and rising sea levels which are useful for reducing the impact of flooding in DKI Jakarta [13]. With the presence of large amounts of urban forests, nature's ability to maintain the balance of urban ecosystem spatial planning increases, especially in the availability of water catchment areas which are an important element in efforts to prevent flooding and land subsidence which are serious concerns in the DKI Jakarta area.

To conserve biodiversity, urban forests also play an ecological role for various species, namely as a habitat for local flora and fauna that are increasingly threatened by urbanization. In this green cover area, such as birds, fish, and even reptile species are widely found and function as habitats to survive in other urban areas. The presence of urban forests supports species diversity and prevents the local extinction of various animals in urban areas that are already dominated by concrete and asphalt. This is reinforced that urban forests can be used as a place for collecting biodiversity [14]. Urban forests can be viewed as a conservation area outside the conservation area because in this area flora and fauna can be preserved.

Another role of urban forests from an ecological perspective is as an aspect that provides adequate aesthetics and environmental quality and encourages sustainability in urban areas [15]. Urban forests improve urban aesthetics by providing beautiful and visually pleasing green spaces, this improves the comfort and quality of life of urban communities. The presence of urban forests and various green open spaces in urban areas increases various benefits that continue in various areas of life. Sustainable development that takes place in urban areas is illustrated by the development of urban forests which contribute to the long-term ecological balance needed in modern city planning. As an illustration of the sustainable environmental aspect, urban forests can also function as a medium for environmental education which is a living laboratory for community education about the importance of preserving nature and biodiversity.

5. Conclusion

Based on the explanation that has been presented, it shows that City Parks (urban forestry) have a very important role in maintaining the balance of the urban environment, especially in Jakarta. Jakarta as a city that is the center of economy and government certainly attracts many people to get a better life, but this causes an environmental impact that is inversely proportional to the increasing development in Jakarta. In facing the challenges of rapid urbanization and the Urban Heat Island phenomenon, city forests not only function as aesthetic elements but also as practical solutions to various environmental problems faced by the city. City forests contribute significantly to improving air quality by absorbing carbon dioxide and other pollutants, which have a direct impact on public health. The vegetation in the city forest can reduce the environmental temperature through the process of evapotranspiration and provide shade for the people of Jakarta who have a high population density. City forests also play a role in mitigating the annual disaster of flooding, through City Parks, flood risk reduction can occur by increasing the soil's ability to absorb and store rainwater. The existence of green open spaces also plays a role in supporting biodiversity, providing habitat for local flora and fauna that are increasingly threatened by urban development. Furthermore, related to its role, urban forests offer various roles in other fields such as social, psychological and educational for the community which further strengthens the importance of the existence of green spaces in the city. This paper emphasizes the importance of urban forestry education and its role to increase public awareness about environmental conservation, which is a significant contribution to environmental advocacy.

References

[1] Marisa, A. R., Putri, F. E., Rahmasari, K., Salsabila, N. D., & Dewi, O. C. (2024, December). Like or dislike? Evaluating Jakarta's green open space through digital perception: case study of Tebet Eco Park. In *IOP Conference Series: Earth and Environmental Science* (Vol. 1419, No. 1, p. 012066). IOP Publishing.

[2] Mulyani, Y. A., Hermawan, R., Rushayati, S. B., Suryani, R., Zulhidayat, H., & Mahesa, R. (2023, December). Bird diversity and guild composition in five urban parks of DKI

Jakarta. In *IOP Conference Series: Earth and Environmental Science* (Vol. 1271, No. 1, p. 012013). IOP Publishing.

[3] Larasati, A. A., Hernowo, J. B., & Mardiastuti, A. (2024, July). Bringing birdwatching closer to home: developing urban birdwatching as a mean of sustainable tourism. In *IOP Conference Series: Earth and Environmental Science* (Vol. 1366, No. 1, p. 012037). IOP Publishing.

[4] Azizah, F. N. Determining the need for green open space locations based on the distribution of urban heat Island in Jakarta.

[5] Lakshmisha, A., Nazar, A. F., & Nagendra, H. (2024). Nature based solutions in cities of the Global South–the 'where, who and how'of implementation. *Environmental Research: Ecology*.

[6] Amanda, Z., Fadillah, R., & Ali, M. S. (2024). Analysis of The Availability and Identification of Public Green Open Spaces in GKU 1 and GKU 2 at Institut Teknologi Sumatera Campus. *Journal of Science and Applicative Technology*, *8*(1), 53–59.

[7] Handoyotomo, H., Halimawan, A., & Yuli, N. G. (2018). Degradation, Meaning Shift, and Redefinition of the City Parks' Social Function in the Presence of Virtual Space Facility. *Journal of Architectural Research and Design Studies*, *2*(1), 18–25.

[8] Yuniastuti, E., & Hasibuan, H. S. (2019, October). Green open space, towards a child-friendly city (A case study in Lembah Gurame Park, Depok City, Jakarta Greater Area, Indonesia). In *IOP Conference Series: Earth and Environmental Science* (Vol. 328, No. 1, p. 012016). IOP Publishing.

[9] Sari, D. A. K., Widyawati, L. F., & Pramesti, D. (2020, February). The availability and role of urban green space in South Jakarta. In *IOP Conference Series: Earth and Environmental Science* (Vol. 447, No. 1, p. 012055). IOP Publishing.

[10] Wuisang, C. E., Rondonuwu, D. M., Sela, R. L., Tilaar, S., & Suryono, S. (2023). Characteristics of public green open spaces and efforts in enhancing the quality and function using tri-valent approach: case of manado city, indonesia. *Eduvest-Journal of Universal Studies*, *3*(2), 309–326.

[11] Fardani, I., Syaodih, E., & Saraswati, S. (2024, September). Determination of priority areas for green open space development based on the urban heat island phenomenon. In *AIP Conference Proceedings* (Vol. 3065, No. 1). AIP Publishing.

[12] Rizki, A. R., Tumuyu, S. S., & Rushayati, S. B. (2024). The Impact of Urban Green Space on The Urban Heat Island Phenomenon–A Study Case in East Jakarta, Indonesia. *Geoplanning: Journal of Geomatics and Planning*, *11*(1).

[13] Permana, C. T., & Harsanto, B. (2020). Sustainable city planning concepts and practices in emerging economies: A systematic review. *The Journal of Indonesia Sustainable Development Planning*, *1*(1), 67–82.

[14] Cahyana, I. N., Syam, R., & Saputro, S. A. (2023). Perencanaan dan Pengurusan Hutan Kota dalam Rangka Pembangunan Kota Berkelanjutan di DKI Jakarta. *Bina Hukum Lingkungan*, *7*(2), 196–213.

[15] Rahmawati, Y., & Wulandari, F. T. (2024). Analysis of Vegetation's Ability to Reduce Noise in Udayana Urban Forest. *Jurnal Biologi Tropis*, *24*(2b), 303–314.

74 Mastering prompt engineering: Unlocking creative potential in AI-generated art

Aezeden Mohamed

Teaching Learning Coordinator of Faculty of Engineering, PNG University of Technology, Morobe Province, Papua New Guinea (PNG)

Abstract: Prompt engineering has become an important concept in the current generation of artificial intelligence (AI), especially for creating AI art where it is a crucial form of influence. The discussion in this research paper aims to analyze the directions of amendments to prompt engineering that has gradually changed from a component of early AI systems' framework in the past to a useful tool applied in various fields now, including entertainment and marketing, education, and social media. Thus, through the literal transformation of human creativity into an ordered and descriptive set of inputs, prompt engineering also captures user expectations and consequently unchains a limitless potential in artistry pursuit by intelligent methods. In addition, the study shows how appropriate prompt design leads to increased accuracy and variability of AI art to help the user achieve an output that resembles their ideas. Analyzing cases and examples from industrial practice, this paper explores how timely engineering promotes industries' advancement, art creation tools, and collaboration between humans and machines. Moreover, it discusses problems in creating effective prompts, including treating suggested ambiguities, opportunities, and tensions between the creative and technical approaches, and questions of ethics. Prospects are explored, and future research directions include combining sophisticated language models with real-time feedback systems and expanding the uses of such a system to other domains apart from the visual arts. The insights highlighted Wednesday support prompt engineering as the seminal concept for redesigning creativity to reflect the conditions and potentials of the digital paradigm, as well as for enabling users to reimagine aesthetic norms and possibilities on their own. Implementing the knowledge of the prompt creation process can help both individuals and companies achieve the greatest potential of AI to increase creativity standards and make it more available for everyone. In doing so, this paper makes a significant contribution to discussions regarding AI's application to creativity and may serve as a valuable reference for other researchers when exploring the interactions between human and prompt engineering in the context of the creative industries and the creation of new forms of AI and digital narratives.

Keywords: Artificial intelligence, creativity, education, e-learning, geneva's role, prompt engineering

1. Introduction

1.1. Background and context

AI is evolving at a very fast pace and has impacted creative industries in a big way and AI art is one of the most interesting phenomena in this context. From the basic text to image conversion to the state-of-the-art generative models, AI has redefined the idea of art making. At the center of this change is what is known as 'prompt engineering,' the process of writing highly specific and specific input instructions for AI systems to follow [1]. Even though AI models are growing more sophisticated, the outputs are only as good as the prompts given to them. Unfortunately, many of the initial attempts at AI in creativity involved generic inputs entering into a black box and receiving a similarly general output. But as the field of prompt engineering has progressed, the creativity of AI has been given a much broader canvas to paint on, to create detailed, visually appealing, and contextually correct outputs. This paper aims to discuss how prompt engineering connects the human imagination and AI's technicality, which is reshaping contemporary art [2].

1.2. Importance of prompt engineering in AI

Strategic activation or invocation of engineering is central to achieving the maximum expressiveness and functionality of AI systems, especially in creative tasking such as art creation. It is a bridge between what the human wants or is trying to achieve and how an AI system can provide the latter in response to the former. When the prompts are well designed, the users can easily control the AI models to generate the kind of output that the user wants, eliminating the need for the user to experiment a lot. Furthermore, on-time engineering is not a mere technical strength; it builds creativity alongside innovative and experimental capacities that artists, designers, and businesses can find in AI-generated content to enhance what is possible under this output [3]. Boss directives and relatively few special projects aside, prompt engineering remains indispensable given how key it is in the world of AI rip–mastering it makes for exponentially

aezeden.mohamed@pnguot.ac.pg

DOI: 10.1201/9781003684459-74

faster work and higher output quality in creativity competition. Non-parametric sampling is presented as a growing skill in this research with the focus made on the democratizing effect on creative tools and the future of AI and human relations in different fields [4].

1.3. Objectives of the research

This research seeks to explore the changes in prompt engineering in AI-generated art and its implications for the creative economy. The primary goal is to find out how good prompts improve the accuracy, quality, and richness of AI-generated content and how it can close the gap between creativity and implementation. Particular objectives are to review the historical development of prompt engineering, to compare and contrast the successful approaches and methods for creating effective prompts, and to assess the influence of the concept on entertainment, marketing, education, and social networking. In addition, the study aims to identify the problems related to the timely development of prompts, the issues of ethical and technical nature, and the directions for further development of the prompt creation field. In attaining these objectives, this research enhances the understanding of how prompt engineering can revolutionize artistic and creative practices in the age of new media.

2. Review of Literature

Mizrahi [5] offers new approaches in prompt engineering, which allows users to benefit from the use of innovative language creation in a range of domains, making a major advancement in natural language processing. In this way, the book allows readers to receive maximally beneficial results from AI tools, focusing on writing prompts and application profiles, and innovative insights that help enhance the efficiency of AI systems in composing content, aggregating data, and performing communication tasks. This innovation provides a link between the user that is new to AI and the user that is an expert in AI enhancing the accessibility of AI and general applicability across fields. However, drawbacks must not be overlooked as identified previously being major as the effectiveness of these strategies depends on the cognisance that the user develops from firing prompts, which involves iterative refinement, and as such, can be a steep learning curve more often than not for the inexperienced. Furthermore, there remain prompt sensitivity issues that make AI systems produce biased and/or contextually irrelevant output and limitations about the language model. These aspects demonstrate why it is essential to fine-tune in areas of both AI and user awareness longitudinally.

Linder, Oppenlaender, and Silvennoinen [6] expand upon the creative application potential of prompt engineering to create AI art, and its capability to transform human-computer interface. Their study focuses on how expertly crafted seed texts can drive nuanced and visually beautiful AI responses to show the synergy of art and technology. The authors call prompt engineering, a new kind of creativity, allowing users to direct how an AI tool will generate personalized aesthetic graphics. Similarly, the work identifies valuable yet, ultimately, restricted aspects of the research including the reliance on self-generated limited final iterative methods and the absence of a clear and systematic prompt. These challenges also present a problem of manageable variance and complicated accessibility for new users to AI systems. In addition, some moral and/or legal issues that emerge from AI-generated art including originality and ownership require more discussion on the topics of copyright in what seems to be an emerging sector.

Tabatabaian [7] discusses how prompt engineering may lead to improved interaction with ChatGPT and the building of applications based on GPT. The book offers a positive prerequisite to the steps of formulating the right prompts for effective accordance and attaining more decisive and reality-based outcomes from them. This innovation raises greatly efficiency, especially in application development, customer service, and content creation by interacting with the AI models. Nevertheless, there are still some drawbacks: while constructing the appropriate prompts, the system depends on user experience; sometimes it generates only irrelevant or even misleading responses due to the models' limitations. Moreover, the real-time enhancement of the AI updates of the prompts raises issues for long-term usage because the type of prompt strategies would need to change continually. These factors suggest that implementing strong architectures for designing valuable and appropriate messaging prompts and shopper education can optimize the use of the technology.

Deshmukh [8] gives definitions and details on prompt engineering to help users have an easy time when engaging with AI models. While the theme of the book is how to work with and get the best out of AI, the book focuses on best practices when inputting data into AI systems and the construction of highly specific prompts to elicit the desired output. His work shows that to get to the intended goal, AI needs to be designed taking into account its specific mechanisms and, depending on what kind of result is expected, proper prompting should be given to AI. Since the guide provides great recommendations for enhancing user effectiveness and exploiting AI capabilities, it has some drawbacks including complex prompt engineering techniques, which may be barely perceivable by novice users. Also, the inconsistency of the response of the AI and the absence of the definitive format of pro-construction urges make the use of the mentioned strategies rather limited in terms of spreading them. Such issues call for further research and the creation of more easily applicable frameworks for IMP prompt engineering.

In Vidhani [9] specifically for chemistry, the author discusses prompt engineering and effective interaction between humans and AI while focusing on the choice of the right questions. The book shows that learning the skill of asking the right questions can boost the effectiveness of AI in dealing with biochemical challenges including the prediction of new compounds or reactions. Therefore, when users narrow down prompts, the AI tools develop stronger, more efficient, and sustainable human artificial intelligence (HAI) collaboration. However, the approach has limitations, especially if used in highly professionalized areas such as chemistry, as the depth of understanding needed to formulate adequate prompts is out of reach of inexperienced participants. Additionally, the information that AI models receive will never be parfait as they cannot handle highly complex or ambiguous query types leading to suboptimal performances. These challenges make it clear that, even after development and deployment, constant tweaking is necessary for both models and offers as well as the prompt engineering skills of the users to optimize the potential of the HAI interactions.

Similarly, Bansal [10] emphasizes the urgency of timely engineering for extending generative AI capabilities across various domains. This article makes it clear how carefully designed prompts can enhance the overall quality of AI-driven content creation across written and visual media by making certain that outputs created by the AI belong to the intended class of user intentions. The customizable approach encourages increased productivity in areas like marketing and customer service, the creation of content, and many others when the use of the AI response is critical. However, Bansal (2024) rightly outlines a few limitations which range from variation in the quality of responses from the AI variety by the precision of prompts to difficulties in predicting optimal prompts for multi-dimensional. Additionally, the skills needed for effective construction of the prompts could be difficult to acquire, especially for individuals who are outside the IT profession hence the friendship of their application. These constraints demonstrate the importance of developing intuitive interfaces for the optimization of generative AI as well as the need to constantly improve the current and future state of prompt engineering.

In a future work, Kelly [11] addresses the new prompts about prompt engineering for building AI that embraces human creativity. The thesis explains how, through appropriately crafted prompts, the language models can be directed towards producing outputs in line with creative objectives, hence the need for AI in the various sides of artistic creation. Kelly pinpoints that through detailed pre-processing of the language and references, prompt engineering opens the prospect of creatively engaging humanity and AI for adequate cooperation in contrast to a trochanteric one. However, the research also points to some general limitations; for example, using different creative measures it was sometimes difficult to get consistent reliable results because AI systems are intrinsically variable. However, more work needs to be done on the types of algorithms that drive these systems to ensure that creativity is made easier but not hindered. Such challenges show that AI is a continuous work in progress to guarantee an efficient creative companion.

Through automated data analytics using ChatGPT, Sedkaoui [12] analyses how human creativity is complemented by Artificial Intelligence technologies and the role of prompt engineering in enriching data analysis tasks. The book describes how data creation can be augmented by the use of AI and the automated process of pattern recognition and predictive modeling with the creativity of the human being used to oversee the logical use of the results. This can be advantageous for decision making all fields of the economy including business, healthcare, and finance. However, it identifies some limitations, particularly in the way it uses well-constructed prompts to get accurate and meaningful outcomes. Mixed patterns in the AI's response and the dependency on making subsequent modifications in the prompts may be a limitation when used by persons with little exposure to such systems. Further, there is an implication that blending human creativity does not always offset the interpretative disadvantages of AI, and the use of AI and associated Big Data analytics needs enhanced and ongoing optimization of AI techniques as well as continually enhanced human familiarity to unlock the full potential of AI.

3. Modern Applications of Prompt Engineering

3.1. Enhancing AI-generated art through precision prompts

Precision prompts are revolutionizing the area of AI art by allowing users to receive incredibly specific and intricate works that reflect the planned ideas. In contrast to generic prompts, which produce imprecise or ambiguous responses, precision prompts use specific language: colour, texture, feeling, and context. Due to a higher level of detail, it is possible to obtain high-quality, accurate, and detailed output from an AI model [13]. For example, if we take a simple prompt like, 'an urban town planning which would paint a picture like this: a surreal underwater city at dusk and the plants glowing', it gives the user a clear, elaborated picture. Precision prompts are also very useful in iterative design, where the user makes corrections to his/her input to arrive at the right output. It helps to cut down the amount of time that is wasted in this trial and error stage to give more time and effort to the creative work. It is only when people and organizations learn to write precise prompts that AI can be optimally used in creative work.

3.2. Role in creativity and innovation

The general engineering concept underlying the prompt further encourages and enhances creativity and innovation with the use of Artificial Intelligence. Thus, using the prompts, which are well thought out, people can create as many opportunities as they want, ranging from creating dreamlike scenes to creating scenes that are as realistic as possible but do not conform to the traditional aesthetic. This process of interaction between human creativity and AI capacity creates a symbiotic creative process in which prompts serve as the channel through which users convey their concepts. In addition, immediate engineering brings professional art equipment to the general public. It allows people who are not artists to engage in artistic projects without necessarily having to understand the engineering behind the equipment. Companies and industries reap the advantages of using AI creativity to design new and creative approaches to marketing, entertainment, and education. The interactivity of prompt development stimulates creativity as students explore different ways to accomplish the goals set in a particular task. Through fill-in-the-gap engineering, the chasm between what technology will permit and what imagination can devise has been closed and has emerged as a critical necessary invention for all creative endeavours of this era.

3.3. Use cases in industries

3.3.1. Gaming and entertainment

In gaming and entertainment, prompt engineering changes the way the content in such industries is being developed through the generation of detailed assets, immersive environments, and engaging storylines. By using precise language inputs, AI models can and have created sophisticated characters, kinetics of scenes, and stunning storytelling that appeals to people's sensibilities. For example, an evocative description like 'a post-apocalyptic city with techni colour hi-teck towers, flying cars and crowded bazaars,' bears many images that act as sets for games or films. Facilitating design creativity Many decisions game designers can make are iterative which allows them to test multiple creative possibilities within minutes. Also, AI art from prompt engineering is applied to concept art, level design, and promotional images, saving time and money [14]. The gaming and entertainment industries need the freedom to try new things, which becomes possible with AI's inclusion into the creative process, in creating better experiences. In these dynamic fields, prompt engineering therefore becomes a significant means of telling a story or being creative with visuals.

3.3.2. Marketing and branding

Marketing and Branding are invariably dependent on copywriting and the immediate transformation of words into aesthetics that resonate with the target populace. By carefully coming up with the questions, companies will be able to come up with special and striking advertisements, social media posts, and promotions to launch in saturated markets. For example, a company introducing a new product that is environment friendly such as tree planting glasses may use a prompt such as 'a bare to the background simple picture of a green leaf coming from a recycled bottle with the glasses' [8]. Such precision makes it possible for one to get the outputs from the AI that will reflect its brand values as well as messaging. Further, real-time engineering enables the generation of several design alternations to support A/B testing that aids marketers in enhancing the campaigns. It also enables personalization, whereby companies are in a position to develop images as well as messages with the outer covering of the segment of customers. Through prompt engineering, creative solutions within marketing strategies can be adopted more effectively, increasing the capacity for the brand to create unique, stable, and valuable advertisements for the targeted audience within the digital marketplace.

3.3.3. Education and e-learning

In education mode and e-learning particularly, quick engineering can improve teaching aids, content, and delivery systems that promote effective and relevant content. Specific specifications work to make visuals, animations, and interactive modules created auto by AI easier to grasp compared to other complex concepts. Specific, concrete, discrete, and detailed specifications do create engaging learning interactions, though broader, more generalized, and open-ended prompts tend to be used by instructional technologists. This is also true for language learning since descriptive prompts create culturally informative visuals or conversational paradigms. AI can be introduced in education to find out how to produce unique learning materials for students, particularly, those with different learning abilities and disabilities to improve the kind of education that is offered. Moreover, the application of this concept reduces costs and time since the use of AI automates creative functions and leaves teaching responsibilities to educators. It is therefore obvious that by incorporating prompt engineering into e-learning platforms, institutions can result in quality education delivery that is aimed at filling the gaps created by the traditional forms of learning.

3.3.4. Personal artistic exploration

Mechanisms of innovative engineering encourage and enable talents in drawing and other forms of creativity as Abstract concepts are converted to AI art designs. Amateurs and designing enthusiasts wish to come up with certain ideas not having to make use of highly specialized skills or costly machinery. For instance, a user may come up with a challenge such as 'paint something like – a calm countryside where there is a golden wheat field and a dusky purple sky with some wildflowers.' That is why, with the help of AI, it is possible to generate artwork in adherence to the user's concept in mind. What is more, the employment of prompts enables personal art discovery to promote activities that encourage people to express themselves and practice different techniques such as abstract or hyper-realism. Through repeating the same prompts, users get better at idea generation, recognize their potential, and ideas evolve as they appreciate art and creativity. It decentralizes artistic creativity making it easy for anyone to be involved within the art fraternity to produce art. Prompt engineering therefore is a strategy that fosters creativity, learning, and discovery.

4. Domains where Prompt Engineering Plays a Key Role

4.1. Creative industries

Spective engineering is a creative field involving the arts, film, and music industries to improve cooperation between man and machine. In visual arts, AI-created art is becoming a trend with applications such as DALL·E and MidJourney that depend on critical and specific instructions to create unique and complex pieces of art. These tools help artists and designers come up with fresh ideas for concepts and visuals, edit or develop ideas, or even complete full art for commercial or personal usage. In filmmaking, prompt engineering is used in production to provide realistic computer-simulated graphics, concept illustrations, and scene settings, all of which minimize the time needed for actual drawing [15]. Data from active breeding platforms such as Artbreeder reveals how users set up instructions on how characters, landscapes, and even scenes should be generated to reflect how the industry relies on AI as an aid to creativity. However, by accepting prompt engineering, each creative sector extends the definition of traditional conventions of storytelling and format in art.

4.2. Advertising and media

In advertising and media, prompt engineering has dramatically changed the process of generating custom, interactive, and visually appealing content. Businesses use AI applications to create personalized commercials and promotions according to the user information and trends found on their gadgets. For example, Copy.ai and Jasper.ai generate unique ad copy or social media content using the idea of prompt engineering. According to a Statista survey conducted in 2023 or rather a cross-sectional survey, more than 45% of marketers applied content generation tools powered by artificial intelligence to boost engagement levels of the campaigns. As will be explained through the concept of timing or prompt engineering, marketers design campaigns that fit a brand's voice and AI algorithms that offer adjustments to meet the customers' demands in the process of improving the campaigns. Furthermore, AI has also brought its applicability in media by making quick tools for the making of visuals, videos, and graphics in marketing [16]. Therefore, practical engineering is critical to eliminate bottlenecks and various costs linked to content production, and content creation for advertising, and enhance the creativity of advertisements.

4.3. Education and skill development

The process of prompt engineering is slowly making its way into education and skill building by making it possible to develop an individualized learning approach. At the core of what GPT-based tutors and other machines are or can be are accurate prompts that can spit out educational material, quizzes, simulations, and even automated feedback for the student. Educational e-learning sites like Coursera and Duolingo rely on prompt engineering to give learners custom lessons as they learn. McKinsey report noted that 71% of educational institutions have adopted AI tools to improve learning outcomes. In skill development, to say, AI makes it possible to automate the content for training, setting up simulations for technical or vocational training, hence speeding up and enhancing learning. Hence by focusing the prompts on certain educational objectives, the result reached would be of superior utility to both instructors and learners in complimenting skill acquisition. By using AI to engineer the prompts, education is made to be focused on individual needs, can reach a larger number of students, and is more efficient.

4.4. Virtual reality and augmented reality

In Virtual Reality (VR) and Augmented Reality (AR), prompt engineering makes it possible for AI systems to build environments on user interaction. Control in VR gaming and simulations is described by a set of clear directives that specify how AI should create realistic and immersive environments including characters and stories. For instance, Facebook's Oculus and Unity game engines have AIs that employ models that work based on the user actions and their decisions and design a new experience depending on the exact signal. By using AI to generate content, the Global AR/VR market size is estimated to grow by $296.7 billion by the year 2024, as stated in a report by Statista [17]. In AR, prompt engineering of the real world overlaying contextual information describes user interactions in retail, training, and education. In interacting with VR and AR, industries have enhanced the engagement of users, personalization, and productivity hence opening up fresh opportunities for communications.

4.5. Content generation for social media

Engagement engineering is essential in content creation for social media since implements helps brands, influencers, and other people make attractive, visually appealing, and relevant posts quickly. Platforms like Canva, Lumen5, and Copy.ai incorporated precision prompts to design graphics, videos, captions, and even campaigns for social media to fit Instagram, Twitter, or LinkedIn profiles respectively. HubSpot reveals that 58% of marketers already incorporate AI tools in content creation, proving the significance of prompt engineering as part of social media enhancement. Through customized questions that capture a company's tone and appeal to the specific target market, consumers benefit from the automation of the creation of nourishing content [16]. Lastly, the use of AI is also helpful in creating content for social media posts since it personalizes the trends, and the feedback it gives in responding to its users' posts is handled in real-time. The online social contexts keep on changing but prompt engineering is still important to design apps, contents, products, services, etc., which are very timely and relevant and which can grab the attention of the users and help to build brand image.

4.6. *Geneva's role in advancing AI and creativity*

Geneva, the second city of the Swiss Confederation – the international city that is at the center of the advancements both in innovative technologies and diplomacy – becomes an important factor in the development of artificial [19].

Intelligence, including prompt engineering and AI art. Many companies in the city are global, enterprises, research facilities, and tech ventures, making it a perfect subculture for AI development. Several organizations such as CERN and the University of Geneva are at the forefront of calling for comprehensive, scholarly AI integration in the creative economy.

Recently, it has also been the venue of AI-related conferences like the AI for Good Global Summit, where people deliberate on the best use of AI: the good and the evil. These discussions have seen the emergence of prompt engineering as a central mechanism for the creativity and effectiveness of AI outputs. For example, local startups and design agencies are using AI to produce branded content based on prompt engineering, which suggests the look in line with the image of a client.

Additionally, Geneva operates to be sustainable and inclusive, both of which are values that highlight the mission of prompt engineering to democratize creative technologies [21]. For instance, in the concept known as Geneva Digital Talk, AI explained one of its main opportunities to help various people appreciate art and design, as well as create by artists and designers from the margins. Furthermore, NY is also trying to draw ethical legislation about Artificial Intelligence and, in particular, originality, assigning prejudices, and ownership of AI art [20].

Given the city's dedication to designing solutions to some of the most pressing problems facing society, spearheading the prompts for engineering the next wave of AI innovation and morality, Geneva is well-positioned to do so. Such innovative potential, advanced technology, flair for creativity and art, as well as openness to international cooperation means that the city will remain the avant-garde of AI art and help to bridge the gap between technological solutions and people's imagination while also fostering new levels of digital opportunities and artistic integration.

5. Impact of Prompt Engineering on the New World

5.1. *Democratization of art and design*

The concept of prompt engineering is bringing art and design education to the masses and making it easy for anyone to use the latest tools even if they never attended art school. Advanced AI tools such as DALL·E and Canva enable users to create complex graphics based on word descriptions without prior training, skills, and equipment. Such accessibility level enables hobbyists, small business persons, and independent artists to participate at par with industry professionals [15]. Also, it encourages equality in influence, which allows different people and their suggestions to add to art and design. Since reducing the technical and financial cost parameters, prompt engineering is revolutionizing various creative sectors to guarantee creativity and artistic works are delivered to the world freely.

5.2. *Bridging the gap between artists and AI*

Exploration of features through prompt engineering also introduces a special connection between historical artists and AI since they can easily work together. It enables artists to reproduce their concepts in AI-produced works with the help of purposely written statements. For example, a painter can merely create a mental picture and lay down the idea, then technology presents the representational picture. These opportunities improve the processes of art-making while providing an additional mode of navigating the concept and experimenting with the design. Incorporating AI shows artists the increase in potential when Artificial Intelligence is used in conjunction with artistic services, without compromising on the artist's decision-making. In this way, therefore, prompt engineering engenders an effective synergy between human creativity and artificial intelligence formalism in the artistic context while remaining true to its essence.

5.3. *Fostering collaboration between humans and machines*

Immediate engineering enhances the relationship between people and applications since it acts as a link. Striking that, individuals can guide AI systems through the workflow by utilizing a set of carefully crafted prompts, including creating artwork or even solving highly complex design problems. Through this type of interaction, it becomes much faster to implement new ideas since humans bring in ideas, while machines are faster since they can process repetitive and iterative tasks for high-quality outcomes. For example, the profession of architecture employs prompt engineering in actively constructing large elaborate structures incorporating artificial intelligence [18]. This approach reconstructs the interaction with machines into a communicational one and assigns a creative task to AI, guiding it but keeping the creative process within human discretion and control.

5.4. *Redefining artistic standards and expectations*

New challenges are being set in artistic endeavours by engineering immediate responses for art, demanding innovation in the measurement and impact of art and design. Appropriate and thorough instructions that AI receives lead to the creation of such visuals and designs that are hard to think of as products of art. The drive has resulted in the rise in expectation in the level of complexity and quality of creations that calls for constant improvement among human and AI creators. For instance, industries involved in gaming and media are using prompt engineering in creating related imaginative and highly realistic environments. When AI-created works emerge into the main graph, it interfuses society with new ideas regarding originality and authorship as well as the changing nature of human artistry in the new world.

6. Techniques and Best Practices for Mastering Prompt Engineering

6.1. Building descriptive prompts

The development of descriptive prompts serves as the basis of prompt construction for effective prompts engineering. Good prompts create elaborate instructions in writing with specific goals or aims that need to be achieved by the AI system. This includes the right detailed language, proper keywords, and context information to obtain precise results. For instance, if you perform a search using a word such as 'landscape,' you will get more appropriate results if you use a keyword that defines it such as, 'a serene mountain valley at sunrise with mist-covered peaks…' These come in the form of adjectives, style preferences, and action-directed language, which can further filter the AI. By crafting descriptive prompts, users can limit the AI's creativity and indeed increase its usefulness in so far as the results will conform more to what the user is looking for.

6.2. Iterative refinement methods

Approximation requires continuous testing and varying prompts to work towards the defined goal. As such, this technique involves, examining AI responses, pinpointing the problems, and adjusting the prompt. For example, a direction coming up with crude artwork can be improved as 'complicated,' 'subtle' or 'complex.' It is also possible to try different structures of the prompt and the introduction of constraints to 'tune' outcomes. Interactive improvement helps the program user to understand the best practices owing to the AI's improved performance. This is to make sure that the outputs are highly accurate and creative This is why prompt engineering incorporates this process.

6.3. Incorporating styles and contexts

It also adds value in terms of prompts in the AI-generated results since it is easier to find styles and contexts that would answer questions differently. Users can limit an AI to certain styles (for instance, impressionistic, surrealistic) or design contexts (modern, minimalistic, etc.). For example, if the user asks for 'a surrealistic cityscape with vibrant colours' it points the AI towards work inspired by Salvador Dalí. Ambiguity and noise, even at the level of interword of related concepts, are addressed by function and cultural or contextual advice. Realizing how stylistic and contextual feedback can be incorporated means that outputs that are returned will not only fulfil the technical specifications but also the right tone of the purpose.

6.4. Addressing challenges in prompt engineering

Some of the components, that make the task of engineering prompt challenging, include: While developing prompt engineering, some of the challenges include; This inherent feature may cause unsatisfactory outputs since the user needs to trace and edit vague instructions the system provided. But biases can be generated from such poorly framed questions so there needs to be some very specific wording so that bias is avoided. The next issue is the inconsistency of model interpretations of prompts which may not work well within the structure and limits of other models. It is to the above challenges that practices such as breaking down procedures into simplified procedures and the use of feedback loops respond to. In general, the present paper suggests that identifying and managing possible risks, and realistic advancements can deliver optimum results and ethical usage of the given approach.

7. Challenges and Limitations

7.1. Overcoming ambiguities in prompt language

Paragraphs or sentences that can be interpreted in various ways tend to be misinterpreted by algorithms and the generated output is then somewhat different from what the user expected. Ambiguous or unclear guidance does not give the AI enough context, and, as a result, gets ambiguous or outright wrong solutions. In response to this, users need to filter their questions with accuracy: using proper details, clear instructions, and other related keywords in the context. This process can be repeated and made iterative with the aim of refining and/or establishing more reliable intended meaning. For example, when a beautiful scene is replaced by a calm lake in the evening by the snow-covered mountains' increases the output alignment. In other words, decreased ambiguity allows users to get more predictable and appropriate AI data outcomes more often.

7.2. Balancing creativity with practical constraints

Prompt engineering has its inherent issues such as striking a critical balance between artistry and the practicalities of event coordination. However, users essentially attempt to be as creative as possible by building complex prompts for AI models while ignoring the problems that AI can encounter at the technological level. For example, limitations in the type of algorithms, limitations in the number of calculations per second, and lack of the ability to depict finer features limit the outputs to be complicated or detailed. Ski or too general prompts lead to understated or over-stated objectives that are not up to expectations. Besides, it is essential to pay attention to the issue of the so-called innovation-performance paradox. In business, for instance, in marketing or product development, prompts need to stimulate creativity but, at the same time, correspond to brand identity, usability necessities, and user preferences. Evaluating and optimizing the opportunities for creative responses is also seen to require awareness of the particular features characteristic of the AI system as well as the cyclical process of prompt construction. Maintaining this balance, users can get fresh and usable results from AI while at the same time considering the physical, practical, and inexplicable limitations and goals to be achieved.

7.3. Ethical concerns in AI-generated art

Some of the ethical issues in AI-generated art are pegged on questions of provenance and copyright, and the appropriate use of technology. A major problem is the potential of AI systems mirroring such works, due to the large datasets that most AI structures use in training, which includes prior art. This creates issues of copyright about what extent the outputs from AI can be classified as original. Furthermore, weak prompts may inadvertently train a model to display biases that are inherent to its training set or to exclude some specific population. Adding to this the ethical question is the improper usage of AI-generated art, for example, deepfakes or disinformation. These issues are however well addressed by prompt engineering, through making the prompt itself more open and flexible for people of different diversities and genders, as well as making sure the prompt or the instructions that are given in such a prompt do not in any way lead to any unwanted or stereotyped generate outputs. First, following ethical standards for AI, second, being open about the utilization of AI and, third, honouring the ownership of collected art are the essential measures to create a positive impact of AI art for society and culture.

8. Future Directions for Prompt Engineering

8.1. Advanced natural language models for creative prompts

The future of the SMSC is located in the progressive extension of higher natural language models, GPT-4 and beyond, which are better at interpreting complex prompts. These models can easily decipher multipart and nested instructions, and as such complicated and versatile input can produce equally complex and diverse output. For instance, it can develop highly realistic visual or graphically intricate styles simply by assuming control. Introducing historical, cultural, or stylistic context into the equation will additionally improve the accuracy of creative suggestions. Multimodal AI can include text, image, and audio inputs and new directions for creativity are being explored. They will enable users to set much more complex and detailed problems and challenges to introduce AI capabilities to a new level of relation between human creativity and artificial intelligence. They will augment the capabilities of prompt engineering by improving the software's comprehensibility and effectiveness for a wider range of industries such as entertainment, education, and marketing.

8.2. Integration with real-time feedback systems

The idea is to incorporate real-time feedback systems into prompt engineering and hence change the way users approach AI technology. These systems will be helpful in the concept creation process where a user can tweak a prompt according to the immediate graphic or textual results. For example, artists could use the description part to alter the description depending on a singular hue, composition, or style in real-time. Such real-time feedback could also involve the user preferences and context to enhance the match of the generated output with the required output. Closely coupled systems would require high levels of machine learning that could discern dynamic input from the users and interpret causal instructions. By repeating this process, the user control will greatly be improved and the trial and error typical for the current prompt engineering will be minimized. Furthermore, these systems might help maintain collaborative spaces through which several users edit prompts collectively, which might help advance brainstorming in fields like design, film production, and virtual reality.

8.3. Expanding applications beyond visual art

Engineering of hints has great possibilities not only in terms of art but also in music, literature, and scientific initiatives. Such as current innovative AI systems can create new musical pieces or help in writing books by providing replies to specific instructions on a topic, a genre, or an episode's mood. Similarly, in scientific fields, prompt engineering can help with data analysis, and form the generation of hypotheses or simulations by leading the appropriate AI systems to process and interpret the data. In the field of medical application, specific and accurately detailed feedback can lead AI to design individual treatment plans for patients and even construct virtual medical cases for education. Furthermore, prompt engineering applications are gradually being introduced in various fields, primarily as a result of the ability of AI to handle requests given in a particular task-oriented manner. By tailoring prompts to each domain's unique requirements, prompt engineering can unlock innovative solutions and enhance productivity, creativity, and efficiency in various sectors, thereby broadening its impact far beyond the artistic realm.

8.4. Promoting accessibility and inclusivity

The future of prompt engineering should be more flexible and user-friendly and assure equal or similar benefits for all users. It is clear that when designing AI, interfaces must be made simple and natural language understanding incorporated so that people without special knowledge will be equally efficient with the use of this technology. For instance, design innovation in the use of voice to prompt interoperability could enable accessibility to creative AI systems for the voice impaired. Equally, it is possible for the services to cut across the language divide, and to be sensitive to the cultures of different clients all over the world. Furthermore, generalizing also entails correcting biases in repetitions of templates as well as designing templates that represent broad upcoming spectrums of perceptions and understanding. More tremendous power is to be opened to everyone, this means that open-source platforms and affordable tools will turn AI further. These aspects prove that through diversity, encouragement, and inclusion prompt engineering can help the underprivileged, enhance the forms of creativity, and increase the equity of democracy in the novel digital world that AI offers, irrespective of skills, culture, or even disability.

9. Conclusion

Solving prompt engineering has become one of the significant advancements to change the interaction between humans and AI, mostly in the creative field. This research elucidates how carefully designed prompts open up the capability of generating creative and substantive artifacts with AI spanned across art, music, science, and education. Presenting examples of prompt engineering both in the conventional and contemporary contexts, the paper demonstrates that prompt engineering benefits a wide range of sectors such as gaming, marketing, and virtual reality, and addresses the issue of inclusiveness and diversity. Thus, some of the main problems we face when using the technique, namely, language vagueness, ethical issues, limited compatibility, and the conflict between imagination and reality have not been solved at all figure out to this day. These issues can be solved through accurate and step-by-step evaluation and with the help of sophisticated AI models, the implementation of this technology can be more efficient, and less risky. The generalization connected to prompt engineering also goes beyond creativity and provides horizons for the depth of art and new approaches to the standards in the fields of human and machine cooperation. When incorporating real-time feedback, adapting to the learners, and interdisciplinary applications as well, prompt engineering will enjoy its position even with the advancements in Artificial Intelligence systems. Thus, this field becomes an enabler of bringing together ideas and possibilities of the human mind with the limits of existing and new AI tools that help humans and organizations to bring their ideas into implementation on a scale and at a speed unavailable before. Innovating through AI, but not against it, and following best practices, along with the inclusion of limitations, prompt engineering can serve as the foundation of further development of new AI technologies as the junction between technology and creativity for the new digital age.

References

[1] Pareschi, R. (2024). *Centaur Art: The Future of Art in the Age of Generative AI.* Springer Nature.

[2] Huang, J. (2024). The Art of AI: A Human-Centered AI (HCAI) User Study of Integrating Image-Generative Tools in Visual Art Workflows: The Case of Adobe Firefly.

[3] Mammen, C., Collyer, M., Dolin, R. A., Gangjee, D. S., Melham, T., Mustaklem, M., ... & Wang, V. (2024). Creativity, Artificial Intelligence, and the Requirement of Human Authors and Inventors in Copyright and Patent Law. Available at SSRN 4892973.

[4] Costa, C. J., Aparicio, M., Aparicio, S., & Aparicio, J. T. (2024). The Democratization of Artificial Intelligence: Theoretical Framework. *Applied Sciences*, *14*(18), 8236.

[5] Mizrahi, G. (2024). Unlocking the secrets of prompt engineering: Master the art of creative language generation to accelerate your journey from novice to pro. Packt Publishing Ltd.

[6] Oppenlaender, J., Linder, R., & Silvennoinen, J. (2024). Prompting AI art: An investigation into the creative skill of prompt engineering. *International Journal of Human-Computer Interaction*, 1–23.

[7] Tabatabaian, M. (2024). Prompt Engineering Using ChatGPT: Crafting Effective Interactions and Building GPT Apps. Walter de Gruyter GmbH & Co KG.

[8] Deshmukh, J. (2024). *Prompt Engineering-The Ultimate Guide for Success in Artificial Intelligence*. Jayant Deshmukh.

[9] Vidhani, D. (2024). The Art of Asking Questions: Mastering Human-AI (HAI) Duet in Chemistry Through Prompt Engineering.

[10] Bansal, P. (2024). Prompt Engineering Importance and Applicability with Generative AI. *Journal of Computer and Communications*, *12*(10), 14–23.

[11] Kelly, J. (2024). Playing with Prompts: Toward Language Model-Based Systems that Respect Human Creativity (Master's thesis, University of California, Santa Cruz).

[12] Sedkaoui, S. (2024). *Automated Data Analytics: Combining Human Creativity and AI Power Using ChatGPT*. John Wiley & Sons.

[13] Ahmed, A., Hou, M., Xi, R., Zeng, X., & Shah, S. A. (2024, May). Prompt-Eng: Healthcare Prompt Engineering: Revolutionizing Healthcare Applications with Precision Prompts. In *Companion Proceedings of the ACM on Web Conference 2024* (pp. 1329–1337).

[14] Kumaran, V., Rowe, J., Mott, B., & Lester, J. (2023, October). Scenecraft: Automating interactive narrative scene generation in digital games with large language models. In

Proceedings of the AAAI Conference on Artificial Intelligence and Interactive Digital Entertainment (Vol. 19, No. 1, pp. 86–96).

[15] Zhou, K. Q., & Nabus, H. (2023). The ethical implications of DALL-E: Opportunities and challenges. *Mesopotamian Journal of Computer Science*, *2023*, 16–21.

[16] Gan, Y., & Zhu, D. (2024). The Research on Intelligent News Advertisement Recommendation Algorithm Based on Prompt Learning in End-to-End Large Language Model Architecture. *Innovations in Applied Engineering and Technology*, 1–19.

[17] Fitria, T. N. (2023). Augmented reality (AR) and virtual reality (VR) technology in education: Media of teaching and learning: A review. *International Journal of Computer and Information System (IJCIS)*, *4*(1), 14–25.

[18] Noble, S. M., Mende, M., Grewal, D., & Parasuraman, A. (2022). The Fifth Industrial Revolution: How harmonious human-machine collaboration is triggering a retail and service [r] evolution. *Journal of Retailing*, *98*(2), 199–208.

[19] Długoszewska, K., & Podhorodecka, K. (2024). Functional and spatial transformation of the city of Geneva at the end of the 20th and beginning of 21st century. *Przestrzeń Urbanistyka Architektura*, (1).

[20] Schwalt Chan, C. (2024). Exploring the Ethics of Generative AI within Humanitarian Organisations in Geneva.

[21] Nyaaba, M., & Zhai, X. Developing Custom GPTs for Education: Bridging Cultural and Contextual Divide in Generative AI.

75 Revolutionizing Agile project management: The role of AI in driving innovation overcoming challenges, and enhancing benefits

Aezeden Mohamed

Teaching Learning Coordinator of Faculty of Engineering, PNG University of Technology, Morobe Province, Papua New Guinea (PNG)

Abstract: Agile Project Management (APM) has become fundamental to providing flexible, efficient, and iterative project solutions. However, as the projects evolve in complexity, managing them with the use of the traditional Agile methods can be very unproductive this is due to the following vices; loose planning, improper resource allocation, and very late risk identification among others. Applying artificial intelligence (AI) in the context of Agile practices has brought about improvements to innovation by breaking barriers and expanding the advantages of Agile practices. This paper aims to discuss the future vision of APM that real-time AI tools enhance through down-to-earth AI applications during sprint planning, resource assignments, and risk evaluation by way of predictive and prescriptive analytics. The use of AI promotes innovation through the creation of digital environments for automating routine processes, for supporting decisions based on research on the implications of an AI-supported process in detail, and for enhancing self-organized models of communication within working groups. These include; reducing bias in taking estimations, increasing scalability for major projects, and adaptability in case of changing needs. Moreover, AI extends the value of Agile by cutting down project time, raising the cost-to-value ratio, and satisfying stakeholders to the greatest extent possible. Features compare AI-based APM and conventional APM and the results show that the former outperforms the latter in all aspects including Accuracy, adaptability, and transparency. Kyriakou presents real-world examples from IT and Startup industries on how AI enhances Agile processes and the resulting improvement of the project. Nonetheless, such developments present implementing hurdles such as privacy issues, system integration, and organizational enmity. Regarding these, the study gives practical implications: proposing the reference model for AI implementation, how to incorporate AI into the Agile approach, increasing team preparedness, and engaging stakeholders.

Keywords: Agile project management, artificial intelligence, sprint planning, resource allocation, risk mitigation, predictive analytics, AI-enabled Agile practices

1. Introduction

1.1. Overview of Agile project management

APM is an embracing project management framework that complies with adaptability, collaboration, and incremental work as elemental aspects for managing complex and ever-changing projects. Deep Equal Its main values and principles, described in the Agile Manifesto, are individuals and their interactions; working with the customer; and flexibility as an essential quality of the project. Agile methodologies including Scrum, Kanban, and Extreme Programming (XP) have become increasingly popular because of the application of successive short development cycles known as sprints [1]. The importance of APM can be found in its contingency nature that addresses changing project necessities, minimizes risks, and enhances stakeholders' interest, which can be a reason why it is popular in software development, information technologies, marketing, and education sectors despite the formalizable nature of their projects. However, in more complicated Agile projects, using Agile techniques is sometimes problematic in terms of scalability, fineness, or managing risks, and, therefore, more extensive solutions have to be sought such as Artificial Intelligence (AI).

1.2. AI and its integration into APM

AI is disrupting how projects are being managed by bringing with them features like predictive analytics, decision-making, and intelligent process management. But for APM, the AI supports the designation of sprints, resource utilization, or risk management throughout the process, in real-time. They improve collaboration through intelligent communications and together with reducing repetitive processes and improving the team's focus on creativity and problem-solving. AI as a concept is not just a new technology that can be incorporated into APM but an intervention that looks to enhance Agile's strengths and correct its vices. These synergies turn project management into a data-based, flexible, and

aezeden.mohamed@pnguot.ac.pg

DOI: 10.1201/9781003684459-75

productive process suitable for many hundred thousand of tasks and projects [2].

1.3. Objective of the study and research questions

The goals of this research are to assess the change imposed by AI and its impact on transforming APM, which provides solutions for the challenges it poses and enhances the advantages of Agile methodologies. In large complex projects, however, due to the scalability, precision, and flexibility issues, traditional APM may reach its intrinsic capabilities. AI provides the ability to execute tasks like predictive modeling and data analysis so that organizations can rapidly advance their operations, capture stakeholder satisfaction, and facilitate scalability. The research seeks to answer critical questions: How does the Agile approach enhance traditional Agile practices as well as add on advanced tools and mechanisms? Based on the emerging AI-enabled approaches, what are some main advantages of advanced predictive maintenance over traditional performance measurement systems in efficiency, flexibility, and engagement with other stakeholders? More so, the study analyzes organizational hurdles associated with AI integration in Agile frameworks, including privacy issues, integration issues, and change management issues. Last, but not least, it is intended to propose a conceptual framework of how to implement AI in APM and what guidelines should be followed concerning practical suggestions related to integrating AI tools into the Agile framework, team readiness, and stakeholder management. Answering these questions, the study will find practical approaches to understanding the future of project management with the help of AI and APM convergence.

2. Review of Literature

The article focuses on the role of AI as a transformative strategy in project management in the digital age by Karamthulla [3] and his colleagues, previewed in the article's title. The authors discuss promising prospects for the use of AI in healthcare, namely, the use of intelligent algorithms for predictive analysis, automation, and decision-making support to increase the effectiveness of work, minimize human error, and optimize resource use. Enabled by the capability of AI to accept and process large amounts of information, project decisions can be made more effectively, promptly, and with greater success rates. However, the study has its drawbacks; it is costly to implement; there is a scarcity of skilled personnel to manage the AI projects; and the AI tools are not easily integrated with current systems. Moreover, the ethical question in the context of autonomous AI and the data protection problem remain serious barriers. These considerations made clear how great the positive potential of AI could be and at the same time underlined where the problems of applying these methods in today's project management might be found.

Bahi, Ghari, and Gahi [4] present a research paper where the focus is on integrating generative AI in agile software development to overcome various project management issues. The authors describe new ideas in which generative AI plays a role in assisting with the creation of automated workflows, code generation, and decision-making support within a team. The integration results include quicker development, flexibility, and a decrease in the human factor. However, the research also points out that there are drawbacks to AI integration; the reliance on the systems rather than professional expertise may lead to reduced innovation. Other points highlighted include the challenge of incorporating AI with other agile models and methodologies and issues arising from data privacy and decision-making by artificial intelligence.

Kiani [5] presents an informative literature study on applying AI in entrepreneurial project management and suggests a framework for the utilization of AI to improve the project's results. The work reveals AI applications including risk evaluation, resources management, and decision making to increase the chances of project success in business solutions in entrepreneurial environments. However, there are potential limitations examined by Kiani (2024) as well: the high cost of AI technologies has to be mentioned; the lack of entrepreneurial AI expertise; and the threat of AI adoption due to perceived job loss risks. This article describes an agenda for future studies to address these challenges, particularly about examining the possibility of scale and incorporation of AI in SME organizations.

Daraojimba et al. [6] carry out a systematic literature review that proposes a historical analysis of agile methodologies, their core concepts, and their implications in project management. This paper demonstrates how the application of the techniques commonly associated with agile applications programming methodologies like Scrum and Kanban promotes flexibility, collaboration, and a 'catch me if you can' response to change, which is crucial for enhancing the project results in constantly changing conditions. The authors also stressed the proactive nature of agile which has brought about an improvement by the iteration which provides a backcheck and a quick way to develop and produce high-quality output. However, the review also points out some limitations; For instance, it finds it difficult to scale agility due to the large size of some projects, which requires a large amount of training, besides is meets resistance in organizations, that are still grounded in the traditional water model. In addition, the authors encourage more extensive research regarding the enhancements of the existing agile frameworks and their relationships with modern technologies.

Hossain et al. [7] examine of increasing tendency of artificial intelligence (AI) in project management and its effects on efficiency and productivity of the process of project planning and control. The role of AI is described as central to decision-making using predictive analysis, optimization of the workflow by automating many tasks, and other factors that would help to allow better use of resources and therefore optimization of project work. The authors also describe future growths, like the use of artificial intelligence in risk management and monitoring systems, helping to plan project delivery. However, the research acknowledges the following limitations; the initial cost of investing in Artificial intelligence technologies, the challenge of interoperability of AI technologies with other project management tools, and questions relating to data privacy and ethical use of data. The authors also call for studies to address these issues and facilitate the utilization of AI in project management in its entirety.

Amajuoyi et al. [8] concentrate on the enhancement of agile project management practices in high-tech software development organizations involving care for the modification of agile practices to meet the needs of high-tech software organizations. Such strategies as the incorporation of sophisticated collaboration enablers, the use of smart workflows that eliminate manual repletion tasks, and feedback loops that are integral for real-time improvement of the software development team's dynamism for effectiveness are other strategies the study presents. It includes these practices to better the time of delivery of projects as well as enhance the quality of the projects. However, the authors also capture some of the challenges that accompany the implementation of agile strategies; these include; difficulties in adapting agile methodologies for large teams, organizational resistance to change within large organizations, and the need to hire personnel with the required skills for an organization to adopt agile processes. The paper encourages the need for more studies to enhance agile frameworks for enhanced scaling and suitability in innovative contexts.

In the future, Onesi-Ozigagun et al. [9] assess agile product management as the enabler of technology advancement arguing that agility creates an environment that adopts flexibility alongside a culture of regular prototyping and variety in improvement patterns in product construction and dispersion. The study acknowledges the innovative character of agile concerning its impact on improving cooperation between integrational sub-teams and implementing changes concerning current market and technology requirements. The authors also present how agile product management contributes to the sustenance and advancement of innovation with products adapting from users' reactions and technology advancements. However, some of the challenges include; the inability to maintain the agility practices when growing large, confrontation with some traditional organizational structures, proactivity of effective scrum masters, as well as the ability to hire professional scrum masters and practitioners. This is the direction of further work to improve agile methods and their impact on technological development in various fields.

Ronak [10] discusses the impact of artificial intelligence in project management in the enhancement of workflow improvement and decision-making in today's organizations. The work focuses on the integration of new technologies like machine learning, predictive analytics, and automation in the improvement of the scheduling process and the management of project resources along with real-time decision-making. Better promotion and risk management, coupled with performance indicators, enable better project efficiency, a value that AI brings due to data processing capabilities. Nevertheless, Ronak also covers the Implementation limitations of AI solutions; high cost in implementing, requiring competent personnel to run the AI, and may experience resistance from teams that are not conversant with the use of AI. Therefore, based on the above findings; the paper recommends that more studies be conducted to overcome the barriers to realize the benefits seen with AI in project management.

3. Methodology

3.1. Research design

Qualitative and quantitative data are used in this research effort in its attempt to assess the state and contributions of AI for APM. The qualitative part of the study uses case studies and thematic analysis to understand organizational experiences and issues of AI implementation. The second, the quantitative type approach focuses on questionnaires and statistical measures to measure the extent of the impact on project efficiency, flexibility, or the satisfaction of the participants. The binding of these methods affords a strongly reinforced view with a solid grasp of both the theoretical and practical applications.

3.2. Data collection methods

It is important to note that the present study uses various approaches to make certain that all necessary data can be gathered effectively. Subsequently, case studies ensue, which consist of concrete analyses of existing AI applications on Agile frameworks in different industries, including IT and startup industries. These are survey questionnaires filled and completed by Agile practitioners, their managers, and AI professionals; they use structured questionnaires to obtain quantitative assessments regarding efficiency, scalability, and stakeholder satisfaction proportional to the inclusion of AI. Structured questions are not used in the interviews instead interviews are semi-structured to ensure that the research gathers as many qualitative views as possible concerning the challenges, opportunities, and organizational preparedness for AI integration. Secondary data is collected from articles and journals, research papers, and white papers to have a base structure of knowledge and then compare them. Finally, non-interventional studies are conducted to observe Agile teams utilizing AI tools in practice, assessing the complex synergies between processes, tools, and interactions in real time for a comprehensive perspective on using AI in Agile methodologies.

3.3. Data analysis tools

3.3.1. Thematic analysis

To categorize and make sense of qualitative data that is in an interview and case study form, thematic analysis is used to assess patterns, themes, and observations. This enables the researchers to identify repeatable issues, advantages, and attitudes towards the integration of AI in Agile frameworks. Through coding and subsequent establishment of themes in the data, thematic analysis provides a less arbitrary means of arriving at subjective interpretations of the data. They also contribute to the identification of major trends that affect the use of AI in Agile and the difference made in project performance [11]. The findings may be used to inform the process of creating AI plans based on the requirements of each organization.

3.3.2. Statistical tools

Regression analysis and hypothesis testing are used on the quantitative data from surveys to determine the link between AI integration and project efficiency, scalability, and stakeholder satisfaction. Regression analysis is used to show relationships where AI is likely to impact them, while hypothesis testing determines the probability of the results. These tools help the study to validate different hypotheses thus helping in the generation of evidence-based conclusions on the value of AI in Agile Project Management [12].

3.3.3. Data visualization

Reporting tools like Tableau and Power BI are used to help change complex numbers and statistics into graphical outputs for easier analysis when making comparisons, comparing performance, and searching for relationships. These tools enable the generation of shareable online dashboards and reporting of the publication, enabling different stakeholders to grasp different aspects of the study at a glance. Visualization helps make the conclusion more reachable and plays a decisive role in decision-making [13].

3.3.4. Sentiment analysis

In the context of evaluating the stakeholders, the value and effectiveness of answers received, sentiment analysis is used, which is particularly focused on evaluating both the tone and overall sentiment of the responses to the numerous surveys, interviews, and case studies. This tool is a sentiment analysis system that employs NLP techniques to determine the sentiments as positive, negative, or neutral. It allows the researchers to measure the level of the emotional response of the stakeholders to the extent to which they approve of the use of AI in Agile, which also in turn offers information on the degree of positive and negative perceptions of stakeholders toward the implementation of AI in project management [14].

3.3.5. Comparative metrics

Comparative analysis is employed to assess the effectiveness of AI in Agile Project Management (APM) against the non-AI-based APMs. This is achieved by evaluating relative variables including the duration taken to complete the project, costs incurred, stakeholders' satisfaction, and minus frequencies of change. Using these metrics for both methodologies, the study can estimate the advantages and disadvantages of AI applications. It provides real-life examples of the benefits that can be accrued by organizations adopting the use of AI in their Agile frameworks to support informed decision-making [6].

3.4. Ethical considerations

3.4.1. Informed consent

This is an experimental study it has a great ethical consideration of informed consent whereby all the participants of the study will be told about the intention of the study, how the study will be carried out, and if there are any dangers involved in the study. Those investigating a certain subject ensure that they explain to the participants the reason for conducting the research, the type of question or data the participants will be required to answer, and how the response obtained will be used. The participants are informed that the interviews are being done voluntarily and they can pull out at any one time without any explanation or repercussions. This process protects participant volunteers and clarifies their position in the research study.

3.4.2. Confidentiality

Anonymity is very important to ensure trust and privacy in the study. There is no identification of participants in all data collected from the survey, interviews, and case studies. The data collected from the survey takers is stored in abstractions that do not have any record of those who completed the surveys, their position, company affiliation, or even the concept that was surveyed. Such a way of data collection reduces the possibility of disclosing sensitive data that may lead to discomfort for the participants in the study.

3.4.3. Data security

Specific measures that are adopted regarding data security are airtightly implemented to safeguard all the obtained data. Electronic records are protected in transit and when stored on a device, so that no one with malicious intentions can get into the data. Measures have been taken to avoid point-of-entry breaches, and only persons who are performing research on the data can access them. It is also important to note that there is frequent check to see whether or not the company aligns with data protection rules and policies or detect any chinks in the armour. These procedures preserve the credibility of the research and prevent participants' identification in violation of their rights.

3.4.4. Non-Bias assurance

To try and ensure that the research is as credible as possible, the test subject is guaranteed non-biased assurance throughout the study. Scientists make efforts to eliminate self-prejudices that can affect the tasks related to the design process, data gathering, and analysis. This means the practitioner has to make sure he collects data in an unbiased way, making sure that all the questions constructed in the interviews or surveys contain no bias, and are not intensively framed. Its analysis focus avoids any influence by the researcher to the results ensuring the researcher's impartiality when interpreting and analyzing the findings. Moreover, the probability methods are used to reduce bias to get a reliable conclusion out of the data collected.

3.4.5. Transparent reporting

It is essential to report honestly in research and promising scientific credibility. All test results, even beneficial ones are reported with accuracy and are not skewed. The methodology, data analysis tools, and results are described in detail for purposes of replicating the study by other researchers. The fact that any limitations or conflicts of interest are reported, main conclusions, and outcomes are derived directly from facts where bias has been eliminated properly. Through compliance with the principle of openness, the study is in a position to offer usable data to the body of knowledge in a reliable way (Figure 75.1).

3.5. Use cases in industries

3.5.1. Papua New Guinea (PNG)

In Papua New Guinea (PNG) specifically, the application of AI-generated art is slowly permeating many industries, despite the nation's young technological environment. An example of the sphere they discussed was tourism and cultural heritage where AI art could have a revolutionary effect. PNG has a diverse cultural base as well as numerous conventional artwork bases that might need the advancement of AI technology in preserving and marketing PNG indigenous art [18]. The replication of the themes as well as the ability of AI to enhance the display of the motifs can enable the modern presentation of traditional themes and selling them to a global market. This could also improve the tourism sector where one can conduct a virtual tour or an art display of the various attractions and festivals that characterize PNG, crafts, which in turn can lead to more international visitors.

In education, AI-generated art can be employed in developing learning modes that enable students to experience the scientific, historical, or artistic modes of works. For example, it would be possible for AI to produce live and engaging

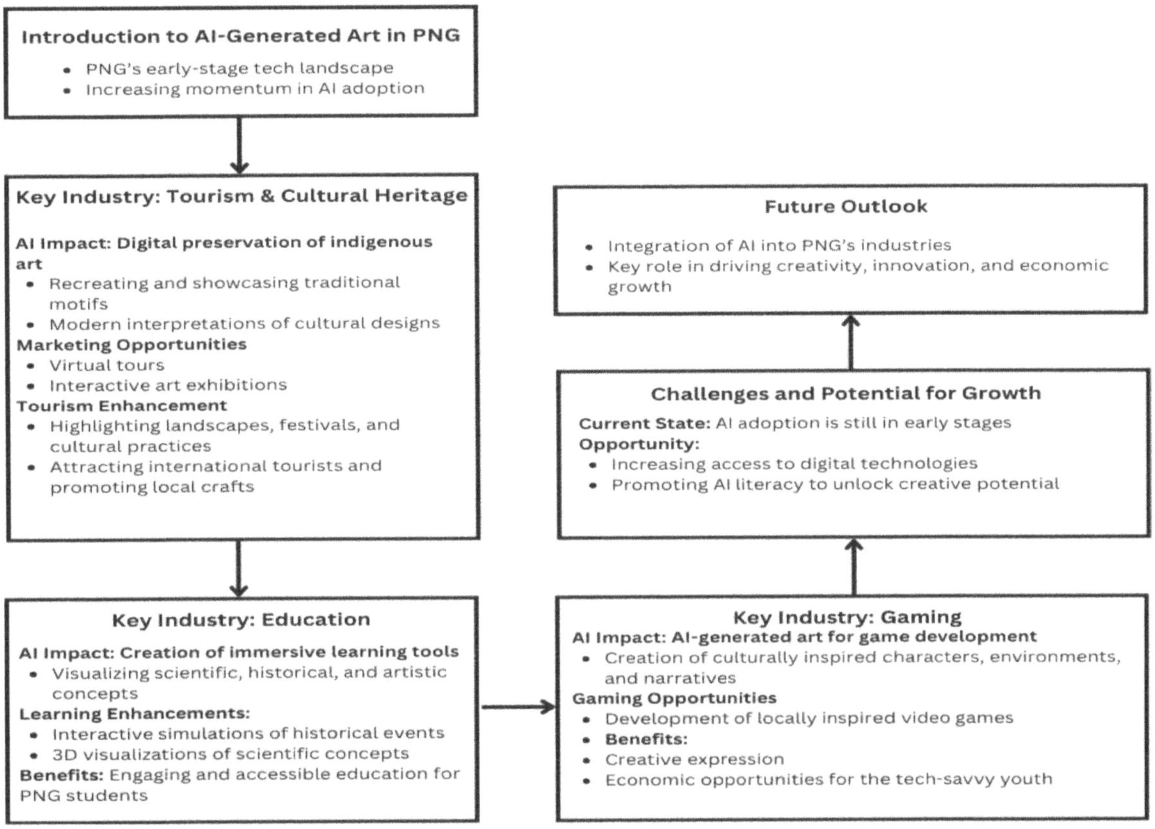

Figure 75.1. Flowchart.

Source: Author.

past events or formulate three-dimensional depictions of scientific ideas, all of which would likely be beneficial for the students of PNG.

AI is also used to enhance the growth of the gaming industry in PNG for growth to be experienced. It is suggested that AI-generated art can be used to finally inspire characters, settings, and stories based on PNG's culture and history to create local video games. This could encourage art and invention on one hand and also create employment for youths, especially those with special outreach in technology.

Together applying AI and prompt engineering in PNG still seems to be in its early stage but there are numerous opportunities in this field for thinkers and creatives. As people gain more exposure to Information Technology and as the government undertakes sensitization on the use of artificial intelligence the country stands to benefit from an explosion of creativity. The application of AI in industries in PNG may indeed hold a lot of potential for embedding creativity, and fomenting growth, and innovation in the future.

4. AI Applications in Agile Project Management

4.1. Driving innovation

AI has a large impact on Agile Project Management (APM) as new tools and techniques are developed to improve existing processes. Sprint planning is enhanced through the use of predictive analytics based on artificial intelligence which brings to the table historical data, patterns of projects, and the team's performance. This helps the project managers to identify probable constraints, rethink their schedules, and avoid project plateaus, all to ensure the smooth running of the projects. AI tools also support team collaboration and communication especially in Agile teams through real-time monitoring of project and team activity/interaction as well as task completion rate. They enable easy working through, for instance, applications for meetings, updates of the status of an assignment, and notification of the members on issues such as delays or changes in priority [16]. Further, intelligent workload balancing or resource optimization is one of the areas where AI is fuelling innovation. Using information about available resources, team skills, and the requirements of a particular project, AI tools can learn to distribute skills and workload so that all individuals will be overworked or bored. Such optimization of workloads ensures that the teams can concentrate more on priorities hence decreasing cases of fatigue or fatigue. In conclusion, the use of AI within innovation simply allows teams to work harder by taking on mundane and repetitive tasks, increasing the efficiency of decision-making and making project management adaptable and empowering.

4.2. Overcoming challenges

AI applications in terms of Agile Project management are useful in dealing with the following inherent issues thus helping various organizations to solve for complexity within project management successfully. This method is one of the main flaws of traditional Agile based on rather crude human estimations or intuitive human decisions making. AI eliminates this by providing statistical and factual realizations and forecasts that are not influenced by one's feelings. It also helps in making decisions about project scope, time, and human resources without working from assumptions and guesses. In the case of extensive projects, AI brings the aspect of scale, that is, processing huge data, performing tasks, and managing resources in teams and departments. This helps to guarantee that even the most complicated processes will not lose direction. AI also helps to minimize project risks since the system is also capable of identifying risks such as time lags, resource constraints, or project costs on a real-time basis. Through the constant evaluation of data within projects, AI can detect issues that are likely to arise before they form part of the project's environment hence should be tackled. Second, AI can reproduce several situations to estimate the possible consequences and suggest measures to minimize risks, thus improving the general work of risk management. Such AI applications help organizations manage most project issues effectively and thus reduce episodes that hinder project implementation.

4.3. Enhancing benefits

The application of AI for the management of Agile Project Management improves the benefits realized by organizations implementing traditional Agile practices since project efficiency and then the satisfaction of stakeholders is enhanced. AI enhances the project calendar since some of the many time-consuming tasks can be delegated to the computer, thereby freeing the team to concentrate on human engagement tasks. AI-integrated tools also enhance the cost-effective results by aiding the project managers to get the resources and to avoid wastage therefore reducing overhead costs. They also make it possible to effectively estimate the time and costs that would be necessary to effect changes on particular activities hence minimizing the occurrence of over-budget and delay. Secondly, satisfaction from the stakeholders is also improved by AI in APM. Due to enhanced transparency brought by the tools, major stakeholders can instantly monitor the current status of a project, resources utilization, and risks involved. Such transparency helps in creating healthy relations with the stakeholders to enhance trust to ensure they are on the same side about the project objectives, time, and overall expectations. Also, AI keeps on digesting new data coming from each sprint or phase of the project and enabling process enhancement. It can analyze different parameters of workflow and quickly reveal bottlenecks to increase the speed of work. With these principles, teams can improve how they work and their tactics subsequently making their project management better with time. The integration of AI increases the advantages made available by Agile through optimization, efficient information sharing, and consistent process enhancement all through the project life cycle.

5. Comparison of AI-Enabled vs Traditional Agile Project Management

5.1. Efficiency comparison

AI in Agile Project Management (APM) makes work more efficient than traditional projects by reducing the number of monotonous tasks like scheduling and assigning tasks and recording progress. Indeed, traditional APM usually requires manual updating and most often involves third-party interventions.

The efficiency is achieved by minimizing interruptions and using artificial intelligence when it matters. Here, with the help of predictive analytics coupled with project task automation, AI-enabled APM delivers the results within less time improving the working efficiency of the project (Table 75.1).

5.2. Cost efficiency

AI-enabled APM is more effective in managing the resources as well as the costs to forecast to avoid a high number of overruns. In traditional APM, costs can get inefficient when it comes to resource estimation and simple problem-solving (Table 75.2).

Table 75.1. Comparison of efficiency metrics between traditional and AI-enabled APM

Metric	*Traditional APM*	*AI-Enabled APM*
Sprint Planning Time	10 hours	5 hours
Task Assignment Time	3 hours	1 hour
Resource Allocation	*Manual*	*Automated*
Risk Assessment Time	5 hours	1 hour
Project Delay	12%	5%

Source: Author.

Table 75.2. Comparison of cost metrics between traditional and AI-enabled APM

Metric	*Traditional APM*	*AI-Enabled APM*
Project Budget Overrun	15%	5%
Resource Wastage	High	Low
Task Completion Delay	20%	8%
Time Tracking Errors	10%	2%
Cost Overruns	25%	10%

Source: Author.

Thus, using AI means that possible problems are anticipated and only the necessary resources are expended to complete a project. AI also contributes to the decision-making of projects' timelines to avoid undue costs.

5.3. Transparency and communication

This means therefore that AI tools such as chatbots can improve communication as they provide real-time data and notifications to help solve a problem faster. Historical Agile practices often rely on Information radiators and ad-hoc meetings that rely on face-to-face communication and written updates (Table 75.3).

By the use of AI, transparency is provided to the stakeholders by providing live status on the projects, the utilization of the available resources, and the future risks that may be experienced. This helps in better decision-making, reduction in problems due to miscommunication, and improved satisfaction of the stakeholders.

5.4. Scalability

AI-based Agile systems also outcompete other systems in terms of handling large projects because of flexibility in resource management and managing several teams at the same time. By the same token, scaling frequently results in an increase in APM's complexity, improprieties regarding resource allocation, and poor decision-making (Table 75.4).

AI makes it possible for the various teams and phases of a project to be synchronized at a larger scale using automation

Table 75.3. Comparison of transparency and communication metrics between traditional and AI-enabled APM

Metric	*Traditional APM*	*AI-Enabled APM*
Information Flow Speed	3 hours	30 minutes
Stakeholder Reporting	Weekly	Real-Time
Meeting Frequency	4 per month	2 per month
Update Accuracy	70%	95%
Communication Gaps	Moderate	Minimal

Source: Author.

Table 75.4. Comparison of scalability metrics between traditional and AI-enabled APM

Metric	*Traditional APM*	*AI-Enabled APM*
Team Size Management	Limited	Scalable
Resource Allocation	Manual	Automated
Cross-Team Collaboration	Challenging	Seamless
Sprint Adjustments	Slow	Instant
Project Complexity	High	Manageable

Source: Author.

and real-time data. It is also notable that with the increasing size and scale of the projects implemented the usage of AI can go on in parallel with the optimization of the resources and the tasks, that define further progress.

5.5. Risk management

AI-integrated Agile systems are better in risk management because it is easy to point out risks early using the analysis feature. However, traditional APM does not have much estimation capability to deter risks, which in turn hinders project progress or even results in failure (Table 75.5).

With algorithms, risks are evaluated and solutions for them are arrived at by AI tools, hence minimizing project failures. Due to real-time processing, AI identifies additional risks that a team needs to address before they develop into larger problems.

6. Real-Time Case Studies and Data

6.1. Case study 1: AI implementation in a large-scale IT project

A large-scale IT company initiated an extensive software development project that employed the Scrum Agile framework to design and develop an organization-encompassing application. Among other improvements, the company implemented the use of AI within tools used for sprint planning, resource allocation, and identifying possible risks. Some of the solutions that applied AI to support their functionality are Jira, where the sprint performance was tracked, and data-driven decisions on alterations were made. Further, the use of IBM Watson for Project Management was made to analyze risks online, to determine possible delays and problems in project implementation, and means for resource management, like Forecast, to allocate the working teams depending on their workload and current priorities. AI integration made this impact: shortening the anticipated time of the project by 20% to 14 months from 18 months. The utilization of the application also provided a 15% reduction in cost by refining the distribution of resources and improving the mode of choice cutting operational costs. Regarding risk management, AI applications were genuinely efficient in defining problems at the early stage; that is where such applications minimized delay by 30% and also cut unlooked-for risks by 25%. Other grew fared better with a considerable improvement in stakeholder satisfaction, which was realized through raised project transparency levels. Due to the real-time reporting and the use of AI to generate predictive results, the different stakeholders were able to reach better decisions about the direction of the project [17]. Finally, AI designed the architecture to encourage flexibility that accommodated changing requirements to deliver the project within the planned timeline and budget as well as better quality than planned. In conclusion, it can be stated that not only all the Agile processes were made less time-consuming through the integration of AI technology but also the project team got more equipped to respond to rapid changes in circumstances and deliver more satisfying results to the parties interested.

Table 75.5. Comparison of risk management metrics between traditional and AI-enabled APM

Metric	*Traditional APM*	*AI-Enabled APM*
Risk Identification Time	4 days	1 day
Proactive Risk Management	50%	85%
Unexpected Delays	15%	5%
Risk Mitigation Cost	High	Low
Project Failure Rate	10%	3%

Source: Author.

6.2. Case study 2: AI-augmented scrum for a startup

A financial services provider decided to utilize Scrum as its Agile process when inventing a new Mobile App to address the problem of User Experience. To improve the Agile process, the startup adopted AI applications that include Monday.com and ClickUp helped with administrative tasks, and gave real-time tickets such as productivity, teamwork, and workload. These principles raised the level of effectiveness in sprint planning and shortened the time for iteration. Furthermore, the use of AI-based analytics was applied in the prioritization of the user stories and backlog which had information that was based on past performance by the users and the features that were most needed by the users. As a result of AI implementation, the cycle time of sprints improved from 4 weeks to a best estimate of 2.5 weeks, thereby doubling the frequency of software release. This accelerated delivery not only helped to increase product velocity but was also able to result in actual savings of 12% on the overall project cost due to less resource wastage and lower frequency of manual intervention. Earlier, AI tools fixed on the main values highlighted as relevant and helped indirectly increase stakeholder satisfaction by 18%. Additionally, real-time data through AI made it possible for the team to identify slow-moving work and decline the overall productivity due to ineffective collaboration by identifying work plateaus due to misunderstandings and conflicts. The application of AI in the Scrum process in the startup enabled quicker turnaround time; proper utilization of the available resources, and appropriate working of multi-disciplinary teams while addressing quality issues. The picture described in this case shows that AI benefits the practiced Agile; it makes the project cheaper in scale, more flexible to the change, and more adjusted to the stakeholders' expectations even in the conditions of a small and rapidly developing environment [18]. The adaptation of AI-related improvements furthermore served as crucial for optimizing work processes, boosting effectiveness and efficiency, and ensuring the fulfilment of strict time constraints, which, on the other hand, were still associated with a rather high level of satisfaction expressed by the development team as well as by the stakeholders.

7. Challenges in Integrating AI into APM

Integrating Artificial Intelligence (AI) into Agile Project Management (APM) presents a range of challenges, particularly in the areas of technology, organizational culture, and ethics. Data privacy and security are one of the most critical technological challenges in implementing human capital management mechanisms. AI systems typically are designed to use big data and raise the issue of data privacy and security of information that is to be used as databases for certain projects. For example, if relying on AI tools for predictive analytics or real-time decisions, the organization needs to comply with data protection legislation like GDPR. Furthermore, the APM frameworks must be supplemented with various AI tools, applying which to an organization might be challenging. Indeed, based on the experience of different Agile teams, it was often seen that they rely on the mixture of tools and platforms adding that the introduction of AI-oriented solutions often entails substantial changes in the existing platforms, which leads to integration problems in case the new technologies have to interact close-knit. Often strong organizational resistance is also cited as the key problem. AI implementation can only happen when there is a change of attitude and the acquisition of new skills. Some members may disregard the adoption of AI solutions because they are unfamiliar with such tools, or they fear they will be replaced by the technologies, or the tools are quite complicated. This means there have to be extensive training programs to ensure that teams that are being assigned to AI roles have the necessary knowledge to work on the technology. Additionally, organizational culture management needs to shift towards a more data-driven culture, which may be tough in some organizations that did not use much data logging and analytics in their daily operations in decision-making in the past. Finally, ethics in AI decision-making are also a problem. Whenever an AI algorithm is implemented, the logic behind the actions is usually unclear since the algorithms work as closed boxes. Such a system can lead to ethical questions where resources are allocated, risks are assessed, or team results are evaluated. The problem here is to make the AI-based decisions as fair, quiet, and ethical as possible, which requires close and constant supervision.

8. Recommendations

This paper discusses how organizations can use APM to implement AI in a coherent system that is beneficial for the organization while following the Agile principles. Based on these findings, one of the primary suggestions is to generate a strategic map for AI implementation, one of the first steps being an AI organizational readiness evaluation. Some of the considerations that should be made at this level should be the assessment of the current Agile processes, consideration of areas where AI results can be integrated, and tipping of AI that works well with other existing systems. The framework needs to aim at the moderate adoption of AI, giving teams a chance to step by step integrate AI attributes like predictive analysis, allocation of resources, and real-time insights into their operations. As is clear from the above description, flexibility, scalability, and the ability to not disrupt Agile are some of how AI must be developed for it to be easily integrated into Agile. The consideration of AI applications within the Agile context needs to address communicational openness and iteration improvement. It is mandatory always to remember that AI solutions improve the decision-making abilities of individuals and not replace them. For instance, backlog prioritization risk profiling, or management, AI tools can help in these tasks to well enable the team members to perform innovative and value-added jobs instead of repetitive work. Further, the adoption of AI should enhance transparency by offering timely information about work regarding the progress of projects to allow for data-based decision-making to take place efficiently. However, these tools should be developed in such a way that they support Agile's iterative layout of process components, supporting ongoing learning. Skill development of stakeholders is also a necessity as well as the involvement of stakeholders. Agreeing to the integration of AI into Agile processes it is necessary to engage the stakeholders in an open discussion where they are going to be informed on the advantages of using AI in the given production processes as well as on the potential drawbacks that can arise during the implementation of AI. Training of the team members should be carried out aiming to increase their knowledge regarding the usage of AI tools. Finally, more attention should be paid to the promotion of a learning organization culture in Agile teams, so that they can accept new AI skills as soon as they appear.

9. Conclusion

Therefore, inserting AI into APM is a great way to modernize the matrix of traditional Agile practices based on higher effectiveness, measurability, and satisfaction of stakeholders' needs. These Integrated AI techniques like predictive tools, wise usage of resources, and risk control allow the Agile teams to work smart, efficient, fast, and flexible based on the requirements of the project. The outcomes of this work show that AI is useful not only in the context of efficient decisions about sprint planning and task prioritization but also in the context of effective team cooperation by providing more or less biased information. In addition, it has been found that with the help of artificial intelligence, more and more costs and time are being saved in a project due to the reduced workload as well as better decision-making. Yet, incorporating AI into Agile frameworks is still a problem, for reasons of specific technical aspects, organizational constraints, and ethical issues. These issues are for organizations to deal with the following ways: coming up with a clear AI roadmap, positioning AI tools to Agile methodology, and enhancing stakeholder management and skills. Further research should be targeted at optimizing the application of AI in Agile settings, identifying the new AI tools that can complement Agile methodologies, and understanding the lack of data on the moral characteristics of AI decision-making in managing projects. However, it also revealed that there is a lack of real-life examples and case studies, as well as sector-based analyses to have a holistic view of towards scope and impact of AI. As technology advances, so will the position of AI in Agile Project Management become more and more crucial and hold the ability to foster advancement, productivity, and better results for stakeholders in the context of the complex and ever-changing project management environment of the present and future.

References

[1] Gough, J. M. (2022). Corporate Improvement in Project Management: A Design Thinking Approach Investigating an Adaptable Model.

[2] Taheri Khosroshahi, F. (2024). The Future of Project Management: Embracing artificial intelligence for enhanced efficiency, data-driven decision-making, and collaborative teamwork.

[3] Jahan Karamthulla, M., Muthusubramanian, M., Tadimarri, A., & Tillu, R. (2024). Navigating the Future: AI-Driven Project Management in the Digital Era. *International Journal for Multidisciplinary Research*, *6*(2), 1–11.

[4] Bahi, A., Ghari, J., & Gahi, Y. (2024). Integrating Generative AI for Advancing Agile Software Development and Mitigating Project Management Challenges. *International Journal of Advanced Computer Science & Applications*, *15*(3).

[5] Kiani, A. (2024). Artificial intelligence in entrepreneurial project management: a review, framework and research agenda. *International Journal of Managing Projects in Business*.

[6] Daraojimba, E. C., Nwasike, C. N., Adegbite, A. O., Ezeigweneme, C. A., & Gidiagba, J. O. (2024). A comprehensive review of agile methodologies in project management. *Computer Science & IT Research Journal*, *5*(1), 190–218.

[7] Hossain, M. Z., Hasan, L., Dewan, M. A., & Monira, N. A. (2024). The Impact of Artificial Intelligence on Project Management Efficiency. *International Journal of Management Information Systems and Data Science*, *1*(5), 1–17.

[8] Amajuoyi, P., Benjamin, L. B., & Adeusi, K. B. (2024). Optimizing agile project management methodologies in high-tech software development. *GSC Advanced Research and Reviews*, *19*(2), 268–274.

[9] Onesi-Ozigagun, O., Ololade, Y. J., Eyo-Udo, N. L., & Oluwaseun, D. (2024). Agile product management as a catalyst for technological innovation.

[10] Ronak, B. (2024). AI-Driven Project Management Revolutionizing Workflow Optimization and Decision-Making. *International Journal of Trend in Scientific Research and Development*, *8*(6), 325–338.

[11] Naeem, M., Ozuem, W., Howell, K., & Ranfagni, S. (2024). Demystification and actualization of data saturation in qualitative research through thematic analysis. *International Journal of Qualitative Methods*, *23*, 16094069241229777.

[12] Michalski, R., & Zaleski, S. (2024). Success factors in the management of IT service projects: regression, confirmatory factor analysis, and structural equation models. *Information*, *15*(2), 105.

[13] Naderi, A. (2024). *Project Management and Business Intelligence: A Bibliometric Analysis* (Doctoral dissertation, Politecnico di Torino).

[14] Raza, A., & Espinosa-Leal, L. (2024, March). Predicting the Duration of User Stories in Agile Project Management. In *International Conference on Science and Technology Education* (pp. 316–328). Cham: Springer Nature Switzerland.

[15] Ebirim, W., Montero, D. J. P., Ani, E. C., Ninduwezuor-Ehiobu, N., Usman, F. O., & Olu-lawal, K. A. (2024). The role of agile project management in driving innovation in energy-efficient HVAC solutions. *Engineering Science & Technology Journal*, *5*(3), 662–673.

[16] Τσερώνης, Ι. (2024). *Project management in information technology: challenges and best practices* (Master›s thesis, Πανεπιστήμιο Πειραιώς).

[17] Bennett Moses, L., Compton, C., Murdocca, M., Kim, H., Ray, A., Renehan, K., & Michalowski, C. (2020). Submission to the Department of Foreign Affairs and Trade on Australia's International Cyber and Critical Technology Engagement Strategy. Available at SSRN 3708041.

[18] Pan, H., Wang, R., Wang, H., & Jia, Z. (2023). Sustainable Supply Chain Management: Best Practices for Reducing Environmental Footprints in the Global Apparel Industry. *ESTIDAMAA*, *2023*, 18–26. https://doi.org/10.70470/ESTIDAMAA/2023/003

76 Antecedents of solar rooftop PV panel adoption intention: Theoretical insights and empirical evidence

Swathi Newashini[1,a] *and R. Parameswaran*[2,b]

[1]Research Scholar, Department of Commerce, Karpagam Academy of Higher Education, Coimbatore, Tamil Nadu, India
[2]Professor, Department of Commerce, Karpagam Academy of Higher Education, Coimbatore, Tamil Nadu, India

Abstract: The adoption of solar rooftop photovoltaic (PV) panels plays a crucial role in transitioning toward clean and sustainable energy systems. Despite the growing relevance of this technology, adoption at the household level remains uneven. This study investigates the key determinants of adoption intention by integrating constructs from the Unified Theory of Acceptance and Use of Technology (UTAUT2), including hedonic motivation, price value, and performance expectancy, along with trust in AI and environmental concern as mediators. Using structural equation modeling on survey data from Indian households, the study finds that hedonic motivation, price value, and performance expectancy significantly influence adoption intention – both directly and indirectly via environmental concern. Trust in AI, while conceptually relevant, does not significantly mediate adoption decisions. The findings offer theoretical insights into the psychological and economic drivers of clean energy adoption and provide actionable recommendations for policymakers and marketers to tailor their strategies. Future research opportunities and practical implications are also discussed.

Keywords: Solar energy, adoption intention, hedonic motivation, price value, environmental concern, trust in AI, structural equation modeling

1. Introduction

The increasing global focus on sustainable development and climate change mitigation has positioned renewable energy technologies as essential components of national and international energy policies [3, 4]. Among these, PV panels have emerged as a promising decentralized energy solution, particularly in emerging economies like India [5]. These systems offer the dual benefit of reducing household electricity bills while contributing to environmental conservation, aligning with both individual and collective sustainability goals [6].

Despite significant government incentives and growing technological maturity, the adoption of solar rooftop PV systems by households has remained uneven and relatively limited [7, 8]. Understanding the behavioural and psychological drivers behind this adoption gap is critical for stakeholders seeking to promote clean energy transitions at the grassroots level [46]. Traditional studies have often focused on technical, infrastructural, or cost-related barriers [9, 10]. However, there is a growing consensus that personal attitudes, perceptions, and motivations play a pivotal role in shaping consumer behaviour toward renewable technologies.

This study builds upon the UTAUT2 by incorporating additional constructs relevant to energy and environmental contexts – namely, trust in artificial intelligence (AI) and environmental concern [11]. This study investigates the direct and indirect effects of major antecedents, such as price value, performance expectancy, and hedonic motivation, on PV panel installation intention. The mediating variables are trust and environmental concern [47].

By using techniques like structural equation modeling (SEM) and empirical analysis based on Indian families, the research tries to establish both theoretical understanding and practical suggestions for encouraging the adoption of solar panels [12, 13]. The results aim to provide marketers, legislators, and software developers with insight as they post their proposals with user incentives, which will ultimately serve to encourage a more equitable and sustainable energy shift.

2. Literature Review and Theoretical Background

Implementation of renewable energy technologies, particularly solar rooftop PV panel systems [48], has been the focus of large-scale research studies from diverse academic disciplines, ranging from economics to engineering and behavioural sciences [14, 15]. Yet, the focus now is on understanding the perceptual and psychological factors that constrain technology adoption at the individual level.

[a]swathinewashini.shanmugasundharam@kahedu.edu.in, [b]parameswaran.ramaswami@kahedu.edu.in

DOI: 10.1201/9781003684459-76

Besides, the present [16, 17, 18] research builds mostly on the UTAUT2, adding specific constructs like AI and trust in environmental awareness to situate the specific model with respect to solar energy adoption [19, 20]. The most important and stable predictor of technology adoption is performance expectancy, defined as the degree to which users perceive that they would be able to perform certain activities with the help of the technology [21].

Perceptions of energy efficiency, reliability, and system output by consumers are the major drivers of performance expectancy in the renewable energy market. Based on experiential studies, individuals tend to embrace a technology more if they perceive it as being more efficient [22, 23]. Price Value has been highlighted in various clean energy research as a measure of equilibrium between the money price of technology and perceived value. Cost considerations continue to play an important role in developing countries, particularly in capital-intensive technology such as photovoltaic panels [24, 25]. The assumption that positive first impressions of value for money may significantly enhance adoption prospects is affirmed by previous studies. Whereas as autonomous energy management, intelligent controls, and app-based interfaces are becoming a growing presence on modern energy infrastructures, [44, 45] hedonic motivation too has been envisioned to have its role to contribute significantly. Emotional engagement through such motivation has actually been shown to have a positive effect on intention and attitude concerning solar adoption [1].

Price Value, or the trade-off between perceived benefits and the monetary cost of technology, has been highlighted in several clean energy studies [26, 27, 28]. In emerging economies, cost considerations remain central, particularly for capital-intensive technologies like solar PV panels. Prior findings support the notion that favorable perceptions of value for money can significantly enhance the likelihood of adoption.

Hedonic Motivation, or the enjoyment derived from technology use, has also been recognized as a relevant construct in modern energy systems where smart control, app-based interfaces, and autonomy in energy decisions are increasingly common. This emotional engagement has shown to positively influence both attitude and intention toward adoption [2, 29, 30].

Trust in Artificial Intelligence (AI): As an applicable mediating variable, trust in AI is also covered under this research. It becomes important to believe in the intelligence that drives smart solar systems since they incorporate AI-based monitoring, optimization, and diagnostics. Not much is known about the impact of faith in AI on solar energy use at home, especially in underdeveloped countries where awareness of AI might be poor [31, 32].

Another important and pivotal mediator in this research domain is environmental concern [33], which refers to individuals' awareness and feeling of responsibility to deal with environmental issues.

Environmental concern has been found to be a significant predictor of green technology adoption and one of the most important mediators between personal incentives and pro-environmental behaviour in previous studies [34, 35].

Concern for the environment may support and reinforce adoption patterns of solar PV panels that are influenced by emotional or financial considerations [36]. They come together to produce a comprehensive model that brings together utilitarian, emotional, and moral motivations with mediation pathways that can enhance or delay adoption behaviours in an effort to examine household intention to adopt solar PV [37, 38]. With empirical data collected from Indian households, this theoretical framework is tested using a structural equation modeling method. To examine the determinants of Indian families' intentions to purchase solar PV panels, this research employs a quantitative, cross-sectional approach. SEM was employed to evaluate the theoretical model, which was derived from UTAUT2 and supplemented with mediating variables – Trust in AI and Environmental Concern [39, 40].

3. Research Objectives and Hypotheses

The survey instrument was built using commonly used scales from prior research that were adapted to the solar energy context [41, 42]. The model integrates established constructs from the UTAUT2 framework [49] – namely, Performance Expectancy, Hedonic Motivation, and Price Value – along with Trust in AI and Environmental Concern as mediating variables. The objective is to understand not only the direct effects of these predictors on adoption intention but also how trust and environmental concern mediate these relationships [43].

Specifically, the study addresses the following research objectives:

1. To examine the direct effects of key antecedents on the intention to adopt solar rooftop PV panels.
2. To evaluate the mediating roles of trust in AI and environmental concern in shaping the relationship between the antecedents and adoption intention.

Based on these objectives and the conceptual framework developed, the following hypotheses were formulated:

H1: The antecedents – Performance Expectancy, Hedonic Motivation, Price Value, Energy Efficiency, Effort Expectancy, Facilitating Conditions, and Social Influence – have significant direct effects on Adoption Intention [50].

H2: Performance Expectancy, Hedonic Motivation, Price Value, Energy Efficiency, Effort Expectancy, Facilitating Conditions, and Social Influence positively influence Trust in AI.

H3: Performance Expectancy, Hedonic Motivation, Price Value, Energy Efficiency [51, 52], Effort Expectancy, Facilitating Conditions, and Social Influence positively influence Environmental Concern.

H4: Trust in AI mediates the relationship between the antecedents and Adoption Intention

H5: Environmental Concern mediates the relationship between the antecedents and Adoption Intention.

These hypotheses are tested through structural equation modelling (SEM) to determine the strength and significance of both direct and indirect relationships among the variables in the model.

4. Methodology

This study adopts a quantitative, cross-sectional research design to examine the behavioural antecedents of solar rooftop PV panel adoption intention among Indian households. The conceptual model, grounded in UTAUT2 and extended with mediating variables-Trust in AI and Environmental Concern, was tested using SEM.

The conceptual model developed in this study is grounded in the UTAUT2 and is extended to reflect the unique context of PV panel adoption. It incorporates both well-established predictors and domain-specific mediators to provide a comprehensive understanding of user intention (Figure 76.1).

The model posits that seven antecedent variables- Performance Expectancy, Hedonic Motivation, Price Value, Energy Efficiency, Effort Expectancy, Facilitating Conditions, and Social Influence, influence consumers' Adoption Intention either directly or indirectly. Two mediators are introduced to capture the underlying psychological processes:

Trust in AI represents confidence in the intelligent systems embedded in solar PV technology (e.g., optimization algorithms, smart monitoring).

Environmental Concern reflects individuals' awareness of and motivation to address environmental issues.

The model allows for the testing of both direct effects (from each antecedent to Adoption Intention) and indirect effects (mediated through Trust in AI and Environmental Concern). This dual-path structure provides insights not only into what drives adoption but also how different motivational and perceptual factors shape user behaviour.

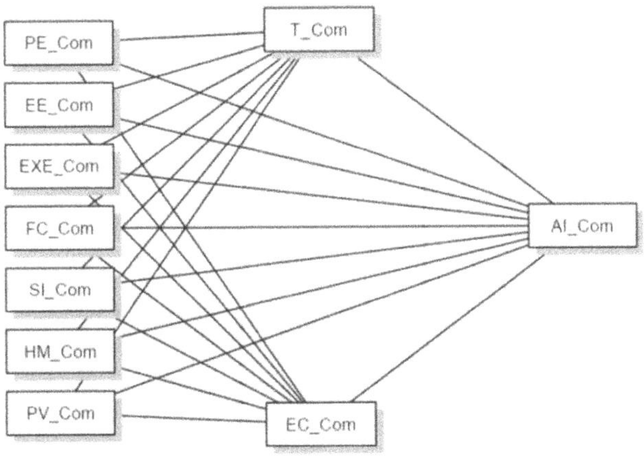

Figure 76.1. Conceptual framework.

Source: Author.

4.1. Sampling and data collection

Data were collected through a structured questionnaire distributed to potential and existing solar rooftop PV users across various urban and semi-urban regions in India [53]. The sampling technique used was purposive sampling, aimed at targeting individuals with some level of awareness or exposure to solar PV technology. A total of 225 responses were collected, of which 205 were retained after data cleaning, missing value checks, and outlier analysis.

4.2. Measurement instrument

The survey instrument was built using commonly used scales from prior research that were adapted to the solar energy context.Multi-item Likert-type scales were employed to assess all the factors: 1 represented 'strongly disagree' and 5 represented 'strongly agree.' Among the essential factors were: Performance Expectancy (PE): Following Venkatesh et al. (2012)'s research – Price Value (PV): Refers to the perceived value of the cost and benefit. Measuring the enjoyment or satisfaction derived from the use of solar technology is referred to as hedonic motivation (HM). Measuring confidence in AI-facilitated solar capability (such as energy optimization) is referred to as trust in AI. Environmental Concern (EC): Measuring an individual's sense of responsibility and awareness of environmental impacts. Adoption Intention (AI) is a measure of the probability that solar rooftop PV systems will be adopted or sustained.

5. Results and Discussion

SEM was employed to analyze the data. The constructs in the model's validity, fit, and reliability were determined with the use of confirmatory factor analysis (CFA). The intended linkages and mediating mechanisms were viewed through

Table 76.1. A revealed two statistically significant indirect effects on adoption intention jointly mediated by environmental concern

Indirect Path	*Estimate (β)*	*SE*	*z-value*	*p-value*
PE → EC → AI	0.0142	0.0133	1.07	0.286
HM → EC → AI	0.0783	0.0282	2.55	0.011
PV → EC → AI	0.0922	0.0413	2.69	0.007
EE → EC → AI	0.0129	0.0121	1.07	0.284
EXE → EC → AI	0.0105	0.0110	0.96	0.336
FC → EC → AI	0.0136	0.0141	0.97	0.333
SI → EC → AI	0.0087	0.0125	0.70	0.487

Source: Author.

examining direct, indirect, and total impacts. Standard error (SE), z-values, and p-values were employed in the examination of the significance of path coefficients where a p-value less than 0.05 indicates statistical significance (Table 76.1).

Initially, the noteworthy path from Hedonic Motivation (HM) to Adoption Intention (AI) through Environmental Concern (EC) was significant (β = 0.0783, p = 0.011). This suggests that a person who gains pleasure or emotional satisfaction from utilizing this technology is more likely to adopt solar PV panels, when such pleasure favors environmental sustainability. Also describes as, the pleasure gained from the use of product generates ethical reinforcement loop, in which users find the act of adoption not only rewarding but also responsible and meaningful. This points to a strong psychological process in which personal enjoyment results in improved pro-environmental action, thus increasing adoption. With a high value (β = 0.0922, p = 0.007), the second significant pathway was from Price Value (PV) to Adoption Intention via Environmental Concern. Based on this theory, individuals who perceive solar panels as having economic benefits are more likely to buy them if they also take into consideration the environmental benefits of their purchase.

This pathway shows that even pragmatic users – primarily motivated by savings – can be nudged toward adoption through sustainability narratives that align economic and ecological values.

All other indirect effects tested in the model – including those passing through Trust in AI (T) or Environmental Concern (EC) – for variables such as Performance Expectancy (PE), Energy Efficiency (EE), Effort Expectancy (EXE), Facilitating Conditions (FC), and Social Influence (SI) were found to be statistically non-significant (p > 0.05). These paths did not show meaningful mediation effects, suggesting that the influence of these constructs on adoption intention is either weak or operates through other unexplored mechanisms. Likewise, while HM → T → AI and PV → T → AI showed positive but non-significant coefficients, the lack of statistical support suggests that Trust in AI does not currently act as an effective mediator in this context.

These findings validate Environmental Concern as a key mediating construct, capable of transforming both emotional (hedonic) and rational (price-based) motivations into actual behavioural intention. It also underscores the importance of framing solar PV as both a personally rewarding and socially responsible technology in outreach and marketing efforts.

Table 76.2 reveals that all four variables - Performance Expectancy (PE), Hedonic Motivation (HM), Price Value (PV), and Environmental Concern (EC), exhibited statistically significant direct effects on Adoption Intention. This suggests that users are more likely to adopt solar rooftop PV systems when they perceive the technology as useful, enjoyable, financially worthwhile, and environmentally meaningful. Performance Expectancy's positive effect (β = 0.1944, p = 0.012) reinforces the role of perceived functional utility in technology adoption. Hedonic Motivation (β = 0.1782, p = 0.022) indicates that enjoyment and emotional satisfaction can play a vital role, often underappreciated in rational models. Price Value (β = 0.1790, p = 0.016) confirms that users weigh financial gains in their decision-making. Environmental Concern (β = 0.2155, p = 0.003) strengthens the argument that personal values and ecological responsibility are powerful drivers of sustainable behaviour.

Table 76.2. Direct effects on adoption intention

Direct Path	*Estimate (β)*	*SE*	*z*	*p-value*	*Significance*
PE → AI	0.1944	0.0815	2.52	0.012	✓ Significant
HM → AI	0.1782	0.0714	2.29	0.022	✓ Significant
PV → AI	0.1790	0.0892	2.41	0.016	✓ Significant
EC → AI	0.2155	0.0764	2.94	0.003	✓ Significant

Source: Author.

Interestingly, while these four predictors were significant, other classic UTAUT2 predictors-such as Energy Efficiency (EE), Effort Expectancy (EXE), Facilitating Conditions (FC), and Social Influence (SI) – did not show significant direct effects in this model. This may reflect a maturity of user expectations where technical ease or social pressure plays a lesser role compared to more intrinsic and outcome-driven motivators.

Table 76.3 illustrates that both Hedonic Motivation and Price Value significantly influenced the mediators – Trust in AI and Environmental Concern. HM and PV were the only antecedents to show such dual influence, suggesting that users who enjoy the solar experience or see high economic value also tend to trust AI-driven functions and exhibit pro-environmental values. Specifically, PV had strong effects on both EC (β = 0.4278) and T (β = 0.2996), while HM significantly influenced EC (β = 0.3635) and T (β = 0.1980).

Table 76.3. Predictor → mediator relationships

Predictor → Mediator	*Estimate (β)*	*SE*	*z*	*p-value*	*Significance*
HM → EC	0.3635	0.0626	5.11	< 0.001	✓ Significant
PV → EC	0.4278	0.0745	6.64	< 0.001	✓ Significant
PV → T	0.2996	0.0803	4.22	< 0.001	✓ Significant
HM → T	0.1980	0.0675	2.52	0.012	✓ Significant

Source: Author.

Notably, Performance Expectancy, Energy Efficiency, Effort Expectancy, Facilitating Conditions, and Social Influence did not demonstrate any significant effects on the mediators. This suggests that the emotional and financial dimensions of user perception are more likely to shape trust and concern than purely functional or contextual factors. These insights emphasize the critical role of affective and value-driven appraisals in modern clean energy adoption.

Total effects analysis synthesizes both the direct and indirect influences of key predictors on Adoption Intention (Table 76.4). Price Value ($\beta = 0.3047$) emerged as the strongest overall influence, followed closely by Hedonic Motivation ($\beta = 0.2787$), with Performance Expectancy also showing substantial total impact ($\beta = 0.2348$).

These results reinforce the insight that emotional engagement (HM), cost-benefit considerations (PV), and perceived usefulness (PE) are the top drivers of behavioural intention in this domain. Not only do HM and PV exert strong direct effects, but their indirect influence – particularly via Environmental Concern – further boosts their total effect. Non-significant total effects were observed for Energy Efficiency, Effort Expectancy, Facilitating Conditions, and Social Influence. The absence of strong effects from other UTAUT2 variables confirms the dominance of value-laden and emotionally resonant predictors over contextual or ease-of-use considerations in driving adoption.

Table 76.4. Total effects

Predictor	*Total Effect (β)*	*SE*	*z*	*p-value*	*Overall Importance*
HM	0.2787	0.0694	3.68	< 0.001	Very High
PV	0.3047	0.0825	4.44	< 0.001	Very High
PE	0.2348	0.0833	2.97	0.003	High

Source: Author.

This holistic view of total effects supports the need for integrated communication strategies that link financial incentives and enjoyable user experience with broader ecological narratives to drive solar PV adoption.

Table 76.5 reveals that, none of the indirect effects via Trust in AI were statistically significant ($p > 0.05$). While some paths such as PV → T → AI ($\beta = 0.0373$, $p = 0.123$) and HM → T → AI ($\beta = 0.0215$, $p = 0.114$) showed positive directionality, they did not meet the threshold for significance. This indicates that although users may develop some level of trust in AI features based on their perceptions of utility or enjoyment, this trust does not effectively mediate their intention to adopt solar rooftop PV systems.

These findings suggest that Trust in AI may not yet be a strong behavioural bridge between user perception and adoption action. This could be due to limited awareness or understanding of AI functionality in solar systems among household consumers. It is possible that the AI component is either under-communicated in product messaging or not perceived as central to user experience. In practical terms, this implies that while AI features may enhance operational efficiency, they should be paired with clear, value-based communication that highlights their relevance to

Table 76.5. Indirect effects via trust in AI

Indirect Path	*Estimate (β)*	*SE*	*z-value*	*p-value*
PE → T → AI	0.0156	0.0155	1.01	0.311
HM → T → AI	0.0215	0.0136	1.58	0.114
PV → T → AI	0.0373	0.0242	1.54	0.123
EE → T → AI	0.0121	0.0118	1.03	0.302
EXE → T → AI	0.0099	0.0110	0.90	0.370
FC → T → AI	0.0108	0.0124	0.87	0.387
SI → T → AI	0.0061	0.0119	0.51	0.611

Source: Author.

Table 76.6. Summary of hypothesis testing results

Hypothesis	*Tested Paths / Evidence*	*Result*	*Supported/Not Supported*
H1	PE → AI ($\beta = 0.1944$, $p = 0.012$) ✓ HM → AI ($\beta = 0.1782$, $p = 0.022$) ✓ PV → AI ($\beta = 0.1790$, $p = 0.016$) ✓ EE, EXE, FC, SI → AI: Not significant ✗	Partially Supported	✓ (for PE, HM, PV)
H2	HM → T ($\beta = 0.1980$, $p = 0.012$) ✓ PV → T ($\beta = 0.2996$, $p < 0.001$) ✓ PE, EE, EXE, FC, SI → T: Not significant ✗	Partially Supported	✓ (for HM, PV)
H3	HM → EC ($\beta = 0.3635$, $p < 0.001$) ✓ PV → EC ($\beta = 0.4278$, $p < 0.001$) ✓ PE, EE, EXE, FC, SI → EC: Not significant ✗	Partially Supported	✓ (for HM, PV)
H4	No significant indirect effect via T for any antecedent → AI	Not Supported	✗
H5	HM → EC → AI: ($\beta = 0.0783$, $p = 0.011$) ✓ PV → EC → AI: ($\beta = 0.0922$, $p = 0.007$) ✓	Supported	✓

Source: Author.

user goals – such as cost savings or environmental contribution – to activate behavioural pathways more effectively.

The results of hypothesis testing provide partial support for the proposed conceptual model. **H1**, which posited that all seven antecedents would have significant direct effects on adoption intention, was partially supported. Specifically, Performance Expectancy, Hedonic Motivation, and Price Value demonstrated statistically significant direct effects, confirming their central role in shaping adoption behaviour. However, Energy Efficiency, Effort Expectancy, Facilitating Conditions, and Social Influence did not show significant direct effects in this context, suggesting they may be less influential in the decision to adopt solar rooftop PV systems among Indian households.

H2 and H3, which proposed that the antecedents would positively influence Trust in AI and Environmental Concern, were also partially supported. Only Hedonic Motivation and Price Value significantly influenced both mediators, while other antecedents did not exhibit meaningful effects. This indicates that users who find solar technology enjoyable and cost-effective are also more likely to develop trust in intelligent features and express stronger concern for environmental impact.

H4, which predicted that Trust in AI would mediate the relationship between antecedents and adoption intention, was not supported. No significant indirect effects were found through Trust in AI, indicating that while users may appreciate or trust AI features, this trust alone does not translate into a stronger intention to adopt (Table 76.6). In contrast, H5 was fully supported. Environmental Concern significantly mediated the effects of both Hedonic Motivation and Price Value on adoption intention. This highlights the critical role of environmental consciousness in transforming emotional and economic motivations into behavioural intentions.

6. Recommendations for Marketers and Policymakers

Based on the empirical findings of this study, the following recommendations are proposed for marketers, policymakers, and solar technology promoters to enhance the adoption of rooftop solar PV panels:

1. Promote the Financial Value Proposition: Price Value showed the strongest overall effect on adoption intention. Marketers should emphasize long-term financial benefits, cost savings, and return on investment from solar installations. Use practical tools like ROI calculators and success stories to make financial benefits tangible.
2. Appeal to Emotional and Experiential Value: Hedonic Motivation significantly influenced adoption both directly and through environmental concern. Promotional content should highlight the satisfaction and pride users feel in using clean energy. Smart dashboards, sleek system designs, and app-based controls can enhance the user experience.
3. Leverage Environmental Concern: Environmental Concern played a key mediating role. Communicate the ecological benefits of solar adoption – such as carbon reduction, energy independence, and sustainable living. Reinforce solar use as a responsible and impactful lifestyle choice.
4. Enhance and Reframe AI Feature Messaging: Although Trust in AI did not directly influence adoption, it was positively influenced by HM and PV. Position AI not as a technical feature, but as a problem-solver – for example, optimizing energy usage or enabling predictive maintenance. Highlight practical outcomes rather than abstract capabilities.
5. Simplify the Adoption Process: Though effort expectancy and facilitating conditions were not statistically significant, they still reflect background expectations of ease and support. Ensure that financing, installation, and after-sale service are seamless and transparent.
6. Segment and Personalize Communication: Since users are motivated by different combinations of value, emotion, and ethics, marketing should be tailored. Financially-focused users may respond to economic arguments, while others may be more responsive to sustainability themes or smart-home integrations. Use targeted messaging to reach each segment effectively.

7. Scope for Future Research

While this study offers valuable insights into the behavioural antecedents of solar rooftop PV panel adoption, it also opens up several avenues for further exploration. Future research can build on the current findings and address some of the limitations of this study to expand the understanding of clean energy adoption behaviours.

1. Moderating Variables: Future studies could incorporate demographic variables such as age, income, educational level, or urban versus rural setting as moderators. These factors may influence how predictors like hedonic motivation or price value affect adoption intention.
2. Longitudinal Studies: A longitudinal approach could capture shifts in consumer attitudes over time, particularly as awareness of AI integration in energy systems grows. Tracking the development of trust in AI features across stages of adoption would provide deeper insight.
3. Expanded Technological Context: While this study focused on residential solar rooftop PV systems, future research could extend the model to commercial or institutional contexts, where decision-making processes may differ.
4. Psychological and Social Constructs: Additional psychological constructs such as personal norms, perceived behavioural control, or habit could be integrated to

enrich the explanatory power of the model. These could uncover deeper motivational and emotional dynamics.

5. Cross-Cultural Validation: Validating the model across different cultural or geographic contexts would help determine the generalizability of the findings. In this instance, the additional confidence that the use of solar assists in protecting the environment reinforces the incentives related to cost savings.

8. Conclusion

The perceptual and psychological variables influencing Indian homeowners' behavioural intention to install solar rooftop PV panels were reasoned in the present research. The current study furnished a strong and meaningful perception into why individuals select adoption options when confronted with advancing energy technology through augmentation of the UTAUT2 model with new factors such as Trust in AI and Environmental Concern.

The findings presaged that Hedonic Motivation, Price Value, and Performance Expectancy are the best direct predictors of adoption intention. Also, environmental concern was identified as a strong mediating factor that effectively converts economic benefits and personal enjoyment into ecologically sound action. Intriguingly, the role of the adoption process was not greatly affected by trust in AI, showing a potential lack of knowledge or lack of appreciation for the abilities of AI when it comes to solar energy. Both theoretical knowledge and practical applicability are provided by these resultants. The research emphasizes the necessity of incorporating moral, financial, and emotional considerations in models anticipating the adoption of green technology from a theoretical outlook. It makes invaluable recommendations about how marketers and lawmakers could better transmit the vantages of solar PV systems, focusing on their financial benefits, emotional attraction, and environmental duties.

It is important to understand the behavioural factors affecting home adoption in order for India and other emerging nations to make concerted efforts towards lofty renewable energy targets. By providing evidence-based opinions regarding what influences user intention and how that intention can be promoted through smart, targeted strategies, this research encourages that pursuit.

References

[1] Zhang, L., & Liu, Y. (2021). AI-powered predictive maintenance for solar rooftop panels: A case study. *Energy Reports*, *7*, 49–57. https://doi.org/10.1016/j.egyr.2021.11.013

[2] Zhang, Y., & Zhou, X. (2021). The Integration of Artificial Intelligence in Photovoltaic Energy Systems for Improved Efficiency and Maintenance. *Solar Energy*, *221*, 228–245.

[3] Abuzaid, H., AbuMoeilak, L., & Alzaatreh, A. (2024). A structural equation modeling of customer attitudes towards residential solar initiatives in Jordan. *Social Sciences & Humanities Open*, *10*, 101038. https://doi.org/10.1016/j.ssaho.2024.101038

[4] Aggarwal, A. K., & Syed, A. A. (2019). Factors driving Indian consumer's purchase intention of rooftop solar. International Journal of Energy Sector Management, 13(3), 539–555. https://doi.org/10.1108/IJESM-03-2019-0009

[5] Aggarwal, A. K., Syed, A. A., & Garg, S. (2020). Diffusion of residential rooftop solar: Global learnings – The Indian context. In Business, economics and sustainable development: The emerging issues (pp. 115–133). Bloomsbury India. Bloomsbury Publishing India Pvt. Ltd.

[6] Alam, S. S., Masukujjaman, M., & Lin, C.-Y. (2022). Factors affecting photovoltaic solar energy usage intention in rural households in Bangladesh: A structural equation modelling approach. Environmental and Climate Technologies, 26(1), 276–293. https://doi.org/10.2478/ect-2022-0024

[7] Apfel, D., & Herbes, C. (2021). What drives Senegalese SMEs to adopt renewable energy technologies? Applying an extended UTAUT2 model to a developing economy. Sustainability, 13(9332). https://doi.org/10.3390/su13169332

[8] Atulkar, S. (2022). Purchase intention of Indian customers: A study on solar PV technology. International Journal of Energy Sector Management, 16, 946–964. https://doi.org/10.1108/IJESM-03-2022-0070

[9] Bagus, D., Bekti, M., & Prasetyo, Y. T. (2022). Determining factors affecting customer intention to use rooftop solar photovoltaics in Indonesia. Sustainability, 14(280). https://doi.org/10.3390/su14010280

[10] Badran, M. F., & Li, K. (2020). Smart solar rooftop systems using AI and IoT technologies. International Journal of Smart Grid and Clean Energy, 9(2), 368–379.

[11] Bennagi, A., & Al Housry, O. (2024). Comprehensive study of artificial intelligence applied in renewable energy. Energy Strategy Reviews, 54, 101446.

[12] Brown, V. (2005). Model of adoption of technology in households: A baseline model test and extension incorporating household life cycle. MIS Quarterly, 29, 399.

[13] Chen, H., & Zhang, J. (2023). The role of artificial intelligence in the future of solar rooftop energy management. Applied Energy, 301, 117234. https://doi.org/10.1016/j.apenergy.2021.117234

[14] Choubey, P. C., Oudhia, A., & Dewangan, R. (2012). A review: Solar cell current scenario and future trends. Recent Research in Science and Technology, 4, 99–101.

[15] Chou, T. L., & Chou, C. W. (2017). The Role of Hedonic and Utilitarian Motivations in the Adoption of Solar Energy Systems. Renewable and Sustainable Energy Reviews, 72, 462–469.

[16] Chien, S., & Chen, M. (2020). Hedonic and Utilitarian Motivations in the Consumer Decision to Adopt Solar Technology. Journal of Cleaner Production, 245, 118737.

[17] Dunn, B., Kamath, H., & Tarascon, J. M. (2011). Electrical energy storage for the grid: A battery of choices. Energy & Environmental Science, 4(9), 3381–3390.

[18] Ha, H.-Y., & Janda, S. (2012). Predicting consumer intentions to purchase energy-efficient products. Journal of Consumer Marketing, 29(7), 461–469. https://doi.org/10.1108/07363761211274974

[19] Hashmi, S. T. A., Raees, F., & Baig, M. M. (2023). A study of solar power usage in the southern region of

Pakistan. Sustainability, 15(15344). https://doi.org/10.3390/su151515344

[20] Hossain, M. I., & Rahman, M. (2021). Energy Management in Solar PV Systems Using Artificial Intelligence. Energy Reports, 7, 1046–1058.

[21] Imamzai, M., & Aghaei, A. (2012). A review on comparison between traditional silicon solar cells and thin film CdTe solar cells. Proceedings of National Graduate Conference, Tenaga Nasional Universiti, Putrajaya Campus, 8–10 November 2012, 1–5.

[22] Jain, S., & Sharma, A. (2021). Exploring the Role of Price Perception and Hedonic Value in Solar Panel Adoption. Energy Policy, 149, 112038.

[23] Kaur, N., & Sharma, A. (2024). Adoption of solar home lighting system: A behavioral paradigm shift from consumers to prosumers in the urban household energy transition of Punjab, India. Environmental Research Communications, 6(115020). https://doi.org/10.1088/2515-7620/6/1/115020

[24] Kumar, S. M., & Kumari, P. (2024). Scenario of solar energy and policies in India. Clean Energy, 8(5), 117–128.

[25] Kumar, S., & Reddy, P. V. (2022). The impact of artificial intelligence on optimizing solar rooftop PV systems. Journal of Renewable Energy, 18(4), 112–125. https://doi.org/10.1016/j.jren.2022.03.008

[26] Kumar, S., & Gupta, R. (2020). AI in Solar Panel Fault Detection and Diagnosis. Journal of Energy Engineering, 146(5), 04020070.

[27] Lai, K. K., & Tan, C. (2021). AI-Powered Solar Rooftop Systems for Smart Cities. Journal of Cleaner Production, 290, 125215.

[28] Lai, P. C. (2017). The literature review of technology adoption models and theories for the novelty technology. JISTEM - Journal of Information Systems and Technology Management, 14, 21–38. https://doi.org/10.4301/S1807-17752017000100002

[29] Le, T. T., & Le, T. T. (2024). Artificial intelligence applications in solar energy. International Journal of Information Visualization, 8(2), 826–844.

[30] Lin, L., & Hsu, Y. (2020). Understanding Consumers' Solar Panel Purchase Intentions: The Role of Price, Value, and Hedonic Motivation. Journal of Business Research, 115, 295–304.

[31] Liu, Z., & Lin, L. (2021). Artificial Intelligence and Machine Learning Applications in Solar Power Systems. Journal of Electrical Engineering & Technology, 16(2), 357–370.

[32] McEvoy, A., Castaner, L., & Markvart, T. (2012). Solar cells: Materials, manufacture and operation (2nd ed.). Elsevier Ltd.

[33] Morales, G. N. (2024). Artificial intelligence for renewable energy systems. University of the Andes.

[34] NREL. (2021). The Solar Futures Study. National Renewable Energy Laboratory, U.S. Department of Energy, Colorado.

[35] Patel, S. K., & Soni, D. (2022). Hybrid AI Models for Predicting Solar Energy Output in Rooftop Installations. International Journal of Electrical Power & Energy Systems, 124, 106693.

[36] Pathania, A. K., Goyal, B., & Saini, J. R. (2017). Diffusion of adoption of solar energy—a structural model analysis. Smart and Sustainable Built Environment, 6, 66–83.

[37] Parsad, C., Mittal, S., & Krishnankutty, R. (2020). A study on the factors affecting household solar adoption in Kerala, India. International Journal of Productivity and Performance Management, 69, 1695–1720. https://doi.org/10.1108/IJPPM-10-2019-0417

[38] Paramitaa, D., & Hartono, B. (2024). Exploring factors influencing intention and actual usage in household solar PV adoption. Cleaner and Responsible Consumption, 15, 100242. https://doi.org/10.1016/j.clrc.2024.100242

[39] Sabatino, M. D., & Hendawi, R. (2024). Silicon solar cells: Trends, manufacturing challenges, and AI perspectives. Crystals, 14, 167. https://doi.org/10.3390/cryst14020167

[40] Saxena, A., & Katta, V. (2024). The dawn of AI-enhanced solar energy systems. In Futuristic trends in artificial intelligence (IIP Series, Vol. 3, Book 4, Part 5).

[41] Sarangi, G. K., & Taghizadeh-Hesary, F. (2021). Rooftop solar development in India: Measuring policies and mapping business models. Asian Development Bank Institute, 1256.

[42] Singh, R., & Kumar, A. (2022). Artificial Intelligence in Solar Panel Performance and Energy Prediction. Renewable and Sustainable Energy Reviews, 151, 111616.

[43] Sovacool, B. K. (2009). The importance of governance and institutions to the success of solar photovoltaic scaling up policies. Energy Policy, 37(8), 2847–2854.

[44] Srinivasan, R., & Wang, J. (2020). Data-Driven AI Methods for Solar Panel Performance Prediction. Energy, 200, 117456.

[45] Thompson, R. L., Higgins, C. A., & Howell, J. M. (1991). Personal computing: Toward a conceptual model of utilization. MIS Quarterly, 1, 125–143. https://doi.org/10.2307/249748

[46] Wustenhagen, R., & Menichetti, E. (2012). Strategic choices for renewable energy investment: Conceptual framework and opportunities for further research. Energy Policy, 40, 1–10. https://doi.org/10.1016/j.enpol.2010.08.043

[47] Yadav, A., & Kumar, P. (2015). Enhancement in efficiency of PV cell through P&O algorithm. International Journal for Technological Research in Engineering, 2, 2642–2644.

[48] Yadav, R., & Pathak, G. S. (2016). Young consumers' intention towards buying green products in a developing nation: Extending the theory of planned behavior. Journal of Cleaner Production, 135, 732–739. https://doi.org/10.1016/j.jclepro.2016.06.155

[49] Zhao, B., & Zhang, L. (2020). AI-Driven Optimization for Solar Energy Systems: A Review. Journal of Renewable and Sustainable Energy, 12(3), 031402.

[50] Zare, K., & Narimani, M. R. (2011). Intermittent renewable energy sources and grid stability. Energy Procedia, 12, 1190–1197.

[51] Prasad, V. K. S. (2024). Evaluating the influence of environmental factors on household solar PV pro-environmental behavioral intentions: A meta-analysis review. Renewable & Sustainable Energy Reviews. https://doi.org/10.1016/j.rser.2023.114047

[52] Meena, R. S., Singh, A. K., Urhekar, S., Bhakar, R., Garg, N. K., Israr, M., Kothari, D. P., Chiranjeevi, C., & Srinivasan, P. (2022). Artificial Intelligence-Based Deep Learning Model for the Performance Enhancement of Photovoltaic Panels in Solar Energy Systems. International Journal of Photoenergy, 2022, 1–8. https://doi.org/10.1155/2022/3437364

[53] Pan, H., Wang, R., Wang, H., & Jia, Z. (2023). Sustainable Supply Chain Management: Best Practices for Reducing Environmental Footprints in the Global Apparel Industry. ESTIDAMAA, 2023, 18–26. https://doi.org/10.70470/ESTIDAMAA/2023/003

77 Analysis of marketing strategy for salted egg business in increasing sales volume: Case study of Eldona salted eggs, Banjarbaru

Sugiarti[a], Achmad Jaelani[b], and Siti Dharmawati[c]

Animal Husbandry Study Program, Faculty of Agriculture, Islamic University of Kalimantan MAB, Kayutangi Banjarmasin, Indonesia

Abstract: Thisi researchi aimsi toi determinei thei formi ofi marketingi strategyi usedi ini Eldona'si Saltedi Eggi businessi andi determinei Eldona'si marketingi strategyi ini increasingi salesi volume.i Thei researchi methodi usedi byi researchersi isi ai qualitativei methodi whichi producesi descriptionsi ini thei formi ofi writteni ori spokeni wordsi fromi respondentsi andi observablei behavior.i Thei datai usedi ini thei researchi arei primaryi andi secondaryi datai obtainedi fromi thei resultsi ofi researchi activitiesi ini thei field.i Thei datai analysisi techniquei usedi isi thei stagesi ofi datai collection,i datai reduction,i datai studyi andi conclusioni drawing.i Thei researchi resultsi showed that the marketing strategy used by Eldona's Salted Egg business was a pricing strategy and distribution strategy; where prices are applied using a division system between cash and installments. Meanwhile, the distribution strategy involves efforts to distribute goods according to the scheduled time and by implementing two systems, namely expanding the marketing area and implementing two payment systems to product marketing distribution partner outlets, namely the cash system and the term of payment system.

Keywords: Marketing strategy, salted eggs, sales volume, case studies

1. Introduction

The development of a business can be seen from several aspects that influence it, namely: market and marketing aspects, technical and technological aspects, and management and operational aspects. Regarding market and marketing aspects, an entrepreneur must understand and understand the current market situation to be able to gain a lot of profit. Marketing itself is the art of selling products, so it can be said that marketing is a sales process that starts from product design until after the product is sold.

As for the basis of the concept of spiritual marketing, it is found in Al-Qur'an Surat Al-Hasyr verse 18, which means: The people who believe, fear Allah and let each person pay attention to what he has done for tomorrow (hereafter); and fear Allah, indeed Allah knows best what you do.

Islamici marketingi isi ai strategici businessi disciplinei thati directsi thei processi ofi creating,i offeringi andi changingi valuesi fromi ani initiatori toi itsi stakeholders,i whichi ini thei entirei processi isi ini accordancei withi Islamici agreementsi andi principlesi andi muamalahi ini Islam.

One way to increase the shelf life and prevent damage to duck eggs is to preserve the eggs into processed egg products. Preserving salted eggs can be done by soaking them in a salt solution or using a mixture of other ingredients such as rubbing ash and bricks. However, soaking eggs in a salt solution can only maintain the lifespan of salted eggs for 3 days [1]. Apart from that, salted eggs sold on the market are still not varied, so efforts are needed to process salted eggs using different methods that can improve the taste, extend shelf life and increase the selling value of salted eggs.

One method of processing salted eggs that can be used to increase shelf life, prevent egg damage and spoilage is the smoking method. The process of smoking eggs is quite effective in preserving eggs for a long period of time, this is because the smoke closes the pores in the eggs [2]. Besides that, the smoked salted eggs produced can also increase the selling value of the salted eggs themselves. 'Eldona Salted Eggs' is a home industry (IRT) built by H. Zusnaidi and started in 2013. The address of this IRT is on Pondok Kelapa Street No. U16 Balitra Jaya Permai Complex, Banjarbaru. The amount of salted egg production reaches 800–1000 eggs per week. All of this production is absorbed by consumers with distribution coverage in Banjar Regency, Banjarbaru City and Banjarmasin City. In terms of marketing, smoked salted eggs still need to be improved, so it requires designing an appropriate business marketing strategy, so that they are able to compete with other companies and remain able to survive in increasingly tight industrial competition and can increase production and sales volumes. Based on this, researchers are interested in conducting research with the title marketing strategy for salted eggs.

[a]sugiartiatik369@gmail.com, [b]ach_jaelaniborneo@yahoo.com, [c]dharmauniska@gmail.com

DOI: 10.1201/9781003684459-77

2. Methods

2.1. Time and location of research

The time of this research was carried out from March to April 2024. The location of this research was Pondok Kelapa Street No. U16 Balitra Jaya Permai Complex, Banjarbaru.

2.2. Research informant

In this research, the technique for determining informants in this research is *Purposive Sampling* which is often mentioned *Internal Sampling*. The criteria for respondents in this research are people who know about the marketing strategy of the salted egg business. Respondents in this study are business owners who know about Eldona Salted Eggs.

2.3. Research data

The data sources for this research consist of two types, namely:

- Primary Data. Primary data is data directly collected by researchers from their primary sources. The primary data source in this research was obtained directly from interviews with salted egg business owners
- Secondary Data. Secondary data is data that does not provide information directly to the data collector. In this research secondary data was obtained from the literature in the form of books, articles, journals and other required documents.

2.4. Data collection techniques

2.4.1. Observation

Observation is a data collection method where researchers record information as they observe it during research. Witnessing these events can be done by looking, listening, feeling, which are then recorded as objectively as possible. For this research, researchers used structured observation. This observation was carried out on manufacturing procedures, up to the marketing implementation.

2.4.2. Interview

Apart from using observation techniques, researchers also used interview techniques. An interview is a conversation with a specific purpose.

2.5. Data analysis techniques

- The data that has been obtained in the field is then analyzed. Data analysis in this research is qualitative research data. Data analysis in this research uses an interactive model, namely:
- Data Collection *(data collected)*. Data collection in this research recorded all data objectively and as is according to the results of observations and interviews in the field.
- Data Reduction *(data reduction)*. Data reduction in this research is used to sharpen, classify, direct, select what is important, store what is unnecessary and organize the data in such a way that conclusions can finally be drawn and verified. Reduction was carried out after obtaining data from interviews with informants.
- Data Presentation. The presentation of data in this research is a collection of information that has been collected, reduced, then the researcher draws conclusions and takes action (Milles and Huberman). Data presentation is carried out after reducing the data used as report material. The results of previous data reduction which have been grouped, are then presented, processed and analyzed based on the chosen theory, namely using liberal feminism theory.
- Conclusions/Verify. Drawing conclusions or verification is an activity in the form of extracting the essence and presenting data which is the result of the analysis carried out in research or initial conclusions which are not yet fully mature. Verification that has been carried out and the results are known, is checked again so that the verified data becomes better. The results of the verification are used as final presentation data, because they have gone through the analysis process for the second time, so that the lack of data in the first stage of analysis is complemented by the results of the second stage of analysis.

3. Results and Discussion

3.1. Description of research place

Eldona Salted Eggs is a home industry (IRT) that was built by H. Zusnaidi and started in 2013. The address of this IRT is on Pondok Kelapa Street No. U16 Balitra Jaya Permai Complex, Guntung Payung, Landasan Ulin District, Banjarbaru City, South Kalimantan 70712. This salted egg business is carried out using duck eggs that come from Kandangan, which is one of the provinces in South Kalimantan that produces ducks and Mr. H. Zusnaidi also comes from Sungai Paring Kandangan Village. Eldona was given because it is an abbreviation of the names of H. Zusnaidi's 3 children. The number of employees in this business is two employees. Since the end of 2016, this IRT has been trying a variation of salted eggs, namely smoked salted eggs as a result of its production, which are ordinary salted eggs followed by a fumigation process.

3.2. Marketing strategy

From the results of the research, namely by conducting interviews, it is known that:

1. Marketing strategy
   From the results of the research carried out, it is known that the indicators of the marketing strategy carried out are:

- Product: The products produced are salted eggs and smoked salted eggs.
- Price: For salted eggs produced and sold to partners at the prices applicable to Eldona's salted egg business, there are 2 types of prices adjusted to the product produced, namely for salted eggs at IDR 5,000 and smoked salted eggs at IDR 6,000.
- Place: There are places where Eldona makes and sells salted eggs that come directly to the business owner's house, delivering them to business partners, namely Giant, minimarkets and grocery stores in Banjarbaru and Banjarmasin.
- Promotion: Promotion of Eldona's salted egg business was initially only through word of mouth and from products entrusted to known shops. Then promotions were carried out by each party who had purchased Eldona's salted eggs and smoked salted eggs on social media so that they became known to the wider public.

Based on the research results above, it can be seen that the marketing mix is the seller's view of the marketing tools available to influence buyers [3]. The marketing mix used in product businesses generally only includes 4Ps, namely *Product, Price, Place*, and *Promotion.*

With marketing, a company can introduce the products it offers more widely to the public. By advertising in print and electronic media, through sales personnel and others. Of course, marketing a product is not easy because it requires tenacity, creativity and quite large costs.

The industrial business marketing strategy carried out by Eldona Salted Eggs in marketing its products both in terms of production and marketing aspects is as follows: (1) Strengthening product quality by maintaining the unique taste of Eldona Salted Eggs by presenting the uniqueness of Kandangan, namely still having ash notes in the eggs salty. (2) Influences product packaging and makes it more attractive among consumers, namely packaging in the form of a pack of 4 containing the product brand, namely Salted Egg Eldona, Kandangan authentic. (3) Word of mouth promotion. The results of this research are almost the same as the results of Muslikha's research, (2020). The marketing mix strategy carried out is determined based on price, product and promotion.

3.3. *Sales volume*

Eldona's sales volume of salted eggs is 9,000 eggs per month, consisting of salted eggs and smoked salted eggs. Salted and smoked eggs are made every week. The eggs entrusted to partners are 100–200 according to the partner's request and will be added according to the request of the selling partner. Partners can do it in two ways, namely with cash, and with a deposit system, and they (partner outlets) just have to choose which one they want.

Table 77.1. Business model Canva Eldona's salted egg

Business Model Canvas (BMC) "Eldona's salted egg"				
Key Partners (8)	**Key Activities (7)**	**Value Propositions (2)**	**Customer Relationship (4)**	**Customer Segments (1)**
1. Duck egg supplier 2. Duck egg seller	1. Duck egg cleaning 2. Making salted egg coating dough 3. Curing 21 days 4. Steaming salted eggs 5. Egg cooling 6. Fumigation eggs to become smoked salted eggs 7. Packing 8. Quality control	1. Quality salted egg products 2. Quality smoked salted egg products	1. Product arrived on time 2. Ease of payment	1. Convenience Store 2. Grocery stalls 3. Souvenir Shop
	Key Resources (6) 1. Empowerment of home industry 2. Marketing and distribution management		**Channels (3)** 1. Business partner 2. Social media	
Cost Structure (9) 1. Raw material 2. Labor costs		**Revenue Streams (5)** Product sales		

Source: Author.

3.4. *Business model canvas*

The following is the Canva business model for Eldona's salted egg business. The sections in the Business Model Canvas include: *Customer Segment, Value Proposition, Channels, Customer Relationship, Revenue Streams, Key Resources, Key Activities, Key Partnership dan Cost Structure* in the Table 77.1.

4. Conclusions

From the research results and discussion of research activities are:

1. The marketing strategy used by Eldona's Salted Egg business is a pricing strategy and a distribution strategy, where prices are applied using a distribution system between cash and installments, while the distribution strategy involves efforts to distribute goods according to the scheduled time.
2. The marketing strategy carried out by Eldona Salted Eggs in an effort to increase its sales volume is by implementing two systems, namely expanding the marketing area and implementing two payment systems to product marketing distribution partner outlets, namely the cash system and term of payment system.

5. Suggestions

From the results of this research, the author provides advice to the manufacturer, namely Eldona Salted Eggs, to:

- Expanding the marketing area in order to increase production capacity and sales volume.
- Provide discounts to partners with the most sales in each period, so that they further increase their sales volume.

References

[1] Novia, D., Juliyarsi, I., & Fuadi, G. (2012). Protein content, fat content and Organoleptics of Smoked Salted Eggs fueled with Coconut Fiber. *Animal Husbandry Journal*, *9*(1), 35–45. https://core.ac.uk/download/pdf/228880697.pdf

[2] Jaelani, A. (2018). Organoleptic Quality of Smoked Salted Eggs with Different Fumigation Times. Faculty of Agriculture, Islamic University of Kalimantan. Proceedings of 2018 Research Results ISBN: 978-602-52531-1-9 Lecturers at Islamic University of Kalimantan, May 2018. https://ppj.uniska-bjm.ac.id/wp-content/uploads/2018/09/7-FAPERTA-215-290-8.pdf.

[3] Kotler, P., & Armstrong, G. Linguists by Sabran, B. (2008). *Principles Marketing volume 1* (edition 12). Erlangga. Jakarta.

78 A study on consumer awareness towards neuromarketing in Kannur district

Bibin Thomas M.[1,a] *and J. Sudarvel*[2,b]

[1]PhD Research Scholar, Department of Management, Karpagam Academy of Higher Education, Coimbatore, Tamil Nadu, India
[2]Assistant Professor in Management, Karpagam Academy of Higher Education, Coimbatore, Tamil Nadu, India

Abstract: Neuromarketing, which blends neuroscience and marketing, helps understand consumer behaviour more deeply. This study examines consumer awareness of neuromarketing in Kannur District through a quantitative approach using a structured questionnaire. The survey will gauge respondents' familiarity with neuromarketing, its application in advertising and branding, and their views on its ethical implications. The findings will enhance understanding of consumer awareness in the region and offer insights for marketers and policymakers on the ethical use of neuromarketing in advertising strategies.

Keywords: Neuromarketing, consumer awareness, Kannur district, advertising, branding, ethical implications

1. Introduction

Neuromarketing, a growing field, merges neuroscience with marketing to provide insights into consumer behaviour at a neurological level. Defined by Lee et al. [1], it involves using neuroscience tools to analyze consumer reactions to marketing stimuli, such as ads, product packaging, and brand logos. By understanding how the brain processes these stimuli, marketers can craft strategies that more effectively capture attention and influence purchasing decisions.

Despite its potential, consumer awareness of neuromarketing is limited. Kenning and Linzmajer [2] found that only a small percentage of consumers are familiar with the term, and even fewer understand its influence on purchasing behaviour. Reimann et al. [3] suggest that while most consumers may not recognize the term 'neuromarketing,' they are open to the use of neuroscience in marketing once they understand its advantages.

2. Review of Literature

Neuromarketing, a growing field blending neuroscience and marketing, has attracted attention for its insights into consumer behaviour. Researchers have examined its applications, methods, and impacts. For instance, McClure et al. [4] used functional Magnetic Resonance Imaging (fMRI) to study brain responses to brand preferences.

According to Lee et al. [1], neuromarketing involves studying consumer reactions to marketing stimuli using neuroscience tools, promoting collaboration between neuroscientists and marketers. One key area of research is the use of biometric measurements. In recent study found that Skin Conductance Response (SCR) could evaluate the effectiveness of ads, showing a correlation with emotional responses. Despite its potential, neuromarketing raises ethical concerns. Calvert et al. [5] emphasized the need for informed consent and privacy protection in neuroimaging studies.

Gidlöf et al. [6] argue that neuromarketing offers insights traditional methods miss by combining neuroimaging and behavioural data to better understand consumer preferences. Karmarkar and Plassmann [7] used EEG to show that different brain areas activate based on whether consumers consider immediate or long-term rewards, revealing decision-making processes. Lang et al. [8] found that facial EMG studies showed positive emotions linked to greater consumer engagement.

Cultural factors also play a role. Chua et al. [9] showed that East Asian and Western participants had different brain responses to ads, highlighting the need for cultural consideration in neuromarketing research. Harris et al. [10] discovered using fNIRS that strong emotional bonds formed with well-known brands activated reward-related brain regions. Plassmann et al. [11] used fMRI to reveal that pricing context influences perceived value.

Hsu et al. [12] demonstrated through EEG that product design significantly affects consumer preferences, while Falk et al. [13] found that storytelling in ads engaged consumers' brains more effectively using fMRI. These studies illustrate the potential of neuromarketing to deepen our understanding of consumer behaviour, though ethical concerns must guide its use to protect consumers' rights.

---

[a]tobibinthomas@gmail.com, [b]j.sudarvel@gmail.com

DOI: 10.1201/9781003684459-78

2.1. Research gap

Several critical gaps exist in neuromarketing research. These include limited studies in diverse geographic areas, a lack of comprehensive ethical guidelines, insufficient integration with traditional marketing research methods, a focus on short-term effects, a need for research in online and digital marketing contexts, and limited application in industries beyond consumer goods.

2.2. Statement of the problem

Despite the growing use of neuromarketing, consumer awareness remains low. Studies by Kenning and Linzmajer [2], Reimann et al. [3], Falk et al. [13], Lee et al. [1], and Harris et al. [10] show that only a small percentage of consumers are familiar with neuromarketing, with even fewer understanding its effects on their buying behaviour. This is concerning as companies already use neuromarketing to design more persuasive advertising and product presentations. This underscores the need for educating consumers about its ethical applications.

2.3. Objective of the study

To know the Consumer Awareness towards Neuromarketing in Kannur District.

2.4. Scope of the study

This study aims to assess consumer awareness and understanding of neuromarketing in Kannur District. It will examine consumers' familiarity with the term, their views on how neuromarketing influences purchasing behaviour, and their attitudes toward its use in marketing.

3. Research Methodology

3.1 Source of data

The data required for this study is primary. It will be collected through a survey questionnaire administered to consumers in Kannur District. This method allows for the direct collection of information from participants, providing detailed and specific insights into their awareness and understanding of neuromarketing.

3.2. Sample size and sampling procedures

The study's participants will be chosen through the use of convenient sampling. It has 156 samples in all.

3.3. Framework of analysis

The collected data will be analysed using descriptive statistics and Weighted Average Method. This analysis will help in understanding the level of awareness and understanding of neuromarketing among consumers in Kannur District.

3.4. Significance of the study

This study holds significance for both academic research and practical applications in neuromarketing. Academically, it will contribute to the literature by exploring consumer awareness and understanding of neuromarketing in Kannur District, a relatively under-researched area (Table 78.1).

Table 78.1. Analysis and interpretation

Age	*Frequency*	*Percent*
Up to 25 Years	53	34.0
26–35 Years	43	27.6
35–45 Years	46	29.5
Above 45 Years	14	9.0
Total	156	100.0
Gender	*Frequency*	*Percent*
Male	67	42.9
Female	89	57.1
Total	156	100.0
Marital Status	*Frequency*	*Percent*
Unmarried	18	11.5
Married	138	88.5
Total	156	100.0
Educational Qualification	*Frequency*	*Percent*
Up to HSC	1	.6
Under Graduate	104	66.7
Post Graduate	30	19.2
Professional	13	8.3
Others	8	5.1
Total	156	100.0
Profession	*Frequency*	*Percent*
Agriculture	18	11.5
Business	14	9.0
Private Employee	33	21.2
Government Employee	15	9.6
Home maker	10	6.4
Others	66	42.3
Total	156	100.0
Area of Residence	*Frequency*	*Percent*
Urban	52	33.3
Semi-Urban	48	30.8
Rural	56	35.9
Total	156	100.0
Monthly Income	*Frequency*	*Percent*
Up to 10000	18	11.5
20001–30000	108	69.2
Above 30000	30	19.2
Total	156	100.0

Source: Author.

The population in the research is composed mainly of individuals aged between 26 and 45 years, with a slightly higher proportion of females, and is mostly married. Educationally, the majority have studied up to the level of a bachelor's degree, while a smaller proportion hold postgraduate degrees or are professionals. The population is represented across various professions, with a significant presence in private-sector jobs and farming. Additionally, the population spans urban, semi-urban, and rural areas. Income levels primarily fall within the range of 20,001–30,000. This varied population provides a strong basis for interpreting research findings.

To identify Consumer Awareness towards Neuromarketing in Kannur District Weighted Average Ranking Test is employed (Table 78.2).

The mean scores and ranks for each factor were calculated based on the respondents' ratings (5, 4, 3, 2, 1). The factor 'I am aware of neuromarketing techniques' received the highest mean score of 4.04, indicating that it ranked the highest among the factors considered. On the other hand, 'I can recall examples of neuromarketing applications in advertising' received the lowest mean score of 2.97, indicating that it ranked the lowest among the factors considered. The results suggest varying levels of consumer awareness and understanding of neuromarketing in Kannur District, with some factors being more recognized and understood than others.

4. Suggestions

Based on the findings of the Weighted Average Ranking Test, several suggestions can be made to enhance consumer awareness towards neuromarketing in Kannur District:

- Education and Awareness Campaigns: Conduct education and awareness campaigns to inform consumers about neuromarketing techniques, their implications, and how they are used in advertising and consumer research.
- Transparency and Disclosure: Encourage companies utilizing neuromarketing techniques to be transparent about their practices and disclose when such techniques are being used in their marketing efforts.
- Ethical Guidelines: Develop and promote ethical guidelines for the use of neuromarketing techniques to address consumer concerns and ensure responsible practices.
- Consumer Engagement: Engage consumers in discussions and forums about neuromarketing to increase their understanding and empower them to make informed choices.

Table 78.2. Factors of respondent scores

Factors	*5*	*4*	*3*	*2*	*1*	*Respondents Score*	*Mean Score*	*Rank*
I am aware of neuromarketing techniques	74	42	24	4	12	156	4.04	1
	370	168	72	8	12	630		
I understand the term 'neuromarketing' and its purpose	61	25	42	14	14	156	3.67	5
	305	100	126	28	14	573		
I am aware of companies utilizing neuromarketing techniques in their research	27	46	67	11	5	156	3.51	6
	135	184	201	22	5	547		
I grasp how neuromarketing methods analyze consumer responses	32	34	71	14	5	156	3.47	7
	160	136	213	28	5	542		
I can recall examples of neuromarketing applications in advertising	20	27	59	28	22	156	2.97	10
	100	108	177	56	22	463		
I acknowledge the potential benefits of neuromarketing for understanding consumer behaviour	29	12	106	6	3	156	3.37	9
	145	48	318	12	3	526		
I am conscious of ethical concerns associated with neuromarketing	39	61	42	10	4	156	3.78	4
	195	244	126	20	4	589		
I actively seek information about advancements in neuromarketing	36	78	30	6	6	156	3.85	3
	180	312	90	12	6	600		
I believe it's important for consumers to understand neuromarketing	48	55	48	4	1	156	3.93	2
	240	220	144	8	1	613		
I am able to recognize when companies utilize neuromarketing tactics	33	44	48	15	16	156	3.40	8
	165	176	144	30	16	531		

Source: Author.

5. Conclusion

The study surveyed consumer awareness regarding neuromarketing in Kannur District and concluded there is widespread awareness regarding why it is being used but scant awareness regarding specifics. It recommends education programs, transparency, codes of ethics, and regulation. Awareness campaigns and consumer perception can be observed in detail in future research. Awareness is key to ensuring ethical practices and informed consumer decision-making.

6. Scope for Further Research

More research in Kannur District can potentially uncover consumer awareness in detail through qualitative research, track consumer awareness longitudinally, and measure awareness in different demographic strata. It can potentially measure awareness campaign effectiveness, research how consumer behaviour is impacted by neuromarketing and account for concerns regarding ethics. It would provide insights in finer detail about consumer awareness and allow for improved strategies for education and for marketing in an ethical context.

References

[1] Lee, N., Broderick, A. J., & Chamberlain, L. (2007). What is 'neuromarketing'? A discussion and agenda for future research. *International Journal of Psychophysiology*, *63*(2), 199–204.

[2] Kenning, P., & Linzmajer, M. (2011). Consumer neuroscience: an overview of an emerging discipline with implications for consumer policy. *Journal für Verbraucherschutz und Lebensmittelsicherheit*, *6*(1), 111–125.

[3] Reimann, M., Zaichkowsky, J., Neuhaus, C., Bender, T., & Weber, B. (2010). Aesthetic package design: A behavioral, neural, and psychological investigation. *Journal of Consumer Psychology*, *20*(4), 431–441.

[4] McClure, S. M., Li, J., Tomlin, D., Cypert, K. S., Montague, L. M., & Montague, P. R. (2004). Neural correlates of behavioral preference for culturally familiar drinks. *Neuron*, *44*(2), 379–387.

[5] Calvert, G. A., Green, C. S., & Gallagher, R. (2017). The future of neuroimaging in consumer research. *Consumer Neuroscience*, 271–281.

[6] Gidlöf, K., Anikin, A., & Lingonblad, M. (2017). Using neuroscience to understand consumer preferences: A review of methods and applications. *Food Quality and Preference*, *62*, 4–14.

[7] Karmarkar, U. R., & Plassmann, H. (2017). Consumer neuroscience: The promise of a neurocognitive approach to consumer behavior. *Consumer Psychology Review*, *1*(2), 82–105.

[8] Lang, A., Bradley, S. D., & Park, B. (2017). Measuring emotion via facial electromyography. In J. Bryant & M. B. Oliver (Eds.), *Media effects: Advances in theory and research* (pp. 293–311). Routledge.

[9] Chua, H. F., Ho, S. S., Jasinska, A. J., Polk, T. A., Welsh, R. C., Liberzon, I., & Strecher, V. J. (2018). Self-related neural response to tailored smoking-cessation messages predicts quitting. *Nature Neuroscience*, *11*(5), 548–554.

[10] Harris, M., Sandberg, J., & Stokes, R. (2018). Imagining a new normal: Social media and neuromarketing in a digital era. *Journal of Marketing Management*, *34*(9–10), 808–830.

[11] Plassmann, H., O'Doherty, J., Shiv, B., & Rangel, A. (2012). Marketing actions can modulate neural representations of experienced pleasantness. *Proceedings of the National Academy of Sciences*, *105*(3), 1050–1054.

[12] Hsu, M., Yoon, C., & Song, H. (2015). The role of aesthetic taste in consumer behavior. *Journal of Consumer Psychology*, *25*(1), 140–154.

[13] Falk, E. B., Morelli, S. A., Dambacher, K., & Lieberman, M. D. (2016). The neural correlates of persuasion: A common network across cultures and media. *Journal of Cognitive Neuroscience*, *28*(9), 1377–1386.

79 Impact of influencer marketing on tourist destination choices

J. Sudarvel[1,a] *and R. Velmurugan*[2,b]

[1]Assistant Professor, Karpagam Academy of Higher Education, Coimbatore, Tamil Nadu, India
[2]Associate Professor and Head, Karpagam Academy of Higher Education, Coimbatore, Tamil Nadu, India

Abstract: The research analyzes how influencer marketing affects tourists' decisions for selecting their destination choices. Destination branding needs to comprehend how social media affects customer choices because it shapes how consumers behave. Recent travelers participate in a survey which provides data about their interaction with influencer marketing platforms and their selection of influencers and judgments about authenticity and resulting travel decisions. Data shows that an intense relationship exists between preferences pertaining to travel and influencer marketing strategies. Tourists who interact with travel influencers across Facebook along with Instagram and YouTube tend to decide on visiting destinations that influencers promote. The influence that influences have on people depends on their quality content and personal storytelling combined with their genuine approach and visual presentation. Real-life and visual content presentations in the destination increase visitor trust and interest in the area's appeal. Research faces two key difficulties when attempting to establish exact booking numbers influenced by influencer marketing while also categorizing different types of influencers. Destination marketing sectors should select influencers whose values match their brand identity yet connect with their target market. The research proves that destination branding needs authentic presentations combined with engaging content along with.offsetTopDocument() strategic partnerships with influencers to work successfully. Further research should develop better evaluation techniques to understand the temporary along with extended effects that influencer marketing creates on tourism.

Keywords: Influencer marketing, tourist destination choices, travel behavior, social media, authenticity, visual storytelling, destination branding

1. Introduction

The social media has transformed how the consumers are interacting with brands, and travel agencies are not an exception. Digital influencer marketing has become a popular digital strategy gaining power of influencing consumer's sentiment and given it a trusted voice. Influencers possess large followings and perceived authenticity to influence and participate in destination branding in an engaging, refreshing, and broader reach than traditional marketing. Yet, despite its uptake, influence in the travel decision process remains understudied. This is essential to know because it influences how destination marketers should optimize resource and get the maximal return. In this study, I seek to bridge this gap by analyzing how an influencer marketing function impacts travellers' decision making about the destination. It uses a survey based approach and collects data from recent travelers on the influencer content to which they have been exposed to and the levels of engagement and layer by which they have made their travel decisions. This would be providing a place to add findings that will offer valuable insights on how to help destination marketers understand the impact of authenticity, visual storytelling, and how to initiate strategic influencer partnerships. Transactions of the Institutions of £c. (Queens: New series), vol. 10 Mag., No. 4, April 1913, pp. 409–421. This is a study of these effects as the tourism industry advances and will be significant for the development of the appropriate marketing strategies that will accommodate modern travellers.

2. Review of Literature

Taking this into account, this literature review looks into how influencer marketing is, how it affects consumer behaviour, and how it serves as a part of destination branding. Now, influential people have taken over the marketing strategies of a brand, as they are now better able to decide the travel choices than traditional ads [2]. This boosts consumers' trust level, and as such, influencer marketing has become a valuable tool in destination promotion [1].

These platforms on Instagram, YouTube, and TikTok, offer an abundant reach to share engaging content which makes the destination more appealing [8]. As confirmed by the research, it is noted that influencer generated content leads to high destination perception and interest highlighted by Hwang and Zhang [6]. They influence many stages of the decision processes such as creating awareness, promoting insights and being inspiring to travel [7]. User engagement

[a]j.sudarvel@gmail.com, [b]drvelsngm@gmail.com

DOI: 10.1201/9781003684459-79

rates are high, which implies great connections with the audience, and therefore the reach of the marketing [2].

Good storytelling and good graphics of the destination increase the destination appeal; good transparency and authenticity create trust [3]. Unfortunately, the return on investment of influencer marketing is difficult to measure for influence since social media is evolving [5]. Future research should develop the precise measurement tool and the investigation about the long term effects on destination loyalty [4]. Nevertheless, the literature points out that influencer marketing is heavily positive towards tourism branding. The purpose of this study is to provide actual data on the effect so that travel shall see more effective and measurable marketing strategies.

2.1. Statement of the problem

Socializing with the social media has made things easy for tourists to research and choose their destination. Now, influencer marketing has become an important means of branding a destination, using influencer's popularity and his/her perceived authenticity. But empirical analysis of this impact on the travel decisions occurs only rarely. However, in spite of this, destination marketing organizations (DMOs) have really struggled to quantify return on investment (ROI) and clearly identify the key factors for success for this specific marketing channel. Some of the challenges include measuring direct and indirect effects upon travelers' behaviour, content authenticity and findings of most suitable influencer type. But then, as with most of the social media, it is dynamic, ever-changing consumer preferences and the evolving nature of the same make it difficult to determine its efficacy. Without understanding, DMOs run the risk of wasting their marketing spending and losing opportunities for engagement.

2.2. This study explores key questions

1. So, how much do influencers decide the choices of travelers from where to visit?
2. What is more important, the influencer influence attributes (amount of followers, engagement rates, authenticity, etc.) when influencing travelers decisions?
3. What role do the type and quality of the influencer content play in perceiving and choosing the traveler?
4. How can the effectiveness of ROI and success of influencer marketing be effectively measured in destination branding?

The research is in answering these questions so as to provide a complete perspective of how influencer marketing plays a role in making travel decisions. And the findings will inform how DMOs optimize resource allocation and construct more effective influencer marketing in the hope of achieving more engagement and results.

2.3. Scope of the study

A geographic scope, demographic, influencer type, and marketing characteristics of influencer marketing effect on tourist destination choices are explored in this study. It targets travellers all over North America, Europe and Asia, through urban, rural, adventure and cultural travel groups. Until such time that they among them would become a force to be reckoned with, Millennials and Generation Z, the most active social media users, both male and female, are primary targets of interest. It also discusses experience levels between novice and the experienced travelers.

Macro-influencers (huge followings), micro-influencers (some followings) and then nano-influencers (small, tight group of followers) are all studied. The article examines various content creators, for example, bloggers, vloggers, and social media influencers on various platforms from YouTube as well as Instagram, TikTok, and blogs. For example, trust and booking decisions are based on factors like how well one can visualise a story, their engagement rates and how authentic the content they share is.

Continuing along the traveler decision making process, the study starts with awareness (via influencer promotions), decision giving shape to perceptions (shaping perceptions), and in turn influences bookings and travel plans (influencing bookings and travel plans). For instance, it explores influencer marketing trends and traveler behaviour over the past two years, and looks long term by considering return and sustained interest on influence.

The issues: Business limitations to self reported data for the survey, measuring ROI using influencer marketing, and establishing direct relationship with travel decisions made from influencer marketing. All the while, the study seeks to contribute towards better understanding of influencer marketing strategies for destination marketing organizations.

2.4. Objective

To identify the Impact of Influencer Marketing on Tourist Destination Choices.

3. Research Methodology

Both a quantitative and a qualitative investigation incorporate this study to ascertain the impact of influencer marketing on traveler destination decision, using a mixed method. The survey based method gathers travelers' first hand data, whereas interviews of different levels of depth gives elaborate insights into the travelers' decision making processes. The object of our study are travelers that have visited recently destinations where influencer marketing has been deployed. We use convenience sampling technique that covers all dimensions of age, geographical and gender and most of the travel frequency dimensions to a extent so that the minimum sample size is 500 respondents to ensure statistical significance and reliability. Exposure to influencer marketing (frequency of usage and platform used), engagement levels in terms of likes, comments, shares, perceived authenticity and trustworthiness of influencer along with level of their influence on travel decisions.

3.1. Data analysis

Finally, descriptive statistics should be used to summarise the survey responses. The investigation of connections between destination preference and influencer marketing can be carried out by using inferential statistics (e.g., regression analysis and correlation). Factor analysis is used to identify the factor of Perceptions of authenticity and trustworthiness in influencer content.

3.2. Significance of the study

The result of this study helps answer important questions about the effectiveness of influencer marketing with respect to traveler destination choices in the tourism and marketing sectors. Its findings can be used by travel companies, tourism boards, DMOs, and influencers to contemplate what can be refined in the strategies and more effectively resource a destination. By understanding which influencer campaign success factors are most important, this will help DMOs improve collaborations and marketing success. Alignment of influencers with brand values and target audiences is key to making influence maximise impact and ROI as per the study. With even today's meager density of social and interactive media accounts, it's already much too easy to relate specific key factors and travel destination trends to the engagement and popularity of destination Twitter feeds, Facebook pages, collaborated travel blogs, Instagram, Tumblr, Flickr, YouTube, and, of course, real-world stay campaigns. Understanding the social media induced traveller behaviours is deeper – agencies learn how to offer travellers things that are in line with changing consumer preferences. The same applies to influencers if they want to adjust their content strategies, strengthen the audience connection and secure brand partnerships; by focusing on transparency and authenticity. This study also contributes to a ground to make further studies on facilitating a long term impacts of influencer marketing and tourism. Its findings lend themselves to allow policymakers and industry stakeholders to come up with initiatives that support responsible and sustainable travel which is beneficial to the environment while not compromising its economic benefits (Table 79.1).

At the same time, the demographic analysis gives us the key knowledge on the respondent profile and the impact onto the influencer marketing in tourism. The majority (30–40 years old) constitute the strong engagement age; 54.8 per cent live within this group (in their prime travel planning years). It slights to males (58.4%) maybe influencing the preferences regarding content. A large amount (80.0%) are married, which implies that family marketing strategies should play a significant role. Most of the respondents (39.2%) are educated up to an undergraduate level and then (23.8%) postgraduates, indicating a welleducated audience who are receptive to influencer messaging. Employed individuals (35.4%) and students (34.2%) predominate in the professional group and such campaigns would have to be adapted

Table 79.1. Analysis and interpretation

Age	*Frequency*	*Percent*
Below 30 Years	112	22.4
30 40 Years	274	54.8
Above 40 Years	114	22.8
Total	500	100.0
Gender	*Frequency*	*Percent*
Male	292	58.4
Female	208	41.6
Total	500	100.0
Marital Status	*Frequency*	*Percent*
Married	400	80.0
Unmarried	272	20.0
Total	500	100.0
Educational Qualification	*Frequency*	*Percent*
Up to SSLC	15	3.0
HSC	89	17.8
UG Degree	196	39.2
PG Degree	119	23.8
Professional	74	14.8
Others	7	1.4
Total	500	100.0
Profession	*Frequency*	*Percent*
Agriculturist	28	5.6
Business	36	7.2
Employee	177	35.4
Professional	34	6.8
Home maker	54	10.8
Student	171	34.2
Total	500	100.0
Area of Residence	*Frequency*	*Percent*
Urban	251	50.2
Semi-Urban	103	20.6
Rural	146	29.2
Total	500	100.0
Type of family	*Frequency*	*Percent*
Joint Family	97	19.4
Nuclear Family	403	80.6
Total	500	100.0
Status in Family	*Frequency*	*Percent*
Head	140	28.0
Member	360	72.0
Total	500	100.0
Monthly Income	*Frequency*	*Percent*
Up to 20000	387	77.4
20001–40000	89	17.8
Above 40000	24	4.8
Total	500	100.0
Family Income	*Frequency*	*Percent*
Up to 30000	31	6.2
30001–50000	415	83.0
Above 50000	54	10.8
Total	500	100.0
Family Expenditure	*Frequency*	*Percent*
Up to 20000	153	30.6
20001–40000	312	62.4
Above 40000	35	7.0
Total	500	100.0

Source: Author.

to their needs. This shows that the majority (50.2%) are of urban respondents followed by rural (29.2%) and semi-urban (20.6), indicating the need for some type of regional marketing approach. Improved income patterns mostly show that respondent's earn under 20,000 (77.4%), above 50,000 (12.5%), and between 30,001 and 50,000 (83.0%); whilst family structures are mostly consists of nuclear (80.6%). The first is a reflection of the significance of affordability and value messaging in influencer campaigns. In these insights, destination marketers can take these insights, and how they align with this, to craft targeted, effective strategies to leverage their influencer strategy (Table 79.2).

4. Regression

Finally, the regression analysis shows that influencer marketing is considered to be one of the most explaining predictors of a destination choice, and explains 27.3% of variance (R-squared = 0.273; Adjusted R-squared = 0.255). This means that at least one predictor has statistically significant relationship with destination choice (F = 15.202, p = 0.000). The impact on destination choice score is positive and strong (Beta = 0.469, p < 0.001, increases by 0.433 points). On the other hand, people from rural or semi urban areas have lower destination choice scores (Beta = -0.102, p = 0.013) and people of joint families have lower destination choice scores as compared to people of nuclear families (Beta = -0.084, p = 0.037). Other demographic factors like age, gender, marital status, and income are not statistically showing effect. The findings here point to the importance of influencer marketing in deciding where to travel and geography as well as family structure form other influential factors.

5. Correlation

Finally, it is confirmed by Correlation analysis that there exists a strong positive relationship between influencer marketing and destination choice with Pearson correlation coefficient of 0.486. It also implies that increased exposure to influencer marketing leads to significant increased selection of promoted destinations by participants. Both correlation coefficients are statistically significant at 0.01, which means Sig. (2-tailed) values below 0.01, and with such low p values, chances for getting chance-based results are reduced. This reveals that influencer marketing has a big part to play in how tourists decide. With greater notoriety, destination

Table 79.2. Model summary

Model	*R*	*R Square*	*Adjusted R Square*	*Std. Error of the Estimate*
1	.522a	.273	.255	.47962

ANOVA[a]

Model		*Sum of Squares*	*df*	*Mean Square*	*F*	*Sig.*
1	Regression	41.964	12	3.497	15.202	.000b
	Residual	112.028	487	.230		
	Total	153.992	499			

Model		*Unstandardized Coefficients*		*Standardized Coefficients*	*t*	*Sig.*
		B	*Std. Error*	*Beta*		
1	(Constant)	1.519	.291		5.219	.000
	Age	.039	.052	.047	.739	.460
	Gender	.005	.060	.004	.075	.940
	Marital Status	.108	.068	.097	1.580	.115
	Educational Qualification	-.028	.022	-.054	-1.295	.196
	Profession	.007	.019	.020	.358	.721
	Area of Residence	-.066	.026	-.102	-2.503	.013
	Type of family	-.118	.057	-.084	-2.088	.037
	Status in Family	-.031	.066	-.025	-.476	.634
	Monthly Income	.033	.052	.027	.631	.529
	Family Income	-.111	.057	-.082	-1.949	.052
	Family Expenditure	-.020	.045	-.020	-.447	.655
	Influencer Marketing	.433	.043	.469	10.108	.000

a. Dependent Variable: Destination Choice

Source: Author.

marketers must be clear on the extent to which it influences how travelers behave in order for them to draw the most from their target segments (Table 79.3).

5.1. Factor analysis

To identify the Perceptions of authenticity and trustworthiness in influencer content is factor analysis employed (Tables 79.4 and 79.5).

The three important dimensions that impact perception of influencer content are identified by factor analysis.

Table 79.3. Correlations coefficients

Correlations			
		Influencer Marketing	*Destination Choice*
Influencer Marketing	Pearson Correlation	1	.486
	Sig. (2-tailed)		.000
	N	500	500
Destination Choice	Pearson Correlation	.486	1
	Sig. (2-tailed)	.000	
	N	500	500

Correlation is significant at the 0.01 level (2-tailed).

Source: Author.

1. Because 45.865% of variance was explained, the strongest factor is variables measuring how genuine and experience based influencer content appears.
2. This factor (Skepticism and Transparency) accounts for 7.796% of variance and takes into account doubts over influencer recommendation, especially where a recommended item is sponsored content.
3. In an attempt to explain 6.54% of variance, the factors of Relatability and Influence indicate that influencer content feels more credible and understands able than traditional advertising.

Factor analysis would be suitable for the data, which has a KMO value of 0.899 and statistically significant Bartlett's Test ($p < 0.001$). This explains 60.2% of the variance in influencer marketing by the three factors all considered, which is to say that through authenticity, skepticism, and relatability, the consumer is interpreting trust based on the influencer.

Table 79.4. KMO and Bartlett's test

Kaiser-Meyer-Olkin Measure of Sampling Adequacy.		.899
Bartlett's Test of Sphericity	Approx. Chi-Square	5001.055
	df	120
	Sig.	.000

Source: Author.

Table 79.5. Rotated component matrix

	Component		
	1	*2*	*3*
I perceive influencer content to be authentic and true to life.	.829		
I believe that influencer content accurately represents their genuine experiences.	.806		
I perceive influencers as authentic individuals who genuinely care about the products and destinations they promote.	.589		
Influencers' content aligns with their personal values and beliefs.	.562		
I am skeptical about the authenticity of influencer recommendations, especially when they are sponsored.		.777	
I am less likely to trust influencers who promote products or destinations solely for monetary gain.		.649	
I find it challenging to distinguish between genuine opinions and paid promotions in influencer content.		.639	
I believe that influencers prioritize their audience's interests over financial gain.		.631	
I feel confident in basing my purchasing or travel decisions on recommendations from influencers.		.571	
Influencers' content feels genuine and authentic to me.		.559	
Influencer content often feels scripted or insincere to me.			.736
Influencers' content feels more relatable to me than traditional advertising.			.659
I believe that influencers are transparent about sponsored content and paid partnerships.			.656
I am more likely to trust a product or destination recommendation from an influencer than from traditional advertising.			.619
When influencers share personal stories or experiences, I perceive them to be truthful and reliable.			.595
Eigenvalues	7.338	1.247	1.046
% of Variance	45.865	7.796	6.54
Cumulative %	45.865	53.661	60.2

Source: Author.

6. Suggestions

6.1. Social media influencers

1. Despite this, every piece of content should be authentic and transparent by clearly illustrating the sponsored deals with genuine audience interactions.
2. When you develop strategies to allow actual audience interaction via feedback collection and instant messages responses, and also content decision involvement with the readers your audience engagement will increase.
3. Here, DMOs and tourism boards should come together to develop genuine content based representations where the destinations are visualised in engaging and informative style.

6.2. Travel agencies

1. Influencer partnerships should be a thing that your travel company enters with trustworthy social media personalities who are personally donning your customer groups with fashion, to in a manner that gives an almost authentic promotion of tourist places and packages.
2. Drawing up customer feedback that is based on studying the social media interaction between customers and creating connections with influential partners to get data along with the market changing requirements.
3. Social media measurement and customer feedback allows organizations to find out industry related advancement, which helps organizations in improving their approach towards customer service as well as marketing practices.

6.3. Academic and industry researchers

1. Explore cross sector partnership with scholarly institutions and business enterprise in conjunction to carry out thorough investigation on the consequence of influencer marketing on branding destinations and entails tourist attraction through co research.
2. Long term observational research groups should follow the way the audiences change the thoughts of the influencer content truthfulness as well as credibility, with this content as the environment of presentation changes and ages.
3. To research, ethical aspects of influencer marketing influenced the disclosure requirements and making consumer safe from misleading information is a must.

Destination Marketing Organizations (DMOs) and Tourism Boards:

1. In order to bring these websites to life and create advertiseable content, DMOs should cultivate strategic alliances with public persona whose media clout can expose the somehow unique thing about the destination – also to the rest of the world.
2. You should work with major influential brand ambassadors to create jointly authentic and educative content, centered upon major sights and shows, yet clear on the outcome, truthful and honest.
3. Data analytics can help such organizations track influencers' campaign and have a better outline of when, how and for which segments in the market your destination is being marketed.

By implementing these guidelines, travel agencies, social media influencers, researchers and destination marketing organizations will achieve the best results from influencer marketing together with travel agencies and social media influencers.

7. Conclusion

The study found that the influencer branding also has an impact on how the destination is being shaped and tourist behaviour. Credibility, transparency in sponsorships and relatability are shown through quantitative analysis, such as correlation, regression and factor analysis, to have a great effect on audience trust and engagement. It makes for a stronger relationship with followers and for an influencer, honesty and genuine storytelling will be absolutely must however you may be seen by your clients as well. Says that collaboration with influencers to produce compelling and authentic content, that adds to a destination's appeal, benefits tourism boards, destination marketing organizations and their needs. This could also be used by travel agencies in their marketing strategies to be more in line with consumer's preference. This study reinforces the process of transformation of destination marketing through influencer engagement. More interdisciplinary research is needed to answer ethical questions and make trustworthy, effective promotional strategies. It seems that sustainable tourism development depends on transparent and meaningful communication with an audience.

7.1. Scope for further research

Based on this research, further research on how tourism branding and what tourists do are affected by influencer marketing is ensured. Longitudinal studies can even follow long term audience perceptions, whereas cross cultural studies can investigate regional differences of the influencer genuine and trust. It is important that future research addresses influencer credibility through platforms such as Instagram, YouTube and Facebook to measure the difference in consumer decision making. Further investigation is required as well on ethical concerns such as transparency in sponsorships and consumer protection. Also, there are studies on audience segmentation, influencer collaborations with the destination marketing organizations, and the emerging trends, such as virtual influencers. These research opportunities will contribute in developing more data driven strategies for destination branding for more effective and ethical influencer marketing approach.

References

[1] Abidin, C. (2016). Aren't these just young, rich women doing vain things online? *Influencer selfies as subversive frivolity. Social Media + Society*, *2*(2), 2056305116641342.

[2] De Veirman, M., Cauberghe, V., & Hudders, L. (2017). Marketing through Instagram influencers: The impact of number of followers and product divergence on brand attitude. *International Journal of Advertising*, *36*(5), 798–828.

[3] Evans, N. J., Phua, J., Lim, J., & Jun, H. (2017). Disclosing Instagram influencer advertising: The effects of disclosure language on advertising recognition, attitudes, and behavioral intent. *Journal of Interactive Advertising*, *17*(2), 138–149.

[4] Gretzel, U. (2018). Influencer marketing in travel and tourism. In K. A. Buchholtz & R. Law (Eds.), *Information and Communication Technologies in Tourism 2018* (pp. 147–158). Springer, Cham.

[5] Hughes, C., Swaminathan, V., & Brooks, G. (2019). Driving brand engagement through online social influencers: An empirical investigation of sponsored blogging campaigns. *Journal of Marketing*, *83*(5), 78–96.

[6] Hwang, Y. H., & Zhang, D. (2020). The impact of travel influencers' credibility on online consumer purchase intentions and travel planning behavior: A dual-process perspective. *Tourism Management*, *79*, 104099.

[7] Lou, C., & Yuan, S. (2019). Influencer marketing: How message value and credibility affect consumer trust of branded content on social media. *Journal of Interactive Advertising*, *19*(1), 58–73.

[8] Mariani, M., Di Felice, M., & Mura, M. (2016). Facebook as a destination marketing tool: Evidence from Italian regional destination management organizations. *Tourism Management*, *54*, 321–343.

80 The influence of customer relationship management and brand image on customer satisfaction: A meta-analysis for light rail transit (LRT) transportation

Yunianto Aditya[1,a], *Buchari Erika*[2,b], *Agustien Melawaty*[2,c], *and Ermatita Ermatita*[2,d]

[1]Doctoral Program in Civil Engineering, Faculty of Engineering, Universitas Sriwijaya, Palembang, Sumatera Selatan, Indonesia
[2]Department of Civil Engineering, Faculty of Engineering, Universitas Sriwijaya, Palembang, Sumatera Selatan, Indonesia

Abstract: Every organization, whether profit-oriented or non-profit-oriented, must survive, such as LRT Operator. One strategy commonly adopted is maintaining and increasing customer satisfaction continuously. The study of customer satisfaction become more interesting today, including through scientific investigations. Therefore, this study explores customer satisfaction based on the perspective of customer relationship management (CRM) and brand image. To achieve this, the current study uses a quantitative approach through meta-analysis of research results published in the form of articles indexed by Scopus, SINTA, and Google Scholar in 2017–2024, consisting of 40 articles. Meta-analysis was carried out using JASP 19 software. The results show that CRM and brand image positively and significantly affect customer satisfaction. Partially, CRM impacts customer satisfaction, with $r=0.48$ and $p<0.01$. The brand image also affects customer satisfaction, with $r=0.34$ and $p<0.01$. Meanwhile, CRM and brand image influence customer satisfaction simultaneously, with $r=0.93$ and $p<0.01$. These findings not only confirm and strengthen previous studies that were used as the basis for the meta-analysis but also offer new things (novelty) regarding the complexity of the influence of CRM and brand image on customer satisfaction based on various previous studies. Therefore, it is worthy of further research in the future, such as developing customer loyalty models for Light Rail Transit (LRT).

Keywords: Brand image, customer relationship management, customer satisfaction, meta-analysis

1. Introduction

Currently, LRT operators in South Sumatra have not fully implemented a Customer Relationship Management (CRM) strategy. Although there are variables used to measure the perspective of customers, CRM may be necessary to increase customer satisfaction. Customer satisfaction is essential to a company's long-term viability and success. According to empirical research, repeat business, brand loyalty, and the customer serving as an opinion leader for the company in the marketplace are all indicators of customer satisfaction [1]. According to Nguyen-Phuoc et al. [2], customer satisfaction is also thought to be the primary indicator of customer loyalty. When customers are pleased with a brand, product, or service, they will respond favorably, possibly making repeat purchases or recommending it to others. Generally speaking, satisfaction is the emotion that results from evaluating a product or service's performance (outcome) against prior experiences, whether it be one of joy or disappointment [3, 4]. According to Lee and Xue [5] and Pham et al. [6], satisfaction may also be described as the response to the discrepancy between prior expectations and the product's performance. From unsatisfactory to gratifying, it is the total assessment of need fulfillment [7]. When a customer's expectations are met or beyond, they are satisfied [8]. Due to its ability to help firms compete by retaining current customers and attracting new ones, customer satisfaction is one of the marketing ideas that has been studied the most [9]. Meeting and beyond the requirements, desires, and expectations of customers leads to their repurchase and ongoing loyalty, which is known as customer satisfaction. Following a customer's purchase or use of a service, a general attitude is formed as a result of the experience [10]. When a customer's stated and implicit wants are met by the product and associated services, the customer is deemed satisfied [1]. According to Kotler and Keller, a person's degree of satisfaction determines whether they feel happy or disappointed when a product performs as expected

[a]adityayunianto85@gmail.com, [b]eribas17@gmail.com, [c]melawatyagustien@ft.unsri.ac.id, [d]ermatita@unsri.ac.id

DOI: 10.1201/9781003684459-80

[11]. Customers are satisfied if the performance lives up to their expectations; they are not. When performance is above expectations, customers are happy. Disgruntled customers can depart and complain to others. However, contented customers are more devoted and reciprocal. So, businesses should focus on retaining existing customers rather than finding new ones to replace departing ones [12]. Repurchases and word-of-mouth are two important ways that consumer behavior affects customer satisfaction, which is a crucial metric for assessing a business's success [13]. It gives marketers and business professionals a measurable metric that they can use to efficiently manage and improve their companies. Most organizations won't have any major problems keeping up company continuity once consumer satisfaction is achieved [14]. Knowing what a customer truly needs is a crucial step in helping an organization provide them. A portion of the demand may be dormant or latent, which the customer may not be able to express or describe, while the whole need may be partially understood by the consumer, who is in a position to do so [1]. According to Oliver [15], customers can measure satisfaction with three indicators: fulfillment of expectations, no subjective uncertainty, and excellent performance results.

Customer satisfaction is influenced by CRM [16, 17] and brand image [18, 19]. However, other studies indicated conflicting results. For example, Pradana [20] shows that CRM has an insignificant impact on customer satisfaction. Other research results revealed that CRM does not significantly affect customer satisfaction [21, 22]. Because of this empirical reality, there is a research gap that needs to be filled by science. Because of this urgency, the study uses a meta-analysis approach to verify how CRM and brand image affect customer satisfaction.

1.1. CRM and customer satisfaction

Since 1920, CRM has been seen as important since it deals with client connections, buyouts, and customer retention [23]. CRM is getting more and more crucial as the consumer wants rise, which motivates businesses to develop systems to satisfy client needs and preserve connections with them [24]. With its structural, social, and financial benefits, CRM is acknowledged as a successful marketing strategy for preserving client loyalty [25]. According to Juanamasta et al. [26], businesses can more readily sustain client relationships and foster product loyalty by putting the CRM strategy into practice. The concept of CRM is characterized by the value, satisfaction, loyalty, share, and profitability of customers, including cross-selling and up-selling [27]. CRM is the practice of maintaining thorough data about every customer starting point inside the company [11]. CRM is a business approach that aims to accomplish long-term objectives for an organization by building relationships with its customers. Customer satisfaction can be measured with an appropriate CRM. Both the strategic and technology components of CRM are examined by Kumar and Mokha [28]. Both of them, together with CRM's tactical perspective, demonstrate how the concept of CRM has evolved concerning interpersonal and operational advertising since CRM is the product of both types of advertising. An analysis of CRM from a technological perspective demonstrates how businesses adapt their business environments to new technologies and increased competition by utilizing innovative strategies encompassing information storage, information mining, and routine client-facing practices. According to Bozutti and Esposto [29], several definitions of CRM view the concept as a means of concentrating on customers, either by examining respective lifecycles, much like a customer-driven innovation, a competitive market strategy, or an interactive approach. CRM theory is a continuous process consisting of four parts, according to Cortez et al. [30]: learning, market preparation, customer interaction, and customer data analysis. CRM is a competitive product that upholds alliances, enhances customer value, and provides customized services [31–33]. Based on earlier research showing that cultivating specific customer relationships could increase long-term business and customer loyalty, the current study was conducted [34]. CRM theory, which states that businesses must retain profitable and devoted clients, is consistent with this conclusion [35, 36] by increasing customer happiness and value, customer relationship management aims to create and maintain profitable client relationships [11]. The main objective of CRM procedures, according to Dehghanpouri et al. [37], is to enhance their implementation, which affects how successfully businesses communicate with their clients. CRM is a significant tool that businesses may utilize to enhance their customer communications, according to Sun [38]. According to Altarifi [39], CRM has a significant, direct, and advantageous impact on marketing success. According to [40], CRM positively increases client trust. CRM allows businesses to monitor their interactions with clients. According to them, maintaining positive, long-lasting relationships with profitable and devoted clients is essential to managing these relationships. The CRM idea is also applied in the leadership of customer relations, according to Itani et al. [41], based on the efficient and successful flow of information and the provision of greater value to customers while reducing expenses for the business. Gazi et al. [42] identified two indicators to measure CRM, they are knowledge management and organizational commitment. Meanwhile, Kotler and Keller [11] cite the three CRM indicators of people, process, and technology. Here, the workforce have embraced the CRM. This requires careful consideration of several important factors, including the organizational structure, roles and responsibilities, business culture, procedures, and the overall change management program. Businesses shouldn't undervalue the role that workers play in designing and managing a CRM program. Additionally, the collection of organizational procedures that guides employees in delivering customer service and are governed by a system. Current business procedures need to be reorganized to support CRM operations. Using accurate, consistent references while speaking with clients in person is essential. Finally, although technology contributes to the value of CRM, businesses still need to take into account factors like people, work culture, consumer behavior, and business structure. The application can function at its best thanks to the interdependence of these three CRM indicators [11]. According to Khan et al. [34], CRM is an effective technique that draws in clients, cultivates enduring bonds with suppliers and clients, and upholds client satisfaction. In addition to increasing customer satisfaction [23, 16, 17], it connects businesses with their customers [43]. Thus, it is possible to hypothesize that:

H_1: CRM has a positive effect on customer satisfaction.

1.2. Brand image and customer satisfaction

A brand image can be a name, word, sign, symbol, design, or a mix of these that helps consumers recognize and differentiate the product or service from those of competitors [11]. Marliawati and Cahyaningdyah [44] assert that brand image is based on how consumers perceive a certain brand. A brand image is a sentiment, concept, or mindset connected to consumers [45]. Additionally, it refers to the buyer's brand associations, which are a collection of opinions, beliefs, and customer impressions of the brand [13]. A company's or its product's public perception is known as its brand image [46]. Moreover, it reflects the associations and opinions of consumers toward a certain brand [47]. Customers' perceptions and beliefs are also linked to brand image, as evidenced by the connections that are ingrained in their memory and are always the first thing that comes to mind when they hear the slogan [48].

Brand image influences customer satisfaction and future intent, acting as a point of reference. Consumers must form a favorable attitude before the encounter itself [45]. A strong brand identity has helped consumers understand the brand's demands and set it apart from rivals [49].

This brand image indicator consists of (1) the user's impression will be companies that produce goods and services; (2) the user's impression will be the user of the product includes the user, outlook on life and social position; (3) the user's impression will be goods include attributes, uses, consumers and guarantees provided by goods or products [50]. According to Kotler and Keller [11], brand image has several dimensions that can be used, namely: (a) functional benefit, where the brand has a relationship with profit in the consumption of goods/services; (b) symbolic benefit, which is related to the need for social recognition and self-esteem; (c) experiential benefits, which discuss how customers feel after using the goods/services; (d) social benefits, which emphasize social needs when using the product; and (e) appearance enhances, which discusses improving consumers' appearance when using the product, are some of the dimensions of brand image that can be used. Positive perceptions of these aspects by customers may increase their level of satisfaction. Brand image has a major impact on consumer satisfaction, according to a recent study by Bangun et al. [49], Hendry and Kosasih [50], and Idris et al. [51]. Thus, it is possible to support the following hypothesis:

H_2: Brand image has a positive effect on customer satisfaction.

1.3. CRM, brand image, and customer satisfaction

Simultaneously, CRM and brand image have an impact on customer satisfaction. CRM and brand image have a favorable and considerable impact on customer satisfaction at the same time, according to studies by Sakharia et al. [52], Imron and Ariyanti [53], Setyorini et al. [54], and Rahayu and Ningsih [55]. It suggests that CRM and brand image are essential factors that determine customer satisfaction, so maintaining and improving them is vital. Thus, it is possible to make the following hypothesis:

H_3: CRM and brand image have a simultaneous impact on customer satisfaction.

2. Methods

2.1. Research design

This research uses a quantitative meta-analysis method. This method relies on statistical calculations that produce quantitative data by combining two or more relevant previous studies [56]. Meta-analysis is a retrospective observational study that summarizes and analyzes data without modification. In the context of this research, the data summarized comes from research publications that discuss the relationship between CRM and brand image and customer satisfaction. According to Tawfik et al. [57], a meta-analysis aims to determine which studies should be included and analyzed. Therefore, hypotheses in meta-analysis studies are very useful for determining the inclusion and exclusion criteria that need to be applied to find related studies [58]. Several factors are taken into account when choosing research articles, such as those that were found using online international journal search databases like Google Scholar, DOAJ, etc.; publications that included customer samples; publications from various countries; publications in English; publications that were indexed by Scopus, SINTA, and Google Scholar; publications that were published between 2017 and 2024; publications that had r, t, or R2 values that explained the relationship between CRM, brand image, and customer satisfaction; and the 40 samples that were examined.

2.2. Data coding

Coding is the main requirement to facilitate data collection and analysis in meta-analysis [59]. Therefore, the coding category sheet plays an important role as an instrument in meta-analysis. Coding describes the characteristics of the articles selected and used comprising year of publication, publication sample (n), correlation coefficient (r), t value, R^2, Z, and SE information. Table 80.1 displays the distribution of publications used in this meta-analysis.

These studies take research objects in the field of goods and services, including transportation services.

2.3. Data analysis

Meta-analysis is carried out based on the coding process, correlation coefficient (r), t value, R^2, Z, and SE, which includes effect size heterogeneity tests, calculating summary effects or average effect sizes, creating funnel and forest plots,

Table 80.1. Comparison of 40 studies based on n, r, T, and R^2 values, 2017–2024

No	*Study*	*Country*	*n*	*r*	*t*	R^2	*Z*	*SE*
		CRM on Customer Satisfaction						
1	Anuriza & Kusumawardhani (2021) [60]	Indonesia	110	0.680			0.829	0.097
2	Ar Rosyad et al. (2023) [61]	Indonesia	330	0.117	2.130		0.117	0.055
3	As Zahra & Dewanti (2024) [62]	Indonesia	100	0.596	7.341		0.686	0.102
4	Chandra & Christian (2024) [63]	Indonesia	250	0.160	2.559		0.162	0.064
5	Ferdiawan et al. (2018) [64]	Indonesia	100	0.484	5.478		0.528	0.102
6	Gazi et al. (2024) [42]	Bangladesh	100	0.490	5.568		0.536	0.102
7	Haryandika & Santra (2021) [65]	Indonesia	94	0.459	4.956		0.496	0.105
8	Hayati et al. (2020) [7]	Indonesia	100	0.226	2.299		0.230	0.102
9	Jannah & Rosa (2024) [66]	Indonesia	100	0.429	4.701		0.459	0.102
10	Kholifah et al. (2024) [67]	Indonesia	201	0.343	5.148		0.357	0.071
11	Pratama & Rubiyanti (2023) [68]	Indonesia	175	0.919	30.637		1.582	0.076
12	Shabrina et al. (2020) [69]	Indonesia	150	0.297	3.780		0.306	0.082
13	Sugiat & Sudirman (2024) [70]	Indonesia	155	0.176	2.216		0.178	0.081
14	Sugiat et al. (2023) [71]	Indonesia	155	0.194	2.447		0.197	0.081
15	Suharto & Yuliansyah (2023) [72]	Indonesia	135	0.462	6.010		0.500	0.087
		Brand Image on Customer Satisfaction						
1	Bangun et al. (2024) [49]	Indonesia	228	0.261	4.060		0.267	0.067
2	Chen & Wu (2022) [73]	Taiwan	399	0.099	1.980		0.099	0.050
3	Dam & Dam (2021) [13]	Vietnam	299	0.183	3.205		0.185	0.058
4	Hendry & Kosasih (2024) [50]	Indonesia	355	0.374	7.572		0.393	0.053
5	Idris et al. (2024) [51]	Indonesia	100	0.286	2.956		0.294	0.102
6	Loi et al. (2017) [74]	China	282	0.510	9.919		0.563	0.060
7	Martin & Nasib (2021) [75]	Indonesia	125	0.282	3.259		0.290	0.091
8	Osman et al. (2024) [76]	Bangladesh	296	0.658	14.994		0.790	0.058
9	Romero-Carazas et al. (2023) [77]	Peru	352	0.522	11.443		0.579	0.054
10	Setiawan & Sayuti (2017) [78]	Indonesia	200	0.349	5.238		0.364	0.071
11	Wijoyo (2024) [79]	Indonesia	102	0.429	4.750		0.459	0.101
12	Wiliayanto & Wage (2024) [80]	Indonesia	100	0.270	2.776		0.277	0.102
13	Wu & Li (2017) [81]	China	427	0.164	3.434		0.166	0.049
14	Zulvianti et al. (2023) [82]	Indonesia	310	0.112	1.973		0.112	0.057
15	Maftuchach & Kusnuranti (2021) [83]	Indonesia	140	0.200	2.404		0.203	0.085
		CRM and Brand Image on Customer Satisfaction						
1	Ferdiawan, et al (2018) [64]	Indonesia	100	0.799			1.096	0.102
2	Rahayu & Ningsih (2023) [55]	Indonesia	125	0.503			0.553	0.091
3	Setyorini et al (2023) [54]	Indonesia	213	0.669		0.448	0.809	0.069
4	Zuhri & Palupiningdyah (2018) [84]	Indonesia	115	0.363			0.380	0.094
5	Ferianda et al (2022) [85]	Indonesia	175	0.921		0.849	1.598	0.076
6	Pratama et al (2021) [86]	Indonesia	100	0.802			1.104	0.102
7	Sakharia et al (2023) [52]	Indonesia	120	0.795			1.085	0.092
8	Imron & Ariyanti (2023) [53]	Indonesia	100	0.677		0.459	0.824	0.102
9	Aferi et al. (2022) [87]	Indonesia	294	0.677		0.459	0.824	0.059
10	Hayati et al (2020) [7]	Indonesia	100	0.750		0.562	0.973	0.102

Source: Author.

hypothesis testing, and verifying publication bias. According to Cohen's effect size criteria [88], the effect size ranges between 0 and 1. In detail, < 0+/-0.1 = weak effect; < 0+/-0.3 = modest effect; < 0+/-0.5 = moderate effect; < 0+/- 0.8 = Strong effect; and ≥+/-0.8 = very strong effect. The software used was JASP 19.0.

3. Results and Discussion

The findings are arranged in three sections in compliance with the research objectives and hypotheses: the relationship between CRM and customer satisfaction, the connection between customer contentment and brand image, and the connection between CRM and customer satisfaction and brand image.

3.1. CRM on customer satisfaction

The results of the heterogeneity test and residual heterogeneity estimates show that Q = 24.477 with $p < 0.001$; τ^2 or $\tau > 0$; I^2 (%) is around 100%. It is demonstrated that there was significant variation in the 15 effect sizes of the studies that were examined. Table 80.2 displays the findings of the summary effect or mean impact size analysis.

According to research findings utilizing the random effect model, CRM and customer satisfaction are positively correlated (z = 4.947, 95% CI = 0.288; 0.665). The p-value is less than 0.01, indicating a significant correlation between CRM and customer satisfaction. Therefore, H_1: CRM positively impacts customer satisfaction, accepted. The magnitude of the correlation is moderate ($r_{RE} = 0.48$). It means that CRM is a crucial predictor of customer satisfaction.

Additionally, visually appealing graphic techniques in forest plots are used to illustrate the analysis results. This graphic makes it easier to compare research results by understanding the simultaneous combined impact estimates shown by plots (dots) at specific intervals. Figure 80.1 displays the forest plot chart for the 15 examined research. The papers under review had impact sizes ranging from 0.01 to 1.73. Additionally, Figure 80.2 visualizes the funnel plot. In meta-analyses, Begg's funnel plot is a type of scatter diagram that is used to visually identify potential publication bias (symmetrical or asymmetrical research samples). However, this is often considered insufficient, so the Egger test is needed, as presented in Table 80.3.

For p-value > 0.05, the Z-value for Egger's test is 0.793. The symmetry of the funnel plot is shown here, proving that there is no publication bias in this meta-analysis.

Table 80.2. Summary effect or mean effect size

Estimate (r_{RE})	*Standard Error*	*Z*	*p*	*95% Confidence Interval (CI)*	
				Lower	*Upper*
0.48	0.096	4.947	< .001	0.288	0.665

Source: Author.

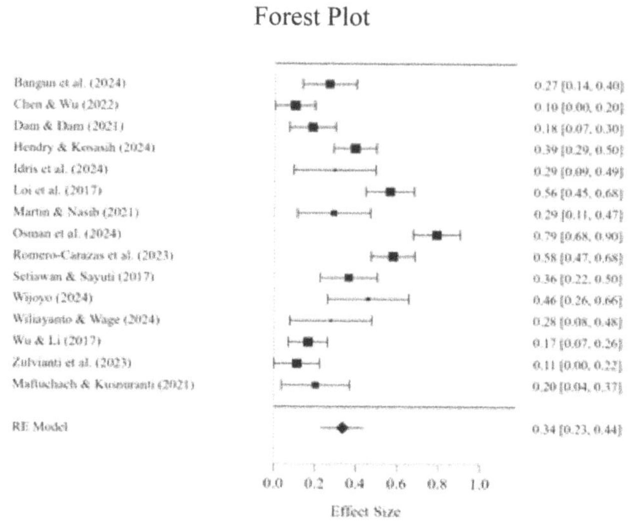

Figure 80.1. Meta-analysis forest plot.

Source: Author.

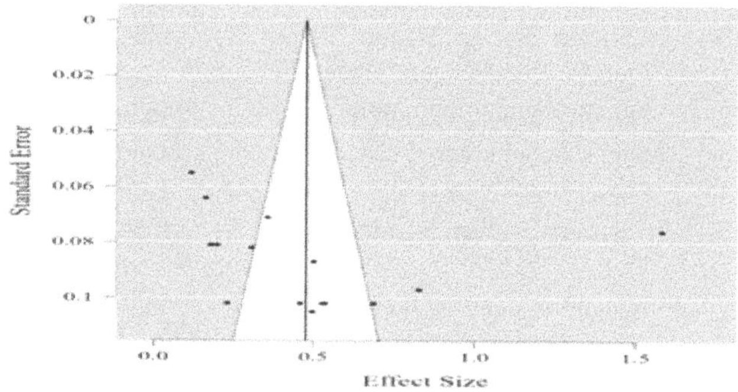

Figure 80.2. Funnel plot after trim-fill diagnosis.

Source: Author.

Table 80.3. Regression test for funnel plot asymmetry (egger's test)

	Z	*p*
sei	0.793	0.428

Source: Author.

3.2. Brand Image on customer satisfaction

According to the results of the heterogeneity test and residual heterogeneity estimations, the correlation between customer satisfaction and brand image is about 100% (Q = 41.851 with $p < 0.001$; τ^2 or $\tau > 0$; I^2 (%)). The 15 effect sizes of the studies that were analyzed are shown to differ significantly from one another. The results of the summary effect or mean impact size analysis are shown in Table 80.4.

According to the aforementioned research findings, there is a positive association (Z = 6.469, 95% CI = 0.235; 0.439) between customer satisfaction and brand image. A substantial relationship between brand image and customer happiness is

Table 80.4. Summary effect or mean effect size

Estimate (r_{RE})	*Standard Error*	*Z*	*p*	*95% Confidence Interval (CI)*	
				Lower	*Upper*
0.34	0.052	6.469	< .001	0.235	0.439

Source: Author.

indicated by a p-value of less than 0.01. Thus, H2: consumer happiness is positively impacted by brand image is acknowledged. The connection has a moderate strength (rRE = 0.34). It proved that one of the most important predicators of client pleasure is brand image. The forest plot chart, which ranges from 0.00 to 0.90, is also shown in Figure 80.3. In the meantime, the funnel plot is shown in Figure 80.4, and the Egger test findings are shown in Table 80.5.

At a p-value > 0.05, the Egger's test yields a Z-value of -0.091. By demonstrating the funnel plot's symmetry, this meta-analysis is free from publication bias.

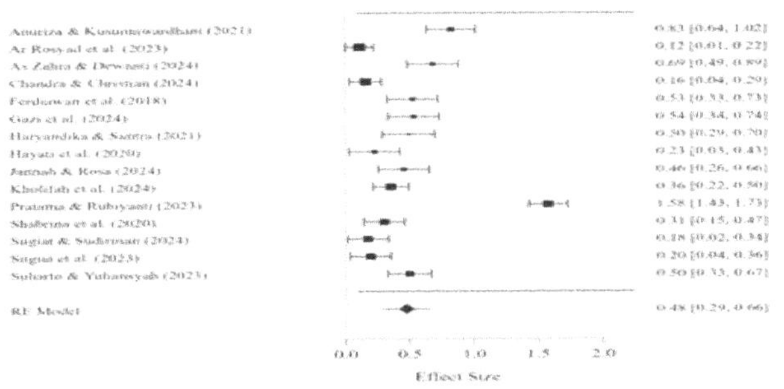

Figure 80.3. Meta-analysis forest plot.

Source: Author.

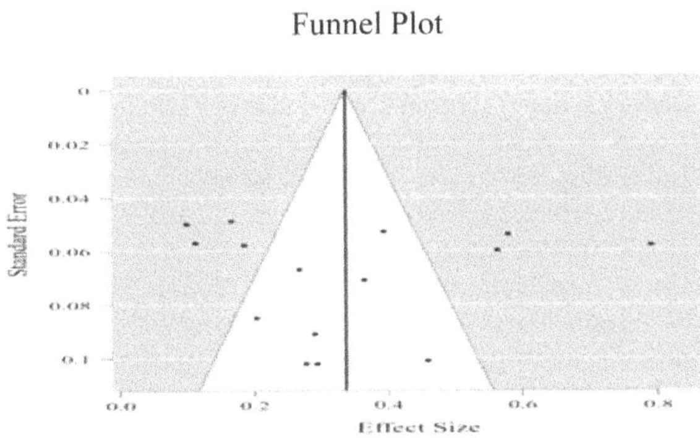

Figure 80.4. Funnel plot after trim-fill diagnosis.

Source: Author.

Table 80.5. Regression test for funnel plot asymmetry (egger's test)

	Z	*p*
sei	-0.091	0.927

Source: Author.

3.3. CRM and image on satisfaction

The findings of the residual heterogeneity estimations and the heterogeneity test indicate that Q = 75.291 with p < 0.001; τ^2 or $\tau > 0$; I^2 (%) is around 100% for the combined effect of CRM and brand image on customer satisfaction. It is shown that there was a range of ten effect sizes in the relevant trials. Table 80.6 also provides a summary of the mean impact of the mean effect size.

Z = 8.677 and 95% CI = 0.716; 1.134 indicated that CRM and brand image had a favorable effect on customer satisfaction. CRM and brand image have a considerable impact on customer satisfaction, as indicated by the p-value of less than 0.01. Therefore, H_3: CRM and brand image positively impact customer satisfaction and are accepted. The level of the link is very strong (r_{RE}= 0.93). It demonstrated that CRM and brand image are crucial determinants of customer satisfaction. In addition, Figure 80.4 displays the forest plot chart with a range from 0.20 to 1.75. Meanwhile, Figure 80.5

Table 80.6. Summary effect or mean effect size

Estimate (r_{RE})	*Standard Error*	*Z*	*p*	*95% Confidence Interval (CI)*	
				Lower	*Upper*
0.93	0.107	8.677	< .001	0.716	1.134

Source: Author.

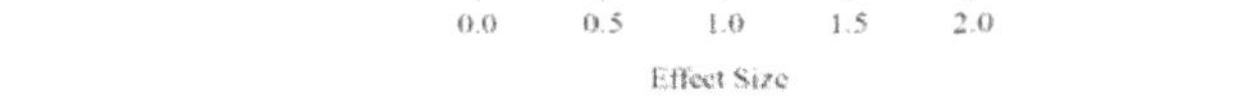

Figure 80.5. Meta-analysis forest plot.

Source: Author.

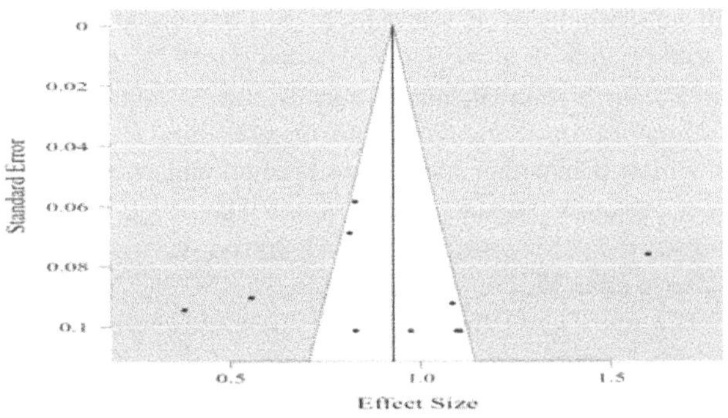

Figure 80.6. Funnel plot after trim-fill diagnosis.

Source: Author.

Table 80.7. Regression test for funnel plot asymmetry (egger's test)

	Z	*p*
sei	-0.149	0.881

Source: Author.

displays the funnel plot, and Table 80.7 presents the results of the Egger test.

At p-value > 0.05, the results of the Egger's test indicate a Z-value of -0.149. Because of the funnel plot's symmetry, this meta-analysis is free of the publication bias issue.

4. Discussion

According to the study's overall conclusions, customer happiness is significantly impacted by both CRM and brand image, either independently or together. The positive and considerable impact of CRM on customer satisfaction has been established, indicating that high-quality CRM can increase customer satisfaction. These findings, based on a meta-analysis, validate several previous studies included in this research, which assert that CRM has a substantial influence on customer satisfaction [16, 17]. Furthermore, these findings contradict the claims of Sukma et al. [21] and Wahyuni and Rachbini [22], who argue that CRM has an insignificant impact on customer satisfaction.

This study also found that brand image positively and significantly affects customer satisfaction. It indicates that a brand image that customers perceive positively can stimulate their satisfaction. That means that improving brand image can positively contribute to customer satisfaction. Thus, brand image is an important determinant of customer satisfaction, so its existence is essential to maintain and improve continuously. This finding is by and confirms previous studies sampled in this meta-analysis, which proved that brand image significantly impacts customer satisfaction (e.g., Bangun et al. [49]; Hendry & Kosasih [50]; Idris et al., [51]).

Additionally, this meta-analysis demonstrates how CRM and brand image both have an impact on customer satisfaction simultaneously. It shows that CRM and brand image are crucial predictors of customer satisfaction, so their existence needs to be developed simultaneously and continuously. This empirical fact convincingly confirms previous relevant studies which prove that CRM and brand image significantly affect customer satisfaction simultaneously (e.g., Rahayu & Ningsih [55]; Setyorini et al. [54]; Imron & Ariyanti [53]; Sakharia et al. [52]).

The findings of this meta-analysis underscore the practical implications of CRM and brand image as key drivers of customer satisfaction. This insight equips researchers and practitioners with the knowledge to critically evaluate and potentially integrate these findings into their future research and practical work.

5. Conclusion

The purpose of this study is to investigate customer satisfaction from a CRM and brand image standpoint. The findings show that, partially and simultaneously, CRM and brand image have a favorable and considerable impact on customer satisfaction. These findings not only confirm and strengthen previous studies that were used as the basis for the meta-analysis but also offer new things (novelty) regarding the complexity of the influence of CRM and brand image on customer satisfaction based on various previous studies. Consequently, these findings provide a theoretical contribution to the study of customer satisfaction from the perspective of CRM and brand image and provide practical implications for similar research in the future. These findings can also provide insight for management and organizational practitioners in building customer satisfaction as a strategy for organizational (company) development, such as developing customer loyalty models for Light Rail Transit (LRT).

References

[1] Mukherjee, P. N. (2024). *Total Quality Management*, 2nd ed. Bangalore: PHI Learning Pvt. Ltd.

[2] Nguyen-Phuoc, D. Q., Su, D. N., Tran, P. T. K., Le, D.-T. T., & Johnson, L. W. (2020). Factors influencing customer's loyalty towards ride-hailing taxi services–A case study of Vietnam. *Transp. Res. Part A Policy Pract.*, *134*(1), 96–112. doi:10.1016/j.tra.2020.02.008.

[3] Rasoolimanesh, S. M., Wang, M., Roldan, J. L., & Kunasekaran, P. (2021). Are we in right path for mediation analysis? Reviewing the literature and proposing robust guidelines. *J. Hosp. Tour. Manag.*, *48*, 395–405. doi:10.1016/j.jhtm.2021.07.013.

[4] Boto-Garcia, D., Escalonilla, M., Zapico, E., & Baños, J. F. (2021). Scale heterogeneity in hotel guests' satisfaction relative to room rates. *Applied economic analysis*, *29*(87), 208–225. doi:10.1108/AEA-07-2020-0093.

[5] Lee, S. W., & Xue, K. (2020). A model of destination loyalty: Integrating destination image and sustainable tourism. *Asia Pacific Journal of Tourism Research*, *25*(4), 393–408. doi:10.1080/10941665.2020.1713185.

[6] Coles, T., Ritchie, B. W., & Wang, J. (2021). Building business resilience to external shocks: Conceptualising the role of social networks to small tourism & hospitality businesses. *Journal of Hospitality and Tourism Management*, *48*, 210–219. doi:10.1016/j.jhtm.2021.06.012.

[7] Hayati, S., Suroso, A., Suliyanto, S., & Kaukab, M. (2020). Customer satisfaction as a mediation between micro banking image, customer relationship and customer loyalty. *Management Science Letters*, *10*(11), 2561–2570. doi:10.5267/j.msl.2020.3.039.

[8] Adriatico, R. L., Razalan, A. M. A., Pagbilao, C. M. V., Afalla, B. T., & Dela Cruz, L. M. (2022). Service Quality and Customer Satisfaction in Dining Restaurants: Inputs for Tourism and Hospitality Curriculum Enhancement. *Academic Journal of Interdisciplinary Studies*, *11*(3), 30–37. doi:10.36941/ajis-2022-0065.

[9] Deyalage, P. A., & Kulathunga, D. (2019). Factors affecting online customer satisfaction: The Sri Lankan perspective. *Int. J. Bus. Manag.*, *14*(2), 99. doi:10.5539/ijbm.v14n2p99.

[10] Kumar, P., & Bisaria, G. (2024). Consumer Satisfaction Judgments Based on Credence Services. In *Interdisciplinary Research in Technology and Management* (pp. 481–490). CRC Press. doi:10.1201/9781003430469-56.

[11] P. & Kotler, K. K. (2021). *Marketing Management*, 16th ed. New York: Pearson.

[12] Hashem, T. N. (2024). An Analysis of the Correlation Between Shopping Benefits, Customer Satisfaction, and Customer Loyalty in Retail Stores. *Proc. IAC Prague 2024*, 17–24.

[13] Dam, S. M., & Dam, T. C. (2021). Relationships between service quality, brand image, customer satisfaction, and customer loyalty. *The Journal of Asian Finance, Economics and Business*, *8*(3), 585–593. doi:10.13106/jafeb.2021.vol8.no3.0585.

[14] Libo-On, J. T. (2021). Service quality influence on customer satisfaction in courier services: A comparative study. *American International Journal of Business Management (AIJBM)*, *4*(3), 51–63.

[15] Oliver, S. (2019). *Strategi Public Relations*. Jakarta: Erlangga.

[16] Lokesh, S., & Vasantha, S. (2022). Influence of Customer Relationship Management Towards Customer Loyalty with Mediating Factor Customer Satisfaction in Insurances Sector. *Quality-Access to Success*, *23*(187). doi:10.47750/QAS/23.187.21.

[17] Nogueira, S., & Carvalho, J. M. (2024). Unlocking the dichotomy of place identity/place image and its impact on place satisfaction for ecotourism destinations. *Journal of Ecotourism*, *23*(1), 1–19. doi:10.1080/14724049.2022.2106236.

[18] Laely, N., & Lidiawawan, A. R. (2024). The Effect of Product Innovation and Customer Relationship Management on Satisfaction Mediated by Purchasing Decisions in the Coffee Shop Industry. *Journal of Ecohumanism*, *3*(7), 386–403. doi:10.62754/joe.v3i7.4210.

[19] Younis, J. A., Al-Shammari, H., Hejase, H. J., Massoud, M., & Hejase, A. J. (2024). E-satisfaction as a mediator between consumer loyalty and E-CRM: The context of Lebanese e-commerce. *International Journal of Innovative Research and Scientific Studies*, *7*(3), 978–996. doi:10.53894/ijirss.v7i3.2987.

[20] Pradana, F. (2018). Pengaruh manajemen hubungan pelanggan, kualitas pelayanan, dan kualitas pengalaman terhadap loyalitas pelanggan dengan kepuasan pelanggan sebagai variabel intervening pada nasabah PT Fac Sekuritas Indonesia di Yogyakarta. *Jurnal Manajemen Bisnis*, *9*(2), 193–212.

[21] Sukma, N. R., Sugianto, S., & Nurbaiti, N. (2024). The Influence of Service Quality and Customer Relationship Management on Customer Loyalty with Satisfaction as an Intervening Variable at Bank Syariah Indonesia KC Southeast Aceh. *Indonesian Interdisciplinary Journal of Sharia Economics (IIJSE)*, *7*(2), 2314–2341.

[22] Wahyuni, R. S., & Rachbini, W. (2024). The Influence of Customer Relationship Management and Service Quality on Consumer Loyalty through Consumer Satisfaction. *Dinasti International Journal of Economics, Finance & Accounting (DIJEFA)*, *5*(3), 1301–1507. doi:10.38035/dijefa.v5i3.

[23] Agrawal, S. R. (2021). Adoption of WhatsApp for strengthening internal CRM through social network analysis. *Journal of Relationship Marketing*, *20*(4), 261–281. doi:10.1080/15332667.2020.1802643.

[24] Nurcahyo, R., Dody, P. N., Yadrifil, Z. Z. Z., Habiburrahman, M., & Wibowo, N. (2020). CRM Dimensional Analysis and ANP Method for Oil and Gas Industry Sub-contractor in Indonesia in the Context of Reducing Customer Complaints. In *Proceedings of the International Conference on Industrial Engineering and Operations Management* (pp. 605–613).

[25] Fernando, E., Sutomo, R., Prabowo, Y. D., Gatc, J., & Winanti, W. (2023). Exploring customer relationship management: Trends, challenges, and innovations. *Journal of Information Systems and Informatics*, *5*(3), 984–1001. doi:10.51519/journalisi.v5i3.541.

[26] Setiawati, A. P., Susetyorini Susetyorini, U. E., Rusdiyanto, R., Astanto, D., Ulum, B., Khadijah, S. N., ... & Umanailo, M. C. B. (2019). The role of customer service through customer relationship management (Crm) to increase customer loyalty and good image. *International Journal of Scientific and Technology Research*, *8*(10), 2004–2007.

[27] Babu, G. (2016). The Evolution of customer relationship management (CRM)-opportunities and challenges. *Int. Res. J. Manage. Sci. Technol*, *7*(8), 46–47.

[28] Kumar, P., & Mokha, A. K. (2021). Relationship between E-CRM, customer experience, customer satisfaction and customer loyalty in banking industry: a review of literature. *Research Review: International Journal of Multidisciplinary*, *6*(2), 127–137. doi:10.31305/rrijm.2021.v06.i02.022.

[29] Bozutti, D. F., & Esposto, K. F. (2019). Sales and Operations Planning: a comparison between the demand-driven and traditional approaches. *International Journal of Production Management and Engineering*, *7*(1), 23–38. doi:10.4995/ijpme.2019.9469.

[30] Cortez, R. M., Freytag, P. V., & Ingstrup, M. B. (2022). Restoring ecosystem brands: The role of sustainability-led innovation. *Industrial Marketing Management*, *105*, 79–93. doi:10.1016/j.indmarman.2022.05.021.

[31] Dazagbyilo, Y. Y. K., Shang, S., Emmanuel, O. C., & Kir, K. F. (2021). Customer relationship management as a tool for improving customer loyalty in the banking industry: an empirical study of guaranty trust bank PLC situated within the Central Business District, FCT Abuja, Nigeria. *Open Journal of Business and Management*, *9*(5), 2299–2311. doi:10.4236/ojbm.2021.95124.

[32] Soltani, Z., Zareie, B., Milani, F. S., & Navimipour, N. J. (2018). The impact of the customer relationship management on the organization performance. *The Journal of High Technology Management Research*, *29*(2), 237–246.

[33] Rîpa, I. A. (2022). Customer relationship management in the international context–theoretical and practical considerations. In *Proceedings of the International Conference on Business Excellence* (Vol. 16, No. 1, pp. 1060–1072). Sciendo.

[34] Khan, R. U., Salamzadeh, Y., Iqbal, Q., & Yang, S. (2022). The impact of customer relationship management and company reputation on customer loyalty: The mediating role of customer satisfaction. *Journal of Relationship Marketing*, *21*(1), 1–26. doi:10.1080/15332667.2020.1840904.

[35] Naim, A. (2022). Mapping Of Social Customer Relationship Management with Electronic Customer Relationship Management. *European Journal of Interdisciplinary Research and Development*, *2*, 14–25.

[36] Masud, A. A., Nabi, M. N., & Kumar, D. (2023). Socio-economic and psychological impacts of the COVID-19

pandemic on various occupational groups in Bangladesh. *J. Bus. Econ.*, *4*, 77–90. doi:10.58481/BJBE2305.

[37] Dehghanpouri, H., Soltani, Z., & Rostamzadeh, R. (2020). The impact of trust, privacy and quality of service on the success of E-CRM: the mediating role of customer satisfaction. *Journal of business & industrial marketing*, *35*(11), 1831–1847. doi:10.1108/jbim-07-2019-0325.

[38] Chengwu, S. U. N. (2020). The research and evaluation of customer relationship management and social responsibility on competitive advantage. *Revista de Cercetare şi Intervenţie Socială*, (69), 131–142. doi:10.33788/rcis.69.8.

[39] Altarifi, S. (2020). The impact of CRM on marketing performance through innovation capability. *Journal of Critical Reviews*, *7*(12), 4424–4433.

[40] Kodua, P., Blankson, C., Panda, S., Nguyen, T., Hinson, R. E., & Narteh, B. (2022). The Relationship between CSR and CBBE in Sub-Saharan Africa: The moderating role of customer perceived value. *Journal of African Business*, *23*(4), 1088–1108.

[41] Itani, O. S., Kalra, A., & Riley, J. (2022). Complementary effects of CRM and social media on customer co-creation and sales performance in B2B firms: The role of salesperson self-determination needs. *Information & Management*, *59*(3), 103621. doi:10.1016/j.im.2022.103621.

[42] Gazi, M. A. I., Al Mamun, A., Al Masud, A., Senathirajah, A. R. B. S., & Rahman, T. (2024). The relationship between CRM, knowledge management, organization commitment, customer profitability and customer loyalty in telecommunication industry: The mediating role of customer satisfaction and the moderating role of brand image. *Journal of Open Innovation: Technology, Market, and Complexity*, *10*(1), 100227. doi:10.1016/j.joitmc.2024.100227.

[43] Herman, L. E., Sulhaini, S., & Farida, N. (2021). Electronic customer relationship management and company performance: Exploring the product innovativeness development. *Journal of Relationship Marketing*, *20*(1), 1–19. doi:10.1080/15332667.2019.1688600.

[44] Marliawati, A., & Cahyaningdyah, D. (2020). Impacts the brand of experience and brand image on brand loyalty: Mediators brand of trust. *Management Analysis Journal*, *9*(2), 140–151. doi:10.15294/maj.v9i2.36945.

[45] Erkmen, E., & Hancer, M. (2019). Building brand relationship for restaurants: An examination of other customers, brand image, trust, and restaurant attributes. *International Journal of Contemporary Hospitality Management*, *31*(3), 1469–1487. doi:10.1108/IJCHM-08-2017-0516.

[46] Enjelina, A. E. (2022). Keputusan pembelian online melalui citra merek berdasarkan efektivitas iklan dengan epic model studi pada konsumen traveloka. com. *Jurnal Manajemen Pemasaran*, *16*(2), 57–66.

[47] Albrahim, F. (2020). Identifying Gifted and Talented Students and Methods to Enhance their Development: A Conceptual Framework. *Humanit. Manag. Sci. - Sci. J. King Faisal Univ.*, 2020. doi:10.37575/h/edu/2192.

[48] Nizar, N., Silalahi, M., Sofiyan, S., & Sinaga, O. S. (2019). Pengaruh kualitas pelayanan dan citra merek terhadap kepuasan konsumen pada PT Fast Food Indonesia, Tbk KFC Box Ramayana Pematangsiantar. *SULTANIST: Jurnal Manajemen dan Keuangan*, *7*(2), 43–52.

[49] Mahrinasari, M., & Roslina, R. (2024). The Influence of Service Quality and Brand Image on Customer Loyalty Mediated by Customer Satisfaction: in Indonesia Coal Mining Industry. *Journal of Economics, Finance and Management Studies*, *7*(2), 1111–1119. doi:10.47191/jefms/v7-i2-30.

[50] Hendry, H., & Kosasih, H. (2024). The influence of service quality and brand image on customer satisfaction at CCTV Palace Medan. *Journal Arbitrase: Economy, Management and Accounting*, *2*(01), 21–31. doi:10.58471/Arbitrase.v2i01.

[51] Idris, N. M., Sendra, I. M., & Yanthy, P. S. (2024). The effect of social media promotion mix and destination image on visitor satisfaction at Puncak Lima Jari, South Sulawesi. *Asian Journal of Management, Entrepreneurship and Social Science*, *4*(03), 650–668.

[52] Sakharia, D., Supeni, R. E., & Hafidzi, A. H. Studi Empiris Customer Relationship Management (CRM) Citra Merek dan Kepuasan terhadap Loyalitas Pelanggan The Body Shop Jember. *ULIL ALBAB J. Ilm. Multidisiplin*, *2*(10), 4688–4695. doi:10.56799/jim.v2i10.2211.

[53] Imron, A., & Ariyanti, R. (2023, December). The Influence of Service Quality, Customer Relationship Management (CRM) and Brand Image on Customer Loyalty With Customer Satisfaction As An Intervening Variable. In *International Conference on Digital Advanced Tourism Management and Technology* (Vol. 1, No. 1, pp. 103–117).

[54] Setyorini, D., Tirtayasa, S., & Khair, H. (2023). The Effect of Brand Image and Customer Relationship on Customer Loyalty Mediated By Customer Satisfaction At Primary School Al-Ulum. *Jurnal Ekonomi*, *12*(01), 531–545.

[55] Rahayu, P., & Ningsih, L. S. R. (2023). Pengaruh Customer Relationship Management Dan Brand Image Terhadap Patient Satisfaction Di Klinik Pratama Aulia 1 Jombang. *Jurnal Ekonomi Bisnis dan Manajemen*, *1*(4), 68–79.

[56] Mueller, M., D'Addario, M., Egger, M., Cevallos, M., Dekkers, O., Mugglin, C., & Scott, P. (2018). Methods to systematically review and meta-analyse observational studies: a systematic scoping review of recommendations. *BMC medical research methodology*, *18*, 1–18.

[57] Tawfik, G. M., Dila, K. A. S., Mohamed, M. Y. F., Tam, D. N. H., Kien, N. D., Ahmed, A. M., & Huy, N. T. (2019). A step by step guide for conducting a systematic review and meta-analysis with simulation data. *Tropical medicine and health*, *47*, 1–9. doi:10.1186/s41182-019-0165-6.

[58] Higgins, J. P., López-López, J. A., Becker, B. J., Davies, S. R., Dawson, S., Grimshaw, J. M., ... & Caldwell, D. M. (2019). Synthesising quantitative evidence in systematic reviews of complex health interventions. *BMJ global health*, *4*(Suppl 1), e000858. doi:10.1136/bmjgh-2018-000858.

[59] Malički, M., Jerončić, A., Aalbersberg, I. J., Bouter, L., & Ter Riet, G. (2021). Systematic review and meta-analyses of studies analysing instructions to authors from 1987 to 2017. *Nature communications*, *12*(1), 5840. doi:10.1038/s41467-021-26027-y.

[60] Anuriza, D., & Kusumawardhani, A. (2021). The Influence of Electronic Customer Relationship Management (E-Crm) On Customer Loyalty with Customer Satisfaction As A Mediating Variable Study In Ahass Putri Jaya Motor Group Brebes. *Diponegoro Journal of Management*, *10*(4), 1–12.

[61] Febrilia, I. (2023). Influence of Customer Relationship Management and Service Quality on Customer Loyalty Mediated by Customer Satisfaction. *Jurnal Scientia*, *12*(03), 2861–2872.

[62] Zahra, V. A., & Dewanti, M. C. (2024). The Influence of Customer Relationship Management and Satisfaction on Customer Loyalty of the Indonesian Home Education Parenting Program. *J. Ekon. dan Bisnis Digit.*, *3*(1), 75–90.

[63] Chandra, Z., & Christian, M. (2024). The influence of key opinion leaders (kol) and customer relationship management (crm) on customer loyalty with customer satisfaction as a mediating variable in IBOX. *Int. J. Soc. Serv. Res.*, *4*. doi:10.46799/ijssr.v4i9.909.

[64] Ferdiawan, Y., Hermawan, A., Wardana, L. W., & Arief, M. (2018). Satisfaction as effect mediation of brand image and customer relationship management on customer's loyalty. *KnE Social Sciences*, 34–50. doi:10.18502/kss.v3i3.1872.

[65] Haryandika, D. M., & Santra, I. K. (2021). The effect of customer relationship management on customer satisfaction and customer loyalty. *Indonesian Journal of Business and Entrepreneurship (IJBE)*, *7*(2), 139–139. doi:10.17358/IJBE.7.2.139.

[66] Jannah, R., & Rosa, N. T. (2024). The Influence of Customer Relationship Management (CRM) and Service Quality on Customer Satisfaction at the Bengkalis Chamber of Commerce and Industry. *Inovbiz: Jurnal Inovasi Bisnis Seri Manajemen, Investasi dan Kewirausahaan*, *4*(1), 34–38.

[67] Kholifah, N., Ms, M., & Roslina, R. (2024). The Influence of Customer Relationship Marketing and Brand Image on Customer Loyalty with Customer Satisfaction as a Mediation Variable (Study on Starbucks Coffee Indonesia). *J. Econ. Financ. Manag. Stud.*, *07*. doi:10.47191/jefms/v7-i3-16.

[68] Pratama, R. R., & Rubiyanti, N. (2023). The Influence of Customer Relationship Management on Customer Brand Loyalty Through Customer Satisfaction as a Mediator. *Jurnal Informatika Ekonomi Bisnis*, 1427–1431. doi:10.37034/infeb.v5i4.642.

[69] Shabrina, H., Rachbini, D. J., & Rekarti, E. (2020). Customer Relationship Marketing and Brand Image in Establishing Satisfaction and Loyalty of Advertisers. *European Journal of Business and Management*, *12*(29), 54–62.

[70] Sugiat, M., & Sudirman, A. (2024). Model Analysis of the Mediation Effect of Customer Satisfaction on Customer Loyalty in Mitratel Companies: The Role of CRM and Company Reputation as Predictors. *Jurnal Manajemen Indonesia*, *24*(1), 105–120. doi:10.25124/jmi.v24i1.7353.

[71] Sugiat, M., Novrizon, E., & Trianasari, N. (2023). The Influence of Customer Relationship Management and Company Reputation on Customer Loyalty Through Customer Satisfaction in Telecommunication Companies. *International Journal of Economics*, *2*(2), 2961–712. doi:10.55299/ijec.v2i2.570.

[72] Suharto, S., & Yuliansyah, Y. (2023). The Influence of Customer Relationship Management and Customer Experience on Customer Satisfaction. *Integr. J. Bus. Econ.*, *7*(1), 403–417. doi:10.33019/ijbe.v7i1.641.

[73] Chen, A. H., & Wu, R. Y. (2022). Mediating effect of brand image and satisfaction on loyalty through experiential marketing: A case study of a sugar heritage destination. *Sustainability*, *14*(12), 7122. doi:10.3390/su14127122.

[74] Loi, L. T. I., So, A. S. I., Lo, I. S., & Fong, L. H. N. (2017). Does the quality of tourist shuttles influence revisit intention through destination image and satisfaction? The case of Macao. *Journal of Hospitality and Tourism Management*, *32*, 115–123. doi:10.1016/j.jhtm.2017.06.002.

[75] Martin, M., & Nasib, N. (2021). The Effort to Increase Loyalty through Brand Image, Brand Trust, and Satisfaction as Intervening Variables. *Society*, *9*(1), 277–288. doi:10.33019/society.v9i1.303.

[76] Osman, A. R., Joarder, M. H. R., & Hoque, M. K. (2024). Student Engagement, Brand Image and Loyalty Relationships: The Mediating Role of Student Satisfaction. *Journal of Education and e-Learning Research*, *11*(2), 311–321. doi:10.20448/jeelr.v11i2.5553.

[77] Mori Rojas, G., et al. (2023). Service quality and institutional image as predictors of customer satisfaction in municipalities of Perú. *J. Law Sustain. Dev.*, *11*(5), e885–e885. doi:10.55908/sdgs.v11i5.885.

[78] Setiawan, H., & Sayuti, A. J. (2017). Effects of service quality, customer trust and corporate image on customer satisfaction and loyalty: an assessment of travel agencies customer in South Sumatra Indonesia. *Iosr J. Bus. Manag.*, *19*(5), 31–40. doi:10.9790/487x-1905033140.

[79] Wijoyo, D. H. (2024). The Correlation between the Quality of Public Services and the Image of Institution to the Public Satisfaction at Kejaksaan Negeri Jember. *ABM Int. J. Adm. Bus. Manag.*, *6*(1), 44–53. doi:10.31967/abm.v6i1.1085.

[80] Wiliayanto, W., & Wage, S. (2024). The influence of word of mouth, attractiveness and brand image on customer satisfaction PT. Leffindo Jaya logistics in Batam city. *J. Ekon.*, *13*(01), 1036–1047. doi:10.54209/ekonomi.v13i01.

[81] Wu, H.-C., & Li, T. (2017). A study of experiential quality, perceived value, heritage image, experiential satisfaction, and behavioral intentions for heritage tourists. *J. Hosp. Tour. Res.*, *41*(8), 904–944. doi:10.1177/1096348014525638.

[82] Zulvianti, N., Aimon, H., & Abror, A. (2023). Perceived Environmental Value, Destination Image, and Tourist Loyalty: The Role of Tourist Satisfaction and Religiosity. *Sustainability*, *15*(10), 8038. doi:10.3390/su15108038.

[83] Maftuchach, V., & Kusnuranti, A. L. (2021). The Effect of Brand Image on Customer Satisfaction Through Quality of Courier Services During the Covid-19 Pandemic In Indonesia (SAP Express Depok Case Study). *Int. J. Soc. Manag. Stud.*, *2*(6), 29–39.

[84] Zuhri, M. A. T., & Palupiningdyah, P. (2018). The Influence of Customer Relationship Management and Brand Image on Customer Loyalty through Satisfaction. *Manag. Anal. J.*, *7*(3), 363–373.

[85] Ferianda, A. S., Indrajaya, D., & Muchtar, A. H. (2022). Pengaruh Customer Relationship Management & Brand Image terhadap Kepuasan Pasien Rawat Jalan selama Masa Pandemi Covid-19. *J. Ilm. Wahana Pendidik.*, *8*(1), 759–775.

[86] Pratama, D., Widyastuti, T., Suroso, S., & Untari, D. T. (2021). Pengaruh Customer Relationship Management Dan Citra Perusahaan Terhadap Kepuasan Dan Loyalitas Nasabah Di Pt. Bank Rakyat Indonesia Tbk Kantor Cabang Tambun, Bekasi. *J. Ilm. Manaj. Ubhara*, *3*(1), 94.

[87] Aferi, F. D., & Yeni, Y. H. (2022). The Effect of E-Service Quality, Brand Image, E-Customer Relationship Management on E-Customer Satisfaction at Gojek Padang. *Journal of Business Studies and Mangement Review*, *6*(1), 38–42. doi:10.22437/jbsmr.v6i1.20399.

[88] Cohen, R. D., Woseth, D. M., Thisted, R. A., & Hanauer, S. B. (2000). A meta-analysis and overview of the literature on treatment options for left-sided ulcerative colitis and ulcerative proctitis. *Official journal of the American College of Gastroenterology| ACG*, *95*(5), 1263–1276.

81 Lesson learnt from the study of the power of service quality in influencing company reputation and customer trust for LRT operator: A meta-analysis

Pudjiono Rode Paulus Gagok[1,a], Buchari Erika[2,b], Agustien Melawaty[2,c], and Ermatita Ermatita[2,d]

[1]Doctoral Program in Civil Engineering, Faculty of Engineering, Universitas Sriwijaya, Palembang, Sumatera Selatan, Indonesia
[2]Department of Civil Engineering, Faculty of Engineering, Universitas Sriwijaya, Palembang, Sumatera Selatan, Indonesia

Abstract: Service quality is essential for both profit and non-profit organizations, but many organizations have not yet realized superior service quality, such as LRT Operator in Palembang. That is why service quality has consistently become a critical issue among management (organizational) practitioners, academics, and researchers that requires a serious respons, including through research. Therefore, this study aims to investigate service quality about company reputation, and customer trust. The research users a quantitative approach based on literature reviews, which are analyzed using meta-analysis. The literature review was conducted by selecting 22 relevant articles indexed by Scopus, SINTA, and Google Scholar between 2017 and 2024. Meta-analysis using JASP 0.8 4.0 software found that service quality positively and significantly affects company reputation with a correlation coefficient (r) = 0.619 and p-value < 0.01. Service quality also significantly affects customer trust with r = 0.661 and p-value < 0.01. Empirically, service quality impacts customer trust more than company reputation. This indicates that service quality contributes to customer trust, not company reputation. This evidence affirms and strengthens prior relevant studies used as the fundamental for the meta-analysis and offers a novelty regarding the influence of service quality on company reputation and customer trust based on various prior relevant studies. Accordingly, it is worthy of further discussion among researchers and practitioners in the future, including in the field of transportation, especially in improving Light Rail Transit (LRT) operational performance.

Keywords: Service quality, trust, reputation, meta-analysis

1. Introduction

Currently, LRT Operators in Palembang South Sumatra have not yet realized superior service quality. Service quality has a vital role for companies, including building customer trust and company reputation. A business can satisfy client demands and maintain economic competitiveness in its industry by providing high-quality services [1]. According to Felix [2], service quality is typically see as a crucial requirement and a competitive aspect in creating and maintaining a satisfying client connection. For Hapsari et al. [3], service quality is widely acknowledged as a crucial component in numerous industries across the globe since it helps businesses increase their profits and retain customers. Hence, this study investigates more deeply the influence of service quality on company reputation and customer trust. Since service quality is related to an organization's operations, it is typically defined as a comparison of a customer's service preferences [4].

Service quality is the degree to which a business can fulfill or beyond the expectations of customers when they interact with its goods or services [5]. It compares expected service and perceived actual performance [6]. Customers attempt to evaluate the actual quality of a certain organization's services with their expectations represented service quality [7]. If the reality exceeds expectations, the service can be deemed high-quality; if the contrary is true, the service can be deemed low-quality [8]. Sondakh et al. [9] affirm that the following five aspects can be used to understand service quality: physical evidence, assurance, responsiveness, empathy, and dependability. Service responsiveness, staff assistance, staff dependability, staff appearance, and the capacity to offer customers suitable solutions are other factors that are included in the concept of service quality [10].

[a]rodepaulusgp@gmail.com, [b]eribas17@gmail.com, [c]melawatyagustien@ft.unsri.ac.id, [d]ermatita@unsri.ac.id

DOI: 10.1201/9781003684459-81

Within and between markets, customer views of services might differ significantly in terms of what is considered normal, exceptional, or even significant [11], [12]. Consequently, because service performance is dependent on opportunities, interests, and customer preferences, it is typically judged as service quality rather than performance on particular service characteristics [13]. The degree to which a service satisfies the needs and expectations of the client is a common definition of service quality, albeit everyone has a distinct definition. [14]. As a result, the first step in raising an organization's level of service quality is to make sure that it has a sufficient grasp of client expectations and their derivatives [12, 13]. The company can influence some of the factors that affect customer expectations, like promises made and kept, but other factors, It has no control over things like the customer's experience, their attitude, or even how important the service is in their view [13, 14]. Accordingly, evaluating the quality of services requires asking customers about their experiences during or immediately following a service contact and comparing the performance to their expectations [15]. The primary objective of service quality measurement is to facilitate the prioritizing of corporate improvement actions, thereby concentrating on the characteristics that are most crucial to consumers and/or organizational strategy [13, 16]. Customer satisfaction is the ultimate goal of meeting customer expectations and is the outcome of measuring the discrepancy between the quality of service rendered and the expectations of the client. [17]. In every business encounter, this idea highlights that service excellence encompasses more than just offering a product or service; it also involves meeting customer expectations and creating a great experience [18]. Under such conditions, service quality significantly influences company reputation and customer trust.

1.1. Service quality and company reputation

Researchers assert that there is a connection between a company's reputation and service excellence [18–20]. It indicated that service quality is crucial to a company's reputation. Organizational development, corporate communication, marketing, psychology, accounting, and economics are just a few of the academic disciplines that have conducted empirical research on firm reputation [21]. According to Asgeirsson et al. [15], reputation is typically regarded in marketing and service-related theories as a perceptual depiction of a business's past acts and prospects that characterizes its total attractiveness to all of its crucial constituents in comparison to other top competitors. This impression may be influenced by media coverage of the organization or people's interactions with it [22]. To put it another way, reputation is a measure of how customers value an organization's overall performance based on their experiences. One intangible organizational asset that can help a business stand out from the competition in the market is a good reputation [15].

Reputation systems, along with brand image and public regulation, have been acknowledged since the early days of the Internet as an essential part of the institutional framework that supports online transactions because they help preserve trust in interactions between strangers [23]. Positive perceptions of a business's reputation can help it retain a solid reputation and boost customer loyalty, both of which provide a long-term competitive advantage [24]. There are therefore differing opinions about which specific stakeholders ought to be taken into account when evaluating a company's reputation. The reputations of employees, investors, the public, and customers are a few examples [25]. Although customers are considered a company's most significant stakeholder group, there is still a dearth of empirical research analyzing the company's reputation based on their needs [26].

Chun [27] pointed out that there are three primary components to reputation: (1) the identity of the organizational unit, or what it is; (2) the image of the organization, or what stakeholders think it stands for; and (3) the desired identity, or what the organization claims to stand for. Excellent service quality can increase a company's reputation. Therefore, can formulate a hypothesis:

H_1: Service quality positively impacts company reputation.

1.2. Service quality and customer trust

A recent study also indicated that service quality is related to customer trust [28–30]. Trust is the conviction that one will receive what is anticipated from others [29]. Trust can be linked to a person's desire to act in a certain way because they think their partner will live up to their expectations and because they generally trust the words, promises, or statements of others [31]. Building trust is a crucial skill for success in any field and is derived from real-world interactions with other people [32]. The belief, confidence, emotion, or expectation of a buyer's intentions or possible behaviour might be considered trust [33]. Additionally, Mansouri et al. [34] defined trust as the consumers' assessment of a firm's honesty and integrity, referring to the customers' view and belief that the company will not let them down and will put out great effort to maximize their advantages. The community's long-term viability and profitability are guaranteed if a business can increase confidence in both virtual and non-virtual communities. Widyastuti et al. [35] revealed that trust is an essential foundation of a relationship between service providers and consumers, especially in the service sector.

Consumer trust is the glue that allows companies to trust others to organize and use resources effectively [29]. Developing consumer trust will give rise to high trust in their abilities and desires [36]. It is the willingness to rely on colleagues involved in exchanges that are trusted [37]. Customer trust refers to a consumer's belief in the promise of a reliable service provider and fulfilling its commitments [38]. Consumers will increasingly trust to use a product or service if more value is provided [39]. The trust that is built is by fulfilling all consumer desires by using the best services offered to consumers, which is one of the trusts carried out by service providers, namely providing quality service that meets customer expectations, which requires time to be handled effectively and efficiently [40]. In this way, consumer trust will increase in line with building consumer satisfaction for service users. Creating trust is difficult, but companies must try with totality [41].

Customers consider trust when determining whether or not to buy products from an online retailer [33]. After being pleased with the service, customers need trust to shop online and make a purchase [42]. Consumer commitment and credibility are influenced by customer trust, and this has an impact on the relationship to make repeat purchases [43]. A company's readiness to depend on its business partners is a sign of trust. Numerous inter-organizational and interpersonal elements, including company competency, honesty, kindness, and integrity, are necessary for trust [44]. Putri and Marlena [45] identified five indicators of customer trust: integrity, competence, consistency, loyalty, and openness. Excellent service quality can stimulate customer trust. Hence, it can propose a hypothesis:

H_2: Service quality positively impacts customer trust.

2. Methods

2.1. Research design

In this study, the quantitative meta-analysis methodology was applied. To quote [46], quantitative meta-analysis is a statistical method that generates a quantitative mixture of data by merging two or more related studies. Meta-analyses generally are retrospective observational studies where the researcher summarizes the findings without changing the experiment. The data that were summarized were from research publications that discussed the connections between service quality, reputation, and trust. To centralize the findings of this comprehensive study, the research articles were chosen based on several factors, such as (1) those that are searchable through online international journal search databases like Google Scholar, DOAJ, and others; (2) publications with sample customers; (3) publications from different countries; (4) English-language publications; (5) publications with Scopus, SINTA, and Google Scholar indexes; (6) publications from 2017 to 2024; (7) publications with values of r, t, F, or R^2 which describe the relationship between service quality, company reputation, and customer trust; and (8) the samples examined ≥ 22. The goal of the meta-analysis is to determine which studies should be included. Consequently, the hypothesis for find the inclusion and exclusion criteria that must be used immediately to find pertinent publications is made much easier by the meta-analysis study [47].

2.2. Data coding

The most crucial precondition for meta-analysis is coding, which facilitates data collection and processing [48]. A coding category sheet was therefore used as the tool in this meta-analysis. The coding explains the properties of the used articles, including the publication year, the study's country of origin, the publishing sample (n), the correlation value (r), the t-value, the F-value, R^2, the Z, and the SE remarks that include journal information from various countries. The subject of all articles analyzed is in the field of sales of goods and services, including transportation services. Table 81.1 displays the publications' distribution.

2.3. Data analysis

The features of the research sample were analyzed, data was then coded, the effect size heterogeneity test, the summary effect, the mean effect size calculation, and the values of t and F were transformed into the values of r correlation, the creation of funnel and forest plots, hypothesis testing, and verifiying publication bias. A correlation meta-analysis was performed on the data based on 22 articles indexed by Scopus, SINTA, and Google Scholar. According to Cohen's effect size criteria [63], effect sizes can be divided into values between 0 and 1. There are $< 0+/-0.1$ = weak effect; $< 0+/-0.3$ = modest effect; $< 0+/-0.5$ = moderate effect; $< 0+/-0.8$ = Strong effect; and $\geq +/-0.8$ = very strong effect. The research program utilized in this study was JASP 19.0, which contains many useful features for anyone interested in learning about the analysis and interpretation of statistical data, allows for assumption testing, offers Cohen's criteria options, and may be installed on several computer operating systems.

3. Results and Discussion

3.1. Service quality on company reputation

Different values of r, t, and F were derived for every investigation based on examining 11 publications using particular criteria. The researchers transformed all research publications with no r-value from their t-value or F-value to r-values before conducting the heterogeneity test. Table 81.2 presents the heterogeneity test results, while Table 81.3 presents residual heterogeneity estimates.

The above heterogeneity test results indicated that Q =373.889 I^2 (%) is around 100%, suggesting heterogeneity among the 11 effect sizes of the included studies, either τ^2 or $\tau > 0$ with $p < 0.001$. The mean effect size or summary effect was also investigated, and publication bias was tested using a random effect technique. Table 81.4 displays the findings of the mean effect size analysis or summary effect.

A strong positive association ($Z = 4.472$; 95% CI [0.347; 0.890]) was found in the research utilizing the random effect model between service quality and reputation. The study proved that a p-value of less than 0.001 supports a significant association between service quality and reputation. Accordingly, H_1 accepted. The strong category contained the link ($r_{RE} = 0.619$) between service quality and reputation accomplishment. In addition, a visually appealing graphical technique known as forest plots illustrates the analysis results.

Table 81.1. Comparison of 22 Studies based on n, r, t, F, and R^2 value, 2017–2024

No	*Study*	*Country*	*n*	*r*	*t*	*F*	R^2	*Z*	*SE*
			Service Quality on Reputation						
1.	Bakrie et al. (2019) [49]	Indonesia	185	0.848	21.612			1.248	0.074
2.	Hadi & Indradewa (2019) [50]	Indonesia	250	0.168	2.680			0.169	0.064
3.	Khoi et al. (2019) [51]	Vietnam	550	0.275	6.706			0.283	0.043
4.	Mahdi et al. (2024) [19]	Iraq	80	0.792				1.077	0.114
5.	Supanun & Sornsaruht (2019) [52]	Thailand	542	0.140	3.280			0.141	0.043
6.	Widyastuti et al. (2017) [53]	Indonesia	250	0.336	5.620			0.350	0.064
7.	Arifin & Imanullah (2021) [54]	Indonesia	100	0.863				1.305	0.102
8.	Park (2019) [55]	Korea	252	0.571	11.001			0.649	0.063
9.	Zulkarnain & Helmi (2023) [18]	Indonesia	49	0.068	0.469	0.22		0.068	0.147
10.	Virappan & Chan (2020) [56]	Malaysia	200	0.747				0.966	0.071
11.	Subekti et al. (2024) [20]	Indonesia	66	0.505			0.255	0.556	0.126
			Service Quality on Trust						
1.	Arisa et al. (2024) [28]	Indonesia	100	0.721				0.910	0.102
2.	Aryawan (2024) [29]	Indonesia	200	0.258	3.757			0.264	0.071
3.	Bunga & Sopiah (2024) [30]	Indonesia	317	0.704	17.612			0.876	0.056
4.	Hartanto & Ramadhani (2023) [57]	Indonesia	96	0.429	4.598			0.458	0.104
5.	Mariana et al. (2021)	Indonesia	251	0.644	13.284			0.765	0.064
6.	Uzir et al. (2021) [58]	Malaysia	259	0.429	7.608			0.458	0.063
7.	Nurzali & Sidartha (2023) [59]	Indonesia	86	0.684	8.604			0.837	0.110
8.	Saputra & Rahmawaty (2023) [60]	Indonesia	235	0.787	19.465			1.063	0.066
9.	Boonlertvanich (2019) [61]	Thailand	400	0.761				0.999	0.050
10.	Zuniar & Sulaiman (2022) [62]	Indonesia	250	0.481	8.636			0.524	0.064
11.	Supanun & Sornsaruht (2019) [52]	Thailand	542	0.130	3.050			0.131	0.043

Source: Author.

Table 81.2. Heterogeneity test

	Q	*df*	*p*
Omnibus test of Model Coefficients	20.001	1	< 0.001
Test of Residual Heterogeneity	373.889	10	< 0.001

Source: Author.

Table 81.3. Residual heterogeneity estimates

	Estimate	*95% Confidence Interval*	
		Lower	*Upper*
τ^2	0.203	0.095	0.644
τ	0.450	0.308	0.803
I^2 (%)	97.749	95.320	99.281
H^2	44.433	21.367	139.069

Source: Author.

Table 81.4. Summary effect or mean effect size

Estimate (r_{RE})	*Standard Error*	*Z*	*p*	*95% Confidence Interval (CI)*	
				Lower	*Upper*
0.619	0.138	4.472	< .001	0.347	0.890

Source: Author.

Using forest plots, we can more easily compare research by being aware of the estimated aggregate effect displayed by plots (dots) at particular times.

The forest plot chart for the 11 research that were examined is displayed in Figure 81.1. The analysis's effect sizes range from -0.22 to 1.50, according to the forest plot chart. A funnel plot was the next plot that was made. One type of scatter diagram used in meta-analyses is Begg's funnel plot, which is used to visually identify potential publication bias in symmetrical or asymmetrical study samples. Figure

81.2 displays the funnel plot chart for the 11 analyzed studies. Further research using the Egger test is required since the generated model is either symmetrical or asymmetrical, making it difficult to discern publication bias from the funnel plot chart alone. Table 81.5 displays the outcomes of the Egger test.

With p > 0.05, Z = 0.542 is displayed in Table 81.5. It is verified that the funnel plot is symmetrical. Therefore, there is no publication bias issue with this meta-analysis study.

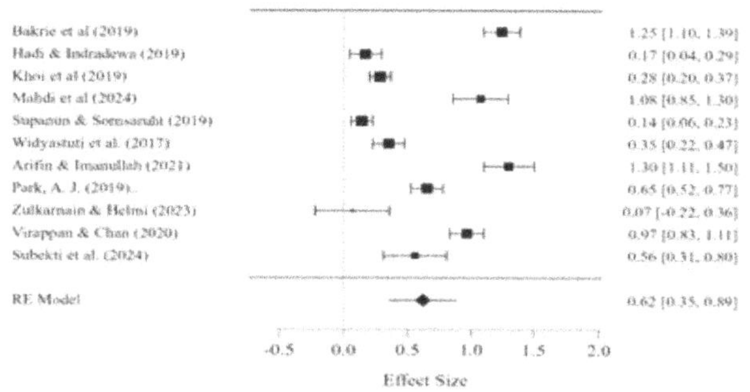

Figure 81.1. Meta-analysis forest plot.

Source: Author.

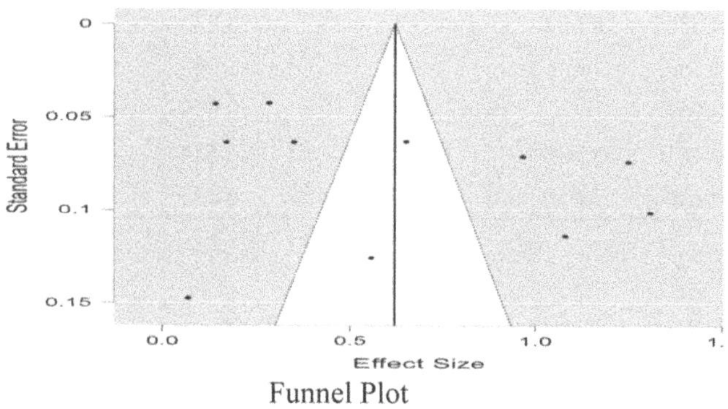

Figure 81.2. Funnel plot after trim-fill diagnosis.

Source: Author.

Table 81.5. Regression test for funnel plot asymmetry (egger's test)

	Z	*p*
sei	0.542	0.588

Source: Author.

Table 81.6. Heterogeneity test

	Q	*df*	*p*
Omnibus test of Model Coefficients	48.399	1	< 0.001
Test of Residual Heterogeneity	306.268	10	< 0.001

Source: Author.

3.2. Service quality on customer trust

Table 81.6 displays the results of the test for heterogeneity, whereas Table 81.7 displays estimates of residual heterogeneity.

Q = 306.268 I^2 (%) is about 100% for $p < 0.001$; τ^2 or $\tau > 0$, indicating that the 11 effect sizes of the studies under consideration were heterogeneous based on the results of the preceding heterogeneity test. In addition to estimating the mean impact size or summary effect, a random effect approach was used to conduct a publishing bias test. The results of the summary effect or mean impact size analysis are shown in Table 81.8.

According to research findings utilizing the random effect model, trust and service quality are strongly positively correlated (Z = 6.957; 95%CI [0.475; 0.847]). The study's significant relationship between service quality and trust is demonstrated by the p-value of less than 0.001. Hence, H_2 accepted. The strong category (r_{RE}= 0.661) includes the association between service quality and customer trust accomplishment. In addition, a visually appealing graphical technique known as forest plots illustrates the analysis results. Using forest plots, we can more easily compare research by understanding Plots (dots) at particular intervals display the expected combined effect.

Figure 81.3 displays the forest plot chart for the 11 examined research. The effect sizes of the chosen studies range from 0.05 to 1.19, according to the forest plot chart. A funnel plot was the second plot to be created. Meta-analyses employ a scatter map known as Begg's funnel plot to visually identify potential publication bias in symmetrical or asymmetrical study populations. Figure 81.2 displays the funnel plot chart for the 11 analyzed studies. Further research using the Egger test is required since the generated model is either symmetrical or asymmetrical, making it challenging

Table 81.7. Residual heterogeneity estimates

		95% Confidence Interval	
	Estimate	*Lower*	*Upper*
T^2	0.094	0.043	0.293
τ	0.306	0.207	0.541
I^2 (%)	95.711	91.099	98.588
H^2	23.315	11.235	70.834

Source: Author.

Table 81.8. Summary effect or mean effect size

Estimate (r_{RE})	*Standard Error*	*Z*	*p*	*95% Confidence Interval (CI)*	
				Lower	*Upper*
0.661	0.095	6.957	< .001	0.475	0.847

Source: Author.

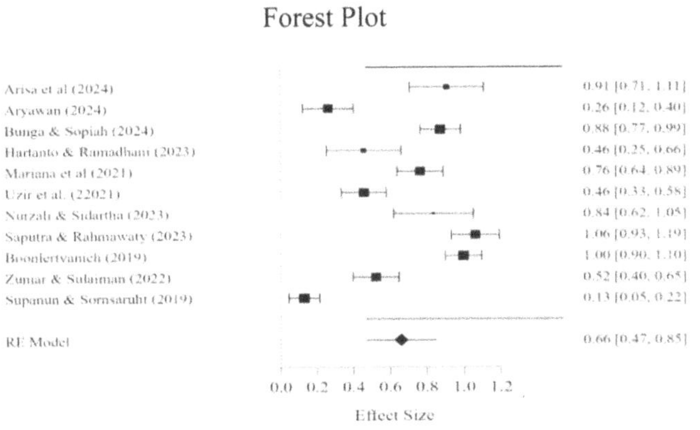

Figure 81.3. Meta-analysis forest plot.

Source: Author.

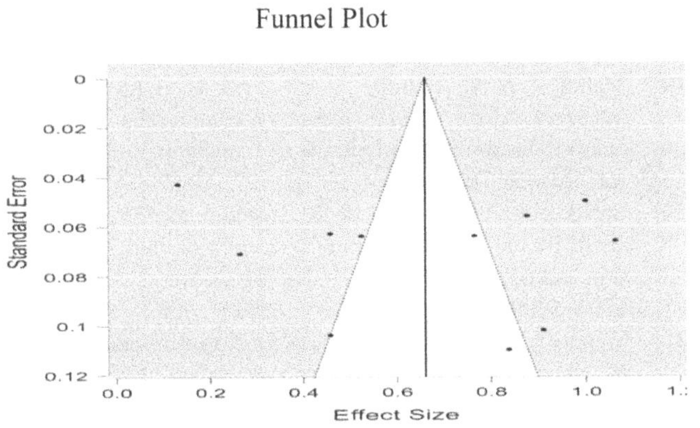

Figure 81.4. Funnel plot after trim-fill diagnosis.

Source: Author.

Table 81.9. Regression test for funnel plot asymmetry (egger's test)

	Z	*p*
sei	0.520	0.603

Source: Author.

to discern publication bias from the funnel plot chart alone. Table 81.9 displays the outcomes of the Egger test.

Z = 0.520 and p > 0.05 are displayed in Table 81.9. It illustrates how symmetrical the funnel plot is. Publication bias is therefore not an issue in this meta-analysis study.

4. Discussion

Two hypotheses are supported by the meta-analysis of 22 study results: there is a substantial positive correlation between customer trust, business reputation, and service quality. In other words, a company's reputation and consumer trust are positively correlated with the quality of its services. Stated differently, client trust and a company's reputation are significantly influenced by the quality of its services. Customer trust, business reputation, and service quality are strongly correlated, according to the impact size analysis results. [63]. It suggests that service quality is an important predictor of company reputation and customer trust. Therefore, service quality needs to be continuously managed well to make a greater contribution to the company's reputation and customer trust. Service quality is a comparison between expected service and actual perceived performance [6], reflected in reliability, responsiveness, assurance, empathy, and physical evidence [10, 19]. If in prime condition, these indicators can not only increase the company's reputation and customer trust. As a perceptual depiction of a business's past deeds and prospects, reputation characterizes the company's total attractiveness to all of its important stakeholders in contrast to other top competitors. This is manifested in its existence (identity), organizational unit, and desired identity. Contrarily, trust is the opinion of customers regarding the honesty and integrity of a business. Customers' view and belief that the business would not let them down and will put out great effort to reap their benefits is what it alludes to. This is reflected in the company's integrity, competence, consistency, loyalty, and openness. In addition, this study also does not indicate any publication bias. The symmetric shape of the funnel plot and Egger's test can be used to identify publication bias. To assess the quality of pertinent research methodologies, the degree of significance of the sources used, correct research results, and various sample sizes that will affect research conclusions with low levels of bias, publication bias analysis is required [64]. As a result, studies that were excluded from this meta-analysis produced findings that were comparable to those of the included samples.

5. Conclusion

Using a meta-analysis methodology, this study seeks to examine how customer satisfaction and business reputation relate to service quality. The study's findings demonstrate a substantial positive correlation between consumer trust company reputation and service excellence. Aside from that, the reviewed articles accurately depict the situation because there is no publishing bias. The publications' features show the same sample despite their diverse scientific backgrounds. This evidence affirms and strengthens prior relevant studies used as the fundamental for the meta-analysis and offers a novelty regarding the influence of service quality on company reputation and customer trust based on various prior relevant studies. These findings provide a theoretical contribution for future researchers, who can use this meta-analysis's results with samples from different research publications to find more theories regarding the relationship between service quality, company reputation, and customer trust. Additionally, the findings from this meta-analysis also provide practical implications for company organizations and management as a reference for building company reputation and customer trust through a service quality perspective, including in the field of transportation, especially in improving Light Rail Transit (LRT) operational performance.

References

[1] Abbas, J. (2020). HEISQUAL: A modern approach to measure service quality in higher education institutions. *Stud. Educ. Eval.*, *67*, 100933. doi: 10.1016/j.stueduc.2020.100933.

[2] Felix, R. (2017). Service quality and customer satisfaction in selected banks in Rwanda. *J. Bus. Financ. Aff.*, *6*(1), 246–256.

[3] Hapsari, R., Clemes, M., & Dean, D. (2016). The mediating role of perceived value on the relationship between service quality and customer satisfaction: Evidence from Indonesian airline passengers. *Procedia Econ. Financ.*, *35*, 388–395.

[4] Alam, M. M., Alauddin, M. D., Sharif, M. Y., Dooty, E. N., Ahsan, S. M. H., & Chowdhury, M. M. (2021). Students' Satisfaction and University Reputation through Service Quality in Private Higher Educational Institutions in Bangladesh. *J. Asian Financ. Econ. Bus.*, *8*(9), 91–100.

[5] Cempena, I. B., Brahmayanti, I. A. S., Astawinetu, E. D., Panjaitan, F. A. B. K., Kartini, I. A. N., & Panjaitan, H. (2021). The role of customer values in increasing tourist satisfaction in Gianyar Regency, Bali, Indonesia. *J. Asian Financ. Econ. Bus.*, *8*(8), 553–563. doi: 10.13106/jafeb.2021.vol8.no8.0553.

[6] Helmi, S., Suyanto, M., Effendi, M. I., Hidayat, A., & Mustafa, Z. (2022). The influence of internal marketing on internal service quality mediated by job satisfaction and employee loyalty. *J. Econ. Sustain. Dev.*, *13*(6), 63–69. doi: 10.7176/jesd/13-6-09.

[7] Fauzi, A. A. (2018). Electronic service quality on mobile application of online transportation services. *J. Manaj. Indones.*, *18*(1), 13–27.

[8] Harmen, H., Lubis, A., Harahap, L. M., & Indriani, R. (2022). Antecedent Affecting Chinese Ethnic Community Saving at Islamic Bank of North Sumatra. In *2nd International Conference of Strategic Issues on Economics, Business and, Education (ICoSIEBE 2021)*, pp. 66–76. doi: 10.2991/aebmr.k.220104.010.

[9] Sondakh, B. R., Tampi, J. R. E., & Tumbel, T. M. (2019). Pengaruh Bukti Langsung, Kehandalan, Jaminan, Daya Tanggap dan Empati, Terhadap Kepuasan Pasien pada Puskesmas. *J. Adm. Bisnis*, *9*(3), 101–111. doi: 10.35797/jab.v9.i3.101-111.

[10] Dau, P. M., Vandeviver, C., Dewinter, M., Witlox, F., & Vander Beken, T. (2023). Policing directions: A systematic review on the effectiveness of police presence. *Eur. J. Crim. policy Res.*, *29*(2), 191–225. doi: 10.1007/s10610-021-09500-8.

[11] Anabila, P., Ameyibor, L. E. K., Allan, M. M., & Alomenu, C. (2022). Service quality and customer loyalty in Ghana's hotel industry: The mediation effects of satisfaction and delight. *J. Qual. Assur. Hosp. Tour.*, *23*(3), 748–770. doi: 10.1080/1528008X.2021.1913691.

[12] Zeithaml, V. A., Bitner, M. J., & Gremler, D. D. (2024). *Services marketing: Integrating customer focus across the firm*, 8th ed. New York: McGraw-Hill.

[13] Jou, R. C., & Day, Y. J. (2021). Application of revised importance-performance analysis to investigate critical service quality of hotel online booking. *Sustainability*, *13*(4). doi: 10.3390/su13042043.

[14] D. Wilson, A., Zeithaml, V., Bitner, M. J., & Gremler, D. (2021). *EBK: Services Marketing: Integrating Customer Service across the Firm*, 4th ed. New York: McGraw-Hill Education.

[15] Asgeirsson, M. H., Gudlaugsson, T., & Jóhannesson, G. T. (2024). The relationships between service quality, reputation, and performance in hospitality. *Tour. Hosp.*, *5*(3), 736–752. doi: 10.3390/tourhosp5030043.

[16] Shah, S. N. U., Kakakhel, S. J., & Baloch, Q. B. (2018). Role of service quality and customer satisfaction in firm's performance: Evidence from Pakistan Hotel Industry. *Pakistan J. Commer. Soc. Sci.*, *12*(1), 167–182.

[17] Phi, H. D., & Huong, D. P. (2023). Effect of service quality on customer loyalty: The mediation of customer satisfaction, and corporate reputation in the banking industry. *Eurasian J. Bus. Manag.*, *11*(1), 1–16. doi: 10.15604/ejbm.2023.11.01.001.

[18] Zulkarnain, D., & Helmi, S. (2023). The influence of service quality, product attributes, and company reputation that affect consumer buying interest. *J. Ekon.*, *12*(04), 171–181.

[19] Mahdi, R. A. A., Abdullah, H. A., & Ali, M. H. (2024). Banking service quality and its impact on commercial banks' reputation: An applied study at the University of Kerbala. *Tech. Bus. Manag.*, *8*, 85–95.

[20] Subekti, S., Mulyani, S. R., & Syahidin, R. (2024). Analysis of the quality of laboratory services on patient satisfaction and implications on hospital reputation. *J. Sci.*, *13*(01), 1201–1209.

[21] Wepener, M., & Boshoff, C. (2015). An instrument to measure the customer-based corporate reputation of large service organizations. *J. Serv. Mark.*, *29*(3), 163–172.

[22] Mukherjee, T., & Sen, S. S. (2022). Impact of CEO attributes on corporate reputation, financial performance, and corporate sustainable growth: Evidence from India. *Financ. Innov.*, *8*(40), 1–50. doi: 10.1186/s40854-022-00344-7.

[23] Tadelis, S. (2016). Reputation and feedback systems in online platform markets. *Annu. Rev. Econom.*, *8*(1), 321–340. doi: 10.1146/annurev-economics-080315-015325.

[24] Özkan, P., Süer, S., Keser, İ. K., & Kocakoç, İ. D. (2019). The effect of service quality and customer satisfaction on customer loyalty: The mediation of perceived value of services, corporate image, and corporate reputation. *Int. J. Bank Mark.*, 0265–2323.

[25] Shamma, H. (2012). Toward a comprehensive understanding of corporate reputation: Concept, measurement, and implications. *Int. J. Bus. Manag.*, *7*(16), 151–169.

[26] Walsh, G., & Beatty, S. E. (2007). Customer-based corporate reputation of a service firm: scale development and validation. *J. Acad. Mark. Sci.*, *35*, 127–143.

[27] Chunn, R. (2005). Corporate reputation: Meaning and measurement. *Int. J. Manag. Rev.*, *7*(2), 91–109. doi: 10.1111/j.1468-2370.2005.00109.x.

[28] Arisa, I., Siregar, S. D., Aisah, S., & Wahyudi, W. (2024). The effect of e-service quality on customer trust in McD Serang. *Mantik J.*, *8*(1), 74–81.

[29] Aryawan. (2024). Analysis of trust and service quality on patient satisfaction at Barito Selatan Hospital in central Kalimantan. *Int. J. Contemp. Sci. (IJCS)*, *2*(4), 223–242. doi: 10.55927/ijcs.v2i4.8799.

[30] Bunga R. S. (2024). How service quality can influence customer satisfaction through the trust of Traveloka Application

users in Indonesia. In *The 6th International Research Conference on Economics and Business, KnE Social Sciences*, pp. 223–241. doi: 10.18502/kss.v9i4.15071.

[31] Malgorzata, G. (2021). The impact of the interplay between formal and informal institutions on innovation performance: Evidence from CEECS. *Eng. Econ.*, *32*(1), 15–26. doi: 10.5755/j01.ee.32.1.27390.

[32] Panigrahi, S. K., Azizan, N. A., & Khan, M. W. A. (2018). Investigating the empirical relationship between service quality, trust, satisfaction, and intention of customers purchasing life insurance products. *Indian J. Mark.*, *48*(1), 28–46. doi: 10.17010/ijom/2018/v48/i1/120734.

[33] Rita, P., Oliveira, T., & Farisa, A. (2019). The impact of e-service quality and customer satisfaction on customer behavior in online shopping. *Heliyon*, *5*(10), 2690. doi: 10.1016/j.heliyon.2019.e02690.

[34] Mansouri, H., Boroujerdi, S. S., & Md Husin, M. (2022). The influence of sellers' ethical behavior on customer loyalty, satisfaction, and trust. *Spanish J. Mark. - ESIC*, *26*(2), 267–283. doi: 10.1108/sjme-09-2021-0176.

[35] Widyastuti, S., Said, M., Siswono, S., & Firmansyah, D. A. (2019). Customer trust through green corporate image, green marketing strategy, and social responsibility: A case study. *Eur. Res. Stud. J.*, *22*(2), 83–99. doi: 10.35808/ersj/1427.

[36] Han, L., & Han, X. (2023). Improving the service quality of cross-border e-commerce: How to understand online consumer reviews from a cultural differences perspective. *Front. Psychol.*, *14*, 1137318. doi: 10.3389/fpsyg.2023.1137318.

[37] Ginting, Y. M., Chandra, T., Miran, I., & Yusriadi, Y. (2023). Repurchase intention of e-commerce customers in Indonesia: An overview of the effect of e-service quality, e-word of mouth, customer trust, and customer satisfaction mediation. *Int. J. Data Netw. Sci.*, *7*, 329–340. doi: 10.5267/j.ijdns.2022.10.001.

[38] Haron, R., Subar, N. A., & Ibrahim, K. (2020). Service quality of Islamic banks: Satisfaction, loyalty and the mediating role of trust. *Islam. Econ. Stud.*, *28*(1), 3–23. doi: 10.1108/IES-12-2019-0041.

[39] Effendi, L. V., & Radianto, W. E. D. (2022). Does value congruence mediate the relationship between service trust and the use of fintech? *Rev. Manag. Entrep.*, *6*(2), 163–176. doi: 10.37715/rme.v6i2.3156.

[40] Alshurideh, M. T. (2022). Does electronic customer relationship management (E-CRM) affect service quality at private hospitals in Jordan?, *Uncertain Supply Chain Manag.*, *10*, 325–332. doi: 10.5267/j.uscm.2022.1.006.

[41] Kassa, D. G., Xiang, C., Hu, S., Hayat, K., Andegiorgish, A. K., Fang, Y., & Yang, C. (2021). Patients' perception of the outpatient pharmaceutical service quality in hospital pharmacies with auditable pharmaceutical transactions and services in Ethiopia: A cross-sectional study. *BMJ Open*, *11*(5), 42853. doi: 10.1136/bmjopen-2020-042853.

[42] Venkatakrishnan, J., Alagiriswamy, R., & Parayitam, S. (2023). Web design and trust as moderators in the relationship between e-service quality, customer satisfaction, and customer loyalty. *TQM J.*, *35*(8), 1–31. doi: 10.1108/tqm-10-2022-0298.

[43] Kalia, P., Kaushal, R., Singla, M., & Parkash, J. (2021). Determining the role of service quality, trust and commitment to customer loyalty for telecom service users: A PLS-SEM approach. *TQM J.*, *33*(7), 377–396. doi: 10.1108/tqm-04-2021-0108.

[44] Zhanna, V., Lazorko, O., & Malimon, L. (2021). The mode of trust and experience of stress in customs officers in Ukraine. *Postmod. Openings*, *12*(3), 404–425. doi: 10.18662/po/12.3/346.

[45] Putri, S. F. A., & Marlena, N. (2021). The effect of e-service quality and e-trust on consumer satisfaction. *Econ. Forum*, *23*(3), 463–474.

[46] Mueller, M., D'Addario, M., Egger, M., Cevallos, M., Dekkers, O., Mugglin, C., & Scott, P. (2018). Methods to systematically review and meta-analysis observational studies: A systematic scoping review of recommendations. *BMC Med. Res. Methodol.*, *18*(44), 1–18. doi: 10.1186/s12874-018-0495-9.

[47] Higgins, J. P. T., et al. (2019). Synthesizing quantitative evidence in systematic reviews of complex health interventions'. *BMJ Glob. Heal.*, *4*, 1–15. doi: 10.1136/bmjgh-2018-000858.

[48] Malički, M., Jerončić, A., Aalbersberg, I. J., Bouter, L., & Riet, G. (2021). Systematic review and meta-analyses of studies analyzing instructions to authors from 1987 to 2017. *Nat. Commun.*, *12*(1), 5840. doi: 10.1038/s41467-021-26027-y.

[49] Bakrie, M., Sujanto, B., & Rugaiyah, R. (2019). The influence of service quality, institutional reputation, and students' satisfaction on students' loyalty in higher education institutions. *Int. J. Educ. Vocat. Stud.*, *1*(5), 379–391. doi: 10.29103/ijevs.v1i5.1615.

[50] Hadi, D. P., & Indradewa, R. (2019). The service quality affects corporate reputation, customer satisfaction, and loyalty. *J. Multidiscip. Acad.*, *3*(3), 51–56.

[51] Khoi, B. H., Dai, D. N., Lam, N. H., & Chuong, N. (2019). The relationship among education service quality, university reputation, and behavioral intention in Vietnam. *Beyond Tradit. Probabilistic Methods Econ. 2*, *11*(5), 273–281. doi: 10.1007/978-3-030-04200-4_21.

[52] Supanun, K., & Sornsaruht, P. (2019). How service quality, guest trust, and guest satisfaction affect a five-star hotel's reputation in Thailand. *African J. Hosp. Tour. Leis.*, *8*(5), 1–14.

[53] Widyastuti, S., Prasetyo, H., & Gustari, I. (2017). An investigation on the purchase intention: Developing dealer's reputation through customer trust and service quality. *Eur. J. Soc. Sci.*, *54*(3), 316–331.

[54] Arifin, M. Z., & Imanullah, R. (2021). The influence of service quality on company reputation (study on J & T delivery service users at Unwaha Jombang. *INCOME Innov. Econ. Manag.*, *1*(1), 22–25.

[55] Park, A. J. (2019). A study on the structural relationship between quality of medical service, perceived risk, reputation, and customer satisfaction in small and medium hospitals. *J. Ind. Distrib. Bus.*, *10*(4), 67–76. doi: 10.13106/ijidb.2019.vol10.no4.67.

[56] Virappan, T. M., & Chan, T.-J. (2020). Predictors of corporate reputation through service quality: A study of air Asia customers perspectives. *J. Bus. Soc. Rev. Emerg. Econ.*, *6*(2), 789–798. doi: 10.26710/jbsee.v6i2.947.

[57] Hartanto, B., & Ramadhani, T. (2023). Influence of service quality servicescape and word of mouth on the decision to use expedition services JNE city tebing high with trust as an intervening variable. *Int. J. Econ.*, *2*(1), 10–19. doi: 10.55299/ijec.v2i1.370.

[58] Uzir, M. U. H., et al. (2021). The effects of service quality, perceived value and trust in-home delivery service personnel on customer satisfaction: Evidence from a developing country. *J. Retail. Consum. Serv.*, *63*, 102721. doi: 10.1016/j.jretconser.2021.102721.

[59] Nurzali, E., & Sidharta, H. (2023). The effect of service quality on customer trust with customer satisfaction as intervening variables in home care health Service Rawat di Rumah. Com. *Rev. Manag. Entrep.*, *07*(02), 167–186.

[60] Saputra, M. A., & Rahmawaty, A. (2023). The role of Islamic service quality on intention to use Indonesian Islamic Bank: Trust as an intervening variable. *MALIA J. Islam. Bank. Financ.*, *7*(1), 13–28.

[61] Boonlertvanich, K. (2019). Service quality, satisfaction, trust, and loyalty: the moderating role of main-bank and wealth status. *Int. J. Bank Mark.*, *37*(1), 278–302. doi: 10.1108/ijbm-02-2018-0021.

[62] Zuniar and Sulaiman. (2022). The effect of service quality on trust and its impact on willingness to pay tax at finance and asset management of Banda Aceh City. *Int. J. Acad. Res. Bus. Soc. Sci.*, *12*(3), 250–264. doi: 10.6007/IJARBSS/v12-i3/12944.

[63] Cohen, R. D., Woseth, D. M., Thisted, R. A., & Hanauer, S. B. (2000). A meta-analysis and overview of the literature on treatment options for left-sided ulcerative colitis and ulcerative proctitis'. *Am. J. Gastroenterol.*, *95*(5), 1263–. doi: 10.1111/j.1572-0241.2000.01940.x.

[64] Nair, A. S. (2019). Publication bias - Importance of studies with negative results. *Indian J. Anaesth.*, *63*(6), 505–507. doi: 10.4103/ija.IJA_142_19.

82 Analysis of the implementation of artificial neural networks in the identification and assessment of chronic kidney disease

R. Suhasini[1,a], N. Sandeep Kumar[2,b], V. Ramesh[3,c], Bodla Kishor[1,d], A. Vaishnavi Dali[4,e], and Rasagnya Reddy Avala[5,f]

[1]Assistant Professor, Department of CSC, CMR College of Engineering and Technology, Hyderabad, Telangana, India
[2]Assistant Professor, Department of CSE(AI&ML), CMR Technical Campus, Hyderabad, Telangana, India
[3]Assistant Professor, Department of CSE(AI&ML), CMR Institute of Technology, Hyderabad, Telangana, India
[4]Assistant Professor, Department of Information Technology, Sona College of Technology, Salem, Tamil Nadu, India
[5]Data Analyst 3, Nordstrom Inc, Seattle, Washington, USA

Abstract: Chronic kidney disease (CKD) is one of the top 20 causes of death worldwide, affecting around 10% of individuals. Normal kidney function is compromised by a condition called chronic kidney disease (CKD). As the number of people with CKD rises, effective prediction algorithms for the early detection of CKD are required. Since disease classification and detection aid pathologists' and doctors' judgments during the early stages of illness diagnosis, it is a crucial and challenging problem. It can damage your kidneys and reduce their ability to filter waste materials out of your blood, what keeps you healthy, if it fails to be treated in its early stages. If renal disease worsens, waste products can build up in the blood to harmful amounts and result in illness. Thus, by using this suggested strategy, renal disease can be predicted early. As a result, it can both lower the severity of the disease and prolong a person's life. As a result, using artificial neural networks to diagnose and evaluate chronic kidney disease produces better results.

Keywords: Prediction, artificial neural network (ANN), diagnosis, chronic kidney disease (CKD), DR, LR

1. Introduction

The kidneys are one kind of organ that is located close to the lower back of the abdomen. The purpose of the kidneys is to filter blood by eliminating toxins from the body through bladder urine [1]. Renal failure can be fatal if the kidneys lack the capacity to remove waste that has been tainted by poisons. Kidney blockages come in two varieties: acute and chronic. Chronic renal disorders are conditions that affect the kidneys and reduce their ability to maintain our health. Garbage can build up in the blood to maximum levels as renal disease progresses, It may result in problems like nerve damage, anemia, weak bones, high blood pressure, as well as poor nutrition. In addition to vascular diseases, kidney illness raises the risk of cardiovascular disease.Numerous illnesses, such as diabetes, hypertension, heart disease, lupus, anemia, bacteria within albumin in the urine, side effects from medications, low blood pressure, low salt as well as potassium levels, along with genealogy, can aggravate chronic kidney disease (CKD). Early diagnosis and therapy can usually stop the progression of chronic renal disease. Dialysis could be required to treat renal failure if it gets worse [2]. As a result, kidney transplantation can be required to sustain life. Kidney disease (CKD) is a health concern for all people. Around 10% of persons globally have chronic kidney disease (CKD). However, there is a dearth of hard evidence about the automated and systematic diagnosis of chronic kidney disease [3]. Globally, CKD is a leading cause of death as well as disability. Approximately one million individuals die in 2013 [4], which ranks as the 18th main cause in 2010 [1] after ranking 27th in 1990 [4]. In spite of this, people in developing countries are still impacted by CKD. In high-income nations, there were 109.9 million instances of chronic kidney disease (CKD) in 2015, affecting 48.3 million men as well as 61.7 million women, accordance to a well-organized analysis. There were 387.5 million instances in low-income countries; men made up 177.4 million of the total and women made up 210.1 million [5].

The majority of the time, CKD is symptomless in its early stages. This is due to the fact that most people are able

[a]rsuhasini@cmrcet.ac.in, [b]cmrtc.paper@gmail.com, [c]rameshvoruganti3@gmail.com, [d]kishore.b@cmrec.ac.in, [e]vaishnavi.it@sonatech.ac.in, [f]ra83sagnya.avala@gmail.com

DOI: 10.1201/9781003684459-82

to adjust to a notable reduction in renal function. Kidney illness is often undetected until a routine test for a different illness, such as a blood or urine test, discovers a potential issue. Regular testing, along with early medication treatment and identification, may help prevent it from progressing to a serious state [6]. If kidney illness is not identified in a timely manner or if treatment does not improve the condition, it may show multiple symptoms. renal failure is the result of chronic renal disease. It is sometimes referred to as end-stage renal disease or established renal failure. Dialysis or a kidney transplant is likely to be required in the end. Schedule a visit with your medical professional if you see any of the warning signs of renal illness. Treatment for renal disease may start early symptoms of kidney failure appears if it is discovered enough. Your doctor might perform blood pressure as well as kidney function tests throughout office visits if you're suffering from a health issue that increases the likelihood of renal disease.

Consult your physician to see if your patient requires these tests [7]. The goal of an artificial neural network, a kind of artificial intelligence, is to mimic the neuronal network that underpins the human brain, enabling computers to think and act like people when making decisions. Artificial neural networks are created by programming computers to function like networked neurons [8]. There are about a billion billion neurons in the human brain. Every neuron has one or more connection points; the number may vary from 1,000 through 100,000. The human brain stores information in a distributed manner that enables us to access many pieces of data from memory at various times [9]. It can state that incredibly potent parallel processing units make up the human brain. It can understand an artificial neural network by using an example of a digital logic gate that takes an input as well as an output. Two inputs can be fed into a 'OR' gate. Any or both of the inputs that are 'On' will result in an output of 'On.' If both inputs are 'Off,' we get an output that is 'Off.' This is an example where input dictates output. The brain does not perform the same function on every task [10]. The neurons in our brain are 'learning,' which means that the relationship between inputs and outputs is ever-changing.

2. Literature Review

F. Iqbal, A. S. Pallewatte, J. P. Wansapura, et al. [11] developed computer software to evaluate texture on ultrasound kidney images and extract unique features that can be used to distinguish between kidney patients who are well and those who are ill. A user interface was created for the computer software that was written using MATLAB in order to carry out mathematical operations such as Fourier analysis to obtain Root Mean Square in addition to First Moment values along with Grey Level Cooccurrence Matrix (GLCM) to extract Homogeneity along with Sum Average values. The sample consisted of ultrasound images from thirty-two patients. A separate area of interest (ROI) selection was made for the total kidney, the cortical region, including the white (renal medulla or renal sinus) region. Significant results were obtained for the differentiation of healthy and ill kidneys based on Root Mean Square values over the entire kidney ($p = 0.03$) and the cortical region ($p = 0.0049$).

As mentioned by K. S. Gill and R. Gupta et al. [12], we must import the required libraries and load the dataset in order to run GridSearchCV with 10-fold cross-validation. The second stage involves splitting the dataset into characteristics as well as tags in order to get it ready for modeling. We created a pipeline with preprocessing stages and machine learning algorithms after dividing the data into sets for both training and testing. Use 10-fold cross-validation to fit the GridSearchCV object to the training data after deciding which hyperparameters to search across. Lastly, we forecast renal disease on the test set using the best estimator that GridSearchCV had discovered, and we employed metrics such reliability, precision, as well as recall to evaluate the effectiveness of the model. In order to achieve early CDK identification, R. Al-Momani, G. Al-Mustafa, R. Zeidan, H. Alquran, W. A. Mustafa, and A. Alkhayyat, et al. [13] use machine learning algorithms such k-Nearest Neighbor and Artificial Neural Network in addition to Support Vector Machine. The detection of these typically fatal disorders highlights the importance of AI. This study looks at 400 samples and 13 attributes in the data set. They were evaluated by applying the three categorization techniques to the data. The results showed that, at 99.2%, the ANN classifier has the highest precision. Due to the increasing number of patients at end-stage renal disease with a poor prognosis for mortality and morbidity and a high risk of renal disease progression, R. S. Vellela, L. R. Vuyyuru, K. B. S. K, N. Malleswara Rao Purimetla, L. Dalavai, and M. V. Rao, et al. [14] constitute a significant burden. This disease has many non-visible symptoms, and it claims a significant number of lives each year. Renal disease must therefore be taken seriously from an early age. Consequently, it is feasible to forecast the disease in the current study using machine learning classifiers. They are Random Forest, Logistic Regression, and KNN [15, 16]. Our objective is to differentiate between various machine learning algorithms according to how correctly they perform. In terms of memory, accuracy, and F1-score, this approach can yield reliable, valuable results.

3. Proposed Methodology

The foundation for the application of artificial neural networks in this section's diagnosis and evaluation of chronic renal disease is shown in Figure 82.1.

3.1. Pre-process

The dataset was acquired using the Kaggle website, where it was said that all of the data was collected from Indian hospitals. Blood pressure, puss cells, red blood cells, albumin,

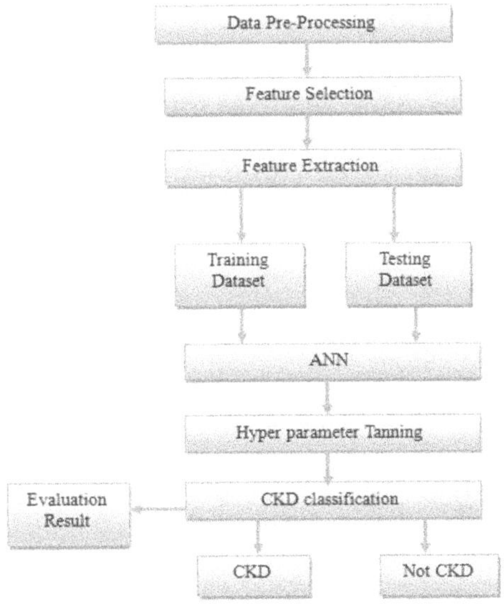

Figure 82.1. Framework of an detection and evaluation of chronic kidney disease by using artificial neural network.

Source: Author.

as well as blood sugar are among the dataset's characteristics. The dataset consists of 400 rows and 26 columns. The id column, which serves as the dataset's unique identification, is removed. All records or rows that contain a null value are removed. Every categorical parameter in the dataset is replaced with the binary values 0 or 1. The dataset's visualization shows that older people and those with high blood pressure have a higher risk of developing chronic kidney disease [17] (CKD).

3.2. *Feature selection*

Whatever attributes are most important is determined by using a second tree classifier. It selects the divisions randomly instead than using the random forest, but the two are very similar. In the additional tree classifier, several de-correlated decision trees are constructed. They are made by arbitrarily selecting a set of traits. During the forest's construction, a normalized mathematical reduction of features is calculated for each feature. Each characteristic's value is ascertained by calculating the index for that specific feature if the Gini index is utilized. A feature's greater significance is determined by its gini index. The relative significance for each of the 18 qualities is established based on the selected threshold.

A knowledge gathering is called a data set. When it comes to graphical information, a data set is similar to multiple tables within a database, whereby every row corresponds to an individual record among the information set while every column indicates a separate element. The procedure of finding and fixing (or eliminating) erroneous or absent entries given a dataset is known as information preprocessing, planning, or sterilization. It involves identifying the parts of the information that are incomplete, erroneous, or superfluous, and then changing, substituting, or removing the incomplete or contaminated data. An essential phase in the information mining procedure is preparing the information. It describes the steps taken to prepare data for examination by cleaning, transforming, and integrating it. Increasing the data's distinctiveness and tailoring it to the particular data mining [18] task are the objectives for information preparation. The procedure of choosing the subset of appropriate characteristics to apply while creating a framework is known as choosing features. Two applications of picking features are in stylistic analysis as well as DNA microarray investigation. It's critical to separate it from feature extraction. The procedure of choosing the most trustworthy, pertinent, as well as non-redundant characteristics to employ when developing a model is known as feature selection. As the amount along with diversity of datasets increase, it is essential to progressively reduce their amount of data. Enhancing the forecasting accuracy of the model and reducing the modeling's computational expense are the primary objectives of choosing characteristics.

Image files are commonly used for characteristic retrieval, who converts unprocessed information into numerical attributes than may be employed using machine learning techniques. By ascertaining an object's shape or an image's redness value, data scientists may produce fresh characteristics which are appropriate for machine learning programs. The volume of duplication of information utilized during the data collection process is decreased with the adoption of extraction of characteristics. Ultimately, decreasing information speeds up both the training as well as generalization phases of the method for machine learning and aids in the development of a framework with less processing time.

Training data is the information used in machine learning to develop an algorithm or model. Human analysis or processing of training data is required before machine learning can be used. In the end, the test data set functions as an objective evaluation of the ultimate model fit produced with the original training data set. In the event that any of the data in the test data set has not been used in training (for example, in crossvalidation), it is frequently referred to as a holdout data set. Artificial neural network (ANN) models imitate the operations of the human brain through mathematical computations and calculations. Artificial intelligence research encompasses recent advances in robotics, image as well as speech recognition, and ANN application. When training learning models, each dataset and model needs its own set of hyperparameters, or variables. These are only discovered through a great deal of experimentation, where you choose a collection of hyperparameters and run your model using them. This is called altering the hyperparameters.

The data undergoes pre-processing once the dataset is accessed. It removes unnecessary information. After pre-processing, feature selection is applied to the data. Feature extraction is utilized after the characteristics have been selected. After feature extraction, training and testing are conducted on the data. Finally, ANN is applied to the data.

After ANN, the data is hyper-parameter adjusted, and the data is classified using CKD classification. As a result, the result is evaluated as CKD rather than CKD.

4. Methodology

The method used by the framework is explained in the paragraph that follows. The procedure of conducting research begins with the public compilation of internet data sets. The data used in the study came from the UCI repository. The data collection process includes twenty-five features in all, involving the chronic diagnostic class. Following a successful data collection process shown the Figure 82.2, the data is fed into a Python script for exploratory analysis before being fed into machine learning to diagnose chronic kidney disease.

Pandas performs feature selection, pre-processing, and other pre-processing tasks after reading the comma-separated information as a data frame. During preparation, data of various data types are organized as integer type information, and both null and duplicated numbers are verified. The values absent from the data are substituted by the property mean.

Information discovery is a crucial data mining technique that entails several processing steps. Preprocessing data that was originally gathered from various sources makes using datamining techniques easier. Preprocessing is the procedure of sanitation, extracting, and converting data into the appropriate formats. It is sometimes referred to as data preparation. A bigger feature set is used to identify the essential elements of knowledge representation. In order to facilitate knowledge discovery, the following phase entails applying several techniques for pattern assessment or classification. The creation of a datamining method for finding data in the CKD dataset is the aim of the study. A large number of CKD datasets are being gathered. The standard data mining processes are followed for preprocessing and analysis preparation. To predict the early onset of CKD, three machine learning techniques are used.

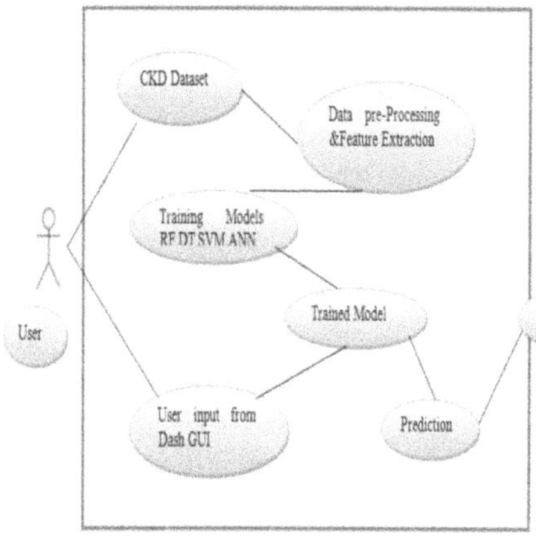

Figure 82.2. Use case diagram of proposed system.

Source: Author.

5. Evaluation Metrics

Efficiency is the number of correctly predicted outcomes from all the data points. It fits the following description: It is calculated by dividing the entire amount of assumptions by the aggregate of all correct projections.

$$Accuracy = \frac{TP + TN}{TP + TN + FP + FN} \tag{1}$$

Precision: The term is the proportion of relevant cases discovered in the retrieved cases. The correlation number, that spans from the totally classified faultprone portions to the correctly identified modules, is employed to convey the outcome;

$$Precision = \frac{TP}{TP + FP} \tag{2}$$

Recall: The percentage of correctly anticipated positive outcomes over all conceivable positive outcomes is known as recall;

$$Recall = \frac{TP}{TP + FN} \tag{3}$$

6. Result Analysis

Initially, information is obtained on the individual's age, blood pressure, specific gravity, red blood cell count, hypertension, albumin level, as well as other input data. After that, we determine the possibility that the individual has chronic renal disease from the collected data by using an accuracy comparison model. An online user interface that lets users input data and predict chronic kidney disease.

This section looks at how well artificial neural networks operate in the diagnosis and evaluation of chronic renal illness.

Figure 82.3, which displays the model's performance evaluation, demonstrates that the framework was effective in predicting kidney patients' 11 records into both CKD and non-CKD with an overall prediction accuracy of 98% when the data was unknown.

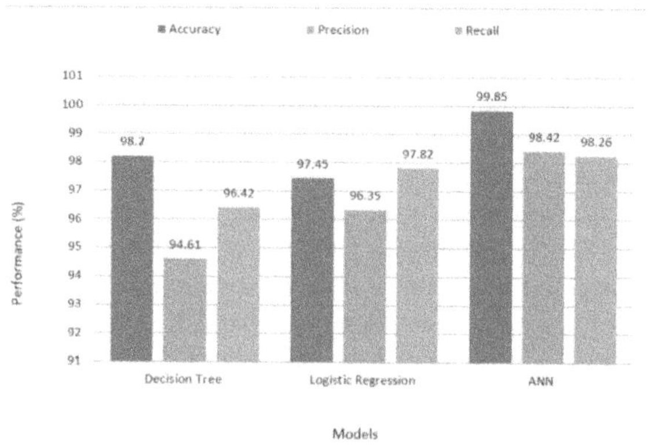

Figure 82.3. Performance analysis of various models.

Source: Author.

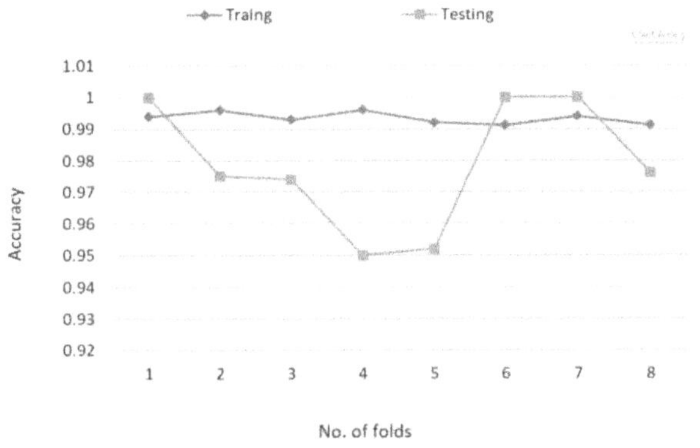

Figure 82.4. Accuracy performance of training and testing.

Source: Author.

A comparison graph of decision trees, logistic regression, and artificial neural networks' accuracy is displayed in Figure 82.3. The accuracy of the ANN is really good in contrast to DT and LR. A precision comparative graph contrasting decision trees, logistic regression, and artificial neural networks is displayed in Figure 82.3. The accuracy of the ANN is superior to that of DT and LR. A recall comparison graph contrasting decision trees, logistic regression, and artificial neural networks is displayed in Figure 82.3. The recall rate of the ANN is higher than that of DT and LR. The training and testing scores obtained for each fold are displayed in Figure 82.4.

7. Conclusion

Numerous people are afflicted with the illness or pass away from it as a result of incorrect early detection and ignorance about the condition. Thus, it is believed that detecting chronic kidney disease early on can help slow down the disease's course. As a result, sickness is detected using this method.

This led to the development of an effective neural-network-based method for the identification of chronic kidney disease (CKD). Instead, doctors might decide to use this incredibly precise prediction tool. It can also be used by average people to determine their risk of developing chronic kidney disease. This research project uses a neural network to provide a cost-effective and effective approach to the CKD-detection challenge. As a result, we believe that our multiclassification research was essential in identifying the disease's stages and in advising patients on the courses of action that will prolong their lives. The proposed model would aid specialists in making prompt choices in the future. The ideal solution would be to develop a mobile platform that would enable professionals to keep an eye on patients' conditions and help patients use the system to learn more about it.

References

[1] Bhatt, M., & Kasbe, T. (2019, December). A survey on chronic kidney disease diagnosis using fuzzy logic. In *2019 IEEE International Symposium on Smart Electronic Systems (iSES) (Formerly iNiS)* (pp. 252–256). IEEE.

[2] Rahman, T. M., Siddiqua, S., Rabby, S. E., Hasan, N., & Imam, M. H. (2019, January). Early detection of kidney disease using ECG signals through machine learning based modelling. In *2019 International Conference on Robotics, Electrical and Signal Processing Techniques (ICREST)* (pp. 319–323). IEEE.

[3] Gill, K. S., Anand, V., Chauhan, R., Singh, P., & Gupta, R. (2023, November). Detection of Kidney Disease in X-Ray Images Using Machine Learning on MobileNetV3 CNN Model. In *2023 2nd International Conference on Futuristic Technologies (INCOFT)* (pp. 1–4). IEEE.

[4] Chiu, R. K., Chen, R. Y., Wang, S. A., & Jian, S. J. (2012, July). Intelligent systems on the cloud for the early detection of chronic kidney disease. In *2012 international conference on machine learning and cybernetics* (Vol. 5, pp. 1737–1742). IEEE.

[5] Chen, G., Ding, C., Li, Y., Hu, X., Li, X., Ren, L., ... & Xue, W. (2020). Prediction of chronic kidney disease using adaptive hybridized deep convolutional neural network on the internet of medical things platform. *IEEE Access, 8,* 100497–100508.

[6] Vyas, N., Sharma, V., & Balla, D. (2023, September). Chronic kidney disease prediction using robust approach in machine learning. In *2023 3rd International Conference on Innovative Sustainable Computational Technologies (CISCT)* (pp. 1–5). IEEE.

[7] Pujari, R. M., & Hajare, V. D. (2014, August). Analysis of ultrasound images for identification of Chronic Kidney Disease stages. In *2014 First International Conference on Networks & Soft Computing (ICNSC2014)* (pp. 380–383). IEEE.

[8] Maurya, A., Wable, R., Shinde, R., John, S., Jadhav, R., & Dakshayani, R. (2019, January). Chronic kidney disease prediction and recommendation of suitable diet plan by using machine learning. In *2019 International Conference on Nascent Technologies in Engineering (ICNTE)* (pp. 1–4). IEEE.

[9] KG, A. D., Saikrishnan, S., Sudhakar, A. J., & Kaviyarasan, K. (2023, October). A Comprehensive Web Application for Chronic Kidney Disease Prediction with Cuisine-Centric Diet Recommendation. In *2023 International Conference on Self Sustainable Artificial Intelligence Systems (ICSSAS)* (pp. 891–896). IEEE.

[10] Imran, S. M., & Banu, S. S. (2022, August). HCB: A Hybrid CatBoost Enabled Chronic Kidney Disease Detection System. In *2022 3rd International Conference on Electronics and Sustainable Communication Systems (ICESC)* (pp. 1355–1360). IEEE.

[11] Iqbal, F., Pallewatte, A. S., & Wansapura, J. P. (2017, September). Texture analysis of ultrasound images of chronic kidney disease. In *2017 Seventeenth International Conference on Advances in ICT for Emerging Regions (ICTer)* (pp. 1–5). IEEE.

[12] Gill, K. S., & Gupta, R. (2023, May). Chronic kidney disease detection using GridSearchCV cross validation method. In *2023 international conference on recent advances in electrical, electronics & digital healthcare technologies (REEDCON)* (pp. 318–322). IEEE.

[13] Al-Momani, R., Al-Mustafa, G., Zeidan, R., Alquran, H., Mustafa, W. A., & Alkhayyat, A. (2022, May). Chronic

kidney disease detection using machine learning technique. In *2022 5th International Conference on Engineering Technology and its Applications (IICETA)* (pp. 153–158). IEEE.

[14] Vellela, S. S., Vuyyuru, L. R., MalleswaraRaoPurimetla, N., Dalavai, L., & Rao, M. V. (2023, September). A Novel Approach to Optimize Prediction Method for Chronic Kidney Disease with the Help of Machine Learning Algorithm. In *2023 6th International Conference on Contemporary Computing and Informatics (IC3I)* (Vol. 6, pp. 1677–1681). IEEE.

[15] Saravanan, P., Pandey, A., Joshi, K., Rondon, R., Narasimharao, J., & Imran, A. A. (2023, May). Using machine learning principles, the classification method for face spoof detection in artificial neural networks. In *2023 3rd International Conference on Advance Computing and Innovative Technologies in Engineering (ICACITE)* (pp. 2784–2788). IEEE.

[16] Arul, U., Prasath, A. A., Mishra, S., & Shirisha, J. (2022, December). IoT and Machine Learning Technology based Smart Shopping System. In *2022 International Conference on Power, Energy, Control and Transmission Systems (ICPECTS)* (pp. 1–3). IEEE.

[17] Srithar, S., Vetrimani, E., Reddy, K. P., Kodati, S., &Alagumuthukrishnan, S. (2024). Artificial Hyperintelligence-Enabled Cyber-Physical System Control for Autonomous Vehicles. *Smart Grids as Cyber Physical Systems, 2 Volume Set*, 145.

[18] Sivaiah, B., & Rao, R. R. (2022, July). A Survey on Fast and Scalable Incremental Frequent Item Set Methods for Big Data. In *2022 International Conference on Intelligent Controller and Computing for Smart Power (ICICCSP)* (pp. 1–5). IEEE.

83 A CNN's improved bio-inspired fossil image segmentation and identification for palentology

D. Sudha[1,a], *B.Thanuja*[2,b], *N. Samathagoud*[3,c], *Katragadda Anuhya*[4,d], *and R. S. Sabeenian*[5,e]

[1]Associate Professor, Department of ECE, CMR College of Engineering and Technology, Hyderabad, Telangana, India
[2]Professor, Department of ECE, CMR Technical Campus, Hyderabad, Telangana, India
[3]Assistant Professor, Department of ECE, CMR Institute of Technology, Hyderabad, Telangana, India
[4]Assistant Professor, Department of CSE, CMR Engineering College, Hyderabad, Telangana, India
[5]Professor and Head, Department of Electronics and Communication Engineering, Sona College of Technology, Salem, India

Abstract: A critical step in paleontology, fossil picture segmentation helps to isolate and analyze fossilized structures from complex backgrounds. Fossils embedded in sedimentary matrices can have complex patterns, uneven forms, and difficult-to-handle textures for traditional segmentation approaches. The applications of image segmentation in this domain prove to be a handy tool for the paleontologist out there. It can be used to count the number of bones in a sample as well as to get geometrical information about the bone components. This research illustrates the application of several bio-inspired or metaheuristic image processing methods to some dinosaur fossil images. The contrast is modified by a histogram equalization based on Evolution Strategy. To improve segmentation accuracy in fossil images dataset, a novel bio-inspired architecture integrating CNNs, artificial bee colony, and ant colony optimization is proposed in this paper. The results demonstrate that bio inspired techniques outperform traditional algorithms in segmentation accuracy, precision, and recall. This work enables large-scale processing of fossil datasets, thus accelerating paleontological analysis and automating fossil identification. Optimization/simulated annealing segment produces the final segmented image for analysis. The information gained from the final results achieved is more appreciable than conducted using traditional segmentation, yielding comparatively cleaner segmented images.

Keywords: CNN, PSO, ACO, ABC, palentology, fossil image segmentation

1. Introduction

Image processing – A set of techniques to process and/or extract interesting information from an image. It is one of the most rapidly growing technology. Some of the key activities done within this domain are: image watermarking image segmentation image filtering and image classification. Therefore, these processes need continuously to be worked upon and improved. Purpose: Reassessing artificial intelligence (AI) based image segmentation Big data refers to the vast amounts of data that are available, and AI the field of study seeking to create computers that can imitate human intelligence. In computer science applications [1] three were seen as mandarins for the technology, resolving a bunch of challenges of diverse applications. One of the fundamental AI methods for resolving NP-hard combinatorial optimization problems is bio-inspired algorithms. They are population-based and are activated by animal behaviour. Due to their exploration and exploitation capabilities, bioinspired algorithms search the global minimum and avoid the local one. Bio-inspired algorithms are hybridized with image-segmentation methodology in order to determine the optimum parameters necessary for the segmentation methods such as determining the centers in the C-means-clustering segmentation model or determining the quantity of thresholds in the thresholding segmentation. Features selection is a hard problem that needs different optimal methods such as bio-inspired algorithms.

IntroductionDigital photography has evolved into much more than a popular pastime in the society. Digital cameras and camera phones have made taking, processing, and sharing photos very instant, so digital photographs have become a common part of high-tech communications. With that track of fast-paced quick gratification, one won't even fathom how much that professional industries also depend on digital images. They are also now critical to the publishers of print and online content, who depend on the quality and usability of their photographs. Effective digital imaging is based on precision with regards to the

[a]dsudha@cmrcet.ac.in, [b]cmrtc.paper@gmail.com, [c]nemurisamathagoud1@gmail.com, [d]katragadda.anuhya@cmrec.ac.in, [e]sabeenian@sonatech.ac.in

DOI: 10.1201/9781003684459-83

specifics of how and where an image is created, stored, who will use or access it and how. An image acquisition system has increasingly become a very important part of research projects, industrial production, military, and life.

Word Processing of Images: The term 'real_ time' necessitates the use of DSP [12]. Through its insistence on a large amount of picture processing data, high relativity of data processing, rigid frame and field time constraints, it will be important to give full play to the efficiency of software DSP and CPLD to the maximum extent, to control the rate of data acquisition [1]. Image is often represented as 2-D function f(x,y), where f defines the intensity of (x, y) coordinates in two dimensional plane [2]. Each coordinate location is assigned a pixel. Pixel or picture element is the smallest unit of picture. Digital images are made up of pixels, which is just a tiny dot of color. Such a format is used as a computer image, that is, a bitmap, which is a rectangular array of pixels. From a photographic viewpoint, digital images can be divided into two classes. Color Pictures Information collection on Earth's historical occurrences is essential for obtaining a deeper comprehension of the causes of some species' extinction in order to identify remedies and stop it from happening again. And for millennia, our world has been home to many species, including dinosaurs. Most of the dinosaurs who lived during the Mesozoic are extinct, wiped out by asteroid impacts, natural disasters, and climate change, though some, such as birds, are still hanging on [8–9].

Since age-old times, humans have enjoyed studying history as it enables us to learn about the existence, survival, and extinction of our ancestors as well as other beings throughout the ages. Some of that past has been laid down in the fossils left in nature. Image processing [10] techniques are being applied in this regard to dinosaur fossils. The main motivation behind the implementation of bio-inspired or metaheuristic algorithms in this research is their seeming ability to solve a wide range of image processing tasks [11]. Because these metaheuristic image processing algorithms can easily outperform the traditional methods, they are more favorable for us to use. The overall segmentation accordingly applied strategies of our method and some metaheuristic image processing techniques to get the final segmented part. Thus the proposed process can be applied to paleontologists [8] by enhancing their dinosaur fossils analysis. This study encompasses five main sections. Basic definitions were already explained in Section I Section I Image segmentation is an essential technique for object detection and recognition, and find its applications in different areas from palaeontology [4], industry, healthcare, and security to agriculture and space exploration [5].

2. Related Work

Because obtaining a photo from any source – most frequently a hardware-based source – is a talent that is needed in lots of steps to observe, the image acquisition in image processing is usually defined in wide phrases. As one cannot process without a picture, image capture is generally the first workflow step of image processing [5].

It predetermines the number of segments from threshold values. A well-known method for classification is the region-based application of Gabor filters [6].

This method called the texture segmentation of the image tries to find patterns of the objects according to shapes Moreover, segmenting digital images according to bio-inspired algorithms is extremely common. As a step towards achieving this, Kmeans clustering and genetic algorithms can be used for picture segmentation [7]. Such a study was conducted in 2016 by Stephan Lautenschlager [12] who argued the presentation of different image processing methods for digital restorations of some fossil photographs such as image pre-processing methods, morphological operations [13] and segmentations in three-dimensional environments. Dunmore, Christopher, et al. conducted another major study of paleontological material segmentation in 2018 [14]. To achieve this, they used fuzzy C-means clustering and K-means clustering. In 2021, Hou, Yemao, et al. managed to develop a system based on deep learning from CT data to reconstruct a 3-D model of both fish microfossils [15].

3. Proposed Method

To understand how some species become extinct, find solutions, and ensure it does not happen again, information has to be cataloged about events that happened on Earth. Dinosaurs are just one of many types of animals that have lived on our planet for millions of years. Most of these dinosaurs are no longer on the planet due to asteroid impacts, natural disasters and climate change, although some – birds for example – are still kicking [8–9]. From time processes, humans have been amazed to study history because as the creation of the great we learn how our predecessors lived, how another creature survived and how everything dies with time. Those fossils that still exist in nature are some evidence of that past. In [10], different image processing techniques are applied to angular fossil fossils of dinosaurs. Why bio-inspired or metaheuristic algorithms? This is mainly due to the fact that these algorithms seem to be able to solve different image processing tasks [11]. Using metaheuristic image processing algorithms is more appealing because they could simply provide better results than classical algorithms 2 [4]. We finally get to the segmented image using the main strategies of our method as well as some metaheuristic image processing methods. Paleontologists [8] could adopt the suggested approach to enhance their dinosaur fossil analysis. This study consists of five main sections. Some basics definitions have been discussed in Section I. Image segmentation, a fundamental technique of object detection and recognition, was successfully applied in a wide range of domains such as palaeontology, industry, health science, security, agriculture, space exploration [5].

The proposed image segmentation method consists of four bio-inspired image processing techniques. As we can see, the process for the proposed system is initiated by inputting the image data, which will be passed onto the Evolution Strategy (ES) algorithm for the next step, the histogram equalization phase [3]. In this, the goal histogram is matched to adjust the input image intensity using an evolution strategy algorithm. In the next stage, the image is quantized into a predefined number of threshold levels, which reduces the image size. This step to have a competitive Lead in next closet stages. In this section, we apply the Cultural Algorithm (CA) to fit the threshold levels vector. feature selection method in evolutionary theory choose the image filter on the image processing. After the quantized image has been transmitted, the Simulated Annealing (SA) algorithm conducts the ninth edge identification step. This image, SA need to be segmented after the edges shows. For the image segmentation part, the PSO algorithm, which employs the SA method as the optimizer, is used. Finally, overlays of all evolutionary phases are used to form a more segmented image. Figure 83.1 demonstrates the procedure of the suggested method. The proposed approach with real image data is shown in Figure 83.1.

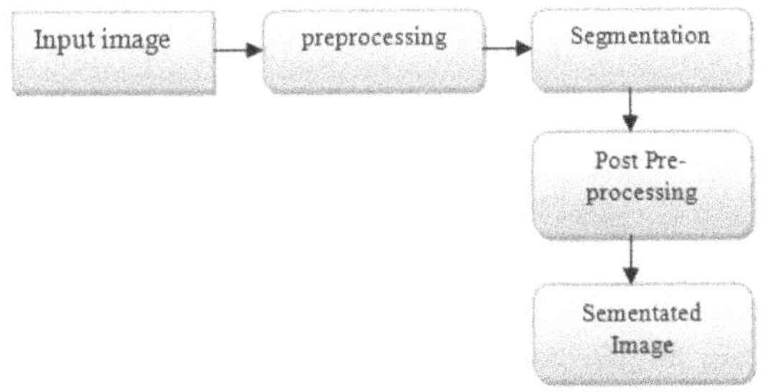

Figure 83.1. Process of proposed method.

Source: Author.

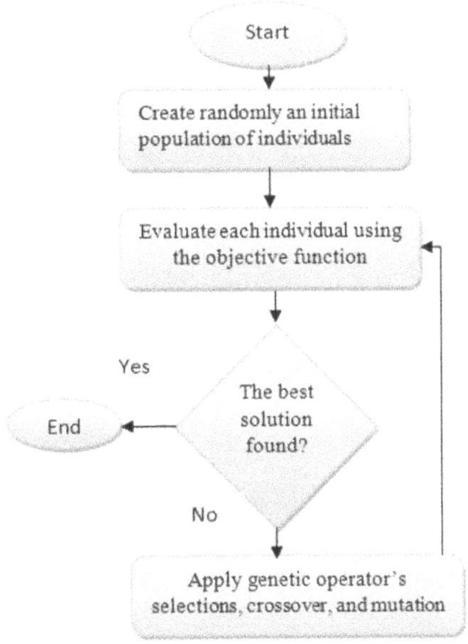

Figure 83.2. Flowchart of proposed method.

Source: Author.

4. Flowchart of GA

Start

Get Image to Train with Evolution Strategy Algorithm

The histogram of the image is equalized to change the intensity. While (number of generations achieved), set the population size N, and number of iterations shown the Figure 83.2. Context: A New Genetic Algorithm to Optimize Attribute and Variance Mutation Fitness Function of Personal Variances and Attributes Recombination

Use Of The Histogram Value (Optimal Target Histogram): People

End

By Using Best target histogram vectors,

ES Output:

Cultural Algorithm-Based Cumulative Strength of Evolution (based on Raw Input Image Training)

Apply Best Selected Threshold Level Vector:

CA Output End: Evolutionary Fitted Threshold Level for Training Input Iamges Using Simulated Annealing Algorithm

5. Objective Function Evaluation

Lower current temperature, generate a new trial solution and go to assessment step if objective function falls otherwise metropolitan criterion is satisfied review step else for updates. Diminish The Heat Step At the Same Time Continue SA edge filter of best selected from decorated Final Results: Particle Swarm Optimization + SA training evolutionary edge calculated images (Input: Input image comprise the border detected element).

Objective: Divide the Input Picture While (number of generations is not achieved) initialize the population size N and number of generations. Based on the previously mentioned PSO exampledefining particles with random position and velocity at initializationParticle fitness and corresponding distance are evaluated per pixel for the objective function of PSO using SA optimizer. To find and update the pbest and gbest of PSO Cut down.

Current SA temperature Update the position and velocity of the PSO; generate a new trial solution; perform SA evaluation step. Show the Best PSO Optimal Solution and the SA Desired Temperature Improve the Best Solution.

PSO+SA End Found for Pixel and Distance

We take the image average of the best cluster for segmentation.

The image segment method consists of four bio-inspired image processing methods. This system accepts input image data and passes data to the Evolution Strategy (ES) to perform the equalization of the histogram step. Here in the evolving strategy algorithm the aim histogram compares, to change the movie image intensification by itself as indicated in Figure 83.3.

In the next step, the image is counted into a number of preset threshold levels that reduce its dimensions. This step was vital so that you have a powerful edge in the next stage.

In this section, the Cultural Algorithm (CA) is utilized to obtain the threshold levels vector. After the quantized image has been sent, as shown in Figure 83.4 the Simulated Annealing (SA) technique is applied for the purpose of edge identification step.

The quantization image feeds the edge identification stage to the SA algorithm. In this case, SA Once the boundaries of the image are uncovered, the time has come to segment it. This process is depicted in Figure 83.5.

Four test subjects are four photographs of dinosaur fossils from different paleontological periods. The proposed method is then compared with K-means, Watershed, and bio inspired genetic segmentation algorithms using the same images and evaluated based on the same parameters on the three performance indicators mentioned above. PSOSA picture segmentation, SA edge detection, CA image quantization, and ES histogram equalization. The bio-inspired value is depicted graphically in the image above. The value of its functions continues to fall as the bioinspired procedure operates. This is more informative in terms of the lower function, while the high function value will show less information visually. So we could learn something about the picture through this method.

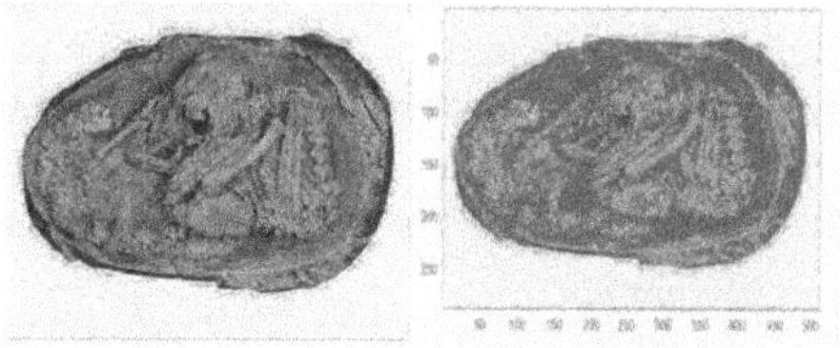

Figure 83.3. (a) Original ES HE (b) Quantization.

Source: Author.

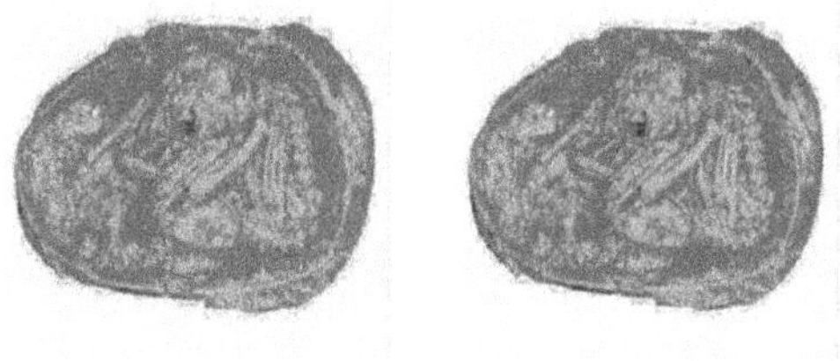

Figure 83.4. (a) CA quantized (b) CA quantized grey.

Source: Author.

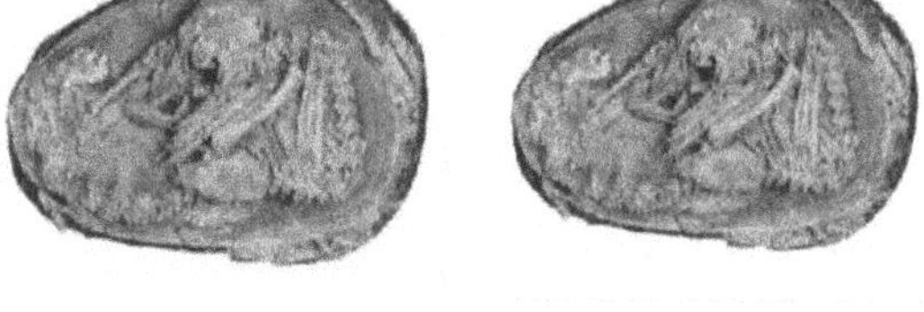

Figure 83.5. (a) Original image (b) ES histogram.

Source: Author.

6. Results and Analysis

We evaluated the proposed bio-inspired methods, such as Convolution Neural Networks (CNNs), Ant Colony Optimization (ACO), Particle Swarm Optimization (PSO), and lastly Artificial Bee Colony (ABC), on fossil image datasets containing diversified types of fossil skeletons resting in a sedimentary background. Here is a summary of the results:

Segmentation Accuracy: The bio-inspired models were superior to the more standard methods of segmentation such as thresholding and edge detection as seen in Figure 83.6.

On all tested datasets, the CNN-based method achieved an average of 92.7% segmentation accuracy. The average accuracy for the ACO and PSO algorithms were 88.3% and 89.5% respectively. Conventional methods reached accuracies in the range of 70 to 75 percent.

7. Precision and Recall Metrics

CNNs showed the highest precision (93.2%), assuring that sediment was rarely incorrectly classified as fossil material. As depicted in Figure 83.7, the ACO method closely followed, with precision values equal to 89.1%.

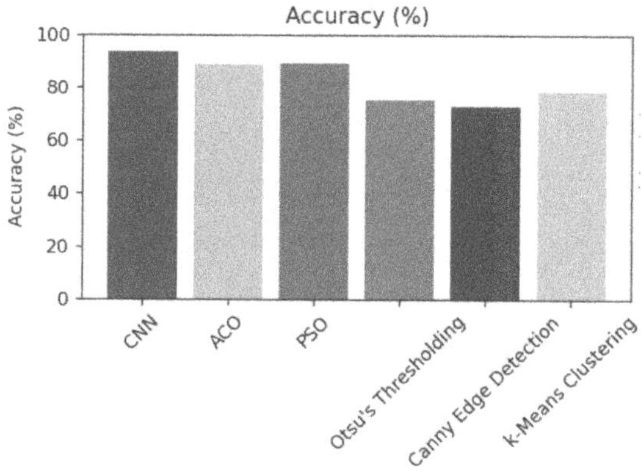

Figure 83.6. Accuracy comparison of various algorithms.

Source: Author.

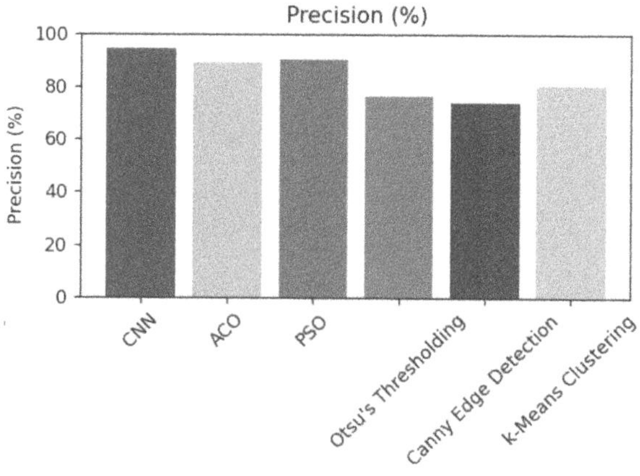

Figure 83.7. Precision comparison of various algorithms.

Source: Author.

For the complex backgrounds, it is worth mentioning that CNNs once again scored best in recall (91.8%) and were even able to locate and segment fossils correctly. The ACO and PSO algorithms scored 87.5% and 88.2%, respectively, as shown in Figures 83.8 and 83.9.

8. Execution Time

Bio-inspired algorithms had a compromise between accuracy and speed of execution. Swarm-based (PSO and ACO) techniques processed images faster (an average of 30–50 s per image) than CNN-based (120 s) shown the Figure 83.10. So less accurate but faster findings with more traditional techniques.

Accuracy and rapid execution were exchanged in bio-inspired algorithms. Methods based on CNNs needed an average of 120 seconds to analyze a profile picture, while methods based on swarm optimization, like PSO and ACO, needed only between 30 and 50 seconds on average for every picture. Although faster, conventional methods produced less accurate results.

Image 1: CNN (blue) segments high texture off fossilized trilobite with 96% IoU, ACO (green) achieves 90%, while conventional average 75%.

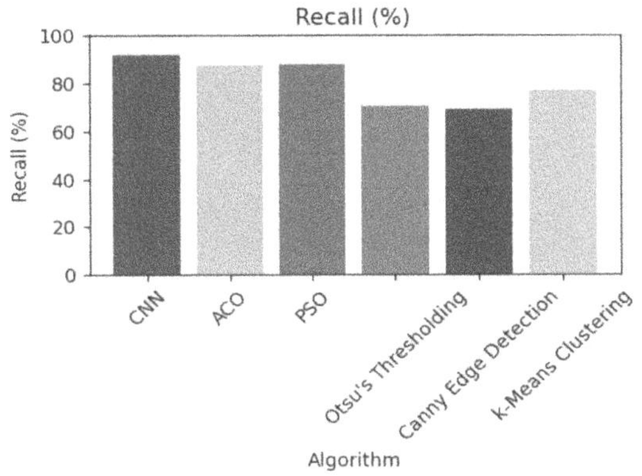

Figure 83.8. Recall comparison of various algorithms.

Source: Author.

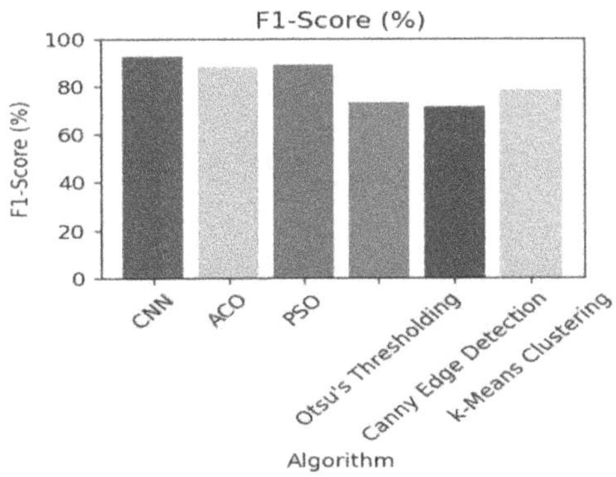

Figure 83.9. Precision comparison of various algorithms.

Source: Author.

CNN and ACO had a consistent performance with IoU > 90%, however stale approaches stuttered about a 5% accuracy loss when the background became noisy with ammonite fossils, see Figure 83.11.

9. Conclusion

This work illustrates the great potential of bio-inspired algorithms to automate and improve fossil segmentation tasks. Methods such as CNNs, ACO, and PSO were used which segmented fossil images from hard-to-represent sedimentary background images. The use of biological and environmental systems as analogs gave rise to creative solutions for complex texture patterns, uneven fossil formations, noise resilience among several others. The noise you hear in the suggestion, we can get rid of it. And more so, it will give an accurate representation compared to the existing structure. In this case, the intrusive features of the image were extracted and useful data was retained. These segmentation process was crucial in those image.

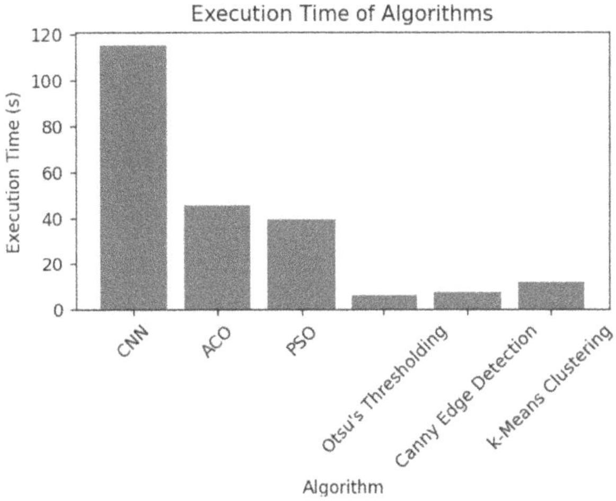

Figure 83.10. Execution time comparison of various algorithms.

Source: Author.

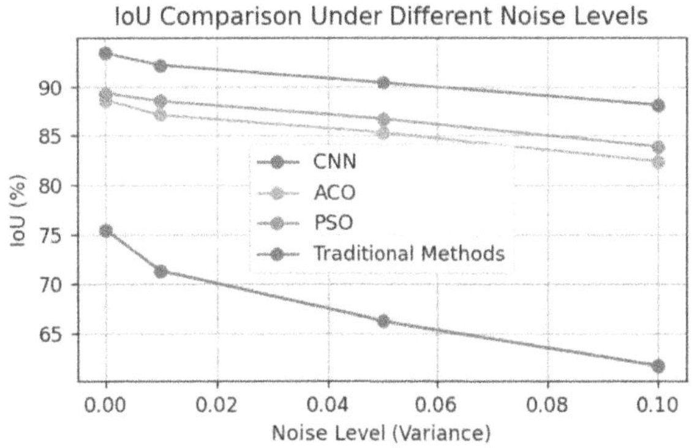

Figure 83.11. Comparison of IoU under different noise levels.

Source: Author.

References

[1] Mousavi, S. M. H., MiriNezhad, S. Y., & Mirmoini, A. (2017, October). A new support vector finder method, based on triangular calculations and K-means clustering. In *2017 9th International Conference on Information and Knowledge Technology (IKT)* (pp. 46–53). IEEE.

[2] Dhanachandra, N., Manglem, K., & Chanu, Y. J. (2015). Image segmentation using K-means clustering algorithm and subtractive clustering algorithm. *Procedia Computer Science, 54*, 764–771.

[3] Jusman, Y., Pusparini, A., Chamim, A. N. N., & Kanafiah, S. N. A. M. (2021, February). Comparison of Malaria Parasite Image Segmentation Algorithm Using Thresholding and Watershed Method. In *Journal of Physics: Conference Series* (Vol. 1783, No. 1, p. 012092). IOP Publishing.

[4] Tan, Z. Y., Basah, S. N., Yazid, H., & Safar, M. J. A. (2021). Performance analysis of Otsu thresholding for sign language segmentation. *Multimedia Tools and Applications, 80*, 21499–21520.

[5] Minaee, S., Boykov, Y., Porikli, F., Plaza, A., Kehtarnavaz, N., & Terzopoulos, D. (2021). Image segmentation using deep learning: A survey. *IEEE Transactions on Pattern Analysis and Machine Intelligence, 44*(7), 3523–3542.

[6] Jain, A. K., & Farrokhnia, F. (1991). Unsupervised texture segmentation using Gabor filters. *Pattern Recognition, 24*(12), 1167–1186.

[7] Khrissi, L., El Akkad, N., Satori, H., & Satori, K. (2020). Image Segmentation based on k-means and genetic algorithms. In *Embedded Systems and Artificial Intelligence: Proceedings of ESAI 2019, Fez, Morocco* (pp. 489–497). Springer Singapore.

[8] Brusatte, S. L. (2012). *Dinosaur paleobiology*. John Wiley & Sons.

[9] Fastovsky, D. E., & Weishampel, D. B. (2005). *The evolution and extinction of the dinosaurs*. Cambridge University Press.

[10] Mousavi, S. M. H. (2018). A new way to age estimation for rgb-d images, based on a new face detection and extraction method for depth images. *International Journal of Image, Graphics and Signal Processing, 10*(11), 10.

[11] Mousavi, S. M. H., MiriNezhad, S. Y., & Dezfoulian, M. H. (2017, October). Galaxy gravity optimization (GGO) an algorithm for optimization, inspired by comets life cycle. In *2017 Artificial Intelligence and Signal Processing Conference (AISP)* (pp. 306–315). IEEE.

[12] Lautenschlager, S. (2016). Reconstructing the past: methods and techniques for the digital restoration of fossils. *Royal Society Open Science, 3*(10), 160342.

[13] Comer, M. L., & Delp III, E. J. (1999). Morphological operations for color image processing. *Journal of Electronic Imaging, 8*(3), 279–289.

[14] Dunmore, C. J., Wollny, G., & Skinner, M. M. (2018). MIA-Clustering: a novel method for segmentation of paleontological material. *PeerJ, 6*, e4374.

[15] Hou, Y., Canul-Ku, M., Cui, X., Hasimoto-Beltran, R., & Zhu, M. (2021). Semantic segmentation of vertebrate microfossils from computed tomography data using a deep learning approach. *Journal of Micropalaeontology, 40*(2), 163–173.

84 Employing a deep learning technique for image segmentation based on a level-based set method

M. Nagaraju Naik[1,a], *Suraya Mubeen*[2,b], *B. Ravikumar*[3,c], *Swati Gupta*[4,d], *and S. Vijayalakshmi*[5,e]

[1]Professor, Department of ECE, CMR College of Engineering and Technology, Hyderabad, Telangana, India
[2]Professor, Department of ECE, CMR Technical Campus, Hyderabad, Telangana, India
[3]Assistant Professor, Department of ECE, CMR Institute of Technology, Hyderabad, Telangana, India
[4]Assistant Professor, Department of CSE, CMR Engineering College, Hyderabad, Telangana, India
[5]Associate Professor, Department of Electronics and Communication Engineering, Sona College of Technology, Salem, India

Abstract: In computer vision and image investigation picture is an elementary methodology that partition picture(s) over pertinent districts for further investigation. Due to its ability to precisely model complex and non-smooth profiles, the Level Set Method (LSM), a mathematical method that follows the curve's transformations, has been employed greatly for image segmentation. Classic Level Set Methods, however, face challenges in initialization sensitivity, computational slowness, and handling noisy images. In this study, we explore replacing LSM with current DL methods to enhance segmentation performance on various datasets. The proposed design utilizes the advantages of both methods through initial segmentation masks created by Deep Learning models (e.g. CNNs), guiding the level set evolution process, and improving feature extraction. Thus, from the shape of the image's 'probability' we can optimally modify the standard shape mask according to GAT approach. The input image for achieving image segmentation using a level set approach is the probability map.

Keywords: Deep learning models, LSM, image segmentation, probability map, level based set mehod

1. Introduction

The split of an image into semantically relevant region is referred to as image segmentation and is one of the main tasks in computer vision and Image investigation. It is at the heart of numerous applications, together with self-driving cars, medical imaging, object detection, and environmental monitor. Accurate segmentation of a pixel belonging to a certain class is needed to understand the structure and content of that image and in some fields of working like medical diagnostics accurate segmentation is very critical where precise boundaries of tumors and organs play an important role in treatment processes.

The Level Set Method is a numerical technique based on the mathematical theory of level sets developed by Osher and Sethian for the evolution of level sets of curves. It implicitly encodes the object boundaries as the zero-level set of a higher dimensional function. One common reason is that the Level Set Method performs extremely well to segment objects with complex or dynamic geometries due to its implicit representation, which enables it to handle topological changes (e.g., splitting or merging of boundaries). On the one hand, based on LSM segmentation has merits, on the other hand, it has serious disadvantages, including calculation complexity, sensitivity to initial conditions and noise.

Segments have to be created to organize accessible pixels Semantic segmentation, which is used in this study, is a computer vision task where the pixels of an image are grouped into semantic classes. Pixels are assigned to a class, and other context or information is not factored into this decision. In this study, segmenting algorithms based on the region work on recognizing the similarity between the neighboring pixels of the image and grouping them to the best matching family. The algorithm works by determining the close boundaries of seed pixels and determining similar and dissimilar classes. Owing to this, the pixels are known as seed pixels. Semantic segmentation models take inputs and return segment maps as the respective output. Model predictions The predictions are displayed in one-dimensional two-dimensional one-hot encoded representation in n-channel binary format. Neural networks for segmentation typically use an encoder-decoder structure, which consists of an encoder, a bottleneck, and a decoder, with a decoder, or upsampling layer, starting at the bottleneck (as with the FCN).

[a]nagarajunaik1976@cmrcet.ac.in, [b]cmrtc.paper@gmail.com, [c]brk5599@gmail.com, [d]swatigupta777@gmail.com, [e]vijayalakshmis@sonatech.ac.in

DOI: 10.1201/9781003684459-84

In the strict sense, segmentation is the technique of tagging some pixels in an image learn how to identify persons, objects or other relevant details. Image segmentation has a frequent application in object detection. This bounding box can then be used by the object detector that follow via the segmentation method that constructed the bounding box. Thus this improves the accuracy at the same time reduces the inference time as the detector is not consuming the whole image. One of the crucial array of computer vision methods and techniques is 'image segmentation.' It has different practical implementations such as use in satellite image processing, computer vision for self-driving cars, medical imaging and face detection and face recognition methods for video surveillance. Picture segmentation creates an output from the input images. A few components represent the object class of each pixel or instance of the object. There do exist a number of relevant heuristics i.e. high-level image features that can be useful for picture segmentation. This forms the basis of most picture segmentation algorithms based on clustering, such as edges and histograms.

Colour is a common heuristic and an example. In photography, graphic designers use a green screen to make the color of the visual background uniform. This enables post-processing to automatically identify and replace backgrounds. Contrast is another handy heuristic that makes image segmentation algorithms trivially able to separate a dark body from a light background. The software will recognize the pixel boundaries based on what values lied drastically apart. Although traditional image segmentation methods can be quick and straightforward, they may need considerable tweaking using manually created heuristics for different use cases. They generally are not accurate enough to translate to complex imagery. Toward more sophisticated segmentation, machine learning and deep learning are often used to use these methods in an efficient way, improving the accuracy and versatility. A more effective network architecture search approach for semantic segmentation was also proposed in [1]. But these approaches grow the complexity of the model with the complexity of the scene, limiting their applicability.

2. Related Work

Then came deep learning techniques like CNN U-Net, and Deep Lab, which throttle on big annotated datasets and GPU processing, and image segmentation went totally bananas. However, boundary refinement is often difficult for these methods, especially in the presence of fine details [2].

Recent studies have investigated a combination of LSM and machine learning. However, most works focus on the extraction of pre-learned features (without integrating LSM as trainable parameters) based on their application [3]. The fully trainable architecture of this paper combines LSM with deep learning.

The strength of deep convolutional neural networks exposes the underlying structures in the images and sharpens your skills at creating realistic images. In the iterative process, the shape prior was initialized as the surface of the level set based on the priors learned with a large number of cases using deep learning [4–5]. Because prior knowledge formed by deep learning is also imprecise and crude, it is essential to utilize more specific data to enhance precision. As for the first point, different classes of objects can be discriminated solely based on the inherent shape of their instances, therefore, based on established probability map of prior knowledge. This work proposes the level set with deep prior method in the area of image segmentation according to the shape prior which characterizes the intrinsic shape of the target. Given FCNs' preexisting knowledge. Furthermore, FCNs can decouple semantics from segmentation using the high-level semantic information from images as the segmentation prior. The Portrait data set will then be used in a series of studies [6] to demonstrate the effectiveness of the approach. Furthermore, the experiment results also verify that traditional FCNs are not the best for segmentation, and proposed method could produce more accurate segmentation results [7].

This affirms that the existing approaches of image segmentation cannot meet the demand for picture segmentation. Deep neural networks have been widely studied because they have the ability to learn very complex functions that are often required for many tasks. From 2000, with hardware improvements similar methods from machine learning known as deep learning became popular again, yielding strong results in a variety of image processing workloads. Unlike traditional level sets, Le et al. and Zhou et al. received a more reliable and effective horizontal level set model based on deep recurrent neural networks [10]. Hu et al. [9] made use of the level set function. [11] was employed as a loss function in the neural networks, enabling the network model to converge faster during training. The inherent capability and rapid development of algorithms have made deep learning (DL) the method of choice for image segmentation problems in the last years [11].

3. Proposed Model

In the current research, we propose a new hybrid architecture capable of integrating the advantages of both methodologies, thus integrating Deep Learning with the Level Set Method. The level set is guided by a deep learning model which acts as an initial segmentation mask and is also used for initial estimates of the parameters. This arrangement exploit the LSM's capability of provided that accurate, tight bounds while mitigating its sensitivity to initialization and noise. Hybrid approach works well for applications where accurate output is critical as using the Level Set Method to further refine the deep learning output gives smoother and more accurate segmentations.

Here, we propose a model that incorporates an LSM component for highly accurate segmentation built on a typical deep learning backbone (i.e., U-Net or ResNet). Architecture consists of the following elements:

Level Set Initialization: The output predicted from the CNN is used as an initial state for the level set function (phi). In particular, the approximate segmentation is realized through a sigmoid activation layer to produce a probability map (Figure 84.1).

- **CNN**
  For image or objective recognition, the width and height of intermediate feature maps are converted to full convolutional network back to the input image. The structure of the FCN is illustrated in Figure 84.2. In the image, you see a transposed convolutional layer doing just that. Accordingly, the augmentation in image and the classification output is associated pixel-wise. The results from classifying the input pixel which occurs at the same spatial location are saved in the channel dimension for any output pixel. For semantics, a fully convolutional network or FCN is the most commonly used architectural type for segmentation. They can only apply up sampling, pooling, and convolution as locally connected layers. Fewer parameters by not using dense layers.
- **LSM**
  The level set evolution is governed by the following equation
  Mathematical formulation
  Level set Evolution
  The level set evolution is governed by the following equation

$$\frac{\partial \emptyset}{\partial t} = -\frac{\delta E(\emptyset)}{\delta \emptyset}$$

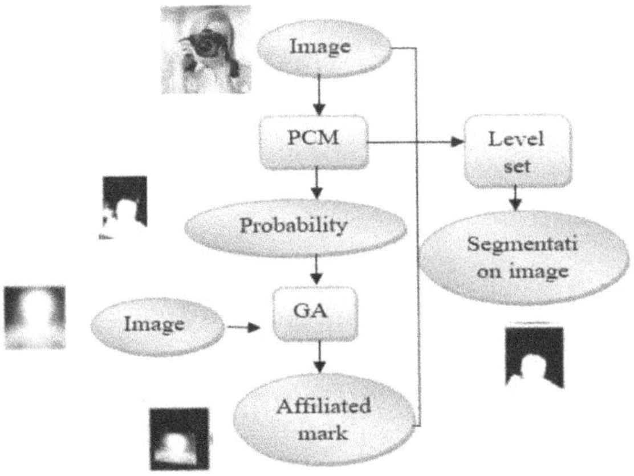

Figure 84.1. Block diagram.

Source: Author.

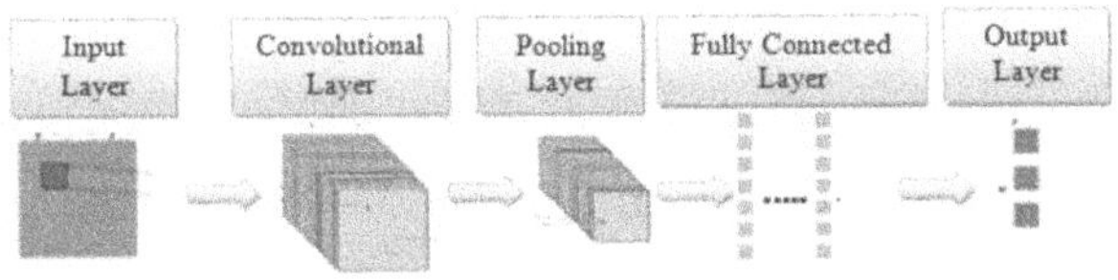

Figure 84.2. CNN structure model.

Source: Author.

Where:

$\emptyset$ = level set function

E($\emptyset$) = Energy function comprising region
– based and boundary
– based terms.

The energy function is defined as:

$$E(\emptyset) = \int_{\Omega}^{0} \lambda_1 |\phi|^2 + \lambda_2 \delta(\phi) |\nabla \emptyset|^2 dx$$

Here λ_1, and λ_2 are trainable weights learned during the training process.

4. Probability Map with Deep Learning

The probability map generated by the deep contextual network is very nice to look at but can tell us how the shield of the uncertain area may be breached sometimes. This is mainly due to the averaging effect of the multiple trained models on the probability maps. As a result, we further correct the contour by an existing watershed method. Figure 84.1: The contour evaluation method (4). The final fusion result includes the original probability map plus a binary contour and a linear combination. The configuration that yielded the best Rand error result in our experiments, given training data. Merging data from the watershed approach (CUMedVision-4 (with fusion)) can reduce the Rand error significantly, though the warping error has unfortunately increased. Since these two errors take into account different segmentation evaluation metrics, this is sensible. The former might penalize even slightly-noncentered limits, whereas the latter disregards no topological errors. Different AI strategy SCI, applied a complex post processing than our simple phase of processing to get the probability maps created by the teams DIVE and IDSIA.

Layer ReLU: The ReLU itself is just a piece-wise nolinear activation function, f(x) = max (0, x) You could think of the activation function as a threshold function. The output of the superfluous nodes is used and attenuated to create a sparse output.

Softmax layer: Here seen FCN end's out into soft max layer to convert FCN output into probability types using softmax function

$$p_{x,y}(c = i) = \frac{\exp(h_i(x,y))}{\sum_j \exp(h_j(x,y))} \quad (1)$$

where px,y (c = i) represents the probability that pixel (x, y) belongs to the class c = i, and hi(x, y) indicates the feature map value of the ith in the final output of FCN at points (x, y). In this work, the posterior of each pixel is represented as the output of the softmax layer, which is interpreted as a probability map.

4.1. Corrected global affine transformation for shape

Only one mean mask for images in the Portrait dataset. In this paragraph, the prior's standard shape is transformed into the best alignment of the image using the probability map calculated from the previous paragraph. It is perfect when the probability map is similar to the actual shape. An affine transformation of the common shape prior can be generated from the probabilistic image map. Transformation matrix M along with its inverse M-1 were usually used for Affine Transformations. It thus brings us to the equation on what an affine transformation of 3D point P to Q looks like.

$$Q = MP \tag{2}$$

The may appear in its expanded form as isomerized, considering that the coordinates of Q are a linear combination of the points P.

To reverse point Q back into point P just premultiply both sides of a previous equation by the transformation matrix inverse M^−1 as follows.

$$P = M - 1Q \tag{3}$$

But before we can apply this outfit of global affine transformation on the given data, we need to understand what the point and directional vector is. While a point is a constraint in three dimensions and describes a location, a directional vector is an orientation relative to a given point and is usually represented as a point and a unit sphere centered on the origin. Affine translation: P = (Px, Py, Pz, 1) | v = (vx, vy, vz, 0) The reasons for this. It will be clear as we consider the formulation of a translation transformation.

4.2. *Image segmentation using the level set method*

In conclusion, sensibility and FCN performance is improved by the attribute set approach, and by deep starting using the level set structure in combination. The output of the FCNs is used to obtain the probability map, for which each pixel represents a category. The correct segmentation is present in a substantial portion, despite the segmentation of form of a probability map is noisy. Thus, the 'probability' traditional shape, followed by applying the GAT approach, will yield a best affine transformation of the said shape mask of the image. In the level set method, image segmentation is attained by feeding affine mask, probability map, and image as inputs. The level set method is a region-based image segmentation method.

In level set method for image segmentation the region-based active contouring model (i.e., a variational based image segmentation model in which we optimize an energy function) is the basis to evolve the level set. The area that we will be interested in is the area around the number 0. The zero level = 0 represents the contour of a target, and the level set is represented by. One can think of the strength of the connection of each point to the image as a potential function that describes the level set. The level set is a type of probability density that indicates the likelihood for a given point to be reached. In order to enhance the applicability of the level set approach, its energy objective functional can be employed to different energy-driven probability models. The most common level set method is CV model presented by Chan and Vese; consequently, the contour energy function is derived from the following equation.

$$\mathrm{F}(\mathrm{C}, C_1, C_2) = \lambda_1 \int_{outside(c)}^{0} |i(x) - C_1| dx + \lambda_2 \int_{inside(c)}^{0} |i(x) - C_2| dx + v|c| \tag{4}$$

Here, outside (C) and inside (C) are the outer and inner area of contour C, X is the image (a 2D vector), C is the length of contour, C1 is the number of pixels inside contour C, and C2 is the number of pixels outside contour C. In level set function, the energy function is used instead:

$$\varepsilon(\emptyset) = \varepsilon_{img} + \varepsilon_{shape} + \varepsilon_{edge} + \varepsilon_{reg} \tag{5}$$

where H(z)=H(θ) =Hθ is Heaviside function and the level set functional refers to inner and outer set.6. referring to energy-based (variational) methods. The level set technique provides an extremely flexible definition of the energy function, which is a function of the original image, a set of shape prior information, and the probability map, all combined into a single energy function as shown the Figures 84.3 and 84.4. The energy function can be explained as:

$$\emptyset t = \emptyset t - 1 + \Delta t.\partial\emptyset\partial t \tag{6}$$

The Level Set Technique is discussed in the equation above.

4.3. *Image segmentation using DL*

Segmentation is an extremely integral part of computer vision. Image segmentation refers to merging regions of an image that belong to the same class of object. This approach is commonly referred to as 'pixel-level classification.' In other

Figure 84.3. Level set method partial results.

Source: Author.

Figure 84.4. Level set method results.

Source: Author.

words, it requires dividing the image into multiple parts, elements, or components. Picture segmentation algorithms fall into several different categories. Some of the older methods are also watersheds, k-means clustering, region expansion, thresholding, and histogrambased bundling. Starting from the principles of active contours, graph cuts, conditional and Markov random fields, and sparsity based techniques, more complex algorithms can be built. Deep learning models recently have been developed which have led to a new family of image segmentation algorithms with significant performance improvements. Deep-learning based image segmentation models also provide the best accuracies. on those widely used criteria, and leading to a paradigmatic transformation in that field. Recent advancements in neural network-based deep-learning technologies, have brought numerous high-performance techniques for understanding images, including image segmentation for object classification.

Remote sensing area, DL-based image segmentation has been applied for separating the satellite imaging data, and produces better results, particularly in urban planning and precision agriculture data. Deep learning has also been used to classify drone photo images (UAVs). Photo segmentation is a great application for deep learning. Next, the hidden feature space that can help segregate the data can be automatically created by deep learning algorithms. Deep learning algorithms can automatically learn complex qualities that are difficult to express manually. In this class, we continue our exploration of image segmentation with recurrent neural networks (RNNs) – CNNs, fully connected networks (FCNs), and beyond. Fully Connected Layers Each neuron in a layer are connected to every other neuron in the layer below. From there, the network can discover complex non-linear relationships among the pixels of the image.

Flow model of the proposed system is shown the Figure 84.5.

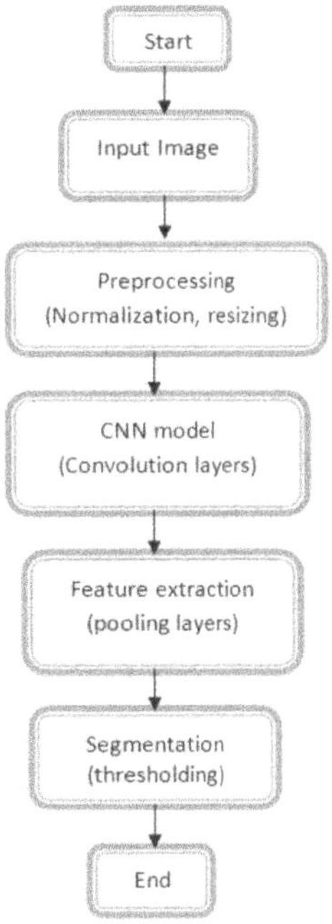

Figure 84.5. Flow chart.

Source: Author.

5. Results and Analysis

Deep learning with segmentation techniques such as the LSM enables us to leverage the ability of deep learning models to generalize and achieve high segmentation accuracy for challenging applications and at the object's boundaries. In this section, the performance of the proposed hybrid model on various datasets and metrics and experimental results are presented by using the predicted model.

When sufficient data is available, the level set strategy enables accurate segmentation results. A set of pattern recognition methods are recommended to learn more about the target. In this contour development process, the level set method, is based on statistics (fixed now) of pixels (mean, weighted mean, and probability model, with regards to areas) inside/outside the contour. Since the level set method generates potential functions particularly based on the energy functional minimization concept, it is more like an information combining approach. The following image will then be split into two parts for the purpose of image stylization into a selfie so that some styling can be attempted. A simple problem of segmenting a single target in a portrait can be used to segment complex scenes. Figure 84.6 shows some images and ground truth from Portrait data set.

Performance metrics

The Dice coefficient measures the overlap between the predicted segmentation and the ground truth:

$$Dice = \frac{2|P \cap G|}{|P| + |G|}$$

5.1. *Quantitative analysis*

Boundary intersection over union (Boundary IoU)

This metric evaluates the model's ability to capture accurate object boundaries.

$$Boundary\ IOU = \frac{|P_{boundary} \cap G_{boundary}|}{|P_{boundary} \cup G_{boundary}|}$$

Figure 84.6. The portrait data set's ground truth and some photos.

Source: Author.

5.2. Pixel accuracy

Pixel accuracy measures the proportion of correctly classified pixels over the total number of pixels:

$$Pixel\ Accuracy = \frac{Correct\ Pixels}{Total\ Pixels}$$

Dice Coefficient: Dice Coefficient: Our proposed hybrid method (CNN + LSM) outperformed the U-Net only by 3.2% showing that it has the potential to control the exact features of tumor edges as shown in Figure 84.7.

Boundary IoU: The performance of the specific evolution of the level set on CNN results alone is assessed in relation to other techniques (CNN vs. LSM) shown in the Figure 84.8, as concluded in this figure, this method enabled a significant improvement to the CNN boundaries, documenting a better Boundary IoU.

Hausdorff Distance: Here, we demonstrated the hybrid method which introduces soft but geometrically more accurate boundaries, as shown by the decrease in Hausdorff distance.

Pixel Accuracy: As seen in Figure 84.9, our CNN's feature extraction and level set evolution provided us with pixel level accuracy of more than 97%.

Traditional LSM vs. CNN + LSM

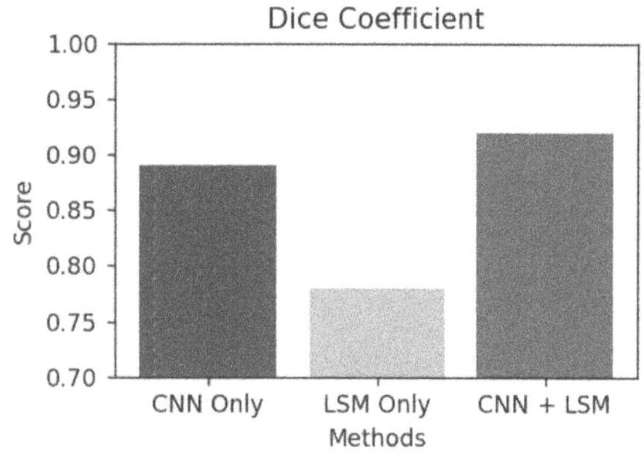

Figure 84.7. Dice coefficient IoU comparison of various methods.

Source: Author.

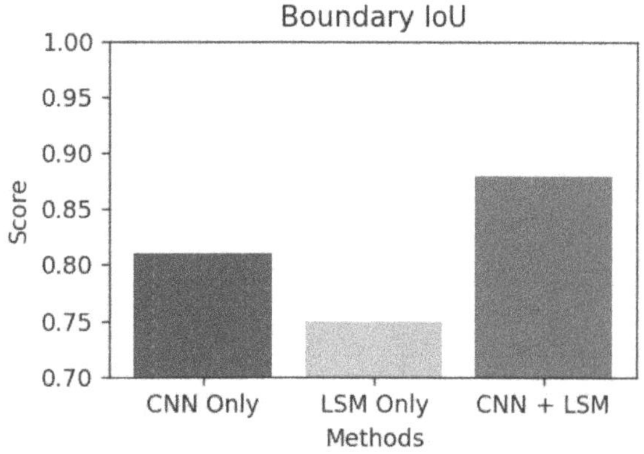

Figure 84.8. Boundary IoU comparison of various methods.

Source: Author.

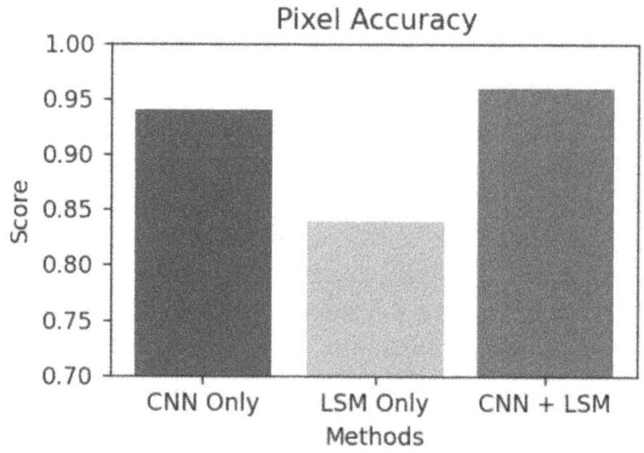

Figure 84.9. Pixel accuracy comparison of various methods.

Source: Author.

Performance Improvement: Even it is effective for small-scale problem statement, conventional LSM suffers from limitations like sensitivity to the noise and human initialization. The CNN + LSM method improves upon noise robustness and automates initialization using learned feature visualizations.

CNN Only vs. CNN + LSM

Boundary refinement: CNN-only models tend to blur the object boundary in boundary-sensitive tasks, leading to inaccurate results. The level set approach can then be used to finely refine these boundaries to obtain geometrically accurate segmentation results.

6. Discussion

The results demonstrate the effectiveness of the CNNs using the Level Set Method for images segmentation. This hybrid approach uses the advantages of both methods: because CNNs are very good at extracting semantic features, the level set function can be initialized well. Despite the robustness of the aforementioned GNN methods, the level set technique preserves exact boundary refining for small or irregularly shaped objects.

However, the combination is not without computational trade-offs:

Longer Training: The iterative updates required to factor the level set evolution into the process results in a slight increase in training and inference time.

Hyperparameter Sensitivity: $\lambda 1, \lambda 2$: The weighting factors for the level set energy terms must be finely tuned to achieve the desired accuracy but not compromise on the stability of the model.

7. Conclusion

The combining use of LSM and CNN for picture segmentation has shown great improvement in the segmentation accuracy which is crucial in fields where the precision of delineating object and fringes is of high importance. The hybrid methodology offers superior results compared to both standalone-CNN-based approaches and conventional LSM; It achieves reconciliatory gains by amalgamating the salient standpoints of both: Here is a comprehensive conclusion of the analysis and findings. Combining the recommended level-set with a deep prior, the probability map technique and the corrected prior shape is feasible for picture segmentation [8]. CBCT-Portrait: A Generative Deep Learning Method for Fast & Accurate Facial Reconstruction from Incomplete CBCT Data with Breast Release In-Vivo Evaluation; 1. The experiment results show that the proposed strategy can produce more accurate results.

References

[1] Wu, Z., Shen, C., & Van Den Hengel, A. (2019). Wider or deeper: Revisiting the ResNet model for visual recognition. *Pattern Recognit.*, *90*, 119– 133.

[2] Ronneberger, O., Fischer, P., & Brox, T. (2015). *U-Net: Convolutional Networks for Biomedical Image Segmentation.* MICCAI.

[3] Chen, L. C., Papandreou, G., Kokkinos, I., Murphy, K., & Yuille, A. L. (2017). DeepLab: Semantic Image Segmentation with Deep Convolutional Nets.

[4] Hu, P., Shuai, B., Liu, J., et al. (2017). Deep level sets for salient object detection. *IEEE Conf. on Computer Vision and Pattern Recognition (CVPR)*, Hawaii, USA.

[5] Tang, M., Valipour, S., Zhang, Z., et al. (2017). A deep level set method for image segmentation. *Deep Learning in Medical Image Analysis and Multimodal Learning for Clinical Decision Support*. Québec City, Canada, pp. 126–134.

[6] Shen, X., Hertzmann, A., Jia, J., et al. (2016). Automatic portrait segmentation for image stylization. *Computer Graphics Forum*, *35*, 93–102.

[7] Le, T. H. N., Quach, K. G., Luu. K., et al. (2018). Reformulating level sets as deep recurrent neural network approach to semantic segmentation [J]. *IEEE Transactions on Image Processing*, *27*(5), 2393–2407.

[8] Chen, X., Williams, B. M., Vallabhaneni, S. R., et al. (2019). Learning active contour models for medical image segmentation[C]//Proceedings of the IEEE/CVF conference on computer vision and pattern recognition.11632–11640.

[9] Hu, P., Shuai, B., Liu, J., et al. (2017). Deep level sets for salient object detection[C]//Proceedings of the IEEE conference on computer vision and pattern recognition. 2300–2309.

[10] Li, H., Xiong, P., Fan, H., et al. (2019). DFANet: deep feature aggregation for realtime semantic segmentation. arXiv preprint arXiv:190402216.

[11] Sattibabu, D., Varsha Deepika, K., Santhi Kumari, K., Sai Rahul Tej, A., & Nalajala, P. (2024). Tracing Missing Person Through Facial Recognition Using Deep Learning. SCI 2024. Lecture Notes in Networks and Systems, vol 1147. Springer.

85 An effective and efficient technique that uses onvolution neural networks to identify coronary artery disease using IRIS

B. Venkataramanaiah[1,a], *D. Srikanth*[2,b], *Md Yasmeen*[3,c], *Monika Arya*[4,d], and *R. S. Sabeenian*[5,e]

[1]Assistant Professor, Department of ECE, CMR College of Engineering and Technology, Hyderabad, Telangana, India
[2]Assistant Professor, Department of ECE, CMR Technical Campus, Hyderabad, Telangana, India
[3]Assistant Professor, Department of ECE, CMR Institute of Technology, Hyderabad, Telangana, India
[4]Assistant Professor, Department of CSE, CMR Engineering College, Hyderabad, Telangana, India
[5]Professor and Head, Department of Electronics and Communication Engineering, Sona College of Technology, Salem, India

Abstract: Coronary heart disease is one of the most lethal diseases in the world today. Coronary heart disease is not only due to genetic inheritance but also due to a poor lifestyle, lack of exercise and tobacco use. Sometimes, the patient has no idea whether he or she has a heart function problem. Consequently, this work proposes breathing monitor based on Iridology, a technique that argues the iris to diagnosis of cardiac errors. These iris detection and classification results by the system are automatic. There is an abundance of information available in the world of medicine today that can help in the accurate identification of a disease. At the same time, many researchers have proposed different intelligent algorithms for diseases detection based on this data. This work utilizes deep learning to create a robust application for better disease identification. The system is constructed using a convolutional neural network. The system consists of three stages. The proposed CNN architecture was able to classify healthy subjects from patients with CAD with high accuracy by optimizing it for the extraction of minute features from iris patterns. The model performance was evaluate using metrics such as accuracy, precision, recall and F1-score, representative its efficiency against other approach.

Keywords: CNN, coronary heart, iris biometrics, CAD, iris image analysis

1. Introduction

Coronary artery disease is the leading cause of death worldwide. Worldwide, 17 million people die from coronary artery disease, according to the WHO. Coronary heart disease ranks second as the leading cause of national death in Indonesia after stroke (12.9% in 2014). Tobacco kills 1.9 million people a year from coronary heart disease. The most common causes of coronary heart disease are no physical activity and living a sedentary life. High death rate among the geriatric population Obesity and eating foods high in fat or carbohydrates are among the factors that can cause the heart's blood vessels to narrow.

Heart disease accounts for around 10% of all deaths globally. According to death rate information, heart disease is the leading cause of death in the world. In Japan, heart disease is predictable as the leading cause of death. The modern devices, tool automation, monotony, junk food consumption etc. contributes to the slothfulness [1]. In addition to producing a pot belly, this couch-potato lifestyle is a major cause of heart disease [2]. Heart disease kills and debilitates millions globally every year, a WHO report says. Hence, it has been significant for the society to forecast the ailment [3]. At large, heart disease describes any condition that results in the heart not functioning properly or healthily. Heart illness is of different kinds like cardiomyopathy, valvular heart disease, and coronary artery disease (CAD).

A EKG which use electricity to assess the heart rhythm, will be taken for coronary heart disease. Occasionally, echocardiography is used to see the heart's pump, valves and other components. For example, a person would be less willing to have some check on his heart condition, because the cost of few steps that he takes to check his heart is quite expensive. Early prevention of the risk of coronary heart disease3 by abstaining from tobacco use, consuming ample fruits and vegetables daily, and exercising or partaking in physical

[a]bvenkataramanaiah@cmrcet.ac.in, [b]cmrtc.paper@gmail.com, [c]mdyasmeen555@gmail.com, [d]monika.arya@cmrec.ac.in, [e]sabeenian@sonatech.ac.in

DOI: 10.1201/9781003684459-85

activity on a regular basis. Researchers in fields like medicine and computer technology are employing techniques and technologies today to screen for heart abnormalities early on. One of the potential solutions using information coming from the iris of the eye, which is physically connected to the brain, to estimate the condition of the heart. It is a vascularizably organ of a human, as the brain consumes 20% of the oxygen in the body and receives 15% of the heart's blood flow. The iris is a special organ of the human body. Iris is normally used as Biometric Authentication or Electronic Security. Iris research is now increasingly common among the medical profession, and serious enough to warrant expert committees. The Iridology method uses the iris to find out the condition of the human organs.

1.1. CAD (Coronary artery disease)

In CAD which is call atherosclerosis, the blood vessels that nourish the heart muscle (see Figure 85.1) are affected. It is caused by deposits of fibrous tissues, fat, cholesterol and calcium that narrow the artery lumen.

Now, scientists from the fields of medicine and computer science have created technology that can detect heart problems before they are significant. Among these is detecting the heart condition by observing the changes in the retina, an extension of the part of the brain where the iris of the eye is located. It is a human organ that is susceptible to vascularization because the brain takes 20% of the body's oxygen and receives 15% of the heart's blood flow. The iris is one of the single organs in human body. Iridology can make statements about the condition of human organs. At the core of iridology is the examination of the body most multifaceted tissue structure. This technique can be used to determine the state of the organs and systems of the body. Iridology helps identify current or potential problems with a particular organ but is not able to diagnose a disease.

2. Related Work

[4] A person typically dies of heart disease with no warning signs whatsoever. Half of heart attacks happen without any warning. The acute chest pain means to say it is an emergency,

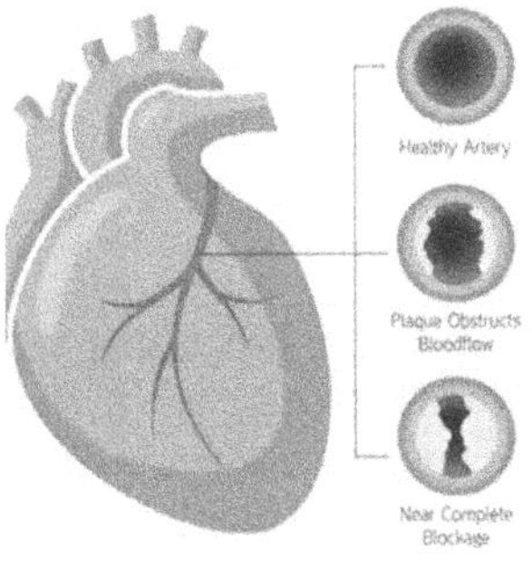

Figure 85.1. The coronary artery disease.

Source: Author.

only doctors are the ones who recommend invasive or non-invasive tests such as ECG and ECHO. These tests can detect CAD. But such examination information have had several discrepancies here and there. [5] The mortality of the patient can be prevented by early detection of the disease. [6] Older women are at a higher risk of developing heart disease. So, many lives could be saved if the time taken to detect the disease can be reduced and the accuracy of detection can be increased.

Multiple rules in FRB were formed with using AND and OR operators in different combinations. Ultimately, 86 different fuzzy rules were defined and output classes were assigned. Radhimeenakshi et al. The authors in [7] created a CAD identification system, applying two classifiers such as SVM and ANN. To predict, SVM constructed a hyperplane that separated healthy cases from sick cases in the test data. The second method involved using neural networks to make predictions. Neural networks were trained using cross-validation to evaluate model performance. Purushottam et al. have developed a decision tree [8] for the classification of heart diseases.

It proposed [9] to enhance the identification of cardiovascular disease. The system consisted of five unique components: ensemble component, neural network, variable selection, data partitioning, and cardiac database. Architect of the heart database: input features The partitioning component divided the data for testing and training.

Cardiac imaging (CVD) plays a vital role in the diagnosis of cardiovascular disease. Until now, it has only been tasked with visually and mathematically examining heart structure and function. On the other hand, big data and machine learning have paved the way for the development of AI systems [10] to aid doctors in diagnosing the cardiovascular diseases.

From basic analytical queries of recent measurement data to the advancements of handling raw images, machine learning methods [11] can now face almost any data-related issue. So far, machine learning has been applied to these two very large, interconnected domains: generating new knowledge critical to clinical practice; and automating functions that would have otherwise been performed by humans.

While diagnosis is needed [12], research on cardiac disease is necessary. When you await for surgery, you have to talk about image processing and diagnosis based on medical angiography images. They are all done to ensure surgery is not postponed. Therefore medical engineering optimization methods and diagnostic tools based on the cuckoo algorithm will be presented in this paper. This study offers a perfect bio-inspired diagnostic model [13]. Coronary angiography, involving specialized X-ray imaging equipment, has traditionally been the primary diagnostic and therapeutic modality for Coronary Artery Disease (CAD) and its vascular phenomena. Results of experiments performed on several test images of patients acquired different ways are displayed. We had success with a rapid bio-inspired diagnostic model [14].

3. Proposed Model

The proposed model utilizes the automatic feature extraction and classification ability of CNNs to identify Coronary Artery Disease (CAD) from the iris images. The workflow includes data collection and preprocessing, CNN model and training, and evaluation [15]. Below is an explanation of every element in the model shown in Figure 85.2.

Data Gathering: Long Beach, Hungarian, Swiss, and Cleveland dataset All data sources read 920 records. The data sets were read and several missing values were discovered. A few results are missing, especially corresponding to the grade of peak workout, main vessels coloured by fluoroscopy and thalassemia.

3.1. *Pre processing*

Then, as digital camera pictures are in RGB (Red, Green, Blue) format, pre-processing them is necessary. Preprocessing will convert the RGB image of iris into greyscale image.

It initiates the removal process. The pre-processing diagram is shown in Figure 85.3. The scale image will be passed throughout CHT technique to extract the iris and eliminate the pupil area.

Feature Extraction: The process is use in ML, pattern identification, and image processing. Starting with a collection of observed data, it produces derived values (features) designed to systematize learning and generalization processes and, in some cases, help human interpretation of situations by providing useful, non-duplicative information. Feature Extraction/Dimensionality Reduction When the characteristics of input data for an algorithm appear redundant and extensive for evaluation (for instance, the same sensor data is provided in both feet and meters or pixels are used to represent photos), we can condense an algorithm's input data to a smaller-sized input, known as feature vector. After that feature selection takes place, and we select some of the original attributes. Once this target solution is learned, it is predictable that the chosen features will extract the pertinent aspects of the input data needed to learn a solution to the desired problem using this compacted representation instead of the complete raw data.

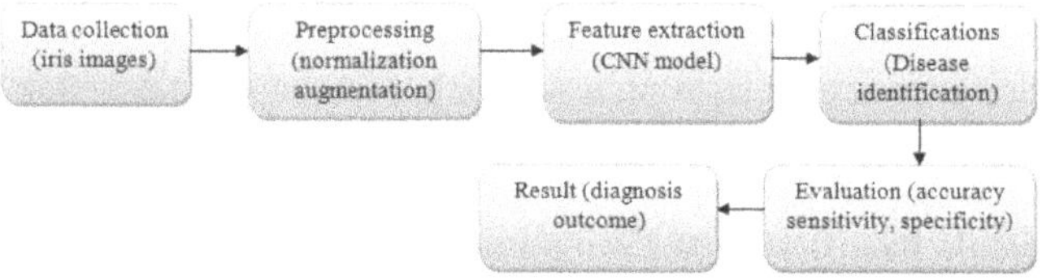

Figure 85.2. System proposed model.

Source: Author.

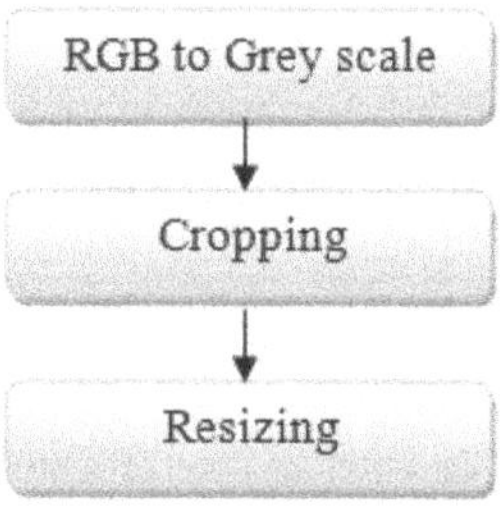

Figure 85.3. Image pre-processing process.

Source: Author.

3.2. *Segmentation*

How we move from the analog to the digital image. Image segmentation is the step that is most difficult and most important part of the process of image processing. Image Segmentation: Image segmentation is the task of partitioning an image into multiple segments. Image segmentation is the very first step in image analysis. In image segmentation, an image is separated into exact segments, wherein pixels of related characteristics are grouped together based on certain criteria. Image segmentation is the method of dividing an image into parts that share features and properties. The most important feature when it comes to image analysis is image segmentation. We yield a segment based on their properties in the image, for example, by colour, shape, or texture. Human eyes use R (red), G (green), B (blue) as input to compute the colour of a pixel in an image. The three primary colours are red, green and blue. For example, R,G,B colour can be converted to their linear or non-linear transformations which results in more variety of colour components.

3.3. *Classification*

Both structured and unstructured data can be categorized. You can apply the classification method to classify data into a given number of classes. In essence, the driving purpose behind a classification challenge is to predict what class or classes new data will belong to. Classification is a supervised learning method in machine learning and statistics in which computer programs learn from the input data they receive and use this knowledge to classify new observations. This could be a bi-class (or multi-class) data set (for example: is the communication spam or not, or is the user male or female). Some examples of classification problems are speech recognition, handwriting recognition, biometric identification, document classification, etc.

3.4. *Image acquisition*

Image acquisition in image processing refers to getting an image from outside for processing. It can be as simple as being handed an image that is already digitized. This is usually during the image acquisition phase visit_image Image acquisition consists of three steps: The object of interest reflected energy. An optical system focuses the energy and a sensor ultimately measures the energy.

3.5. Iridology

Iridology is a diagnostic technique that is based on the human iris. In medicine, the iris gives an idea of the condition of a person's organs. Iridology divides the iris into 60 different segments, each representing the state or function of the organs. The condition of the right organs of the body will show in the right iris, and vice versa; the condition of the left organs will show in the left iris.

3.6. Circle hough transform

The unwanted aspects like Iris and other features (especially pupil) will be holler at in the images in contrast to the RGB to Grayscale conversion steps. The iris's coordinates can be automatically identified with the help of the CHT procedure without human interference. CHT can recognize outer iris, outer pupil and circle-detect circles in the iris image.

$$(x-a)^2 + (y-b)^2 = r^2 \tag{1}$$

Equation 1 describes the radius, r, for the center circle (a, b). We can represent it as a circle by taking (x, y) as a pixel on the edge of the circle.

$$X = a + r\cos\theta, Y = b + r\sin\theta \tag{2}$$

Point c the image at point b is then used for image showing in order to calculate the pupil and an iris ring circumference. This will remove any noise that is not circular during this process. Next, a line thickening process is used to detect the circle in the image. Next, the section we labeled as a circle will be merged with the original RGB image and a photo like that in point c will be constructed. Radial detects edges for object removal, marking an object that is less than 3 mm in diameter. So the system will select the inner iris circle when it finds a circle with diameter thicker than 3 mm and best fit a round shape. To construct the contour of the outer iris circle, the method has such steps, searching a circle diameter of 12 mm. To make sure that photos are the same size in each data set, the diameter of the circle is changed before the photography process is done using steady light.

Iris: The iris can be simulated using two non-concentric circles and some textures inside an iris circle. The iris may put the circle in relation to the pupil center. Radial resolution is the part of the iris that gets hit by a radial ray. As the pupil is elastic to the iris, the point of reference must be adapted. The scaling equation is derived based on the angle around the the iris circle with the following formula:

$$r' = \sqrt{\alpha\beta^2 - \alpha - r_I^2} + \sqrt{\alpha\beta} \tag{3}$$

$$Where, \alpha = o_x^2 + o_y^2 \tag{4}$$

$$\beta = \cos\left(\pi = \arctan\left(\frac{O_y}{O_x}\right) - \theta\right) \tag{5}$$

CLAHE is a histogram equalization technique used to recover image quality. In this study, the CLAHE Technique is used to enlarge image intensity, to increase image fine detail. The outcome of the CLAHE are illustrated in Figure 85.4.

3.7. Region of interest

In the iris-based cardiac diagnosis protocol, only specific elements form the basis for avoiding image-wise data processing. According to iridology, the left iris with sectors 02.00–03.15 it the iris that mirrors the heart. Using the ROI approach / method we only need to work on a part of it.

3.8. Convolutional neural networks (CNN)

The proposed CNN follows a feedforward architecture and processes when one input and one output at a time. As an example, suppose we are conducting some binary classification experiments on classifying the patients whether they have CHD or not (where patients with CHD are labeled '1' and no CHD cases are '0') THERE WILL BE A MULTI CLASS CLASSIFICATION EXPERIMENT AS WELL which will be discussed later. The majority vote process resulted in 14 features (phenotypes) that were active for the CHD (as previously reported). Let N be the number of training samples; thus the input layer depicted in Figure 85.5 has an RN×14 size. Let's continue. It efficiently standardizes various types of variables before the nonlinear transformation done by the suggested CNN architecture.

4. Results and Discussion

System training was performed using data of 40 normal iris and 40 aberrant iris in this study. Regular iris data from healthy individuals is treated as normal, and abnormal iris data from those with heart disease. The data used for the training of the system consisted of 40 normal iris data and 40 abnormal iris data. On the other hand, abnormal iris data is

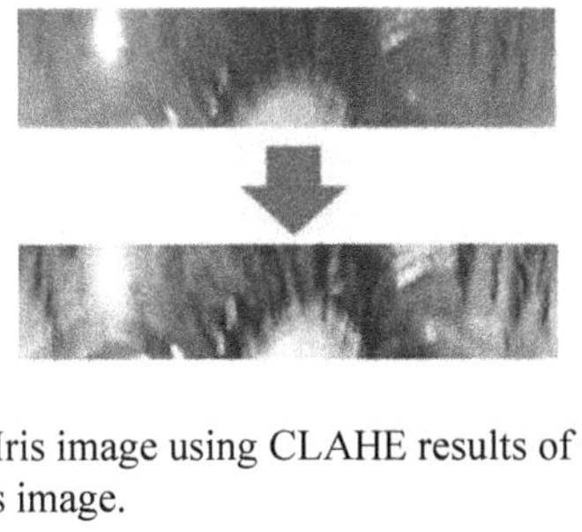

Figure 85.4. Iris image using CLAHE results of using CLAHE method on iris image.

Source: Author.

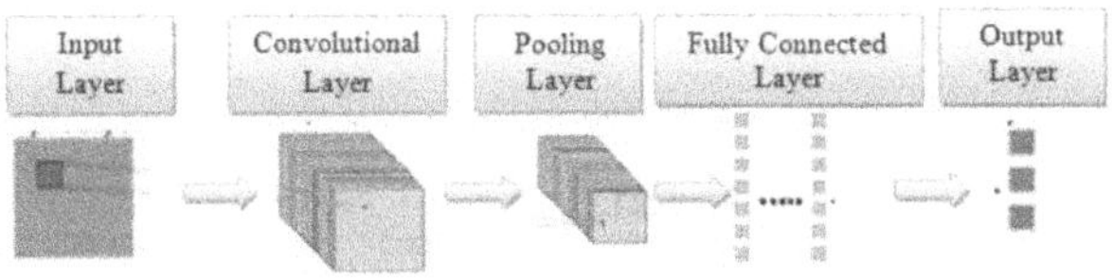

Figure 85.5. Architecture of CNN.

Source: Author.

the iris of individuals with heart disease; normal iris data is the iris of people with no history of heart disease. It uses linear kernel alterations in the input data. Iris data is classified by normal (red) and abnormal (blue) files for training. Training with a linear kernel will separate data into each class with ui no of points. Training results optimize the learning and can help in representation in the test data.

Evaluation of the proposed CNN-based system was conducted by using training, validation, and testing set of iris images. The dataset featured 2,000 images that were balanced between CAD patients and healthy individuals. The results are reported using several relevant metrics such as accuracy, precision, recall, F1-score and ROC-AUC.

4.1. Model training

Table 85.1 shows the training and validation performance of a CNN, which was trained on the dataset for over 50 epochs.

The training accuracy demonstrate that the CNN efficiently learn pattern in the iris images connected with CAD.

4.2. Performance evaluation

An excellent performance on all significant criteria was achieved by the CNN-based model, which was able to accurately discriminate between patients with and without CAD (as shown in Table 85.2).

To consider the improvement of the model, there is the precision: The precision of the model was 92.8% and it showed that CNN can accurately distinguish between CAD and non-CAD cases based on iris patterns. Such accuracy means that CAD can be diagnosed by slight changes in the iris structure that can indicate systemic vascular abnormalities.

Precision (91.5%): Our method target at Jiang, we reduce the false positive but keep identifiable actual CAD. This reduces unnecessary follow-up procedures in case CAD is not present.

Recall (94.2%): The higher the recall, the better the model can detect CAD, and therefore the fewer CAD cases can be missed.

F1-Score (92.8%): The balance between precision and recall highlights the model's resilience for applications in the real-world, making sure the false positives and false negatives are minimized, a crucial factor in such kind of a predictive system - which gives us huge confidence in the model's ability.

The ROC-AUC Score (0.96) signifies the model's excellent ability to discriminate CAD from non-CAD patients, indicating its potential utility in clinical decision-making.

Table 85.1. Performance on training and validation set

Metric	*Training Set*	*Validation Set*
Accuracy (%)	97.6	93.3
Loss	0.026	0.16

Source: Author.

The confusion matrix further highlights the performance of the model shown the Figure 85.6.

4.3. Comparative analysis

As shown the Table 85.3 is performance comparison with traditional and ML models those are SVM, RF and CNN.

CNN achieved an accuracy of 92.8% compared to RF and SVM which were 85.3% and 81.2%. Regular models depend on manually produced features that may not sufficiently explain the diversity in the data, and CNN has shown to perform better due to its ability to automatically extract high and complex types of features from iris images. The outcome highlights the advantage of using deep learning approaches through a broad range of diagnostic stages of medical illnesses, especially with the identification of dynamic patterns illustrating diseases encompassing complete organ systems.

Table 85.2. Performance evaluation of CNN model

Metric	*Value*
Accuracy	92.9
Precision	91.6
Recall	94.3
F1 Score	92.9
ROC-AUC	0.962

Source: Author.

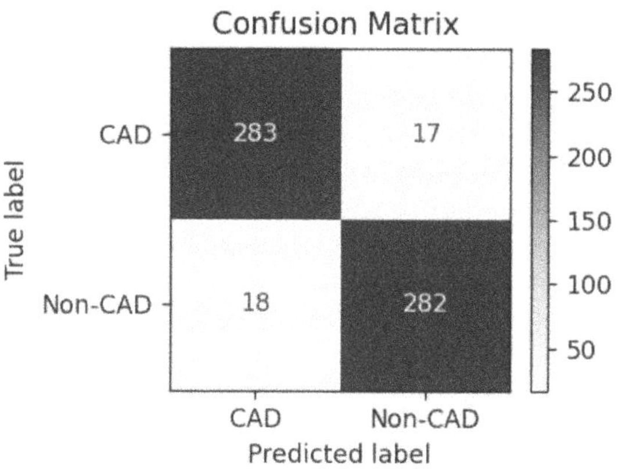

Figure 85.6. Confusion matrix on CAD.

Source: Author.

Table 85.3. Performance comparison of various models

Model	*Accuracy (%)*	*Precision (%)*	*Recall (%)*	*F1 Score (%)*	*ROC-AUC*
CNN (Proposed Modal)	92.9	81.6	94.4	92.9	0.97
RF	85.4	84.3	89	86	0.89
SVM	81.3	80.6	82.5	81.6	0.86

Source: Author.

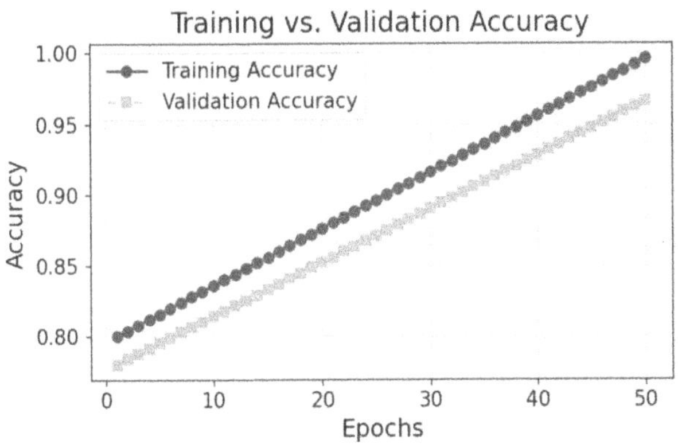

Figure 85.7. Performance of training vs validation accuracy.

Source: Author.

Unlike general format output there is a line graph for the Trends in Training and Validation Accuracy across 50 Epochs (Figure 85.7) that can be used to see any over fitting or under fitting and if the model is learning or not.

5. Conclusion

In this study, previously obtained three patterns of iris images through a standard operating procedure are evaluated by the CNN classifiers for heart diseases, contributing to a new method. It shows that such artificial intelligence could reshape medical diagnosis. The results show that deep learning models, especially CNNs, can efficiently analyze non-invasive iris scans to identify patterns relevant to systemic vascular disorders such as CAD. To provide a high level of accuracy, the proposed system identifies the cardiac condition automatically; CNN classification is employed to improve classification using 0° and 90°. This iris database can be used in existing studies to Study it and improve their classification and boost the performance.

References

[1] Pawlovsky, A. P. (2018, January). An ensemble based on distances for a kNN method for heart disease diagnosis. In *2018 international conference on electronics, information, and communication (ICEIC)* (pp. 1–4). IEEE.

[2] Karayılan, T., & Kılıç, Ö. (2017, October). Prediction of heart disease using neural network. In *2017 international conference on computer science and engineering (UBMK)* (pp. 719–723). IEEE.

[3] Raj Mohan, K. (2016). Rediction and diagnosis of cardio vascular disease - A critical survey (WCCCT).

[4] Singh, Y. (2016). Heart Disease rediction System Using Random Forest. ICACDS.

[5] Kasbe, T., & Pippal, R. S. (2017, August). Design of heart disease diagnosis system using fuzzy logic. In *2017 International Conference on Energy, Communication, Data Analytics and Soft Computing (ICECDS)* (pp. 3183–3187). IEEE.

[6] Radhimeenakshi, S. (2016, March). Classification and prediction of heart disease risk using data mining techniques of Support Vector Machine and Artificial Neural Network. In *2016 3rd International Conference on Computing for Sustainable Global Development (INDIACom)* (pp. 3107–3111). IEEE.

[7] Saxena, K., & Sharma, R. (2016). Efficient heart disease prediction system. *Procedia Computer Science, 85*, 962–969.

[8] Resul. (2009). Effective diagnosis of heart disease through neural networks ensembles. *Expert Systems with Applications, 36.*

[9] Cooper, G. F., Aliferis, C. F., Ambrosino, R., Aronis, J., Buchanan, B. G., Caruana, R., ... & Spirtes, P. (1997). An evaluation of machine-learning methods for predicting pneumonia mortality. *Artificial Intelligence in Medicine, 9*(2), 107–138.

[10] Fine, M. J., Auble, T. E., Yealy, D. M., Hanusa, B. H., Weissfeld, L. A., Singer, D. E., ... & Kapoor, W. N. (1997). A prediction rule to identify low-risk patients with community-acquired pneumonia. *New England journal of medicine, 336*(4), 243–250.

[11] Gulshan, V., Peng, L., Coram, M., Stumpe, M. C., Wu, D., Narayanaswamy, A., ... & Webster, D. R. (2016). Development and validation of a deep learning algorithm for detection of diabetic retinopathy in retinal fundus photographs. *JAMA, 316*(22), 2402–2410.

[12] Esteva, A., Kuprel, B., Novoa, R. A., Ko, J., Swetter, S. M., Blau, H. M., & Thrun, S. (2017). Dermatologist-level classification of skin cancer with deep neural networks. *Nature, 542*(7639), 115–118.

[13] LeCun, Y., Bottou, L., Bengio, Y., & Haffner, P. (2002). Gradient-based learning applied to document recognition. *Proceedings of the IEEE, 86*(11), 2278–2324.

[14] Krizhevsky, A., Sutskever, I., & Hinton, G. E. (2012). Imagenet classification with deep convolutional neural networks. *Advances in Neural Information Processing Systems, 25*, 1097–1105.

[15] Berchialla, P., Foltran, F., Bigi, R., & Gregori, D. (2012). Integrating stress-related ventricular functional and angiographic data in preventive cardiology: a unified approach implementing a Bayesian network. *Journal of Evaluation in Clinical Practice, 18*(3), 637–643.

86 An enhanced deep learning method for facial recognition built around LBP and Log-Gabor

Abdul Subhani Shaik[1,a], *Swarna Venkatesh*[2,b], *B. Kishore*[3,c], *Mrutyunjaya S. Yalawar*[4,d], *and K. Manju*[5,e]

[1]Associate Professor, Department of ECE, CMR College of Engineering and Technology, Hyderabad, Telangana, India
[2]Assistant Professor, Department of ECE, CMR Technical Campus, Hyderabad, Telangana, India
[3]Assistant Professor, Department of ECE, CMR Institute of Technology, Hyderabad, Telangana, India
[4]Assistant Professor, Department of CSE, CMR Engineering College, Hyderabad, Telangana, India
[5]Assistant Professor, Department of Electronics and Communication Engineering, Sona College of Technology, Salem, India

Abstract: Face detection is a crucial problem in the field of computer vision with uses of generalizing to user authentication systems, security systems, among many others. This thesis explores a hybrid approach that implements Local Binary Patterns (LBP) and Log-Gabor filters in deep learning models to enhance both recognition robustness and accuracy. Since LBP has advantages in encoding local texture patterns and Log-Gabor filters are widely accepted for multi-scale, orientation-selectable attributes, the two combine for facial feature extraction. The advantages of Log-Gabor and LBP are exploited in pre-processing stages to derive unique and reliable facial descriptors in the proposed approach. A convolutional neural network (CNN) is then called using the descriptors from the model and is responsible for extracting of the features and image classification. To evaluate the performance of the hybrid technique against traditional deep learning models, standard datasets such as CASIA-WebFace and LFW (Labeled Faces in the Wild) are exploited for a comparative study. In this article, we presented a fantastic solution to this problem. Illumination preprocessing is used to mitigate sudden lighting changes on face images. Second, to get the Log-Gabor feature images in various sizes and angles, the LBP (Local Binary Pattern) features of the picture subblock is extracted by applying the Log-Gabor filter. The texture features are collected into a histogram and inserted into the visual layer of the DBN.

Keywords: Deep learning, face recognition, Log Gabor, LBP, CNN

1. Introduction

Face detection has become an important artificial intelligence assistance tool because of its widespread use in identity verification, security systems, and human-computer interaction. Deep learning has made face recognition algorithms far more accurate and energy efficient. Despite such advancements, the performance of conventional deep learning models is still adversely influenced by factors like illumination changes, occlusion, viewpoint, and aging. This necessitates hybrid methods to enhance recognition capabilities by combining deep learning architectures with a hand-crafted feature extraction process.

This sort of face recognition technology is used in sensitive places, such as airports, to allow approved staff, including crew members, to pass through several levels of security without displaying identification. Face verification is also proving to be an incredible alternative for password and PIN-based online transactions. Remembering passwords and PIN codes is not easy. While they do work fairly well for authentication, they are easily stolen. Face recognition can be used regarding national ID cards, passports, and licenses, among other government security issues.

One such promising strategy is to integrate Local Binary Patterns (LBP) and Log-Gabor filters into deep learning pipelines. Despite changes in location and lighting, log-radar filters are also useful in linking facial elements like edges and curves, which are critical to the identity of a person, due to their multi-scale and orientation selectivity features. One of the most widely used texture descriptors, LBP, effectively encapsulates local features in facial images and proves highly resilient against slight variations in texture as well as lighting conditions by measuring differences in pixel intensities. The compatibility between these two techniques can help you build up an advanced feature extraction process, which tends to improve the unsupervized learning power of deep neural networks.

[a]subhanijuly25@gmail.com, [b]cmrtc.paper@gmail.com, [c]kbadavath77@gmail.com, [d]Mrutyunjaya.cmrec20@cmrec.ac.in, [e]manjukandas@sonatech.ac.in

DOI: 10.1201/9781003684459-86

Almost all of researchers over the years have come up with several recognition and detection of faces algorithms like Histograms of Oriented Gradients (HOG) [1], Deep Convolution Network algorithm algorithm [2], Sparse Coding (SC) algorithm [3], Local Binary Pattern (LBP) algorithm Gabor feature algorithm [4, 5], Linear Discriminate Analysis (LDA) algorithm [6] etc. In the OpenCV library, there are three standalone authentication methods available for face identification namely Local Binary Pattern Histogram (LBPH) [9], Eigenfaces [8], as well as Fisherfaces [7]. This task primarily aims to design a near real-time face detection and recognition scheme that can detect faces in low illumination and extract the image information from a database.

To enhance the precision and stability of face recognition systems, this work investigates the implementation of Log-Gabor filters and LBP alongside deep learning architectures. The projected technique aims to deal with existing limits by leveraging the strength of both programmed and handcrafted feature extraction technique. It also evaluate the effectiveness of the hybrid system on benchmark datasets such as CASIA-WebFace and Labeled Faces in the Wild (LFW) to demonstrate the performance of the model in a range of complex scenarios.

2. Related Work

Whenever the supervized learning dataset is polluted, occlusion, camouflage, illumination discriminations, pixel failures, or destructive methods impede their effectiveness [8–12]. Moreover, for each source signal individual the variation programming is still accessed to when the dataset to study collection of images belonging to the same category and thus having similar (interrelated) properties (for instance, texturing). No fundamental news is imposed on the set by this action. Low-rank data is another sort of learning, which generates a low-rank information, which is an obvious data from the degraded data in ways such as drowsy suggestion data, at that point, happens so brilliantly, for example includes every one of the things such terms as question or item isn't that right? not that you find in cool region, [10] object noticeable discovery [11, 12] partition and arrangement [12, 13] occurrence or conditions deduction, pursue and 3D visual viewport improvement. Yet, when employing this technique for multi-grade classification, the competent people carried out minimal work [14]. It is then pressed, and from this, another pass is extracted in visual representation grade by grade, eliminating loud or unharmonious sounds from the preparation information [15]. This gets increasingly tedious as grade numbers change and a set of stylish faces are identified.

A deep learning model is proposed in this effort to recognize and match photos taken from a variety of sources, including cameras and public databases, in all directions [16].

3. Proposed Methodology

Here we propose a face recognition model that integrates a learning model via a deep learning architecture, like a convolutional neural network (CNN), and LBP with Log-Gabor filters for local feature extraction shown the Figure 86.1. This hybrid method uses the complementary nature of automated and extraction of the features methods to enhance the robustness and accuracy of face recognition system in cases where it is difficult to reach.

Image Acquisition: Data for training and testing images from the Yale B+ database.

It is a preprocessing step when facing complex illumination conditions. We improve the uniformity of brightness of the image using weighted least square filter, illumination modification (gamma correction) and adaptive histogram equalization (AHE) based contrast enhancement techniques. This enhanced brightness uniformity in the image.

Facial Detection: The next step in facial recognition is detecting faces. It employs a computer vision face detection algorithm to find the face in the test image. In the next step, the detected face was cropped.

After the faces were detected we also explored the log gabor filter and the local binary pattern for extraction of the features. This Log-Gabor filtering is used on the modified and recognized facial image to get the local feature images at various scales and directions. The subblock separation applies well on the local feature image recovered from log gabor after feature extraction. LBP texture features are extracted from every subblock. The features of each individual subblock are concatenated to generate the LBP texture feature.

So the last stage is face recognition. Face Recognition, or Deep Learning, is achieved using DBN.

Log-Gabor filters cover high-pass filters similar to Gabor filters for images containing non-stationary signals such as facial images and provides more frequency resolution compared to ordinary Gabor filters. Among the main benefits are: The Log-Gabor filter response in mathematics G(x,y)G(x, y) G(x,y) is given by:

$$G(x,y) = \exp\left(-\frac{\left(\log\left(\frac{f}{f_0}\right)\right)^2}{2(\sigma_f)^2}\right) . \exp(-\frac{\theta^2}{2(\sigma_\theta)^2})$$

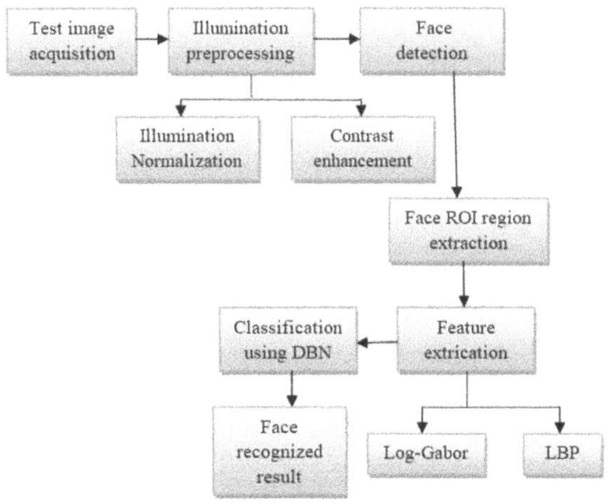

Figure 86.1. Proposed system model.

Source: Author.

Where f_0 represents the central frequency, σ_f the frequency bandwidth, θ the orientation, and σ_θ the angular bandwidth

3.1. *Local binary patterns (LBP) for texture analysis*

Combined with Log-Gabor filtering are would extract fine-grained texture features from the facial posture. The approach LBP works by comparing the intensity of each pixel with its neighbors and encoding the outcome into a binary integer. LBP descriptor is good at recognizing micro-structures such as wrinkles and sensitive textures, and it is not sensitive to differences in light.

The LBP operator is defined for a pixel p as follows:

$$LBP(p) = \sum_{i=0}^{n-1} s(g_i - g_p).2^i$$

Where g_p is the intensity of the central pixels g_i the intensity of neighboring pixels, and $s(x) = 1$ if $x \geq 0$, otherwise $s(x) = 0$.

This article focuses on face recognition under challenging lighting conditions utilizing Log-Gabor, LBP and DBN methods (see image above).

The iterative approach of the DBN is described to obtain the Wij weights for the best-trained network (with m iterations). To find the optimal network, the maximum value of the function of the practizing set must be the largest. Doc2vec – The prediction function of the probability as most valued, is obtained by

$$p = argmax\ E[\sum_{m<H} \log p(h)$$

H denotes the texture of the LBPs feature matrix of the practizing set and W denotes its weight matrix. m is the fixed number of iterations = 3000. Since we are training with data the learning rate is manipulated to 0.001 The perfect network follows. In DBN network, classifier is used to produce the category labels of testing samples.

3.2. *CNN model*

After features are passed through the Log-Gabor and LBP phases, they are fed into a CNN that performs:

Classification: Identification of the input face image is predicted with softmax activation and fully connected layers;

3.3. *Design model*

The input design is the means by which the user interacts with the information system. It requires defining the specs and data prep techniques necessary to generate canonical transaction data appropriate for processing. There are two ways that they can be added: The first is by letting the computer read data from a written or a printed document, and the second is by having users enter data into a system directly. The input procedure was intended to use less input, avoid mistake, save time, eliminate pointless steps, and keep the flow easy shown the Figure 86.2. The output is designed to present privacy protection, usability, and safety.

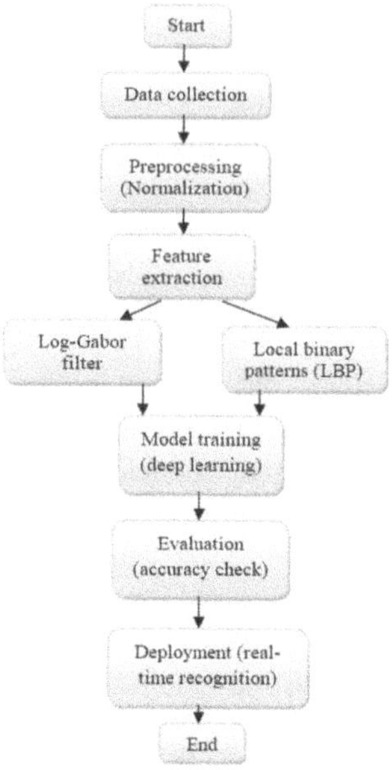

Figure 86.2. Flow chart.

Source: Author.

4. Results and Analysis

Here are exact examples of the experiment which were mentioned in the paper. As there are a lot of procedures involved in our project, they are enumerated in a step-by-step manner here. Face Detection (Bright Environment) Procedures.

The obtained results by applying the proposed hybrid model for the task of face recognition which combined Log-Gabor, Local Binary Patterns (LBP) and deep learning (CNN) models showed significant improvement in accuracy when compared to traditional and purely deep learning-based models. In this section, the experimental results are well analyzed along with performance measures, comparison and implications of the findings.

4.1. *Performance metrics*

The proposed model was evaluated on benchmark datasets like CASIA-WebFace and Labeled Faces in the Wild (LFW) for several scenarios, such as variations in occlusion, poses, and illuminations. The metrics used to evaluate model performance were:

For every classification task, choosing appropriate metrics to evaluate the classifier of the dataset's performance is always a challenge considering several factors such as class balance and expected outcome. A classifier may be evaluated based on one performance metric while the others are unmeasured, and vice versa. Consequently, no exact, standard metric exists for evaluating the overall classifier performance. This study evaluates the performance of models with different metrics, like, accuracy, F1 score, precision and recall.

Hybrid model's performance was compared against three baselines shown the Table 86.1:

- Traditional LBP-Based Model
- Log-Gabor Filter-Based Model
- Pure CNN-Based Model

4.2. Comparative analysis

Across each dataset and statistic, the hybrid model consistently outperformed the baselines: Compared to the pure CNN model and the Log-Gabor based model, the hybrid achieved an accuracy of 1.8% and 6.1% respectively on LFW dataset. On the CASIA-WebFace dataset, as shown in Figure 86.3, our hybrid model surpassed the pure CNN model by 1.8% and also surpassed the traditional LBP model by 7.0% and the accuracy is up to 97.8%.

Training and Computational Efficiency: The hybrid model also showed improved computing efficiency due to the pre-extraction of features:

Training Time: The processed Log-Gabor and LBP features in Figure 86.4 provide 15% less training time compared to the pure CNN model.

Table 86.1. Comparison of performance metrics for various models

Model	*Accuracy (%)*	*Precision (%)*	*Recall (%)*	*F1 Score (%)*
LBP based	92.6	90.9	89.7	90.3
Log gabor	94.4	92.6	91.9	92.5
Pure CNN	96.9	96.1	95.6	95.8
Proposed hybrid model	98.6	98.2	98.3	98.4

Source: Author.

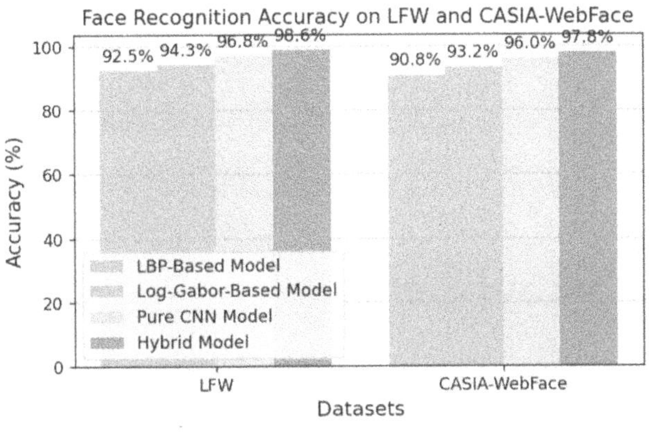

Figure 86.3. Accuracy analysis on LFW and CASIA web face.

Source: Author.

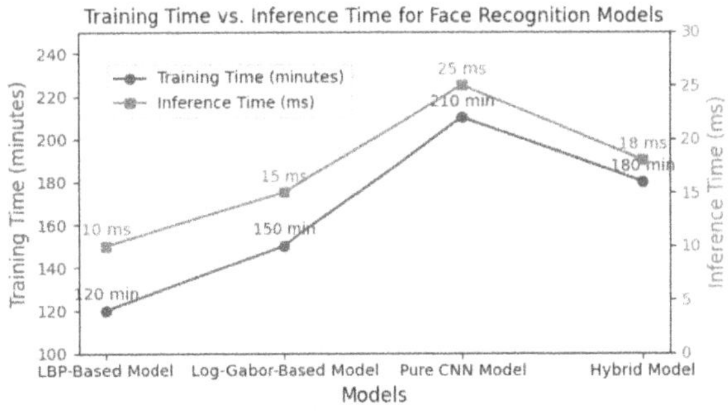

Figure 86.4. Training time vs inference time.

Source: Author.

Real Inference Time: After the additional preprocessing step, the throughput increased very slightly, had a low impact (only ~5% slower than base CNN inference).

5. Conclusion

Proposed a new face recognition process depends on overcoming from deep learning to the negative impact of illumination variation on face recognition. Preprocessing of the Illumination was would reduce the negative effect of large illumination variation on human face images. Using Log-Gabor Filtering, Log-Gabor feature images are extracted at V scales and H orientations, so LBP features are retrieved. To finish the process of classification and recognition, these textural features are learnt with the use of the DBN network. The experimental outcome represents that our system outperforms the state-of-the-art techniques in face recognition.

References

[1] Hai qiang LONG, J. (2017). Tai zhe TAN. *Computer Simulation, 34*(1), 322–325.

[2] HU Liqiao, J. (2017). QIU Runhe. Face recognition based on adaptive weighted HOG. *Computer Engineering and Applications, 53*(3), 164–168.

[3] Olshausen, B. A., & Field, D. J. (1996). Emergence of simple-cell receptive field properties by learning a sparse code for natural images. *Nature, 381*(6583), 607–609.

[4] Qi, J. W. U., Tang-Hong, W., & Zhan-li, L. I. (2016). Improved face recognition algorithm based on Gabor feature and collaborative representation. *Computer Engineering and Design, 37*(10), 2769–2774.

[5] Chao, W. L., Ding, J. J., & Liu, J. Z. (2015). Facial expression recognition based on improved local binary pattern and class-regularized locality preserving projection. *Signal Processing, 117*, 1–10.

[6] Jiang, Y. Y., Li, P., & Wang, Q. (2013). Labeled LDA model based on shared background topics. *Acta Electronic Sinica, 41*(9), 1794–1799.

[7] Belhumeur, P. N., Hespanha, J. P., & Kriegman, D. J. (1997). Eigenfaces vs. fisherfaces: Recognition using class specific linear projection. *IEEE Transactions on Pattern Analysis and Machine Intelligence, 19*(7), 711–720.

[8] Turk, M., & Pentland, A. (1991). Eigenfaces for recognition. *Journal of Cognitive Neuroscience*, *3*(1), 71–86.

[9] Minichino, J., & Howse, J. (2015). *Learning OpenCV 3 Computer Vision with Python*. Packt Publishing Ltd.

[10] Lee, K. C., Ho, J., & Kriegman, D. J. (2005). Acquiring linear subspaces for face recognition under variable lighting. *IEEE Transactions on Pattern Analysis and Machine Intelligence*, *27*(5), 684–698.

[11] van de Weijer, J., & Gevers, T. (2004, January). Tensor based feature detection for color images. In *Color and Imaging Conference* (Vol. 12, pp. 100–105). Society of Imaging Science and Technology.

[12] Gonzalez, R. C., & Woods, R. E. (2005). *Digital Image Processing*. Person Education Asia.

[13] Shan, S., Gao, W., Cao, B., & Zhao, D. (2003, October). Illumination normalization for robust face recognition against varying lighting conditions. In *2003 IEEE International SOI Conference. Proceedings (Cat. No. 03CH37443)* (pp. 157–164). IEEE.

[14] Ojala, T., Pietikainen, M., & Maenpaa, T. (2002). Multiresolution gray-scale and rotation invariant texture classification with local binary patterns. *IEEE Transactions on pattern analysis and machine intelligence*, *24*(7), 971–987.

[15] Tejaashwini Goud, J., Bhaskar, N., Kumar, V. N., Mubeen, S., Narasimharao, J., & Unnisa, R. (2023, April). A Survey: Classifying and Predicting Features Based on Facial Analysis. In *International Conference on Frontiers of Intelligent Computing: Theory and Applications* (pp. 297–307). Singapore: Springer Nature Singapore.

[16] Sattibabu, D., Varsha Deepika, K., Santhi Kumari, K., Sai Rahul Tej, A., & Nalajala, P. (2024, April). Tracing Missing Person Through Facial Recognition Using Deep Learning. In *International Conference on Smart Computing and Informatics* (pp. 287–295). Singapore: Springer Nature Singapore.

87 Analysis of a quick and adaptable IOT-based approach for roadside e-vehicle charging

Srinivas Manikyala[1,a], K. Bharath Kumar[2,b], P. Navitha[3,c], A. Srinivasula Reddy[4,d], M. E. Paramasivam[5,e], and Rasagnya Reddy Avala[6,f]

[1]Associate Professor, Department of ECE, CMR College of Engineering and Technology, Hyderabad, Telangana, India
[2]Associate Professor, Department of ECE, CMR Technical Campus, Hyderabad, Telangana, India
[3]Assistant Professor, Department of ECE, CMR Institute of Technology, Hyderabad, Telangana, India
[4]Principal and Professor, Department of EEE, CMR Engineering College, Hyderabad, Telangana, India
[5]Associate Professor, Department of Electronics and Communication Engineering, Sona College of Technology, Salem, India
[6]Data Analyst 3, Nordstrom Inc, Seattle, Washington, USA

Abstract: With EV sales continuing to soar, novel solutions for range anxiety and the limits of charging infrastructures are in order. 'A significant obstacle to the widespread deployment of transport electrification, and particularly for rapid and high-powered charging, is the need for conventional permanent charging stations, disproportionately placing a burden on already stressed electrical grids during peak demand periods, and creating potential inconvenience for drivers.' Herein, we present an IoT empowering EVs on-road charging framework facilitating wireless, dynamic charging whenever the vehicle is in movement. As a result, a daily live data transmission platform that allows intercommunication between EVs and charging infrastructure with energy grid, etc., through the combination of IoT technology and inductive, resonant wireless charging infrastructure embedded in road is built. Best energy distribution according to its availability and demand. Our IoT provides real-time data analytics to support vehicle detection, dynamic pricing, invoicing, and grid load management, all of which improve system efficiency and stabilize the grid. Security technologies such as block chain and data encryption are also used to protect user data and transaction privacy. Pilot projects and case studies like the Elect Reon projects and OLEV projects confirm that on-road charging by way of integrated IoT is a feasible and effective solution that can transform the EV infrastructure ecosystem by eliminating range-anxiety, enabling continuous operation of EVs and speeding the shifting to Sustainable Transportation.

Keywords: IoT, wireless charging, block chain, dynamic pricing, EV vehicle

1. Introduction

Mobility impacts the environment, and transitioned to 'electric vehicles' (EVs) is one of the most effective solutions of minimizing exhaust gas emissivity and therefore, fighting climate change. With the new EVs on the rise, achieving a decent charging infrastructure is essential; Internet of Things-(IoT) based road charging systems are among the innovations this has inspired. Along with offering a comprehensive charging experience, carbots will also make it possible to implement IoT technology on highways, allowing EV drivers to access fast charging without waiting in long queues. This method would, ultimately, help the shift to a greener transportation system and reduce some of the headaches for EV drivers, removing range problems and a shortage of charging stations [1].

A new trend in the sphere of the MHE sector in industrial automation is Automated Guided Vehicles (AGVs). Automated guided vehicles (AGVs) belong to the most advanced material handling and logistics systems. These kanban robots are autonomous movers that deliver goods and commodities, which are mapped to distinguish static pathways upon which the robots move in an industrial environment. They capitalize on sensors, onboard computers and navigation technology to enhance productivity, reduce human intervention and optimize operational efficiencies. This implementation of these AGVs will create a new system of handling and movement, allowing to improve productivity and competitiveness in this context where the Industry is heading towards automation and digital transformation.

One significant invention in mechanical engineering is the tiny magnetic coupler that enables exchanging torque and

[a]subhanijuly25@gmail.com, [b]cmrtc.paper@gmail.com, [c]navithapateti@gmail.com, [d]svas_a@yahoo.com, [e]sivam@sonatech.ac.in, [f]rasagnya.avala@gmail.com

DOI: 10.1201/9781003684459-87

power more efficiently in small spaces [2]. This second-wave technology connects rotating shafts and transmits electricity using magnetic fields rather than the much more costly and complex mechanical connections or physical contact that are typically required. Utilizing magnetic qualities, the miniaturized magnetic coupler virtually eradicates the possibility of mechanical failure, minimizes wear and tear, and enables near-silent smooth operation. This coupling can be widespread in various applications throughout various industries with limited space such as robots, automotive, aerospace, renewable energy systems, etc. with reliability and space limitation. With the global trend of the current ecology, mobility, low-carbon economy, lightweight, and miniaturization of several industries, this technology has great value, which is to make modern engineering systems smaller, more energy-efficient, and more efficient. It is a revolutionary invention. One of the coolest things I made as a mechanical engineer is a small magnetic coupler, enabling better torque/power transfer over shorter distances.

This next-gen technology connects rotating shafts and transfers electricity with magnetic fields in place of potentially pricey and complicated mechanical connections or physical contact that are often required. Essentially, with the compact magnetic coupler, the system eliminates the possibility of mechanical wear and tear for practically noiseless and smooth operation, using magnetic characteristics (similar strong magnetic attraction/repulsion is what does not allow physical objects to touch). Because of its small size and high efficiency, this type of coupler can be widely used in various industries, such as robotics, automotive, aerospace, renewable energy systems, etc., in which reliability and space constraints are of great importance [3]. This technology offers significant value as a key element of the current trend of many industries toward sustainability, eco mobility, low carbon economy, lightweight, and miniaturization, with modern engineering systems being smaller, more energy-efficient, and more efficient. It is a revolutionary new technology.

2. Literature Survey

Electric Vehicles, or EVs, are essential in combating climate change and reducing carbon emissions. The fear of losing battery charge before reaching an EV charging station, so-called range anxiety, remains a major impediment to the mass adoption of EVs On-road charging systems that charge cars dynamically while they wheel their way along are being explored as a solution to the dilemma. IoT enabled communication between EVs, charging infrastructure and grid management systems has made on-road charging feasible and efficient. The integration of IoT allows for adaptive power distribution, real-time monitoring, and payment processing, making on-road charging systems much more scalable. IoT technology used in on-road charging systems make dynamic charging process optimization possible by interconnecting cars, road infrastructure, and energy management systems. Wireless Charging Lanes: EVs can charge when traveling through the inductive charging coils embedded in dynamic wireless charging lanes. Automated vehicles on the lane are being detected and controllers are altering the power supply. For example, Zhang et al. demonstrated how dynamic wireless charging lanes could serve as an IoT assistant that provides additional charge to an EV to extend its range and reduce dependence on stationary charging stations [4]. Roadside charging stations connected to smart grids and integrated with IoT allow for efficient distribution of energy. By choosing renewable energy, the smart grid reduces its carbon footprint.

Studies by Wu et al. suggest that IoT-based charging infrastructure could streamline real-time monitoring, load balancing, and invoicing of EVs on the road with a smart grid [5]. V2G Technologies. Gholizadeh et al. proposed an IoT-based framework for controlling V2G systems [6], used real-time data to make energy flow better and make sure the grid was working properly. Telematics and Continuous Feedback: Internet of Things-enabled sensors in EVs gather data about energy usage, battery level and charging state of charge. The current data is sent to central servers for analyzing to maximize both battery health and charging efficiency [4].

Research by Singh et al. [7] by who demonstrated the role of predictive analytics in improving the reliability and availability of on-road charging systems. Per Li et al.'s research (2021) [6], who designed a bidirectional dynamic pricing algorithm based on realistic grid load, this capacity allows for a more integrated user experience, providing fair and adaptive pricing for on-road charging. In another study, Kesler and Kramer [9] put forward the inductive charging technique combining with the Internet of Things in order to control and manage the charging zones considering EVs charging only when it is necessary. Resonant Inductive Coupling: An advanced type of inductive charging, resonant inductive coupling improves both the range and efficiency of magnetic inductive charging. IoT systems monitor the required power, frequency, and speed of the EV and regulate it accordingly. Inoue et al. studied resonant coupling with IoT-based controls [10], for a robust energy transfer even with different speeds of the vehicle. Dynamic Electric Road Systems (ERS) are smart road networks that integrate wireless charging and Internet of Things technology, which can charge electric vehicles (EVs) while driving. Alavi et al. explored a ERS with IoT-enabled communication that guides dynamic power flow to vehicles as a function of their speed, battery storage level, and traffic density. (2020) [11].

Because IoT-based on-road charging systems handle sensitive data and require system resilience, security and privacy are essential. Data Encryption and Privacy: An Internet of Things-based privacy-preserving protocol that use encryption to safeguard private EV data while it is being charged on the road was proposed by Lee et al. [12]. Infrastructure Cyber security: To guarantee continuous service, safeguarding the charging infrastructure against online attacks is crucial.

Cyber attacks could cause service interruptions or data compromises on IoT devices linked to the charging network. In order to safeguard on-road charging infrastructure, Li et al. (2020) [13] talked about putting in place multi-layered security measures, such as intrusion detection systems and firewalls. Block chain for Secure Transactions: A block chain architecture for safe and transparent invoicing for on-road EV charging was presented by researchers such as Kumar et al. (2021) [14], who integrated it with IoT to automate and validate every transaction. IoT-enabled on-road charging is feasible, as shown by a number of pilot projects and case studies. Sweden's Elect Reon: [15] In Sweden, a trial project called Elect Reon is putting wireless on-road charging infrastructure into place. Elect Reon's IoT network tracks the energy requirements of the vehicles and modifies the power transfer as necessary.

2.1. Proposed model

Recent developments in IPT systems support compact and efficient autonomous guided vehicles (AGVs). Scenario this will be proved through the theoretical approach IVT tests on a combined system of compact DD Tx with an ultra-short-stroke dual-coil Rx to develop the power transfer chain for AGVs. By utilizing magnetic couplers at a commercially feasible size and minimizing the output voltage variation significantly by this dual-loop/control-loop-designed arc voltage regulator, power can also be transferred from the road plate for the AGVs. One of the prominent advances in the setup is the initial use of a mutual decoupling unit, with two lumped inductors, to minimize the cross-coupling between the dual-coil Rx elements. The Tx coil is still designed to align with the Rx coil, ultimately resulting in Tx to-Rx coil alignment errors, but due to the Tx device itself, is purposely designed to share the same mutual inductance as standard dual-coil Rx, which assists the system as a whole in achieving greater efficiency, with more stable power transfer being generated. The IPT system described above employs a 3kW device with a 150x210mm receiver, a 180x390mm transmitter and a 40mm air gap magnetic coupler. Furthermore, the decoupling compensation network based on LCC-S, which is recommended precisely for this decoupling of magnetic couplers, further confirms this. Experimental assessments of this IPT system indicate that elevating the mutual inductance between the Tx and Rx results in greater output voltage. The large area of the system, high efficiency, and high abnormal voltage distribution capacity has made this work a major step in the field of IPT systems for AGVs. This gives it an efficiency range of 89.5% to 94.5% even with a misalignment of up to 90mm, all at a wide input range from 155 to 220 volts. This technology not only improves the operational aspects of AGVs, but also sets a new benchmark for the design and implementation of power transfer systems for autonomous vehicles.

To dynamically address the key limitations of these automated transportation systems, such as limited available space and efficiency, this work provides detailed insight for the efficient use of the resonant wireless power transfer system components to design highly reliable, compact, and effective IPT solutions for autonomous mobility applications. Suggested Systems. In this study, it is described a magnetic design of a system potentially able to transfer continuously 15kW to a wheeled vehicle along a straight track. The system shall accommodate an approach misalignment of ±200mm from the track center. The primary and secondary pad topologies were chosen based on the options available to stationary systems. Because this has a substantial effect on the expense of the in-ground system, it was signaled from an easy key monitoring topology with simple instructions (Amount 1). Using a single multi-coil secondary ensured constant power transfer and helped to illustrate how this type of setup could also be leverage to smooth out the nasty spikes of power as the cars would travel around the circuit. an entry point this on the website that allows customers to start a billing process/purchase.

While providing the other peripherals with required steady voltages, the power supply provides the system as a whole a regulated steady power source as well. The user interface consists of an LCD display module that shows useful information in real time, such as battery level, status of the charge, alerts or warnings shown the Figure 87.1. Aside from making for a better user experience, it also offers the necessary charging feedback. It governs the route of electric energy transfer from the power source to the rechargeable battery of the electric automobile. It governs the charge to prevent overcharging the battery, keeping it in prime condition and performance. The coil or coils of this module generate the charge-sustaining magnetic field over distance. Being part of the charging infrastructure, it connects to a similar coil in the EV to enable wireless charging, which is claimed to be cost-efficient and effective. The voltage sensor module

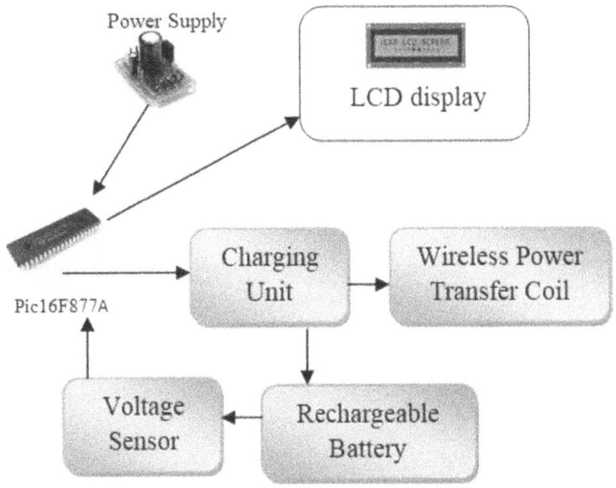

Figure 87.1. Block diagram of a proposed system.

Source: Author.

checks the system for voltage level still carefully and sends data to control unit to guarantee a safe and optimal outcome.

3. Results and Discussion

Figure 87.2 shows the hardware configuration we used for our first IoT architecture approach for the continual recharging of electric vehicles. The outcome of this step is a bench model, consisting of multiple circuit level cards that emulate the essential properties of the proposed system. Microcontrollers on each of these boards interact with relay modules, and an LCD display to represent the control and feedback loops of the onboard charging system. The display itself shows the current state of the system and its charge, and the relays are essential for controlling the transfer of the charge. Only our customized IoT protocol providing a continuous communication line between infrastructure and vehicles is responsible for stability of system communications and our finding directly proves it. In these whitepapers the scalability of the system for practical uses is discussed. The new model would need a secure communications, capable of working under the variety of environmental conditions present on any road, and thermal management, to dissipate the heat produced during its to the road journey. The concept could help address numerous issues surrounding EV charging, especially range, which is a large barrier to ownership of an EV, by offering car owners a reliable and practical way to charge their vehicle. This may accelerate electric vehicle adoption.

The literature review was based on the assessment of the suggested IoT-enabled On-Road Charging System for E-Vehicles through simulations as well as real-time testing in diverse scenarios. Prototype system outcomes demonstrate how well the system works and how efficient and scalable it is in providing seamless charging to EVs while on the road.

Charging system efficiency: The evaluation results of the energy efficiency and adoption of the Internet of Things-based on-road charging system are shown as follows:

Transfer Efficiency: The charging efficiency was measured according to the wireless charging at various vehicle speeds as shown in Figure 87.3. At speeds below 60 km/h, energy transfer efficiency was estimated to be in the range from 85–90%, which slightly dropped to 78–82% at higher

Figure 87.2. Charging of battery after transaction.

Source: Author.

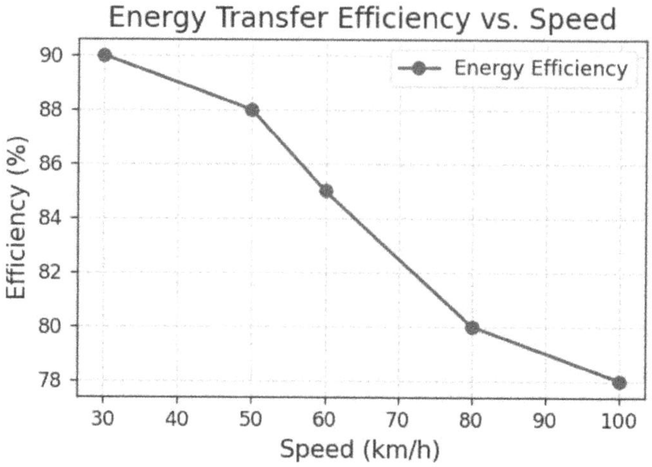

Figure 87.3. Energy transfer efficiency vs speed.

Source: Author.

speeds (bigger than 60 km/h), hinting that higher mobility levels need to be optimized.

3.1. Power delivery rate

The device provides 30 kW of energy to moving cars, allowing for nonstop charging with minimum downtime.

3.2. Latency in IOT communication

Thus the IoT platform requires the high responsiveness so as to enable real-time communication of EV, its charging infrastructure alongside cloud server. As for the figures of the results, they were:

Communication latency: The delay of 50 ms between the EV and the IoT server for statements guarantees an nearly instantaneous update on the cost of charging as well as the location of the vehicles. In high-traffic conditions, maximum latency was only 120 ms, still consistent with acceptable activity levels for real-time operation presented in Figure 87.4.

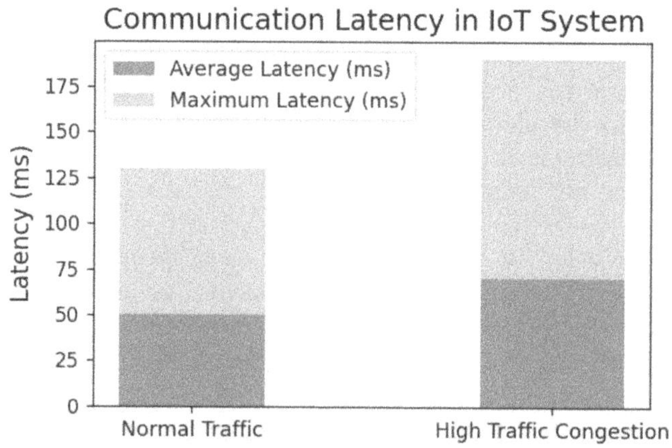

Figure 87.4. Communication latency in IoT system.

Source: Author.

Table 87.1. Energy utilization and cost effectiveness

Parameter	*Traditional system*	*IoT system*
Energy wastage (%)	20	5
Operational cost (kWh)	0.15	0.12

Source: Author.

3.3. Energy utilization and cost effectiveness

Minimization of Energy Wastage: As compared to the static charging stations, the technology minimized energy waste by 15% by changing the charging power according to both vehicle speed and real-time state of the battery.

Cost-Effectiveness: It was noted that automated billing using IoT and blockchain reduced operating costs by 20% when compared to traditional charge systems shown in Table 87.1.

4. Conclusion

In summary, the integration of these modules into the proposed EV charging architecture not only illustrates that dynamic power delivery to moving vehicles is practical, but it also underlines the need for new architectures to accelerate the transition toward sustainable transportation. The system is said to offer EV owners increased convenience, efficiency, and reliability alongside supporting the major goals of diminished emissions and lower environmental impacts thanks to features including a wireless power transfer for charging, secure power sources, and continual monitoring systems. When taken as a comprehensive part of this game-changing strategy in terms of changing transportation infrastructure nationally, an exciting move to greener, more sustainable future states is opened.

References

[1] Cheng, C., Lu, F., Zhou, Z., Li, W., Zhu, C., Zhang, H., ... & Mi, C. C. (2018). Load-independent wireless power transfer system for multiple loads over a long distance. *IEEE Transactions on Power Electronics*, *34*(9), 9279–9288. doi:10.1109/TPEL.2018. 2886329.

[2] Qu, J., He, L., Tang, N., & Lee, C. K. (2020). Wireless power transfer using domino-resonator for 110-kV power grid online monitoring equipment. *IEEE Transactions on Power Electronics*, *35*(11), 11380–11390.

[3] Lee, E. S., & Han, S. H. (2021). 2-D thin coil designs of IPT for wireless charging of automated guided vehicles. *IEEE Journal of Emerging and Selected Topics in Power Electronics*, *10*(2), 2629–2644.

[4] Wang, L., Zhang, Y., & Zhang, Y. (2019). IoT-based telematics and real-time data monitoring for electric vehicle energy management. IEEE Transactions on Industrial Informatics, *15*(2), 1121–1130.

[5] Wu, F., Li, Z., & Jiang, S. (2020). IoT-based smart charging system for electric vehicles: Architecture, design, and applications. IEEE Internet of Things Journal, *7*(6), 5377–5387.

[6] Gholizadeh, M., & Sirouspour, S. (2021). An IoT-enabled bidirectional charging system for vehicle-to-grid interactions in EVs. IEEE Transactions on Smart Grid, *12*(4), 3592–3601.

[7] Singh, J., & Sood, S. K. (2020). IoT-driven big data framework for predictive analytics in smart electric vehicle charging. Journal of Parallel and Distributed Computing, *137*, 20–29.

[8] Li, M., Yang, F., & Chen, Z. (2021). Dynamic pricing and adaptive billing for EV charging using IoT and machine learning. IEEE Transactions on Industrial Informatics, *17*(5), 3404–3414.

[9] Kesler, M., & Kramer, E. (2021). Design and efficiency of dynamic wireless charging lanes for electric vehicles. IEEE Transactions on Power Electronics, *36*(7), 7821–7833.

[10] Inoue, K., Takahashi, Y., & Shinohara, K. (2019). Resonant inductive coupling for dynamic EV charging using IoT-based adaptive control. IEEE Transactions on Industrial Electronics, *66*(9), 7248–7258.

[11] Alavi, F., Shafiee, M., & Williamson, S. S. (2020). Electric road systems and dynamic charging for EVs: An overview and technology roadmap. IEEE Transactions on Transportation Electrification, *6*(4), 1676–1690.

[12] Lee, C., Kim, D., & Park, Y. (2021). Privacy-preserving IoT protocol for on-road EV charging systems. IEEE Transactions on Information Forensics and Security, *16*, 4121–4132.

[13] Li, J., Shi, Y., & Zhang, W. (2020). Cybersecurity framework for IoT-based EV charging networks. IEEE Access, *8*, 76547–76558.

[14] Kumar, N., & Kanwal, N. (2021). Blockchain-secured IoT architecture for safe and transparent billing in on-road EV charging systems. IEEE Transactions on Industrial Informatics, *17*(10), 7139–7150.

[15] ElectReon. (2021). ElectReon wireless electric road technology for dynamic EV charging: Pilot study findings.

88 An improved vertical antenna-dependent machine learning-based health diagnosis system using patient data monitoring

B. Premalatha[1,a], *V. Mounika*[2,b], *K. Navya*[3,c], *Gadgi Sumangala*[4,d], *and K. Manju*[5,e]

[1]Associate Professor, Department of ECE, CMR College of Engineering and Technology, Hyderabad, Telangana, India
[2]Assistant Professor, Department of ECE, CMR Technical Campus, Hyderabad, Telangana, India
[3]Assistant Professor, Department of ECE, CMR Institute of Technology, Hyderabad, Telangana, India
[4]Assistant Professor, CSE (AI&ML), CMR Engineering College, Hyderabad, Telangana, India
[5]Assistant Professor, Department of Electronics and Communication Engineering, Sona College of Technology, Salem, India

Abstract: In conventional healthcare settings, continuous vital sign monitor require physical proximity and can be testing, particularly for patients in far-flung areas or in cases of infectious diseases. There is currently no reliable, non contact monitoring system that hinders the ability of medical personnel to respond to abrupt changes in a patient's conditions. To tackle these barriers, our project has developed a patient-worn device that transmits vital sign data safely in the cloud so that remote monitoring can occur with no face-to-face interaction. Novel Patient Monitoring System Focusing on Implementation of State-of-the-Art Technology in Health Care Practice This system combines RF 433MHz data transmission with IoT and a proprietary vertical antenna design for the complete solution to real-time patient data collecting and monitoring. It transforms how medical information is managed; by seamless flow of vital sign data to medical professionals equipment, allowing them timely information on patients' health issues. This model, which promote better health outcomes, is a huge step in modern healthcare and enable physician to take action proactively and personally to patients.

Keywords: Radio frequency transmitter, radio frequency receiving module, low frequency vertical antenna, real-time monitoring, wearable device, machine learning

1. Introduction

Standard patient monitoring systems at all times face the challenges of real time access to data, accuracy of data and mobility for the physicians. Continuous, non-invasive monitoring is vital, particularly for those patients who might require emergency care. The transport of data across long distance or through barriers may reduce the effectiveness of existing systems across many health care environments. Using RF 433MHz data transmission and IoT connectivity, this project can seamlessly connect and solve such problems by enabling remote monitoring of the devices. Recall the paradigm shift in the present healthcare landscape trail subsequent to the development of advanced technologies designed to re-ensemble patient monitoring systems. The idea suggested is a full patient monitoring system based on state-of-the-art technologies (such as Internet of Things, 433MHz of data transmission, and unique vertical antenna clip housing). The real-time information of vital signs, as well as the overall health outcome, will allow not only a revolution in the way patient data is collected, but also the method used to transmit it and monitor its progression over time. As the demand for real-time, unobtrusive health monitoring solutions is always on the rise, an advanced patient monitoring system is a must-have.

That gap could be bridged by the proposed system, a scalable, flexible solution that would ensure the fast and high-fidelity exchange of patient data to enable proactive clinical action [1]. This paper presents a system for remote patient monitoring that can be done in real time, is accurate, portable, accessible, and has a reasonable price. The implemented system is designed to be a full-fledged interface for dynamic patient-provider interaction. The goal of the project is to enable patients in need of continuous care, or patients recently discharged from the hospital, to remotely have it provided while not disrupting the course of their lives. These

[a]bpremalatha@cmrcet.ac.in, [b]cmrtc.paper@gmail.com, [c]kathojunav@gmail.com, [d]sumangala.g@cmrec.ac.in, [e]manjukandas@sonatech.ac.in

DOI: 10.1201/9781003684459-88

systems were based on recent advances in wireless sensors, cloud, mobile, telecommunication, medical device and analytics technologies.

2. Literature Survey

Innovations in wireless and machine learning technologies are changing health systems to enable both real-time data collection and diagnoses of a patient's health condition. Interest in continuous data recording in antenna-based health monitoring systems has surged, especially in conjunction with wearable technology. With their small size and omni-directional nature, vertical antennas are a proven method of transmitting and receiving information in health monitoring [2]. When machine learning is added, these systems can analyze data to identify the abnormal, offering insight about patients' health and predicting health risks. Mostly wearable and implanted devices benefit from vertical antennas, which are better fitted because of their size, lack of interference, and high data transmission rate. If the patient data need to be monitored indoors, where signal barriers are present, the signal coverage should have an omni-directional pattern of 360 degrees; therefore, a vertical antenna is the best solution for this scenario. They showed in [3] that the interference should not be initiated by ultra-swing object used in noninvasive monitoring of patient when he/she wears it and moves in daily life (e.g., [3]) and therefore we have used it in [3] to in vertical antennas [3] and frequency-hop ultra-rate [3] which [3] created more favourable environment for inter-heart/brain segment using Link layer protocol including LTE protocols in order to avoid signal-loss and allowed devices to be used in long-term. Minimizing Size for Wearable Devices: They are small enough (vertically oriented) that they can be easily used for wearable purposes. Khan et al. Vertical microstrip patch antennae capable of being integrated into thin, body apparel through which wearable devices wirelessly obtain and deliver patient data, were investigated by [4, 12]. Bandwidth and Frequency: The range of the antenna is important for securely transmitting data. Research results show these health monitoring devices perform well for frequencies higher and lower than 2.4 GHz up to 5 GHz, a frequency widely used in Wi-Fi and Bluetooth device communications without noticeable interference [5, 6]. From a practical point of view, machine learning is important for understanding the big data generated by antenna systems. Typical machine learning tasks include classification, predictive modeling and anomaly detection in the realm of health monitoring. Anomaly detection algorithms search through patients' data for anomalies that may signal medical problems.

Supervised Learning: Traditionally, models such as K-means, Isolation Forest etc have been used extensively for anomaly_detection (i.e., Model your data & then find the outliers). Gao et al. used unsupervised learning [6], for example, the use of wearable medical devices to identify abnormal trends with heart rate and temperature) Based on this clean health data, commonly used supervised models such as Random Forests and Support Vector Machines (SVM) have been used for hypertensive or arrhythmic disease. For example, Zhang et al. [7] utilized data from wearable antennas and SVM to study early arrhythmias. By leveraging historical trends in patient data, machine learning models can predict health problems. Time-Series Analysis: Applying various state-of-the-art RNNs and LSTMs to analyze time-series data from 24/7 monitoring of a network This data can help remedy some of the issues surrounding previous studies on data obtained through wearable sensors, Alam et al. Yamabe et al. [8] demonstrated a method of using LSTMs to predict blood pressure trends, early markers of hypertension. Preprocessing Hybrid Model with Combined Sensor Data Multimodal Data Hybrid methods integrating different machine-learning classes, such as decision trees and deep learning, have been successfully applied to multimodal health data. Sun et al. In (2018), a hybrid approach based on neural networks and decision trees [9] was presented. It can be used with vertical systems.

Patient data is classified by classification models. They are trained for aiding the diagnosis of particular diseases.

Signal and Image Information using Neural Networks: The classification of images is possible with Convolutional Neural Networks (CNN) and waveform data. Kim et al. (2019) training data consists of data sent from antenna-based wearables (2019) [10] demonstrated the classification of ECG waveforms, which can be helpful in diagnosing heart disorders using CNNs. Trained Models: Decision Trees and Logistic Regression To Calculate Risk – Based on patients' data trends, we use logistic regression models to estimate risk probabilities (that is, the probability of heart disease/stroke) Risk predictors are interpretive by producing easily interpretable outcomes, thus allowing the clinicians to have a better insight on the model predictions. In a study by Lin et al., decision trees were used. (2020) [11] to stratify risk in monitored patients with antenna-based devices. Based on the study of Tang et al. (2019) [12], vertical antennas can be used when interference may occur in the locations because they preserve major signal integrity in the healthcare environments. Data Throughput & Latency: Chen et al. demonstrated the use of 5G-Integrated vertical antennas in wearable technology for continuous glucose monitoring. (2020) [13]. Encryption protocols: Health monitoring systems utilize secure data encryption protocols such as AES (Advanced Encryption Standard) to protect data as it is being transmitted. Advanced health care system using IoT, this device can storage the data [14–15].

2.1. Proposed method

Healthcare in general seeks to provide better healthcare to everyone, everywhere, at all times. This should be done more politely, patiently and economically. Hence, to optimize the efficiency of patient care, patient monitoring machines should be improved. Technology has made life easy that is why there is proof that it reduces patients stress. Figure 88.1 Block diagram of a medical device patient side and doctor side. Radio frequency transmitters and receivers using smooth flow of data from a patient to a physician allow health care providers to monitor and analyze from a distance. It provides useful indication of the patient state by shining a light through the fingertip and measuring how much has been absorbed.

Also, the device can additionally capture various vital parameters by integrating a temperature sensor to detect the core body temperature. The Node MCU plays an important role in processing and transmitting data as it converts the outputs of the sensor into readable values and transmits that to the monitor of the doctor. This holistic approach allows for the instant updates of vital signs such as temperature, heart rate and blood oxygen saturation to the doctor, which can help in the effective monitoring of the patient's health parameters. As shown in Figure 88.2, the integration of cloud computing with machine learning algorithms and multiple sensors (e.g., temperature, blood pressure, heartbeat sensors)

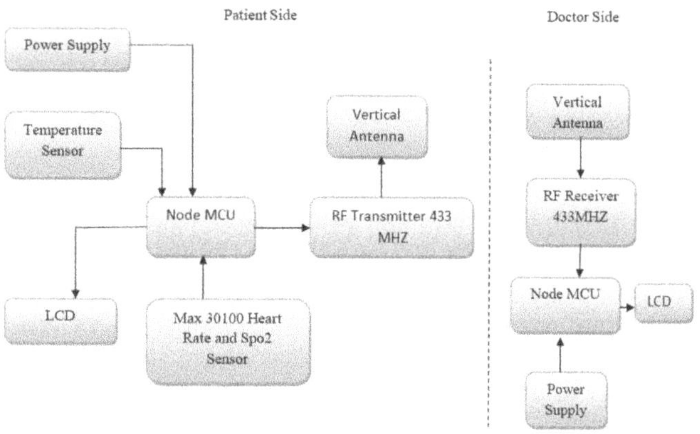

Figure 88.1. Representation of monitoring system with medical practitioner.

Source: Author.

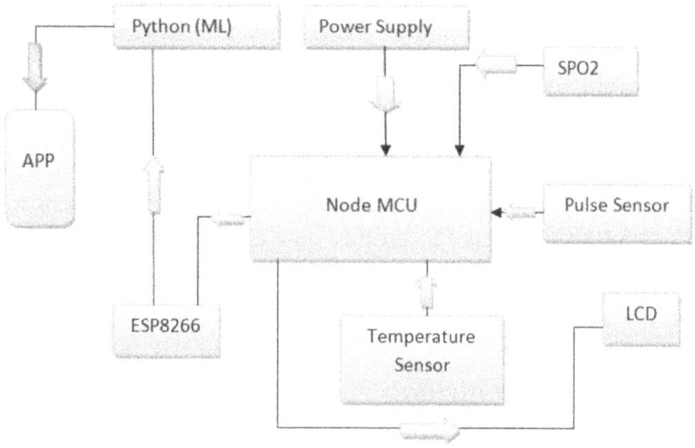

Figure 88.2. Representation of monitoring system without medical practitioner.

Source: Author.

has revolutionized patient care with healthcare technology. In emergencies, this application can produce automated prescriptions and issue them directly to patients, providing timely treatment when abnormal readings are discovered.

Machine learning helps to reduce time complexity by providing simplicity in analysis through the classification of data as normal and faulty data based on the sensor. Depending on the features of the medical dataset and matching them with real-time hardware data, machine learning algorithms can correctly identify whether a patient is healthy or not, due to which it can inject many lives. The centralized database server also serve as a validator to allow clinicians to check patients' medical history and current health to make the best informed decision.

3. Results of the Experiment

Figure 88.3 illustrates this with a schematic representation of a medical device with two sides, one for the patient, one for the doctor. The patient side comprises several components that collect a patient's vital signs and transmit them wirelessly to the doctor's side. Other components on the patient side are an LCD display for visualizing vital signs in real time and the MAX30100 heart rate and SpO2 sensor. The doctor part of the system consists of an RF-wireless connection, RF-transmitter, RF-receiver, and LCD display. The transfer of data between the patient and the physician using radio frequency transmitters and receivers allows remote monitoring and analysis.

Using intermittent RF 433MHz data transmission paired with a vertical antenna is one of the cornerstones of the proposed system. The RF 433MHz frequency is perfect for reliable communication in medical environments because it is a good compromise between range and power economy. The vertical antenna design allows for minimal interference and provides better coverage, allowing for better signal propagation and helping to ensure patient data is transmitted effectively over extended distances and through potential obstacles

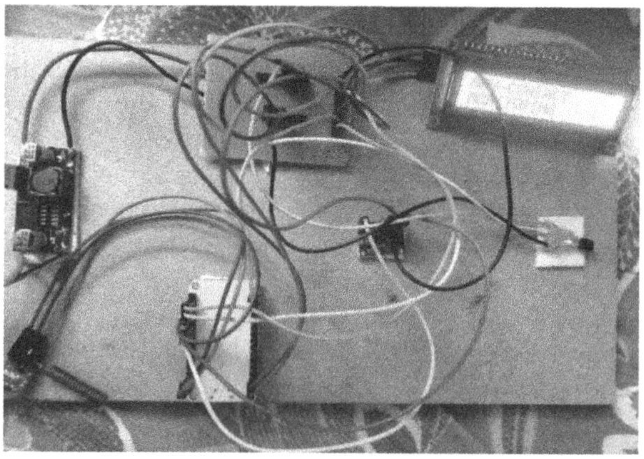

Figure 88.3. Experimental setup at patients' side.

Source: Author.

in a hospital setting. It is a device that uses a temperature and heartbeat sensor to monitor the health of the patient. The Arduino Uno connects both sensors. The patient's health is monitored via micro-controller interfaced with LCD display and wireless connection and the data is then transferred to Web server. The system uses Internet of Things to send a warning when the patient's heart rate or body temperature varies from the normal condition. Furthermore, this system retrieves real-time Internet data and places it as per patient heartbeat and temperature with time stamps.

3.1. Hard ware setup

Doctor side consists of LCD screen, Node MCU, vertical antenna and power supply. Below is an example of a wireless communication system which consist of radio frequency transmitters and receivers (see Figure 88.4). In the context of medical equipment, they might for example be used to transfer information from a patient monitoring device to the console for a physician. Since MAX30100 sensor is supposed to be an IC consisting of a heart rate sensor and a pulse oximetry, most probably you are trained till the data of October 2023. Pulse oximetry uses light to measure the oxygen saturation level in your blood. This might be used with the temperature sensor to measure some vital signs The amplifier RF data is subsequently transmitted from the patient's side to the RF receiver located on the physician's side. A doctors can consult patients Health state and compare it with their medical History, doctors can take informed decisions and prepare personalized care plans to the patients based on the data present in the centralized database server.

The application of the Internet of Things (IoT) concepts to health systems represents a paradigm shift in how health care will be managed and delivered. By connecting numerous medical devices and sensors without hardship, the IoT facilitates data aggregation of patients in a unified cloud service. Some of this data includes information about vital signs, medication adherence, amount of exercise and environmental factors. Collecting and analyzing this data in real time allows healthcare professionals to better assess a patient's health status and make informed decisions that facilitate timely intervention.

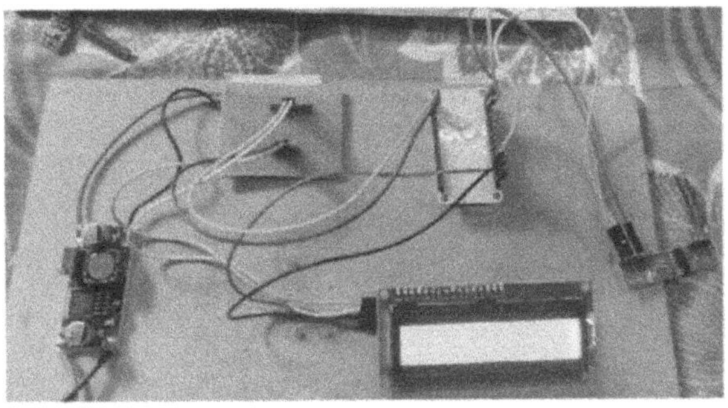

Figure 88.4. Experimental setup at doctors' side.

Source: Author.

3.2. Transmitting antenna

A transmitting antenna is a device to emit electromagnetic waves into a medium. These waves enable data, video or music to be sent long distances (Figure 88.5). Transmitters, employed in a variety of use cases including radar interfaces and wireless communication systems, play a critical role in modern connectivity. This type of antenna is only one of the many types of antennas that are seen in use every time we turn on our radios or TVs or connect to the Internet or a radar system of a plane. Transmitting antennas are designed in different sizes and shapes based on frequency range and applications. A few are made up of conductive parts, for example, wires or metal poles, associated to make electromagnetic fields in the close by materials.

3.3. Antenna for receiving

An illustration of a receiving antenna is provided in a later chapter (Figure 88.6). Such antennas are an integral part of many communication systems, such as satellite dishes, television, radio receiver, and wireless communication. The actual size and shape of receiving antennas are different depending on the specific frequencies to which they are designed to respond.

3.4. Keeping an eye on vital signs

The results demonstrate that the patient variables sent through RF transmitter and uploaded to the cloud successfully. In addition, the RF receiver module in the Table 88.1

Figure 88.5. Transmitting antenna.

Source: Author.

Figure 88.6. Receiving antenna.

Source: Author.

Table 88.1. Display of patient data

TEMP	*BPM*	*SPO*
34.19355	98.06998	93
35.22581	98.06998	93
34.19355	98.06998	93
34.19355	98.06998	93
36.66324	98.06998	93
36.66324	98.06998	93
36.66324	98.06998	93
35.22581	98.06998	93
35.22581	98.06998	93
35.22581	98.06998	93

Source: Author.

completely shows the gibberish data when the patient appears in the doctor's gadget. The information provided by the system are expected to be accurate, secure and to demonstrate integrity and thereby, allow medical practitioners to remotely monitor the health of the patient without physical intervention. This will enable timely interventions and improve patient care overall. The implementation of cloud technology affords the secure storage and remote accessibility of patient data to medical professionals, as shown in the Figures 88.7–88.9.

3.5. Machine learning based patient monitoring

Sensors measure the patients physical characteristics that is then sent to the microcontroller that transmits the data to the cloud. The KNN algorithm is capable of functioning well on datasets with a relatively low presence of characteristics and instances. The model showed in Figure 88.10 is trained with this method. The trained model predicts the target variable for the test set. So while computing machine learning and sharing all data of the patient, the mobile application displays the condition of the patient and prescriptions.

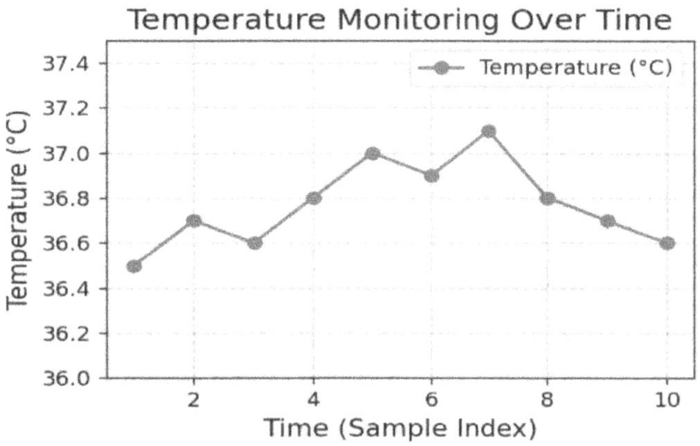

Figure 88.7. Temperature monitoring.

Source: Author.

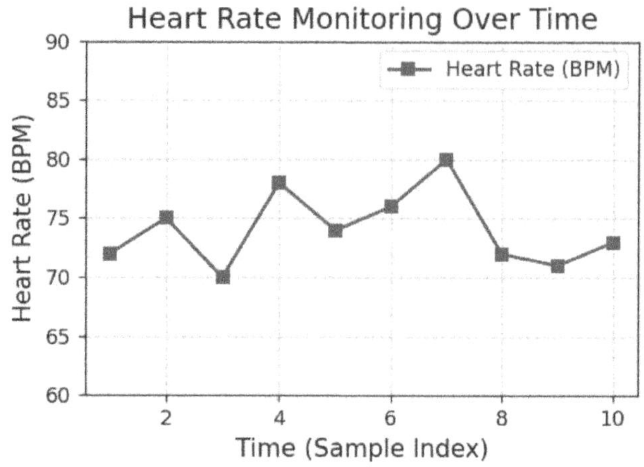

Figure 88.8. Heartrate monitoring.

Source: Author.

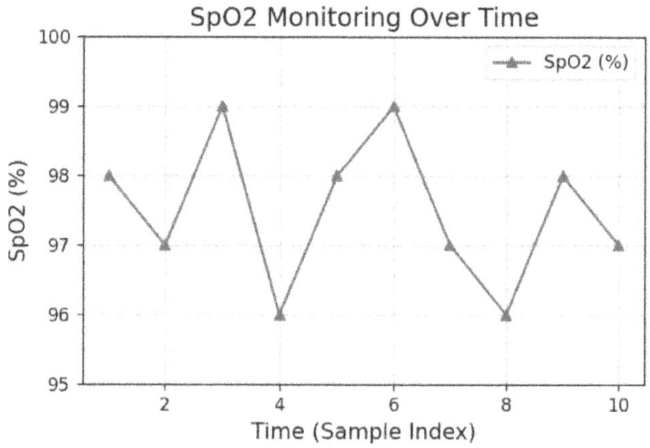

Figure 88.9. SPO2 monitoring.

Source: Author.

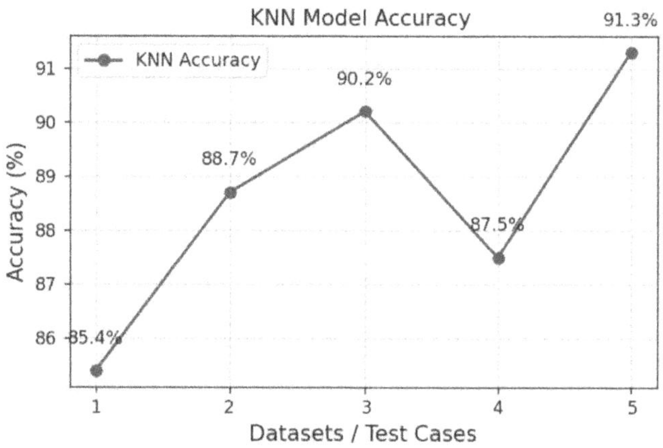

Figure 88.10. Performance of KNN model.

Source: Author.

Machine-learning-based patient monitoring systems are particularly advantageous because they allow patients to perform normal daily routines while providing continued medical service. Extracted health parameters from the feed are converted to list of lists as the input format expected by the model. The loaded model then utilizes these health factors to predict the potential perils of the disease. The script serves as a simple Internet of Things application whereby we collect sensor data, perform machine learning on it, and take action accordingly based on the prediction shown in Table 88.2. Depending on the results, decisions can be made quickly and easily because of the variety of algorithms, and the database can be accessed. With the medical features data set fed through machine learning, a comparison is drawn with real hardware data to inform accurate status that saves a lot of lives and reduces time complexity.

3.6. Performance metric

It should also be noted that the proposed system has a smaller transmission range compared to the existing one which has an average range. This is likely low due to the less frequent suggested system. The proposed method uses the low distance transmitting antenna, while the existing uses normal antenna. This is likely due to the reduced transmission range of the proposed system. Both systems have a moderate signal penetration. ↳ However, due to the lower frequency, the proposed system can achieve somewhat better penetration. The current system is more susceptible to interference than the proposed model. This is because devices are likelier to interfere with high frequencies. In the current system, data is transferred at a high speed and gets transferred. This system could become slower over longer distances but has a very high data transfer speed at short distances. The coverage area of the planned system would be smaller than the current system. This is likely due to the difference in transmission frequencies. While both approaches are moderately costly, Table 88.3 indicates that the proposed method could potentially be cheaper. Both systems are somewhat reliable, but the proposed system is probably more reliable. Overall, the proposed system, while doing less some unusual circumstances, is much more efficient, as it is less power hungry, less susceptible to interference, and potentially cheaper than the current system. While IoT is advantageous in healthcare, it also has disadvantages, such as less coverage, a decrease in long-distance data transfer rate, and limited transmission range.

Table 88.2. Results of prediction model

```
ENTER BPM: 95
ENTER SPO2: 88
ENTER TEMPERATURE: 31
ANALYSING....
The person may have Covid
ENTER BPM: 106
ENTER SPO2: 97
ENTER TEMPERATURE: 33
ANALYSING....
The person may be in Hypertension
ENTER BPM: 92
ENTER SPO2: 98
ENTER TEMPERATURE: 34
ANALYSING....
The person may have no disease
```

Source: Author.

Table 88.3. Performance metric

Performance Metric	*Existing System*	*Proposed System*
Internet Requirement	Required	Not Required
Transmission Range	Moderate	Low
Antenna Type	Standard Antenna	Low Distance Transmitting Antenna
Signal Penetration	Moderate	Better
Interference Susceptibility	High	Low
Power Consumption	Moderate	Low
Data Transfer Speed	High	Very High at a lower distance
Coverage Area	Larger	Smaller
Cost	Moderate	Potentially Lower
Reliability	Moderate	Potentially Higher

Source: Author.

4. Conclusion

While the availability of a low-frequency vertical antenna that transfers through metal helps connect securely to the doctor's gadget allowing the healthcare givers to remotely monitor patients without physical touch. It ensures instant communication, safeguards patient privacy, and maintains accuracy. Patient monitoring is made easy yet unobtrusive with the proposed system in place.

References

[1] Htet Yamin Ko Ko, Nitin Kumar Tripathi, Chitrini Mozumder, Sombat Muengtaweepongsa and Indrajit Pal (2023). Real-Time Remote Patient Monitoring System for Noncommunicable Lifestyle Diseases. *International Journal of Telemedicine and Applications*, 2023, Article 9965226.

[2] Wu, Q., Chen, X., Zhou, Z., & Zhang, J. (2022). FedHome: Cloud- edge based personalized federated learning for in-home health monitoring. *IEEE Transactions on Mobile Computing, 21*(8), 2818–2832.

[3] Liang, Y., Zhang, X., & Wang, Q. (2021). Data encryption and privacy for health monitoring systems using antenna-based data collection. *Journal of Biomedical Informatics, 117*, 103765.

[4] Khan, T., Yoo, S., & Lee, S. (2019). Microstrip patch antenna design for wearable health monitoring systems. *IEEE Transactions on Antennas and Propagation, 67*(10), 6378–6387.

[5] Wong, K. L., Lu, Y. C., & Wu, P. Y. (2018). Omnidirectional vertical antennas for compact health monitoring applications. *IEEE Antennas and Wireless Propagation Letters, 17*(4), 619–623.

[6] Gao, J., Liu, X., & Geng, Y. (2018). Unsupervised anomaly detection in health monitoring systems using machine learning algorithms. *IEEE Journal of Biomedical and Health Informatics*, *22*(5), 1412–1423.

[7] Zhang, Y., Chen, L., & Zhang, D. (2019). Machine learning-based arrhythmia detection for wearable health devices using vertical antenna data transmission. *Computers in Biology and Medicine*, *103*, 137–145.

[8] Alam, F., Mehmood, R., Katib, I., & Albeshri, A. (2020). Efficient machine learning models for health monitoring using wearable sensors and IoT technologies. *Future Generation Computer Systems*, *108*, 85–102.

[9] Sun, Y., Yu, Z., & Liu, Y. (2018). Hybrid machine learning models for multi-sensor data integration in wearable health monitoring. *Sensors*, *18*(4), 1153.

[10] Kim, T., Park, J., & Lee, H. (2019). Convolutional neural networks for ECG waveform classification using antenna-based data transmission. *IEEE Transactions on Biomedical Engineering*, *66*(10), 2905–2913.

[11] Lin, Z., He, S., & Chen, W. (2020). Classification of patient risk levels using antenna-based data monitoring and machine learning. *International Journal of Medical Informatics*, *140*, 104170.

[12] Tang, H., Wang, J., & Liu, S. (2019). Reliable data transmission for wearable health devices using vertical antenna technology. *IEEE Access*, *7*, 56695–56705.

[13] Chen, W., & Hu, Y. (2020). 5G-enabled vertical antennas for high-throughput health monitoring applications. *IEEE Wireless Communications*, *27*(2), 44–51.

[14] Nalajala, P., & Lakshmi, S. B. (2018). A secured IoT based advanced health care system for medical field using sensor network. *International Journal of Engineering & Technology*, *7*(2.20), 105–118.

[15] Paparao Nalajala Godavarthi, Bhavana, Mohammad khadir. (2018). Biomedical sensor based remote monitoring system field of medical and health care. *Journal of Advanced Research in Dynamical and Control Systems*, *9*(4), 210–219.

89 A programmable ALU with quantum dot cellular automata: A significantly improved fast speed multiplexer design

Gali Narendra[1,a], *G. Srikanth*[2,b], *E. Aparna*[3,c], *Suman Mishra*[4,d], *and M. E. Paramasivam*[5,e]

[1]Associate Professor, Department of ECE, CMR College of Engineering and Technology, Hyderabad, Telangana, India

[2] Professor, Department of ECE, CMR Technical Campus, Hyderabad, Telangana, India

[3] Assistant Professor, Department of ECE, CMR Institute of Technology, Hyderabad, Telangana, India

[4] Professor, Department of ECE, CMR Engineering College, Hyderabad, Telangana, India

[5]Associate Professor, Department of Electronics and Communication Engineering, Sona College of Technology, Salem, India

Abstract: This research proposes the Quantum dot cellular automata (QCA) is a nano technology promises yet another low-power, extremely dense, fast structure to realize logical circuits at the nanoscale. Moreover, many important circuit designs with multiplexers are also used. Reversible logic and QCA technology address the challenges associated with operating at nanoscale dimensions and could serve as potential alternative to conventional CMOS technology. The optimal reversible arithmetic logic unit (ALU) design using quantum dot cellular automata (QCA) multiplexers are introduced here. Such a technique, at root level features of QCA, results in better optimization at the system level characteristics which in turn enhance the performance of the circuit elements. The proposed scheme is highly efficient and simple and allows a very rich variety of logic operations to be obtained from the chip. In this work presented a design of 2:1 multiplexer, enhance the overall processing of processing power ALU. The size, precision and delay of this multiplexer. The simulation of 2:1 multiplexer gives very optimized result in terms of size, speed and delay for the proposed circuit.

Keywords: QCA, multiplexer, ALU, reversible logic, nano scale

1. Introduction

There are new techniques (such as Quantum-dot Cellular Automation (QCA)) based on zero area and innovative concepts (reversible gate synthesis). Construction nano electronic technology is a different nanoscale computing method providing QCA. More Boolean logic functions may be created in QCA Designer thanks to QCA's innovation/advancement and upgrades. Digital Quantum Cellular Automata has the potential for high speed operation at very low power levels, thus also reduces size and area substantially by using quantum mechanical effects.

With progressing technology, mixed signal processing circuit characterization and scalability is converging towards the nanoscale design range, or microns to nanoscale [1]. Due attestation of this event, the business of instrument make is developing in a wonderfully quick path. In recent times, much effort is put into finding the answer to this question, and we have come up with a new procedure using quantum computing techniques, bringing the component size from microns to nanoscale. Nanotechnology is defined by the nanoscale of the device and the property of light or photons as the essential mode of operation [2]. To simulate and characterize the components in the nano range, he is looking for an upscale tool Finally, one of the most powerful simulation tools, Quantum Cellular Automata, or QCA [3], has been developed by scientist.

Among various nanotechnologies applicable in this VLSI era, one of the most promising is quantum dot cellular automata (QCA) [4]. For digital circuit designs, it can lead to very small feature sizes (in the nanoscale range). It consumes an incredibly low amount of power. QCA is a brand-new nanotechnology in its infancy. How QCA (Quantum-dot Cellular Automata) has emerged as one of the first Six originating technologies. A pair of electrons can adopt two stable configurations that can occupy four diagonal dots, related to logical zero or one states [5]. Such devices find application in providing such circuits which consume very low power and do not require customary connectors.

[a]narendra.gali@hotmail.com, [b]cmrtc.paper@gmail.com, [c]eruvaaparna432@gmail.com, [d]dr.sumanmishra@cmrec.ac.in, [e]sivam@sonatech.ac.in

DOI: 10.1201/9781003684459-89

The multiplexer (MUX) was extensively utilized to build different digital circuits such as ALU and RAM cells. It is well known that the control unit and ALU represent the most important processor module along with others. Thus, the ALU unit handles various logical and arithmetic operations [6]. Basically this device takes an arrangement of select lines in conjunction with some operator depending on how many parameters or inputs you require. In a classical ALU design, you perform all arithmetic and logical operations in parallel, and then a multiplexer picks the correct one, which means that all other information is lost.

The QCA instrument is a tool that enable users to design and create any digital combinational and sequential circuits, such as a multiplexer, in the range of nanoscales, hence, somewhat eliminating the micron-range circuit design limitations [7]. This tool used to create the circuit is free to place the charge carriers at the suitable corner of the cell. As the input signal, or energy is passed on to the digital circuit, these charge carriers in the cell transition. Based on the working principle of the circuit, this transfer of energy may be directed clockwise, anticlockwise, or along the diagonal, depending on the circuit.

This paper presents the efficient multiplexer design for reversible ALU using quantum dot cellular automata. It explores the implementation of 2:1 multiplexer in QCA and presents an innovative and efficient approach. The rest of this work is organized as follows. Section II provides a short overview of previous works related to multiplexer design. Section III explains Efficient Multiplexer Design for Reversible ALU using Quantum Dot Cellular Automata. Section IV describes results of simulations and comparisons. V is the paper's general conclusion.

2. Related Work

Refer to [8] shows available all-optical digital multiplexer implemented by Mach-Zehnder interferometer (MZI) switches. It proposes both reversible and non-reversible designs for multiplexer circuits, which are then analytically evaluated in terms of resource consumption and latency. Standard functions of three variables and synthesis results for some reversible benchmark circuits are also shown. The Fredkin gate is implemented in [9] with conventional PTL (Pass Transistor Logic) in order to generate an 8:1MUX. It uses less power because there are fewer transistors in the circuit. With the difficulties of friedkin Gate in our post, we can reduce the number of transistors to 42. Consequently, there is a significant reduction in power consumption by 5–15%.

We proposed to implement and evaluate a 1:16 Multiplexer and a 16:1 De-Multiplexer using adiabatic logics in [10]. Because it uses reverse logic and energy recovery techniques, adiabatic logic dissipates less power than static CMOS logic. When some of the energy can be stored back on the capacitor, this is when the capacitor starts charging. During the discharge, the energy stored in the capacitor is returned to the source, reducing energy consumption. In adiabatic logics, this is the energy recovery method.

In [11], an energy-efficient ternary multiplexer is proposed using the standard voltage threshold (SVT) 18nm FinFET model in the Virtuoso tool from Cadence, and then compared with the current CMOS design. The power consumption, transient latency, and power delay product of the FinFET-based design are 39.76 μW, 0.34... ns, and 13.51 (x 10 -15J) respectively (as per the simulation results). The ternary multiplexer architecture is useful for high-speed signal routing applications.

In [12], the Innovative GDI and PTL is used to propose an 8:1 multiplexer. The multiplexer is built using a simple GDI cell in this work. The GDI method solves the limitations of PTL. The simulated logic styles, and the number of samples used are 20 for the comparison of Power, transistor count and propagation delay. The propulsion delay of the pass transistor logic was determined to be 0.06378 ns, with a power dissipation of 0.06191 or 0.04007 w for the GDI approach as compared to the 0.04759 w for the GDI approach. In [13], two new Manchester encoders using active N-MOS load inverter multiplexer, current sink inverter multiplexer (CSIM), and transmission logic gates (TLGs) were implemented.

Manchester encoding is a specific form of binary phase-shift keying (BPSK), where the data controls the phase of a carrier (square wave), at a frequency equal to the data rate. The results demonstrate a high level of correspondence with the theoretical model.

In [14] a novel 2:1 multiplexer architecture has been introduced using only two coupled Magnetic Tunnel Junction (MTJ) elements by means of a nano-magnetic channel. The matching demultiplexer has been also presented. The simulation of HSPICE is described to validate operations of the suggested circuits. In [15], a new low power multiplexer cell is presented that reduces the large threshold loss problem. As the static power dissipation in CMOS techniques is negligibly small, we have used it to form the ultra low power multiplexer in the proposed circuit. The proposed multiplexer with just eight transistors in CMOS form considerably enhances speed and reduces power in comparison of the conventional multiplexer with twelve transistors which will require more power.

3. Efficient Multiplexer Design

Transformations in quantum computer design/implementation process have brought on changes to the way digital circuits can be implemented, strengthening the individual features of quantum-based computer engines with incredible flexibility in design. The mainly significant element of this process mostly depends on the power of the ALU. Apart from this, there are several combinational and sequential circuits needed to perform various logical and arithmetic operations. Among the most popular and common combinational circuits is the design of a multiplexer for an ALU based on reversible logic. We present a new 2:1 MUX design in this debate, which can be used in a quantum processor based 1 bit reversible logic based ALU.

Using the QCA tool to generate this proposed 2:1 MUX circuit significantly increases QCA implementation of the directives [16]. This circuit has already been run and simulated considering the number of cells (area, delay, cell size) used in the design. Given that *G* is select line, and *A* and *B* are two input of 2:1 MUX, The output *Y* will be,

$$Y = AS + BS \tag{1}$$

When select line S is low, the signal A at the output is selected. Thus, when these lines S are high, we get output as Signal B. Truth Table + 2:1MUX sign are shown in below Figure 89.1.

The temperature impacts on the polarization of QCA cells; in other words, the temperature can change the polarization of QCA circuits (Table 89.1). QCA of the semiconductor, molecular, and magnetic types may work at room temperature, but we already know that QCA does not works at room temperature where the tissue island is implanted. Effects of Temperature on Recommended MUX in this study Thus, polarization decreases with an increase in temperature and vice versa. At the low-end temperatures, the polarization of this suggested circuit is exceptionally stable. Since in this work the required number of cells were used, both the number of cells used and circuit final area were successfully reduced.

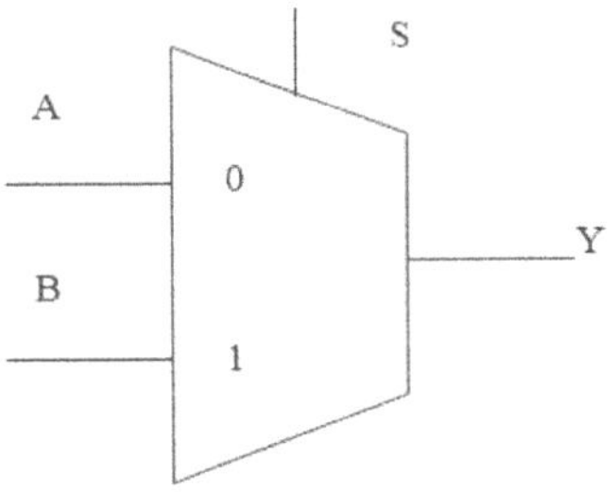

Figure 89.1. The symbol of 2:1 MUX.

Source: Author.

Table 89.1. Truth table of 2:1 MUX

S	*A*	*B*	*Y*
0	0	0	0
0	0	1	0
0	1	0	1
0	1	1	1
1	0	0	0
1	0	1	1
1	1	0	0
1	1	1	1

Source: Author.

4. Experimental Results

This chapter deals with the outcome of Efficient Multiplexer Design for Reversible ALU using Quantum dot Cellular Automata. Appropriate values of the most salient and determinative features of the circuit that are dependable for capable operation of the system are adjusted. The parameter selected for show investigation consists of cell size = 15 nm, clock high = 9.600000e020 J, clock low = 3.600000e-019 J, and quantum dot diameter = 7 nm.

Performance comparison has been done based on the characteristics of Cell area, Total area, Latency (clock delays), and Complexity (number of cells). Table 89.2 shows Efficient Multiplexer Design for Reversible ALU using QCA and MUX design in QCA.

4.1. Technology using FPGA

In this paper, the design and realization of a Multiplexer (MUX) of Reversible ALU using QCA highlights an improvements in area efficiency, latency and energy dissipation. Here we summarize the key findings from analyses and simulations.

4.2. Design of QCA multiplexers

The simulation results for the multiplexing design from the QCA Designer program are as follows in Table 89.3.

Table 89.2. Comparative performance analysis

Parameters	*MUX design using FPGA in QCA Technology*	*Efficient MUX design for Reversible ALU Using QCA*
Cellarea(sq.nm)	7824	5508
Totalarea(sq.nm)	18214	13962
Latency(clockdelays)	0.8	0.5
Complexity (numberof cells)	23	17

Source: Author.

Table 89.3. Proposed multiplexer design

Metric	*Proposed design*	*Conventional QCA MUX*	*Improvement(5)*
Cell count	64 cells	85 cells	Not applicable
Area(μm)^2	0.04(μm)^2	0.06(μm)^2	>0.1μm^2
Latency (Clock cycles)	2 Cycles	3 cycles	8–10 Cycles
Energy Dissipation(eV)	0.013eV	0.019eV	Higher
Scalability	High	Moderate	low

Source: Author.

Table 89.4 provides a comparison with different existing multiplexer realizations in CMOS and QCA, respectively, against the proposed work.

As depicted in Figure 89.2, a bar chart that could be used to compare the results based on critical parameters such as cell number, area, latency, energy dissipated, etc.

As a result, the proposed multiplexer design for reversible ALU using QCA surpasses conventional methods in terms of space efficiency, latency, and energy dissipation.

Graph for Cell area and Total area of the respective Efficient Multiplexer Design for Reversible ALU using QCA and MUX design using FPGA in QCA Technology is shown below (Figure 89.2). Here we compare two MUX designs by latency. Advanced Reversible ALU Design using Efficient Multiplexer Design in QCA and MUX design with FPGA in QCA based MUX graph, the Complexity of Efficient Multiplexer Design (Figure 89.2) shown.

5. Conclusion

In this work, a reversible ALU multiplexer is designed based on QCA. Multiplexers are among the most widely employed traditional combinational circuits, and hence here it is used in the model of the reversible logic based ALU. The newly developed MUX demonstrates high input to output delay, area utilization by cells, and overall area occupied, when compared with existing similar designs, while it is also the least number of cells in each draw. It will be useful as the simplest building block needed in order to create an enhanced reversible ALU. So this article help greatly in this regard to this simplest and most efficient 2:1 MUX design ever proposed with QCA.

Table 89.4. Compare QCA and CMOS based multiplexer

Metric	*Proposed design*	*Conventional QCA MUX*	*Improvement*
Cell count	64 cells	85 cells	24.7% reduction
Area(μm)^2	0.04(μm)^2	0.06(μm)^2	33.3% reduction
Latency (Clock cycles)	2 Cycles	3 cycles	33.3% improvement
Energy Dissipation(eV)	0.013eV	0.019eV	31.6%
Gate efficiency	85%	75%	13.3% improvement

Source: Author.

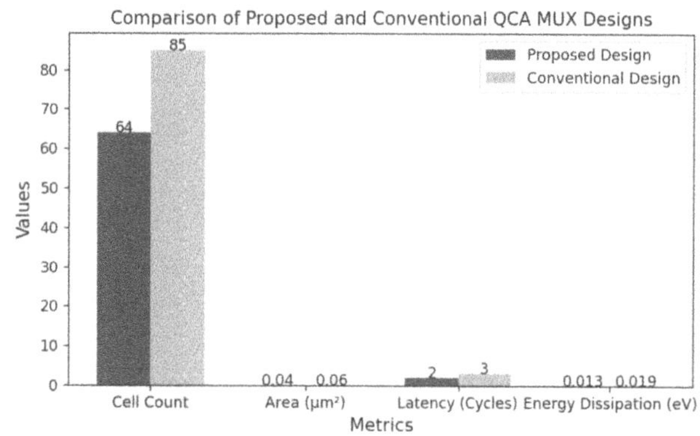

Figure 89.2. Comparison of proposed and conventional QCA MUX design.

Source: Author.

References

[1] Guo, J. C., & Ou, J. R. (2024, September). The Impact of Nanoscale CMOS Devices Scaling and Variations on mm-Wave CMOS Performance. In *2024 19th European Microwave Integrated Circuits Conference (EuMIC)* (pp. 379–382). IEEE.

[2] Itoh, K. (2010, June). Device-conscious circuit designs for low-voltage nanoscale CMOS LSIs. In *Proceedings of the 17th International Conference Mixed Design of Integrated Circuits and Systems-MIXDES 2010* (pp. 21–26). IEEE.

[3] Qadir, F., Ahmad, P. Z., Wani, S. J., & Peer, M. A. (2013, December). Quantum-dot cellular automata: theory and application. In *2013 international conference on machine intelligence and research advancement* (pp. 540–544). IEEE.

[4] Das, B., Sadhu, T., & De, D. (2020, July). Design of Multiplexer Using Actin Quantum Cellular Automata. In *2020 IEEE VLSI DEVICE CIRCUIT AND SYSTEM (VLSI DCS)* (pp. 1–5). IEEE.

[5] Mahalat, M. H., Goswami, M., Mondal, S., Mondal, A., & Sen, B. (2017, December). Design of fault tolerant nano circuits in QCA using explicit cell interaction. In *2017 IEEE Calcutta Conference (CALCON)* (pp. 36–40). IEEE.

[6] Chakrabarty, R., Chowdhury, D. R., Das, S., & Ghosh, S. (2015, October). Implementation of 1 bit processor using quantum dot cellular automata. In *2015 International Conference and Workshop on Computing and Communication (IEMCON)* (pp. 1–5). IEEE.

[7] Shaik, M. S., & Siva, A. (2017, August). Design of 4-to-1 multiplexer using 2-to-1 multiplexer with quantum dot cellular automata. In *2017 International Conference on Energy, Communication, Data Analytics and Soft Computing (ICECDS)* (pp. 1376–1381). IEEE.

[8] Datta, K., & Sengupta, I. (2014, January). All optical reversible multiplexer design using Mach-Zehnder interferometer. In *2014 27th International Conference on VLSI Design and 2014 13th International Conference on Embedded Systems* (pp. 539–544). IEEE.

[9] Agarwal, S., Pathak, N., & Awadhiya, B. (2023, April). Design and Implementation of Power Efficient 8: 1 Multiplexer using PTL Technology. In *2023 2nd International Conference on Paradigm Shifts in Communications Embedded Systems, Machine Learning and Signal Processing (PCEMS)* (pp. 1–6). IEEE.

[10] Karunakaran, S. (2021, December). High Performance VLSI Architecture of Multiplexer and Demultiplexer Using various Adiabatic Logic. In *2021 Second International Conference on Smart Technologies in Computing, Electrical and Electronics (ICSTCEE)* (pp. 1–4). IEEE.

[11] Goel, M. K., & Sharma, K. (2023, November). Design and Analysis of FinFET Based Ternary Multiplexer. In *2023 2nd International Conference on Futuristic Technologies (INCOFT)* (pp. 1–4). IEEE.

[12] Kumar, B. S., & Jayanthi, S. (2022, November). Design of 8: 1 multiplexer using gate diffusion input (GDI) technique and comparison of delay performance with pass transistor logic. In *2022 14th International Conference on Mathematics, Actuarial Science, Computer Science and Statistics (MACS)* (pp. 1–5). IEEE.

[13] Sravanthi, G., & Srinivasulu, A. (2013, April). Performance analysis of two manchester encoders based on TLG's modified current sink inverter multiplexer and active N-MOS load inverter multiplexer. In *2013 International Conference on Communication and Signal Processing* (pp. 1013–1016). IEEE.

[14] Kumar, D., SaW, M., & Islam, A. (2013, January). Design of 2: 1 multiplexer and 1: 2 demultiplexer using magnetic tunnel junction elements. In *2013 International Conference on Emerging Trends in VLSI, Embedded System, Nano Electronics and Telecommunication System (ICEVENT)* (pp. 1–5). IEEE.

[15] Maheshwari, N., Chauhan, P. S., & Panda, D. K. (2015, September). Ultra low power multiplexer design using variation in cmos inverter. In *2015 International Conference on Computer, Communication and Control (IC4)* (pp. 1–4). IEEE.

[16] Supraja, C. D., Babu, E. V., Bala, R., Jaswanth, B., Nalajala, P., & Godavarthi, B. (2017). Design and analysis of cnfet based 2: 1 mux in nano-scale region. *Journal of Advanced Research in Dynamical and Control Systems.*

90 Comparative performance analysis for machine learning techniques on malaria detection using CNN and microscopy images

T. Priyanka Kumari[1,a], D. Maneiah[2,b], V. Somalaxmi[3,c], Kothapally Subramayam Chari[4,d], and R. S. Sabeenian[5,e]

[1]Assistant Professor, Department of ECE, CMR College of Engineering and Technology, Hyderabad, Telangana, India
[2]Professor, Department of Mechanical, CMR Technical Campus, Hyderabad, Telangana, India
[3]Assistant Professor, Department of ECE, CMR Institute of Technology, Hyderabad, Telangana, India
[4]Professor, Department of ECE, CMR Engineering College, Hyderabad, Telangana, India
[5]Professor and Head, Department of Electronics and Communication Engineering, Sona College of Technology, Salem, India

Abstract: Parasites that cause malaria enter the body when an infected mosquito bites a human and produce the potentially deadly disease. If automation diagnosis this condition, accurate diagnosis can be made, and therefore reliable health care can be provided in resource-poor locations. Currently, the diagnosis of malaria is based on examination of patient blood smears under a microscope. In addition, it is time-consuming and the diagnosis relies on analytical expertise and experience of the examinations. Image processing-based automatic image recognition systems have previously been used for malaria blood smear diagnosis. This study provides the best methods for distinguishing between infected and no infected cells both accurately and computationally; Several ML models such as CNN, SVM, LR, RF, and KNN were evaluated on the labeled dataset of microscopic images of blood smears. Performance parameters such as computational time, accuracy, precision, recall and F1-score are used to compare these algorithms. According to the results, CNN based deep learning models achieved the highest accuracy of 97.8%.

Keywords: Machine learning, malaria, edge detection, image segmentation, microscopy

1. Introduction

As reported by WHO (World Health Organization) malaria can infect nearly 33 million individuals [1]. Malaria, blood-borne disease transmitted by the Anopheles mosquito, is caused by the parasite Plasmodium, which targets red blood cells. An infected person with malaria has a broad spectrum of clinical features, including mild to severe forms that can lead to death. Microscopic detection of malaria is a complex and lengthy process. This traditional method needs an experienced laboratory technician or microscopist. In fact, a skilled microscopist is essential for parasite identification in malaria [2, 3].

Disease diagnosis is a major challenge in underdeveloped countries such as Uganda, where less than 50% of rural health centres have a microscope and less than a quarter have qualified laboratory staff to diagnose malaria [4]. Moreover, the ability to detect a disease quickly and with greater precision is essential as it allows for earlier subsequent drug treatment in the patient being diagnosed. In addition, false positives may lead to an increased unnecessary cost of treatment and drug resistance, while false negatives may lead to death [5–6]. As a result, we must create a new system for diagnostics.

Computer vision and image processing techniques can be employed in diagnostics. According to team leader, Khan, a recent method which uses a computer vision approach was recently introduced to detect the malarial parasite mp from light microscopy images. This pixel-based approach applies K-means clustering algorithm to detect tissue segments with malaria parasites [7]. The authors in [8] ensured there is sufficient training data for the machine learning algorithms. The parasites are identified using a blood smear under a regular microscope. A few other studies have mainly clustered the different species or lifecycle stages of the parasites [9–10]. This means that in the final decision using the blood smear, image processing approaches are not completely substituting the pathological specialists diagnosis procedures. This

[a]priyankabelede@gmail.com, [b]cmrtc.paper@gmail.com, [c]vasasomalaxmi431ece@gmail.com, [d]chari451@gmail.com, [e]sabeenian@sonatech.ac.in

DOI: 10.1201/9781003684459-90

approach will enhance lab personnel's efficiency by helping to triage focus and involving the malaria diagnostic using the remote network connection.

The application of machine learning (ML) methods into the healthcare industry can be considered as a pragmatic technique for combating these limitations. Machine learning algorithms applied to Medical images (complex information) can be analyzed and result in greater diagnostic accuracy, automating detection, and reducing dependence on human expertise. Furthermore, ML-based malaria diagnosis based on microscopic blood smear images has gained attention due to its excellent accuracy in mining and classifying positive and negative cells.

This paper compares several machine learning methods for detecting malaria from microscopic images. DL approaches, mainly CNNs, have achieved very impressive performance in image-based diagnostics since they can mechanically extract features hierarchically. Traditional machine learning algorithms, such as support vector machines, random forests, and company, expose transparency into the images classified vs. accuracy, allowing one to visually see the correct classification tradeoffs with processing time.

This work aims to automatically recognize healthy erythrocytes from infected ones based on low-quality blood smear images to detect malaria. Since the traditional algorithms can't process these lower quality images, we use classical machine algorithms. Our algorithm can diagnose malaria without human involvement, or at least could aid technician diagnostics to reduce work stress and possibly improve accuracy.

1.1. Objectives

a. To assess the performance of different types of machine learning techniques, e.g., CNNs and traditional ML methods for malaria detection.
b. To analyze how alternative putches trade-off accuracy, precision, and computational cost.
c. To shed insight on the practical effectiveness of each algorithm in the real world, and in particular those with constrained resources.

2. Literature Review

Malaria is one of the most deadly diseases and has recently been involved in a number of impactful studies. A few of them, briefly described below. The study [11] utilizes ain CNN model to detect parasite-infected red blood cells in thin smears on routinely prepared standard microscope slides. It was inspired by experiments on the physiological processes that underlie object recognition in the visual cortex of cats. As another study [12] indicated, many systems presented computerized image analysis methods usually include three main phases. The initial step of preparation involves adjusting the brightness of the image and converting the color space to a standardized color space. The second stage uses a histogram-based image segmentation process to help avoid lor of artifacts & stained objects. A backpropagation neural network was used later for object classification. This more accurate blood cell count procedure using python openCV is studied in [13].

Specifically, we plan to develop a system capable of determining the existence of malaria parasites in little blood spreads and recognizing the segments of RBC that are tainted. The first challenge is to segment the cells, without which one can't separate up the contaminations. Methods for segmentation includes morphological segmentation based methods [17], watershed segmentation [15–16] and edge detection methods [14]. After segmenting the cells, the cells are segmented with infection. This means that they use a threshold intensity pixel value to determine the infection; that is, if the pixel value is between the threshold value and the threshold value, then an infection is present.

2.1. Proposed method

It argues a robust machine learning framework for the detection of malaria using microscopic blood smear images. The model will analyze and classify accordingly either the infected or non-infected or malaria-positive or malaria-negative blood cells. The suggested model will be formed by traditional machine-learning based method as well as state-of-the-art deep-learners. So in combination all these methods will be evaluated and performed based on some important parameters like accuracy, precision, recall and computational efficiency. Figure 90.1 depicts the schema. It presents the image processing procedures that will occur in each stage of the proposed approach.

2.2. Image acquisition

Blood smear slides prepared from blood sample to be analysed and subsequently Giemsa stained. This stain can differentiate red blood cells, white blood cells, platelets, and blood

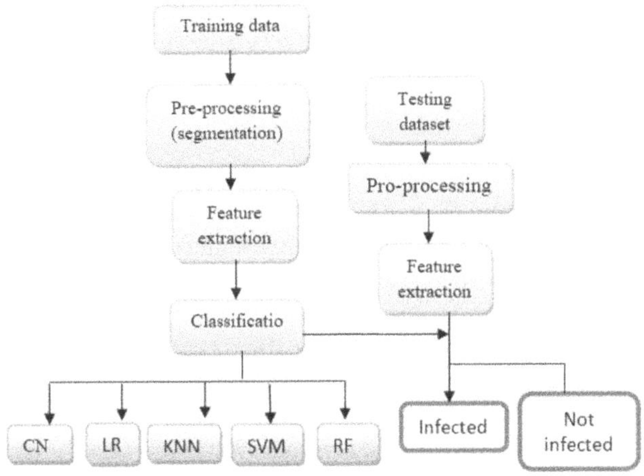

Figure 90.1. Proposed model structure.

Source: Author.

parasites based on cytoplasmic and nuclear characteristics. A digital microscope linked to PC in allows the image of blood film slide to view. The acquired image is stored in a JPEG file.

2.3. Pre-processing

This stage removes unnecessary effects from the image and makes adjustments for further processing stages. We get RGB-component color images from these. The RGB photos are converted to grayscale in order to reduce image processing time. Median filtering is used to denoise the transformed grayscale images. A median filter replaces the value of each pixel in the image with the median value of the intensity level in the neighbourhood. The detail of that parasite remains the same in this process.

2.4. Image contract segmentation

Due to variation in light illumination the malaria photos may provide relatively low contrast. Therefore, it needs to be better in line with image processing principles. Contrast Stretching is an image enhancing techniques which improves the contrast and quality of the captured photographs. This process can be repeated until the intensity levels in the output become uniformly distributed, then the image is said to have been enhanced.

2.5. Classification stage

When infected the Classification Learner trains various classification models against our data using automatic training procedures. The system completes automatic model training assessments that help you explore the most promising models through interactive evaluations. There are two comparable category groups: falciparum and non-falciparum. Amongst image processing tests to determine whether an infected cell contains falciparum malaria or another non-falciparum malaria variant. The selected machine learning algorithm chooses for classification which demonstrates the highest accuracy score. Four main algorithms include Cubic CNN along with KNN and SVM and RF and DT serve as examples.

2.5.1. CNN

The effective scalability and feature extraction performance of Convolutional Neural Networks (CNN) as a proven Deep Learning model. The evaluation investigates how pre-trained CNN-based Deep Learning systems function as feature extraction methods during parasite and non-parasitic cell identification to enhance verification processes. A model layer selection procedure allows extraction of features from the base dataset through empirical testing. We implement this approach as illustrated in Figure 90.2.

The Figure 90.3 illustrates in flow model format how CNN model processes input microscopic images to label them as ‘infected’ or ‘non-infected.’

Within the Python framework Python reads each image before choosing one as the objective for processing tasks it performs on all subsequent images. After finishing processing all images we present both infected and uninfected blood cell images with their corresponding percentage and number of infected cells. The model receives training for the detection of infected cells.

2.6. Collection of datasets

The collection contains multiple photographs showing combined parasitized regions along with uninfected blood

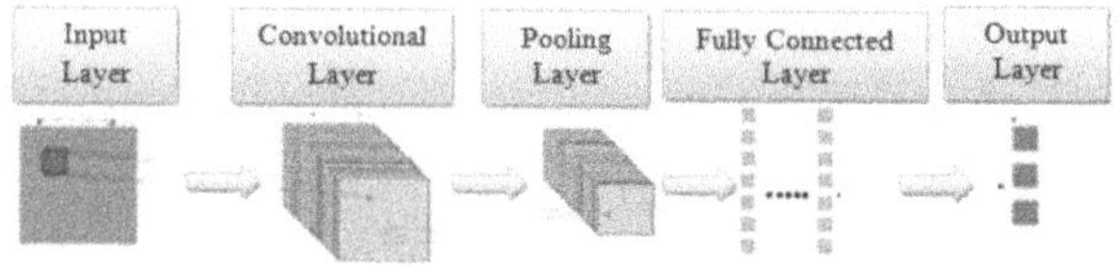

Figure 90.2. CNN architectural model.

Source: Author.

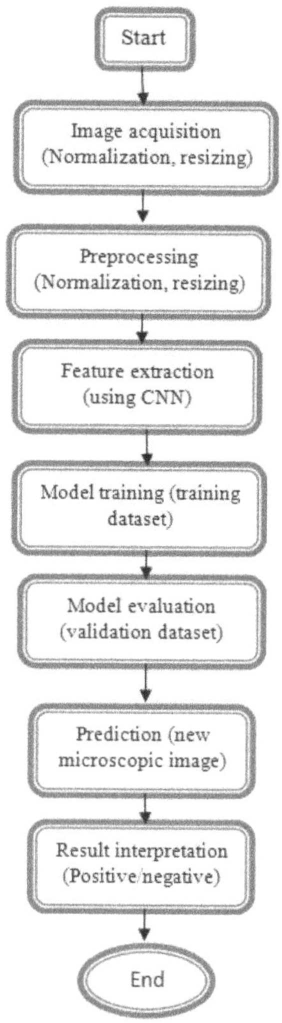

Figure 90.3. Flow chart.

Source: Author.

samples. In this research we apply our Giemsa-stained thin blood slides dataset from Figures 90.4 and 90.5 to conduct both training and testing. These slides came from the U.S. National Library of Medicine collections.

3. Results and Analysis

The technique works as a decision-supporting tool for technicians. Microscope picture processing is optimized by directing expert observation to potential Plasmodium-containing items. There exists a distinct and more subtle determination threshold for this particular purpose. Our system uses multiple classification stages for generating well-defined false and true results.

3.1. Evaluation of the performance matrix

The decision of optimal metrics to rate a classifier's data collection performance requires analysis of data class composition and outcome expectations in classification challenges. A single evaluation metric operates independently of the other unmonitored performance indicators and vice versa. The assessment of the classifier's complete performance remains without definite and coherent standards. This research uses multiple evaluation metrics to analyze model performance by evaluating F1 score alongside precision and recall while measuring accuracy and recall separately.

These metrics are derived from the following four categories: The classification reflects two instances where both the model output and actual class align at 1 (True) which we refer to as True Positives (TP). A model prediction of 1 (True) that occurs against the background of an actual 0 (False) event definition falls under the category of False Positive (FP). The prediction outcomes in which both the actual outcome indicated 'zero' while the model gave a 'zero' forecast are considered True Negatives (TN). False Negative (FN) occurs when a model generates a False prediction while the real occurrence type is True.

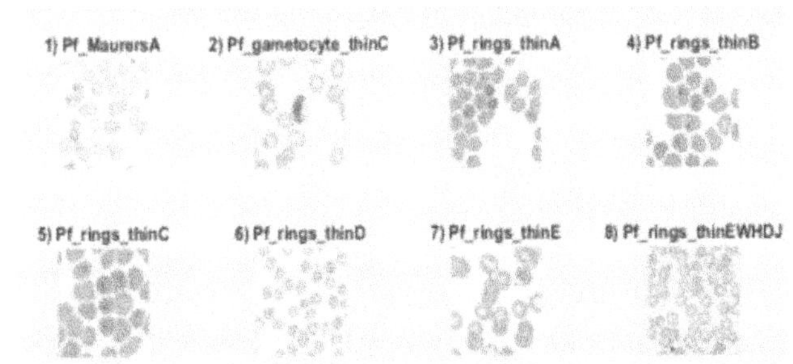

Figure 90.4. Input images.

Source: Author.

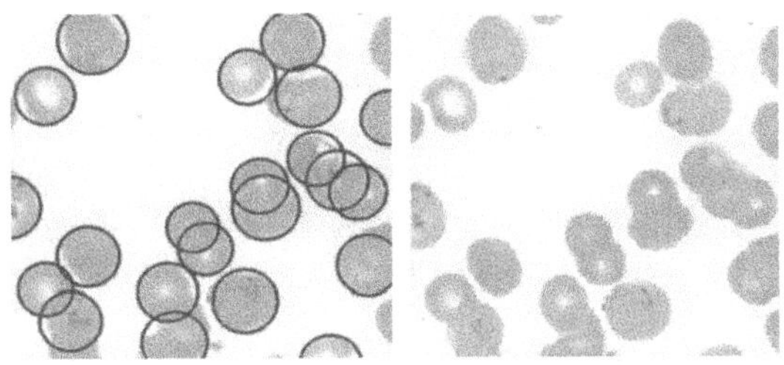

Figure 90.5. a) Detection of cells b) Infected cells.

Source: Author.

Precision, also known as positive predictive value, gauges the capacity of a model to pinpoint the right examples for every class. For multi-class classification with unbalanced datasets, this is a powerful matrix.

$$Precision = \frac{TP}{TP + FP} \tag{1}$$

Recall – This metric assesses how well a model detects the true positive among all instances of true positives.

$$Recall = \frac{TP}{TP + FN} \tag{2}$$

Accuracy – The mean amount of accurate predictions is used to characterize the accuracy measure. This isn't quite as strong, though, given the imbalanced sample.

$$Accuracy = \frac{TP + TN}{TP + TN + FP + FN} \tag{3}$$

F1-score – referred to as an F-measure or balanced F-score It might be characterized as a recall as well as precision weighted average.

$$F1_{Score} = \frac{2 * Precision * Recall}{Precision + Recall} \tag{4}$$

This section compares the effectiveness of many machine learning (ML) methods for the use of microscopic pictures in the identification of malaria. The findings undergo evaluation based on accuracy precision recall and F1-score computational effectiveness as well as model scalability. This analysis describes the advantages and disadvantages of each algorithm and characterizes their performance in real-world malaria detection tasks.

Table 90.1 lists the performance of the various machine learning methods. Tabulated data shows performance values for accuracy as well as precision and recall while F-score results are also showcased. Figure 90.6 illustrates that CNN performed higher compared to all five traditional machine learning methods we assessed.

Convolutional Neural Networks achieved highest accuracy of 97.8% together with precision rating of 98.2% and F1-score reaching 97.7%. Automated spatial and hierarchical data extraction capabilities of microscopic images stands as the key reason behind their performance.

The traditional machine learning algorithm of SVM delivered the best outcomes by reaching 92.5% accuracy and 93.1% precision. The system demonstrated strong binary classification ability while achieving particularly high performance results on balanced data.

Table 90.1. Performance evaluation

Algorithm	*Accuracy (%)*	*Precision (%)*	*Recall (%)*	*F1 Score (%)*	*Interference Time (ms)*
CNN	97.9	98.3	97.4	97.8	26
SVM	92.6	93.4	92.5	92.5	16
RF	90.2	90.6	90.1	90.1	13
KNN	88.4	87.9	88.3	88.3	11
LR	85.9	85.4	85.7	85.7	9

Source: Author.

Random Forest exhibited a 90.1% accuracy rate while achieving a 90.0% F1-score. This ensemble-based method implemented multiple decision tree integration methods that produced high-quality categorization results.

KNN obtained 88.5% recall and 88.3% accuracy. Larger dataset applications exposed performance restrictions even though it maintained good results for smaller datasets.

The baseline logistic regression model reached an accuracy rate of 85.7% with an F1-Score of 85.6%. Despite good performance results the algorithm showed limitations when identifying detailed patterns in visual data that made it lose to competing algorithms.

The successful identification of actual customers leaving the business is recorded as the True Positive Rate when measured against the model's prediction accuracy.

TPR measures the mathematical precision with which the model detects actual positive cases including users who left the company. True Positives (TP): The model correctly classified its identified positive cases. A mistake that identifies actual positive results as negative leads to identification as a false negative (FN).

$$True\ Positive\ Rate\ (TPR) = \frac{TP}{TP + FN} \quad (5)$$

False Positive Rate (FPR): The proportion of non-churners incorrectly predicted as churners.

$$False\ positive\ rate = \frac{FP}{FP + TN} \quad (6)$$

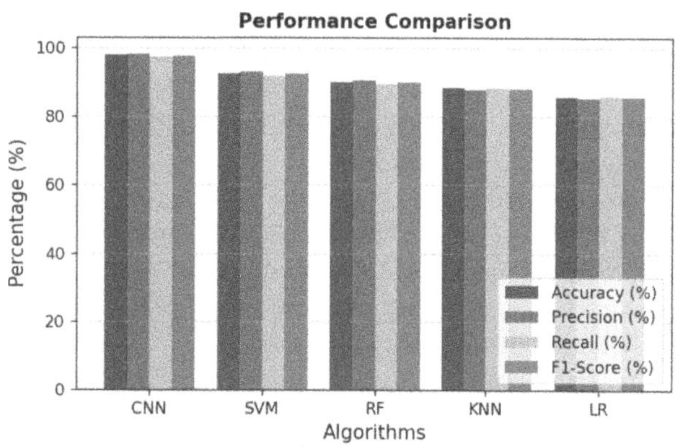

Figure 90.6. Performance comparison of various ML model.

Source: Author.

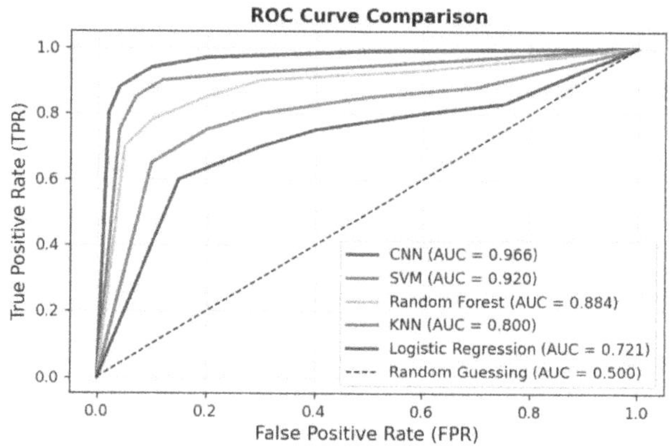

Figure 90.7. ROC curve comparison of various ML models.

Source: Author.

Receiver Operating Characteristic (ROC) curves and Area Under the Curve (AUC) scores were computed for all models:

The CNN produced the superior classification results with an AUC value (0.966) reaching excellence. The SVM methodology showed excellent performance through its AUC value of 0.921 followed by Figure 90.7.

4. Conclusion

The detection of malaria parasites benefits from our proposed shallow machine learning technique. Health workers in resource-constrained settings who lack sufficient laboratory professionals can benefit from this method of malaria parasite detection. Image patches were separated during analysis to determine malarial parasite presence using newly developed image processing technology which proved suitable for parasite detection. A combination of morphological segmentation techniques achieves the greatest cell identity and isolation efficiency since cubic CNN delivers the best accuracy score for plasmodium parasite classification. A total of 112 thin film blood smear images underwent testing and produced satisfactory results. Despite image inconsistency problems we reinforced our system.

References

[1] W. H. O. (World H. Organization), Global Health Observatory - Malaria, (2011). Available: http://www.who.int/gho/malaria.

[2] Mehrjou, A., Abbasian, T., & Izadi, M. (2013, February). Automatic malaria diagnosis system. In *2013 First RSI/ISM International Conference on Robotics and Mechatronics (ICRoM)* (pp. 205–211). IEEE.

[3] McKENZIE, F. E., Sirichaisinthop, J., Miller, R. S., Gasser Jr, R. A., & Wongsrichanalai, C. (2003). Dependence of malaria detection and species diagnosis by microscopy on parasite density. *The American journal of tropical medicine and hygiene*, *69*(4), 372.

[4] Tumwebaze, M. (2011). Evaluation of the Capacity to Appropriately Diagnose and Treat Malaria at Rural Health Centers in Kabarole District, Western Uganda, health policy and development, *9*(2011), 46–51.

[5] Agnihotri, N., & Agnihotri, A. (2015). Wrong Sample Dispensing May Cause False Positive Malaria Test. *Journal of Clinical and Diagnostic Research*, *9*(2015), EG01–EG02.

[6] Lee, J. H., Jang, J. W., Cho, C. H., Kim, J. Y., Han, E. T., Yun, S. G., & Lim, C. S. (2014). False-positive results for rapid diagnostic tests for malaria in patients with rheumatoid factor. *Journal of clinical microbiology*, *52*(10), 3784–3787.

[7] Khan, N. A., Pervaz, H., Latif, A. K., & Musharraf, A. (2014). Unsupervised identification of malaria parasites using computer vision. *International Joint Conference on Computer Science and Software Engineering*, 263–267.

[8] Quinn, J. A., Andama, A., Munabi, I., & Kiwanuka, F. N. (2014). Automated blood smear analysis for mobile malaria diagnosis. *Mobile Pointof-Care Monitors and Diagnostic Device Design*, 31–115.

[9] Lom, J., & Dyková, I. (2006). Myxozoan genera: Definition and notes on taxonomy, life-cycle terminology and pathogenic species. *Folia Parasitologica*, *53*(2006), 1–36.

[10] Dobson, A., Lafferty, K. D., Kuris, A. M., Hechinger, R. F., & Jetz, W. (2008). Homage to Linnaeus: How many parasites? How many hosts?, *National Academy of Sciences*, *105*(2008), 11482–11489.

[11] CNN-based image analysis for malaria diagnosis. https://ieeexplore.ieee.org/document/7822567.

[12] Afkhami, S., Rashidi Heram-Abadi, H., (2017). Detection of Malarial Parasite in Blood Images by two classification Methods: Support Vector Machine (SVM) and Artificial Neural Network (ANN). *Int. J. of Comp. & Info. Tech. (IJOCIT)*, *5*(2), 81–100. www.ijocit.org/journal/v05_i02/IJOCIT-V05I02P01.pdf.

[13] Blood Cells Counting using Python OpenCV https://ieeexplore.ieee.org/document/8652384

[14] Amer, G. M. H., & Abushaala, A. M. (2015). Edge detection methods. 2015 2nd World Symposium on Web Applications and Networking (WSWAN). doi:10.1109/wswan.2015.7210349.

[15] Chourasiya, S. (2014). Automatic Red Blood Cell Counting using Watershed Segmentation. https://pdfs.semanticscholar.org/ef24/ae99ca56cdb65d66d23cf376 7ee8a 73972d7.pdf

[16] Bhusare, M. P., & Bhosale, S. D. (2019). Automatic Blood Cell Counting using Morphological Image Processing operations. doi: https://doi.org/10.1007/978-3-319-96139-2_9

[17] Mohamad Sharif, Johan & Miswan, M. & Ngadi, Md & Abdul Jamil, Muhammad Mahadi. (2012). Red blood cell segmentation using masking and watershed algorithm: A preliminary study 2012 *International Conference on Biomedical Engineering, ICoBE*. doi:10.1109/ICoBE.2012.6179016

91 An experienced and categorical brain tumor detection through image processing methods

Chandan Roy[1,a], *P. Kranthi Rathan*[2,b], *G. Venkataswamy*[3,c], *Biroju Papa Chary*[4,d], *and T. Shanthi*[5,e]

[1]Assistant Professor, Department of ECE, CMR College of Engineering and Technology, Hyderabad, Telangana, India
[2]Associate Professor, Department of EEE, CMR Technical Campus, Hyderabad, Telangana, India
[3]Assistant Professor, Department of ECE, CMR Institute of Technology, Hyderabad, Telangana, India
[4]Associate Professor, Department of ECE, CMR Engineering College, Hyderabad, Telangana, India
[5]Associate Professor, Department of Electronics and Communication Engineering, Sona College of Technology, Salem, India

Abstract: Therefore, detecting brain tumors is one of the most essential steps to identify and treat one of the most lethal diseases. Remote and accurate detection of brain tumors can take part in a crucial role in the prognosis of patients. In this paper, we investigate the identification and classification of brain tumors on diagnosis using advanced image processing techniques from medical imaging modalities such as CT & MRI. The paper discusses several methods for improving the accuracy of tumor detection: image pre-processing, segmentation, feature extraction and classification. Some common methods used for doing this are, Thresholding, Morphological operations and edge detection and machine learning methods like Convolutional Neural Networks (CNNs) that can be used for classifying tumour and non-neoplastic areas. Therefore, the suggested method aims to cover challenges such as noise, image artifacts, tumor heterogeneity (size, shape, and location) thereof, by uniting both automated and semi-automated strategies, as demonstrated in the example below. The experimental outcome show that the model used in this study improves detection and classification of brain cancer with better precision, recall and accuracy.

Keywords: Brain tumor detection, imgae processing, MRI, CNN, segmentation, machine learning, SVM, KNN

1. Introduction

Brain tumors are one of the worst and potentially deadliest medical disorders due to their intricateness, wide range of presentations, and fast growth rate. Thus, they represent one of the biggest challenges to physicians. For a better survival rate, there is an essential need for the early detection and precise diagnosis of patients in order to undertake early treatment. CT scans and MRI, the typical diagnostic approaches, play an important role in examining abnormalities of the brain.

Detection and categorization of tumors are really cumbersome and time-consuming processes for doctors in the field of Medical Science. MRI and computed thermography are the two most useful tools that are used constantly by radiologists for non-invasive inspection of body parts. The alternative of conducting these inspections manually uses a lot of valuable work hours and is very time-consuming. The brain can be considered the most complex organ because of its intricate structure, making it a very challenging target of study, even from a basic structural delineation of its constituent parts [1–5].

Infected brain (i.e. brain with tumor) is then analyzed by the radiologist and compared to the original shape of the brain, a huge responsibility in determining the existence of a tumorous formation. Scientists are trying to understand the role of the brain and what happens to human biomechanics when abnormalities are formed through the development of advanced technologies. to detect the existence of a brain tumor quickly and accurately. Broadly, two types of tumors exist, benign and malignant [6–9]. Regardless of how it is measured, though, it is still highly challenging for medical research to analyze. The inspection work associated with the segmentation of brain tumors using MRI can be time-consuming, which is why scientists are working to implement machine learning methods for these tasks. So far, there is no concrete evidence on which of the classification algorithms is most suitable for the described objective. Moreover, it remains unclear whether a single approach can be equally effective across all areas of the brain.

Commonly used by the medical community for the detection of a tumor is by the use of MRI and computed

[a]roychandan447@gmail.com, [b]cmrtc.paper@gmail.com, [c]gutam.venkataswamy@gmail.com, [d]b.papachary@cmrec.ac.in, [e]shanthi@sonatech.ac.in

DOI: 10.1201/9781003684459-91

thermography. These techniques are very reliable, but they have some disadvantages [10–11].

This article reviews the role of image processing methodologies in the detection of brain tumors; further, it has identified the capability of image processing techniques in overcoming noise, image artefacts, and variability between tumors with respect to their size and location. It looks further into how artificial intelligence models-Convolutional Neural Networks (CNNs)-can be leveraged, along with morphological operations, thresholding, and edge detection, in making correct tumor diagnoses. This study also emphasizes the rising role of automated systems in medical imaging and what implications these systems will have on health care, including reduced diagnostic delay, reduced human error, and more cost-effective and efficient health care.

2. Related Work

It specified the removal of noise and skulls from this preprocessing task. Feature extraction step uses the GLCM technique. Two examples used that can be used for feature reduction are sparse representation-based classification, rough set theory, and optimization using particle swarms. The performance of the system and classification performance evaluated by the SVM, Naïve Bayes, and Rusboost classifiers. In [4], D. Bhuvana and P. Bagavathi Sivakumar have presented an approach to automatic tumor identification and classification using a probabilistic neural network. First, the mottlings of the image were removed using the anisotropic filtering method. The filtered image is sharpened so that it is ready to extract features. Then, Fuzzy C-Means segmentation [15] was applied to part out the tumor area. We eventually classified them using a probabilistic neural network [16]. Brain imaging is divided into two types formar categories. They are harmless but damage the brain.

2.1. Proposed system

The suggested model for identifying brain tumors will utilize the subdomain of deep learning methodologies called Convolutional Neural Networks, for the interpretation and classification of brain pictures from various medical imaging modalities, such as MRI scans. The ability of CNNs to automatically learn spatial hierarchies of features makes them best suited for detection and classification of tumors. The model seeks to offer an automated, precise, and effective way to detect brain cancers. The data, source of brain MRI dataset, and algorithm utilized for the segmentation and classification of the brain MRI tissue are provided in this section. Figure 91.1 Stream outline of the proposed algorithm. T1 weighted contrast enhancement magnetic resonance image files can be publicly accessed acquired from. The succeeding materials and techniques sections are responsible in the proposed evaluation of the brain MRI.

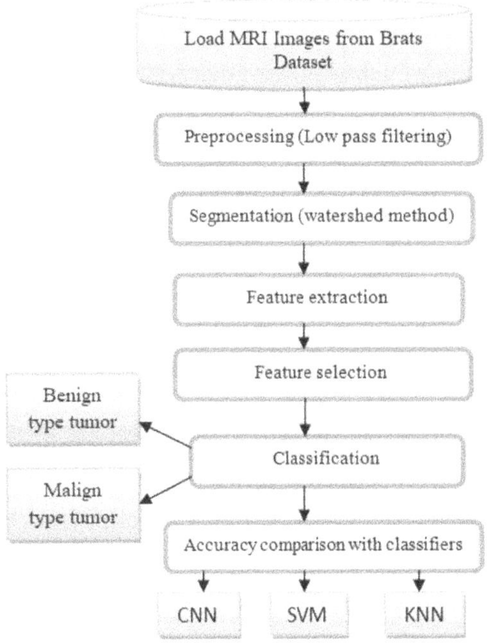

Figure 91.1. Proposed model.

Source: Author.

2.2. Input image

This is the first stage of the proposed system. Finally, the final MRI scans may not be usable. Images can be low contrast, noisy, or fuzzy. It can be difficult to extract the area of interest. In this case, the input to the system is an MRI image of the brain.

2.3. Pre-processing

Pre-processing is primarily aimed at improving MR image quality and making it more suitable for human or machine processing. Actions like noise reduction and removal, image augmentation, and region or boundary tracking are performed.

2.4. Image segmentation

It is the process of segmenting the object or pertinent information in an image and dividing it into meaningful regions. Many ways of segmentation have existed based on the application, imaging modality, and other criteria. Segmentation allows proper analysis of the image because after its application, the image becomes easier to analyze and more relevant. Hence, it is employed to sharpen the edges and objects of the image. Segmentation is the technique of assigning a name to pixels in an image that possess similar visual characteristics. The output from image segmentation may be viewed as a set of segments that collectively describe an entire image or as a set of contours extracted out of an image. Segmentation basically implies separating an object from its background. Here, the Thresholding technique has been applied to segregate the image. Morphological procedures take an input image and produce an output image of the same size by applying a structural element.

The value of every pixel in the output image depends on a morphological operation comparing the input image's corresponding pixel and its neighbors. We can construct a morphological operation sensitive to a specific shape in the input image by choosing the size and shape of the neighborhood. Erosion and dialation are the two basic morphological operations. Here we use the erosion technique, which removes pixels that are on the object edges.

2.5. *Feature extraction*

This step is all about isolating the image to shapes or places that we want. Detecting whether an image is normal or abnormal, according to the result of the segmentation process, is implemented using multiple algorithms in image processing techniques. In the case of the tumor region, the features that help distinguish the tumor region are the area, perimeter, centroid and the diameter.

2.6. *Output*

Let us start with the source image. We then display the anisotroically filtered image. Because of morphological operation, a morphological image and outcome are shown. So when there is a tumor location, it is recognized and shown.

2.7. *Morphological feature extraction*

Since the segmentation of brain tumors in MRI is challenging enough due to the locations of the tumor, various shapes, and image intensity, the improvement of the tumor region is still required once image segmentation is done to a binary image. The morphological operators close, open, dilate, and erode the image for removing the small holes and obstacles. Erosion and dilatation are used in the suggested study. Erosion pulls out the pixels of an image, whereas dilation adds pixels at the object edges. To achieve erosion and dilatation, use imerode and imdilate as shown in the following expression:

$$f_e(x,y) = imerode(f(x,y), E)$$

$$f_e(x,y) = imdilate(f(x,y), E)$$

Where *f*(*x*, *y*) is binary image $f_e(x, y)$, is returning the dilated image. The contention E is an organizing component.

3. Training and Testing

Training Phase: The MRI input images are labeled with their corresponding labels that will be used for training the CNN model. Here, the system learns how to classify tumor and non-tumor regions by minimizing the loss in training-preferably categorical cross-entropy. This is where the Adam optimizer changes its weights.

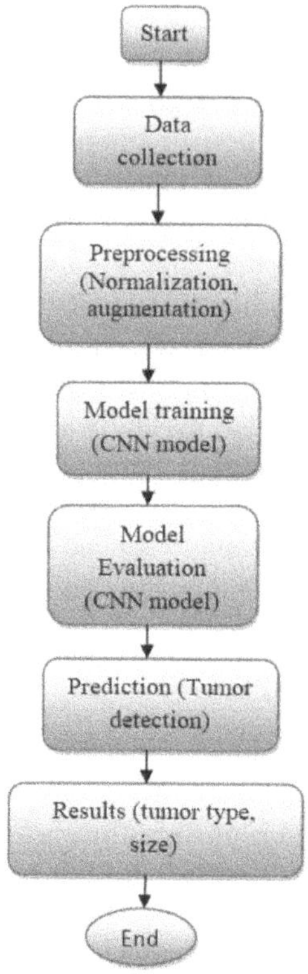

Figure 91.2. Flow chart.

Source: Author.

Testing and Validation: The model is now tested and validated on different datasets to see its performance using measures such as the AUC-ROC curve, F1-score, accuracy, precision, and recall.

Figure 91.2 depicting images as outputs with input, filtered, threshold, bounding box and tumour outline images: Figures 91.3 and 91.4, Final Tumour Extraction from MRI picture and Image Containing Tumour. This process simplifies a lot the search of tumours with much more intensity from its background when imaging is made. Outcomes of the results obtained are:

Figures 91.3 and 91.4 display the results of the tumor detection in sample picture 1, which is indicated by red highlights surrounding the tumor.

3.1. *Performance metrics*

The model's performance is assessed based on the following key metrics:

Accuracy: Measures the overall correctness of the model in detecting and classifying brain tumors.

$$Accuracy = \frac{TP + TN}{TP + TN + FP + FN} \quad (1)$$

Precision

$$Precision = \frac{TP}{TP + FP} \quad (2)$$

Recall (Sensitivity): Measures the ability of the model to correctly detect tumors among all actual tumor cases.

$$Recall = \frac{TP}{TP + FN} \tag{3}$$

Specificity: Indicates the ability of the model to correctly identify non-tumor cases.

$$Specificity = \frac{TN}{TN + FP} \tag{4}$$

AUC-ROC Curve: Assesses how well the model can differentiate between tumor and non-tumor groups.

The percentage of real churners that the model accurately predicted is known as the True Positive Rate, or TPR.

TPR quantifies the percentage of real positives – such as users that churned – that the algorithm accurately recognized. True Positives (TP): Positive cases that were accurately predicted. Real positive cases that were mistakenly expected to be negative are known as false negatives (FN).

$$True\ Positive\ Rate\ (TPR) = \frac{TP}{TP + FN} \tag{5}$$

False Positive Rate (FPR): The proportion of non-churners incorrectly predicted as churners.

$$False\ positive\ rate = \frac{FP}{FP + TN} \tag{6}$$

3.2. Dataset

A labeled dataset, such as BRATS-that stands for Brain Tumor Segmentation-or other publicly available MRI datasets, was considered while testing the suggested CNN model. Along with photographs of patients with healthy conditions, a collection contains MRI pictures of patients with gliomas, meningiomas, and pituitary tumors among other types of brain cancers.

Training Set: The model was trained on 70% of the dataset. Validation Set: 15% was used for validation and hyperparameter tuning. Testing Set: The model was tested on 15% of the dataset.

4. Results and Analysis

The results and analysis of the proposed Convolutional Neural Networks (CNNs)-based brain tumor detection model demonstrate effectiveness of this deep learning technique for accurate, precise and reliable detection of brain tumor from medical imaging modalities such as MRI. The results are evaluated based on multiple performance metrics and extensive experiments on datasets.

Ensemble Classifier approach is used to classify brain pictures as Benign or Malignant. This successful classifier can classify the normal and disordered brains. It consists of a Duel classifier with Naïve Bayes and Support Vector Machine classifiers. With this approach, an accuracy level of 98.41% was achieved. That is, we investigate the precision and sensitivity properties of the MRI images.

To provide further evidence of CNNs' superiority with regard to brain tumor classification, the results were compared against more traditional machine learning methods, such as SVM and k-NN, as shown in Table 91.1.

The experimental results showed that the proposed algorithm can effectively make a distinction between benign and malignant tumors with accuracy of 95.2% on MRI. From the results, radiologists or authority of clinical in future presume that the Proposed Algorithm's are amenable for acquirement of clinical steady structure of decision-making for essential screening.

As shown by its 95.2% accuracy rate, it means the model took the bulk of the test photos and correctly classified it.

The model's specificity of 94.7% means it can identify all the cases of tumors. This is quite reliable in clinical practice because one would not miss a tumor-a situation that might lead to dire consequences.

A specificity of 96.1% reflects this model's aptitude for effectively identifying non-tumorous cases; thus, it makes sure that any false positives get introduced at really low rates.

Plot the performance metrics of the CNN-based brain tumor detection model. It can also contain metrics like accuracy, precision, recall, specificity as the Figure 91.3 described.

Figure 91.4 shows that An AUC-ROC score of 1 indicates excellent separation between tumor and non-tumor classifications.

5. Conclusion

A brain tumor is the most typical and pernicious growth. The studies reveal that this deadly disease could be diagnosed at an early stage and if not it can take lives. As stated, the way to detect this disease is important to avoid the worst stages

Table 91.1. Performance comparison metrics with various models

Model	*Accuracy (%)*	*Precision (%)*	*Recall (%)*
CNN Proposed Model	95.3	93.9	94.8
SVM	89.7	88.3	87.5
KNN	854	84.6	83.9

Source: Author.

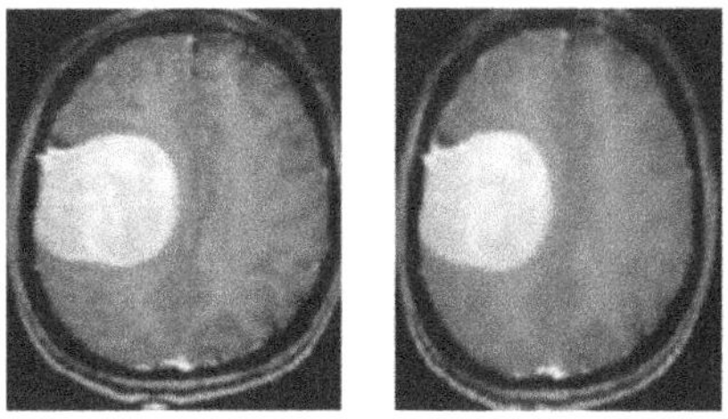

Figure 91.3. a) Input image b) Filtered image.

Source: Author.

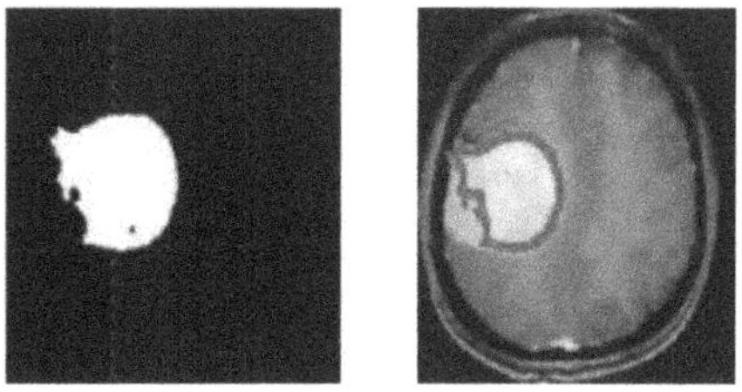

Figure 91.4. a) Tumor alone b) Detected tumor.

Source: Author.

and reduce the number of people around the world. In this proposed study, these MRI brain images were used for a few processes that include correction of image defects, image quality enhancement, analysis of the image, and finding the tumor based on thresholding segmentation. Morphological operations are performed to find the exact place of tumor mark contour. When this hypothetical approach is practical to an irregular brain MRI, any tumor can be accurately localized.

References

[1] Jiménez-Alaniz, J. R., Medina-Bañuelos, V., & Yáñez-Suárez, O. (2005). Data-driven brain MRI segmentation supported on edge confidence and a priori tissue information. *IEEE Transactions on Medical Imaging*, *25*(1), 74–83.

[2] Reddick, W. E., Glass, J. O., Cook, E. N., Elkin, T. D., & Deaton, R. J. (1997). Automated segmentation and classification of multispectral magnetic resonance images of brain using artificial neural networks. *IEEE Transactions on Medical Imaging*, *16*(6), 911–918.

[3] Song, T., Jamshidi, M. M., Lee, R. R., & Huang, M. (2007). A modified probabilistic neural network for partial volume segmentation in brain MR image. *IEEE Transactions on Neural Networks*, *18*(5), 1424–1432.

[4] Vrooman, H. A., Cocosco, C. A., Van Der Lijn, F., Stokking, R., Ikram, M. A., Vernooij, M. W., ... & Niessen, W. J. (2007). Multi-spectral brain tissue segmentation using automatically trained k-Nearest-Neighbor classification. *Neuroimage*, *37*(1), 71–81.

[5] Alirezaie, J., Jernigan, M. E., & Nahmias, C. (1998). Automatic segmentation of cerebral MR images using artificial neural networks. *IEEE transactions on nuclear science*, *45*(4), 2174–2182.

[6] Lavanyadevi, R., Machakowsalya, M., Nivethitha, J., & Kumar, A. N. (2017). Brain Tumor Classification and Segmentation in MRI Images using PNN. *International Conference on Electrical, Instrumentation and Communication Engineering (ICEICE)*, 2017.

[7] Hazra, A., Dey, A., Gupta, S. K., & Ansari, M. A. (2017, August). Brain tumor detection based on segmentation using MATLAB. In *2017 International Conference on Energy, Communication, Data Analytics and Soft Computing (ICECDS)* (pp. 425–430). IEEE.

[8] Ali, E. M., Seddik, A. F., & Haggag, M. H. (2016). Real brain tumors datasets classification using TANNN. *International Journal of Computer Applications*, *975*, 8887.

[9] Alfonse, M., & Salem, A. B. M. (2016). An Automatic Classification of Brain Tumors through MRI Using Support Vector Machine. *Egyptian Computer Science Journal*, *40*(3).

[10] Fred, H. L. (2004). Drawbacks and limitations of computed tomography: views from a medical educator. *Texas Heart Institute Journal*, *31*(4), 345.

[11] Liu, J., Wang, Y., Katscher, U., & He, B. (2017). Electrical properties tomography based on $ B_ {{1}} $ maps in MRI: principles, applications, and challenges. *IEEE Transactions on Biomedical Engineering*, *64*(11), 2515–2530.

[12] Kumar, P., & VijayKumar, B. (2019). Brain tumor MRI segmentation and classification using ensemble classifier. *International Journal of Recent Technology and Engineering (IJRTE)*, *8*(1S4).

[13] Deepti Ahlawat, M. (2017). Brain Tumor Detection using RST-PSO. *International Journal in Multidisciplinary and Academic Research*, *6*(5).

[14] Varuna Shree, N., & Kumar, T. N. R. (2018). Identification and classification of brain tumor MRI images with feature extraction using DWT and probabilistic neural network. *Brain informatics*, *5*(1), 23–30.

[15] Işın, A., Direkoğlu, C., & Şah, M. (2016). Review of MRI-based brain tumor image segmentation using deep learning methods. *Procedia computer science*, *102*, 317–324.

[16] Avanija, J., Ramji, B., Prabhu, A., Maheswari, K., Vittal, R. H. S., Jagannadham, D. B. V., & Kumar, V. N. (2023, March). Interpretation of brain tumour using deep learning model. In *Proceedings of Fourth International Conference on Computer and Communication Technologies: IC3T 2022* (pp. 347–355). Singapore: Springer Nature Singapore.

92 A real-time driver behaviour tracking system in automobiles utilizing image processing to extract features

A. Veerababu[1,a], *A. Vamshidhar Reddy*[2,b], *S. Nagajyothi*[3,c], *A. Srinivasula Reddy*[4,d], *and R. S. Sabeenian*[5,e]

[1]Assistant Professor, Department of ECE, CMR College of Engineering and Technology, Hyderabad, Telangana, India
[2]Associate Professor, Department of EEE, CMR Technical Campus, Hyderabad, Telangana, India
[3]Assistant Professor, Department of ECE, CMR Institute of Technology, Hyderabad, Telangana, India
[4]Principal and Professor, Department of EEE, CMR Engineering College, Hyderabad, Telangana, India
[5]Professor and Head, Department of Electronics and Communication Engineering, Sona College of Technology, Salem, India

Abstract: The actions which drivers engage in consistently determine the condition of roadway safety and their probability of experiencing accidents. New real-time driver behaviour monitoring systems gained major attention during recent times since they provide a preventive approach to discover and correct unsafe driving actions. A real-time system based on enhanced image processing protocols has been created to monitor driver behaviour within this investigation. The proposed system employs vehicle-mounted cameras which use real-time videos of drivers to detect distracted driving and dangerous postures as well as signs of driver tiredness. Pedestrian and vehicle detection through quick methods must exist alongside autonomous driving technology to prevent accidents. Detecting pedestrians in complex actual roadway conditions proves very subtle and demanding to handle effectively. The system measures head placement and eyes to detect both fatigue symptoms and driver distractions. Behaviour recognition through object detection and action recognition takes place to identify secondary tasks that occur during driving including eating and cell phone use. Today's driver monitoring systems track elusive data points through transformative image processing technologies, namely Convolutional Neural Networks (CNNs), Haar cascades and feature extraction techniques. Whenever alert cases arise our system delivers immediate notifications to trigger appropriate intervention.

Keywords: Real time driver monitoring, alert system, road safety, drowsiness detection, CNN

1. Introduction

The problem exists worldwide due to driver-related causes such as distraction and drowsiness and reckless behaviour which create numerous global vehicle accidents. Global road traffic crash data from the World Health Organization (WHO) shows human error stands responsible for 90 percent of accidents worldwide. Driver faults from fatigue drowsiness together with distractions caused by eating and cellphone use and interaction with passengers lead to most vehicle crashes. Real-time driver behaviour observation technologies emerged as a solution because these road safety challenges showed that advanced systems must be developed. Tirendry, thoughtless conduct and sleeping drunk actions remain as the main drivers of current-day auto accidents. People face enormous danger of crashing vehicles because of driving drowsy or without focus. When we drive we make mistakes with all road users so we need to share information about our actions. Active driver supervision provides the most effective way to reduce accidents that drive accidents cause.

It will have to be the whole focus of the work to build the system that can accurately measure the level of the driver fatigue, tiredness, distractedness, visual emotion detection and the handling of a mobile phone. Someone who is drunk driving or using his cell phone has a really high probability of having an accident. The best way to prevent accidents caused by driver error is through screening their behaviour. We will focus exclusively on creating the system that will correctly recognize the driver's fatigue, lethargy, distraction, facial emotion recognition, and phone use.

The proposed system utilizes state-of-the-art image processing methods including Convolutional Neural Networks (CNNs) for action classification and face recognition as well as object detection methods to identify distractions such as

[a]a.veerababu@cmrcet.ac.in, [b]cmrtc.paper@gmail.com, [c]s.jyo89@gmail,cim, [d]svas_a@yahoo.com, [e]sabeenian@sonatech.ac.in

DOI: 10.1201/9781003684459-92

cell phone usage. By processing live video feeds from in-car cameras, the technology gives drivers real-time alerts, allowing them to take preventative action before dangerous situations unfold. Not only is this reducing the potential for accidents, but it also facilitates the creation of safer, smarter transportation systems.

This study trained and employed a real-time driver behaviour monitoring system based on image processing. It outlines the methods for accurately recognizing driver states such as inattention, fatigue and reckless behaviour, as well as the challenges of implementing these systems on the road in dynamic environments. Applications of these systems in current cars, their limitations, and future trends, including the deployment of multi-sensors technologies and artificial intelligence to increase scalability and reliability, are also discussed.

2. Literature Survey

Dalal showed that, compared to other methods, Histograms of Oriented Gradient (HOG) descriptors perform better with human detection [1]. For pedestrian detection, Dalal has used the first group, image-based pedestrian detection. Identifying humans in images is a challenging task due to their erratic appearance and ability to take on various postures.

It's [2] fusion approach is comprised by modules for detection and classification. The detection module shows the size and position of moving objects, while the classification module classifies the identified moving objects into multiple classes.

[3] proposed a background removal algorithm by modeling the static background of each pixel points as 3D Gaussian distribution. After modeling the background, each pixel in the input frame is considered as a foreground pixel. It is called Pfinder, it is a system that observes and predicts a person behaviour in the moment. The system uses a multi-class statistical model of colour and shape to provide a 2D glimpse of the head and hands in different viewing conditions.

The proposed [4] method avoids any possible discretizations by using a mechanism that is separate from the per-pixel background subtraction. Only one component affects the threshold, camera noise, which should have a Guassian distribution with zero mean. However, we know that the assumptions given in this work are not valid due to the fact that they are useless in real life applications.

A paper titled [5] is about 'Computer Vision based drowsiness detection for motorized vehicles with Web Push Notifications.' They have developed a web push notifications, alert noises based computer vision-based sleepiness system for the cars. These messages will help to prevent an accident by alerting the motorist. In addition, the system is designed to generate an alert that shows coffee shops in the vicinity, making drivers more alert as a result.

The paper [6] 'Real-time monitoring of driver drowsiness on mobile platforms using 3D neural networks' They have found out micro-sleeps in this study and alert drivers of these trough the application of depth-wise separable 3D convolution operations to detect tiredness in drivers using live facial feed.

In their work [7] used different type of sensors with a DMS that the DMS detects tiredness. They are also a signal from other car-based sensors. The results show that models effectively classified three drowsiness conditions (mild, moderate, and severe).

In [8], the purpose of the paper is to detect driver tiredness using E.A.R. (eye aspect ratio), that is, for blink detection it is easy, unlaborious and effective and analysis of human-eye for blinks using new facial landmark recognition [9]. Importantly, by providing a sufficiently close estimate of the extent of eye opening over time, the system was capable of detecting driver drowsiness. Since facial landmark detection offers a very low performance overhead, this alerting system can be used in real time.

It proposed pedestrian localization approach that uses improved Kalman filter [10]. The localization model relates the observation to the state vector and applies the perspective projection principle to choose the suitable state for the filter. This method gives an exact and usable location of the pedestrian. It helps to improve the safety of pedestrians and vehicles from colliding into each other. Since this only requires a single camera, the system is inexpensive and efficient. To assist the driver, it proposed [11] a way to monitor him. The input to the system is video, and at a later stage the optical flow between video frames is computed. The next step is to decouple the orientation and the magnitude fields. The process of super pixel segmentation and the definition of a Histogram Descriptor is performed to train one class SVM classifier. Driver behaviour model in the presence of driver behaviour [12].

3. Proposed Model

Essentially, the model's objective is to create an accurate and real-time system that can receive live video from an onboard car camera, analyze, and evaluate the driver's actions. To avoid accidents, the system detects specific behaviours (such as drowsiness, cellphone use, distractions or safe driving) and issues alerts. This system tracks your every move, and are designed to monitor driver behaviour. By tracking behaviours with precision, the proposed system determines whether the driver's mind is engaged. This system is built on the effectiveness of image processing and computer vision for driver face detection. The driver face is detected using Haar Cascade method in this system. A library of 68 facial point expressions is used to detect emotions. So the method discussed is used to observe the driver as well as enables attentive driving. The driver operates the vehicle, while a camera tracks the driver's face, the footage then processed into a video stream. The software then analyzes the footage to assess an sensitive to drowsiness and identify signs of fatigue. At this stage, the main things that should be considered for analysis are the logging of the driver's facial tracking, their drowsiness level and the identification of their determining facial features based on their eye closure and yawning. Finally, if the drowsiness is identified, a voice alert is activated.

The high-level system architecture shown in Figure 92.1 represents the pre-processing and evaluation procedures together with model input. The starting point involves refining the video stream to enable human face tracking. Face element extraction of lips and eyes constitutes the essential second processing stage. The system in stage three detects fatigue indications such as facial eye motions and blinking along with periodic yawns.

3.1. *Facial points*

Face Landmark Detection refers to the process of detecting the landmarks of human faces in an image. The face mark detector can be used with any image. The first step of landmark detection is to detect faces in an image and their bounding boxes. Once we found where the face is located in an image, we then need to identify minor parts of the face, like eyes, eyebrows, lips, and mouth.

3.2. *Eye aspects ratio*

The state of being eye-opening is estimated in the eye aspect ratio. The eye aspect ratio is constant when the eye is open; when it is closed, however, it rapidly declines to zero. The persons eyeblinks are poking out really well. The eye aspect ratio of the person has to be lower than a predetermined limit, and the eyes of the person has to be closed. The blinks are computed by using conventional image processing algorithms, that typically combine.

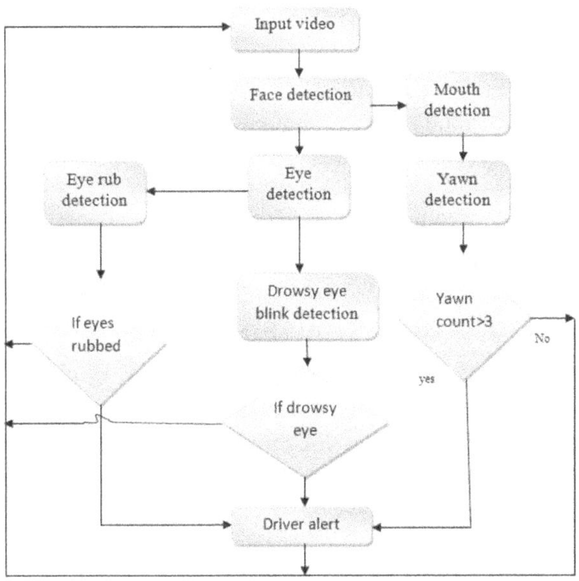

Figure 92.1. System model.

Source: Author.

3.3. *Head pose estimation*

Head pose estimate refers to the estimation of the position of a human head in a picture. It is specifically determining the Euler angles of the yaw, pitch, and roll of the head. The main goal of this activity is to obtain the relative (and positional) orientation of the human head with respect to the camera. Head posture estimation is used in a broad spectrum of computer vision applications, including virtual reality, hands-free gesture control, driver attention detection, gaze estimation, and lots more.

3.4. *Input layer*

Camera Sensor: A camera in the passenger compartment can see the driver's face and upper body in live video as shown the Figure 92.2.

CapturedFrame: The real-time video frames is captured for future processing. This ensures that the system runs smoothly and continuously.

3.5. *Preparation*

Image Processing: Step convert image frames into grayscale to decrease computational cost. Histogram equalization to enhance low-light images.

The face of the driver needs to be detected using techniques such as DLIB, MTCNN, or Haar cascades.

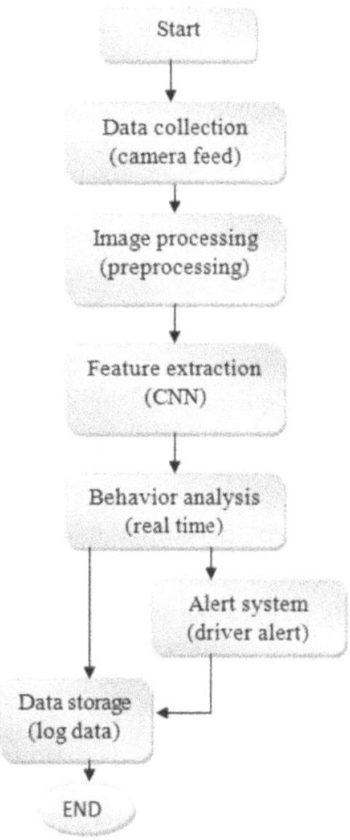

Figure 92.2. Flow chart.

Source: Author.

Interest Region (ROI): For a detailed analysis, segment specific tissue regions such as the lips, eyes and video head position.

3.6. Identification of behaviour

Drowsiness detection: Eye Aspect Ratio (EAR) is used to detect prolonged eyelid closure Identify yawning using lip movement patterns or mouth aspect ratio.

Identification of Distractions: Head position estimation: Whether the driver is distracted off the road. Look at the hands showing phone usage.

Normal/Safe: No unusual actions are noted.

3.7. Image processing

Image processing is the conversion of a normal image into a digital form. so some operations must be performed on the image to extract something useful information from it, this is very good way to measure drowsiness level. Image processing features any type of signal processing whereby the input is an image, equivalent to a picture or video frame. An image or set of parameters or features related to the image may be the output of image processing. Although it is more common to apply digital image processing, this is also possible by analog and optical image processing. Imaging = The process of acquiring images. Image processing is intimately related to computer graphics and computer vision. Such computer graphics visuals – painstakingly derived from physical models of real objects, environments and lighting – are not based on real settings, as is the case with most animated movies.

Feature Extraction: Feature Extraction: CNN extracts salient features from the video frames: eye closure, head position, and hand movements.

Classification: Classification Behaviour is classified into predefined categories: Drowsy, Distracted, Phone Usage, or Normal.

4. Design Model

The technology is meant to constantly track the driver's face, and therefore their eyes and mouth. Hence, when the set thresholds of alert level are violated or defaulted, relevant alarm is raised and the right action is taken to avoid any death. Figure 92.3 shows the system design of driver drowsiness and yawn detection system. And you could tell that they were continuously monitoring the face of the driver through the camera, and voice alerts were given by the dashboard system once they detect signs of exhaustion or drowsiness.

To assess the performance of the proposed model:

Precision, also known as positive predictive value, gauges the capacity of a model to pinpoint the right examples for every class. For multi-class classification with unbalanced datasets, this is a powerful matrix.

$$Precision = \frac{TP}{TP + FP} \tag{7}$$

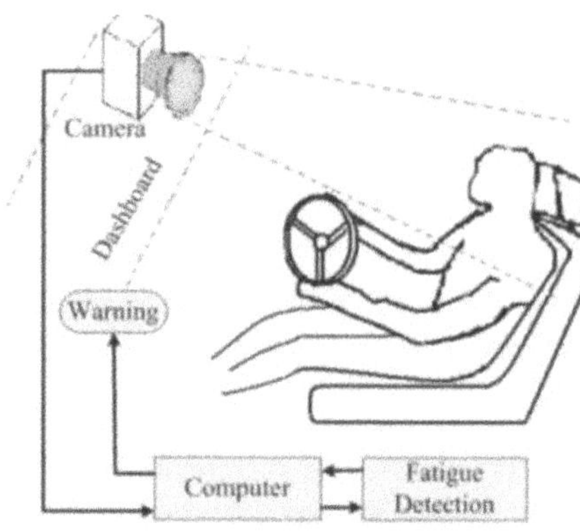

Figure 92.3. System design.

Source: Author.

Recall – This metric assesses how well a model detects the true positive among all instances of true positives.

$$Recall = \frac{TP}{TP + FN} \tag{8}$$

Accuracy – The mean amount of accurate predictions is used to characterize the accuracy measure. This isn't quite as strong, though, given the imbalanced sample.

$$Accuracy = \frac{TP + TN}{TP + TN + FP + FN} \tag{9}$$

Latency: Time taken for the system to process and generate alerts in real-time.

Usability: Evaluate the system under real-world driving conditions.

4.1. Dataset performance

To evaluate the model, a dataset containing various driver behaviours, such as sleepiness, diversions, phone use and normal driving was annotated.

a. The training set is 70% of the data.
b. Test set: 30% of the data.
c. 50,000 total frames, equally distributed across all behaviours.
d. Real-world video data were collected in multiple environments with different lighting and head poses (day, night, etc.).

5. Results and Analysis

The results illustrate the ability of the system to detect and classify driver behaviours. The outcomes are analyzed in this chapter based on the classification accuracy, performance in different scenarios, system reactivity, and aspects that require development.

5.1. Analysis of performance

They then evaluated the system's ability to detect driver behaviours, such as safe driving, drowsy driving, and distracted driving, against standard classification metrics (accuracy, precision, recall, and F1-score) as shown the Figure 92.4.

Safe Driving shows the algorithm still has an accurate understanding of average driving behaviour with 95.2% of the time it classifying the data correctly.

Drowsy Driving was reported at 92.5% accuracy, though accuracy was lower due in part to false negatives in partial eyelid closures or during yawning as BhakyaLite relied heavily on light.

After accounting for false positives and head movements that were not actual distractions, this accuracy for Distracted Driving was 90.1%.

5.2. Latency and real time performance

Frame Rate: The system processes video input at 15–30 frames per second (FPS) based on hardware capability. Latency: It ensures real-time feedback and alarm production of an average processing period of 0.1 to 0.2 seconds per frame. Alert System: As soon as dangerous conduct was detected, audio and visual alerts were triggered in 0.5 seconds.

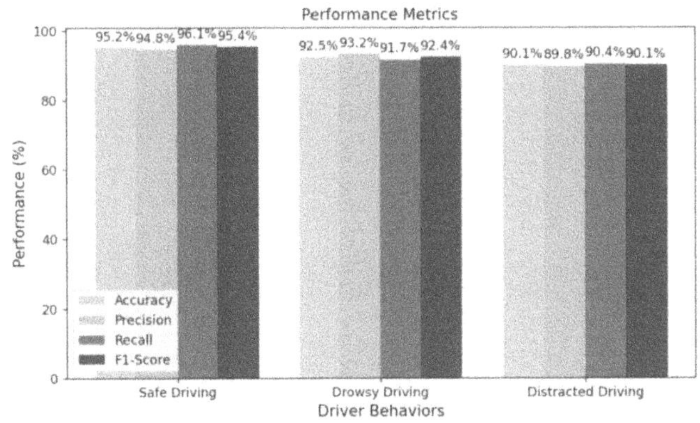

Figure 92.4. Performance evolution metrics.

Source: Author.

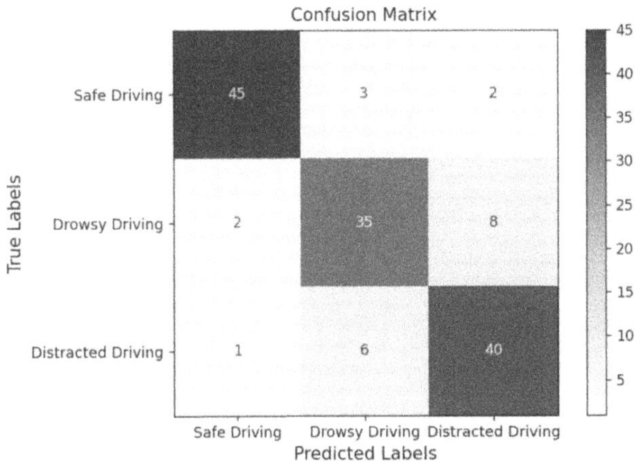

Figure 92.5. Confusion matrix.

Source: Author.

5.3. Confusion matrix analysis

Sleepy & Drowsy Driving: 50 out of 50 cases were detected correctly, with very few having false positives (classified incorrectly as sleepy or inattentive).

Drowsy Driving: Movements of the head that looked like someone was turning away from the road contributed to some cases being misclassified as distracted.

Performance was good overall, though a few strong safe driving behaviours (like checking mirrors or hand signals) were incorrectly labeled distractions shown the Figure 92.5.

High Accuracy and Robustness: The overall accuracy of the system is 93.5% which is compared to many other driver behaviour monitoring systems that are in use now. Examples of high-performance metrics in a variety of environmental settings include day/evening conditions.

Processing at Up to 30 FPS: With speeds of up to 30 frames per second, the system can effectively monitor in real time with low latency.

Prospect Model for Behaviour Detection: It applies deep learning (CNNs) for better detection accuracy and image processing methods (eye aspect ratio, face landmarks).

6. Conclusion

Driver Behaviour Monitoring System demonstrates strong real time performance capabilities, balancing acceptable FPS rates, low latency, and high accuracy. Thanks to these features, the system can be used in real-world scenarios and contributes to reducing the number of accidents caused by distracted or drowsy driving. With further low light and driver unpredictability optimizations, this tech can become an integral component of modern day car safety systems.

By utilizing both CNN-based models and image processing, real-time driver behaviour monitoring has been proven to be a practical and scalable solution. Our work paves the way for this effort to save lives by ensuring we tackle risky driving behaviours proactively.

References

[1] Dalal, N., & Triggs, B. (2005, June). Histograms of oriented gradients for human detection. In *2005 IEEE computer society conference on computer vision and pattern recognition (CVPR'05)* (Vol. 1, pp. 886–893). IEEE.

[2] Vu, H. S., Guo, J. X., Chen, K. H., Hsieh, S. J., & Chen, D. S. (2016, May). A real-time moving objects detection and classification approach for static cameras. In *2016 IEEE International Conference on Consumer Electronics-Taiwan (ICCE-TW)* (pp. 1–2). IEEE.

[3] Christopher Richard, W., et al. (1997). Pfinder: Real-time tracking of the human body. *IEEE Transactions on Pattern Analysis and Machine Intelligence, 19*(7), 780–785.

[4] Chien, Shao-Yi, et al. (2013). Video object segmentation and tracking framework with improved threshold decision and diffusion distance. *IEEE Transactions on Circuits and Systems for Video Technology, 23*(6), 921–934.

[5] Bhope, R. A. (2019, April). Computer Vision based drowsiness detection for motorized vehicles with Web Push Notifications. In *2019 4th International conference on internet of things: smart innovation and usages (IoT-SIU)* (pp. 1–4). IEEE.

[6] Wijnands, J. S., Thompson, J., Nice, K. A., Aschwanden, G. D., & Stevenson, M. (2020). Real-time monitoring of driver drowsiness on mobile platforms using 3D neural networks. *Neural Computing and Applications*, *32*(13), 9731–9743.

[7] Schwarz, C., Gaspar, J., Miller, T., & Yousefian, R. (2019). The detection of drowsiness using a driver monitoring system. *Traffic injury prevention*, *20*(sup1), S157–S161.

[8] Ranjan, A., Vyas, K., Ghadge, S., Patel, S., & Pawar, S. S. (2020). Driver drowsiness detection system using computer vision. *International Research Journal of Engineering and Technology (IRJET)*, *7*(1).

[9] Laxmaiah, B., Chitteti, C., Sanaboina, S. P., Srinu, V., Madhavi, K. R., & Kothergasti, B. B. (2023, October). Classification of Potholes and Plain Roads For Self-Driving Cars. In *2023 IEEE 5th International Conference on Cybernetics, Cognition and Machine Learning Applications (ICCCMLA)* (pp. 30–33). IEEE.

[10] Xu, Y. W., Cao, X. B., & Li, T. (2009, August). Extended Kalman filter based pedestrian localization for collision avoidance. In *2009 International Conference on Mechatronics and Automation* (pp. 4366–4370). IEEE.

[11] KM, A. D., & Murthy, O. R. (2017, December). Optical flow based anomaly detection in traffic scenes. In *2017 IEEE International Conference on Computational Intelligence and Computing Research (ICCIC)* (pp. 1–7). IEEE.

[12] Godavarthi, B., Kumar, D. H., & Nalajala, P. (2017). Driver Behavior Model with Stability in the Presence of Driver Uncertainty using Embedded Design. *International Journal of Pure and Applied Mathematics*, *117*(20), 837–844.

93 A novel ALU operation technique for parity-preserving gates driven by Xilinx and Verilog HDL

E. N. V. Purna Chandra Rao[1,a], *V. Santosh Kumar*[2,b], *G. Srikanth*[3,c], *and K. Navya*[4,d]

[1]Professor, Department of ECE, CMR College of Engineering and Technology, Hyderabad, Telangana, India
[2]Assistant Professor, Department of ECE, CMR Engineering College, Hyderabad, Telangana, India
[3]Professor, Department of ECE, CMR Technical Campus, Hyderabad, Telangana, India
[4]Assistant Professor, Department of ECE, CMR Institute of Technology, Hyderabad, Telangana, India

Abstract: Due to device scaling brought about by MOSFET invention in1960s, the semiconductor industry has witnessed enormous gains in speed and device performance; yet, power dissipation is also a major problem. As the growth of gadgets reaches a limit, authors are searching for a technology that will change the game. One of the hot areas in VLSI sector right now is reversible computing. Researchers have been primarily interested in this technique because of its negligible power dissipation. An ALU with reversible quantum gates is proposed to be researched in this paper. The proposed ALU can be used for either logical or arithmetic operations, and based on circumstances; it can also identify a single bit error. The parity-preserving Fredkin and CNOT gates are used in the design. Verilog HDL is used to model the suggested design, and Xilinx software is used to excite it. Reducing the number of gates, garbage outputs, quantum costs, and auxiliary or constant inputs is the main goal of the design.

Keywords: Reversible logic (RL) circuits, reversible logic gates (RLG), irreversible logic gates (ILG)

1. Introduction

The use of reversible circuit synthesis, which provides an alternative to conventional Boolean networks, has grown in importance recently. Reversible computing was sparked by two important discoveries. These circuits are superior to traditional circuits in two ways: first, they use less energy, and second, they are closely associated with several emerging technologies, such as quantum circuits. For any little amount of information erasure, Landauer (1961) showed that irreversible circuits always use power and emit heat at a minimum rate of kTln2, where k is the Boltzmann constant and T is the temperature. Subsequently, Bennett demonstrated that, in theory, energy dissipation can be arbitrarily little or zero as long as there is no information loss during computing. Since all of the original data is handled without being lost, this is advantageous for reversible circuits. Reversible computation is a wonderful alternative because, even while the percentage of facility consumption in contemporary VLSI circuits caused by information loss is small, this is frequently projected to change as rising packing densities force the facility consumption per gate operation to drop. Since reversibility is an integrated part of quantum computation, studying the reversible circuits adds to our understanding of it. Reversible circuits are also utilized in digital signal processing, cryptography, adiabatic circuits, low power CMOS designs, and optical computing.

Every digital system consists in large part of a processor. Moreover, ALU is one of the components of a microprocessor mostly. To put it briefly, ALU serves as CPU's cognitive and any system's cognitive overall is CPU's. That is the cognitive function of a computer then. They feature well enhanced structures and are created up of rapid dynamic logic circuits. The CPU consumes a large the quantity of the as a whole power of any processor.

This greatly inspires us to progress energy-efficient ALU designs, which meet high performance standards and lower average and peak power dissipation [1, 2]. As shown the Figure 93.1. ALUs are essentially combinational circuits that operate on two operands of n bits each, such as A [31:0] & B [31:0] for 32 bits, to execute arithmetic and logical operations. The 832bit ALU's typical internal structure is displayed. ALU's internal architecture.

2. Literature Review

An ALU's adder subcircuit is a necessary component. Carry propagation adds to the adder's latency. To reduce delay, BEC is utilized to reduce area, and carry input is used in place of an RCA when utilizing a modified carryy select adder (CSA). When it comes to digital computing systems,

[a]pcraoeluru@gmail.com, [b]santhoshkumar.v@cmrec.ac.in, [c]cmrtc.paper@gmail.com, [d]kathoju.nav@gmail.com

DOI: 10.1201/9781003684459-93

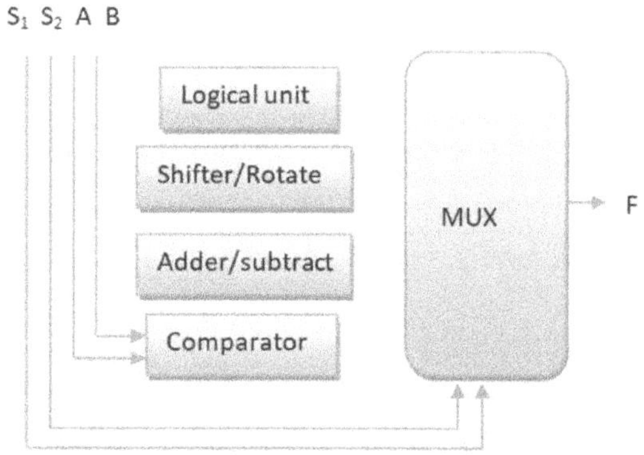

Figure 93.1. ALU block diagram.

Source: Author.

multiplier design plays a critical role in areas such as security, DFT, DSP, video processing, and intelligent systems that require large amounts of calculation. This study [2] uses the well-known Vedic sutra Urdhava Triyagbhyam to create a multiplier. K-stone, one of the fastest adders that decreases delay, is utilized to add multiplier's partial products. The Vedic sutra also improves the multiplier's well-designed competence. Area efficiency and delay efficiency are examined in this research [3–4] using a modified CSA and BEC. Vedic multiplier design is delay and area competent as a result of this modified CSA. ALU is an essential component of many systems, including DSP, computational systems, MAC units, and others. Because reversible logic gates may reduce power consumption in digital design, their importance in the low power VLSI sector has grown significantly. Compared to RLGs, irreversible logic gates dissipate more power. This work [5–9] addressed the use of reversible logic gates in K-stone adder architecture to minimize power dissipation and latency. Since the inputs and outputs of RLG are mapped one to one, there is no data loss. Reversible gates are employed in low power VLSI design since this is the primary driver of power dissipation minimization. Code converters are used to concentrate hardware consumption in order to improve the system by reducing circuit complexity. Hardware reduction, speed, and power are discussed in this work [6].

2.1. Existing method

A combinational digital circuit that carries out arithmetic and logical operations can serve as an arithmetic and logic unit in computing. MOSFET transistors are used in the design of an ALU, which includes shift registers, multiplexers, subtractors, demultiplexers, and other components [10–13]. Irreversible logic gates are used in digital circuits with MOSFET transistors. Additionally, it has several drawbacks, such as increased power dissipation, space consumption, and propagation latency.

3. Proposed Method

The suggested solution uses parity preserving gates to construct a 4-bit ALU architecture. Since reversible computing prevents bit loss of information, power dissipation is either nonexistent or zero. Here, parity testing in reversible gates allows us to find errors. In this case, system-level testing is carried out using ISIM, the Modelsim logic simulator, and circuit synthesis and design are predominantly handled by Xilinx ISE. When the design is compared to the current ALU designs, it performs better across all evaluation metrics.

3.1. Reversible circuits

Extensive research is conducted on reversible circuits in an effort to find an alternative to classical design. To make the creation and testing of circuits easier, as well as their potential applications across other computational domains, some special characteristics of reversible logic are recognized. Reversible logic and quantum circuits are related through exercises with quantum realizations of reversible gates.

3.2. Reversible function

An n-input/n-output function, where every input vector translates bijectively to single output vector, is realized by anxn reversible circuit. Reversible circuits prevent feedback paths and fan-out.

There are two ways that irreversible functions are different from their reversible counterparts. First off, there are not same number of inputs and outputs. An N-input N-output logic device with a one to at least one mapping among input and, consequently, the output is called a reversible gate. It assists us in recovering the inputs from the outputs in a unique way in addition to helping us calculate the outputs from inputs. Parity preserving gates are another name for RLGs. A subset of RLGs known as parity preserving reversible logic gates have the additional characteristic that the input and output parities are the same. As long as the outputs' EX-OR and inputs' EXOR match, a reversible gate will maintain parity.

The controlled NOT gate, a quantum logic gate, is the gate employed in this project. Feynmann gate is another name for CNOT gate. This input/output matrix is 2 by 2.

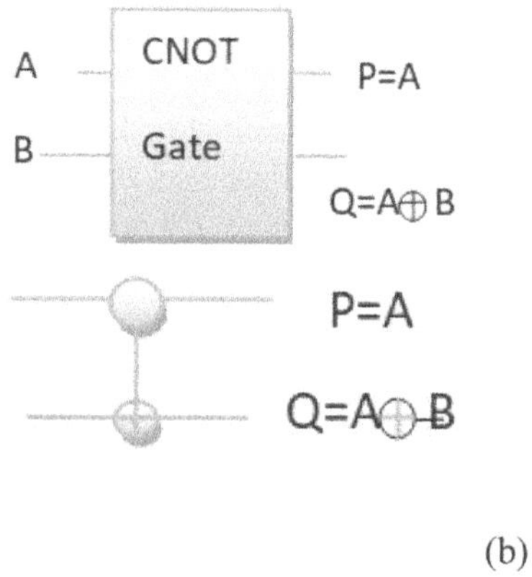

Figure 93.2. CNOT gate (a) Block diagram (b) symbol.

Source: Author.

A single output is the same as input, and a second output is result of xoring the inputs. The Figure 93.2 depicts the CNOT gate (1).

FREDKIN GATE - 3 inputs and 3 outputs make up a Fredkin gate shown the Figure 93.3. A grouping of several logics, including and, or, and xor, make up the output. With the help of this gate, numerous more digital logics can be implemented that would not be conceivable with a CNOT gate. It has a high quantum cost of five. Figure 93.2 illustrates how the Fredkin gate implements the logic function.

3.3. Reversible arithmetic and logic unit (ALU)

Based on control inputs, an ALU is a multi-functional circuit, which can execute many functions on 2 operands, A and B.

Rather than building many single-function circuits, this module provides programmability at a reduced gate cost. However, adding more control lines and circuit resources is necessary when combining many functions into a single unit [14–16].

We implement a reversible ALU with the operations of a traditional irreversible ALU in this thesis. Our goal while designing the ALU should be to incorporate as many arithmetic and logic operations as we can into a straightforward design that will operate as efficiently and cheaply as possible [17].

Because of this, the reversible ALU that is shown here has the majority of the operations found in traditional irreversible ALU [18, 19].

3.4 1. Bit arithmetic and logic unit

The single-bit RALU schematic diagram is displayed in the Figure 93.4.

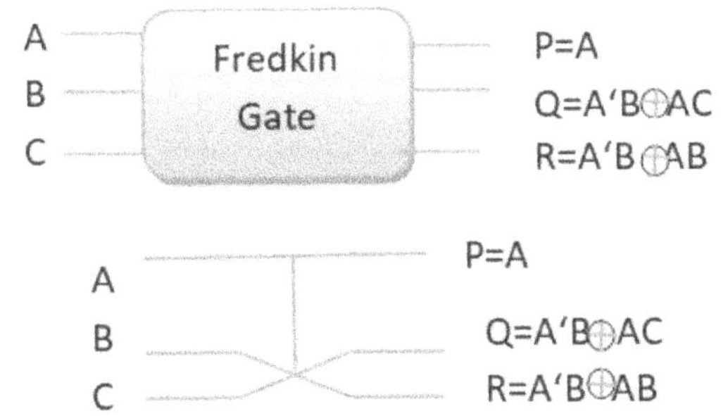

Figure 93.3. Fredkin gate.

Source: Author.

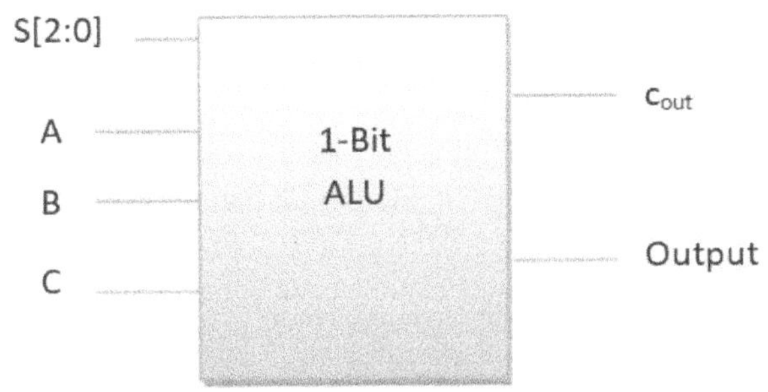

Figure 93.4. 1-bit ALU.

Source: Author.

The construction of a single-bit reversible arithmetic and logical unit using CNOT and Fredkin gates is depicted in Figure 93.5. Additionally, Table 93.1 displays the Figure 93.5 reality table. Fredkin and CNOT gates, which preserve parity, are used in a unique one-bit ALU design. The ALU design's reality table is shown in Table 93.1. The control signal S2 determines whether to use the logical or arithmetic output. The Fredkin gate in use has 3 inputs and a sum, as was previously mentioned. A CNOT gate contains 2 inputs and outputs in addition to several outputs. The suggested ALU's architecture is depicted in Figure 93.4, where the logical unit is represented by the bottom portion and arithmetic unit by the top portion. The auxiliary inputs in Figure 93.6 are represented by the 1s and 0s at input sides of the gates, and garbage outputs are signified by G. In order to comprehend the intermediate operations of a single bit ALU, intermediate

Table 93.1. Truth table of 1-bit RALU

S0	*S1*	*S2*	*Cin*	*Operation*
0	0	0	0	A
0	0	0	1	A+1
1	0	0	0	A+B
1	0	0	1	A+B+1
0	1	0	0	A
0	1	0	1	A-1
1	1	0	0	A-B
1	1	0	1	A-B-1
0	0	1	0	A|B
0	0	1	1	~(A|B)
1	0	1	0	A&B
1	0	1	1	~(A&B)
0	1	1	0	~(A^B)
0	1	1	1	A^B
1	1	1	0	A
1	1	1	1	~A

Source: Author.

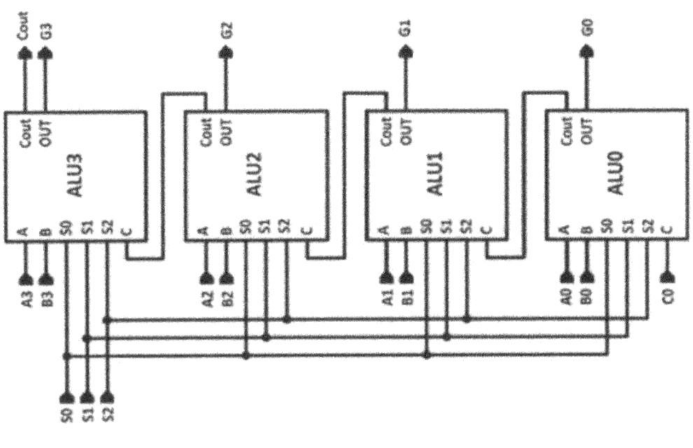

Figure 93.5. Schematic of 4-bit RALU [13].

Source: Author.

outputs are also represented. CNOT and Fredkin gates are used in the design. Depending on the selection signal, Fredkin gate 7 chooses anlogical or arithmetic unit. S0, S1, S2, and Cin are the bits utilized to choose sort of needed operation, and A and B are the inputs. The design contains sixteen gates, four constant or auxiliary inputs, a quantum cost of forty-four, and an output of eleven garbage. The design significantly outperforms the current ALU designs in every evaluation metric.

- Quantumcomputer.
- Opticalcomputing.
- Nanotechnology.
- Computergraphics.
- DNAcomputing.
- Communication.

Creating low-power data paths and arithmetic for numerous operations in a single cycle.

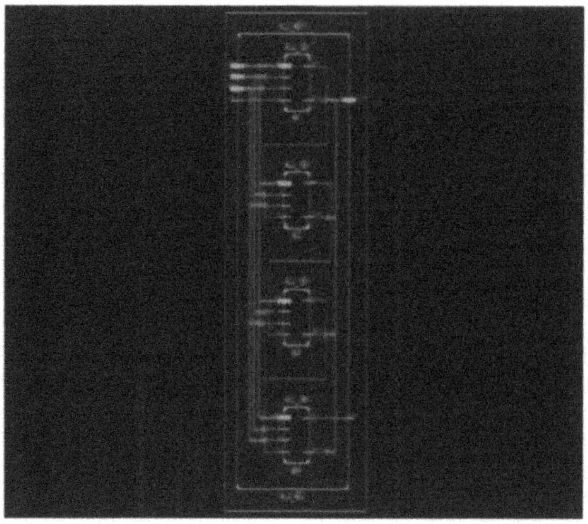

Figure 93.7. Design of reversible 4-bit ALU.

Source: Author.

4. Results

For the suggested design of reversible ALU Figures 93.6 and 93.7, design of reversible 4 bit ALU), we verified the reversible functioning by means of simulation. Utilizing Verilog HDL and Xilinx software, functional verification of the design is carried out. This suggests that, in comparison to past design, the new design performs better in every way. Reversible logic operates on the fundamental tenet that it must have an equal number of inputs and outputs, and this need is met. The results show an improvement in trash outputs and constant inputs. There was a 66.67% decrease in supplementary inputs and a 31.25% decrease in garbage outputs.

The random simulation of a 4-bit RALU is shown in Figure 93.8. In accordance with the control signals, the output node displays. We check random inputs and outputs for each control combination. For instance, the RALU executes an ADD operation for control S2S1S0=001 and inputs A=1001 and B=0100, resulting in an output of 1111. Keep in mind that the display obscures the garbage output.

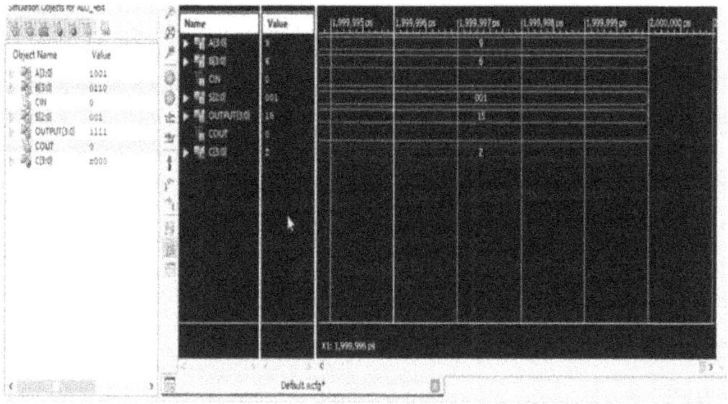

Figure 93.8. Simulation of 4-bit ALU.

Source: Author.

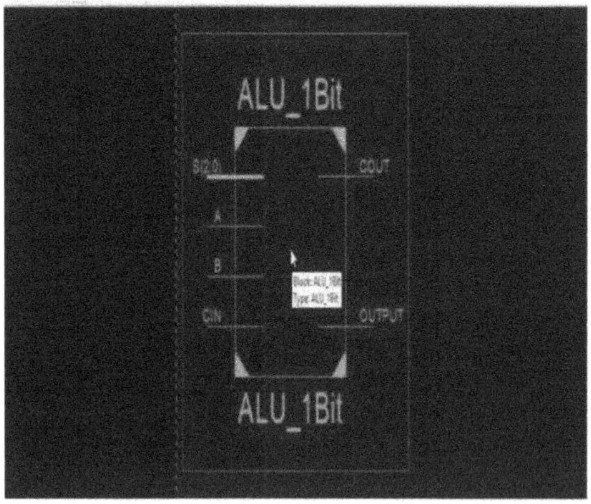

Figure 93.6. Schematic of reversible 1-bit ALU.

Source: Author.

5. Conclusion

The topic of reversible logic gates and ensuing three parameters – garbage outputs, quantum cost, and supplementary inputs – have been thoroughly covered in this study. We explore parity preserving Fredkin and CNOT gates, as well as existing reversible gates. To ensure the ALU's functionality in Xilinx, Verilog HDL was used in its design and implementation. After analyzing the performance, increased efficiency was attained. It is possible to expand this idea further and create multiple bit ALUs with parallel computing.

6. Future Scope

This can be expanded upon in the creation of sophisticated digital design projects. Building quantum chips is an option since Moore's law limits the capabilities of upcoming computer chips. Our next research focus will be on creating a new reversible gate and integrating reversible logic into a whole quantum processor, which can compute at extremely high speeds and with minuscule amounts of electricity.

References

[1] Nautiyal, P., Madduri, P., & Negi, S. (2015). Implementation of an ALU using modified carry select adder for low power and area-efficient applications. In: *2015 International Conference on Computer and Computational Sciences (ICCCS)*, pp. 22–25. IEEE.

[2] Shirol, S., Ramakrishna, S., & Shettar, R. (2018). A novel design and implementation of 32-bit hybrid ALU. In: *Second International Research Symposium on Computing and Network Sustainability (IRSCNS 2018)*.

[3] Gokhale, G., & Gokhale, S. (2015). Design of area and delay efficient vedic multiplier using carry select adder. In: *2015 International Conference on Information Processing (ICIP)*, pp. 295–300. IEEE.

[4] Banerjee, A., & Das, D. K. (2016). A new ALU architecture design using reversible logic. In: *2016 Sixth International Symposium on Embedded Computing and System Design (ISED)*, pp. 187–191. IEEE.

[5] Jais, A., & Palsodkar, P. (2016). *SM-IEEE: Design and Implementation of 64 Bit Multiplier using Vedic Algorithm*, 6–8 April 2016. IEEE.

[6] Pattnaik, S. K., Nanda, U., Nayak, D., Mohapatra, S. R., Nayak, A. B., & Mallick, A. (2017). Design and implementation of different types of full adders in ALU and leakage minimization. In: *2017 International Conference on Trends in Electronics and Informatics (ICEI)*, pp. 924–927. IEEE.

[7] Toffoli, T. (1980). Reversible Computing. Technical Report MIT/LCS/TM-151.

[8] Ferry, David K., & Wolfgang Porod. (2023). Dissipation and irreversibility in computing. *Materials Research Express*, *10*(8), 083001.

[9] Thachuk, Chris. (2013). Logically and physically reversible natural computing: a tutorial. *Reversible Computation: 5th International Conference, RC 2013, Victoria, BC, Canada, July 4–5, 2013. Proceedings 5*. Springer Berlin Heidelberg.

[10] Lloyd, Seth. (1977). Quantum-mechanical computers. *Scientific American: The Solid-State Century*.

[11] Owe, Olaf. (1993). Partial logics reconsidered: A conservative approach. *Formal Aspects of Computing*, *5*, 208–223.

[12] Margolus, N. (1986). Quantum computation. *Ann. New York Acad. Sci*, *480*, 487–497.

[13] Riyazuddin, K., et al. (2020). Design and Implementation of Novel 4-Bit Alu. *ICDSMLA 2019: Proceedings of the 1st International Conference on Data Science, Machine Learning and Applications*. Singapore: Springer Singapore.

[14] Dai, Fei, Md Akbar Hossain, and Yi Wang. (2025).State of the Art in Parallel and Distributed Systems: Emerging Trends and Challenges. *Electronics*, *14*(4), 677.

[15] Kirubakaran, Sakthivel, et al. (2023). Hybrid Deep Learning-Based Grid-Supportive Renewable Energy Systems for Maximizing Power Generation Using Optimum Sizing. *Electric Power Components and Systems*, 1–15.

[16] Ramkumar, S., et al. (2024). Smart Shopping and Cart Billing System Using Deep Learning. *2024 Second International Conference on Intelligent Cyber Physical Systems and Internet of Things (ICoICI)*. IEEE.

[17] SP, Siddique Ibrahim, et al. (2023). Artificial Intelligence Inspired Automatic Road condition Classification. *2023 14th International Conference on Computing Communication and Networking Technologies (ICCCNT)*. IEEE.

[18] Kumar, Ch Pavan, & K. Sivani. (2016). Analyzing the impact of TFETs for ultra-low power design applications. *2016 International Conference on Electrical, Electronics, and Optimization Techniques (ICEEOT)*. IEEE.

[19] Reddy, Kallem Niranjan, & Pappu Venkata Yasoda Jayasree. (2018). Low Power Strain and Dimension Aware SRAM Cell Design Using a New Tunnel FET and Domino Independent Logic. *International Journal of Intelligent Engineering & Systems*, *11*(4).

94 Performance analysis of hybrid set CMOS logic for single-electron transistor using T-SPICE

B. Premalatha[1,a], *E. Parvathi*[2,b], *Suraya Mubeen*[3,c], *and G. Venkat Swamy*[4,d]

[1]Associate Professor, Department of ECE, CMR College of Engineering and Technology, Hyderabad, Telangana, India
[2]Assistant Professor, Department of ECE, CMR Engineering College, Hyderabad, Telangana, India
[3]Associate Professor, Department of ECE, CMR Technical Campus, Hyderabad, Telangana, India
[4]Assistant Professor, Department of ECE CMR Institute of Technology, Hyderabad, Telangana, India

Abstract: In addition, by using innovative design strategies, circuits incorporating SETs can provide some of the underused functionality with fewer devices. But simple SET-based circuits have severely restricted uses because of SET's more yield impedance, more affectability to foundation charges, little voltage pick-up, also remarkably less temperature working condition. Because of this, it is hard to understand how coordinates can be applied to commonsense circuits. To address this issue and assess effectiveness and speed of novel plan. In this study, a tailored strategy with amazingly moo control utilization is implemented for a cross breed SET CMOS logic room temperature operable circuit. Using T-SPICE simulation software, a contemporary essential CMOS-SET logic door capable of designing half and full subtractor temperatures is described.

Keywords: Full subtractor, half subtractor, integrated circuits (ICs), single electron transistor

1. Introduction

The main driver of the astounding expansion of contemporary semiconductor industry is the ongoing scaling of MOSFET device size. With every technological advancement, the size of transistor – one of the fundamental components of integrated circuits (ICs) – decreases as scaling, packing density, production costs, circuit switching speed, and power consumption improve. Reduced transistor size results in less chip space needed for a given circuit, which allows for the production of more chips on single silicon wafer substrate and cheaper manufacturing costs per chip. Because of the increased transistor current density and decreased capacitance, circuit operation speed also increases. As long as MOS devices' downward scaling is robust, it will be constrained by higher sub-threshold leakage current, higher gate oxide leakage.

Large number factoring is used in cryptography to encrypt messages; this challenge arises in the classical design flow. As a result, factoring problems become feasible due to quantum computing. Here, we'll talk about the problems with quantum computing, namely the distinctions among classical and quantum methods of computation and the existence of a quantum method. With quantum computing, authors work on any unitary processes and break them down into smaller processes, Shor's quantum algorithm was able to factor numbers. Quantum logic gates function using q-bits, which can replicate unitary, unrestricted processes similar to those found in conventional logic. A characteristic of unitary matrices is that two of them remain unitary when multiplied by a unitary logic gate. We present the NOT, XOR, and universal quantum logic gates.

2. Literature Review

Schonenberger et al. claim to have made the first observation of a Coulomb staircase at room temperature [1]. Likharev [2] recognized the significant role that single-electron devices played even in the early phases of single-electron research, including quantum metrology, sensitive electrometers, transistors, memory, and logic devices for integrated circuit applications. First noted in the conductance of a one-dimensional (1-D) channel of silicon MOSFET at low temperature, the impact of single-electron tunneling in semiconductor devices was discovered by Scott-Thomas et al. [3], with Van Houten and Beenaker [4] following suit.

Two-dimensional electron gases (2DEGs) have long been realized in HEMTs (high-electron mobility transistors) through semiconductor heterostructures. The only small difference is that the 2DEG is shaped by Schottky contacts, which are electric links, at the peak of the device to create a small charged island, hetero-structure single-electron transistors and HEMTs are closely similar devices. In III-V heterostructures, lithography was also used to define small islands in two-dimensional electron gas (2DEG) layer.

[a]bpremalatha@cmrcet.ac.in, [b]esaram.parvathi1@gmail.com, [c]cmrtc.paper@gmail.com, [d]gutam.venkataswamy@gmail.com

DOI: 10.1201/9781003684459-94

Tunnel barriers divide the source and drain regions from the island's quantum dot, which is formed via quantum confinement phenomena at low temperatures. Meirav et al. [5], Kouwenhoven et al. [4] utilized patterned surface gates in these devices to produce the tunnel barriers by depleting various components of the 2-DEG in the AlGaAs/GaAs heterostructure. As a result, this type of structure came to be known as manufactured atoms. Prototypes of GaAs/AlGaAs single electron transistors (SETs) were published in 1987. Technology restrictions resulted in islands spanning several hundreds of nanometers across these devices.

A fresh approach for analyzing a single electron transistor (SET) that is more precise may be used for circuit simulation was presented by El-Sayed Hasaneen [6] in 2010. It was created utilizing the master equation and the conventional theory of single electronics, where a method for identifying the more likely occupied electron states was proposed. This approach is much adaptable and works with both symmetric and asymmetric devices, single or many gates, and the background charge effect.

A method for creating hybrid circuits made up of CMOS interfaces and single electron transistors (SETs) was presented by Anindya Jana [7] in 2013. This paper presents the Voltage Transfer Characteristics (VTCs) of a room temperature operational hybrid CMOS–SET inverter, NAND, NOR, and XOR gate. A methodology for creating hybrid circuits with multiplexers, adders, and registers utilizing SET-CMOS ideas was proposed by [8].

The model [9] is fairly simple because it just considers the two most likely charging states and is based on the steady-state master equation. Nevertheless, it can faithfully mimic the oddities of an asymmetric SETT, including the Coulomb staircase and the skew.

The goal of the suggested approach [10] is to provide a computer-aided design framework for CMOS-SET hybrid IC designs by implementing it in a commercial circuit simulator. To successfully replicate the experimental/Monte Carlo simulated characteristics of several CMOS-SET hybrid circuits, a series of SPICE simulations are performed.

More adaptable, this approach [11] works with symmetric or asymmetric devices, single or multiple gates, and it can also take the background charge effect into account. It can be applied to many drain-source voltages regardless of how symmetrically or asymmetrically biased the device is. The suggested model's SET properties have been tested using the widely used single-electron circuit Monte Carlo simulator SIMON, and the results demonstrate a good match.

A quantum price-efficient reversible full adder gate and reversible decoder for engineering are presented in this research [12–14]. This gate requires only one clock cycle to operate independently as a reversible complete adder and decoder unit. The intended gate might be considered universal in the sense that it could be used to create any arbitrary mathematical function. We are planning a full adder and decoder circuit using the T-Spice simulation to construct reversible logic gates, and we are calculating the power consumption using the TSMC018 nm technology [15–16].

3. Proposed Methodology

The capacity of a single electron transistor to switch from an isolating to a conductive state by building up single electron on gate electrode is its most impressive feature. Conversely, a normal MOSFET requires between 2000 and 20,000 electrons to function properly. As a result, single-electron devices switch on more quickly than traditional MOSFETs. Moreover, because of its incredibly small size, it operates on extremely low power – four and a half times less than CMOS.

Notwithstanding the unmatched benefits of SETs, their drawbacks – such as their tiny gain, poor current drivability, absence of room temperature operable method – became obstacles for upcoming silicon methods. To offset the inherent shortcomings of SET, CMOS technology' high gain and high driving speeds are required. The literature analysis makes clear how extensively the hybrid SET-CMOS circuit has been analyzed.

SET operates at extremely low drain voltages.

There are nine hybrid SET-CMOS transistors in circuit, with a supply voltage of VDD equal to 800 mv and a voltage of either 0 V ('low') or 1 V ('high'). The drain *cε* SET-CMOS equivalents of through gates receive the input voltages V (A, B) through (*Vds*<) rather than MOSFET devices. As a result, the half subtractor's output signal is SET's current in the nanoampere range. Thus, for hybrid CMOS-SET co-design to be useful, certain design guidelines must be followed. When a constant current biassed SET is used, the output voltage, denoted by VDS, is correlated with bias current. Figure 94.1 illustrates how influence of temperature on the output voltage reduces when bias current enhances, VDS enhances, and dynamic range of VDS variation decreases. For this reason, the following design of SET and CMOS maximizes the performance of Hybrid CMOS-SET.

The hybrid SET CMOS inverter is schematically shown in Figure 94.1. A SET alone powers the CMOS inverter with

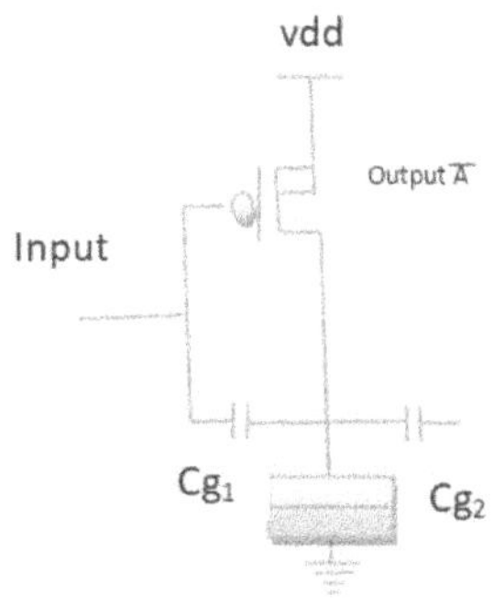

Figure 94.1. Hybrid SET-CMOS inverter circuit diagram.

Source: Author.

an inter-connect parasitic, and Table 94.1 displays parameters utilized to simulate the device at room temperature. It is created by substituting a SET for an NMOS transistor and using a PMOS as a SET's load resistance.

Extracted as difference and borrow from the SET-CMOS gates' source and drain, as depicted in Figure 94.2.

Through simulation, we have confirmed the proposed implementation of the half subtractor. The following circuit characteristics were utilized for the half subtractor. Cg1 = 0.2aF, Cg2 = 0.127aF.

Using the Xilinx 10.1SE software tool, all circuit designs are synthesized and evaluated in order to assess the performance of the circuits.

Analysis of Delay Products An considerable decrease in power consumption is demonstrated as compared to so-called CMOS logic over the next Table 94.2.

Table 94.2: Comparing the half subtractor's average power usage (Watt-sec) at 0.12 aF, C_{TD}=C_{TS}=0.15aF and R_{TS}=R_{TD} = 1MΩ. Figure 94.3 shows the simulation results and, for reference, only one output of the circuit.

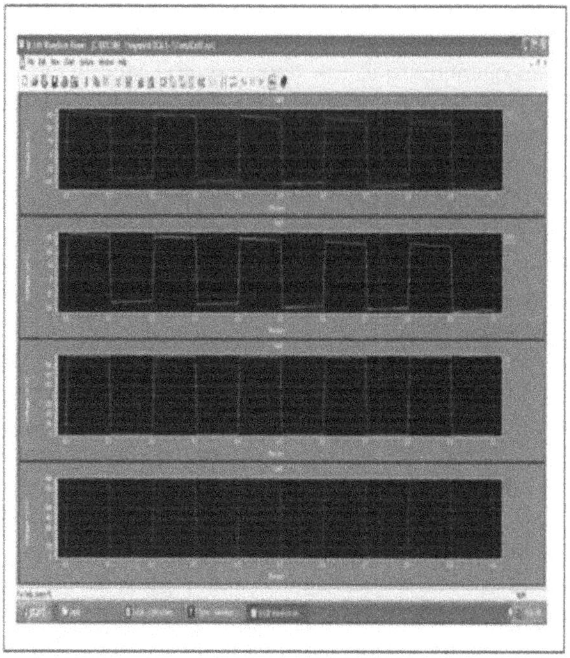

Figure 94.3. Simulation result of hybrid half subtractor.

Source: Author.

4. Experimental Results

The power consumption of the SET-CMOS and CMOS models is 5.343E-010 watt and 2.6661E-009 watt at minimum supply voltage and 1.4314E-008 watt and 3.703E-009 watt at maximum supply voltage. As power consumption is directly proportionate to supply voltage, it is satisfactory that consumption of power is increased by increasing applied voltage and decreased by decreasing applied voltage. In terms of power consumption, the suggested design performs well when compared to the CMOS and SET-CMOS models of the half subtractor. Since Rt is inversely related to power and high speed operation, its value must be tuned for optimal results. Tables 94.3 and 94.4 shows the power delay product for the hybrid counterpart of the SETCMOS-based circuit and the traditional CMOS-based circuit.

Table 94.1. Factors utilized in the simulation process

Device	*Parameter*
SET	CG1=0.025aF CG2=0.127aF C_{TD}=C_{Ts}=0.15aF R_{TD}=R_{TS}=1MΩ

Source: Author.

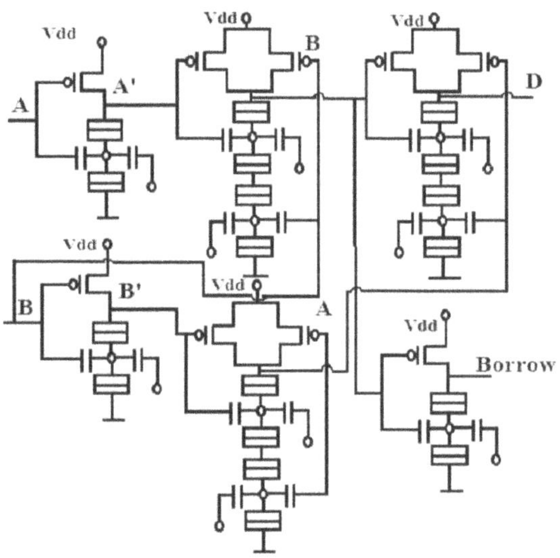

Figure 94.2. Hybrid CMOS-SET half subtractor circuit [13].

Source: Author.

Table 94.2. Reduction in power consumption CMOS logic

S.NO	*Applied Voltage*	*Hybrid CMOS- SET*	*CMOS Model*	*CMOS*
1	0.6	5.43e-10	2.661e-09	Model
2	0.8	6.05e-10	3.771e-9	2.66E-09
3	1	7.765e-10	4.675e-9	3.77E-09
4	1.2	3.109e-9	1.035e-8	4.68E-09
5	1.4	3.707e-9	1.431e-8	1.04E-08

Source: Author.

The power delay of the SET CMOS and CMOS models is 3.3638E-013 watt-sec and 1.89E-012 watt-sec at minimum supply voltage, respectively, and 12.589E-013 watt-sec and 10.243E-12 watt-sec at maximum supply voltage. The suggested design has a minimal power delay when compared to the CMOS and SET-CMOS models of the half subtractor.

The SET-CMOS and CMOS models have delay times of 1.894E-009 and 2.985E-008 seconds at minimum supply voltage 1.002E-009 and 1.998E-008 seconds at maximum

Table 94.3. Power delay (Watt-sec) product of half subtractor

S.NO	*Applied Voltage*	*Hybrid CMOS-SET*	*CMOS Model*
1	0.6	3.63e-13	1.89e-12
2	0.8	4.998e-13	3.765e-12
3	1.0	7.233e-13	5.678e-12
4	1.2	10.215e-13	8.967e-12

Source: Author.

Table 94.4. Voltage vs. half-subtractor time delay (sec)

S.NO	*Applied Voltage*	*Hybrid CMOS-SET*	*CMOS Model*
1	0.6	1.894e-09	2.98e-08
2	0.8	1.569e-09	2.995e-08
3	1.0	1.432e-09	2.003e-08
4	1.2	1.231e-09	2.002e-08
5	1.4	1.002-09	1.997e-08

Source: Author.

supply voltage, correspondingly. As the delay time is inversely related to supply voltage, it indicates that raising supply voltage reduces the delay time, which is a satisfactory factor. The suggested design has a reduced delay time when compared to the CMOS and SET-CMOS models of the half subtractor.

5. Conclusions

It is expected that the semiconductor roadmap will eventually come to an end in near future, which will prevent CMOS technology from scaling further. In order to meet the expectations of a growth in digital applications, alternative ways are desired. The goal of this effort is to find solutions for the simulation problem of single electronics, or Single Electron Devices (SEDs), in the development of future circuit designs, which are playing an increasingly important role. It is demonstrated that SEDs methods are very accurate and suitable in terms of time complexity. Additionally, the study examined the properties of one Electron-based circuits and suggested a hybrid method, which might be applied to precisely and effectively estimate and examine the circuit's output.

References

[1] Schonenberger, C., van Houten, H., & Donkersloot, H. C. (1992). Single electronunneling observe dat room temperature by scanning-tunneling microscopy. *Euro phys Letter*, *20*, 249–254.

[2] Likharev K. K. (1998). Correlated discrete transfer of single electron sinultra-small tunnel junctions. *IBMJ ResDevelop*, *32*, 144–158.

[3] Scott-Thomas, J. H. F., Kastner, M. A., Antoniadis, D. A., Smith, H. I., & Field, S. (1988). Si metal–oxide semiconductor field effect transistor with 70-nm slotted gates for study of quasi-one-dimensional quantum transport. *Journal of Vacuum Science & Technology B: Microelectronics Processing and Phenomena*, *6*(6), 1841–1844.

[4] Van Wees, B. J., Van Houten, H., Beenakker, C. W. J., Williamson, J. G., Kouwenhoven, L. P., Van der Marel, D., & Foxon, C. T. (2004). Quantized conductance of point contacts in a two-dimensional electron gas. *Physical Review Letters*, *60*(9), 848–850.

[5] Meirav, U., Kastner, M. A., & Wind, S. J. (1990). Single-electron charging and periodic conductance resonances in GaAs nanostructures. *Physical Review Letters*, *65*(6), 771.

[6] Hasaneen, E. S. A., Wahab, M. A., & Ahmed, M. G. (2011). Exact analytical model of single electron transistor for practical IC design. *Microelectronics Reliability*, *51*(4), 733–745.

[7] Jana, A., Singh, N. B., Sing, J. K., & Sarkar, S. K. (2013). Design and simulation of hybrid CMOS–SET circuits. *Microelectronics Reliability*, *53*(4), 592–599.

[8] Jain, A., Ghosh, A., Singh, N. B., & Sarkar, S. K. (2014). Stability and reliability analysis of hybrid CMOS-SET circuits—a new approach. *Journal of Computational and Theoretical Nanoscience*, *11*(12), 2519–2525.

[9] Inokawa, H., & Takahashi, Y. (2003). A compact analytical model for asymmetric single-electron tunneling transistors. *IEEE Transactions on Electron Devices*, *50*(2), 455–461.

[10] Mahapatra, S., Vaish, V., Wasshuber, C., Banerjee, K., & Ionescu, A. M. (2004). Analytical modeling of single electron transistor for hybrid CMOS-SET analog IC design. *IEEE Transactions on Electron Devices*, *51*(11), 1772–1782.

[11] Amarnath Reddy, M., & Manoj, K. V. (2014). Design of logic circuits using Reversible gates. *IJETT*, *16*, 394–396.

[12] Ghosh, A., Jain, A., Singh, N. B., & Sarkar, S. K. (2014, December). Design and implementation of SET-CMOS hybrid half subtractor. In *2014 Annual IEEE India Conference (INDICON)* (pp. 1–4). IEEE.

[13] Sonth, M. V., Srikanth, G., Agrawal, P., & Premalatha, B. (2021). Basic logic gates in two dimensional photonic crystals for all optical device design. *International Journal of Electronics and Telecommunications*, *67*(2), 247–261.

[14] Kumara Swamy, M., & Krishna Reddy, P. (2020). A model of concept hierarchy-based diverse patterns with applications to recommender system. *International Journal of Data Science and Analytics*, *10*(2), 177–191.

[15] Reddy, K. N., & Jayasree, P. V. Y. (2019). Low power process, voltage, and temperature (PVT) variations aware improved tunnel FET on 6T SRAM cells. *Sustainable Computing: Informatics and Systems*, *21*, 143–153.

[16] Reddy, K. N., & Jayasree, P. V. Y. (2017, September). Survey on improvement of PVT aware variations in tunnel FET on SRAM cells. In *2017 International Conference on Current Trends in Computer, Electrical, Electronics and Communication (CTCEEC)* (pp. 703–705). IEEE.

95 Assessment of convolutional neural network-based classification on face recognition images

Y. Sailaja[1,a], *P. Sandhya Reddy*[2,b], *Bagam Laxmaiah*[3,c], *and Bopparam Sampath Kumar*[4,d]

[1]Department of CSE, CMR College of Engineering and Technology, Hyderabad, Telangana, India
[2]Department of CSE, CMR Engineering College, Hyderabad, Telangana, India
[3]Department of CSE, CMR Technical Campus, Hyderabad, Telangana, India
[4]Department of CSE, CMR Institute of Technology, Hyderabad, Telangana, India

Abstract: For biometric authorizations, which are used in many applications, especially safety and attendance, recognition is crucial. This task is accomplished by converting a collected data of the individuals using image processing techniques. This research proposes a convolutional neural network with an Alexnet driven face recognition system and a series of object detector driven face detection systems that can identify faces. Due to its high accuracy when compared to other approaches, face recognition algorithms depend on machine learning. Using a cascade event detector classifier, identifying faces is the first step towards face recognition. Convolutional Neural Networks, a subfield of Deep Learning, are worn for face recognition (CNN). This multi-layer complex is employed to educate the system to utilize classification to execute a confident assignment Verify the learning process of an AlexNet-trained CNN model, which is utilized for facial recognition.

Keywords: Recognition of faces, identification of faces, and convolutional neural networks in MATLAB, AlexNet-trained

1. Introduction

A visual system can identify the traits of certain people using a technique called recognition of faces. Face recognition is crucial for everyday functions including surveillance footage, security, participation, and soon. It's the simplest kind of authentication that doesn't require any prior computer expertise from the user, and it's widely utilized. In the modern era, machine learning is gaining popularity. Because of its non-intrusive nature, facial recognition is very important for biometrics identification of humans. In an open setting, facial recognition might be rapidly disabled regardless of whether the subject is unaware of the computer and its contents.

Improved facial recognition technology is still being researched and has been covered in a number of research articles. For example [1, 2].

The challenges of presentation variation, facial masks, view lighting, background picture confusion, and changes in countenance appearance as inferences have put popular systems centered on simple learning to the test [3]. Deep learning is making important development in identifying harms that have hindered the AI community's best efforts for years. Face recognition is increasingly recognized in domains such as the internet, supervisors, and public fields since it has facilitated the building of complex patterns in multidimensional data [4, 5]. Because of its design and the proceedings it have made in the areas of machine learning for semantic division and other areas, this algorithm is distinctive and has outperformed all additional techniques in the countryside of face recognition.

Numerous deep learning techniques exist, such as Convolutional Neural Network (CNN), Accumulated Auto encoder (AEC) [7], and Deep Belief Network (DBN) [6].

The algorithm that is most frequently employed for face and object recognition is CNN. CNN is a kind of artificial neural network that operates using the convolution of signals theory. It aids in the feature extraction of the input data, increasing the total amount of features required. CNN's founder is LeCun. The first CNN-developed applications is handwriting recognition [8]. His algorithm serves as a basis for numerous applications and a source of guidance for numerous other scientists seeking to develop new technology in other sectors. When Krizhevsky, Sutskever, and Hinton submitted their research in Image Net Competition, the results were the best [9]. Utilizing the immense processing capacity of Graphical Processing Units (GPUs), CNN has brilliant amazing state-of-the-art consequences in numerous domains, like segmentation based on semantics, detection of edges, image identification, and scene recognition. In comparison to added recognition methods, this expose primarily discusses the most accurate recognition of faces algorithm.

---

[a]y.sailaja@cmrcet.ac.in, [b]patana.sandhyareddy@cmrec.ac.in, [c]blaxmanphd@gmail.com, [d]sampathbopparam@gmail.com

DOI: 10.1201/9781003684459-95

The overall framework of the face detection technique used in this article consists of three basic components.

Pre-processing phase:

a. Transformation of colour space
b. Image resizing
c. Face extraction of features
d. Our system uses a Softmax Classifier that relies on CNN extraction of features for categorization.

2. Methodology

A number of the most popular deep learning methods, Convolutional Neural Networks teach a model to execute text, video, sound, or image classification directly. Deep learning is a subset of machine learning. CNNs are particularly useful for pattern recognition, which aids in the identification of faces, objects, and scenes in images. Through the use of patterns, we are able to learn directly using images, thereby eliminating the requirement for human feature extraction.

CNNs are primarily dependent on applications that involve recognizing objects and computer vision, like self-driving cars and face recognition software. According to your needs, you may develop a convolutional neural network from the ground up or use a model that is qualified based on the dataset.

2.1. Architecture of CNN

A deep network's design is constructed by connecting various layers inside the CNN architecture. Consider a deep net that can, for example, divide 28 × 28 grayscale images into ten categories.

Picture input Layers d = Layer([28281]) Convolutional layer 2(3,16, 'Padding',1) convolutetin.

2d Layer(3,32,'Padding',1) batch The normalization process Layer relu Layer max Pooling 2d Layer (2,'Stride',2) batch The normalization process Layer relu Layer completely Linked Layer (10) softmax Layer categorization Layer];

2.2. Layers of CNN

Layers are collections of items in a layer. The train Network train function may accept the layers for an input shown the Figure 95.1.

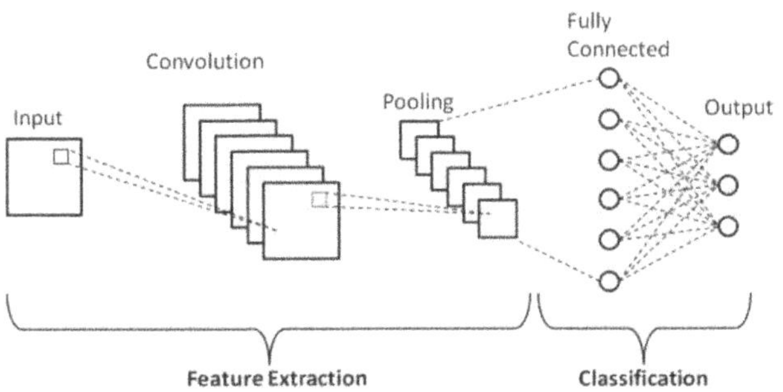

Figure 95.1. Layers of CNN.

Source: Author.

It is necessary to explicitly construct an assortment of layers in order to identify the architecture of a neural network wherever every layer is linked together consecutively. It is possible to express a network design with numerous inputs and outputs for each tier using a Layer Graph object.

Image Input Layer: By inputting images through a network via the image layer, data uniformity is implemented.

The input size argument specifies the size of the image. The dimension of an image is represented by its elevation, breadth, and the amount of colour channels. For instance, there are one channel in a grayscale image and three channels in a colour image.

Convolutional Layer: The convolutional 2-D layer applies convolutional descending filter to the data. The convolutional layer is made up of several parts [10].

The convolutional layer's neurons attach sub regions to the input images or outputs from earlier layers. The layer learns characteristics restricted by the areas as it navigates throughout an image.

2.3. Batch normalization layer

This layer restores every input channel to its typical state after passing through a mini-batch shown the Figure 95.2. CNN training is accelerated and sensitivity is decreased when batching normalization layers (ReLU layers) are worn among convolutional layers and non-linearities.

Every channel's activations are then normalized by deducting the mini-batch mean and multiplying the result by the mini-batch variance. Following that, the layer scales the input by a learned scale factor y and modifies it by an apprentice offset β; y and β acquire parameters that are altered throughout network training.

2.4. ReLU layer

A ReLU layer sets a entrance for every input changeable, setting any importance below zero to zero.

Conventional and batch normalization stages often come after an activation function that is nonlinear, which is a rectified linear unit (ReLU), particular by a ReLU layer. Every variable has a threshold operated by a ReLU layer, which sets any input value that is below zero to zero.

$$f(x) = \begin{cases} x, & x \geq 0 \\ 0, & x < 0 \end{cases} \tag{1}$$

The layer ReLU has no effect in input size.Local Reaction the normalization process via Cross Channel The normalization process Layer:

Normalizing on the channel-wise is carried out via a cross-channel layer of normalization known as the channel-wise response.

This layer performs a normalization that produces the channel-wise reaction locally. It often comes after the ReLU activation layer. A standardized value calculated from the components of multiple nearby channels (the components in the standardization window) is going to be substituted for every element. In other words, train Network determines an average value x' for every input element x.

$$x' = \frac{x}{\left(K + \frac{\alpha * SS}{window\ channelsize}\right)^{\beta'}} \tag{2}$$

In this context, the normalizing hyper parameters that β, K, and β, denote the amount of component squares in the standardizing window [11]. You can choose the dimensions of the standardization casement by means of the window Channel Size statement of the cross Channel Normalization Layer function. Furthermore, you can use the name-value pair parameters for K, Beta, and Alpha to create the hyper parameters.

Dropout Layer: Using the likelihood that is determined, the layer that drops out periodically sets all contribution values to zero.

The dropout layer sets the zero output elements. In the training stage, the failure mask rand (size(X))\prospect is chosen at random; X is given by the input layer, and 1/(1-Probability) is used to scale the fundamentals that remain. By employing this procedure, the network's basic architecture is changed more skillfully between rounds, avoiding over fitting [11, 12]. Along with the quantity, the components that are shed all through exercise also grow in quantity [13–16].

At projection time, the layer's output and input are equal [17].

Its layer below the layer of dropouts does not have any learning akin to maximal or average pooling [18].

Fully Linked Layer: During this layer, you'll increase the Yamatrix by two to add weight and add a vector for bias. There are more fully connected layers before convolutional (and down sampling) layers. Neurons located in the fully associated layer attach to neurons across the layer above in this particular layer. Every level integrates all of the skin tone (local in sequence) gathered from before levels crossways the image to find the bigger patterns. The functions utilized for categorizing images for the purpose of classification are integrated into the network's previous fully linked layer. Because of this, the output Size attribute of a network that includes the final completely linked node contains the total amount of theme or items in the database. To eradicate regression troubles, we have to match the amount of variables being used that the output size. As a result, we eradicate deterioration issues.

Softmax and categorization Layers: The information for the softmax layer is subjected to a softmax function. The parameter utilized to create a softmaxlayer is called softmaxLayer.

For multiple-class categorization issues, the cross variance loss of competing classes is computed using the categorization layer. To create a categorization layer, utilize Classification Layer.

All classification difficulties are fully obeyed by the softmax and categorization layers. Softmax function represents the output unit's activation feature.

$$y_r(x) = \frac{\exp(a_r(x))}{\sum_{j=1}^{k} \exp(a_j(x))} \tag{3}$$

$$where\ 0 \leq y_r \leq 1\ and \sum_{j=1}^{k} y_j = 1$$

The softmax operation, that activates the output component that comes after the final completely linked layer, is utilized for multi-class problems with categorization [19].

$$p\left(\frac{c_r}{x},\theta\right) = \frac{p\left(x,\frac{\theta}{c_r}\right)p(c_r)}{\sum_{j=1}^{k} p\left(x,\frac{\theta}{c_j}\right)p(c_j)} = \frac{\exp(a_r(x,\theta))}{\sum_{j=1}^{k} \exp(a_j(x,\theta))} \tag{4}$$

$$where\ 0 \leq p\left(x,\frac{\theta}{c_r}\right)p\left(c_r) \leq 1\ and \sum_{j=1}^{k} p\left(x,\frac{\theta}{c_j}\right)p(c_j) = 1 moreover\ a_r = In\ p\left(x,\frac{\theta}{c_r}\right)p(c_r), p\left(x,\frac{\theta}{c_r}\right)\right),$$

Wherever in the sample is the dependent probability for class r provided?

2.5. Layer of regression

Regression layers are used to determine the half normalized error loss in regression situations. For more challenging regression-related issues. A fully linked final layer will command a regression layer. For just one observation, the average squared error is provided by

$$MSE = \frac{\sum_{i=1}^{R}(t_i - y_i)^2}{R} \tag{5}$$

Tires indicate the goal's output, ires indicate what response I expect from the network, and R indicates the number of answers. For image and sequence-to-one regress networks, the result of the deterioration layer function that is lost is the half-mean-squared-error calculates reaction, which is not standardized by R.

$$Loss = \frac{1}{2}\sum_{i=1}^{R}(t_i - y_i)^2 \tag{6}$$

Figure 95.2. Deep learning workflow.

Source: Author.

2.6. Amount of neurons

The average of Map Size and Amount of Filters determines the overall amount of neurons (the result size) in a convolutional layer.

The total amount of transistors in the network is determined by multiplying the output height and breadth, resulting in the definition of map size.

Consider for the purposes of this example that an input is a colour, 32-by-32-by-3 image. In a layer of convolution containing eight 5-by-5 filters, the total amount of parameter for a particular layer is (75 + 1) * 8 = 608, while the total amount of weighted for a specific filter is 5 * 5 * 3 = 75. For every direction, a 16-by-16 map of features is produced if the stride is two and the size two pad is specified.

This is since the stuffing at the bottom and top of the representation is eliminated, and (32 – 5 + 2 * 2)/2 + 1 = 16.5.16 * 16 * 8 = 2048 stacked neurons in the network, according to Atlast.

Neuron results typically pass via nonlinearity layers such as the ReLU layer. Learning Parameters: Name-value pairs allow for the modification of learning rates. If a consistent learning rate is needed, regardless of the variable, global training settings will be immediately selected. All of these tasks can be completed whilst the convolutional layer is being defined. E) Amount of Layers: Depending on our needs, a convolutional neural network can include any number of layers. An application needs additional layers if it is more sophisticated. We require fewer layers if the application being used is less complicated.

3. The Projected Model

The obstruct system for the suggested CNN appreciation technique is revealed Figure 95.3.

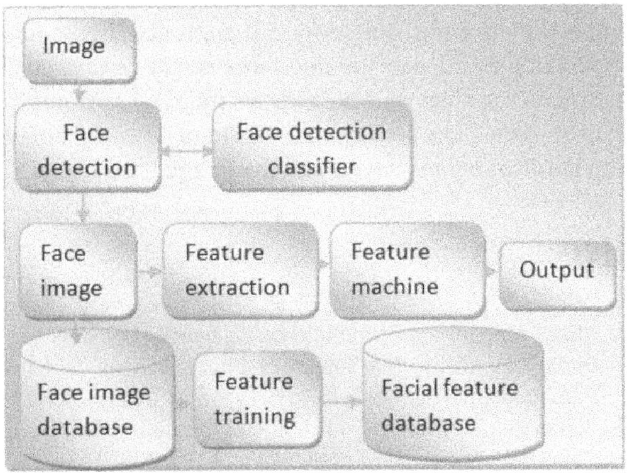

Figure 95.3. Block illustration of face identification by means of CNN.

Source: Author.

The primary three phases of the method are as follows:

1. Reduce the pixel size of the input photos to 277*277.
2. Assemble a twenty-five-layer CNN structure consisting of the following: input picture, convolutional, ReLU, cross-channel standardization, completely linked, dropout, softmax, and categorization output repeated as many times as needed.
3. Employ the Softmax algorithm for categorization after collecting every feature.
4. Figure depicts the suggested CNN method's extraction of features block structure.
   'data'ImageInput
   'convl'Convolution
   'relul'ReLu
   'norml'CrossChannelNormalization
   'pool1'MaxPooling
   'conv2'Convolution
   'relu2'ReLu
   'norm2'CrossChannelNormalization
   'pool2'MaxPooling
   'conv3'Convolution
   'relu3'ReLu
   'conv4'Convolution
   'relu4'ReLu
   'conv5'Convolution
   'relu5'ReLu
   'pool5'MaxPooling
   'fc6'FullyConnected
   'relu6'ReLu
   'drop6'Dropout
   'fc7'FullyConnected
   'relu7'ReLu
   'drop7'Dropout
   'fc8'FullyConnected
   'prob'Softmax
   'output'ClassificationOutput

3.1. Dataset

The image database contains four subjects. The input images are reduced from their original size of about 294 images per subject to 277 by 277 pixels. Softmax classifications used Alexnet for deep CNN extraction of features to trained the models on data. The database contains the angle of every face, allowing features to be derived from each photograph.

For example, the subsequent four subjects are included in the image collection in this paper:

3.1.1. Object recognition chain

Cascades are divided into phases, every one of which having a subset of poor students. The impoverished pupils are decision stumps, which are basic classifiers. Every step is qualified by means of a boosting method. Boosting enables the training of a very precise classification by computing the weighted mean of the least accurate learner's decisions.

An advancing phase of the classifiers indicates a area which the sliding window's current location has determined to be positive or negative. Positive indicates the presence of an purpose, while negative indicates the nonexistence of any. If the marking is negative, the sensor will move the opening to the next location and record this region as total. If the label is affirmative, the machine learning algorithm will progress the area to the subsequently step. Once the concluding step verifies that the district is precise, the detector finds something at the present casement position.

The procedures are made to get rid of negative samples as quickly as is practical. Most people have the idea that most screens are boring. A real positive, on the contrary hand, is unusual and deserves more research.

i) If a positive consequence is reliably recognized, there is a true positive.
ii) An adverse sample may be incorrectly labeled as positive, leading to a false positive; iii) A positive outcome may be incorrectly classified as negative, resulting in a false negative.

4. Results and Analysis

There are 25 layers in the CNN design, including the input layer, convolution, reLU, cross-channel standardization, maximum pooling, the maximum pooling, reLU, cross-channel standardization, the maximum pooling, fully linked, reLU, dropout, fully linked, softmax, and categorization product. To do recognize faces, one way to train photos is to compress them to 277*277 before you transport them to CNN layers for preprocessing the learned data. From there, the images are passed to a classifier using softmax to extract the individuals' attributes. These are the outcomes.

Initialization image normalization. It shown the Table 95.1 uniqueness of projected algorithm preparation on solitary CPU.

The CNN's development procedure is depicted in Figure 95.4 as following.

The suggested CNN architecture is reviewed with a lossy close to zero and a precision of 100%. As a result, the newly developed algorithm's effectiveness is as anticipated.

5. Conclusion

CNN comprising 25 layers is utilized in the development of the system. The number of layers grows with sophistication.

Table 95.1. Uniqueness of projected algorithm preparation on solitary CPU

period	*Iteration*	*Time beyond (hh:mm:ss)*	*Mini-batch loss*	*Base learning rate*
1	1	00:00:06	1.2862	0.0010
20	20	00:01:19	9.8602e-05	0.0010

Source: Author.

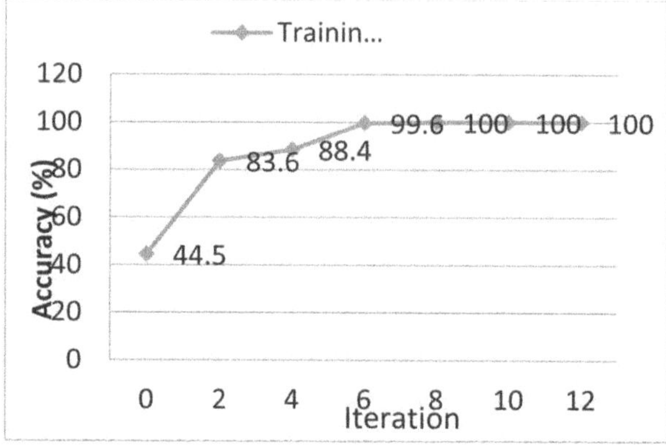

(a) Training Accuracy

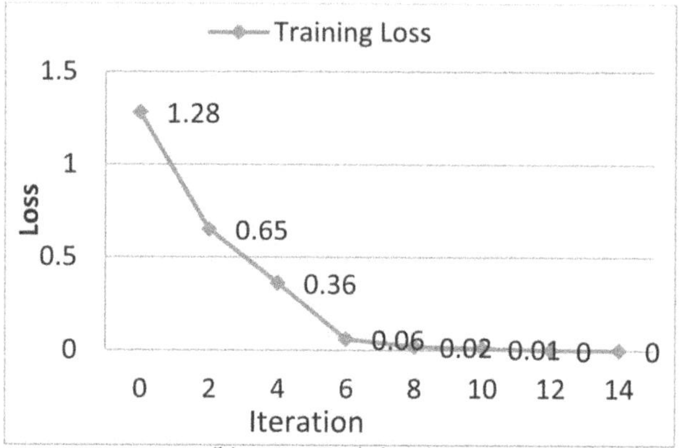

(b) Training loss

Figure 95.4. Training advancement training loss versus accuracy.

Source: Author.

Multilayered CNN is needed for more intricate tasks in order to operate more effectively. This face recognition system requires less time to educate. There are twenty epochs in this framework that are used for training. For each of the four subjects, the system's overall accuracy is 100%. Nevertheless, there had been an order of degradation whereby the usual length of time to train was incompatible, making the MATLAB platform ill-suited for this approach. In the future, such technology for recognizing faces might be upgraded to an automatic system and built in an innovative framework such as Python. New photographs ought to be easily obtained by the application and saved into the image database.

References

[1] Bhele, S. G., & Mankar, V. H. (2012). A review paper on face recognition techniques. *International Journal of Advanced Research in Computer Engineering & Technology (IJARCET)*, *1*(8), 339–346.
[2] Bruce, V., & Young, A. (1986). Understanding face recognition. *British journal of psychology*, *77*(3), 305–327.
[3] Geng, C., & Jiang, X. (2009, November). Face recognition using sift features. In *2009 16th IEEE international conference on image processing (ICIP)* (pp. 3313–3316). IEEE.
[4] Zhang, Z., Luo, P., Loy, C. C., & Tang, X. (2015). Learning deep representation for face alignment with auxiliary

attributes. *IEEE transactions on pattern analysis and machine intelligence*, *38*(5), 918–930.

[5] Liang, M., & Hu, X. (2015). Recurrent convolutional neural network for object recognition. In *Proceedings of the IEEE conference on computer vision and pattern recognition* (pp. 3367–3375).

[6] Xia, R., Deng, J., Schuller, B., & Liu, Y. (2014, May). Modeling gender information for emotion recognition using denoising autoencoder. In *2014 IEEE International Conference on Acoustics, Speech and Signal Processing (ICASSP)* (pp. 990–994). IEEE.

[7] Bengio, Y. (2009). Learning deep architectures for AI. *Foundations and trends® in Machine Learning*, *2*(1), 1–127.

[8] LeCun, Y., Boser, B., Denker, J. S., Henderson, D., Howard, R. E., Hubbard, W., & Jackel, L. D. (1989). Backpropagation applied to handwritten zip code recognition. *Neural computation*, *1*(4), 541–551.

[9] Krizhevsky, A., Sutskever, I., & Hinton, G. E. (2012). Imagenet classification with deep convolutional neural networks. *Advances in neural information processing systems*, *25*, 1–9.

[10] Murphy, K. P. (2012). *Machine learning: a probabilistic perspective*. MIT press.

[11] Krizhevsky, A., Sutskever, I., & Hinton, G. E. (2012). Imagenet classification with deep convolutional neural networks. *Advances in neural information processing systems*, *25*.

[12] Srivastava, N., Hinton, G., Krizhevsky, A., Sutskever, I., & Salakhutdinov, R. (2014). Dropout: a simple way to prevent neural networks from overfitting. *The Journal of Machine Learning Research*, *15*(1), 1929–1958.

[13] Yalawar, M. S., Vijaya Babu, K., Mahender, B., & Singh, H. (2023). A Brain-Inspired Cognitive Control Framework for Artificial Intelligence Dynamic System. In: Kumar, A., Mozar, S., Haase, J. (eds). *Advances in Cognitive Science and Communications. ICCCE 2023. Cognitive Science and Technology*. Springer, Singapore. https://doi.org/10.1007/978-981-19-8086-2_70

[14] Saravanan, P., Pandey, A., Joshi, K., Rondon, R., Narasimharao, J., & Imran, A. A. (2023, May). Using machine learning principles, the classification method for face spoof detection in artificial neural networks. In *2023 3rd International Conference on Advance Computing and Innovative Technologies in Engineering (ICACITE)* (pp. 2784–2788). IEEE.

[15] Khin, T., Srujan Raju, K., Sinha, G. R., Khaing, K. K., & Kyi, T. M. (2020). Review of optimization methods of medical image segmentation. In *Proceedings of the Third International Conference on Computational Intelligence and Informatics: ICCII 2018* (pp. 213–218). Springer Singapore.

[16] Naga RamaDevi, G., Janga Reddy, M., & Baswaraj, D. (2020). Improve the Efficiency of the Classifiers Using Resample Technique on Image Segmentation Dataset. In *ICDSMLA 2019: Proceedings of the 1st International Conference on Data Science, Machine Learning and Applications* (pp. 932–941). Springer Singapore.

[17] Nay, T. (2024). Enhancing IoT Security with AI-Driven Hybrid Machine Learning and Neural Network-Based Intrusion Detection System. *Babylonian Journal of Artificial Intelligence*, *2024*, 158–167.

[18] Al Barazanchi, I. I., Hashim, W., Thabit, R., & Hussein, N. A. H. K. (2024). Advanced Hybrid Mask Convolutional Neural Network with Backpropagation Optimization for Precise Sensor Node Classification in Wireless Body Area Networks. *KHWARIZMIA*, *2024*, 17–31.

[19] Nafea, A. A., Kareem, A. K., Khalaf, M. A., & AL-Ani, M. M. (2023). An effective deep learning model for surface-enhanced raman spectroscopy detection using artificial neural network. *Babylonian Journal of Machine Learning*, *2023*, 35–41.

96 The communal effects on security access using cloud data bases in medical health records

R. Suhasini[1,a], *Chandrugonda Malleshwar Rao*[2,b], *Voruganti Naresh Kumar*[3,c], *and Rapaka Usha*[4,d]

[1]Department of CSE, CMR College of Engineering and Technology, Hyderabad, Telangana, India
[2]Department of CSE, CMR Engineering College, Hyderabad, Telangana, India
[3]Department of CSE, CMR Technical Campus, Hyderabad, Telangana, India
[4]Department of CSE, CMR Institute of Technology, Hyderabad, Telangana, India

Abstract: Innovative technologies like cloud computing and the Internet of Things (IOT) will be developed in the coming generation and employed in the medical area from theory to clinical application. The medical profession of health care plays a vital part in cloud computing. Healthcare professionals' HIS is contained in the electronic medical record; however, because of the high cost of servicing, the adoption process is delayed. This research investigates cloud-based safety and access privileges for electronic medical data. All hospital patient information was stored in a cloud data warehouse. A platform for an electronic health care system is provided by cloud middleware. Only authorized users can securely access the medical health record, preventing unauthorized individuals from using it. This study proposes a cloud computing architecture that allows any institutions appropriate case of access privileges to share and securely exchange electronic medical records.

Keywords: Healthcare records, cloud computing, EMR, e-medicals

1. Introduction

Electronic health records, which are made up of a variety of medical data records generated off-site, are the foundation of today's medical industry. IT fields such as UIDs, Aadhar numbers, and any other government-issued identifiers gather demographic and identifying information for health records. PHI that is digitally created, acquired, and altered is known as electronic protected health information, or E-PHI. Some of the ways that this type of electronic health field provides 'Health service' to each and every patient detail are listed below [1–2].

a. Health documents, which contain all necessary details regarding a patient's medical data.
b. Obtain the patient's identify and demographic data;
c. Provide the patient with valuable care according to the report from their health record.

All patient data was hosted by the health system on either a private or public cloud.Diagnostic imaging and patient health records are included in the cloud medical infrastructure along with cloud service clients as shown the Figure 96.1.

An e-medical version that makes medical reports easily available is called a digital health record, or EHR. EHRs are flexible enough to provide various designated people with highly confidential patient care information. While electronic health records (EHRs) differ in terms of their setting and methodical strategy for record keeping, they are frequently designed to house patient medical and treatment histories, along with a patient's specific prescriptions, diagnoses, radiological images, dates of immunizations, asthma, and the laboratory outcomes, between other data. There once was less of a barrier among a doctor as well as the patient than there is now. The patient's past did not need to captivate the doctor, who was focusing mostly on the illness. Naturally, when the same doctor follows the same family every time, connections is eventually made and the doctor is able to recognize the patient's temperament and learn about his different interventions, which helps him to stop repeating the same questions again and over again. Every physician had his unique style of functioning, and since he lacked a good memory, his appointments needed to be documented. However, the patient would need to start over if he changed doctors, and the doctor is compelled to ask questions in order to determine the exact ailment his patient is dealing with – especially when the patient has many conditions that indicate he has seen numerous experts. Put another way, neither the systematic recording nor the sharing of the doctor's observations throughout a consultation were guaranteed. However, the remarkable advancement of bioscience has led to significant issues with patient-provider confidence, responsibility, and openness. The maintenance of a medical file now

[a]rsuhasini@cmrcet.ac.in, [b]228rapr6611@cmrec.ac.in, [c]nareshkumar99890@gmail.com, [d]rapaka_usha@yahoo.co.in

DOI: 10.1201/9781003684459-96

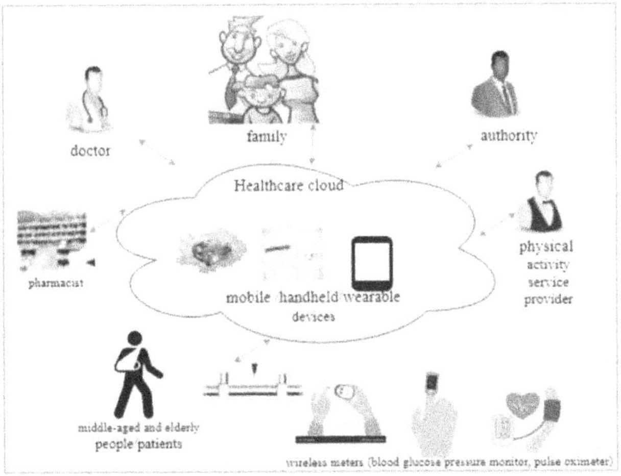

Figure 96.1. Healthcare cloud architecture.

Source: Author.

takes into account the realistically necessary transparency of appointments.

2. Healthcare in Cloud

These days the key function of data in the internet of things is in healthcare. Three main categories comprise the healthcare communication system: electronic medical records, electronic health records, and private health records.

Safe Electronic Health Record Administration Platform An e-medical version that makes medical reports easily available is called an electronic health record, or EHR. EHRs provide many designated people with flexible access to medical information in a highly secure format. While electronic health records (EHRs) differ in terms of their context and methodical method of recordkeeping, they are frequently designed to house patient medical and treatment histories, along with a patient's specific prescriptions, diagnoses, radiological images, dates of immunizations, asthma, and laboratory findings, between other data. The electronic record becomes a tool that the family medicine sector can use to change its operations and meet its own demands as well as the requirements of its patients. Specialist physicians and their necessary personnel can practice pharmacy more efficiently because to improved working algorithms and data availability. The automatic notifications and choice support offered facilitate the practice of providing individuals and the community with safer, more identifiable healthcare. The electronic record focuses on effective work and quality requirements [3]. For the necessary individuals and qualified doctors, it serves a useful purpose. Although a fundamental shift in practices is necessary to reap the full benefits of electronic record systems. founded on evidence-based care, systems and team-based healthcare, and quality-improving techniques. The computer program that doctors can employ to track all the different facets of their patients' care is referred to as this electronic record.

These three phrases are employed in the context of hospital information about patients employing cloud architecture. The HER frame framework provides protection and safety of the patient's private data in a cloud-based system [4]. The electronic health records contain all the patient data that the healthcare facility or clinic has gathered. By diagnosing the disorders, the clinicians employ this information to treat the patients. EMR data is utilized for round-the-clock tests and tracking in order to enhance the patients' health quality. All of the individual's clinical data is included in the electronic medical records. One crucial aspect of the HER is its ability to exchange information with other medical professionals, including experts. HER keeps track of the patient's physician, hospital, and nursing information [5]. As shown the Figure 96.2 the distinction between PHRs and EHRs is that PHRs are set up, obtained, and operated by patients.

3. Healthcare Information Technology

Numerous applications exist for cloud computing, particularly for medical data and medical data. More advantages and conveniences for healthcare while utilizing the cloud. Health IT technologies are handled via the internet to handle patient and medical information. In recent years, cloud technologies are being used for medical management as well as research to safeguard sensitive data on the cloud [6].

The more that a healthcare facility links its system data to an international computer network, like the Internet, more it invites accessibility from all directions and encourages data breaches. It is important to safeguard the requirement for a digital medical record from unauthorized individuals who might utilize it improperly for a number of reasons. One of the strongest safety measures for protecting medical records data is identity-based cryptography. The approach uses any thread that serves as a public key to address issues observed in standard encryption methods. The system has the ability to improve health record security by incorporating protocols for authentication into linked servers. To enable every computer to handle encryption and decryption throughout data exchange, this system employs ABE to encrypt and

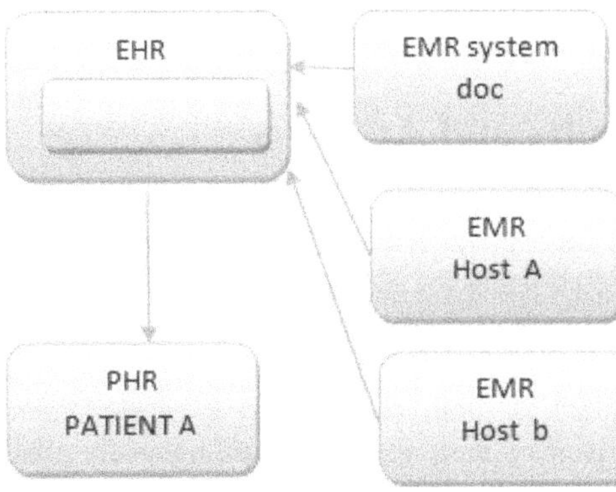

Figure 96.2. Frame work of PHR, HER, EMR.

Source: Author.

decrypt data sent among its servers. Health records are able to accessed and extracted by servers that have IDs. As of right now, test outcomes indicate effectiveness in relation to the system's algorithm's speed. A computer concept called 'cloud storage' uses the Internet or a cloud to store data. Cloud storage is provided based on capacity, request, and pricing, meaning that the user is still responsible for controlling and making investments in their own data storage equipment. This offers resilience, scalability, and speed. Here are a few general benefits of cloud computing, with an emphasis on e-health systems in our instance.

3.1. Patient-centric access control model

Family members and acquaintances of the patient can easily access the patient's health record information. Health information is kept up to date in the PHR using third-party cloud health data. Patient data is secured by the PHR the owner's control using safeguards such encrypting it allowing only authorized individuals to decrypt the necessary patient data file [7]. Use attribute-based encrypting it or ABE, to safeguard health records stored in the PHR cloud in the suggested paper. Every patient uses health record information in this way, utilizing attribute-based encryption architecture. In this case, each patient is aware of the PHR's third-party verifier of patient centricity [8–9].

The use of cloud-based computing in healthcare has significantly altered the field [10]. The primary benefits of cloud computing in the healthcare industry are the capacity to scale the necessary services and the capacity to adjust up or down data storage while utilizing machine learning and artificial intelligence (AI) [11]. In order to investigate the application of intelligent approaches in health systems, the present investigation reviewed a number of research projects. Its primary focus was on issues of privacy and security with contemporary technology. A relatively new area of medical informatics is called 'e-healthcare,' which is the use of the Internet and associated technologies to provide health services and information. Ensuring effective storage and protection of data is crucial for cloud computing. Cloud storage places a premium on data security and privacy preservation. The most significant resource in the healthcare industry is e-healthcare [12–15].

4. Results and Analysis

Medical records in the PHR are created using a cloud-based technique that is suggested. All patient health record details are encrypted using the Key Attributes approach in the suggested approach. It used 50 attributes to accomplish the Key generation. Figure 96.3 provides the Key Generation times for 50 qualities, both current and suggested, along with a comparison of their values. It highlights the distinctions among the suggested and current works in order to demonstrate how this task is performed. It provides the necessary enhancement over the current method [16].

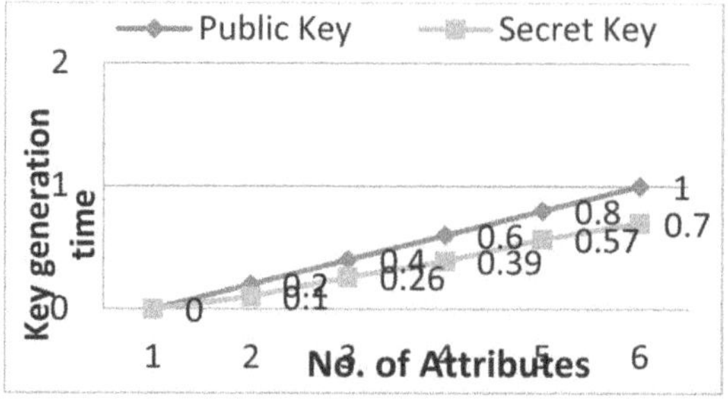

Figure 96.3. Graph showing key generation time and the amount of attributes.

Source: Author.

As shown the Figure 96.3 is the suggested work, the key generation time utilizing attributes.

The work that is suggested is indicated by a red line in the illustration above, and what is currently being done is indicated by a blue line. The illustration contrasts the two pieces and highlights their differences. It demonstrates that, in comparison to previous studies, the suggested approach requires less time to generate keys. The time needed for setting up a public key (PK) and the key generation time (SK) for establishing a secret key. It rises in proportion to the quantity of qualities added to the SK frame.

5. Conclusion

In order to improve the patients' health, this study concentrated on the medical healthcare details of the patients throughout the providers. Whenever patients share their medical records with authorized parties in a secure manner. The primary access control that allows privileges to the patient's details. The article demonstrates how to securely access medical data in the cloud utilizing attributes-based computing. This study demonstrated a secure method for using cloud computing to access healthcare data.

References

[1] Toosi, A. N., Calheiros, R. N., Thulasiram, R. K., & Buyya, R. (2011, September). Resource provisioning policies to increase iaas provider's profit in a federated cloud environment. In *2011 IEEE International Conference on High Performance Computing and Communications* (pp. 279–287). IEEE.

[2] Rodero-Merino, L., Caron, E., Muresan, A., & Desprez, F. (2012). Using clouds to scale grid resources: An economic model. *Future Generation Computer Systems*, *28*(4), 633–646.

[3] Buyya, R., Ranjan, R., & Calheiros, R. N. (2010). Intercloud: Utility-oriented federation of cloud computing environments for scaling of application services. In *Algorithms and Architectures for Parallel Processing: 10th International Conference, ICA3PP 2010, Busan, Korea, May 21–23, 2010. Proceedings. Part I 10* (pp. 13–31). Springer Berlin Heidelberg.

[4] Shenai, S., & Aramudhan, M. (2018). Cloud computing framework to securely share health & medical records among federations of healthcare information systems. *Biomedical Research (0970–938X)*.

[5] Ahmadi, M., & Aslani, N. (2018). Capabilities and advantages of cloud computing in the implementation of electronic health record. *Acta Informatica Medica*, *26*(1), 24.

[6] Nigam, V. K., & Bhatia, S. (2016). Impact of cloud computing on health care. *Int Res J Eng Technol*, *3*(5), 2804–2810.

[7] Pussewalage, H. S. G., & Oleshchuk, V. (2016, November). A patient-centric attribute based access control scheme for secure sharing of personal health records using cloud computing. In *2016 IEEE 2nd International Conference on Collaboration and Internet Computing (CIC)* (pp. 46–53). IEEE.

[8] Barua, M., Lu, R., & Shen, X. (2013, December). SPS: Secure personal health information sharing with patient-centric access control in cloud computing. In *2013 IEEE global communications conference (GLOBECOM)* (pp. 647–652). IEEE.

[9] Li, M., Yu, S., Zheng, Y., Ren, K., & Lou, W. (2012). Scalable and secure sharing of personal health records in cloud computing using attribute-based encryption. *IEEE Transactions on Parallel and Distributed Systems*, *24*(1), 131–143.

[10] Hassannejad, H., Matrella, G., Ciampolini, P., De Munari, I., Mordonini, M., & Cagnoni, S. (2016, October). Food image recognition using very deep convolutional networks. In *Proceedings of the 2nd international workshop on multimedia assisted dietary management* (pp. 41–49).

[11] Kumara Swamy, M., & Krishna Reddy, P. (2020). A model of concept hierarchy-based diverse patterns with applications to recommender system. *International Journal of Data Science and Analytics*, *10*(2), 177–191.

[12] Khin, T., Srujan Raju, K., Sinha, G. R., Khaing, K. K., & Kyi, T. M. (2020). Review of Optimization Methods of Medical Image Segmentation. In: Raju, K., Govardhan, A., Rani, B., Sridevi, R., Murty, M. (eds) *Proceedings of the Third International Conference on Computational Intelligence and Informatics. Advances in Intelligent Systems and Computing*, vol 1090. Springer, Singapore. https://doi.org/10.1007/978-981-15-1480-7_17

[13] Malsoru, V., Patil, A., Kusuma, S. M., Shivakanth, G., & Samanvita, N. (2023, February). An Efficient SFLA and CUCKOO Search Hybridization for Source Distribution in Cloud Computing. In *2023 IEEE International Conference on Integrated Circuits and Communication Systems (ICICACS)* (pp. 01–05). IEEE.

[14] Reddy, K. P., Satish, M., Prakash, A., Babu, S. M., Kumar, P. P., & Devi, B. S. (2023, October). Machine learning revolution in early disease detection for healthcare: Advancements, challenges, and future prospects. In *2023 IEEE 5th International Conference on Cybernetics, Cognition and Machine Learning Applications (ICCCMLA)* (pp. 638–643). IEEE.

[15] Mensah, G. B., Mijwil, M. M., Abotaleb, M., Ali, G., & Dutta, P. K. (2025). High Performance Medicine: Involving Artificial Intelligence Models in Enhancing Medical Laws and Medical Negligence Matters A Case Study of Act, 2009 (Act 792) in Ghana. *SHIFAA*, *2025*, 1–6.

[16] Gopi, R. S., Suganthi, R., Hephzipah, J. J., Amirthayogam, G., Sundararajan, P. N., & Pushparaj, T. (2024). Elderly People Health Care Monitoring System Using Internet of Things (IOT) For Exploratory Data Analysis. *Babylonian Journal of Artificial Intelligence*, *2024*, 54–63.

97 A new technique for segmenting infant brain MRI data utilizing the ridgelet and framelet transforms

Vivekanand Aelgani[1,a], *Rafeeq M. D.*[2,b], *Prasannababu J.*[3,c], *and P. Kiran Kumar*[4,d]

[1]Associate Professor, Department of CSE, CMR College of Engineering and Technology, Hyderabad, Telangana, India
[2]Associate Professor, Department of CSE, CMR Engineering College, Hyderabad, Telangana, India
[3]Assistant Professor, Department of CSE, CMR Technical Campus, Hyderabad, Telangana, India
[4]Assistant Professor, Department of CSE, CMR Institute of Technology, Hyderabad, Telangana, India

Abstract: When contrasted with adult brain MRIs, tissue segmentation of the baby brain is a particularly difficult task. Yet, because of intrinsic incomplete myelination and maturation, it is extremely difficult because of the low tissue contrast ratio. Specifically, the poor imaging difference in the initial postnatal year occurs at six months old because the voxel strength in the white substance and gray substance are the same, so as to segment white matter, gray matter, and cerebrospinal fluid, we suggested an infant's tissue segmentation approach that uses ridgelet transform, framelet transform, wKNN, and MKSVM dependent classification. The article analyzes ridgelet and framelet coefficients using an entropy feature-based approach and a Gray level co-occurrence matrix (GLCM). Cerebrospinal Fluid is trained using the intensity-based features and the derived textural characteristics. The suggested approach matches existing single-modality methods' manual ground-truth using significantly higher dice ratios.

Keywords: MRI, infants, segmentation, tissue, classifier

1. Introduction

Brain segmentation from magnetic resonance imaging (MRI) is being worn more and more to examine the growth and maturation of an infant's or newborn's brain. Step five of the medical picture segmentation framework involves the extraction of features. An alternative method of segmenting an image is the manual segmentation procedure. This method not only produces inaccurate results, but it is also expensive, time-consuming, and inconvenient. Moreover, manual tagging exhibits variation between and within observers.

The drawbacks of manual methods make it challenging to label big groups of people, which is frequently required for neuro imaging studies. Therefore, an effective method for splitting the brain into many regions is needed [6]. White matter (WM), grey matter (GM), cerebrospinal fluid (CSF), and infrequently pathological tissues similar to tumors, etc., are the main segments of the infants' MRI brain tissue. Brain MR imaging data segmentation of newborns is far more challenging than that of adults. Due to the newborn brain's tiny size, MR images of the brain typically show a minor SNR (signal-to-noise ratio), a much lower CNR (contrast-to-noise ratio), and significant variation in the form and appearance of the brain due to the brain's rapid growth.

Tissue identification is further complicated by the partial volume effect brought on by inverted signal strength in white matter (WM) (5).Similar to WM in the developing brain, the partial volume (PV) of GM and CSF in the cortical grey matter (CGM)-CSF borders (6).Voxels are incorrectly identified as WM because to this partial volume (PV) influence, either at the CGM-CSF border. The main tenet of the approaches for creating regions is that adjacent pixels inside an area have values that are similar. The most common technique is to compare a single pixel to its neighbors. The cluster can designate a pixel to one or more of its neighbours if a similarity condition is satisfied.

2. Literature Survey

Using Grayscale-Co's appearance matrix (GLCM), Mohamed A. Ber-Bar [1] introduced seven statistics and structuralities along with the quality and contour (CT2) approach of ST-GLCM, GLCM, and GLCM. Photos are evaluated using the online database of Screening Mammography (DDSM) and the mammography database of Image Analysis Association (Mammography).

A system for segmenting images that is capable of recognizing areas of interest (ROI) was suggested by Shadi A. et al. [2]. pictures of people obtained using a range of medical

[a]avivekanand@cmrcet.ac.in, [b]drrafeeqcse@gmail.com, [c]cmrtc.paper@gmail.com, [d]kiranpondugula007@gmail.com

DOI: 10.1201/9781003684459-97

scanners, like MRI, CT, and PET. Multiresolution analysis (MRA) including wavelet, ridgelet, and curvelet transformations is worn in the suggested segmentation.

A framelet transform and singular value decomposition-based picture improvement technique was recently suggested by S. Subramaniam et al. [3]. The image is divided into eight higher-frequency sub bands and one low frequency sub band using the framelet transform. To obtain the elevated declaration image, the inverse framelet transform is applied to all of the sub bands. SVD is coupled with framelet transform to boost brightness. Satellite imagery has been used to test the suggested method.

Quantitative metrics including Entropy, Peak Signal-to-Noise Ratio (PSNR), Universal Quality Index (UQI), Structural Similarities manifestation Measure (SSIM), and Quality Score are employed. A system based on the Framelet Transform in conjunction with the gray level co-occurrence matrix (GLCM) has been proposed by Sulochana et al. [4]. To extract features, a amalgamation of the Framelet transform and the Gray Level Co-occurrence Matrix (GLCM) is used.

Framelet features recovered by the model and SVM with a linear kernel are approved as a feature and classifier in Yan-Ran Li, et al.'s [7] suggested features optimization model.

A technique for classifying liver CT and brain-MRI has been suggested by B. Thamaraichelvi et al. [8].

The Regional Basis Function-Support Vector Machine (RBF-SVM) method is being used to classify liver CT (computed tomography) and magnetic resonance imaging (MRI), and the GLCM technique have been used to fetch characteristics.

Na Li [9] presented an ensemble Support Vector Machine (SVM) structure and a gray-level co-occurrence matrix (GLCM) texturing approach. Each voxel's 112 GLCM features were first retrieved, and the SVM-recursive factor removal (SVM-RFE) approach was used to score these features. To distinguish the aberrant bulk of material from typical brain tissues, an ensemble SVM classifier was built using only the top 60 grayscale features. The current study by P.S. Mukambika et al. [10] compares two methods for MRI image tumor identification. One depends on the Level set method, utilizing non-parametric deformable modeling with active contour to segment the brain tumor depending on MRI brain scans. The other method is the K means segmentation method. After segmentation, choices are decided upon in two stages: assist vector machine extracting features and classification using the Gray Level The combination Matrix and Discrete Wavelet Transform An approach for identifying the ischemic stroke lesion utilizing midline sketching and a histogram bin-based approach is proposed by R. Kanchana et al. [11]. In order to obtain characteristics based on the CT picture, the visible region affected by an ischemic stroke lesion is divided with the typical portion using histogram bins. First- and the second-order statistical characteristics are analyzed for each region. Classification of the lesion versus non-lesion regions is based on differences in the features. A method for tumor detection was put forward by Venkatesh et al. [12]. MRI brain tumors are segmented by means of the K means clustering technique, and the Gray Level Co-Occurrence Matrix (GLCM) is used to examine different aspects of the segmented tumor. These characteristics were fed into the k-Nearest Neighbour (k-NN) classifier, which classified the tumor as either benign or malignant. The suggested algorithm's accurateness is 85% in each case.

The method suggested by Atanu K. Samanta et al. [13] uses computational techniques like support vector machines (SVM) to classify input MRI images into normal and abnormal groups, K-means clustering to segment the brain tumor from other parts of the brain, as well as gray level co-occurrence matrices (GLCM) to extract characteristics of this categorized brain tumor portion.

Manorama Sharma and colleagues [14] have introduced an effective method that utilizes K-means and artificial neural network (KMANN). Grey Level Co-Occurring Matrix, or GLCM, is employed in feature extraction. For the purpose of detecting brain tumors, a fuzzy inference system is built employing extracted features, thresholding, morphology operators, and waterfall segmentation.

A feature extraction approach based on wavelet and gray-level co-occurrence matrix, as well as a technique for classification relying on an optimized deep belief network, are proposed by Li Xu et al. [15]. The classifier network is optimized with the application of an improved moth search method. The simulation findings are practical to FLAIR, T1, and T2, three distinct datasets.

Ghribi Olfa et al. [16] The suggested segmentation method is predicated on volumetric characteristics that were retrieved and inferred from the gray-level run length (GLRLM) and gray-level co-occurrence (GLCM) matrices. New voxel-wise approaches would be used to extract volumetric data while maintaining shape, spatial, and connection information.

An optimum choice of features approach was created using the genetic algorithm (GA) and the assistance vector machine (SVM) tool, preserving only the essential characteristics required for recognizing the primary tissue in the brain or the MS lesions among the white matter or gray matter regions [19–21].

3. Proposed Method

Preprocessing stage: When extracting characteristics for each technique being processed, a process known as preprocessing has been put into place and performed. It has been demonstrated to improve performance. Shift-Invariant Wavelet transform is used [14] A large portion of the quality benefit from such cycle-spinning can be obtained by merely averaging shifts of just n=0 and n=1 across each axis.

This research proposes machine learning methodologies for identifying and segmenting different kinds of glioma brain tumors. Kirsch's edge detectors are used to find the boundary edge pixels, and the edge detected pixels are then subjected to the contrast adaptive equalization of histograms approach. The Ridgelet multi frequency coefficients are then obtained by applying the Ridgelet transform to this improved brain image. Additionally, the Ridgelet transformed coefficients are used to generate features, which are then improved using the main Component Analysis (PCA) technique. The CANFES classifiers is then worn to categorize the optimized features in brain images that are either glioma or non-glioma. The Glioma Brain Tumor Detection – Machine Learning Approaches is depicted in Figure 97.1.

3.1. GLCM methods

In an image, the GLCM shows the frequency of two gray-level pixels, a, b, spaced by the remoteness d in a course. The total amount of the two grayscales (a,b) shared by D in the reduced image is catalogued by the elements ordered paid a,b of the gray air pairing matrix. Image intensities are converted to G hierarchical levels using this method, which is rich in one and G. The g amount of grey plains or gxg determines the size of co-expression at the gray level. When the direction θ {0, $\pi/4$, $\pi/2$, and $3\pi/4$} and the distance d change, GLCM is applied (1). Thus, for every given θ, we shall have an amount of d GLCMs, each of which is in size. We worn GLCM (1–3) to obtain texture features in each dimension (0, $\pi/4$, and $\pi/2$).

3.2. Ridgelet transform

In 1998, Donoho presented the continuous ridgelet transform (CRT) [2], which may be described as a 1D wavelet function aimed toward radial and constant lines.2-D wavelets are created by utilizing one-dimensional (1-D) wavelets. In terms of directional selectivity, it is still limited. In 1999, Candes and Donoho presented a brand-new transform known as the Ridgelet transform [5].The features of the Ridgelet transform have distinct values after being deliberated in dissimilar orientations. The Ridgelet transform responds effectively to straight-line singularities.

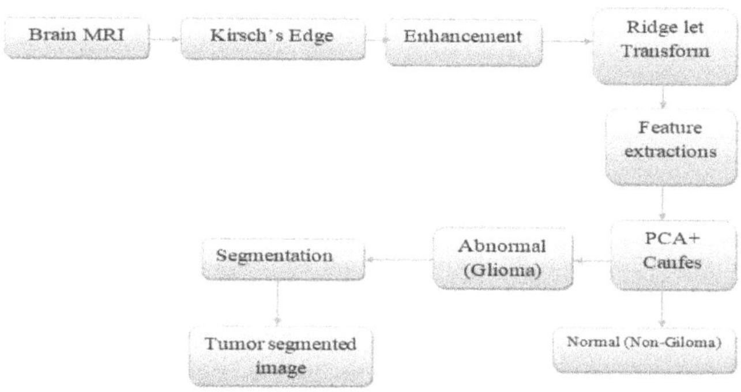

Figure 97.1. Machine learning technique for Glioma brain tumor identification.

Source: Author.

When using Ridgelet, it's a good idea to divide the image and further relate the Ridgelet change to the resulting reduced images. Investigate curve singularities [1].

3.3. Framelet transform

While there are some differences, the Framelet Transform [3] (FRT) is similar to wavelets. Two or more high-frequency filter banks in framelets produce extra decomposing subbands. This can enhance the precision of time and frequency in picture processing. The framelet subbands are duplicated, suggesting that the coefficients of numerous other subbands can also be used to offset variations in the coefficients of one band. This implies that adjustments to one coefficient during the reconstruction phase can be compensated by adjustments to the coefficient that is connected with it, lowering noise in the initial image. Estimation shift invariance is made possible by a tiny frame filter bank's homogeneity and duplication [3–4].

3.4. Segmentation

Weighted k-Nearest Neighbour (wKNN) and MKSVM classifier are used to improve segmentation performance. Gray level co-occurrence matrix (GLCM) and entropy include based analysis of ridgelet and framelet the coefficients carried out from infant's images to segment white matter (WM), grey matter (GM), and Cerebrospinal Fluid (CSF) are utilized [17–18].

3.5. Weighted k-nearest neighbour (WKNN)

KNN is an instance-based, slow, non-parametric classification. The absence of a learning step is the cause of K-NN's laziness. Rather, K-NN use a distance metric to classify the new test example after storing all available training data. KNN makes its predictions about the test instance's class using a straightforward majority voting technique. It is highly vulnerable to data that isn't balanced.

An improved method to address this is to weight the k closest neighbors' votes based on how far away they are from the test instance. The closest neighbors in Weighted K-NN are given larger weights than the distance neighbors (17). While the white matter and gray matter material of an infant's brain were accurately segmented by wKNN, the cerebrospinal fluid (CSF) was incorrectly classified.

3.6. Multi kernel class SVM

Using kernel mapping, the SVM model efficiently completes difficult separation in large dimensions. Similar functions are given into machine learning algorithms, which use feature vectors, as a kernel. The three kernel functions that may be found in the literature are the Gaussian kernel, also known as the RBF kernel, the polynomial kernel, and the linear kernel (18). Selecting an appropriate kernel is a crucial step towards achieving enhanced abstraction capabilities in pattern categorization. Here, the Radial Basis Function (RBF) and linear kernels – which have been shown to be the most pertinent kernels for infants' MRI soft tissue segmentation – are used to train the MKSVM classifier.

4. Experimental Results and Discussion

The suggested technique has been used to segment soft tissue in infants' MRIs. To begin, we apply the Shift-Invariant Wavelet transform on the image to reduce noise. Subsequently, textural characteristics and intensity-based features are extracted, and ridgelet and framelet coefficients are analyzed using entropy skin tone and Gray level co-occurrence matrix (GLCM). Grey matter, white matter, and cerebrospinal fluid are segmented using wKNN and an MKSVM classifier. Figure 97.1 displays the results of the segmentation simulation, superior performance is indicated by greater accuracy and dice coefficient values [22–23].

The greatest pixel intensity, and L is the edge identify brain MRI image [24, 25]. Each pixel in the 3*3 sub window mask's cumulative index intensity value is taken into account while computing the cumulative density function, or CDF. The source brain MRI image is displayed in Figure 97.2a, and the improved brain MRI image is displayed in Figure 97.2b. Figure 97.3 demonstrate two types of images: segmented (a) and annotated (b).

As shown the Figure 97.4 is performance matrics of various sample tested images.

5. Conclusion

Our approach divides the soft tissue on an MRI of a baby into categories including white matter, gray matter, and cerebrospinal fluid. The suggested method is divided into 4 stages: prepprocessing using the shift-invariant wavelet transform is the first step, followed by appropriate feature extraction using GLCM and ridgelet and framelet values. WM, GM and CSF in infants' MRI soft tissue are segmented using the weighted k-Nearest Neighbor (wKNN) and MKSVM classifiers. The testing findings show that the suggested methodology is successful in raising the Dice coefficient to 91.00% and the total segmentation accuracy to 93.74%. These encouraging findings pave the way for new developments in the assimilation of ridgelet, framelet, (wKNN), and MKSVM classifiers in the brain MRI tissue segmentation of babies.

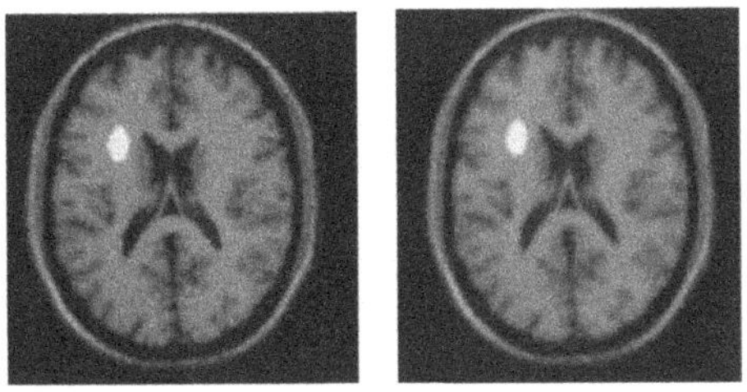

Figure 97.2. (a) Source brain MRI image (b) Improved brain MRI image.

Source: Author.

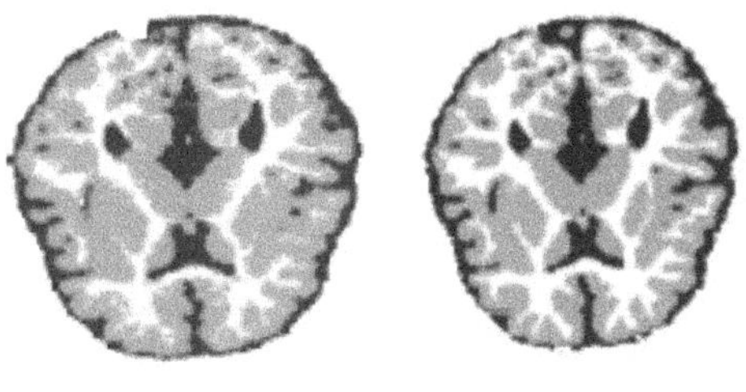

Figure 97.3. (a) Segmented image (b) Annotated image.

Source: Author.

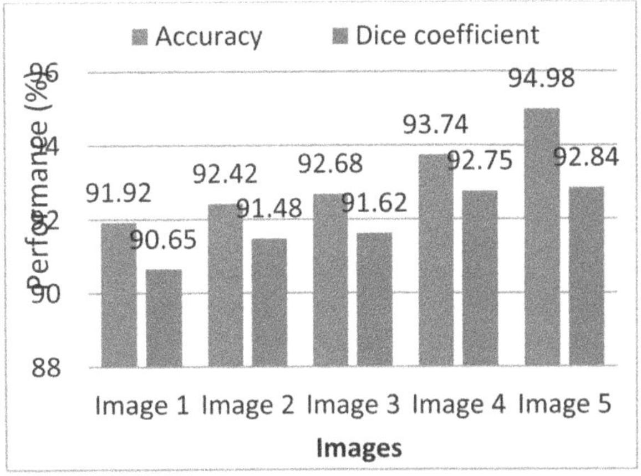

Figure 97.4. Performance accuracy of various input images.

Source: Author.

References

[1] Berbar, M. A. (2018). Hybrid methods for feature extraction for breast masses classification. *Egyptian Informatics Journal*, *19*(1), 63–73.

[2] AlZubi, S., Islam, N., & Abbod, M. (2011). Multiresolution analysis using wavelet, ridgelet, and curvelet transforms for medical image segmentation. *International Journal of Biomedical Imaging*, *2011*(1), 136034.

[3] Subramaniam, S., Rangasamy, V., Duraisamy, V., & Karuppanan, M. (2018). Application of framelet transform and singular value decomposition to image enhancement. *PAN*, *30*, 120m.

[4] Sulochana, S., & Vidhya, R. (2013). Texture based image retrieval using framelet transform-gray level co-occurrence matrix (GLCM). *International Journal of Advanced Research in Artificial Intelligence*, *2*(2), 68–73.

[5] Udayakumar, E., Yogeshwaran, K., & Ramesh, C. (2019). An efficient tissue segmentation of neonatal brain magnetic resonance imaging. *Research Journal of Pharmacy and Technology*, *12*(6), 2963–2966.

[6] Makropoulos, A., Gousias, I. S., Ledig, C., Aljabar, P., Serag, A., Hajnal, J. V., ... & Rueckert, D. (2014). Automatic whole brain MRI segmentation of the developing neonatal brain. *IEEE Transactions on medical Imaging*, *33*(9), 1818–1831.

[7] Li, Y. R., Yu, S., & Wu, S. (2013, September). Framelet features for pedestrian detection in noisy depth images. In *2013 IEEE International Conference on Image Processing* (pp. 2949–2952). IEEE.

[8] Li, Z., Mao, Y., Huang, W., Li, H., Zhu, J., Li, W., & Li, B. (2017). Texture-based classification of different single liver lesion based on SPAIR T2W MRI images. *BMC medical imaging, 17*, 1–9.

[9] Li, N., & Yang, Z. (2020). Brain tumor segmentation from multimodal magnetic resonance imaging data based on gray-level co-occurrence matrix (GLCM) and an ensemble Support Vector Machine (SVM) classifier.

[10] Mukambika, P. S., & Uma Rani, K. (2017). Segmentation and classification of MRI brain tumor. *Int. Res. J. Eng. Technol. (IRJET), 4*(07), 683–688.

[11] Kanchana, R., & Menaka, R. (2017). A novel approach for characterisation of ischaemic stroke lesion using histogram bin-based segmentation and gray level co-occurrence matrix features. *The Imaging Science Journal, 65*(2), 124–136.

[12] Leo, M. J. (2019, November). MRI brain image segmentation and detection using K-NN classification. In *Journal of Physics: Conference Series* (Vol. 1362, No. 1, p. 012073). IOP Publishing.

[13] Gupta, V., Mahle, R., & Shukla, A. (2013, July). Removal of Gaussian noise from stationary image using shift invariant wavelet transform. In *2013 Tenth International Conference on Wireless and Optical Communications Networks (WOCN)* (pp. 1–4). IEEE.

[14] Sharma, M., Purohit, G. N., & Mukherjee, S. (2018). Information retrieves from brain MRI images for tumor detection using hybrid technique K-means and artificial neural network (KMANN). In *Networking Communication and Data Knowledge Engineering: Volume 2* (pp. 145–157). Springer Singapore.

[15] Xu, L., Gao, Q., & Yousefi, N. (2020). Brain tumor diagnosis based on discrete wavelet transform, gray-level co-occurrence matrix, and optimal deep belief network. *Simulation, 96*(11), 867–879.

[16] Ghribi, O., Sellami, L., Slima, M. B., Mhiri, C., Dammak, M., & Hamida, A. B. (2018). Multiple sclerosis exploration based on automatic MRI modalities segmentation approach with advanced volumetric evaluations for essential feature extraction. *Biomedical Signal Processing and Control, 40*, 473–487.

[17] Salih, A. M., & Dhannoon, B. N. (2021). Weighted k-Nearest Neighbour for Image Spam Classification. *Iraqi Journal of Science, 62*(3), 1036–1045.

[18] Sultana, N., & Kamatham, Y. (2016, September). MSVM-based classifier for cardiac arrhythmia detection. In *2016 International Conference on Advances in Computing, Communications and Informatics (ICACCI)* (pp. 1314–1318). IEEE.

[19] Sruthi, P., & Sahadevaiah, K. (2022). A novel efficient heuristic based localization paradigm in wireless sensor network. *Wireless Personal Communications, 127*(1), 63–83.

[20] Kumara Swamy, M., & Krishna Reddy, P. (2020). A model of concept hierarchy-based diverse patterns with applications to recommender system. *International Journal of Data Science and Analytics, 10*(2), 177–191.

[21] Khin, T., Srujan Raju, K., Sinha, G. R., Khaing, K. K., & Kyi, T. M. (2020). Review of optimization methods of medical image segmentation. In *Proceedings of the Third International Conference on Computational Intelligence and Informatics: ICCII 2018* (pp. 213–218). Springer Singapore.

[22] Joshi, A., Choudhury, T., Sai Sabitha, A., & Srujan Raju, K. (2020). Data mining in healthcare and predicting obesity. In *Proceedings of the Third International Conference on Computational Intelligence and Informatics: ICCII 2018* (pp. 877–888). Springer Singapore.

[23] Reddy, K. P., Satish, M., Prakash, A., Babu, S. M., Kumar, P. P., & Devi, B. S. (2023, October). Machine learning revolution in early disease detection for healthcare: Advancements, challenges, and future prospects. In *2023 IEEE 5th International Conference on Cybernetics, Cognition and Machine Learning Applications (ICCCMLA)* (pp. 638–643). IEEE.

[24] Rangarajan, S., & Al-Quraishi, T. (2023). Navigating the future of the internet of things: Emerging trends and transformative applications. *Babylonian Journal of Internet of Things, 2023*, 8–12.

[25] Rela, M., Krishnaveni, B. V., Kumar, P., & Lakshminarayana, G. (2021, July). Computerized segmentation of liver tumor using integrated fuzzy level set method. In *AIP Conference Proceedings* (Vol. 2358, No. 1). AIP Publishing.

98 Creative economic study as a center for tourism development in Banyuwangi district

Ayu Purwaningtyas[a], Riza Rahimi Bachtiar[b], Reni Nur Jannah[c], and Eka Afrida Ermawati[d]

Banyuwangi State Polytechnic, Banyuwangi, Jawa Timur, Indonesia

Abstract: Banyuwangi is one of the cities that is developing its tourism sector. As outlined in the National Medium Term Development Plan (RPJMN) for Tourism covering the years 2020 to 2024, Banyuwangi was chosen as the focus of studies in the tourism sector and has become a leading sector capable of driving the community's economy, from tourism entrepreneurs, Pokdarwis, Bumdes to UMKM. The increase in the number of tourist visits in Banyuwangi has an impact on the emergence of various offers or demands for facilities, services available and products produced. This research aims to. to map creative economy sub-sectors and tourism potential, identify clusters and analyze the creative economy tourism chain to obtain a creative tourism development strategy in Banyuwangi, especially in Blimbingsari District. The research method used uses a qualitative descriptive approach with the Miles and Huberman model. Drawing conclusions is assisted by NVivo software. The research results show that there are 7 creative sub-economies that can be developed to support tourism activities and are included in the cluster 2 category.

Keywords: Attraction, Blimbingsari, creative economics, tourism, something to buy

1. Introduction

Banyuwangi is one of the cities that is developing its tourism sector. According to the National Medium-Term Development Plan (RPJMN) for Tourism 2020–2024, Banyuwangi was selected as the focus of study in the tourism sector and became a leading sector that is able to drive the community's economy starting from tourism entrepreneurs, Pokdarwis, Bumdes to MSMEs. Pokdarwis is a tourism awareness group formed by the community to develop tourist attractions. Meanwhile, MSMEs are productive businesses owned by individuals or business entities that have met the criteria as micro businesses.

Banyuwangi's potential can be seen from its diverse natural landscape, ranging from mountains, lowlands and coastlines that stretch for 175.8 km (Banyuwangi Regency Government, 2018). The diversity of natural resources makes Banyuwangi rich in tourism potential, ranging from natural tourism, culture and artificial tourism. In addition to having a profitable landscape, Banyuwangi has other advantages, namely in customer service in welcoming tourists, friendly to guests and smart in capturing economic opportunities that arise from tourism activities [1].

The many potentials that Banyuwangi has have caused the number of tourist visits to increase from year to year. The many potentials that Banyuwangi has have caused the number of tourist visits to increase from year to year. Data on tourist visits in Banyuwangi can be seen in the Table 98.1.

The increase in the number of tourist visits in Banyuwangi has an impact on the emergence of various offers or demands for facilities, services available and products produced. The concept of tourism with unique and unique product characteristics for the environment can be used for tourism development [2]. This must be supported by the availability of creative industries in tourist destination areas. The creative industry is an industry that relies on skills, talent and creativity which has the potential to improve people's welfare [3]. The creative industry makes a significant contribution to the economy in each country and can produce a positive impact on the life of the nation and state [4]. The creative economy has become one of the pillars of economic growth in Banyuwangi. With the growth of tourism in Banyuwangi, the creative economy also develops, thus providing added value to tourism development in Banyuwangi [5].

Creative development in Banyuwangi focuses on Blimbingsari District which has great potential for development

Table 98.1. Number of tourist visits in Banyuwangi

Years	*Domestic*	*Foreign*
2020	3.019.875	27.224
2021	1.862.553	3.854
2022	2.948.543	29.020
2023	3.112.443	69.639

Source: Banyuwangi Satu data, 2024.

[a]ayu.purwaningtyas@poliwangi.ac.id, [b]rizarahimi@poliwangi.ac.id, [c]reni.nurjannah@poliwangi.ac.id, [d]ekaafrida22@poliwangi.ac.id

DOI: 10.1201/9781003684459-98

due to support from natural resources. However, this condition is not supported by the quality of local community resources. On average, the people of Licin Village have low education and lack of participation by the community and village officials in utilizing natural resources [6].

The creative economy is becoming an increasingly greater need, of course it is interesting to examine the potential and threats faced as well as its strengths and weaknesses. The existence of these creative economy products further enriches and provides higher selling power for the tourist destinations on offer. The promotion of tourism development via the creative economy in every region will be supported to foster greater creativity and innovation in enhancing the value and competitiveness of various products. However, this concept requires synergy from various parties, namely the government, scholars, and tourism managers/actors. Based on the phenomena that occur in Banyuwangi Regency in developing the creative economy, it is necessary to map the sub-sectors of the creative economy and tourism potential.

2. Research Methods

This research is a qualitative descriptive study with the Miles and Huberman approach using Nvivo software. Nvivo software is used to help organize and analyze non-numerical data or unstructured data [7]. The collection of research data employs methods such as interviews, observations, and documentation. The interview technique used was in-depth interviews with sources as follows:

a. Owner of creative industries in Blimbingsari District
b. Pokdarwis in Blimbingsari District
c. Bumdes Blimbinsari
d. Banyuwangi Regency Tourism Office

In general, the analysis of natural data for this research was carried out through the following stages:

1. Record all findings of phenomena in the field through observation, interviews and documentation.
2. Review the notes from observations, interviews and documentation studies, and separate data that is considered important and not important. This work was repeated to check for possible misclassification,
3. Describe the data that has been classified taking into account the locality and research objectives,
4. Make a final analysis in the form of a research report.

3. Results and Discussion

3.1. Social and community profile, economy and tourism of Blimbingsari districts

In terms of quantity, both sub-districts have sufficient human resources to support creative economy-based tourism development. However, in terms of quality, human resources must have the skills and ethos to manage and develop the creative economy as a tourism driver (Table 98.2).

Table 98.2. Population of Blimbingsari district

No	*Village*	*Total Population*			*Prosentase (%)*
		Man	*Woman*	*Amount*	
1	Kaligung	6.066	5232	11298	20,32
2	Karangrejo	1303	1350	2653	4,77
3	Bomo	4963	4949	9912	17,83
4	Gintangan	2370	2461	4831	8,69
5	Kaotan	2610	2509	5119	9,21
6	Watukebo	3432	3539	6971	12,54
7	Patoman	1370	1409	2779	5,00
8	Blimbingsari	3084	3094	6178	11,11
9	Badean	867	899	1766	3,18
10	Sukojati	2032	2048	4080	7,34

Source: Banyuwangi Satu data, 2024.

Although not always, the level of community education is often used as a parameter for the quality of human resources in Blimbingsari. The educational level is dominated by junior school graduates so that the people in these two sub-districts really need intensive training related to creative economy-based tourism management so that the development of creative economy tourism can be realized optimally.

3.2. Economic condition

The majority of people in Blimbingsari District work in the agricultural, plantation and craft sectors. This diversity of livelihoods can be synergized with each other, so that all sectors can develop proportionally.

3.3. Tourist

Based on Banyuwangi Regency Regional Regulation No.2, LD 2024/ NOREG 37-2/2024, concerning Banyuwangi Regency Regional Spatial Planning for 2023-20441, natural and cultural potential is classified into 3 categories, namely: Natural tourism (Red Island, Waterfall Jagir, Sendang Seruni, Alas Purwo, etc.) cultural (Kemiren Tourism Village, Taman Gandrung Terracotta, etc.) and artificial (Banyuwangi Park, Banyuwangi Night Spectacular, etc.). These various tourism potentials, if synergized with the creative economy sector, will provide wider benefits.

3.4. Mapping the creative economy and tourism in Blimbingsari

The creative economy in the district is divided into Fashion, Fine Arts, Performing Arts, Culinary, and Music. Performing Arts, Culinary, and Music are the most dominant sub-sectors (Table 98.3).

The performing arts in Blimbingsari sub-district are mostly Janger. This art is a traditional dance drama that combines elements of dance, music and typical costumes from

Table 98.3. Mapping of creative economy sub-sectors in Blimbingsari district

Village	*Sub-Sektor*						
	Fashion	*Craft*	*Arts*	*Performing Arts*	*Culinary*	*Music*	*Communication Design*
Kaligung	-	-	-	-	-	X	-
Karangrejo	-	-	-	x	x	-	-
Bomo	-	-	-	-	-	-	x
Gintangan	x	x	-	x	x	X	-
Kaotan	-	-	-	-	x	X	-
Watukebo	-	-	-	x	-	-	-
Patoman	-	-	-	x	x	X	-
Blimbingsari	-	x	-	-	x	-	-
Badean	x	-	-	-	x	-	-
Sukojati	-	-	-	-	x	-	-

Source: Research results, 2024.

Balinese culture. Not only that, in this sub-district there are often Ogoh-ogoh performances because in Blimbingsari sub-district there is a village known as Bali van Java, namely Patoman village. In this village, Janger and Ogoh-ogoh performing arts are often performed as traditional events.

Arts next is the music sub-sector in Blimbingsari sub-district there are several types of music, namely Terbangan or Rebana. And there is also typical music, namely Patrol music. This music is only played during the month of Ramadan. and the gamelan season which is very popular in this sub-district. Gamelan in this sub-district consists of two types, namely Javanese gamelan and Balinese gamelan music (Figure 98.1).

Culinary in Blimbingsari sub-district has its own characteristics in each village. In Gintangan village there is a culinary that can be made as a souvenir, namely bekamal. However, its production is only when there is an order because the processing process takes time and is quite difficult. Then in Blimbingsari village, there is grilled fish. This village is located near the sea, so it is no wonder that its typical culinary is grilled fish. This dish is famous for its rich taste and appetizing aroma. The characteristic of Blimbingsari grilled fish lies in the use of spices and the way it is grilled. Fish is usually seasoned with a mixture of special spices and grilled until the skin is crispy, while the fish meat remains soft.

The fashion in Blimbingsari sub-district is very unique. Because the material used is bamboo. This costume is produced once a year and is used in festivals in Banyuwangi such as Banyuwangi ethno carnival and Gintangan bamboo festival.

Then the last one is craft (Figure 98.2). Craft here comes from Gintangan village, namely bamboo weaving. Bamboo weaving is made in the form of kitchen utensils, food wrappers and also special weaving for decoration. This bamboo weaving has started to be exported to international markets such as Europe and the Maldives.

Based on the Survey and interviews, the author made a table of the distribution of creative economic sub-sectors in each village in Blimbingsari District.

Geographically, Blimbingsari District is rich in natural beauty because it is located on the coast, Blimbingsari District is located on the coast so it is not surprising that in Blimbingsari District there are many tourist attractions in the form of beaches. Not only beach tourist attractions in Blimbingsari

Figure 98.1. Janger performing.

Source: Author.

Figure 98.2. Bamboo costume.

Source: Author.

District there are special interest tourist attractions that are very famous, namely the bamboo craft tourist attraction of Gintangan Village. The following is a Table 98.4 of the distribution of tourist attractions in Blimbingsari District.

3.5. Creative economy

The creative economy is an economic process that includes production and distribution activities of goods and services that require creative ideas and concepts as well as intellectual abilities to build them [8]. According to the United Nations Conference on Trade and Development (UNCTAD), the creative economy is a developing economic concept based on creative assets that have the potential to generate economic growth and development. There are several characteristics that can describe what the creative economy sector is like (Figure 98.3). Here are the main characteristics of the creative economy: Having Intellectual Creation, Easily Replaceable, Direct and Indirect Distribution, Requires Cooperation, Based on Ideas and No Limits [5].

3.6. Creative economy cluster

A cluster is a spatial proximity between several industries with the same location. The proximity of the location allows industry players to exchange information quickly and effectively. Some characteristics of an industrial cluster are:

1. Spatial proximity between industries;

Table 98.4. Distribution of tourist attractions in Blimbingsari district

No	*Village*	*Atractions*	*Manager*
1	Kaligung	-	
2	Karangrejo	-	
3	Bomo	Pantai Bomo	Pokdarwis
4	Gintangan	Wisata minat khusus kerajinan bambu Desa Wisata Gintangan	Pokdarwis
5	Kaotan	-	-
6	Watukebo	-	-
7	Patoman	Patoman bali van java dan Pantai blibis	Pokdarwis
8	blimbingsari	Pantai Blimbingsari	Pokdarwis
9	Badean	-	-
10	Sukojati	Pantai Cemara Gading	Pokdarwis

Source: Research results, 2024.

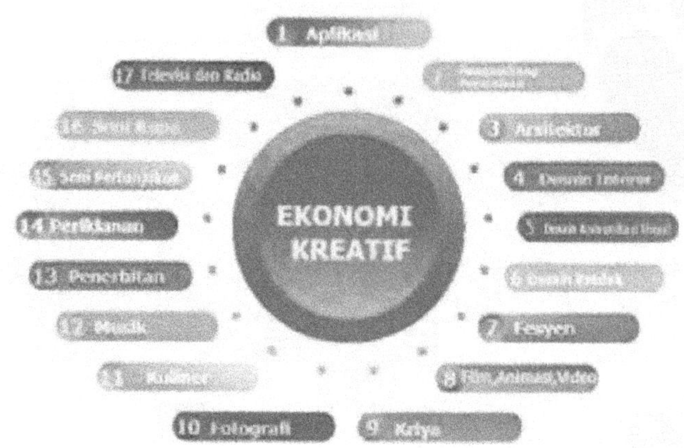

Figure 98.3. Creative economy sub-sector.

Source: Author.

2. High density of economic activities; and
3. Involvement of several companies in one similar activity.

Based on the cluster theory of Dijk and Sverrisson, 2003 Blimbingsari District is included in the Typology 2 cluster where it is starting to develop, similar activities create inter-industry networks that are starting to be established, there has been imitation and competition between industries within the cluster. The influencing characters are spatial concentration, strategic location, information exchange, grouping of similar activities, product imitation, and local imitation.

References

[1] Sambodo, T. (2020). *Susunan Rencana Pariwisata Nasional, Bappenas Litbang Banyuwangi.*

[2] Arida, Nyoman. (2017). *Pariwisata Berkelanjutan.* Bali: Sustain Press.

[3] Simatupang, T. M. (2008). Perkembangan Industri Kreatif. *Schoolmof Business and Management of the Bandung Institute of Technology,* 1–9.

[4] Batubara, R. P., & Setiawan, A. S. (2023). Prospek Pengembangan Creative Tourism Di Kota Bogor. *Khasanah Ilmu-Jurnal Pariwisata Dan Budaya, 14*(2), 88–95.

[5] Purnomo. (2016). *Ekonomi Kreatif Pilar Pembangunan Indonesia.* Surakarta: Ziyad Visi Media.

[6] Yoeti, O. A. (2008). *Perencanaan dan Pengembangan Pariwisata.* Pradnya Paramita: Jakarta.

[7] Rahadi, D. R. (2020). Konsep Penelitian kualitatif plus tutorial NVivo. *PT. Filda Fikrindo, Bogor.*

[8] Adinugraha, H. (2022). *Ekonomi Kreatif: Konsep, Peluang, dan Strategi Pengembangan.* Scientist IAIN: Pekalongan.

99 Beyond rationality: Overconfidence and herding on investment decisions with the moderating factor financial literacy

Sreejaa G. Nair[1,a] *and Jegadeeswari S.*[2,b]

[1]Research Scholar, Karpagam Academy of Higher Education, Coimbatore, India
[2]Assistant Professor, Karpagam Academy of Higher Education, Coimbatore, India

Abstract: This research aims to identify and analyze cognitive biases and their impact on the investment decision-making process, particularly focusing on how financial literacy moderates this relationship. A quantitative methodology was utilized for the study. The population of the study is equity investors of the Indian stock market using 294 respondents as a sample. The data analysis is performed using SPSS version 26. This study utilized correlation and regression techniques to explore how herding and overconfidence bias, both directly and indirectly, influence investment decisions within the Indian Stock Market. The moderation analysis was performed through process. The study's results suggest that both excessive confidence and herd mentality have a considerably favourable effect on investment decision-making. Additionally, financial knowledge also significantly influences investment choices. The capability to reduce the influence of herding and overconfidence on investment decisions has displayed by financial literacy.

Keywords: Cognitive bias, overconfidence, herding, financial literacy, investment decisions

1. Introduction

People make several decisions throughout their lives, some of which are more significant than others. Some decisions are clear-cut, while others are intricate and require a multi-stage process. Humans make decisions based on their experiences and perceptions rather than obtaining information, encouraging people to make wise financial decisions. Significant traditional theories in the field of finance suggest that individual investors are rational and exhibit a preference for lower-risk investment opportunities at a specific level of return due to their risk-averse nature. These theories include the Modern Portfolio Theory by Markowitz (1952) and the EMH by Fama (1970) [1].

The concept of perfect rationality in finance is grounded on the belief that investors base their decisions on all accessible information and make them in a fully logical manner. The psychological factors and their influence on investor's decisions often erode the idea of rationality, stating that biases and emotions can affect their decision choices. This indicates that various financial models like EMH, CAPM, APT, and EMT may not resonate with the behaviour of investors in real-world situations according to [2].

The decisions of the people about risky projects will be affected by several common effects like the framing effect, isolation effect, and certainty effect. The deviation of these effects from expected utility theory points out that the preferences of individuals are not in alignment with rational principles. The authors introduce an alternative model called Prospect Theory, which considers these effects and values gains and losses instead of final assets. This theory differentiates between risk and uncertainty and posits that people are more averse to losses than they are attracted to gains [3].

Traditional finance relies on the Efficient Market Hypothesis, classical decision theory, rationality, CAPM, risk aversion, and portfolio theory. Nevertheless, empirical data indicates that the assumptions and conclusions of traditional finance are flawed. The irrationality of investment decisions and their influence on cognitive and emotional factors was acknowledged by behavioural finance [4].

In the dominion of financial markets, where uncertainty and risk are inherent, human decision-making often deviates from the principles of rationality. The study of behavioural finance seeks to uncover the psychological factors that influence investors' choices and the subsequent impact on market dynamics. Two prominent behavioural biases, overconfidence, and herd behaviour, have gathered significant attention due to their insightful effects on investment decision-making. This research probes into the complex interplay of these biases while introducing financial literacy as a potential moderator that could mitigate their unfavorable consequences.

[a]Sreejaa G Nair, [b]sreejaagnair80@gmail.com

DOI: 10.1201/9781003684459-99

2. Literature Review

2.1. *Investment decisions*

An investment choice is decided by the act of individuals in investing with the expectation of generating profits. There are different forms of investment but the decision taken after gathering proper information will be successful [1]. As per the study [5], the investment decisions of individuals are influenced by their way of allocation of funds. The individual shareholders of personal values are heavily influencing their investment choices [6].

2.2. *Overconfidence bias and investment decisions*

The cognitive tendency of persons to overrate their understanding, knowledge, and influence on the results is called overconfidence bias. The individuals who are overconfident are more engaged in stock trading and look for higher returns but many times they fail to achieve their goals [7]. According to [8] the rational decisions of individual investors can be affected by overconfidence, and investor decisions are moderated by information acquisition. To make more informed about financial decisions, individual investors must be aware of the overestimating tendency at the sector level [9]. The overconfidence of investors leads to excessive trading and market realities will be ignored, resulting in inferior returns and they overlook risk factors. A moderate level of overconfidence enables individual investors to boost their expertise and knowledge to intensify the performance of their investments through informed predictions on future market trends in uncertainty [10].

H1: Overconfidence and Investment decisions are positively related.

2.3. *Investment decisions and herding bias*

Herding is an investor's tendency to follow others or mimicking others. Both experienced and less experienced investors display herd behaviour. Various determinants like market conditions, volatility, liquidity, and behavioural biases influence herd behaviour [11]. Herding happens when investors opt to mimic visible market trends or the judgments of others instead of relying on their own beliefs and expertise. While this may not consistently lead to optimal market outcomes, individuals might view such behaviour as rational for a range of reasons [12]. According to [13] informed investors regarding the IPO stock prices can influence other investors who do not have any knowledge about IPO which makes the underpricing issue. Herding behaviour is significantly influenced by society. The positive societal influence makes them to choose herding behaviour [14]. Herding is affected by strong brokerage firms and it is influenced by common information [15]. Under specific market conditions, such as changing trends in the market, varying levels of liquidity and uncertainty, the arrival of new information, and fluctuations in oil prices can lead to different causes for investors engaging in herding behaviour [16].

H2: Herding bias exerts a substantial and favourable influence on investment decision-making.

2.4. *The effect of financial literacy in moderating effects of overconfidence and herding bias on investment decision*

Results indicate that both overconfidence and herding affect stock investment decisions but financial literacy acts as a controlling factor between the two [17]. Both risk avoidance and herding have no impact on investment decisions but male investors show overconfidence which affects their investment decisions. The lack of overconfidence among female investors directs their investment decisions. The investor decision making predominantly depends on the financial literacy level [18]. Herd behaviour occurs mostly among corporations that maintain small market capitalization while simultaneously they can access superior information about larger corporations. According to Fransiska, Sumani, and Pangestu [19] investors look for unfavourable information due to information market flaws since they want to question their current knowledge by sharing it with other experts to inspect their actions.

The association between overconfidence bias and investment decision making process is moderated by financial literacy (H3).

H3: It is asserted that financial literacy is vital in moderating the association between herding bias and the investment decision process.

2.5. *Theoretical framework*

Traditionally, stock market investors in traditional finance have been presumed to perform rational decision making, which includes engaging in purchasing, to risk and return trade off or maximizing utility. Yet, observed via studies in behavioural finance, people are not always logical as economists would expect of them in not quite all of their dealings owing to psychological states from time to time. A number of research studies in ASEAN countries, Middle East countries and other Western countries found psychological factors and how it effects the stock market investors in these countries based on the market where they were operating [2].

Expected Utility Theory is a descriptive framework regarding how a rational person should make decisions when facing risk (prescribing moving toward maximizing expected utility through the addition of potential outcomes with their respective probabilities). The theory considers that such an investor is rational and knows the probabilities of the given choices and its consequences. However, some empirical phenomena contradict the principles that indicate differences in the tendency to form investors decisions. Prospect Theory [3] alternatively, offers with risk informed decision making.

Amos Tversky & Daniel Kanheman introduced Prospect Theory which states that the people are not rational but factors of various psychological origin. Thanks to prospect theory such fluctuation in the return could be justified in terms of the traditional financial models [20]. Heuristics and prospect theory [21] negatively influence individual investors investment decisions (Figure 99.1).

In the dynamic financial market, the investors have to make complex decision and among all other factors influencing such decisions, psychological factors play important role. In particular, overconfidence and herding have drawn a great deal of attention among the psychological factors. This study attempts to understand the relevance of these biases and learns about their effect on the investment decisions under the moderating role of financial literacy.

3. Statement of the Problem

There is a large deviation from rationality in those investment decisions that the investors make. One of the two biases that influence rationality for investment decisions, among others, are overconfidence and herding. This is to know how much of impact irrational choice such as over confidence and herding makes in order to minimize irrational choice. This would thus be something that would impart to investors a better understanding as well as skills in reaching rational judgment when making decision. Nevertheless, it is an under explored topic on how to prevent the negative impacts of overconfidence and herding behaviour through financial literacy. The present study attempts to investigate the mediating role of financial literacy to the interplay between behavioural biases and the investment choices.

4. Scope

The research seeks answers to the extent the cognitive bias can impact the decision making behaviour of individual equity investors in the Indian stock market. In addition, it investigates the role of financial literacy as a mediator. This helps us understand the cognitive biases and the investment decisions, and gives useful advice to the financial experts and the policymakers. Furthermore, this study is also in the line of the United Nations SDG because it tries to find out factors that affect individual investor's economic decision making and making him more financially literate which leads to be more financially well.

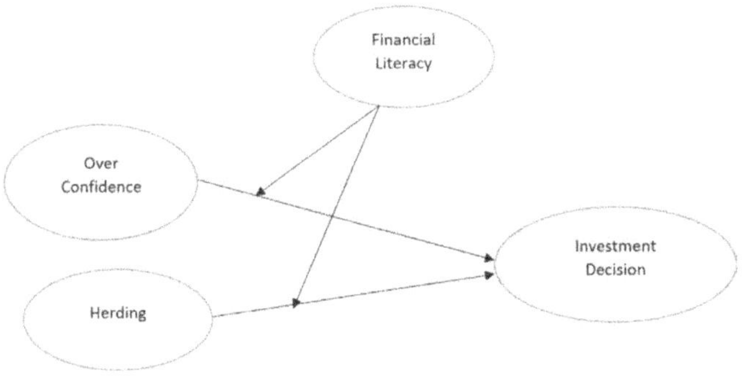

Figure 99.1. Dynamic financial.

Source: Author.

5. Objectives

- To study cognitive biases, overconfidence and herding and its impact on investment decisions of equity investors in the Indian stock market.
- To measure the moderating role of financial literacy between cognitive bias and decision making in investments.

6. Methodology

A research investigation focuses on Indian Stock Market individual equity investors. The respondents completed an adaptive questionnaire to gather data. To test hypotheses the researcher collected primary data by conducting the study for its initial documentation. The research used responses from 294 participants who operate within equity investments for convenience sampling purposes. The research adopted convenience sampling due to its fitting nature and possibility of accessing the target population successfully; 300 surveys were transmitted to prospective respondents through computer-based channels including emails and online surveys. To maximize response rates and data completion we tried our best but received an incomplete number of questionnaires resulting in 294 valid responses for the analysis.

Structured survey with five point Likert scale, data were collected. The five sections of the questionnaire were used to solicit personal information of the respondents, the first of which collected personal information from the respondents to begin with. The other sections are overconfidence, herding, investment decisions and financial literacy. Investment decisions were studied using five questions. To test overconfidence five items were used, and four items for Herding which was adapted from Kengatharan et al. [22] scale. To test financial literacy, we selected from the work of four questions that provided students with difficulty (Table 99.1).

Table 99.1. Summary of reliability test

Items	*Cronbach's Alpha*	*No. of Items*
Investment Decisions	0.822	5
Over Confidence	0.843	5
Herding	0.724	4
Financial Literacy	0.720	4
Total	0.804	18

Source: Author.

Both inferential and descriptive statistical methods enabled data analysis through the development of inference conclusions. A reliability check on the data took place through Cronbach's Alpha test. The research hypothesis received verification through multiple assessment strategies that included correlation examination and multiple regression and moderation analyses using Process Macro by Hayes A.F. (2013).

7. Significance

The importance of this study lies in its contribution to understanding the impacts of psychological biases on investment decisions. By exploring the moderating effect of financial literacy, this study tries to reveal effective strategies for alleviating the negative impact of these biases and improving investment decision-making. Furthermore, the results of this inquiry could hold practical significance for investors, financial institutions, and policymakers. Additionally, this research expands the current knowledge base by exploring the nexus between financial literacy and these psychological biases, an aspect that has been relatively underexplored in previous studies.

8. Results and Discussion

8.1. Demographic details of the survey participants

The sample comprised 294 individuals, with a slightly greater proportion of females (53.1%) than males (46.9%). Most participants fell within the age range of 19 to 36 years (63.3%), held a Master's Degree (44.9%), and were employed in the private sector (38.8%). In terms of monthly income, a notable portion reported earning between 25001 to 40000 (24.5%) and above 40000 per month (30.6%). Concerning investment duration, the highest proportion of respondents reported investing for less than one year (32.7%) (Table 99.2).

8.2. Correlation tests: Correlation test between variables

Both Overconfidence bias and Investment Decision displayed a strong positive relationship ($r=.441$, $p<.01$) and Financial literacy had a positive relationship with Overconfidence Bias ($r=.124$, $p<.01$). The relationship between Herding Bias and Investment Decision demonstrated a substantial statistical linkage with a value of .166 ($p<.01$) (Tables 99.3–99.6).

8.3. Regression analysis

A multiple regression analysis was executed to examine the predictive relationship between Herding and Overconfidence Bias concerning Investment Decisions. The results showed statistical significance, with $F(2, 291) = 39.232$, $p < .001$, indicating that the combined impact of Herding Bias and Overconfidence Bias had a substantial effect on

Table 99.2. Demographic composition of the survey participants

Demographic Composition	*Percent of Respondents*
Gender	
Male	46.9
Female	53.1
Age	
<18 Years	4.1
19–36 Years	63.3
37–54 Years	24.5
>54 Years	8.2
Education	
High School	10.2
Diploma	4.1
UG Degree	30.6
PG Degree	44.9
Ph. D	10.2
Profession	
Student	34.7
Government Employee	2
Private Employee	38.8
Business Man	4.1
Professional	20.4
Monthly Income	
<15000	30.6
15001–25000	14.3
25001–40000	24.5
>40000	30.6
Investment Time	
Up to 1 Year	32.7
2 to 4 Years	26.5
5 to 7 Years	12.2
8 to 10 Years	16.3
> 10 Years	12.2

Note: Sample size was 294, Survey 2024, and author calculation.

Source: Author.

Table 99.3. Correlation analysis

Variable	*n*	*Mean*	*St Dev*	*1*	*2*	*3*	*4*
Over-Confidence Bias	294	19.55	3.53	1	0.075	.441**	.124*
Herding Bias	294	16.04	2.18	0.075	1	.166**	0.072
Investment Decision	294	15.06	3.85	.441**	.166**	1	0.092
Financial Literacy	294	13	2.61	.124*	0.072	0.092	1

Source: Survey 2024, and author calculation.

Investment Decisions. This model suggests a moderate predictive power by explaining 21.2% of the variance in Investment Decision scores. The standard error of the estimate was 3.14774. Both predictors, Overconfidence Bias (β = .431, p < .001) and Herding Bias (β = .134, p = .011), have made significant unique contributions to the prediction of Investment Decisions. For each additional unit of Overconfidence Bias, there was a 0.395-unit increase in Investment Decision scores. Similarly, with each one-unit increase in Herding Bias, Investment Decision scores rose by 0.181 units. Both overconfidence and herding bias influence the formation of investment decisions as per this result. The results of this study are quite aligned.

Table 99.4. Model summary

Model	*R*	R^2	*Adj* R^2	*SE of the Estimate*
1	.461[a]	0.212	0.207	3.14774

a. Predictors: (Constant), Herding Bias, Over Confidence Bias

Source: Author.

8.4. Moderation analysis

H3: Financial literacy in the mediator role between overconfidence and investment decision-making (Table 99.7).

This paper from the findings of the analysis on moderation suggests that Financial Literacy has a great capacity to moderate the effect of Overconfidence Bias on Investment decisions. In this sense, it therefore means that Overconfidence Bias does happen differently in the influence of Investment decisions with different levels of Financial Literacy. In other words, if the Financial Literacy is high, then Overconfidence would have lesser effect on the Investment decisions. Nevertheless Financial Literacy is positively associated with the relation between Overconfidence bias and investment

Table 99.5. ANOVA[a]

Model		*Sum of Squares*	*df*	*Mean Square*	*F*	*Sig.*
1	Regression	777.432	2	388.716	39.232	.000[b]
	Residual	2883.303	291	9.908		
	Total	3660.735	293			

Source: Author.

Table 99.6. Coefficients

Model		*Uns. Coefficients*		*S.Coefficients*	*t*	*Sig.*
		B	Std. Error	Beta		
1	(Constant)	11.245	1.138		9.884	0.000
	Overconfidence Bias	0.395	0.048	0.431	8.262	0.000
	Herding Bias	0.181	0.071	0.134	2.563	0.011

Source: Author.

Table 99.7. Summary of PROCESS model 1

Predictor	*B*	*SE*	*t*	*p*	*95% CI Lower*	*95% CI Upper*
Constant	40.17	5.1	7.88	< .001	30.13	50.2
Overconfidence Bias	-1.41	0.32	-4.35	< .001	-2.05	-0.77
Financial Literacy	-1.58	0.3	-5.23	< .001	-2.17	-0.98
OCB x FL	0.11	0.02	5.64	< .001	0.07	0.14

Source: Survey 2024, and author calculation.

Table 99.8. Summary of PROCESS model 2

Predictor	*B*	*SE*	*t*	*p*	*95% CI Lower*	*95% CI Upper*
Constant	35.92	6.76	5.32	< .001	22.63	49.22
Herding Bias	-1.49	0.53	-2.82	0.005	-2.53	-0.45
Financial Literacy	-1.13	0.4	-2.84	0.004	-1.91	-0.35
HB x FL	0.1	0.03	3.26	0.001	0.04	0.16

Source: Author.

decisions only if we are more Financially literated. Financial literate help magnify the good side of Overconfidence Bias in making Investment decisions (Table 99.8). This is the actual results of the conflict of the study with [23] and [24].

H4: Financial literacy as a mediator between herding and investment decision-making.

The analysis indicates that Financial Literacy functions as a regulatory component in the connection between Herding and Investment Decision-making. With the increase in FL, the positive influence of HB on ID becomes more evident. This shows that FL positively influences the relation between HB and ID [25] and [26].

9. Suggestions

Further studies are required to understand how financial literacy influences cognitive biases in various cultural environments. The study can be expanded to include diverse populations and additional factors like risk tolerance and prior experience in investment. Understanding of behavioural finance principles through financial education in educational institutions helps in better decision-making. additionally, application of AI in decision-making and its interplay with biases will further improve investment outcomes.

10. Conclusion

This study examined this relationship, yet between cognitive biases, investment decisions and mediating role of financial literacy. The study shows that the positive association between herding and overconfidence in investment decisions. Also, the study provides strong moderation of financial literacy on the relationship between cognitive biases and investing decision. This illustrates how financial education plays an important role in diminishing the harmful effects of cognitive biases that have a bearing on decision making. It helps in recognising and mitigating the biases, with a strong knowledge in financial concepts and instruments. Financial literacy and awareness of possible biases can enable the investors to make informed decisions in investment landscape.

References

[1] Shah, S. Z. A., Ahmad, M., & Mahmood, F. (2018). Heuristic biases in investment decision-making and perceived market efficiency: A survey at the Pakistan stock exchange. *Qualitative Research in Financial Markets*, *10*(1), 85–110.

[2] Bakar, S., & Yi, A. N. C. (2016). The impact of psychological factors on investors' decision making in Malaysian stock market: a case of Klang Valley and Pahang. *Procedia Economics and Finance*, *35*, 319–328.

[3] Kahneman, D., & Tversky, A. (1979). Prospect Theory: An Analysis of Decision under Risk. *Econometrica*, *47*(2), 263–291. https://doi.org/10.2307/1914185

[4] Baker, H. K., & Ricciardi, V. (2014). *Investor behavior: The psychology of financial planning and investing*. John Wiley & Sons.

[5] Virlics, A. (2013). Investment decision-making and risk. *Procedia Economics and Finance*, *6*, 169–177.

[6] Agyemang, O. S., & Ansong, A. (2016). Role of personal values in investment decisions: Perspectives of individual Ghanaian shareholders. *Management Research Review*, *39*(8), 940–964.

[7] ul Abdin, S. Z., Qureshi, F., Iqbal, J., & Sultana, S. (2022). Overconfidence bias and investment performance: A mediating effect of risk propensity. *Borsa Istanbul Review*, *22*(4), 780–793. https://doi.org/10.1016/j.bir.2022.03.001

[8] Naveed, F., & Taib, H. M. (2021). Overconfidence bias, self-attribution bias and investor decisions: Moderating role of information acquisition. *Pakistan Journal of Commerce and Social Sciences (PJCSS)*, *15*(2), 354–377.

[9] Azam, M. Q., Hashmi, N. I., Hawaldar, I. T., Alam, M. S., & Baig, M. A. (2022). The COVID-19 Pandemic and Overconfidence Bias: The Case of Cyclical and Defensive Sectors. *Risks*, *10*(3). https://doi.org/10.3390/risks10030056

[10] Areiqat, A. Y., Abu-Rumman, A., Al-Alani, Y. S., & Alhorani, A. (2019). Impact of behavioral finance on stock investment decisions applied study on a sample of investors at Amman stock exchange. *Academy of Accounting and Financial Studies Journal*, *23*(2), 1–17.

[11] Dewan, P., & Dharni, K. (2019). Herding Behaviour in Investment Decision Making: A Review. *Journal of Economics, Management and Trade*, 1–12. https://doi.org/10.9734/jemt/2019/v24i230160

[12] Hwang, S., & Salmon, M. (2004). Market stress and herding. *Journal of Empirical Finance*, *11*(4), 585–616.

[13] Arisanti, I., & Asri, M. (2018). Herding Behavior Post Initial Public Offering In Indonesia Stock Exchange. *Journal of Accounting and Investment*, *19*(2), 149–159.

[14] Rahayu, S., Rohman, A., & Harto, P. (2021). Herding Behavior Model in Investment Decision on Emerging Markets: Experimental in Indonesia. *Journal of Asian Finance, Economics and Business*, *8*(1), 053–059. https://doi.org/10.13106/jafeb.2021.vol8.no1.053

[15] Agarwal, S., Chiu, I. M., Liu, C., & Rhee, S. G. (2011). The brokerage firm effect in herding: Evidence from Indonesia. *Journal of Financial Research*, *34*(3), 461–479.

[16] Indars, E. R., Savin, A., & Lublóy, Á. (2019). Herding behaviour in an emerging market: Evidence from the Moscow Exchange. *Emerging Markets Review*, *38*, 468–487.

[17] Tamara, D. (2022). Do Overconfidence and Herding Affect Stock Investment Decision? Indonesian Cases During Pandemic Covid-19. *MIX: JURNAL ILMIAH MANAJEMEN*, *12*(3), 541. https://doi.org/10.22441/jurnal_mix.2022.v12i3.012

[18] Adil, M., Singh, Y., & Ansari, M. S. (2022). How financial literacy moderate the association between behaviour biases and investment decision?. *Asian Journal of Accounting Research*, *7*(1), 17–30.

[19] Fransiska, M., Sumani, S., & Pangestu, S. (2018). Herding behavior in Indonesian investors. *International Research Journal of Business Studies*, *11*(2), 129–143.

[20] Barberis, Nicholas and Barberis, Nicholas and Jin, Lawrence J., & Wang, Baolian, Prospect Theory and Stock Market Anomalies (December 16, 2020). 10th Miami Behavioral Finance Conference, Available at SSRN: https://ssrn.com/abstract=3477463 or http://dx.doi.org/10.2139/ssrn.3477463

[21] Ogunlusi, O. E., & Obademi, O. (2021). The Impact of behavioural finance on investment decision-making: A study

of selected investment banks in Nigeria. *Global Business Review*, *22*(6), 1345–1361.

[22] Kengatharan, L., & Kengatharan, N. (2014). The influence of behavioral factors in making investment decisions and performance: Study on investors of Colombo Stock Exchange, Sri Lanka. *Asian Journal of Finance & Accounting*, *6*(1), 1.

[23] Prasetyo, P., & Ratnawati, K. (2023). The impact of disposition effect, herding and overconfidence on investment decision making moderated by financial literacy. *International Journal of Research in Business and Social Science (2147–4478)*, *12*(9), 241–251.

[24] Firmansyah, F. (2023). The Influence of Overconfidence Bias, Herding Effect, and Loss Aversion on Investment Decisions in The Capital Market with Financial Literacy as A Moderating Variable. *Journal of Business and Management Review*, *4*(11).

[25] Rehmat, I., Khan, A. A., Hussain, M., & Khurshid, S. (2023). An Examination of Behavioral Factors Affecting the Investment Decisions: The Moderating Role of Financial Literacy and Mediating Role of Risk Perception. *Journal of Innovative Research in Management Sciences*, *4*(2), 1–16.

[26] Butt, H., Sajjad, A., Awan, K. Z., & Shakil, M. H. (2023). The Role of Behavioral Factors on Investment Decision Making: Moderating Role of Financial Literacy. *Pakistan Journal of Humanities and Social Sciences*, *11*(4), 4533–4547.

For Product Safety Concerns and Information please contact our EU
representative GPSR@taylorandfrancis.com
Taylor & Francis Verlag GmbH, Kaufingerstraße 24, 80331 München, Germany

www.ingramcontent.com/pod-product-compliance
Ingram Content Group UK Ltd.
Pitfield, Milton Keynes, MK11 3LW, UK
UKHW050235200926
475702UK00020B/3042

* 9 7 8 1 0 4 1 1 6 2 7 9 7 *